Encyclopedia of Metalloproteins

Robert H. Kretsinger • Vladimir N. Uversky
Eugene A. Permyakov
Editors

Encyclopedia of Metalloproteins

Volume 3

M–R

With 1109 Figures and 256 Tables

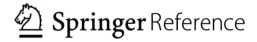

Editors
Robert H. Kretsinger
Department of Biology
University of Virginia
Charlottesville, VA, USA

Vladimir N. Uversky
Department of Molecular Medicine
College of Medicine
University of South Florida
Tampa, FL, USA

Eugene A. Permyakov
Institute for Biological Instrumentation
Russian Academy of Sciences
Pushchino, Moscow Region, Russia

ISBN 978-1-4614-1532-9   ISBN 978-1-4614-1533-6 (eBook)
ISBN 978-1-4614-1534-3 (print and electronic bundle)
DOI 10.1007/ 978-1-4614-1533-6
Springer New York Heidelberg Dordrecht London

Library of Congress Control Number: 2013931183

© Springer Science+Business Media New York 2013
This work is subject to copyright. All rights are reserved by the Publisher, whether the whole or part of the material is concerned, specifically the rights of translation, reprinting, reuse of illustrations, recitation, broadcasting, reproduction on microfilms or in any other physical way, and transmission or information storage and retrieval, electronic adaptation, computer software, or by similar or dissimilar methodology now known or hereafter developed. Exempted from this legal reservation are brief excerpts in connection with reviews or scholarly analysis or material supplied specifically for the purpose of being entered and executed on a computer system, for exclusive use by the purchaser of the work. Duplication of this publication or parts thereof is permitted only under the provisions of the Copyright Law of the Publisher's location, in its current version, and permission for use must always be obtained from Springer. Permissions for use may be obtained through RightsLink at the Copyright Clearance Center. Violations are liable to prosecution under the respective Copyright Law.
The use of general descriptive names, registered names, trademarks, service marks, etc. in this publication does not imply, even in the absence of a specific statement, that such names are exempt from the relevant protective laws and regulations and therefore free for general use.
While the advice and information in this book are believed to be true and accurate at the date of publication, neither the authors nor the editors nor the publisher can accept any legal responsibility for any errors or omissions that may be made. The publisher makes no warranty, express or implied, with respect to the material contained herein.

Printed on acid-free paper

Springer is part of Springer Science+Business Media (www.springer.com)

Preface

Metal ions play an essential role in the functioning of all biological systems. All biological processes occur in a milieu of high concentrations of metal ions, and many of these processes depend on direct participation of metal ions. Metal ions interact with charged and polar groups of all biopolymers; those interactions with proteins play an especially important role.

The study of structural and functional properties of metal binding proteins is an important and ongoing activity area of modern physical and chemical biology. Thirteen metal ions – sodium, potassium, magnesium, calcium, manganese, iron, cobalt, zinc, copper, nickel, vanadium, tungsten, and molybdenum – are known to be essential for at least some organisms. Metallo-proteomics deals with all aspects of the intracellular and extracellular interactions of metals and proteins. Metal cations and metal binding proteins are involved in all crucial cellular activities. Many pathological conditions are correlated with abnormal metal metabolism. Research in metallo-proteomics is rapidly growing and is progressively entering curricula at universities, research institutions, and technical high schools.

Encyclopedia of Metalloproteins is a key resource that provides basic, accessible, and comprehensible information about this expanding field. It covers exhaustively all thirteen essential metal ions, discusses other metals that might compete or interfere with them, and also presents information on proteins interacting with other metal ions. *Encyclopedia of Metalloproteins* is an ideal reference for students, teachers, and researchers, as well as the informed public.

Acknowledgements

We extend our sincerest thanks to all of the contributors who have shared their insights into metalloproteins with the broader community of researchers, students, and the informed public.

About the Editors

Robert H. Kretsinger Department of Biology, University of Virginia 395, Charlottesville, VA 22904, USA

Robert H. Kretsinger is Commonwealth Professor of Biology at the University of Virginia in Charlottesville, Virginia, USA. His research has addressed structure, function, and evolution of several different protein families. His group determined the crystal structure of parvalbumin in 1970. The analysis of this calcium binding protein provided the initial characterization of the helix, loop, helix conformation of the EF-hand domain and of the pair of EF-hands that form an EF-lobe. Over seventy distinct subfamilies of EF-hand proteins have been identified, making this domain one of the most widely distributed in eukaryotes.

Dr. Kretsinger has taught courses in protein crystallography, biochemistry, macromolecular structure, and history and philosophy of biology, and has served as chair of his department and of the University Faculty Senate.

His "other" career as a sculptor began, before the advent of computer graphics, with space filling models of the EF-hand (www.virginiastonecarvers.com). He has also been an avid cyclist for many decades.

Vladimir N. Uversky Department of Molecular Medicine, College of Medicine, University of South Florida, Tampa, FL 33612 USA

Vladimir N. Uversky is an Associate Professor at the Department of Molecular Medicine at the University of South Florida (USF). He obtained his academic degrees from Moscow Institute of Physics and Technology (PhD in 1991) and from the Institute of Experimental and Theoretical Biophysics, Russian Academy of Sciences (DSc in 1998). He spent his early career working mostly on protein folding at the Institute of Protein Research and Institute for Biological Instrumentation, Russia. In 1998, he moved to the University of California, Santa Cruz, where for six years he studied protein folding, misfolding, protein conformation diseases, and protein intrinsic disorder phenomenon. In 2004, he was invited to join the Indiana University School of Medicine as a Senior Research Professor to work on intrinsically disordered proteins. Since 2010, Professor Uversky has been with USF, where he continues to study intrinsically disordered proteins and protein folding and misfolding processes. He has authored over 450 scientific publications and edited several books and book series on protein structure, function, folding and misfolding.

About the Editors

Eugene A. Permyakov, Institute for Biological Instrumentation, Russian Academy of Sciences, Pushchino, Moscow Region, Russia

Eugene A. Permyakov received his PhD in physics and mathematics at the Moscow Institute of Physics and Technology in 1976, and defended his Doctor of Sciences dissertation in biology at Moscow State University in 1989. From 1970 to 1994 he worked at the Institute of Theoretical and Experimental Biophysics of the Russian Academy of Sciences. From 1990 to 1991 and in 1993, Dr. Permyakov worked at the Ohio State University, Columbus, Ohio, USA. Since 1994 he has been the Director of the Institute for Biological Instrumentation of the Russian Academy of Sciences. He is a Professor of Biophysics and is known for his work on metal binding proteins and intrinsic luminescence method. He is a member of the Russian Biochemical Society. Dr. Permyakov's primary research focus is the study of physico-chemical and functional properties of metal binding proteins. He is the author of more than 150 articles and 10 books, including *Luminescent Spectroscopy of Proteins* (CRC Press, 1993), *Metalloproteomics* (John Wiley & Sons, 2009), and *Calcium Binding Proteins* (John Wiley & Sons, 2011). He is an Academic Editor of the journals *PLoS ONE* and *PeerJ*, and Editor of the book *Methods in Protein Structure and Stability Analysis* (Nova, 2007).

In his spare time, Dr. Permyakov is an avid jogger, cyclist, and cross country skier.

Section Editors

Sections: Physiological Metals: Ca; Non-Physiological Metals: Pd, Ag

Robert H. Kretsinger Department of Biology, University of Virginia, Charlottesville, VA, USA

Section: Physiological Metals: Ca

Eugene A. Permyakov Institute for Biological Instrumentation, Russian Academy of Sciences, Pushchino, Moscow Region, Russia

Sections: Metalloids; Non-Physiological Metals: Ag, Au, Pt, Be, Sr, Ba, Ra

Vladimir N. Uversky Department of Molecular Medicine, College of Medicine, University of South Florida, Tampa, FL, USA

Sections: Physiological Metals: Co, Ni, Cu

Stefano Ciurli Laboratory of Bioinorganic Chemistry, Department of Pharmacy and Biotechnology, University of Bologna, Italy

Sections: Physiological Metals: Cd, Cr

John B. Vincent Department of Chemistry, The University of Alabama, Tuscaloosa, AL, USA

Section: Physiological Metals: Fe

Elizabeth C. Theil Children's Hospital Oakland Research Institute, Oakland, CA, USA
Department of Molecular and Structural Biochemistry, North Carolina State University, Raleigh, NC, USA

Sections: Physiological Metals: Mo, W, V

Biswajit Mukherjee Department of Pharmaceutical Technology, Jadavpur University, Kolkata, India

Sections: Physiological Metals: Mg, Mn

Andrea Romani Department of Physiology and Biophysics, Case Western Reserve University, Cleveland, OH, USA

Sections: Physiological Metals: Na, K

Sergei Yu. Noskov Institute for BioComplexity and Informatics and Department for Biological Sciences, University of Calgary, Calgary, AB, Canada

Section: Physiological Metals: Zn

David S. Auld Harvard Medical School, Boston, Massachusetts, USA

Sections: Non-Physiological Metals: Li, Rb, Cs, Fr

Sergei E. Permyakov Protein Research Group, Institute for Biological Instrumentation, Russian Academy of Sciences, Pushchino, Moscow Region, Russia

Sections: Non-Physiological Metals: Lanthanides, Actinides

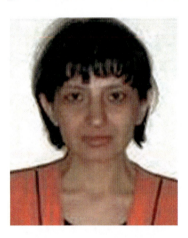

Irena Kostova Department of Chemistry, Faculty of Pharmacy, Medical University, Sofia, Bulgaria

Sections: Non-Physiological Metals: Sc, Y, Ti, Zr, Hf, Rf, Nb, Ta, Tc, Re, Ru, Os, Rh, Ir

Chunying Chen Key Laboratory for Biological Effects of Nanomaterials and Nanosafety of CAS, National Center for Nanoscience and Technology, Beijing, China

Sections: Non-Physiological Metals: Hg, Pb

K. Michael Pollard Department of Molecular and Experimental Medicine, The Scripps Research Institute, La Jolla, CA, USA

Sections: Non-Physiological Metals: Al, Ga, In, Tl, Ge, Sn, Sb, Bi, Po

Sandra V. Verstraeten Department of Biological Chemistry, University of Buenos Aires, Buenos Aires, Argentina

List of Contributors

Satoshi Abe Department of Biomolecular Engineering, Graduate School of Bioscience and Biotechnology, Tokyo Institute of Technology, Yokohama, Japan

Vojtech Adam Department of Chemistry and Biochemistry, Faculty of Agronomy, Mendel University in Brno, Brno, Czech Republic

Central European Institute of Technology, Brno University of Technology, Brno, Czech Republic

Olayiwola A. Adekoya Pharmacology Research group, Department of Pharmacy, Institute of Pharmacy, University of Tromsø, Tromsø, Norway

Paul A. Adlard The Mental Health Research Institute, The University of Melbourne, Parkville, VIC, Australia

Magnus S. Ågren Department of Surgery and Copenhagen Wound Healing Center, Bispebjerg University Hospital, Copenhagen, Denmark

Karin Åkerfeldt Department of Chemistry, Haverford College, Haverford, PA, USA

Takashiro Akitsu Department of Chemistry, Tokyo University of Science, Shinjuku-ku, Tokyo, Japan

Lorenzo Alessio Department of Experimental and Applied Medicine, Section of Occupational Health and Industrial Hygiene, University of Brescia, Brescia, Italy

Mamdouh M. Ali Biochemistry Department, Genetic Engineering and Biotechnology Division, National Research Centre, El Dokki, Cairo, Egypt

James B. Ames Department of Chemistry, University of California, Davis, CA, USA

Olaf S. Andersen Department of Physiology and Biophysics, Weill Cornell Medical College, New York, NY, USA

Gregory J. Anderson Iron Metabolism Laboratory, Queensland Institute of Medical Research, PO Royal Brisbane Hospital, Brisbane, QLD, Australia

Janet S. Anderson Department of Chemistry, Union College, Schenectady, NY, USA

João Paulo André Centro de Química, Universidade do Minho, Braga, Portugal

Claudia Andreini Magnetic Resonance Center (CERM) – University of Florence, Sesto Fiorentino, Italy

Department of Chemistry, University of Florence, Sesto Fiorentino, Italy

Alexey N. Antipov A.N. Bach Institute of Biochemistry Russian Academy of Sciences, Moscow, Russia

Tayze T. Antunes Kidney Research Center, Ottawa Hospital Research Institute, University of Ottawa, Ottawa, ON, Canada

Varun Appanna Department of Chemistry and Biochemistry, Laurentian University, Sudbury, ON, Canada

Vasu D. Appanna Department of Chemistry and Biochemistry, Laurentian University, Sudbury, ON, Canada

Cristina Ariño Department of Analytical Chemistry, University of Barcelona, Barcelona, Spain

Vladimir B. Arion Institute of Inorganic Chemistry, University of Vienna, Vienna, Austria

Farukh Arjmand Department of Chemistry, Aligarh Muslim University, Aligarh, UP, India

Fabio Arnesano Department of Chemistry, University of Bari "Aldo Moro", Bari, Italy

Joan L. Arolas Proteolysis Lab, Department of Structural Biology, Molecular Biology Institute of Barcelona, CSIC Barcelona Science Park, Barcelona, Spain

Afolake T. Arowolo Department of Biochemistry, Microbiology & Biotechnology, Rhodes University, Grahamstown, Eastern Cape, South Africa

Nebojša Arsenijević Faculty of Medical Sciences, University of Kragujevac, Centre for Molecular Medicine, Kragujevac, Serbia

Samuel Ogheneovo Asagba Department of Biochemistry, Delta State University, Abraka, Delta State, Nigeria

Michael Aschner Department of Pediatrics, Division of Pediatric Clinical Pharmacology and Toxicology, Vanderbilt University Medical Center, Nashville, TN, USA

Center in Molecular Toxicology, Vanderbilt University Medical Center, Nashville, TN, USA

Center for Molecular Neuroscience, Vanderbilt University Medical Center, Nashville, TN, USA

The Kennedy Center for Research on Human Development, Vanderbilt University Medical Center, Nashville, TN, USA

Michael Assfalg Department of Biotechnology, University of Verona, Verona, Italy

William D. Atchison Department of Pharmacology and Toxicology, Michigan State University, East Lansing, MI, USA

Bishara S. Atiyeh Division of Plastic Surgery, Department of Surgery, American University of Beirut Medical Center, Beirut, Lebanon

Sílvia Atrian Departament de Genètica, Facultat de Biologia, Universitat de Barcelona, Barcelona, Spain

Christopher Auger Department of Chemistry and Biochemistry, Laurentian University, Sudbury, ON, Canada

David S. Auld Harvard Medical School, Boston, MA, USA

Scott Ayton The Mental Health Research Institute, The University of Melbourne, Parkville, VIC, Australia

Eduard B. Babiychuk Department of Cell Biology, Institute of Anatomy, University of Bern, Bern, Switzerland

Petr Babula Department of Natural Drugs, Faculty of Pharmacy, University of Veterinary and Pharmaceutical Sciences Brno, Brno, Czech Republic

Damjan Balabanič Ecology Department, Pulp and Paper Institute, Ljubljana, Slovenia

Wojciech Bal Institute of Biochemistry and Biophysics, Polish Academy of Sciences, Warsaw, Poland

Graham S. Baldwin Department of Surgery, Austin Health, The University of Melbourne, Heidelberg, VIC, Australia

Cynthia Bamdad Minerva Biotechnologies Corporation, Waltham, MA, USA

Mario Barbagallo Geriatric Unit, Department of Internal Medicine and Medical Specialties (DIMIS), University of Palermo, Palermo, Italy

Juan Barceló Lab. Fisiología Vegetal, Facultad de Biociencias, Universidad Autónoma de Barcelona, Bellaterra, Spain

Khurram Bashir Graduate School of Agricultural and Life Sciences, The University of Tokyo, Tokyo, Japan

Partha Basu Department of Chemistry and Biochemistry, Duquesne University, Pittsburgh, PA, USA

Andrea Battistoni Dipartimento di Biologia, Università di Roma Tor Vergata, Rome, Italy

Mikael Bauer Department of Biochemistry and Structural Biology, Lund University, Chemical Centre, Lund, Sweden

Lukmaan Bawazer School of Chemistry, University of Leeds, Leeds, UK

Carine Bebrone Centre for Protein Engineering, University of Liège, Sart–Tilman, Liège, Belgium

Institute of Molecular Biotechnology, RWTH–Aachen University, c/o Fraunhofer IME, Aachen, Germany

Konstantinos Beis Division of Molecular Biosciences, Imperial College London, London, South Kensington, UK

Membrane Protein Lab, Diamond Light Source, Harwell Science and Innovation Campus, Chilton, Oxfordshire, UK

Research Complex at Harwell, Harwell Oxford, Didcot, Oxforsdhire, UK

Catherine Belle Département de Chimie Moléculaire, UMR-CNRS 5250, Université Joseph Fourier, ICMG FR-2607, Grenoble, France

Andrea Bellelli Department of Biochemical Sciences, Sapienza University of Rome, Rome, Italy

Gunes Bender Department of Biological Chemistry, University of Michigan Medical School, Ann Arbor, MI, USA

Stefano Benini Faculty of Science and Technology, Free University of Bolzano, Bolzano, Italy

Stéphane L. Benoit Department of Microbiology, The University of Georgia, Athens, GA, USA

Tomas Bergman Department of Medical Biochemistry and Biophysics, Karolinska Institutet, Stockholm, Sweden

Lawrence R. Bernstein Terrametrix, Menlo Park, CA, USA

Marla J. Berry Department of Cell & Molecular Biology, John A. Burns School of Medicine, University of Hawaii at Manoa, Honolulu, HI, USA

Ivano Bertini Magnetic Resonance Center (CERM) – University of Florence, Sesto Fiorentino, Italy

Department of Chemistry, University of Florence, Sesto Fiorentino, Italy

Gerd Patrick Bienert Institut des Sciences de la Vie, Universite catholique de Louvain, Louvain-la-Neuve, Belgium

Andrew N. Bigley Department of Chemistry, Texas A&M University, College Station, TX, USA

Luis M. Bimbo Division of Pharmaceutical Technology, University of Helsinki, Helsinki, Finland

Ohad S. Birk Head, Genetics Institute, Soroka Medical Center Head, Morris Kahn Center for Human Genetics, NIBN and Faculty of Health Sciences, Ben Gurion University, Beer Sheva, Israel

Ruth Birner-Gruenberger Institute of Pathology and Center of Medical Research, Medical University of Graz, Graz, Austria

Cristina Bischin Department of Chemistry and Chemical Engineering, Babes-Bolyai University, Cluj-Napoca, Romania

Florian Bittner Department of Plant Biology, Braunschweig University of Technology, Braunschweig, Germany

Jodi L. Boer Department of Biochemistry and Molecular Biology, Michigan State University, East Lansing, MI, USA

Judith S. Bond Department of Biochemistry and Molecular Biology, Pennsylvania State University College of Medicine, Hershey, PA, USA

Martin D. Bootman Life, Health and Chemical Sciences, The Open University Walton Hall, Milton Keynes, UK

Bhargavi M. Boruah CAS Key Laboratory of Pathogenic Microbiology and Immunology, Institute of Microbiology, Chinese Academy of Sciences, Beijing, China

Graduate University of Chinese Academy of Science, Beijing, China

Sheryl R. Bowley Division of Hemostasis and Thrombosis, Beth Israel Deaconess Medical Center, Harvard Medical School, Boston, MA, USA

Doreen Braun Institute for Experimental Endocrinology, Charité-Universitätsmedizin Berlin, Berlin, Germany

Davorka Breljak Unit of Molecular Toxicology, Institute for Medical Research and Occupational Health, Zagreb, Croatia

Leonid Breydo Department of Molecular Medicine, Morsani College of Medicine, University of South Florida, Tampa, FL, USA

Mickael Briens UPR ARN du CNRS, Université de Strasbourg, Institut de Biologie Moléculaire et Cellulaire, Strasbourg, France

Joan B. Broderick Department of Chemistry and Biochemistry, Montana State University, Bozeman, MT, USA

James E. Bruce Department of Genome Sciences, University of Washington, Seattle, WA, USA

Ernesto Brunet Dept. Química Orgánica, Facultad de Ciencias, Universidad Autónoma de Madrid, Madrid, Spain

Maurizio Brunori Department of Biochemical Sciences, Sapienza University of Rome, Rome, Italy

Susan K. Buchanan Laboratory of Molecular Biology, National Institute of Diabetes and Digestive and Kidney Diseases, US National Institutes of Health, Bethesda, MD, USA

Gabriel E. Büchel Institute of Inorganic Chemistry, University of Vienna, Vienna, Austria

Živadin D. Bugarčić Faculty of Science, Department of Chemistry, University of Kragujevac, Kragujevac, Serbia

Melisa Bunderson-Schelvan Department of Biomedical and Pharmaceutical Sciences, Center for Environmental Health, The University of Montana, Missoula, MT, USA

Jean-Claude G. Bünzli Center for Next Generation Photovoltaic Systems, Korea University, Sejong Campus, Jochiwon–eup, Yeongi–gun, ChungNam–do, Republic of Korea

École Polytechnique Fédérale de Lausanne, Institute of Chemical Sciences and Engineering, Lausanne, Switzerland

John E. Burke Medical Research Council, Laboratory of Molecular Biology, Cambridge, UK

Torsten Burkholz Division of Bioorganic Chemistry, School of Pharmacy, Saarland State University, Saarbruecken, Germany

Bruce S. Burnham Department of Chemistry, Biochemistry, and Physics, Rider University, Lawrenceville, NJ, USA

Ashley I. Bush The Mental Health Research Institute, The University of Melbourne, Parkville, VIC, Australia

Kunzheng Cai Institute of Tropical and Subtropical Ecology, South China Agricultural University, Guangzhou, China

Iván L. Calderón Laboratorio de Microbiología Molecular, Universidad Andrés Bello, Santiago, Chile

Glaucia Callera Kidney Research Center, Ottawa Hospital Research Institute, University of Ottawa, Ottawa, ON, Canada

Marcello Campagna Department of Public Health, Clinical and Molecular Medicine, University of Cagliari, Cagliari, Italy

Mercè Capdevila Departament de Química, Facultat de Ciències, Universitat Autònoma de Barcelona, Cerdanyola del Vallés (Barcelona), Spain

Fernando Cardozo-Pelaez Department of Pharmaceutical Sciences, Center for Environmental Health Sciences, University of Montana, Missoula, MT, USA

Bradley A. Carlson Molecular Biology of Selenium Section, Laboratory of Cancer Prevention, National Cancer Institute, National Institutes of Health, Bethesda, MD, USA

Silvia Castelli Department of Biology, University of Rome Tor Vergata, Rome, Italy

Tommy Cedervall Department of Biochemistry and Structural Biology, Lund University, Chemical Centre, Lund, Sweden

Sudipta Chakraborty Department of Pediatrics and Department of Pharmacology, and the Kennedy Center for Research on Human Development, Vanderbilt University Medical Center, Nashville, TN, USA

Henry Chan Department of Molecular Biology, Division of Biological Sciences, University of California at San Diego, La Jolla, CA, USA

N. Chandrasekaran Centre for Nanobiotechnology, VIT University, Vellore, Tamil Nadu, India

Loïc J. Charbonnière Laboratoire d'Ingénierie Moléculaire Appliquée à l'Analyse, IPHC, UMR 7178 CNRS/UdS ECPM, Strasbourg, France

Malay Chatterjee Division of Biochemistry, Department of Pharmaceutical Technology, Jadavpur University, Kolkata, West Bengal, India

Mary Chatterjee Division of Biochemistry, Department of Pharmaceutical Technology, Jadavpur University, Kolkata, West Bengal, India

François Chaumont Institut des Sciences de la Vie, Universite catholique de Louvain, Louvain-la-Neuve, Belgium

Juan D. Chavez Department of Genome Sciences, University of Washington, Seattle, WA, USA

Chi-Ming Che Department of Chemistry, State Key Laboratory of Synthetic Chemistry and Open Laboratory of Chemical Biology of the Institute of Molecular Technology for Drug Discovery and Synthesis, The University of Hong Kong, Hong Kong, China

Elena Chekmeneva Department of Chemistry, University of Sheffield, Sheffield, UK

Di Chen The Developmental Therapeutics Program, Barbara Ann Karmanos Cancer Institute, and Departments of Oncology, Pharmacology and Pathology, School of Medicine, Wayne State University, Detroit, MI, USA

Hong-Yuan Chen National Key Laboratory of Analytical Chemistry for Life Science, School of Chemistry and Chemical Engineering, Nanjing University, Nanjing, China

Jiugeng Chen Laboratory of Plant Physiology and Molecular Genetics, Université Libre de Bruxelles, Brussels, Belgium

Sai-Juan Chen State Key Laboratory of Medical Genomics, Shanghai Institute of Hematology, Rui Jin Hospital Affiliated to Shanghai Jiao Tong University School of Medicine, Shanghai, China

Zhu Chen State Key Laboratory of Medical Genomics, Shanghai Institute of Hematology, Rui Jin Hospital Affiliated to Shanghai Jiao Tong University School of Medicine, Shanghai, China

Robert A. Cherny The Mental Health Research Institute, The University of Melbourne, Parkville, VIC, Australia

Yana Chervona Department of Environmental Medicine, New York University Medical School, New York, NY, USA

Christopher R. Chitambar Division of Hematology and Oncology, Medical College of Wisconsin, Froedtert and Medical College of Wisconsin Clinical Cancer Center, Milwaukee, WI, USA

Hassanul Ghani Choudhury Division of Molecular Biosciences, Imperial College London, London, South Kensington, UK

Membrane Protein Lab, Diamond Light Source, Harwell Science and Innovation Campus, Chilton, Oxfordshire, UK

Research Complex at Harwell, Harwell Oxford, Didcot, Oxforsdhire, UK

Samrat Roy Chowdhury Department of Pharmaceutical Technology, Jadavpur University, Kolkata, West Bengal, India

Stefano Ciurli Department of Agro-Environmental Science and Technology, University of Bologna, Bologna, Italy

Stephan Clemens Department of Plant Physiology, University of Bayreuth, Bayreuth, Germany

Nansi Jo Colley Department of Ophthalmology and Visual Sciences, UW Eye Research Institute, University of Wisconsin, Madison, WI, USA

Gianni Colotti Institute of Molecular Biology and Pathology, Consiglio Nazionale delle Ricerche, Rome, Italy

Giovanni Corsetti Division of Human Anatomy, Department of Biomedical Sciences and Biotechnologies, Brescia University, Brescia, Italy

Max Costa Department of Environmental Medicine, New York University Medical School, New York, NY, USA

Jos A. Cox Department of Biochemistry, University of Geneva, Geneva, Switzerland

Adam V. Crain Department of Chemistry and Biochemistry, Montana State University, Bozeman, MT, USA

Ann Cuypers Centre for Environmental Sciences, Hasselt University, Diepenbeek, Belgium

Martha S. Cyert Department of Biology, Stanford University, Stanford, USA

Sabato D'Auria National Research Council (CNR), Laboratory for Molecular Sensing, Institute of Protein Biochemistry, Naples, Italy

Verónica Daier Departamento de Química Física/IQUIR-CONICET, Facultad de Ciencias Bioquímicas y Farmacéuticas, Universidad Nacional de Rosario, Rosario, Argentina

Charles T. Dameron Chemistry Department, Saint Francis University, Loretto, PA, USA

Subhadeep Das Division of Biochemistry, Department of Pharmaceutical Technology, Jadavpur University, Kolkata, West Bengal, India

Nilay Kanti Das Department of Dermatology, Medical College, Kolkata, West Bengal, India

Rupali Datta Department of Biological Sciences, Michigan Technological University, Houghton, MI, USA

Benjamin G. Davis Chemistry Research Laboratory, Department of Chemistry, University of Oxford, Oxford, UK

Dennis R. Dean Department of Biochemistry, Virginia Tech University, Blacksburg, VA, USA

Kannan Deepa Biochemical Engineering Laboratory, Department of Chemical Engineering, Indian Institute of Technology Madras, Chennai, Tamil Nadu, India

Claudia Della Corte Unit of Liver Research of Bambino Gesù Children's Hospital, IRCCS, Rome, Italy

Simone Dell'Acqua REQUIMTE/CQFB, Departamento de Química, Faculdade de Ciências e Tecnologia, Universidade Nova de Lisboa, Caparica, Portugal

Dipartimento di Chimica, Università di Pavia, Pavia, Italy

Hakan Demir Department of Nuclear Medicine, School of Medicine, Kocaeli University, Umuttepe, Kocaeli, Turkey

Sumukh Deshpande Institute for Biocomplexity and Informatics, Department of Biological Sciences, University of Calgary, Calgary, AB, Canada

Alessandro Desideri Department of Biology, University of Rome Tor Vergata, Rome, Italy

Interuniversity Consortium, National Institute Biostructure and Biosystem (INBB), Rome, Italy

Patrick C. D'Haese Laboratory of Pathophysiology, University of Antwerp, Wilrijk, Belgium

José Manuel Díaz-Cruz Department of Analytical Chemistry, University of Barcelona, Barcelona, Spain

Saad A. Dibo Division Plastic and Reconstructive Surgery, American University of Beirut Medical Center, Beirut, Lebanon

Pavel Dibrov Department of Microbiology, University of Manitoba, Winnipeg, MB, Canada

Adeleh Divsalar Department of Biological Sciences, Tarbiat Moallem University, Tehran, Iran

Ligia J. Dominguez Geriatric Unit, Department of Internal Medicine and Medical Specialties (DIMIS), University of Palermo, Palermo, Italy

Delfina C. Domínguez College of Health Sciences, The University of Texas at El Paso, El Paso, TX, USA

Rosario Donato Department of Experimental Medicine and Biochemical Sciences, University of Perugia, Perugia, Italy

Elke Dopp Institute of Hygiene and Occupational Medicine, University of Duisburg-Essen, Essen, Germany

Melania D'Orazio Dipartimento di Biologia, Università di Roma Tor Vergata, Rome, Italy

Q. Ping Dou The Developmental Therapeutics Program, Barbara Ann Karmanos Cancer Institute, and Departments of Oncology, Pharmacology and Pathology, School of Medicine, Wayne State University, Detroit, MI, USA

Ross G. Douglas Zinc Metalloprotease Research Group, Division of Medical Biochemistry, Institute of Infectious Disease and Molecular Medicine, University of Cape Town, Cape Town, South Africa

Annette Draeger Department of Cell Biology, Institute of Anatomy, University of Bern, Bern, Switzerland

Gabi Drochioiu Alexandru Ioan Cuza University of Iasi, Iasi, Romania

Elzbieta Dudek Department of Biochemistry, University of Alberta, Edmonton, AB, Canada

Todor Dudev Institute of Biomedical Sciences, Academia Sinica, Taipei, Taiwan

Henry J. Duff Libin Cardiovascular Institute of Alberta, Calgary, AB, Canada

Evert C. Duin Department of Chemistry and Biochemistry, Auburn University, Auburn, AL, USA

R. Scott Duncan Vision Research Center and Departments of Basic Medical Science and Ophthalmology, School of Medicine, University of Missouri, Kansas City, MO, USA

Michael F. Dunn Department of Biochemistry, University of California at Riverside, Riverside, CA, USA

Serdar Durdagi Institute for Biocomplexity and Informatics, Department of Biological Sciences, University of Calgary, Calgary, AB, Canada

Kaitlin S. Duschene Department of Chemistry and Biochemistry, Montana State University, Bozeman, MT, USA

Ankit K. Dutta School of Molecular and Biomedical Science, University of Adelaide, Adelaide, South Australia, Australia

Naba K. Dutta Ian Wark Research Institute, University of South Australia, Mawson Lakes, South Australia, Australia

Paul J. Dyson Institut des Sciences et Ingénierie Chimiques, Ecole Polytechnique Fédérale de Lausanne (EPFL) SB ISIC-Direction, Lausanne, Switzerland

Brian E. Eckenroth Department of Microbiology and Molecular Genetics, University of Vermont, Burlington, VT, USA

Niels Eckstein Federal Institute for Drugs and Medical Devices (BfArM), Bonn, Germany

David J. Eide Department of Nutritional Sciences, University of Wisconsin-Madison, Madison, WI, USA

Thomas Eitinger Institut für Biologie/Mikrobiologie, Humboldt-Universität zu Berlin, Berlin, Germany

Annette Ekblond Cardiology Stem Cell Laboratory, Rigshospitalet University Hospital, Copenhagen, Denmark

Jean-Michel El Hage Chahine ITODYS, Université Paris-Diderot Sorbonne Paris Cité, CNRS UMR 7086, Paris, France

Alex Elías Laboratorio de Microbiología Molecular, Departamento de Biología, Universidad de Santiago de Chile, Santiago, Chile

Jeffrey S. Elmendorf Department of Cellular and Integrative Physiology and Department of Biochemistry and Molecular Biology, and Centers for Diabetes Research, Membrane Biosciences, and Vascular Biology and Medicine, Indiana University School of Medicine, Indianapolis, IN, USA

Sanaz Emami Department of Biophysics, Institute of Biochemistry and Biophysics (IBB), University of Tehran, Tehran, Iran

Vinita Ernest Centre for Nanobiotechnology, VIT University, Vellore, Tamil Nadu, India

Miquel Esteban Department of Analytical Chemistry, University of Barcelona, Barcelona, Spain

Christopher Exley The Birchall Centre, Lennard-Jones Laboratories, Keele University, Staffordshire, UK

Chunhai Fan Laboratory of Physical Biology, Shanghai Institute of Applied Physics, Shanghai, China

Marcelo Farina Departamento de Bioquímica, Centro de Ciências Biológicas, Universidade Federal de Santa Catarina, Florianópolis, SC, Brazil

Nicholas P. Farrell Department of Chemistry, Virginia Commonwealth University, Richmond, VA, USA

Caroline Fauquant iRTSV/LCBM UMR 5249 CEA-CNRS-UJF, CEA/Grenoble, Bât K, Université Grenoble, Grenoble, France

James G. Ferry Department of Biochemistry and Molecular Biology, Eberly College of Science, The Pennsylvania State University, University Park, PA, USA

Ana Maria Figueiredo Instituto de Pesquisas Energeticas e Nucleares, IPEN-CNEN/SP, Sao Paulo, Brazil

David I. Finkelstein The Mental Health Research Institute, The University of Melbourne, Parkville, VIC, Australia

Larry Fliegel Department of Biochemistry, University of Alberta, Edmonton, AB, Canada

Swaran J. S. Flora Division of Pharmacology and Toxicology, Defence Research and Development Establishment, Gwalior, India

Juan C. Fontecilla-Camps Metalloproteins; Institut de Biologie Structurale J.P. Ebel; CEA; CNRS; Université J. Fourier, Grenoble, France

Sara M. Fox Department of Pharmacology and Toxicology, Michigan State University, East Lansing, MI, USA

Ricardo Franco REQUIMTE FCT/UNL, Departamento de Química, Faculdade de Ciências e Tecnologia, Universidade Nova de Lisboa, Caparica, Portugal

Stefan Fränzle Department of Biological and Environmental Sciences, Research Group of Environmental Chemistry, International Graduate School Zittau, Zittau, Germany

Christopher J. Frederickson NeuroBioTex, Inc, Galveston Island, TX, USA

Michael Frezza The Developmental Therapeutics Program, Barbara Ann Karmanos Cancer Institute, and Departments of Oncology, Pharmacology and Pathology, School of Medicine, Wayne State University, Detroit, MI, USA

Barbara C. Furie Division of Hemostasis and Thrombosis, Beth Israel Deaconess Medical Center, Harvard Medical School, Boston, MA, USA

Bruce Furie Division of Hemostasis and Thrombosis, Beth Israel Deaconess Medical Center, Harvard Medical School, Boston, MA, USA

Roland Gaertner Department of Endocrinology, University Hospital, Ludwig-Maximilians University Munich, Munich, Germany

Sonia Galván-Arzate Departamento de Neuroquímica, Instituto Nacional de Neurología y Neurocirugía Manuel Velasco Suárez, Mexico City, DF, Mexico

Livia Garavelli Struttura Semplice Dipartimentale di Genetica Clinica, Dipartimento di Ostetrico-Ginecologico e Pediatrico, Istituto di Ricovero e Cura a Carattere Scientifico, Arcispedale S. Maria Nuova, Reggio Emilia, Italy

Carolyn L. Geczy Inflammation and Infection Research Centre, School of Medical Sciences, University of New South Wales, Sydney, NSW, Australia

Emily Geiger Department of Biological Sciences, Michigan Technological University, Houghton, MI, USA

Alayna M. George Thompson Department of Chemistry and Biochemistry, University of Arizona, Tucson, AZ, USA

Charles P. Gerba Department of Soil, Water and Environmental Science, University of Arizona, Tucson, AZ, USA

Miltu Kumar Ghosh Department of Pharmaceutical Technology, Jadavpur University, Kolkata, West Bengal, India

Pramit Ghosh Department of Community Medicine, Medical College, Kolkata, West Bengal, India

Saikat Ghosh Department of Pharmaceutical Technology, Jadavpur University, Kolkata, West Bengal, India

Hedayatollah Ghourchian Department of Biophysics, Institute of Biochemistry and Biophysics (IBB), University of Tehran, Tehran, Iran

Jessica L. Gifford Department of Biological Sciences, Biochemistry Research Group, University of Calgary, Calgary, AB, Canada

Danuta M. Gillner Department of Chemistry, Silesian University of Technology, Gliwice, Poland

Mario Di Gioacchino Occupational Medicine and Allergy, Head of Allergy and Immunotoxicology Unit (Ce.S.I.), G. d'Annunzio University, Via dei Vestini, Chieti, Italy

Denis Girard Laboratoire de recherche en inflammation et physiologie des granulocytes, Université du Québec, INRS-Institut Armand-Frappier, Laval, QC, Canada

F. Xavier Gomis-Rüth Proteolysis Lab, Department of Structural Biology, Molecular Biology Institute of Barcelona, CSIC Barcelona Science Park, Barcelona, Spain

Harry B. Gray Beckman Institute, California Institute of Technology, Pasadena, CA, USA

Claudia Großkopf Department Chemicals Safety, Federal Institute for Risk Assessment, Berlin, Germany

Thomas E. Gunter Department of Biochemistry and Biophysics, University of Rochester School of Medicine and Dentistry, Rochester, NY, USA

Dharmendra K. Gupta Departamento de Bioquímica, Biología Celular y Molecular de Plantas, Estación Experimental del Zaidin, CSIC, Granada, Spain

Nikolai B. Gusev Department of Biochemistry, School of Biology, Moscow State University, Moscow, Russian Federation

Mandana Haack-Sørensen Cardiology Stem Cell Laboratory, Rigshospitalet University Hospital, Copenhagen, Denmark

Bodo Haas Federal Institute for Drugs and Medical Devices (BfArM), Bonn, Germany

Hajo Haase Institute of Immunology, Medical Faculty, RWTH Aachen University, Aachen, Germany

Fathi Habashi Department of Mining, Metallurgical, and Materials Engineering, Laval University, Quebec City, Canada

Alice Haddy Department of Chemistry and Biochemistry, University of North Carolina, Greensboro, NC, USA

Nguyêt-Thanh Ha-Duong ITODYS, Université Paris-Diderot Sorbonne Paris Cité, CNRS UMR 7086, Paris, France

Jesper Z. Haeggström Department of Medical Biochemistry and Biophysics (MBB), Karolinska Institute, Stockholm, Sweden

James F. Hainfeld Nanoprobes, Incorporated, Yaphank, NY, USA

Sefali Halder Department of Pharmaceutical Technology, Jadavpur University, Kolkata, West Bengal, India

Boyd E. Haley Department of Chemistry, University of Kentucky, Lexington, KY, USA

Raymond F. Hamilton Jr. Department of Biomedical and Pharmaceutical Sciences, Center for Environmental Health, The University of Montana, Missoula, MT, USA

Heidi E. Hannon Department of Pharmacology and Toxicology, Michigan State University, East Lansing, MI, USA

Timothy P. Hanusa Department of Chemistry, Vanderbilt University, Nashville, TN, USA

Edward D. Harris Department of Nutrition and Food Science, Texas A&M University, College Station, TX, USA

Todd C. Harrop Department of Chemistry, University of Georgia, Athens, GA, USA

Andrea Hartwig Department Food Chemistry and Toxicology, Karlsruhe Institute of Technology, Karlsruhe, Germany

Robert P. Hausinger Department of Biochemistry and Molecular Biology, Michigan State University, East Lansing, MI, USA

Department of Microbiology and Molecular Genetics, 6193 Biomedical and Physical Sciences, Michigan State University, East Lansing, MI, USA

Hiroaki Hayashi Department of Dermatology, Kawasaki Medical School, Kurashiki, Japan

Xiao He CAS Key Laboratory for Biomedical Effects of Nanomaterials and Nanosafety & CAS Key Laboratory of Nuclear Analytical Techniques, Institute of High Energy Physics, Chinese Academy of Sciences, Beijing, China

Yao He Institute of Functional Nano & Soft Materials, Soochow University, Jiangsu, China

Kim L. Hein Centre for Molecular Medicine Norway (NCMM), University of Oslo Nordic EMBL Partnership, Oslo, Norway

Claus W. Heizmann Department of Pediatrics, Division of Clinical Chemistry, University of Zurich, Zurich, Switzerland

Michael T. Henzl Department of Biochemistry, University of Missouri, Columbia, MO, USA

Carol M. Herak-Kramberger Unit of Molecular Toxicology, Institute for Medical Research and Occupational Health, Zagreb, Croatia

Christian Hermans Laboratory of Plant Physiology and Molecular Genetics, Université Libre de Bruxelles, Brussels, Belgium

Griselda Hernández New York State Department of Health, Wadsworth Center, Albany, NY, USA

Akon Higuchi Department of Chemical and Materials Engineering, National Central University, Jhongli, Taoyuan, Taiwan

Department of Reproduction, National Research Institute for Child Health and Development, Setagaya–ku, Tokyo, Japan

Cathay Medical Research Institute, Cathay General Hospital, Hsi–Chi City, Taipei, Taiwan

Russ Hille Department of Biochemistry, University of California, Riverside, CA, USA

Alia V. H. Hinz Department of Chemistry, Western Michigan University, Kalamazoo, MI, USA

John Andrew Hitron Graduate Center for Toxicology, University of Kentucky, Lexington, KY, USA

Miryana Hémadi ITODYS, Université Paris-Diderot Sorbonne Paris Cité, CNRS UMR 7086, Paris, France

Christer Hogstrand Metal Metabolism Group, Diabetes and Nutritional Sciences Division, School of Medicine, King's College London, London, UK

Erhard Hohenester Department of Life Sciences, Imperial College London, London, UK

Andrij Holian Department of Biomedical and Pharmaceutical Sciences, Center for Environmental Health, The University of Montana, Missoula, MT, USA

Richard C. Holz Department of Chemistry and Biochemistry, Loyola University Chicago, Chicago, IL, USA

Charles G. Hoogstraten Department of Biochemistry and Molecular Biology, Michigan State University, East Lansing, MI, USA

Ying Hou Key Laboratory for Biomechanics and Mechanobiology of the Ministry of Education, School of Biological Science and Medical Engineering, Beihang University, Beijing, China

Mingdong Huang Division of Hemostasis and Thrombosis, Beth Israel Deaconess Medical Center, Harvard Medical School, Boston, MA, USA

David L. Huffman Department of Chemistry, Western Michigan University, Kalamazoo, MI, USA

Paco Hulpiau Department for Molecular Biomedical Research, VIB, Ghent, Belgium

Amir Ibrahim Plastic and Reconstructive SurgeryBurn Fellow, Massachusetts General Hospital / Harvard Medical School & Shriners Burn Hospital, Boston, USA

Mitsu Ikura Ontario Cancer Institute and Department of Medical Biophysics, University of Toronto, Toronto, Ontario, Canada

Andrea Ilari Institute of Molecular Biology and Pathology, Consiglio Nazionale delle Ricerche, Rome, Italy

Giuseppe Inesi California Pacific Medical Center Research Institute, San Francisco, CA, USA

Hiroaki Ishida Department of Biological Sciences, Biochemistry Research Group, University of Calgary, Calgary, AB, Canada

Vangronsveld Jaco Centre for Environmental Sciences, Hasselt University, Diepenbeek, Belgium

Claus Jacob Division of Bioorganic Chemistry, School of Pharmacy, Saarland State University, Saarbruecken, Germany

Sushil K. Jain Department of Pediatrics, Louisiana State University Health Sciences Center, Shreveport, LA, USA

Peter Jensen Department of Dermato-Allergology, Copenhagen University Hospital Gentofte, Hellerup, Denmark

Klaudia Jomova Department of Chemistry, Faculty of Natural Sciences, Constantine The Philosopher University, Nitra, Slovakia

Raghava Rao Jonnalagadda Chemical Laboratory, Central Leather Research Institute (Council of Scientific and Industrial Research), Chennai, Tamil Nadu, India

Hans Jörnvall Department of Medical Biochemistry and Biophysics, Karolinska Institutet, Stockholm, Sweden

Olga Juanes Dept. Química Orgánica, Facultad de Ciencias, Universidad Autónoma de Madrid, Madrid, Spain

Sreeram Kalarical Janardhanan Chemical Laboratory, Central Leather Research Institute (Council of Scientific and Industrial Research), Chennai, Tamil Nadu, India

Paul C. J. Kamer School of Chemistry, University of St Andrews, St Andrews, UK

Tina Kamčeva Laboratory of Physical Chemistry, Vinča Institute of Nuclear Sciences, University of Belgrade, Belgrade, Serbia

Laboratory of Clinical Biochemistry, Section of Clinical Pharmacology, Haukeland University Hospital, Bergen, Norway

ChulHee Kang Washington State University, Pullman, WA, USA

Kazimierz S. Kasprzak Chemical Biology Laboratory, Frederick National Laboratory for Cancer Research, Frederick, MD, USA

Jane Kasten-Jolly New York State Department of Health, Wadsworth Center, Albany, NY, USA

Jens Kastrup Cardiology Stem Cell Laboratory, Rigshospitalet University Hospital, Copenhagen, Denmark

The Heart Centre, Cardiac Catheterization Laboratory, Rigshospitalet University Hospital, Copenhagen, Denmark

Prafulla Katkar Department of Biology, University of Rome Tor Vergata, Rome, Italy

Fusako Kawai Center for Nanomaterials and Devices, Kyoto Institute of Technology, Kyoto, Japan

Jason D. Kenealey Department of Biomolecular Chemistry, University of Wisconsin, Madison, WI, USA

Bernhard K. Keppler Institute of Inorganic Chemistry, University of Vienna, Vienna, Austria

E. Van Kerkhove Department of Physiology, Centre for Environmental Sciences, Hasselt University, Diepenbeek, Belgium

Kazuya Kikuchi Division of Advanced Science and Biotechnology, Graduate School of Engineering, Osaka University, Suita, Osaka, Japan

Immunology Frontier Research Center, Osaka University, Suita, Osaka, Japan

Michael Kirberger Department of Chemistry, Georgia State University, Atlanta, GA, USA

Masanori Kitamura Department of Molecular Signaling, Interdisciplinary Graduate School of Medicine and Engineering, University of Yamanashi, Chuo, Yamanashi, Japan

Rene Kizek Department of Chemistry and Biochemistry, Faculty of Agronomy, Mendel University in Brno, Brno, Czech Republic

Central European Institute of Technology, Brno University of Technology, Brno, Czech Republic

Nanne Kleefstra Diabetes Centre, Isala clinics, Zwolle, The Netherlands

Department of Internal Medicine, University Medical Center Groningen, Groningen, The Netherlands

Langerhans Medical Research Group, Zwolle, The Netherlands

Judith Klinman Departments of Chemistry and Molecular and Cell Biology, California Institute for Quantitative Biosciences, University of California, Berkeley, Berkeley, CA, USA

Michihiko Kobayashi Graduate School of Life and Environmental Sciences, Institute of Applied Biochemistry, The University of Tsukuba, Tsukuba, Ibaraki, Japan

Ahmet Koc Department of Molecular Biology and Genetics, Izmir Institute of Technology, Urla, İzmir, Turkey

Sergey M. Korotkov Sechenov Institute of Evolutionary Physiology and Biochemistry, The Russian Academy of Sciences, St. Petersburg, Russia

Peter Koulen Vision Research Center and Departments of Basic Medical Science and Ophthalmology, School of Medicine, University of Missouri, Kansas City, MO, USA

Nancy F. Krebs Department of Pediatrics, Section of Nutrition, University of Colorado, School of Medicine, Aurora, CO, USA

Zbigniew Krejpcio Division of Food Toxicology and Hygiene, Department of Human Nutrition and Hygiene, The Poznan University of Life Sciences, Poznan, Poland

The College of Health, Beauty and Education in Poznan, Poznan, Poland

Robert H. Kretsinger Department of Biology, University of Virginia, Charlottesville, VA, USA

Artur Krężel Department of Protein Engineering, Faculty of Biotechnology, University of Wrocław, Wrocław, Poland

Aleksandra Krivograd Klemenčič Faculty of Health Sciences, University of Ljubljana, Ljubljana, Slovenia

Peter M. H. Kroneck Department of Biology, University of Konstanz, Konstanz, Germany

Eugene Kryachko Bogolyubov Institute for Theoretical Physics, Kiev, Ukraine

Naoko Kumagai-Takei Department of Hygiene, Kawasaki Medical School, Okayama, Japan

Anil Kumar CAS Key Laboratory for Biomedical Effects of Nanoparticles and Nanosafety, National Center for Nanoscience and Nanotechnology, Chinese Academy of Sciences, Beijing, China

Graduate University of Chinese Academy of Science, Beijing, China

Thirumananseri Kumarevel RIKEN SPring-8 Center, Harima Institute, Hyogo, Japan

Valery V. Kupriyanov Institute for Biodiagnostics, National Research Council, Winnipeg, MB, Canada

Wouter Laan School of Chemistry, University of St Andrews, St Andrews, UK

James C. K. Lai Department of Biomedical & Pharmaceutical Sciences, College of Pharmacy and Biomedical Research Institute, Idaho State University, Pocatello, ID, USA

Maria José Laires CIPER – Interdisciplinary Centre for the Study of Human Performance, Faculty of Human Kinetics, Technical University of Lisbon, Cruz Quebrada, Portugal

Kyle M. Lancaster Department of Chemistry and Chemical Biology, Cornell University, Ithaca, NY, USA

Daniel Landau Department of Pediatrics, Soroka University Medical Centre, Ben-Gurion University of the Negev, Beer Sheva, Israel

Albert Lang Department of Molecular and Cell Biology, California Institute for Quantitative Biosciences, University of California, Berkeley, Berkeley, CA, USA

Alan B. G. Lansdown Faculty of Medicine, Imperial College, London, UK

Jean-Yves Lapointe Groupe d'étude des protéines membranaires (GÉPROM) and Département de Physique, Université de Montréal, Montréal, QC, Canada

Agnete Larsen Department of Biomedicine/Pharmacology Health, Aarhus University, Aarhus, Denmark

Lawrence H. Lash Department of Pharmacology, Wayne State University School of Medicine, Detroit, MI, USA

David A. Lawrence Department of Biomedical Sciences, School of Public Health, State University of New York, Albany, NY, USA

Laboratory of Clinical and Experimental Endocrinology and Immunology, Wadsworth Center, Albany, NY, USA

Peter A. Lay School of Chemistry, University of Sydney, Sydney, NSW, Australia

Gabriela Ledesma Departamento de Química Física/IQUIR-CONICET, Facultad de Ciencias Bioquímicas y Farmacéuticas, Universidad Nacional de Rosario, Rosario, Argentina

John Lee Department of Biochemistry and Molecular Biology, University of Georgia, Athens, GA, USA

Suni Lee Department of Hygiene, Kawasaki Medical School, Okayama, Japan

Silke Leimkühler From the Institute of Biochemistry and Biology, Department of Molecular Enzymology, University of Potsdam, Potsdam, Germany

Herman Louis Lelie Department of Chemistry and Biochemistry, University of California, Los Angeles, CA, USA

David M. LeMaster New York State Department of Health, Wadsworth Center, Albany, NY, USA

Joseph Lemire Department of Chemistry and Biochemistry, Laurentian University, Sudbury, ON, Canada

Thomas A. Leonard Max F. Perutz Laboratories, Vienna, Austria

Alain Lescure UPR ARN du CNRS, Université de Strasbourg, Institut de Biologie Moléculaire et Cellulaire, Strasbourg, France

Solomon W. Leung Department of Civil & Environmental Engineering, School of Engineering, College of Science and Engineering and Biomedical Research Institute, Idaho State University, Pocatello, ID, USA

Bogdan Lev Institute for Biocomplexity and Informatics, Department of Biological Sciences, University of Calgary, Calgary, AB, Canada

Aviva Levina School of Chemistry, University of Sydney, Sydney, NSW, Australia

Huihui Li School of Chemistry and Material Science, Nanjing Normal University, Nanjing, China

Yang V. Li Department of Biomedical Sciences, Heritage College of Osteopathic Medicine, Ohio University, Athens, OH, USA

Xing-Jie Liang CAS Key Laboratory for Biomedical Effects of Nanoparticles and Nanosafety, National Center for Nanoscience and Nanotechnology, Chinese Academy of Sciences, Beijing, China

Patrycja Libako Faculty of Veterinary Medicine, Wroclaw University of Environmental and Life Sciences, Wrocław, Poland

Carmay Lim Institute of Biomedical Sciences, Academia Sinica, Taipei, Taiwan

Department of Chemistry, National Tsing Hua University, Hsinchu, Taiwan

Sara Linse Department of Biochemistry and Structural Biology, Lund University, Chemical Centre, Lund, Sweden

John D. Lipscomb Department of Biochemistry, Molecular Biology, and Biophysics, University of Minnesota, Minneapolis, MN, USA

Junqiu Liu State Key Laboratory of Supramolecular Structure and Materials, College of Chemistry, Jilin University, Changchun, China

Qiong Liu College of Life Sciences, Shenzhen University, Shenzhen, P. R. China

Zijuan Liu Department of Biological Sciences, Oakland University, Rochester, MI, USA

Marija Ljubojević Unit of Molecular Toxicology, Institute for Medical Research and Occupational Health, Zagreb, Croatia

Mario Lo Bello Department of Biology, University of Rome "Tor Vergata", Rome, Italy

Yan-Chung Lo The Genomics Research Center, Academia Sinica, Taipei, Taiwan

Institute of Biological Chemistry, Academia Sinica, Taipei, Taiwan

Lingli Lu MOE Key Laboratory of Environment Remediation and Ecological Health, College of Environmental & Resource Science, Zhejiang University, Hangzhou, China

Roberto G. Lucchini Department of Experimental and Applied Medicine, Section of Occupational Health and Industrial Hygiene, University of Brescia, Brescia, Italy

Department of Preventive Medicine, Mount Sinai School of Medicine, New York, USA

Bernd Ludwig Institute of Biochemistry, Goethe University, Frankfurt, Germany

Quan Luo State Key Laboratory of Supramolecular Structure and Materials, College of Chemistry, Jilin University, Changchun, China

Jennene A. Lyda Department of Pharmaceutical Sciences, Center for Environmental Health Sciences, University of Montana, Missoula, MT, USA

Charilaos Lygidakis Regional Health Service of Emilia Romagna, AUSL of Bologna, Bologna, Italy

Jiawei Ma Key Laboratory for Biomechanics and Mechanobiology of the Ministry of Education, School of Biological Science and Medical Engineering, Beihang University, Beijing, China

Jian Feng Ma Plant Stress Physiology Group, Institute of Plant Science and Resources, Okayama University, Kurashiki, Japan

Megumi Maeda Department of Biofunctional Chemistry, Division of Bioscience, Okayama University Graduate School of Natural Science and Technology, Okayama, Japan

Axel Magalon Laboratoire de Chimie Bactérienne (UPR9043), Institut de Microbiologie de la Méditerranée, CNRS & Aix-Marseille Université, Marseille, France

Jeanette A. Maier Department of Biomedical and Clinical Sciences L. Sacco, Università di Milano, Medical School, Milano, Italy

Robert J. Maier Department of Microbiology, The University of Georgia, Athens, GA, USA

Masatoshi Maki Department of Applied Molecular Biosciences, Graduate School of Bioagricultural Sciences, Nagoya University, Nagoya, Japan

R. Manasadeepa Department of Pharmaceutical Technology, Jadavpur University, Kolkata, West Bengal, India

David J. Mann Division of Molecular Biosciences, Department of Life Sciences, Imperial College London, South Kensington, London, UK

G. Marangi Istituto di Genetica Medica, Università Cattolica Sacro Cuore, Policlinico A. Gemelli, Rome, Italy

Wolfgang Maret Metal Metabolism Group, Diabetes and Nutritional Sciences Division, School of Medicine, King's College London, London, UK

Bernd Markert Environmental Institute of Scientific Networks, in Constitution, Haren/Erika, Germany

Michael J. Maroney Department of Chemistry, Lederle Graduate Research Center, University of Massachusetts at Amherst, Amherst, MA, USA

Brenda Marrero-Rosado Department of Pharmacology and Toxicology, Michigan State University, East Lansing, MI, USA

Christopher B. Marshall Ontario Cancer Institute and Department of Medical Biophysics, University of Toronto, Toronto, Ontario, Canada

Dwight W. Martin Department of Medicine and the Proteomics Center, Stony Brook University, Stony Brook, NY, USA

Ebany J. Martinez-Finley Department of Pediatrics, Division of Pediatric Clinical Pharmacology and Toxicology, Vanderbilt University Medical Center, Nashville, TN, USA

Center in Molecular Toxicology, Vanderbilt University Medical Center, Nashville, TN, USA

Jacqueline van Marwijk Department of Biochemistry, Microbiology & Biotechnology, Rhodes University, Grahamstown, Eastern Cape, South Africa

Pradip K. Mascharak Department of Chemistry and Biochemistry, University of California, Santa Cruz, CA, USA

Anne B. Mason Department of Biochemistry, University of Vermont, Burlington, VT, USA

Hidenori Matsuzaki Department of Hygiene, Kawasaki Medical School, Okayama, Japan

Jacqueline M. Matthews School of Molecular Bioscience, The University of Sydney, Sydney, Australia

Andrzej Mazur INRA, UMR 1019, UNH, CRNH Auvergne, Clermont Université, Université d'Auvergne, Unité de Nutrition Humaine, Clermont-Ferrand, France

Paulo Mazzafera Departamento de Biologia Vegetal, Universidade Estadual de Campinas/Instituto de Biologia, Cidade Universitária, Campinas, SP, Brazil

Michael M. Mbughuni Department of Biochemistry, Molecular Biology, and Biophysics, University of Minnesota, Minneapolis, MN, USA

Joseph R. McDermott Department of Biological Sciences, Oakland University, Rochester, MI, USA

Megan M. McEvoy Department of Chemistry and Biochemistry, University of Arizona, Tucson, AZ, USA

Astrid van der Meer Interfaculty Reactor Institute, Delft University of Technology, Delft, The Netherlands

Petr Melnikov Department of Clinical Surgery, School of Medicine, Federal University of Mato Grosso do Sul, Campo Grande, MS, Brazil

Gabriele Meloni Division of Chemistry and Chemical Engineering and Howard Hughes Medical Institute, California Institute of Technology, Pasadena, CA, USA

Ralf R. Mendel Department of Plant Biology, Braunschweig University of Technology, Braunschweig, Germany

Mohamed Larbi Merroun Departamento de Microbiología, Facultad de Ciencias, Universidad de Granada, Granada, Spain

Albrecht Messerschmidt Department of Proteomics and Signal Transduction, Max-Planck-Institute of Biochemistry, Martinsried, Germany

Marek Michalak Department of Biochemistry, University of Alberta, Edmonton, AB, Canada

Faculty of Medicine and Dentistry, University of Alberta, Edmonton, AB, Canada

Isabelle Michaud-Soret iRTSV/LCBM UMR 5249 CEA-CNRS-UJF, CEA/Grenoble, Bât K, Université Grenoble, Grenoble, France

Radmila Milačič Department of Environmental Sciences, Jožef Stefan Institute, Ljubljana, Slovenia

Glenn L. Millhauser Department of Chemistry and Biochemistry, University of California, Santa Cruz, Santa Cruz, CA, USA

Marija Milovanovic Faculty of Medical Sciences, University of Kragujevac, Centre for Molecular Medicine, Kragujevac, Serbia

Shin Mizukami Division of Advanced Science and Biotechnology, Graduate School of Engineering, Osaka University, Suita, Osaka, Japan

Immunology Frontier Research Center, Osaka University, Suita, Osaka, Japan

Cristina Paula Monteiro Physiology and Biochemistry Laboratory, Faculty of Human Kinetics, Technical University of Lisbon, Cruz Quebrada, Portugal

Augusto C. Montezano Kidney Research Center, Ottawa Hospital Research Institute, University of Ottawa, Ottawa, ON, Canada

Pablo Morales-Rico Department of Toxicology, Cinvestav-IPN, Mexico city, Mexico

J. Preben Morth Centre for Molecular Medicine Norway (NCMM), Nordic EMBL Partnership, University of Oslo, Oslo, Norway

Jean-Marc Moulis Institut de Recherches en Sciences et Technologies du Vivant, Laboratoire Chimie et Biologie des Métaux (IRTSV/LCBM), CEA–Grenoble, Grenoble, France

CNRS, UMR5249, Grenoble, France

Université Joseph Fourier–Grenoble I, UMR5249, Grenoble, France

Isabel Moura REQUIMTE/CQFB, Departamento de Química, Faculdade de Ciências e Tecnologia, Universidade Nova de Lisboa, Caparica, Portugal

José J. G. Moura REQUIMTE/CQFB, Departamento de Química, Faculdade de Ciências e Tecnologia, Universidade Nova de Lisboa, Caparica, Portugal

Mohamed E. Moustafa Department of Biochemistry, Faculty of Science, Alexandria University, Alexandria, Egypt

Amitava Mukherjee Centre for Nanobiotechnology, VIT University, Vellore, Tamil Nadu, India

Biswajit Mukherjee Department of Pharmaceutical Technology, Jadavpur University, Kolkata, West Bengal, India

Balam Muñoz Department of Toxicology, Cinvestav-IPN, Mexico city, Mexico

Francesco Musiani Department of Agro-Environmental Science and Technology, University of Bologna, Bologna, Italy

Joachim Mutter Naturheilkunde, Umweltmedizin Integrative and Environmental Medicine, Belegarzt Tagesklinik, Constance, Germany

Bonex W. Mwakikunga Council for Scientific and Industrial Research, National Centre for Nano–Structured, Pretoria, South Africa

Department of Physics and Biochemical Sciences, University of Malawi, The Malawi Polytechnic, Chichiri, Blantyre, Malawi

Chandra Shekar Nagar Venkataraman Condensed Matter Physics Division, Materials Science Group, Indira Gandhi Centre for Atomic Research, Kalpakkam, Tamil Nadu, India

Hideaki Nagase Kennedy Institute of Rheumatology, Nuffield Department of Orthopaedics, Rheumatology and Musculoskeletal Sciences, University of Oxford, London, United Kingdom

Sreejayan Nair University of Wyoming, School of Pharmacy, College of Health Sciences and the Center for Cardiovascular Research and Alternative Medicine, Laramie, WY, USA

Manuel F. Navedo Department of Physiology and Biophysics, University of Washington, Seattle, WA, USA

Tim S. Nawrot Centre for Environmental Sciences, Hasselt University, Diepenbeek, Belgium

Karel Nesmerak Department of Analytical Chemistry, Faculty of Science, Charles University in Prague, Prague, Czech Republic

Gerd Ulrich Nienhaus Institute of Applied Physics and Center for Functional Nanostructures (CFN), Karlsruhe Institute of Technology (KIT), Karlsruhe, Germany

Department of Physics, University of Illinois at Urbana–Champaign, Urbana, IL, USA

Crina M. Nimigean Department of Anesthesiology, Weill Cornell Medical College, New York, NY, USA

Department of Physiology and Biophysics, Weill Cornell Medical College, New York, NY, USA

Department of Biochemistry, Weill Cornell Medical College, New York, NY, USA

Yasumitsu Nishimura Department of Hygiene, Kawasaki Medical School, Okayama, Japan

Naoko K. Nishizawa Graduate School of Agricultural and Life Sciences, The University of Tokyo, Tokyo, Japan

Research Institute for Bioresources and Biotechnology, Ishikawa Prefectural University, Ishikawa, Japan

Valerio Nobili Unit of Liver Research of Bambino Gesù Children's Hospital, IRCCS, Rome, Italy

Nicholas Noinaj Laboratory of Molecular Biology, National Institute of Diabetes and Digestive and Kidney Diseases, US National Institutes of Health, Bethesda, MD, USA

Aline M. Nonat Laboratoire d'Ingénierie Moléculaire Appliquée à l'Analyse, IPHC, UMR 7178 CNRS/UdS ECPM, Strasbourg, France

Sergei Yu. Noskov Institute for Biocomplexity and Informatics, Department of Biological Sciences, University of Calgary, Calgary, AB, Canada

Wojciech Nowacki Faculty of Veterinary Medicine, Wroclaw University of Environmental and Life Sciences, Wrocław, Poland

David O'Connell University College Dublin, Conway Institute, Dublin, Ireland

Masafumi Odaka Department of Biotechnology and Life Science, Graduate School of Technology, Tokyo University of Agriculture and Technology, Koganei, Tokyo, Japan

Akira Ono Department of Material & Life Chemistry, Faculty of Engineering, Kanagawa University, Kanagawa-ku, Yokohama, Japan

Laura Osorio-Rico Departamento de Neuroquímica, Instituto Nacional de Neurología y Neurocirugía Manuel Velasco Suárez, Mexico City, DF, Mexico

Patricia Isabel Oteiza Departments of Nutrition and Environmental Toxicology, University of California, Davis, Davis, CA, USA

Takemi Otsuki Department of Hygiene, Kawasaki Medical School, Okayama, Japan

Rabbab Oun Strathclyde Institute of Pharmacy and Biomedical Sciences, University of Strathclyde, Glasgow, UK

Vidhu Pachauri Division of Pharmacology and Toxicology, Defence Research and Development Establishment, Gwalior, India

Òscar Palacios Departament de Química, Facultat de Ciències, Universitat Autònoma de Barcelona, Cerdanyola del Vallès (Barcelona), Spain

Maria E. Palm-Espling Department of Chemistry, Chemical Biological Center, Umeå University, Umeå, Sweden

Claudia Palopoli Departamento de Química Física/IQUIR-CONICET, Facultad de Ciencias Bioquímicas y Farmacéuticas, Universidad Nacional de Rosario, Rosario, Argentina

Tapobrata Panda Biochemical Engineering Laboratory, Department of Chemical Engineering, Indian Institute of Technology Madras, Chennai, Tamil Nadu, India

Lorien J. Parker Biota Structural Biology Laboratory, St. Vincent's Institute of Medical Research, Fitzroy, VIC, Australia

Department of Biochemistry and Molecular Biology, Bio21 Molecular Science and Biotechnology Institute, The University of Melbourne, Parkville, VIC, Australia

Michael W. Parker Biota Structural Biology Laboratory, St. Vincent's Institute of Medical Research, Fitzroy, VIC, Australia

Department of Biochemistry and Molecular Biology, Bio21 Molecular Science and Biotechnology Institute, The University of Melbourne, Parkville, VIC, Australia

Marianna Patrauchan Department of Microbiology and Molecular Genetics, College of Arts and Sciences, Oklahoma State University, Stillwater, OK, USA

Sofia R. Pauleta REQUIMTE/CQFB, Departamento de Química, Faculdade de Ciências e Tecnologia, Universidade Nova de Lisboa, Caparica, Portugal

Evgeny Pavlov Department of Physiology & Biophysics, Faculty of Medicine, Dalhousie University, Halifax, NS, Canada

V. Pennemans Biomedical Institute, Hasselt University, Diepenbeek, Belgium

Harmonie Perdreau Centre for Molecular Medicine Norway (NCMM), Nordic EMBL Partnership, University of Oslo, Oslo, Norway

Alice S. Pereira Departamento de Química, Faculdade de Ciências e Tecnologia, Requimte, Centro de Química Fina e Biotecnologia, Universidade Nova de Lisboa, Caparica, Portugal

Eulália Pereira REQUIMTE, Departamento de Química e Bioquímica, Faculdade de Ciências da Universidade do Porto, Porto, Portugal

Eugene A. Permyakov Institute for Biological Instrumentation, Russian Academy of Sciences, Pushchino, Moscow Region, Russia

Sergei E. Permyakov Protein Research Group, Institute for Biological Instrumentation of the Russian Academy of Sciences, Pushchino, Moscow Region, Russia

Bertil R. R. Persson Department of Medical Radiation Physics, Lund University, Lund, Sweden

John W. Peters Department of Chemistry and Biochemistry, Montana State University, Bozeman, MT, USA

Marijana Petković Laboratory of Physical Chemistry, Vinča Institute of Nuclear Sciences, University of Belgrade, Belgrade, Serbia

Le T. Phung Department of Microbiology and Immunology, University of Illinois, Chicago, IL, USA

Roberta Pierattelli CERM and Department of Chemistry "Ugo Schiff", University of Florence, Sesto Fiorentino, Italy

Elizabeth Pierce Department of Biological Chemistry, University of Michigan Medical School, Ann Arbor, MI, USA

Andrea Pietrobattista Unit of Liver Research of Bambino Gesù Children's Hospital, IRCCS, Rome, Italy

Thomas C. Pochapsky Department of Chemistry, Rosenstiel Basic Medical Sciences Research Center, Brandeis University, Waltham, MA, USA

Ehmke Pohl Biophysical Sciences Institute, Department of Chemistry, School of Biological and Biomedical Sciences, Durham University, Durham, UK

Joe C. Polacco Department of Biochemistry/Interdisciplinary Plant Group, University of Missouri, Columbia, MO, USA

Arthur S. Polans Department of Ophthalmology and Visual Sciences, UW Eye Research Institute, University of Wisconsin, Madison, WI, USA

K. Michael Pollard Department of Molecular and Experimental Medicine, The Scripps Research Institute, La Jolla, CA, USA

Charlotte Poschenrieder Lab. Fisiología Vegetal, Facultad de Biociencias, Universidad Autónoma de Barcelona, Bellaterra, Spain

Thomas L. Poulos Department of Biochemistry & Molecular Biology, Pharmaceutical Science, and Chemistry, University of California, Irivine, Irvine, CA, USA

Richard D. Powell Nanoprobes, Incorporated, Yaphank, NY, USA

Ananda S. Prasad Department of Oncology, Karmanos Cancer Center, Wayne State University, School of Medicine, Detroit, MI, USA

Walter C. Prozialeck Department of Pharmacology, Midwestern University, Downers Grove, IL, USA

Qin Qin Wise Laboratory of Environmental and Genetic Toxicology, Maine Center for Toxicology and Environmental Health, Department of Applied Medical Sciences, University of Southern Maine, Portland, ME, USA

Thierry Rabilloud CNRS, UMR 5249 Laboratory of Chemistry and Biology of Metals, Grenoble, France

CEA, DSV, iRTSV/LCBM, Chemistry and Biology of Metals, Grenoble Cedex 9, France

Université Joseph Fourier, Grenoble, France

Stephen W. Ragsdale Department of Biological Chemistry, University of Michigan Medical School, Ann Arbor, MI, USA

Frank M. Raushel Department of Chemistry, Texas A&M University, College Station, TX, USA

Frank Reith School of Earth and Environmental Sciences, The University of Adelaide, Centre of Tectonics, Resources and Exploration (TRaX) Adelaide, Urrbrae, South Australia, Australia

CSIRO Land and Water, Environmental Biogeochemistry, PMB2, Glen Osmond, Urrbrae, South Australia, Australia

Tony Remans Centre for Environmental Sciences, Hasselt University, Diepenbeek, Belgium

Albert W. Rettenmeier Institute of Hygiene and Occupational Medicine, University of Duisburg-Essen, Essen, Germany

Rita Rezzani Division of Human Anatomy, Department of Biomedical Sciences and Biotechnologies, Brescia University, Brescia, Italy

Marius Réglier Faculté des Sciences et Techniques, ISM2/BiosCiences UMR CNRS 7313, Aix-Marseille Université Campus Scientifique de Saint Jérôme, Marseille, France

Oliver-M. H. Richter Institute of Biochemistry, Goethe University, Frankfurt, Germany

Agnes Rinaldo-Matthis Department of Medical Biochemistry and Biophysics (MBB), Karolinska Institute, Stockholm, Sweden

Lothar Rink Institute of Immunology, Medical Faculty, RWTH Aachen University, Aachen, Germany

Alfonso Rios-Perez Department of Toxicology, Cinvestav-IPN, Mexico city, Mexico

Rasmus Sejersten Ripa Cardiology Stem Cell Laboratory, Rigshospitalet University Hospital, Copenhagen, Denmark

Cluster for Molecular Imaging and Department of Clinical Physiology, Nuclear Medicine and PET, Rigshospitalet University Hospital, Copenhagen, Denmark

Marwan S. Rizk Deptartment of Anesthesiology, American University of Beirut Medical Center, Beirut, Lebanon

Nigel J. Robinson Biophysical Sciences Institute, Department of Chemistry, School of Biological and Biomedical Sciences, Durham University, Durham, UK

João B. T. Rocha Departamento de Química, Centro de Ciências Naturais e Exatas, Universidade Federal de Santa Maria, Santa Maria, RS, Brazil

Juan C. Rodriguez-Ubis Dept. Química Orgánica, Facultad de Ciencias, Universidad Autónoma de Madrid, Madrid, Spain

Harry A. Roels Louvain Centre for Toxicology and Applied Pharmacology, Université catholique de Louvain, Brussels, Belgium

Andrea M. P. Romani Department of Physiology and Biophysics, School of Medicine, Case Western Reserve University, Cleveland, OH, USA

S. Rosato Struttura Semplice Dipartimentale di Genetica Clinica, Dipartimento di Ostetrico-Ginecologico e Pediatrico, Istituto di Ricovero e Cura a Carattere Scientifico, Arcispedale S. Maria Nuova, Reggio Emilia, Italy

Barry P. Rosen Department of Cellular Biology and Pharmacology, Florida International University, Herbert Wertheim College of Medicine, Miami, FL, USA

Erwin Rosenberg Institute of Chemical Technologies and Analytics, Vienna University of Technology, Vienna, Austria

Amy C. Rosenzweig Departments of Molecular Biosciences and of Chemistry, Northwestern University, Evanston, IL, USA

Michael Rother Institut für Mikrobiologie, Technische Universität Dresden, Dresden, Germany

Benoît Roux Department of Pediatrics, Biochemistry and Molecular Biology, The University of Chicago, Chicago, IL, USA

Namita Roy Choudhury Ian Wark Research Institute, University of South Australia, Mawson Lakes, South Australia, Australia

Jagoree Roy Department of Biology, Stanford University, Stanford, USA

Kaushik Roy Division of Biochemistry, Department of Pharmaceutical Technology, Jadavpur University, Kolkata, West Bengal, India

Marian Rucki Centre of Occupational Health, Laboratory of Predictive Toxicology, National Institute of Public Health, Praha 10, Czech Republic

Anandamoy Rudra Department of Pharmaceutical Technology, Jadavpur University, Kolkata, West Bengal, India

Giuseppe Ruggiero National Research Council (CNR), Laboratory for Molecular Sensing, Institute of Protein Biochemistry, Naples, Italy

Kelly C. Ryan Department of Chemistry, Lederle Graduate Research Center, University of Massachusetts at Amherst, Amherst, MA, USA

Lisa K. Ryan New Jersey Medical School, The Public Health Research Institute, University of Medicine and Dentistry of New Jersey, Newark, NJ, USA

Janusz K. Rybakowski Department of Adult Psychiatry, Poznan University of Medical Sciences, Poznan, Poland

Ivan Sabolić Unit of Molecular Toxicology, Institute for Medical Research and Occupational Health, Zagreb, Croatia

Kalyan K. Sadhu Division of Advanced Science and Biotechnology, Graduate School of Engineering, Osaka University, Suita, Osaka, Japan

Anita Sahu Institute of Pathology and Center of Medical Research, Medical University of Graz, Graz, Austria

P. Ch. Sahu Condensed Matter Physics Division, Materials Science Group, Indira Gandhi Centre for Atomic Research, Kalpakkam, Tamil Nadu, India

Milton H. Saier Jr. Department of Molecular Biology, Division of Biological Sciences, University of California at San Diego, La Jolla, CA, USA

Jarno Salonen Laboratory of Industrial Physics, Department of Physics, University of Turku, Turku, Finland

Abel Santamaría Laboratorio de Aminoácidos Excitadores, Instituto Nacional de Neurología y Neurocirugía Manuel Velasco Suárez, Mexico City, DF, Mexico

Luis F. Santana Department of Physiology and Biophysics, University of Washington, Seattle, WA, USA

Hélder A. Santos Division of Pharmaceutical Technology, University of Helsinki, Helsinki, Finland

Dibyendu Sarkar Earth and Environmental Studies Department, Montclair State University, Montclair, NJ, USA

Louis J. Sasseville Groupe d'étude des protéines membranaires (GÉPROM) and Département de Physique, Université de Montréal, Montréal, QC, Canada

R. Gary Sawers Institute for Microbiology, Martin-Luther University Halle-Wittenberg, Halle (Saale), Germany

Janez Ščančar Department of Environmental Sciences, Jožef Stefan Institute, Ljubljana, Slovenia

Marcus C. Schaub Institute of Pharmacology and Toxicology, University of Zurich, Zurich, Switzerland

Sara Schmitt The Developmental Therapeutics Program, Barbara Ann Karmanos Cancer Institute, and Departments of Oncology, Pharmacology and Pathology, School of Medicine, Wayne State University, Detroit, MI, USA

Paul P. M. Schnetkamp Department of Physiology & Pharmacology, Hotchkiss Brain Institute, University of Calgary, Calgary, AB, Canada

Lutz Schomburg Institute for Experimental Endocrinology, Charité – University Medicine Berlin, Berlin, Germany

Gerhard N. Schrauzer Department of Chemistry and Biochemistry, University of California, San Diego, La Jolla, CA, USA

Ruth Schreiber Department of Pediatrics, Soroka University Medical Centre, Ben-Gurion University of the Negev, Beer Sheva, Israel

Ulrich Schweizer Institute for Experimental Endocrinology, Charité-Universitätsmedizin Berlin, Berlin, Germany

Ion Romulus Scorei Department of Biochemistry, University of Craiova, Craiova, DJ, Romania

Lucia A. Seale Department of Cell & Molecular Biology, John A. Burns School of Medicine, University of Hawaii at Manoa, Honolulu, HI, USA

Lance C. Seefeldt Department of Chemistry and Biochemistry, Utah State University, Logan, UT, USA

William Self Molecular Biology & Microbiology, Burnett School of Biomedical Sciences, University of Central Florida, Orlando, FL, USA

Takashi Sera Department of Applied Chemistry and Biotechnology, Graduate School of Natural Science and Technology, Okayama University, Okayama, Japan

Aruna Sharma Laboratory of Cerebrovascular Research, Department of Surgical Sciences, Anesthesiology & Intensive Care medicine, University Hospital, Uppsala University, Uppsala, Sweden

Hari Shanker Sharma Laboratory of Cerebrovascular Research, Department of Surgical Sciences, Anesthesiology & Intensive Care medicine, University Hospital, Uppsala University, Uppsala, Sweden

Honglian Shi Department of Pharmacology and Toxicology, University of Kansas, Lawrence, KS, USA

Xianglin Shi Graduate Center for Toxicology, University of Kentucky, Lexington, KY, USA

Satoshi Shinoda JST, CREST, and Department of Chemistry, Graduate School of Science, Osaka City University, Sumiyoshi-ku, Osaka, Japan

Maksim A. Shlykov Department of Molecular Biology, Division of Biological Sciences, University of California at San Diego, La Jolla, CA, USA

Siddhartha Shrivastava Rensselaer Nanotechnology Center, Rensselaer Polytechnic Institute, Troy, NY, USA

Center for Biotechnology and Interdisciplinary Studies, Rensselaer Polytechnic Institute, Troy, NY, USA

Sandra Signorella Departamento de Química Física/IQUIR-CONICET, Facultad de Ciencias Bioquímicas y Farmacéuticas, Universidad Nacional de Rosario, Rosario, Argentina

Amrita Sil Department of Pharmacology, Burdwan Medical College, Burdwan, West Bengal, India

Radu Silaghi-Dumitrescu Department of Chemistry and Chemical Engineering, Babes-Bolyai University, Cluj-Napoca, Romania

Simon Silver Department of Microbiology and Immunology, University of Illinois, Chicago, IL, USA

Britt-Marie Sjöberg Department of Biochemistry and Biophysics, Stockholm University, Stockholm, SE, Sweden

Karen Smeets Centre for Environmental Sciences, Hasselt University, Diepenbeek, Belgium

Stephen M. Smith Departments of Molecular Biosciences and of Chemistry, Northwestern University, Evanston, IL, USA

Małgorzata Sobieszczańska Department of Pathophysiology, Wroclaw Medical University, Wroclaw, Poland

Young-Ok Son Graduate Center for Toxicology, University of Kentucky, Lexington, KY, USA

Martha E. Sosa Torres Facultad de Quimica, Universidad Nacional Autonoma de Mexico, Ciudad Universitaria, Coyoacan, Mexico DF, Mexico

Jerry W. Spears Department of Animal Science, North Carolina State University, Raleigh, NC, USA

Sarah R. Spell Department of Chemistry, Virginia Commonwealth University, Richmond, VA, USA

Christopher D. Spicer Chemistry Research Laboratory, Department of Chemistry, University of Oxford, Oxford, UK

St. Hilda's College, University of Oxford, Oxford, UK

Alessandra Stacchiotti Division of Human Anatomy, Department of Biomedical Sciences and Biotechnologies, Brescia University, Brescia, Italy

Jan A. Staessen Study Coordinating Centre, Department of Cardiovascular Diseases, KU Leuven, Leuven, Belgium

Unit of Epidemiology, Maastricht University, Maastricht, The Netherlands

Maria Staiano National Research Council (CNR), Laboratory for Molecular Sensing, Institute of Protein Biochemistry, Naples, Italy

Anna Starus Department of Chemistry and Biochemistry, Loyola University Chicago, Chicago, IL, USA

Alexander Stein Hubertus Wald Tumor Center, University Cancer Center Hamburg (UCCH), University Hospital Hamburg-Eppendorf (UKE), Hamburg, Germany

Iryna N. Stepanenko Institute of Inorganic Chemistry, University of Vienna, Vienna, Austria

Martin J. Stillman Department of Biology, The University of Western Ontario, London, ON, Canada

Department of Chemistry, The University of Western Ontario, London, ON, Canada

Walter Stöcker Johannes Gutenberg University Mainz, Institute of Zoology, Cell and Matrix Biology, Mainz, Germany

Barbara J. Stoecker Department of Nutritional Sciences, Oklahoma State University, Stillwater, OK, USA

Edward D. Sturrock Zinc Metalloprotease Research Group, Division of Medical Biochemistry, Institute of Infectious Disease and Molecular Medicine, University of Cape Town, Cape Town, South Africa

Minako Sumita Department of Biochemistry and Molecular Biology, Michigan State University, East Lansing, MI, USA

Kelly L. Summers Department of Biology, The University of Western Ontario, London, ON, Canada

Raymond Wai-Yin Sun Department of Chemistry, State Key Laboratory of Synthetic Chemistry and Open Laboratory of Chemical Biology of the Institute of Molecular Technology for Drug Discovery and Synthesis, The University of Hong Kong, Hong Kong, China

Claudiu T. Supuran Department of Chemistry, University of Florence, Sesto Fiorentino (Florence), Italy

Hiroshi Suzuki Department of Biochemistry, Asahikawa Medical University, Asahikawa, Hokkaido, Japan

Q. Swennen Biomedical Institute, Hasselt University, Diepenbeek, Belgium

Ingebrigt Sylte Medical Pharmacology and Toxicology, Department of Medical Biology, University of Tromsø, Tromsø, Norway

Yoshiyuki Tanaka Graduate School of Pharmaceutical Sciences Tohoku University, Sendai, Miyagi, Japan

Shen Tang Department of Chemistry, Georgia State University, Atlanta, GA, USA

Akio Tani Research Institute of Plant Science and Resources, Okayama University, Kurashiki, Okayama, Japan

Pedro Tavares Departamento de Química, Faculdade de Ciências e Tecnologia, Requimte, Centro de Química Fina e Biotecnologia, Universidade Nova de Lisboa, Caparica, Portugal

Jan Willem Cohen Tervaert Clinical and Experimental Immunology, Maastricht University, Maastricht, The Netherlands

Tiago Tezotto Departamento de Produção Vegetal, Universidade de São Paulo/ Escola Superior de Agricultura Luiz de Queiroz, Piracicaba, SP, Brazil

Elizabeth C. Theil Children's Hospital Oakland Research Institute, Oakland, CA, USA

Department of Molecular and Structural Biochemistry, North Carolina State University, Raleigh, NC, USA

Frank Thévenod Faculty of Health, School of Medicine, Centre for Biomedical Training and Research (ZBAF), Institute of Physiology & Pathophysiology, University of Witten/Herdecke, Witten, Germany

David J. Thomas Pharmacokinetics Branch – Integrated Systems Toxicology Division, National Health and Environmental Research Laboratory, U.S. Environmental Protection Agency, Research Triangle Park, NC, USA

Ameer N. Thompson Department of Anesthesiology, Weill Cornell Medical College, New York, NY, USA

Department of Physiology and Biophysics, Weill Cornell Medical College, New York, NY, USA

Rüdiger Thul School of Mathematical Sciences, University of Nottingham, Nottingham, UK

Jacob P. Thyssen National Allergy Research Centre, Department of Dermato-Allergology, Copenhagen University Hospital Gentofte, Hellerup, Denmark

Milon Tichy Centre of Occupational Health, Laboratory of Predictive Toxicology, National Institute of Public Health, Praha 10, Czech Republic

Dajena Tomco Department of Chemistry, Wayne State University, Detroit, MI, USA

Hidetaka Torigoe Department of Applied Chemistry, Faculty of Science, Tokyo University of Science, Tokyo, Japan

Rhian M. Touyz Kidney Research Center, Ottawa Hospital Research Institute, University of Ottawa, Ottawa, ON, Canada

Institute of Cardiovascular & Medical Sciences, BHF Glasgow Cardiovascular Research Centre, University of Glasgow, Glasgow, UK

Chikashi Toyoshima Institute of Molecular and Cellular Biosciences, The University of Tokyo, Tokyo, Japan

Lennart Treuel Institute of Applied Physics and Center for Functional Nanostructures (CFN), Karlsruhe Institute of Technology (KIT), Karlsruhe, Germany

Institute of Physical Chemistry, University of Duisburg–Essen, Essen, Germany

Shweta Trivedi Department of Animal Science, North Carolina State University, Raleigh, NC, USA

Thierry Tron iSm2/BiosCiences UMR CNRS 7313, Case 342, Aix-Marseille Université, Marseille, France

Chin-Hsiao Tseng Department of Internal Medicine, National Taiwan University College of Medicine, Taipei, Taiwan

Division of Endocrinology and Metabolism, Department of Internal Medicine, National Taiwan University Hospital, Taipei, Taiwan

Tsai-Tien Tseng Center for Cancer Research and Therapeutic Development, Clark Atlanta University, Atlanta, GA, USA

Samantha D. Tsotsoros Department of Chemistry, Virginia Commonwealth University, Richmond, VA, USA

Petra A. Tsuji Department of Biological Sciences, Towson University, Towson, MD, USA

Hiroshi Tsukube JST, CREST, and Department of Chemistry, Graduate School of Science, Osaka City University, Sumiyoshi-ku, Osaka, Japan

Sławomir Tubek Institute of Technology, Opole, Poland

Raymond J. Turner Department of Biological Sciences, University of Calgary, Calgary, AB, Canada

Toshiki Uchihara Laboratory of Structural Neuropathology, Tokyo Metropolitan Institute of Medical Science, Tokyo, Japan

Takafumi Ueno Department of Biomolecular Engineering, Graduate School of Bioscience and Biotechnology, Tokyo Institute of Technology, Yokohama, Japan

Christoph Ufer Institute of Biochemistry, Charité – Universitätsmedizin Berlin, Berlin, Germany

İrem Uluisik Department of Molecular Biology and Genetics, Izmir Institute of Technology, Urla, İzmir, Turkey

Balachandran Unni Nair Chemical Laboratory, Central Leather Research Institute (Council of Scientific and Industrial Research), Chennai, Tamil Nadu, India

Vladimir N. Uversky Department of Molecular Medicine, University of South Florida, College of Medicine, Tampa, FL, USA

Joan Selverstone Valentine Department of Chemistry and Biochemistry, University of California, Los Angeles, CA, USA

Marian Valko Department of Chemistry, Faculty of Natural Sciences, Constantine The Philosopher University, Nitra, Slovakia

Faculty of Chemical and Food Technology, Slovak Technical University, Bratislava, Slovakia

J. David van Horn Department of Chemistry, University of Missouri-Kansas City, Kansas City, MO, USA

Frans van Roy Department for Molecular Biomedical Research, VIB, Ghent, Belgium

Department of Biomedical Molecular Biology, Ghent University, Ghent, Belgium

Marie Vancová Institute of Parasitology, Biology Centre of the Academy of Sciences of the Czech Republic and University of South Bohemia, České Budějovice, Czech Republic

Jaco Vangronsveld Centre for Environmental Sciences, Hasselt University, Diepenbeek, Belgium

Antonio Varriale National Research Council (CNR), Laboratory for Molecular Sensing, Institute of Protein Biochemistry, Naples, Italy

Milan Vašák Department of Inorganic Chemistry, University of Zürich, Zürich, Switzerland

Claudio C. Vásquez Laboratorio de Microbiología Molecular, Departamento de Biología, Universidad de Santiago de Chile, Santiago, Chile

Oscar Vassallo Department of Biology, University of Rome Tor Vergata, Rome, Italy

Claudio N. Verani Department of Chemistry, Wayne State University, Detroit, MI, USA

Nathalie Verbruggen Laboratory of Plant Physiology and Molecular Genetics, Université Libre de Bruxelles, Brussels, Belgium

Sandra Viviana Verstraeten Department of Biological Chemistry, IQUIFIB (UBA-CONICET), School of Pharmacy and Biochemistry, University of Buenos Aires, Argentina, Buenos Aires, Argentina

Ramon Vilar Department of Chemistry, Imperial College London, South Kensington, London, UK

John B. Vincent Department of Chemistry, The University of Alabama, Tuscaloosa, AL, USA

Hans J. Vogel Department of Biological Sciences, Biochemistry Research Group, University of Calgary, Calgary, AB, Canada

Vladislav Volarevic Faculty of Medical Sciences, University of Kragujevac, Centre for Molecular Medicine, Kragujevac, Serbia

Anne Volbeda Metalloproteins; Institut de Biologie Structurale J.P. Ebel; CEA; CNRS; Université J. Fourier, Grenoble, France

Eugene S. Vysotski Photobiology Laboratory, Institute of Biophysics Russian Academy of Sciences, Siberian Branch, Krasnoyarsk, Russia

Anne Walburger Laboratoire de Chimie Bactérienne (UPR9043), Institut de Microbiologie de la Méditerranée, CNRS & Aix-Marseille Université, Marseille, France

Andrew H.-J. Wang Institute of Biological Chemistry, Academia Sinica, Taipei, Taiwan

Jiangxue Wang Key Laboratory for Biomechanics and Mechanobiology of the Ministry of Education, School of Biological Science and Medical Engineering, Beihang University, Beijing, China

Xudong Wang Department of Pathology, St Vincent Hospital, Worcester, MA, USA

John Wataha Department of Restorative Dentistry, University of Washington HSC D779A, School of Dentistry, Seattle, WA, USA

David J. Weber Department of Biochemistry and Molecular Biology, University of Maryland School of Medicine, Baltimore, MD, USA

Nial J. Wheate Faculty of Pharmacy, The University of Sydney, Sydney, NSW, Australia

Chris G. Whiteley Graduate Institute of Applied Science and Technology, National Taiwan University of Science and Technology, Taipei, Taiwan

Roger L. Williams Medical Research Council, Laboratory of Molecular Biology, Cambridge, UK

Judith Winogrodzki Department of Microbiology, University of Manitoba, Winnipeg, MB, Canada

John Pierce Wise Sr. Wise Laboratory of Environmental and Genetic Toxicology, Maine Center for Toxicology and Environmental Health, Department of Applied Medical Sciences, University of Southern Maine, Portland, ME, USA

Pernilla Wittung-Stafshede Department of Chemistry, Chemical Biological Center, Umeå University, Umeå, Sweden

Bert Wolterbeek Interfaculty Reactor Institute, Delft University of Technology, Delft, The Netherlands

Simone Wünschmann Environmental Institute of Scientific Networks in Constitution, Haren/Erika, Germany

Robert Wysocki Institute of Experimental Biology, University of Wroclaw, Wroclaw, Poland

Shenghui Xue Department of Biology, Georgia State University, Atlanta, GA, USA

Xiao-Jing Yan Department of Hematology, The First Hospital of China Medical University, Shenyang, China

Xiaodi Yang School of Chemistry and Material Science, Nanjing Normal University, Nanjing, China

Jenny J. Yang Department of Chemistry, Georgia State University, Atlanta, GA, USA

Natural Science Center, Atlanta, GA, USA

Vladimir Yarov-Yarovoy Department of Physiology and Membrane Biology, Department of Biochemistry and Molecular Medicine, School of Medicine, University of California, Davis, CA, USA

Katsuhiko Yokoi Department of Human Nutrition, Seitoku University Graduate School, Matsudo, Chiba, Japan

Vincenzo Zagà Department of Territorial Pneumotisiology, Italian Society of Tobaccology (SITAB), AUSL of Bologna, Bologna, Italy

Carla M. Zammit School of Earth and Environmental Sciences, The University of Adelaide, Centre of Tectonics, Resources and Exploration (TRaX) Adelaide, Urrbrae, South Australia, Australia

CSIRO Land and Water, Environmental Biogeochemistry, PMB2, Glen Osmond, Urrbrae, South Australia, Australia

Lourdes Zélia Zanoni Department of Pediatrics, School of Medicine, Federal University of Mato Grosso do Sul, Campo Grande, MS, Brazil

Huawei Zeng United States Department of Agriculture, Agricultural Research Service, Grand Forks Human Nutrition Research Center, Grand Forks, ND, USA

Cunxian Zhang Department of Pathology, Women & Infants Hospital of Rhode Island, Kent Memorial Hospital, Warren Alpert Medical School of Brown University, Providence, RI, USA

Chunfeng Zhao Institute for Biocomplexity and Informatics and Department of Biological Sciences, University of Calgary, Calgary, AB, Canada

Anatoly Zhitkovich Department of Pathology and Laboratory Medicine, Brown University, Providence, RI, USA

Boris S. Zhorov Department of Biochemistry and Biomedical Sciences, McMaster University, Hamilton, ON, Canada

Sechenov Institute of Evolutionary Physiology and Biochemistry, Russian Academy of Sciences, St. Petersburg, Russia

Yubin Zhou Department of Chemistry, Georgia State University, Atlanta, GA, USA

Division of Signaling and Gene Expression, La Jolla Institute for Allergy and Immunology, La Jolla, CA, USA

Michael X. Zhu Department of Integrative Biology and Pharmacology, The University of Texas Health Science Center at Houston, Houston, TX, USA

Marcella Zollino Istituto di Genetica Medica, Università Cattolica Sacro Cuore, Policlinico A. Gemelli, Rome, Italy

M

Mad Cow Disease

▸ Copper and Prion Proteins

Magnesium

▸ Magnesium and Cell Cycle

Magnesium (Mg²⁺)

▸ Magnesium and Vessels

Magnesium and Cell Cycle

Jeanette A. Maier
Department of Biomedical and Clinical Sciences L. Sacco, Università di Milano, Medical School, Milano, Italy

Synonyms

Cancer; Cell division; Ion channels; Magnesium; Senescence

Definition

Regulation of cell proliferation by extracellular mitogens is governed through receptor-mediated signals which ultimately converge on the cell cycle machinery driven by cyclins and cyclin-dependent kinases (CDKs) and counteracted by CDK inhibitors (CKIs). Magnesium (Mg) is involved in the regulation of eukaryotic cell proliferation. In response to mitogens, intracellular Mg increases in the G1 and S phases of the cell cycle and this event correlates with the enhancement of protein synthesis and the onset of DNA synthesis. Later in the cell cycle, Mg contributes to the formation of the mitotic spindle and cytokinesis. Accordingly, low extracellular Mg retards cell proliferation by accumulating the cells in the G1 phase of the cell cycle through the upregulation of CKIs and the downregulation of cyclins. The relevance of Mg in cell cycle progression is underscored by the evidence that the silencing of its transporter Transient Receptor Potential Melastatin-7 inhibits cell growth.

It is noteworthy that perturbations of Mg homeostasis have been reported in cancer, characterized by uncontrolled cell growth, and in cellular senescence, which is the irreversible arrest of cell growth.

Overview

Cell proliferation is a complex, highly regulated process which requires a huge number of molecules and interrelated pathways (Alberts et al. 2002). It is triggered by growth factors and hormones or by

V.N. Uversky et al. (eds.), *Encyclopedia of Metalloproteins*, DOI 10.1007/978-1-4614-1533-6,
© Springer Science+Business Media New York 2013

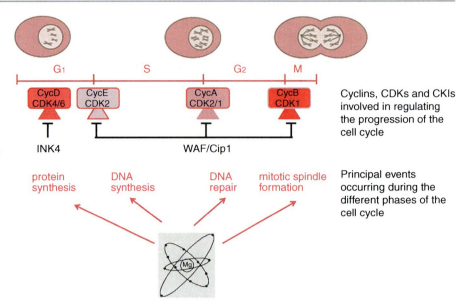

Magnesium and Cell Cycle, Fig. 1 Magnesium is involved in the regulation of cell cycle progression. The different phases of the cell cycle, their dependence upon cyclins/CDKs, and the principal events occurring in the single phases are summarized. The sites on the inhibitory effects of Waf/Cip1 and INK4 are also indicated

signaling from components of the extracellular matrix, all activating the transcription of genes that control cell cycle entry and progression to ultimately achieve cell division. These events are in strict coordination with a metabolic response that includes the increased uptake of various substrates, the increased synthesis of protein and RNA, and, in the end, DNA replication. Because of its pivotal role in maintaining tissue homeostasis and eventually driving tissue regeneration and repair, the cell cycle is rigorously controlled by activators, inhibitors, and also by sensors of damage to DNA and chromosomes. A critical, rate-limiting step of the cell cycle is the transition from the G1 to S phase, since upon passing this restriction point the cells are irreversibly committed to DNA replication. Briefly, at the G1/S transition, cyclin-dependent kinases (CDK)-4 and -6 acquire catalytic activity by binding to cyclin D and CDK2 by interacting with cyclin E. Cyclin A-CDK2 and cyclin A-CDK1 are active in the S phase, while cyclin B-CDK1 is fundamental for G2/M transition (Fig. 1). CDKs phosphorylate target proteins crucial for cell cycle progression and are opposed by two families of CDK inhibitors (CKI), namely, INK4 and WAF/Cip1. A particular attention should be devoted to the p21 and p27 members of the WAF/Cip1 family, which play a central role as negative regulator of cell cycle progression.

Surveillance mechanisms are constantly active during cell cycle to ensure that cells bearing damaged DNA do not complete replication. In particular, the first checkpoint is at the G1/S transition and monitors the integrity of DNA before duplication, while the G2/M checkpoint controls DNA after replication so that the cells can safely divide (Alberts et al. 2002).

Magnesium: A Versatile Cation Involved in the Control of Cell Proliferation

Magnesium (Mg) is the most abundant intracellular divalent cation, serving a wide range of metabolic, structural, and regulatory functions. Mostly complexed with ATP and other molecules with negatively charged moieties, it is involved in a wide variety of biochemical reactions activating a vast array of enzymes (Wolf and Trapani 2008), thus participating in all the major metabolic processes as well as in redox reactions. Being associated with ATP, Mg is crucial for all trans-phosphorylation reactions through the formation of ATP-Mg complexes which anchor substrates to the active sites of enzymes. These reactions are fundamental in signal transduction pathways. Recently, Mg has been shown to act as an intracellular second messenger coupling cell-surface receptor activation to intracellular effectors in T lymphocytes (Li et al. 2011). Mg homeostasis in the cells is tightly regulated by precise control mechanisms operating at the level of

Mg influx and efflux across the plasma membrane, and at the level of intracellular Mg buffering and organelle localization (Wolf and Trapani 2008).

Because of its versatility, Mg contributes to the regulation of cell proliferation. In the beginning, Mg is required for the activation of the kinases which transduce the mitotic signal from cell-surface receptors to the nucleus. Then Mg is necessary to sustain protein synthesis and is involved in DNA replication. In the end, Mg has a role in the formation of the mitotic spindle (Wolf and Trapani 2008).

Extracellular Magnesium and the Regulation of Cell Proliferation

Evidence from prokaryotes and lower eukaryotes reinforces the view that magnesium plays a role in the control of the cell cycle (Walker 1986). Of particular interest is the evidence that Mg is essential for cell division in yeast, a single-cell eukaryote widely used as a model system for studying the cell cycle. Specifically, Mg concentration is the transducer for size and, consequently, time related control of yeast cell cycle. Accordingly, limited Mg supply constrains yeast growth by accumulating cells in G0/G1 and decreasing the cells in the S phase. Low Mg also influences the formation and breakdown of the mitotic spindle (Walker 1986).

Also in mammalian cells, Mg is crucial in sustaining cell growth. Indeed, intracellular Mg concentration is higher in growing cells than in quiescent cells. Upon exposure to growth factors, intracellular Mg increases fractionally in the G1 and S phases of the cell cycle. This increase correlates with the enhancement in the rate of protein synthesis and the onset of DNA synthesis. In particular, the rise of intracellular Mg activates mTOR kinase, thus speeding the initiation of translation, and accelerating the passage of cells through the G1 to the S phase (Rubin 2005). The changes in Mg and protein synthesis have to be maintained throughout the G1 period to produce their full effect.

Lowering Mg concentration in the medium and, therefore, within the cells reduces all the early reactions including protein synthesis, which is followed by a significant reduction in the rate of DNA synthesis (Wolf and Trapani 2008; Rubin 2005). It is also noteworthy that DNA polymerases and ligases require Mg-ATP. In the G2 phase of the cell cycle, DNA repair eventually occurs through the intervention of specific enzymes, some of which are known to require Mg to properly function. In addition, chromatin condensation and decondensation have also been linked to alterations in intracellular Mg levels. Moreover, Mg has an important role in tubulin assembly regulation and, therefore, in mitotic spindle formation (Fig. 1). On these bases, it is not surprising that low concentrations of extracellular Mg inhibit the growth of different types of cultured mammalian cells (Table 1) by decreasing the number of cells in the S phase and promoting their accumulation mainly in the G0/G1 and also in G2/M phases of the cell cycle (Wolf and Trapani 2008). It is noteworthy that Mg deprivation downregulates cyclin D and the transcription factor E2F, which promote cell cycle progression by activating CDKs and transcription of S phase–specific genes, respectively. In addition, growth inhibition correlates with the upregulation of p27 and p21 which inhibit the progression from the G1 to the S phase. In particular, p21 prevents the activation of cyclin E/CDK2 and also interacts with the proliferating cell nuclear antigen PCNA, thus blocking DNA synthesis by DNA

Magnesium and Cell Cycle, Table 1 *Mammalian cells growth-inhibited by low extracellular Mg.* Different cell types have been cultured in medium containing low Mg and revealed how Mg controls cell cycle progression. For some cells, growth-inhibition by Mg deficiency has relevant clinical implications

Epithelial cells	
Breast epithelial MCF7 and HC 11 cells	
Renal epithelial NRK52E cells	
Leukocytes	
HL-60 leukemic cells	
J774.E macrophages	
D10.G4.1 T lymphocytes	
Vascular cells	
Smooth muscle cells	Implications in atherosclerosis
Macrovascular endothelial cells	
Microvascular endothelial cells	Implications in angiogenesis
Endothelial progenitors	
Bone and cartilage cells	
Osteoblasts	Implications in osteoporosis
Chondrocytes	
Fibroblasts	

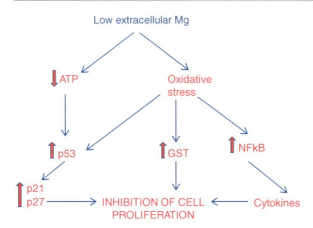

Magnesium and Cell Cycle, Fig. 2 Low Mg inhibits mammalian cell proliferation. The mechanisms reported in the literature as involved in retarding cell growth are shown

polymerase δ. The increase of p21 and p27 levels by low Mg is mainly due to their transcriptional activation by transcription factors activated by various extrinsic stress signals, such as oxidative stress and low ATP. Interestingly, in renal epithelial cells, Mg deficiency decreases ATP content, an event which induces p53 that, in turn, upregulates p21. In endothelial cells, culture in low Mg induces oxidative stress which triggers the activation of the transcription factor nuclear factor-kB (NF-kB). NF-kB orchestrates the inflammatory response and, therefore, modulates the release of growth factors and cytokines. In these cells, the activation of NF-kB leads to the upregulation of the prototypical inflammatory cytokine interleukin (IL)-1α, which is also a potent inhibitor of endothelial proliferation (Castiglioni and Maier 2011). Also the detoxifying enzyme glutathione-S-transferase (GST), which is activated under Mg restriction, is involved in the modulation of cell growth. In mammary epithelial cells, low Mg increases GST activity which, apart from being an antioxidant, controls cell proliferation by modulating JNK and Janus kinase/Stat (Wolf and Trapani 2008). Figure 2 summarizes present knowledge about the mechanisms underlying Mg deficiency–dependent inhibition of mammalian cell growth.

It is relevant to point that the inhibition of cell proliferation is reversible upon return to physiologic concentrations of Mg. In addition, some normal diploid cells proliferate faster than controls when exposed to very high concentrations of extracellular Mg.

Magnesium Transporters and Cell Proliferation

That Mg importantly contributes to cell division in yeast is also demonstrated by the fact that loss-of-function mutations of the yeast Mg transporter Alr1 reduce Mg uptake and induce a growth defect that is suppressible by an excess of Mg (Romani 2011).

Similar results were obtained in mammalian cells. Transient Receptor Potential Melastatin (TRPM)-6 and −7 are the first molecularly defined components of the mammalian Mg transport machinery. They are unique bifunctional molecules which consist of an ion channel fused to a protein kinase (Romani 2011). TRPM7, ubiquitously expressed, is vital for cellular Mg homeostasis and crucial for cell survival. It has been shown to be particularly important to ensure a rapid Mg entry during the G1 phase of the cell cycle, when most cell constituents are doubled, and, therefore, additional Mg is needed to interact with newly synthesized molecules (Ryazanova et al. 2010). Accordingly, TRPM7 has been implicated in regulating cell proliferation. In various cell types apart from human umbilical vein endothelial cells, silencing TRPM7 induces cell cycle arrest by increasing the number of cells in the G1 and G2 phases. This failure of G1/S and G2/M progression can be overridden by Mg supplementation, thereby indicating that growth inhibition is based on the channel function of TRPM7.

Implications of Magnesium Regulation of Cell Proliferation: Cancer and Senescence

Cancer and senescence represent the two opposite extremes of disturbed cell proliferation. In replicative senescence, cells stop growing, while in cancer cells begin to grow out of control.

Cellular senescence is an irreversible growth arrest in the G1 phase of the cell cycle which limits the lifespan of mammalian cells and prevents unlimited cell proliferation. This process, first glimpsed in cell culture, is now confirmed also in vivo as a vital mechanism that constrains the malignant progression of many tumors. Replicative senescence in human cells can invariably be induced by repeated serial passaging in vitro, which leads to telomere shortening. Telomeres are distinctive DNA-protein structures that cap chromosomes and are considered cellular timekeepers. Alternatively, various types of stressful conditions, such as DNA damage and oxidative stress, can induce premature senescence. It has been demonstrated that long-term culture of human fibroblasts and endothelial cells in Mg-deficient medium accelerates cellular senescence (Killilea and Maier 2008). The cells become enlarged, elongated, and vacuolated and express senescence-associated β-galactosidase activity. The increased amounts of p21 explain the loss of replicative capacity associated with cell dysfunction. Endothelial cells in low Mg also overexpress interleukin (IL)-1α, which is considered a marker of endothelial senescence. Interestingly, silencing IL-1α extends their lifespan although it does not lead to the acquisition of an immortal phenotype.

In fibroblasts from low Mg conditions, telomeres are shorter compared to standard Mg cultures from the same passaging time. Telomere shortening, which is known to contribute to accelerated replicative senescence, seems to be driven by oxidative stress in fibroblasts cultured in Mg-deficient media (Killilea and Maier 2008).

Opposite to senescence, cancer is the uncontrollable growth of abnormal cells. As mentioned above, in normal diploid cells, the total concentration of Mg increases throughout the G1 and S phases of the cell cycle and, accordingly, low extracellular Mg markedly inhibits their proliferation (Castiglioni and Maier 2011). On the contrary, neoplastic cells are rather refractory to Mg deprivation and extremely avid for Mg. They accumulate Mg even when cultured in low Mg containing media, partly because of an impairment of Na-dependent Mg extrusion and the overexpression of TRPM7 (Castiglioni and Maier 2011). High intracellular Mg is likely to provide a selective advantage for the neoplastic cells because it acts a second messenger and activates a vast array of enzymes among which DNA polymerases, ribonucleases, adenylyl cyclases, phosphodiesterases, guanylate cyclases, ATPases, and GTPases which are implicated in the metabolism of nucleic acids and proteins as well as in signal transduction. In particular, Mg is necessary for the activity of glucokinase, phosphofructokinase, phosphoglycerate kinase, and pyruvate-kinase, enzymes of the glycolysis which is known to be the preferential pathway utilized by neoplastic cells to produce energy also in the presence of oxygen.

A peculiarity of tumor cells is their limitless proliferative potential. It is therefore relevant that Mg is required to activate telomerase, a specialized DNA polymerase that extends telomeric DNA and counters the progressive telomere erosion associated with cell duplication and, therefore, with cell senescence.

Magnesium and Proliferation In Vivo

The effects of Mg on cell proliferation in vivo have been studied in relation to tumor development. Mg deficiency seems to be linked to increased risk of some types of cancers in humans (Castiglioni and Maier 2011). In experimental models, some studies indicate that Mg exerts a protective effect in the early phases of chemical cancerogenesis. In rodents, Mg deficiency inhibits tumor growth at its primary site. In rats, Mg deprivation reduces tumor growth by limiting glutathione synthesis for which Mg is an obligatory cofactor (Castiglioni and Maier 2011). In mice subcutaneously injected with Lewis lung carcinoma, mammary adenocarcinoma, and colon carcinoma cells, a low-Mg-containing diet leading to hypomagnesemia inhibits primary tumor growth, an effect which is promptly reverted by reintroducing Mg in the diet. However, Mg-deficient mice develop more metastases than controls. To explain how hypomagnesemia inhibits the development of primary tumors while enhancing metastatization, some cellular and molecular aspects have been investigated. Regarding primary tumors, Mg deficiency directly inhibits tumor cell growth by downregulating cyclin B and D3, crucial for the progression through the cell cycle, and by upregulating p21, p27, and Jumonji, all involved in breaking cell proliferation (Castiglioni and Maier 2011). Mg deficiency also affects tumor development indirectly, because it promotes inflammation and DNA oxidative damage (Mazur et al. 2007) and it impairs angiogenesis, one of the hallmarks of cancer. Oxidative stress and inflammation generate genetic instability and, therefore, increase the risk of mutations. Since Mg is an essential cofactor in some enzymatic systems involved in DNA repair, under low Mg availability mutations may become permanent, thus generating the neoplastic cell. The persistence of oxidative stress and inflammation together facilitate further mutations which render the cell immortal and self-sufficient in terms of proliferation and, eventually, invasive and metastatic, thereby capable of colonizing distant organs.

Conclusions

All data available indicate that Mg homeostasis is fundamental for cell proliferation. While the evaluation of the effects of Mg on the regulation of cell growth in vivo is rather complicated, evidence in cultured cells allows to conclude that the increase of intracellular Mg in the G1 phase of the cell cycle is necessary for progression to S phase and that Mg is fundamental in orchestrating all the events leading to cytokinesis.

Cross-References

▶ Magnesium and Cell Cycle
▶ The Effect of Selenium on Cell Proliferation
▶ Magnesium in Eukaryotes
▶ Magnesium in Health and Disease

References

Alberts B, Johnson A, Lewis J, Raff M, Roberts K, Walter P (2002) Molecular biology of the cell. Garland Science, New York
Castiglioni S, Maier JA (2011) Magnesium and cancer: a dangerous liaison. Magnes Res 24:92–100
Killilea D, Maier JA (2008) A connection between magnesium deficiency and aging: new insights from cellular studies. Magnes Res 21:77–82
Li FY, Chaigne-Delalande B, Kanellopoulou C, Davis JC, Matthews HF, Douek DC, Cohen JI, Uzel G, Su HC, Lenardo MJ (2011) Second messenger role for Mg revealed by human T-cell immunodeficiency. Nature 475:471–476
Mazur A, Maier JA, Rock E, Gueux E, Nowacki W, Rayssiguier Y (2007) Magnesium and the inflammatory response: potential physiopathological implications. Arch Biochem Biophys 458:48–56
Romani AM (2011) Cellular magnesium homeostasis. Arch Biochem Biophys 512:1–23
Rubin H (2005) The membrane, magnesium, mitosis (MMM) model of cell proliferation control. Magnes Res 18:268–274
Ryazanova LV, Rondon LJ, Zierler S, Hu Z, Galli J, Yamaguchi TP, Mazur A, Fleig A, Ryazanov AG (2010) TRPM7 is essential for Mg homeostasis in mammals. Nat Commun 1:109
Walker GM (1986) Magnesium and cell cycle control: an update. Magnesium 5:9–23
Wolf FI, Trapani V (2008) Cell (patho)physiology of magnesium. Clin Sci 114:27–35

Magnesium and Inflammation

Andrzej Mazur[1], Patrycja Libako[2],
Wojciech Nowacki[2] and Jeanette A. Maier[3]
[1]INRA, UMR 1019, UNH, CRNH Auvergne,
Clermont Université, Université d'Auvergne, Unité de Nutrition Humaine, Clermont-Ferrand, France
[2]Faculty of Veterinary Medicine, Wroclaw University of Environmental and Life Sciences, Wrocław, Poland
[3]Department of Biomedical and Clinical Sciences
L. Sacco, Università di Milano, Medical School,
Milan, Italy

Synonyms

Acute Phase; Calcium; Immunity; Low-grade inflammation; Oxidative stress; Phagocytes

Definition

Inflammation is an automatic response to harmful internal or external agents, acknowledged as a type of nonspecific immune response, which aims to defend and repair injured organs and tissues. Inflammation is a complex reaction which consists of various interconnected events involving blood vessels and leukocytes. In particular, phagocytes and vascular endothelial cells are the most important cellular players in orchestrating this process. Even though inflammation is fundamentally a protective response, its deregulation and loss of control is potentially harmful. Classically, the inflammatory response is described as an acute response with activation of nonspecific immune system and production of wide variety of pro- and anti-inflammatory mediators. Nowadays, the importance of a chronic low-grade inflammation is highlighted as implicated in diverse pathophysiological conditions (e.g., diabetes, obesity, hepatic steatosis, metabolic syndrome, and debilitating aging). Western lifestyle, characterized by sedentarity and unhealthy diet deficient in fibers, vitamins, minerals, trace elements, and bioactives of vegetal origin (e.g., polyphenols, carotenoids), is likely to contribute to generate a pro-inflammatory condition. With regard to their multiple biological and chemical properties, the aforementioned nutrients and bioactives may have an important impact on the inflammatory process-related events (e.g., oxidative stress) and subsequent onset of pathological conditions.

Overview

Mg is an essential cation, serving vital functions in all the cells of the body. Consequently, its crucial role for the maintaining adequate immune response is not surprising. For a long time, it was acknowledged that Mg deficiency leads to the decline in the specific immune response but induces an inflammatory response. On the contrary, Mg-rich preparations are known to exert an anti-inflammatory action. In humans, this relationship between Mg and inflammation is mainly based on their association in cohorts, but direct clinical observations are scarce. Therefore, the fundamental knowledge of the role of Mg in inflammation is based on experimental in vivo and in vitro studies. The modulatory action of Mg on inflammation is based on multiple mechanisms at the cellular and molecular levels. In addition, its crucial role in maintaining Ca^{2+} homeostasis has to be underscored.

Mg Deficiency–Induced Inflammation: Evidences from Experimental Models

Experimental dietary Mg depletion within a short time induces Mg deficiency in different animal models. The rapidity and severity of deficiency depends on the Mg content in the diet, and on nutritional and pathophysiological states of animals (Mazur et al. 2007; Nielsen 2010). Young growing animals are particularly prone to develop severe deficiency because of their lower Mg stores in the bone when compared to older ones. Severe depletion leads to a very fast and significant drop in magnesemia. It has been well recognized for a long time that Mg deficiency induces inflammatory features in addition to characteristic neuromuscular symptoms. These inflammatory signs are particularly remarkable in rats which present characteristic allergy-like symptoms, i.e., erythema, hyperemia, and edema, most likely related to histamine release from mast cells. In other species (mice, hamsters, pigs, rabbits, quails, and guinea pigs), some inflammatory features were also observed but not as ample as in rats. Pathological and hematological examinations support the occurrence of general inflammation. Mg-deficient rats show an enlarged spleen, marked leukocytosis (principally neutrophils and eosinophils), and leukocyte infiltration in numerous tissues. This latter could be facilitated by the tissue-damaging effect of Mg deficiency. Once attracted within the injured tissues, neutrophils may contribute to aggravate the damage by producing reactive oxygen species (ROS). The presence in the plasma of deficient animals of pro-inflammatory cytokines, neuropeptides (i.e., substance P), inflammatory mediators, and positive acute phase proteins completes the portrait of the inflammatory features. Obviously, the time course and sternness of these changes are related to the severity and duration of Mg deficiency.

The pathological outcomes of Mg deficiency–induced inflammation are numerous, i.e., hyperlipemia, blood pressure alteration, hyperalgesia, immune system alterations, cardiac damages, thrombosis, and enhanced response to nonspecific immune stimuli. Several studies have underlined the importance of the oxidative stress generated in Mg deficiency in promoting inflammation. Therefore, any condition leading to oxidative stress (among which, high-fructose diet) worsens Mg deficiency outcomes and, on the contrary, any measure protecting against oxidative stress (e.g., estrogens and drugs with antioxidant properties) attenuates injuries by Mg deficiency (Kramer et al. 2009; Mazur et al. 2007).

A question arises about the specificity of Mg deficiency in inducing inflammation. Some experimental evidences support a specific role of Mg deficiency as an inducer of inflammatory reaction. It should be noted that in these animals, no bacteria and no lipopolysaccharide in the blood and organs have been detected. In addition, restoring normal Mg levels by supplementing Mg to Mg-deficient rats alleviates inflammatory symptoms. On the basis of these data, it could be anticipated that the deficiency of Mg is able per se to induce inflammation.

One of the most remarkable issues of the experimental Mg deficiency is an enhancement of the nonspecific immune response to various immune stresses, e.g., sepsis, endotoxemia, sensitization with ovalbumin or haptens. This increased responsiveness related to low Mg condition was confirmed ex vivo on macrophages and neutrophils isolated from Mg-deficient animals and exposed to immune challenge.

Relationship Between Mg Status and Inflammation in Humans

The direct induction of inflammation by Mg deprivation was not clearly observed in humans. The available studies on the experimental deficiency in men are limited and do not provide information on inflammatory parameters nor symptoms. Because of the adverse cardiac effects of Mg deficiency induced by low Mg diet, for ethical reasons, further results on experimental deficiency on volunteers are not expected. In large populations, it is feasible that Mg deficiency resulting from a low Mg intake is subclinical and not severe enough to induce clinical symptoms. In patients, Mg deficiency is predominantly the result of diseases leading to urinary or intestinal Mg loss. However, in these patients, it is difficult to reach conclusions about the specific contribution of Mg deficiency to inflammation because of the complex pathogenesis and pathophysiology of the diseases.

From epidemiological studies, there is a growing body of evidence indicating the association between low Mg intake or status and low-grade inflammation. Several studies have shown that Mg intake was inversely related to serum concentration of the acute phase proteins, such as C-reactive protein (CRP). At the moment, serum CRP level, determined with a high-sensitivity assay, is the best used inflammatory marker for low-grade inflammation. The persistent low-grade inflammation may be found in apparently healthy subjects and is frequently encountered in subjects with type-2 diabetes, obesity, nonalcoholic steatohepatitis (NASH), and metabolic syndrome. People with these metabolic disorders often present a low Mg status. Interestingly, it is now commonly accepted that low-grade chronic inflammation plays a key role in the initiation and development of these metabolic diseases (Baker et al. 2011). On the bases of these associations, it could be hypothesized that low Mg status contributes to these metabolic diseases and their manifestations (e.g., hyperlipemia, blood pressure elevation, endothelial dysfunction, and thrombosis tendency), at least in part, through its pro-inflammatory effect. Finally, a low Mg status could contribute to the increased risk of CVD. Interestingly, Mg supplementation (300 mg/day) was efficient to attenuate elevated serum CRP in patients with heart failure (Almoznino-Sarafian et al. 2007). However, in men, a low Mg status alone is unlikely to induce inflammation, but it plays a key role in potentiating existing low-grade inflammatory processes (King 2009).

Hypotheses on the Mechanisms Responsible for the Link between Mg and Inflammation

A question arises about the exact role played by low Mg in inflammatory processes. Even though present knowledge is not sufficient to propose an accurate and detailed picture of the events and their schedule occurring under Mg deficiency, several hypotheses can be elaborated about the role of Mg in inflammation. The possible involved mechanisms are presented in Table 1.

One of the issues to be considered is that changes in Mg concentration affect Ca^{2+} homeostasis, and Ca^{2+} is a pivotal second line messenger in the immune system. In early 1980 Iseri and French (1984) defined Mg as nature's physiological calcium blocker. The blocking effect of Mg has been mainly explained as the result of a nonspecific antagonism of Ca channels.

The role of Na^{2+}/Ca^{2+} exchanger in respiratory burst in neutrophils, which produce reactive oxygen species (ROS) in response to pro-inflammatory stimuli, was investigated (Simchowitz et al. 1990). A Na^{2+}/Ca^{2+} exchange carrier is recognized to be highly involved in Ca^{2+} fluxes in cells. Additionally, the rise in cytosolic Ca^{2+} is itself sufficient to prime/initiate oxygen metabolites' generation by immune cells. It has been documented that in calcium-buffering conditions (extracellular concentration of Ca^{2+} was 1.0 mM), the generation of ROS by human neutrophils is inhibited by a wide spectrum of extracellular tri- and divalent cations including Mg. Moreover, divalent cations reduce the increase of cytoplasmatic Ca^{2+} after stimuli, maintaining its levels as low as in the absence of extracellular Ca^{2+}. The "blocking hypothesis" assumes that the polyvalent ions inhibit the influx of Ca^{2+} by being a rival of calcium ions for binding to the translocation site of Na^{2+}/Ca^{2+} counter-transport system. These in vitro data have been supported by in vivo studies in animals, showing that under Mg-restricted conditions, nifedipine (organic Ca^{2+} antagonist/blocker) prevents the adverse effects of Mg deficiency.

More recently, it was hypothesized that the neuropeptide-mediated inflammation is the first

Magnesium and Inflammation, Table 1 Potential mechanistic relationships between Mg and inflammation

Mg acts as calcium antagonist
Mg stabilizes membranes, an important issue in the control of degranulation of mast cells or basophils
Mg regulates the activation of N-Methyl-D-aspartic acid (NMDA) receptors. Mg deficiency activates NMDA
Mg deficiency increases the production of neuromediators
Low Mg determines cellular dysfunction, including oxidative stress, which leads to the activation of inflammatory pathways by NFκB

repercussion of developing Mg deficiency. Indeed, elevated levels of substance P were found in early Mg deficiency in rats (Kramer et al. 2009). Substance P is a tachykinin neuropeptide with a wide range of biological functions, very well known for its involvement in neurogenic inflammation and pro-inflammatory cytokines' production.

The increased sensitivity of phagocytic cells to immune stimuli indicates their preactivation – "priming." The priming effect might be defined as the state of cell where former exposure to a so-called priming factor influences (most often amplifies) forceful response to a stimuli. Priming is considered as reversible phenomenon but the priming's profile depends on the priming agent applied. The wide spectrum of priming mediators has been identified, e.g., substance P, interferon-γ, TNF-α which are at the origin of the further potentiation of the phagocyte response to immune stimuli.

On this basis, it is easy to notice that in Mg deficiency, several conditions are encountered for phagocyte priming. As shown in Fig. 1 an enhanced entry and storage of intracellular Ca in low extracellular Mg conditions offers an excellent condition for the enhanced phagocyte response. Summarizing, it could be hypothesized that changes in extracellular Mg concentration alter activity of phagocytes, directly by its calcium antagonism and indirectly by being an important modulator of immune-inflammatory processes.

Endothelial Cell: An Important Player in the Relationship Between Mg and Inflammation

If leukocytes are the protagonists of inflammation, the endothelial cells are certainly costars, being crucial for the initiation and perpetuation of local inflammation. Indeed, the endothelium is a target and also a source of inflammatory mediators and cytokines. Diverse mechanical, chemical, and immunological injuries lead to the so-called activation of the endothelium, which corresponds to the acquisition of a pro-inflammatory phenotype. While quiescent endothelial cells express an anticoagulant and anti-adhesive phenotype, activated endothelial cells undergo functional changes, such as pro-adhesive and pro-coagulant properties, and morphological alterations, leading to increased vascular leakage and leukocyte transmigration (Maier 2012).

Mg deficiency is one of the activators of the endothelial cells in vivo and in vitro. Dietary Mg is inversely associated with plasma concentrations of soluble VCAM (vascular cell adhesion molecule)-1, one of the markers of endothelial activation. In addition, in experimental models as well as in humans, low Mg status is associated with increased levels of CRP, which has been shown to possess a direct pro-inflammatory effect on endothelial cells (Pasceri et al. 2000). It is therefore feasible that high CRP cooperates with low Mg in inducing the activation of the endothelium in vivo.

Studies on cultured endothelial cells have unquestionably advanced our understanding of the molecular pathways involved in low-Mg-induced endothelial activation. It is oxidative stress that triggers the cascade of events leading to the acquisition of a pro-inflammatory phenotype. Early after exposure to low Mg, endothelial cells produce an excess of free radicals which rapidly activate the transcription factor NFκB (Maier 2012) through the canonical pathway. This implies the rapid degradation of the inhibitor IκB by the proteasome, the release of NFκB subunits, their translocation to the nucleus, and binding to the cognate DNA motifs in target genes. Cytokines, adhesion molecules, and enzymes catalyzing the synthesis of inflammatory mediators are all targets of NFκB which therefore orchestrates the molecular events resulting in vasodilation, vasopermeabilization, leukocyte recruitment, and diapedesis.

Interestingly, Mg deficiency activates cyclooxygenases in endothelial cells, thus leading to an increased production of prostacyclin (PGI2), a potent vasodilator. Exposure of endothelial cells to low Mg concentrations also induces the synthesis of nitric oxide (NO) through the activation of nitric oxide

Magnesium and Inflammation,

Fig. 1 Proposed action of the low extracellular magnesium on the calcium homeostasis and availability to biological processes in the basal condition and under immune stimulation. It is hypothesized that in phagocytic cells in Mg-deficient condition more [Ca^{2+}]i is available from the internal store for the response after stimulation *(Illustration réalisée grâce à Servier Medical Art)*

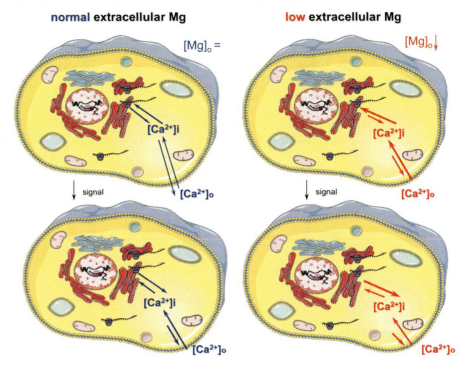

synthases. In addition, human endothelial cells exposed to low Mg concentrations synthesize increased amounts of interleukin (IL) 1α, tumor necrosis factor (TNF) α, RANTES, IL-8, and GM-CSF. IL-1α and TNFα are the prototypical pro-inflammatory cytokines. Central in mediating some of low Mg effects on endothelial cells is IL-1α, a target of NFκB and an inducer of NFκB itself. It is through the induction of IL-1α that Mg deficiency retards endothelial growth by delaying the transit through the G1 and S phases of the cell cycle. Moreover, IL-1α induces various chemokines, among which are RANTES and IL-8, and adhesion molecules in vascular endothelial cells, thus favoring the recruitment, adherence, and diapedesis of leukocytes. In particular, IL-1α is responsible for the induction of VCAM in human endothelial cells cultured in low Mg. RANTES directs leukocyte attraction while IL-8 is critical for chemotaxis and adhesion of monocytes to the endothelial cells. Low concentrations of Mg also upregulate GM-CSF which induces the differentiation, survival, proliferation, migration, and metabolism of macrophages/granulocytes, and also the migration and growth of endothelial cells.

It is therefore feasible to propose the following scenario. Upon exposure to low Mg, EC release PGI2 and NO which contribute to the vascular phase of inflammation by determining vasodilatation and increase of vascular permeability, thus leading to edema. Later, endothelial synthesized inflammatory cytokines enhance the expression of adhesion molecules on the endothelial surface while chemokines recruit leukocytes to the site of inflammation, thus resulting in leukocyte diapedesis and accumulation in the site of inflammation.

Endothelial activation is a welcome response in local defense against pathogens and in tissue repair. However, long-term maintenance of this phenotype contributes to the pathogenesis of vascular diseases.

Anti-inflammatory Action of Mg In Vivo

Fairly few experimental studies have been performed on the anti-inflammatory properties of increasing Mg in vivo. As discussed above, studies have shown that Mg repletion in experimental deficiency alleviates inflammatory symptoms. Other data from experimental and clinical studies, in apparently non-deficient conditions, also suggest the anti-inflammatory action of oral, parenteral, or local Mg treatment. In rats, Mg administration prevented paw edema in adjuvant-induced arthritis, mitigated lung injury in endotoxemia, and inhibited placental inflammation after *i.p.* LPS. In humans, Mg supplementation (300 mg/day) in patients with heart failure attenuated elevated serum CRP (Almoznino-Sarafian et al. 2007). Similarly, Mg supplementation vs. placebo decreased plasma CRP in adults with baseline values of CRP > 3.0 mg/L (Nielsen 2010). However, in ex vivo study on the human whole blood exposed to LPS stimulation, only a weak effect of high Mg was observed. This supports the possibility that a low Mg status plays a role in enhancing the inflammatory response and suggests that anti-inflammatory efficiency of Mg supplementation depends on Mg status.

Some observations also support that high Mg exerts an anti-inflammatory action when administered locally. One of the first observations of the local administration of Mg was published in 1920 and reported usefulness of Mg sulfate solution in the treatment of inflammatory conditions of the conjunctiva and surface of cornea.

The efficiency of Dead Sea water and mud rich in Mg in treatment of psoriasis and atopic dermatitis has been known for a long time. Results showing that Mg ions inhibit the antigen-presenting capacity of human epidermal Langerhans cells make Mg possible compound, contributing to the beneficial effect of this topical treatment.

Conclusion

The available data strongly support an essential role for Mg in controlling inflammation. At present, an ultimate mechanism for such action is not identified, and certainly, there are several targets for Mg at the cellular and molecular levels. Nevertheless, the control of intracellular Ca^{2+} is an important and primary mechanism by which Mg regulates cellular processes related to inflammation. Interestingly, Mg deficiency not only induces but also potentiates inflammatory response induced by the immune stress. This suggests that, even if in humans a low Mg status is not sufficient to induce clinically discernible inflammation, low Mg background is favorable for the induction of inflammatory processes. The concomitant occurrence of metabolic disorders (e.g., diabetes, obesity, metabolic syndrome), low-grade inflammation, and inadequate Mg status would point to Mg as a leading player in the development of these disturbances and subsequent pathological issues.

Considering the potential role of Mg in inflammatory responses, further understanding of the role and underlying mechanisms of the pro-inflammatory effect of the low Mg status assumes great importance. The human genome sequencing permitted the recent explosion in the genetics of the cellular Mg regulation. The discovery of new Mg transporters, channels, and sensors will shortly contribute to a better understanding of the intimate mechanisms of Mg contribution to the inflammatory processes.

Cross-References

▶ Calcium in Health and Disease
▶ Magnesium and Cell Cycle
▶ Magnesium in Biological Systems
▶ Magnesium in Health and Disease
▶ Magnesium Metabolism in Type 2 Diabetes Mellitus
▶ Zinc and Immunity

References

Almoznino-Sarafian D, Berman S, Mor A et al (2007) Magnesium and C-reactive protein in heart failure: an anti-inflammatory effect of magnesium administration? Eur J Nutr 46:230–237
Baker RG, Hayden MS, Ghosh S (2011) NF-κB, inflammation, and metabolic disease. Cell Metab 13:11–22
Iseri LT, French JH (1984) Magnesium: nature's physiologic calcium blocker. Am Heart J 108:188–193
King DE (2009) Inflammation and elevation of C-reactive protein: does magnesium play a key role? Magnes Res 22:57–59
Kramer JH, Spurney C, Iantorno M et al (2009) Neurogenic inflammation and cardiac dysfunction due to hypomagnesemia. Am J Med Sci 338:22–27
Maier JA (2012) Endothelial cells and magnesium: implications in atherosclerosis. Clin Sci (Lond) 122:397–407
Mazur A, Maier JAM, Rock E et al (2007) Magnesium and the inflammatory response: potential physiopathological implications. Arch Biochem Biophys 458:48–56
Nielsen FH (2010) Magnesium, inflammation, and obesity in chronic disease. Nutr Rev 68:333–340
Pasceri V, Willerson JT, Yeh ET (2000) Direct proinflammatory effect of C-reactive protein on human endothelial cells. Circulation 102:2165–2168
Simchowitz L, Foy MA, Cragoe EJ Jr (1990) A role for Na+/Ca^{2+} exchange in the generation of superoxide radicals by human neutrophils. J Biol Chem 265:13449–13456

Magnesium and Vessels

Augusto C. Montezano[1], Tayze T. Antunes[1], Glaucia Callera[1] and Rhian M. Touyz[1,2]
[1]Kidney Research Center, Ottawa Hospital Research Institute, University of Ottawa, Ottawa, ON, Canada
[2]Institute of Cardiovascular & Medical Sciences, BHF Glasgow Cardiovascular Research Centre, University of Glasgow, Glasgow, UK

Synonyms

Calcium (Ca^{2+}); Magnesium (Mg^{2+}); $[Mg^{2+}]_i$ = Intracellular free magnesium concentration; Potassium (K^+); Sodium (Na^+); Vascular remodeling (change in vascular structure); Vascular smooth muscle cell (VSMC); Vasoconstriction (contraction of vessels); Vasodilation (relaxation of vessels)

Definition

Magnesium modulates mechanical, electrical, and structural functions of vascular cells. Small changes in extracellular Mg^{2+} levels and/or intracellular free Mg^{2+} concentration may have significant effects on vascular tone, contractility, reactivity, and endothelial function. Thus, Mg^{2+} may be important in the physiological regulation of the vascular system and in the pathophysiology of cardiovascular disease. Emerging evidence indicates a possible role for Mg^{2+} in diseases associated with vascular damage and impaired endothelial function, such as atherosclerosis, hypertension, and vascular calcification.

Overview

Magnesium is a chemical element. It is abundant in the earth's crust and is also found in all living organisms. In the human body, magnesium is the fourth most abundant ion. In plants, magnesium is the metallic ion at the center of chlorophyll, and hence, green vegetables are a rich nutritional source of magnesium. All enzymes that require phosphate for catalytic function have an obligatory need for magnesium and enzymes that utilize nucleotides to synthesize DNA and RNA require magnesium. Hence, magnesium is essential for all living cells and organisms. At the molecular level, magnesium regulates contractile proteins; modulates transmembrane transport of calcium (Ca^{2+}), sodium (Na^+), and potassium (K^+); acts as an essential cofactor in the activation of ATPase; controls metabolic regulation of energy-dependent cytoplasmic and mitochondrial pathways; and influences DNA and protein synthesis. Mg^{2+} catalyzes over 320 cellular enzymes. Most important among these are enzymes that hydrolyze and transfer phosphate groups, including those associated with reactions involving adenosine triphosphate (ATP) (Grubbs and Maguire 1987). Since ATP is essential for glucose utilization, fat, protein, nucleic acid, and coenzyme synthesis, muscle contraction, methyl group transfer, and many other reactions, alterations in magnesium metabolism can potentially have significant effects on multiple cellular functions.

Small changes in extracellular Mg^{2+} levels ($[Mg^{2+}]_e$) and/or intracellular free Mg^{2+} concentration ($[Mg^{2+}]_i$) have important effects on cardiac excitability and on vascular tone, contractility, and reactivity. Thus, Mg^{2+} may be physiologically important in blood pressure and cardiac regulation whereas changes in Mg^{2+} levels could contribute to cardiovascular disease (Altura et al. 1991). Epidemiological and experimental studies support a role for Mg^{2+} deficiency in the pathogenesis of cardiovascular diseases, with reports demonstrating, in large part, an inverse correlation between body Mg^{2+} levels and cardiovascular risk factors (hypertension, atherosclerosis, diabetes) (Adamopoulos et al. 2009; Kesteloot and Joossens 1988). However, some studies failed to show a relationship between Mg^{2+} and cardiovascular risk (Khan et al. 2010).

Magnesium Metabolism

The normal dietary adult intake of Mg^{2+} is about 300 mg/day, of which approximately 40% is absorbed (Fine et al. 1991). Mg^{2+} absorption depends on the dietary Mg^{2+} content and on body Mg^{2+} status. In patients consuming a low dietary Mg^{2+} or those in negative Mg^{2+} balance, relative Mg^{2+} absorption is increased. Mechanisms controlling Mg^{2+} absorption are unclear, but vitamin D metabolites, parathyroid hormone, calcitonin, insulin, and vasoactive intestinal peptide may be important.

Renal Mg^{2+} excretion, which reflects the dietary intake and amount of Mg^{2+} absorbed from the gastrointestinal tract, is filtered at the glomerulus and reabsorbed along the nephron (especially the ascending limb) (Ferrè et al. 2011). The loop of Henle is the major regulator of Mg^{2+} reabsorption and urinary Mg^{2+} excretion (San-Cristobal et al. 2010). It is at this site that major disturbances of Mg^{2+} reabsorption occur, e.g., furosemide-induced hypomagnesaemia. Hormones regulating renal Mg^{2+} excretion include vasopressin, aldosterone, and thyroid hormone, which increase magnesuria, and parathyroid hormone, insulin, and vitamin D which reduce magnesuria (Glaudemans et al. 2010).

Mg^{2+} is distributed in three major compartments of the body: ~65% in the mineral phase of bone, ~34% in muscle, and ~1% in plasma and interstitial fluid. The intracellular Mg^{2+} stores are important in maintaining plasma levels. Unlike plasma calcium, where 40% is protein bound, only about 20% of plasma Mg^{2+} is protein bound. Consequently, changes in plasma protein concentrations have less effect on plasma Mg^{2+} than on plasma calcium.

Regulation of Intracellular Mg^{2+} in Vascular Cells

For Mg^{2+} to significantly modulate intracellular events, Mg^{2+} itself must be regulated within the cell. Despite the fact that Mg^{2+} is the most abundant cytosolic divalent cation, little is known about intracellular Mg^{2+} homeostasis and mechanisms controlling $[Mg^{2+}]_i$ are poorly understood. Mg^{2+} is a dynamic ion that moves between compartments and across membranes. Mg^{2+} enters cells along a concentration gradient and through recently identified Mg^{2+} transporters, including MagT1, TRPM6, and TRPM7, and is extruded from cells via Mg^{2+} exchangers, such as the Na^+-Mg^{2+} exchanger and Mg^{2+}-ATPase (Quamme 2010). Mg^{2+} exchangers and TRPM6/7 are functionally present in vascular cells and are regulated by vasoactive peptides such as angiotensin II, aldosterone and bradykinin (Touyz and Schiffrin 1996; Yogi et al. 2010; Callera et al. 2009; Touyz et al. 2006). Within cells, Mg^{2+} is compartmentalized in mitochondria, endoplasmic/sarcoplasmic reticulum, and nuclei and is highly regulated by intracellular buffering processes as well as by hormones and biochemical metabolites, such as isoproterenol, glutathione, bradykinin, angiotensin II, vasopressin, aldosterone, norepinephrine, and epinephrine (Yogi et al. 2009, 2010). Intracellular Mg^{2+} is tightly regulated and because Mg^{2+} is a critical component in multiple biochemical reactions, even small changes in vascular smooth muscle cell $[Mg^{2+}]_i$ can have significant effects on signaling pathways that influence contraction and relaxation.

Magnesium and Vascular Smooth Muscle Function

The direct vascular effect of Mg^{2+} was first suggested in the early 1900s when it was observed in clinical studies that Mg^{2+} salt infusion lowers blood pressure via a reduction in peripheral vascular resistance in spite of a slight increase in myocardial contractility. Experimental studies support these clinical observations and confirm that acute Mg^{2+} administration induces hypotension through vasodilatory actions. Vasorelaxation and increased blood flow induced by Mg^{2+} have been observed in various vascular beds. Increased extracellular Mg^{2+} concentration improves blood flow, decreases vascular resistance, and increases capacitance function of the peripheral, renal, coronary, and cerebral blood vessels, while decreasing the extracellular Mg^{2+} concentration exerts opposite effects (Heidarianpour et al. 2011; Barbagallo et al. 2010).

Mechanisms whereby elevations in extracellular Mg^{2+} concentration decrease vascular resistance are not fully understood, but vasorelaxation induced by local application of Mg^{2+} occurs immediately in a dose-dependent manner indicating that the beneficial vascular effects of Mg^{2+} are mediated via direct actions on the vasculature (Gilbert D'Angelo et al. 1992). In vitro studies demonstrated that the increase in extracellular Mg^{2+} concentration inhibits calcium-induced contraction of isolated vessels in calcium-free, high potassium physiological salt solution, indicating that external Mg^{2+} may block calcium influx. Reducing basal vascular tone and spontaneous mechanical activity by raising the extracellular Mg^{2+} concentration is reversed by a concentration-dependent elevation in extracellular calcium. When extracellular Mg^{2+} concentration is below the physiological level, the elevation in mechanical activity and tone is reduced as extracellular calcium is lowered.

The exact molecular mechanisms whereby Mg^{2+} impacts on vascular tone are unknown, but Mg^{2+} probably influences $[Ca^{2+}]_i$, which plays an important role in regulating endothelial, vascular smooth muscle, and

Magnesium and Vessels, Fig. 1 Increased extracellular Mg^{2+} concentration improves blood flow, decreases vascular resistance, and increases capacitance function of the peripheral, renal, coronary, and cerebral blood vessels while decreasing the extracellular Mg^{2+} concentration exerts opposite effects. These effects of Mg^{2+} are mediated via direct actions on the vasculature by regulating $[Ca^{2+}]i$

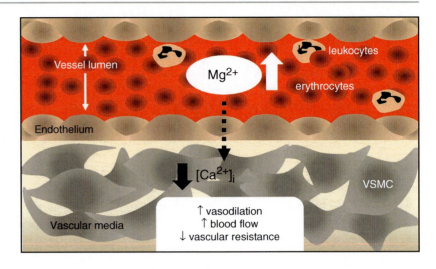

cardiac function. In vascular smooth muscle cells, Mg^{2+} acts extracellularly by inhibiting transmembrane calcium transport and calcium entry and decreasing contractile actions of vasoactive agents or intracellularly as a calcium antagonist, thereby modulating the vasoconstrictor actions of increased $[Ca^{2+}]_i$ (Altura et al. 1987) (Fig. 1).

Mg^{2+} also influences vascular tone by modulating effects of vasoconstrictor and vasodilator agents (Fig. 2). An increase in the extracellular Mg^{2+} concentration attenuates the vasoconstrictor actions and potentiates the vasorelaxant properties of many vasoactive agents. These effects may be related to altered binding of agonists to their specific cell membrane receptors and/or to modulation of $[Ca^{2+}]_i$. Mg^{2+} may also influence production of certain vasoactive agents such as endothelin-1 (ET-1), a potent vasoconstrictor, and prostacyclin (PGI_2), a vasodilator (Laurant and Berthelot 1996). In Mg^{2+}-deficient rats, plasma ET-1 levels are elevated, whereas in Mg^{2+}-supplemented rats, plasma ET-1 levels are reduced. These changes could influence vascular tone as the vasoconstrictor activity and sensitivity to ET-1 are modulated depending on extracellular Mg^{2+} concentration. Increased Mg^{2+} attenuates ET-1-induced contraction whereas reduced Mg^{2+} levels augment ET-1-stimulated contractile responses. Mg^{2+} also influences production and release of vasodilators. Elevation of extracellular Mg^{2+} levels stimulates endothelial release of PGI_2 from human umbilical arteries and from cultured umbilical vein endothelial cells. These effects may be particularly important in $MgSO_4$ treatment of eclampsia and preeclampsia.

Administrating $MgSO_4$ to preeclamptic patients induces significant changes in nitric oxide production which has a major role in modulating vasculature changes in preeclampsia (Pryde and Mittendorf 2009).

Another possible mechanism whereby Mg^{2+} influences vascular function and contractility is via its antioxidant actions. There is increasing evidence that the vasculature is a rich source of reactive oxygen species, which directly influences vascular smooth muscle cell contraction and growth (Weglicki et al. 2011). Mg^{2+} has antioxidant properties that could attenuate damaging actions of reactive oxygen species on the vasculature, thereby preventing increased vascular tone and contractility. These effects may be particularly important in hypertension, diabetes, and atherosclerosis, where generation of reactive oxygen species is increased and $[Mg^{2+}]_i$ is reduced. Low Mg also promotes vascular inflammation by increasing vascular smooth muscle cell activation of ceramide and NFκB and by stimulating production of cytokines.

Magnesium and the Endothelium

In addition to the direct modulatory action that Mg^{2+} has on vascular smooth muscle, it modulates endothelial function (Maier 2012). The vascular endothelium plays a fundamental role in the regulation of vasomotor tone by releasing nitric oxide (NO), ET-1, cyclooxygenase-derived prostanoid(s) such as prostacyclin (PGI_2), and endothelial-derived hyperpolarizing factor. An acute reduction of extracellular Mg^{2+} leads to a transient vasodilation followed by a sustained contraction. When there is endothelial damage, low Mg^{2+} induces a sustained contraction without the

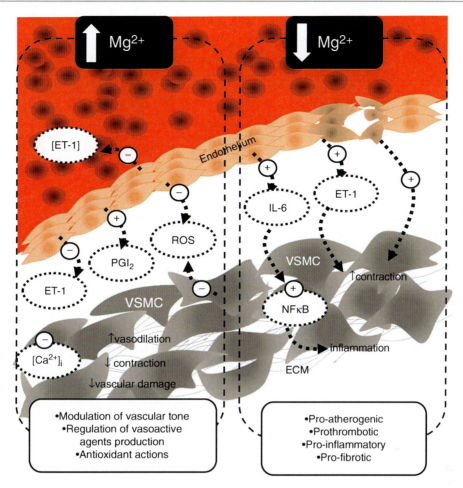

Magnesium and Vessels, Fig. 2 Increase in the extracellular Mg^{2+} concentration attenuates the vasoconstrictor actions and potentiates the vasorelaxant properties of many vasoactive agents. Mg^{2+} decreases contractile actions of vasoactive agents by regulating $[Ca^{2+}]i$. Mg^{2+} may also influence production of vasoactive agents such as endothelin-1 (*ET-1*) and prostacyclin (*PGI$_2$*). Another possible mechanism whereby Mg^{2+} influences vascular function and contractility is via its antioxidant actions. These antioxidant properties may attenuate damaging actions of reactive oxygen species (*ROS*) on the vasculature, thereby preventing increased vascular tone and contractility. In addition to influencing endothelium-dependent vasodilation, low $[Mg^{2+}]$ promotes atherosclerosis and thrombosis through its pro-atherogenic, prothrombotic, and pro-inflammatory effects. Low $[Mg^{2+}]$ also promotes vascular inflammation by increasing vascular smooth muscle cell activation of NFκB and by stimulating production of cytokines. In the presence of an injured endothelium, low $[Mg^{2+}]$ has a direct constrictor effect on vascular smooth muscle. The effects may be particularly important in hypertension, diabetes, and atherosclerosis, where generation of reactive oxygen species is increased and $[Mg^{2+}]$ intracellular is reduced. *ECM* extracellular matrix, *VSMC* vascular smooth muscle cell

transient vasorelaxation phase. These findings suggest that Mg^{2+} could have a dual effect in the regulation of vascular reactivity, depending on the integrity of the endothelium. An intact endothelium prevents against the detrimental effects of acute hypomagnesemia, whereas in the presence of an injured endothelium, as is the case in many cardiovascular diseases, the compensatory vasodilatory effect is absent and low Mg^{2+} has a direct constrictor effect on vascular smooth muscle. The endothelium-dependent relaxation observed after acute Mg^{2+} withdrawal seems to be related to Ca^{2+}-dependent release of NO.

In addition to influencing endothelium-dependent vasodilation, low endothelial cell Mg^{2+} promotes atherosclerosis and thrombosis through its pro-atherogenic, prothrombotic, and pro-inflammatory effects.

At the molecular levels, these processes involve generation of reactive oxygen species, and activation of NFκB and IL-6-sensitive pathways.

Conclusions

Magnesium regulates vascular tone and reactivity by modulating intracellular Ca^{2+}, Na^+, K^+, and pH_i, all of which are important factors in the processes that regulate vascular contraction and relaxation. In addition, Mg^{2+} influences production of vasoconstrictor and vasodilator agents, modulates redox-sensitive processes, and influences molecular events associated with vascular inflammation and calcification. For Mg^{2+} to effectively exert its biological actions, Mg^{2+} itself must be regulated. Mg^{2+} is a mobile cation that moves rapidly between intracellular organelles and across membranes through recently identified exchangers and transporters, such as TRPM6 and TRPM7. Future research will elucidate the exact role of these transporters in the regulation of vascular function and the implications in cardiovascular disease.

Cross-References

▸ Calcium in Biological Systems
▸ Calcium in Health and Disease
▸ Calcium Signaling
▸ Magnesium and Cell Cycle
▸ Magnesium in Eukaryotes
▸ Magnesium in Health and Disease
▸ Magnesium, Physical and Chemical Properties
▸ Magnesium and Inflammation
▸ Magnesium in Biological Systems

References

Adamopoulos C, Pitt B, Sui X et al (2009) Low serum magnesium and cardiovascular mortality in chronic heart failure: a propensity-matched study. Int J Cardiol 136:270–277

Altura BM, Altura BT, Carella A, Gebrewold A, Murakawa T, Nishio A (1987) Mg^{2+}-Ca^{2+} interaction in contractility of vascular smooth muscle: Mg^{2+} versus organic calcium channel blockers on myogenic tone and agonist-induced responsiveness of blood vessels. Can J Physiol Pharmacol 65:729–745

Altura BM, Zhang A, Altura BT (1991) Magnesium, hypertensive vascular diseases, atherogenesis, subcellular compartmentation of Ca^{2+} and Mg^{2+} and vascular contractility. Miner Electrolyte Metab 9(4–5):323–336

Barbagallo M, Dominguez LJ, Galioto A, Pineo A, Belvedere M (2010) Oral magnesium supplementation improves vascular function in elderly diabetic patients. Magnes Res 23(3):131–137

Callera GE, He Y, Yogi A et al (2009) Regulation of the novel Mg^{2+} transporter transient receptor potential melastatin 7 (TRPM7) cation channel by bradykinin in vascular smooth muscle cells. J Hypertens 27:155–166

Ferrè S, Hoenderop JG, Bindels RJ (2011) Insight into renal Mg^{2+} transporters. Curr Opin Nephrol Hypertens 20(2):169–176

Fine KD, Santa Ana CA, Porter JL (1991) Intestinal absorption of magnesium from food supplements. J Clin Invest 88:396–400

Gilbert D'Angelo EK, Singer HA, Rembold CM (1992) Magnesium relaxes arterial smooth muscle by decreasing intracellular Ca^{2+} without changing intracellular Mg^{2+}. J Clin Invest 89:1988–1994

Glaudemans B, Knoers NV, Hoenderop JG, Bindels RJ (2010) New molecular players facilitating Mg(2+) reabsorption in the distal convoluted tubule. Kidney Int 77(1):17–22

Grubbs RD, Maguire ME (1987) Magnesium as a regulatory cation: criteria and evaluation. Magnesium 6:113–127

Heidarianpour A, Sadeghian E, Gorzi A, Nazem F (2011) The influence of oral magnesium sulfate on skin microvasculature blood flow in diabetic rats. Biol Trace Elem Res 143(1):344–350

Kesteloot H, Joossens JV (1988) Relationship of dietary sodium, potassium, calcium, and magnesium with blood pressure. Belgian interuniversity research on nutrition and health. Circulation 12:594–599

Khan AM, Sullivan L, McCabe E, Levy D, Vasan RS, Wang TJ (2010) Lack of association between serum magnesium and the risks of hypertension and cardiovascular disease. Am Heart J 160(4):715–720

Laurant P, Berthelot A (1996) Endothelin-1-induced contraction in isolated aortae from normotensive and DOCA-salt hypertensive rats: effect of magnesium. Br J Pharmacol 119:1367–1374

Maier JA (2012) Endothelial cells and magnesium: implications in atherosclerosis. Clin Sci 122(9):397–407 (Lond)

Pryde PG, Mittendorf R (2009) Contemporary usage of obstetric magnesium sulfate: indication, contraindication, and relevance of dose. Obstet Gynecol 114(3):669–673

Quamme GA (2010) Molecular identification of ancient and modern mammalian magnesium transporters. Am J Physiol Cell Physiol 298(3):C407–C429

San-Cristobal P, Dimke H, Hoenderop JG, Bindels RJ (2010) Novel molecular pathways in renal Mg^{2+} transport: a guided tour along the nephron. Curr Opin Nephrol Hypertens 19(5):456–462

Touyz RM, Schiffrin EL (1996) Angiotensin II and vasopressin modulate intracellular free magnesium in vascular smooth muscle cells through Na^+-dependent protein kinase C-pathways. J Biol Chem 271(40):24353–24358

Touyz RM, He Y, Montezano AC et al (2006) Differential regulation of transient receptor potential melastatin 6 and 7 cation channels by ANG II in vascular smooth muscle cells from spontaneously hypertensive rats. Am J Physiol Regul Integr Comp Physiol 290:R73–R78

Weglicki WB, Chmielinska JJ, Kramer JH, Mak IT (2011) Cardiovascular and intestinal responses to oxidative and nitrosative stress during prolonged magnesium deficiency. Am J Med Sci 342(2):125–128

Yogi A, Callera G, Tostes R, Touyz RM (2009) Bradykinin regulates calpain and proinflammatory signaling through TRPM7-sensitive pathways in vascular smooth muscle cells. Am J Physiol Regul Integr Comp Physiol 296(2): R201–R207

Yogi A, Callera GE, Antunes TT, Tostes RC, Touyz RM (2010) Vascular biology of magnesium and its transporters in hypertension. Magnes Res 23(4):S207–S215

Magnesium Binding Sites in Proteins

Shen Tang[1] and Jenny J. Yang[1,2]
[1]Department of Chemistry, Georgia State University, Atlanta, GA, USA
[2]Natural Science Center, Atlanta, GA, USA

Synonyms

Aspartic acid: Aspartate, Asp, D; Binding site: Binding motif; Calcium: Ca^{2+}, Ca^{2+} ion; Dissociation constant: K_d; EF-hand III: the third EF-hand; Glutamatic acid: Glutamate, Glu, E; Magnesium: Mg^{2+}, Mg^{2+} ion; Millimolar: mM

Definition

Metal coordination chemistry: The spatial geometry of the binding ligands electrostatically interacting with a metal ion.

Binding ligands: Atoms or residues directly interacting with a metal ion.

Continuous/discontinuous binding site: A continuous binding site is formed by sequentially close binding ligands. A discontinuous binding site is formed by binding ligands that are sequentially separated, but spatially close.

EF-hand protein: a continuous Ca^{2+}-binding motif featured with helix-loop-helix secondary structure, and the metal-binding ligands are conserved within the loop region.

Biological Functions of Mg^{2+}

Mg^{2+} is one of the most abundant divalent physiological metals. Similar to intracellular Ca^{2+} concentration, magnesium has a concentration around 25 and 100 mM intracellularly in prokaryotic and eukaryotic cells, respectively. The free intracellular Mg^{2+} concentration is reported around 0.3–0.6 mM with a negligible concentration gradient between plasma membranes, tightly controlled by large amounts of Mg^{2+} chelators, such as ATP and parvalbumin. In contrast, the free Ca^{2+} concentration inside different subcellular organelles varies from sub-micromolar in nucleus and cytosol to sub-millimolar in endoplasmic reticulum and Golgi. Such significant differences in homeostasis between Mg^{2+} and Ca^{2+} is likely a result of large differences in metal coordination chemistry, binding affinity, selectivity, and chemical properties of intracellullar chelating molecules. The Mg^{2+} homeostasis between intracellular and extracellular prokaryotic cells is regulated by several plasma membrane anchored Mg^{2+} transporters of which the high-resolution X-ray crystal structures have been solved recently. The Mg^{2+} transporters have also been found in squid giant axons and barnacle muscle cells in a Na^+- or ATP- dependent way.

Metal Coordination Chemistry

Mg^{2+} has a radius of 0.6 Å that is significantly smaller than that of Ca^{2+} (0.95 Å). The resulting great charge/radius ratio of Mg^{2+} to Ca^{2+} dictates the main differences in coordination properties between Mg^{2+} and Ca^{2+}. Different from the pentagonal bipyramidal geometry of EF-hand-binding site of Ca^{2+} surrounded by seven oxygen atoms with average Ca^{2+}-O distance about 2.4 Å, up to six oxygen atoms octahedrally coordinate Mg^{2+} with average Mg^{2+}-O distance 2.05 Å (Fig. 1a, b).

Usually, at the physiological pH range, Mg^{2+} is hydroxylated with six H_2O molecules with a large hydration energy to form a complex with a large radius of approximately 5 Å. In addition to protein ligands, Mg^{2+} also strongly interacts with phosphate ligands from nucleotides such as ATP and DNA due to its high charge/radius ratio. Coordination number and spatial distribution of water molecules surrounding the Mg^{2+} influence its binding thermodynamics with the protein. Mg^{2+} and its associated water or phosphate

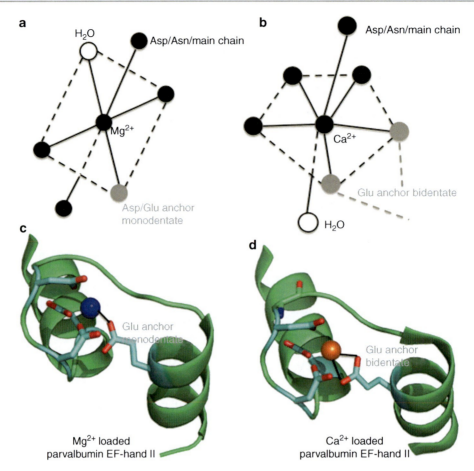

Magnesium Binding Sites in Proteins, Fig. 1 Comparison of coordination difference between Ca^{2+}-binding site and Mg^{2+}-binding site. (**a**) Octahedral coordination of Mg^{2+}-binding site. Mg^{2+} (*black circle*) locates in the center, surrounded by one H_2O (*empty circle*), four protein ligands (*black circle*), and one Asp or Glu as an anchor (*gray circle*) in which one side chain carboxylate oxygen atom monodentately interact with Mg^{2+}. (**b**) Pentagonal bipyramidal coordination of Ca^{2+}-binding site. Ca^{2+} (*black circle*) located in the center, surrounded by one H_2O, four protein ligands (*black circle*), and one Glu as an anchor (*gray circle*) in which two side chain carboxylate oxygen atoms bidentately interact with Mg^{2+}. (**c**) Mg^{2+}-loaded EF-hand II of parvalbumin (PDB code: 4PAL). The Glu (*cyan stick*) in position 12 of the binding loop monodentately (*black line*) interacts with Mg^{2+} (*blue circle*). (**d**) Ca^{2+}-loaded EF-hand II of parvalbumin (PDB code: 4CPV). The Glu (*cyan stick*) in position 12 of the binding loop bidentately (*black line*) interacts with Ca^{2+} (*orange circle*)

complex binds protein with three to five coordinating oxygen atoms from protein, and the side chain carboxylate oxygen plays a major role in the protein-binding ligands. The classification of several representative Mg^{2+}-binding protein sites based on the number of side chain carboxylate is given in Table 1. Mg^{2+} often regulates numerous enzymes through water-mediated interaction. Further, the stable octahedron of Mg^{2+} and higher surface charge also lead to 10^4 times slower exchange rate between the Mg^{2+} and water compared with Ca^{2+}. Because of this, Mg^{2+} cannot function as a fast trigger metal inside of cells.

In addition to maintaining the three-dimensional conformation of DNA and RNA, and interacting with phospholipids to influence the membrane fluidity and permeability, Mg^{2+} binds to diversified classes of proteins such as DNA/RNA polymerases, reverse transcriptases, telomerases, regulating numerous important cellular processes and a wide range of enzymes involved in the metabolism of phosphorylation, or the glycolytic and tricarboxylic acid pathways. The types of Mg^{2+}-binding proteins can be classified into several categories based on the arrangement of the ligand residues, binding

Magnesium Binding Sites in Proteins, Table 1 Classification of representative Mg^{2+}-binding proteins based on the number of side chain carboxylate

PDB entry	Resolution (Å)	Protein	Ligands
Sites with one carboxylate			
4PAL	1.8	Parvalbumin	D53
1INP	2.3	Insitol Phosphatase	E79, I155
1IDO	1.7	I-Domain from Integrin Cr3	E314, S142, S144, T209
Sites with two carboxylates			
1RDD	2.8	Ribonuclease	D10, E48, G11
1WDC	2.0	Scallop myposin	D28, D30, F34
5ICB	1.5	Bovine calbindin	D54, D58, N56, E60
Sites with three carboxylates			
1MPM	2.6	Maltoporin maltose complex	D78A, D78B, D78C
Sites with four carboxylates			
4PAL	1.8	Parvalbumin	D90, D92, D94, E101, M96

coordination properties, metal-binding specificity, and ligand types such as with cofactors (ATP and other nucleotide triphosphates) in a wide range of enzymes.

Metal Selectivity and Affinities of Metalloproteins

The selective binding of Mg^{2+} to metalloproteins rather than other physiological metals is attributed to several factors. For monovalent metals, the high charge to radius ratio of Na^+ and K^+, for example, compared to Mg^{2+}, assists the strong electrostatic attraction to binding sites consisting of several negatively charged ligands Asp/Glu, even though the free metal concentration of these monovalent metals is more than 100-fold higher than that of Mg^{2+}.

Hard metal Mg^{2+} is prone to bind oxygen mainly from the side chain carboxylates of Asp and Glu. Physiological, divalent soft metals, such as Zn^{2+} and Cu^{2+}, prefer to bind nitrogen from the side chain of His. This prevents Zn^{2+} and Cu^{2+} from competing with the major Mg^{2+}-binding sites, though Zn^{2+} was found to bind a different site from Ca^{2+} in Ca^{2+}-binding protein such as calmodulin. Moreover, the approximately 10^6 lower concentration of free Zn^{2+} and Cu^{2+} than Mg^{2+} at 10^{-4} M further reduces their likelihood to occupy the abundant Mg^{2+}-binding sites. The several-magnitude concentration difference of free metals also assists the selectivity of metalloproteins between Mg^{2+} and Ca^{2+}, as the latter is at 10^{-6} to 10^{-7} M. In addition, different from monodentate ligand interaction between carboxylate oxygen and Mg^{2+}, Glu binds Ca^{2+} bidentate fashion, enhancing the metal selectivity of the EF-hand to Ca^{2+} due to the frequent occupancy of Ca^{2+} at position 12 of the binding loop. The minor perturbation of the binding loop observed during Mg^{2+} interaction prevents further interaction between Mg^{2+}-loaded calmodulin and target peptides, affecting calmodulin's biological functions, due to a dramatic, global conformational change triggered by Ca^{2+} binding in general.

The size and the total charges of the binding cavity also influence the metal-binding affinity, as the binding constants (K_a) of Mg^{2+} to numerous intracellular EF-hand proteins are usually lower than 10^4 M^{-1}, in contrast to 10^7 M^{-1} of Ca^{2+}. The experimental K_a of N- and C- domain calmodulin are 3.5×10^6 M^{-1}, 2.0×10^7 M^{-1} to Ca^{2+}, respectively, but 2.7×10^3 M^{-1} and 5.8×10^2 M^{-1} to Mg^{2+}, respectively. The structural basis of Ca^{2+}/Mg^{2+} metal selectivity of EF-hand proteins will be discussed latter.

Major Types of Mg^{2+}-Binding Protein Sites

The major types of Mg^{2+}-binding protein sites can be classified as follows: (1) the conserved, continuous, non-EF-hand peptide containing multiple acidic residues, (2) EF-hand-binding motif and Ca^{2+}/Mg^{2+} selectivity, (3) the discontinuous binding site formed by sequentially distant residues, (4) protein motif binding multiple Mg^{2+} ions, (5) Mg^{2+}-binding sites assembled by different subunits/several monomers, and (6) hybrid of phosphate and protein ligand-binding site.

The Conserved Continuous non-EF-Hand Peptide Containing Multiple Acidic Residues

Different from Ca^{2+}-binding proteins, in which the continuous binding motifs can be identified by primary sequence due to well-established theories and methods (Yang et al. 2002), it is difficult to define general Mg^{2+}-binding motifs, and only a few short continuous Mg^{2+}-binding sequences with high sequence homology have been discovered among the Mg^{2+}-regulated enzymes, such as -NA*DFDGD*-, and -*GDD*- mainly identified in different RNA polymerases, -*D*-NSLYP- and -K-NS(L/V)YG-, found in DNA polymerase, and -YX*DD*- or -LX*DD*- motifs in reverse transcriptase and telomerase, respectively (Dudev and Lim 2007). The acidic residues in bold among these sequences directly binds Mg^{2+}, similar to high preference of D/E as Ca^{2+}-binding ligands. The relatively shorter sequence with fewer binding ligands of Mg^{2+} binding site in comparison to Ca^{2+} binding site is plausibly due to constant hydrolysis of Mg^{2+} and fewer coordination number of Mg^{2+} binding site. A conserved Mg^{2+}-binding sequence ^{281}EFMPELKWS^{289} plays an important role in metal selectivity of CorA channel (Moomaw and Maguire 2010). The mutation of E281 and K287 to other residues does not affect the metal selectivity, so the electrostatics of these two residues do not play significant roles in the biological function of CorA channel. However, the mutation of E285 will substantially impair the function of the CorA channel, though the variant with Ala or Lys in this position maintained highest activity, suggesting the minimal contribution of electrostatic of this residue. Though these results are distinct from the traditional theory of the strong electrostatic interaction between divalent metal ion and acidic residues, it is hypothesized that this conserved loop mainly interacts with the hydrolyzed Mg^{2+} ion but not the dehydrolyzed Mg^{2+} ion.

EF-Hand-Binding Motif and Ca^{2+}/Mg^{2+} Selectivity

Endogenous Ca^{2+}-binding proteins are mainly EF-hand proteins, classified as Ca^{2+} sensors, like calmodulin, transferring the chemical signals (intracellular Ca^{2+} concentration change) to diversified biological responses (interacting numerous targeting peptides through Ca^{2+}-binding-induced conformational change), and Ca^{2+} buffers, like parvalbumin, with strong binding affinity to control the free Ca^{2+} concentration, without experiencing global conformational change.

Due to the similarity of coordination chemistry between Mg^{2+} and Ca^{2+} and high physiological Mg^{2+} concentration, Mg^{2+} is often observed to bind endogenous calcium-binding proteins. Thus, calcium-binding motifs such as EF-hand-binding motifs are often classified as calcium specific, magnesium specific, or mixed type.

Mg^{2+} competes for Ca^{2+}-binding sites to some degree, especially in intracellular environment where both of them are at millimolar total concentration. EF-hand is a Ca^{2+}-binding motif with helix-loop-helix structure originally reported in the X-ray structure of parvalbumin which is intracellularly abundant. Other Ca^{2+}-binding proteins such as calmodulin, troponin C, and the regulatory domain of scallop myosin are reported to have this motif in high-resolution structure and further define the geometry of EF-hand. The canonic EF-hand is featured by a continuous 12 amino acid loop containing multiple D/E occupying positions 1, 3, 5, 9, and 12 as Ca^{2+}-binding ligands flanked with two helices. Special computational algorithms have been designed to predict the EF-hand-based Ca^{2+}-binding sites, based on the structural study of this motif. The alternation of amino acid sequences enables EF-hands to bind Ca^{2+} with K_d from 10^{-4} to 10^{-9} M (Clapham 2007); even they have a similar secondary structure. Recent studies demonstrated that a non-Ca^{2+}-binding protein fused this isolated 12 amino acid loop originating from the third EF-hand of calmodulin that can bind Ca^{2+} with K_d around 10^{-5} to 10^{-4} M even without two helices, though these helices can influence the Ca^{2+}-binding capacity and maintain the secondary structure of EF-hand. This short loop wraps Ca^{2+} or Mg^{2+} ion to form a binding cavity in pentagonal bipyramid and octahedral coordination, respectively, with different binding affinities. One of the key factors to determine the coordination is the amino acid at 12 position, as Glu with long side chain favorably forms bidentate with Ca^{2+} whereas Asp interacts with Mg^{2+} in a monodentate fashion. In general, EF-hand motif that exhibits high Ca^{2+}/Mg^{2+} selectivity featured with glutamate occupied this position. Moreover, the Ca^{2+}/Mg^{2+} metal selectivity and binding affinity of this motif can determine the biological functions of some EF-hand proteins, such as calmodulin in cytosol and parvalbumin in muscle.

Calmodulin has four canonical EF-hand motifs coupled into two pairs distributed in N and C domains, respectively, with hydrogen bonds formed between the loops and generating positive cooperativity within each pair. The calmodulin interaction with Mg^{2+}/Ca^{2+} has been explored by nuclear magnetic resonance (NMR) heteronuclear single quantum coherence (HSQC) titration, and the results suggest (Ohki et al. 1997): (1) Ca^{2+} triggered global conformational change, Mg^{2+} binding only influences the local loop. (2) Mg^{2+} went to the same location as Ca^{2+}, but possibly binds different ligands. (3) Mg^{2+} has preference to binding EF-hand loop I and IV, for loop II and III, only high concentration of Mg^{2+} at the ratio of 1:10 (protein: Mg^{2+}) can trigger significant conformational change. (4) Mg^{2+} binding will influence calmodulin binding affinity to its target peptide. (5) The presence of Mg^{2+} will not influence the binding of Ca^{2+}, as Ca^{2+} can still induce conformational change in the presence of Mg^{2+}. (6) The presence of Mg^{2+} will occupy more than 90% of loop I of calmodulin and slow down the k_{off} value of Ca^{2+}. The metal selectivity between Ca^{2+}/Mg^{2+} of EF-hand proteins is mainly determined by the sequence of loops. The biological function of calmodulin is specially carried by interacting Ca^{2+} as only Ca^{2+}-loaded calmodulin experiencing significant conformational change to interact with numerous targeting peptides.

Parvalbumin is a 12 kd endogenous calcium buffer protein with two canonical EF-hands, however, no cooperativity among the binding sites was demonstrated though hydrogen bonds were reported to exist between the two antiparallel binding loops. Different from calmodulin, the metal selectivity of parvalbumin between Ca^{2+}/Mg^{2+} is achieved by different binding coordination. The Ca^{2+} ion monodentately binds to four oxygen ligands at positions 1, 3, 5, 9 of the binding loop and bidentately binds to glutamate at position 12; however, Mg^{2+} only binds monodentately to the glutamate. The metal-binding specificity of Ca^{2+} or Mg^{2+} is determined by glutamate or asparate in position 12 of the loop, respectively. The D/E replacement at position 12 determining the metal selectivity has been supported by computational simulation, based on E101D mutant (Cates et al. 2002). The in-depth study of coordination difference of Mg^{2+}/Ca^{2+} binding is due to high-resolution X-ray crystal structures of metal-loaded parvalbumin which have been solved. Moreover, parvalbumin does not interact with peptides, so neither Ca^{2+} nor Mg^{2+} binding can trigger significant conformational change (Fig. 1c, d).

Calbindin D9k has one pseudo EF-hand, and in contrast to parvalbumin and calmodulin featured with Glu in position 12 of the binding loop. Asp in this position decreases the metal selectivity between Ca^{2+} and Mg^{2+}, as the sidechain of Asp is too short to coordinate Ca^{2+} with bidentate, leaving an unsealed binding cavity compared to Ca^{2+}-loaded EF-hand (Fig. 2a). Similar to calbindin D9k, position 12 of RLC loop of scallop myosin is also Asp, providing a more preferential site for Mg^{2+}.

For the other endogenous EF-hand proteins, Mg^{2+} binding also helps to stabilize the conformation of protein from the molten globule apo state in the absence of Ca^{2+}. Calcium- and integrin-binding (CIB) protein, calcium-binding protein 1 (CaBP1), and guanylyl cyclase-activating protein 1 (GCAP1) are still able to respond to sub-micromolar Ca^{2+} in the presence of millimolar Mg^{2+}.

The Discontinuous Binding Site Formed by Sequentially Distant Residues

A highly conserved Mg^{2+}-binding motif containing only three discontinuous residues (DED) has been discovered in the RNase H family (Babu and Lim 2010), supported by the Mg^{2+}-loaded X-ray crystal structures of three RNase H from *Escherichia coli* (1rdd), human immunodeficiency virus type 1 (HIV-1) (1o1w), and moloney murine leukemia virus (MMLV) (2hb5). The secondary structure of each individual residue is beta-sheet (first D), alpha-helix (E), and a connection between alpha-helix and beta-sheet (second D). A distinct equilibrium of bidentate and monodentate interaction between the second D and Mg^{2+} was discovered in the MMLV RNase H, whereas the other side chain carboxylates of DED motif from E. coli, HIV-1 and the first D and E from that of MMLV monodentately bound Mg^{2+} (Fig. 2b).

Protein Motif Binding Multiple Mg^{2+} Ions

Two distinct protein motifs binding multiple Mg^{2+} ions were discovered in a recent report of high-resolution structure of bacterial Mg^{2+} transporter MgtE (Hattori et al. 2009). A discontinuous sequence containing three glutamates (E216, 255, and 258) and two aspartates (D259 and 418) with their side chain carboxylates coordinating two Mg^{2+} ions formed the first metal-protein-binding site. The second site was composed by side chain carboxylates of five aspartates

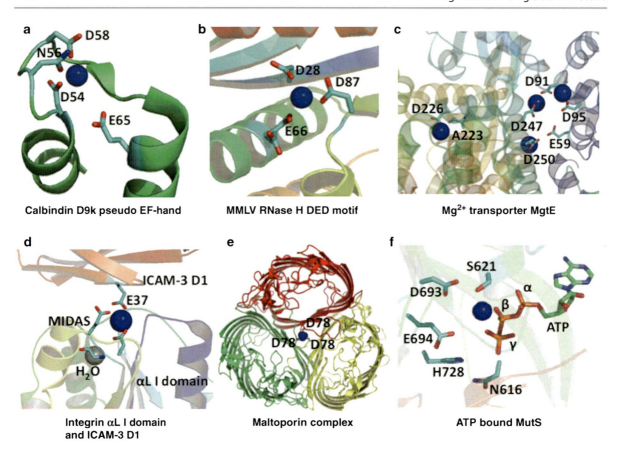

Magnesium Binding Sites in Proteins, Fig. 2 Major class of Mg^{2+}-binding sites. (**a**) Mg^{2+}-loaded pseudo EF-hand of calbindin D9k (PDB code: 3ICB). The E65 in position 12 of the binding loop is too distant to interact with Mg^{2+}. (**b**) Discontinuous Mg^{2+}-binding site of RNase H from moloney murine leukemia virus (MMLV) featured with DED residues (PDB code: 2hb5). (**c**) Multiple Mg^{2+} ions-binding site in Mg^{2+} transporter MgtE (PDB code: 2ZY9). Four Mg^{2+} ions clustered within one domain, and three Mg^{2+} ions partially share the same binding ligands. (**d**) Mg^{2+}-binding site formed by metal ion-dependent adhesion site (MIDAS) of integrin αL I domain and E37 of its ligand ICAM-3 (PDB code: 1T0P). (**e**) Mg^{2+}-binding site is assembled by three maltoporin monomers (*red*, *green*, and *yellow*), and each monomer provides D78 (*blue stick*) as a binding ligand (PDB code: 1MPM). (**f**) Hybrid phosphate and protein-binding site from DNA mismatch repair protein (MutS) (PDB code: 1W7A). The binding residues (*blue stick*) from protein are S621, D693, and E694, and the β-, γ-triphosphate (*orange stick*) of ATP bind Mg^{2+}

(D91, 95, 226, 247, and 250), one glutamate (E59) and mainchain hydroxyl from glycine 136 and alanine 223, binding four Mg^{2+} ions (Fig. 2c). Though some binding ligands were shared by more than one Mg^{2+} ion, the octahedral coordination geometry of Mg^{2+} was preserved by the other octahedron metal Co^{2+} and Ni^{2+} binding these two same sites of MgtE in new crystal structures. Such multiple divalent metals clustered within a high negatively charged, dense cavity is very similar to calcium-binding proteins such as C2 domain of protein kinase c-beta, and cadherins with more than two Ca^{2+} ions bind to the protein.

Mg^{2+}-Binding Site Assembled by Different Subunits

Integrins, a family of cell adhesion molecules noncovalently interacting with diversified ligands to integrate the extracellular and intracellular informations, plays important roles in immunology. In vertebrates, the integrins are heterodimers containing 1 alpha and 1 beta subunit, and in total, 24 different alpha beta pairs are formed by 18 alpha and 8 beta subunits. The Mg^{2+}-binding sites in integrins have been reported in crystal structures of several subfamilies and are crucial to influence integrins binding their ligands.

One universal Mg^{2+}-binding site called metal ion-dependent adhesion site (MIDAS) was initially discovered on the surface of alpha subunit (alphaM I domain) of integrin CR3, and the high-resolution crystal structure showed that MIDAS was formed by a continuous binding loop featured by DXSXS, and two Asp from the other part of the protein. A single Mg^{2+} octahedrally coordinated with three side chain hydroxyl oxygen atoms of S142, S144, and T209, two water oxygen atoms and one carboxylate oxygen atom of E314 of a neighboring A domain with short bond distance of around 2.0 A. After the first shell was fully occupied by six ligands, D140 and D242 did not directly coordinate Mg^{2+} but formed hydrogen bonds with the binding ligands and maintained in the second shell. One carboxylate oxygen atom from both D140 and D242 bound to the same water molecule and the other one bound to the side chain hydroxyl oxygen atom, respectively (D140 to S142 and D242 to S144). However, this open-form alphaM I domain MIDAS can switch to close-form via replacing E314 with D242 as Mg^{2+}-binding ligand, maintaining the same number of negatively charged ligand between these two forms. The site-directed mutagenesis of the second shell-binding ligand D140A and D242A not only abolished the affinity of the recombined protein to divalent metal, but also reduced its binding to ligand iC3b and NIF. A similar affinity decrement was observed in the presence of divalent metal chelators, EDTA, without introducing mutations, suggesting the crucial role of Mg^{2+} in the protein-protein interaction for achieving the biological functions of integrins.

Mg^{2+}-binding site also influences the binding between integrin and ligands from other subfamilies. Instead of forming Mg^{2+}-binding sites within integrin, the Mg^{2+} coordination ligands can come both from the integrin and its ligands. High-resolution crystal structures showed that Mg^{2+} bound to MIDAS on the surface of αL I domain, and the carboxylate group of E37 from the ligand ICAM-3 also directly coordinated Mg^{2+} (PDB codes: 1T0P) (Fig. 2d). A similar Mg^{2+}-binding site of lymphocyte function-associated antigen-1 (LFA-1) was identified, and the site-directed mutagenesis of this identified Mg^{2+}-binding site decreased the binding affinity between LFA-1 and its ligand ICAM-1, suggesting that the Mg^{2+} ions, rather than particular residues, play an important role of Integrin and ligand binding. Moreover, E34 of ICAM-1 directly coordinated Mg^{2+}, as an analogy to E37 of ICAM-1 in binding αL I domain.

Moreover, the crystal structure of maltoporin complex showed an assembled Mg^{2+}-binding site formed within a small triangle space among three parallel beta-can maltoporin monomer with three D78 as binding ligands provided by each monomer (Fig. 2e).

Hybrid Phosphate and Protein-Ligand-Binding Site

Magnesium-binding sites of protein also widely contain the binding ligands from the phosphates and nucleotides functioned as cofactors. One example is the DNA mismatch repair protein (MutS) which plays an important role in recognizing and repairing mismatched DNA base pairs and inserting or deleting short DNA sequences. Mg^{2+} influences DNA mismatch repair protein (MutS) molecular switch, as it can accelerate the ATP-binding process and enhance the binding affinity, influencing the switching of DNA mismatch repair. These conclusions are supported by the mutation experiments that the impaired Mg^{2+}-binding site will prevent the fast switching and DNA mismatch repair (Lebbink et al. 2010).

Magnesium-binding ligands of MutS are a hybrid of phosphate from ATP or ADP and the side chain hydroxyl group of S621, and four water molecules directly coordinated with Mg^{2+} in the first binding shell. The carboxylate oxygen atoms of E694 and D693 form hydrogen bonds with water molecule and S621 in the second shell to stabilize the binding site (Fig. 2f). The non-Mg^{2+}-binding crystal structure of site-directed mutagenesis of D693N suggests the significance of this stabilization to the Mg^{2+} binding.

Computational Algorithm to Predict and Design Mg^{2+}/Ca^{2+}-Binding Sites

Based on rapid understanding of 3D structures of Mg^{2+}-binding motifs and large data bank of structures available, the prediction of Mg^{2+}-binding domains has evolved from initial based on amino acid sequence alignment to current 3D structure homology based computational algorithms (Dudev and Lim 2007), which overcome the limited Mg^{2+}-binding sequence patterns available and the difficulties of identifying discontinuous Mg^{2+}-binding motifs. The optimized prediction parameters rely on the common features of existing

Mg^{2+}-binding cavities, and the concept to describe Mg^{2+}-binding ligands by the first and second shells is also used in computational algorithms to predict Ca^{2+}-binding sites independently (Wang et al. 2010). Such knowledge is important for future design therapeutics to modulate biological functions of Mg^{2+}-binding proteins.

Acknowledgments We are sincerely grateful for Natalie White's helpful editing and Yunmei Lu's insightful advices.

Cross-References

▶ Calbindin D_{28k}
▶ Calcium and Viruses
▶ Calcium-Binding Protein Site Types
▶ Calmodulin
▶ EF-Hand Proteins
▶ Integrins
▶ Magnesium in Biological Systems
▶ Metal-Binding Proteins
▶ Parvalbumin

References

Babu CS, Lim C (2010) Protein/solvent medium effects on Mg^{2+}-carboxylate interactions in metalloenzymes. J Am Chem Soc 132(18):6290–6291
Cates MS, Teodoro ML et al (2002) Molecular mechanisms of calcium and magnesium binding to parvalbumin. Biophys J 82(3):1133–1146
Clapham DE (2007) Calcium signaling. Cell 131(6):1047–1058
Dudev M, Lim C (2007) Discovering structural motifs using a structural alphabet: application to magnesium-binding sites. BMC Bioinformatics 8:106
Hattori M, Iwase N et al (2009) Mg^{2+}-dependent gating of bacterial MgtE channel underlies Mg^{2+} homeostasis. EMBO J 28(22):3602–3612
Lebbink JH, Fish A et al (2010) Magnesium coordination controls the molecular switch function of DNA mismatch repair protein MutS. J Biol Chem 285(17):13131–13141
Moomaw AS, Maguire ME (2010) Cation selectivity by the CorA Mg^{2+} channel requires a fully hydrated cation. Biochemistry 49(29):5998–6008
Ohki S, Ikura M et al (1997) Identification of Mg^{2+}-binding sites and the role of Mg^{2+} on target recognition by calmodulin. Biochemistry 36(14):4309–4316
Wang X, Zhao K et al (2010) Analysis and prediction of calcium-binding pockets from apo-protein structures exhibiting calcium-induced localized conformational changes. Protein Sci 19(6):1180–1190
Yang W, Lee HW et al (2002) Structural analysis, identification, and design of calcium-binding sites in proteins. Proteins 47(3):344–356

Magnesium in Biological Systems

Robert H. Kretsinger
Department of Biology, University of Virginia, Charlottesville, VA, USA

Synonyms

Definition of dissociation constant; Ratio of off rate to on rate

Definition

The Mg^{2+} ion is often bound to phosphate groups of substrates and/or phosphates attached to proteins as posttranslational modifications. Magnesium is also bound to nominal calcium-binding sites of calcium-modulated proteins when $[Ca^{2+}]_{cytosol}$ is $<10^{-7}$ M.

Overview

The Mg^{2+} ion has a similar electron configuration, $1s^2\ 2s^2\ 2p^6$, to that of the Ca^{2+} ion, $1s^2\ 2s^2\ 2p^6\ 3s^2\ 3p^6$; however, its ionic radius is only 0.65 Å as compared with 0.85 Å for Mn^{2+} and 0.99 Å for Ca^{2+}. Its charge density $\sim (.99/.65)^3$ is three times greater than that of calcium and its affinity for electronegative ligands, almost always oxygen in biological systems, is much greater. The magnesium–oxygen bond length is ~2.05 Å. When magnesium is octahedrally coordinated by six oxygen atoms, the oxygen–oxygen distance is $2.05 \times 2^{1/2} = 2.9$ Å, optimal van der Waals contact.

Magnesium does not function as a charge carrier; it exerts its effects by its interactions, primarily with nucleic acids and with proteins. It also binds specifically to the heme group of several chlorophylls, primarily chl-a (P700) of photosystem I and chl-b (P680) of photosystem II. About half of human's total magnesium is incorporated into bone; how it interacts with hydroxylapatite is not known. Both magnesium and calcium interact with the head groups of phospholipids in membrane; however, details of stoichiometries and affinities are not known.

Magnesium has a relatively high affinity and selectivity for the oxygen atoms of phosphate. It is one of the usual counterions for RNA and DNA, along with potassium and polyamines, where it often coordinates two non-bridging phosphates, thereby forming a characteristic ten-atom ring, a "magnesium clamp."

In contrast, many proteins, especially those involved in cell signaling, bind calcium with $\sim 10^4$ higher affinity than they bind magnesium. In quiescent cells, with $[Ca^{2+}]_{cytosol} < 10^{-7}$ M and $[Mg^{2+}]_{cyt} \sim 10^{-3}$ M, these calcium-modulated proteins are in the apo- or magnesi-form. Following a pulse of informational calcium, they assume the calci-form, the next step in information transduction.

The dissociation constant is the ratio of the off-rate to the on-rate: K_d (M) = k_{off} (s^{-1})/k_{on} ($M^{-1}s^{-1}$). The rate-limiting dehydration of $Ca(H_2O)_7^{2+}$ is fast, $\sim 10^{8.0}$ s^{-1}, while that of $Mg(H_2O)_6^{2+}$ is slow, $\sim 10^{4.6}$ s^{-1}. This difference reflects the loose pentagonal bipyramidal versus the tight octahedral packing of the oxygen ligands about the central cation. Magnesium must be (partially) dehydrated before it can bind to the protein or to RNA. The greater affinity of most proteins for calcium relative to magnesium derives primarily from this difference in k_{on}.

Many other proteins – intracellular and extracellular, eukaryotic and prokaryotic – have been characterized with magnesium bound. In most cases, this is assumed, but not proven, to be the functional counterion. The concentrations of $[Mg^{2+}]_{cyt}$ and $[Mg^{2+}]_{extracellular}$ are similar, but not identical, for different cells under different conditions. Several channels and pumps are involved in maintaining magnesium homeostasis.

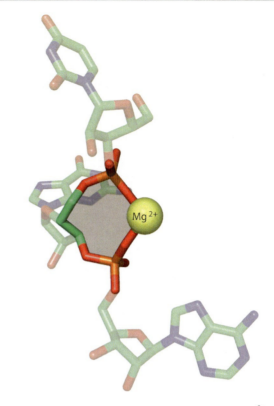

Magnesium in Biological Systems, Fig. 1 A Mg^{2+} ion coordinates the negatively charged, non-bridging oxygen atoms of two adjacent phosphates groups of RNA, thereby forming a ten-atom ring (Courtesy of L. D. Williams, School of Chemistry and Biochemistry, Georgia Institute of Technology, Atlanta, GA, U.S.A.). This so-called magnesium clamp stabilizes the polymer and restricts its flexibility

Magnesium–Nucleic Acid Interactions

Magnesium, as well as potassium and polyamines, are the "diffuse countercations" that balance the negative charges of the phosphates of DNA and various RNA's. In contrast, magnesium is usually, but not always, the "specifically associated" cation that coordinates two or three adjacent or distant, non-bridging phosphate oxygen atoms. In ribosomal RNA and ribozymes, it is coordinated by oxygens, frequently from adjacent phosphates, thereby forming a stabilizing, ten-member ring, a "magnesium clamp" (Fig. 1) (Hsiao et al. 2009).

Magnesium is the preferred counterion for many phosphorylated compounds, most notably ATP and ADP. It is also essential in many enzymatic reactions that involve nucleic acids; its replacement with other divalent or monovalent cations results in greatly diminished activity. It often bridges an oxygen atom of a phosphate group with one or several oxygens from the protein.

Magnesium Binding to Calcium-Modulated Proteins

Several families of calcium-modulated proteins have been characterized in detail, including annexins and those containing the C2 domain (Permyakov and Kretsinger 2010). The family most extensively studied consists of proteins that contain 2–12, tandem EF-hands, sometimes spliced to other domains. Most of the crystal and NMR structures are in the calci-form;

however, several have been determined in the magnesi-form and are inferred to reflect the structure of the protein in the quiescent cell with $[Ca^{2+}]_{cyt} < 10^{-7}$ M and $[Mg^{2+}]_{cyt} \sim 1.0^{-3}$ M. The calcium-binding loop of the EF-hand changes to accommodate the six coordinate magnesium instead of the seven coordinate calcium. This often involves the reorientation of the side chain of Glu at position 12 in the loop so that its carboxyl group coordinates the Mg^{2+} ion with only one oxygen (monodentate) instead of two oxygens (bidentate) as with the Ca^{2+} ion. Also the calcium-binding loop may contract because the Mg–O distance is ~ 2.05 Å; whereas, the Ca–O distance is ~ 2.3 Å. If this contraction cannot be realized, then one of the oxygen (of the protein) ligands may be lost and be replaced by water (Gifford et al. 2007).

Proteins with Magnesium Bound

Several publications refer to "300" magnesium-binding proteins. There are scores of protein homolog families in the Protein Data Bank reported to have magnesium bound. Most of these Mg^{2+} ions are coordinated by six oxygen atoms in octahedral configuration; however, three cautions obtain.

First, these statements of the number of magnesium-binding proteins do not explore whether each of the "300" represents a distinct homolog family, or whether some are paralogs with slightly different characteristics, or whether they are orthologs with different names in different organisms. Often one does not know whether all, or only some, of the orthologs bind magnesium.

It is difficult to establish that magnesium is the functional counterion in vivo. In vitro studies often reveal that several cations – e.g., Mn^{2+}, Zn^{2+}, or Ca^{2+} – activate these enzymes to similar levels. Such evaluations are even more difficult for proteins that are not enzymes. There are few magnesium-specific dyes or indicator proteins, such as aequorin for calcium, that bind magnesium specifically and can be used in vivo. Neither magnesium dyes, such as "mag Fura-2," nor magnesium electrodes are highly sensitive or selective and are not often used. The only radioactive isotope, ^{28}Mg, emits a β particle and has a half-life of only 21 h; it is very expensive. Most magnesium determinations are made by atomic absorption spectroscopy, which accurately gives total magnesium but does not distinguish the free Mg^{2+} ion from bound magnesium.

Third, often the metal content of the protein to be crystallized, and/or of the crystallization medium, was not determined. The electron density observed in the crystal structure, solved with high-resolution data, is sixfold coordinate by oxygen and the "metal," oxygen distance is ~ 2.05 Å; hence, the density is inferred to reflect an Mg^{2+} ion. The only other cation that would be six coordinate and have a similar bond distance, ~ 2.20 Å is manganese, whose ionic radius of 0.80 Å does not usually accommodate seven coordinating oxygens. If there is any question, the Mn^{2+} ion could easily be identified by its anomalous dispersion; K edge = 6.539 keV, 1.896 Å (f″ = +4.0 electrons, f′ = −8.0 electrons) from x-ray diffraction data measured at a synchrotron beam line. Magnesium cannot be so identified; its K edge (1.30 keV, 9.500 Å) is far beyond the "optics" of contemporary synchrotrons.

Many of the actions of magnesium reflect its role as a cofactor for a wide range of enzymes. It activates many of the enzymes involved in the metabolism of phosphorylated compounds. When bound to ATP in the active site of an enzyme, not all of the normal, six coordination positions for ligands are filled by interaction with either the protein or with ATP. One or more water molecules remain coordinated to the Mg^{2+} ion. The function of its binding to the phosphoryl groups of ATP appears in many cases to be activation of the phosphate ester toward hydrolysis (reviewed by Maguire and Cowan 2002). Examples include:

Mandelate racemase binds an Mg^{2+} ion with three carboxyl groups. Mandelate displaces two waters from the coordination sphere of the Mg^{2+} ion and binds it with a carboxyl group and a hydroxyl group of the protein. The magnesium holds the substrate in an orientation so that the catalytic abstraction and addition of a hydrogen atom by His 297 or Lys 166 is precisely executed.

Topoisomerases play a central role among enzymes involved in processing the main chain of DNA, (Sissi and Palumbo 2009). They participate in essential cellular processes, such as DNA replication, transcription, and chromosome condensation; this requires that the enzymes be able to regulate the ensuing topological changes produced in the nucleic acid. Topoisomerases produce these changes in the conformation of DNA by the unwinding or by supercoiling the double helix, thereby releasing the torsional strain imposed by

DNA processing. The topoisomerase-mediated cleavage process consists of a nucleophilic attack of a Tyr located in the catalytic pocket on a phosphodiester bond of the DNA backbone. The transphosphorylation reaction produces a covalent protein, nucleic acid linkage, and a free hydroxyl group at the split deoxyribose group. Topoisomerases often require cofactors for full catalytic activity. ATP regulates the conformational changes required for enzyme action through its binding and subsequent hydrolysis. Topoisomerase I and topoisomerase II require magnesium to relax supercoiled DNA. They have a conserved domain about 100 residues long, called Toprim, which has an invariant Glu and a DxD motif. This Glu is located in a sharp turn connecting a β-strand to an α-helix. The three acidic residues are near one another and available for concerted interactions. Given its chemical nature and the requirement for magnesium, this triad motif has been proposed to be the Me^+ ion(s) binding element in the catalytic core. In the topoisomerase II family, the DxD motif is extended to a conserved DxDxD pattern that offers an additional carboxylic site to generate a structural and electronic network for coordination and correct positioning of the catalytically relevant Mg^{2+} ion(s).

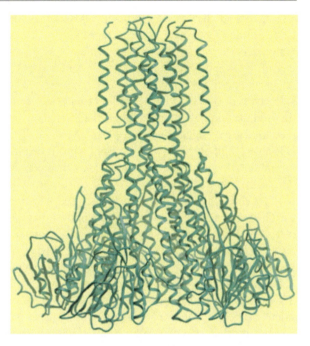

Magnesium in Biological Systems, Fig. 2 Structure of the magnesium transporter CorA of *Thermatoga* (PDB file 2BBJ) (Courtesy of E. A. Permyakov, Institute for Biological Instrumentation, Russian Academy of Sciences, Moscow, Russia). CorA is a homo-pentamer; each monomer has two transmembrane helices. The central pore is formed by the five TM1 helices (residues 293–315); these are enclosed by five TM2 "stem" helices (326–346), which extend 100 Å into the cytosol

Magnesium Channels and Pumps

The concentration of free Ca^{2+} ions in cells varies over several decades in tenths of seconds. Organisms evolved strategies for keeping cytoplasmic concentrations of $[Mg^{2+}]_{cyt}$ near constant ~1.0 mM, in spite of fluctuations of magnesium concentration outside and of calcium inside the cell. The Gram-negative bacterium, *Salmonella enterica*, has three magnesium transporters – CorA, MgtA, and MgtB – and a regulatory system – PhoP/PhoQ – that respond to the external levels of magnesium (Chamnongpol and Groisman 2002).

CorA is the major magnesium transporter in *Salmonella*; homologs have been identified in over 30 organisms from eubacteria, archea, and eukaryotes. Although CorA can mediate both influx and efflux, influx is believed to be its primary physiological function. Besides magnesium, CorA can also transport cobalt, nickel, and iron, although with a reduced efficacy. CorA of *Thermatoga* is pentameric and funnel shaped as seen in the crystal structure (Fig. 2).

Each monomer contains two transmembrane helices; the pore is formed by the TM1 helix (residues 293–315), which is kinked at Pro303 and Gly312. The narrow opening of the pore is at the membrane interface; the wider mouth is within the cell. It is enclosed by the TM2 helix (residues 326–346); this "stem" helix extends about 100 Å into the cytosol, the longest continuous helix found in proteins, except the coiled-coil of tropomyosin. The rest of the N-terminal part of CorA forms a separate domain outside the funnel. It consists of an αβα sandwich comprising two sets of three α-helices ($\alpha_1\alpha_2\alpha_3$ and $\alpha_4\alpha_5\alpha_6$) on either side of a seven-stranded antiparallel β-sheet ($\beta_2\uparrow\beta_1\downarrow\beta_3\uparrow\beta_7\uparrow\beta_6\downarrow\beta_5\uparrow\beta_4\downarrow$). The relatively long α_5- and α_6-helices, which extend back toward the membrane, are called "willow" helices as they hang down like the branches of a willow tree. The loop between β_5 and β_6 extends back toward the membrane surface. The tips of this loop and the tips of the willow helices have many Asp's and Glu's. This high concentration of negative charge is counterbalanced by a unique ring of

Lys's, termed the "basic sphincter." Asn314 at the extracellular membrane surface appears to block completely the entrance to the ion channel. The diameter of the channel varies from about 6 Å to about 2.5 Å, it is too narrow for a hydrated Mg^{2+} ion to pass. The side chain of Met302 constricts the channel to about 3.3 Å in the middle of the membrane; additional obstacles to ion movement are Leu294 and Met291. The side chains of these hydrophobic residues protrude into the interior of the channel, narrowing its diameter to 2.5 Å. Both blockage by Asn314 and constriction by Met302 show that this is a closed state of CorA. It is assumed that the negatively charged willow helices can pull the positive charge of the basic sphincter away from the central axis of CorA at the level of Leu294 and Met291. This would allow the stalk helix to act as a lever to rotate or otherwise move the TM1 α-helix, which forms an extension of the membrane. Together, these movements would have the combined effect of opening both the Leu294 barrier and the Met291, Asn314 barrier at both ends of the channel.

MgtA and MgtB are P-type ATPases that primarily mediate magnesium influx. They can also mediate nickel influx at a lower capacity.

PhoP/PhoQ is a two-component regulatory system that mediates adaptation to magnesium-limiting environments (Chamnongpol and Groisman 2002). Growth in micromolar concentrations of magnesium promotes transcription of PhoP-activated genes; in contrast, growth in millimolar concentrations of magnesium represses transcription of PhoP-activated genes. Residues G93, W97, H120, and T156 in PhoQ are required for a wild type response to magnesium; its binding to the periplasmic domain seems to regulate several activities of the PhoQ protein.

In mammals, magnesium homeostasis is strictly controlled and depends on a balance between intestinal absorption and renal excretion (Voets et al. 2004). The kidney provides the most sensitive control of magnesium balance. About 80% of the total serum magnesium is filtered through the glomerular membrane. The final excretion of magnesium to the urine is determined by the active reabsorption of magnesium in the distal convoluted renal tubules. They contain a magnesium-permeable channel, TRPM6 (transient receptor potential channel, melastatin subtype 6), that is specifically localized to the apical membrane of magnesium-reabsorbing tubules in the kidney and to the brush border membrane of the magnesium absorptive cells in the duodenum. The tight regulation of the TRPM6-induced current by intracellular magnesium provides a feedback mechanism for regulation of influx and implies that intracellular magnesium buffering and extrusion mechanisms strongly impact channel functioning. Parvalbumin and calbindin D_{28K}, which are co-expressed with TRPM6, might function as intracellular magnesium chelators in the distal convoluted renal tubule.

Chlorophyll

Richard Martin Willstätter received the Nobel Prize in 1915 for his studies of chlorophylls and specifically the demonstration that magnesium is an integral part of the molecule. It can be removed under acid conditions. The porphyrin ring remains intact and the functioning molecule can be reconstituted by addition of magnesium. The Mg^{2+} ion just fits in the plane of the four surrounding nitrogen atoms; its +2 charge is balanced by the −2 charge distributed over the ring. Waters are coordinated at the other two vertices of the octahedron. Chlorophyll is quite specific in its requirement for magnesium, as opposed to other divalent cations, e.g., Mn^{2+}, Fe^{2+}, Zn^{2+}, and Ca^{2+}.

As noted, magnesium is essential for many enzymes because it functions as the preferred counterion for phosphoryl groups. By analogy magnesium is essential for the proteins of photosystems I and II. It seems that neither proteins nor biochemists have evolved or designed a protein that directly binds the Mg^{2+} ion in a rigid ring with four nitrogens.

Cross-References

▶ Calcium in Biological Systems
▶ EF-hand Proteins and Magnesium
▶ Magnesium
▶ Magnesium and Cell Cycle
▶ Magnesium in Eukaryotes
▶ Magnesium in Health and Disease
▶ Magnesium in Plants
▶ Magnesium, Physical and Chemical Properties
▶ Potassium in Biological Systems

References

Chamnongpol S, Groisman EA (2002) Mg^{2+} homeostasis and avoidance of metal toxicity. Mol Microbiol 44:561–571

Gifford JL, Walsh M, Vogel HJ (2007) Structures and metal-ion-binding properties of the Ca^{2+}-binding helix–loop–helix EF-hand motifs. Biochem J 405:199–221

Hsiao C, Tannenbaum E, van Deusen H, Hershkovitz E, Pering G, Tannenbaum AR, Williams LD (2009) Complexes of nucleic acids with group I and II cations. In: Hud NV (ed) Nucleic acid – metal ion interactions. RSC Biomolecular Sciences, London

Maguire ME, Cowan JA (2002) Magnesium chemistry and biochemistry. Biometals 15:203–210

Permyakov EA, Kretsinger RH (2010) Calcium binding proteins. Wiley, Hoboken

Sissi C, Palumbo M (2009) Effects of magnesium and related divalent metal ions in topoisomerase structure and function. Nucleic Acids Res 37:702–711

Voets T, Nilius B, Hoefs S, van der Kemp AW, Droogmans G, Bindels RJ, Hoenderop JG (2004) TRPM6 forms the Mg^{2+} influx channel involved in intestinal and renal Mg^{2+} absorption. J Biol Chem 279:19–25

Magnesium in Eukaryotes

Andrea M. P. Romani
Department of Physiology and Biophysics,
School of Medicine, Case Western
Reserve University, Cleveland, OH, USA

Synonyms

Mg^{2+} (magnesium ions); $[Mg^{2+}]_i$ (free magnesium concentration)

Definition

The concept that cellular Mg^{2+} content remains stable has been replaced in recent times by the notion that the cell can dynamically modulate total Mg^{2+} content as a result of hormonal and metabolic stimuli, with minimal variation in free cytoplasmic Mg^{2+} concentration. This notion suggests the operation of abundant and powerful transport mechanisms in the cell membrane as well as in the membrane of cellular organelles, namely, mitochondria, endoplasmic reticulum, and Golgi, and supports the idea that changes in cellular Mg^{2+} content regulate enzymes located predominantly within cellular organelles.

Overview

Magnesium is the second most abundant cellular cation after potassium. High concentrations of total and free magnesium ion (Mg^{2+}) have been measured within mammalian cells through a variety of techniques. These concentrations are essential to regulate numerous cellular functions and enzymes, including ion channels, metabolic cycles, and signaling pathways. Despite significant progress, our understanding of the regulation of cellular Mg^{2+} homeostasis is far from incomplete. In the absence of metabolic and hormonal stimuli, the majority of mammalian cells exhibits a slow Mg^{2+} turnover with limited changes in cytoplasmic $[Mg^{2+}]_i$. As a consequence, it has been assumed that cellular Mg^{2+} concentration does not change significantly over time but it is constantly maintained at a level more than adequate for its role of cofactor for various cellular enzymes, and there has been limited interest in developing techniques and methodologies to measure changes in cellular Mg^{2+} content. This assumption has drastically changed in the last two decades as an increasing body of evidence indicates that metabolic or hormonal stimuli elicit major Mg^{2+} fluxes in either direction across the plasma membrane of mammalian cells leading to changes in total and free Mg^{2+} content within the cytoplasm and cellular organelles alike. This has resulted in the identification of several Mg^{2+} entry mechanisms, and it has indicated a major regulatory role for Mg^{2+} in a variety of cellular functions. The interest to understand how Mg^{2+} regulates biological functions has also promoted the development of new techniques to measure in a more reliable manner variations in total and free Mg^{2+} content within mammalian cells.

Cellular Mg^{2+} Distribution

Total Mg^{2+} concentration ranging between 17 and 20 mM in the majority of mammalian cell types (Table 1 in Romani and Scarpa 1992; Wolf et al. 2003), with concentrations between 16 and 18 mM being measured within mitochondria, nucleus, and endoplasmic or sarcoplasmic reticulum (Fig. 1). Only a small precentual

Magnesium in Eukaryotes, Fig. 1 Cellular Mg^{2+} distribution: The cartoon summarizes the principal cellular pools involved in Mg^{2+} compartmentation within a eukaryotic cell, and the total and free Mg2+ concentrations measured within each of these compartments

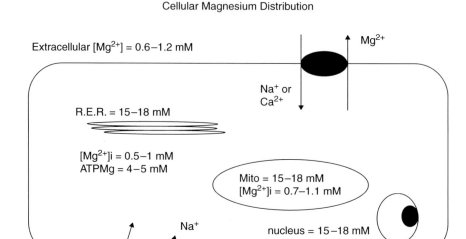

fraction is free within the cytoplasm or the organelles' lumen. For example, free $[Mg^{2+}]$ in the matrix of cardiac and liver mitochondria ranges between 0.8 and 1.2 mM, that is, a value very similar to that measured in the cytoplasm of cardiac myocytes and hepatocytes, or in the extracellular space (Romani and Scarpa 1992). A similar $[Mg^{2+}]_i$ is assumed to be present in the nucleus and the endo-(sarco)-plasmic reticulum, although technical limitations (i.e., elevated luminal Ca^{2+} concentration with three orders of magnitude higher affinity for magnesium fluorescent dyes than Mg^{2+} itself) have prevented proper Mg^{2+} determination.

Cytoplasmic Mg^{2+} represents the best detectable pool of Mg^{2+} within the cell. The majority of Mg^{2+} ($\sim$4–5 mM) in this pool is in the form of a complex with ATP, phosphonucleotides, and phosphometabolites, cytosolic $[Mg^{2+}]_i$ ranging between 0.5 and 1 mM (i.e., less than 5% of total cellular Mg^{2+} content). Overall, this distribution supports the presence of a very limited chemical Mg^{2+} gradient across the cell membrane, and across the membrane of cellular organelles.

Little is known about the ability of cellular proteins to bind Mg^{2+}. Aside from hemoglobin, calmodulin, troponin C, parvalbumin, and S100 protein, no information is available as to whether other cytoplasmic or intra-organelle proteins can bind Mg^{2+} (Romani 2007). Moreover, the physiological relevance of Mg^{2+} binding by proteins has been called into question by the report that no hypomagnesemia or detectable changes in tissue Mg^{2+} homeostasis are observed in parvalbumin *null* mice.

Mg^{2+} Entry Mechanisms

Eukaryotic cells stably maintain Mg^{2+} content below the predicted electrochemical equilibrium potential of $\sim$50 mM (Flatman 1984) even when significant inward and outward gradients are imposed across the cell membrane. This observation strongly supports the notion that eukaryotic cells possess powerful and abundant Mg^{2+} transport mechanisms that maintain total and *free* Mg^{2+} content within the measured levels. The main transport mechanisms involved in regulating cellular Mg^{2+} homeostasis are summarized in Fig. 2. These mechanisms can be cursorily divided into channels, predominantly involved in promoting Mg^{2+} accumulation, and exchangers, by and large involved in favoring Mg^{2+} extrusion.

Channels: Magnesium entry through channels was first observed in prokaryotes and protozoan. More recently, several channels or channel-like structures responsible for Mg^{2+} entry have been identified in eukaryotic cells as well. Exhibiting a relatively high specificity for Mg^{2+}, these channels are predominantly located in the cell membrane, but also in the mitochondrial membrane or in the Golgi cysternae (Fig. 2). The interested audience is referred to severeal recent reviews (Touyz et al. 2006; Schmitz et al. 2007; Quamme 2010) for a more detailed description of the

Magnesium in Eukaryotes, Fig. 2 Mg^{2+} transport mechanisms: The cartoon summarizes the main transport mechanisms responsible for Mg^{2+} extrusion and entry in eukaryotic cells (see text for more details). *NMX* Na/Mg exchanger, *SLC41* solute carrier family 4, *TRPM6/7* Transient Receptor Potential, Melastatin subfamily, isoform 6 or 7, *MMgT* Membrane Mg^{2+} transporter

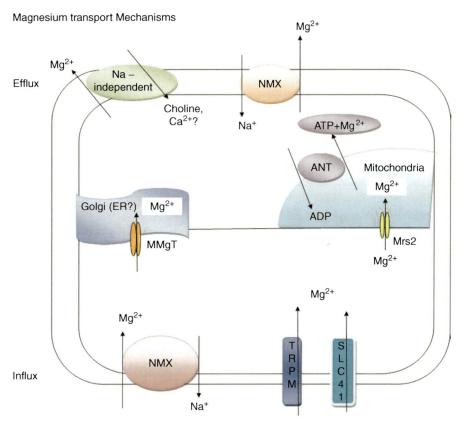

intrinsic characteristics of each of the Mg^{2+} entry mechanisms summarily mentioned in this entry.

TRPM Channels: TRPM7 and TRPM6 were the first Mg^{2+} channels identified in mammalian cells. These two channels share ~70% homology, and show a very distinct localization. TRPM7 is ubiquitous and therefore more in control of Mg^{2+} homeostasis in individual cells. In contrast, TRMP6 is specifically localized in the colon and the distal convolute tubule of the nephron, a distribution that strongly emphasizes the role of this channel in controlling whole body Mg^{2+} homeostasis via intestinal absorption and renal resorption. Both channels combine a channel structure with an alpha-kinase domain at the C-terminus, thereby the term "*chanzyme*" used to indicate these channels and their dual biological actvitiy. Despite the homology in structure and sequence, these two channels present significant differences.

TRPM7 inward current is markedly enhanced by protons (acidic pH), which compete with Ca^{2+} and Mg^{2+} for binding sites within the channel pore, and increase in PIP2 level, which decreases TRPM7 run-down (Romani 2007; Schmitz et al. 2007). Activation of TRPM7, however, only occurs in the presence of physiological [Mg^{2+}]$_i$, within the cell as reduction in [Mg^{2+}]i below its physiological levels results in PLC-mediated inactivation of the channel activity. A functional TRPM7 is also required for sustained phosphoinositide-3-kinase-mediated signaling and cell growth, to the point that TRPM7 deficient cells exit cell cycle and enter quiescence (Romani 2011).

The alpha-kinase domain of TRPM7 phosphorylates serine and threonine residues within an alpha-helix (Schmitz et al. 2007). Not necessary to activate the channel, this domain undergoes massive autophosphorylation, which promotes target recognition and enhances substrate phosphorylation rate (Romani 2011; Schmitz et al. 2007). Presently, only annexin I, myosin IIA heavy chain, and calpain have clearly been identified as phosphorylation substrates for the TRPM7 kinase domain (Romani 2011; Schmitz et al. 2007), highlighting a double role of TRPM7 in regulating Mg^{2+} homeostasis and also cell adhesion, contractility, or inflammation, based upon the cell type. Removal of the kinase domain results in embryonic lethality in homozygous mice, and hypomagnesemia in heterozygous mice (Romani 2011). Embryonic stem cells lacking TRPM7 kinase domain show an arrest in proliferation that can be rescued by Mg^{2+} supplementation. Hence, it has been proposed that TRPM7 regulates the rate of protein synthesis based upon Mg^{2+} availability by phosphorylating eEF2 via eEF2 cognate kinase (eEF2-k), a direct target of TRPM7 kinase-mediated phosphorylation (Romani 2011).

TRPM6 null mice have also been developed. While heterozygous $Trpm6^{+/-}$ mice present a modest 10% decrease in plasma Mg^{2+} level, the majority of the homozygous $Trpm6^{-/-}$ mice die by embryonic day 12.5 unless a high Mg^{2+}- diet is administered to the dams (Romani 2011; Schmitz et al. 2007).

TRPM6 also possesses an alpha-kinase domain at its C-terminus and, as in the case of TRPM7, removal of the kinase domain does not abolish channel activity but affects its regulation by cellular free Mg^{2+} or Mg*ATP complex (Schmitz et al. 2007), and the ability of TRPM6 to phosphorylate downstream targets. Presently, with the exception of TRPM7 itself, no specific substrate for the TRPM6 kinase has been identified. Hence, it is undefined whether TRPM7 and TRMP6 phosphorylate similar or different substrates within the tissues in which they are expressed (Schmitz et al. 2007).

In terms of expression and regulation, TRPM6 is quite different from TRPM7. For one, in vivo expression and activity of TRPM6, but not TRPM7, are extremely sensitive to changes in estrogens (17β-estradiol) level and dietary Mg^{2+} intake as both upregulate TRPM6 mRNA in colon and kidney in a selective manner (Schmitz et al. 2007). TRPM6 activity is inhibited by cellular ATP in a manner similar to that reported for TRPM7. The site of inhibition resides in a conserved ATP-binding motif within the alpha-kinase domain. The effect of ATP, however, does not depend on alpha-kinase autophosphorylation activity (Romani 2011; Schmitz et al. 2007). Lastly, TRPM6 activity is negatively modulated by RACK1 (*r*eceptor for *a*ctivated protein *k*inase *C*) and positively regulated by protein kinase C activation (Schmitz et al. 2007). Interestingly, RACK1 binds TRPM6 in the same region involved in estrogen regulation.

Recent evidence indicates EGF as an autocrine/paracrine magnesiotropic hormone, activating TRPM6 at the apical domain of the cell through ERK1/2 signaling and inducing cellular Mg^{2+} accumulation. The relevance of EGF in TRPM6/Mg^{2+} reabsorption is underscored by the clinical evidence that cancer patients undergoing treatment with antibodies anti-EGFR present massive renal Mg^{2+} wasting (Romani 2011).

Several other putative Mg^{2+} entry mechanisms have been recently identified. They include:

Claudins: Claudin-16, originally termed paracellin-1, is a member of the claudin family, which comprehends a group of tight junction proteins with four transmembrane spans coordinated by two extracellular loops, and with both C- and N-termini on the cytoplasm side. More than 20 mutations affecting trafficking or permeability of claudin-16 have been currently identified (Romani 2011).

Claudin-16 mediates paracellular Ca^{2+} and Mg^{2+} fluxes throughout the nephron by increasing paracellular Na^+-permeation, which, in turn, generates a positive potential within the lumen of the nephron that acts as driving force for Mg^{2+} and Ca^{2+} reabsorption. It is worth nothing that claudin-16 expression is positively modulated by the Mg^{2+} concentration present in the extracellular medium (Romani 2011). Claudin-16 can form head-to-head complexes with claudin-19 at the level of the tight junction, with enhanced cation selectivity.

MagT1: This protein does not present a significant homology to prokaryotic Mg^{2+} transporters but it exhibits some similarities with the oligosaccharide transferase complex OST3/OST6 that regulates protein glycosylation in yeast (Quamme 2010). Highly expressed in liver, heart, kidney, and colon, MagT1 appears to possess high specificity for Mg^{2+} (Quamme 2010). Limited information is available about N33, a second member of the MagT family, which transports Mg^{2+} with lower specificity than MagT1. Because of its high selectivity for Mg^{2+}, MagT1 is considered essential to regulate Mg^{2+} homeostasis in eukaryotes as knocking out MagT1 or its human homolog TUSC3 markedly reduces cellular Mg^{2+} content.

Mrs2: Mrs2 shows several regions of structural homology and a similar two transmembrane domains topology to the bacterial transporter CorA. Mutant yeasts lacking Mrs2 show a decrease in total mitochondrial and matrix free Mg^{2+} level, and they can be rescued by CorA fused to the mitochondrial N-terminus leader sequence of Mrs2 (Romani 2011). Cells deprived of Mrs2 present depressed mitochondrial complex I expression, reduced mitochondrial Mg^{2+} level, and major morphological changes. Overexpression of Mrs2 results instead in a marked increase in matrix free Mg^{2+}, thus supporting an essential role of Mrs2 in regulating mitochondrial Mg^{2+} homeostasis. Functionally, Mrs2 operates as a $\Delta\psi$-dependent channel (Romani 2011).

MMgTs: MMgT1 and MMgT2 (*m*embrane Mg^{2+} *t*ransporter 1 and 2) are located in the Golgi complex and post-Golgi vesicles in which they may contribute to the regulation of Mg^{2+}-dependent enzymes involved in protein assembly and glycosylation (Quamme 2010). However, their transport via Golgi vesicles to the cell membrane or other cellular destinations cannot be excluded. Their short amino acid sequence (131 for MMgT1, and 123 for MMgT2) suggests that these proteins form homo- or hetero-oligomeric channels to favor Mg^{2+} permeation. Both MMgT1 and MMgT2 lack specificity for Mg^{2+} as they can also transport other cations (Quamme 2010).

SLC41: This family of Mg^{2+} transport mechanisms includes three members that are distantly related to prokaryotic MgtE channel. SLC41A1 was the first member of the family to be identified. Broadly distributed, this protein can be markedly upregulated in tissues like kidney following prolonged exposure to low Mg^{2+} diet (Quamme 2010). Functional expression of mouse SLC41A1 in *X.Laevis* oocyte indicates that this protein can transport not only Mg^{2+} but also other cations (Romani 2011). Controversy exists as to whether SLC41A1 operates as a channel or an electrogenic antiporter to promote Mg^{2+} influx, or it operates as a carrier and actually promotes Mg^{2+} efflux.

A second isoform labeled SLC41A2 has been identified. It shares ~70% homology with SLC41A1 and can transport other divalent cations in addition to Mg^{2+} (Quamme 2010). Widely expressed in mammalian tissues, SLC41A2 expression is not affected by low Mg^{2+} diet, and hydrophobicity analysis indicates that the C- and N- termini are located on different sites of the cell membrane, supporting an 11 transmembrane segments configuration (Romani 2011).

As for SLC41A3, the third isoform of the family, no study has investigated in detail its operation and cation selectivity.

ACDP2: The human *ACDP* gene family comprises four isoforms. ACDP1 is restricted to the brain. ACDP2 is more widely expressed, but it is absent in skeletal muscles. ACDP3 and ACDP4 are both ubiquitous, but have the highest expression in the heart [1]. Termed *a*ncient *c*onserved *d*omain *p*rotein because all isoforms share one domain phylogenetically conserved from bacteria to man, these proteins are >50% homologous to the prokaryotic CorC transporter in bacteria. This protein can transport a variety of divalent cations including Mg^{2+} (Romani 2011). Mg^{2+} transport via ACDP2 is voltage dependent and does not require the presence of extracellular Na^+ or Cl^-. *ACDP2* gene also becomes overexpressed following exposure to Mg^{2+} deficient diet (Quamme 2010).

NIPA: The *NIPA1* gene is so called for *n*on-*i*mprinted in *P*rader-*W*illi/*A*ngelman syndrome (Quamme 2010; Romani 2011). Yet, it is unknown whether this syndrome presents alteration in Mg^{2+} homeostasis. Located among a set of 30 genes linked to the disease, *NIPA1* has also been implicated in autosomal dominant *h*ereditary *s*pastic *p*araplegia. The human and mouse genomes contain four members of the NIPA family, termed NIPA1 trough NIPA4, with an overall similarity of ~40%. NIPA1 and NIPA2, but not NIPA3 and NIPA4, can operate as Mg^{2+} transporters carrying Mg^{2+} in a saturable fashion, with different K_m and specificity, and only NIPA2 shows high specificity for Mg^{2+} (Quamme 2010).

Huntingtin: Both *H*untingtin-*i*nteracting *p*rotein 14 (HIP14) and its related protein HIP14-like (HIP14L) are significantly upregulated (~3-fold) by low-extracellular Mg^{2+} (Quamme 2010). HIP14 also possesses a cytoplasmic DHHC cysteine-rich domain that confers palmitoyl-acyltransferase activity to the protein, and gives it the ability to palmitoylate membrane components. Mg^{2+} accumulation via HIP14 and HIP14L appears to be electrogenic, voltage-dependent, and saturable (Quamme 2010). Although primarily located in Golgi and post-Golgi vesicles, the widespread tissue distribution and intracellular localization of HIP14 (nuclear and perinuclear regions, Golgi complex, mitochondria, microtubules, endosomes, clathrin-coated and non-coated vesicles, and plasma membrane) potentially implicates this protein in numerous cellular processes.

Extrusion Mechanisms

Mg^{2+} extrusion occurs through the operation of two exchange mechanisms termed Na^+-dependent and Na^+-independent Mg^{2+} exchanger, respectively (Fig. 2) based upon their specific electrochemical requirements. Because neither mechanism has been cloned as yet, information about their operation, abundance, and tissue specificity remains largely circumstantial or indirect.

Na-dependent Exchanger (Na$^+$/Mg^{2+} Exchanger): Observed originally in chicken red blood cells, this mechanism has been reported to operate in almost all the mammalian cell types examined (see Romani and Scarpa 2000 for a list). Inhibited by amiloride, this Na$^+$-dependent Mg^{2+} extrusion mechanisms is activated via cAMP-dependent phosphorylation irrespective of the modality by which cellular cAMP is increased (stimulation of β-adrenergic, glucagon, or PGE2 receptors, administration of forskolin or cell-permeant cyclic-AMP analogs, etc.).

While Mg^{2+} extrusion through this exchanger requires the presence of a physiological extracellular Na$^+$ concentration, it is debated whether the exchanger operates on electroneutral (2Na$^+_{in}$:1 Mg$^{2+}_{out}$) or electrogenic (1Na$^+_{in}$:1 Mg$^{2+}_{out}$) bases (Romani 2011). The extrusion of Mg^{2+} via this exchanger appears to be coupled to the outward movement of Cl$^-$ ions, the electrogenic exchange becoming electroneutral in the absence of Cl$^-$ (Romani 2011). The Cl$^-$ extrusion has been interpreted as an attempt to equilibrate charge movement across the cell membrane but it is unclear whether Cl$^-$ is extruded through the Na$^+$/Mg^{2+} exchanger directly or through Cl$^-$ channels activated by the exchanger.

Operating with a K_m for Na$^+$ between 15 and 20 mM, this exchanger is inhibited by amiloride, imipramine, or quinidine (Romani 2011). The limited specificity of these agents does not clarify as to whether they inhibit the Na$^+$/Mg^{2+} exchanger directly, or indirectly by operating on other transport mechanisms including Na$^+$ and K$^+$ channels, which, in turn, alter the cell membrane potential and the driving force for Mg^{2+} transport across the plasma membrane.

Na$^+$-independent: In the absence of extracellular Na$^+$, or in the presence of amiloride, imipramine, or quinidine, which all block the operation of the Na$^+$/Mg^{2+} exchanger, Mg^{2+} can still be extruded via a Na$^+$-independent mechanism. It is unclear whether this transport mechanism operates as a cation antiporter or a sinporter for cations and anions as it appears to utilize Ca^{2+}, Mn^{2+}, HCO$_3^-$, Cl$^-$, or choline to promote Mg^{2+} extrusion. It is also uncertain whether this pathway can be activated by hormonal stimulation.

Regulation of Mg^{2+} Transport and Homeostasis

The majority of mammalian cells retain their basal Mg^{2+} content virtually unchanged under resting conditions even when a major transmembrane gradient is artificially imposed (Romani and Scarpa 2000). Following the addition of different hormones, large amounts of Mg^{2+} can move in either direction across the cell membrane, resulting in changes in total and free Mg^{2+} content in the serum, within the cell and also within cellular organelles (Romani and Scarpa 2000).

Mg^{2+} Extrusion: Hormones like catecholamine or glucagon induce Mg^{2+} extrusion from various tissues and cell types (see Romani and Scarpa 2000 for a list) via an increase in cellular cAMP level. While the Mg^{2+} extrusion elicited by these hormones depletes to a varying extent the Mg^{2+} pools present within the cytoplasm and the cellular compartments, the physical extrusion of Mg^{2+} across the cell membrane primarily occurs via the Na$^+$/Mg^{2+} exchanger (Romani 2011), although a contribution of the Na$^+$-independent Mg^{2+} extrusion mechanism cannot be completely excluded. Following the removal of extracellular Na$^+$ or pre-treatment with amiloride or imipramine, Mg^{2+} extrusion is abolished almost completely, and a sustained rise in cytosolic free [Mg^{2+}]$_i$ is observed, suggesting that blocking the Na$^+$-dependent Mg^{2+} transport mechanism prevents Mg^{2+} from being extruded across the cell membrane but not from being released into the cytoplasm from binding/buffering sites and/or cellular pool(s) (Romani 2011).

Phenylephrine administration also promotes Mg^{2+} extrusion via alpha$_1$-adrenergic stimulation. The co-stimulation of α_1- and β-adrenergic receptor are not alternative but rather additive and complementary processes to induce Mg^{2+} extrusion from liver cells. This event is of particular relevance when the two classes of adrenergic receptors are stimulated by mix-adrenergic agonists such as epinephrine or norepinephrine, resulting in the mobilization of an amount of Mg^{2+} equivalent to the sum of the amounts of Mg^{2+} mobilized by the separate stimulation of α_1- and β-adrenergic receptors (Romani 2011).

It is interesting to note that catecholamine, glucagon, and adrenergic agonists like isoproterenol or phenylephrine all activate glycogenolysis and promote release of hepatic glucose into the bloodstream within a time frame similar to that of Mg^{2+} extrusion, both processes being inhibited in the presence of amiloride or imipramine (Romani 2011). A role of Mg^{2+} in regulating glucose homeostasis is supported by the observation that several glycolytic enzymes, including hexokinase, phosphofructokinase, phosphor-glycerate mutase, phosphoglycerate kinase, enolase, and pyruvate kinase, show activation at low and inhibition at high Mg^{2+} concentrations (Romani 2011).

Magnesium extrusion has also been observed following metabolic treatments that decrease cellular ATP content, the main Mg^{2+} buffering component. Cyanide, mitochondrial uncouplers, fructose, ethanol, or hypoxia are just some of the agents or conditions that affect cellular ATP level and Mg^{2+} homeostasis (Romani 2011). All these agents, in fact, decrease ATP content by preventing the mitochondrial electron chain from generating ATP (cyanide or uncouplers), by acting as an ATP trap (fructose), or by altering the redox state of pyridine nucleotide within the cytoplasm and the mitochondrion (ethanol). Because ATP represents the major Mg^{2+} buffering component within the cell (Fig. 1), a decrease in its content or its degradation into ADP or AMP results in an increase in dissociation of Mg^{2+} from the binding and an increase in cytosolic free $[Mg^{2+}]_i$, which ultimately results in Mg^{2+} extrusion from the cell.

Mg^{2+} *Accumulation*: Mammalian cells can also accumulate large amounts of Mg^{2+} as a result of hormonal stimulation. Administration of hormones like carbachol, vasopressin, angiotensin-II, or insulin to various cell types (Romani and Scarpa 2000) results in the inhibition of cAMP-mediated Mg^{2+} extrusion and/or the reversal of Mg^{2+} extrusion into Mg^{2+} accumulation. In addition to inhibiting cAMP production, several of the hormones indicated above activate protein kinase C (PKC) as part of their cellular signaling.

Protein kinase C: Several lines of evidence support a role of PKC in mediating Mg^{2+} accumulation, and downregulation of PKC by PMA or its inhibition by calphostin or staurosporine abolishes the process. A point of inconsistency is provided by the observation that administration of phenylephrine, which activates PKC signaling in addition to IP_3 and Ca^{2+} signaling, does not elicit Mg^{2+} accumulation but induces Mg^{2+} extrusion. These results raise the possibility that different PKC isoforms are activated under one condition but not the other. This seems to be the case following ethanol administration to liver cells, a condition in which Mg^{2+} accumulation is inhibited and PKCε translocation to the cell membrane is prevented (Romani 2011). Interestingly, this PKC isoform has the highest affinity for Mg^{2+} among all PKC isoenzymes, with a K_m ~1 mM, close to the physiological free $[Mg^{2+}]_i$ measured in the cytoplasm of hepatocytes and other mammalian cells (Romani 2011).

Role of MAPKs: The involvement of MAPKS in modulating Mg^{2+} accumulation is supported by the evidence that pharmacological inhibition of ERK1/2 and p38 MAPKs abolishes PKC-mediated Mg^{2+} accumulation and affects cyclin activity, preventing progression through the cell cycle (Romani 2011). This effect may occur via changes in nuclear functions directly regulated by Mg^{2+}, or changes in nuclear signaling by ERK2, which depends on Mg^{2+} level to properly dimerize, translocate, and activate specific nuclear targets (Romani 2011).

Role of EGF: EGF has also been implicated in promoting Mg^{2+} accumulation by controlling TRPM6 channel expression and operation in the apical domain of renal epithelial cells. Point mutations in EGF sequence limit TRPM6 functioning and cellular Mg^{2+} accumulation. The modulation of TRPM6 expression appears to occur via ERK1/2 signaling coupled to *a*ctivator *p*rotein-1 (AP-1). Further evidence that EGF regulates Mg^{2+} homeostasis is provided by the clinical observation that anti-EGF antibodies used in colon cancer treatment induce Mg^{2+} wasting and hypomagnesemia (Romani 2011).

Physiological Role of Intracellular Mg^{2+}

The data discussed previously support the notion that Mg^{2+} is a key regulatory cation for enzymes, phosphometabolites, channels, and signaling components including adenylyl cyclase (Table 1). The regulation of these cellular components occur at Mg^{2+} concentrations between 0.5 and 1 mM, which are well within the fluctuations in free $[Mg^{2+}]_i$ measured in the cytoplasm of various cells including hepatocyte (Grubbs and Maguire 1987).

Ca^{2+}- *and* K^+-*Channels*: Various Ca^{2+} channels are modulated by intracellular free Mg^{2+} including L-type $Ca2+$ channels, T-type Ca^{2+}-channel, store-operated Ca^{2+} channels (SOC), and store-operated calcium release-activated Ca^{2+} (CRAC) channels (Agus and Morad 1991). In most cases, Mg^{2+} acts directly on the phosphorylated channel or on dephosphorylation rate of the channel rather than to changes in cAMP concentration or cAMP-dependent phosphorylation (Agus and Morad 1991).

Magnesium in Eukaryotes, Table 1 Mg^{2+}-regulated enzymes and cell functions in eukaryotic cells

Cell membrane proteins and enzymes	5′-nucleotidase
	Aquaporin 3
	Ca^{2+}-ATPase
	Ca^{2+}-channels
	G-proteins
	K^+-channels
	K^+/H^+-ATPase
	Na^+/K^+-ATPase
	NBCe1-B
Cytosolic signaling components	Adenylyl cyclase
	ERK1/2
	IRAK4
	PKCε
Glycolytic enzymes	Glycogen phosphorylase
	Phospho-glucomutase
	Phosphorylase kinase
Endoplasmic reticulum functions	cADPr-induced Ca2+ release
	Glucose 6-Phosphatase
	IP3-induced Ca2+ release
	RyR-induced Ca2+ release
	SERCA pumps
Mitochondrial enzymes and fucntions	Acetyl-CoA synthetase
	Adenine Nucleotide Translocase
	α-ketoglutarate-dehydrogenase
	Isocitrate carrier
	Mitochondrial respiration
	Mitohcndrial permeability transition pore
	Voltage-dependent anion channel
Nuclear fucntions	Cyclin D
	Cyclin E
	Phosphoribosylpyrophosphate transferase (m3)
	RNAses

The table reports a few examples out of more than 180 enzymes and cell functions regulated by magnesium (Romani and Scarpa 1992).

Potassium channels are also regulated by cytosolic Mg^{2+}, which blocks the outward current of inwardly rectifying K^+ channels without affecting the inward currents. In the case of K_v channels, intracellular Mg^{2+} slows down the kinetic of activation of the K_v channel, causing also inward rectification at positive membrane potentials and a shift in voltage-dependent inactivation. Intracellular Mg^{2+} also modulates large-conductance (BK type) Ca^{2+}-dependent K^+ channels by blocking the pore of the channels in a voltage-dependent manner (Romani 2011).

The inhibitory effect of Mg^{2+} on K^+ channels also occurs in mitochondria, in which Mg^{2+} within the matrix appear to modulate gating and conductance of mitochondrial K_{ATP} channels, which are essential to promote mitochondrial recovery and cell survival under ischemia/reperfusion conditions.

Mitochondrial Channels and Dehydrogenases: Mitochondria represent one of the major cellular Mg^{2+} pools, with a concentration of Mg^{2+} ranging between 14 and 16 mM (Romani 2011). In the absence of changes in matrix Ca^{2+}, changes in matrix Mg^{2+} stimulate the activity of succinate and glutamate dehydrogenases but not α-ketoglutarate and pyruvate dehydrogenase. This observation is consistent with the changes in mitochondrial Mg^{2+} content observed during transition from state 3 to state 4.

Within the mitochondria, Mg^{2+} also regulates the inner membrane anion channel (IMAC) and the permeability transition pore opening.

The *IMAC channel* is essential to maintain mitochondrial volume in conjunction with the K^+/H^+ antiporter. Matrix Mg^{2+} will maintain the IMAC channel in a closed state, a conformation that would allow fine modulation by small changes in pH and proton distribution under physiological conditions, and will retain an optimal $\Delta\psi$ across the mitochondrial membrane, essential to preserve proper organelle function and intra-mitochondrial Mg^{2+} content (Romani 2011).

The permeability transition pore (PTP) opens in the inner mitochondrial membrane following a decrease in mitochondrial $\Delta\psi$ or ATP level, or an increase in Ca^{2+} content. Mitochondrial Mg^{2+} prevents PTP opening by acting from the mitochondrial matrix as well as by maintaining creatine kinase in an active state and tightly associated to the mitochondrial membrane. Both binding and activity state of creatine kinase are Mg^{2+}-dependent, and removal of Mg^{2+} from the extra-mitochondrial environment results in a decline in creatine kinase activity and PTP opening (Romani 2011).

Reticular G6Pase: The Endoplasmic Reticulum (ER) represents another major Mg^{2+} pool within the cell, with a total concentration estimated to be between 14 and 18 mM (Romani 2011). No information is available about the mechanisms by which Mg^{2+} enters and exits the organelle and is buffered within the ER lumen. Aside from protein synthesis, experimental evidence suggests that cytosolic and luminal Mg^{2+} concentration can limit Ca^{2+} uptake into the ER/SR and its release from the organelle via IP_3 and ryanodine receptor (RyR) (Romani 2011). Cytosolic Mg^{2+} can also regulate the activity of reticular glucose 6-phosphatase (G6Pase) in liver cells by affecting the glucose 6-phosphate (G6Pi) transport component of the G6Pase enzymatic complex. This effect is biphasic, with an optimal stimulatory effect at ~0.5 mM $[Mg^{2+}]_i$ and an inhibitory effect at higher Mg^{2+} concentrations (Romani 2011). A similar modulatory effect of Mg^{2+} on G6Pase hydrolysis rate is observed in microsomes isolated from livers of animals exposed to Mg^{2+} deficient diet (Romani 2011). It is presently undetermined whether Mg^{2+} exerts a similar modulating effect on other reticular enzymatic activities.

Cell pH and Volume: Cells exposed to cyanide, fructose, hypoxia, or ethanol undergo cellular acidification, decrease in cellular ATP content, and Mg^{2+} extrusion. This extrusion is the consequence of a decrease in ATP buffering capacity and binding affinity within the cytoplasm. Recent evidence indicates that physiological $[Mg^{2+}]_i$ inhibits by ~50% the current generated by the electrogenic Na^+-HCO_3^- cotransporter NBCe1-B, and increasing *free* $[Mg^{2+}]$ to 3 mM completely abolished NBCe1-B current. This regulatory effect is exerted by Mg^{2+} and not Mg*ATP, and occurs at the N-terminus of the transporter (Romani 2011). Increasing cellular Mg^{2+} content also enhances aquaporin 3 expression through cAMP/PKA/CREB signaling, as well as MEK1/2 and MSK1. As aquaporin 3 is highly expressed in brain, erythrocytes, kidney, and skin, in addition to the gastrointestinal tract, the modulating effect of Mg^{2+} on aquaporin 3 expression in these tissues may be highly relevant for various physiological and pathological conditions including brain swelling following traumatic injury.

Cell Cycle: Cell cycle, cell proliferation, and cell differentiation have all been associated with the maintenance of an optimal Mg^{2+} level. Under conditions in which cellular Mg^{2+} accessibility is reduced, cell proliferation, cell differentiation, and cell cycle progression become markedly impaired. The mechanisms by which a decrease in cellular Mg^{2+} content affects these cellular processes revolve around MAPKs and p27 signaling, increased oxidative stress, and decreased Mg*ATP levels (Romani and Scarpa 1992). Cells maintain Mg*ATP concentration at a level optimal for protein synthesis. Hence, any decrease in this parameter will have major repercussion on proper cell functioning. In addition, extracellular Mg^{2+} levels regulate integrins signaling, de facto modulating the interaction among cells, and between cells and extracellular matrix (Wolf et al. 2009).

Conclusions

Recent years have registered a significant advancement in understanding the mechanisms that regulate cellular Mg^{2+} homeostasis. Although the field still lags behind in knowledge as compared to other ions such as Ca^{2+}, H^+, K^+, or Na^+, the identification of Mg^{2+} channels and transport mechanisms and a better comprehension of the signaling pathways and conditions regulating Mg^{2+} transport are providing new tools to address essential questions about the relevance of Mg^{2+} for various cell functions under physiological and pathological conditions.

Cross-References

▶ Calcium in Biological Systems
▶ EF-hand Proteins and Magnesium
▶ Magnesium
▶ Magnesium in Health and Disease
▶ Magnesium in Plants
▶ Magnesium, Physical and Chemical Properties
▶ Potassium in Biological Systems

References

Agus ZS, Morad M (1991) Modulation of cardiac ion channels by magnesium. Annu Rev Physiol 53:299–307
Flatman PW (1984) Magnesium transport across cell membranes. J Membr Biol 80:1–14
Grubbs RD, Maguire ME (1987) Magnesium as a regulatory cation: criteria and evaluation. Magnesium 6:113–127
Quamme GA (2010) Molecular identification of anceitn and modern mammalian magnesium transporters. Am J Physiol 298:407–429
Romani A (2007) Regulation of magnesium homeostasis and transport in mammalian cells. Arch Biochem Biophys 458:90–102
Romani A (2011) Cellular magnesium homeostasis. Arch Biochem Biophys 512:1–23
Romani A, Scarpa A (1992) Regulation of cell magnesium. Arch Biochem Biophys 298:1–12
Romani A, Scarpa A (2000) Regulation of cellular magnesium. Front Biosci 5:D720–D734

Schmitz C, Deason F, Perraud A-L (2007) Molecular components of vertebrate Mg2+ homeostasis regulation. Magnes Res 20:6–18

Touyz RM, He Y, Montezano ACI, Yao G, Chubanov V, Gudermann T, Callera GE (2006) Differential regulation of transient receptor potential melastatin 6 and 7 cation channels by Ang-II in vascular smooth muscle cells from spontaneously hypertensive rats. Am J Physiol 290: R73–R78

Wolf FI, Torsello A, Fasanella S, Cittadini A (2003) Cell physiology of magnesium. Mol Asp Med 24:11–26

Wolf FI, Cittadinin AR, Maier JA (2009) Magnesium and tumors: ally or foe?. Cancer Treat Rev 35:378–382

Magnesium in Health and Disease

Cristina Paula Monteiro[1] and Maria José Laires[2]
[1]Physiology and Biochemistry Laboratory, Faculty of Human Kinetics, Technical University of Lisbon, Cruz Quebrada, Portugal
[2]CIPER – Interdisciplinary Centre for the Study of Human Performance, Faculty of Human Kinetics, Technical University of Lisbon, Cruz Quebrada, Portugal

Synonyms

Glutathione, gamma-glutamyl-cysteinyl-glycine, GSH; [Mg2+]i, intracellular free magnesium; Oxygen free radicals, reactive oxygen species

Definitions

Magnesium deficiency results from insufficient intake of magnesium.

Magnesium depletion results from deregulation of factors controlling magnesium status such as intestinal magnesium hypoabsorption, urinary leakage, reduced magnesium bone uptake and mobilization, hyperglucocorticism, insulin-resistance, and adrenergic hyporeceptivity.

Primary magnesium deficit originates from two etiological mechanisms: deficiency and depletion.

Secondary magnesium deficit results from various pathologies and/or treatments: non-insulin-dependent diabetes mellitus, alcoholism, or ingestion of hypermagnesuric diuretics, etc.

Overview

Magnesium is an essential cation playing a crucial role in many physiological functions. It is critical in energy-requiring metabolic processes, in protein synthesis, membrane integrity, nervous tissue conduction, neuromuscular excitability, muscle contraction, hormone secretion, and in intermediary metabolism. Serum magnesium concentration is maintained within a narrow range by the small intestine and kidney both of which increase their fractional magnesium absorption under conditions of magnesium deprivation. If magnesium depletion is prolonged, the bone store helps to maintain serum magnesium concentration by exchanging part of its content with extracellular fluid. The abundance of magnesium within cells is consistent with its relevant role in regulating tissue and cell functions. Imbalances of magnesium are common and are associated with a great number of pathological situations responsible for human morbidity and mortality. A large part of the population may have an inadequate magnesium intake inducing a magnesium deficiency that, associated to alterations in magnesium metabolism, may induce magnesium deficit. Magnesium deficit is frequently found in industrialized countries and may contribute to neuromuscular, cardiovascular, renal, and other diseases.

Early Magnesium Research

Magnesium is found in all tissues and may affect many functions in the body. Its multiple physiological actions have been discovered, thanks to the numerous convergent efforts of multinational research.

One may consider the recognition by Grew, in 1695, of magnesium sulfate as one of the essential constituents of Epsom salts, as marking the entry of magnesium in medicine. It was considered as an internal remedy and purifier of the blood. After several centuries of research, Leroy in 1926 recognized magnesium as an essential ion. The physiological properties of this cation in the rat were described by MacCollum and Greenberg in the 1930s who showed the multiple effects of lack of magnesium intake on development, reproduction, neuromuscular apparatus, and humoral balance. Flink, in 1956, carried out the earliest clinical study clearly identifying the syndrome of magnesium deficiency in man. Aikawa published his first book about magnesium in 1963, but the understanding of magnesium biology was still highly rudimentary due to the lack of suitable analytical methods. In 1968, Wacker published a review on magnesium metabolism, which was a classic reference for several years. Meanwhile, Seelig in the United States, and Durlach in Europe were beginning to carry out clinical studies of magnesium metabolism. A series of clinical case reports in the early 1960s helped focus attention on the role of hypomagnesemia in various malabsorptive states and stimulated efforts to study magnesium depletion and its consequences under controlled conditions (Rayssiguier et al. 2000).

Magnesium Physiology and Biochemistry

Role of Magnesium in Metabolic Functions

Magnesium, the second most abundant intracellular cation, is essential for the optimal function of a diversity of life-sustaining processes. It is a cofactor for more than 300 enzymes, participating in the metabolism of carbohydrates, lipids, proteins, and nucleic acids, in the synthesis of hydrogen transporters, and particularly in all reactions involving the formation and use of adenosine triphosphate (ATP). Magnesium is also known to alter both receptor sites and ion movements across the cell membrane. By forming complexes with phospholipids, magnesium stabilizes all the biological membranes, reducing their fluidity and permeability. Thus, magnesium is an important modulator of intracellular ion concentrations. In magnesium deficit, cytoplasmic concentrations of calcium and sodium increase, whereas those of potassium and phosphorus decrease. Simultaneously, the membrane depolarizes. These alterations may be the result of magnesium's direct effect on sodium, calcium, or potassium channels, or the indirect result of its effect on enzymes in the cell membrane that are involved in active transport, e.g., (Na^+K^+)–ATPase. Magnesium also regulates the levels of lipid and phosphoinositide-derived second messengers.

Within the cell, magnesium affects the function of organelles, such as sarcoplasmic reticulum, primarily by its ability to alter calcium flux, or mitochondria, by altering their membrane's permeability to protons, which leads to alterations in the coupling of oxidative phosphorylation to electron transport chains, thus affecting the efficiency of ATP production.

Magnesium also acts as a calcium antagonist. In the neuromuscular system, it reduces the electric excitability of the neurons, inhibits the release of acetylcholine by the nerve endings at the neuromuscular junction, and blocks the effect of N-methil-D-aspartate, an excitatory neurotransmitter of the central nervous system. In muscle contraction, both stimulation and the activity of the calcium transport system in the sarcoplasmic reticulum membranes depend on the presence of Mg^{2+}. Troponin contains four calcium-binding sites, two of which have a high affinity for calcium and bind Mg^{2+} competitively. These calcium-magnesium-binding sites do not seem to be directly involved in any rapid muscle twitching mechanism, but play a structural role in muscle physiology. Magnesium bound to these sites may maintain the protein in a particular conformational state regardless of the fluctuation in calcium concentration (assuming that the structural changes induced by magnesium and calcium are essentially the same). This conformation may be a prerequisite for calcium activation via binding at the calcium-specific sites. Magnesium also acts as a vasodilator, like other calcium antagonists, and inhibits the coagulation process.

Compelling evidence shows that magnesium is directly related to proliferation in normal cells as magnesium stimulates DNA and protein synthesis.

These and many other cellular functions make magnesium a crucial element in living beings. As a result, magnesium acts as a regulator of many physiologic functions in various apparata including neuromuscular, cardiovascular, immune, and hormonal systems. It is easy to understand why disruptions in magnesium metabolism may be a factor in the development of pathological conditions such as neuromuscular, cardiovascular, and renal diseases, pre-eclampsia, or diabetes (Beyenbach 1995; Reinhart 1990).

Magnesium Distribution and Regulation

In biological systems, magnesium ions exist in three different states: bound to proteins, complexed to anions, and free (Mg^{2+}). Only free magnesium has biological activity.

The adult human body contains approximately 24 g (1 mol) of magnesium in cells versus 280 mg in extracellular fluids. The skeleton represents the body's largest magnesium store (about 60% of total magnesium), divided into two subcompartments. The slow bone compartment corresponds to firmly apatite-bound magnesium that cannot be mobilized even under conditions of extreme depletion. The second is the mobile compartment comprising magnesium that is absorbed to the surface of mineral crystals. This fraction can be increased by increasing magnesium supply, and it is mobilized during hypomagnesemia. Thus, bone functions as a large magnesium reservoir, helping to stabilize its concentration in serum. About one fourth of total magnesium is located in skeletal muscle; the nervous system and other organs with high metabolic rates, namely, liver, myocardium, digestive tube, kidney etc., account for the remaining 15%. In serum, about one third of circulating magnesium is bound to proteins, mainly albumin; the remaining two thirds is ultrafiltrable, being approximately 92% free and 8% complexed to citrate, phosphate, and other compounds. The mean serum magnesium concentration is about 0.8 mmol/L. The concentration of magnesium in red blood cells is approximately 2.5 mmol/L and is genetically controlled, with the oldest cells having the least magnesium (Laires et al. 2004).

Regulation of the intracellular magnesium compartmentalization by the organism is an important evolutionary development. The intracellular free magnesium ($[Mg^{2+}]i$) is the component essential for the regulation of a large number of cellular functions and can be controlled within a narrow range by a large number of membrane transporters. Several specific mechanisms have been described to regulate cellular magnesium homeostasis. Their modality of action, however, has not been fully elucidated (Touyz et al. 2001; Yago et al. 2000).

Magnesium Balance: Absorption, Excretion, and Homeostasis

As for other minerals, magnesium homeostasis is preserved through control of intestinal absorption and urinary losses.

Under physiological conditions, 30–50% of magnesium uptake is absorbed mainly by the small intestine. The net magnesium absorption results from the balance between the amount of magnesium released into the digestive tract by the digestive fluids and the amount of magnesium absorbed both from the ingested food and from the digestive fluids. There are two mechanisms of absorption: one active and saturable and another passive. Passive absorption occurs via a paracellular pathway following a favorable electrochemical gradient as a function of water and solute movements and appears to be proportional to dietary intake. Regulated active transport operates only under conditions of low-magnesium intake. Magnesium concentration in the digestive tube is the most important determinant of the amount of magnesium absorbed. Nevertheless, food elements capable of influencing magnesium absorption should be taken into account in the investigation of cases of low-magnesium intake. Substances increasing magnesium solubility favor its absorption, in opposition to the action of substances that form insoluble complexes. Recent data show that no competition exists between magnesium and calcium for absorption, as calcium supplementation does not decrease magnesium absorption. Phosphates can inhibit magnesium absorption through the formation of insoluble magnesium complexes in the intestine. Several studies have popularized the concept according to which diets rich in fiber have unfavorable effects on magnesium absorption. However, from recent studies it is clear that the effect of fiber on magnesium absorption depends on the nature (solubility and fermentability) and the amount of fiber and on the associated compounds in the meal such as phytates. Drinking water may be an important source of magnesium as it may be better absorbed than magnesium in food (Rayssiguier et al. 2000).

Analyzing magnesium intake and urinary excretion, one can observe that on average about one third of the dietary magnesium is eliminated in the urine. When magnesium intake is severely restricted in humans with normal kidney function, magnesium output becomes much smaller. Supplementing a normal magnesium intake increases urinary magnesium excretion without altering normal serum levels, as long as renal function is normal and the amounts of magnesium administered are not excessive. Hence, the kidney conserves magnesium in response to a deficiency, and renal excretion increases in proportion to the load presented to the kidney. Approximately 70–80% of the serum magnesium is filtered through the glomerular membrane but only 20–30% of the filtered magnesium is reabsorbed along the proximal tubule. The primary site for magnesium reabsorption is the thick ascending limb of the loop of Henle, where about 65% of the filtered magnesium is reabsorbed. At this site, magnesium reabsorption is associated with NaCl cotransport. The increase in renal excretion that accompanies increased magnesium uptake seems to be a response to increased plasma magnesium.

The decreased urinary excretion that is observed in deficit situations seems to result from increased magnesium reabsorption in the loop of Henle. Factors decreasing NaCl reabsorption in the thick ascending limb of the loop of Henle, like osmotic diuretics, increase magnesium excretion.

Several studies indicate that glucagon, parathyroid hormone, calcitonin, and insulin increase magnesium reabsorption in the loop of Henle and proximal tubule. On the contrary, hypercalciuria and hypophosphatemia decrease tubular reabsorption. Metabolic acidosis also increases magnesium urinary excretion. Increases in catecholamine levels from any cause can induce hypomagnesemia that seems to be associated with an increase in magnesium excretion in the urine. Female sex hormones significantly affect magnesium metabolism. Serum magnesium levels in women of active reproductive age have been reported to be lower than the levels in males of equal age. In addition, serum magnesium levels in women taking hormonal contraceptives are lower than those in age-matched controls. A decrease in urinary magnesium excretion has been observed in postmenopausal women treated with estrogens in cyclical combination with gestagens as compared with a group of postmenopausal women receiving placebo treatment. This may depend on the fact that female sex hormones favor magnesium uptake into bone as part of the mineralization process. Hypomagnesemia, which frequently develops during the second half of pregnancy, is probably caused by increased urinary magnesium losses.

In chronic magnesium deficit, magnesium decreases in plasma and tissues, and bone magnesium is mobilized to restore normal magnesemia. According to some authors, vitamin D plays a minor role in magnesium homeostasis, while for others it has a direct action on magnesium intestinal absorption.

Thus, the mechanisms leading to hypomagnesemia may include decreased intake or absorption, internal redistribution, and/or increased renal or nonrenal loss (Laires et al. 2004; Rayssiguier et al. 2000).

Dietary Magnesium Intake

A large number of balance studies have been performed over the years in an effort to obtain quantitative data on magnesium requirements. These studies involved the measurement of daily dietary intake along with subtraction of daily excretory losses via urine and feces, yielding a positive or negative "balance." Based on long-term balance studies, Seelig and Durlach, in the early 1960s, recommended a daily magnesium intake of 6 mg/kg/day. More recently, other authors have agreed with Seelig and Durlach's recommendation. However, when analyzing the results of the reported surveys in several populations, it is concluded that this value is frequently not reached in developed countries.

This insufficient intake has been attributed to a wide variety of factors, including a drop in the consumption of grain products, agricultural techniques of accelerated growth decreasing magnesium fixation by plants, use of magnesium-poor soil fertilizers, use of pesticides (inhibition of absorption), refining of foods resulting in large losses of magnesium, especially in sugars and grains, and boiling of vegetables, which causes major losses to water, etc.

As marginal magnesium intake may condition hypomagnesemia, which is observed in several pathological situations, more attention needs to be paid to magnesium intake. In some cases, the need for supplementation should be considered. However, the use of supplementation should be carefully evaluated, as renal disturbances may lead to magnesium excess and pathological consequences (Durlach 1989).

Magnesium Deficit

Etiological Mechanisms of Magnesium Deficit

Primary magnesium deficit originates from two etiological mechanisms: deficiency and depletion. Deficiency is due to insufficient intake, and depletion is due to deregulation of factors controlling magnesium status such as intestinal magnesium hypoabsorption, urinary leakage, reduced magnesium bone uptake and mobilization, hyperglucocorticism, insulin-resistance, and adrenergic hyporeceptivity. Secondary magnesium deficit results from various pathologies and treatments: non-insulin-dependent diabetes mellitus, alcoholism, ingestion of hypermagnesuric diuretics, etc. Magnesium deficit may contribute to neuromuscular, cardiovascular, renal, and other pathologies.

Consequences of Chronic Magnesium Deficit

The assessment of a patient's magnesium status is problematic because there are no easily performed tests that reliably predict the intracellular concentration. The patient should be tested for clinical and paraclinical signs of neuromuscular hyperexcitability. The most significant is Chvostek's sign: careful auscultation of the heart reveals a systolic click and a mitral valve murmur in magnesium deficiency. Paraclinical examination involves electromyogram (EMG), electroencephalogram (EEG), and echocardiogram (ECC), after which mitral valve prolapse is often identified. Clinical assessment also includes measurement of magnesium in plasma and red blood cells, plasma calcium levels, and urinary magnesium.

When analyzing the consequences of magnesium deficit in man, the best-documented clinical aspects are the nervous alterations resulting from primary chronic magnesium deficiency. Nervous hyperexcitability due to chronic insufficient magnesium intake results in a nonspecific clinical pattern with an associated central and peripheral neuromuscular symptomatology, analogous to what has been described in latent tetany, chronic fatigue, and asthenia. Central nervous hyperexcitability may be associated with decreased energy and cationic gradients, along with disturbances in calcium distribution.

Cardiovascular manifestations of magnesium deficit have also been described. They mainly include modifications of EMG, and arrhythmias. Nevertheless, these disturbances are similar to those described for potassium deficit. Magnesium deficit is often in the origin of a mitral dyskinesia (prolapse). Besides disturbances in the nervous and cardiac system, magnesium deficit can influence fiber and collagen structures, lipid profile, and the endocrine system. Other consequences of magnesium deficit are an increase in lipid peroxidation, due to the increase in reactive oxygen species, and alterations in immune system (Rayssiguier et al. 2000).

Magnesium Deficit and Oxidative Stress

Oxidative stress is a state where the production of reactive oxygen species (ROS) in the body transcends the antioxidant defense capacity.

Several physiological processes are involved in ROS generation and propagation such as mitochondrial electron transport; hemoglobin oxidation to methemoglobin; autoxidation of catecholamines to adrenochrome; prostaglandin synthesis; or phagocytosis. Most pathological conditions also promote ROS generation. For example, situations involving hypoxia can induce the increase of cytosolic calcium concentrations promoting proteolytic modification of xanthine dehydrogenase to xantine oxidase as well as the accumulation of hypoxanthine. When reperfusion occurs, xanthine oxidase promotes the transformation of hypoxanthine to xantine, and ultimately to uric acid, with superoxide anion (O_2^-) formation. When trauma occurs, it may lead to extravasation of blood and introduction of free iron and copper into tissues. Some O_2^- originated in mitochondria may leak into the cytosol, where the free iron and copper can promote its transformation into the highly reactive hydroxyl radical (OH). Inflammatory injury, with the infiltration and activation of monocytes, produces a spectrum of free radicals. The formed ROS, especially OH, can modify a wide range of biomolecules including DNA, proteins, or polyunsaturated fatty acids in the cell membrane. If the rise in oxygen free radicals exceeds the protective capacity of cell's antioxidant defense systems, it can lead to free radical mediated injury and consequently to loss of cell membrane integrity and tissue damage. The efficiency of the antioxidant defense systems relies on adequate dietary vitamin (C, E, and β-carotenes) and mineral (Cu, Zn, Se, Mn, etc.) intake and on endogenous production of antioxidants such as antioxidant enzymes and glutathione (GSH) (Halliwell and Gutteridge 1989).

Although magnesium is not directly involved in antioxidant defense systems, its deficiency favors the increase in ROS generation as magnesium has an important role in the inhibition of ROS-induced cell injury. Magnesium inhibits catecholamines release. Additionally, catecholamine methylation, which prevents their oxidation, is catalyzed by catechol-O-methyl transferase, a magnesium-dependent enzyme. Magnesium also participates in the de novo synthesis of GSH, important in the restoring of GSH cellular levels.

In magnesium-deficient rodents, it has been demonstrated that ROS formation and the concentration of thiobarbituric acid reactive substances (TBARS), reflecting lipid peroxidation, are increased. These changes were associated with structural damage in skeletal muscles affecting sarcoplasmic reticulum and mitochondria. The authors suggested that magnesium deficiency induces lipid and protein damage and impairs intracellular calcium homeostasis by altering the integrity of the sarcoplasmic reticulum membrane. However, the mechanism by which magnesium deficiency potentiates injury is not totally clear. Besides the disturbance in calcium homeostasis, other explanations include increased cytokine concentrations, alterations in iron metabolism, or decreased endogenous antioxidant capacity. The lack of magnesium is also associated with a depletion of antioxidant defense systems including glutathione peroxidase activity, a selenium-dependent enzyme (Halpern and Durlach 1996; Rayssiguier et al. 2001).

Magnesium supplementation to patients with oxidative stress diseases has induced antioxidant protection resulting in decreased lipid peroxidation.

Evaluating magnesium status is important because magnesium deficit may render the individual more susceptible to the occurrence of cellular damage, including that associated to oxidative stress.

Conclusions and Perspectives

The authors have focused on reviewing the major aspects of magnesium physiology which may explain the physiopathology of several pathological situations where magnesium deregulation is observed. Several gaps in the knowledge of magnesium metabolism and its regulation still challenge the investigators. In the future, it is expected that research on human genome and the proposed mechanisms for magnesium transport will allow the formulation of informed hypothesis regarding the role of magnesium in the pathological conditions mentioned above.

In the upcoming years, controlled randomized clinical trials will also be necessary to characterize the role of magnesium in physiopathological conditions, and to determine whether magnesium supplementation can improve the symptomatology of neuromuscular, cardiovascular, renal, and other diseases.

Cross-References

▶ Magnesium and Cell Cycle
▶ Magnesium in Biological Systems

References

Beyenbach KW (1995) The physiology of intracellular magnesium. In: Vecchiet L (ed) Magnesium and physical activity. The Parthenon Publishing Group, New-York, pp 93–116

Durlach J (1989) Recommended dietary amounts of magnesium: Mg RDA. Magnes Res 2(3):195–203

Halliwell B, Gutteridge JMC (1989) Oxygen free radicals in biology and medicine, 2nd edn. Clarendon, Oxford

Halpern MJ, Durlach J (eds) (1996) Current research in magnesium. Jonh Libbey, London

Laires MJ, Monteiro CP, Bicho M (2004) Role of cellular magnesium in health and human disease. Front Biosci 9:262–276

Rayssiguier Y, Durlach J, Boirie Y (2000) Métabolisme du magnésium et son rôle en pathologie. In: Encyclopédie Médico-Chirurgicale, Endocrinologie-Nutrition, vol 10-357-A-10. Editions Scientifiques et Médicales Elsevier SAS, tous droits réservé, Paris

Rayssiguier Y, Mazur A, Durlach J (eds) (2001) Advances in magnesium research: nutrition and health. John Libbey, Eastleigh

Reinhart RA (1990) Magnesium metabolism. Wis Med J 89(10):579–583

Touyz RM, Mercure C, Reudelhuber TL (2001) Angiotensin II type I receptor modulates intracellular free Mg2+ in renally derived cells via Na+ -dependent Ca2+ -independent mechanisms. J Biol Chem 276(17):13657–13663

Yago MD, Manas M, Singh J (2000) Intracellular magnesium: transport and regulation in epithelial secretory cells. Front Biosci 5:D602–D618

Magnesium in Plants

Christian Hermans, Jiugeng Chen and
Nathalie Verbruggen
Laboratory of Plant Physiology and Molecular
Genetics, Université Libre de Bruxelles,
Brussels, Belgium

Synonyms

Mineral deficiency; Mineral homeostasis; Mineral nutrition; Photosynthetic organisms

Definitions

- Biofortification is a strategy to increase nutritional value (minerals and vitamins concentrations) of crops, either through conventional selective breeding, or through genetic engineering.
- *M*itochondrial *R*NA *S*plicing 2 (MRS2)/ *Mag*nesium *T*ransporter (MGT) is a family of Mg^{2+} channels in *Arabidopsis*.

- Sucrose phloem loading is the entry of sucrose in the sieve-element companion cell (SECC) mediated by H$^+$/sucrose symporter (in plant species where SECC is isolated symplamically from mesophyll tissue).

Importance of Magnesium in Agronomy and for Human Health

Magnesium Deficiency in Plant Production Systems

Magnesium deficiency in plants is a widespread problem, affecting productivity and quality in agriculture, horticulture, and forestry. In arable land, the incidence of Mg deficiency symptoms is increasing, due to intensive harvesting, amplified rotations per site, and unbalanced application of fertilizers lacking in secondary elements (e.g., calcium, magnesium). In forest regions, Mg deficiency is widespread and associated with the "new type forest decline" or "crown thinning" in Europe and the northeastern part of North America and the "upper mid crown yellowing" of conifers in New Zealand.

Handbooks exist to distinguish between mineral deficiencies disorders in major crop plants (Bennett 1997). Signs of Mg deficiency in most plants usually manifest belatedly as a chlorophyll breakdown between the veins (Fig. 1), usually associated with Mg concentrations below 0.2% dry weight in leaves (Marschner 2003). Chloroses make their appearance first on the oldest and recently expanded leaves, and systematically progress from them toward the youngest ones.

Magnesium Distribution in Soil and Interaction with the Rhizosphere

The distribution of Mg in soil is divided into three: non-exchangeable, exchangeable, and water-soluble pools. (a) By far, the largest fraction is the non-exchangeable one present in ferromagnesian minerals such as olivine (forsterite: $2MgO \cdot SiO_2$) and secondary clay minerals such as phyllosicates (chlorite: $Mg_3[Si_4O_{10}](OH)_2 \cdot Mg_3(OH)_8$). Additionally, parent material may contain Mg-rich rocks such as dolomite ($CaCO_3 \cdot MgCO_3$). In arid or semiarid regions, soils may contain large amount of kieserite ($MgSO_4$). (b) The exchangeable Mg held on the surface of clay and organic matter particles is usually in the order of 5% of the total soil Mg. (c) In the water-soluble fraction, Mg is present in fairly high concentrations (between 2 and 5 mM) and is accessible to roots. As a general rule, an agricultural soil is considered to have a relative low, medium, or high level when its total Mg content is in the range of 0–50, 50–150, or more than 150 mg Mg kg^{-1} soil, respectively.

Because Mg has the highest hydrated radius among divalent cations, it sorbs less strongly to soil colloids than other ones. It is therefore more prone to leaching in sandy and well-drained soils, which is considered as the most important factor reducing Mg availability for roots. Mg deficiency can be brought by depletion in the soil reserve but can also be induced by a failure of the roots to assimilate Mg^{2+}, due to imbalanced competitive inhibition of uptake by other ions. High levels of Ca^{2+}, K$^+$, Mn^{2+}, or NH$_4^+$ or low pH strongly depresses Mg^{2+} uptake by plants (Marschner 2003).

Impact of Low Plant Magnesium Nutritional Value on Human and Animal Diet

Magnesium is the fourth-most common cation in the human body and it literally drives our fuel sources as it regulates over 300 biochemical reactions and the chelation to nucleotidyl phosphate forms. Hypomagnesaemia is an electrolyte disorder affecting virtually every organ system and causing arrhythmia, sudden cardiac death, muscle dysfunction, migraines, attention deficit disorder, etc. (▶ Magnesium Metabolism in Type 2 Diabetes Mellitus, ▶ Magnesium in Health and Disease). In addition, chronic Mg inadequacy can promote or exacerbate age-related diseases. Studies have shown that nearly two-thirds of the US population does not meet the dietary reference intake for Mg. There are also indications of genetic inheritance in hypomagnesaemia susceptibility, primarily due to defects in Mg intestinal absorption and renal-wastage. Mineral salts (MgCl$_2$) and organic complexes and chelates (Mg-citrate, Mg-taurate, Mg in chlorophyll structure) are the most readily absorbed forms by the body. However, absorption may be impaired by a variety of factors such as phytates, oxalates, and nonfermentable fibers. Eating Mg-rich foods or taking oral Mg supplements can prevent chronic diseases associated with Mg deficiency.

In most instances, elevated Mg^{2+} content improves the nutritional quality of plants for animal and human

Magnesium in Plants, Fig. 1 *Magnesium deficiency symptoms observed in Arabidopsis thaliana.* Comparison of control (**a**) and Mg-deficient (**b**) rosettes. Close-up of young mature leaves of control (**c**) and Mg-deficient (**d**) plants. Scale bars: a, b: 2 cm and c, d: 1 cm. Plants were grown hydroponically for 5 weeks. At day 0, plants were fed with Mg-saturated and Mg-depleted solutions. Pictures were taken 28d after omitting Mg from the nutrient solution

diet (Shaul 2002). It is estimated that half of the Mg intake of humans is from foods of plant origin. Biofortification offers a solution to overcome human malnutrition by increasing edible crop nutritional value. Therefore, expanding knowledge on the mechanisms underlying Mg homeostasis is necessary to increase plant Mg content.

Magnesium in Plant Physiology

Roles of Magnesium in Photosynthesis
Structural Role in Chlorophyll Antenna
Plants require magnesium to harvest solar energy and to drive photochemistry. Because of its tendency to bind six ligands in a regular octahedral arrangement, resulting in strong electrophilic axial coordination, Mg is able to occupy a central position in chlorophyll, the pigment responsible for light absorption. It is estimated that the magnesium structural pool, associated with chlorophyll, represents between 15% and 20% of the total leaf content. The ATP-dependent insertion of Mg^{2+} into protoporphyrin IX is catalyzed by Mg^{2+}-chelatase. In the degradation pathway, chlorophyllase most likely removes the phytol group, producing chlorophyllides, whereas Mg-dechelatase removes Mg atoms from chlorophyllides. In addition to its central role in chlorophyll, Mg participates in the cation-mediated grana stacking of thylakoid membranes.

Physiological Significance of the Fluctuations in Mg^{2+} and pH Levels in the Stroma and Lumen
Proton pumping from the stroma into thylakoids in the light results in acidification of the thylakoid lumen. The accompanied electrical gradient across the thylakoid membrane is compensated by concomitant fluxes of Cl^-, K^+, and Mg^{2+} from the cytosol and the lumen. The increase of stromal Mg^{2+} levels by light transition is well documented, and is reported to be in the range of 1–5 mM, depending on the assay conditions. The modulation of stromal Mg^{2+} concentrations by light transition has an important regulatory effect on the activity of enzymes in Calvin cycle. For example, increased stromal Mg^{2+} concentration activates fructose 1,6-bisphosphatase and sedoheptulose

1,7-biphosphatase, and the ribulose 1,5-bisphosphate carboxylase (Rubisco). A CO_2 molecule (not the substrate that will be fixed) reacts with an uncharged ε-NH_2 group of a lysine within the active site of Rubisco. The resulting carbamate derivative then rapidly binds to Mg^{2+} to yield the activated complex (Taiz and Zeiger 2010) (▶ Magnesium Binding Sites in Proteins).

Importance of Mg in the Cell Energy Budget

The concentration of Mg^{2+} in the metabolic pool (cytoplasm and chloroplast) of leaf cells is assumed to be in the range of 2–10 mM (Igamberdiev and Kleczkowski 2001). However, free Mg^{2+} concentration in the cytosol is lower as Mg^{2+} is largely complexed with inorganic pyrophosphate (PPi) or with nucleotidyl tri- and diphosphates (ATP and ADP). For example, in mung bean root tips, it is estimated that about 90% of the cytoplasmic ATP is complexed to Mg^{2+}. The nucleotidyl phosphate Mg-chelated forms are the active conformation and the substrate for proton pumps such as H^+-ATPase and vacuolar pyrophosphatase. The energy-rich compounds Mg-ATP and Mg-ADP balance with the free Mg^{2+} pool under the control of adenylate kinase (Igamberdiev and Kleczkowski 2001).

Magnesium-Promoted Activity of Enzymes in Metabolic Biochemistry

Many metabolic cycles are mediated by magnesium-dependent enzymatic reactions, falling into two general classes. First, enzymes may bind magnesium-substrate complexes, interacting primarily with the substrate and showing little interaction with Mg^{2+} (e.g., isocitrate lyase). Alternatively, Mg^{2+} directly binds enzymes (direct interaction to side-chains or indirect interaction through metal-bound waters) and alters their structure and/or serves a catalytic role. The catalytic metal may be delivered as a chelate complex of the substrate, or the enzyme makes an authentic complex with the Mg^{2+} cofactor. Typically, coordinating ligands are carboxylates, although occasionally an amide carbonyl can form the backbone.

Importance of Mg in Nucleic Acid Structure

The roles played by Mg in the nucleic acids structure and metabolism are treated somewhere else in more details (▶ Magnesium). Magnesium is engaged in cation-π interactions with the bases of DNA and RNA to stabilize their conformation. Most functional RNA molecules are compact and tightly folded into a unique conformation. Magnesium is the most common ion involved in RNA folding. Interaction may exist between water ligands of hexahydrated Mg sphere and RNA base and backbone substituents to stabilize specific RNA motifs. Additionally, direct contacts exist between Mg^{2+} and RNA.

All biomolecular machines that catalyze RNA splicing are dependent on Mg^{2+} for both folding and chemical catalysis. Among them, the group II autocatalytic introns ribozymes are true metalloenzymes that require Mg^{2+} specifically for folding, substrate binding, and for the living group during chemical catalysis.

Physiological Roles Revealed Through Studies of Magnesium Limitation in Plants

Few reports exist concerning the effects of an Mg shortage on physiological processes in plants. They describe an activation of antioxidative mechanisms, an impairment in sugar metabolism, or a decline in photosynthetic CO_2 fixation rate and stomatal conductance. Transcriptional profiling of the model plant *Arabidopsis thaliana* is providing non-biased approaches to the study of Mg deficiency targets. Recent studies implicate several circadian clock genes in the response to Mg limitation. The involvement of hormones (e.g., abscisic acid and ethylene) in the regulation of Mg responsive genes is also apparent from those studies.

Magnesium Involvement in Antioxidative Mechanisms

Magnesium deficiency is largely reported to induce an oxidative stress (Cakmak and Kirkby 2008). Mg-deficient leaves are highly photosensitive and develop severe chlorosis, particularly under high light conditions. Before distinct depression in growth and in chlorophyll content, an enhancement of the mechanisms against toxic reactive oxygen species (ROS) is observed in leaves upon Mg depletion. In particular, an increase of ascorbate and SH-compounds, higher activities of superoxide dismutase, ascorbate peroxidase, dehydroascorbate reductase, and glutathione reductase are often observed. Enhanced production of ROS may result from restricted consumption of photoreductants in CO_2 fixation.

Mg Involvement in Sugar Metabolism

Reports exist concerning the effects of Mg depletion on the impairment of sugar partitioning in various species (Hermans et al. 2006). Actually, plants are photoautotrophic organisms building complex carbohydrates components (sugars) from atmospheric CO_2. However, plants also contain sink organs (typically roots, fruits, and immature leaves), which are heterotrophic tissues and therefore require a carbon import from photosynthesizing organs. Sink and source organs are separated from each other in space but connected together through the phloem vasculature. Upon Mg depletion, it is proposed that either a decrease in sucrose (the carbon currency for long-distance transport) export is responsible for the inhibition of sink organs' growth or a decrease in the metabolic activity in sink organs causes inhibition of sucrose export (Hermans et al. 2006). In the first situation, the high susceptibility of sucrose partitioning to low Mg levels may be related to the carrier-mediated uptake of sucrose in the conducting complex of phloem (event v in Fig. 2). Mg is possibly involved in sucrose phloem loading because it interacts with nucleotidyl triphosphates fuelling the H^+-ATPases which creates the proton motive force energizing the sucrose symporters (in type IIb phloem loading strategy plants). Likewise, pyrophosphatases, playing a role in the long-distance transport of sugars, also require Mg for pyrophosphate hydrolysis.

Non-biased Global Transcriptomic Approaches to Identify Early Magnesium Deficiency Targets

The transcriptome changes in response to Mg availability were recently identified in *Arabidopsis thaliana* (Hermans et al. 2010 and companion paper). A thorough description of the transcriptomic responses (within hours or days before the outbreak of visual symptoms) is opening new routes to decipher the cascade of Mg deficiency responses. Mg starvation triggers a different temporal response in organs, with a higher number of genes being differentially regulated in the root after 8 h of depletion and in young mature leaves after 28 h. Unlike other mineral deficiencies, putative Mg transporters genes are not induced at the organ level (post-transcriptional induction cannot be excluded). Interestingly, the expression of circadian clock component genes is altered in roots, while abscisic acid signaling is triggered in leaves. One-week Mg starvation impacts on the abundance of 33% of the transcripts in leaves and less than 2% of the transcripts in roots. That result confirms the visual observation that Mg starvation affected the shoot more than the root in *Arabidopsis*. In long-standing Mg-deficient leaves, the amplitude in the peak expression of clock-associated genes is altered but not their phase. Ethylene could also play a key role in the Mg starvation response, as the expression levels of several genes encoding enzymes in the C_2H_4 biosynthetic pathway are enhanced and Mg-deficient plants produce twice as much gas than Mg-supplied plants.

Magnesium Homeostasis in Plants

General Definitions of Nutrients Uptake Mechanisms

Mineral nutrients absorbed from the soil solution by roots are transported to the aerial biomass by water transpiration stream moving through the xylem vessels. Prior to xylem loading, minerals have to reach the stele. To reach that objective, ions can diffuse or be carried passively by water flow in the apoplast of the root cortex (event I in Fig. 2). Alternatively, ions can move from cell to cell by simple diffusion through plasmodesmata within the symplast (event II in Fig. 2). At the boundary between stele and cortex, a thickening of the endodermis cells, known as the Casparian strip, blocks the entry of water and ions into the stele via the apoplasmic pathway. Therefore, further radial transport necessarily involves a symplasmic pathway, and ions have to enter the symplast at a certain point before entering the stele. The cellular plasma membrane contains different types of transporter proteins: channel proteins, ATP-powered pumps, and cotransporters. *Channel* proteins facilitate the diffusion of water and ions down energetically favorable gradients. Active transport utilizes *carrier-type* proteins that are energized directly by ATP hydrolysis to move ions across the plasma membrane against chemical and electrical gradients, such as H^+-ATPases and Ca^{2+}-ATPases. *Cotransporter* proteins (symporters and antiporters – also called exchangers) move a given substance against its own concentration gradient and indirectly use the electric potential and/or chemical gradient of secondary substances energy, usually a proton gradient.

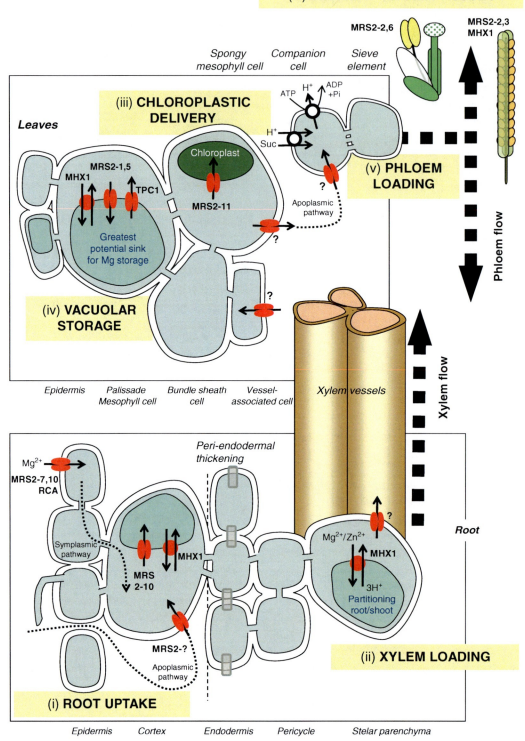

Magnesium in Plants, Fig. 2 (continued)

Characteristics of Mg Transport

Magnesium is distinctive from potassium or calcium because it has a smaller ionic radius and a higher charge density. It binds water molecules three to four orders of magnitude more tightly than any other biological divalent cation does. The inner coordination sphere contains those atoms in contact with the metal ion and the outer sphere contains those in contact with the inner sphere ligand atoms. Within the inner coordination sphere, hexa-aquated magnesium ion ($Mg[H_2O]_6^{2+}$) has an octahedral liganded state and binds six oxygen atoms from water molecule ligands which surrounds the cation. Globally, around 300 molecules of water compose the hydration sphere of Mg^{2+} (by comparison only 50 in calcium).

Magnesium Transport in Plants

During the last decade, a few reports have been published on Mg transport systems in plants. Potential Mg^{2+} transport systems are the homologues of the bacterial CorA proteins. In *Arabidopsis thaliana*, a family of ten genes and one pseudo-gene have been annotated as *MITOCHONDRIAL RNA SPLICING 2* (*MRS2*) or *MAGNESIUM TRANSPORTERS* (*MGT*) because they share the CorA Gly-Met-Asn (GMN) signature motif (Waters 2011 and references therein). Even if all the family members can complement yeast *mrs2* mutants, the roles of MRS2 are still largely unidentified *in planta*. Also, *MAGNESIUM/PROTON EXCHANGER 1* (*MHX1*) encodes a vacuolar transporter, which is present in root and shoots and exchanges protons with Mg^{2+}, Zn^{2+}, and Fe^{2+} ions (Shaul 2002). The implication of *AtMHX1* in metal homeostasis still has to be elucidated, as upon overexpression of that gene, no major change is detected in the mineral content of any organ.

Figure 2 presents a schematic model of Mg movement in plant organs and summarizes current knowledge on Mg transport systems. (i) *Root Uptake*. Mg^{2+} from the soil solution can diffuse or be carried passively by water flow in the apoplast of the root cortex (apoplasmic pathway). The exchange through the plasma membrane with Mg^{2+} present in the cytoplasm of the cortical cells is almost two orders of magnitude slower than the exchange between soil solution and cortical apoplast. Alternatively, ions can enter epidermis or cortex cells through carriers and move from cell to cell by simple diffusion through plasmodesmata within the symplast (symplasmic pathway). Very little is known about the transport systems engaged in Mg^{2+} uptake from the soil solution into the root. It is assumed that entry of magnesium into root cells is mediated by permeable channels. One possible way of Mg^{2+} entry could be through RCA calcium channels at the plasma membrane of root cells, which are permeable to a wide variety of monovalent and divalent cations (Shaul 2002). Another possibility is through the AtMRS2/MGT family members, likely to function as Mg^{2+} channels. Most of those genes follow partly overlapping tempo-spatial activity patterns, but six members are expressed in root tissues, indicating a possible role in Mg^{2+} uptake. For example, after germination, *MRS2-7* and *MRS2-10* expression is restricted to the root (except the root tip for *MRS2-10*) and the proteins localize to the plasma membrane and endoplasmic reticulum, respectively. Interestingly, *mrs2-7* knockout mutants are severely retarded upon low Mg supply only (Gebert et al. 2009). For Mg storage, the vacuole is considered the main organelle and an Mg^{2+}/H^+ exchange activity is observed in tonoplast vesicles isolated from roots. Because of the peri-endodermal thickening, further radial transport in root necessarily involves a symplastic pathway, and Mg^{2+} has to enter the symplast at a certain point before entering the stele. Transfer of Mg^{2+} from the root symplast to the xylem occurs against its electrochemical gradient. (ii) *Xylem loading*. Mg^{2+} absorbed from the soil solution by roots is transported to the aerial biomass by water transpiration stream moving through the xylem vessels. AtMHX1 is an antiporter, which might control the partitioning of Mg^{2+} between plant organs, determining the proportion of cations to be stored in the

Magnesium in Plants, Fig. 2 *Schematic model of magnesium movement in plant organs.* Mg^{2+} is taken up from soil solution into root cells, transported to shoots, and recycled to sink organs. Mg^{2+} transporters of the MRS2/MGT family (Gebert et al. 2009; Waters 2011) and MHX1 (Shaul 2002) seem to play a central role in the processes of root uptake, xylem loading, delivery to chloroplasts, vacuolar storage, phloem loading, and delivery to sink tissues. MRS2 proteins have been localized to the plasma membrane, mitochondria, chloroplast, and tonoplast of various types of cells. Further description is given in the text vacuoles of the xylem parenchyma cells, and the amount available in the cytosol for xylem loading (Shaul 2002). (iii) *Chloroplastic delivery*. MRS2-11 could play a role in Mg^{2+} uptake and translocation into chloroplast. *MRS2-11* is expressed in stomata guard cells and spongy mesophyll cells and the protein localizes to the chloroplast envelope membrane. (iv) *Vacuolar storage in leaves*. MRS2-1 and MRS2-5 are important for Mg accumulation in the vacuoles of palisade mesophyll cells. It is shown that enrichment of those transporters in the mesophyll enables the accumulation of greater vacuolar Mg^{2+} than in the epidermis. Mg^{2+} could be released from vacuoles through cation channels, including the SV channel TPC1 (*t*wo-*p*ore *c*hannel). (v) *Phloem loading*. Mg is relatively phloem mobile and is recycled from old to developing tissues. However, it is not currently clearly identified which transporters mediate Mg loading inside and unloading outside the vasculature tissues. Note that an apoplasmic phloem loading strategy IIb scheme is presented here. (vi) *Delivery to sink tissues*. MRS2-2 and MRS2-6 play an essential role for pollen development and male fertility. They have both high expression levels in floral parts or pollen. Hemizygous *mrs2-2* and *mrs2-6* mutants have defects in pollen formation but homozygous mutants are lethal. *AtMRS2-3* and *AtMHX1* co-localize with major chromosomal quantitative trait loci (QTLs) controlling seeds Mg concentration. Expressions of *MRS2-4* and *MRS2-10* are increased in senescing rosette leaves, which are tissues that supply Mg to developing seeds through remobilization.

How the activity of Mg transporters is regulated is not well understood, but to date, there is little evidence of regulation at the transcriptional level (Hermans et al. 2010). However, when *Arabidopsis* is grown in low Ca solution, several *MRS2* genes have increased expression (Waters 2011, and references therein).

Benefits from Magnesium Homeostasis Research in Plants

Historical trends in the concentrations of mineral elements (and particularly Mg) in edible crop tissues from developed countries have declined over the last half century (White and Broadley 2009). Determining how plants regulate Mg uptake from the rhizosphere, as well as transport and allocate it to organs will have significant implications for plant nutrition and human health. With the knowledge of genes governing Mg homeostasis in plants, it will be possible to develop biofortification strategies through conventional breeding or genetic engineering. The end-goal would be to enrich edible parts of crops, which would offer humans improved Mg sources to overcome mineral malnutrition.

Cross-References

▶ Magnesium
▶ Magnesium Binding Sites in Proteins
▶ Magnesium in Biological Systems
▶ Magnesium in Health and Disease
▶ Magnesium Metabolism in Type 2 Diabetes Mellitus

References

Bennett WF (1997) Nutrients deficiencies & toxicities in crop plants. APS Press/The American Phytopathological Society, St Paul

Cakmak I, Kirkby EA (2008) Role of magnesium in carbon partitioning and alleviating photooxidative damage. Physiol Plant 133:692–704

Gebert M, Meschenmoser K, Svidová S, Weghuber J, Schweyen R, Eifler K, Lenz H, Weyand K, Knoop V (2009) A root-expressed magnesium transporter of the MRS2/MGT gene family in *Arabidopsis thaliana* allows for growth in low-Mg^{2+} environments. Plant Cell 21:4018–4030

Hermans C, Hammond JP, White PJ, Verbruggen N (2006) How do deficiencies of essential mineral elements alter biomass allocation? Trends Plant Sci 11:610–617

Hermans C, Vuylsteke M, Coppens F, Craciun A, Inzé D, Verbruggen N (2010) The early transcriptomic changes induced by magnesium deficiency in *Arabidopsis thaliana* reveal the perturbation of the circadian clock in roots and the triggering of ABA-responsive genes. New Phytol 187: 119–131. Companion paper New Phytol 187: 132–144

Igamberdiev AU, Kleczkowski LA (2001) Implications of adenylate kinase-governed equilibrium of adenylates on contents of free magnesium in plant cells and compartments. Biochem J 360:225–231

Marschner H (2003) Mineral nutrition of higher plants, 2nd edn. Academic, London

Shaul O (2002) Magnesium transport and function in plants: the tip of the iceberg. Biometals 15:309–323

Taiz L, Zeiger E (2010) Plant physiology, 5th edn. Sinauer Associates, Sunderland

Waters BM (2011) Moving magnesium in plant cells. New Phytol 190:510–513

White PJ, Broadley MR (2009) Biofortification of crops with seven mineral elements often lacking in human diets – iron, zinc, copper, calcium, magnesium, selenium and iodine. New Phytol 182:49–84

Magnesium Metabolism in Type 2 Diabetes Mellitus

Mario Barbagallo and Ligia J. Dominguez
Geriatric Unit, Department of Internal Medicine and Medical Specialties (DIMIS), University of Palermo, Palermo, Italy

Overview

The link between magnesium (Mg) deficiency and diabetes mellitus is well known. Diabetes mellitus is frequently associated with both extracellular and intracellular Mg depletion. Several epidemiologic studies have recognized a high prevalence of hypomagnesemia in subjects with type 2 diabetes, especially in those with poorly controlled glycemic profiles (Barbagallo and Dominguez 2007). Because of the lack of sensitivity of total serum magnesium (MgT), a depletion in intracellular (Mgi) and serum ionized Mg (Mg-ion) can be found in many subjects with MgT still in the normal range. Resnick et al. (1993) measured concurrently MgT, Mg-ion, and Mgi levels in the same subjects using ^{31}P-NMR and ISE Mg-selective electrode, and found that both Mgi and Mg-ion (but not MgT) were significantly reduced in diabetic subjects, and that a close direct relationship was present between the ionized extracellular and the intracellular Mg measurements (Resnick et al. 1993).

Mg deficiency in diabetes, which may take the form of a chronic latent Mg deficit rather than an overt clinical hypomagnesemia, may have clinical importance because Mg ion is a crucial cofactor for many enzymatic reactions involved in a myriad of metabolic processes.

At the cellular level, cytosolic free Mgi levels are consistently reduced in subjects with type 2 diabetes mellitus. Using gold standard NMR techniques, significantly lower steady-state Mgi and reciprocally increased intracellular calcium (Cai) levels have been shown in old and young type 2 diabetic subjects, when compared with young nondiabetic subjects (Barbagallo et al. 2000). An impairment of cellular Mg uptake mechanism, and a decrease in cellular ATP level, may contribute, at least in part, to explain the decrease in cellular Mg content observed under diabetic conditions.

Mechanisms of Magnesium Deficiency in Diabetes

Among the mechanisms that may favor Mg depletion, the most important are a low Mg intake and an increased Mg urinary loss, while absorption and retention of dietary Mg seems not to be impaired in patients with type 2 diabetes.

With regard to low Mg intake, changes in dietary habits in the western world have resulted in daily Mg intake close to, or even below, the minimum recommended daily allowances. An inverse correlation between dietary Mg and the incidence of type 2 diabetes has been suggested (see below). A diet deficient in Mg is associated with a significant impairment of insulin-mediated glucose uptake and with a considerable increased risk of developing glucose intolerance and diabetes (Barbagallo et al. 2007).

Diabetes is associated with renal Mg and calcium wasting. Hyperglycemia and hyperinsulinemia may both have a role in the increased urinary Mg excretion contributing to Mg depletion. Hyperglycemia, which is a hallmark of lack of good metabolic control, may have a role in urinary Mg wasting. Plasma Mg levels were found inversely correlated with urinary Mg excretion rate and with fasting blood glucose values, suggesting that tubular reabsorption of Mg is decreased in the presence of severe hyperglycemia (Barbagallo et al. 2007). The increased renal Mg transporter abundance may represent a compensatory adaptation for the increased load of Mg to the distal tubule. An improved metabolic control corrects the hyperglycemia-associated hypercalciuria and hypermagnesiuria, reverses the increase of Mg transporter abundance, and is associated with reduced urinary Mg losses. Hyperinsulinemia, which is present in insulin resistant states, may contribute per se to the urinary Mg depletion and the reduced sensitivity to insulin, and may itself affect Mg transport. As an additional factor, the use of loop and thiazide diuretics, often prescribed in diabetic patients with hypertension and/or cardiovascular diseases, also promotes Mg wasting.

Magnesium Deficiency and Insulin Resistance

Mg is a necessary cofactor in over 300 enzymatic reactions and specifically in all phosphorylation processes. Thus, intracellular Mg is a critical cofactor for enzymes involved in carbohydrate metabolism, and because of its role as part of the activated Mg-ATP complex required for all of the rate-limiting enzymes of glycolysis, Mg regulates the activity of all enzymes involved in phosphorylation reactions. Mg concentration is critical in the tyrosine-kinase-mediated phosphorylation of the insulin receptor as well as the phosphorylation by the other protein kinases participating in the insulin signaling. Magnesium also regulates all the enzymes involved in ATP and phosphate transfer, such as the Ca-ATPases in the plasma membrane and the endoplasmic reticulum. Mg deficiency may result in disorders of tyrosine kinase activity of the insulin receptor, event related to the development of post-receptorial insulin resistance and decreased cellular glucose utilization, whereby the lower the basal Mg, the greater the amount of insulin required to metabolize the same glucose load, indicating a decreased insulin sensitivity. Measurements of intracellular free Mg concentrations using ^{31}P-NMR have revealed that cellular concentrations of Mgi are in the 100–300 μM range, which is close to the dissociation constant of many enzymes systems using ATP or phosphate transfer. Because tissue Mg uptake is regulated by insulin, impairment of this process by insulin resistance could either cause or exacerbate intracellular Mg deficiency (Barbagallo et al. 2007) (Fig. 1).

A deficient Mg status may both be a secondary consequence or may precede and cause insulin resistance and altered glucose tolerance, and even diabetes. Inflammation and oxidative stress have been proposed to be the link between Mg deficit and insulin resistance/metabolic syndrome. More generally, chronic hypomagnesaemia and conditions commonly associated with Mg deficiency, such as type 2 diabetes mellitus and aging, are associated with increased free radical formation and subsequent damage to cellular processes (Barbagallo and Dominguez 2010). Antioxidant therapies with vitamin E and glutathione have been shown to improve insulin sensitivity and whole body glucose disposal and their action is mediated, at least in part, by their specific action to increase intracellular Mg and improve cellular Mg homeostasis. This is consistent with a role of oxidative stress and inflammation in the Mg deficiency-associated development of insulin resistance, vascular remodeling, atherosclerosis, type 2 diabetes, and cardio-metabolic syndrome.

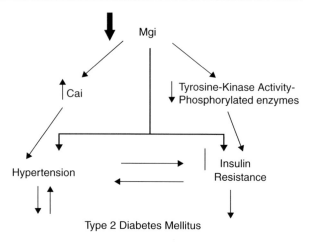

Magnesium Metabolism in Type 2 Diabetes Mellitus, Fig. 1 Overall hypothesis in which intracellular magnesium deficiency may mediate the relationship between insulin resistance, hypertension, and type 2 diabetes

Mgi depletion in diabetes has been shown to be clinically and pathophysiologically significant; intracellular Mg levels have been found to quantitatively and inversely predict the fasting and post glucose levels of hyperinsulinemia, peripheral insulin sensitivity, and systolic and diastolic blood pressures (Barbagallo et al. 2007). Specifically, (a) fasting insulin levels; (b) the integrated insulinemic response to a standard oral glucose tolerance test; (c) the steady-state plasma glucose response to insulin infusion and indices of peripheral insulin sensitivity derived from euglycemic hyperinsulinemic clamps, all have been found to be inversely related to Mg levels, whether measured as Mgi in situ in brain, free or total Mg in peripheral red cells, or even as circulating Mg. Furthermore, inverse relations have been observed between steady-state fasting levels of Mgi and: (a) fasting blood glucose, (b) blood pressure, (c) glycated hemoglobin (HbA1c), and (d) the glycemic and insulinemic responses to oral glucose loading in normal, hypertensive, and diabetic subjects. In other words, the lower the Mgi, the higher is the blood pressure, and the more hyperinsulinemic the response to oral glucose loading becomes (Barbagallo et al. 2007).

Aging is frequently associated with insulin resistance and glucose intolerance. A continuous age-dependent fall of Mgi levels in peripheral blood cells of healthy elderly subjects is present, these alterations being indistinguishable from those occurring, independently of age, in essential hypertension or type 2 diabetes (Barbagallo et al. 2000). In other terms, essential hypertension and/or type 2 diabetes appear to determine an acceleration of the usual age-dependent Mg depletion, suggesting that these ionic changes may be clinically significant, underlying the predisposition in elderly subjects to hypertension and metabolic diseases (Barbagallo et al. 2000; Barbagallo and Dominguez 2010).

The link between Mg deficiency and the development of insulin resistance and type 2 diabetes is strengthened by the observation that several treatments for diabetes appear to increase Mg levels. Metformin raises Mg levels in the liver and pioglitazone (a thiazolidinedione antidiabetic agent that increases insulin sensitivity) has positive actions on Mg metabolism both in vitro, and in vivo, stimulating free Mg concentration in adipocytes and increasing serum Mg. Other insulin-mimeting substances such as vanadate and IGF-1 are also associated with a direct effect to stimulate Mgi levels (Barbagallo and Dominguez 2007).

The relevance of altered cellular Mg metabolism to tissutal insulin sensitivity suggests a possible role of Mg in contributing to the clinical coincidence of Mg depletion with clinical conditions of insulin resistance such as hypertension, metabolic syndrome, type 2 diabetes as well with the increased incidence of each of these conditions with age, a condition itself characterized by a tendency to Mg depletion. Thus, Mgi depletion can directly promote tissutal insulin resistance and altered vascular tone, thus helping to understand the mechanisms underlying the clinical association among these apparently disparate conditions.

Altogether, independent of the cause of poor plasma and intracellular Mg content, Mg depletion seems to contribute to an impairment of insulin sensitivity. Mg deficiency, which may take the form of a chronic latent Mg deficit rather than clinically evident hypomagnesemia, may have crucial importance because of the key role of Mg as a cofactor in many enzymatic reactions regulating glucose metabolism. Thus, a deficient Mg status may not just be a secondary consequence of type 2 diabetes but may precede and contribute itself to the development of insulin resistance and altered glucose tolerance, and even type 2 diabetes. We have suggested a role for Mg deficit as a possible unifying mechanism of conditions associated to "insulin resistance," including type 2 diabetes mellitus, metabolic syndrome, and essential hypertension (Barbagallo and Dominguez 2007; Barbagallo et al. 2007).

Mg Deficiency and Diabetes Complications

Mg deficiency has been proposed as a factor implicated in the pathogenesis of diabetes complications. Cellular ionic alterations are related to the cardiovascular structural modifications often present in diabetes. A significant relation was found between fasting levels of Mgi levels and cardiovascular structural indices. In diabetic subjects, even in the absence of elevated blood pressure, suppressed Mgi levels are associated with cardiac hypertrophy, and specifically with increased echocardiographically measured posterior wall thickness and left ventricular mass index in both diabetic and/or hypertensive subjects. Similarly, aortic distensibility values determined by magnetic resonance imaging in normal and hypertensive humans were closely and positively related to concomitantly measured levels of Mgi in situ in brain and skeletal muscle tissue by ^{31}P-NMR magnetic resonance spectroscopic techniques: the more suppressed the Mgi, the stiffer (less distensible) the aorta (Barbagallo et al. 2007). In children and adolescent patients with type 1 diabetes, serum Mg levels have been associated with early atherosclerosis, and a significant relationship was found between serum Mg levels with intima-media thickness, and functions of common carotid artery, accepted as markers of early carotid atherosclerosis. In patients with type 2 diabetes mellitus, low circulating Mg levels have been associated also with a more rapid decline of renal function. To confirm the potential role of Mg deficits in the pathogenesis of diabetic vascular complications, in an experimental model of diabetes in rats, Soltani has suggested a potential role for oral Mg supplementation in the prevention of the vascular complications of the disease, as well as in the prevention of the pathological changes in the aorta and pancreas of diabetic rats (Soltani et al. 2005).

Magnesium Deficiency May Predispose to Diabetes

Lower Mg levels may not only be a consequence, but may also predispose to the development of diabetes. In a national population-based cross-sectional nutrition survey in Taiwan, hypomagnesemia was associated with an increased risk of diabetes. Thus, the risk of diabetes was elevated 3.25 times at plasma Mg levels lower than <0.863 mmol/L. Not only serum Mg, but also lower Mg intake has been suggested to predispose to diabetes. In a large cohort of young American adults participating in the Coronary Artery Risk Development in Young Adults (CARDIA) study, during the 20-year follow-up, Mg intake was inversely longitudinally associated with the incidence of diabetes in this young American population, after adjustment for potential confounders. This inverse association may be explained, at least in part, by the inverse correlations of Mg intake with systemic inflammation and insulin resistance (Kim et al. 2010).

The hypothesis that alterations of Mg metabolism would induce insulin resistance and predispose to diabetes mellitus is confirmed by data in both, experimental animals and humans, showing that dietary-induced Mg deficiency is correlated with insulin resistance. A Mg-deficient diet in sheeps has been found to be associated with a significant impairment of insulin-mediated glucose uptake while Mg supplementation delayed the development of diabetes in a rat model of diabetes. Higher Mg intake is associated with lower fasting insulin concentrations among women without diabetes, and a significant negative correlation is present between total dietary Mg intake and insulin response to oral glucose tolerance test. Rats fed a low Mg diet showed a significant increase in blood glucose and triglyceride levels. The effects of dietary-induced Mg deficiency on glucose disposal, glucose-stimulated insulin secretion, and insulin action on skeletal muscle were studied in rats, which were fed a low Mg-containing diet. Mg depletion provoked a deleterious effect on glucose metabolism due to an impairment of both insulin secretion and action. The insulin resistance observed in skeletal muscle of Mg-deficient rats was linked, at least in part, to a defective tyrosine kinase activity of insulin receptors.

Mg Supplementation in the Management of Diabetes

While deficiencies of both dietary Mg and serum Mg content were associated with an increased risk to develop glucose intolerance and diabetes mellitus, there are potential benefits supporting the use of Mg supplementation in persons who have diabetes or risk factors for diabetes. Increased Mg intake is associated with decreased risk of developing type 2 diabetes in populations. In a prospective study of almost 85,000 women, the relative risk of developing diabetes for women in the highest quintile of Mg consumption was 0.68 when compared with women in the lowest quintile (Lopez-Ridaura et al. 2004). In the Women's Health Study, a cohort of 39,345 US women aged ≥ 45 years with no previous history of cardiovascular disease, cancer, or type 2 diabetes was followed for an average of 6 years; in 918 participants, confirmed incident cases of type 2 diabetes were documented, and a significant inverse association was found between Mg intake and the risk of developing type 2 diabetes, supporting a protective role of higher intake of Mg in reducing the risk of becoming diabetic.

Thus, the use of Mg supplements could be a potential tool for the prevention of type 2 diabetes and metabolic syndrome, a hypothesis that needs to be confirmed by specific and well-designed trials with Mg. However, the effects of Mg supplements on the metabolic profile of diabetic subjects are controversial, benefits having been found in some, but not all clinical studies. Oral Mg supplementation is contraindicated in patients with significant renal impairment. Differences in baseline Mg status and metabolic control may explain the differences among these studies. Mg may mediate the favorable impact of whole grains on insulin sensitivity. While the benefits of oral Mg supplementation on glycemic control have yet to be clearly demonstrated in diabetic patients, Mg supplementation has been shown to improve insulin sensitivity. Based on current knowledge, clinicians have good reason to believe that Mg repletion may play a role in delaying type 2 diabetes onset and potentially in warding off its devastating complications, i.e., cardiovascular disease, retinopathy, and nephropathy. In a recent small clinical trial, Mg supplementation was able to restore altered endothelial function in elderly diabetic subjects

(Barbagallo et al. 2010). Mg supplementation improves the insulin-mediated glucose uptake measured by euglycemic insulin clamp, with a significant relationship between the parallel increase in plasma and erythrocyte Mg concentration and the progressive increase in insulin sensitivity (Barbagallo et al. 2007). In a clinical trial specifically conducted among type 2 diabetics with clinical hypomagnesemia (index of an already advanced Mg deficit), oral Mg supplementation had beneficial effects on fasting and postprandial glucose levels and on insulin sensitivity (Rodriguez-Moran and Guerrero-Romero 2003), and in a study from Taiwan, the risk of dying from diabetes was inversely proportional to the level of Mg in the drinking water.

Among nondiabetic, apparently healthy subjects, there is some evidence of relatively small but significant beneficial effects of Mg supplements on insulin sensitivity. Altogether, Mg supplementation in diabetic patients with Mg deficiency corrects the deficit in intracellular free Mg levels, improves insulin sensitivity, and may protect against diabetic complications. The positive effects of a high intake of Mg on systemic inflammation and insulin resistance may help to explain the favorable effects on the risk of developing diabetes.

The fact that most but not all diabetic subjects have a Mg deficiency and that no large clinical trial has been specifically focused on subjects with a Mg deficit, diagnosed with an accurate and reliable technique, may help to explain the discrepancy between the unclear role of supplemental Mg on glycemic control in diabetics, and the significant impact on diabetes risk in prospective epidemiologic studies. Future prospective studies are needed to support the potential role of dietary Mg supplementation as a possible public health strategy to reduce diabetes risk in the population.

Cross-References

▸ Calcium in Health and Disease
▸ Cellular Electrolyte Metabolism
▸ Magnesium and Inflammation
▸ Magnesium in Biological Systems
▸ Magnesium in Health and Disease
▸ Potassium in Biological Systems
▸ Potassium in Health and Disease

References

Barbagallo M, Dominguez LJ (2007) Magnesium metabolism in type 2 diabetes mellitus, metabolic syndrome and insulin resistance. Arch Biochem Biophys 458:40–47

Barbagallo M, Dominguez LJ (2010) Magnesium and aging. Curr Pharm Des 16:832–839

Barbagallo M, Gupta RK, Dominguez LJ et al (2000) Cellular ionic alterations with age: relation to hypertension and diabetes. J Am Geriatr Soc 48:1111–1116

Barbagallo M, Dominguez LJ, Resnick LM (2007) Magnesium metabolism in hypertension and type 2 diabetes mellitus. Am J Ther 14:375–385

Barbagallo M, Dominguez LJ, Galioto A et al (2010) Oral magnesium supplementation improves vascular function in elderly diabetic patients. Magnes Res 23:131–137

Kim DJ, Xun P, Liu K et al (2010) Magnesium intake in relation to systemic inflammation, insulin resistance, and the incidence of diabetes. Diabetes Care 33:2604–2610

Lopez-Ridaura R, Willett WC, Rimm EB et al (2004) Magnesium intake and risk of type 2 diabetes in men and women. Diabetes Care 27:134–140

Resnick LM, Altura BT, Gupta RK et al (1993) Intracellular and extracellular magnesium depletion in type 2 (non-insulin-dependent) diabetes mellitus. Diabetologia 36:767–770

Rodriguez-Moran M, Guerrero-Romero F (2003) Oral magnesium supplementation improves insulin sensitivity and metabolic control in type 2 diabetic subjects: a randomized double-blind controlled trial. Diabetes Care 26:1147–1152

Soltani N, Keshavarz M, Sohanaki H et al (2005) Oral magnesium administration prevents vascular complications in STZ-diabetic rats. Life Sci 76:1455–1464

Magnesium, Physical and Chemical Properties

Fathi Habashi
Department of Mining, Metallurgical, and Materials Engineering, Laval University, Quebec City, Canada

Magnesium is a silvery white light metal. It is a typical metal in that when it loses its outermost electrons it attains the electronic structure of the inert gases; hence, it is a reactive metal. It is produced commercially by electrolysis of magnesium chloride melts and by metallothermic reduction of magnesium oxide with silicon. The largest single use of magnesium is as an alloying element in aluminum alloys.

Other uses include its use as a reducing agent in the production of titanium, zirconium, uranium, beryllium, and hafnium. Magnesium enters in the structure of chlorophyll; hence, it is essential for plants.

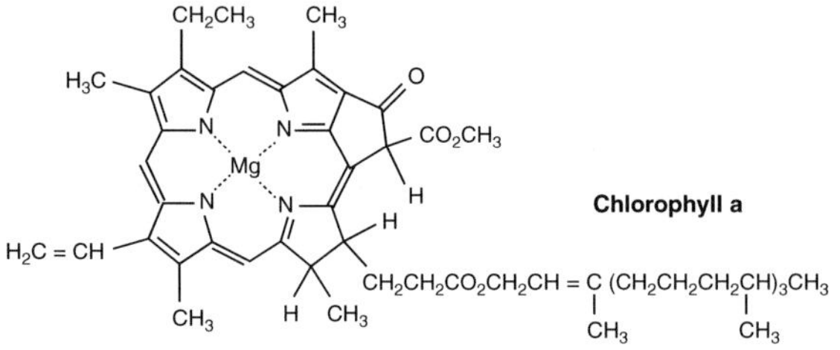

Chlorophyll a

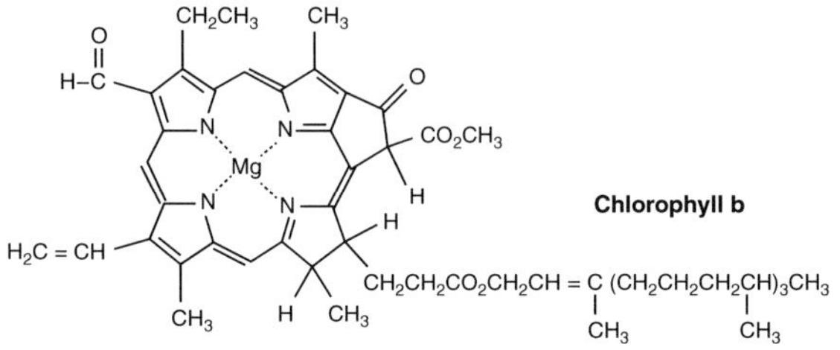

Chlorophyll b

Physical Properties

Atomic number	12
Atomic weight	24.31
Relative abundance in Earth's crust, %	2.09
Density (solid), g/cm^3	
At 20°C	1.738
At 600°C	1.622
Density above 650°C (liquid)	$1.834 - 2.647 \times 10^{-4} T$
Crystal structure	Hexagonal
Lattice constants at 20°C, nm	$a = 0.32$
	$c = 0.52$
Melting point, °C	650 ± 2
Boiling point, °C	$1{,}107 \pm 10$
Latent heat of fusion, MJ/kg	0.37
Latent heat of evaporation, MJ/kg	5.25
Heat of combustion, MJ/kg	25.1
Specific heat, J kg^{-1} K^{-1}	
At 20°C	1,030
At 600°C	1,178
Electrical resistivity at 20°C, mW cm	4.45
Thermal conductivity at 25°C, W m^{-1} K^{-1}	155
Linear coefficient of thermal expansion, K^{-1}	
At 20°C	25.2×10^{-6}
At 20–300°C	$27–28 \times 10^{-6}$

(continued)

Standard redox potential, V	−2.372
Electrical resistivity at 20°C, nΩ·m	43.9
Thermal conductivity, W·m^{-1}·K^{-1}	156
Thermal expansion at 25°C, μm·m^{-1}·K^{-1}	24.8
Young's modulus, GPa	45
Shear modulus, GPa	17
Bulk modulus, GPa	45
Poisson ratio	0.290
Mohs hardness	2.5

Chemical Properties

Magnesium burns in air with an intense white flame forming a mixture of oxide and nitride. It is readily dissolved by most organic and inorganic acids but resistant to alkali hydroxide solutions, hydrofluoric acid, fluorine, and fluorine compounds due to the formation of protective hydroxide and fluoride films. Magnesium occurs in divalent form in all of its compounds. The bromide, iodide, sulfate, and nitrate salts are water soluble. Some of the water-soluble salts are highly hygroscopic and form crystals with a high water content. Magnesium fluoride, oxide, hydroxide, phosphate, and carbonate are sparingly soluble or insoluble in water.

The most important magnesium compounds are magnesium oxide, which is used for refractory magnesia bricks, and magnesium chloride, which is electrolyzed to produce magnesium metal. Naturally occurring magnesium carbonate (magnesite, $MgCO_3$) is mainly burned to give sintered magnesia for the production of magnesia bricks. Seawater is an important source of magnesium oxide and magnesium chloride. Calcined dolomite ($CaCO_3 \cdot MgCO_3$) is used as a raw material for the silicothermic production of magnesium metal. Naturally occurring magnesium silicates include asbestos and talc. Materials for electrical insulation are also based on magnesium silicates. A series of refractory ceramics are based on magnesium. Magnesium chloride can be decomposed by oxygen to form magnesium oxide and chlorine.

Organomagnesium compounds are also well known, the most important of these being the Grignard reagents which play an important role in synthetic organic chemistry.

References

Andereassen K et al (1997) In: Habashi F (ed) Handbook of extractive metallurgy. Wiley, Weinheim, pp 981–1038

Habashi F (2006) A history of magnesium. In: Pekguleryuz MO, Mackenzie LWF (eds) Magnesium technology in the global age. Canadian Institute of Mining, Metallurgy, and Petroleum, Montreal, pp 31–42

Magnesium: Mg^{2+}, Mg^{2+} Ion

▶ Magnesium Binding Sites in Proteins

Magnetic Resonance Cell Imaging

▶ Labeling, Human Mesenchymal Stromal Cells with Indium-111, SPECT Imaging

Major Intrinsic Proteins

▶ Aquaporins and Transport of Metalloids

Major Intrinsic Proteins and Arsenic Transport

▶ Arsenic and Aquaporins

Major Intrinsic Proteins and Boron Transport

▶ Boron and Aquaporins

Major Intrinsic Proteins and Selenite Transport

▶ Selenium and Aquaporins

Major Intrinsic Proteins and Silicon Transport

▶ Silicon and Aquaporins

Malignant Neoplasm

▶ Lanthanides and Cancer

Malignant Tumor

▶ Lanthanides and Cancer

Mallory Body-Like Desmin-Related Myopathy (MB-DRM)

▶ Selenium and Muscle Function

Manganese and Catalases

Sandra Signorella, Claudia Palopoli, Verónica Daier and Gabriela Ledesma
Departamento de Química Física/IQUIR-CONICET, Facultad de Ciencias Bioquímicas y Farmacéuticas, Universidad Nacional de Rosario, Rosario, Argentina

Synonyms

Antioxidant enzymes; Hydrogen-peroxide oxidoreductase enzymes

Definition

Catalases are enzymes that protect cells from deleterious effects caused by hydrogen peroxide by decomposing it into water and dioxygen. Manganese catalases are a class of manganese-containing catalases that are widespread among prokaryotes, which act as the frontline molecular defense against hydrogen peroxide. All manganese catalases share a bridged binuclear manganese cluster that serves as the active site to perform the two-electron disproportionation of hydrogen peroxide, cycling between the reduced $Mn_2(II,II)$ and oxidized $Mn_2(III,III)$ states during turnover. A combination of spectroscopic and reactivity studies show that the ligands bridging the manganese ions and the protein environment surrounding the dimanganese center of manganese catalases play a key role in the catalytic mechanism for decomposing hydrogen peroxide efficiently.

Introduction

Manganese is an abundant element and constitutes a plentiful and freely available resource. One of the major biological roles of Mn is related to oxygen metabolism, as the cofactor for antioxidant defense enzymes (Wu et al. 2004). The special role played by manganese in the redox buffering of living cells, results of its unrivaled repertoire of redox chemistry under highly oxidizing conditions. Owing to the difficulty in oxidizing Mn^{2+} in simple aqueous environments, it is almost certain that this form is the one available to organisms. In aqueous solution, Mn^{2+} resists oxidation beyond 1 V; however, the oxidation potential is well different when manganese is coordinated by protein ligands in a less hydrophilic environment. Therefore, once taken up by an organism, Mn will tend to remain as Mn(II) unless it is captured by ligands, mostly bound to oxygen and nitrogen donor atoms. The number and type of ligands, the local charge, more than one manganese ions clustered together, are among the factors that introduce a way of tuning the redox potential of the metal center to face oxidative conditions (Signorella et al. 2008). This redox buffering function of Mn is reflected in the active site of specialized enzyme systems that allow the cell to create an effective defense against oxidative challenges. Manganese catalases (MnCATs) are one of these antioxidant enzymes (Yoder et al. 2000), and serve as the frontline molecular defense against hydrogen peroxide in a number of organisms.

Hydrogen peroxide is one of the most frequently occurring reactive oxygen species in the biosphere. It appears as a by-product of aerobic metabolism in respiratory and photosynthetic electron-transport chains and as product of enzymatic activity. Excess of hydrogen peroxide and hydroxyl radical, its decomposition product formed in a Fenton-type reaction (see below), are harmful for almost all cell components. Thus, the rapid and efficient removal of H_2O_2 is essential to all aerobically living prokaryotic and eukaryotic cells.

The disproportionation of H_2O_2 is a thermodynamically favorable process ($\Delta G° = -233.56$ kJ mol^{-1}) but proceeds slowly in the absence of catalyst. Nature has developed catalase enzymes to catalyze the redox disproportionation of hydrogen peroxide into dioxygen and water at reasonable rates (Wu et al. 2004). Although the majority of known catalase enzymes are iron-heme-based enzymes that are spread among bacteria, archaea, and eukarya, several organisms utilize dinuclear manganese-containing enzymes to disproportionate hydrogen peroxide. The non-heme manganese-containing catalases appear to be widespread among prokaryotes and have been isolated from bacteria (*Thermus thermophilus, Thermoleophilum album, Salmonella enterica, Thermus sp., and Lactobacillus plantarum*), a hyperthermophilic archeon (*Pyrobaculum caldifontis*), and some cyanobacteria.

All known MnCAT enzymes share an unusual bridged binuclear manganese cluster that serves as the active site to perform the efficient two-electron disproportionation of hydrogen peroxide (Reaction 3), interconverting between reduced (Reaction 1) and oxidized (Reaction 2) states during turnover.

$$H_2O_2 + Enzyme_{red} + 2\ H^+ \longrightarrow Enzyme_{ox} + 2\ H_2O \quad (1)$$

$$H_2O_2 + Enzyme_{ox} \longrightarrow Enzyme_{red} + 2H^+ + O_2 \quad (2)$$

$$2\ H_2O_2 \longrightarrow 2\ H_2O + O_2 \quad (3)$$

At pH 7, the electrochemical potentials (vs. NHE: normal hydrogen electrode) for the two-electron O_2/H_2O_2 and H_2O_2/H_2O couples are +0.28 and +1.35 V, respectively. To efficiently catalyze the H_2O_2 disproportionation, the protein environment controls the reduction potential of the dimanganese active site to a value much lower than that of the Mn^{3+}/Mn^{2+} couple (1.54 V). Besides, the presence of two Mn ions provides with

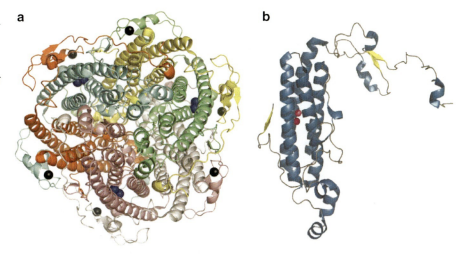

Manganese and Catalases, Fig. 1 (a) Ribbon diagram of MnCAT from *L. plantarum* indicating Ca^{2+} ions as *black dots* and manganese ions as *blue dots*; (b) Single subunit of the hexameric protein and the elements of secondary structure: α-helix (*blue*), β-sheet (*yellow*). Based on PDB ID 1JKU

a further tool to allow the protein to exchange more than one electron, such as required by the two-electron redox half-reactions (1) and (2). This is an important issue because the one-electron reduction of H$_2$O$_2$ leads to the formation of hydroxyl radicals, the reactivity of which is impossible to control and can cause severe damage to macromolecules. The one-electron reduction of H$_2$O$_2$ to hydroxyl radicals (E(H$_2$O$_2$/OH$^•$) = +0.38 V, pH 7) occurs in the presence of small traces of redox-active metal ions via a Fenton-type reaction (Eqs. 4 and 5). Thus, MnCATs have developed a precise catalytic mechanism to disproportionate H$_2$O$_2$ without formation of hydroxyl radicals, where the ligands bridging the manganese ions and the protein environment surrounding the dimanganese center should play a key role (de Boer et al. 2007).

$$H_2O_2 + M^{n+} + H^+ \rightarrow M^{(n+1)+} + OH^• + H_2O \quad (4)$$

$$M^{(n+1)+} + {}^1/_2 H_2O_2 \rightarrow M^{n+} + {}^1/_2 O_2 + H^+ \quad (5)$$

Structural Characterization of MnCAT Enzymes

The three-dimensional structures of MnCAT enzymes isolated from *T. thermophilus* and *L. plantarum* were determinated by X-ray diffraction analysis. Both structures comprise six identical subunits (~30 kDa), where each monomer is based on a motif of four anti-parallel α-helices and a binuclear manganese center in the active site. A structural calcium ion seems to crosslink side chains arising from two subunits and this fact is a distinctive attribute of the protein fold for the MnCAT from *L. plantarum*, absent in the enzyme from *T. thermophilus* (Fig. 1).

The crystal structures of native catalase from *T. thermophilus* and its chloride-bound derivative were refined at 1 Å resolution, and suggest that the dimanganese active site is located in the space between the four helices, deeply immersed in a hydrophobic cavity (Antonyuk et al. 2000). The asymmetric unit possesses two independent bundles of four helices, and each of them has two different conformations for the dimanganese site, in such a way that four independent dimanganese center structures exist. Two channels lead to the active center of the molecule: one of them spans 22 Å direct to the molecule surface, and the other, 16 Å long, connects with the principal channel of the hexamer. In the active site (Fig. 2, left), the two manganese ions are each coordinated to one histidine residue (His73 for Mn1 and His188 for Mn2) and one glutamate residue (Glu36 for Mn1 and Glu155 for Mn2) as terminal ligands, and are bridged by a μ$_{1,3}$-carboxylate ligand (Glu70) and two solvent-derived μ-oxygen bridges (W1 and W2, either aqua, hydroxo, or oxo ligands).

The arrangement of the amino acid residues in the active center of native MnCAT from *T. thermophilus* and its chloride complex revealed that the side chains of residues Lys162 and Glu36 are flexible and exist in two related conformations. The two conformations vary in the ligation mode of carboxylate from Glu36, being a bidentate chelate in one conformation and

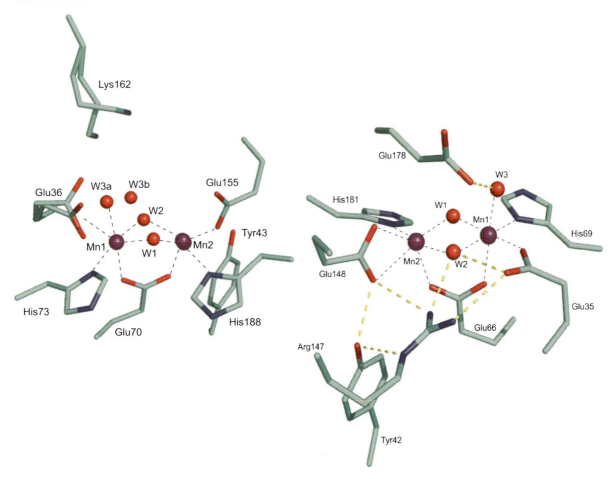

Manganese and Catalases, Fig. 2 Crystal structures of the dimanganese active sites of native MnCAT from *T. thermophilus* (*left*, two conformations of Glu36 and Lys162 are shown) and *L. plantarum* (*right*). Hydrogen-bonding network are indicated by *yellow dashed lines*. Figures based on PDB ID 2V8U and 1JKU, respectively

monodentate in the other (with the vacant coordination site occupied by water W3a). Therefore, there is a six-coordinate Mn1 site (with either bidentate Glu36 or monodentate Glu36 + H_2O), and a five-coordinate Mn2 site (with Glu155 bound via one oxygen). The inhibited-by-chloride enzyme shows that two chloride ions substitute both μ-oxygen bridges, increasing the Mn···Mn distance from 3.13 Å in the native enzyme to 3.30 Å in the chloride-bridged derivative.

A structural correlation between the MnCAT from *T. thermophilus* and *L. plantarum* shows that the dimanganese cores keep the same major features. In the MnCAT from *L. plantarum*, the dimanganese core is embedded in the center of a coiled-coil α-helical domain, with the two manganese ions triply bridged by a $\mu_{1,3}$-carboxylate from Glu66 (Fig. 2, right) and two solvent-derived ligands that were modeled as μ-oxo and μ-hydroxo (W1 and W2) in the resting $Mn_2(III)$ state (Barynin et al. 2001). Furthermore, each manganese ion is bound to one histidine (His69 for Mn1 and His181 for Mn2) and one carboxylate from glutamate residues (monodentate Glu35 for Mn1 and bidentate Glu148 for Mn2). The two manganese subsites are six-coordinate, with the upper-axial position of Mn1 completed by a third solvent molecule (W3), where substrate binding is presumed to occur, as will be discussed below.

The dimanganese cluster is surrounded by a web of hydrogen bonds spread over the outer sphere protein environment, as shown in Fig. 2 for MnCAT from *L. plantarum*. A nonligating glutamate residue (Glu178) is suspended above the cluster and hydrogen bonded to the apical water (W3). Arg147 anchors the hydrogen bonds network by means of a series of bonds: from guanidinium NH2 group to Glu148 and the bridging hydroxide (W2); and from guanidinium NH1 to Glu35. Also, an outer sphere tyrosine residue (Tyr42) acts as a hydrogen bond acceptor to Nε of Arg147 and as a donor to Glu148.

In the azide complex of the Mn_2(III) form of the protein from *L. plantarum*, the azide anion replaces the terminal water on the Mn1 subsite, probably in a mode similar to H_2O_2. The X-ray structure of the azide-bound derivative reveals that the Mn(μ-oxygen)$_2$Mn core flattens somewhat, lengthening the intermetal distance from 3.03 to 3.19 Å.

The outer sphere interactions between tyrosine and glutamate ligand (Tyr43 and Glu155 in *T. thermophilus*; Tyr42 and Glu148 in *L. plantarum*) are also preserved within this family of enzymes; nevertheless other details of the spatial arrangement around the active site are fairly different. The nonligating glutamate Glu178 that lies above the cluster in *L. plantarum* is missed in *T. thermophilus* and functionally replaced by a lysine residue (Lys162). The functional replacement of lysine residue by glutamate modifies the charge balance in the cluster, and the neutralizing effect of arginine (Arg147) is not present in the *T. thermophilus* structure. The absence of residues corresponding to Glu178 and Arg147 turns the active site of *T. thermophilus* into a more spacious one (when compared to the site in *L. plantarum*) and opens up a new channel to the surface of the protein. This enlargement allows the access of a diversity of large anions to the manganese complex in *T. thermophilus* that together with their relaxed steric constraints have implications on the biological functions: *T. thermophilus* enzyme may have functions other than peroxide decomposition (catalase and peroxidase); meanwhile *L. plantarum* enzyme exerts a restricted access to H_2O_2 and specifically serves as a catalase.

Analysis of multiple-sequence alignment discloses that an essential outer-sphere tyrosine (Tyr43 in MnCAT from *T. thermophilus* and Tyr42 in MnCAT from *L. plantarum*) is highly conserved within the manganese catalase family, and this residue seems to be essential to preserve the hydrogen-bonding network. Site-directed mutagenesis was used to prepare recombinant Tyr42Phe MnCAT from *L. plantarum* (Whittaker et al. 2003). Its structure is almost identical to that of the wild type, but the absence of the hydroxyl group belonging to tyrosine residue perturbs the structure of the dimetallic active site, disrupting one of the solvent bridges (W2). In fact, the lack of the phenolic hydroxyl group in the Tyr42Phe mutant eliminates the hydrogen bonding contact with carboxylate of Glu148 residue and contributes to the increase of the intermetallic distance (3.33 Å in the mutant), by shifting the Mn2 atom toward the carboxylate of Glu148. In addition, the absence of Arg147–Tyr42 hydrogen bond raises the Mn–O–Mn angle and destabilizes the W2 bridge, generating an open form of the cluster (Fig. 3). The reduced catalytic activity of the mutant (Section "Reactivity of MnCAT") demonstrates that the outer-sphere Tyr42 possesses a key role on the stability and reactivity of the dimanganese cluster in MnCAT.

Spectroscopic Characterization of the diMn Active Site

A combination of electron paramagnetic resonance (EPR), UV-visible, and X-ray absorption spectroscopies has shown that MnCATs can exist in at least four oxidation states: reduced Mn_2(II,II), mixed valence Mn(II)Mn(III), oxidized Mn_2(III,III) and superoxidized Mn(III)Mn(IV), without evidence to date for a Mn_2(IV,IV) state.

X-ray absorption near-edge structure (XANES) spectroscopy is sensitive to all of the Mn in the sample, regardless of spin state, and has been particularly useful for assigning the oxidation states of MnCAT (Yoder et al. 2000). By this technique, it was found that the as-isolated enzyme contains a mixture of the Mn_2(III,III) and Mn_2(II,II) states with a small portion of superoxidized Mn(III)Mn(IV) state also present. In addition, XANES data provided direct evidence that during the catalytic decomposition of H_2O_2, the enzyme cycles between the reduced Mn_2(II,II) and oxidized Mn_2(III,III) forms, which are present in approximately a 2:1 ratio, and confirmed the observation that reduced Mn_2(II,II) state formed when catalase was treated with H_2O_2 in the presence of inhibitory concentrations of Cl^- or F^-.

X-band (9 GHz) EPR spectroscopy has also proven useful for identifying the Mn_2(II,II), Mn(II)Mn(III), and Mn(III)Mn(IV) oxidation states of the enzyme (Wu et al. 2004). In biological systems, manganese is always in a high-spin state, because the amino acids

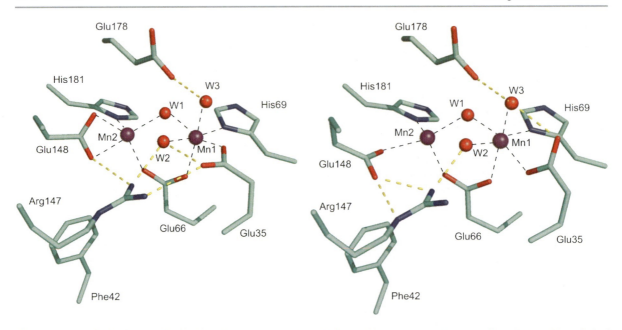

Manganese and Catalases, Fig. 3 Crystal structure of the dimanganese active site of Tyr42Phe MnCAT from *L. plantarum*, showing conformations A (*left*) and B (*right*). Hydrogen-bonding networks are indicated by *yellow dashed lines*. Based on PDB ID 1O9I

lead to only weak ligand fields. Indeed, in MnCATs all the Mn oxidation states are paramagnetic, the coupling between the metal spins is antiferromagnetic, and the EPR signals originate from the exchange coupling between the two Mn ions. The first evidence for the presence of Mn_2(II,II)-active site in MnCATs just emerged from EPR investigations. The reduced form of the enzyme shows a broad, weak, and complex EPR signal that has been attributed to the $S = 1$ and $S = 2$ excited states of a weakly antiferromagnetically coupled Mn_2(II,II) center. This signal is quite sensitive to the anions that are present in solution, as the consequence of anions binding to Mn(II), a result later confirmed by X-ray diffraction studies.

In particular, relatively strong exchange coupling between the two Mn ions in the mixed-valence Mn(II)Mn(III) and superoxidized Mn(III)Mn(IV) states of catalase produces well-isolated ground state doublets, with resolved hyperfine couplings from the two ^{55}Mn nuclei that lead to the detection of multiline EPR signals. The mixed-valence Mn(II)Mn(III) derivative has an 18-line signal that is difficult to saturate and is detectable only at low temperatures, attributed to an antiferromagnetically coupled $S = ½$ ground state. The superoxidized Mn(III)Mn(IV) form is characterized by a 16-line signal which is easily saturated at low temperature. This signal is typical of the di-μ-oxo bridged Mn(III)Mn(IV) core, and is in agreement with extended X-ray absorption fine structure (EXAFS) data from which a Mn⋯Mn distance of 2.7 Å was measured. For superoxidized MnCAT from *L. plantarum* and *T. thermophilus* further magnetic coupling between the electron spin and neighboring ^{14}N (not resolved in the EPR spectra) were detected by electron spin echo envelope modulation (ESEEM) spectroscopy (Stich et al. 2010). ESEEM spectra at 9.77 and 30.67 GHz showed modulations consistent with direct coordination of the imidazole nitrogen of the two histidines to Mn(IV) and Mn(III) ions, respectively. The ESEEM spectra are not sensitive to the addition of CN^- or N_3^-, suggesting that these anions do not bind Mn in the superoxidized enzyme.

The reduced enzyme from *L. plantarum* and *T. thermophilus* lacks significant UV-visible absorption; however, the oxidized and superoxidized forms have significant electronic transitions. The oxidized MnCAT exhibits characteristic absorptions centered at near 470 nm with a shoulder at around 500 nm (extinction coefficients close to 300 M^{-1} cm^{-1} per Mn ion) and a tail between 600 and 800 nm, typical of Mn(III) complexes. The electronic spectrum of the superoxidized state of MnCATs is blueshifted and has higher absorptivity. When azide is added to the oxidized form of MnCAT from *L. plantarum*, an intense new band is observed at 370 nm. This band was assigned to the azide → Mn charge-transfer transition and was used to monitor the ligand binding at room temperature.

Reactivity of MnCAT

MnCATs dismutate H_2O_2 at high rates. Kinetic studies indicate a rate-limiting step that is first-order in enzyme. Saturation kinetics is observed with respect to substrate, and kinetic parameters summarized in Table 1 (Wu et al. 2004) were obtained for MnCAT from *L. plantarum*, *T. thermophilus*, and *T. album* by fit of rate data to the Michaelis-Menten equation. The Michaelis constants (K_M) differ by 20-fold, while the catalytic efficiencies (k_{cat}/K_M) exhibit only a 5-fold variation with species. The turnover numbers of MnCAT, although quite fast, are significantly slower than those of heme-CATs that have turnover numbers close to the diffusion-limited rate and are also more efficient (higher k_{cat}/K_M ratio) than MnCAT. However, MnCAT is, for the experimentalist, a more convenient scavenger of H_2O_2 rather than the heme-CAT because it is stable to freezing and thawing, tolerates moderate heating, and is insensitive toward many compounds which inhibit the heme-CAT.

The activity of MnCAT is dependent on the pH and the oxidation state of the dimanganese center (de Boer et al. 2007). For the *L. plantarum* and *T. thermophilus* enzymes the activity was found to occur over a pH range of 5–12, with activity nearly independent of pH in the 7–10 pH range, and falling to zero at more extreme pH values. Based upon magnetic studies on MnCAT from *T. thermophilus* in the pH range 6.5–9.5, two different pH-dependent bridging modes for the $Mn_2(III,III)$ state have been proposed: a strong coupled $Mn(\mu\text{-oxo})_2Mn$ closed form in equilibrium with a weakly coupled $H_2OMn(\mu\text{-oxo})MnOH$ open form that appears at pH 6.5. These results may explain the loss of catalytic activity at pH < 5 upon protonation of the bridges and formation of the open form of the enzyme, and support the notion that the catalytically active $Mn_2(III,III)$ state of the enzyme must contain a closed cluster, associated with an antiferromagnetically coupled electronic ground state mediated by a pair of solvent bridges. In line with this, the active form of Tyr42Phe MnCAT from *L. plantarum* is much more sensitive to pH than the wild type enzyme (Whittaker et al. 2003). This mutant has less than 5% of wild type MnCAT activity at neutral pH and much higher K_M for H_2O_2 ($\approx$1.4 M) at neutral pH than at pH 10 (K_M 220 mM). As shown in Fig. 2, in the wild type enzyme, Tyr42 is involved in a web of hydrogen bonds that contributes to stabilize the diMn core with a pair of solvent bridges over a wide pH range. Replacement of Tyr42 residue disrupts the hydrogen-bonded web (Fig. 3), the double solvent bridged diMn core is not further stabilized and efficient CAT activity and substrate affinity are only achieved at high pH, where the broken bridge (W2) appears to be rescued by exogenous hydroxide, regenerating the bis-bridged $Mn_2(III,III)$ closed form of the native enzyme and restoring the catalytic function.

The mixed valence state does not form during enzymatic turnover, but can be induced artificially as an intermediate during reduction with hydroxylamine and other one-electron reductants (Wu et al. 2004). If this is allowed to occur in the presence of peroxide, the intermediate is quantitatively oxidized to the catalytically inactive di-μ-oxo $Mn_2(III,IV)$ form. Although the superoxidized state can be fully reactivated, it occurs so slowly (half-time $\approx$ 30 min) as to render the enzyme effectively unavailable during the cell's lifetime.

When the enzyme is in the reduced $Mn_2(II,II)$ state it can be inhibited by the reversible binding of several anions, including halides and oxoanions. Consequently, when H_2O_2 decomposition is performed in the presence of either Cl^- or F^- the enzyme is trapped in the reduced state, a fact confirmed by X-ray studies that showed the replacement of the μ-oxygen bridges by the μ-chloride ones. Unlike the pH-independent reactivity, anion binding typically shows strong pH dependence, increasing at lower pH. The higher affinity of the reduced enzyme for the halide explains the observation that it is the reduced enzyme that is trapped during turnover with halide.

Although the MnCATs were originally identified on the basis of their insensitivity to azide and cyanide, more recent work has shown that they are in fact inhibited by both azide and cyanide, albeit at much higher concentration than the heme-CAT. Besides, in the case of MnCAT from *L. plantarum*, azide inhibition is competitive with peroxide, implying that azide and peroxide share a common binding site. In contrast to halides, X-ray studies have shown that azide binds to Mn1 in the oxidized $Mn_2(III,III)$ state of the enzyme

Manganese and Catalases, Table 1 Kinetic parameters for dismutation of H_2O_2 by MnCATs

Catalase	k_{cat} (s^{-1})	K_M (mM)	k_{cat}/K_M (M^{-1}s^{-1})
L. plantarum	2.0×10^5	350	5.7×10^5
T. thermophilus	2.6×10^5	83	3.1×10^6
T. album	2.6×10^4	15	1.7×10^6

from L. plantarum as a terminal ligand, by replacing the labile water molecule in the native enzyme (Barynin et al. 2001). Since azide has frontier molecular orbitals similar to those of H_2O_2, the interaction of this anion with catalase is probably the same as the mode of initial peroxide binding to the active enzyme.

Role of Bridging Ligands and Mn Environment on CAT Activity

The highly efficient dismutation of H_2O_2 by MnCAT requires a two-electron redox cycle, made possible at the dinuclear site, which operates as a unit and suppresses discrete one-electron redox processes that yield catalytically inactive mixed valence Mn(II)Mn(III) and Mn(III)Mn(IV) states. Mechanistic studies performed on biomimetic model compounds, together with biochemical, structural, and spectroscopic studies on MnCATs, have provided some clues on the role played by the ligands bridging the manganese ions and the protein environment surrounding the dimanganese center in the enzymatic mechanism.

The fact that the two Mn ions of the active site of MnCAT possess the same NO_5 coordination sphere (section "Structural Characterization of MnCAT Enzymes") provides a symmetrical environment that stabilizes the homovalent diMn core and facilitates the observed redox activity based on shuttling between $Mn_2(II,II)/Mn_2(III,III)$ states during disproportionation of H_2O_2. Evidence on the effect of the cluster symmetry on the oxidation states employed during catalysis has been obtained from studies on CAT activity of diMn model complexes of symmetrical and asymmetrical ligands (Signorella et al. 2008): the former employ homovalent states to disproportionate H_2O_2, while the latter involve mixed-valent states.

One of the most important structural features of MnCAT is the carboxylate bridge which, besides acting as a bridging ligand, is believed to be critical to H_2O_2 disproportionation (de Boer et al. 2007).

While μ-oxo bridges facilitate communication between the two Mn centers, carboxylate bridges increase the Mn⋯Mn separation and electronically shield the diMn center, thus promoting two-electron $Mn_2(II,II)/Mn_2(III,III)$ over one-electron $Mn_2(II,II)/Mn(II)Mn(III)/Mn_2(III,III)$ processes. In addition to MnCATs, several other dinuclear metalloproteins that participate in two-electron redox chemistry, such as hemerythrin, methane monooxygenase, and ribonucleotide reductase, contain carboxylate bridges. In all these cases, nature has selected the carboxylate bridge to separate and decouple the redox centers ensuring minimal formation of harmful mixed-valence species.

It has been demonstrated that among complexes with the same terminal ligands, the number of bridging acetate/oxo/aqua groups directly correlates with Mn oxidation states and CAT activity. In many manganese model complexes, carboxylate and aqua bridging ligands favor the lower oxidation states whereas μ-oxo bridging ligands favor the higher oxidation states. Arginase, whose physiological role is to catalyze the hydrolysis of L-arginine to L-ornithine, is an example of this kind of correlation. The active site of arginase consists of two Mn(II) ions triply bridged by two aspartates and one water molecule. Although able to dismutate H_2O_2, arginase is five orders of magnitude less efficient than MnCAT. The low CAT activity of arginase reflects the role of bis-carboxylate/aqua bridges in stabilizing the $Mn_2(II,II)$ redox state.

Studies of model systems have shown that the presence of the bridging glutamate or carboxylate bends the Mn_2O_2 core, activating the oxido groups for protonation and exchange with solvent (Dubois et al. 2008). The histidines in MnCAT could be further contributing to this activation, labilizing the solvent bridge to accommodate the substrate hydrogen peroxide during the reduction step of the disproportionation (Stich et al. 2010). Evidence for the communication between the trans oxido bridge and the active site histidines was obtained by X-band ESEEM on MnCAT from T. thermophilus where the Lys162 residue (Fig. 2, left) had been reductively methylated. Methylation of Lys162 induced significant changes in the spectral features that were assigned to the histidine bound to the Mn ion. Lys162 is able to hydrogen bond directly with Glu36 that is postulated to be critical in proton transfer from bound hydrogen peroxide, and, through a water molecule, to the μ-*oxygen* bridge (W2) that is trans to the Mn-bound histidines. The loss of these interactions

Manganese and Catalases, Scheme 1 Molecular mechanism of catalytic decomposition of H_2O_2 by MnCAT

upon methylation of Lys162 disrupts the active site, leading to a change in the coordination geometry of Mn1 and protonation of the μ-oxo bridges. Also, non-ligating carboxylates participate in hydrogen bonds and have been shown to be important in base-assisted proton-coupled electron transfer for CAT activity.

Whereas carboxylate bridge of MnCAT is stable, oxo/hydroxo/aqua ligands are labile with respect to dissociation during catalysis. The lability of ligands, in particular water, is central to activity and decreases with increasing oxidation state of the Mn ions. Detailed mechanistic studies performed on diMn complexes of binucleating ligands have shown that the presence of a labile H_2O molecule bound to one of the Mn ions facilitates terminal H_2O_2 binding through H_2O/H_2O_2 exchange (Signorella et al. 2008).

Mechanism of H_2O_2 Disproportionation by MnCAT

As mentioned in the introduction, the overall *catalase* reaction includes the degradation of two molecules of hydrogen peroxide to water and molecular oxygen (Reaction 3), and it has been well established that the active site cycles between the $Mn_2(II,II)$ and $Mn_2(III,III)$ states during turnover. Therefore, Eqs. 1 and 2 may be rewritten as:

$$H_2O_2 + Mn_2(II, II) - CAT(2H^+) \rightarrow Mn_2(III, III) - CAT + 2H_2O \quad (6)$$

$$H_2O_2 + Mn_2(III, III) - CAT \rightarrow Mn_2(II, II) - CAT(2H^+) + O_2 \quad (7)$$

This "ping-pong" mechanism can be described by a two-stage two-electron redox process, where the reduced dimanganese cluster favors peroxide reduction to water while the oxidized form catalyzes peroxide oxidation, releasing oxygen by proton-coupled electron transfer, without temporal order in these reduction and oxidation stages.

The observations described in the precedent sections provide the basis for mechanistic insights, illustrated in Scheme 1. In this mechanism, two different coordination modes of H_2O_2 to the dimanganese center are proposed for the oxidative and reductive half-reactions, where the solvent bridges serve as proton storage sites and dictate the coordination chemistry of the dimanganese cluster for maximum catalytic efficiency.

Initial peroxide binding to the Mn_2(III,III) form of the enzyme is proposed to occur at a terminal site on one of the Mn centers by displacement of the labile water ligand with concomitant protonation of the μ-oxo bridge. Subsequent reduction of the Mn_2(III,III) core accompanied by a second proton transfer from the terminally bound substrate results in the formation of the μ-aquo Mn_2(II,II) enzyme and release of O_2. The second equivalent of H_2O_2 binds to the Mn_2(II,II) form of the enzyme as a bridging $\mu_{1,1}$-hydroperoxo. The $\mu_{1,1}$-bridging mode polarizes the O–O bond and may even be polarized further by protonation of hydrogen peroxide, thus facilitating heterolytic O–O bond cleavage coupled to cluster reoxidation to Mn_2(III,III) with loss of water, to close the catalytic cycle. It is interesting to note that in the oxidative half-reaction, as the Mn_2(III) cluster is already oxidized, the bridging geometry may be unproductive and should decrease the catalytic efficiency of the enzyme.

The presence of donor/acceptor residues near the active site (e.g., Glu178 in MnCAT from *L. plantarum* and Lys162 in *T. thermophilus*) facilitates proton transfers between nonadjacent atoms in the complex and is central for efficient catalysis. Biomimetic models which possess the same bridging motif as MnCAT but lack proton acceptor near the dimanganese core are less efficient than the enzyme to disproportionate H_2O_2 (Palopoli et al. 2011), emphasizing the importance of these residues in proton transfers during catalysis.

Cross-References

▶ Magnesium in Biological Systems
▶ Magnesium, Physical and Chemical Properties
▶ Manganese and Photosynthetic Systems
▶ Manganese, Interrelation with Other Metal Ions in Health and Disease
▶ Mercury and Low Molecular Mass Substances
▶ Metals and the Periodic Table
▶ Peroxidases

References

Antonyuk S, Melik-Adamyan V, Popov A et al (2000) Three-dimensional structure of the enzyme dimanganese catalase from *Thermus thermophilus* at 1 Å resolution. Crystallogr Rep 45:105–116

Barynin V, Whittaker M, Antonyuk S et al (2001) Crystal structure of manganese catalase from *Lactobacillus plantarum*. Structure 9:725–738

De Boer J, Browne W, Feringa B et al (2007) Carboxylate-bridged dinuclear manganese systems – from catalases to oxidation catalysis. Comptes Rendus Chimie 10:341–354

Dubois L, Pécaut J, Charlot M et al (2008) Carboxylate ligands drastically enhance the rates of oxo exchange and hydrogen peroxide disproportionation by oxo manganese compounds of potential biological significance. Chem Eur J 14:3013–3025

Palopoli C, Bruzzo N, Hureau C et al (2011) Synthesis, characterization and catalase activity of a water soluble diMn(III) complex of a sulphonato-substituted Schiff base ligand: an efficient catalyst for H_2O_2 disproportionation. Inorg Chem 50:8973–8983

Signorella S, Tuchagues J-P, Moreno D et al (2008) Catalase activity of diMn(III) complexes with the $[Mn_2(\mu-O_2C_2H_3)(\mu-OL)(\mu-OX)]^{3+}$ core (L = polydentate ligand; X = CH_3 or OC_2H_3). Structural features that control catalysis. In: Hughes J, Robinson A (eds) Inorganic biochemistry research progress. Nova Science, New York

Stich T, Whittaker J, Britt R (2010) Multifrequency EPR studies of manganese catalase provide a complete description of proteinaceous nitrogen coordination. J Phys Chem B 114:14178–14188

Whittaker M, Barynin V, Igarashi T et al (2003) Outer sphere mutagenesis of *Lactobacillus plantarum* manganese catalase disrupts the cluster core. Eur J Biochem 270:1102–1116

Wu A, Penner-Hahn J, Pecoraro V (2004) Structural, spectroscopic, and reactivity models for the manganese catalases. Chem Rev 104:903–938

Yoder D, Hwang J, Penner-Hahn J (2000) Manganese catalases. In: Sigel A, Sigel H (eds) Metal ions in biological systems, vol 37. Marcel Dekker, New York

Manganese and Photosynthetic Systems

Alice Haddy
Department of Chemistry and Biochemistry,
University of North Carolina, Greensboro, NC, USA

Synonyms

Mn in photosystem II; Mn cluster in photosynthesis

Definition

Mn and photosynthetic systems refers to the role of Mn in the biochemical processes of organisms that carry out photosynthesis as part of their life cycle.

Photosynthetic systems include many forms of bacteria and all types of algae and higher plants. However, only those that produce molecular oxygen during photosynthesis require Mn as an essential cofactor. In this context, Mn has an important role in the oxidation-reduction reactions that are involved in converting light energy into chemical energy that is used by all living systems.

Role of Manganese

Manganese is well known to be an essential metal for photosynthesis, because it is indispensible for oxygen evolution. The only significant natural source of atmospheric O_2 is through oxidation of H_2O during biological photosynthesis, and manganese is the key component of the catalytic site at which this occurs. This feature of photosynthesis is therefore responsible for the oxygen that respiring creatures require for survival and so plays a major role in maintaining life on earth as it is known. Oxygen production takes place in those higher photosynthetic forms that carry out electron transfer through the use of two light-absorbing membrane-bound protein complexes, photosystems I and II. These photosynthetic organisms include higher plants, algae, and cyanobacteria (also known as blue-green algae). Lower photosynthetic forms that make use of a single photosystem, including various types of photosynthetic bacteria, do not produce oxygen.

Location and Function of Manganese in Oxygen Evolution

The site of catalytic O_2 production from H_2O is a Mn_4CaO_5 cluster of photosystem II (PSII), which is embedded within the thylakoid membrane of chloroplasts and cyanobacteria (for reviews on PSII, see McConnell 2008; Barber 2002). PSII is one of three major membrane-embedded protein complexes that participate in the light reactions of photosynthesis. It serves as the starting point for electron transfer, with H_2O as the electron donor. The energetic driving force for water oxidation is derived from the absorption of light energy at the PSII reaction center, which promotes electron transfer through several redox-active components. The electron transfer components are oriented within PSII such that the oxidation of H_2O takes place on the inside of the thylakoid membrane. Oxidation of water to molecular oxygen proceeds according to the half reaction:

$$2H_2O \rightarrow O_2 + 4e^- + 4H^+.$$

Concurrent with the formation of O_2 is the release of four protons to the lumen or inside of the thylakoid membrane and the donation of four electrons to the electron transfer chain. Electrons extracted from water are withdrawn via the light-absorbing reaction center toward the plastoquinone-binding site, which is located near the stroma or outside of the thylakoid membrane. Oxidation of water is thereby physically separated from the electron acceptor half reaction, in which plastoquinone is reduced to hydroquinone. The protons released to the inside of the membrane contribute to the transmembrane proton gradient, which is used for phosphorylation of ADP to ATP by the ATP synthase, while the electrons are transferred through subsequent protein complexes to their eventual end point in the reduction of $NADP^+$ to NADPH. The Mn_4CaO_5 cluster is therefore responsible for a catalytic process in which O_2 is the by-product of electron donation from H_2O to the electron transfer chain that ultimately produces the high-energy molecules ATP and NADPH.

Manganese is absolutely required for photosynthetic oxygen production and no other ion has been found to be functional in its place. (The calcium within the Mn_4CaO_5 cluster can, on the other hand, be replaced by strontium with partial effectiveness.) The versatility of manganese in terms of available oxidation states accounts in part for its usefulness in the site of oxygen production. Manganese can take on a wide range of oxidation states from 0 to 7+, with the highest states acting as powerful oxidants. Although nature provides many examples of compounds employing Mn^{II}, Mn^{III}, Mn^{IV}, Mn^{VI}, and Mn^{VII}, the first three oxidation states are those that are most relevant biochemically. Mn^{II} is the most stable oxidation state because of its d^5 electron configuration, but it is more labile to cluster formation than Mn^{III} or Mn^{IV}. It can be found as a monomer in proteins where its role is likely to involve providing an octahedral coordination environment, as do Ca^{2+} and Mg^{2+}. Mn^{III} and Mn^{IV} are able to form strong bonds with O-donating ligands (oxides and hydroxides). Mn^{III} is highly oxidizing and in aqueous environment tends to disproportionate into Mn^{II} and Mn^{IV}. Mn^{IV} is relatively stable and is often found as MnO_2 in aqueous environments. In the proper ligation environment, the Mn^{III} and

Mn^{IV} species are able to form stable mixed-valence clusters, which can participate in multi-electron transfer reactions. Such is the case for the photosynthetic Mn_4CaO_5 cluster, which cycles through a series of oxidation states that are mainly combinations of Mn^{III} and Mn^{IV}. Iron, the transition metal with abilities closest to those of manganese, forms redox-active clusters from Fe^{II} and Fe^{III} species, but it is evidently unable to fulfill the role required for water oxidation chemistry. This may be partly because Fe^{II} is not able to associate as strongly with O-donating ligands as Mn^{III} and because clusters with a high proportion of Fe^{II} are more labile (Armstrong 2008).

The oxygen evolving Mn_4CaO_5 cluster of PSII can be compared with a few other Mn-containing enzymes, most notably manganese superoxide dismutase, which removes O_2^- or HO_2 by dismutation into O_2 and H_2O_2, and manganese catalase, which removes H_2O_2 by disproportionation into H_2O and O_2. These have in common with PSII a role in water/oxygen chemistry in which catalysis is carried out by an oxidation state cycle that includes Mn^{III}. However, there are forms of superoxide dismutase and catalase that function with iron or other metals at the catalytic center, whereas photosystem II strictly requires manganese at the catalytic center. No biological catalytic center other than the Mn_4CaO_5 cluster of photosystem II is able to carry out conversion of H_2O into O_2 and no forms with alternative metal-requirements are known to exist.

Oxidation State Cycle of the Manganese Cluster

The characteristics of the oxidation state cycle of the Mn_4CaO_5 cluster during oxygen production has been the focus of much research attention over the past four decades, because this information is expected to lead to an understanding of the mechanism of this energetically difficult reaction. Early on, the oxidation states were identified as S_0 through S_4 based on measurements of oxygen yield in response to a series of carefully spaced light flashes (Fig. 1). It was found that in dark-adapted samples, the oxygen yield cycles with the flash, with peaks in yield occurring on the third flash, then on subsequent fourth flashes. This behavior was explained by the S-state cycle or Kok cycle (named after Bessel Kok), in which the absorption of a photon by the reaction center of PSII promotes an increase in oxidation state by one step, from S_i to S_{i+1}. Once the S_4 state is reached, a molecule of oxygen is produced and released, returning the system to the S_0 state.

The explanation for the first peak occurring on the third rather than the fourth flash is that the S_1 state is the most stable state reached after a period of dark adaptation. Thus, the oxidation state cycle of the Mn_4CaO_5 cluster consists of five formal oxidation states, with transitions to higher oxidation states accompanied by electron transfer until dioxygen is produced concurrent with return to the lowest oxidation state. This model of the oxygen evolution catalytic cycle has continued to be used over the decades since first proposed, with subsequent additions and modifications that have served to clarify it.

The multistep catalytic cycle of the Mn_4CaO_5 cluster succeeds in breaking down the oxidation of water into electron transfer steps with energy demands that are manageable within a protein environment (Armstrong 2008; Brudvig 2008). The half reaction for the oxidation of $H_2O(2H_2O \rightarrow O_2 + 4H^+ + 4e^-)$ has a standard reduction potential, E_0, of -1.23 V ($25°C$). A more realistic value would be one adjusted for a biological pH, thus at pH 7 the potential would be -0.82 V. Taking into account that this value represents an average per electron, then the complete four-electron oxidation of two water molecules corresponds to a potential of -3.27 V. This quite large negative potential indicates the difficulty of driving the reaction forward. The potential provided by the absorption of a photon at P680, the reaction center of PSII, is around $+1.25$ V, while the oxidizing potential of Tyr Z, the immediate electron acceptor of Mn_4CaO_5 oxidation, is believed to be about $+1$ V for the phenoxyl form. Thus, the Mn_4CaO_5 cluster is able to accumulate enough oxidizing potential over the four photon absorption/electron transfer steps to carry out water oxidation. While oxidation of the Mn_4CaO_5 cluster takes place in single electron steps, the final step in the formation of O_2 during the S_4 to S_0 transition does not. The completion of water oxidation in this step, with concurrent reduction of the manganese cluster, is believed to take place in either two two-electron steps or a single concerted four-electron step based on theoretical considerations. Peroxide intermediates have not been detected and are not believed to have a role.

Coupling in the Manganese Cluster

The oxidation states and coupling of manganese within the Mn_4CaO_5 cluster during the S-state cycle have been examined extensively using electron paramagnetic resonance (EPR) spectroscopy and X-ray

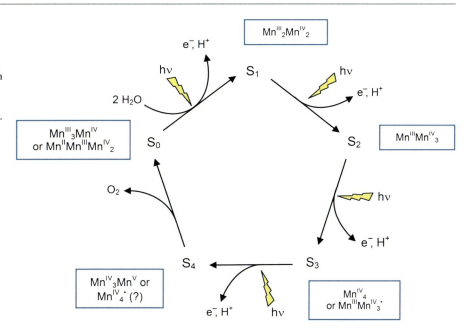

Manganese and Photosynthetic Systems, Fig. 1 Oxidation state or S-state cycle of the Mn_4CaO_5 cluster of photosystem II, with likely oxidation states of the individual manganese ions given (see text for discussion). H_2O is shown to enter the cycle at the S_0 state, but it is probably able to enter or exchange at all steps. Proton release and electron release are shown to occur concurrently for simplicity, although this may not be the case for each step

absorption near edge structure (XANES) methods. EPR spectroscopy can reveal the overall spin state of a paramagnetic species, which indicates the oxidation state or likely coupling; XANES can reveal the oxidation state of a metal species or the average oxidation state in the case of a cluster. These studies of the S-states and how they change during the catalytic cycle have been important in supporting postulates of the mechanism of water oxidation. XANES analysis, in combination with spin state information from EPR, has led to a fairly accurate estimate of the individual manganese oxidation states during the catalytic cycle of oxygen evolution (Sauer et al. 2008; Dau and Haumann 2008), although some questions remain. The most extensively studied oxidation states are S_1 and S_2 because of their ease of preparation. There is general agreement that the manganese cluster is present as $Mn_2^{III}Mn_2^{IV}$ in the S_1 state and as $Mn^{III}Mn_3^{IV}$ in the S_2 state. The S_0 state is likely to be due to a $Mn_3^{III}Mn^{IV}$ cluster or possibly a $Mn^{II}Mn^{III}Mn_2^{IV}$ cluster. The origin of the S_3 state has been less clear with some evidence that supports oxidation of the Mn^{III}, resulting in a Mn_4^{IV} cluster, and other evidence that supports oxidation of a nearby amino acid residue, resulting in a $Mn^{III}Mn_3^{IV}$ cluster coupled with an organic radical (represented as $Mn^{III}Mn_3^{IV\bullet}$). As a highly transient state, there is extremely little evidence to support a specific assignment for the S_4 state, but suggested possibilities include a $Mn_3^{IV}Mn^V$ cluster or a Mn_4^{IV} cluster coupled with an organic radical (represented as $Mn_4^{IV\bullet}$).

Throughout most of the S-state cycle, the Mn^{III} and Mn^{IV} ions of the Mn_4CaO_5 cluster are thought to be coupled antiferromagnetically, resulting in the lowest possible spin state. Using EPR spectroscopy, the S_0 and S_2 states have been shown to have net electron spin of $S = ½$ in the ground state, while the S_1 and S_3 states have net spin of $S = 0$ in the ground state (Haddy 2007; Peloquin and Britt 2001). This information leads to postulates of how the ions within the Mn_4CaO_5 cluster must be coupled. In the case of the S_2 oxidation state, for example, an EPR signal appears with a multiline pattern at a g-factor of 2.0, which is characteristic of an $S = ½$ state. As the first EPR signal to be observed from the manganese cluster, it helped to confirm the presence of manganese at the site of oxygen evolution. The spin state of $S = ½$ for this signal can arise from a $Mn^{III}Mn_3^{IV}$ cluster if the Mn^{III} (d^4, $S = 2$) is antiferromagnetically coupled with a Mn^{IV} (d^3, $S = 3/2$), leading to a spin of $S = ½$, while antiferromagnetic coupling of the two remaining Mn^{IV} results in no additional contribution to the total spin. The S_0 oxidation state also shows a spin $S = ½$ state leading to an EPR signal with a different multiline pattern than the S_2 state EPR signal. The S_1 and S_3 oxidation states arise from clusters in which the manganese ions are coupled so as to cancel like electron spins, resulting in a total electron spin of $S = 0$. While no EPR signal is

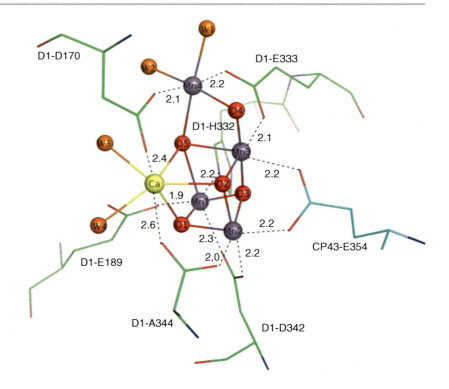

Manganese and Photosynthetic Systems, Fig. 2 Structure of the Mn_4CaO_5 cluster where formation of oxygen from water takes place, as determined by X-ray diffraction at 1.9 Å resolution (From Kawakami et al. 2011)

observed from an $S = 0$ state, these two S-states have been characterized by EPR signals arising from higher spin states ($S = 1$, $S = 2$, etc.) within the same spin manifold as the $S = 0$ state. Higher spin states from alternative coupling schemes have also been shown to be present for several of the S-states under certain conditions and these are probably involved in the progress of the catalytic cycle. For example, the S_2 state shows a second EPR signal arising from an $S = 5/2$ spin state, which is characterized by a broad EPR signal with a g-factor of 4.1 and appearing at lower magnetic field than the $S = ½$ multiline signal.

Structure of the Manganese Cluster

Since the first crystal structure of PSII was determined in 2001, X-ray diffraction (XRD) studies have contributed greatly to understanding the structure of the Mn_4CaO_5 cluster and its nearby protein environment (Umena et al. 2011; Kawakami et al. 2011). These studies have been carried out using PSII purified from thermophilic cyanobacteria, since it is very stable to crystallization. The resulting structures are considered to be applicable to all species because of the high level of conservation of the core proteins of PSII, although extrinsic subunits are known to vary across divergent species. The resolution of the structures has improved from 3.8 Å in the first structure, in which the Mn_4CaO_5 cluster appeared as a "papaya-shaped" blob of electron density, to a resolution of 1.9 Å in the most recently determined structure as of 2011 (Umena et al. 2011). As resolution increased, the identity of amino acid residues ligating the cluster became more certain, the coordinated calcium ion became visible, and activating chloride ions near the cluster were resolved. Once 1.9 Å resolution was reached, the individual atoms of the cluster, as well as surrounding H_2O molecules, were clearly resolved for the first time. XRD is difficult to use for PSII in any oxidation state other than the dark-adapted, stable S_1 state and concerns have been raised about reductive damage due to X-ray radiation. Another method that has contributed significantly to the present understanding of the Mn_4CaO_5 structure is extended X-ray absorption fine structure (EXAFS) analysis, an extension of XANES that can measure with high precision the distances between Mn ions and nearby atoms. This method, which does not require crystallization, has the advantage of being usable with PSII from a variety of species and in a wider range of S-states. In addition, the Mn ions are not prone to X-ray damage. However, the method is unable to resolve the directions of distance vectors unless the sample has been crystallized or otherwise oriented. Some studies have combined XRD and EXAFS data, resulting in a highly accurate picture of the Mn_4CaO_5 structure.

The Mn_4CaO_5 cluster, as determined in the S_1 state, is arranged such that three manganeses (designated Mn1, Mn2, and Mn3) and the calcium are coordinated through μ-oxobridges, thus this part of the cluster can be viewed as a near-cubane of Mn_3CaO_4 (Fig. 2) (Umena et al. 2011; Kawakami et al. 2011). However, the bonds to the calcium are longer than those to the three manganeses, thereby distorting the symmetry. The fourth manganese (Mn4), which is referred to as the distal or "dangler" manganese, is ligated to the calcium and two manganeses through μ-oxo bridges. Altogether the structure resembles a "distorted chair" with the Mn_3CaO_4 cubane forming the seat and the distal Mn plus one O atom forming the back of the chair. One O atom, designated O5, has the special role of ligating three manganeses and the calcium and is, therefore, thought to be relatively unstable. The metal ions of the cluster are ligated to the protein through amino acids mainly from the D1 or PsbA subunit (one of the core subunits that contains the electron transfer components of PSII), plus two amino acids from the CP43 or PsbC subunit. These include D1-Glu189, D1-His332, and D1-Asp342 to Mn1; D1-Asp342, D1-Ala344 (the C-terminal residue), and CP43-Glu354 to Mn2; D1-Glu333 and CP43-Glu354 to Mn3; D1-Asp170 and D1-Glu333 to Mn4; and D1-Asp170 and D1-Ala344 to Ca. Thus, most of the coordinating amino acid residues serve as bidentate ligands. Four water molecules are directly coordinated to the cluster, with two ligated to Mn4 and two to the Ca atom. Given the complete ligation for the cluster, it appears that all four manganeses have six ligands each while the calcium has seven ligands.

Although manganese is required for only one main function in photosynthesis, its role is critical for both photosynthesis and supporting life on earth. With its unique oxidation properties and ability to bind oxygen, it plays a critical and evidently irreplaceable role in the production of molecular oxygen. The mechanism of this unusual four-electron oxidation reaction will continue to be the focus of much research.

Cross-References

▸ Magnesium in Biological Systems
▸ Magnesium, Physical and Chemical Properties
▸ Manganese and Catalases

References

Armstrong F (2008) Why did nature choose manganese to make oxygen? Philos Trans R Soc B 363:1263–1270
Barber J (2002) Photosystem II: a multisubunit membrane protein that oxidises water. Curr Opin Struct Biol 12:523–530
Brudvig GW (2008) Water oxidation chemistry of photosystem II. Philos Trans R Soc B 363:1211–1219
Dau H, Haumann M (2008) The manganese complex of photosystem II in its reaction cycle – basic framework and possible realization at the atomic level. Coord Chem Rev 252:273–295
Haddy A (2007) EPR spectroscopy of the manganese cluster of photosystem II. Photosynth Res 92:357–368
Kawakami K, Umena Y, Kamiya N, Shen J-R (2011) Structure of the catalytic, inorganic core of oxygen-evolving photosystem II at 1.9 Å resolution. J Photochem Photobiol B 104:9–18
McConnell IL (2008) Substrate water binding and oxidation in photosystem II. Photosynth Res 98:261–276
Peloquin JM, Britt RD (2001) EPR/ENDOR characterization of the physical and electronic structure of the OEC Mn cluster. Biochim Biophys Acta 1503:96–111
Sauer K, Yano J, Yachandra V (2008) X-ray spectroscopy of the photosynthetic oxygen-evolving complex. Coord Chem Rev 252:318–335
Umena Y, Kawakami K, Shen J-R, Kamiya N (2011) Crystal structure of oxygen-evolving photosystem II at a resolution of 1.9 Å. Nature 473:55–59

Manganese in Biological Systems

Ebany J. Martinez-Finley[1,2], Sudipta Chakraborty[3] and Michael Aschner[1,2,4,5]
[1]Department of Pediatrics, Division of Pediatric Clinical Pharmacology and Toxicology, Vanderbilt University Medical Center, Nashville, TN, USA
[2]Center in Molecular Toxicology, Vanderbilt University Medical Center, Nashville, TN, USA
[3]Department of Pediatrics and Department of Pharmacology, and the Kennedy Center for Research on Human Development, Vanderbilt University Medical Center, Nashville, TN, USA
[4]Center for Molecular Neuroscience, Vanderbilt University Medical Center, Nashville, TN, USA
[5]The Kennedy Center for Research on Human Development, Vanderbilt University Medical Center, Nashville, TN, USA

Definition

Manganese is an essential mineral, required in the diet, yet toxic in excess. Its atomic number is 25; it is found in nature and has many industrial uses. Manganese functions as a cofactor in many enzymes and metalloproteins; of these the most widely known is manganese superoxide dismutase (MnSOD), whose primary function is detoxification of superoxide free radicals.

Manganese in the Environment

Manganese (Mn) is found naturally in the environment, accounting for about 0.1% of the earth's crust, making it the 12th most abundant element and the 5th most abundant metal. Mn is released into the environment through soil erosion, leading to an average concentration of 0.02 $\mu g/m^3$ in ambient air. However, both organic and inorganic forms of Mn can be found at high levels in various industrial settings, leading to air concentrations of 0.2–0.3 $\mu g\ Mn/m^3$.

Surface waters in the United States contain a median concentration of 16 μg Mn/L, with mean concentrations in public water drinking supplies ranging from 4 to 32 μg Mn/L. Groundwater may naturally contain 5–150 μg Mn/L. High Mn concentrations in groundwater have been documented in Bangladesh where the local water drinking supply contain up to 8.31 mg/L. In general, water sources found in urban areas contain higher levels of Mn due to industrial run-off. While water Mn levels are moderate within city limits, residents utilizing wells as their primary source of drinking water and rural residents can experience significant Mn concentrations. Mn particulates can be found in dust and affect those who live in mining towns or areas near industrial factories that may be using Mn in their products.

Human Health Impact

The World Health Organization estimates that adults consume between 0.7 and 10.9 mg of Mn per day (WHO 1996). Normal ranges of Mn are 4–15 μg/L in blood, 1–8 μg/L in urine, and 0.4–0.85 μg/L in serum. Due to its ubiquitous presence in the environment, exposure to low levels in water, air, and soil is inevitable but levels are negligible compared to dietary levels. Dietary intake is the primary source of exposure to Mn and includes foods such as grains, beans, leafy green vegetables, nuts, and teas. Because of this vegetarians are likely to have a higher dietary intake of Mn. It is also available as a nutritional supplement as well as a component in most infant formulas (ATSDR 2008). One dietary source to infants that may result in excess Mn is total parenteral nutrition given to some neonates. This nutritional solution is usually supplemented with a trace element solution that contains Mn. However, infants lack proper excretory functions and have little control over absorption into the intestine, which may cause Mn toxicity (Aschner and Aschner 2005). Methcathinone drug abusers constitute another group of people with adverse effects associated with Mn due to their exploitation of the oxidant properties of potassium permanganate in fabrication of the psychostimulants. A Parkinsonian syndrome has been described in these illicit-drug users most likely resulting from permanganate contamination secondary to poorly controlled processes of synthesizing the drug.

The primary source of toxic exposures is inhalation of contaminated air. Groups most at-risk for considerable exposure are those occupationally exposed or those who are downwind of a manganese source. Workplace exposure to Mn-containing fumes or dusts can be a concern in the iron and steel, ferromanganese, dry-cell battery, welding and smelting industries. Although the number of United States Mn mines is currently minimal, mining can lead to concentrations as high as 1.5–450 mg Mn/m^3, with dry-cell battery factories producing up to 18 mg Mn/m^3. A significant concern also involves the amount of Mn in steel. However, one of the highest sources of occupational Mn exposure is associated with welding. Studies have found that welders are exposed to high Mn fumes. These exposures have been linked to higher prevalence of Parkinsonian symptoms. Asymptomatic welders exposed to Mn-containing welding fumes show reduced [^{18}F]fluoro-L-DOPA (FDOPA) uptake using PET imaging compared to control subjects, indicating dysfunction within the nigrostriatal dopaminergic (DAergic) system (Criswell et al. 2011). This is the same brain region compromised in Parkinson's disease, implicating a potential connection between Mn toxicity and Parkinsonism. MMT (methylcyclopentadienyl manganese tricarbonyl, or MCMT) combustion represents another concern in the workplace, for example, garage mechanics are exposed to high levels of MMT, resulting in exposure with a mean 0.45 $\mu g Mn/m^3$. However, it is difficult to determine how much of the Mn found in these workplaces is arising from MMT alone.

The Environmental Protection Agency (EPA) has established that lifetime drinking water exposure of 0.3 mg/L is not expected to produce adverse health effects and, in children, exposure to 1 mgMn/L for 1 or 10 days in drinking water does not cause adverse effects. The inhalation reference concentration (RfC) for Mn is 0.00005 mg/m^3 and is based on neurobehavioral function in humans. The oral reference dose (RfD) for Mn is 0.14 mg/kg/day and is based on central nervous system effects. The Food and Drug Administration (FDA) regulates the concentration of Mn in bottled water at 0.05 mg/L and the Occupational Safety and Health Administration (OSHA) regulates workplace air at 5 mg/m^3 Mn in air averaged over an 8-h work day (ATSDR 2008).

Organic and Inorganic Manganese

Organic forms of Mn are used as fungicides, fuel-oil additives, smoke inhibitors, and anti-knock additives in gasoline. They include MMT and mangafodipir (MnDPDP). Mangafodipir is a Mn-containing contrast agent which is intravenously delivered to enhance contrast in magnetic resonance imaging (MRI) of the liver. It contains the paramagnetic Mn along with the chelating agent fodipir (dipyridoxyl diphosphate, DPDP). MMT and mangafodipir toxicities are thought to be mediated by excess Mn and not other organic moieties.

Inorganic forms include manganese chloride ($MnCl_2$; Mn (II);Mn^{2+}), manganese sulfate ($MnSO_4$; Mn (II)), manganese acetate (MnOAc; Mn (III); Mn^{3+}), manganese phosphate ($MnPO_4$), manganese carbonate ($MnCO_3$), and the manganese dusts manganese dioxide (MnO_2; Mn (IV);Mn^{4+}) and trimanganese tetraoxide (Mn_3O_4). Inorganic Mn is found in many factory settings, as it is a major component of steel, dry-cell batteries, fireworks, ceramic and glass manufacturing, and various leather and textile industries. The most common forms found in the environment and in the workplace are in the Mn (II), Mn (III), and Mn (IV) oxidation states. Inorganic forms, such as the pigment known as manganese violet (manganese ammonium pyrophosphate complex), are found in cosmetics and in certain plants (ATSDR 2008). In addition to soil erosion, Mn can be released into the air and waterways via industrial waste run-off, as well as seepage from Mn-containing man-made products found in landfills (ATSDR 2008). The available human and animal studies suggest that inorganic Mn exposure does not cause significant injury to the heart, stomach, blood, muscle, bone, liver, kidney, skin, or eyes, and there is no evidence that it causes cancer in humans (ATSDR 2008). However, ingestion of potassium permanganate (Mn in the VII oxidation state) may lead to severe corrosion at the point of contact.

Fate of Mn in the Human Body

Primary routes of exposure include inhalation, ingestion, and dermal contact. Mn is found in all tissues and fluids. The highest levels of Mn are found in the liver, pancreas, and kidney and the lowest in bone and fat, in humans most tissue concentrations range between 0.1 and 1 μgMn/g wet weight (ATSDR 2008). Following dietary absorption (most commonly in Mn II form), Mn enters the portal circulation from the gastrointestinal tract and binds to plasma alpha2-macroglobulin or albumin. It is then delivered to the liver to be secreted in the bile although some may also be oxidized to Mn III by ceruloplasmin. Transferrin is a plasma carrier protein for Mn (III), and Mn (III) can be transported into neurons via transferrin receptor-mediated endocytosis. Following oral exposure, humans excreted Mn with whole-body retention half-times of 13–37 days (ATSDR 2008). High iron diets have been shown to suppress Mn absorption and those poor in iron show increased Mn uptake (ATSDR 2008). In addition, deficiencies in other essential nutrients, such as magnesium, may increase the toxicity of Mn. In regard to organic Mn exposure, MMT has been shown in animal studies to be eliminated very slowly with a half-time of 55.2 h compared to elimination of $MnCl_2$ with a half-time of 4.56 h. Because the oral pathway is not the primary exposure pathway for mangafordipir there is insufficient data on disposition and excretion following oral exposure.

Absolute amounts of Mn absorbed from Mn-containing dusts are unknown. Following inhalational exposure Mn can be directly transported to the brain through olfactory or trigeminal presynaptic nerve endings in the nasal mucosa before metabolism in the liver (ATSDR 2008). Smaller particles can also directly enter the blood stream through the gastrointestinal epithelial lining following mucociliary elevator clearance from the respiratory tract (Aschner and Dorman 2006). Particle transport from the lungs has not been quantified in humans but the half-lives of elimination in animal models ranged from 3 h to 1 day (ATSDR 2008). Mn is shuttled to the liver and conjugated to bile and then passed to the intestine to be excreted in the feces. At this point some of the Mn in the intestine is reabsorbed through the enterohepatic circulation (ATSDR 2008). Mn is primarily excreted in the feces within a few days. Mn can also be detected in small amounts in urine, sweat, and breast milk (ATSDR 2008). Mn can readily cross both the blood–brain and placental barriers. Sixty percent of Mn dusts, $MnCl_2$ or MnO_4, were cleared via the feces 4 days following inhalational exposure (ATSDR 2008). In intratracheal instillation rat studies excretion of 50% of the dose happened within 3–7 days and monkeys exposed to aerosolized $MnCl_2$ excreted most of the manganese with a half-time of 0.2–0.36 days (ATSDR 2008). In the monkey studies part of the $MnCl_2$ remained deposited in the brain and lung and clearance occurred with half-times 12–250 days (ATSDR 2008). There is insufficient data on absorption and excretion of Mn following inhalation exposure to organic manganese.

Absorption, distribution, and excretion following dermal exposure have not been documented. Other delivery methods are not discussed because they are not the primary routes of exposure except for mangafodipir which is commonly injected. Infusion of mangafodipir is rapidly dephosphorylated to manganese dipyridoxyl monophosphate which is then dephosphorylated to manganese dipyridoxyl ethylenediamine (MnPLED). This metabolite is measured in the blood within 18 min and up to 1.3 h later. Transmethylation with zinc can occur and yield ZnPLED (ATSDR 2008). In humans, mangafodipir is excreted in the feces via the bile and incompletely cleared from the body within 24 h, with ~8% retained after 1 week (ATSDR 2008).

Mn forms tight complexes with other substances and therefore, free plasma and tissue concentrations are very low (Aschner et al. 2007). Because of this it is difficult to adequately assess overexposure as plasma levels do not reliably reflect current or cumulative exposures. In addition, because Mn is a normal part of the diet the baseline is variable and overexposure can only be quantified as Mn levels over the variable baseline. Prediction of toxicity is also difficult due to the rapid clearance of Mn and its excretion in fecal matter as opposed to urine.

Role of Mn in Proteins/Enzymes

Mn serves as an important cofactor and activator of several pertinent enzymes that are necessary for proper functioning. Through the action of these Mn-dependent enzymes, processes that are dependent on Mn include proper development, digestion, reproduction, regulation of cellular energy, immune function, as well as antioxidant defenses (Aschner and Aschner 2005). Although Mn can be found in many oxidation states (II, III, and IV) in biological systems, Mn^{2+} is the primary state as an activator of many metal-enzyme complexes, as well as the component of some metalloenzymes. Mn activates these enzymes either through an intermediate interaction with some substrate (like ATP), or by binding to the protein itself and causing a conformational change to produce the enzymatic activity (ATSDR 2008). Some metalloenzymes that contain Mn include glutamine synthetase, pyruvate carboxylase, and arginase. Glutamine synthetase is found in the liver, kidneys and is localized to astrocytes in the brain and is necessary for nitrogen metabolism via the formation of glutamine from ammonia and glutamate, allowing for the astrocyte-specific protection of neurons from glutamate excitotoxicity. The majority of brain Mn is found bound to this enzyme, with two Mn(II) cation-binding sites in every enzyme molecule. For this reason, dietary Mn deficiency can lead to epileptic seizures. Moreover, pyruvate carboxylase is necessary for carbohydrate synthesis from pyruvate. In situations of Mn deficiency, lowered activity of this enzyme could compromise proper energy metabolism (Takeda 2003). Arginase is responsible for the generation of urea, and requires Mn^{2+} for its pH-sensing ability that is necessary for its proper functionality. Two Mn^{2+} cations are required to work with water and arginine in the enzyme's active site. In situations of Mn deficiency, the activity of arginase can drastically decrease, leading to deficient ammonia elimination and dire functional implications.

Mn can also activate several enzyme families, including transferases, involved with the transfer of glycosyl groups, kinase reactions, DNA and RNA polymerases, and sulfotransferases. Mn is also involved in phosphatase, ligase, hydrolase, oxidoreductase, peroxidase, and isomerase activities (Weatherburn 2001). An interesting aspect of Mn's biochemistry is that it mimics the chemistry of other metals in the same divalent state, including Zn^{2+}, Ca^{2+}, and especially Mg^{2+}. In many of these enzyme families, Mg^{2+} substitution for Mn^{2+} does not alter the enzyme's activity.

Mn^{3+} is not found as prevalently as Mn^{2+}, but is found in the essential antioxidant enzymes manganese catalase and manganese superoxide dismutase (MnSOD). Catalase plays a significant role in combating oxidative stress, as it converts hydrogen peroxide into oxygen and water. Meanwhile, MnSOD or SOD2 represents one of three SOD enzymes in humans and is involved in the dismutation of superoxide in mitochondria to decrease oxidative stress. Both of these antioxidant enzymes break down oxidants using Mn^{3+} found in their reactive catalytic centers. Mn^{3+} is also the state that can bind to transferrin, the iron-binding glycoprotein known to regulate iron levels in the blood. This metalloprotein complex can be essential to Mn in the context of neuronal import, as Mn can be transported into neurons via transferrin receptor-mediated endocytosis (Takeda 2003).

There are also several Mn-containing proteins that have no enzymatic function. Lectins require both Mn^{2+} and Ca^{2+} in order to bind carbohydrates. Integrins represent other proteins that require divalent cations such as Mn^{2+} for proper functioning. These proteins are found on the plasma membrane and regulate adhesion of cells to the extracellular matrix. Thus, they are important for immune response, embryogenesis, and proper maintenance of tissue structure. This interaction between integrins and other protein ligands is dependent on the presence of Mn^{2+} or Mg^{2+} (Weatherburn 2001).

Neurotoxicity of Mn

Although inhalational exposure can adversely affect the respiratory system resulting in increased susceptibility to infections and manganic pneumonia, the brain is the primary target site for Mn toxicity. The threshold level of Mn for production of neurological effects in humans has not been established. Entry of Mn to the brain occurs through three known pathways: (1) from the nasal mucosa to the brain olfactory bulb through olfactory neural connections, (2) from the blood through capillary endothelial cells of the blood–brain barrier, and (3) from the blood through the cerebral spinal fluid via the choroid plexuses (ATSDR 2008).

Transport mechanisms responsible for delivery of Mn to the brain include facilitated diffusion, active transport, transferring-mediated transport, divalent metal transporter-1 mediated transport, ZIP8 and transferrin-dependent transport and through store-operated calcium channels (Aschner et al. 2007). Mechanisms associated with Mn neurotoxicity in the CNS include impairment of transport systems, enzyme activities, and receptor functions. Manganese enhances the autoxidation or turnover of intracellular catecholamines producing free radicals, reactive oxygen species, and toxic metabolites while also depleting cellular antioxidant defense mechanisms, all of which likely contribute to its toxicity in the brain. The human striatum, globus pallidus, and substantia nigra are the primary target sites (Aschner et al. 2007).

Exposure to high levels of Mn can lead to behavioral changes, including slow and clumsy movements, tremors, difficulty walking and facial muscle spasms collectively referred to as "manganism." Symptoms may not manifest until a period of months and in some cases years following the exposure. Other less severe symptoms of exposure that precede symptoms of "manganism" include slowed hand movements, irritability, aggressiveness, and hallucinations. Excess Mn in the body preferentially accumulates in the basal ganglia of the brain and can be detected by magnetic resonance imaging (MRI), positron emission tomography (PET), and single photon emission computed tomography (SPECT).

Although clinically "manganism" presents similar to Parkinson's disease (PD), there are several key features that aid in distinguishing the two. Patients exposed to high Mn levels have a propensity to fall backward when pushed, less frequent resting tremor, more frequent dystonia, a "cock-walk," hypokinesia and tremor differing from PD patients, and they also do not respond to pharmacotherapy targeting the dopamine system. Absence of Lewy bodies is also indicative of manganism. Studies indicate that chronic exposure to Mn in certain occupations may accelerate PD or the risk of acquiring PD (Aschner et al. 2007).

Treatment of Mn Poisoning

Several therapeutic strategies have been employed in Mn poisoning, but no single treatment has been successful at relieving all associated symptoms although a combinatorial treatment plan that encompasses all aspects of Mn toxicity may be promising. An initial approach for relieving Mn toxicity is to reduce the total intake of Mn via diet. It has been shown that other essential minerals can impact Mn toxicity, especially iron (Fe) and calcium (Ca). This likely occurs due to shared cellular transport systems, such that a deficiency in iron can lead to increased Mn absorption via competition at the transporter level. Therefore, maintaining or restoring ample iron or calcium levels could aid in reducing potential toxicity by decreasing the amount of Mn absorbed. However, further investigation into whether dietary iron or calcium supplementation actually reduces Mn toxicity is necessary, as it remains unknown whether sufficient iron or calcium stores are beneficial when Mn has already been absorbed.

A more extensively researched treatment for severe cases of Mn toxicity is chelation therapy. This treatment option uses metal chelators in an attempt to scavenge the metal ions already found within the system. The most common chelator is ethylenediaminetetraacetic acid, or EDTA. Although the idea of a metal chelator seems favorable in reducing the body's total Mn burden, studies have found that improvement upon treatment with EDTA chelators is highly inconsistent, with no success in ameliorating all symptoms permanently. This is thought to occur due to the structural moieties that are responsible for chelating activity, but result in decreased lipophilic characteristics, which compromise its ability to cross the blood–brain barrier. Thus, although chelation therapy may decrease Mn concentrations in the blood and urine, alleviation of symptoms is minimal due to its inability to counter the neuronal accumulation of Mn in severe cases of Mn poisoning.

As excess Mn exposure can induce a Parkinsonian illness known as manganism, treatment with L-DOPA has also been tested. Although some of the motor symptoms associated with manganism were reversed by L-DOPA treatment, there are serious side effects. Additionally, treatment with anti-Parkinsonian drugs may not be effective due to the preferential accumulation of Mn in specific brain regions, such as the globus pallidus, that differs from that of targeted cell death in the substantia nigra seen in PD. Other studies have assessed antioxidant treatment options, such as vitamin E, but there is no evidence thus far that such treatment can improve symptoms after Mn exposure (ATSDR 2008). As manganism is an irreversible condition, no evidence currently exists for a treatment that offers more than just symptomatic control on a more permanent basis for cases involving prolonged Mn exposure.

The anti-tuberculosis drug para-aminosalicylic acid (PAS) has been a newer avenue for potential treatment options. Although the mechanism behind its therapeutic actions in Mn poisoning remains unknown, it has shown the most promise in maintaining improvement of symptoms associated with several cases of Mn poisoning. In a 17-year follow-up study, a patient exposed to high levels of Mn for 21 years exhibited all signs of having occupational manganism. However, after a 4-month period of PAS treatment administered 4 days of the week, the patient showed substantial improvement. Furthermore, this improvement appeared to be maintained in a follow-up conducted 17 years after the original treatment paradigm. The mechanism behind these protective effects remains unknown, although it is postulated that PAS could be a Mn chelator, or that it gains its protective effects from its salicylate moiety that confers anti-inflammatory attributes to PAS (Jiang et al. 2006). However, it is still unclear whether PAS has the ability to cross the BBB in order to provide these protective effects. Such promising evidence in maintaining improvement of symptoms upon prolonged Mn exposure warrants blinded, randomized studies.

Another area of research has focused on differences in oxidation states and their effects on cellular Mn uptake and subsequent toxicity. It has been shown that Mn^{2+} is more readily cleared than Mn^{3+}. Moreover, Mn^{3+} is a much more potent oxidizing agent, as Mn has been shown to readily oxidize dopamine in this state, leading to higher concentrations of damaging dopamine quinone products. Thus, it could be beneficial in finding a way to selectively decrease the generation of Mn^{3+}, allowing for decreased Mn^{3+}-induced toxicity and faster elimination of Mn^{2+}. However, the consequences of selectively changing the oxidation state of Mn are unclear, as some metalloenzymes are dependent on this oxidation state of Mn for proper enzymatic function (ATSDR 2008). Although oxidative stress is thought to be the primary mechanism behind Mn-induced neurotoxicity, further examination is necessary to understand how to reduce Mn-induced oxidation without affecting the vital roles of Mn for normal cellular function.

Conclusions

Because Mn is an essential nutrient, small amounts in the diet are required for health maintenance; it is exposure to high levels that causes toxicity. The primary route of exposure is ingestion, although occupational exposures are primarily inhalational. The absorption and excretion of Mn is tightly controlled in order to maintain stable tissue Mn levels for essential reactions as Mn is both a constituent of metalloenzymes and an enzyme activator. Preferential uptake by the brain plays a role in the distinct neurological effects of Mn.

Acknowledgment These authors wish to acknowledge funding by Grants from the National Institutes of Environmental Health (R01 ES010563 and T32 ES007028).

Cross-References

▶ Blood Clotting Proteins
▶ Iron, Physical and Chemical Properties
▶ Magnesium, Physical and Chemical Properties
▶ Manganese, Interrelation with Other Metal Ions in Health and Disease
▶ Mitochondria
▶ Manganese Toxicity

References

Agency for Toxic Substance and Disease Registry (2008) Toxicological profile for manganese. U.S. department of health and humans services, public health service, centers for disease control, Atlanta

Aschner JL, Aschner M (2005) Nutritional aspects of manganese homeostasis. Mol Aspects Med 26(4–5):353–362

Aschner M, Dorman DC (2006) Manganese: pharmacokinetics and molecular mechanisms of brain uptake. Toxicol Rev 25(3):147–154

Aschner M, Guilarte TR, Schneider JS, Zheng W (2007) Manganese: recent advances in understanding its transport and neurotoxicity. Toxicol Appl Pharmacol 221:131–147

Criswell SR, Perlmutter JS, Videen TO, Moerlein SM, Flores HP, Birke AM, Racette BA (2011) Reduced uptake of [^{18}F] FDOPA PET in asymptomatic welders with occupational manganese exposure. Neurology 76(15):1296–1301

Jiang YM, Mo XA, Du FQ, Zhu XY, Gao HY, Xie JL, Liao FL, Pira E, Zheng W (2006) Effective treatment of manganese-induced occupational Parkinsonism with p-aminosalicylic acid: a case of 17-year follow-up study. J Occup Environ Med 48(6):644–649

Takeda A (2003) Manganese action in brain function. Brain Res Rev 41:79–87

Weatherburn DC (2001) Chapter 8. Manganese-containing enzymes and proteins. In: Handbook on metalloproteins. Bertini I, Sigel A, Sigel H (eds.), Marcel Dekker, New York, pp 196–246

World Health Organization (WHO) (1996) Guidelines for drinking water quality, vol 2, Health criteria and other supporting information. WHO, Geneva, pp 275–278

Manganese Toxicity

Marcello Campagna[1], Roberto G. Lucchini[2,3] and Lorenzo Alessio[2]
[1]Department of Public Health, Clinical and Molecular Medicine, University of Cagliari, Cagliari, Italy
[2]Department of Experimental and Applied Medicine, Section of Occupational Health and Industrial Hygiene, University of Brescia, Brescia, Italy
[3]Department of Preventive Medicine, Mount Sinai School of Medicine, New York, USA

Synonyms

Adverse health effects related to Mn

Definition

The term manganese (Mn) toxicity refers to a condition in which the Mn concentration in the body can produce adverse health effects. The degree to which Mn can harm humans or animals. The Mn quality of being poisonous. The degree to which Mn can damage an organism.

Mn Deficiency

Manganese (Mn) is an essential element for humans and animals like iron, zinc, and copper. Many people are likely below the estimated safe and adequate daily Mn intake. Mn is needed for normal prenatal and neonatal development of bone mineralization, protein and energy metabolism, metabolic regulation, cellular protection from damaging free radical species, and the formation of glycosaminoglycans. Mitochondrial superoxide-dismutase, pyruvate carboxylase, and liver arginase are known as Mn metalloenzymes.

Several enzyme systems: transferases, decarboxylases, hydrolases, dehydrogenases, synthetases, and lyases have been reported to interact with or depend on Mn for their catalytic or regulatory function. Mn has been shown to stimulate the synthesis of chondroitin sulfate, an important constituent of the cartilage and connective tissue. The US Food and Nutrition Board of the National Research Council established ESADDI (Estimated Safe and Adequate Daily Dietary Intake) levels as follows: 0.3–0.6 mg/day for infants from birth to 6 months; 0.6–1.0 mg/day for infants from 6 months to 1 year; 1.0–1.5 mg/day for children from 1 to 3 years; 1.0–2.0 mg/day for children from 4 to 6 years and 7 to 10 years; and 2.0–5.0 mg/day for adolescents (>11 years) and adults (NRC 1989).

Neurotoxicity

When Mn exceeds the homeostatic range, Mn becomes a neurotoxic agent with well-known clinical manifestations that can result in severe parkinsonian signs and symptoms, and psychiatric features. This syndrome has been described as "manganism" and can be determined by exposure generally higher than 2 mg/m^3, but susceptible individuals may develop the clinical pictures also at 1 mg/m^3 (WHO 1981). Although manganism, which typically follows acute, high-level exposure, is the most obvious clinical manifestation of Mn neurotoxicity, subclinical and subfunctional declines, related to motor coordination of fine movements, cognitive and behavioral functions, have been documented in the context of lower level exposure. Chronic lifetime exposure to very low levels is currently hypothesized as a possible risk factor for the onset of Parkinson's disease (Lucchini et al. 2009). These three manifestations may be associated with a continuum of dysfunction (Martin 2006).

Clinical Manganism

Sir James Couper (1837) is credited with the first clear description of the adverse neurological effects of Mn in five Scottish men employed in grinding Mn dioxide ore in 1837. At that time, the industrial use of Mn was limited, and in Couper's case series, the application was to generate "bleaching powder." Although not referenced by Couper, his report was only 20 years after James Parkinson's seminal description of the "shaking palsy." Several decades after Couper's first description, Mn began to be used much more widely in what remains by for its most common application, as a metal essential to the manufacture of steel alloy. Although symptoms may overlap, manganism is a distinct entity from Parkinson's disease at many levels. Shared features include generalized rigidity and bradykinesia. However, unlike the festinating gait in Parkinson's disease, the gait abnormality of manganism is a cock-walk with an associated foot dystonia, such that patients walk on the balls of the feet with the heels elevated above the ground. Dystonia has also been documented in other locations, unlike Parkinson's disease. In manganism, tremor is less prominent, postural, of higher frequency and lower amplitude. Manganism patients are more prone to fall backward. Manganism patients do not show a sustained response to dopamine replacement and functional imaging studies using fluorodopa-labeled positron emission tomography (PET) scans fail to show the pattern of reduced striatal uptake which is uniformly present in Parkinson's disease. Finally, the condition has been noted to progressively deteriorate following removal from exposure.

Subclinical Neurotoxicity

At lower exposure levels, less severe, subtle, preclinical, neurobehavioral effects have been widely reported in various occupational settings. Most studies have been cross-sectional, neurobehavioral investigations designed to compare the test performance of a Mn-exposed group of workers with nonexposed matched controls. The main effects are represented by dose-related impairment of motor function and coordination, but cognitive and behavioral deficits have been observed as well. Several studies have also addressed the existence of neurobehavioral effects due to nonoccupational exposure to Mn. Adult populations residing in the vicinities of ferroalloy plants and mines have been studied to assess Mn-related toxicity on motor and cognitive functions. Dose-effect association between Mn exposure and several neurobehavioral tests examining motor and cognitive functions and mood were carried in South Quebec, Mexico, Brazil and Italy. The main exposure route in these populations is from inhalation. Environmental exposure to Mn can also affect more sensitive individuals like children. Exposure through drinking water has been related to impairment on IQ levels in Bangladesh, USA, and Quebec (Lucchini and Kim 2009).

Mn-Induced Parkinsonism

It has also been suggested that Mn exposure may play a role in the development of idiopathic Parkinson disease by acting as an environmental trigger able to accelerate the onset of the neurodegerative damage (Martin 2006). Welders have been particularly studied for a possible increase of parkinsonism due to Mn exposure. An increased frequency of parkinsonian disturbances has been shown in case control studies on large groups of welders in the USA. The parkinsonian features in welders seem not to be different from idiopathic Parkinson disease (PD), except for a younger age of onset and a tendency to familiarity. Epidemiological studies conducted in Norway, Italy, and Canada have shown increased prevalence of Parkinson's disease and parkinsonism in the vicinity of industrial sites causing emission of Mn dust (Lucchini and Kim 2009).

The brain needs Mn during the early phases of development as a constituent of important metalloenzymes such as arginase, glutamine synthetase, pyruvate carboxylase, and superoxide dismutase. Mn exposure can start before birth from the maternal exposure through inhalation and ingestion of food items that may contain higher Mn concentration from environmental pollution. Therefore, excessive concentration of Mn may cause an overload that is potentially harmful for the fetus. Postnatal exposure can also be relevant due to a relatively high concentration of Mn in formulas. In order to provide Mn to the developing brain, the intestinal absorption of this element is high, whereas the excretion rate is low due to the incomplete development of the biliary pathway, responsible for Mn elimination. Mn exposure can continue during childhood and adulthood from both environmental and occupational exposure. According to each life stage, different absorption routes and different potentials for increased exposure may occur, consistently during an entire lifetime or during discrete periods, leading to a final total body burden that may result in neurotoxicity. The concept of lifetime exposure is an important toxicological aspect to be considered for substances like Mn, which are characterized by a cumulative mechanism of action (Lucchini and Zimmerman 2009). Cumulative exposure can result in delayed, long-term toxicity. According to the principle of "fetal programming" of the brain, prenatal exposure may be of concern for late-onset neurodegenerative effects. The occurrence of delayed effects can be explained by the mechanisms of transport across the blood–brain barrier. Taken together, a change in exposure scenario from occupational settings to the general environment represents an important background to be considered when approaching the relationship between current Mn exposure and the subsequent neurotoxic effects. Currently, overt, clinical manganism is very rare, while low-level prolonged Mn exposure remains of concern. Inhaled Mn that can be transported directly to the brain through nasal deposition and transport along olfactory neurons, especially when transported by ultrafine particles such as welding fumes. Changes in olfactory threshold and odor identification have been shown in Mn-exposed workers and children and the same tests are recognized as predictive of Parkinson's disease.

Mechanism of Mn Neurotoxicity

Mn-induced damage seems to occur at the postsynaptic level to the nigrostriatal system, predominantly in the globus pallidus. Mn may enhance the autoxidation or turnover of various intracellular catecholamines such as dopamine, leading to an increased production of free radicals, reactive oxygen species, and other cytotoxic metabolites. It may disturb also the gamma-Aminobutyric acid (GABA) regulation as well as glutamatergic transmission. It also seems to impair the cellular antioxidant defense mechanism. Oxidative stress generated through mitochondrial perturbation may be a key event in the demise of the affected central nervous system cells. Studies with primary astrocytes cultures revealed that they are a critical component in the defenses against Mn-induced neurotoxicity. Enhanced oxidative stress may take place particularly in catecholaminergic (i.e., dopamine) cells. Some experimental evidence suggests that the mechanisms of Mn toxicity may depend on the oxidation state of Mn. However, both the trivalent (MnIII) and divalent (MnII) forms have been demonstrated to be neurotoxic, but it is important to note that the oxidation of catechols is more efficient with Mn(III), than with Mn(II) or Mn(IV). Formation of Mn(III) may occur by oxidation of Mn(II) by superoxide. Both, Mn(III) and Mn(II) can cross the blood-brain barrier, although it is suggested that Mn(III) is predominantly transported bound to the protein transferrin, whereas Mn(II) may enter the brain independently of such a transport mechanism. A large portion of Mn is bound to Mn metalloproteins, especially glutamine synthetase in astrocytes. A portion of Mn probably exists in the synaptic vesicles in glutamatric neurons and Mn is dynamically coupled to the electrophysiological activity of the neurons. Mn released into the synaptic cleft may influence synaptic transmission.

The involvement of oxidative stress has been demonstrated by significant changes in intracellular antioxidant levels, including increased glutathion and catalase levels due to Mn exposure and the observed attenuation of Mn cytotoxicity when antioxidants are used. When Mn is inside the cell, it preferentially accumulates into the mitochondria, mainly as Mn^{2+} via the Ca^{2+} uniporter. Inside the mitochondria, Mn can disrupt oxidative phosphorylation and increase reactive oxygen species (ROS) production.

The ability of Mn to enhance oxidative stress is due to the transition of its oxidative state +2 to +3, increasing its pro-oxidant capacity. Experimental evidence suggests that pro-oxidant activity of Mn^{2+} is dependent on trace amounts of Mn^{3+}. Superoxide produced in the mitochondrial electron transport chain may catalyze this transition through a set of reactions similar to those mediated by superoxide dismutase and thus lead to the increased oxidant capacity of the metal.

Additionally, basal ganglia are regions with high oxidative activity, possibly favoring the oxidation of Mn^{2+} to Mn^{3+} in those specific areas. Until now there are controversies about the Mn chemical specie able of producing more damage; however, Mn^{3+} is considered the most toxic specie. A reasonable hypothesis to explain Mn neurotoxicity is that Mn^{3+} is responsible for the dopaminergic toxicity. In this case, Mn^{2+} would undergo cyclic reactions with Mn^{3+} that would lead to aminochrome production in the presence of dopamine. Aminochrome is the main dopanine (DA) oxidation product.

The selectivity of Mn for mitochondria of neurons from basal ganglia has been suggested. Long-term Mn-intoxication effects are similar to those produced by other substances that also affect the mitochondria in the same brain structures, for instance carbon monoxide and cyanide. Parkinsonism and dystonia are common as sequelae of Mn, cyanide, and carbon monoxide (CO) intoxications. Why Mn is able to produce the same effects that those substances is not known but there may be common mechanisms of injury; for example all of them are able to produce alterations in oxidative phosphorylation and both Mn and CO induce mitochondrial permeability transition (mTP) and ROS production (ATSDR 2008).

Mn impact on energetic metabolism is significant. It can inhibit some tricarboxylic acid cycle enzymes and reduce electron transport chain activity, leading to decreased adenosine-5′-triphosphate (ATP). Additionally, it is known that Mn decreases astrocytic glutamate uptake and reduces the expression of astrocytic glutamate transporter (GLAST) and glutamate transporter 1 (GLT-1). As Mn alters ATP production and glutamate uptake in astrocytes, basal ganglia neurons could be susceptible of excitotoxic damage. Decreased expression of glutamine transporters has been also reported in Mn-exposed cultured astrocytes, contributing to the disruption of the glutamate-glutamine cycle.

Changes in glutamic acid decarboxylase (GAD) and GABA content can be found as a consequence of Mn exposure. It has been proposed that loss of GAD-positive cells in the striatum and globus pallidus of Mn-treated rats may be caused by the loss of chemical, electrical, or physical support because of neurite dysfunction encountered in mesencephalic cultures exposed to Mn, resulting in profound changes in cytoskeleton and neurite length, or may be an independent neurotoxic event. Hence, manganism features can be due to the effect of the metal on the interactions among dopaminergic, gabanergic, and glutamatergic neurotransmission systems.

Individual Susceptibility Factors

Besides the mechanisms of direct toxicity in the basal ganglia, the importance of interindividual variability is quite evident in all Mn exposure conditions. Genetic mutations of two genes play an important role in rendering some individuals more at risk for parkinsonism when chronically exposed to Mn. The ubiquitin E3 ligase parkin, which is associated with early onset of Parkinson's disease, can also protect from Mn toxicity. A genetic interaction has also been observed between α-synuclein, an important protein for Parkinson's disease, and PARK9, a yeast ortholog of the human gene ATP13A2, which is also important for Parkinson's disease. Yeast PARK9 protects the cells from Mn toxicity. Mutations in these genes may be ultimately important for the expression of DMT1, therefore influencing the transport of Mn (Lucchini et al. 2009).

Interactions between Mn and iron have been pointed out as important for Mn absorption and transport across biological barriers. Anemic condition can increase the Mn uptake and should be considered as a potential cause of hypersensitivity to Mn exposure.

Another condition able to increase the risk of parkinsonism in Mn-exposed individuals may be represented by subclinical impairment of liver function. Mn is almost totally excreted via the biliary system, and therefore any impairment of this pathway is potentially able to cause Mn overload due to insufficient elimination from the body. This is well known in the case of cirrhotic patients showing Mn-related abnormalities, where the "liver encephalopathy" may be partially explained by the excessive Mn in the brain. Milder liver abnormalities may become highly important under conditions of lifetime Mn exposure. A possible role played by subclinical liver impairment may render Mn-exposed individuals at higher risk for parkinsonism. Imbalance of copper and zinc is also important in Mn exposure and was already observed in exposed workers and nonhuman primates. These metals are both important in cellular redox reactions; therefore, a dysregulation of their homeostasis may potentiate the cellular damage resulting from reactive oxygen species. Ceruloplasmin was found to influence the disposition and neurotoxicity of Mn. Therefore, underlying impaired liver function likely influences the neurotoxicity of Mn through a number of pathways.

Pulmonary Toxicity

Lungs are the second targets organs affected by the exposure to Mn. Mn inhalation can result in adverse effects to the respiratory system. Exposure to high levels of Mn dusts (especially Mn dioxide [MnO2] and Mn tetroxide [Mn3O4]) can indeed cause an inflammatory response in the lung, which can result in damage of lung function. Lung toxicity can cause increased susceptibility to infections such as bronchitis and can result in *manganic pneumonia* that is frequently associated to exposures to high levels of dusts containing other metals. These effects have been reported primarily in workers exposed to high levels of Mn dusts in the workplace. Some recent surveys indicate although that people who live and attend school next to ferroMn industries have an increased risk of respiratory effects. Inhaled Mn in nonsoluble forms can impair resistance to respiratory infections. This assumption has been supported by experimental and epidemiological studies. An increased morbidity and mortality rate from pneumonia has been found among workers with high exposure to Mn dusts in the old literature Clinically, Mn pneumonia has been characterized as an acute alveolar inflammation with marked dyspnea. A typical finding was that antibiotics were often without effect. In vitro studies have demonstrated the cytotoxic action of Mn, including the inhibition of activities of alveolar macrophages. Overall, the pulmonary toxicity of Mn seems to be related to exposure intensity and also to the chemical form of Mn with the less soluble oxides being more dangerous for the respiratory system (Šaric and Lucchini 2007).

Reproductive Toxicity

Few studies examined male fertility, with conflicting results likely due to difference in solubility of the different chemical forms of Mn involved. In these studies, however, the birth of a live baby to the man's wife was used as the outcome. This method is influenced by the wife's fertility, and by many social, sexual, and health factors. More supportive evidence for reproductive effects of Mn was derived recently from infertility clinic clients, where blood Mn was inversely related to sperm motility, concentration, and morphology. Adverse reproductive effects may be determined by changes in serum prolactin as a result of Mn interference with the dopaminergic system, which is a tonic inhibitor of prolactin secretion. Studies on ferroalloy workers and welders found a significant positive association between blood Mn and serum prolactin. Raised prolactin concentrations might have an effect on male fertility and sexual function, as hyperprolactinaemia in men may be associated with decreased sex drive, impotence, and reduced sperm production (Šaric and Lucchini 2007).

Other Toxic Effects

Mn seems not to cause significant damages to the stomach, blood, muscle, bone, kidney, skin, or eyes. Only studies conducted on animals have demonstrated potential adverse coronary effects due to high levels of exposure to Mn. Studies in humans conducted on male exposed workers pointed out a greater incidence of low diastolic blood pressure and a significantly increased incidence in sudden death mortality nevertheless other confounding factors were not completely excluded.

References

ATSDR (2008) Toxicological profile for Mn. US Department Of Health and Human Services, Public Health Service, Agency for Toxic Substances and Disease Registry, Atlanta

Couper J (1837) On the effects of black oxide of Mn when inhaled into the lungs. Brit Ann Med Pharmacol 1:41–42

Lucchini RG, Kim Y (2009) Health effects of Mn. In: Vojtisek M, Prakash R (eds) Metals and neurotoxicity. Society for Science and Environment, Jalgaon, pp 119–147. ISBN/ISSN: 81-85543-09-7

Lucchini RG, Zimmerman NJ (2009) Lifetime cumulative exposure as a threat for neurodegeneration: need for prevention strategies on a global scale. Neurotoxicol 30(6):1144–1148

Lucchini RG, Martin C, Doney B (2009) From Manganism to Mn-induced parkinsonism: a conceptual model based on the evolution of exposure. Neuro Mol Med 11(4):311–321

Martin C (2006) Mn neurotoxicity: connecting the dots along the continuum of dysfunction. Neurotoxicol 27:347–349

NRC (1989) Recommended dietary allowances. National Research Council, Washington, DC, pp 231–235

Šaric M, Lucchini R (2007) Mn. In: Nordberg GF, Fowler BA, Nordberg M, Friberg L (eds) Handbook on the toxicology of metals, 3rd edn. Elsevier, Amsterdam, pp 645–674

WHO (1981) Mn. Environmental Health Criteria 17. World Health Organization, Geneva

Manganese, Interrelation with Other Metal Ions in Health and Disease

James C. K. Lai[1] and Solomon W. Leung[2]
[1]Department of Biomedical & Pharmaceutical Sciences, College of Pharmacy and Biomedical Research Institute, Idaho State University, Pocatello, ID, USA
[2]Department of Civil & Environmental Engineering, School of Engineering, College of Science and Engineering and Biomedical Research Institute, Idaho State University, Pocatello, ID, USA

Synonyms

Interdependence between manganese and other metal ions; Manganese-metal ion interactions

Definition

Manganese and its interrelation with other metal ions in health and disease refers to how the level and metabolism of manganese in an organ are related to the levels and metabolism of other metal ions in the same organ in a healthy individual and how disease states exert modulatory effects on such interrelations.

Introduction

To understand the "state of the art" regarding manganese (Mn) and its interrelation with other metal ions in health and disease, one needs to appreciate the advantages and the limitations of the approaches that critically drive the advances of this multidisciplinary field. As have been emphasized earlier, "Ideally, if one can demonstrate directly the mechanisms that causally link the metabolism of manganese to those of other metals, then one can define the physiological ... situations or conditions that show *interdependence* between Mn and other metal ions. However, largely because of methodological limitations as well as limitations of experimental approaches, the data in the literature and those of ongoing studies are better or more accurately defined as depicting the *interrelations* between Mn and other metal ions. This conceptual distinction should be borne in mind when one assesses the advances in the topic areas to be discussed..." (Lai et al. 2000).

The scope of this entry is limited to a summary discussion of the interrelations between Mn and other elements after chronic exposure of animals or humans to Mn because distributions of Mn in brain regions and in peripheral organs usually reflect chronic rather than acute exposure to this metal. Several aspects of this topic not covered in this entry because of space limitations have been discussed in some detail in one or another of our previous publications (Chan et al. 1983; Lai et al. 1984, 1985a, c, 1999, 2000; Leung et al. 2011; Wright et al. 2011).

Interdependence Between Manganese and Other Metals in Absorption

Because manganese (Mn) is an essential trace metal, intake of this metal is usually via food intake and intake via the drinking water (Underwood 1977; Lai et al. 1984, 2000). Respiratory exposure of Mn to humans is rare except under poorly controlled industrial environments such as manganese mining, steel mills, and metal soldering in a confined space.

Consequently, the major findings to date employing animal models concerning the interdependence between Mn and other metals are centered upon intestinal absorption and subsequent organ distribution. Several generalizations are evident: intestinal Mn absorption depends on dietary iron (Fe). Low-dietary Fe enhances Mn absorption whereas high-dietary Fe lowers Mn absorption (Lai et al. 1985c, 2000). Similarly, high-dietary Mn intake results in lowered Fe absorption (Lai et al. 1985c, 2000). Furthermore, inclusion of essential and nonessential trace metals in the drinking water of rats and mice can alter accumulation of essential trace metals (including Mn) in their peripheral organs (see Lai et al. 2000 for discussion).

Interrelations Between Manganese and Other Metals in Several Peripheral Organs

A developmental rat model had been developed so that the organ distribution of multiple metals and other elements could be systematically analyzed based on dietary intake. Furthermore, by exposing rats in utero to manganese (added as the chloride form in the drinking water) and continuously until they were employed for analysis, the effects of Mn on the distributions of other trace metals and electrolytes in various organs during development and aging can be systematically investigated employing versatile techniques such as instrumental neutron activation analysis (INAA) (Chan et al. 1983; Lai et al. 1984, 1999, 2000). Those series of studies constitute the largest and most complete sets of data in the Mn-metal interaction literature (Chan et al. 1983; Lai et al. 1985a, 1985c, 1999, 2000; Leung et al. 2011; Wright et al. 2011).

The relative distribution of trace metals in different peripheral organs subsequent to their intestinal absorption reflects the amount of metal-binding molecules (e.g., proteins and nucleic acids) in each organ and the organ's storage capacity (Lai et al. 2000). In rats, the rank order of the relative abundance of trace metals in peripheral organs is Fe $\gg$ Zn > Cu > Al > Mn > Se. The rank order of Fe level in peripheral organs is spleen $\gg$ liver > lung > heart $\approx$ kidney. The rank order of Zn level is liver $\gg$ kidney > spleen $\approx$ lung $\approx$ heart. The rank order of Cu level is kidney $\gg$ liver $\approx$ heart > spleen > lung. The rank order of Mn level is liver $\gg$ kidney $\gg$ spleen $\approx$ heart > lung whereas that for Se is kidney $\geq$ liver > spleen > lung = heart. However, the rank order of the level of the nonessential metal Al is spleen > liver > heart > kidney $\approx$ lung.

In adult (i.e., 120-day-old) rats, chronic Mn treatment induces dose-related differential elevation in Mn in heart, kidney, spleen, and liver (Lai et al. 1985c, 2000). Associated with the differential elevation of Mn in the peripheral organs is a shift in the patterns of organ distribution of trace metals (Lai et al. 1985c, 2000). For example, in Mn-treated rats, Fe levels are increased in spleen and lung whereas Zn level is decreased in liver and Cu level is decreased in spleen, compared to corresponding levels in untreated rats (Lai et al. 1985c, 2000). However, the organ distributions of several other metals are also affected by chronic manganese exposure. (Because of space limitation, those findings will not be discussed here: see Lai et al. 1985c, 2000 for additional details.)

Based on the findings discussed above, two generalizations can be arrived at: (1) Dietary intake of trace metals and electrolytes and their subsequent organ distribution is under regulatory control even though the mechanisms mediating the control are far from being understood. (2) Increasing the levels of Mn in dietary intake (e.g., via addition of manganese salt in the drinking water) ultimately alters the patterns of organ distribution of trace metals.

Regional Differences in Distribution of Trace Metals in Brain

In the adult rat, whole brain levels of trace metals differ and the rank order of levels of several trace metals is Fe > Zn > Al > Cu > Se $\approx$ Mn (Lai et al. 1985c, 2000). Additionally, brain levels of several trace metals (e.g., Fe, Cu, Mn, and Se) are altered in the aged rat brain compared to the corresponding levels in the young adult rat brain (Lai et al. 1985c, 2000). On the other hand, each trace metal has its own, almost unique, brain regional distribution.

In development, various regions of the mammalian brain undergo structural changes concomitant with increases in glial cell numbers and migrations of growing and maturing neurons (Lai et al. 2000). Although not yet fully understood, these developmental structural and cellular changes ultimately give rise to each trace metal exhibiting its characteristic regional distribution in the adult brain (Lai et al. 1985c, 2000).

For example, in the adult rat, Fe level is higher in hippocampus, hypothalamus, pons and medulla, and cerebellum than in midbrain, striatum, and cerebral cortex. Zn level is higher in hippocampus and cerebral cortex than in the other regions whereas Cu level is highest in hypothalamus, intermediate in striatum, midbrain, and cerebellum but lowest in hippocampus, pons and medulla, and cerebral cortex. Se level is highest in hippocampus, midbrain, and hypothalamus, intermediate in cerebellum and cerebral cortex, but lowest in striatum and pons and medulla. Mn level is highest in midbrain, cerebellum, and pons and medulla, intermediate in cerebral cortex and hypothalamus, but lowest in striatum and hippocampus. The level of the nonessential trace metal, Al, unlike those of the essential metals (e.g., Fe, Zn, Cu, Mn, Se) is lower in cerebral cortex than those in other regions.

Distributions of electrolytes are also somewhat different from those of trace metals (Lai et al. 1985c, 2000). In the adult rat, Ca level is lower in cerebral cortex and midbrain than in the other regions. Mg level is also lower in cerebral cortex than those in other regions.

Interrelations Between Manganese and Other Metals in Brain

To facilitate the investigation of the interrelations between manganese and other trace metals and electrolytes in brain and other organs, a model of chronic and life-span exposure of rats to manganese had been developed (Lai et al. 1984, 1985b). Chronic Mn exposure was carried out by adding $MnCl_2.4H_2O$ into the drinking water given to the rats. Four groups of rats were studied: the control group was given just the double-distilled water; group A was given 1 mg of $MnCl_2.4H_2O$ per mL of double-distilled water; group B was given 10 mg of $MnCl_2.4H_2O$ per mL of double-distilled water; and group C was given 20 mg of $MnCl_2.4H_2O$ per mL of double-distilled water (Lai et al. 1984, 1985b). The rats were exposed to Mn in utero via their mothers' circulation; postnatally, they were exposed to Mn via their mothers' milk; and from around weaning (i.e., 21/22 days postnatal) onward, they were directly exposed to Mn in the drinking water (Lai et al. 1984, 1985b). When the rats attained full adulthood (i.e., at 120 days of age), they were euthanized and their organ removed (Lai et al. 1984, 1985b).

Each rat brain was dissected into the following seven regions: cerebellum, cerebral cortex, hippocampus, hypothalamus, midbrain, pons and medulla, and striatum (Lai et al. 1984, 1985b). The contents of trace metals and other elements in each brain region were determined employing instrumental neutron activation analysis (INAA) (Chan et al. 1983; Lai et al. 1985c, 1999, 2000). By combining the versatile technique of INAA with the animal model of chronic and life-span exposure to Mn, a comprehensive metallomic database had been collected of levels of trace metals and other essential and nonessential elements in all the brain regions in the four groups of rats (one control and three Mn-treated groups, namely, A, B, and C) (Chan et al. 1983; Lai et al. 1985c, 1999, 2000). The data were taken from the comprehensive metallomic database and then analyzed so as to allow the interpretation of how Mn can interact with trace and major elements in discrete rat brain regions (Leung et al. 2011; Wright et al. 2011). The level of each trace metal or element in each brain region in the Mn-treated rat was compared with the corresponding level (set as 100%) in the same region in the untreated (i.e., control) rat and the difference ranked as % increase or % decrease relative to the value in the control animal. For ease of visual comparison, the rank orders of increases or decreases of levels were color-coded (Tables 1 and 2) (Leung et al. 2011; Wright et al. 2011).

As evident from the rank-order patterns of the regional distributions of trace metals (Table 1) and electrolytes (Table 2) in the seven brain regions in the rats treated with the three doses of Mn, several generalizations regarding how Mn interacts with other trace metals and other elements can be arrived at. (1) Because the distributions of trace metals and other elements in the adult rat brain show region-specific differences, their patterns of distribution in the untreated normal rat brain are significantly altered by Mn treatment and such alterations can be generally correlated with the dose of Mn administered (Tables 1 and 2). (2) Mn treatment leads to selective, region-specific increases in the accumulation of some trace and major elements (Tables 1 and 2). (3) Mn treatment results in selective, region-specific decreases in the accumulation of other trace and major elements (Tables 1 and 2). (4) By contrast, Mn treatment gives rise to both increases and decreases in the accumulation of some other trace and major elements, and these fluctuations in their accumulation are dependent on

Manganese, Interrelation with Other Metal Ions in Health and Disease, Table 1 Changes in levels of trace metals and other elements in brain regions in adult female rats after chronic treatment with Mn compared with corresponding levels in untreated (i.e., control) rats. The Mn-treated groups consisted of group A (1 mg of $MnCl_2.4H_2O$ per mL of double-distilled water), group B (10 mg of $MnCl_2.4H_2O$ per mL of double-distilled water), and group C (20 mg of $MnCl_2.4H_2O$ per mL of double-distilled water). The data were derived from the comprehensive metallomic database. The level of each trace metal or element in each brain region in the Mn-treated rat was compared with the corresponding level (set as 100%) in the same region in the untreated (i.e., control) rat and the difference ranked as % increase or % decrease relative to the value in the control animal. For ease of visual comparison, the rank orders of increases or decreases of levels were color-coded as follows: Those levels showing % higher than corresponding values in control rats were color-coded as: *red*, 0–25%; *purple*, 25–50%; and *brown*, >50%. Those levels showing % decrease compared with corresponding values in control rats were color-coded as: *yellow*, 0% to −25%; *green*, −25% to −50%; and *blue*, >−50%. Those levels that were the same as corresponding values in control rats were coded in *black* and those levels that were not detectable (i.e., below the limits of detection) by the INAA technique were not color-coded (i.e., *white*)

both brain region in question and the dose of Mn administered (Tables 1 and 2).

As may be expected, upon chronic Mn treatment, Mn levels show the largest region-specific increases among all the trace metals that were investigated and showed increases (Table 1). Among the trace metals examined showing region-specific increases in their levels induced by Mn treatment, the rank order of their relative changes compared with each other is Mn > Hg > Cu > Se (Table 1). Because all these trace metals are neurotoxic when taken in excess, this finding suggests that increased brain regional Mn accumulation consequent to Mn treatment results in increased regional accumulation of Hg (a nonessential toxic metal), Cu, and Se, ultimately leading to a cumulative toxicity due to all three metals and perhaps even neurodegeneration.

Among the trace metals and other elements showing region-specific decreases in their levels induced by Mn treatment, the rank order of their relative changes compared with each other is F > V > Br > Al > Mo (Table 1). The functional roles of F, V, Br, and Mo in mammalian brain are poorly defined. Nevertheless, Al, a nonessential trace metal in mammals, is known to be neurotoxic (Lai et al. 2000). Thus, this finding suggests that Mn accumulation consequent to Mn treatment induces region-specific decreases in Al accumulation, leading to a lowering of neurotoxicity due to Al.

Among the trace metals and other elements showing region-specific fluctuations (i.e., both trends of increases and decreases) in their levels induced by Mn treatment, the rank order of their relative changes compared with each other is Cr > Zn > Co > Rb > I (Table 1). The functional roles of Cr, Rb, and I in mammalian brain are not defined. Zn is an essential trace metal: its decreases and increases in brain regional accumulation induced by Mn treatment could lead to, respectively, deficiency and toxicity states (Lai et al. 2000). On the other hand, the only known role of Co in mammals is that it is a component in vitamin B12 and is required in ultra-trace quantities.

Manganese, Interrelation with Other Metal Ions in Health and Disease, Table 2 Changes in levels of iron and electrolytes in brain regions in adult female rats after chronic treatment with Mn compared with corresponding levels in untreated (i.e., control) rats. The Mn-treated groups consisted of group A (1 mg of $MnCl_2 \cdot 4H_2O$ per mL of double-distilled water), group B (10 mg of $MnCl_2 \cdot 4H_2O$ per mL of double-distilled water), and group C (20 mg of $MnCl_2 \cdot 4H_2O$ per mL of double-distilled water). The data were derived from the comprehensive metallomic database. The level of each trace metal or element in each brain region in the Mn-treated rat was compared with the corresponding level (set as 100%) in the same region in the untreated (i.e., control) rat and the difference ranked as % increase or % decrease relative to the value in the control animal. For ease of visual comparison, the rank orders of increases or decreases of levels were color-coded as follows: Those levels showing % higher than corresponding values in control rats were color-coded as: *red*, 0–25%; *purple*, 25–50%; and *brown*, >50%. Those levels showing % decrease compared with corresponding values in control rats were color-coded as: *yellow*, 0% to −25%; *green*, −25% to −50%; and *blue*, >−50%. Those levels that were the same as corresponding values in control rats were coded in *black* and those levels that were not detectable (i.e., below the limits of detection) by the INAA technique were not color-coded (i.e., *white*)

Region	Ca A	Ca B	Ca C	Cl A	Cl B	Cl C	Fe A	Fe B	Fe C	K A	K B	K C	Mg A	Mg B	Mg C	Na A	Na B	Na C
Cerebellum	green	yellow	green	red	red	red	brown	red	red	yellow	yellow	yellow	yellow	yellow	yellow	red	red	red
Cerebral Cortex	yellow	red	red	red	red	red	red	red	red	red	red	red	red	blue	red	red	red	red
Hippocampus	green	yellow	yellow	red	red	red	brown	brown	purple	red	red	red	green	red	red	red	yellow	red
Hypothalamus	red	red	red	purple	purple	purple	red	red	red	red	red	red	red	purple	red	purple	purple	purple
Midbrain	brown	brown	brown	red	red	red	red	red	red	red	red	red	red	red	red	red	red	red
Pons & Medulla	brown	brown	brown	yellow	yellow	yellow	purple	yellow	yellow	yellow	yellow	yellow	red	red	green	green	blue	green
Striatum	red	red	red	purple	red	red	brown	brown	brown	red	red	red	red	red	purple	purple	red	red

However, it is neurotoxic when taken in excess. Thus, the decreases and increases in brain regional accumulation of Co induced by Mn treatment could give rise to, respectively, vitamin B12 deficiency and Co neurotoxicity.

Brain Fe level approaches those of major elements and electrolytes such as Ca (Lai et al. 1985c, 2000). Fe interacts with Mn in intestinal absorption (see above): thus, consistent with this notion is the observation that in Mn-treated rats, the largest increase in Fe levels are noted in hypothalamus, striatum, and cerebellum (Table 2), and these are the same regions that show the highest accumulation of Mn (Table 1). Because Fe is neurotoxic when taken and/or accumulated in excess, this observation suggests that the increase in Fe accumulation could accentuate the neurotoxicity of the accumulated Mn, possibly leading to enhanced neurodegeneration (Lai et al. 2000).

Similar to Fe, two electrolytes (namely, Na and Cl) show region-specific increases in their levels induced by Mn treatment, the rank order of their relative changes compared with each other is Fe > Na > Cl (Table 2). This finding suggests that increases in Na and Cl accumulation induced by Mn treatment could result in altered ionic balance in neural cells.

Unlike the trace metals, none of the electrolytes exhibit predominantly region-specific decreases in their levels induced by Mn treatment (Table 2).

Three other electrolytes exhibit region-specific fluctuations (i.e., both trends of increases and decreases) in their levels induced by Mn treatment, the rank order of their relative changes compared with each other is Ca > K > Mg (Table 2). It is interesting to note that Mn treatment results in increased Ca accumulation in midbrain, pons and medulla, and striatum but decreased accumulation in cerebellum and hippocampus (Table 2), suggesting in Mn-treated rats, midbrain, pons and medulla, and striatum are the regions that are prone to calcification and Ca-mediated neurodegeneration (Lai et al. 2000). On the other hand, the Mn-treatment-induced fluctuations in regional K may lead to altered neuronal excitability and conduction, while the fluctuations in regional Mg could result in some alterations in key protein functions (e.g., receptor-gaited ion channels and protein kinases) in which Mg plays important roles.

In addition to the findings discussed above, chronic and life-span Mn treatment also modulates on the region-specific distributions of trace metals and electrolytes during development and aging. Some of those findings have been discussed previously and will not be reiterated here (see Lai et al. 1985c, 1999, 2000 for detailed discussion).

Taken together, the region-specific interactions between Mn and other trace metals and other elements

(Tables 1 and 2) may have pathophysiological implications in neurodegenerative and other neurological diseases (see below).

Interrelations Between Manganese and Other Metals in Brain in Human Health and Diseases

While Mn can interact with trace elements (e.g., F, V, Br, Al, Mo) leading to region-specific decreases in their accumulation, the health impact of the Mn-induced decreases is largely unknown as the functional roles of these trace elements in brain, with the exception of Al, are unknown. Of some health concern are the Mn interactions where brain regional levels of trace metals become increased: these trace metals include Mn, Hg, Cu, Se, Fe, and Zn. The available evidence suggests that the accumulation of these metals may have pathophysiological implications in several neurodegenerative and other neurological diseases and the implications are briefly summarized below.

High brain levels of Mn, Ca, and Al are found in Alzheimer's disease (AD). In AD patients, plasma levels of Al, Cd, Hg, and Se are increased (see Lai et al. 2000 for discussion). Fe is also elevated in multiple regions of the AD brain. Fe in the cores and rims of senile plaques of amygdala of AD patients is elevated and Fe (like Cu and Zn) can accelerate aggregation of β–amyloid peptide, a hallmark of AD neuropathology (see Lai et al. 2000 for discussion). Additionally, Zn level is increased in different regions of the AD brain (see Lai et al. 2000 for discussion). Consequently, Fe, Zn, Al, and Cu have been proposed to be responsible for inducing cell death, especially neuronal cell death, in the AD brain because of (a) formation of reactive oxygen species, (b) other metal-mediated cytotoxic effects, or (c) a combination of mechanisms (see Lai et al. 2000 for discussion).

Chronic ► Mn toxicity in humans shows signs and symptoms that are reminiscent of those noted in Parkinson's disease (PD) and dystonia (Lai et al. 2000). Some but not all studies have demonstrated an increased brain level of Mn in PD. However, increased levels of Fe and Zn are found in the substantia nigra of PD patients.

In several neurological diseases other than AD and PD, brain metabolism of Mn and other metals are also disturbed. Pick's disease is associated with increases in brain Mn, Fe, and Na and decreases in brain Cr and Se. In amyotrophic lateral sclerosis (ALS) brain, levels of Al and Ca are elevated. These findings suggest Mn may interact with other metals in Pick's disease and in ALS (see Lai et al. 2000 for discussion).

In some metabolic encephalopathies (e.g., dialysis encephalopathy, chronic renal failure without dialysis, and hepatic coma), brain Al is elevated. Thus, Al neurotoxicity is implicated in these encephalopathies. There is also good evidence that Mn level is elevated in globus pallidus of patients with liver cirrhosis, thereby implicating Mn in brain in hepatic encephalopathy (see Lai et al. 2000 for discussion).

Conclusions and Prospects for Future Investigation

There have been significant advances in our understanding of manganese and its interrelations with other metal ions in health and disease. The advances in the last 25 years have been built upon the skillful exploitation of the versatile technique of instrumental neutron activation analysis (INAA) in conjunction with a rat model of chronic and life-span exposure to Mn in constructing and assembling a comprehensive metallomic database containing the distributions of over 20 trace and major elements in seven discrete brain regions and in multiple peripheral organs in rats during their life span (i.e., from early postnatal development until aging). This comprehensive database also contains the distributions of some 20 trace and major elements in seven discrete brain regions and in multiple peripheral organs in rats continuously exposed to three doses of Mn over their life span. The gradual mining and analyses of selected sets of data within this comprehensive metallomic database have allowed us to arrive at a number of fundamental conclusions as to how Mn interacts with other trace and major elements in various regions of the rat brain and in various peripheral organs.

(1) Mn interacts with multiple trace elements (especially Fe) in the process of intestinal absorption. (2) Each trace and major element has its own, almost unique, brain regional distribution. (3) The patterns of distributions of trace and major elements in the untreated rat brain show region-specific differences, and these patterns are significantly altered by Mn treatment in an approximately dose-related manner. (4) The patterns of brain regional distributions of trace and

major elements in the adult rats and the modulations by chronic Mn treatments thereon have implications in health and disease, especially in regard to several neurodegenerative (e.g., Alzheimer's disease and Parkinson's disease) and other neurological diseases. (5) However, the cellular and molecular mechanisms underlying how Mn interacts with other trace and major elements in brain and other organs are almost totally unknown. (6) Nevertheless, the continued mining and systematic analyses of discrete segments of the comprehensive metallomic database are beginning to facilitate the formulation of discrete, testable hypotheses regarding such cellular and molecular mechanisms and lead to novel and productive research. Thus, these endeavors promise to break new grounds in our understanding of how trace and major elements interact in biological systems.

Acknowledgments We wish to acknowledge the contributions of Drs. Alex W.K. Chan and Margaret J. Minski (now retired), both formerly of Imperial College, University of London, UK, and Dr. Louis Lim of Institute of Neurology, University of London, UK, and their role in either designing or producing or supporting the creation of the comprehensive metallomic database. JCK Lai wishes to thank the Worshipful Company of Pewterers, London, UK, for their generous support to allow him to initiate the project and create the metallomic database. We thank Mr. G.L. Wright for his timely help with mining and analyzing some sets of the data. This study is supported, in part, by a US Army Medical Research and Material Command Project Grant (Contract W81XWH-07-2-0078).

Cross-References

▶ Manganese Toxicity

References

Chan AWK, Minski MJ, Lai JCK (1983) An application of neutron activation analysis to small biological samples: simultaneous determination of thirty elements in rat brain regions. J Neuro Method 7:17–328

Lai JCK, Leung TKC, Lim L (1984) Differences in the neurotoxic effects of manganese during development and aging: some observations on brain regional neurotransmitter and non-neurotransmitter metabolism in a developmental rat model of chronic manganese encephalopathy. Neurotoxicology 5:37–48

Lai JCK, Chan AWK, Minski MJ, Lim L (1985a) Roles of metal ions in brain development and aging. In: Gabay S, Harris J, Ho BT (eds) Metal ions in neurology and psychiatry. Alan Liss, New York

Lai JCK, Leung TKC, Lim L (1985b) Effects of metal ions on neurotransmitter function and metabolism. In: Gabay S, Harris J, Ho BT (eds) Metal ions in neurology and psychiatry. Alan Liss, New York

Lai JCK, Chan AWK, Minski MJ, Leung TKC, Lim L, Davison AN (1985c) Application of instrumental neutron activation analysis to the study of trace metals in brain and metal toxicity. In: Gabay S, Harris J, Ho BT (eds) Metal ions in neurology and psychiatry. Alan Liss, New York

Lai JCK, Minski MJ, Chan AWK, Leung TKC, Lim L (1999) Manganese mineral interactions in brain. Neurotoxicology 20:433–444

Lai JCK, Minski MJ, Chan AWK, Lim L (2000) Interrelations between manganese and other metal ions in health and disease. In: Sigel A, Sigel H (eds) Metal ions in biological systems, vol 37. Marcel Dekker, New York

Leung SW, Chan A, Minski M, Lai JCK (2011) Comparison of elemental distribution in rat's brain after lifelong treatment with excessive Mn^{2+} in drinking water and the health implications. In: Proceedings of international water convention, Singapore, 4–8 July 2011, Article IWA-6218R1, pp 1–10

Underwood EJ (1977) Trace elements in human and animal nutrition, 4th edn. Academic, New York

Wright GL, Lai JCK, Chan A, Minski M, Leung SW (2011) Influence of metallomic distribution in brain by prolong consumption of contaminant in drinking water. Proceedings of International Water Convention, Singapore, 4–8 July 2011, Article IWA-6228R1, pp 1–21

Manganese, Physical and Chemical Properties

Fathi Habashi
Department of Mining, Metallurgical, and Materials Engineering, Laval University, Quebec City, Canada

Manganese is a silver-gray metal, resembling iron. It is hard and brittle, and its primary uses in a metallic form are as an alloying, desulfurizing, and deoxidizing agent for steel, cast iron, and nonferrous metals. It is a transition metal; hence, it is less reactive than the typical metals and more reactive than the less-typical metals. Manganese is an essential trace nutrient in all forms of life.

Physical Properties

Atomic number	25
Atomic weight	54.94
Relative abundance in the Earth's crust, %	0.10

(*continued*)

Manganese, Physical and Chemical Properties, Table 1 Phase modifications of manganese

Property	α-Mn	β-Mn	γ-Mn	δ-Mn	Molten Mn
Crystal structure	Complex cubic	Complex cubic	Face-centered cubic	Body-centered cubic	
Lattice parameter, nm	0.8894	0.6290	0.3774	0.3080	
Transition temperature, °C	α ⇌ β	β ⇌ γ	γ ⇌ δ	δ ⇌ Liquid	Liquid ⇌ Gas
	727	1,095	1,104	1,244	2,032
Density, g/cm^3	7.44	7.29	7.21		7.55–1.2 × 10^{-3} t (1,244–1,500°C)
Volume increase on transition, %	α ⇌ β	β ⇌ γ	γ ⇌ δ	δ ⇌ Liquid	
	3.4	0.8	0.8	2.9	
Resistivity, Ω cm	150–260 × 10^{-6}	90 × 10^{-6}	40 × 10^{-6}		
Coefficient of linear expansion, K^{-1}	22.3 × 10^{-6}	24.9 × 10^{-6}	14.8 × 10^{-6}	41.6 × 10^{-6}	
Heat of transition, J/mol	2,240	2,282	1,800	14,655	219,901
Specific heat, J g^{-1} K^{-1}	0.477	0.482	0.502		
Enthalpy, J/mol	4,999		5,112		
Entropy, J mol^{-1} K^{-1}	32.0		32.3		173.8

The vapor pressure of manganese
α-Mn (298–1,000 K): $C_p = 23.9 + 14.2 \times 10^{-3} T - 1.55 \times 10^5 T^{-2}$
β-Mn (1,000–1,374 K): $C_p = 34.9 + 2.765 \times 10^{-3} T$
γ-Mn (1,374–1,400 K): $C_p = 44.80$
δ-Mn (1,410 K–mp): $C_p = 47.31$
Liquid Mn (mp–bp): $C_p = 46.06$

Density at room temperature, g·cm^{-3}	7.21
Density at melting point, g·cm^{-3}	5.95
Atomic radius, pm	127
Melting point, °C	1,246
Boiling point, °C	2,061
Heat of fusion, kJ·mol^{-1}	12.91
Heat of vaporization, kJ·mol^{-1}	221
Molar heat capacity, J·mol^{-1}·K^{-1}	26.32
Electrical resistivity at 20°C, μΩ·m	1.44
Thermal conductivity, W·m^{-1}·K^{-1}	7.81
Thermal expansion at 25°C, μm·m^{-1}·K^{-1}	21.7
Young's modulus, GPa	198
Bulk modulus, GPa	120
Mohs hardness	6.0
Brinell hardness, MPa	196
Standard electrode potential, V	1.134

Manganese exists in four allotropic forms: α, β, γ, and δ. Data of the different phases are given in Table 1.

Chemical Properties

Manganese dissolves in acids with liberation of hydrogen and formation of Mn (II) salts. Hot concentrated sulfuric acid dissolves manganese with evolution of SO_2, and nitric acid with evolution of hydrogen, nitrogen, and dinitrogen monoxide. At normal temperatures, pure manganese is not attacked by oxygen, nitrogen, or hydrogen. At high temperatures it reacts violently with oxygen, sulfur, and phosphorus. It is therefore a powerful agent for deoxidizing and desulfurizing metals, whereby it is converted to a divalent oxide or sulfide. Reaction with chlorine forms manganese (II) chloride, and with SO_2 the products are MnS and MnO.

Manganese exists in the following oxidation states: +7, +6, +5, +4, +3, +2, +1, −1, −2, −3. The oxides are acidic, basic, or amphoteric depending on the oxidation state. Manganese compounds where manganese is in oxidation state +7, which are restricted to the unstable oxide Mn_2O_7 and compounds of the intensely purple permanganate anion MnO_4^-, are powerful oxidizing agents. Compounds with oxidation states +5 (blue) and +6 (green) are strong oxidizing agents and are vulnerable to disproportionation. The most stable oxidation state for manganese is +2, which has a pale pink color, and many manganese (II) compounds are known, such as manganese (II) sulfate, $MnSO_4$, and manganese (II) chloride, $MnCl_2$.

This oxidation state is also seen in the mineral rhodochrosite [manganese (II) carbonate]. The +2 oxidation state is the state used in living organisms for essential functions; other states are toxic for the human body.

Manganese (IV) oxide, MnO_2, is amphoteric. The oxidation states Mn (II) and Mn (VII) are of industrial importance, forming a large number of useful salts.

References

Craven PM et al (1997) Manganese. In: Habashi F (ed) Handbook of extractive metallurgy. Wiley-VCH, Weinheim, pp 1813–1860

Manganese-Metal Ion Interactions

▶ Manganese, Interrelation with Other Metal Ions in Health and Disease

Marine Sponge Skeletons

▶ Silicateins

MAs: Mugineic Acid Family Phytosiderophores

▶ Iron Proteins, Plant Iron Transporters

MATE: Multidrug and Toxic Compound Extrusion (MATE)

▶ Iron Proteins, Plant Iron Transporters

Materials

▶ Tellurium in Nature

Matrilysin

▶ Zinc Matrix Metalloproteinases and TIMPs

Matrix Metallopeptidases

▶ Silicosis

Matrixin

▶ Zinc Matrix Metalloproteinases and TIMPs

MCD, Magnetic Circular Dichroism

▶ Zinc Aminopeptidases, Aminopeptidase from Vibrio Proteolyticus (Aeromonas proteolytica) as Prototypical Enzyme

MCR

▶ Methyl Coenzyme M Reductase

Mechanisms of Action of Gold(III) Compounds

▶ Gold(III) Complexes, Cytotoxic Effects

Melarsoprol

▶ Arsenic in Therapy

Membrane Proteins

▶ Sodium-Hydrogen Exchangers, Structure and Function in Human Health and Disease

Membrane Proteins with Transport Activity for Silicic Acid

▶ Silicon Transporters

Membrane Trafficking

▶ C2 Domain Proteins

Membrane-Bound Methane Monooxygenase

▶ Particulate Methane Monooxygenase

Meprin a Complex Peptidase

▶ Zinc Meprins

Mercury (Hg^{2+}) Permeases/Porters

▶ Mercury Transporters

Mercury and Alzheimer's Disease

Boyd E. Haley[1] and Joachim Mutter[2]
[1]Department of Chemistry, University of Kentucky, Lexington, KY, USA
[2]Naturheilkunde, Umweltmedizin Integrative and Environmental Medicine, Belegarzt Tagesklinik, Constance, Germany

Synonyms

Biomedical plausibility of exposure and mechanism of action; Mercury exacerbation of Alzheimer's disease

Definition

Most of the prominent neurological illnesses that affect many people in developed countries, such as Alzheimer's Disease (AD), Parkinson's (PD), Amyotrophic Lateral Sclerosis (ALS) and Multiple Sclerosis (MS), are still without a proven etiology even though millions of research dollars have been spent many people in developed countries. The difficulty of identifying the possible causal or exacerbating factors for these diseases indicates that the factors involved are complex and a simple exposure/causation to a single toxicant/microbial model is not likely and that synergistic toxicities may be involved. Autopsy studies have shown that first signs of neuronal damage (neurofibrillary tangles) occur in the brain 40-50 years before clinical manifestation of Alzheimer's disease. The slow development of the neuronal damage involved in AD before the first detectable biological signs of deterioration and the first clinical manifestation of AD also makes any causation model difficult to prove. Further, if multiple genetic differences can induce different susceptibilities to toxicity from the same level of exposure, the determination of causation is even more difficult. Mercury, due to its neurotoxic properties, should at least exacerbate any neurological illness that involves synapse deterioration, oxidative stress and inflammation, and inhibition of biochemical reactions that are sensitive to sulfhydryl reactive agents. Experimental studies in animals and in-vitro systems not only confirmed the well-known toxicology of mercury, but also reproduced the pathological signs of Alzheimer's disease quite accurately: hyperphosphorylation of tau protein, degeneration of microtubules and the increased formation of beta-amyloid protein.

Specific Problems Associated with Mercury Toxicity That Prevent Its Recognition as Associated with Neurological Illnesses like Alzheimer's Disease

Mercury is well known as the most toxic, non-radioactive metal, with a well-described neurotoxicology. Widespread use of inorganic mercury started around 1830, when dental amalgams became rapidly popular. From this source, 1.2 to 27.0 μg of Hg are taken up per day, and 1.0 to 22.0 μg are retained. In contrast, the intake of organic Hg is only about 2.4 μg per week, if consuming one fish meal per week. Mercury from amalgam is taken up via the lungs, and 80% of it is absorbed. It is highly diffusible and crosses the blood-brain barrier or bilayers of cells and cell organelles easily. Hg reacts immediately with enzymes, glutathione, tubulin, ion channels or transporters, inhibiting their activities. A very low level (180nM) of Hg, which is much less than the levels found in brains of individuals with dental amalgam (Mutter 2011), lead to neurofibrillary tangels and secretion of β-Amyloid in human neurons. It decrease glutathione levels (GSH) and increase oxidative and nitrosative stress, which lead to cytotoxicity (Olivieri et al. 2000). Hg also potently interferes with neural stem cell development, which could contribute to reduced cortical and hippocampal neuronal density in AD. Elevated levels of homocysteine (HCY) in AD have been reported, and the rate of cognitive decline is correlated with HCY levels. HCY is normally converted to methionine by the vitamin B12 and methyl-folate-dependent enzyme methionine synthase, which is highly sensitive to cellular redox status. Mercury inhibits effectively methionine synthethase in human neuronal cells in concentrations which are normally found in post-mortem brain of individuals with dental amalgam.

Synergistic Toxicities. It is widely believed that the old adage "the dose makes the toxin" covers what is needed to associate a known toxin with a biomedical abnormality. Therefore, some of the research that has been used to refute the association of mercury with AD development have major methodical failures (Mutter et al. 2010; Mutter 2011). However, the actual toxicity of mercury is very strongly associated with synergistic toxic effects with other known toxic metals such as lead and cadmium. It has been reported that a LD-1 level of mercury when mixed with the LD1 of lead produces a high toxic mixture with a LD-100 value. That means, all animals die instead of the expected two percent if the toxicities were additive (for review see Mutter 2011). Because humans are different with their diets and life stile, the body burden of synergistic acting environmental toxins, like pesticides, lead, arsenic and cadmium varies greatly. Therefore, their toxic response to an identical exposure of mercury is different. Smokers are known to be higher in cadmium than nonsmokers and this represents a group that would be more sensitive to mercury toxicity. In sum, not only the level of exposure to mercury is a measure of its toxicity, but rather the amount retained in the body and the toxicity caused by the combined presence of other synergistic toxins.

Genetic Susceptibilities. The most single recognized genetic risk factor for AD are the inherited forms of apolipoprotein E (APO-E). It is well documented that APO-E2 reduces the risk of AD by 50% and APO-E4 increases the risk with the most common form, APO-E3, in between. All three of these brain-produced APO-E forms do an adequate job of transporting oxidized cholesterol from the brain into the cerebrospinal fluid (CSF) where it is cleared into the blood for removal by the liver. There is no widely accepted explanation why APO-E2 decreases the risk for AD while the e4-genotype increases it by 15-fold relative to the e3 genotype. It is important to notice that all three ApoE forms consist of 299 amino-acids, and the only differences are that ApoE e4 has an arginine in position 112 and 158, where ApoE e2 has two cysteines, and ApoE e3 one arginine and one cysteine. Average age of onset depends on the combination of APO-E genotypes and therefore the number of cysteines:

Number of cysteines	Genotype combinations		Average age of AD onset
Four	APO-E2	APO-E2	>90
Three	APO-E2	APO-E3	>90
Two	APO-E2	APO-E4	80–90
	APO-E3	APO-E3	80–90
One	APO-E3	APO-E4	70–80
Zero	APO-E4	APO-E4	<70

APO-E proteins are commonly called "housekeeping proteins" in that they are made to be excreted from the brain and the body in the process of cleaning the brain of oxidized lipids and cholesterol. One theory explaining how APO-E2 protects the brain is based on a "counter current" scheme where APO-E2 on its way out of the brain into the CSF has the potential to bind to two mercury atoms in the brain/CSF and carry them into the blood, removing these mercury atoms from the central nervous system (CNS) and later from the body (Pendergrass and Haley 1996). Obviously, APO-E4 has reduced mercury binding capacity compared to APO-E2. However, the presence of the two sulfhydryl groups on E2 also increases the antioxidant capacity in regards to the ability to absorb hydroxyl free radicals. This could also reduce brain inflammation and cell damage but the oxidation of the cysteines on E2 has never been reported.

A recent publication reports on the association of polymorphisms of glutathione enzymes and selenoproteins with mercury biomarkers (Goodrich et al. 2011). The capability to recover GSH from oxidized glutathione (GSSG) and the ability to synthesize new reduced GSH is dependent on the stability of the different enzymes to the toxic effects of mercury. The authors state that the polymorphisms in this area may influence the elimination of mercury in the urine and hair or mercury retention following exposure. This further adds to the concept that excretion ability has as much to do with resulting toxicity as does the exposure level. This would be especially true with a chronic low level exposure.

Since blood, urine, feces, and hair levels of mercury are dependent both on exposure and excretion rates (or retention effects) it is obvious that merely observing the mercury levels in one biopsy material is not sufficient enough to determine exposure. For example, measuring the mercury in the birth hair of autistic versus healthy children and plotting this against the number of dental amalgams in the birth mother showed the fallacy of this approach (Holmes et al. 2003). The mercury in the birth hair of healthy children increased appropriately with the mother's increased amalgam count. In contrast, the autistic children did not show any increase in birth hair mercury even in mothers with the highest level of amalgam fillings indicating these children had a problem excreting mercury in utero.

Biochemical Abnormalities in Alzheimer's Disease Brain That May Be Mimicked by Mercury Exposure

Brain Nucleotide Binding Protein Abnormalities. It is well known that neurons and related synaptic junctions are deteriorated in AD brain. Also, there are numerous hypothesis as to what causes this deterioration in AD brain and the most common are the loss of specific biochemical activities or the induction of toxic protein such as beta-amyloid. The problem with these hypotheses is that there is no identified toxin or infection that sets off the aberrant biochemistry in the brain. Since over 95% of AD cases are not genetically caused, it is likely that susceptibility to environmental toxins are involved.

Early studies comparing the nucleotide-binding proteins in AD brain to those found in normal brain homogenates showed that tubulin (Pendergrass and Haley 1996) and creatine kinase (CK) (David et al. 1998) nucleotide-binding sites were dramatically reduced in AD brain. These studies were performed using a technique called "radioactive nucleotide photoaffinity labeling" that allows the detection and quantification of proteins which bind specific nucleotides. This entails using radioactive nucleotide photoaffinity probes to radiolabel binding proteins in brain tissue. For the GTP-binding site of tubulin $[^{32}P]8N_3GTP$ was used, and for the ATP binding site of CK $[^{32}P]8N_3ATP$ was used. Separation of treated tissue proteins by polyacrylamide gel electrophoresis (PAGE) allowed identification of the radiolabeled proteins and quantification of the relative number of binding sites for tubulin or CK. An example is shown in Fig. 1 where normal and clinically diagnosed AD brain tissues were compared for their tubulin GTP-binding sites. The three putative AD samples with normal type binding were later found not to have the AD diagnostic NFT marker.

A decrease in binding site radiolabel indicates an inactivation of the protein or a decrease in expression of the protein. This indicates some toxic events has affected these proteins. A decrease in binding site detection indicates an inactivation of the protein or a decrease in expression of the protein. With tubulin and CK quantification of protein, copy number by immunoassay showed that there was not any significant decrease in copy numbers of tubulin or CK but that nucleotide-binding sites were unavailable for binding. Figure 2 shows an example of tubulin which had similar levels of total tubulin in both normal and AD brain. It also shows that tubulin in AD brain partitions into the particulate fraction whereas in normal brain a large fraction (about 70%) of the tubulin remains in the soluble fraction. This indicates some toxic event has affected these proteins and their solubility. In normal brain, both tubulin and CK are soluble proteins which primarily remain in the supernatant on high speed centrifugation, but both partitioned primarily into the insoluble fraction on centrifugation in the AD samples. This indicates a major structural change has occurred that did not change the molecular size of the protein subunits. Most likely, the structural changes involved hydrophobic interactions since solubilizing with lipid-detergent in the PAGE process released the intact proteins.

CK was well known to have a very reactive sulfhydryl that controlled its activity, and it was proven that

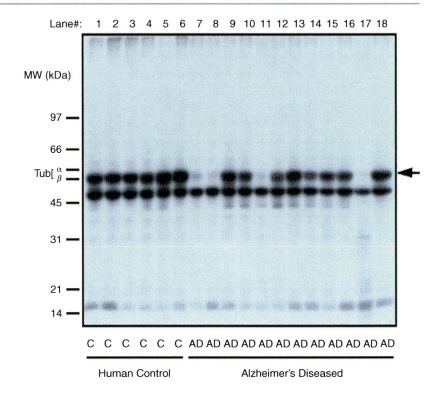

Mercury and Alzheimer's Disease, Fig. 1 An autoradiogram made from a PAG on which brain proteins have been separated showing the $[^{32}P]8N_3GTP$ bioavailable tubulin GTP-binding sites in control versus clinically diagnosed AD brain homogenates

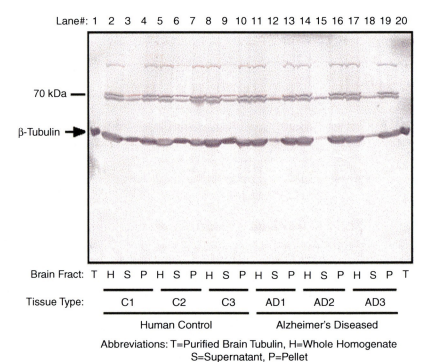

Mercury and Alzheimer's Disease, Fig. 2 Immunostaining for beta-tubulin shows the loss of beta-tubulin from the soluble fraction into the particulate fraction of human brain homogenates in AD brain compared to the normal human brain (compare levels in lanes 3, 6, and 9 to levels in lanes 12, 15, and 18)

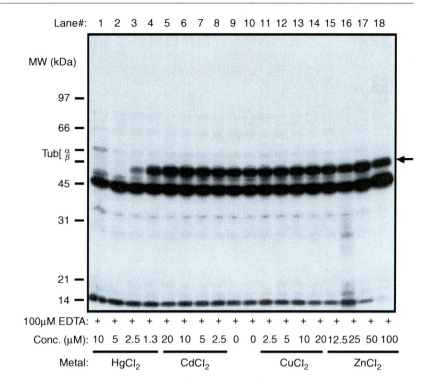

Mercury and Alzheimer's Disease, Fig. 3 An autoradiogram made from a PAG on which normal brain proteins were separated after radiolabeling with [^{32}P] 8N$_3$GTP in the presence of different metal and excess EDTA. This represents an example of one of the screening tests investigating the effects of various heavy metals on tubulin viability

its nucleotide substrate, ATP, protected this sulfhydryl from inhibition by thiol reactive compounds and, later, that this thiol group was in a peptide within the active site (Olcott and Haley 1994). Also, the decrease in the GTP-binding sites of tubulin involved in the polymerization of tubulin into microtubulin was the most pronounced observation in comparing AD to normal brain since tubulin is a major brain protein (Pendergrass and Haley 1996). Studies were done to determine if any known toxins could cause the specific inhibition of tubulin GTP binding and cause its abnormal polymerization, resulting in partitioning into the insoluble fraction on centrifugation.

Initially, several metals toxic to tubulin activity were tested by the simple process of adding them to normal brain homogenates before photoaffinity testing as described above. Since brain tissues contain organic acids with chelation capabilities (such as citrate), the metal tests were done in the presence of excess EDTA (ethylene diamine tetraacetic acid) to insure that any metal effects were of exceptionally high affinity and not easily prevented by normal cellular chelation ability. A sample of one such experiment is shown in Fig. 3. The results were very clear, only mercury of all metals tested, specifically prevented the interaction of tubulin with the GTP photoaffinity probe, even in the presence of excess EDTA. Also, other proteins binding this affinity probe were not affected by the mercury addition, just as appears in AD brain homogenates (Pendergrass and Haley 1996). It is well known that mercury compounds, as well as other factors, can cause the abnormal polymerization of tubulin which would explain its partitioning into the insoluble fraction. Also, the amount of Hg^{2+} needed to cause these abnormal tubulin properties are below the range of Hg^{2+} concentrations reportedly found in human brain. Supporting the synergistic concept, mercury was especially effective in the presence of other divalent metals such as zinc, copper, lead, and aluminum and some of these metals are known to be elevated in AD brain (Thompson et al. 1988). The data in Fig. 4 shows the effect of 10 and 20 μM zinc on the toxic effects of mercury on GTP binding to normal brain tubulin. The presence of 10 μM zinc increased the toxic effects of 0.625 μM mercury on brain tubulin from about 3% to 40%. Importantly, the incubation with both zinc and mercury were in the matter of minutes, not days or weeks.

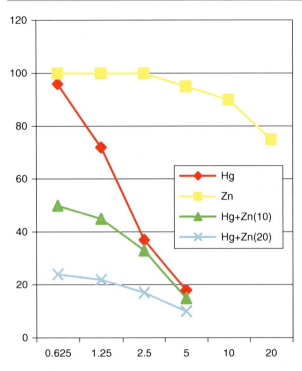

Mercury and Alzheimer's Disease, Fig. 4 The synergistic effects of nontoxic levels of zinc on mercury's toxic effects on the GTP-binding site of tubulin in normal brain

The results of any homogenate test may not occur in a living animal. Therefore, tests were done to determine if exposure of rats to mercury vapor, known to result in Hg^{2+} in the brain, would cause the same abnormal toxic effects on tubulin seen in adding Hg^{2+} to homogenates (Pendergrass et al. 1997). Two experiments showed a 41% and 75% reduction of the GTP-binding capability of rat tubulin on exposure to mercury vapor. Results of one such study is shown in Fig. 5. This was very similar to the effect seen in AD brain and confirmed the homogenate results. It may also be an explanation of the mechanism of mercury vapor causing dementia of the "Mad Hatter" and "Danbury Shakes" syndromes and implies that exposure to mercury vapor would likely exacerbate the AD syndrome in any afflicted individual.

Research on animal neurons demonstrated that mercury at low nanomolar concentrations could strip the axons of tubulin leaving behind neurofibrils that represent the initial stage of neurofibrillary tangles (NFTs). NFTs are a major pathological, diagnostic marker for AD (Leong et al. 2001). Other metals, like lead, cadmium, copper, manganese were not able to produce NFTs. Mercury vapor and inorganic mercury are

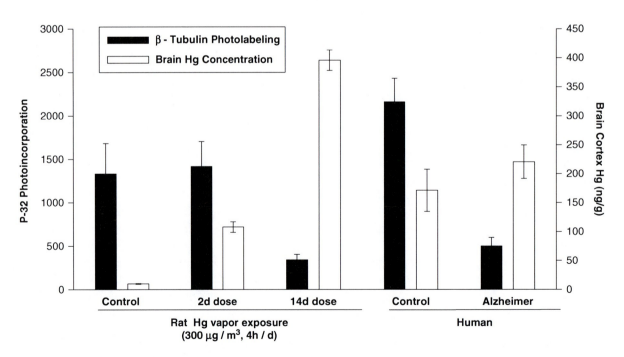

Mercury and Alzheimer's Disease, Fig. 5 A comparison of brain mercury increase and a correlated decrease in brain tubulin GTP-binding sites in rats exposed to mercury vapor. A comparison of normal and AD human brain data shows a similar effect

available environmental toxins and may represent a potential contribution to the development of AD.

In support of the homogenate and animal studies (Olivieri et al. 2000), studies using neuroblastoma cells demonstrated that exposure to sublethal doses (36×10^{-9} M) of Hg^{2+} caused a significant drop in GSH, an increased phosphorylation of the microtubulin protein Tau and secretion of beta-amyloid protein. All of these biochemical changes are observed in AD brain and the latter two are uniquely observed in AD brain tissues and are widely considered to be diagnostic, pathological markers of the disease. Beta-amyloid protein makes up the "amyloid plaques" that was one of the first diagnostic markers reported for AD brain pathology. A very strong component of AD researchers believe that amyloid protein is the cause of AD while others have speculated that abnormal phosphorylation of Tau is involved in the pathology. Therefore, mercury exposure at nanomolar levels causes neuroblastoma cells to suffer oxidative stress (low GSH levels), produce a protein that is diagnostic for AD and believed to be involved directly in AD. This leads to the authors' conclusion that mercury would have to be considered as causal for AD (Olivieri et al. 2000).

In summary, research has shown that mercury is found in human brains at the level that could inhibit two specific brain proteins (tubulin and creatine kinase) and affect their particulate/soluble partitioning as observed by experimental additions of Hg^{2+} to normal brain homogenates and also seen in AD brain homogenates without experimental Hg^{2+} added. Further, other experiments using neurons in culture can lead to the production of neurofibrillary tangles, decreased glutathione levels, elevation of beta-amyloid protein, and abnormal phosphorylation of protein Tau, all very highly related to observations also seen in AD brain tissue. Most of these observations specifically require mercury and are not caused by other metals. This is highly suggestive, but is not conclusive evidence, that mercury is involved in the etiology of AD. It definitely implies that exposure to mercury would be an exacerbating factor for those who do become afflicted with AD.

Cross-References

▶ Aluminum, Biological Effects
▶ Biomarkers for Cadmium
▶ Cadmium and Health Risks
▶ Lead and Alzheimer's Disease
▶ Lead Nephrotoxicity
▶ Mercury and DNA
▶ Mercury and Immune Function
▶ Mercury and Lead, Effects on Voltage-Gated Calcium Channel Function
▶ Mercury and Low Molecular Mass Substances
▶ Mercury Nephrotoxicity
▶ Mercury Neurotoxicity
▶ Mercury Toxicity
▶ Mercury Transporters
▶ Mercury, Physical and Chemical Properties
▶ Metallothioneins and Lead
▶ Metallothioneins and Mercury
▶ Zinc Metallothionein

References

David S, Shoemaker M, Haley B (1998) Abnormal properties of creatine kinase in alzheimer's disease brain: correlation of reduced enzyme activity and active site photolabeling with aberrant cytosol-membrane partitioning. Mol Brain Res 54:276–287

Goodrich JM, Wang Y, Gillespie B, Werner R, Franzblau A, Basu N (2011) Glutathione enzyme and selenoprotein polymorphisms associate with mercury biomarker levels in michigan dental professionals. Toxicol Appl Pharmacol. doi:10.1016/j.taap. 2011.09.014

Holmes AS, Blaxill MF, Haley B (2003) Reduced levels of mercury in first baby haircuts of autistic children. Int J Toxicol 22:1–9

Leong CCW, Syed NI, Lorscheider FL (2001) Retrograde degeneration of neurite membrane structural integrity and formation of neruofibillary tangles at nerve growth cones following in vitro exposure to mercury. NeuroReports 12(4):733–737

Miu AC, Benga O (2006) Metals in Alzheimer's disease. J Alzheimers Dis 10(2–3):133

Mutter J, Curth A, Naumann J, Deth R, Walach H (2010) Does inorganic mercury plays a role in Alzheimer's Disease? A systematic review and an integrated molecular mechanism. J Alzheimers Dis 22:357–374

Mutter J (2011) Is dental amalgam safe for humans? The opinion of the scientific committee of the europaen commission. J Occup Med Toxicol 6:2

Olcott M, Haley B (1994) Identification of two peptides from the ATP-binding domain of creatine kinase. Biochemistry 33:11935–11941

Olivieri G, Brack Ch, Muller-Spahn F, Stahelin HB, Herrmann M, Renard P, Brockhaus M, Hock C (2000) Mercury induces cell cytotoxicity and oxidative stress and increases beta-amyloid secretion and tau phosphorylation in SHSY5Y neuroblastoma cells. J Neurochem 74:231–231

Pendergrass JC, Haley BE (1996) Inhibition of brain tubulin-guanosine 5'-triphosphate interactions by mercury: similarity to observations in Alzheimer's diseased brain. In: Sigel H, Sigel A (eds) Metal ions in biological systems, vol 34, Mercury and its effects on environment and biology, Chapter 16. Marcel Dekker, New York, pp 461–478

Pendergrass JC, Haley BE, Vimy MJ, Winfield SA, Lorscheider FL (1997) Mercury vapor inhalation inhibits binding of GTP to tubulin in rat brain: similarity to a molecular lesion in Alzheimer's disease brain. Neurotoxicology 18(2):315–324

Thompson CM, Markesbery WR, Ehmann WD, Mao YX, Vance DE (1988) Regional brain trace-element studies in Alzheimer's disease. Neurotoxicology (Spring) 9(1):1–7

Mercury and DNA

Hidetaka Torigoe[1], Yoshiyuki Tanaka[2] and Akira Ono[3]
[1]Department of Applied Chemistry, Faculty of Science, Tokyo University of Science, Tokyo, Japan
[2]Graduate School of Pharmaceutical Sciences Tohoku University, Sendai, Miyagi, Japan
[3]Department of Material & Life Chemistry, Faculty of Engineering, Kanagawa University, Kanagawa-ku, Yokohama, Japan

Synonyms

Double-stranded DNA; Mercury ion; Specific binding; T:T mismatched base pair

Definition

Mercury ion has the ability to specifically bind with T:T mismatched base pair in double-stranded DNA. The specific binding is quite important not only for the biological meaning of metal ion-DNA interaction but also for its application to the development of various fields, such as environmental, life, and material sciences.

Introduction

The interactions between metal ions and natural nucleic acids have attracted considerable interest for their involvement in structure formation and folding of nucleic acids, such as triplex, quadruplex, and RNA folding, and their possible roles in catalytic activity of nucleic acids, such as catalytic cofactors in ribozymes. The interactions between metal ions and nucleic acids including both natural and artificially designed molecules are also important for their wide variety of potential applications in nanotechnology including the design of biomolecular nanomachines and nanodevices. This essay describes the binding between the Hg^{2+} ion and T:T mismatched base pair in duplex DNA to form T-Hg-T metal-mediated base pair (Fig. 1) (Ono et al. 2011) and the potential applications of T-Hg-T metal-mediated base pair to the development of various fields, such as environmental, life, and material sciences.

Binding of the Hg^{2+} Ion with Large DNA Polynucleotides

The Hg^{2+} ion has been known to bind selectively with nucleobases rather than with the phosphate or sugar groups in DNA (Saenger 1984). Due to the high toxicity of mercury, interactions between the Hg^{2+} ions and DNA have been widely examined. In the 1960s, binding of the Hg^{2+} ions with natural large DNA polynucleotides was studied by UV absorption spectra and Hg^{2+} titration experiments. In 1961, Yamane and Davidson reported that protons were released when the Hg^{2+} ions bound with natural large DNA polynucleotides (Reference 55 cited in Ono et al. 2011). In 1963, Katz proposed the possibility of the formation of a 1:2 complex between the Hg^{2+} ions and thymine bases in large double-stranded DNA polynucleotides, $d(AT)_n:d(AT)_n$, with the release of protons (Reference 56 cited in Ono et al. 2011). On the other hand, in 1974, Kosturko et al. reported that a covalent and linear T-Hg-T bond was observed in the crystal structure revealing a 1:2 complex of the Hg^{2+} ions and 1-methylthymine (Reference 60 cited in Ono et al. 2011). Thus, the binding of the Hg^{2+} ions with large DNA oligonucleotides may trigger the dissociation of A:T base pairs, followed by the formation of T-Hg-T metal-mediated base pairs.

Binding of the Hg^{2+} Ion with DNA Oligonucleotides

Double-stranded DNA oligonucleotides with T:T mismatched base pair can be prepared by annealing

Mercury and DNA, Fig. 1 Formation of T-Hg-T metal-mediated base pair by the binding between the Hg^{2+} ion and T:T mismatched base pair in duplex DNA. The specific binding of the Hg^{2+} ion significantly stabilizes naturally occurring T:T mismatched base pair duplex DNA. Following this finding, we can prepare duplex DNAs containing metal-mediated base pairs at the desired sites as well as novel double helical architectures consisting of only T-Hg-T base pairs (Reprinted with permission from Miyake Y et al. (2006) MercuryII-mediated formation of thymine−HgII−thymine base pairs in DNA duplexes. J Am Chem Soc. Copyright (2006) American Chemical Society)

synthesized single-stranded DNA oligonucleotides. Thus, it is possible to examine the binding of the Hg^{2+} ions with the T:T mismatched base pair double-stranded DNA oligonucleotides (Fig. 1). In 1996, Kuklenyik and Marzilli analyzed the binding of the Hg^{2+} ions with a hairpin duplex DNA, d(GCGCTTTTGCGC) (Reference 61 cited in Ono et al. 2011). The Hg^{2+} ions formed an intrastrand T-Hg-T cross-linking between the first and fourth thymine bases in the loop region (underlined) of the hairpin duplex DNA. The Hg^{2+}-hairpin adduct was formed at a molar ratio of 1:1.

Thermal denaturation method has been widely used to examine the thermal stability of nucleic acid structures. The absorbance at 260 nm of a solution containing a duplex DNA is monitored as the solution is gradually heated. In 2006, by the thermal denaturation method, Ono and Tanaka's group found that the thermal stability of a linear duplex DNA containing the T:T mismatched base pair, $d(A_{10}TA_{10}):dT_{21}$, was increased by the addition of the Hg^{2+} ions at a molar ratio of 1:1 (Reference 10 cited in Ono et al. 2011). The thermal stability of the Hg^{2+} ion-mediated T-Hg-T base pair was comparable to that of the Watson-Crick A-T base pair. Other metal ions known to bind with DNA, such as Mg^{2+}, Ca^{2+}, Mn^{2+}, Fe^{2+}, Fe^{3+}, Co^{2+}, Ni^{2+}, Cu^{2+}, Zn^{2+}, Ru^{3+}, Pd^{2+}, Ag^{+}, Cd^{2+}, and Pb^{2+}, did not induce any notable effects on the thermal stability of the T:T mismatched base pair duplex DNA. The stabilizing effect of the Hg^{2+} ion on the T:T mismatched base pair was highly specific.

To examine the base pair specificity of formation of T-Hg-T base pair, in 2010, by the thermal denaturation method, Torigoe and Ono's group examined the effect of the Hg^{2+} ions on the thermal stability of a series of duplex DNAs with 16 different base pairs, 5′-d(GCCCTGCCTGTC**X**CCCAGATCACTG)-3′/3′-d(CGGGACGGACAG**Y**GGGTCTAGTGAC)-5′ (**X:Y** = A:A, A:C, A:G, A:T, C:A, C:C, C:G, C:T, G:A, G:C, G:G, G:T, T:A, T:C, T:G, and T:T) (Reference 13 cited in Ono et al. 2011). They found that the thermal stability of only the duplex DNA with T:T mismatched base pair (**X:Y** = T:T) was specifically increased by the addition of the Hg^{2+} ions at a molar ratio of 1:1. The thermal stability of the duplex DNAs with perfectly matched (**X:Y** = A:T, C:G, G:C and T:A) or with other mismatched base pairs (**X:Y** = A:A, A:C, A:G, C:A, C:C, C:T, G:A, G:G, G:T, T:C, and T:G) did not significantly change by the addition of the Hg^{2+} ions. Only the duplex DNA with T:T mismatched base pair was specifically stabilized by the addition of the Hg^{2+} ions. Combining these results, it is concluded that the combination of the Hg^{2+} ion and the duplex

DNA with T:T mismatched base pair is highly specific for the stabilization of the complex consisting of metal ion and duplex DNA.

Nuclear Magnetic Resonance (NMR) Study of the Binding Between the Hg^{2+} Ion and T:T Mismatched Base Pair

Nuclear magnetic resonance (NMR) spectroscopy has been widely used to determine structures and behaviors of biopolymers, such as proteins and nucleic acids in solution (Billeter et al. 2008). In 2006, Tanaka and Ono's group analyzed the ^{1}H-NMR spectra of a duplex DNA containing two consecutive T:T mismatched base pairs, 5′-d(CGCGTTGTCC)-3′:3′-d (GCGCTTCAGG)-5′, in the absence and presence of the Hg^{2+} ions (Tanaka and Ono 2009; Reference 10 cited in Ono et al. 2011). They found that the imino proton resonances of the two consecutive T:T mismatched base pairs disappeared by the addition of 2.0 molar equivalents of the Hg^{2+} ions. This result indicated that the imino protons of the two consecutive T:T mismatched base pairs in the Hg^{2+}-free duplex were substituted with the two Hg^{2+} ions, which was consistent with the results of Yamane and Davidson in 1961 (Reference 55 cited in Ono et al. 2011) and Katz in 1963 (Reference 56 cited in Ono et al. 2011), described above. They also found that the exchange rates of the Hg^{2+} association and dissociation processes were slow relative to the timescale of the NMR measurements. Once the Hg^{2+} ion bound with the T:T mismatched pair, the Hg^{2+} ion remained for a long time and did not readily transfer to the neighboring T:T mismatched pair, indicating the tightness of the Hg^{2+} ion binding with the T:T mismatched pair. T:T mismatched base pairs in DNA duplexes tightly captured the Hg^{2+} ions to form highly stable T-Hg-T base pairs (Fig. 1).

^{15}N-NMR has been used to study hydrogen bond formation and metalation of nucleobases in ^{15}N-enriched DNA. In 2007, to demonstrate formation of the specific N3-Hg-N3 linkage in the T-Hg-T base pair, Tanaka and Ono's group investigated the ^{15}N NMR spectra of the complex between the Hg^{2+} ions and a duplex DNA containing two consecutive T:T mismatched base pairs, 5′-d(CGCGTTGTCC)-3′:3′-d (GCGCTTCAGG)-5′, labeled with ^{15}N at the N3 position of thymine bases (Tanaka and Ono 2009;

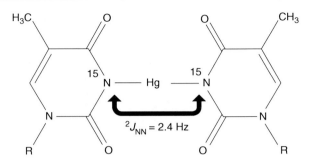

Mercury and DNA, Fig. 2 ^{15}N-^{15}N J coupling constant ($^2J_{NN}$) across the Hg^{2+} ion in T-Hg-T base pair. This observation supports formation of the N3-Hg-N3 hydrogen bond in the T-Hg-T base pair

Reference 12 cited in Ono et al. 2011). To prepare the ^{15}N-labeled DNA, they synthesized N3-labled thymidine nucleoside phosphoramidite and introduced it into DNA oligonucleotide with an automated DNA synthesizer by using the solid-phase cyanoethyl phosphoramidite method. They observed ^{15}N-^{15}N J coupling across the Hg^{2+} ion with a J-coupling constant ($^2J_{NN}$) of 2.4 Hz (Fig. 2). This observation would be evidence supporting the formation of the N3-Hg-N3 linkage in the T-Hg-T base pair. In 2007, Bagno and Saielli theoretically predicted the ^{15}N-^{15}N J coupling constant across the Hg^{2+} ion to be 1.7 Hz by relativistic DFT calculations (Reference 64 cited in Ono et al. 2011), which was consistent with the value observed by ^{15}N-NMR (Fig. 2). Taken together, it is concluded that the Hg^{2+} ion specifically binds with N3 positions of two thymine bases in place of imino protons and bridges two thymine bases to form the T-Hg-T base pair in duplex DNA.

Electrospray Ionization Mass Spectrometry (ESI-MS) Study of the Binding Between the Hg^{2+} Ion and T:T Mismatched Base Pair

Electrospray ionization mass spectrometry (ESI-MS) has been a major tool to investigate reaction mechanisms both in solution and in the gas phase (Reference 10 cited in Ono et al. 2011; Huang et al. 2011). In 2010, Anichina et al. examined the interaction of the Hg^{2+} ions with DNA oligonucleotides rich in thymine by ESI-MS/MS (Reference 43 cited in Ono et al. 2011). Specific interactions were observed for the Hg^{2+} ions with the thymine-rich sequences, in which simultaneous bonding between two thymine bases was indicated. This result is consistent with formation of the T-Hg-T base pair, in which the Hg^{2+} ion covalently coordinated to two thymine bases by replacing two N3 imino protons of the bases. The ESI-MS/MS measurements indicate that the Hg^{2+} ion prefers thymine bases over the other binding sites in DNA oligonucleotides both in solution and in the gas phase.

Isothermal Titration Calorimetry (ITC) Study of the Binding Between the Hg^{2+} Ion and T:T Mismatched Base Pair

Isothermal titration calorimetry (ITC) has been used to study thermodynamic properties of biomolecular complex formation (Ladbury 2010). In 2010, Torigoe and Ono's group examined the thermodynamic properties of the binding between the Hg^{2+} ion and T:T mismatched base pair in a duplex DNA, 5′-d(GCCCTGCCTGTCTCCCAGATCACTG)-3′/3′-d(CGGGACGGACAGTGGGTCTAGTGAC)-5′ (T:T mismatched base pair in bold) by ITC (Reference 13 cited in Ono et al. 2011). They found that the Hg^{2+} ion specifically bound with the T:T mismatched base pair at a molar ratio of 1:1, with a binding constant of 10^6 M^{-1}. The observed molar ratio of 1:1 was consistent with the NMR results described above. Binding with duplex DNA has been previously examined for many metal ions, such as Cr^{3+}, Cr^{6+}, Tl^+, Fe^{2+}, Fe^{3+}, Al^{3+}, Mn^{2+}, and Cd^{2+}. The observed magnitude of the binding constant of 10^6 M^{-1} was significantly larger than those previously reported for the nonspecific binding between metal ions and duplex DNA. In addition, negative enthalpy change and positive entropy change were observed for the specific binding between the Hg^{2+} ion and the T:T mismatched base pair. Because both the observed negative enthalpy change and positive entropy change were favorable for the specific binding between the Hg^{2+} ion and the T:T mismatched base pair, formation of the T-Hg-T base pair in the duplex DNA was driven by both the negative enthalpy change and positive entropy change. As described above in the ^{15}N-NMR results of Tanaka and Ono's group in 2007 (Reference 12 cited in Ono et al. 2011), N3-Hg-N3 bond in the T-Hg-T base pair may have covalent characteristics, and the bond formation should facilitate formation of the T-Hg-T base pair by negative enthalpy change. The negative enthalpy change observed by ITC may be mainly driven by the negative binding enthalpy change from formation of the N3-Hg-N3 bond. On the other hand, structure of the T-Hg-T base pair in the duplex DNA suggests that the mercury atom in the T-Hg-T base pair may be sheltered from water molecules. Water molecules assembled around the Hg^{2+} ions may become free when the Hg^{2+} ions are inserted into the T:T mismatched base pair, which should accelerate formation of the T-Hg-T base pair by positive entropy change via dehydration of the Hg^{2+} ions. The positive entropy change observed by ITC may mainly result from the positive dehydration entropy change from the release of structured water molecules surrounding the Hg^{2+} ions. Taken together, a possible scheme is proposed for the specific binding between the Hg^{2+} ion and the T:T mismatched base pair (Fig. 3). The Hg^{2+} ion surrounded by structured water molecules can be dehydrated with significant contribution of the positive dehydration entropy change. The protons at the N3 position of the two thymine bases in the T:T mismatched base pair can then be released. The dehydrated Hg^{2+} ion can bind with the two deprotonated thymine bases to form the N3-Hg-N3 bond with significant contribution from the negative binding enthalpy change. Thymine deprotonation could be brought about at neutral pH, below the pH at which thymine becomes deprotonated (the pKa of the proton at N3 position of thymine is 9–10), because initial Hg^{2+} ion binding to O4 and/or O2 positions of the canonical tautomer of thymine can acidify the proton at N3 position. The deprotonation could be brought about by Hg-OH, which may be undoubtedly present at physiological pH, because $[Hg(H_2O)_x]^{2+}$ is a strong Lewis acid, followed by metal migration from O4 and/or O2 to N3 and reprotonation of O4 and/or O2.

Application of T-Hg-T Base Pair Formation to Mercury Sensors

Pollution with heavy metal ions may have severe effects on human health and the environment (Fu and Wang 2011). Mercury contamination is widespread and arises from a variety of artificial sources, such as the combustion of solid waste and fuels. When mercury enters the marine environment, bacteria convert inorganic mercury into methylmercury, which may be neurotoxic and a cause of mercury-related diseases. Thus, development of mercury sensors to detect mercury ions in the environment is quite important to increase understanding of mercury pollution.

Mercury and DNA, Fig. 3 Possible scheme for the specific binding between the Hg^{2+} ions and the T:T mismatched base pair. The Hg^{2+} ions surrounded by structured water molecules can be dehydrated, and the protons at the N3 position of the two thymine bases in T:T base pair can be released. The dehydrated Hg^{2+} ions can then bind with the two deprotonated thymine bases to form N3-Hg-N3

i) Dehydration of Hg^{II} ion
ii) Deprotonation of T bases
iii) Binding of dehydrated Hg^{II} ion with deprotonated T bases

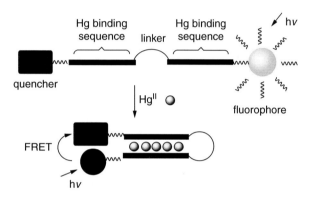

Mercury and DNA, Fig. 4 Schematic representation of the hairpin structure induced in 5′-quencher-oligonucleotide-fluorophore-3′ by Hg^{2+} ion-mediated T-Hg-T base pair formation, which results in the fluorescence resonance energy transfer (FRET)-mediated quenching of fluorophore emission

In 2004, Ono and Togashi developed a novel DNA-based mercury sensor that selectively detects the Hg^{2+} ions in aqueous solution by the specific binding between the Hg^{2+} ion and T:T mismatched base pair (Fig. 4) (Reference 14 cited in Ono et al. 2011). Fluorescence resonance energy transfer (FRET) has been widely used to study a variety of biological phenomena that produce changes in molecular structure. The novel mercury sensor was based on the FRET analyses. The mercury sensor was a single-stranded DNA oligonucleotide composed of two parts, the thymine-rich mercury-binding sequence at both end regions and the linker sequence (underlined) in the middle region, and labeled with a fluorophore (6-fluorescein) at the 3′-end and a quencher (dabcyl) at the 5′-end, (5′-quencher-d(TTCTTTCTTCCCCTTGTTTGTT)-fluorophore -3′) (Fig. 4). Addition of the Hg^{2+} ion formed the T-Hg-T base pairs between the Hg^{2+} ion and two thymine bases from two mercury-binding sequences, which may give rise to a hairpin structure. As a result of the decreased distance between the fluorophore and the quencher in the hairpin structure, fluorescence intensity may be decreased due to the FRET-mediated quenching between the fluorophore and the quencher (Fig. 4). In fact, the intensity of fluorophore emission decreased as the concentration of the Hg^{2+} ions increased. A linear correlation between the fluorescence intensity and the concentration of the Hg^{2+} ions was observed in the concentration range of 40–100 nM. The sensitivity of this DNA-based mercury sensor is greater than those of the previously reported small molecule-based mercury sensors.

In 2007, Lee et al. developed another novel mercury sensor using DNA-functionalized gold nanoparticles, which was also based on the specific binding between the Hg^{2+} ion and T:T mismatched base pair (Reference 85 cited in Lin et al. 2011). They prepared two types of gold nanoparticles, each functionalized with different thiolated-DNA sequences (probe A: 5′-HS-C_{10}-A_{10}-**T**-A_{10}-3′, probe B: 5′-HS-C_{10}-T_{10}-**T**-T_{10}-3′), which are complementary except for a single T:T mismatch (shown in bold). When the duplex was not formed from probes A and B, aggregates may not be formed in the mixture of the two types of gold nanoparticles. When the Hg^{2+} ion specifically bound with the T:T mismatch to form the duplex from probes A and B, aggregates of gold nanoparticles may be formed with a concomitant red-to-purple color change. The presence

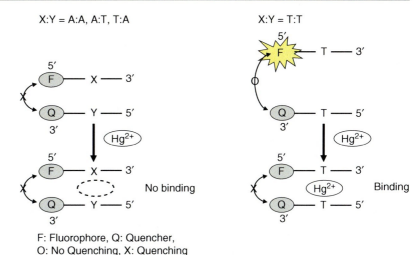

Mercury and DNA, Fig. 5 Schematic representation of the working principles of the probe composed of 5′-end-fluorophore(F)-labeled and 3′-end-quencher(Q)-labeled strands to detect single nucleotide polymorphisms (SNPs) by T-Hg-T base pair formation. The detailed principle is shown in the text

of the Hg^{2+} ion was detected by color change of gold nanoparticle solution. The limit of the Hg^{2+} ion detection for this system was approximately 100 nM.

Application of T-Hg-T Base Pair Formation to Detection of Single Nucleotide Polymorphisms (SNPs)

Single nucleotide polymorphisms (SNPs) are base differences in the human genome (Bichenkova et al. 2011). These differences are favorable markers for genetic factors including those associated with risks of complex diseases (such as cancers, cardiovascular diseases, and diabetes) and individual responses to drugs. When two duplex DNAs with different types of SNPs are mixed and reannealed, the two novel heteroduplexes containing mismatched base pairs are formed in addition to the two initial perfectly matched homoduplexes. Heteroduplex analysis recognizing the newly formed mismatched base pairs is useful for SNP detection. Various strategies to detect the mismatched base pairs, such as nondenaturing gel electrophoresis and addition of small organic molecules to recognize mismatched base pairs, were devised due to the potential applications of SNPs. However, they were not always convenient and accurate.

In 2011, Torigoe et al. proposed a novel strategy to detect the mismatched base pairs by the specific binding between the Hg^{2+} ion and T:T mismatched base pair (Fig. 5) (Reference 20 cited in Ono et al. 2011). The novel strategy is based on the FRET analyses described above. Addition of the Hg^{2+} ions to T:T mismatched base pair heteroduplex DNA labeled with a fluorophore (6-carboxyfluorescein) (F) at the 5′-end of one strand and a quencher (dabcyl) (Q) at the 3′-end of the complementary strand assembled both strands of the heteroduplex DNA due to formation of the T-Hg-T base pair (Fig. 5). The assembly of both strands decreased the distance between F and Q attached with each strand, which may result in the decrease in the intensity of fluorophore emission (Fig. 5). Thus, the addition of the Hg^{2+} ions to dual-labeled T:T mismatched base pair heteroduplex DNA decreased the fluorescence intensity due to the FRET-mediated quenching between F and Q. In contrast, due to lack of the specific binding between the Hg^{2+} ion and the perfectly matched or the other kinds of mismatched base pairs, no significant decrease in the intensity of fluorophore emission and no FRET-mediated quenching was observed by the addition of the Hg^{2+} ions to the dual-labeled perfectly matched homoduplex DNAs and the dual-labeled other kinds of mismatched base pair heteroduplex DNAs (Fig. 5). The intensity of fluorophore emission of only the dual-labeled T:T mismatched base pair heteroduplex DNA was specifically decreased by the addition of the Hg^{2+} ions. It is proposed that addition of the Hg^{2+} ions could be a convenient and accurate strategy to detect T:T mismatched base pairs in heteroduplex DNA. This novel strategy might make the heteroduplex analysis easy and eventually lead to better SNP detection.

In 2008, Lin et al. proposed another novel strategy to detect SNPs based on the specific binding between

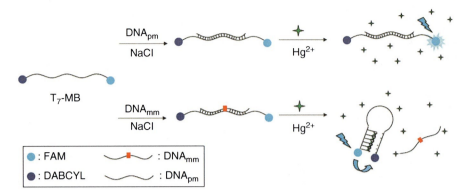

Mercury and DNA, Fig. 6 Schematic representation of the working principles of the fluorophore (FAM) and quencher (DABCYL)-labeled single-stranded probe (T_7-MB) to detect single nucleotide polymorphisms (SNPs) by T-Hg-T base pair formation. Target DNA (DNA_{pm}) is perfectly matched with the probe, T_7-MB. SNP-containing target DNA (DNA_{mm}) forms a mismatched base pair with the probe, T_7-MB. The detailed principle is shown in the text

the Hg^{2+} ion and T:T mismatched base pair (Fig. 6) (Reference 193 cited in Yang and Zhao 2010). They prepared a DNA oligonucleotide probe (T_7-MB) containing a stem of a pair of 7-mer thymidine bases that interact with the Hg^{2+} ion and a loop of 19-mer DNA bases that recognize target DNA sequence. T_7-MB was labeled with a fluorophore (FAM: carboxyfluorescein) at the 5′-end and a quencher (DABCYL: dabcyl) at the 3′-end. The dual-labeled T_7-MB is a random-coil structure that may change into a folded structure by addition of the Hg^{2+} ions due to formation of the T-Hg-T base pair. As a result of the decreased distance between FAM and DABCYL, fluorescence intensity may become weaker due to the FRET-mediated quenching between FAM and DABCYL. When the binding between target DNA (DNA_{pm}) and the loop of the dual-labeled T_7-MB without forming any mismatched base pair is stronger than that between the Hg^{2+} ion and the 7-mer thymidine bases in the stem, a duplex DNA consisting of DNA_{pm} and T_7-MB may form rather than the folded structure of T_7-MB. In this case, FAM and DABCYL of T_7-MB may be separated far apart, resulting in strong fluorescence intensity. In contrast, when the binding between SNP-containing target DNA (DNA_{mm}) and the loop of the dual-labeled T_7-MB with forming a mismatched base pair is weaker than that between the Hg^{2+} ion and the 7-mer thymidine bases in the stem, the folded structure of T_7-MB may form. In this case, fluorescence intensity may be weak due to the FRET-mediated quenching between FAM and DABCYL. This novel strategy could be convenient and accurate to detect SNPs and eventually lead to better SNP detection.

Incorporation of T-Hg-T Base Pair by DNA Polymerases

DNA polymerase is an enzyme that helps catalysis in the polymerization of deoxyribonucleotides into a DNA strand. It has been considered that the number and strength of hydrogen bonds in a base pair may determine the efficiency and fidelity of DNA polymerases. In 2010, Urata's group examined a primer-extension reaction of DNA polymerases (Klenow fragment, KOD Dash, and Taq DNA polymerases) in the presence of the Hg^{2+} ions (Fig. 7) (Reference 27 cited in Ono et al. 2011). They found that, in the presence of the Hg^{2+} ions, the DNA polymerases incorporated thymidine 5′-triphosphate (TTP) at the site opposite thymine in a template strand through formation of the T-Hg-T base pair and made a phosphodiester bond to elongate the primer strand. The DNA polymerases recognized the unusual metal-coordinated type of the T-Hg-T base pair and elongated the primer. These results suggest a potential mechanism for the mutagenic activity of the Hg^{2+} ions. Further, these results open possibility for the metal-ion-mediated enzymatic incorporation of a variety of artificial bases into oligonucleotides to expand the genetic alphabet, and for the functional switching of DNA oligonucleotides through modification involving metal ions.

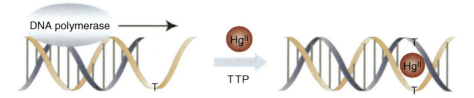

Mercury and DNA, Fig. 7 The incorporation of T-Hg-T metal-mediated base pair by DNA polymerases. In the presence of Hg^{2+} ions, DNA polymerases incorporated thymidine 5′-triphosphate (TTP) at the site opposite thymine in a template strand and made a phosphodiester bond to elongate the primer strand. The T-Hg-T unusual metal-mediated base pair was recognized by the DNA polymerases, which went on to synthesize the full-length product

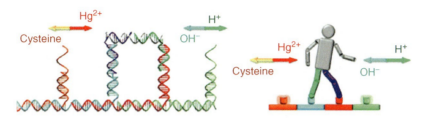

Mercury and DNA, Fig. 8 The assembly of "bipedal walker" using DNA constructs. The DNA-based nanomachine is activated by Hg^{2+}/cysteine trigger. The forward "walking" is activated by Hg^{2+} ions, using the T-Hg-T base pair formation as the DNA translocation driving force. The backward "walking" is activated by cysteine, trigger that destroys the T-Hg-T base pair (Reprinted with permission from Wang Zhen-Gang et al. (2011) DNA machines: bipedal walker and stepper. Nano Lett. Copyright (2011) American Chemical Society)

DNA-Based Nanomachines by T-Hg-T Base Pair Formation and Destruction

The interactions between metal ions and nucleic acids have wide variety of potential applications in nanotechnology including the design of biomolecular nanomachines and nanodevices. In 2011, Wang et al. designed DNA-based nanomachines, a "bipedal walker" (Fig. 8) and a "bipedal stepper," activated by Hg^{2+}/cysteine triggers (Reference 29 cited in Ono et al. 2011). The bipedal walker (Fig. 8) was activated on a DNA template consisting of four nucleic acid footholds. The forward "walking" of the DNA on the template track was activated by the Hg^{2+} ions using the specific binding between the Hg^{2+} ions and T:T mismatched base pairs to form the T-Hg-T base pairs as the DNA translocation driving forces. The backward "walking" was activated by cysteine, trigger that destroyed the T-Hg-T base pairs. Similarly, the bipedal stepper was activated on a circular DNA template consisting of four tethered footholds. With the Hg^{2+}/cysteine triggers, clockwise or anticlockwise stepping was demonstrated.

Conclusion

The Hg^{2+} ion specifically binds with N3 positions of two thymine bases in place of imino protons and bridges two thymine bases to form the T-Hg-T base pair in duplex DNA. The combination of the Hg^{2+} ion and the T:T mismatched base pair is highly specific for the stabilization of the complex consisting of metal ion and duplex DNA. The binding constant of 10^6 M^{-1} for the specific binding is significantly larger than that for the nonspecific binding between metal ions and duplex DNA. The T-Hg-T base pair formation has the ability to be applied to various fields, such as mercury sensor, detection of single nucleotide polymorphisms, enzymatic incorporation by DNA polymerases, and DNA-based nanomachines. Oligonucleotides for the T-Hg-T base pair formation involving natural bases are commercially available without time-consuming synthesis of chemical modification and are, therefore, able to be used in many laboratories. The T-Hg-T base formation has the potential to be used for further applications in various scientific and technological fields.

Cross-References

▶ Gold and Nucleic Acids
▶ Mercury and Low Molecular Mass Substances
▶ Mercury Toxicity
▶ Mercury, Physical and Chemical Properties

References

Bichenkova EV, Lang Z, Yu X et al (2011) DNA-mounted self-assembly: new approaches for genomic analysis and SNP detection. Biochim Biophys Acta 1809:1–23

Billeter M, Wagner G, Wüthrich K (2008) Solution NMR structure determination of proteins revisited. J Biomol NMR 42:155–158

Fu F, Wang Q (2011) Removal of heavy metal ions from wastewaters: a review. J Environ Manage 92:407–418

Huang MZ, Cheng SC, Cho YT et al (2011) Ambient ionization mass spectrometry: a tutorial. Anal Chim Acta 702:1–15

Ladbury JE (2010) Calorimetry as a tool for understanding biomolecular interactions and an aid to drug design. Biochem Soc Trans 38:888–893

Lin Y-W, Huang C-C, Chang H-T (2011) Gold nanoparticle probes for the detection of mercury, lead and copper ions. Analyst 136:863–871

Ono A, Torigoe H, Tanaka Y et al (2011) Binding of metal ions by pyrimidine base pairs in DNA duplexes. Chem Soc Rev 40:5855–5866

Saenger W (1984) Principles of nucleic acid structure. Springer, New York

Tanaka Y, Ono A (2009) Structural studies on the mercuryII-mediated T–T base-pair using NMR spectroscopy. In: Hadjiliadis N, Sletten E (eds) Metal complex-DNA interactions. Wiley, New York

Yang Y, Zhao L (2010) Sensitive fluorescent sensing for DNA assay. TrAC Trends Anal Chem 29:980–1003

Mercury and Immune Function

K. Michael Pollard
Department of Molecular and Experimental Medicine, The Scripps Research Institute, La Jolla, CA, USA

Synonyms

Hg

Definition

Mercury is a heavy, silver-white, highly toxic metallic element that is liquid at room temperature. It is used in barometers, thermometers, pesticides, pharmaceutical preparations, reflecting surfaces of mirrors, dental fillings, and compact fluorescent light (CFL) bulbs. Different forms of mercury have been used to study its effects on the immune system including mercuric chloride, monomethylmercury, thimerosal, mercury vapor, and mercury containing dental amalgam.

Introduction

This entry begins with a brief discussion of the effects of mercury on the immune system. For more details the reader should refer to the cited literature, in particular the reviews by Pollard et al. 2010, and Schiraldi and Monestier 2009. Following these introductory remarks is a detailed description of experiments examining the interaction between mercury and the nucleolar protein fibrillarin, and the role that this interaction plays in mercury-induced autoimmunity. It should be noted that the interaction between mercury and fibrillarin does not generate a classical metalloprotein (which is generally defined as a protein that contains a metal ion cofactor which is required for the protein's biological activity) as there is no evidence that fibrillarin requires mercury for it biological function as a ribosomal RNA methyltransferase. Rather the interaction between mercury and fibrillarin generates a metal-protein complex which is immunologically different from mercury-free fibrillarin. Additionally, proteolysis of the mercury-fibrillarin complex is specific to this xenobiotic and has significant differences to mercury-free fibrillarin. These findings show that interaction with mercury changes both the molecular and immunological properties of fibrillarin which may explain why fibrillarin is the dominant target of the autoantibody response elicited by mercury exposure.

Mercury and Immunity

The effects of mercury on the immune system span a spectrum of responses ranging from immunosuppression to autoimmunity (Pollard and Hultman 1997; Schiraldi and Monestier 2009). While the mechanisms responsible for mercury-mediated effects on the immune system are not entirely clear, there is evidence that genetics and the intensity and duration of mercury exposure play significant roles. Although the role that metalloproteins play in the effects of mercury on the immune system has not been extensively investigated, there is evidence that mercury does cause immune disregulation at the molecular level.

Mercury ions are highly reactive and bind primarily to thiol groups, although hydroxyl, carboxyl, and phosphoryl groups may also bind mercury. Such reactivity may explain the ability of mercury to cause the aggregation of immunologically important cell surface receptors including CD3, CD4, CD45, and Thy-1 on T cells (Pollard and Hultman 1997). Mercury-mediated cell surface receptor aggregation leads to intracellular signaling and tyrosine phosphorylation of a number of nuclear proteins which influence cellular responses ranging from lymphocyte proliferation to cell death (Pollard and Hultman 1997). Mercury-induced lymphocyte activation and proliferation may occur via perturbation of CD95 (Fas)-mediated apoptotic cell death, thereby enhancing cell survival. The effect is argued to proceed through disruption of both the death-inducing signaling complex (DISC) and death receptor–mediated caspase-3 activation due to the ability of low concentrations of mercury to dissociate preassembled Fas receptor complexes required for DISC formation. This mechanism suggests that mercury exposure allows autoreactive T cells to survive Fas-Fas ligand–mediated cell death (Pollard et al. 2010; Schiraldi and Monestier 2009). Other studies have suggested that mercury may act via attenuation of T-cell receptor (TCR) signaling, allowing self-reactive cells to escape elimination during T-cell selection. Experiments have shown that mercury can inhibit TCR activation of Ras and ERK MAP kinase which are associated with negative selection of T cells. Failure of Ras activation is due to lack of phosphorylation of LAT (linker for activation of T cells) and the lack of activation of the LAT reactive tyrosine kinase ZAP-70. The failure of ZAP-70 activation is due to mercury-mediated inhibition of phosphorylation of lymphocyte specific protein tyrosine kinase (Lck). This mechanism would hinder early steps in TCR signal transduction (Pollard et al. 2010). These effects highlight mercury's capacity to impinge on molecular pathways of T-cell survival and selection, and identify proteins that may interact with mercury. However studies characterizing these mercury–protein interactions have not yet been done.

Mercury exposure can result in immune suppression leading to greater sensitivity to infectious diseases. Both humoral and cell-mediated immune responses can be affected. While the mechanism of mercury-mediated immune suppression remains unclear, animal studies have suggested that genetic background may be a major factor (Pollard and Hultman 1997). The importance of genetics in regulating the effects of mercury on the immune system has been clearly shown in animal studies as different inbred animal strains display dissimilar degrees of responsiveness.

Mercury-Induced Autoimmunity

In stark contrast to immunosuppression, mercury exposure can stimulate the immune system causing hypersensitivity, autoimmunity, and nephrotic syndrome (Pollard and Hultman 1997; Schiraldi and Monestier 2009). Exposure to mercury in an occupational setting has been associated with increased risk of the autoimmune disease systemic lupus erythematosus (SLE) (Cooper et al. 2010). Exposure to mercury during gold mining results in the production of autoantibodies and pro-inflammatory cytokines while mercury exposure from skin care products leads to membranous nephropathy (Pollard et al. 2010; Schiraldi and Monestier 2009). The human populations at risk of mercury-induced autoimmunity (HgIA) are sizable and diverse. For example, mercury is used by artisanal miners in 55 countries and is a major source of mercury contamination in the associated rural communities. It is believed that between 10 and 15 million people, including women and children, are directly involved in artisanal gold mining (Veiga et al. 2006). Mercury containing skin care products are widely used, with one commercially successful product available in 35 countries (Pollard and Hultman 2007). Although regulated in the USA, use of mercury containing skin-lightening creams has been documented in Hispanic women on the Texas-Mexico border. Use of mercury containing skin-lightening creams accounted for reportable levels of urinary mercury (≥ 20 μg/L) in 9 of 13 individuals identified during a screen of 1,840 adult New Yorkers (McKelvey et al. 2011).

Animal model studies have revealed considerable insight into the mechanisms of HgIA (Pollard et al. 2010; Schiraldi and Monestier 2009). In mice, HgIA can be induced by subcutaneous injection or oral ingestion of inorganic mercury, inhalation of mercury vapor, and implantation of mercury containing dental amalgam. Disease severity depends upon the murine strain under investigation. Several strains, particularly the DBA/2, are nonresponsive while others, such as the SJL/J and B10.S, exhibit immune cell activation, lymphadenopathy, elevated immunoglobulin levels, autoantibodies, and deposition of immune complexes in kidney and other tissues (Pollard et al. 2010). These features of murine mercury-induced autoimmunity (mHgIA) are consistent with the systemic autoimmunity of SLE.

Genetic studies have shown that the absence of some genes (e.g., interleukin (IL)-4) appears to have little effect on disease expression, while others influence specific facets of the mHgIA phenotype (e.g., β-2-microglobulin, CD86.). Deficiency of some genes results in suppression of disease (e.g., interferon (IFN)-γ, CD28, and CD40 ligand), while lack of others exacerbate disease (e.g., Daf1). Such studies reveal that mHgIA and idiopathic murine SLE share a number of effector genes in common including those for IFN-γ, CD28, CD40 ligand, and Daf1 (Pollard et al. 2010; Schiraldi and Monestier 2009).

Mercury exposure of SLE-prone mice (e.g., NZBWF1, BXSB, and MRL-$Fas^{+/+}$) leads to acceleration of the underlying systemic autoimmune disease. Studies in female BXSB mice show that this acceleration is dose dependent and has features of idiopathic disease. Furthermore, tissue mercury levels of BXSB mice exposed to 0.4 μg $HgCl_2$/injection fell within the range found in non-occupationally exposed humans; yet, these mice have an accelerated disease phenotype including earlier appearance of anti-chromatin autoantibodies and proteinuria. Thus, tissue levels of mercury consistent with environmental exposure can be associated with exacerbations of autoimmunity in genetically susceptible hosts.

The autoantibody responses elicited by exposure to mercury are primarily against nuclear components especially chromatin and the small nucleolar ribonucleoprotein (snoRNP) component fibrillarin but can include antibodies to other components of snoRNP complexes, as well as to other nuclear and cytoplasmic antigens (Pollard and Hultman 1997; Pollard et al. 2010; Schiraldi and Monestier 2009). The most characteristic autoantibody response is the genetically restricted autoantibody response against fibrillarin, a nucleolar protein component of the box C/D small nucleolar ribonucleoprotein (snoRNP) complexes involved in methylation of, mainly, ribosomal RNA. Although induced in specific mouse strains following mercury exposure, anti-fibrillarin autoantibodies have also been found in human patients with SLE and a subset of scleroderma patients (Pollard et al. 1997). Human and murine anti-fibrillarin antibodies are both under the control of major histocompatibility complex (MHC) genes and share recognition of a conserved conformational antigenic determinant (Pollard and Hultman 1997). Autoantibodies in scleroderma and mHgIA also recognize other protein components of snoRNP complexes, particularly those of the U3 snoRNP, suggesting that RNA/protein complexes may be the source of immunogenic material for the anti-snoRNP response in both mercury-induced and idiopathic responses against snoRNPs (Pollard et al. 2010).

Mercury-Fibrillarin Interaction

The mechanism by which a simple chemical such as mercuric chloride elicits an autoantibody response that predominantly targets a nucleolar protein remains uncertain. The most likely possibility is that mercury interacts directly with fibrillarin and/or fibrillarin-containing snoRNPs resulting in potentially immunogenic material. Studies have been done examining the interaction between mercury and fibrillarin (Pollard et al. 1997, 2000). At mercury concentrations, which result in cell death, there is a loss of fibrillarin antigenicity and a change in the migration of 34-kDa fibrillarin to a faster migrating form of 32 kDa under nonreducing SDS-PAGE (Pollard et al. 1997). Exposure of isolated nuclei or in vitro produced fibrillarin to mercuric chloride results in the same increase in the faster migrating 32-kDa form. Prior incubation of isolated nuclei with iodoacetamide followed by addition of mercuric chloride results in the 34-kDa band only showing that alkylation of cysteines inhibited the ability of mercury to modify the migration of fibrillarin under nonreducing SDS-PAGE. To confirm that the 32-kDa band found after incubating isolated nuclei with mercuric chloride is due to interaction of mercury with cysteines, nuclei were incubated with mercuric chloride followed by iodoacetamide. Using nonreducing SDS-PAGE, it was found that there is retention of material at 32 kDa, indicating that mercury was protecting cysteines from alkylation and suggesting that the cysteines of fibrillarin are directly involved in binding Hg^{2+} ions (Pollard et al. 1997).

To examine the role of cysteines in the mercury-induced modification of fibrillarin, the mouse cDNA sequence of fibrillarin was used to produce cDNA in which cysteines at position 105 and/or 274 were mutated to alanines. Analysis of fibrillarin produced from these mutant clones showed that absence of one or both cysteines does not result in material migrating at 32 kDa regardless of whether mercury is present or not. In contrast, non-mutated fibrillarin incubated with mercuric chloride shows a broad band of material migrating at 32 kDa under nonreducing SDS-PAGE conditions. These observations clearly demonstrate that the mercury-induced modification of fibrillarin requires both cysteines and strongly support the argument that the disulfide-bonded form of fibrillarin migrates in nonreducing SDS-PAGE at 32 kDa. In addition, the presence of mercuric chloride leads to accumulation of the 32-kDa form, possibly through cross-linking of the free sulfhydryls of the cysteines by mercury.

To determine if mercury modification of fibrillarin affects the interaction of anti-fibrillarin autoantibodies and fibrillarin, wild-type and Cys → Ala mutated mouse fibrillarin were used in immunoprecipitation studies (Pollard et al. 1997). All forms of fibrillarin were immunoprecipitated by anti-fibrillarin autoantibodies, indicating that mutation of either or both cysteines does not influence immunoreactivity. In contrast, incubation of wild-type fibrillarin with $HgCl_2$, reduces immunoprecipitation by anti-fibrillarin autoantibodies. Loss of immunoreactivity is not due to interaction of autoantibodies with mercury alone, as mutant fibrillarin is immunoprecipitated by anti-fibrillarin autoantibodies in the presence and the absence of $HgCl_2$. Monoclonal and polyclonal mouse anti-fibrillarin autoantibodies as well as human anti-fibrillarin autoantibodies fail to react with mercury-modified fibrillarin confirming that non-mercury-modified fibrillarin is the form recognized by autoantibodies. This failure of autoantibodies to recognize mercury-modified fibrillarin demonstrates that the anti-fibrillarin response in mHgIA is directed against mercury-free fibrillarin and not against a foreign antigen comprised of a complex of mercury and fibrillarin. Thus, mercury-modified fibrillarin is not a B-cell antigen in mHgIA.

Proteolysis of Mercury

Proteolysis of nuclear autoantigens during cell death results in distinctive cleavage patterns that differ between apoptotic and nonapoptotic cell death. This generation of unique protein fragments during cell death suggests a mechanism for the production of novel antigenic determinants that may lead to autoimmune responses (Pollard et al. 2000). As noted above, mercury-induced cell death is accompanied by modification of the molecular properties of fibrillarin. Moreover, as mercury-protein interaction has been shown to alter protease sensitivity, it is possible that mercury modification of fibrillarin may influence proteolytic processing of a mercury-fibrillarin complex, leading to the production of cryptic T-cell determinants. The observation that lysates of peritoneal cells from $HgCl_2$-exposed mice enhance inflammation in $HgCl_2$-primed mice suggests that mercury-exposed antigen-presenting cells such as macrophages could be a source of antigenic fibrillarin. To test this possibility, the proteolysis of fibrillarin was examined in macrophages following mercury-induced cell death. Addition of radiolabeled fibrillarin to lysates from mercury-killed macrophages led to proteolytic cleavage of fibrillarin that was specific to this xenobiotic. A prominent feature of mercury-induced cell death and proteolysis is the generation of a 19-kDa fragment of fibrillarin that is not found following apoptotic cell death nor is it associated with nonapoptotic cell death due to stimuli other than mercury. Thus, mercury-induced cell death is associated with a protease activity not found in other forms of induced cell death. Proteolysis of fibrillarin lacking cysteines, and therefore unable to bind mercury, also produces the 19-kDa fragment, suggesting that a mercury-fibrillarin interaction is not necessary for abnormal cleavage of fibrillarin. Immunization of mice with the 19-kDa fragment results in anti-fibrillarin autoantibodies that recognized evolutionarily conserved determinants of fibrillarin, a feature of $HgCl_2$-induced anti-fibrillarin autoantibodies. Immunization with full-length (34 kDa) fibrillarin failed to induce autoantibodies. This highlights the potential of xenobiotic-induced cell death to produce novel protein fragments that stimulate self-reactivity that differs from that elicited by full length, native, protein. These experiments suggest that cell death following exposure to an autoimmunity-inducing xenobiotic such as mercury can lead to the generation of novel protein fragments that may serve as sources of antigenic determinants for self-reactive T lymphocytes.

This study (Pollard et al. 2000) and others (Griem et al. 1996) demonstrate that immunization with metal-protein complexes can lead to T-cell determinants that do not include the metal ion. Instead, the presence of the metal ion appears to influence the generation of antigenic determinants so that novel immunogens become available for immune recognition. The ability of mercury-induced cell death to alter the proteolysis of fibrillarin suggests a mechanism whereby an autoimmunity-inducing xenobiotic might generate unique fragments from a self-antigen. The finding that a protein fragment mimicking such a cleavage product can elicit autoantibodies with novel antigenic specificities suggests that altered proteolysis may contribute to the breaking of self-tolerance.

Acknowledgments The author acknowledges funding by grants ES007511 and ES014847 from the National Institute of Environmental Health Sciences, National Institutes of Health, USA.

Cross-References

▶ Lead and Immune Function
▶ Metallothioneins and Mercury

References

Cooper GS, Wither J, Bernatsky S, Claudio JO, Clarke A, Rioux JD, Fortin PR (2010) Occupational and environmental exposures and risk of systemic lupus erythematosus: silica, sunlight, solvents. Rheumatology (Oxford) 49:2172–2180

Griem P, Panthel K, Kalbacher H, Gleichmann E (1996) Alteration of a model antigen by Au(III) leads to T cell sensitization to cryptic peptides. Eur J Immunol 26:279–287

McKelvey W, Jeffery N, Clark N, Kass D, Parsons PJ (2011) Population-based inorganic mercury biomonitoring and the identification of skin care products as a source of exposure in New York City. Environ Health Perspect 119:203–209

Pollard KM, Hultman P (1997) Effects of mercury on the immune system. Met Ions Biol Syst 34:421–440

Pollard KM, Hultman P (2007) Skin-lightening creams are a possible exposure risk for systemic lupus erythematosus: comment on the article by Finckh et al. Arthritis Rheum 56:1721 author reply 1721–1722

Pollard KM, Lee DK, Casiano CA, Bluthner M, Johnston MM, Tan EM (1997) The autoimmunity-inducing xenobiotic mercury interacts with the autoantigen fibrillarin and modifies its molecular and antigenic properties. J Immunol 158:3521–3528

Pollard KM, Pearson DL, Bluthner M, Tan EM (2000) Proteolytic cleavage of a self-antigen following xenobiotic-induced cell death produces a fragment with novel immunogenic properties. J Immunol 165:2263–2270

Pollard KM, Hultman P, Kono DH (2010) Toxicology of autoimmune diseases. Chem Res Toxicol 23:455–466

Schiraldi M, Monestier M (2009) How can a chemical element elicit complex immunopathology? Lessons from mercury-induced autoimmunity. Trends Immunol 30:502–509

Veiga MM, Maxson PA, Hylander LD (2006) Origin and consumption of mercury in small-scale mining. J Clean Prod 14:436–447

Mercury and Lead, Effects on Voltage-Gated Calcium Channel Function

Brenda Marrero-Rosado, Sara M. Fox, Heidi E. Hannon and William D. Atchison
Department of Pharmacology and Toxicology, Michigan State University, East Lansing, MI, USA

Synonyms

Calcium and apoptosis; Calcium and neurotransmitter

Definitions

Ion channel function refers to the opening of a transmembrane protein pore as a result of a stimulus, which causes the influx or efflux of ions to/from the cell. Specifically, voltage-gated Ca^{2+} channels (VGCCs) are more permeable to Ca^{2+} ions and the opening of the pore results from a change in membrane potential. A substantially higher extracellular concentration of Ca^{2+} and negative membrane potential inside the cell drive the rapid influx of Ca^{2+} through activated voltage-gated Ca^{2+} channels.

Prologue

The work described in this entry is the result of research by many investigators in the field, but outside of the referenced citations. These investigators are acknowledged as being crucial to the field. Unfortunately, as required by the publishers of this book, entries were only to cite up to ten review articles or book chapters. Consequently, it was not possible to give credit to those who actually performed the work in the individual studies. This is not meant in any way to discount their contributions, and readers are referred to the cited book chapters or review articles to ascertain the actual authors whose work is described in this entry and cited in the references listed.

Introduction

A wide variety of nonphysiological heavy metals can be found in the environment. Lead (Pb^{2+}) and mercury (Hg^{2+}) are of particular importance due to their ability to cause irreparable damage to the nervous system. The effects of these metals on the central nervous system have been studied for decades. Pb^{2+} poisoning can occur as a result of exposure to Pb^{2+}-based paint, drinking water from sources in which Pb^{2+} plumbing is used, or inhaling dust or gases that contain Pb^{2+}. In adults, inorganic Pb^{2+} is more easily absorbed in the lungs compared to the gastrointestinal tract; children, on the other hand, not only show a higher Pb^{2+} uptake in the gastrointestinal tract, but also have an underdeveloped blood–brain barrier (BBB). The ability of Pb^{2+} to cross the BBB renders children more susceptible to Pb^{2+}-induced encephalopathies, as well, potentially, to developmental defects with long-lasting consequences. Some of these may be subtle, and subject to exacerbation by other environmental exposures, or lifestyle choices. Hg^{2+} is naturally found in the Earth's crust and can be released into the environment by volcanic activity and the burning of Hg^{2+}-containing coal. As with Pb^{2+}, inorganic Hg^{2+} does not cross the adult BBB and, unless damage is severe, clinical signs can to some extent subside once exposure to the heavy metal has decreased. Both metals can also exist in organic form, which increases their permeability through lipid cell membranes and ability to cross the BBB. Consequently, organometals can sometimes induce more severe and lasting neurological effects than do the inorganic forms. However, this is not always the case.

Voltage-gated calcium (Ca^{2+}) channels (VGCCs) are crucial to several fundamental processes in the nervous system including synaptic transmission and plasticity, neuronal growth, and development. VGCCs are one of the most abundant group of surface proteins present in neurons and are, therefore, readily exposed to toxic compounds that can affect their function. VGCCs are particularly affected by Pb^{2+} and Hg^{2+} because of the chemical similarities (i.e., ionic charge, ionic radius, etc.) that these metals share with Ca^{2+} (Atchison 2003).

Voltage-Gated Ca^{2+} Channels

Ca^{2+} plays a crucial role in the regulation of gene expression and neurotransmitter release, synaptic plasticity, and growth cone elongation in neurons. The majority of the Ca^{2+} needed for these functions enters the neuron or its processes from the extracellular fluid, where the concentration (1.5–2.0 mM) is far in excess of the intracellular bulk free concentration (0.1–0.3 μM). Thus, Ca^{2+} entry is an energy-independent process. VGCCs mediate influx of Ca^{2+} into the cell as a result of a depolarization of the plasma membrane. Their specific role, and hence localization in a neuron, is dictated by the properties that characterize them. They are not uniformly distributed throughout the neuron, nor is their subtype distribution even. Two classes of VGCCs exist: the high-voltage-activated (HVA) and low-voltage-activated (LVA) channels; they are grouped according to the degree of depolarization from rest that is needed for their activation. HVA channels are heteromeric proteins consisting of α, β, $α_2δ$, and sometimes γ subunits. LVA VGCCs consist of the $α_1$ subunit. Other subunits have not been identified. While the pharmacological properties of the channel are imparted by the $α_1$ pore-forming subunit, the β, γ, and $α_2δ$ subunits modulate the voltage-dependence of activation/inactivation and the kinetics of channel opening, as well as play other important roles in VGCC localization or function (Benarroch 2010) (Fig. 1). The HVA category, composed of channels that require a high degree of depolarization, is subdivided into L-, N-, P/Q-, and R-type ($Ca_V2.x$) according to ($Ca_V1.x$) their pharmacological and electrophysiological characteristics (Table 1). The LVA category is

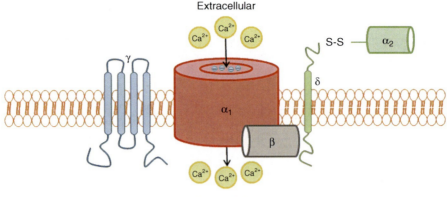

Mercury and Lead, Effects on Voltage-Gated Calcium Channel Function, Fig. 1 *Schematic representation of voltage-gated Ca^{2+} channel (VGCC) quaternary structure.* VGCCs are comprised of five subunits, with the α_1 pore-forming subunit being required for proper function. The α_1 subunit is composed of four homologous domains, each containing six transmembrane segments. Ion selectivity is conferred by a ring of negatively charged glutamate residues (○) which line the pore near the outer vestibule. This subunit not only contains the Ca^{2+} selectivity filter and drug-binding sites, but also plays a role in voltage sensing. The α_1 subunit of certain subtypes of HVA VGCCs interacts with synaptic vesicle release proteins (See Fig. 2.) through the "synprint" site (not shown). The cytoplasmic β subunit interacts with the α_1 subunit at the α-interaction domain (not shown). The β subunit plays a role in trafficking the α_1 subunit to the plasma membrane, stabilizing the final conformation of the α_1 subunit, as well as regulating its activation and inactivation kinetics. The $\alpha_2\delta$ subunit contributes to modulating the kinetics of HVA channels. This subunit is composed of the α_2 subunit located extracellularly and the single transmembrane δ subunit anchoring the protein to the plasma membrane. The role of the γ subunit in the function of Ca^{2+} channels is not completely understood. Co-expression of the α_2 and δ subunits enhances expression of α_1, whereas γ subunit expression is not required for VGCC formation or function in most tissues (Modified from Benarroch 2010)

only composed of the T-type channels ($Ca_V3.x$). Due to their low threshold of activation, T-type VGCCs are responsible for providing neurons Ca^{2+} spikes at resting membrane potentials. Many roles for these Ca^{2+} spikes have been suggested, including pacemaking activity, modulation of Ca^{2+}-dependent ion channels, and neuronal development (Zamponi 2005).

The process of discriminating among extracellular cations is due, in part, to a selectivity filter present at the mouth of the VGCC pore (Fig. 1). A region of four negatively charged glutamate residues projects into the lumen outside the pore. The ability of these residues to bind Ca^{2+} is believed to be crucial for ion discrimination (Atchison 2003; Zamponi 2005). This selectivity is so efficient that, even though the extracellular sodium concentration far exceeds that of Ca^{2+}, VGCCs are still more permeable to Ca^{2+}. However, cations of similar atomic radius (i.e., Na^+, Li^+, K^+, Cs^+, Sr^{2+}, and Ba^{2+}) also can pass through VGCCs when they are substituted for extracellular Ca^{2+} (Zamponi 2005). These cations show a higher single-channel conductance than Ca^{2+} due to their inability to interact with the binding sites present in the channel's vestibule, a characteristic that has been advantageous in studying VGCC function. Toxic metals also interact with and disrupt the function of VGCCs. In particular, Hg^{2+} and Pb^{2+} fall into the category of nonphysiological, environmentally-relevant metals whose neurotoxic effects on Ca^{2+} regulation have been well documented (Atchison 2003).

Effects of Mercury on VGCC Function

Hg, originally known as quicksilver, is a persistent heavy metal naturally found in three forms: (1) the metallic form (Hg^0) exists as a liquid at room temperatures or as a vapor at higher temperatures; (2) the mercurous form (Hg^+) can combine as an inorganic salt with chlorine, sulfur, or oxygen and; (3) the mercuric form (Hg^{2+}) is commonly found associated with carbon. These latter compounds are referred to as organic mercurials (Clarkson and Magos 2006), and include both aryl (phenylmercuric acetate) and alkyl (methylmercury – MeHg) forms. Hg^0 and its chemical derivatives are neurotoxic environmental contaminants of contemporary concern. Hg^0 enters the environment through both natural and anthropogenic

Mercury and Lead, Effects on Voltage-Gated Calcium Channel Function, Table 1 Characteristics of voltage-gated calcium channels

Type	Channel subtype	Antagonist	Electric properties (conductance^a/τ)	α_1-coding gene	Cell/tissue-specific expression	Function
HVA	L	Phenylalkylamines, dihydropyridines, benzothiazapines, and ω-agatoxin IIIA	25 pS/0.5 s	α_{1S} (Ca$_V$1.1)	Skeletal muscle	Excitation-contraction coupling
				α_{1C} (Ca$_V$1.2)	CNS (dendrites and cell bodies)/cardiac muscle	Plasticity, cardiac muscle contraction
				α_{1D} (Ca$_V$1.3)	Cochlea	Sensory transduction
					Striatal medium spiny neurons	Neurotransmitter release
					Substantia nigra pars compacta	Pacemaker activity
					Retina	Sensory transduction
	P/Q	ω-agatoxin IVA (P-type), ω-conotoxin MVIIC (Q-type)	9–20 pS/<116 ms for P-type; <100 ms for Q-type.	α_{1A} (Ca$_V$2.1)	Presynaptic terminals	Neurotransmitter release in neuromuscular junctions and CNS
		FTX (P-type)			Purkinje neurons and cerebellar granule cells	Neurotransmitter release
					Thalamus	Depolarization
	N	ω-conotoxin GVIA	13 pS/50–110 ms (500–800 ms in sympathetic neurons)	α_{1B} (Ca$_V$2.2)	Only expressed in neurons	Neurotransmitter release
		FTX (N-type?)			Nociceptive dorsal root ganglion neurons	Depolarization
	R	SNX 482	14 pS/12 ms	α_{1E} (Ca$_V$2.3)	Presynaptic terminals	Neurotransmitter release
					Hippocampus	Depolarization
LVA	T	Kurtoxin, mibefradil, amiloride	5-11pS/20–50 ms	α_{1G}, α_{1H}, α_{1I} (Ca$_V$3.1–3.3)	Dendrites and cell bodies of some CNS neurons	Rhythmic burst firing and pacemaker activity

τ = time constant of decay of the current. This represents the time required for the current amplitude to decay from its peak value to 63.2% of that value
^{a}The conductance (g) is the reciprocal of resistance (ohms) and reflects the flow of current. Thus, the lower the resistance, the greater the flow of currents – that is, conductance. Originally given units of mhos, g is now given the units of Siemans
? indicates that the results proposed are not confirmed, and thus remain in question (Table modified from Bennaroch 2010 and Shafer 2000)

sources. Natural sources of Hg^0 include release into the atmosphere by the degassing of the Earth's crust, volcanic emissions, and soil erosion. Anthropogenic sources include coal burning and waste combustion as well as mercury and gold mining (Clarkson and Magos 2006). Once released into the atmosphere, Hg^0 can exist as a vapor for a year or longer, allowing it to spread throughout the atmosphere. Thus, point source emissions, such as a volcanic eruption, or high concentration of coal combustion emissions can reach far across the globe. This makes regulation of local Hg air emissions, in many cases, of limited local value inasmuch as the ambient air [Hg^0] may have originated far from the point of measurement, and even at much earlier time points. Atmospheric Hg^0 can be slowly oxidized to Hg$^+$ via a mechanism that is not yet fully understood. The water-soluble Hg$^+$ then returns to the Earth's surface by rainwater. The majority of the oxidized Hg^0 is reduced back to the vapor state and returns to the atmosphere. However, a portion of it is methylated by aquatic methanogenic bacteria to form MeHg which then bioaccumulates up the food chain, with the highest sources being piscivorous fish or mammals. Consequently, humans are primarily exposed to MeHg as a result of consumption of fish and sea mammals (Clarkson and Magos 2006). Although MeHg is considered to be the principal form of Hg relevant to environmental exposures, evidence of exposure to Hg^0 has been increasing due to its use in certain Caribbean religious cults, or Central/South American medicinal practices (*Botanica*). This use, which is typically indoors, can result in retention of the metal in carpets, where it presents a potential route of exposure to infants or young children.

Early studies of mercurials on VGCC function relied on neurochemical measures of uptake of radiolabeled Ca^{2+} ($^{45}Ca^{2+}$) during KCl depolarization in isolated cells in culture or preparations of isolated nerve endings (synaptosomes) prepared from rodent brain. These are produced by density gradient centrifugation of brain homogenates. The resulting structures are pinched off nerve endings, which release neurotransmitter, maintain resting membrane potentials, and exhibit typical ion channel fluxes. Synaptosomes were used for studies of VGCC function due to concerns such as: (1) VGCC phenotype and density varied spatially across the neuron; (2) the identity of distinct VGCC subtypes was uncertain, and (3) the desire to correlate changes in Ca^{2+} influx with effects on neurotransmitter release. The small size of the mammalian nerve terminal (1 μm) precludes the use of intracellular or patch electrode recordings to measure currents. However, the function of several subtypes of VGCC could be studied in isolation using neurochemical flux methods. As such, the synaptosomal model is very convenient for studying the presynaptic effects of metals mediated by VGCCs including Hg^{2+}, Pb^{2+}, and MeHg (Sirois and Atchison 1996). VGCC function is examined, most typically in this model, through flux measurements of $^{45}Ca^{2+}$ in response to KCl depolarization. In these experiments, exposure to the toxic metal was exceedingly brief – 1, or 10 s, and generally limited to the period of depolarization. Such transient exposures minimized possible interaction of the toxicant with other intracellular processes, which may have ultimately been interpreted as a channel-mediated effect (Sirois and Atchison 1996). Ca^{2+} influx into synaptosomes occurs in two kinetically distinct phases following depolarization: the fast phase, which is associated with neurotransmitter release and inactivates rapidly (1–2 s), and the slow phase, which persists for 20–90 s following depolarization and is perhaps the product of a non-inactivating Ca^{2+} channel (Atchison 2003) or reversal of the Ca^{2+}/Na^+ exchanger. Synaptosomes isolated from rat brain were exposed to high concentrations of Hg^{2+} (5–200 μM) to study channel function in isolation for both the fast and slow uptake phases. Hg^{2+} caused a concentration-dependent decrease in total uptake of $^{45}Ca^{2+}$ during KCl depolarization, which suggests that Hg^{2+} inhibited Ca^{2+} influx through VGCCs in some manner (Atchison 2003). The peak effect (∼95% reduction) occurred at ∼200 μM Hg^{2+}. Effects of Hg^{2+} on "fast" and "slow" components were not examined in isolation. Studies performed at the neuromuscular junction (NMJ) of frog sartorius muscle and rat and mouse diaphragm showed that Hg^{2+} disrupted Ca^{2+}-dependent synaptic transmission in an intact synapse. The NMJ is a well-characterized and widely used model in mechanistic studies of metal neurotoxicity (Shafer 2000). When coupled with intracellular recording techniques, isolated NMJ preparations allow for the indirect examination of activity of voltage-sensitive channels through measurements of postsynaptic responses to neurotransmitter release. Release of neurotransmitter can occur either in response to a nerve-evoked depolarizing stimulus or spontaneously. At the NMJ, an end-plate potential (EPP), measured as a graded amplitude depolarization of the postsynaptic membrane, is produced in response to action potential-evoked release of acetylcholine. Given the Ca^{2+}-dependence of evoked transmitter release, the ability of divalent metals to prevent Ca^{2+} entry has been studied extensively at the NMJ, whereby reduction of EPP amplitude has been assumed to be a consequence of block of VGCC function (Atchison 2003). Alternatively, miniature end-plate potentials (MEPPs) result from spontaneous release of single quanta (vesicle) of neurotransmitter; they are measured as small, randomly occurring depolarizations of the postsynaptic membrane, and with a relatively uniform amplitude. Their frequency is correlated with the free $[Ca^{2+}]_i$, but their amplitude is $[Ca^{2+}]$ independent. Since both spontaneous and evoked neurotransmitter release are dependent, in part, upon $[Ca^{2+}]_i$ and a significant portion of it originates extracellularly, a metal which interrupts Ca^{2+} entry into the cell will in turn perturb patterns of transmitter release (Sirois and Atchison 1996). Studies in which Hg^{2+} was applied at the NMJ showed a time- and concentration-dependent decrease in EPP amplitude, suggesting that Hg^{2+} negatively affects VGCC function. In addition, there was an increase in MEPP frequency, which was subsequently decreased to complete block (Atchison 2003). The ability of Hg^{2+} to increase MEPP frequency suggested a mechanism of action that increases the levels of $[Ca^{2+}]_i$ presumably by release from intracellular stores (Atchison 2003).

Subsequent studies were aimed at investigating the effects of Hg^{2+} on specific aspects of VGCC function. These used contemporary single cell electrophysiological methods to study in more depth ion channel function directly and in real time. They typically made use of single cells in culture because of methodological considerations – namely, ease in isolating and recording the VGCC currents of interest. In comparing studies of mercurials in intact tissues versus cells in culture, readers are typically struck by the large difference in concentrations needed to cause effects on VGCC function. Concentrations in intact tissue, such as the NMJ or brain slice, are generally in the low to high μM range (5–100 μM), whereas those needed for equivalent type blocking effects in cells in culture are in the low to sub-μM range (0.1–5 μM). This discrepancy can be explained readily based on pharmacokinetic principles. The ability of mercurials to bind to SH groups makes intact tissue a veritable cornucopia of nonspecific binding sites. Thus, much higher concentrations of mercurials are needed to overcome this impediment in studies of intact tissue. In whole-cell Ca^{2+} current experiments performed in *Aplysia* abdominal ganglia neurons, application of 20 μM Hg^{2+} irreversibly reduced Ca^{2+} current. In primary cultures of rat dorsal root ganglion (DRG) neurons, low μM concentrations of Hg^{2+} resulted in an irreversible reduction of current carried by Ba^{2+}, a Ca^{2+} surrogate, through T-, L-, and N-type Ca^{2+} channels (Atchison 2003). Also, studies in primary hippocampal pyramidal neurons demonstrated that application of Hg^{2+} caused an irreversible inhibition of current through presumed L-type Ca^{2+} channels; however, LVA Ca^{2+} currents were transiently increased (Atchison 2003). Interestingly, similar experiments performed in PC12 cells, an immortalized cell line derived from a rat pheochromocytoma, demonstrated that nM concentrations of Hg^{2+} caused an increase in Ca^{2+} current amplitude (Atchison 2003).

More recent studies have focused on Hg^{2+} effects on individual VGCC subtypes. To do this, a single VGCC isoform is transiently expressed in human embryonic kidney (HEK-293) cells. This transformed cell line lacks native HVA channels, thereby facilitating the ability to study interactions of a mercurial in isolation. (HEK-293 cells do weakly express LVA VGCC, however.) The actions of Hg^{2+} on human N- ($Ca_v2.2$) and R-type ($Ca_v2.3$) VGCCs were studied by transient heterologous expression of recombinant (cloned) channels comprised of neuronal cDNA clones of α_{1E} or α_{1B} along with a constant $\alpha_2\delta$ and β_{3A} in HEK-293 cells. In native neurons, the $\alpha_2\delta$ and β subunit identity may not be immutable. However, by standardizing the identity of these subunits, it was possible to compare the role of mercurials on the α_1 subunit in isolation. Hg^{2+} (0.1–5 μM) caused a time- and concentration-dependent block of both N- and R-type Ba^{2+} current; however, N- but not R- type inhibition was partially reversed when cells were washed with Hg^{2+}-free physiologic saline solution (Atchison 2003). Because HEK-293 cells stably express T-type VGCCs, effects of Hg^{2+} on native T-type VGCCs have also been studied in isolation using this cell line. Whole-cell Ba^{2+} current recordings revealed that Hg^{2+} concentrations ranging from 10 to 100 μM inhibited Ba^{2+} current through these channels. The block was virtually irreversible at lower concentrations and partially reversible at higher concentrations. Hg^{2+} also altered the voltage-dependence and slowed the T-type channel kinetics (Tarabova et al. 2006). This decrease in current of a LVA Ca^{2+} channel is in agreement with the previously mentioned study in primary hippocampal neurons.

MeHg is of particular interest for its prevalence and potent neurotoxic effects. Anthropogenic sources have also been recorded in the past and have been implicated with two distinct episodes of human poisoning in the past 70 years: Minamata Bay (and the less well-known Niigata Bay), Japan, and Iraq (Atchison 2003). Poorly confirmed or unconfirmed episodes have been reported in Asia and North and South America. The clear neurological dysfunction exhibited by Japanese residents of the cities affected by MeHg gave rise to the descriptive title of Minamata disease. Several of the neurological signs presented by these patients have plausible links to alterations in VGCC function. In turn, these episodes of human exposure triggered a raft of studies designed to explore possible interactions of MeHg with VGCCs in isolation. Of particular interest was the fact that, unlike Hg^{2+} and other inorganic divalent cation blockers, MeHg had only a single positive charge. Thus, presumably its interactions with VGCCs should differ from those of Hg^{2+}.

As with other heavy metals suspected of acting on VGCCs, initial studies of actions of MeHg relied upon use of in vitro models at amphibian and mammalian NMJs. This resulted in part from the fact that in the Iraqi exposure to MeHg, an increased incidence of myasthenia gravis-like syndromes was observed, and in part due to the well-characterized nature of the synapse as described above. Somewhat surprisingly, given the chemical differences between the two forms, bath application of MeHg disrupted patterns of nerve-evoked and spontaneous transmitter release in a similar manner to that of Hg^{2+}; there was a decrease in EPP amplitude and an initial increase, then decrease in MEPP frequency to complete block. These effects were also time- and concentration-dependent. However, in contrast to inorganic Hg^{2+}, with MeHg they were irreversible (Atchison 2003). Because the rate of spontaneous neurotransmitter release is related to presynaptic $[Ca^{2+}]_i$, the MeHg-induced increase in MEPP frequency reflects an intracellular effect. This is suggestive either of movement of the neurotoxicant across the plasma membrane, the release of Ca^{2+} directly from internal stores, or the ability to activate metabotropic receptors linked to signaling cascades (i.e., IP_3) and thereby mobilize $Ca^{2+}{}_i$ stores. Treatment of NMJ preparations with Bay K 8644, an L-type VGCC agonist, reduced the latency of MeHg-induced increases in MEPP frequency. These experiments suggest, albeit indirectly, that MeHg might utilize VGCCs to cross the cell membrane into the nerve terminal (Shafer 2000). However, a main attribute that distinguishes MeHg from inorganic Hg^{2+} is the ability of the former to disrupt VGCC function when the channel is at rest. Another important difference is the fact that MeHg acts in a noncompetitive manner, whereas higher $[Ca^{2+}]_e$ can interfere with block of VGCC current caused by Hg^{2+} (Atchison 2003). It has also been hypothesized that MeHg interacts with the channel in additional ways, most likely due to its relatively higher lipophilicity.

Rat brain was also used to study the effects of MeHg on VGCCs. MeHg (5–200 μM) reduced VGCC-mediated $^{45}Ca^{2+}$ uptake in both phases, though the block was more pronounced in the slow phase. The effect was just minimally overcome by increasing $[Ca^{2+}]_e$, suggesting a noncompetitive mode of action (Atchison 2003). This is in contrast to the VGCC blocking effect of inorganic Hg^{2+}, which is highly hindered by an increase in the $[Ca^{2+}]_e$ (Atchison 2003). Even though increasing $[K^+]_e$ concurrently with 100 μM MeHg exposure results in an increased percentage block of synaptosomal $^{45}Ca^{2+}$ influx, VGCC current block occurs irrespective of channel configuration (e.g., closed, activated, inactivated). Prior VGCC activation is not needed for MeHg to induce a decrease in Ca^{2+} influx (Atchison 2003). MeHg also inhibits $^{45}Ca^{2+}$ uptake in synaptosomes isolated from rat cerebellum, so the effect is not brain region-specific. Pretreatment with ω-conotoxin GVIA (1 or 5 μM) or ω-conotoxin MVIIC (0.1 or 0.5 μM) blocked the MeHg-induced inhibition of synaptosomal $^{45}Ca^{2+}$ uptake. These data suggest that MeHg interacts with N- and P/Q-type VGCCs, respectively, in the cerebellum (Sirois and Atchison 1996).

Studies comparing potency and efficacy of block of synaptosomal $^{45}Ca^{2+}$ uptake by mercurials which differ in charge and lipophilicity support the hypothesis that mercurials might enter the cell via VGCCs. Electrically neutral mercurials, such as p-chloromercuribenzoate (PCMB) and dimethyl-mercury, or negatively charged mercurials, p-chloromercuribenzene sulfonate (PCMBS), were unable to block synaptosomal $^{45}Ca^{2+}$ uptake, even at high $[K^+]_e$, indicating that mercurial lipophilicity is not the sole determinant which accounts for blocking capabilities. However, the ability of MeHg to block noncompetitively VGCCs despite channel configuration suggests that lipophilicity is an important factor which contributes to the blocking actions of MeHg (Sirois and Atchison 1996). In contrast to negatively charged or neutral mercurials, those with a positive charge, such as ethylmercury, MeHg, and Hg^{2+}, blocked synaptosomal $^{45}Ca^{2+}$ influx through VGCCs in a voltage-dependent manner, indicating that these ions may interact with the channel's pore (Shafer 2000).

The synaptosomal model has also been used for studies of receptor occupancy involving competition-binding experiments. In these experiments, synaptosomes were pre-incubated with radiolabeled ligands with specificity for distinct VGCC subtypes. Mercurials were introduced as unlabeled "cold" competitors. MeHg reduced the specific binding of $^3[H]$-nitrendipine, an L-type VGCC antagonist (Table 1) at a concentration (100 μM) that also inhibited $^{45}Ca^{2+}$ influx. Specific binding of ^{125}I-ω-conotoxin GVIA in PC12 cells was also reduced by MeHg. Together, these data suggest that MeHg interacts with L- and N-type VGCCs. In particular, mercurials bind to sulfhydryl groups within proteins. Therefore, sulfur-containing residues in the pore of an ion channel represent potential sites for mercurial action. In contrast to Hg^{2+}, which has the ability to bind two sulfhydryl groups, MeHg only binds one. Interactions with sulfur-containing groups in VGCCs may directly alter activation/inactivation of the channel, or disrupt overall channel function indirectly through effects on membrane fluidity. Differences among the physiochemical properties of mercurials can contribute to the differential effects on VGCCs previously discussed (Sirois and Atchison 1996).

Based upon results of synaptosomal flux studies and radioligand-binding assays with acute MeHg exposure, electrophysiological studies were performed to determine how mercurials alter VGCC currents (Atchison 2003). Advantages of electrophysiological techniques include the ability to study: (1) channels in isolation from other cellular events and thus describe more precisely channel kinetics in response to a chemical or electric stimulus, and (2) events resulting from a population of cells down to a single membrane-bound channel using extracellular recording techniques or patch clamp, respectively (Sirois and Atchison 1996). MeHg (1–20 μM) completely and irreversibly blocked whole cell Ba^{2+} current through L- and N-type VGCCs in PC12 cells; onset of current block occurred at a rate which was concentration-dependent. As observed in synaptosomes, block by MeHg was not determined by the configuration of the VGCCs (Atchison 2003). Similar effects on Ba^{2+} current were observed in the highly sensitive cerebellar granule cell following in vitro exposure to MeHg (0.25–1 μM), though it appears that primary cell cultures are generally more sensitive to the effects of MeHg than are transformed cell lines (Sirois and Atchison 1996).

The potency of MeHg as a VGCC blocker is indisputable, with complete block of current often occurring in the low μM range. However, VGCC subtype specificity of MeHg has not been studied extensively in native currents. Previous studies report HVA currents as being more resistant to MeHg (block at 5 and 10 μM), whereas LVA currents are more sensitive with a marked current reduction by 2 μM MeHg. Using heterologous expression of human neuronal VGCC subunits in HEK-293 cells, the effects of mercurials on single VGCC isoforms have been examined; MeHg caused a concentration- (0.125–5.0 μM) and time-dependent reduction in current through the L-type VGCCs, though this block was incomplete, with approximately 25% current remaining in the presence of 5 μM MeHg. Nimodipine, an L-type VGCC antagonist, completely blocked current through the recombinant channel on its own and blocked residual current which remained following 5 μM MeHg exposure. MeHg also induced a concentration- (0.125–5.0 μM) and time-dependent block of current through recombinant N- and R-type VGCCs; this block was complete at the upper range of concentrations examined (Atchison 2003). Thus, MeHg appears to exert subtype-dependent effects on VGCCs.

Effects of Lead on VGCC Function

Studies of the effects of Pb^{2+} on VGCC function were the first to describe interaction of nonphysiological divalent cations on VGCCs. As was the case for other heavy metals, including Hg^{2+}, one of the first pieces of evidence that suggested a deleterious effect of Pb^{2+} on VGCC function came from studies at amphibian NMJs. The pattern of effects seen was reminiscent of that for Hg^{2+}, namely, treatment with Pb^{2+} caused an initial decrease in the amplitude of nerve-evoked EPPs, with a subsequent increase in the frequency of spontaneously occurring MEPPs (Atchison 2003). However, in contrast to the situation with Hg^{2+}, EPP amplitude and MEPP frequency both returned to normal levels once the preparation was washed with a Pb^{2+}-free solution. Due to the dependence of MEPP frequency upon $[Ca^{2+}]_i$, in conjunction with the temporal sequence of events, these data suggest an initial membrane action followed by an intracellular action of Pb^{2+}. Two observations directly support the idea that Pb^{2+} directly blocks Ca^{2+} conductance through VGCCs. At both frog and mammalian NMJs, a higher $[Ca^{2+}]_e$ was required to evoke an EPP in the presence of Pb^{2+} as compared to that needed to evoke an EPP of the same amplitude in a Pb^{2+}-free preparation (Atchison 2003). Flux studies demonstrated a decrease in $^{45}Ca^{2+}$ influx following KCl depolarization of vertebrate synaptosomes exposed to μM concentrations of Pb^{2+} (Atchison 2003; Suszkiw 2004). As with Hg^{2+}, these data imply that Pb^{2+} competes with Ca^{2+} for access to the VGCCs. Furthermore, IC_{50} (the concentration required to cause 50% reduction of the maximal theoretical response) data from HEK-293

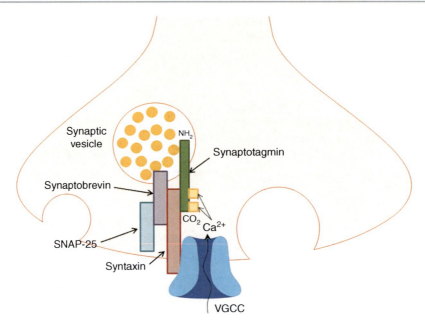

Mercury and Lead, Effects on Voltage-Gated Calcium Channel Function, Fig. 2 *Proteins involved in docking, fusion, and exocytosis of neurotransmitter vesicles.* Syntaxin, synaptobrevin, and SNAP-25 form a complex known as the SNARE complex which links the synaptic vesicle to the plasma membrane and plays a central role in neurotransmitter release. Immunoprecipitation studies have shown that syntaxin coprecipitates with N-type VGCCs, indicating a direct interaction, is found most likely at the so-called "synprint" site. A similar type site exists on P/Q-type channels. However, the SNARE complex itself is not sensitive to increases in $[Ca^{2+}]_i$. Thus, the linkage between SNARE proteins and VGCCs merely provides a site at which synaptic vesicles can dock in close apposition to sites of Ca^{2+} entry. The synaptic vesicle protein, synaptotagmin, acts as the Ca^{2+} sensor and can bind two Ca^{2+} molecules in its C2 domain. This promotes an interaction with syntaxin and regulates the SNARE complex. Synaptotagmin also coprecipitates with N-type VGCCs indicating its ability to interact with the site of Ca^{2+} entry

cells heterologously expressing a single subtype of VGCC demonstrate that subtype-specific susceptibility occurs to the blocking effects of Pb^{2+}; R-type VGCCs are most susceptible ($IC_{50} = 0.1$ μM), followed by L-type ($IC_{50} = 0.38$ μM) and N-type ($IC_{50} = 1.31$ μM) channels (Suszkiw 2004). P/Q-type channels have not been examined. In primary hippocampal neurons on the other hand, data suggest that L-type VGCC currents are more susceptible to block than are N-type (Atchison 2003). These differences may reflect the identity of the auxiliary subunits, which likely differ in native neurons as compared to transfection systems. Therefore, conclusions regarding the relative sensitivity of VGCC subtypes to Pb^{2+}-induced block of current are still controversial.

Two possible mechanisms regarding the Pb^{2+}-induced increase in MEPP frequency have been hypothesized. The first proposed that Pb^{2+} increased levels of $[Ca^{2+}]_i$ through release of Ca^{2+} from intracellular stores. Although there is little evidence of this, pM and nM concentrations of Pb^{2+} can increase protein kinase C (PKC) activity via activation of the enzyme's Ca^{2+}-binding C2 domain. Because PKC has the ability to increase the activity of HVA VGCCs (Benarroch 2010), brief increases in $[Ca^{2+}]_i$ can be observed. In addition, Pb^{2+} activates the Ca^{2+}-binding phospholipase C (PLC), which in turn is able to increase the levels of inositol 1,4,5-trisphosphate (IP_3). $[Ca^{2+}]_i$ could increase as a result of the activation of IP_3 receptors present in the smooth endoplasmic reticulum, which signal this intracellular Ca^{2+} store to release a portion of its contents. The second proposed mechanism involves the entry of Pb^{2+} into the cell followed by a direct interaction with exocytotic components responsible for neurotransmitter release (Fig. 2). One experiment that supports this idea employed staphylococcal α-toxin, which forms pores in the cell membrane and increases its permeability to small extracellular solutes. Bovine adrenal chromaffin cells that were pretreated with this toxin and exposed to Pb^{2+} in a Ca^{2+}-free solution had increased norepinephrine release (Suszkiw 2004). Moreover, because the degree of norepinephrine release did not increase when both cations were present, it is believed that Pb^{2+} and Ca^{2+} share a common intracellular target. Synaptotagmin I, a Ca^{2+}-binding protein that is found in neurotransmitter vesicle membranes, has two Ca^{2+}-binding domains (C2) that are similar to those found in PKC. Under normal circumstances, intracellular Ca^{2+} binds to synaptotagmin's C2 domain and apparently promotes an interaction with syntaxin, a cell-membrane integral protein, and phospholipids (see Fig. 2). This interaction is crucial for neurotransmitter exocytosis. Even though Pb^{2+} is not able to induce a synaptotagmin I/syntaxin interaction, evidence suggests that Pb^{2+} promotes synaptotagmin I/phospholipid binding in a similar manner to Ca^{2+} (Suszkiw 2004). Calmodulin is another intracellular protein that contains a Ca^{2+}-binding domain and can affect spontaneous neurotransmitter release. Pb^{2+} can activate calmodulin and increase the movement of neurotransmitter vesicles from the reserve pool to the readily releasable pool (Suszkiw 2004). These effects are in the same concentration range as that caused by Ca^{2+}.

The effects of Pb^{2+} on the kinetics of VGCCs, if any, are still unknown. As of yet, there are no studies that have focused on single-channel properties of VGCCs after Pb^{2+} exposure. Data from whole-cell Ba^{2+} current recordings do not present a clear picture either. *Aplysia* neuronal VGCCs require higher depolarizations after Pb^{2+}, which suggests that voltage-dependent properties are affected (Atchison 2003). On the other hand, T-type VGCCs of a neuroblastoma cell line, N1E-115, do not show the same change in voltage-dependence of activation or inactivation. There have been reports of an increase in Ca^{2+} current in a small portion of PC12 cells treated acutely and chronically (up to 3 months) with Pb^{2+} (Atchison 2003). The membrane potential at which maximal Ca^{2+} current was observed increased from 0 to +10 mV following Pb^{2+} treatment. Pb^{2+} has also been suggested to impede Ca^{2+}-dependent inactivation of VGCCs intracellularly. Normally, an increase in $[Ca^{2+}]_i$ will cause a reduction in Ca^{2+} current through VGCCs; this mechanism limits open duration of the channel and is believed to be crucial to prevent neuronal excitotoxicity. In vitro experiments, in which Pb^{2+} was applied intracellularly to bovine adrenal chromaffin cells before an increase in $[Ca^{2+}]_i$ showed a reduction in the degree of Ca^{2+} current rundown. A decrease in Ca^{2+} current rundown was also observed in electrophysiological experiments with neurons from the medial septal nucleus of rats that were chronically exposed to Pb^{2+} in their drinking water.

Organic forms of Pb^{2+} were widely used in the twentieth century. One of the most common causes of human exposure to organic Pb^{2+} came from the use of alykl Pb compounds (tetramethyl or tetraethyl Pb^{2+}) in leaded gasoline. Similar to organic Hg^{2+} compounds, organic Pb^{2+} compounds are relatively more lipophilic, able to cross the BBB, and potentially responsible for many of the neurological signs of Pb^{2+} poisoning. Once in the body, tetramethyl-Pb^{2+} is converted into trimethyl-Pb^{2+} (TML) (Gawrisch et al. 1997). The effects of TML on VGCC function have been described in rat primary dorsal root ganglion (DRG) neurons. VGCC currents in a total of 64 DRG neurons were examined, using the patch-clamp configuration. Treatment with 0.5, 1.0, and 5 μM TML caused a 22%, 50%, and 59% reduction of Ba^{2+} current through VGCCs, respectively. Currents were almost completely blocked (85% reduction) following treatment with 50 μM TML. As expected from organic forms of heavy metals, these concentration-dependent effects on VGCC current were irreversible and voltage-independent, in contrast to the effects of inorganic Pb^{2+}. However, a striking difference between TML and MeHg is the dependence of TML on an open channel conformation. When TML is applied followed by a period without stimulation of VGCCs, the initial Ba^{2+} current recorded after this period is of equal amplitude to control currents. The requirement of an open state suggests that TML has a binding site within the channel pore (Gawrisch et al. 1997). It is still unknown whether TML has any intracellular effects.

Conclusions

Even though measures have been taken over the last decades to minimize exposure to these heavy metals, the fact that Hg^{2+} and Pb^2 are naturally found in the Earth's crust makes them a persistent health hazard. Ca^{2+} signaling plays crucial roles within the nervous system. As such, the potential for interaction of these metals with VGCCs – the primary route of Ca^{2+} entry into excitable cells, is significant. As demonstrated in this review, numerous studies have examined this issue and there is no doubt that acute exposures to the metals disrupt VGCC function by these metals. However, important issues remain unresolved. Some studies of the effects of Hg^{2+} and Pb^{2+} on VGCC function have made use of metal concentrations above those normally found in blood samples of affected individuals. Moreover, chronic exposures, which consist of low concentrations over extended periods of time, are the most common manner in which humans are affected. However, there are few studies which have examined this interaction. This is in contrast to the acute treatments that most studies have employed to determine the effects of Hg^{2+} and Pb^{2+} on VGCC function. Consequently, the important issue is how chronic exposure to environmental metals remains untested. Nonetheless, it is evident that both inorganic and organic forms of Hg^{2+} and Pb^{2+} have deleterious effects on neuronal VGCC function, $[Ca^{2+}]_i$ homeostasis, and the process of neurotransmitter release. The complexity of systems affected by the metals have made holistic studies of VGCC function extremely difficult in the whole animal, resulting in the need for use of reductionist systems. The fact that the membrane protein such as VGCCs serve as the "first line of defense" for all cells mandates that membrane proteins will de facto be the first potential targets of exposure to toxicants. For this reason, studies of acute interactions of metals with the channel protein provide a snapshot of prospective initiating events, which may or may not presage later more long-lasting effects of the metals on cellular functions in excitable cells.

Acknowledgements Supported by NIH grants R01ES03299 and R25NS065777. The assistance with word processing and figure illustrations of Jessica Hauptman and Beth Anne Hill is greatly appreciated.

Cross-References

▶ Calcium in Nervous System
▶ Lead, Physical and Chemical Properties
▶ Mercury Neurotoxicity
▶ Mercury, Physical and Chemical Properties

References

Atchison WD (2003) Effects of toxic environmental contaminants on voltage-gated calcium channel function: from past to present. J Bioenerg Biomembr 35:507–532

Benarroch EE (2010) Neuronal voltage-gated calcium channels: brief overview of their function and clinical implications in neurology. Neurology 74:1310–1315

Clarkson TW, Magos L (2006) The toxicology of mercury and its chemical compounds. Crit Rev Tox 36:609–662

Gawrisch E, Leonhardt R, Büsselberg D (1997) Voltage-activated calcium channel currents of rat dorsal root ganglion cells are reduced by trimethyl lead. Toxicol Lett 92:117–122

Shafer TJ (2000) The role of ion channels in the transport of metals into excitable and nonexcitable cells. In: Zalups RK, Koropatnick J (eds) Molecular biology and toxicology of metals, 1st edn. Taylor & Francis, New York

Sirois JE, Atchison WD (1996) Effects of mercurials on ligand- and voltage-gated ion channels: a review. Neurotoxicology 17:63–84

Suszkiw JB (2004) Presynaptic disruption of transmitter release by lead. Neurotoxicology 25:599–604

Tarabova B, Kurejova M, Sulova Z, Drabova M, Lacinova L (2006) Inorganic mercury and methylmercury inhibit the Cav1.3 channel expressed in human embryonic kidney 293 cells by different mechanisms. J Pharmacol Expt Ther 317: 418–427

Zamponi GW (2005) Voltage-Gated Calcium Channels, 1st edn. Kluwer, New York

Mercury and Low Molecular Mass Substances

Miquel Esteban[1], Cristina Ariño[1], José Manuel Díaz-Cruz[1] and Elena Chekmeneva[2]
[1]Department of Analytical Chemistry, University of Barcelona, Barcelona, Spain
[2]Department of Chemistry, University of Sheffield, Sheffield, UK

Synonyms

Interactions of mercury with low molecular mass substances

Definition

Mercury and Low Molecular Mass Substances refer to the main characteristics of the interactions between a mercury ion (either inorganic or organic forms) and a ligand, which can be a pharmaceutical agent, an amino acid, a peptide, a protein, or an ill-defined mixture of substances termed as the Dissolved Organic Matter (DOM). Special attention is paid to thiol-containing compounds because of the extraordinary ability of that group to bind mercury ions.

Overview

Mercury (Hg) is a natural element that can be found in the environment in several forms that will define its physical, chemical, and toxicological properties. According to the World Health Organisation "there is no 'safe' level of mercury." Human activities can release Hg from relatively stable deposits, allowing it to move through air, water, and food chains, and to convert into more toxic compounds. Mercury and its compounds are highly toxic to living organisms, ecosystems, and wildlife. Nowadays Hg pollution is considered to be a global problem, diffuse and chronic. High doses are fatal to humans, but even the low doses can have a serious impact for the neural development, cardiovascular, immune, and reproductive systems. All forms of mercury have toxic effects in a number of organs, especially in the kidneys (Zalups 2000).

At room temperature, elemental mercury (Hg^0) is a liquid metal that, because of its high vapor pressure, can be released into the environment as mercury vapor. Mercury also exists as a cation with an oxidation state of 1+ (mercurous cation, stable form Hg_2^{2+}) or 2+ (mercuric cation, Hg^{2+}). The free mercurous ion Hg_2^{2+} is not stable under environmental conditions, and it tends to convert into Hg^0 and Hg^{2+} (dismutation process), or to form compounds of very low solubility with common ions such as chloride (or other halides) and sulfide. Ions of mercury can form extremely stable complexes or insoluble precipitates. For instance, the solubility product of HgS is of 10^{-52}, while for $Hg_2(OH)_2$ is of $10^{-23.7}$. Within biological systems Hg_2^{2+} or Hg^{2+} ions do not exist as inorganic salts or in an unbound free ionic state.

With respect to organic forms of Hg, methylmercury (CH_3Hg^+) and dimethylmercury a (($CH_3)_2Hg$) are the most frequently encountered compounds in the environment, especially the first one. They are formed mainly as the result of methylation of Hg^{2+} by microorganisms in soil and water. In general R-Hg$^+$ and R-Hg-R compounds can be formed, where R represents any organic ligand.

All mercury forms cause toxic effects in a number of tissues and organs, depending on its chemical form, and the level, duration, and route of exposure.

When considering the effects of mercury ions in biological and environmental systems their bonding characteristics must be taken into account.

Hg^{2+} ion is a "class B" metal, characterized by a "soft sphere" of highly polarizable electrons in its outer shell, according to Pearson classification (Rayner-Canham and Overton 2010). Hg shows a strong affinity for ligands with soft donor atoms such as S, Se, and P, and for the halide ions I$^-$, Br$^-$, and Cl$^-$. Complexes with low coordination numbers, and often with a preference for diagonal two-coordination, are formed. Thiolates are commonly referred to as mercaptans because of their ability to capture Hg^{2+}.

Mercury ions can bind nucleophilic groups on molecules. In particular, they have a greater tendency to bond to reduced sulfur atoms, especially those of endogenous thiol-containing molecules, such as amino acids, peptides, and proteins. Special mention requires cysteine (Cys), homocysteine, glutathione (γGlu-Cys-Gly; GSH), N-acetylcysteine, phytochelatins (PC), metallothioneins (MT), and albumin (Fig. 1).

Hg-thiol compounds are very stable in an aqueous environment inside a very wide pH range. The Hg–ligand formation constants ($log\beta_{pqr}$, p, q, r referred to M, ligand and H$^+$, respectively) for complexes with thiol-containing ligands are very high, usually in the range above 10^{20}, and values of $log\beta_{pqr}$ > 50 have been determined, for instance, for GSH (Oram et al. 1996, and references therein).

In plants and algae, phytochelatins ((γ-Glu-Cys)$_n$-Gly, PC$_n$, n = 2–11), which are synthesized from GSH, immobilize heavy metal ions as Hg in a similar way as MT do in mammalians, and are the basis of phytoremediation (Pilon-Smits 2005).

In contrast, the formation constants for mercury bonding to oxygen- or nitrogen-containing ligands (e.g., carbonyl or amino groups) are some orders of magnitude lower. Hence, in most cases, when considering the biological effects of inorganic or organic Hg, their interactions with thiol-containing compounds are of the greater relevance.

Mercury and Low Molecular Mass Substances, Fig. 1 Chemical structure of cysteine, cystine, glutathione, phytochelatins, and human metallothionein

As it was outlined before in general terms, in the presence of an excess of a low-molecular-weight thiol-compound, mercury ions form diagonal two-coordination compounds.

Interactions of Mercury in Natural Media

Marine Biogeochemical Cycling of Mercury

The inorganic speciation of Hg^{2+} in natural waters and soils is dominated by chloride. Under low-chloride and oxic conditions, organic complexes are dominant. In oxic estuarine and seawater conditions, a progression of chloride complexes is expected in the absence of organic complexing agents (Fitzgerald et al. 2007). The values of the stability constants are highest for Cl^-, OH^-, NH_3, and several orders of magnitude lower for F^-, SO_4^{2-}, and NO_3^-. The latter ligands can be excluded from consideration because they would be only significant under unnaturally high concentrations. Unlike other ligands, Cl^- and OH^- ions are present at sufficiently high concentrations and provide very stable complexes that dominate most natural systems. At pH 4–5, Cl^- concentrations at ppm level are sufficient for all Hg^{2+} to be in the $HgCl_2$ form. Hg-hydroxide complexes ($Hg(OH)_2$, $Hg(OH)^+$) are likely to be the important species in most freshwaters. In aquatic environments containing dissolved sulfide (including some oxic surface waters, where nanomolar levels of sulfide and thiols have been detected), Hg is hypothesized to form Hg-sulfide species (Ravichandran 2004).

Interactions of Mercury with Dissolved Organic Matter

Dissolved Organic Matter (DOM) in aquatic environments consists of a heterogeneous mixture of organic compounds of ill-defined chemical structure. About 20% of DOM consists of carbohydrates, carboxylic acids, amino acids, hydrocarbons, and other identifiable compounds. The remaining 80% of DOM consists of humic substances, which are made up of a complex mixture of residues from the decomposition of plants and animals. Mercury and other trace metals are generally bound at the acid sites in organic matter. The most commonly encountered acidic functional groups in DOM include carboxylic acids, phenols, ammonium ions, alcohols, and thiols. Of these different groups, carboxylic acids and phenols contribute as much as 90% of acidity to organic matter. Despite the abundance of carboxylic acids and other O-containing functional groups in DOM, Hg preferentially binds with thiols and other S-containing groups. Sulfur is a minor constituent in DOM, ranging from ca. 0.5% to 2% by weight. Sulfur in DOM occurs as reduced state (defined as the sum of sulfide, polysulfide, and thiol groups) or as oxidized species (e.g., sulfonate, sulfate), with oxidation states ranging from -2 to $+6$. Of these, only the reduced sulfur sites are expected to be important for mercury binding. However, this has not been confirmed with direct spectroscopic studies as it has been done for soil organic matter, but indirect evidences have been reported recently (Ravichandran 2004, and references cited therein). DOM interacts very strongly with mercury, affecting its speciation, solubility, mobility, and toxicity in the aquatic environment, and this strong binding of Hg by DOM is attributed to coordination of Hg at reduced sulfur sites within the organic matter, which are present at concentrations much higher than those of Hg found in most natural waters. The ability of DOM to enhance the dissolution and inhibit the precipitation of HgS, a highly insoluble solid, suggests that DOM competes with S^{2-} for Hg binding.

Other key role of DOM concerns the photochemical reduction of ionic mercury to elemental mercury (Hg^0) and subsequent reoxidation of Hg^0 to ionic mercury, thus affecting volatilization loss and bioavailability of mercury to organisms. However, global interpretation of these phenomena is quite involved. Some experiments have shown that photolytic reduction of Hg^{2+} is substantially enhanced in the presence of DOM, but it was also observed that Hg^0 production was limited by the amount of Hg^{2+} available in solution, and that high dissolved organic carbon (DOC) concentrations cause reduced light penetration and increased Hg complexation with DOM (Ravichandran 2004, and references therein).

Interactions of Mercury with DOM in Soils

There are direct spectroscopic evidences of the interactions of Hg with soil organic matter. It has been shown by extended X-ray absorption fine structure (EXAFS) spectroscopic studies that Hg in soils is complexed by two reduced organic S groups (likely thiols) at a distance of 2.33 Å in a linear configuration, like in Hg complex with the amino acid cysteine. Furthermore, a third reduced S (likely an organic sulfide) was indicated to contribute with a weaker second shell attraction. When all high-affinity S sites, corresponding to 20–30% of total reduced organic S, are saturated, other groups such as carbonyl-O or amino-N and one carboxyl-O contribute to Hg complexation (Skyllberg et al. 2006, and references therein).

Interactions of Mercury with Amino Acids

Molecular interactions that occur between mercury and amino acids are especially relevant for the amino acid Cys (Fig. 1). $Hg(Cys)_2$ is the predominant Hg(II) complex and, in model systems, it was able to cross cell membranes. Thus, $Hg(Cys)_2$ complex is suggested to mimic cystine (Fig. 1) and use the active cell transport sites usually used for cystine transport through the membrane (Zalups 2000, and references therein).

Interactions with Peptides

In biological systems, GSH (Fig. 1) is the most abundant nonprotein thiol, with intracellular concentrations of between 0.1 and 10 mmol dm^{-3} present in microorganisms, fungi, plant, and animal tissues (Rabenstein 1989). Complex formation between GSH and heavy metal ions is considered as a first key step in biological detoxification processes, prior to vacuolar sequestration or transfer of the heavy metal to MT (in animals) and PC (in plants). GSH has a number of potential coordinating sites, the utilization of which depends on the soft/hard acid properties of the metal ion, the pH of the system, and steric constraints. The average pK_a values of the acidic Glu COOH, Gly COOH, Cys SH, and Glu NH_3^+ groups are 2.06, 3.50, 8.69, and 9.62, respectively. The partially deprotonated GS^- ion is present in solution at physiological pH, while in alkaline solutions of pH 10.5 widely studied in the literature the totally deprotonated GS^{3-} ion is the one predominating (Mah and Jalilehvand 2008).

No crystal structures have been reported of Hg(II)-GSH compounds, although strong complex formation has been described in aqueous solutions by a variety of techniques, among them Nuclear Magnetic Resonance (NMR) and Mass Spectrometry (ESI-MS). While Hg^{2+} is known to coordinate up to four thiolate ligands, there has been no report of a four coordinate $[Hg(GS)_4]^{n-}$ complex.

Recent ^{13}C NMR studies of dilute Hg(II)-GSH solutions at physiological pH by EXAFS yielded two Hg-S bond distances of 2.33 Å and a possible Hg-Hg interaction at 2.9 Å between the $[Hg(GS)_2]^{k-}$ complexes. More recent studies, also by EXAFS, at pH 10.5 and ambient temperature, showed that $[Hg(GS)_2]^{4-}$ and $[Hg(GS)_3]^{7-}$ are the predominant complexes, but $[Hg(GS)_4]^{10-}$, with four sulfur atoms coordinated at a mean Hg-S bond distance of 2.52 Å, is present in minor amounts (<30%) (Mah and Jalilehvand 2008).

Despite the extraordinarily high thermodynamic stability of the covalent bonds formed between mercury ions and the thiol-containing molecules in aqueous solution, these bonds appear to be more labile within the living organisms (Rabenstein 1989). Thiol- and/or other nucleophilic competition and exchange processes seem to be the most likely explanation for the labile nature of Hg-thiol bonding in tissue and cellular compartments.

Interactions with Phytochelatins

Plants are capable of extracting a variety of metal ions from their growth substrates, including Hg. Many studies have shown that plant roots accumulate Hg when they are exposed to Hg-contaminated soils. Laboratory studies have shown that plant roots absorb Hg from solution and that roots accumulate a much greater amount of Hg than do shoots. Hg is able to bind with water channel proteins of root cells causing a physical obstruction to water flow and consequently affect the transpiration in plants. Other toxic symptoms of mercury accumulation include leaf roll and chlorosis, reduced growth in roots and stems, and damage at the water and ionic balance and the activity of several enzymes.

One of the most effective mechanisms for metal detoxification in plants is the synthesis of low-molecular weight proteins and peptides that chelate metal ions in stable complexes. Among them it is worth mentioning phytochelatins ((γ-Glu-Cys)$_n$-Gly; PC_n n = 2–11), which are synthesized under control of the enzyme PC-synthase from glutathione (GSH). The synthesis is induced by some metals such as Cd, Ag, Bi, Pb, Zn, Cu, Hg, and Au. It is known that PC_n can form complexes with Pb, Ag, and Hg in vitro, but in vivo only the complexes with Cd, Ag, and Cu ions were detected (Cobbett and Goldsbrough 2002, and references therein).

Several analytical methods, such as chromatographic separation (gel filtration or HPLC) coupled with UV detection, flame atomic absorption spectrometry, radioactive labeling, and inductively coupled plasma-mass spectrometry (ICP-MS) or electrospray-MS (ESI-MS) or tandem mass spectrometry (ESI-MS/MS), have been used to analyze PC_n and PC_n-metal complexes.

In vitro studies demonstrated that Hg^{2+} is facile to transfer from shorter- to longer chain PC_n. The strength of Hg^{2+} binding to GSH and PC_n follows the order GSH < PC_2 < PC_3 < PC_4.

In vivo poor information on PC_n-Hg complexes is available. Several works concerned with plants subjected to Hg stress showed the increasing of Cys and PC_n and the decreasing of GSH contents as a consequence of PC_n synthesis. In the majority of the studies in vivo only unbound PC_n were identified, but not their Hg-complexes. The Hg accumulation mechanism was studied in *Brassica napus*, and only unbound PC_2 was detected after the addition of the chelator DMPS (2,3-dimercapto-1-propanesulfonic acid (Unithiol)); the presence of polynuclear Hg-PC_n complexes was suggested (Carrasco-Gil et al. 2011 and references therein). However, in vivo oxidized PC_2, PC_3, and PC_4 and their corresponding $HgPC_2$, $HgPC_3$, $HgPC_4$, and Hg_2PC_4 complexes, which were confirmed by their specific isotope distribution, were detected in vivo in *Brassica chinensis* L. by HPLC-ESI-MS/MS (Carrasco-Gil et al. 2011 and references therein).

Recently, a study with alfalfa (*Medicago sativa*), barley (*Hordeum vulgare*), and maize (*Zea mays*) subjected to Hg stress has been carried out. Liquid chromatography coupled with electro-spray/time of flight mass spectrometry (ESI-MS-TOF) showed that Hg was bound to an array of phytochelatins (PC_n) in roots. A total of 28 Hg-containing ions were detected corresponding to 17 different Hg–complexes formed with up to six different biothiols (GC_2 ((γ-Glu-Cys)$_2$), PC_2, hPC_2, hPC_3, PC_4, and hPC_4, where hPC_n are homophytochelatins with a general formula (γ-Glu-Cys)$_n$-Ala).

In barley, three 1:1 Hg–complexes were found, one each with GC_2, PC_2, and PC_4. In maize, Hg formed

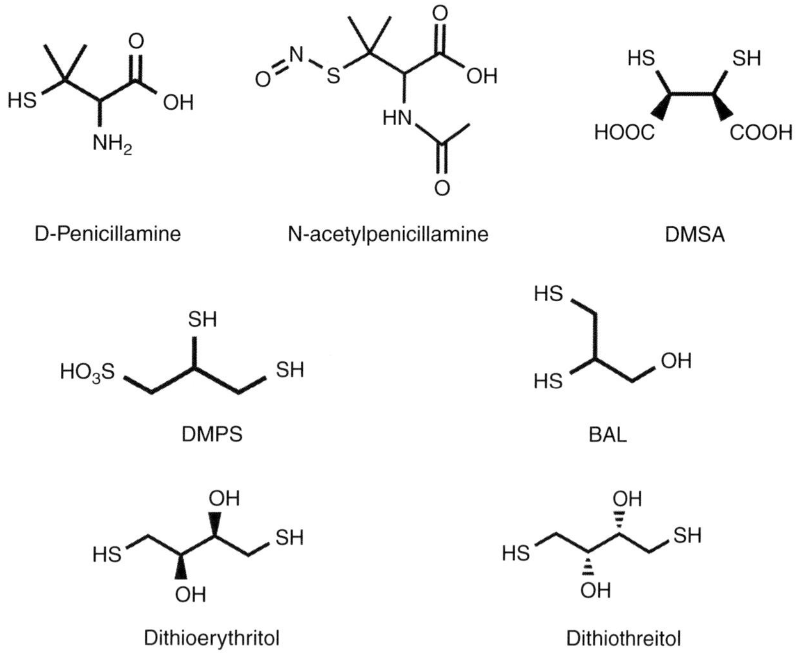

Mercury and Low Molecular Mass Substances,

Fig. 2 Chemical structure of: D-penicillamine; N-acetylpenicillamine; meso-2,3-dimercaptosuccinic (DMSA; Succimer); 2,3-dimercapto-1-propanesulfonic acid (DMPS; Unithiol), 2,3-dimercaptopropanol (BAL); dithioerythritol; and dithiothreitol

only two complexes with GC_2, having 1:1 and 2:2 stoichiometries. In alfalfa, up to 14 different Hg–biothiol complexes were found, each one with GC_2 ($HgGC_2$) and PC_2 ($HgPC_2$), four with hPC_2 ($HghPC_2$, $Hg(hPC_2)_2$, $Hg_2(hPC_2)_2$, and $Hg_3(hPC_2)_3$), three with hPC_3 ($HghPC_3$, $Hg(hPC_3)_2$, and $Hg_2(hPC_3)_2$), two with hPC_4 ($HghPC_4$ and Hg_2hPC_4), and two more with PC_4 ($HgPC_4$ and Hg_2PC_4). Additionally, a complex formed with methyl-Hg and hPC_2 (CH_3-Hg–hPC_2), not previously described in the literature, has been detected in alfalfa. The most abundant Hg-to-ligand stoichiometry found was 1:1. Moreover, some of the Hg–biothiol complexes having thiol groups not bound to Hg were found to occur in oxidized forms (Carrasco-Gil et al. 2011, and references therein).

Interactions with Proteins

Most of the mercury ions present in plasma (after exposure to inorganic mercury) are bound to sulfhydryl-containing proteins, such as albumin, but not for very long. After the exposure, and during the initial hours, there is a rapid decrease in the plasma burden of mercury concurrent with a rapid rate of uptake of inorganic mercury in the kidneys and liver. Because mercuric S-conjugates of small endogenous thiols (e.g., GSH and Cys) are the primary transportable forms of mercury in the kidneys, mercury ions should be transferred from the plasma proteins to these low molecular weight thiols by some ligand-exchange mechanism (Zalups 2000). Endogenous low molecular weight thiols bound to mercury (and other heavy metals) have facilitated entry of the mercury into various cell types via molecular mimicry. "Molecular mimicry refers to the phenomenon whereby the bonding of metal ions to nucleophilic groups on certain biomolecules results in the formation of organo-metal complexes that can behave or serve as a structural and/or functional homolog of other endogenous biomolecules or the molecule to which the metal ion has bonded" (Bridges and Zalups 2005).

Interactions with Thiol-Containing Pharmaceutical Agents

Chelating agents are primarily sulfhydryl-containing compounds such as mono- or dithiol molecules. At the molecular level, the chelation process appears as the competition between the chelating agents and the competing biological ligands. In the past 50 years, there has been substantial progress in understanding, developing, and clinical application of chelating agents used to treat acute and chronic mercury poisonings in humans. It is important to emphasize that a good chelator is usually water soluble, while lipophilic chelators may have a redistribution effect of the mercury to the target organs (Guzzi and La Porta 2008).

Several thiol-containing pharmacological agents have been used against Hg poisoning (Fig. 2): D-penicillamine (β,β,-dimethilcysteine), N-acetylpenicillamine, meso-2,3-dimercaptosuccinic (DMSA; also termed Succimer), 2,3-dimercapto-1-propanesulfonic acid (DMPS; sometimes termed Unithiol), 2,3-dimercaptopropanol (also termed British anti Lewisite (BAL) and dimercaprol), dithioerythritol, and dithiothreitol. The well-known EDTA (Ethylenediaminetetraacetic acid) has also been used, although an important inherent limitation in chelation and elimination of mercury is that EDTA is unable to enter in the cells.

Apart from chelating agents, a number of substances have been proposed as secondary supportive therapy for mercury detoxification: N-acetyl cysteine (NAC), selenium, vitamin E, and choline.

Effectiveness of these agents to remove inorganic and organic mercuric ions from endogenous ligands should be based on the above mentioned ligand-exchange mechanism.

References

Bridges CC, Zalups RK (2005) Molecular and ionic mimicry and the transport of toxic metals. Toxicol Appl Pharmacol 204:274–308

Carrasco-Gil S, Álvarez-Fernández A, Sobrino-Plata J, Millán R, Carpena-Ruiz RO, Leduc DL, Andrews JC, Abadía J, Hernández LE (2011) Complexation of Hg with phytochelatins is important for plant Hg tolerance. Plant Cell Environ 34:778–791

Cobbett C, Goldsbrough P (2002) Phytochelatins and metallothioneins: roles in heavy metal detoxification and homeostasis. Annu Rev Plant Biol 53:159–182

Fitzgerald WF, Lamborg CH, Hammerschmidt CR (2007) Marine biogeochemical cycling of mercury. Chem Rev 107:641–662

Guzzi GP, La Porta CAM (2008) Molecular mechanisms triggered by mercury. Toxicology 244:1–12

Mah V, Jalilehvand F (2008) Mercury(II) complex formation with glutathione in alkaline aqueous solution. J Biol Inorg Chem 13:541–553

Oram PD, Fang X, Fernando Q, Letkeman P, Letkeman D (1996) The formation constants of Mercury(II)-Glutathione complexes. Chem Res Toxicol 9:709–712

Pilon-Smits E (2005) Phytoremediation. Annu Rev Plant Biol 56:15–39

Rabenstein DL (1989) Metal complexes of glutathione and their biological significance. In: Dolphin D, Avramovic O, Poulson R (eds) Glutathione: chemical, biochemical and medical aspects, vol 3, Coenzymes and cofactors. Wiley, New York, pp 147–186

Ravichandran M (2004) Interactions between mercury and dissolved organic matter – a review. Chemosphere 55:319–331

Rayner-Canham G, Overton T (2010) Descriptive inorganic chemistry, 5th edn. Freeman, New York

Skyllberg U, Bloom PR, Qian J, Lin CM, Bleam WF (2006) Complexation of Mercury(II) in soil organic matter: EXAFS evidence for linear two-coordination with reduced sulfur groups. Environ Sci Technol 40:4174–4180

Zalups RK (2000) Molecular interactions with mercury in the kidney. Pharmacol Rev 52:113–143

Mercury Exacerbation of Alzheimer's Disease

▶ Mercury and Alzheimer's Disease

Mercury in Plants

Stephan Clemens
Department of Plant Physiology,
University of Bayreuth, Bayreuth, Germany

Synonyms

Hg in plants

Definition

The entry addresses plant exposure to mercury, uptake of mercury, speciation and distribution, mercury toxicity and detoxification, human health consequences of plant mercury accumulation, and phytoremediation of mercury pollution.

Introduction

Mercury (Hg) has no biological functions in plants. Instead, Hg represents one of the most toxic substances released into the environment. On the CERCLA priority list of hazardous substances, assembled by the US Environmental Protection Agency and the Agency for

Toxic Substances and Disease Registry (www.atsdr.cdc.gov/cercla/07list.html), Hg ranks third behind arsenic and lead. Relevant aspects with respect to the biology of plants are therefore plant exposure to Hg, modes of uptake, accumulation and toxicity, tolerance mechanisms, the fate and speciation of Hg inside a plant, human health consequences of Hg accumulation in crops, and environmental issues such as plant-based remediation of Hg pollution.

Exposure of Plants to Hg

Plants are exposed to Hg in the soil, in water, and in the atmosphere. The latter is a consequence of volatilization due to the unique chemical characteristics of Hg, namely, a low melting point and relatively high vapor pressure. Soil Hg is largely a function of weathering of Hg-containing minerals and of atmospheric deposition. Hg occurs in a range of minerals. The average concentration of Hg in the continental crust is around 0.08 mg kg^{-1}, with sedimentary rocks having higher Hg concentrations than igneous rocks. In soils not affected by Hg emission or contamination, Hg concentrations are usually below 100 μg kg^{-1}. Much higher values can be found near Hg emitting sources or as a consequence of direct application of Hg-containing material to soil. Natural Hg emission exceeds anthropogenic emission by about a factor of 2. Main primary natural source is volcanic activity. However, most natural Hg emission is in fact reemission of previously deposited mercury on land or water surfaces and vegetation (see section "Hg Speciation and Distribution Within Plants") and thus at least partly due to human activities. Main anthropogenic sources include fossil-fuel burning, artisanal small-scale gold mining, metal manufacturing, and the use of Hg in the chlor-alkali industry (Pirrone et al. 2010). Further Hg addition specifically to agricultural soils occurs inadvertently through fertilization with phosphate, sewage sludge, animal manure, etc., or purposely through the application of fungicides, for instance, as seed coat dressings. Hg concentrations in contaminated soil have been reported to reach levels in excess of 100 mg kg^{-1} (McLaughlin et al. 1999). Atmospheric deposition and pollution caused by industrial or agricultural activities also result in Hg contamination of water. Thus, aquatic plants such as macrophytes growing in wetlands or in river sediments are sometimes exposed to comparatively high levels of Hg.

Uptake of Hg

Principal chemical forms of Hg taken up by plants are ionic Hg (mostly Hg^{2+}), organic Hg, and elemental Hg (Hg^0). In the soil, Hg adsorption and desorption processes are predominantly controlled by complexation. Hg hydrolyzes readily but has a strong tendency to form stable complexes with OH-, Cl-, and S^{2-}- or S-containing ligands. The high affinity of Hg to S is the reason for strong interaction with dissolved organic matter. Many factors including pH, redox potential, organic matter content, or inorganic soil constituents influence adsorption. Overall, the concentration of ionic Hg in the soil solution is very low resulting in extremely limited bioavailability of Hg ions for plant uptake when compared to many other metal ions, for example, Cd^{2+} (McLaughlin et al. 1999). Uptake pathways for Hg^{2+} in plant roots have so far not been identified molecularly. It is assumed that similar to other toxic nonessential ions such as arsenate or selenate, transporters mediating the uptake of essential ions do not perfectly discriminate between chemically similar ions and allow entry also of the toxic ions. In case of Hg^{2+}, transporters of metal cations such as Fe^{2+} or Zn^{2+} are hypothesized to account for the uptake (Clemens 2006).

Under anaerobic conditions in soil and in aquatic sediments, methylmercury as the major form of organic mercury is formed by sulfate-reducing bacteria. Methylmercury is a form of Hg that biomagnifies, i.e., accumulates to higher concentrations in organisms at higher trophic levels. It is generally assumed to be more available also for plant uptake, most likely due to its hydrophobicity that allows direct membrane passage (Meagher and Heaton 2005).

In the atmosphere, elemental Hg is mostly present in the gaseous form. This makes it unique among trace metals. Because gaseous elemental mercury shows low reactivity and high stability, its atmospheric residence time is long, enabling transport to areas far remote from Hg emitting sources. Uptake of volatile Hg^0 into leaves is a major factor in the global biochemical cycling of Hg. Entry occurs through stomata, i.e., leaf pores that regulate a plant's gas exchange and mediate CO$_2$ uptake. In addition, a nonstomatal pathway can be detected especially in the dark when stomata are closed. Hg^0 is mildly lipophilic and can therefore diffuse directly through the leaf surface, which is covered by waxes and other hydrophobic polymers in order to reduce water loss. The relative contribution of uptake from soil/water and from the atmosphere into plants has been a major question in understanding the global Hg cycle (Lindberg et al. 2007). The answer depends on the mobility of Hg within plants.

Hg Speciation and Distribution Within Plants

Once taken up into plant cells, both inorganic and organic Hg strongly interact with sulfhydryl groups in biomolecules. This can be demonstrated, for instance, by synchrotron X-ray fluorescence spectroscopy and extended X-ray absorption fine structure analysis of plant issue. These techniques show association with sulfur and binding to organic sulfur compounds in plants exposed to Hg. Analytical tools are only beginning to be developed and exploited to dissect the diversity and dynamics of low molecular weight Hg species formed inside plant cells. The combination of liquid chromatography with electrospray ionization mass spectrometry and inductively coupled plasma mass spectrometry allows detection of a variety of Hg complexes with cysteine or glutathione, two of the major cellular thiols (Krupp et al. 2008).

The mobility of Hg species within plants is very limited. Numerous studies performed with a range of phylogenetically diverse plants such as *Pisum sativum* (pea), *Zea mays* (maize), *Nicotiana tabacum* (tobacco), or Brassica species such as rapeseed agree on low translocation rates from roots to shoots (Patra and Sharma 2000). Only a minute fraction (usually much less than 5%) of total Hg is transported into aboveground tissues following uptake into roots. The transfer factor for Hg is much smaller than for Cd, Cu, or Zn. Respective experimental data have been confirmed under field conditions. Surveys conducted by the US Environmental Protection Agency on soil fertilized with Hg-contaminated sewage sludge showed very low leaf Hg concentrations relative to other metals (McLaughlin et al. 1999). Besides strong binding to thiols, the interaction with negatively charged cell wall polymers contributes to root retention of Hg. Inferable from restricted translocation is that organic Hg-S species and free ionic Hg, which most likely is practically absent from the cytoplasm, are no substrates for efflux transporters such as ATP-driven metal pumps (P-type ATPases) functioning in the loading of metals into the xylem, i.e., the tissue responsible for water and nutrient movement from the root to the shoot. This is again different from Cd, which also strongly interacts with thiols yet can be translocated efficiently via Zn pumps and low-molecular-weight metal chelators (Clemens 2006).

Low translocation implies that most of the Hg accumulated in leaves is originating from atmospheric Hg^0 taken up directly through stomata and the leaf surface. This has been experimentally proven for various plant species. Also within leaves, mobility is low. Most Hg is found in epidermal cells, very little in mesophyll cells. Other than Hg accumulated in root tissue, leaf Hg can be exchanged with the environment, i.e., can be reemitted to the atmosphere as Hg^0 before it becomes oxidized and interacts with sulfhydryl groups in biomolecules. Reemission as opposed to translocation can be demonstrated by feeding of Hg to roots. This does not result in emission from the leaf surface when the experimental setup is well controlled, i.e., volatilization of Hg by soil microbial activity is taken into account (Greger et al. 2005). Most studies agree that only a fraction of total Hg taken up into leaves becomes reemitted. Incorporation into leaf tissue makes plants a net sink for Hg. As a consequence, the shedding of leaves by trees and subsequent decomposition of litter represents a significant input of Hg into forest ecosystems.

Hg Toxicity

Effects of Hg exposure have been investigated in dozens of studies with a wide range of plants under controlled experimental conditions and almost always employing Hg concentrations far above levels encountered in normal or even contaminated sites. The major focus has been on crop plants such as wheat, rice, or *Medicago sativa* (alfalfa). Our knowledge about both acute toxicity and detoxification is mostly derived from such studies while the effects of chronic exposure to concentrations relevant in natural and agricultural environments are poorly understood (Patra and Sharma 2000).

Together with Pb^{2+}, Hg^{2+} has the most pronounced growth-inhibiting effect among metal cations on plants in solution culture (Kopittke et al. 2010). Besides replacement of essential ions in metalloproteins, the strong interaction with sulfhydryl groups is a main reason for the exceptional toxicity of Hg. Little information is available as to the existence and molecular nature of primary target biomolecules (Meagher and Heaton 2005). Often, plastids and the photosynthetic machinery have been reported to represent such targets. However, since many studies have been applying extreme Hg concentrations in vitro, and in view of poor translocation to photosynthetically active cells (see section "Hg Speciation and Distribution Within Plants"), one has to conclude that proteins or peptides particularly vulnerable to inhibition by Hg remain to be identified. Well known, though, is the interaction of Hg^{2+} with plasma-membrane-localized aquaporins, i.e., channel proteins that enhance water permeation through the membrane. In fact, Hg has played an important role in studying the biological function of aquaporins in water transport and in revealing the regulation of water permeability upon environmental stimuli. Hg blocks aquaporins via binding to cysteines located near the channel pore and has therefore been a very useful tool to assess the relative contribution of aquaporins to water movement under various physiological conditions.

A well-established symptom of acute Hg toxicity is oxidative stress even though Hg is not redox active. Numerous studies have shown that plant cells exposed to $HgCl_2$ far above natural levels show an increase in reactive oxygen species, the oxidation of the redox buffer glutathione, and the upregulation of antioxidative systems such as ascorbate peroxidase and glutathione reductase. The underlying mechanism has been suggested to lie in the depletion of the glutathione pool and an inhibition of antioxidative enzymes.

Hg Detoxification

Like other groups of organisms, plants are known to exhibit basal tolerance of Hg, meaning that a variety of mechanisms ensures survival up to a certain level of exposure (Clemens 2006). Loss of such mechanisms in mutants results in hypersensitivity. Antioxidative defenses that are involved in plant responses to practically all forms of abiotic stress certainly contribute to coping with Hg stress. More specific for metal detoxification is the chelation and sequestration of metal ions to suppress interaction with predominantly sulfhydryl groups in essential proteins. Obviously, an efficient way is the synthesis of thiols with high affinity for metals and subsequent transport of thiol-metal complexes out of the cytosol or other sensitive cellular compartments. Binding partners for Hg ions in plant cells are glutathione and glutathione-derived peptides called phytochelatins. Phytochelatins are synthesized in response to exposure to various physiological and nonphysiological metal ions. They are peptides of the general structure (γ-Glu-Cys)$_n$-Gly ($n = 2$–11) and were discovered first in the fission yeast Schizosaccharomyces pombe and then in plants. They are nonribosomally synthesized from glutathione in a transpeptidase reaction, catalyzed by phytochelatin synthases. Phytochelatin synthases are activated by Cd^{2+}, Pb^{2+}, Zn^{2+}, Cu^{2+}, Sb^{3+}, Ag^+, AsO_4^{3-}, and also Hg^{2+}. Thus, in the presence of excess metals, phytochelatins are formed, and they effectively capture metals (Clemens 2006). Plant mutants defective either in glutathione or in phytochelatin synthesis have been shown to be hypersensitive toward Hg. Other ligands have not been proven yet to contribute to basal Hg tolerance in plants.

For the metals/metalloids Cu, Ni, Zn, Cd, As, and Se, naturally selected hypertolerance is known. On metal-rich soil, certain plant species and varieties have evolved that are adapted to toxic metal excess. These so-called metallophytes display an often extreme tolerance toward specific metals. For Hg, such natural variation in tolerance between or within plant species is not documented and most probably does not exist.

Plant Hg and Human Health

Serious incidents of Hg poisoning like the Minamata disease in Japan illustrate the health threat caused by Hg accumulation in food. Agricultural soil is prone to contain more Hg than soil in natural habitats as several agronomic practices result in extra Hg input (see section "Exposure of Plants to Hg"). Still, Hg accumulation in plant-derived food is widely assumed to be of less concern than accumulation of other metals such as Cd simply because of the very limited mobility of Hg within plants. Consumption of fish and other seafood known to bioaccumulate organic Hg continues to be the main source of Hg intake for humans. Recent studies indicate, however, that this could be different when crops are grown on highly Hg-contaminated sites or near Hg emission sources such as Zn smelters. At least for rice grown in paddy fields and therefore exposed to more readily available Hg, available evidence suggests that it could represent the main pathway of methylmercury exposure for humans in certain areas of Asia.

Phytoremediation of Hg Contamination

The cleanup of metal-contaminated soil and water represents a major global task. Phytoremediation refers to the use of plants to remove pollutants from the environment or to render them harmless. As a promising, cost-effective, and environmentally friendly technology for the extraction or at least the stabilization of metal pollutants, phytoremediation has grown tremendously since the 1990s. Principally three approaches have been pursued, the use of plants naturally accumulating metals to extreme levels exceeding those found in normal plants several hundred fold (metal hyperaccumulators), the use of high biomass plants with limited ability to tolerate and accumulate toxic metals, and the genetic engineering of plants with high biomass that can be efficiently grown under various conditions. The early successes following the third approach have been achieved predominantly for Hg (Ruiz and Daniell 2009). Motivated in part by the fact that no plants naturally suitable for Hg phytoremediation are known as well as by the importance of Hg as a pollutant, bacterial Hg detoxification genes have been used to engineer plants suitable for Hg phytoremediation. First, the volatilization of ionic Hg was introduced by expressing a bacterial Hg reductase (merA). This resulted in plants with significantly elevated tolerance of external ionic Hg. A next step was then to target also the most problematic Hg species methylmercury. By combining merA with the expression of an organomercurial lyase (merB) that converts methylmercury to ionic Hg which then is reduced to Hg^0, plants tolerating higher levels of organic mercury were developed. Later refinement of protein localization and expression levels yielded plants accumulating Hg to much higher levels than is normally found in wild-type plants. Thus, a proof of principle has been achieved in different plants. Still, the use of transgenic plants in Hg phytoremediation has yet to reach the stage of application in the field.

Cross-References

▶ Mercury and Low Molecular Mass Substances
▶ Mercury Toxicity
▶ Mercury Transporters
▶ Mercury, Physical and Chemical Properties
▶ Metallothioneins and Mercury

References

Clemens S (2006) Toxic metal accumulation, responses to exposure and mechanisms of tolerance in plants. Biochimie 88:1707–1719

Greger M, Wang Y, Neuschütz C (2005) Absence of Hg transpiration by shoot after Hg uptake by roots of six terrestrial plant species. Environ Pollut 134:201–208

Kopittke PM, Pax F, Blamey C, Asher CJ, Menzies NW (2010) Trace metal phytotoxicity in solution culture: a review. J Exp Bot 61:945–954

Krupp EM, Milne BF, Mestrot A, Meharg AA, Feldmann J (2008) Investigation into mercury bound to biothiols: structural identification using ESI–ion-trap MS and introduction of a method for their HPLC separation with simultaneous detection by ICP-MS and ESI-MS. Anal Bioanal Chem 390:1753–1764

Lindberg S, Bullock R, Ebinghaus R, Engstrom D, Feng X, Fitzgerald W, Pirrone N, Prestbo E, Seigneur C (2007) A synthesis of progress and uncertainties in attributing the sources of mercury in deposition. Ambio 36:19–33

McLaughlin MJ, Parker DR, Clarke JM (1999) Metals and micronutrients – food safety issues. Field Crops Res 60:143–163

Meagher RB, Heaton ACP (2005) Strategies for the engineered phytoremediation of toxic element pollution: mercury and arsenic. J Ind Microbiol Biotechnol 32:502–513

Patra M, Sharma A (2000) Mercury toxicity in plants. Bot Rev 66:379–422

Pirrone N, Cinnirella S, Feng X, Finkelman RB, Friedli HR, Leaner J, Mason R et al (2010) Global mercury emissions to the atmosphere from anthropogenic and natural sources. Atmos Chem Phys 10:5951–5964

Ruiz ON, Daniell H (2009) Genetic engineering to enhance mercury phytoremediation. Curr Opin Biotechnol 20: 213–219

Mercury Ion

▶ Mercury and DNA

Mercury Nephrotoxicity

Lawrence H. Lash
Department of Pharmacology, Wayne State University School of Medicine, Detroit, MI, USA

Synonyms

Inorganic mercury; Mercury-induced nephrotoxicity; Mercury-thiol complexes; Proximal tubular cells

Definitions

Mercury is a metal that exists in the environment in three forms, namely, elemental mercury ($Hg°$), inorganic mercury salts (Hg^+ and Hg^{2+}), and organic mercurials (e.g., dimethyl mercury). When animals or humans are exposed to $Hg°$, it is rapidly oxidized in the body to Hg^+. Hence, only inorganic mercury salts and organic mercurials are relevant forms for understanding mechanisms of toxicity in biological tissues. While inorganic mercury salts have the kidneys as their primary target organ, organic mercurials target both the kidneys and the central nervous system. A critical property of mercuric compounds that is central to understanding how they act is the high affinity for sulfhydryl or thiol groups on proteins and in low-molecular-weight compounds, respectively. The major low-molecular-weight ligands include the amino acid L-cysteine (Cys) and the antioxidant tripeptide glutathione (GSH). Note that typically the term "sulfhydryl" is used to denote the –SH residue on cysteinyl groups of larger peptides or proteins, whereas the term "thiol" is most often used when referring to the –SH group on low-molecular-weight compounds, such as that on Cys, GSH, homocysteine (HCys), or N-acetyl-L-cysteine (NAcCys).

Principles Underlying the Kidneys as a Critical Target for Mercury

There are several key factors that make the kidneys a critical target organ for mercury, in particular Hg^{2+}. First, there are several basic characteristics of renal physiology that make the kidneys targets for many blood-borne chemicals (Lash 2008). These characteristics include the high rate of renal blood flow and glomerular filtration, the presence of numerous plasma membrane transport systems on renal epithelial cells that lead to intracellular accumulation of chemicals from either the renal plasma or interstitial space (nonfiltered chemicals) or the tubular fluid (filtered chemicals), countercurrent flow and other concentrating mechanisms involved in urine formation, and the presence of drug metabolism enzymes in renal epithelial cells that can bioactivate drugs and other chemicals. Additionally, because of high requirements for ATP for transport and biosynthetic processes, the kidneys in general and the proximal tubular (PT) region of the nephron in particular, are especially susceptible to injury from chemicals or pathological conditions that inhibit mitochondrial function and energy metabolism.

The Hg^{2+} ion is one of the major forms of mercury to which biological systems are exposed. Besides the basic physiological processes that render the kidneys, and PT cells in particular, susceptible to Hg^{2+}-induced damage, the physicochemical properties of mercuric-containing chemicals also enhance nephrotoxicity. Perhaps the most important of these properties is the extremely high affinity of Hg^{2+} ions for binding to sulfhydryl (thiol) groups of proteins and certain low-molecular-weight compounds. Consequently, the kidneys are never exposed to the free ion but are exposed to a complex of Hg^{2+} with various sulfhydryl (thiol) groups. The implications for this complexation in terms of understanding how and why Hg^{2+} potently and selectively targets the renal PT cells are described in the next section.

Mercuric-Thiol Complexes as Transport Forms for Inorganic Mercury

The extremely high affinity of Hg^{2+} and organic mercurial ions for sulfhydryl (thiol) groups is an important determinant for both the disposition of the mercuric species and the underlying biochemical and molecular mechanisms of action in the target cell. Consequently, the kidneys are virtually never exposed to free mercuric or organic mercurial ions. Rather, the kidneys are exposed to Hg^{2+} ions in the form of thiol complexes, with each Hg^{2+} ion binding to up to two sulfhydryl (thiol) groups. Thus, the major form of Hg^{2+} in plasma is likely to be that bound to albumin (Fig. 1).

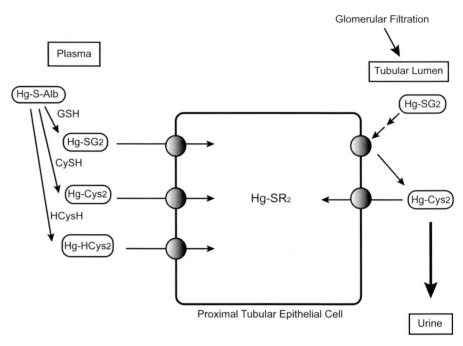

Mercury Nephrotoxicity, Fig. 1 Scheme showing pathways for uptake and intracellular accumulation of mercuric-thiol complexes in the renal proximal tubular (PT) cell. Abbreviations: *Hg-S-Alb* complex of mercuric ion with albumin, *Hg-Cys₂* mercuric-dicysteinyl complex, *Hg-SG₂* mercuric-diglutathionyl complex, *Hg-HCys₂* mercuric-dihomocysteinyl complex, *Hg-SR₂* generic mercuric-thiol complex

Albumin is one of the most abundant proteins and source of sulfhydryl ligands in plasma, with each molecule of the 66 kDa protein containing eight cysteinyl residues. Mercuric ions bound to a large protein, however, cannot permeate the plasma membrane of renal PT cells except by endocytosis. Despite the high affinity of Hg^{2+} ions for the sulfhydryl groups of albumin, exchange reactions can and do occur between these groups and thiol groups of low-molecular-weight compounds as well as sulfhydryl groups of other proteins (Bridges and Zalups 2010; Lash 2010; Zalups and Lash 1994). The most common low-molecular-weight, thiol-containing ligands for Hg^{2+} are GSH, Cys, HCys, and NAcCys.

Although some endocytosis of Hg^{2+}-albumin complexes likely occurs, considerations of general principles of renal cell biology suggest that this mechanism would only account for a minor portion of the intracellular accumulation of Hg^{2+} ions. Rather, it seemed logical that Hg^{2+} ions that bound to low-molecular-weight, thiol-containing ligands could be potential substrates for the large array of organic anion and amino acid transporters that are present on the plasma membranes of the renal PT cell. The concept of "molecular and ionic mimicry," which was originally presented by Clarkson (1993) and expanded upon by Zalups (2000) and Bridges and Zalups (2005), was developed to explain the chemistry by which the various mercuric-thiol complexes are handled by the renal PT cell. For example, comparison of the chemical structure of the disulfide form of the amino acid Cys (i.e., L-cystine or Cys_2) with that of the dicysteinyl conjugate of Hg^{2+} (i.e., Hg-Cys_2) illustrates this concept (Fig. 2).

Moreover, the concept of molecular and ionic mimicry as applied to the biochemistry and physiology of mercuric-thiol complexes is consistent with some key principles and concepts that have developed for understanding the metabolism of drugs and environmental chemicals. Central to this idea is that the vast majority of drugs and environmental chemicals is often metabolized by enzymes that also have endogenous substrates. Thus, although most enzymes have a degree of substrate specificity that includes restrictions on stereochemistry, charge, molecular size, and hydrophobicity among other properties, there is also some range of specificity with regard to some of these properties. Accordingly, some degree of structural modification will not cause a chemical to cease being a substrate for a given enzyme, although kinetic efficiency may certainly change. With respect to plasma membrane transporters, these same principles apply. In fact, substrate specificity for many transporters can be quite broad, thus permitting a good degree of structural

Mercury Nephrotoxicity, Fig. 2 Structure of L-cystine and the mercuric-dicysteinyl complex (Hg-Cys$_2$) illustrating the principle of molecular and ionic mimicry

a L-Cystine.

b Hg-Cys$_2$.

modification to substrates that can still be transported. This broad specificity also illustrates the redundancy present in the array of membrane transporters on renal plasma membranes.

A large number of studies over the past two decades or so have demonstrated that the various mercuric-thiol complexes, which are all organic anions and analogues or molecular mimics of naturally occurring amino acids, are actually transported into the renal PT cell by several known carriers (Table 1) (Bridges and Zalups 2005, 2010; Lash 2010; Zalups 2000; Zalups and Lash 1994). Uptake of mercuric-thiol complexes from the tubular lumen (i.e., those that have undergone glomerular filtration) occurs by the dibasic amino acid transport carrier called System b$^{0,+}$ (Slc7a9). Nonfiltered conjugates can also be transported into the renal PT cell by the action of either the organic anion transporter 1 or 3 (Oat1 or Oat3; Slc22a6 or Slc22a8). Complexes that are accumulated in the renal PT cell may either undergo thiol exchange with intracellular target proteins (see the next section "Intracellular Mechanisms of Mercury-Induced Nephrotoxicity") or may be secreted into the tubular lumen and ultimately into the urine by the action of either the multidrug resistance-associated protein 2 or 4 (Mrp2 or Mrp4; Abcc2 or Abcc4) on the luminal or brush-border plasma membrane (BBM). It should be noted that whereas a role for Mrp2 in the efflux of mercuric-thiol complexes has been unequivocally demonstrated, the function of Mrp4 is only presumed based on its known substrate specificity, which overlaps significantly with that of Mrp2 (Bridges and Zalups 2005, 2010).

The function of these carriers in the uptake into and efflux from the renal PT cell has been demonstrated by the following experimental methods or approaches: (1) consistency of energetics and substrate specificity of transport with those of each carrier; (2) demonstration that intracellular accumulation of Hg^{2+} ions can be diminished by the simultaneous presence of known substrates for specific carrier proteins (i.e., competition experiments); and (3) demonstration that

Mercury Nephrotoxicity, Table 1 Established and postulated membrane transporters for mercuric-thiol complexes in renal plasma membranes

Direction of transport	Transporter	Typical substrates
Uptake at BBM		
	System b$^{0,+}$	Amino acids, including Cys$_2$
Uptake at BLM		
	Oat1, Oat3 (Slc22a6/8)	Exchange of various organic anions with 2-oxoglutarate
Secretion at BBM		
	Mrp2, Mrp4 (?) (Abcc2/4)	ATP-dependent efflux of various S-conjugates, organic anions

BBM brush-border membrane, BML basolateral membrane, Cys$_2$ L-cystine, Mrp multidrug resistance-associate protein, oat organic anion transporter heterologously expressed carrier proteins transport particular mercuric-thiol complexes with saturable kinetics and appropriate energetics and substrate specificity. The latter approach is a particularly powerful method because it involves the selective expression of a purified carrier protein in a model system that contains only the carrier protein of interest without other potential carriers that would confound the results. Model systems include various in vitro preparations such as proteoliposomes, Sf9 insect cells, *Xenopus laevis* oocytes, or Madin-Darby Canine Kidney (MDCK) cells.

A final point about transport of Hg^{2+} ions into the renal PT cell concerns the speciation of the mercury. As noted above, renal PT cells are never exposed to free Hg^{2+} ions but to Hg^{2+} complexed with some thiol ligand. Whereas specific mercuric-thiol complexes, including Hg-GS$_2$, Hg-Cys$_2$, Hg-HCys$_2$, and Hg-NAcCys$_2$, have been synthesized and tested for their abilities to be transported by specific carrier proteins, a key issue concerns which complex is physiologically relevant. Of the four complexes listed above, the Hg-Cys$_2$ complex is likely the predominant one that exists in vivo. Although Hg-GS$_2$ is also likely to exist in vivo, the extremely high renal activity of the enzyme γ-glutamyltransferase (GGT) on the renal BBM will rapidly and efficiently degrade this complex to the Hg-Cys$_2$ complex. Additionally, experimental studies in which the influence of modulation of renal GSH status, including inhibition of GGT activity, support the conclusion that Hg-Cys$_2$ is the speciation that is the most physiologically relevant. While HCys is normally present in plasma at much lower concentrations than either Cys/Cys$_2$ or GSH/GSSG, and thus likely plays a minor role in Hg^{2+} ion transport under normal, physiological conditions, plasma HCys levels may be significantly elevated in various disease states such as cardiovascular disease. This may increase the risk of Hg^{2+}-induced intoxication and nephrotoxicity in those exposed individuals with preexisting disease.

Intracellular Mechanisms of Mercury-Induced Nephrotoxicity

The key to the biochemical and molecular mechanisms of Hg^{2+}-induced nephrotoxicity, as with its pharmacokinetics and transport, is the high affinity of the Hg^{2+} ion for sulfhydryl groups on proteins and other molecules. A schematic illustrating the four major intracellular targets for Hg^{2+} ions in the renal PT cell is shown in Fig. 3. As with any blood-borne or filtered chemical, the first sites in the kidneys that would be exposed to mercuric-thiol complexes are the epithelial cell plasma membranes. Hence, it is not surprising that an early target of Hg^{2+} ions are sulfhydryl-containing membrane proteins such as the (Na^++K^+)-stimulated ATPase, which is found on the basolateral plasma membrane (BLM) (Zalups and Lash 1994). Additional early targets are cytoskeletal proteins such as cytokeratins, actin, and integrins; these are also rich in sulfhydryl groups. Predictably, binding of Hg^{2+} ions to renal plasma membranes leads to inhibition of the function of many membrane proteins, including transporters, and structural changes that lead to increases in membrane permeability, with eventual loss of membrane structure.

Once inside the renal PT cell, the mitochondria are another major site at which binding of Hg^{2+} ions to protein sulfhydryl groups and depletion of GSH by binding to the thiol group of this tripeptide occurs. These events result in inhibition of the mitochondrial electron transport chain, diminished cellular antioxidant defense, and apoptosis. Inhibition of the mitochondrial electron transport chain by numerous toxicants, including Hg^{2+}, is associated with an increase in production and release of reactive oxygen species (ROS), such as superoxide anion and hydrogen peroxide. Hg^{2+} ions may also target the nucleus, leading to DNA damage. Alternatively, the increase in ROS production from the mitochondria may also result in DNA damage. Both direct DNA damage by Hg^{2+} ions, and that which occurs indirectly from the mitochondrial ROS generation, may result in apoptosis. There have also been studies that have shown effects of low-dose exposures to various forms of mercury on signaling pathways, such as the mitogen-activated protein kinase pathway, thus leading to alterations in cell growth and cell death.

Therapeutic Approaches to Mercury-Induced Nephrotoxicity

Experimental and clinical approaches to preventing or counteracting nephrotoxicity caused by exposure to mercuric complexes generally focus on the use of chelators, such as dimercaprol (also known as British anti-lewisite or BAL), meso-2,3-dimercaptosuccinic acid (DMSA), and 2,3-dimercapto-1-propanesulfonic

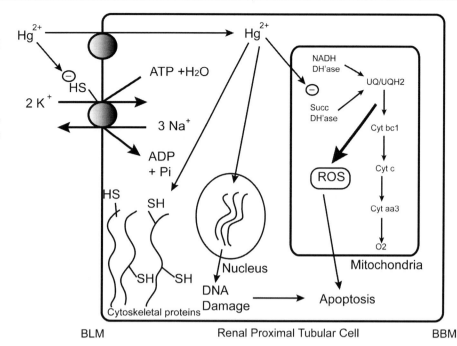

Mercury Nephrotoxicity, Fig. 3 Molecular targets for mercuric ions (Hg^{2+}) in renal proximal tubular cells. Abbreviations: *BBM* brush-border plasma membrane, *BLM* basolateral plasma membrane, *Cyt* cytochrome, *DH'ase* dehydrogenase, *ROS* reactive oxygen species, *–SH* sulfhydryl group, *UQ and UQH$_2$* oxidized and reduced forms of ubiquinone

acid (DMPS) (Aposhian 1983; Aposhian and Aposhian 1990; Aposhian et al 1995). The latter two chelators are more commonly used in clinical applications because of more significant side effects associated with use of dimercaprol.

The basic principle underlying the therapeutic application of chelators is that they possess an extremely high affinity for the Hg^{2+} ion, thus enabling them to pull the metal off of various ligands (mostly thiol and sulfhydryl groups), and the metal-chelator complex is then readily excreted from the target organ. Treatment with chelators can prevent further cytotoxicity due to the metal. Depending on when the chelator is used in relation to the exposure, however, can influence overall efficacy. Hence, chelator use is often accompanied by other supportive measures that seek to preserve or reverse a decline in renal function.

Cross-References

- Bacterial Mercury Resistance Proteins
- Lead Nephrotoxicity
- Mercury and Alzheimer's Disease
- Mercury and DNA
- Mercury and Immune Function
- Mercury and Lead, Effects on Voltage-Gated Calcium Channel Function
- Mercury and Low Molecular Mass Substances
- Mercury in Plants
- Mercury Neurotoxicity
- Mercury Toxicity
- Mercury Transporters
- Mercury, Physical and Chemical Properties
- Metallothioneins and Mercury

References

Aposhian HV (1983) DMSA and DMPS – water soluble antidotes for heavy metal poisoning. Annu Rev Pharmacol Toxicol 23:193–215

Aposhian HV, Aposhian MM (1990) Meso-2,3-Dimercaptosuccinic acid: chemical, pharmacological and toxicological properties of an orally effective metal chelating agent. Annu Rev Pharmacol Toxicol 30:279–306

Aposhian HV, Maiorino RM, Gonzalez-Ramirez D et al (1995) Mobilization of heavy metals by newer, therapeutically useful chelating agents. Toxicology 97:23–38

Bridges CC, Zalups RK (2005) Molecular and ionic mimicry and the transport of toxic metals. Toxicol Appl Pharmacol 204:274–308

Bridges CC, Zalups RK (2010) Transport of inorganic mercury and methylmercury in target tissues and organs. J Toxicol Environ Health Part B 13:385–410

Clarkson TW (1993) Molecular and ionic mimicry of toxic metals. Annu Rev Pharmacol Toxicol 32:545–571

Lash LH (2008) Principles and methods for renal toxicology. In: Hayes AW (ed) Principles and methods in toxicology, 5th edn. CRC Press, Boca Raton

Lash LH (2010) Glutathione, protein thiols, and metal homeostasis. In: Zalups RK, Koropatnick DJ (eds) The molecular biology and toxicology of metals, 2nd edn. Taylor & Francis, London

Zalups RK (2000) Molecular interactions with mercury in the kidney. Pharmacol Rev 52:113–143

Zalups RK, Lash LH (1994) Advances in understanding the renal transport and toxicity of mercury. J Toxicol Environ Health 42:1–44

Mercury Neurotoxicity

Michael Aschner[1,2,3,4], Marcelo Farina[5] and João B. T. Rocha[6]
[1]Department of Pediatrics, Division of Pediatric Clinical Pharmacology and Toxicology, Vanderbilt University Medical Center, Nashville, TN, USA
[2]Center in Molecular Toxicology, Vanderbilt University Medical Center, Nashville, TN, USA
[3]Center for Molecular Neuroscience, Vanderbilt University Medical Center, Nashville, TN, USA
[4]The Kennedy Center for Research on Human Development, Vanderbilt University Medical Center, Nashville, TN, USA
[5]Departamento de Bioquímica, Centro de Ciências Biológicas, Universidade Federal de Santa Catarina, Florianópolis, SC, Brazil
[6]Departamento de Química, Centro de Ciências Naturais e Exatas, Universidade Federal de Santa Maria, Santa Maria, RS, Brazil

Synonyms

Hg; Hydrargyrum; Quicksilver

Definition

A heavy, silvery, toxic univalent and bivalent metallic element. Its atomic number is 80, and it is the only metal that is liquid at ordinary temperatures. The oxidized inorganic [Hg(II)] and organic [CH$_3$Hg(II)] forms of Hg interact with thiol-containing metalloproteins, which mediate their toxicity.

Mercury in the Environment

Mercury (Hg) is a global pollutant and even the most stringent control of its pollution from man-made (anthropogenic) sources will *not* eliminate human exposure to potentially toxic quantities. Ocean sediments represent the largest global repository for Hg and are estimated to contain 10^{17} g of Hg, predominantly in the form of sulfhydryl- sulfur-bound Hg (HgS) (Nriagu 1979). Ocean waters contain approximately 10^{13} g, soils and freshwater sediments 10^{13} g, the biosphere 10^{11} g (mostly in land biota), the atmosphere 10^8 g, and freshwater in the order of 10^7 g. This Hg budget excludes "unavailable" Hg which is found in subterranean repositories, such as mines. Recent estimates of total annual natural and anthropogenic mercury emissions by the US Environmental Protection Agency are about 4,400–7,500 t, with Asia accounting for 53% of the total emissions, Africa (18%), Europe (11%), North America (9%), Australia (6%), and South America (4%). Notably, approximately two thirds of the total Hg emissions are man-made (anthropogenic), predominantly as a result of coal combustion and various industrial uses. Approximately 3% of all global anthropogenic emissions are contributed from US sources, with the power sector accounting for about 1% of the total. These emissions account for approximately 40% of total US man-made Hg emissions.

In nature, Hg exists predominantly in three different molecular species: elemental (Hg°), inorganic (Hg^{2+}), and organic (MeHg). In the environment, Hg released from natural and anthropogenic sources (US EPA 1997) is sustained in the upper sedimentary layers of sea and lake beds, where sulfate-reducing bacteria methylate a portion of the inorganic mercury by the action of microorganisms to form the toxic species, MeHg. Concentrations of MeHg are biomagnified within the food chain, reaching concentrations in fish 10,000–100,000 times greater than in the surrounding water (US EPA 1997; Wiener et al. 2003). The bioaccumulation of Hg in aquatic life is an issue of global human health and ecological risk because Hg input into aquatic systems from atmospheric deposition and terrestrial sources is converted to highly toxic, bioaccumulative MeHg by the action of microorganisms. Nearly all fish have detectable MeHg levels. However, the MeHg enrichment in the aquatic food chain is nonuniform. It depends not only upon the Hg content in the water and bottom sediments, but also on physicochemical properties, such as the water's pH and redox potential, marine species, and their size and age. Anoxia also favors the growth of microorganisms, thus increasing the methylation rate of Hg and the generation of MeHg. Notably, MeHg accumulation in fish represents the major exposure route for human populations around the globe that subside predominantly on fish for their dietary requirements.

Human Health Impact

In the past, Hg compounds have been used as fungicides, antiseptics, vaccine preservatives, disinfectants, laxatives, diuretics, nasal sprays, cosmetics, and other biomedical applications; however, given their propensity to cause ill effects in humans as well as ecological systems, mercurial usage has been greatly curtailed (ATSDR 2003). While human exposures to all forms of Hg have been documented, exposure to the methylated form, MeHg, is the most common and ubiquitous. Accordingly, following a brief discussion on (Hg°) and inorganic Hg toxicity, the remainder of this entry will focus on MeHg neurotoxicity, which in part depends on, or is mediated via its interaction with metalloproteins, such as hemoglobin, the various metallothionein isoforms, and δ-aminolevulinate dehydratase (ALA-D).

Elemental and Inorganic Mercury

Elemental or metallic mercury (Hg°) is the only metal that is liquid at ordinary temperatures. The vapor pressure of Hg at room temperature (25°C) is low and its evaporation is very slow, but it will occur. The evaporation rate depends on the temperature, surface area, and air currents. One gram of mercury with 1 cm^2 surface area and average air currents, at 20°C, will take about 108 weeks to totally evaporate and this process would significantly accelerate at higher temperature. Once heated it volatilizes becoming hazardous to humans.

Acute Hg° vapor exposures cause serious respiratory problems, characterized by dyspnea, associated with increased excitability, and excessive large-scale exposure to Hg° can cause pulmonary fibrosis and death, whereas chronic exposures to Hg° mostly cause disturbances in central nervous system (CNS) characterized by tremors, polyneuropathy, delusions, hallucinations, loss of memory, insomnia, and neurocognitive disorders. Erethism, defined as exposure to Hg° vapor is characterized by bizarre behavior, such as excessive shyness and even aggression. Past exposures to Hg° vapor were occupationally common.

Indeed, the preparation of mercuric nitrate [Hg(II) NO$_3$] is carried out by mixing Hg° with HNO$_3$, which will lead to evaporation of Hg°. Fumes liberated from Hg(II)NO$_3$ mixtures, which were used in the process of curing pelts in some hats, make it impossible for hatters to avoid inhaling it. Such exposures likely led to the well-known symptoms of the Mad Hatter in *Alice in Wonderland*. Contemporary issues with Hg° vapor exposure in the general population are largely limited to release of Hg° from dental amalgams (Clarkson et al. 2003) and from mining gold in developing countries. Exposures in dental personnel result in urinary Hg levels well below those found in people who are occupationally exposed to Hg, and evidence in support of neurotoxicity is limited.

Previous uses of inorganic Hg (Hg^{2+}) include multiple medical products, including topical antiseptic, vermifuges, skin-lightening creams and teething powders. Hg salts are extremely toxic to kidneys, causing severe renal dysfunctions including tubular necrosis and glomerulonephritis (Clarkson and Magos 2006). Acrodynia, characterized by painful extremities and also known as pink disease, can also be induced in response to Hg exposure as evidenced in children exposed to mercurial chloride calomel-containing teething powders. Kawasaki disease, an immune disorder characterized by skin lesions and rashes, peripheral extremity changes, fever, and photophobia, has also been attributed to exposure to inorganic Hg.

Methylmercury

The first case of fatal occupational poisoning by MeHg was recorded in 1863 in researchers synthesizing organic mercurials. Additional four adult cases of occupational MeHg neurotoxicity were reported in 1940. Catastrophic epidemics from environmental MeHg contamination followed in Japan and Sweden. In Iraq, a large outbreak of human MeHg poisoning was traced to consumption of bread prepared from seeds treated with a fungicide containing MeHg. Additional MeHg poisoning outbreaks were reported in Pakistan, Guatemala, and Ghana. Following massive poisoning in Minamata, Japan (late 1950s–early 1960s), the sequelae of in utero MeHg exposure was recognized as congenital Minamata disease. It is characterized by mental retardation, primitive reflexes, coordination disturbances, dysarthria, limb deformation, growth disorder, chorea-athetosis, and hypersalivation (Harada 1995). In contrast, in adults exposed to MeHg the symptoms include paresthesias of the circumoral area and hands and feet, constriction of the visual-field, and ataxia. The neuropathology in adults is characterized by region-specific degeneration of neurons in the visual cortex and cerebellar internal granule cells (Clarkson et al. 2003). Commonly, symptoms commence following a latent phase of weeks or even months between exposure and the appearance of symptoms.

While these effects are inherent to exposure well above the US EPA reference dose (RfD) of 0.1 μg/kg body weight/day (an exposure without recognized adverse effects) it is noteworthy that approximately 8% of US women of child-bearing age have blood Hg concentrations exceeding 5.8 μg/L (level equivalent to the current RfD. In 2000, a National Academy of Sciences (NAS) expert panel reviewed studies of populations consuming large amounts of fish (in the Seychelle Islands, the Faroe Islands, and New Zealand), concluding that the weight of the evidence supported adverse health effects due to MeHg exposure (NAS 2000) and recommended that levels of mercury not exceed 5.0 μg/L in whole blood or 1.0 μg/g in hair, corresponding to a reference dose (RfD) of 0.1 μg/kg body weight/day.

Fate of MeHg in the Human Body

MeHg in food is efficiently absorbed (90%) by the gastrointestinal tract. After ingestion, its distribution to the blood compartment is complete within approximately 30 h, and the blood level accounts for about 7% of the ingested dose. MeHg accumulates predominantly in the red cells where it is bound to cysteinyl residues on the beta-chain of the hemoglobin molecule. Consequently, the fate of MeHg in human body is greatly dictated by its interaction with one of the most abundant metalloproteins found in blood. MeHg is slowly distributed from the blood, with equilibrium between blood and tissues reached at approximately 4 days. About 10% of the MeHg is retained in the brain. MeHg is predominantly excreted (about 90%) by the fecal route. Though most of the MeHg is eliminated through the liver into the bile and gut, it gets reabsorbed, thus establishing an enterohepatic circulation of MeHg. Slow excretion of MeHg via urine has been reported along with breast milk, the latter representing approximately 5% of the MeHg in the maternal blood. The excretion rate of MeHg can be modulated by the composition of the diet.

MeHg has a remarkably high affinity for the anionic form of –SH groups (log K, where k is the affinity constant and is in the order of 15–23). Despite the high thermodynamic stability of the MeHg-SH bond, very rapid exchange of MeHg between –SH groups is known to occur. In cells, MeHg can form a complex with the –SH-containing amino acid cysteine (Cys). The MeHg-S-Cys complex closely mimics the structure of the neutral amino acid, methionine, and is, therefore, a substrate of the neutral amino acid transporter system L. This molecular mimicry is responsible for MeHg uptake into cells. The uptake into the cells can occur by both an active, energy-dependent (e.g., MeHg-Cys) and a passive uptake, depending upon the Hg species. Upon entering the brain, the MeHg bound to low molecular weight thiols can be exchanged with thiols from different proteins, including those from metalloproteins. MeHg readily binds to –SH groups on metallothionines, releasing Zn^{2+} in the process. Intracellular Zn^{2+}, in turn, is known to contribute to MeHg-induced neurotoxicity (Fig. 1).

Neurotoxicity of MeHg

The brain is the primary target site for MeHg. In general, MeHg poisoning results in focal damage in adults, and widespread and diffuse damage in the fetal and neonatal brain (Clarkson et al. 2003), likely reflecting the dynamic nature of the developing CNS (cell division, migration, differentiation, and synaptogenesis). The brain levels of MeHg in the fetus are generally higher than that in the mother, and in newborns or infants, MeHg exposure leads to cerebral palsy-like effects, characterized by ataxic motor and mental symptoms with hypoplastic and symmetrical atrophy of the cerebrum and cerebellum. The neuropathological features include a decreased number of neurons and distortion of cytoarchitecture in the cortical and cerebellar areas. Specifically, these are characterized by incomplete or abnormal migration of neurons to the cerebellar and cerebral cortices, and deranged cortical organization of the cerebrum with heterotopic neurons.

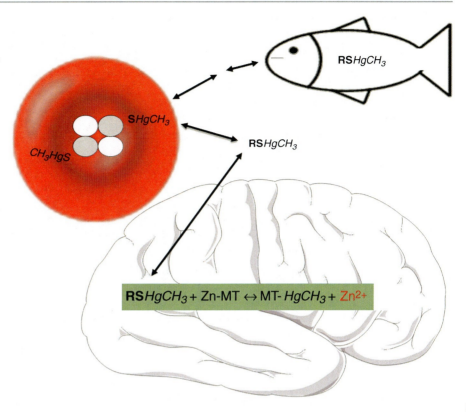

Mercury Neurotoxicity, Fig. 1 Methylmercury derived from fish (and other sources) is bound to molecular weight thiols (RSHgCH$_3$) in blood and brain. RSHgCH$_3$ can be exchanged with thiols from different proteins, including those from metalloproteins. As shown here, MeHg readily binds to –SH groups on metallothionines, releasing Zn^{2+} in the process, the latter contributing to MeHg-induced neurotoxicity

Mechanisms associated with MeHg neurotoxicity in the developing as well as the adult CNS include, but are not limited to: (1) inhibition of macromolecule synthesis (DNA, RNA, and protein), (2) microtubule disruption, (3) increase in intracellular calcium (Ca^{2+}) with disturbance of neurotransmitter function, (4) oxidative stress, and (5) excitotoxicity, secondary to altered glutamate homeostasis. Nonetheless, the primary site of action of MeHg, the genetic bases of its neurotoxicity as well as its specificity to certain cells have yet to be identified. From the above, excessive formation of reactive oxygen species (ROS) and impairments in antioxidant defenses appear to be a major factor in mediating MeHg neurotoxicity (Farina et al. 2010). MeHg exposure is associated with increased brain lipid peroxidation, superoxide and hydrogen peroxide amounts, impaired levels of superoxide dismutase (SOD), glutathione (GSH) reductase and GSH peroxidase activities, as well as a general decrease in GSH levels (Farina et al. 2010).

The molecular mechanism(s) via which MeHg causes oxidative stress in mammalian cells is(are) related to alkylation of the thiol (-SH) and selenol (-SeH) critical group of proteins. Of note, some of these MeHg-target proteins are metalloproteins, including Zn(II)-sulfhydryl-containing enzyme porphobilinogen synthase and the abundant and ubiquitous metallothioneines (Aschner 1996). In effect, MeHg is expected to alkylate thiol groups of the enzyme that are stabilized by Zn^{2+} and to displace Zn(II) from the protein-binding site(s). Another important Zn^{2+}-binding protein that can be targeted by MeHg is metallothionein, which is rich in cysteine residues. Brain metallothionein has a critical role in regulating intracellular Zn^{2+}, and a disruption in Zn^{2+} homeostasis by MeHg (via displacement of Zn^{2+} from its thiol-binding sites in metallothionein) can jeopardize neuronal cell fate (West et al. 2008).

MeHg has also been shown to impair the function of cytoskeleton. The latter regulates many crucial cellular processes, including cell survival, proliferation, differentiation, and migration. Given the high content of sulfhydryl groups (-SH) in microtubules, it is not surprising that binding of MeHg to these organelles causes cytoskeletal breakdown and destruction of mitotic spindles, resulting in cell cycle arrest. Consistent with a microtubule dysfunction are observations on reduced brain size in postmortem brains of infants exposed in utero to MeHg during the Iraqi outbreak.

In addition, MeHg affects the homeostasis of multiple neurotransmitter systems. Synthesis, uptake, release, and degradation of neurotransmitters have all been associated with exposure to MeHg and are mostly documented in glutamatergic, cholinergic, and dopaminergic systems. It is noteworthy that MeHg preferentially accumulates in glial cells, both astrocytes and microglia. The former remove approximately 80% of synaptic glutamate. Exposure to MeHg has been shown to attenuate the astrocytic removal of glutamate, thus sensitizing neurons to extracellularly elevated levels of this excitotoxic amino acid, and activation of neuronal *N*-methyl D-aspartate (NMDA) receptors. A number of studies have also shown that developmental MeHg exposure affects the dopaminergic system, causing reduced levels of dopamine (DA), abnormal DA turnover, and synaptosomal DA uptake.

Treatment of MeHg Poisoning

The body burden of MeHg can be effectively reduced by hemodialysis combined with extracorporeal infusion of chelating agents, such as N-acetylcysteine (NAC) or cysteine and oral administration of 2,3-dimercapto-1-propane sulfonate (DMPS). However, the effectiveness of chelation therapy is unpredictable as occasionally it fails to show efficacy, possibly since binding of MeHg to thiol-containing proteins (including metalloproteins) induces protein conformational changes that preclude its removal by chelating agents.

Conclusions

MeHg is a contemporary environmental neurotoxicant that interacts with different molecular targets. Of particular toxicological significance, its interaction with metalloproteins contributes decisively to dictate either its distribution in the body and its neurotoxicity. Consequently, our knowledge about MeHg neurotoxicity can be considerably advanced if future studies are designed to enhance the understanding of its interaction with different classes of metalloproteins and, particularly, the mechanism which accounts for the ability of MeHg to displace Zn from cerebral metallothionein.

Acknowledgments These authors wish to acknowledge funding by Grants from the Conselho Nacional de Desenvolvimento Científico e Tecnológico (CNPq), Coordenação de Aperfeiçoamento de Pessoal de Nível Superior (CAPES) and from the National Institute of Environmental Health Sciences (USA). The FINEP research Grant "Rede Instituto Brasileiro de Neurociência (IBN-Net)" # 01.06.0842-00 and INCT for Excitotoxicity and Neuroprotection-MCT/CNPq are especially appreciated. MA was supported by NIH Grant R01 ES07331.

Cross-References

▶ Scandium, Physical and Chemical Properties
▶ Silicon, Physical and Chemical Properties
▶ Sodium and Potassium Transport in Mammalian Mitochondria
▶ Sodium Channel Blockers and Activators
▶ Sodium-Coupled Secondary Transporters, Structure and Function
▶ Strontium and DNA Aptamer Folding
▶ Strontium, Calcium Analogue in Membrane Transport Systems
▶ Strontium, Physical and Chemical Properties
▶ Tin, Physical and Chemical Properties

References

Agency for Toxic Substance and Disease Registry (2003) Toxicological profile for mercury. U.S. Department of Health and Humans Services, Public Health Service, Centers for Disease Control, Atlanta

Aschner M (1996) The functional significance of brain metallothioneins. FASEB J 10:1129–1136

Clarkson TW, Magos L (2006) The toxicology of mercury and its chemical compounds. Crit Rev Toxicol 36(8):609–62

Clarkson TW, Magos L, Myers GJ (2003) The toxicology of mercury–current exposures and clinical manifestations. N Engl J Med 349(18):1731–7

Farina M, Rocha JBT, Aschner M (2010) Oxidative stress and methyl mercury-induced neurotoxicity. In: Wang C, Slikker W Jr (eds) Developmental neurotoxicology research: principles, models, techniques, strategies and mechanisms. Wiley, Hoboken, pp 357–385

Harada M (1995) Minamata disease: methylmercury poisoning in Japan caused by environmental pollution. Crit Rev Toxicol 25:1–24

National Academy of Sciences (2000) Toxicologic effects of methylmercury. National Research Council, Washington, DC

Nriagu JO (1979) Global inventory of natural and anthropogenic emissions of trace metals to the atmosphere. Nature 279:409–411

U.S. Environmental Protection Agency (1997) Mercury study: report to congress: vol. 1, Executive summary. EPA-452-R-97-003. Washington, DC: U.S. EPA. http://www.epa.gov/ngispgm3/iris/subst/0073.htm

West AK, Hidalgo J, Eddins D, Levin ED, Aschner M (2008) Metallothionein in the central nervous system: roles in protection, regeneration and cognition. Neurotoxicology 29(3):489–503

Wiener JG, Krabbenhoft DP, Heinz GH, Scheuhammer AM (2003) Ecotoxicology of mercury. In: Hoffman DJ, Rattner BA, Burton GA, Cairns J (eds) Handbook of ecotoxicology. CRC Press, Boca Raton, pp 409–463

Mercury Poisoning

▶ Mercury Toxicity

Mercury Toxicity

Klaudia Jomova[1] and Marian Valko[1,2]
[1]Department of Chemistry, Faculty of Natural Sciences, Constantine The Philosopher University, Nitra, Slovakia
[2]Faculty of Chemical and Food Technology, Slovak Technical University, Bratislava, Slovakia

Synonyms

Chelation therapy; Inorganic mercury; Mercury poisoning; Mercury-induced oxidative stress; Organic mercury

Definition

Mercury is the 80th element on the periodic table and is unique in that it is found in nature in several chemical and physical forms (Stohs and Bagchi 1995; Valko et al. 2005). Hg is a heavy metal and exists as several species: elemental mercury (Hg0, metallic mercury), inorganic mercury compounds (e.g., mercury chloride, $HgCl_2$), and organic mercury (e.g., methylmercury (MeHg) and ethylmercury (EtHg)). Mercury (Hg) is an omnipresent environmental toxin that may cause a serious health risk.

Mercury Toxicity

Mercury poisoning is caused by the ingestion or inhalation of mercury or a mercury compound (Guzzi and La Porta 2008). The chronic form of mercury poisoning (concentrations in the range of 0.7–40 $\mu g/m^3$) results from inhalation of the vapors or dust of mercurial compounds and is characterized by thirst, excessive saliva, irritability, slurred speech, tremors, and other symptoms. Signs of acute mercury poisoning appear within a few minutes and include a metallic taste in the mouth, nausea, vomiting, thirst, abdominal pain, bloody diarrhea, and renal failure that may result in death. The presence of mercury in the body is determined by a urine test.

The most common is exposure to organic mercury; however, all forms of mercury exhibit toxic effects in a number of organs, especially in the kidneys (Zalups 2000). The studies devoted to mercury toxicity became very important after the environmental catastrophes in Minamata (Kumamoto prefecture, Japan) in 1956 and in Niigata prefecture in 1965 (Eto 2000; Takeuchi et al. 1979). Both catastrophes were caused by the release of methylmercury in the industrial wastewater from the local chemical plants. Methylmercury contaminated fish, consumed by local people. Additional methylmercury intoxication occurred in Iraq in the early 1970s. In this case, seed grains were treated with a fungicide containing organic mercury. Several thousand people were hospitalized, and nearly 500 people died from consumption of contaminated bread.

All forms of mercury, including inorganic, organic, and elemental forms of mercury, exhibit toxicological symptoms including neurotoxicity, nephrotoxicity, and gastrointestinal toxicity with ulceration and hemorrhage (Zalups 2000). Generally, organic mercuric compounds exhibit a lesser degree of nephrotoxicity. A variety of mercury-induced changes, including increased oxidative stress, permeability of the blood–brain barrier and disruption of membranes, protein synthesis, microtubule formation, DNA replication, calcium homeostasis, and impairment of synaptic transmission, are being investigated.

Methylmercury and Ethylmercury

While the exact site of absorption is not known, a major portion of the ingested methylmercury (methyl-Hg, MeHg, CH_3Hg^+) by fish enters circulation in the gastrointestinal tract. Conjugates of mercury and low molecular weight ligands such as cysteine and glutathione have been found to be the most likely species of mercury taken up by the basolateral membrane by the organic anion transporter. Since the methylmercuric cation has high affinity for the sulfhydryl groups (SH-), it readily forms complexes with water-soluble molecules (e.g., proteins) or thiol-containing amino acids (Aschner and Aschner 1990).

The proximal convoluted tubule and *pars recta* of the kidney are the most sensitive segments to the toxic effect of mercury (Zalups 2000). The toxic effect in the kidney is given by the bonding properties of mercury compounds which predetermine the molecular interactions between mercury and the target cells. Mercuric ions have a significant affinity toward reduced sulfur atoms, mainly thiol-containing molecules such as glutathione, cysteine, albumin, *N*-acetylcysteine, and others (Guzzi and La Porta 2008; Zalups 2000). While binding constants for Hg bonding to carboxylate oxygens or amino nitrogens are of the order of 10^3–10^8, the binding constants, characterizing Hg-thiol interactions, are approximately 12 orders of magnitude greater ($\sim 10^{15}$–10^{20}). Thus, the toxic effects of organic and inorganic mercury are related to their interactions with SH-containing groups.

MeHg crosses the blood–brain barrier complexed with L-cysteine and in the human body is present in the form of water-soluble complexes. Since methylmercury can enter placental circulation, its levels in the fetal brain were found to be elevated in pregnant women exposed to mercury. Methylmercury is in human body metabolized to inorganic mercury by microflora in the intestines at a slow rate. Urinary excretion of MeHg is very limited; majority of the MeHg is eliminated from the body by demethylation and excretion in the feces.

The ethylmercury (ethyl-Hg or EtHg, $C_2H_5Hg^+$) is a cation and is one of the metabolites of thiomersal, which is used as a preservative in some vaccines given to children. Thiomersal (in the USA known as thimerosal) is an organomercury compound possessing antiseptic and antifungal properties. Vaccines containing thiomersal given at recommended doses exhibited in certain cases local hypersensitivity reaction (Ball et al. 2001). Thiomersal exhibited change in the cell membrane permeability and has been found to induce DNA strand breaks. In addition, apoptosis at micromolar concentrations has been reported. For safety reasons, US and European countries have already removed thiomersal from vaccines.

Exposure to ethylmercury may induce neurodevelopmental disorders. Clinical manifestations of ethylmercury poisoning include speech disorders, vision disorders, tremor, ataxia, spasticity, and other symptoms. Subtle measures of developmental neurotoxicity (as done for methylmercury) have not been evaluated for ethylmercury. Ethylmercury is approximately five times less acutely toxic than methylmercury.

Inorganic Mercury

Inorganic mercury compounds represent another source of intoxication, since such compounds have been used in various industrial and pharmaceutical products. Mercurial chloride is a constituent of various skin creams.

Inorganic mercury ions enter into proximal tubular epithelial cells followed by distribution throughout all intracellular pools. The most common route of exposure to inorganic mercury is ingestion, appearing with clinical manifestations such as diarrhea, vomiting, renal failure, and other symptoms. Acute poisoning of individuals exposed to inorganic mercury might lead to neurologic, renal, and dermatological problems indicated by psychiatric disturbances, flushing, tremor, hypertension, and other problems.

The most critical organ for the ingestion of inorganic mercury appears to be kidney (Afonso and de Alvarez 1960). Acute renal failure has been observed following ingestion of mercuric chloride. This has been accompanied by a significant decrease in urinary protein secretion and increased excretion of albumin and b2-microglobulin. In addition, high levels of serum creatine phosphokinase have been observed. Dermal effects of mercurous chloride in humans include flushing, itching, and swelling. Gastrointestinal problems appeared among patients hypersensitive to mercury. The problems involved abdominal pain, nausea, vomiting, and other symptoms.

Immunotoxic response (Kawasaki disease) has been observed in children exposed to inorganic mercury. Children have a higher absorption rate than adults. The symptoms involved fever, oral lesions, tachycardia, and skin rashes. Urine of patients suffering from Kawasaki disease has shown increased level of mercury.

Elemental Mercury

Mercury vapor is more toxic than liquid mercury. Recent studies conducted on animals and humans have shown that elemental mercury from dental amalgams is another significant source of mercury body burden in humans (Lorscheider et al. 1995; Björkman et al. 1997). Dental amalgams usually contain 43–54% elemental mercury. Amalgam fillings release mercury vapor into the oral cavity. The majority of Hg vapor is absorbed by the lungs, and oxidized Hg is accumulated in the liver, brain, kidney, and cortex. Both acute and chronic exposures to elemental mercury are manifested by cough, fever, tremors, delusions, hallucinations, loss of memory, insomnia, and neurocognitive disorders.

Molecular Mechanism of Mercury Toxicity

As already mentioned above, the most toxic forms of mercury are its organic compounds, such as MeHg and EtHg. Organic mercury compounds quickly diffuse across the biological membranes, thus membrane leakage has been proposed to be the key step in mercury toxicity. The main route of methylmercury transmembrane transport appears to be associated with transport system designed for delivery of amino acids into the cell. MeHg-cysteine conjugate is the principal pathway whereby methylmercury exerts it toxicity. Thus, studies of inhibitors of the amino acid transport system are expected to be useful in preventing disorders triggered by MeHg toxicity.

Calcium homeostasis plays a key role in regulating CNS cell death. A low concentration of MeHg disrupts cell calcium homeostasis by increasing intracellular levels of calcium. In vitro experiments explored that calcium channel blockers have been shown to delay MeHg-induced increase of calcium levels (Levesque and Atchison 1991). In view of this, calcium channel blockers have been shown to prevent the neurological disorders in laboratory animals administrated with MeHg.

In addition to MeHg-induced increase of intracellular calcium in the mouse peritoneal neutrophil, MeHg also potently decreased nitric oxide (NO$^\bullet$) production. The protein and mRNA levels of NO$^\bullet$ synthase induced by lipopolysaccharide were also decreased. Both L-type calcium channel blockers, verapamil and H-89, can antagonize the inhibitory effect of MeHg on NO$^\bullet$ production. These findings lead to the conclusion that MeHg inhibits NO$^\bullet$ production mediated at least in part by calcium-activated adenylate cyclase-cAMP-protein kinase A pathway.

The mitochondria are small intracellular organelles which are responsible for energy production and cellular respiration. Mitochondria accomplish this task through a mechanism called the electron transport chain. Exposure to MeHg under in vivo conditions caused its accumulation inside the mitochondria followed by a series of biochemical changes in this organelle. MeHg induces an impairment of the activity of the enzymes such as cytochrome C oxidase, superoxide dismutase (SOD), and succinate dehydrogenase (SDH), all playing significant roles in the mitochondrial energy metabolism. In addition, it has been demonstrated that administration of mercury in rats resulted in depletion of major cellular antioxidant glutathione followed by increased formation of ROS such as hydrogen peroxide and peroxidation of lipids.

In vitro exposure of isolated rat liver mitochondria to MeHg leads to inhibition of electron transport, phosphorylation and increase of potassium permeability, and decay of the mitochondrial membrane potential (MMP). Decay in MMP results in efflux of calcium from mitochondria and inhibition of mitochondrial calcium uptake (Levesque and Atchison 1991).

Exposure of MeHg to isolated rat brain mitochondria causes ATP-dependent/independent decrease in calcium uptake and increase in calcium release from mitochondria.

MeHg directly affects the mechanism of neurotransmission, involving enzymatic activation of neurotransmitters, postsynaptic events, and uptake/release of neurotransmitters (Atchison 2005). Neurotoxic agents may interfere indirectly with the process of neurotransmission by interacting with Na channels, ATPases, and energy metabolism. Methylmercury triggers spontaneous release of dopamine, acetylcholine, GABA, and serotonin from rat brain synaptosomes (Minema et al. 1989).

Metallothioneins (MTs) are small intracellular proteins (MW of 6–7 kDa) containing numerous cysteine residues. They have the capacity to bind very effectively various metals, including mercury. The detailed description of a binding mode in MeHg-MT complex is rather difficult, given the very complex Me-Hg-thiolate chemistry. MeHg exhibits a clear preference to form linear two-coordinate mercury complexes with thiolate ligands. Several studies reported suppressed cytotoxicity of MeHg by metallothioneins (Hidalgo et al. 2001). Induction of metallothioneins in astrocytes attenuated (and even reversed) the cytotoxicity caused by MeHg. In this case, MeHg is bound by an astrocyte-specific metallothionein, Mt1.

MeHg has been supposed to induce in vivo and in vitro formation of ROS with the consequence of oxidative damage to cells. ROS levels increased following MeHg treatment of neuronal cultures and hypothalamic neuronal cell lines. The formation of these species led to cellular damage and cell death in distinct cell types such as astrocytes and neurons.

Mercury Toxicity and Antioxidants

In vitro and in vivo studies suggested that exposure of laboratory animals to inorganic and organic mercury may to lead to induction of oxidative stress (Zalups 2000). This has been explained by the high affinity of mercuric ions for binding to thiols which in turn leads to depletion of intracellular thiols, especially glutathione and increased formation of hydrogen peroxide and lipid peroxidation in kidney mitochondria. Rats treated with mercuric chloride demonstrated decreased levels of vitamins C and E and glutathione in the kidneys.

Levels of glutathione and catalase activity have been found to correlate proportionally with the MeHg exposure in human population study. Increased levels of glutathione in woman from Amazon area exposed to high levels of MeHg were interpreted by the increased glutathione synthesis and inhibition of glutathione peroxidase activity. These observations may be explained by the adaptive mechanisms for Hg-induced oxidative stress of individuals with a long-term exposure to MeHg.

Animals treated with mercuric chloride and supplemented with vitamin E had lower concentration of mercury in testis than animals fed a diet not supplemented with vitamin E. Thus, vitamin E exhibits a certain level of protection against mercuric chloride-induced reproductive toxicity.

In cerebellar granule cells, antioxidants tocopherols and tocotrienols effectively prevented cell death caused by methylmercury intoxication. MeHg is known to inhibit glutamate transport in astrocytes which in turn leads to increased levels of ROS, and several antioxidants have been shown to be effective in attenuating MeHg neurotoxicity. It has been demonstrated that intrastriatal administration of various concentrations of MeHg induces release of dopamine in rat striatum which has been explained by the interaction of mercury with dopamine transporter. Treatment with cysteine and glutathione suppressed dopamine release induced by MeHg.

Chelation Therapy for Mercury Toxicity

The treatment of mercury toxicity in humans is based on the application of chelation therapy (Aposhian et al. 1995). Chelation therapy most often involves administration of chelating agents to remove the undesirable toxic metals from the body (Fig. 1). A chelator is a molecule that forms multiple interactions with a metal ion or a metalloid ion to form a stable ring complex known as a chelate. The chelate is in most cases subsequently excreted in urine.

Within the past several decades, a major progress has been achieved in development and application of chelating agents to treat both acute and chronic mercury poisonings (Andersen and Molecular 2002). The most commonly used chelators to treat mercury poisoning involve DMPS (2,3-dimercapto-1-propanesulfonic acid) and its sodium salt known as Unithiol, DMSA (dimercaptosuccinic acid), D-penicillamine, and BAL (RS-2,3-disulfanylpropan-1-ol, dimercaprol, British anti-Lewisite) (Fig. 1).

DMPS has been shown to be a useful chelator to prevent fatal damage in fetus following methylmercury intoxication during pregnancy (Rischitelli et al. 2006). Metal complexes with DMPS are not lipid soluble. DMPS is able to penetrate into the kidney cells, mobilize mercury stored in the kidney, and favor urinary excretion (Aposhian et al. 1992). Clinical data proved that DMPS is able to remove a substantial amount of mercury deposited in human tissues within 6 h after administration of the drug. However, postexposure DMPS treatment of mercury vapor in animal models failed to remove the mercury from the tissues of the brain, the most important target organ.

DMSA binds to mercury and then pulls it out of the body. Although DMSA is most often associated with mercury chelation, it also effectively eliminates other heavy metals, including lead (Rischitelli et al. 2006). DMSA is water soluble and therefore is convenient for

Mercury Toxicity, Fig. 1 The structures of chelating agents designed to inhibit mercury poisoning (DMPS, 2,3-dimercapto-1-propanesulfonic acid, DMSA, dimercaptosuccinic acid, BAL, RS-2,3-disulfanylpropan-1-ol)

oral dosing. In animal model studies, subcutaneous treatment with DMSA prevented embryo lethality and fetotoxicity (Domingo 1995). DMSA does not cross the blood–brain barrier and therefore has been shown to be unable to remove mercury from the brain.

D-penicillamine has a preventive effect on developmental toxicity. D-penicillamine was able to prevent morphologic changes in the fetal brain in an animal model of pregnant rats (Aposhian et al. 1992).

BAL (dimercaprol) has been considered the first effective drug against mercury toxicity. While BAL was able to remove inorganic mercury from kidneys, it did not suppress maternal and fetal toxicity against organic mercury exposure.

Acknowledgments This work was supported by Scientific Grant Agency (VEGA Project #1/0856/11) and Research and Development Agency of the Slovak Republic (Contracts No. APVV-0202-10).

Cross-References

- Bacterial Mercury Resistance Proteins
- Mercury and Alzheimer's Disease
- Mercury and DNA
- Mercury and Low Molecular Mass Substances
- Mercury Nephrotoxicity
- Mercury Neurotoxicity
- Mercury Transporters
- Mercury, Physical and Chemical Properties
- Metallothioneins and Mercury

References

Afonso JF, De Alvarez RR (1960) Effects of mercury on human gestation. Am J Obstet Gynecol 80:145–154

Andersen O, Molecular AJ (2002) Mechanisms of in vivo metal chelation: implications for clinical treatment of metal intoxications. Environ Health Perspect 110:887–890

Aposhian HV, Bruce DC, Alter W, Dart RC, Hurlbut KM, Aposhian MM (1992) Urinary mercury after administration of 2,3-dimercaptopropane-1-sulfonic acid: correlation with dental amalgam score. FASEB J 6:2472–2476

Aposhian HV, Maiorino RM, Gonzalez-Ramirez D, Zuniga-Charles M, Xu Z, Hurlbut KM, Junco-Munoz P, Dart RC, Aposhian MM (1995) Mobilization of heavy metals by newer, therapeutically useful chelating agents. Toxicology 97:23–38

Aschner M, Aschner JL (1990) Mercury neurotoxicity: mechanisms of blood-brain barrier transport. Neurosci Biobehav Rev 14:169–176

Atchison WD (2005) Is chemical neurotransmission altered specifically during methylmercury-induced cerebellar dysfunction? Trends Pharmacol Sci 26:549–557

Ball LK, Ball R, Pratt RD (2001) An assessment of thimerosal use in childhood vaccines. Pediatrics 107:1147–1154

Björkman L, Sandborgh-Englund G, Ekstrand J (1997) Mercury in saliva and feces after removal of amalgam fillings. Toxicol Appl Pharmacol 144:156–162

Domingo JL (1995) Prevention by chelating agents of metal-induced developmental toxicity. Reprod Toxicol 9:105–113

Eto K (2000) Minamata disease. Neuropathology 20(Suppl): S14–S19

Guzzi G, La Porta CA (2008) Molecular mechanisms triggered by mercury. Toxicology 244:1–12

Hidalgo J, Aschner M, Zatta P, Vasák M (2001) Roles of the metallothionein family of proteins in the central nervous system. Brain Res Bull 55:133–145

Levesque PC, Atchison WD (1991) Disruption of brain mitochondrial calcium sequestration by methylmercury. J Pharmacol Exp Ther 256:236–242

Lorscheider FL, Vimy MJ, Summers AO (1995) Mercury exposure from "silver" tooth fillings: emerging evidence questions a traditional dental paradigm. FASEB J 9:504–508

Minema DJ, Cooper GP, Greeland RD (1989) Effects of methylmercury on neurotransmitters release from rat brain synaptosomes. Toxicol Appl Pharmacol 99:510–521

Rischitelli G, Nygren P, Bougatsos C, Freeman M, Helfand M (2006) Screening for elevated lead levels in childhood and pregnancy: an updated summary of evidence for the US preventive services task force. Pediatrics 118:1867–1895

Stohs SJ, Bagchi D (1995) Oxidative mechanisms in the toxicity of metal-ions. Free Radic Biol Med 18:321–336

Takeuchi T, Eto N, Eto K (1979) Neuropathology of childhood cases of methylmercury poisoning (Minamata disease) with prolonged symptoms, with particular reference to the decortication syndrome. Neurotoxicology 1:1–20

Valko M, Morris H, Cronin MTD (2005) Metals, toxicity and oxidative stress. Curr Med Chem 12:1161–1208

Zalups RK (2000) Molecular interactions with mercury in the kidney. Pharmacol Rev 52:113–143

Mercury Transporters

Maksim A. Shlykov and Milton H. Saier Jr.
Department of Molecular Biology, Division of Biological Sciences, University of California at San Diego, La Jolla, CA, USA

Synonyms

Hg^{2+} excretion/efflux; Hg^{2+} resistance; Hg^{2+} uptake; Mercury (Hg^{2+}) permeases/porters

Definition

Transmembrane proteins that mediate the transport of Hg^{2+} and its derivatives into and out of the cell.

Introduction

Mercuric ions (Hg^{2+}) and methylmercury (CH_3Hg^+) are major, human-generated, toxic contaminants present in fish and our waterways. The accumulation and amplification of toxic mercury in food chains poses a major risk to all living organisms. Bacteria provide a means of bioremediation by taking up these compounds via membrane potential-dependent sequence divergent members of the mercuric ion (Mer) superfamily and using cytoplasmic mercury reductases to reduce them to volatile, less toxic, elemental mercury (Hg^0) which diffuses out of the cell. Higher eukaryotes pump mercuric ion conjugates into vacuoles or out of cells using ATP-hydrolysis-dependent pumps of the ATP-binding cassette (ABC) functional superfamily. Finally, there are membrane-embedded mercuric ion reductases that eliminate toxic Hg^{2+}, yielding Hg^0.

The Mer Superfamily

One of the ways bacteria detoxify Hg^{2+} is through the use of Mer superfamily mercury uptake channels. The Mer superfamily has recently been expanded to include five different families for which common origins have been established (Mok et al. 2012; see below).

Many publications have dealt with the characterization of mercuric ion transporters and their potential associations with mercuric reductases (Schué et al. 2009; Yamaguchi et al. 2007). Comparisons of the protein phylogenetic trees with the corresponding 16S ribosomal RNA trees led to the conclusion that genes encoding these proteins had undergone extensive horizontal transfer between organisms (Yamaguchi et al. 2007). This observation was not surprising in view of the fact that many of these proteins proved to be plasmid-encoded. Interestingly, not all of these families share a common topology. They exhibit 2, 3, or 4 transmembrane helical segments (TMSs) per polypeptide chain, and some of the 4 TMS homologues have internal duplications of 2 TMS hairpin structures (Mok et al. 2012).

The MerF protein, encoded on plasmid pMER327/419, is an 81 amino acyl residue (aa) polypeptide with 2 putative TMSs (Barkay et al. 2003). It catalyzes uptake of Hg^{2+} in preparation for reduction by mercuric reductase. The MerF gene is found on mercury-resistant plasmids from many Gram-negative bacteria, but the sequence of the protein from these plasmids is the same. Its 2 TMSs show limited sequence similarity with the first 2 TMSs of MerT (TC#1.A.72.3) and MerC (TC# 1.A.72.4). Some members of the MerF family have been designated MerH (Schué et al. 2009).

MerTP permeases catalyze Hg^{2+} uptake into bacterial cells in preparation for its reduction by the MerA mercuric reductase. The Hg^0 produced by MerA is volatile and passively diffuses out of the cell. The *merT* and *merP* genes are found on mercury-resistance plasmids and transposons of Gram-negative and Gram-positive bacteria, but they are also chromosomally encoded in some bacteria. MerT consists of about 130 aas and has three transmembrane helical segments. Evidence for direct interactions between the cytoplasmic face of MerT and the N-terminus of MerA has been presented (Schué et al. 2009). Operon analyses have been reported (Barkay et al. 2003).

MerP is a periplasmic Hg^{2+}-binding receptor of about 70–80 amino acids, synthesized with a cleavable N-terminal leader. It is homologous to the N-terminal heavy-metal-binding domains of the copper and heavy-metal-transporting P-type ATPases (subfamilies 3.A.3.5 and 3.A.3.6 in the Transporter Classification Database (TCDB; www.tcdb.org)). The 3-D structure of MerP from *Ralstonia metallidurans* has been solved to 2 Å resolution (Qian et al. 1998). It is 91 aas long with its leader sequence, is monomeric, and binds a single Hg^{2+} ion. Hg^{2+} is bound to a sequence GMTCXXC found in metallochaperones, mercuric reductases, and metal-transporting ATPases. The fold is $\beta\alpha\beta\beta\alpha\beta$, called the "ferredoxin-like fold." MerT homologues have been identified in which the 3 TMS MerT is fused to a periplasmic MerP "heavy-metal-associated" (HMA) domain via a linker region (see TC# 1.A.72.3.3).

The MerC protein encoded on the IncJ plasmid pMERPH of the *Shewanella putrefaciens* mercuric resistance operon is 137 amino acids in length and possesses 4 putative transmembrane α-helical spanners (TMSs). It has been shown to bind and take up Hg^{2+} ions. *merC* genes are encoded on several plasmids of Gram-negative bacteria and may also be chromosomally encoded. MerC proteins are homologous to other bacterial Hg^{2+} bacterial transporters (Mok et al. 2012; Yamaguchi et al. 2007).

The *merE* gene of transposon Tn21, a pE4 plasmid that contains the *merR* gene of plasmid pMR26 from *Pseudomonas* strain K-62, and the *merE* gene of Tn21 from the *Shigella flexneri* plasmid NR1 (R100) conferred hypersensitivity to CH_3Hg^+ and Hg^{2+}, taking up significantly more CH_3Hg^+ and Hg^{2+} than the isogenic strain (Kiyono et al. 2009). The MerE protein encoded by pE4 was localized in the cell membrane fraction but was not found in the soluble fraction. Kiyono et al. (2009) suggested that the *merE* gene is a broad mercury transporter mediating the transport of both CH_3Hg^+ and Hg^{2+} across the bacterial membrane.

Mok et al. (2012) demonstrated that the five Mer families described above (MerC, MerE, MerH, MerP, and MerT) are related by common descent. Screening of sequences for all five types of Mer proteins and removing false positives yielded GAP scores ranging from 8.6 to 20.4 S.D., strongly suggesting that all Mer proteins belong to one superfamily. This conclusion was confirmed by creating several phylogenetic trees, analyzing sequence motifs, estimating ancestral sequences, and examining protein topologies. These results and the considerations described below, allowed for the placement of all bacterial Mer porters under a single TC#, 1.A.72. TC subclass 1.A refers to channel proteins that span the membrane as α-helices.

Analysis of consensus sequences derived using the MEME program through an 81 amino acyl residue stretch provided further support for homology. The fully conserved residue, cysteine, and the occurrences of only conservative substitutions at many other positions argued against the idea that Mer proteins arose independently from more than one primordial protein. This conclusion was confirmed when it was found that ancestral sequences determined for members of the five families proved to be more similar to each other than were any of the present-day members of these families.

Topological analyses showed that Mer proteins have between 2 and 4 TMSs inclusive. However, two TMSs, 1 and 2, are shared by all Mer proteins, and the conserved motifs are within this region. Motif analysis suggested that TMS 1 plays a more crucial role in protein function while TMS 2 is more important for protein structure. Charge analyses suggested that the N-termini of TMSs 1 and the C-termini of TMSs 2 are in the cytoplasm while the loop between them is extracytoplasmic.

Proteins within the Mer superfamily may function with extracytoplasmic MerP Hg^{2+}-binding receptors as well as with cytoplasmic mercuric ion reductases. These porters probably function as channels, but this has not yet been established experimentally. So far, no secondary active carriers have been shown to have less than 4 TMSs (Saier 2003), providing evidence for an oligomeric channel mechanism. Because the membrane potential is negative inside, and mercuric ions bear two positive charges, a channel-type mechanism allows accumulation against large concentration gradients. The cytoplasmic Hg^{2+} concentrations would then be adequate for rapid reduction of Hg^{2+} to Hg^0 by mercuric ion reductases.

Further analyses of the mechanism and the structures of proteins within the Mer superfamily are of great importance, emphasized by the fact that bacteria possessing these Hg^{2+} transporters, together with cytoplasmic Hg^{2+} reductases, can alleviate the increasing occurrence of mercury poisoning due to practices within the food, mining, and energy industries.

The ATP-Binding Cassette (ABC) Superfamily

In contrast to Mer superfamily transporters which take up Hg^{2+}, known ABC transporters dealing with mercury resistance export Hg^{2+} conjugates. The human multidrug-resistance protein (MRP1) of the ABC superfamily is known to mediate cellular efflux of a wide range of xenobiotics, including anticancer drugs and heavy metals. Using MRP1-overexpressing lung tumor GLC4/Sb30 cells, Vernhet et al. (2000) were able to demonstrate 3.4-, 12.7-, and 16.3-fold increases in resistance over control cells to mercuric ion, arsenite, and arsenate, respectively. GLC4/Sb30 cells remained sensitive to other cytotoxic metals tested such as copper, chromium, cobalt, or aluminum. MK-571, a potent inhibitor of MRP1 activity, almost totally reversed resistance of GLC4/Sb30 cells to mercuric ions and arsenic while it did not significantly alter sensitivity of GLC4 cells to other metal ions. Arsenate-treated GLC4/Sb30 cells were found to poorly accumulate arsenic because of increased MK-571-inhibitable efflux of the metal. Arsenate, however, failed to alter MRP1-mediated transport of known MRP1 substrates such as calcein and vincristine. In conclusion, these findings demonstrated that MRP1 likely handles some, but not all, cytotoxic metals such as arsenic and mercuric ions in addition to antimony, thereby resulting in reduced toxicity of MRP1-overexpressing cells to these compounds (Vernhet et al. 2000).

In mammals, tubular epithelia represent the primary target of Hg^{2+} nephrotoxicity. Although widely investigated, the mechanisms of Hg^{2+} cell uptake, accumulation, and excretion all along the nephron remain largely unknown (Aleo et al. 2005). Native distal tubular-derived Madin-Darby canine kidney (MDCK) cells exposed to subcytotoxic $HgCl_2$ concentrations were used for investigating specific mechanisms involved in the tubular response to toxic metals. Exposed to $HgCl_2$, MDCK cells showed a rapid, but transient, Hg^{2+} accumulation. The metallic cation was found to affect cell density and morphology, proportional to the dose and time of exposure. In parallel, Hg^{2+}-induced upregulation of endogenous MRP1 and MRP2 export pumps, a significant $HgCl_2$-dependent induction of protective cellular thiols, and an increase in glutathione conjugate metabolism were observed (Aleo et al. 2005). The functional suppression of the activity of MRP, obtained following MK-571 treatment, increased the Hg^{2+} cell content and the sensitivity of MDCK cells to $HgCl_2$. The results demonstrate that in MDCK cells, inorganic Hg^{2+} promotes the activation of specific detoxifying pathways that may, at least partly, depend on the activity of MRP transporters. In confirmation of these results, TR-MRP2 knockout rats were found to exhibit greater mercury accumulation in renal and liver tissue, while levels of mercury excretion in these mice were found to be significantly lower (Bridges et al. 2011). Based on results from membrane vesicle experiments, other MRP2 substrates include Cys-S-conjugates of Hg^{2+} and CH_3Hg^+.

The Disulfide Bond Oxidoreductase D (DsbD) Family

The DsbD family is part of the transmembrane 2-electron transfer carrier class in TCDB, and recent evidence points to its inclusion into the rhodopsin superfamily (M. Shlykov and M. Saier, unpublished observations). The transmembrane DsbD protein functions in an electron transport chain (ETC) involving NADPH as the initial electron donor, several thioredoxin proteins, and other Dsb proteins. One DsbD member, MerA of *Streptomyces lividans* (TC# 5.A.1.4.1), uses the two electrons donated by NADPH to reduce Hg^{2+} to the volatile and less toxic Hg^0. Another DsbD homologue, CycZ, is a thiol-disulfide interchange protein containing multiple heavy-metal-associated (HMA) domains including one resembling MerP (TC# 1.A.72.3) and several HMA domains of families 5 and 6 of the P-type ATPase superfamily. CycZ likely binds Hg^{2+} and plays a role in its reduction to a less toxic form.

Conclusions

Mercuric ion (Hg^{2+}) transport allows Hg^{2+} detoxification by at least three mechanisms involving three types of transporters. Bacterial Mer superfamily members accumulate Hg^{2+} and methyl-Hg^+ before reduction to the less toxic Hg^0 which diffuses out of the cell. ABC functional superfamily members pump out Hg^{2+} conjugates, providing a protective function. Finally, MerA Hg^{2+} reductases take electrons from NADPH to generate Hg^0 from Hg^{2+}. These three mechanisms protect cells of various types from mercury toxicity and provide means for bioremediation on both small and large scales.

Acknowledgments Work in the authors' laboratory was supported by NIH grant GM077402.

Cross-References

- Bacterial Mercury Resistance Proteins
- Mercury and Alzheimer's Disease
- Mercury and DNA
- Mercury and Immune Function
- Mercury and Lead, Effects on Voltage-Gated Calcium Channel Function
- Mercury and Low Molecular Mass Substances
- Mercury in Plants
- Mercury Nephrotoxicity
- Mercury Neurotoxicity
- Mercury Toxicity
- Mercury, Physical and Chemical Properties
- Metallothioneins and Mercury

References

Aleo MF, Morandini F, Bettoni F, Giuliani R, Rovetta F, Steimberg N, Apostoli P, Parrinello G, Mazzoleni G (2005) Endogenous thiols and MRP transporters contribute to Hg^{2+} efflux in $HgCl_2$-treated tubular MDCK cells. Toxicology 206:137–151

Barkay T, Miller SM, Summers AO (2003) Bacterial mercury resistance from atoms to ecosystems. FEMS Microbiol Rev 27:355–384

Bridges CC, Joshee L, Zalups RK (2011) MRP2 and the handling of mercuric ions in rats exposed to inorganic and organic species of mercury. Toxicol Appl Pharmacol 251:50–58

Kiyono M, Sone Y, Nakamura R et al (2009) The MerE protein encoded by transposon Tn21 is a broad mercury transporter in *Escherichia coli*. FEBS Lett 583:1127–1131

Mok T, Chen JS, Shlykov MA et al (2012) Bioinformatic analyses of bacterial mercury ion (Hg^{2+}) transporters. Water Air Soil Pollut (Submitted)

Qian H, Sahlman L, Eriksson PO et al (1998) NMR resolution structure of the oxidized form of MerP, a mercuric ion binding protein involved in bacterial mercuric ion resistance. Biochemistry 37:9316–9322

Saier MH Jr (2003) Tracing pathways of transport protein evolution. Mol Microbiol 48:1145–1156

Schué M, Dover LG, Besra GS et al (2009) Sequence analysis of a plasmid-encoded mercury resistance operon from *Mycobacterium marinum* identifies MerH, a new mercuric ion transporter. J Bacteriol 191:439–444

Vernhet L, Allain N, Bardiau C, Anger JP, Fardel O (2000) Differential sensitivities of MRP1-overexpressing lung tumor cells to cytotoxic metals. Toxicology 142:127–134

Yamaguchi A, Tamang DG, Saier MH Jr (2007) Mercury transport in bacteria. Water Air Soil Pollut 182:219–234

Mercury, Physical and Chemical Properties

Fathi Habashi
Department of Mining, Metallurgical, and Materials Engineering, Laval University, Quebec City, Canada

Physical Properties

Mercury is a silvery-white, shiny metal, which is liquid at room temperature.

Atomic number	80
Atomic weight	200.59
Melting point, °C	−38.89
Boiling point at 101.3 kPa, °C	357.3
Density at 0°C, g/cm^3	13.5956
Relative abundance, %	5×10^{-5}
Specific heat capacity at 0°C, J g^{-1} K^{-1}	0.1397
Heat of fusion, j/g	11.807
Heat of evaporation at 357.3°C, kJ/mol	59.453
Thermal conductivity at 17°C, W cm^{-1} K^{-1}	0.082
Thermal expansion coefficient at 0–100°C, K^{-1}	1.826×10^{-4}
Electrical conductivity at 0°C, m Ω^{-1} mm^{-2}	1.063×10^{-4}
Crystal structure	Rhombohedral
Viscosity at 0°C, mPa·s	1.685
Surface tension, N/cm	480.3×10^{-5}

Due to its high surface tension, mercury has the ability to wet metals. When the solubility limit of a metal in mercury is exceeded, mercury still can wet the metal forming a thick silvery amalgam paste. Strictly speaking, an amalgam is a solution of a metal in mercury and a *quasi-amalgam* is a metal wetted by mercury after the solubility limit has been exceeded.

Mercury has a relatively high vapor pressure, even at room temperature. Saturation vapor pressures at 0–100°C are listed in Table 1 (corresponding to a specified mercury content in air).

Concentration of mercury in air can be expressed in two ways:
- Parts per million (by volume), ppm
- Milligrams per cubic meter, mg/m^3

To convert from one system to the other the following relation is used:

One gram equivalent of mercury vapor occupies 22.4 l at NTP.

Atomic weight of mercury vapor $\times\ 10^3$ expressed in milligrams occupies:

$$22.4 \times \frac{298}{273} \text{ liters at } 25°C$$

$$\frac{\text{Atomic weight} \times 10^3}{22.4} \times \frac{273}{298} \text{ mg occupies 1 liter at } 25°C$$

1 ppm (vol.) mercury vapor $= \dfrac{1 \text{ liter polluant}}{10^6 \text{ liters air}}$ at 25°C

$= \dfrac{\frac{\text{Atomic weight} \times 10^3}{22.4}}{10^6} \times \dfrac{273}{298}$ mg/L

$= 0.04 \times$ atomic weight $\times 10^3$ mg/L

$= 0.04 \times$ atomic weight mg/m^3

Maximum allowable concentration of mercury in air is 0.1 mg/m^3 or approximately 0.01 ppm.

The solubility of metals in mercury at room temperature varies greatly (Table 2); the highest is that of indium which is 57%, and the lowest is that of chromium (4×10^{-7}%). Solubility increases with increased temperature. The following types of solubilities can be identified:
- Metals dissolve with formation of compounds at room temperature, e.g., the alkali and alkaline earth and metals; reaction is exothermic.
- Metals having low solubilities and do not form compounds even at high temperature, e.g., aluminum, chromium, cobalt, and iron.
- Metals with low solubilities and form compound only at high temperature, e.g., uranium.

As a result of the high specific gravity of mercury, the solubility of metals in mercury, if expressed in g/L, will be quite high. Thus, a solubility of 1% corresponds to 135.5 g/L.

The ammonium radical behaves as a metal, and ammonium amalgam can be prepared by electrolyzing an ammonium salt solution at low temperature using mercury cathode, or by the reaction of an ammonium salt solution with sodium amalgam:

$$NH_4OH + Na(Hg) \rightarrow NH_4(Hg) + NaOH$$

Ammonium amalgam decomposes readily to ammonia and hydrogen:

$$2NH_4(Hg) \rightarrow Hg + 2NH_3 + H_2$$

The decomposition is accompanied by expansion of the amalgam phase to a large volume due to gas evolution.

When mercury is equilibrated with another metal in the liquid state, mutual solubility is observed, as expected. Thus, at 35°C, the solubility of gallium in mercury is 1.8 at.% and that of mercury in gallium at the same temperature is 32.6 at.% (melting point of gallium is 29.8°C).

Mercury, Physical and Chemical Properties, Table 1 Saturation vapor pressure of mercury at different temperatures

Temperature (°C)	Pressure (Pa)	Mercury content in air (mg/m^3)
0	0.026	2.38
10	0.070	6.04
20	0.170	14.06
30	0.391	31.44
100	36.841	2,404.00

Mercury, Physical and Chemical Properties, Table 2 Solubility of metals in mercury at 20°C

Solubility	Metals	Exceptions	Compound formation with mercury
Highest	Less typical metals	Copper	None forms compounds
Medium	Typical metals	Beryllium	All form compounds except aluminum
		Aluminum	
Least	Transition and inner transition metals		Few form compounds

Chemical Properties

Mercury exists in the oxidation states 0, +1, and +2. Monovalent mercury exists as Hg–Hg bond. Mercury is a relatively noble metal. It dissolves in nitric acid, aqua regia, warm concentrated hydrochloric acid, and sulfuric acid. It is sparingly soluble in dilute HCl, HBr, and HI as well as in cold sulfuric acid. The oxide of mercury, HgO, decomposes at 400–500°C. This effect is utilized in the extraction of mercury from ores. Mercury forms monovalent and divalent compounds with the halogens fluorine, chlorine, bromine, and iodine. It also forms monovalent and divalent compounds with sulfur.

References

Habashi F (1998) Principles of extractive metallurgy, vol 4, Amalgam and electrometallurgy. Métallurgie Extractive Québec, Québec. Distributed by Laval University Bookstore Zone. www.zone.ul.ca

Simon M et al (1998) Mercury. In: Habashi F (ed) Handbook of extractive metallurgy. Wiley-VCH, Weinheim, pp 891–922

Mercury-Induced Nephrotoxicity

▶ Mercury Nephrotoxicity

Mercury-Induced Oxidative Stress

▶ Mercury Toxicity

Mercury-Thiol Complexes

▶ Mercury Nephrotoxicity

Mercury-Thioneins

▶ Metallothioneins and Mercury

Metal Cluster Biosynthesis

▶ Iron-Sulfur Cluster Proteins, Fe/S-S-adenosylmethionine Enzymes and Hydrogenases

Metal Homeostasis

▶ CusCFBA Copper/Silver Efflux System

Metal Nanoconjugates

▶ Palladium, Colloidal Nanoparticles in Electron Microscopy

Metal Neurotoxicity

▶ Lead and Alzheimer's Disease

Metal Resistance

▶ Bacterial Tellurite Processing Proteins
▶ Bacterial Tellurite Resistance

Metal Transporter

▶ Cobalt Transporters

Metal Transporters

▶ Nickel Transporters

Metal-Binding Proteins

▶ Silver, Pharmacological and Toxicological Profile as Antimicrobial Agent in Medical Devices

Metal-Binding Site/Pocket/Cavity/Cleft

▶ Calcium Ion Selectivity in Biological Systems

β_1-Metal-Combining Protein

▶ Chromium(III) and Transferrin

Metal-Dependent Hydrolase

▶ Zinc Amidohydrolase Superfamily

Metal-Induced Oxidative DNA Damage

▶ Lead and Alzheimer's Disease

Metal-Ion-Mediated Protein–Nucleic Acid Interactions

▶ Barium and Protein–RNA Interactions

Metallated DNA

▶ Zinc, Metallated DNA-Protein Crosslinks as Finger Conformation and Reactivity Probes

Metallic Arsenic

▶ Arsenic in Pathological Conditions
▶ Arsenic, Free Radical and Oxidative Stress

Metallocenters of Nickel-Containing Proteins

▶ Nickel-Binding Sites in Proteins

Metallochaperone

▶ Nitrile Hydratase and Related Enzyme

Metallocompound Protein Binding

▶ Platinum- and Ruthenium-Based Anticancer Compounds, Inhibition of Glutathione Transferase P1-1

Metallodrugs in Cancer Treatment

▶ Gallium Nitrate, Apoptotic Effects

Metalloenzymes

▶ Iron-Sulfur Cluster Proteins, Fe/S-S-adenosylmethionine Enzymes and Hydrogenases

Metalloid Resistance

▶ Bacterial Tellurite Processing Proteins
▶ Bacterial Tellurite Resistance

Metalloid Transport

▶ Aquaporins and Transport of Metalloids

Metalloproteomes

▶ Zinc-Binding Proteins, Abundance

Metallothioneins

▶ Silver, Pharmacological and Toxicological Profile as Antimicrobial Agent in Medical Devices

Metallothioneins and Copper

Sílvia Atrian[1], Òscar Palacios[2] and Mercè Capdevila[2]
[1]Departament de Genètica, Facultat de Biologia, Universitat de Barcelona, Barcelona, Spain
[2]Departament de Química, Facultat de Ciències, Universitat Autònoma de Barcelona, Cerdanyola del Vallès (Barcelona), Spain

Synonyms

Copper-thioneins; Thioneins

Definition of the Subject

Metallothioneins (MTs) are cysteine-rich short polypeptides that bind, through metal-thiolate bonds, heavy metal ions. They are ubiquitously present in all living organisms. They serve as main agents for the homeostasis of physiological metals (zinc and copper), and owing to their high coordinating capacity, they protect life systems from deleterious bonding not only of xenobiotic metals but also of excessive Zn^{2+} and Cu^+ overloads. Copper is the active component of metalloproteins involved in electron transfer, oxygen transport, and oxidoreductase reactions so that its availability in cells has to be guaranteed. In turn, intracellular copper levels are tightly controlled to avoid the deleterious effects associated to its excess. Copper binding by metallothioneins has been proposed for both purposes.

General Concepts

Metallothioneins (MTs) are low molecular weight, cysteine-rich proteins which coordinate heavy metal ions through the metal-thiolate bonds established by their highly abundant cysteine residues (average 30%). They constitute a heterogeneous protein superfamily of polymorphic peptides, ubiquitously distributed among all living organisms, i.e., Eukaryota (animals, plants, fungi, and protozoa) and some Prokaryota (cyanobacteria). Apo-MTs (also called thioneins) are random coil peptides that only fold to a definite 3D structure upon metal coordination, this yielding metal-MT complexes containing one or more metal-MT clusters. In vitro coordination of metal ions to MTs, either through direct binding to the apopeptides or as a result of metal ion displacement, is a nonselective process that follows the affinity rules of low-molecular-mass thiolate complexes. However, in organisms, all MTs show a more or less pronounced degree of specificity, this concept referring to the set of determinants – protein sequence features, metal ion availability, and metal-responsive gene expression pattern – that leads a given MT to natively discriminate among metal ions. As a result, each MT peptide is isolated in vivo in the form of a certain metal complex, i.e., complexed to a determined amount of a specific metal ion, and not to others. Consequently, the physiological function of that MT is supposed to be related to the metabolism and/or detoxification of the corresponding metal ion. From a structural perspective, coordination of the favored (cognate) metal ions, results in single, well-folded MT species in which their stoichiometry and folding reflect their kinetic inertness and thermodynamically most stable state. Hence, two genuine categories were considered: Zn-thioneins (or divalent metal ion-binding thioneins) and Cu-thioneins. Later on, the classification was expanded to a stepwise gradation between these two extreme (Zn- and Cu-thionein) characters (Palacios et al. 2011b). According to this classification, copper coordination to MT yields different results when genuine Cu-thioneins, Zn-thioneins, or intermediate forms are considered.

Biological Occurrence of Genuine Cu-thioneins

Genuine Cu-thioneins are MT peptides able to fold into stable, homometallic Cu(I) complexes, which are recovered as such after purification from the corresponding organisms. Genuine Cu-thioneins were first reported in fungi (including yeast), with the *Saccharomyces cerevisiae* (baker's yeast) Cup1 protein the paradigm being among them. Cup1 (reviewed in Dolderer et al. 2009) is a 61-residue polypeptide, posttranslationally processed to a 53-residue mature form that binds up to 8 Cu(I) ions through the metal-thiolate bonds contributed by 12 cysteine residues (the main structural features of the Cu-Cup1 cluster are commented below). It is synthesized by baker's yeast cells in response to copper overload, and therefore it is considered essentially as a detoxifying peptide. Cu-thioneins are well represented among ascomycetes (i.e., *Neurospora crassa*) and basidiomycetes (i.e., *Agaricus bisporus*) fungi, so that Cu-thioneins were once thought to be exclusive of these kinds of lower eukaryotes. However, expansion of the knowledge of MTs showed the inaccuracy of this hypothesis, and currently, genuine Cu-thioneins have been described in other phyla of organisms. Most significant are the MT system of *Drosophila* (Diptera, Insecta, Arthropoda) and the Cu-MT forms of marine Crustacea (Arthropoda) and snails (Gastropoda, Mollusca). *Drosophila* (the fruit fly) organisms synthesize up to 5 MT isoforms, all of them harboring a protein sequence plainly homologous to Cup1, which carry out metabolic copper-related functions, especially in the digestive tract (Atrian 2009). It has recently been shown that pulmonate snail species (*Helix, Cornu*) have developed, together with other isoforms, copper-specific MTs, devoted to handling copper ions related with the homeostasis of hemocyanin, their copper-containing respiratory pigment (Palacios et al. 2011a). Snail MT protein sequence analysis has shown that metal specificity relies on noncysteine amino acid residues since Cd- and Cu-specific isoforms share absolute conservation of the number and position of cysteines. Finally, it is worth noting that specific Cu-thioneins have been also reported in other aquatic organisms, such as Crustacea, in which oxygen is also transported by the blue copper-loaded carrier hemocyanin (Vergani 2009).

Biological Significance of Mammalian MTs as Example of Cu-binding in Zn-thioneins

Besides the cases cited above, no additional genuine Cu-thioneins are known. As a consequence, in all other organisms, MTs of more or less marked Zn-thionein character bind copper ions when required, this yielding mixed ZnCu-MT complexes that are recovered upon purification from native sources. Most illustrative is the case of the mammalian MT system (reviewed in Vasak and Meloni 2011). Mammals synthesize four different types of MTs (MT1 to MT4). MT1 and MT2 are of ubiquitous, metal-inducible synthesis, while MT3 (central nervous system specific) and MT4 (epithelia specific) are constitutively produced. Copper poisoning leads to the synthesis of mixed Zn_xCu_y-MT1/MT2 complexes, with variable amounts of the two metal ions, although Zn_3Cu_7-MT1 has been shown to be the most stable one. This species conserves the two-domain structure typical of the mammalian MTs coordinating divalent metal ions, i.e., an N-terminal or ß domain, encompassing nine cysteines that coordinate three Zn(II) or Cd(II) ions, and a C-terminal or α domain, encompassing 11 cysteines that coordinate four Zn(II) or Cd(II) ions. The Zn_3Cu_7-MT1 keeps one Zn(II) in the ß domain and two Zn(II) in the α domain, which are supposed to maintain the folding in two domains of the whole complex. Eventually, and in the presence of excess copper, Zn_3Cu_7-MT1 is able to evolve in vitro to homometallic Cu_{12}- and Cu_{15}-MT complexes, of uncertain mono- or bi-dominial structure and which are never recovered as native complexes. MT3 and MT4 mammalian isoforms deserve special mention in relation to their function with copper. MT4 shows an increased copper coordination capacity in vivo, as it has been recovered as Cu_{10} complexes. This has been related to the special need of copper imposed by keratin synthesis processes. MT3 was first isolated as a neuronal inhibitory factor downregulated in Alzheimer's disease patients, and consequently it was related to the onset of this pathology. Afterward it has also been associated to other neurological diseases, such as Parkinson's and prion-driven diseases. Since these neuronal disorders are characterized by metal ion disequilibrium in brain cells, the capacity of MT3 for zinc and copper ion coordination, and its relation with the beta-amyloid and alpha-synuclein peptides, has been intensively characterized. MT3 was isolated from native tissues in physiological conditions as a $Zn_{3-4}Cu_4$-MT3 complex, both metal ions separately coordinated in different clusters. The increased ability of this isoform to react and coordinate copper could be on the basis of its protective role toward neuronal degeneration, exerted by two different effects: its capacity to chelate excess metal ions related with neurological disorders and its capacity to react with Cu(II) ions that otherwise would generate redox-associated damage through ROS species formation (reviewed in Vasak and Meloni 2011).

Oxidation State, Coordination Environments, and Structural Aspects of the Cu-MT Complexes

Most of the known Cu-metalloproteins have their Cu (II)/Cu(I) centers surrounded by N-, O-, and S-donor ligands, which generate distinct types of copper sites traditionally used for their classification (e.g., types 1, 2, or 3; Cu_A, or Cu_Z). Hence, Cu(II)/Cu(I) ions are able to play different biological roles in electron transfer, catalytic redox processes, or in the activation of O_2. However, MTs are the only metalloproteins known that contain copper solely bound to S-donor ligands. Furthermore, although MTs are able to "react" with both types of copper ions, Cu(I) and Cu(II), only Cu(I)-MT complexes are recovered from in vivo or in vitro preparations. The inexistence of Cu(II)-MT complexes is due to the well-known capability of Cu(II) to oxidize thiolates with the concomitant formation of Cu(I) and disulphide bridges through the reaction: $Cu^{2+} + 2RS^- \rightarrow Cu^+ + RS-SR$

This is in fact the reaction that takes place when the mammalian Zn-MT3 isoform reacts with Cu(II) ions, and which involves a partial release of the initially bound Zn^{2+} ions. This, together with the Cu(I) ions generated, which are bound by the protein, leads to the formation of the $Zn_{3-4}Cu_4$-MT3 species, characteristic of this isoform (Vasak and Meloni 2011).

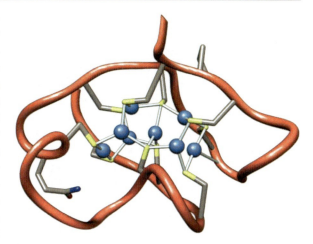

Metallothioneins and Copper, Fig. 1 3D structure of *S. cerevisiae* Cu_8-Cup1, showing the $Cu_8(SCys)_{10}$ cluster (PDB entry 1rju). The Cu(I) ions are shown as *spheres*, the Cys residues as *sticks*, and the S–Cu bonds as *solid lines*

The scarcity of structural data on Cu-MT complexes (see below) limits the information available on the coordination environments about Cu(I) in the Cu-MT species. Overall, all the available data indicate that Cu(I) binding to MTs takes place through its interaction with the S-Cys ligands. Therefore, similar coordination environments to those described in the literature for Cu(I) thiolates could be expected in the Cu-MT complexes: digonal ($n = 2$), trigonal ($n = 3$), or tetrahedral ($n = 4$) $Cu(S-Cys)_n$ environments. The particular case of the unique 3D structure available (1 out of 32 metal-MT structures available in the Protein Data Bank) for a Cu-MT species, which corresponds to a $Cu_8(SCys)_{10}$ cluster, namely Cu_8-Cup1, of the *S. cerevisiae* Cup1 metallothionein, revealed six trigonal-planar and two digonal Cu(I) ions coordinated to the protein cysteines as shown in Fig. 1. This structure was obtained from X-ray diffraction analysis in 2005 (Calderone et al. 2005) after years of hard work. Previously, in 1996, the NMR study of a ^{109}Ag(I) substituted Cu-Cup1 sample revealed that only 10 out of the 12 Cys available were binding 7 M(I) ions (M = Cu or Ag) in a single $M_7(SCys)_{10}$ cluster, where digonal and trigonal metal-coordination sites were proposed. In 2000, an NMR reinvestigation of this Cu_7-Cup1 structure suggested the existence of different possible structures fully compatible with the determined constraints in this single-domain protein. The flexibility of the protein backbone of Cup1, which is a good reflection of the situation in the MT family, seems to be responsible for the disparity of the published structures. It is probably also due to the difficulties, found in other cases, to get crystalline samples or unambiguous NMR data. This has hampered further structural determinations and has limited the knowledge available to that obtained from the NMR techniques in which Cu(I) is unfortunately invisible. This is the case of the Cu_6-NcMT complex of the *Neurospora crassa* metallothionein. Only the 3D structure of the polypeptide backbone has been determined by NMR, which is consistent with the presence of a single-domain protein. Significantly, in the case of mammalian MTs, it has only been possible to obtain spatial information for the in vitro–prepared mixed-metal Zn_xCu_3-αMT1 and Zn_yCu_4-βMT1 independent domains, this revealing important differences with the peptide folding induced by divalent metal ion coordination to these MT moieties. Further Cu(I) addition, eventually leading to homometallic complexes, as well as any Cu-containing full-length MT species, led to a patent loss of defined 3D structures (Dolderer et al. 2007).

Methodologies for the Study of Cu-MTs

There are currently a high number of analytical/spectroscopic techniques to characterize the coordination of the metal-MT sites, even when taking into account the special features of MTs (small size, mainly binding "spectroscopically silent" d^{10} metal ions, and the practical absence of aromatic residues in their amino acid sequences). However, it is important to note that among all the methodological approaches currently applied to the study of MTs, not all of them provide precise, univocal, and valuable information (Capdevila et al. 2012).

The methodologies based on element-specific detection have been the most popular to quantify the protein concentration and its metal content: atomic absorption spectroscopy (AAS), inductively coupled plasma atomic emission spectroscopy (ICP-AES), or inductively coupled plasma mass spectrometry (ICP-MS). Molecular spectroscopies have also been extensively applied to elucidate the properties of MTs on the basis of the ligand-to-metal charge-transfer (LMCT) absorption bands detectable in the UV–vis region, which are related to the metal-MT coordination. In the case of the Cu-MT species, intra- and intermetallic transitions have also been described. Among molecular spectroscopies, UV–vis absorption spectroscopy, circular dichroism (CD), EXAFS, and NMR have been applied in the analysis of Cu binding to MTs, although the first two account for the highest number of published data (Winge 1991). They have provided clues about metal-to-ligand stoichiometries and on the possible coordination geometries. In the particular case of Cu-MTs, a higher number of absorption bands and at higher wavelengths than those found for the Zn(II)- or Cd(II)-MT complexes have been reported.

A particular feature of the Cu-MT aggregates, not shared by the M-MT ones (M = Zn^{II}, Cd^{II}, or other divalent metal ions), is their luminescent properties at room and/or cryogenic temperatures. The emission intensity of the Cu-MT species can be related to the degree of folding of the protein about the metal centers, and the emission profiles recorded during Cu(I) titrations of either apo- or Zn-MT have been interpreted to give information on the domain specificity or occupancy of the added Cu(I).

As mentioned above and due to the nonexistence of an NMR-active copper nucleus, NMR has had a limited application in the study of the Cu-MT systems. To solve this, the substitution of Cu(I) by Ag(I) has been used as a strategy. However, the isomorphous replacement of Cu(I) by Ag(I) has been called into question after having been demonstrated that the differences between their coordination properties are enough to justify differences in the structures of the corresponding complexes. Literature data give support to a comparable behavior of both metal ions only in the case of the formation of heterometallic M_xZn_y-MT1 (M = Cu or Ag; y ≥ 1) species for the mammalian MT1 isoform.

Molecular mass spectrometry, particularly ESI-MS, has been applied to the study of the speciation of the Cu-MT samples, despite the common difficulties associated with the mass determination of Cu-labile species. Even when using high-resolution spectrometers (mainly equipped with TOF analyzers), the Cu-MT and Zn-MT complexes with high metal content are still indistinguishable from the mass spectra, and this hampers the elucidation of the homo- or heteronuclear nature of the complexes. Determination of the metal content of the preparations by ICP may shed light into this subject. Otherwise, recording the mass spectra at different pH values can provide information on the composition of the complexes, as Zn(II) is released from the protein at higher pH values than Cu(I), which remains bound to the protein even at very low pH.

Cross-References

- ▶ Copper-Binding Proteins
- ▶ Copper, Physical and Chemical Properties
- ▶ Zinc in Alzheimer's and Parkinson's Diseases
- ▶ Zinc Metallothionein-3 (Neuronal Growth Inhibitory Factor)

References

Atrian S (2009) Metallothioneins in diptera. In: Sigel A, Sigel H, Sigel RKO (eds) Metallothioneins and related chelators, vol 5, Metal ions in life sciences. RSC Publishing, Cambridge, pp 155–181

Calderone V, Dolderer B, Hartmann H-J, Echner H, Luchinat C, Del Bianco C, Mangani S, Weser U (2005) The crystal structure of yeast copper thionein: the solution of a long-lasting enigma. Proc Natl Acad Sci USA 102:51–56

Capdevila M, Bofill R, Palacios Ò, Atrian S (2012) State-of-the-art of metallothioneins at the beginning of the 21st century. Coord Chem Rev 256:446–62.

Dolderer B, Echner H, Beck A, Hartmann H-J, Weser U, Luchinat C, Del Bianco C (2007) Coordination of three and four Cu(I) to the α- and β-domain of vertebrate Zn-metallothionein-1, respectively induces significant structural changes. FEBS J 274:2349–2362

Dolderer B, Hartmann H-J, Weser U (2009) Metallothioneins in yeast and fungi. In: Sigel A, Sigel H, Sigel RKO (eds) Metallothioneins and related chelators, vol 5, Metal ions in life sciences. RSC Publishing, Cambridge, pp 83–105

Palacios O, Pagani A, Perez-Rafael S, Egg M, Höckner M, Brandstätter A, Capdevila M, Atrian S, Dallinger R (2011a) Shaping mechanisms of metal specificity in a family of metazoan metallothioneins: evolutionary evolution of mollusc metallothioneins. BMC Biol 9:4

Palacios O, Atrian S, Capdevila M (2011b) Zn- and Cu-thioneins: a functional classification for metallothioneins? J Biol Inorg Chem 16:991–1009

Vasak M, Meloni G (2011) Chemistry and biology of mammalian metallothioneins. J Biol Inorg Chem 16:1067–1078

Vergani L (2009) Metallothioneins in aquatic organisms: fish, crustaceans, mollusc and echinoderms. In: Sigel A, Sigel H, Sigel RKO (eds) Metallothioneins and related chelators, vol 5, Metal ions in life sciences. RSC Publishing, Cambridge, pp 199–237

Winge DR (1991) Copper coordination in metallothionein. Method Enzymol 205:458–469

Metallothioneins and Lead

Òscar Palacios and Mercè Capdevila
Departament de Química, Facultat de Ciències, Universitat Autònoma de Barcelona, Cerdanyola del Vallès (Barcelona), Spain

Synonyms

Lead-thioneins; Thioneins

Definition of the Subject

Metallothioneins (MTs) are ubiquitous, cysteine-rich proteins able to bind several kinds of heavy metal ions through their Cys residues. Their function is still under discussion, but they have been related with the homeostasis of essential metals (zinc and copper) as well as with the detoxification of other metal ions (Cd^{2+}, Hg^{2+}, Pb^{2+}, Ag^+, etc.).

The absence of any physiological function related with lead makes necessary its removal from the cellular medium by a potent chelator to form innocuous complexes. MTs have been identified as one of the primary responses of living organisms in the case of intoxication with lead.

Lead is a naturally occurring element that does not break down in either soil or water, although lead compounds are changed by sunlight, air, and water. Lead compounds can be found basically in two different chemical forms: as inorganic ionic compounds (oxide, sulfide, nitrate, halides, etc.), where the Pb^{2+} cation can be more or less easily released, depending on the solubility of the compound, or as alkyl-lead compounds, its organometallic form. Alkyl-lead compounds are man-made compounds in which one or more organic molecules are bound to a lead atom through a carbon atom. The most common alkyl-lead compounds are tetraethyllead and tetramethyllead, which were extensively used as a fuel additive in the past.

The toxicity of lead for the living organisms is perfectly well known, and up to the present, no biological role has been established for this metallic element, which does not exhibit any evidence of essentiality, even at trace levels, for any of its valence states. On the contrary, its interference in the cellular environment with respiratory pigments, in the production of energy, as well as in some membrane functions is well known, as after poisoning, Pb(II) has been found to replace Zn(II) and Ca(II) containing biomolecules. Even at very low concentration, Pb(II) is known to be able to inhibit several enzymes and the biosynthesis of hemoglobin and can cause severe damages to the central nervous system (CNS). Living organisms have developed several mechanisms to survive against this damage in environments polluted with metal ions such as Pb by limiting their intracellular concentrations by their isolation in intracellular organelles or by the formation of innocuous complexes or forms. This role can be performed, in the case of eukaryotes, by three related biomolecules: glutathione, phytochelatins, and metallothioneins (MTs).

There are data in the literature that provides information concerning the participation of MTs in Pb-detoxifying processes. Furthermore, and as happens with other known toxic metal ions (Cd(II), Ag(I), Hg(II), etc.), the induction of the synthesis of MT after administration of Pb(II) to different organisms (mammals, fish, birds, ...) has been reported. Therefore, it cannot be categorically stated that nowadays, there is a lack of information regarding the interaction of Pb(II) with MTs. However, it can be said that the coordination of Pb(II) to MTs is still not well understood. Several facts account for this affirmation. First, there are a limited number of data published on this subject. Second, Pb(II) easily precipitates as $Pb(OH)_2$ or gives rise to hydroxocomplexes at pH values greater than 6. This clearly calls into question the experiments performed at neutral pH – even in the presence of multichelator agents such as MTs. Finally, the complexity and diversity of the Pb(II) coordination chemistry – especially when interacting with biomolecules that can offer several distinct donor ligands (N, O, and S) – have to be also taken into account. All these facts clearly indicate that the Pb-MT interaction must be of a higher complexity than that observed in the Zn(II)-, Cd(II)-, or Cu(I)-MT species. Unfortunately, the limited data concerning Pb(II) binding to MTs are centered on the Pb^{2+} cation, and no information concerning alkyl-lead compounds has been reported. This could probably be due to the lipophilic nature and specific uptake (inhalation and dermal absorption) of the latter, which lead to specific metabolic routes that differ from those related with the Pb^{2+} ion, until its elimination in the urine as diethyllead, ethyllead, and inorganic lead.

Among the few studies reporting the interaction of Pb(II) with MTs, in comparison with those carried out with other metal ions (Capdevila et al. 2012), it has to be noted that most of them were carried out with two of the four mammalian MT isoforms (mainly MT1 and MT2, which are, so far, the most studied MTs since their discovery) and a few more with phytochelatins. It must also be noted that the vast majority of the works published are based on spectroscopic data, i.e., UV–vis absorption and circular dichroism (CD) techniques that report on the average properties of the sample, which clearly limits the significance of the resulting data, and only limited information has been reported about the speciation of the Pb-MT systems.

The first results, based on UV–vis, CD, and magnetic CD, concerning the Pb(II)-MT interaction were reported in the 1980s by Kagi and coworkers (Bernhard et al. 1983). There, the authors already pointed out the clear dependence of the formed Pb-MT species with the working pH and the Pb–MT ratios assayed. In analogy with the initial Zn_7-MT species, formation of a Pb_7-MT single species was proposed, thus assuming that all the Cys residues where involved in Pb(II) coordination, which adopted a tetrahedral environment. Later, Vasak et al. demonstrated, by means of EXAFS measurements, the S coordination in the already assumed Pb_7-MT species, with a Pb-S distance of 2.65 Å (Charnock et al. 1989), which was assigned a coordination number of 2 by other authors. Formation of the Pb_7-MT species was also confirmed by other research groups (Nielson et al. 1985).

Later on, between 1999 and 2001, a series of publications concerning the Pb-MT species appeared, at the same time, from the independent research teams of Prof. Chu and Prof. Ru, both at Peking University. Unfortunately, the Chinese language used mostly prevented its proper dissemination.

By means of spectroscopic data, Prof. Chu's team (He et al. 1999) proposed formation of a single Pb_7-MT2 complex after the addition of lead(II) acetate to mammalian apo-MT2 at neutral pH, this species showing analogous spectroscopic features to those previously reported by Kagi (Bernhard et al. 1983). However, the addition of lead(II) acetate to apo-MT2, now at pH 4.3, was reported to give rise to the formation of a single isostoichiometric Pb_7-MT2 species that, however, showed different spectroscopic features to those observed for the Pb_7-MT2 species formed at pH 7. The complexity of the registered spectroscopic signals leads the authors to propose multiple Pb(II) coordination types, including two coordination, on the basis of previous works with phytochelatins (Mehra et al. 1995). Finally, these authors reported that both Pb_7-MT2 forms cannot interconvert by varying the pH. Special attention must be paid to the fact that these authors were the first to report the determination of the Pb(II) binding constants to MTs by means of isothermal titration calorimetry (Chu et al. 2000). They studied the Pb(II) acetate interaction with rabbit liver Zn_7-MT2 and apo-MT2 at pH 4.7. The results demonstrated that both reactions are spontaneous and exothermic, with binding constants in the same order (1.3×10^5 and 5.3×10^5, respectively), and that the species formed exhibited the same stoichiometry (Pb_7-MT) but a different structure, as previously reported. It should be noted that in all these works, the properties of the acetate anion as a chelating agent were never taken into consideration.

The research team of Prof. Ru (Ji et al. 2000) studied the interaction of Pb(II) with the MT1 and MT2 mammalian isoforms on the basis of the spectroscopic data (mainly UV–vis and CD) and by using distinct metalation states as starting points: apo-MT, Zn_7-MT, and Zn_2Cd_5-MT. Unfortunately, the conclusions they reached were, in a certain way, contradictory. Initially, they reported that MT1 and MT2, in their apo- and Zn_7-MT forms, behave very closely when binding Pb(II) and that Zn(II) ions can be fully displaced from the initial Zn_7-MT species. The same year, another paper by the same authors revealed some differences in Pb(II) binding between apo-MT1 and apo-MT2 (i.e., randomly in the case of MT1 but selectively for apo-MT2 that was favoring one domain over the other). Further studies by this group suggested that Pb(II) was unable to displace Zn(II) from Zn_7-MT1 and Zn_7-MT2 but was able to do this from Zn_2Cd_5-MT1 and Zn_2Cd_5-MT2, although differently for each isoform. Finally, the latest work of Ru et al. concerning the apo-αMT and apo-βMT individual domains, and based only on UV–vis absorption data, proposed the formation of Pb_4-αMT, Pb_7-αMT, and Pb_3-βMT species, with the former being the most stable species.

Some years later, the availability of the electrospray ionization mass spectrometry technique (ESI-MS), allowed a complete analysis of the binding of Pb(II) to the mammalian MT1 isoform and its independent domains, αMT1 and βMT1 (Palacios et al. 2007). Here, the Zn replacement from the initial Zn_7-MT1, Zn_4-αMT1, and Zn_3-βMT1 species was reported to give rise to the formation of several coexisting species, both heteronuclear (Pb, Zn-MT) and homonuclear (Pb-MT), in which the metal content was clearly dependent on several factors: the working pH (neutral or 4.5), the amount of Pb(II) added to the initially Zn-loaded protein solution (a stoichiometric amount or a clear excess), and the protein considered (the whole MT or its individual domains). A common trait in all the experiments was the evolution of the samples with the time (Palacios et al. 2011). Successive ESI-MS spectra (Palacios et al. 2007) revealed the disappearance of the lead-containing species with the concomitant sole detection of Zn-MT species (at pH 7) or of the apo-species (at acidic pH), in a maximum of 48 h after the Pb(II) addition, although the CD fingerprints of the samples continuously grew in intensity (Palacios et al. 2011). These results are consistent with the reported increase in MT synthesis in response to Pb(II) administration to mammals and with the difficulties found in recovering homonuclear Pb-MT species from lead-treated organisms.

An even more limited number of studies than those dedicated to MTs have been devoted to the analysis of the interaction of Pb(II) with phytochelatins (PCs), a special kind of enzymatically synthesized peptides derived from glutathione, and with the general structure (γ-Glu-Cys)$_n$Gly ($n = 2$–11) which occur in plants and some yeasts, which owing to the presence of Cys-X-Cys motifs in their sequence were included in the Class 3 of the first classification of MTs. Special mention must be made here of Mehra et al. (1995), who showed, by means of UV and CD data, that PCs are able to bind Pb(II) ions regardless of the source of lead ($Pb(ClO_4)_2$, $Pb(CH_3COO)_2$, or Pb-GSH). More interestingly, they also showed that Pb(II) forms different complexes (two-, three-, and four-coordinate Pb(II) ions) with PCs, the formation of which clearly depends on the PC chain length, i.e., the number of Cys available. The longest chain assayed, PC_4, gave rise to the proposal of two different coexisting coordination numbers, 2 and 4, thus demonstrating the higher complexity of the Pb(II)-PC system formed when compared with those of Zn(II) or Cd(II).

Currently, and in accordance with the tendency observed during the last decade in the fields of interest of research on MTs (Capdevila et al. 2012), many other studies have been published focusing on the toxicological and environmental aspects of the Pb-MT relationship. In those, the authors investigate the unquestionable role of MTs and PCs in the detoxification of Pb-exposed organisms with the aim of shedding light into the detoxification processes triggered by Pb(II) exposure rather than to the molecular and/or biological aspects of the Pb(II)-MT interaction. Further studies are therefore required to address these latter issues.

Cross-References

▶ Lead and Alzheimer's disease
▶ Lead and Immune Function
▶ Lead Detoxification Systems in Plants
▶ Lead, Physical and Chemical Properties
▶ Zinc Metallothionein

References

Bernhard W, Good M, Vasak M, Kagi JHR (1983) Spectroscopic studies and characterization of metallothioneins containing mercury, lead and bismuth. Inorg Chim Acta 79:154–155

Capdevila M, Bofill R, Palacios O, Atrian S (2012) State-of-the-art of metallothioneins at the beginning of the 21st century. Coord Chem Rev. 256:46–62

Charnock JM, Garner CD, Abrahams IL, Arber JM, Hasnain SS, Henehan C, Vasak M (1989) EXAFS studies of metallothionein. Physica B 158:93–94

Chu D, Tang Y, Huan Y, He W, Cao W (2000) The microcalorimetry study on the complexation of lead ion with metallothionein. Thermochim Acta 352–353:205–212

He W, Chu D, Yang J, Yao D, Shao M (1999) A novel structure motif of metallothionein Pb-MT. Chem J Chinese Univ 2:248–250

Ji Q, Wang L, Zhou Y, Ru B (2000) Study on competition reaction and replacement reaction of metallothioneins with lead ion. Acta Sci Nat Univ Pekin 36:503–508

Mehra RK, Kadati VR, Abdullah R (1995) Chain length-dependent Pb(II)-coordination in phytochelatins. Biochem Biophys Res Commun 215:730–736

Nielson KB, Atkin CL, Winge DR (1985) Distinct metal-binding configurations in metallothionein. J Biol Chem 260:5342–5350

Palacios O, Leiva-Presa A, Atrian S, Lobinski R (2007) A study of the Pb(II) binding to recombinant mouse Zn_7-metallothionein 1 and its domains by ESI TOF MS. Talanta 72:480–488

Palacios O, Atrian S, Capdevila M (2011) Zn- and Cu-thioneins: a functional classification for metallothioneins? J Biol Inorg Chem 16:991–1009

Metallothioneins and Mercury

Òscar Palacios and Mercè Capdevila
Departament de Química, Facultat de Ciències, Universitat Autònoma de Barcelona, Cerdanyola del Vallés (Barcelona), Spain

Synonyms

Mercury-thioneins; Thioneins

Definition of the Subject

Metallothioneins (MTs) are ubiquitous, cysteine-rich proteins able to bind heavy metal ions to form metal-thiolate bonds. Their function is still a matter of debate, but they have been involved in the homeostasis of essential metal ions (Zn^{2+} and Cu^+) as well as in the interaction with toxic metal ions (Cd^{2+}, Hg^{2+}, Pb^{2+}, Ag^+, etc.).

The distribution of mercury and its compounds around the World makes its poisoning a global problem that claims to be solved. The presence of Hg(II) in cells induces the synthesis of several proteins, among them metallothioneins, that act as agents of detoxification in living organisms. Even if MTs have been proposed as biomarkers of the contamination with mercury, the interaction between mercury and MTs is still controversial.

Mercury has a long history of use, both in medicine and as a poison. It is well known that mercury and its compounds are highly toxic, but little is still known about the molecular mechanisms underlying their toxicity and the possible detoxification processes present in living organisms, where metallothioneins (MTs) play a crucial role. To better understand the implications of MTs in mercury detoxification, it is worth mentioning some aspects of mercury chemistry and properties.

The Toxicity of Mercury and Its Compounds

Mercury is a trace element with unique chemical and physical characteristics. It has been known and used for thousands of years. Mercury is a persistent pollutant in all its chemical forms and this has made this element one of the most highly studied of all times. It is the only metal that is liquid at standard temperature and pressure conditions, but it also has one of the narrowest ranges of liquid state than any other metal. Its high vapor pressure in its elemental form is the main reason for the rapid global dispersion from specific localized sources.

Mercury occurs in certain deposits that can be found throughout the world, mostly as cinnabar (mercury(II) sulfide), a dust which is highly toxic by ingestion or inhalation. Mercury poisoning can also result from exposure to water-soluble forms of mercury (such as mercury(II) chloride or methylmercury), inhalation of mercury vapor, or intake of products contaminated with mercury. Unlike what is actually happening with other toxic metals, such as cadmium or lead, and even taking into account that mercury (and most of its compounds) is extremely toxic and must be handled with care, it is still used for several scientific purposes (scientific apparatus and in amalgam material for dental restoration).

Mercury exists in two main oxidation states, I and II, but Hg(II) is the most common and is the main one in nature as well. Mercury, as well as other metals, can be converted into more toxic compounds: its alkyl derivatives (so-called organomercury compounds) by bacteria in the environment, and to a lesser extent through abiotic pathways. Organomercury compounds, the most toxic forms of mercury, are always divalent and usually with a two-coordinate linear geometry. They use to have the HgR_2 (often volatile) or RHgX (normally solids) formula (R = aryl or alkyl, X = halide or acetate). Methylmercury, a generic term for compounds with the formula CH_3HgX, is a dangerous family of compounds, which are found in some mercury-polluted waters. They arise by a process known as biomethylation.

Mercury intake can occur by inhalation and/or absorption through the skin and mucous membranes. It can cause both chronic and acute poisoning. However, mercury and its compounds have been extensively used in medicine, although, fortunately, their use is much less common today than it was in the past. Nevertheless, cinnabar, one of its common ores, is still used in various traditional medicines at low doses.

The way through which mercury accumulates in living organisms must also be considered. Due to the fact that organomercury compounds are stable in water, aquatic life organisms play an essential role in mercury distribution. Fish and shellfish have a natural tendency to concentrate mercury in their bodies, often in the form of methylmercury. Some of the species of fish that are higher on the food chain, such as shark, swordfish, king mackerel, tuna, and tilefish, can contain high concentrations of mercury. Furthermore, and because mercury and methylmercury seem to be fat soluble, they primarily accumulate in the viscera even though they are also found throughout the muscle tissue. When this poisoned fish is consumed by a predator, the mercury level accumulates. Since fish are less efficient at depurating than at accumulating methylmercury, fish-tissue concentrations increase over time. The species higher on the food chain amass body burdens of mercury that can be up to ten times higher than the species they consume. This is the so-called biomagnification process.

Altogether, this points out that the presence of mercury, in any of its chemical forms, is a natural occurring global problem whose effects must be localized in the living organisms.

Mercury in Biological Systems

The presence of Hg(II) in cells induces, in some living organisms, the synthesis of several proteins that can act as detoxifying agents. Among them, the Hg-reductase, metal-regulators proteins (like MerR), and metallothioneins (MT) can be found. Unfortunately, the structural information regarding these metalloproteins is often limited to optical spectroscopy and X-ray absorption.

The chemistry of mercury in biological systems is dominated by coordination to S-rich ligands, namely, the cysteine thiolate groups, in agreement with its preference for soft donor ligands. Therefore, the high affinity of Hg(II) for cysteine residues promotes the irreversible replacement of essential metal ions (Zn, Cu) of cysteine-containing proteins, which accounts for the high toxicity of mercury in living organisms. In several bacteria, resistance to Hg(II) toxicity is based on an ensemble of several proteins, known as the Mer family, which mainly bind Hg(II) through cysteines.

The relationship between the coordination of Cys to Hg(II) and the Hg(II)-thiolate compounds must be useful to better understand the coordination of Hg(II) to metalloproteins. But the chemistry of the Hg(II)-thiolates is particularly complex. Their extensive study has revealed the varied coordination environments that Hg can show: linear, trigonal, and tetrahedral coordination, which have been characterized by means of several techniques. Also the Hg(II)-Hg(II) secondary interactions are common traits of most of these compounds. All this, together with the complexity that the biological systems endow by themselves, make not surprising the disparity of results that can be found in the literature.

Mercury and Metallothioneins

The implication of MTs in the mechanisms of mercury detoxification of a high number of organisms (mammals, fishes, etc.) is a fact widely accepted. Unfortunately, there are a very high number of publications devoted to the toxicological and environmental aspects of the mercury-MT interaction, and only a few have explored the biochemical aspects of this interaction. Most of the former studies assume that MTs act as biomarkers for the mercury contamination of different environments, in spite that not only heavy metals are responsible for the MT synthesis in living organisms (Sigel et al. 2009) and that not always a direct relation between the degree and contamination and the concentration of MT has been found.

When focusing on the most biochemical aspects of the mercury-MTs interactions, it should be mentioned that the number of publications is quite scarce, especially if compared with those devoted to Zn- or Cd-MT (Capdevila et al. 2012). As already occurring with other studies on MTs, most of the literature data about the Hg-MT species has been obtained from only one/two mammalian MT isoforms (MT2 & MT1, the most studied isoforms since their discovery at the end of the 1950s), and mainly with native MT preparations obtained from mammalian organs.

With no doubt, optical spectroscopy (UV–vis and CD) has been the major tool for the study of the mercury-binding properties of mammalian MTs since the first studies. In 1980, based on electronic spectra, a systematic study of the interaction of mammalian MT with metals of group 12 allowed to state that Hg(II) cations, like Zn(II) and Cd(II), are coordinated in the active sites of the corresponding M_7-MT species with tetrahedral geometry about the metal ions (Vasak et al. 1981). These assertions were based on the position of the ligand-to-metal charge transfer bands of lower energy associated with the metal-SCys binding.

The authors observed that these bands were shifted to the red when the metal bound changes from Zn(II) to Cd(II), and the latter to Hg(II). They also observed that an increase of the coordination number about Hg(II) implicates decreases in the energy transfer bands. The semiempirical theory of Jorgensen, which empirically correlates the optical electronegativity of the ligands bound to the metal atoms with the spectra of the corresponding complexes, allowed to associate the presence of an absorption located at *ca.* 300 nm to the presence of Hg(II) ions showing a tetrahedral coordination geometry. In mammalian MTs, a 304 nm band had already been associated to four-coordinated Hg(II) while in yeast MT, Hg(II) with a distorted tetrahedral coordination geometry generated an absorption at 283 nm (Beltramini et al. 1984). This shift to higher energies suggested the possible contribution of bridging cysteines between two metal centers in the mammalian Hg(II)-MT species.

After these initial publications, the main contributions to the spectroscopic study of the Hg-MT species, between the late 1980s and the early 2000s, must be acknowledged to the research team led by Stillman et al. (1992). Through several publications, Stillman and coworkers analyzed the in vitro binding abilities of the rabbit liver MT2 isoform toward Hg(II) at distinct experimental conditions. Therefore, they studied the behavior of Zn_7-MT2, Zn_4-αMT2, apo-MT2 and apo-α MT2 or even Cd_7-MT2 when adding successive Hg(II) equivalents at either pH 7 or 2.4. In additional experiments, several parameters (temperature, counterions, introduction of substances that alter the viscosity of the medium, etc.) were changed. However, the lack of systematization in the design of experiments, the absence of control on variables such as pH – which is often assumed to remain constant – and the assumptions made when interpreting the experimental results difficult a further analysis of published data. In summary, these data resulted inconclusive regarding the aim of determining the behavior of mammalian MT versus Hg(II). The results regarding the interaction of apo-MT2 with Hg(II) at pH 7 and 2.4 indicated that in both cases the already suggested Hg_7-MT species is formed, although it showed different structures depending on the pH. This Hg_7-MT species, only at pH 2.4, can evolve, in the presence of more Hg(II), yielding a Hg_{18}-MT complex. The formation of a third species, Hg_{11}-MT2, was also suggested, although it was only observed after the addition of Hg(II) to

Zn$_7$-MT2 at neutral pH and the authors already stated that its formation was favored in solvents containing 50% ethylene glycol (with high viscosity, which alters the possibility of hydrogen bond formation and reduces the protein mobility) (Stillman 1995).

On the basis of the X-ray absorption studies of the Hg$_7$-MT species obtained by addition of Hg(II) either to apo-MT or to Zn$_7$-MT, Stillman proposed a distorted tetrahedral coordination for Hg(II) with two short (2.33 Å) and two long (3.4 Å) Hg-S distances. The Hg$_{18}$-MT2 species is a special case as it has been the subject of many studies. Therefore, Stillman and coworkers suggested that Hg$_{18}$-MT2 presents a highly chiral monodominial structure where Hg(II) is linearly coordinated to bridging-SCys ligands or showing a coordination geometry consisting of two pseudotetrahedral μ-SCys and two additional Cl$^-$ ligands. A detailed study of this species has shown that its formation strictly depends on several factors: pH, temperature, protein concentration, the presence of Cl$^-$ in the medium, and the MT isoform considered. Interestingly, the authors tried to demonstrate that the formation of Hg$_{18}$-MT takes place with the rabbit liver MT2 isoform but not with the MT1 isoform. In the case of the Hg$_{11}$-MT, a trigonal coordination, even in the absence of X-ray data, was suggested.

The unique data reported concerning Hg(II)-MT species determined by means of ESI-MS are those that describe the coordination of Hg(II) to apo-MT2 at pH 2. Unfortunately, the detection of mixtures of the Hg$_7$-, Hg$_8$-, and Hg$_9$-MT2 species does not give support to the conclusions achieved from the studies based on optical spectroscopy results (Stillman et al. 2000).

As mentioned above, the coordination of Hg(II) to the isolated α domain of rabbit liver MT2 was carried out, but only at pH 7. The data published suggests that the addition of Hg(II) to apo-αMT or to Zn$_4$-αMT, both at neutral pH, leads to the formation of a Hg$_4$-αMT species which shows tetrahedral coordination environment about Hg(II). The addition of further Hg(II) promoted the restructuration of the protein to give rise to the formation of a Hg$_{11}$-αMT species, where Hg(II) was linearly coordinated to the SCys ligands affording a metal-MT aggregate with a very "open" structure, which would explain the low intensity of the recorded CD spectra.

On the basis of the observations reported for the whole MT and for the α fragment cited above, and even without experimental data concerning the interaction of the β domain of MT with Hg(II), Stillman and coworkers hypothesized about the behavior of this MT fragment toward this metal ion. They understood that if the Hg$_7$-MT and Hg$_4$-αMT species contain Hg(II) in a tetrahedral environment in the corresponding Hg$_7$S$_{20}$ and Hg$_4$S$_{11}$ clusters, the β fragment should adopt a folding containing a Hg$_3$S$_9$ cluster with tetrahedral Hg(II). Similarly, the formation of the Hg$_{11}$-MT species, with trigonally coordinated Hg(II), should imply the opening of the α and β domains of the MT, which would lead to the formation of Hg$_6$S$_{11}$ and Hg$_5$S$_9$ aggregates, respectively.

Part of these results were the basis for the Hg-saturation assay proposed for measuring the concentration of MTs in fish (Dutton et al. 1993), as also previously happened with the Cd- and Ag-saturation methodologies. In this case, through the use of ^{203}Hg and assuming the formation of Hg$_7$-MT species, the amount of Hg was measured by the activity of the radionuclide and afterward correlated with the protein concentration. This methodology has been revised by other authors to use nonradioactive, stable Hg for MT measurement in bivalves and fish. It must be noticed here that some authors have suggested that MT can bind up to 20 moles of Hg(II) per MT, depending on the excess mercury added on MT, this supporting again the cited complexity in establishing a definitive Hg-MT stoichiometry.

More recently, in 2004, other authors contributed to the extension of the knowledge on the reaction pathways of Hg(II) toward MT1 (Leiva-Presa et al. 2004a). They investigated the formation of the Hg-MT complexes of MT1 and of its independent α and β fragments at different experimental conditions (pH 7 and 3, by using either HgCl$_2$ or Hg(ClO$_4$)$_2$) by means of the CD and UV-vis spectroscopies. The analysis of the optical data demonstrated that some variables, unconsidered until then, should be taken into account when studying the interaction of MTs with Hg(II), specially "the time elapsed between subsequent additions of Hg(II) to the protein," as well as the nature of the titrating agent and the pH of the solutions. The authors suggested that all these factors drastically affect the structural properties of the Hg-MT species formed, more than their formation by itself. Also, the participation of Cl$^-$ anions as ligands was revealed to be important in the presence of high amounts of Hg(II).

Finally, and in spite of the important toxicity of methylmercury in living organisms, widely spread in the environment, only one work has been devoted to the analysis of the interaction of the MeHg$^+$ cation with MTs (Leiva-Presa et al. 2004b). Also by using optical spectroscopy techniques, the authors reported on the consequences of the addition of MeHgCl to recombinant mouse Zn$_7$-MT1 and their isolated Zn$_4$-αMT1 and Zn$_3$-βMT1 fragments. The results showed that while MeHg$^+$ promotes the total release of Zn(II) from Zn$_7$-MT and Zn$_3$-βMT, with the subsequent unfolding of the corresponding protein, its interaction with Zn$_4$-αMT only promoted a partial Zn(II) release.

Altogether, it becomes difficult to univocally establish the species formed when Hg^{2+} and MeHg$^+$ cations bind to mammalian MTs. No reports about other types of MTs are available and most of the information available comes from optical spectroscopy data that just could be reflecting the average properties of a mixture of species. At this stage, it seems clear that additional studies to gain further information are still needed.

Cross-References

- ► Mercury Toxicity
- ► Mercury Transporters
- ► Mercury, Physical and Chemical Properties
- ► Zinc metallothionein

References

Beltramini M, Lerch K, Vasak M (1984) Metal substitution of Neurospora copper metallothionein. Biochemistry 23:3422–3427

Capdevila M, Bofill R, Palacios O, Atrian S (2012) State-of-the-art of metallothioneins at the beginning of the 21st century. Coord Chem Rev. 256:46–62

Dutton MD, Stephenson M, Klaverkamp JF (1993) A mercury saturation assay for measuring metallothionein in fish. Environ Toxicol Chem 12:1193–1202

Leiva-Presa A, Capdevila M, González-Duarte P (2004a) Mercury(II) binding to metallothioneins. Variables governing the formation and structural features of the mammalian Hg-MT species. Eur J Biochem 271:4872–4880

Leiva-Presa A, Capdevila M, Cols N, Atrian S, González-Duarte P (2004b) Chemical foundation of the attenuation of methylmercury(II) cytotoxicity by metallothioneins. Eur J Biochem 271:1323–1328

Sigel A, Sigel H, Sigel RKO (eds) (2009) Metallothioneins and related chelators. RSC, Cambridge

Stillman MJ (1995) Metallothioneins. Coord Chem Rev 144:461–511

Stillman MJ, Shaw CF III, Suzuki K (eds) (1992) Metallothioneins: synthesis, structure and properties of metallothioneins, phytochelatins, and metal-thiolate complexes. VCH Publishers, New York

Stillman MJ, Thomas D, Trevithick C, Guo X, Siu M (2000) Circular dichroism, kinetic and mass spectrometric studies of copper(I) and mercury(II) binding to metallothionein. J Inorg Biochem 79:11–19

Vasak M, Kagi JHR, Hill HAO (1981) Zinc(II), cadmium(II), and mercury(II) thiolate transitions in metallothionein. Biochemistry 20:2852–2856

Metallothioneins and Silver

Kelly L. Summers[1] and Martin J. Stillman[1,2]
[1]Department of Biology, The University of Western Ontario, London, ON, Canada
[2]Department of Chemistry, The University of Western Ontario, London, ON, Canada

Synonyms

Ag, MT

Definition

Silver has no known natural biological role and has a range of toxicity depending on the form of the metal and the organism involved (Wiley and Brooks 1992). Colloidal silver, which is toxic to microbes, has been used over many centuries as an antiseptic and disinfectant in wound dressings (Klasen 2000a, b). Even though elemental silver is not particularly toxic to humans, silver ions are highly toxic and are thought to be carcinogenic. No studies were found to report cancer in humans after exposure to silver or silver compounds, although subcutaneous imbedding of silver foil induced fibrosarcomas in rats (Oppenheimer et al. 1956).

Metallothioneins (MTs) bind metals in the monovalent group 11 triad (copper, silver, and gold) and in the divalent group 12 triad (zinc, cadmium, and mercury) both in vitro and in vivo (Schmitz et al. 1980; Sutherland and Stillman 2011). One of the proposed roles for MT is in toxic metal detoxification because MT is able to bind a wide range of metals using its 20 cysteinyl thiols. Silver(I) binding to MT has been studied both to aid in the understanding of the biological chemistry of copper(I) and also to understand the role of silver(I) as an antibacterial that aids in wound healing (Klasen 2000a, b). Spectroscopic studies of the d^{10} Cu(I) and Ag(I) are difficult though, due to the lack of chromophores. The properties of Cu(I) make obtaining NMR spectra much more difficult than for metals like Ag(I), Zn(II), or Cd(II) (Narula et al. 1991). Ag(I)-thiolates and Cu(I)-thiolates exhibit orange luminescence; in view of the large Stokes shifts measured, this is most likely phosphorescence (Salgado et al. 2007). Because silver induces MT in vivo, silver binding to MT has been used to mimic that of copper, which is naturally found bound to MT (Cosson 1994).

Silver

Silver, a rare naturally occurring metal, is frequently found deposited as a mineral ore (Howe and Dobson 2002). Silver is a transition or d-block metal which is part of the triad including copper and gold. Copper is an essential metal required for normal functions in most plants and animals. Because copper is essential, a number of proteins, including cuproenzymes, Cu transporters, and Cu chaperones, are required in maintaining Cu homeostasis (Prohaska 2008). Like copper, silver is toxic to microorganisms, although its presence in many higher organisms may suggest that silver also has a biochemical role (Wiley and Brooks 1992). Because copper and silver are much more harmful to microbes than to humans, they have a number of medicinal applications such as antiseptics, analgesics, and antibiotics (Wiley and Brooks 1992; Ratte 1999; Klasen 2000a, b; Casey et al. 2010; Warnes et al. 2012).

An excess of any compound can be toxic, but ionic silver is one of the most toxic heavy metals to bacteria – surpassed only by mercury (Wiley and Brooks 1992; Ratte 1999). Silver complexes, such as silver thiosulfate, silver chloride, and silver sulfide, are less toxic (Ratte 1999; Howe and Dobson 2002). Silver compounds can be slowly absorbed by body tissues and deposited around nerves and in deeper skin layers, causing permanent skin damage (Ratte 1999). Chronic overexposure has been suggested as a cause of cardiac abnormalities, as well as permanent brain and nervous system damage in humans (Ratte 1999).

Metallothioneins

Metallothionein (MT) is a low molecular weight cysteine-rich metal binding protein found in almost all organisms. MT folds into two independent metal-cysteine thiolate clusters induced by metal binding, which requires bridging of the cysteinyl thiols. The two clusters, referred to as the N-terminal β-domain and the C-terminal α-domain, bind tetrahedrally coordinated metals – such as Cd and Zn – in stoichiometries of M_3Cys_9 and M_4Cys_{11}, respectively (Sutherland and Stillman 2011). MT binds the monovalent metals Cu, Ag, and Au in structures that deviate from the M_7Cys_{20} two-domain structure of divalent metals. The most commonly reported structure for Cu(I) is Cu_{12}-MT (Sutherland and Stillman 2011). Figure 1 shows the ESI mass spectrum for rabbit liver metal-free or apo-MT 2a recorded at pH 2. The 5+ manifold maximum of three prominent charge states is characteristic of metallothioneins in general. Deconvolution provides the mass of the parent protein, here with a measured mass of 6,124 Da. More detailed mass spectral data have been published by Stillman's group (Ngu et al. 2010; Sutherland et al. 2012a).

It is proposed that the main physiological role of mammalian MT is in maintaining copper and zinc homeostasis (Richards and Cousins 1975; Kojima and Kägi 1978; Nath et al. 1987); however, synthesis of MT is also induced in tissues (such as the heart, brain, and intestine) to regulate essential metal uptake (Menard et al. 1981; Hamer 1986) or to prevent the absorption of toxic heavy metals such as cadmium (Hamer 1986; Nath et al. 1987) and silver (Liu et al. 1991). Toxic metals either activate MT transcription factors directly or release zinc from MT, which activates the transcription factors (Palmiter 1994; Mayer et al. 2003). The MT transcription factors in turn bind

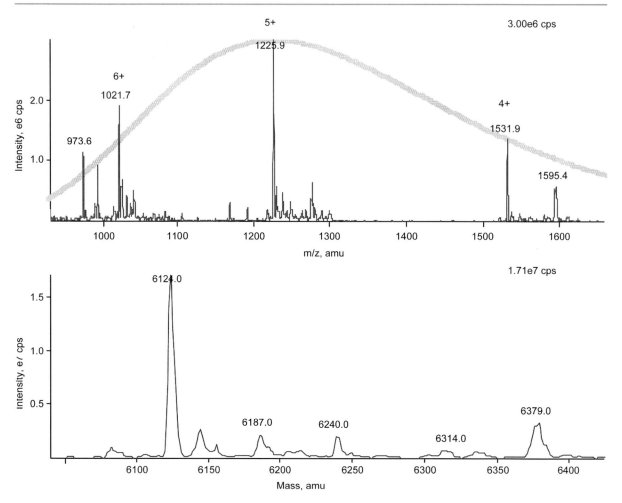

Metallothioneins and Silver, Fig. 1 ESI-MS charge states and deconvoluted spectrum for rabbit liver apo-MT 2a at pH 2.5 (mass 6,124.00 Da) (Reproduced with permission from Stillman 1999)

to recognition sequences and activate transcription of nascent MT (Palmiter 1994; Mayer et al. 2003). Four subisoforms have been reported for human MT, which are localized in different organs (1 and 2 are generally found in all cell types, including large concentrations in the liver; 3 is localized to the brain; and 4 is only found in the skin), which emphasizes the important role that MT plays in regulating both Zn and Cu, as well as the redox properties of the cells.

Finally, while the interactions of a very large number of metals with a range of different metallothioneins have been studied, silver(I) binding to MT has often been used to mimic copper binding because copper(I) is oxygen sensitive. Silver binding to metallothionein has been investigated by titrating Ag(I) into apo-MT 2, Zn_7-MT 1, Zn_4-αMT 1, and Zn_3-βMT 1 using CD, EXAFS, and emission spectroscopy (Zelazowski et al. 1989; Zelazowski and Stillman 1992; Gui et al. 1996). These spectral results identified Ag_{12}-MT, Ag_{18}-MT, Ag_6-αMT, and Ag_6-βMT species. Interestingly, an Ag_{12}-αMT species was also uncovered (Zelazowski et al. 1989; Zelazowski and Stillman 1992). These results deviated from the expected 20 Ag(I) bound singly to the 20 cysteines in βα-MT.

Determination of the Ag/MT Ratio Using Electrospray Ionization Mass Spectrometry

Stillman et al. reported ESI-MS data for a solution of Ag_n-MT 2 (n = 14–19) formed by adding 17 Ag(I) to Zn_7-MT 2 at pH 2.5 (Fig. 2; Stillman et al. 1994).

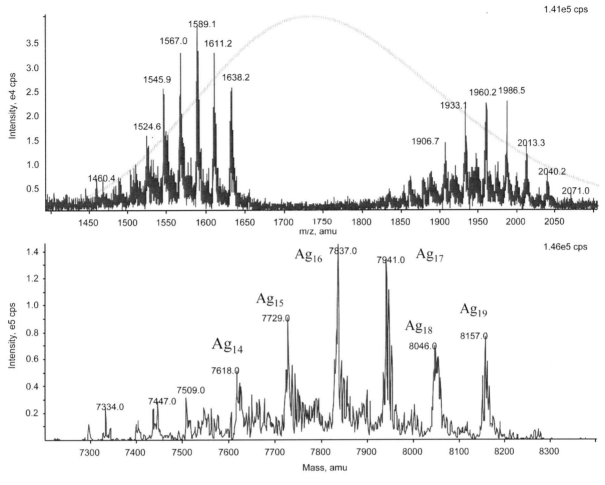

Metallothioneins and Silver, Fig. 2 ESI-MS charge states and deconvoluted spectrum for rabbit liver Ag_n-MT 2a where n = 13–19 as marked. The Ag-MT 2A was formed by adding Ag(I) to Zn_7MT 2A at pH 2.5 (Reproduced from Stillman 1999)

The distribution of Ag(I) in the MT under these reaction conditions is quite reasonable when the more recent data for titrations of MT with Cd(II) (Sutherland and Stillman 2008) and with Zn(II) (Sutherland et al. 2012a) are consulted. The deconvoluted ESI-MS data show the presence of a range of Ag-MT species with maxima for Ag_{16}- and Ag_{17}-MT. Interestingly, there is no evidence for the Ag_{20}-MT species that might be expected if there was linear coordination of each Ag(I) to one of the 20 cysteines. This lack of completely saturated species was also found for Hg-MT, where the Hg_{18}-MT 2 species was considered to be a folded structure. Recently, determination of the NMR spectra for Cd_8-MT 1 suggested that these heavily metalated species are all supermetalated involving a single domain structure (Sutherland et al. 2012b).

Circular Dichroism (CD) Spectroscopy of Silver Metallothionein

CD spectroscopy is an essential technique for identifying metal-MT species with well-defined tertiary structures. A change in molar ratio of metal to protein causes the peptide chain to rearrange to accommodate changes in the number of metal-thiolate bonds required. This in turn changes the tertiary structure of the metalated protein. In studies of cadmium and copper binding, the stoichiometric ratios that caused major changes in the CD spectrum were associated with the formation of specific metal-MT complexes.

There are many reports that describe the coordination and thiol bridging in Ag-MT; not all of them reach the same conclusions. CD spectroscopy has been used extensively to study Ag(I) binding to rabbit liver

Metallothioneins and Silver, Fig. 3 CD spectral data recorded during a titration of rabbit liver apo-MT 2a with Ag(I) at 20°C showing the development of prominent species at the 12 Ag(I) and 17 Ag(I) points. The 6 and 12 Ag(I) spectral inflections are poorly resolved. This research was originally published in the Journal of Biological Chemistry (Zelazowski, A.J., Gasyna, Z. Stillman, M.J. "Silver Binding to Rabbit Liver Metallothionein." *J. Biol. Chem.* 1989; 264, 17091–17099. © The American Society for Biochemistry and Molecular Biology)

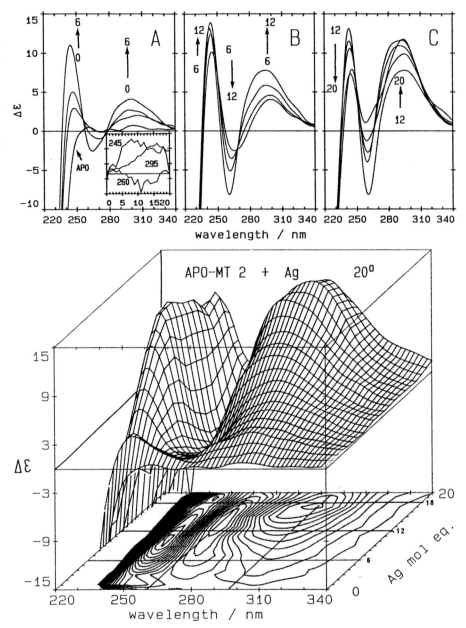

Zn$_7$-MT 1 and 2 and apo-MT 1 and 2, as well as the two isolated fragment domains at a range of temperatures.

Titrations of Ag(I) into rabbit liver apo-MT 2 showed a number of specific species (those identified by spectroscopic signatures), including Ag$_6$-MT, Ag$_{12}$-MT, and Ag$_{18}$-MT (Figs. 3 and 4; Zelazowski et al. 1989). Titrations into the apo-α and apo-β domain fragments showed an Ag$_3$-α species and an Ag$_6$-α species at 20°C and 55°C but only an Ag$_3$-β species at 20°C (Zelazowski et al. 1989). Under these conditions, the initial low pH would ensure that the apo-MT will comprise 20 protonated cysteines, and the

Metallothioneins and Silver, Fig. 4 CD spectral data recorded during a titration of rabbit liver apo-MT 2a with Ag(I) at 55°C showing the development of prominent species at the 12 Ag(I) and 17 Ag(I) points. The 6 and 12 Ag (I) spectral inflections are much better resolved at the higher temperatures, before the steep formation of the Ag_{17}-MT species which indicates a significant change in orientation of the MT peptide as a function of added Ag(I). At 55°C, a spectral inflection at the 6 Ag(I) point can be discerned. This research was originally published in the Journal of Biological Chemistry (Zelazowski, A.J., Gasyna, Z. Stillman, M.J. "Silver Binding to Rabbit Liver Metallothionein." *J. Biol. Chem.* 1989; 264, 17091–17099. © The American Society for Biochemistry and Molecular Biology)

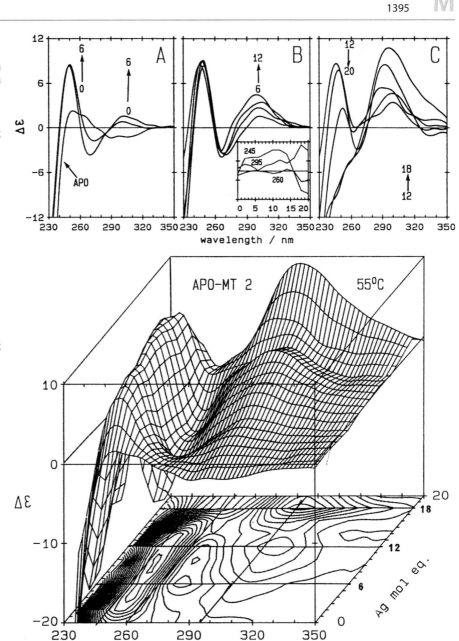

high temperature will provide the thermal energy. The CD spectra were measured at neutral pH.

Interestingly, when Ag(I) was titrated into the zinc-loaded MT (rabbit liver zinc MT 2) at pH 7, only Ag_{12}-MT and Ag_{18}-MT species were observed with the full protein and only the Ag_6-MT 2 species were observed with the domain fragments (Figs. 5 and 6; Zelazowski and Stillman 1992). At 20°C the tetrahedrally coordinated Zn(II) in Zn_7-MT 2 is believed to inhibit formation of the Ag_6- and Ag_{12}-MT observed during Ag(I)

Metallothioneins and Silver, Fig. 5 CD spectral data recorded during a titration of rabbit liver Zn_7-MT 2a at pH 7.4 with Ag(I) at 20°C showing the development of prominent species essentially only at the 17 Ag(I) point (18 is marked). The CD spectrum from the Zn_7-MT with a maximum near 230 nm gradually gives way to the Ag-MT spectrum that maximizes at the $Ag_{17/18}$-MT point (Reproduced with permission of the American Chemical Society, ref Zelazowski and Stillman 1992)

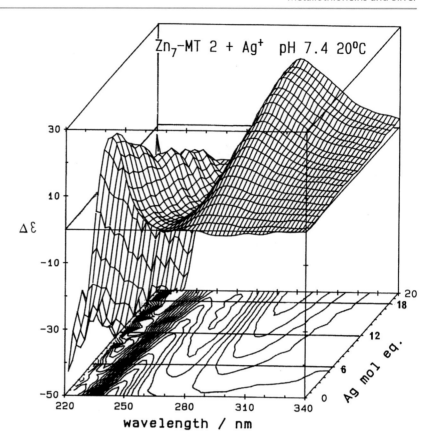

titrations with apo-MT (Zelazowski et al. 1989). Increasing the temperature to 55°C is thought to induce the structural rearrangements necessary to form a stable Ag_{12}-MT species. Ag_{12}-MT 2 is characterized by the CD bands at 263 nm and 314 nm. The stoichiometry of the supermetalated Ag-MT was initially identified as 18 Ag(I) from the CD in 1989. Later spectroscopic data suggested that 17 Ag(I) were bound. The Ag_{17}-MT 1 was identified by the CD bands at 302 nm and 370 nm (Gui et al. 1996).

Emission Spectroscopy of Silver Metallothionein

Emission intensity of Cu-MT was found to be dependent on the Cu(I) content of the protein and was attributed to the copper(I)-thiolate chromophore (Beltramini and Lerch 1981; Stillman 1999). The first emission data for Ag-MT were reported by Zelazowski and coworkers (Fig. 7), who found the emission intensity recorded from frozen glasses at 77 K to be similarly correlated with the stoichiometry of Ag(I) to protein (Zelazowski et al. 1989). The increase in emission intensity as the number of Ag(I) increased was interpreted in terms of the formation of a much tighter tertiary structure, which excluded the solvent, preventing signal quenching.

X-ray Absorption Spectroscopy of Silver Metallothionein

Sulfur K-edge EXAFS can determine the local geometry of the metal-thiolate cluster by measuring the sulfur-metal bond lengths and the coordination number of the protein. The first EXAFS reports for Ag-MT were by

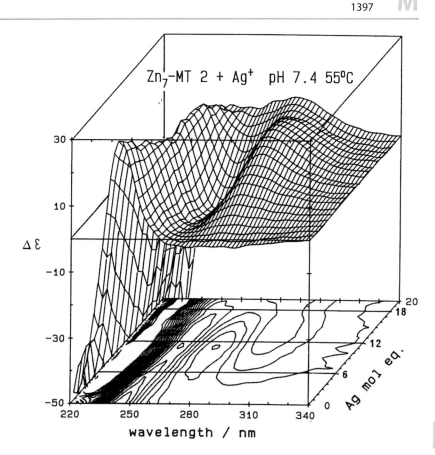

Metallothioneins and Silver, Fig. 6 CD spectral data recorded during a titration of rabbit liver Zn_7-MT 2a at pH 7.4 with Ag(I) at 55 °C showing the development of prominent species essentially only at the 17 Ag(I) point (18 is marked). The CD spectrum from the Zn_7-MT with a maximum near 230 nm gives way to the Ag_{12}-MT spectrum. The 12 Ag(I) spectral inflection is better resolved at the higher temperature, before the formation of the $Ag_{17/18}$-MT species (Reproduced with permission of the American Chemical Society, ref Zelazowski and Stillman 1992)

Hasnain et al. (1987). The Ag-S bond length for Ag_{17}, Cd_2-MT was reported as ca. 2.40 Å with a coordination number of 2 for the Ag(I) based on Ag K-edge EXAFS. The Ag-S bond lengths were later found to be 2.44–2.45 Å in both Ag_{12}-MT and Ag_{17}-MT, which indicated a digonal geometry (Figs. 8 and 9; Gui et al. 1996). In 1996, Gui et al. reported sulfur K-edge EXAFS spectroscopy for Ag_N-MT 1. Gui et al. proposed that 4 bridging and 16 terminal sulfurs are present in Ag_{12}-MT 1 and 14 bridging and 6 terminal sulfurs are present in Ag_{17}-MT 1 (Gui et al. 1996).

Coordination of Ag(I) in Metallothionein

From detailed metalation studies of the individual domain fragments, the stoichiometry of metals like copper and silver have been determined to range from 6 to 7 metals in each domain (Nielson et al. 1985) compared with the 3–4 divalent metals. This higher binding stoichiometry for Cu-MT and Ag-MT suggests that the protein adopts a conformation different from that of Cd or Zn. Cu(I) does not usually form tetrahedral complexes with sulfur ligands (Vortisch et al. 1976), favoring trigonal coordination of the Cu(1) ions with bridging sulfurs. Examples have been characterized in small Cu_4S_6 and Cu_5S_7 cores (Dance 1976; Dance and Calabrese 1976; Griffith et al. 1976). Similarly, Ag(I) may also bind trigonally or diagonally with sulfur ligands (Stillman et al. 1994).

The coordination of Ag(I) in mammalian liver MT was proposed to be mixed (trigonal and digonal) in the commonly observed Ag_{12}-MT and digonal in Ag_{18}-MT (unlike the trigonal coordination of the 12 Cu(I); Stillman et al. 1994). In that case, Ag(I) would not

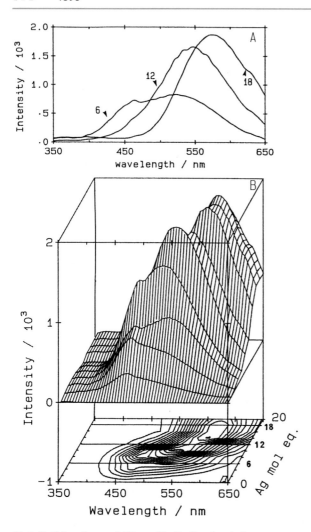

Metallothioneins and Silver, Fig. 7 Total emission spectrum data recorded for a series of separate aliquots of rabbit liver apo-MT 2a increasing molar ratio of added Ag(I) measured at 77 K as a glass showing the development of prominent species at the 12 and 17 Ag(I) points (18 is marked). The individual spectra are shown in (A). This research was originally published in the Journal of Biological Chemistry (Zelazowski, A.J., Gasyna, Z. Stillman, M.J. "Silver Binding to Rabbit Liver Metallothionein." *J. Biol. Chem.* 1989; 264, 17091–17099. © The American Society for Biochemistry and Molecular Biology)

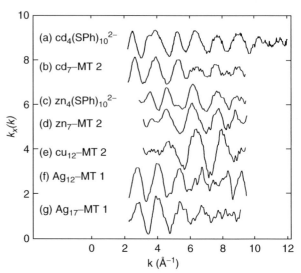

Metallothioneins and Silver, Fig. 8 S K-edge EXAFS recorded at room temperature for a series of metallothioneins and two models. The analysis of the data for the Ag_N-MT is shown in Fig. 9 (Reproduced with permission of the American Chemical Society, ref Gui et al. 1996)

isomorphously replace Cu(I) (Zelazowski and Stillman 1992; Gui et al. 1996; Palacios et al. 2003), although reports of Cu(I) binding to MT have indicated that Cu(I) may also bind in both trigonal and digonal coordination (Merrifield et al. 2002; Calderone et al. 2005). The difference in the CD features of Ag_{12}- and Ag_{17}-MT 1 may be attributed to changes in the tertiary structure that arises from changes in the fraction of bridging and terminal sulfurs, rather than from changes in coordination geometry (Gui et al. 1996).

Salgado et al., reported that the addition of up to 2 molar equivalents of Ag(I) to Cu-MT domain fragments (β and α) at room temperature expands the cluster, but does not extrude the Cu(I) (Salgado et al. 2007). The addition of a third molar equivalent, however, displaces the Cu(I) – as is observed by a shift in the emission band maxima (Salgado et al. 2007). In this metalation experiment, the coordination of Ag(I) was thought to be digonal for up to 3 Ag(I), although a mixture of trigonal and digonal coordination was proposed for stoichiometric species between 4 and 6 Ag(I). Only digonal coordination is then observed at the saturation point of 7 Ag(I) bound to the α and β domains of MT, because no changes in emission properties are observed past that point (Salgado et al. 2007).

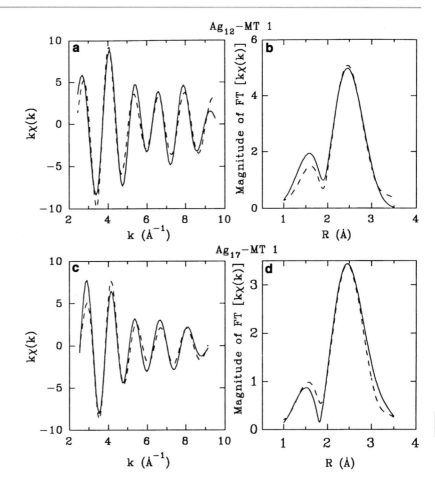

Metallothioneins and Silver, Fig. 9 Room temperature S K-edge EXAFS Fourier filtered (*solid*) and fit (*dotted*) data for rabbit liver Ag_{12}-MT -1 and Ag_{17}-MT -1 and the Fourier transform (*solid*) and fit (*dotted*) in *R*-space. The fits gave Ag-S bond lengths of 2.45 ± 0.02 for Ag_{12}-MT -1 and 2.44 ± 0.03 for Ag_{17}-MT -1 (Reproduced with permission of the American Chemical Society, ref Gui et al. 1996)

Acknowledgements We thank NSERC of Canada for financial support of the research reported from our laboratory.

Cross-References

- Copper, Physical and Chemical Properties
- Magnesium, Physical and Chemical Properties
- Metallothioneins and Copper
- Silver as DISINFECTAnt
- Silver, Burn Wound Sepsis and Healing
- Silver, Neurotoxicity
- Silver, Pharmacological and Toxicological Profile as Antimicrobial Agent in Medical Devices

References

Beltramini M, Lerch K (1981) Luminescence properties of *Neurospora* copper metallothionein. FEBS Lett 127:201–203

Calderone V et al (2005) The crystal structure of yeast copper thionein: the solution of a long-lasting enigma. Proc Natl Acad Sci USA 102:51–56

Casey AL et al (2010) Role of copper in reducing hospital environment contamination. J Hosp Infect 74:72–77

Cosson RP (1994) Heavy metal intracellular balance and relationship with metallothionein induction in the liver of carp after contamination by silver, cadmium and mercury following or not pretreatment by zinc. Biometals 7:9–19

Dance IG (1976) Formation and x-ray structure of hexa(tert-butylthiolato)pentacuprate(I) monoanion. J Chem Soc Chem Commun 2:68–69

Dance IG, Calabrese JC (1976) Crystal and molecular-structure of hexa-(MU2-benzenethiolato)tetracuprate(I) dianion. Inorg Chim Acta 19:L41–L42

Griffith EH, Hunt GW, Amma EL (1976) Adamantane structure in polynuclear Cu4S6 cores – crystal and molecular-structures of Cu_4 $SC(NH_2)_2$ $6(NO_3)_4 \cdot 4H_2O$ and Cu_4 SC $(NH_2)_2$ $9(NO_3)_4 \cdot 4H_2O$. J Chem Soc Chem Commun 12:432–433

Gui Z, Green AR, Kasrai M, Bancroft GM, Stillman MJ (1996) Sulfur K-edge EXAFS studies of cadmium-, zinc-, copper-, and silver-rabbit liver metallothioneins. Inorg Chem 35:6520–6529

Hamer DH (1986) Metallothionein. Annu Rev Biochem 55:913–951

Hasnain SS et al (1987) Chemical structure of metallotheion. In: Kagi JHR, Kojima Y (eds) EXAFS studies of metallothionein. Metallothionein II. Zurich, Birkhäuser, pp 227–236

Howe PD, Dobson S (2002) Silver and silver compounds: Environmental aspects. World Health Organization, Geneva. Available online: http://www.who.int/ipcs/publications/cicad/en/cicad44.pdf

Klasen HJ (2000a) Historical review of the use of silver in the treatment of burns. I. Early uses. Burns 26:117–130

Klasen HJ (2000b) A historical review of the use of silver in the treatment of burns. II. Renewed interest for silver. Burns 26:131–138

Kojima Y, Kägi JHR (1978) Metallothionein. Trends Biochem Sci 3:90–93

Liu J, Kershaw WC, Klaassen CD (1991) The protective effect of metallothionein on the toxicity of various metals in rat primary hepatocyte culture. Toxicol Appl Pharmacol 107:27–34

Mayer GD, Leach DA, Kling P, Olsson P-E, Hogstrand C (2003) Activation of the rainbow trout metallothionein-A promoter by silver and zinc. Comp Biochem Physiol B 134:181–188

Menard MP, McCormick CC, Cousins RJ (1981) Regulation of intestinal metallothionein biosynthesis in rats by dietary zinc. J Nutr 111:1353–1361

Merrifield ME, Huang ZY, Kille P, Stillman MJ (2002) Copper speciation in the alpha and beta domains of recombinant human metallothionein by electrospray ionization mass spectrometry. J Inorg Biochem 88:153–172

Narula SSM, Rajesh K, Winge DR, Armitage IM (1991) Establishment of the metal-to-cysteine connectivities in silver-substituted yeast metallothionein. J Am Chem Soc 113:9354–9358

Nath R, Paliwal VK, Prasad R, Kambadur R (1987) Role of metallothionein in metal detoxification, metal tolerance and metal toxicity. In: Kagi JHR, Kojima Y (eds) Role of metallothionein in metal detoxification and metal tolerance in protein calorie malnutrition and calcium deficient monkeys (Macaca mulatta). Metallothionein II. Zurich, Birkhäuser, pp 631–638

Ngu TT, Dryden MDM, Stillman MJ (2010) Arsenic transfer between metallothionein proteins at physiological pH. Biochem Biophys Res Commun 401:69–74

Nielson KB, Atkin CL, Winge DR (1985) Distinct metal-binding configurations in metallothionein. J Biol Chem 260:5342–5350

Oppenheimer BS, Oppenheimer ET, Danishefsky I, Stout AP (1956) Carcinogenic effect of metals in rodents. Cancer Res 16:439–441

Palacios O, Polec-Pawlak K, Lobinski R, Capdevila M, Gonzalez-Duarte P (2003) Is Ag(I) an adequate probe for Cu(I) in structural copper-metallothionein studies? The binding features of Ag(I) to mammalian metallothionein 1. J Biol Inorg Chem 8:831–842

Palmiter RD (1994) Regulation of metallothionein genes by heavy-metals appears to be mediated by a zinc-sensitive inhibitor that interacts with a constitutively active transcription factor, MTF-1. Proc Natl Acad Sci USA 91:1219–1223

Prohaska JR (2008) Role of copper transporters in copper homeostasis. Am J Clin Nutr 88:826S–869S

Ratte HT (1999) Bioaccumulation and toxicity of silver compounds: a review. Environ Toxicol Chem 18:89–108

Richards MP, Cousins RJ (1975) Mammalian zinc homeostasis – requirement for RNA and metallothionein synthesis. Biochem Biophys Res Commun 64:1215–1223

Salgado MT, Bacher KL, Stillman MJ (2007) Probing structural changes in the alpha and beta domains of copper- and silver-substituted metallothionein by emission spectroscopy and electrospray ionization mass spectrometry. J Biol Inorg Chem 12:294–312

Schmitz G, Minkel DT, Gingrich D, Shaw CF (1980) The binding of gold (I) to metallothionein. J Inorg Biochem 12:293–306

Stillman MJ (1999) Spectroscopic studies of copper and silver binding to metallothioneins. Metal Based Drugs 6:277–290

Stillman MJ, Presta A, Gui Z, Jiang DT (1994) Spectroscopic studies of copper, silver and gold-metallothioneins. Metal Based Drugs 1:375–394

Sutherland DEK, Stillman MJ (2008) Noncooperative cadmium (II) binding to human metallothionein 1a. Biochem Biophys Res Commun 372:840–844

Sutherland DEK, Stillman MJ (2011) The "magic numbers" of metallothionein. Metallomics 3:444–463

Sutherland DEK, Summers KL, Stillman MJ (2012a) Noncooperative metalation of metallothionein 1a and its isolated domains with zinc. Biochemistry 51:6690–6700

Sutherland DEK, Willans MJ, Stillman MJ (2012b) Single domain metallothioneins: supermetalation of human MT 1a. J Am Chem Soc 134:3290–3299

Vortisch V, Kroneck P, Hemmerich P (1976) Model studies on coordination of copper in enzymes.4. Structure and stability of cuprous complexes with sulfur-containing ligands. J Am Chem Soc 98:2821–2826

Warnes SL, Caves V, Keevil CW (2012) Mechanism of copper surface toxicity in Escherichia coli O157:H7 and Salmonella involves immediate membrane depolarization followed by slower rate of DNA destruction which differs from that observed for Gram-positive bacteria. Environ Microbiol 14:1730–1743

Wiley RA, Brooks RR (1992) In: Brooks RR (ed) Noble metals and biological systems: their role in medicine, mineral exploration, and the environment. CRC Press, Boca Raton

Zelazowski AJ, Stillman MJ (1992) Silver binding to rabbit liver zinc metallothionein and zinc alpha-fragment and beta-fragment - Formation of silver metallothionein with Ag(I)-protein ratios of 6, 12, and 18 observed using circular-dichroism spectroscopy. Inorg Chem 31:3363–3370

Zelazowski AJ, Gasyna Z, Stillman MJ (1989) Silver binding to rabbit liver metallothionein. J Biol Chem 264:17091–17099

Metallothioneins as Cadmium-Binding Proteins

▶ Cadmium and Metallothionein

Metallothioneins in Cadmium-Induced Toxicity

▶ Cadmium and Metallothionein

Metallo-β-Lactamase Superfamily

▶ Zinc Beta Lactamase Superfamily

Metal-Protein Interactions

▶ Platinum (IV) Complexes, Inhibition of Porcine Pancreatic Phospholipase A2
▶ Platinum Interaction with Copper Proteins
▶ Platinum-Containing Anticancer Drugs and Proteins, interaction

Metal-Resistant Bacteria

▶ CusCFBA Copper/Silver Efflux System

Metals and the Periodic Table

Fathi Habashi
Department of Mining, Metallurgical, and Materials Engineering, Laval University, Quebec City, Canada

Introduction

Metals are articles of everyday life – a gold ring, an aluminum window, a car made of steel, etc. Nonmetals (except carbon) are hardly seen, e.g., the air we breathe; they are used mainly as compounds or enter in the manufacture of commodities such as fertilizers, plastics, explosives, etc. Metalloids, on the other hand, are mainly used in advanced technology, e.g., transistors, computers, etc. (Fig. 1).

Metals form the major part of the periodic table. Nonmetals include the inert gases; hydrogen, oxygen, nitrogen, fluorine, and chlorine; liquid bromine; and the solid elements carbon, sulfur, phosphorus, and iodine. These elements do not have the properties of a metal. Nonmetals except the inert gases readily share electrons. Their atoms are united together by covalent bond, i.e., atoms that share their outer electrons. They often form diatomic molecules such as H_2, Cl_2, N_2, or larger molecules such as P_4 and S_8, or giant molecules, i.e., a network of atoms of indefinitely large volume such as carbon in form of graphite or diamond. Metalloids have covalent bond like nonmetals, but have intermediate properties between metals and nonmetals (Table 1).

General Properties of Metals

In the solid state, they are composed of crystals made of closely packed atoms whose outer electrons are so loosely held that they are free to move throughout the crystal lattice. This structure explains well their mechanical, physical, and chemical properties.

Mechanical

Since the electrons in the outermost shell are uniform throughout the crystal, the positive ions in the crystal may be moved past one another with relative ease. As is shown in Fig. 2, one group of ions can be changed in position relative to neighboring groups of ions, without changing the internal environment of each positive ion; this makes it possible to change the shape of the crystal without breaking it.

When compared with an ionic crystal, it is evident that the ionic crystal breaks because of the repulsive forces formed as a result of movement. Also, in a covalent bond, where electrons are shared and localized in certain positions, movement causes breaking the bond and results in fracture. Hence, metals have useful mechanical properties. They are malleable, i.e., can be hammered into sheets, and ductile, i.e., can be

Metals and the Periodic Table, Fig. 1 Metals, nonmetals, and metalloids

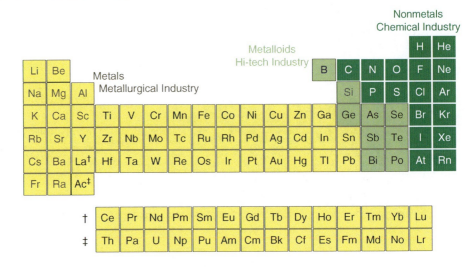

Metals and the Periodic Table, Table 1 Metals, nonmetals, and metalloids

Metals	Metalloids	Nonmetals
Crystalline solids (except mercury) with metallic luster	May be crystalline or amorphous, sometimes have metallic luster	Form volatile or nonvolatile molecules having no metallic luster
Do not readily share electrons; their vapors are monoatomic	Readily share electrons even in the elemental form	Readily share electrons; form diatomic, large, or giant molecules; inert gases are monatomic
Exhibit electrical and thermal conductivity. Electrical resistance usually increases with increased temperature	Low electrical and thermal conductivity	Do not conduct electricity or heat. Electrical resistance decreases with increased temperature
Have high density and useful mechanical properties	Moderate density, no useful mechanical properties	Low density, no useful mechanical properties
Electropositive, form cations, e.g., Cu^{2+}, Na^+, etc.	Sometimes electropositive, sometimes electronegative	Electronegative, form anions, e.g., S^{2-}, Cl^-, etc.
Form basic oxides, e.g., CaO	Form acidic oxides	Form acidic oxides, e.g., SO_2
Deposit on the cathode during electrolysis	Deposit on the cathode	Deposit on the anode, e.g., O_2, Cl_2
Either form no compounds with hydrogen or form unstable compounds usually nonvolatile (metal hydrides)	Form stable compounds with hydrogen, e.g., AsH_3, H_2Se	Form stable compounds with hydrogen, usually volatile, e.g., NH_3, PH_3, H_2S, etc.

drawn into wire. Although mercury is a liquid at room temperature, yet it is considered a metal because when cooled to below its freezing temperature ($-38.87°C$), it has all the characteristics of a metal.

Physical

The physical properties of metals can be readily correlated with the electronic structure model.

Electrical conductivity: Because the electrons in the metallic crystal are not associated with any one atom, then when a voltage difference is applied, they can readily move to the positive electrode. Hence, metals have a high electrical conductivity. As the temperature is raised and the atoms are thermally agitated, the electron movement receives interference. Hence, the electrical conductivity is reduced.

Thermal conductivity: Because the electrons are free to travel from atom to atom in a metal crystal, they are capable of transferring thermal energy. Hence, metals have high thermal conductivity. The thermal transfer through metals generally decreases as the temperature is increased because of interference with thermal vibration of the ions.

Density: Metals usually have a high density because they are arranged in a highly packed crystal structure. There are three main crystal structures for metals:

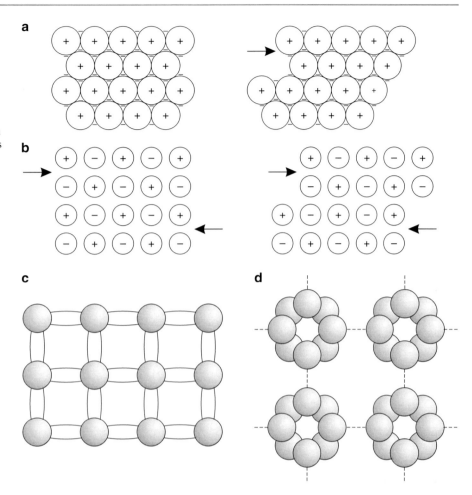

Metals and the Periodic Table, Fig. 2 Effect of displacement of atoms in solids: (**a**) metal; (**b**) ionic crystal; (**c**) covalent-bond crystal, shared areas indicate the fixed positions for electrons; (**d**) molecules in a crystal held together by van der Waals' forces, e.g., forces between S_8 rings

body-centered cubic, face-centered cubic, and hexagonal close-packed (Fig. 3 and Table 2). This is the most efficient way to fill a certain space; a five- or a seven-membered geometry will not fit efficiently (Fig. 4).

The atomic structure of a metallic crystal, whether face-centered cubic or hexagonal close-packed, is also reflected in the actual form of the metal. Thus, precipitated copper crystals are cubic (Fig. 5), while cobalt (Fig. 6) and zinc (Fig. 7) are hexagonal.

Metallic luster: Metals are opaque and capable of reflecting light to a high degree and have a silver to gray metallic color except copper (red) and gold (yellow). This is because the free electrons in the metallic crystal can absorb light energy of a wide range of wavelengths and can re-emit all these radiations. Electrons in metals have all energy levels available.

Metallic vapor: In the elemental state, metals do not share electrons; thus, their vapors are monoatomic.

Chemical

Metals readily lose electrons and therefore tend to form positively charged cations. In solution, these cations can be discharged at the cathode. When oxidized, metals form basic oxides, i.e., when these oxides are dissolved in water, they form basic solutions. For example, calcium forms the oxide CaO which when dissolved in water forms calcium hydroxide, $Ca(OH)_2$, a base. Metals either form no compounds with hydrogen or unstable hydrides, usually nonvolatile.

Classification of Metals

Since metals are those elements capable of losing electrons, they can be divided into typical, less typical, transition, and inner transition metals. This division is a result of their electronic structure (Fig. 8).

Metals and the Periodic Table, Fig. 3 Crystal structure of metals

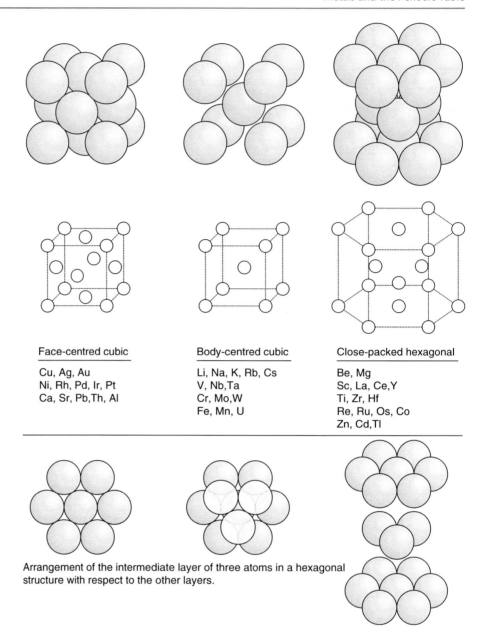

Face-centred cubic

Cu, Ag, Au
Ni, Rh, Pd, Ir, Pt
Ca, Sr, Pb, Th, Al

Body-centred cubic

Li, Na, K, Rb, Cs
V, Nb, Ta
Cr, Mo, W
Fe, Mn, U

Close-packed hexagonal

Be, Mg
Sc, La, Ce, Y
Ti, Zr, Hf
Re, Ru, Os, Co
Zn, Cd, Tl

Arrangement of the intermediate layer of three atoms in a hexagonal structure with respect to the other layers.

Metals and the Periodic Table, Table 2 Crystal structure of metals at ambient conditions

Group			Body-centered cubic	Face-centered cubic	Hexagonal close-packed
Typical			Alkali metals, Ba	Al, Ca, Sr	Be, Mg
Less typical			–	Cu, Ag, Au, Pb	Zn, Cd, Tl
Transition		Vertical	V, Nb, Ta, Cr, Mo, W, Mn	–	Sc, Y, La, Ti, Zr, Hf, Tc, Re
		Horizontal	Fe	Ni	Co
		Vertical/horizontal	–	Rh, Pd, Ir, Pt	Ru, Os
Inner transition		Lanthanides	Eu	Yb	Ce and other lanthanides
		Actinides	U, Np, Pu	Th	

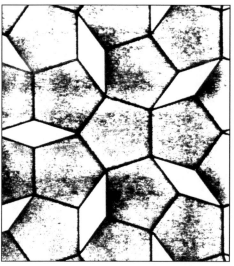
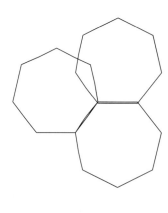

Metals and the Periodic Table, Fig. 4 Geometrical shapes and efficiency of space management

Five-membered geometry   Seven-membered geometry

Metals and the Periodic Table, Fig. 5 Electrodeposited copper in cubic form

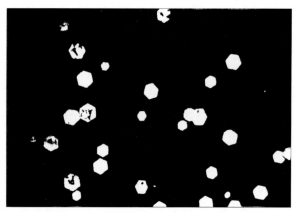

Metals and the Periodic Table, Fig. 6 Photomicrograph of cobalt crystals deposited from aqueous solution

Typical Metals

These are the alkali metals, the alkaline earths, and aluminum. They have the following characteristics:

- They have an electronic structure similar to that of the inert gases with one, two, or three electrons in the outermost shell.
- They have single valency, i.e., they lose their outermost electrons in a single step.
- They are reactive, i.e., react readily with water and oxygen. The driving force for this reactivity is the inclination to achieve maximum stability by attaining the electronic structure of an inert gas. A reactive metal such as aluminum or magnesium may be used as a material of construction because of the protective oxide film that is formed rapidly on its surface.
- They form only colorless compounds.
- Within a certain vertical group, the atomic radius increases with increasing atomic number because of the added electron shells.

Metals and the Periodic Table, Fig. 7 Electrodeposited zinc in hexagonal form (2,000×)

- Within a certain vertical group, the reactivity increases with increasing atomic number because of the ease with which the outermost electrons will be lost since they are further away from the nucleus. Thus, cesium is more reactive than rubidium, and rubidium more than potassium, etc.
- With increasing charge on the nucleus, the electrostatic attraction for the electrons increases and the outermost electrons will not be easily lost; hence, the reactivity decreases. Thus, magnesium is less reactive than sodium, calcium less than potassium, and so on.
- With increased electrostatic attraction for the electrons as a result of increasing charge on the nucleus, the size of the atom decreases. Thus, aluminum has a smaller radius than magnesium, and magnesium smaller than sodium.
- With decreased radius and increased atomic weight, the atom becomes more compact, i.e., the density increases. Thus, aluminum has higher density than magnesium, and magnesium higher than sodium.
- They have appreciable solubility in mercury and form compounds with it except beryllium and aluminum.

Less Typical Metals

These metals are copper, silver, gold, zinc, cadmium, mercury, gallium, indium, thallium, tin, and lead. They differ from the typical metals in that they do not have an electronic structure similar to the inert gases; the outermost shell may contain up to four electrons, and the next inner shell contains 18 instead of 8 electrons as in the inert gas structure. As a result of their electronic configuration, they are characterized by the following:

- The atomic radius is less than the corresponding typical metals in the same horizontal group because the presence of 18 electrons in one shell results in an increased electrostatic attraction with the nucleus. Thus, the atomic radius of copper is less than potassium, silver less than rubidium, and gold less than cesium. However, the atomic radius increases with increased number of electrons in the outermost shell (which is contrary to the typical metals), i.e., the atomic radius of gallium is larger than that of zinc, and zinc is larger than copper. This is demonstrated in Fig. 9: The atomic volume of the typical metals decreases with increased atomic number, while the reverse is true for the less typical metals. The reason for this is the shielding effect of the 18-electron shell, the increased repulsion of the additional electron in the outermost shell and that shell, and also the increased repulsion between the electrons themselves in that shell.
- The outermost electrons will not be easily lost, i.e., these metals are less reactive than their corresponding typical metals for two reasons:
  - There is no driving force to lose electrons since an inert gas electronic structure will not be achieved.
  - There is a stronger electrostatic attraction due to the smaller atomic radius as compared to that of the typical metals.
- Because of the higher atomic weight and the smaller atomic radius, these metals are more dense than their corresponding typical metals.
- Some of these metals show two different valency states, e.g., copper as Cu^I and Cu^{II}, gold as Au^I and Au^{III}, mercury as Hg^I and Hg^{II}, tin as Sn^{II} and Sn^{IV}, and lead as Pb^{II} and Pb^{IV}. This is because of the possibility of removing one or two electrons from the 18-electron shell.
- Few of these metals form colored ions in solution, e.g., Cu^{II} and Au^{III}, or colored compounds, e.g., copper sulfate pentahydrate (blue), cadmium sulfide (yellow), etc. This is due to the possibility of

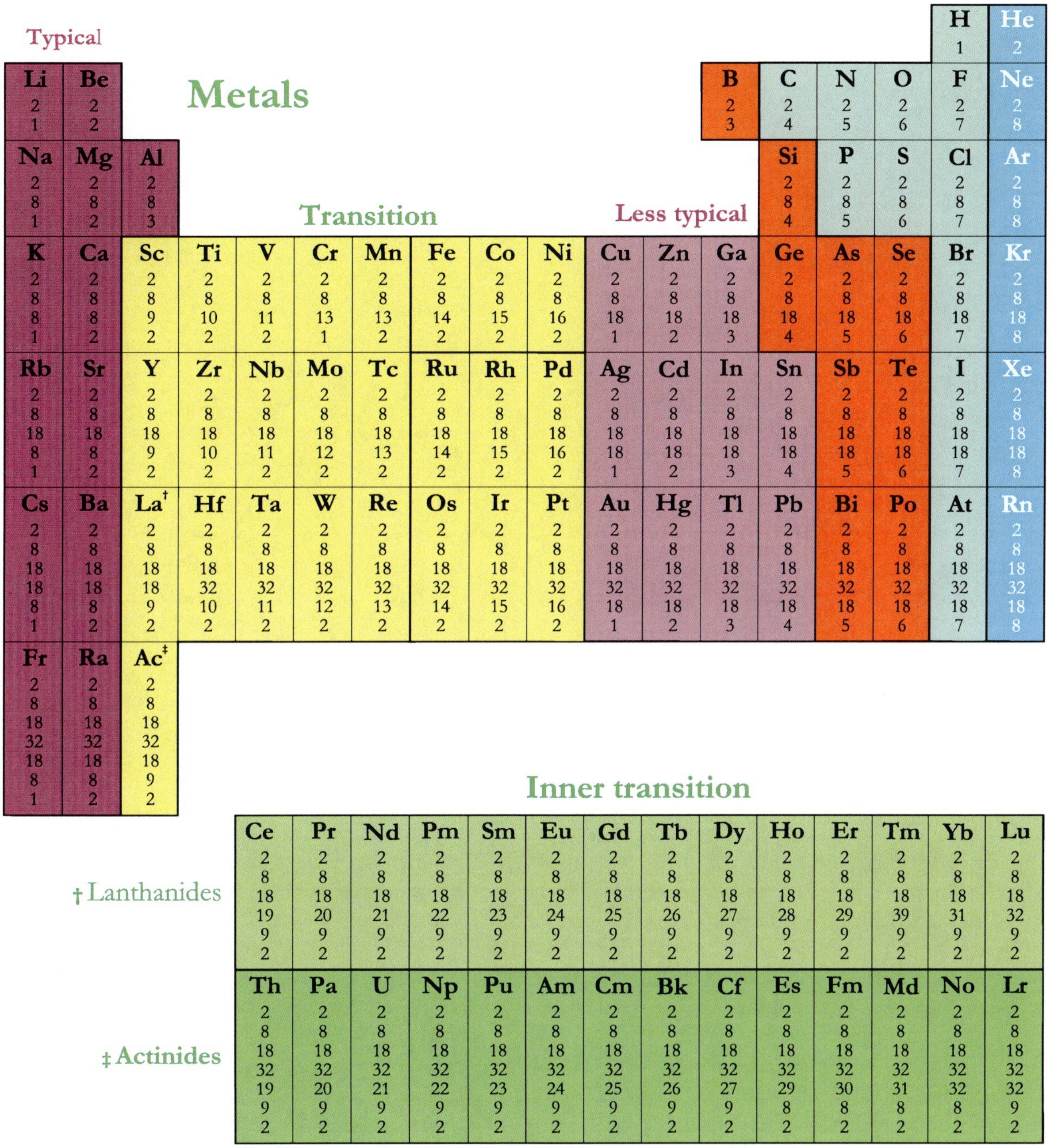

Metals and the Periodic Table, Fig. 8 Electronic configuration of the elements

movement of electrons from the 18-electron shell to a higher level.

- They have the highest solubility in mercury since their electronic structure is similar as that of mercury. Also, they do not form compound with mercury.

Transition Metals

These are the metals in the vertical groups in the periodic table from scandium to nickel. They not only have electronic configuration different from the inert gases but they are characterized by having the

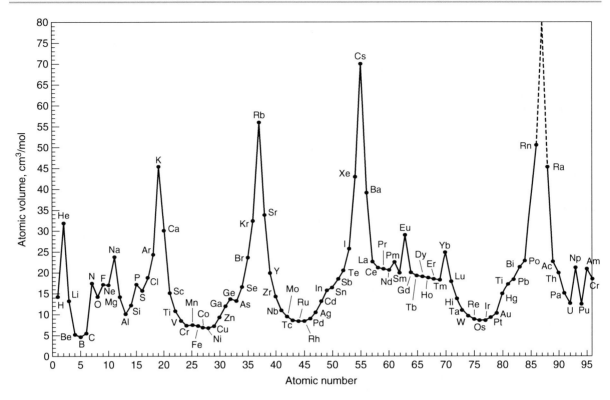

Metals and the Periodic Table, Fig. 9 Atomic volume of the elements

same number of electrons in their outermost shell and a progressively greater number of electrons in the next inner shell. There are, however, some apparent irregularities in the number of electrons in the outermost electron shells. This is due to energy levels, which are determined from spectroscopic measurements (Fig. 10):

- s electrons in the N shell have lower energy than d electrons in the M shell ($4s$ orbitals are filled before $3d$).
- s electrons in the O shell have lower energy than d electrons in the N shell ($5s$ orbitals filled before $4d$).
- d electrons in the O shell have nearly the same energy as f electrons in the N shell, etc.

As their name implies, the transition metals have properties between the typical and less typical metals. They are less reactive than the typical metals because they will not achieve the inert gas structure when they lose their outermost electrons, but they are nevertheless more reactive than the less typical metals. They share the following properties:

- They resemble each other quite closely besides showing the usual group relationships because they have the same number of the outermost electrons.
- They may lose additional electrons from the next lower shell to form ions with higher charges. As a result, they show a variable valence. For example, vanadium exists in +2, +3, +4, and +5 oxidation states, and titanium in +2, +3, and +4.
- The atomic radius of the successive metals in a certain horizontal period decreases slightly as the atomic number rises because when an electron is added to an inner shell, it decreases slightly the size of the atom as a result of increased electrostatic attraction.
- Most of them form colored ions in solution due to electronic transition with the exception of the group Sc, Y, La, and Ac that form only colorless compounds.

1	2	3	4	5	6	7
First shell	Second shell	Third shell	Fourth shell	Fifth shell	Sixth shell	Seventh shell
K	L	M	N	O	P	Q

```
                                              f    d
                                         f    d    p
                                                   s
                                    f    d    p
                                              s
                               d    p
                          d    p    s
                     d    s
                d    p
                p    s
           p    s
      s
s
```

Orbital	s	p	d	f
	□	□□□	□□□□□	□□□□□□□
Maximum number of electrons	2	6	10	14

Metals and the Periodic Table, Fig. 10 Energy states of electrons in various shells. No two electrons have equal energies because of different spin. Atoms that have their outer electrons in the same type of orbital should have similar chemical behavior. Spectroscopic notation: s sharp, p principal, d diffuse, f fundamental

- They form many covalent compounds, e.g., the carbonyls of iron and nickel, the chlorides of titanium, and the oxyacids of chromium, molybdenum, and tungsten.
- They form coordination compounds with ammonia, e.g., the ammines of cobalt and nickel.
- They mostly form borides, carbides, nitrides, and hydrides, which have mostly metallic character.
- They have the lowest solubility in mercury.

The transition metals can be divided into three groups:

- *Vertical similarity transition metals.* These are the vertical groups scandium to manganese. They show similarity in the vertical direction, e.g., Zr–Hf, Nb–Ta, and Mo–W. The group Sc, Y, La, and Ac form colorless compounds and have the same valency (+3).
- *Horizontal similarity transition metals.* This is the group titanium to nickel. They show similarity in the horizontal direction.
    - Their carbides have intermediate properties between the metal-like character of the transition metals and the ionic character of the typical metals. Thus, they have metallic luster and are electrically conductive, but they are attacked by water and dilute acids.
    - They form di- and trivalent compounds.
    - These metals have melting points in the range 1,220–1,800°C.
    - Iron, cobalt, and nickel occur in nature together in the native state in the minerals awaruite, Fe$(Ni,Co)_3$, and josephinite, Fe$(Ni,Co)_2$.
- *Horizontal–vertical transition metals.* This is the platinum metals group where the similarity between the six metals is in the horizontal and vertical direction.
    - They resist corrosion.
    - They occur together in nature in the native state.

Inner Transition Metals

These metals have the same number of electrons in the two outermost shells but a progressively greater number of electrons in the next inner shell. They form two groups: the lanthanides and the actinides.

Summary

The old tradition of numbering the groups in the periodic table has been abandoned and replaced by the following descriptive groups (Fig. 11):

- Monatomic nonmetals
- Covalent nonmetals
- Metalloids
- Typical metals
- Less typical metals
- Transition metals with vertical similarity
- Transition metals with horizontal similarity (Ti to Ni)
- Transition metals with vertical and horizontal similarity (platinum group metals)
- Inner transition metals:
    - The lanthanides
    - The actinides

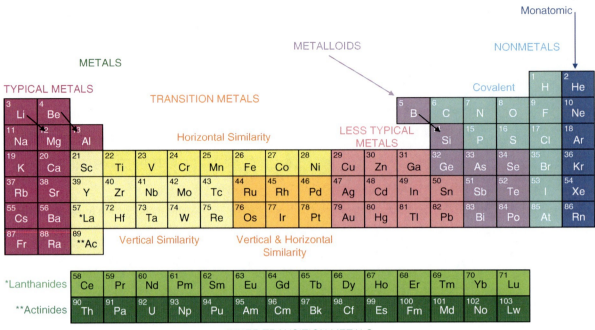

Metals and the Periodic Table, Fig. 11 A periodic table in ten groups

References

Habashi F (2003) Metals from ores. An introduction to extractive metallurgy. Métallurgie Extractive Québec, Sainte-Foy, distributed by Laval University Bookstore

Habashi F (2010) Metals: typical and less typical, transition and inner transition. Found Chem 12(1):31–39

Metals as Antimicrobials

▶ Silver as Disinfectant

Metavanadate

▶ Vanadium in Biological systems

Methionine Amino Peptidase

▶ Cobalt Proteins, Overview

Method of Synthesis

▶ Gold Nanomaterials as Prospective Metal-based Delivery Systems for Cancer Treatment

Methyl Coenzyme M Reductase

Evert C. Duin
Department of Chemistry and Biochemistry, Auburn University, Auburn, AL, USA

Synonyms

2-(Methylthio)ethanesulfonate:N-(7-thioheptanoyl)-3-O-phosphothreonine S-(2-sulfoethyl)thiotransferase; 2-(Methylthio)ethanesulfonic acid reductase; Coenzyme-B sulfoethylthiotransferase; MCR

Definition

Methyl-coenzyme M reductase (MCR, EC 2.8.4.1) catalyzes the (reversible) reduction of methyl-coenzyme

Methyl Coenzyme M Reductase, Fig. 1 *Reaction catalyzed by methyl-coenzyme M reductase (MCR).* $CH_3–S–CoM$, methyl-coenzyme M (2-(methylthio)ethanesulfonate); HS–CoB, coenzyme B (*N*-7-mercaptoheptanoyl-*O*-phospho-L-threonine); CoM–S–S–CoB, heterodisulfide of coenzyme M (2-mercaptoethanesulfonate) and coenzyme B; F_{430}, factor 430

M ($CH_3–S–CoM$) with coenzyme B (H–S–CoB) to methane and the heterodisulfide CoM–S–S–CoB according to (1) (Ragsdale 2003; Duin 2007).

$$CH_3–S–CoM + H–S–CoB \rightleftarrows CH_4 \\ + CoM–S–S–CoB \quad \Delta G^{\circ\prime} = -30 \text{kJ} \cdot \text{mol}^{-1} \quad (1)$$

The structures of these substrates are presented in Fig. 1.

Basic Properties

MCR enzymes have been crystalized from *Methanothermobacter marburgensis*, *Methanosarcina barkeri*, and *Methanopyrus kandleri* (Ermler 2005; Cedervall et al. 2011). The structures showed that the enzyme is a functional dimer of αβγ protomers (Fig. 2). It contains two active site channels that each contain a tightly, but not covalently, bound coenzyme factor 430 (F_{430}). F_{430} is a nickel porphinoid (Fig. 1). The active form of the enzyme has the nickel in the 1+ oxidation state, but it is proposed to go through other states (2+ and 3+) during the catalytic cycle. The channels are lined with the amino acid chains of either subunits α, α′, β, and γ or subunits α′, α, β′, and γ′ (Fig. 3). In or in close vicinity of the active-site channel, five modified amino acids can be found. Four of these modifications are methylations resulting in a 3-methylhistidine, a *S*-methylcysteine, a 5-methylarginine, and a 2-methylglutamine. The fifth modification is the presence of a thiopeptide, the result of the replacement of a carbonyl oxygen of a glycine by a sulfur. The methylations are the result of post-translational modification. It was proposed that the role of these modifications is to make the active-site channel more hydrophobic. The function of the thioglycyl is not clear.

F_{430} is also responsible for the characteristic yellow color of MCR in the so-called MCRsilent (Ni^{2+}) form. Other colors are displayed in the other oxidation states (Table 1). The extinction coefficient of the absorption bands can be used to rapidly determine the enzyme concentration.

Methyl Coenzyme M Reductase, Fig. 2 Schematic presentation of the MCR structure

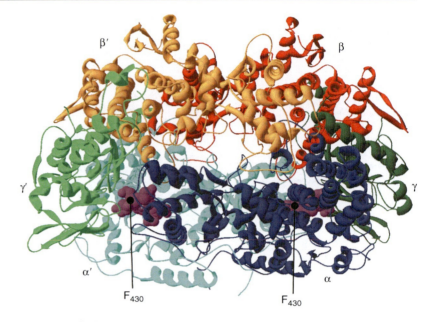

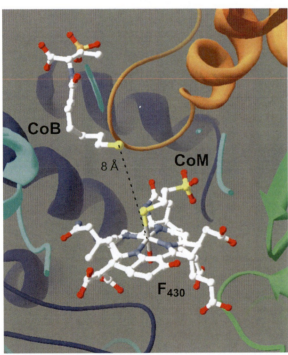

Methyl Coenzyme M Reductase, Fig. 3 *Detail of the active site of MCR in the ox1-silent form, with bound coenzyme M and coenzyme B. The subunit coloring is the same as in Fig. 2*

Physiological Function

MCR has been detected in both methanogenic and methanotrophic archaea (ANME) (Thauer 2010).

Methane production by methanogenic archaea (=methanogenesis) takes place in many anaerobic microbial habitats such as swamps, rice paddies, fresh water sediments and the intestinal tract of animals and insects. Almost all species of methanogenic archaea are able to oxidize H_2 and to use CO_2 as the electron acceptor. Species from several families are also able to use formate, methanol, methylamine, and acetate. Although every pathway starts out different, they all end with the formation of CH_4. The CH_4, however, is a waste product. The concomitant production of the heterodisulfide is much more important for the cell. Most methanogens couple the reduction of the disulfide bond of the heterodisulfide indirectly to ATP synthesis.

Upon breaking the cells of methanogenic archaea and separating the cytosol and membrane fragments, the majority of the MCR activity is present in the cytosol fraction. Immunogold labeling studies in *M. marburgensis* showed that MCR is not equally distributed in the cell but is preferentially localized close to the cell membrane. In *Methanosarcina mazei*, MCR appears to be part of a large membrane associated protein complex called a methanoreductosome.

Anaerobic oxidation of methane (AOM) is a microbial process occurring mainly in anoxic marine sediments. During AOM, methane is oxidized with sulfate as the terminal electron acceptor. This process is mediated by a syntrophic consortium of methanotrophic archaea and sulfate-reducing bacteria.

Methyl Coenzyme M Reductase, Table 1 Spectroscopic and structural properties of different MCR forms

	EPR parameters			Absorption band (nm) (ε, M^{-1} cm^{-1})		Axial ligand	Nickel charge
	g_1	g_2	g_3				
MCRred1a	2.061	2.070	2.250	387	720	–	+1
MCRred1c	2.063	2.068	2.248	387	720	–	+1
MCRred1m	2.061	2.071	2.251	387	748	CH_3–S–CoM (3.9 Å)	+1
MCRred1-silent	–	–	–	423	445 (shoulder)	(–S–CoM)	+2
MCRred2r	2.177	2.234	2.289	416	600, 690	–S–CoM	+1
MCRred2a	2.073	2.077	2.273	416	600, 690	–H	+1
MCRsilent	–	–	–	423 (22,000)	445 (shoulder)	–O–SO_2–CoM–S–S–CoB	+2
MCRox1	2.153	2.168	2.231	420	650	–S–CoM	+3
MCRox1-silent	–	–	–	423	445 (shoulder)	–S–CoM	+2
MCRox2	2.129	2.143	2.226	420	–	nd	+3
MCRox3	2.134	2.140	2.217	nd		nd	+3
MCR–BPS	2.108	2.112	2.219	420	495 (shoulder)	–PS	+3
MCR–BrMe	2.093	2.093	2.216	420	495 (shoulder)	–Me	+3

nd not determined

It is clear that both processes, the anaerobic oxidation of methane and the reduction of sulfate, have to be somehow coupled for methanogenesis to go in reverse, but it is not clear how this is achieved and what compound(s) is (are) exchanged between the two consortia. Recent investigations suggest that AOM is a complete enzymatic reversal of methanogenesis (Thauer 2010).

Activity

Methane formation from methyl-coenzyme M and coenzyme B can be determined under exclusion of air using stoppered serum vials (Duin et al. 2011). The methane that is produced can be detected using gas chromatography with flame ionization detection. The best results are obtained if, in addition to both substrates, aquocobalamin and the reductant titanium(III) citrate are present. The cobalamin will reduce the formed heterodisulfide back to the single thiols, coenzyme M and coenzyme B. This is important since heterodisulfide inhibits the reaction. The specific activity for the *M. marburgensis* isoenzyme I is 70–100 U (μmol·min^{-1}) per mg protein calculated for 1 spin per mol F_{430}. (In this case a spin of 1 means that all nickel in MCR is present as MCRred1 (see below). The spin is related to the metal content, not to the protein content, since enzyme without cofactor should not display any activity.)

MCR is very picky when it comes to substrates. The only other substrates are ethyl-coenzyme M (CH_3CH_2–S–CoM), seleno-methyl-coenzyme M (CH_3–Se–CoM), fluoro-methyl-coenzyme M (CFH_2–S–CoM), difluoro-methyl-coenzyme M (CF_2H–S–CoM), and C_6-coenzyme B (one methylene group shorter than coenzyme B).

The formation of methane is favored under standard conditions. Most of the steps in the hydrogenothropic pathway (conversion of CO_2 into CH_4) have always been considered reversible, except for the reaction catalyzed by MCR. Initially, it was proposed that the oxidation of methane was performed by a new enzyme. MCR, however, was shown to be present in the consortia ANME archaea. These consortia were isolated from biomats that grow on top of methane hydrates at the bottom of the black sea. The high methane concentration in this environment would make the reversible reaction a possibility. Recently it was shown that under certain conditions MCR from *M. marburgensis* (not an ANME) can indeed catalyze the conversion of methane into methyl-coenzyme M (Scheller et al. 2010a). The assay consists of an isotope-labeling experiment in which the formation of $^{13}CH_3$–S–CoM from $^{13}CH_4$ and CoM–S–S–CoB in the presence of $^{12}CH_3$–S–CoM was measured. The experiment is based on three equilibriums:

$$^{13}CH_4 + CoM\text{–}S\text{–}S\text{–}CoB \rightleftharpoons {}^{13}CH_3\text{–}S\text{–}CoM \\ + HS\text{–}CoB \quad \Delta G° = +30 \text{ kJ mol}^{-1} \quad (2)$$

$$^{12}CH_3\text{–}S\text{–}CoM + HS\text{–}CoB \rightleftharpoons {}^{12}CH_4 \\ + CoM\text{–}S\text{–}S\text{–}CoB \quad \Delta G° = +30 \text{ kJ mol}^{-1} \quad (3)$$

$$^{13}CH_4 + {}^{12}CH_3-S-CoM \rightleftarrows {}^{12}CH_4 \\ + {}^{13}CH_3-S-CoM \quad \Delta G° = 0 \text{ kJ mol}^{-1} \quad (4)$$

A specific activity of 11.4 nmol·min^{-1} per mg MCR was found using this procedure.

Different MCR Forms

The spectroscopic and structural properties of all relevant MCR forms are compiled in Table 1. The naming of the different forms of MCR is based on the conditions used to induce these forms. Some of the forms were first discovered in whole cell experiments. For a cell using the hydrogenothropic pathway, H_2, through hydrogenases, is a reductant and the substrate CO_2 can be considered an oxidant. Therefore the signals that were induced by gassing with H_2 were dubbed "red" and the number indicated the order in which they were discovered. In the same way, the signals, which were detected upon incubation with a mixture of N_2 and CO_2, were called "ox" signals. These names are still appropriate since in the MCRred1 form, the Ni in F_{430} is in the 1+ oxidation state. It is also the active form of the enzyme. MCRox1 is an inactive form with the Ni in the 3+ oxidation state. MCRox1, however, can be activated (see below). Both the MCRred1 (Ni^{1+}, d^9) and the MCRox1 (Ni^{3+}, d^7) forms are paramagnetic and are detectable in electron paramagnetic resonance (EPR) spectroscopy. The Ni^{2+} (d^8) from is diamagnetic (S = 1) and is not detectable in EPR spectroscopy, also referred to as "EPR silent" and this form was therefore dubbed MCRsilent. This form is inactive.

The addition of either the substrate methyl-coenzyme M or a coenzyme M analog will stabilize MCR. Both compounds, however, will affect the behavior of MCR in different ways and therefore different forms of MCRred1 are considered. MCRred1c is the form in the presence of coenzyme M. MCRred1m is the form in the presence of methyl coenzyme M. A clear difference in these forms can be detected in their respective EPR signals (Fig. 4 and Table 1). The MCRred1m EPR spectrum shows a more resolved hyperfine-splitting pattern due to the four nitrogen ligand from F_{430} to the nickel. Electron nuclear double resonance (ENDOR) measurements with labeled

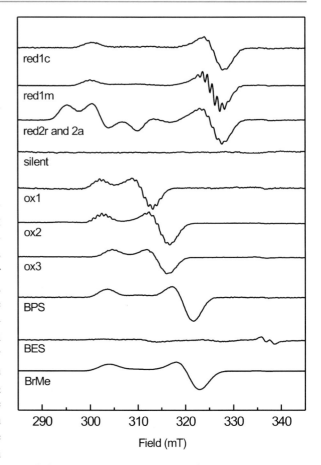

Methyl Coenzyme M Reductase, Fig. 4 *Overview of the different EPR active species in methyl-coenzyme M reductase.* G-values are listed in Table 1

methyl-coenzyme M showed that in this from there is a weak interaction between the Ni(I) center and the thioether sulfur atom of methyl-coenzyme M, which is at a Ni-S distance of 3.94 Å. Removal of methyl-coenzyme M and/or coenzyme M results in the very unstable MCRred1a form ("a" for *a*bsence). Figure 5, shows how the different forms of MCR are related and how they can be converted into each other. In most cases, it required the removal of one compound through a washing procedure (indicated with "−") and the addition of another compound (indicated with "+").

MCRred1c (in the presence of coenzyme M) can be converted into MCRred2 by the addition of coenzyme B. Two forms are induced, MCRred2a ("a" for axial) and MCRred2r ("r" for rhombic) based on their differences in EPR signals (Fig. 4). MCRred2 can be converted into several ox forms: Addition of polysulfide will convert it into MCRox1, addition of sulfite

Methyl Coenzyme M Reductase,
Fig. 5 *Overview of relevant forms of methyl-coenzyme M reductase (MCR) and their interrelationships.* See text for details. The asterisk indicates light-sensitive forms (Taken with permission from Duin et al. 2011)

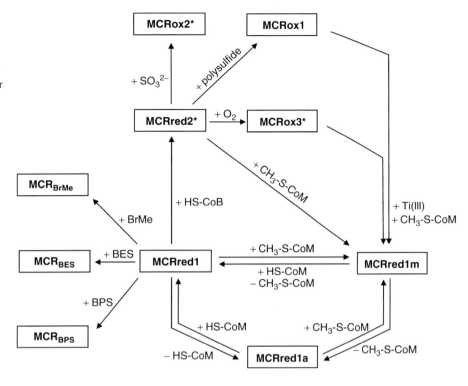

will convert it into the MCRox2 form, and exposure to oxygen will convert it into MCRox3. Both MCRox1 and MCRox3 can be converted back into the active MCRred1m form upon incubation with the reductant Ti(III) citrate. Three of the MCR forms, MCRred2, MCRox2, and MCRox3 are light sensitive (indicated with an asterisk in Fig. 5) and additional "light" forms can be induced.

Absent in this scheme is MCRsilent. All forms will eventually become EPR-silent due to inherent instability of the protein. Not all silent forms are therefore the same. Exposure of oxygen will also turn most forms into an EPR-silent form, including the MCRox forms, albeit very slowly. The only exception to this is MCRred2.

Reaction Mechanism

It appears that the first step in the reaction mechanism is also the rate limiting step, prohibiting the detection of any reaction intermediate when MCRred1 is reacted with the native substrates either separately or together. Therefore most of our knowledge on the mechanism comes from crystallization studies and kinetic studies with substrate analogs and inhibitor compounds.

The most informative is the crystal structure of the MCRox1-silent form (MCRox1 that turned silent upon crystallization). In this form (Fig. 3), coenzyme M coordinates axially with its thiol group to the nickel from the proximal side of F_{430} (Fig. 1, reduced pyrrole rings A, B, C and D clockwise). From the distal side the oxygen of a glutamine residue axially coordinates the nickel, as it does in all crystal structures. Coenzyme B can reach into the active-site channel with its thiol group-containing arm only up to a distance of 8 Å from the nickel. Imagining methyl-coenzyme M in the position where we find coenzyme M in the structure, this would represent a possible starting point of the catalytic cycle. The position of the coenzyme B thiolate is too far from the nickel to be directly involved in the reaction mechanism. The structure of the active site channel would also not allow the coenzyme B to move closer to the nickel ion without a conformational change in the channel.

The next piece of the puzzle comes from ENDOR studies on the MCRred2 form. This form is induced by the addition of coenzyme M and coenzyme B to the MCRred1 form. The EPR spectrum consists of two spectra, an *a*xial spectrum (MCRred2a) and a *r*hombic spectrum (MCRred2r). In the MCRred2a form, the thiolate sulfur of coenzyme M is bound to the nickel, while in the MCRred2a form a hydride species is present. While in all MCR forms the electron spin shows identical coupling constants with the four nitrogen ligands to the nickel ion, this is not the case for MCRred2r, where at least one nitrogen ligand shows a much weaker coupling constant. This indicates a change in the F_{430} electronic structure which is probably correlated to a conformational change in the ring structure. At the same time labeling studies showed that in this form coenzyme B moves about 2 Å closer to the nickel. This means that in the absence of coenzyme B, methyl-coenzyme M can bind in the active site and only interacts weakly with the nickel ion. A very short-lived ternary complex is formed upon binding of coenzyme B. The binding is believed to induce a conformational change in the active-site channel that in the first place allows the thiolate sulfur of coenzyme B to move within 6 Å of the nickel, concomitantly pushing methyl-coenzyme M closer to the nickel. At the same time the F_{430} ring undergoes a conformational change. Many groups have tried to somehow trap this state but for now without success.

Alkylating agents play an important role in the MCR story. Free F_{430} can be easily methylated. The same is true for F_{430} inside MCR but also alkylations have been observed. Bromopropane sulfonate (BPS) is a strong inhibitor of MCR. Addition of this compound to MCRred1 results in a new EPR signal which is due to a Ni(III) species with the propane sulfonate moiety attached to the nickel via the C3 carbon (MCR–BPS or MCR–PS). With bromomethane a relative stable Ni(III)-methyl species is induced (MCR–BrMe). When MCR-BrMe is incubated with coenzyme M, methyl-coenzyme M is formed. As a result the Ni(III)–CH_3 species has been proposed to play an important role in the reaction mechanisms for both methane production and methane oxidation. At the same time density function theory calculations indicate that such a species might not be part of the mechanism since the methyl-Ni(F_{430}) formation is predicted to be endothermic by 98.4 kJ/mol (Chen et al. 2009).

Some inhibitors will also introduce radical species in MCR. When, for example, MCR red1 is incubated with bromoethane sulfonate (BES) the nickel-based signal disappears and a new radical signal is induced (MCR-BES or MCR-ES) (Fig. 4). The relative stability of these radical species indicates that the active-site channel provides a suitable environment for such species to survive for a longer time period suggesting that MCR could make use of a radical-based mechanism.

A typical approach to get around the rate limitation of the first reaction step is to look for substrate analogs that change the energy profile of the reaction and as a result other steps might become rate limiting. Studies have been performed with ethyl-coenzyme M that slows down the reaction by a factor 10, and C_6-coenzyme B that slows down the reaction by a factor 100–1,000. Using this approach the formation of a transient Ni-alkyl species and a transient radical species was detected when MCRred1 was mixed with methyl-coenzyme M and C_6-coenzyme B (Dey et al. 2011). The radical has properties suggestive of a sulfur–based radical and can be assigned to either coenzyme M or coenzyme B or even the nearby thioglycine residue. Additional labeling studies will be needed to discern between these possibilities. Based on these studies the hypothetical mechanism in Fig. 6 was proposed. In the first step (step 1), a nucleophilic attack of Ni(I) on the methyl group of methyl-SCoM takes place to form methyl-Ni(III) and ⁻S–CoB. This is followed by reduction of the methyl-Ni (III) species to methyl-Ni(II) (step 2) and subsequent proton transfer from the HSCoM thiol to generate methane and a CoMS • radical (step 3). Equilibration between the CoMS • radical and CoBS⁻ generates a CoBS • thiyl radical (step 4), and condensation of the CoBS • radical with the thiolate of ⁻SCoM forms a disulfide anion-radical intermediate (step 5). Electron transfer from the disulfide anion radical to Ni(II) completes the catalytic cycle by regenerating active Ni (I)-MCRred1 (step 6).

In addition to the theoretical objections to a Ni(III)–CH_3 species, this type of mechanism might not be able to explain the reversibility of the reaction as recently shown. Based on the conversion of methane into methyl-coenzyme M and ethane into ethyl-coenzyme M (Scheller et al. 2010b) an alternative mechanism was proposed. Figure 7, shows the mechanism for the ethyl-coenzyme M/ethane exchange reaction. In this model, the conformational change in the F_{430} ring is proposed to open up the possibility to form both a hydride-nickel species and a σ-alkane-nickel complex.

The two models are not mutually exclusive. It is important to realize, that for example, the MCRox1

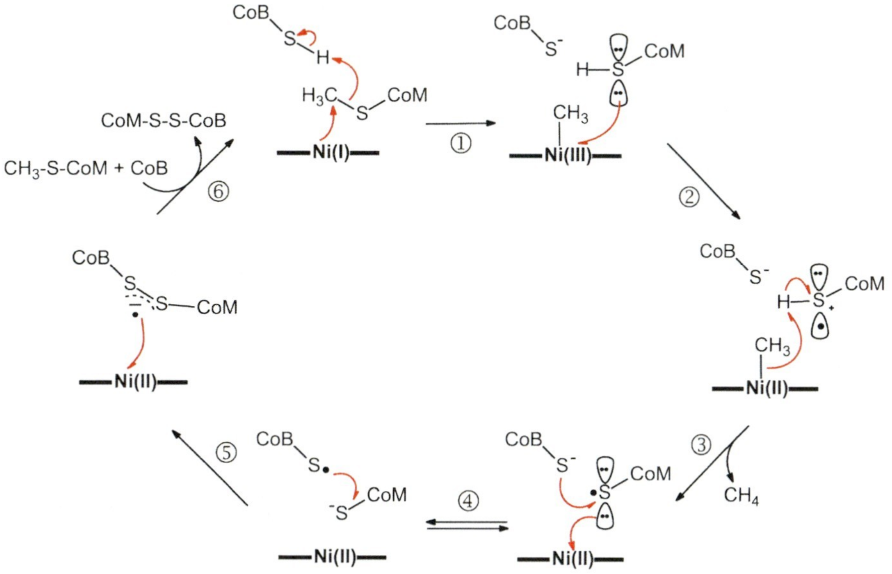

Methyl Coenzyme M Reductase, Fig. 6 *Hypothetical reaction mechanism I*. The ligand of coenzyme F_{430} is represented with bold lines. The mechanism is based on the fact that a methyl-Ni(III) species can easily be generated in the enzyme using strong methylating agents and the detection of the transient formation of the 495 nm band in absorption spectroscopy indicative for a Ni(III)–CH3 species, followed by the transient formation of a radical species detected in EPR-detected rapid-mix, rapid freeze experiments (Adapted from Dey et al. 2011)

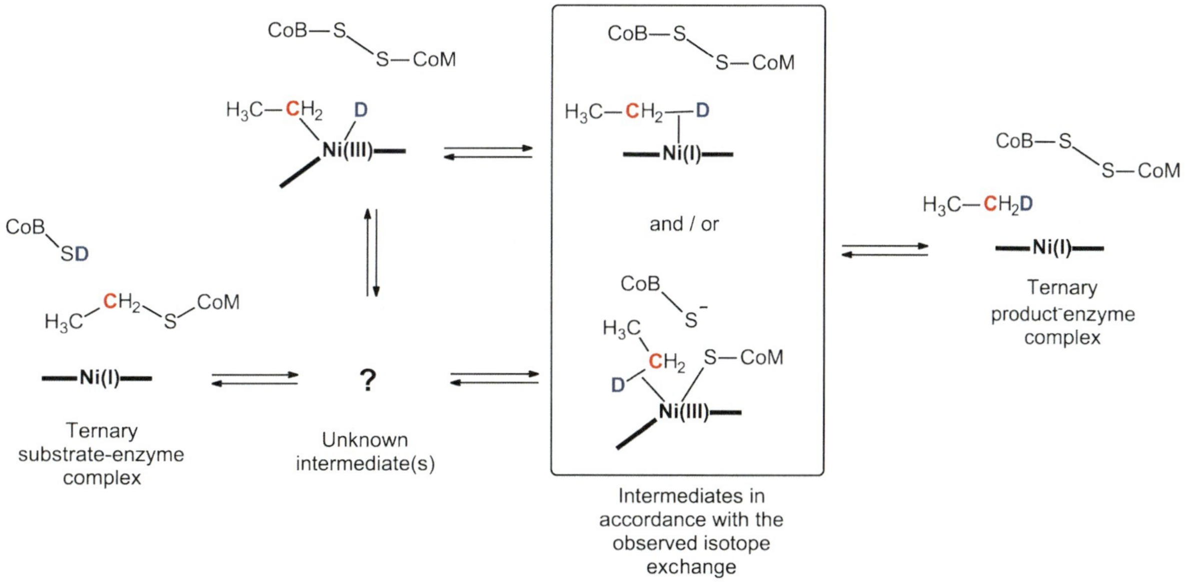

Methyl Coenzyme M Reductase, Fig. 7 *Hypothetical reaction mechanism II*. The mechanism is based on isotope exchange studies using ^{13}C-labeled (in red) ethyl-coenzyme M in deuterated medium. The bend in the Ni(III)-intermediates represents a conformational change in the F_{430} macrocycle (Adapted from Scheller et al. 2010b)

species is formally a Ni(III) species, but a significant amount of the unpaired electron density is found on the axial sulfur ligand. Therefore this species has also been described as a Ni(II)-thiyl species. When the reaction proceeds with the natural substrates it can be expected that some electron density might built up on the a sulfur ligand attached to a Ni(III) species without the formation of a true thiyl radical species as detected with the substrate analogs. The final mechanism is probably going to be a hybrid with elements of both hypothetical mechanisms. Thirty years after the first purification attempts by Ellefson and Wolfe we got a much better understanding about the role and the reaction mechanism of this enzyme, but an essential piece of the puzzle is still missing, as indicated by the question mark in Fig. 7. Key to solving this would be to crystalize the MCRred2 form of the enzyme.

Cross-References

- Nickel in Bacteria and Archaea
- Nickel-Binding Proteins, Overview
- Nickel-Binding Sites in Proteins

References

Cedervall PE, Dey M, Li X et al (2011) Structural analysis of a Ni-methyl species in methyl-coenzyme M reductase from *Methanothermobacter marburgensis*. J Am Chem Soc 133:5626–5628

Chen SL, Pelmenschikov V, Blomberg MRA et al (2009) Is there a Ni-methyl intermediate in the mechanism of methyl-coenzyme M reductase? J Am Chem Soc 131:9912–9913

Dey M, Li X, Kunz RC et al (2011) Detection of organometallic and radical intermediates in the catalytic mechanism of methyl-coenzyme M reductase using the natural substrate methyl-coenzyme M and a coenzyme B substrate analogue. Biochemistry 49:10902–10911

Duin EC (2007) Role of coenzyme F_{430} in methanogenesis. In: Warren MJ, Smith A (eds) Tetrapyrroles: their birth, life and death. Landes Bioscience, Georgetown

Duin EC, Prakash D, Brungess C (2011) Methyl-coenzyme M reductase from *Methanothermobacter marburgensis*. Meth Enzymol 494:159–187

Ermler U (2005) On the mechanism of methyl-coenzyme M reductase. Dalton Trans 3451–3458

Ragsdale SW (2003) Biochemistry of methyl-CoM reductase and coenzyme F_{430}. In: Kadish KM, Smith KM, Guilard R (eds) Porphyrin handbook. Elsevier, San Diego

Scheller S, Goenrich M, Boecher R et al (2010a) The key nickel enzyme of methanogenesis catalyses the anaerobic oxidation of methane. Nature 465:606–609

Scheller S, Goenrich M, Mayr S et al (2010b) Intermediates in the catalytic cycle of methyl coenzyme M reductase: isotope exchange is consistent with formation of a σ-alkane–nickel complex. Angew Chem Int Ed 49:8112–8115

Thauer RK (2010) Functionalization of methane in anaerobic microorganisms. Angew Chem Int Ed 49:6712–6713

2-(Methylthio)ethanesulfonate:N-(7-Thioheptanoyl)-3-O-phosphothreonine S-(2-Sulfoethyl)thiotransferase

- Methyl Coenzyme M Reductase

2-(Methylthio)ethanesulfonic Acid Reductase

- Methyl Coenzyme M Reductase

Metzincin

- Zinc Adamalysins
- Zinc Matrix Metalloproteinases and TIMPs

$[Mg^{2+}]_i$ (Free Magnesium Concentration)

- Magnesium in Eukaryotes

$[Mg^{2+}]_i$, Intracellular Free Magnesium

- Magnesium in Health and Disease

$[Mg^{2+}]_i$ = Intracellular Free Magnesium Concentration

- Magnesium and Vessels

Mg^{2+} (Magnesium Ions)

- Magnesium in Eukaryotes

Microbial Toxic Mercury Resistance

- Bacterial Mercury Resistance Proteins

Micronutrient

- Copper, Biological Functions

Micronutrients

▶ Chromium(III), Cytokines, and Hormones

Millimolar: MM

▶ Magnesium Binding Sites in Proteins

Mimic: Imitate

▶ Monovalent Cations in Tryptophan Synthase Catalysis and Substrate Channeling Regulation

Mineral Deficiency

▶ Magnesium in Plants

Mineral Homeostasis

▶ Magnesium in Plants

Mineral Nutrition

▶ Magnesium in Plants

Mitochondria

▶ Thallium, Effects on Mitochondria

Mitochondrial Sodium and Potassium Channels and Exchangers

▶ Sodium and Potassium Transport in Mammalian Mitochondria

Mixed-Function-Oxidase

▶ Iron Proteins, Mononuclear (non-heme) Iron Oxygenases

MMA^{III} – Monomethylarsenite

▶ Arsenic, Biologically Active Compounds

MMA^{V} – Monomethylarsenate

▶ Arsenic, Biologically Active Compounds

Mn Cluster in Photosynthesis

▶ Manganese and Photosynthetic Systems

Mn in Photosystem II

▶ Manganese and Photosynthetic Systems

Mo Cofactor

▶ Molybdenum and Ions in Prokaryotes

Moco

▶ Molybdenum and Ions in Living Systems

Moco-Containing Enzymes

▶ Molybdenum in Biological Systems

Mode of Action

▶ Arsenic and Primary Human Cells
▶ Gold Complexes as Prospective Metal-Based Anticancer Drugs

Mo-Enzyme

▶ Molybdenum and Ions in Prokaryotes

Molecular Chaperones

▶ Calnexin and Calreticulin

Molybdate

▶ Molybdenum and Ions in Living Systems
▶ Molybdenum and Ions in Prokaryotes

Molybdenum and Ions in Living Systems

Silke Leimkühler
From the Institute of Biochemistry and Biology, Department of Molecular Enzymology, University of Potsdam, Potsdam, Germany

Synonyms

Moco; Molybdate; Molybdoenzyme; Molybdopterin

Definition

Molybdoenzymes are enzymes that harbor either the FeMoco or Moco as catalytical active center. It is a diverse group of enzymes which are involved in the metabolism of nitrogen, sulfur, and carbon compounds. Molybdenum has a direct role in the metabolism of animals and humans.

Background

Although only a minor constituent of the earth's crust, molybdenum is widely bioavailable because of the high solubility of molybdate salts in water; molybdenum is, for example, the most abundant transition metal in seawater. The biological active form is the soluble oxyanion molybdate. Molybdenum was first discovered to be present in molybdenite (MoS_2) by Scheele in 1778–1779 and later isolated by Hjelm in 1782, who named it molybdenum (from Greek molybdos, meaning lead or lead like). The first biological function was proven by Bortels in 1930, who showed that it acts as a catalyst in the fixation of nitrogen by *Arthrobacter chroococum*, and 23 years later, in 1953, it was reported that molybdenum also has a direct role in the metabolism of animals and humans (Coughlan 1983). To date, the biological role of molybdate has been elucidated, and it is known that molybdate can be inserted into two classes of molybdoenzymes: the molybdenum ▶ nitrogenase containing a unique iron–molybdenum cofactor (FeMoco) and the group of molybdenum cofactor (Moco)–containing enzymes in which the metal is always coordinated by a pyranopterin cofactor (named molybdopterin, MPT) (Leimkühler et al. 2011). These mononuclear molybdenum enzymes constitute a large and widely distributed class of proteins, with more than 50 known members in organisms ranging from bacteria and archaea to plants and higher animals (Hille 1996). To date, the five molybdoenzymes identified in humans are xanthine oxidoreductase, sulfite oxidase, aldehyde oxidase, and the newly discovered mARC1 and mARC2, proteins involved in metabolism of several heterocyclic drugs or prodrugs (Hille et al. 2011). Molybdenum, because of its unique chemistry, is the biological catalyst for reactions in which proton and electron transfer and, mostly, oxygen transfer are coupled. The proteins involved in the biosynthesis of the molybdenum cofactor and the role of individual molybdoenzymes found in humans will be explained in detail.

Molybdenum in Biology

Molybdenum is the only 4d transition metal required for biological systems. Molybdenum forms part of the active site of molybdoenzymes that execute key

Molybdenum and Ions in Living Systems, Fig. 1 The three families of enzymes binding Moco. Enzymes of the three classes of molybdoenzymes differ in their coordination of the Mo-atom. The sulfite oxidase family is characterized by a dioxo Mo-center, whereas enzymes of the xanthine oxidase family are characterized by a terminal S-atom covalently bound at the mono-oxo Mo-center, and in enzymes of the DMSO reductase family, the bis-MGD cofactor is coordinated by two dithiolene groups of two MGD cofactors. Moco consists of a molybdenum atom covalently bound to the dithiolate moiety of a tricyclic pterin referred to as molybdopterin (MPT)

transformations in the metabolism of nitrogen, sulfur, and carbon compounds. Molybdenum is used in biology in the form of the MoO_4^{2-} anion and has a chemical versatility that is useful to biological systems: It is redox-active under physiological conditions (ranging between the oxidation states VI and IV); since the V oxidation state is also accessible, the metal can act as transducer between obligatory two-electron and one-electron oxidation–reduction systems, and it can exist over a wide range of redox potentials (Hille 1996). The catalyzed reactions are in most cases oxo-transfer reactions, for example, the hydroxylation of carbon centers and the physiological role is fundamental since the reactions include the catalysis of key steps in carbon, nitrogen, and sulfur metabolism. In the context of availability, its concentration is 10^{-7} mol/l in seawater, which is equal to other transition metals including ▸ iron (Coughlan 1983). Once molybdate enters the cell, it is subsequently incorporated by complex biosynthetic machineries into metal cofactors. These metal cofactors are then incorporated into different enzymes, and these molybdenum enzymes are found in nearly all organisms, with *Saccharomyces* as a prominent eukaryotic exception. There are two distinct types of molybdoenzymes: Molybdenum nitrogenase has a unique molybdenum–iron–sulfur cluster, the $[Fe_4S_3]$-(bridging-S)$_3$-$[MoFe_3S_3]$ center called FeMoco (Schwarz et al. 2009). It catalyzes the reduction of atmospheric dinitrogen to ammonia. All other molybdoenzymes are oxidoreductases that transfer an oxo group or two electrons to or from the substrate (Hille 1996). They have a molybdenum cofactor (Moco) (Fig. 1) in which molybdenum is coordinated to a dithiolene group on the 6-alkyl side chain of a pterin called molybdopterin (MPT) (Leimkühler et al. 2011). In contrast to the multinuclear FeMoco in nitrogenase, the active site of mostly all other well-characterized Moco-containing enzymes is generally mononuclear, with a single equivalent of the metal (carbon monoxide dehydrogenase is so far the only exception). Here, the focus is on the molybdoenzymes identified in humans, which exclusively contain Moco: xanthine dehydrogenase (XDH), aldehyde oxidase (AO), sulfite oxidase (SO), and the mitochondrial amidoxime reducing component (mARC) (Hille et al. 2011). In the redox reactions catalyzed by these molybdoenzymes, electron transfer is linked with proton transfer. Other metal complexes usually do not display the appropriate redox potentials in combination with the appropriate acid–base properties. Most of the other metal centers in biology, including those of iron and ▸ copper, participate only in electron transfer (Coughlan 1983). Moreover, few metals other than molybdenum can effect oxygen transfer.

Reactions, such as Mo$^{(VI)}$ = O → Mo$^{(IV)}$ + O in which oxide transfer is coupled with two-electron transfer, may be significant in the context of reactions catalyzed by AO (RCHO + O → RCOOH), XDH (RH + O → ROH), and SO (SO$_3^{2-}$ + O → SO$_4^{2-}$) (Coughlan 1983). Molybdenum has the ability to exist in various oxidation states under physiological conditions, and to couple oxide or proton transfer with electron transfer which in total makes it the metal of choice for the reactions in which it participates.

Function of Molybdoenzymes in Humans

The mononuclear molybdenum enzymes are categorized on the basis of the structures of their molybdenum centers, dividing them into three families, each with a distinct active site structure and a distinct type of reaction catalyzed (Fig. 1): the xanthine oxidase (XO) family, the SO family, and the dimethyl sulfoxide (DMSO) reductase family (Hille 1996). The XO family is characterized by an MPT-Mo^{VI}OS(OH) core in the oxidized state, with one MPT equivalent coordinated to the metal. The additional sulfido-group is cyanide labile. Enzymes of the SO family coordinate a single equivalent of the pterin cofactor with an MPT-Mo^{VI}O$_2$ core in its oxidized state, and usually an additional cysteine ligand which is provided by the polypeptide. The DMSO reductase family is exclusively found in bacteria and archaea, and all members have two equivalents of the pterin cofactor bound to the metal. The molybdenum coordination sphere is usually characterized by an MPT$_2$-Mo^{VI}S(X) core. The sixth ligand, X, can be a serine, a cysteine, a ▶ selenocysteine, or a hydroxide and/or water molecule. The human molybdoenzymes fall into two of these families: the SO family represented by SO and possibly mARC and the XO family represented by XDH and AO (Hille et al. 2011).

Xanthine Oxidase/Dehydrogenase

Xanthine oxidoreductase (XOR) is a widely distributed enzyme that generally is involved in the later stages of purine catabolism, catalyzing the oxidation of hypoxanthine to xanthine and of xanthine to uric acid, which is finally excreted in the urine (Harrison 2004) (Fig. 2). The enzyme has been studied for more than 100 years. XOR is a complex molybdoflavoprotein, the enzymology of which is now well characterized. Although it is one of the best studied enzymes in vitro, the complexities of its in vivo functions are less well understood. The enzyme is a homodimer, comprising two 150,000 Da subunits, each of which contains one molybdenum, one FAD, and two [2Fe–2S] redox centers (Hille et al. 2011) (Fig. 2). It has a wide range of reducing substrates, including not only hypoxanthine and xanthine but also a broad spectrum of N-heterocycles and aldehydes (Harrison 2004). The simplest representation of the reaction catalyzed by these enzymes is as follows:

$$SH + H_2O + A \rightarrow SOH + AH_2,$$

where SH is the electron donor substrate undergoing hydroxylation, and A is the electron acceptor. The hydroxyl group incorporated into the substrate is derived from water (Hille et al. 2011; Rajagopalan 1988). Electrons are donated from such substrates to the molybdenum site of XOR and are rapidly equilibrated between the redox centers before being passed to NAD$^+$ or to molecular oxygen at the FAD site. Reduction of NAD$^+$ gives NADH, and that of molecular oxygen yields hydrogen peroxide and superoxide anion. The enzyme occurs in two forms, xanthine dehydrogenase (XDH, EC 1.17.1.4) and xanthine

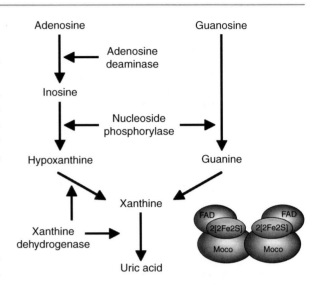

Molybdenum and Ions in Living Systems, Fig. 2 The pathway of purine nucleotide metabolism in humans

oxidase (XO, EC 1.17.3.2). The forms can be interconverted by means of sulfhydryl reagents, whereas proteases convert XDH irreversibly to XO (Hille et al. 2011). Only XDH is capable of reducing NAD^+. Both forms can reduce molecular oxygen, although XO is more effective in this respect. XDH is the predominant enzymatic form found in normal tissues, while the XO form dominates in tissues subjected to injury. The latter reaction results in the production of ▶ reactive oxygen species (ROS) as a metabolic by-product. ROS are produced excessively under certain pathophysiological conditions, and for this reason, the conversion of XDH to XO is of major medical interest. XO has been implicated in many diseases associated with oxygen-radical-induced tissue damage, contributing to postischemic reperfusion injuries as well as aging (Harrison 2004). XOR is a cytoplasmic enzyme and occurs in all mammals and in most tissues, being particularly rich in liver and intestine. Besides its role as a housekeeping gene, XOR is highly expressed in liver and kidney, the main organs involved in purine catabolism and nitrogen elimination. Furthermore, it is found at high levels in a number of other mammalian tissues and organs, including the mammary gland (Harrison 2004). The enzyme is also a major protein component of the milk fat globule membrane (MFGM), which envelops fat droplets in freshly expressed milk (Harrison 2004). It is suggested that in milk fat droplet secretion, XOR may have primarily a structural role as a membrane-associated protein, and thus, XOR provides another example of "gene sharing."

Xanthinuria is the disease in humans involving abnormalities of XDH. It presents itself as three subtypes: classical xanthinuria type I, which is caused by XDH deficiency alone; classical xanthinuria type II, which represents both XDH and AO deficiencies; and Moco deficiency, leading to the concomitant loss of function of all molybdoenzymes (Garattini et al. 2003). It is known that in xanthinuria II, the enzyme responsible for attaching the terminal sulfur ligand at the molybdenum center of the two enzymes is mutated, thus affecting all enzymes of the XO family. The two types of classical xanthinuria can be distinguished by the ability of type I but not type II subjects to convert allopurinol to oxipurinol and N^1-methylnicotinamide to the 2- and 4-pyridones, the latter of which is a substrate exclusively to AO (Garattini et al. 2003). The only symptoms of xanthinuric patients are colics, resulting from hypoxanthine stones in the liver and kidney. This suggests a dispensable role for the enzyme in the homoeostasis of the human organism (Garattini et al. 2003).

Aldehyde Oxidase

While the biochemical function of XOR is well established, the biochemical and physiological functions of AO are still largely obscure. The overall level of similarity between AO and XOR proteins is approximately 50%, which clearly indicates that the two proteins originated from a common ancestral precursor (Garattini et al. 2003). While XORs are represented throughout evolution from bacteria to humans, aldehyde oxidases are thought to be present predominantly in multicellular plants and animals, although reports of the existence of the latter enzymes in certain bacteria are available. In humans, only one gene for AO exists; however, in mouse and other mammals, different homologs of AO were identified. Since the identified proteins were highly homologous to the originally identified AO (named AOX1, the first vertebrate AO identified and characterized), the three related proteins were named aldehyde oxidase homolog 1 (AOH1), 2 (AOH2), and 3 (AOH3) (Garattini et al. 2003). The multiple AO isoforms are expressed tissue specifically in different organisms; however, their existence and expression pattern varies in different animal species. It is believed that the various AO isoforms recognize distinct substrates and carry out different physiological tasks in rodents.

AO is characterized by broad substrate specificity, and this makes it an important enzyme in the metabolism of drugs and xenobiotics (Hille et al. 2011). The enzyme oxidizes aromatic aza-heterocycles containing a –CH = N- chemical function (e.g., phtalazine and purines), aromatic or nonaromatic charged aza-heterocycles with a –CH = N^+- moiety (e.g., N^1-methylnicotinamide and N-methylphthalazinium) or aldehydes, such as benzaldehyde, retinal and vanillin (Fig. 3). Some of the AO substrates are common to XOR, and the relative selectivity of the two enzymes has been systematically reviewed (Garattini et al. 2003). Only limited information is available about the physiological substrates of AO or about the role of this enzyme in the mammalian organism (Garattini et al. 2003). It has been shown

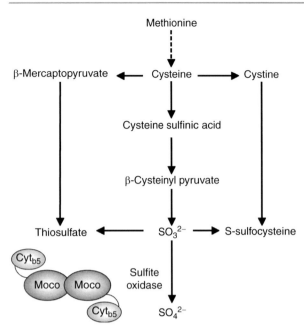

Molybdenum and Ions in Living Systems, Fig. 3 The metabolism of sulfur-containing amino acids in humans

that the enzyme is involved in the metabolism of drugs and xenobiotics of toxicological importance, and it metabolizes N-heterocyclic compounds and aldehydes of pharmacological and toxicological relevance. The protein catalyzes the oxidation of methotrexate, a widely used antineoplastic agent, and is purported to play a role in the oxidation of acetaldehyde, the toxic metabolite of ethanol (Garattini et al. 2003) (Fig. 3). Single monogenic deficits for any AOX1 have not been described yet in mammalia; however, AOX1 is not essential for humans since genetic deficiencies in the Moco sulfurase (MCSF) gene is not associated with pathophysiological consequences (Hille et al. 2011). It is likely, that in animals, AOX1 serves to detoxify exogenously derived unphysiological compounds of wide structural diversity, and it is believed that the absence of AO has symptoms under some unusual circumstances of high intake of such xenobiotics. The electron acceptor of AOX1 is molecular oxygen, and studies on the metabolism of aldehydes demonstrated the production of significant amounts of O_2^- and hydrogen peroxide by AOX1. ROS are thus incorporated in oxygen free radical injury triggered by ischemia–reperfusion injury and various inflammatory diseases (Garattini et al. 2003).

Sulfite Oxidase

SO is a protein which is located in the intermembrane space in mitochondria in mammals. It catalyzes the terminal step in the metabolism of sulfur-containing amino acids cysteine and methionine (Fig. 4). SO is a dimeric enzyme with a molecular mass of 51 kDa per subunit. Each subunit contains one molybdenum cofactor and one cytochrome-b_5-type heme (Fig. 4). In each subunit, the Mo domain and the b_5-type heme domain are linked by a flexible peptide loop of 10 amino acids (Feng et al. 2007). A variation in the heme orientation has been observed which has been interpreted as evidence of domain–domain flexibility. Thus, sulfite oxidase undergoes a conformational change in the electron transfer between the Mo and heme centers during catalysis. Sulfite oxidase has a narrow substrate specificity and does not oxidize compounds other than sulfite (SO_3^{2-}) to any extend. ▶ Cytochrome c is the physiological electron acceptor for the enzyme, with oxygen and ferricyanide serving as alternative oxidizing substrates (Hille et al. 2011):

$$SO_3^{2-} + H_2O + 2(\text{cytc})_{\text{oxidized}} \rightarrow SO_4^{2-} + 2(\text{cytc})_{\text{reduced}} + 2H^+.$$

Oxidation of sulfite to sulfate occurs at the molybdenum center, and the electrons are transferred singly to the heme domain, and ultimately transferred to cytochrome c on the intermembrane space side of the inner mitochondrial membrane. About 1 mole of ATP is synthesized per mole sulfite oxidized. The enzyme also plays an important role in detoxifying exogenously supplied sulfite and sulfur dioxide (Rajagopalan 1988; Feng et al. 2007).

A mutation in the gene for SO leads to isolated SO deficiency, an inherited sulfur metabolic disorder in humans that results in profound birth defects, severe neonatal neurological problems, and early death, with no effective therapies known (Mendel and Schwarz 2011). The inborn error is characterized by dislocation of ocular lenses, mental retardation, and, in severe cases, attenuated growth of the brain. These severe neurological symptoms result from either point mutations in the SO protein itself (so-called isolated SO deficiency, in which only SO activity is affected), or the inability to properly produce the Moco, which results in deficiencies in all Mo-containing enzymes (so-called Mo cofactor deficiency, see below)

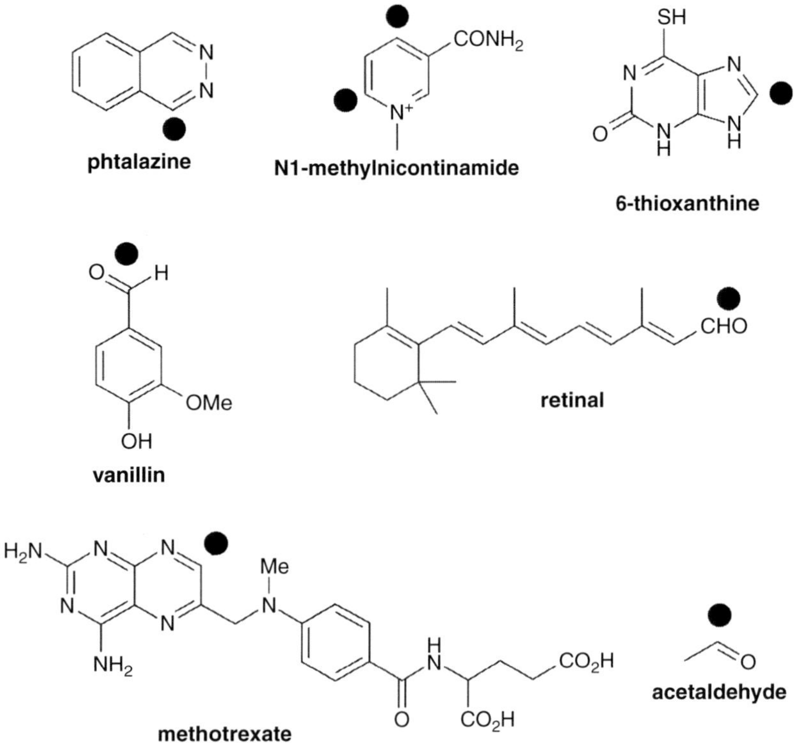

Molybdenum and Ions in Living Systems, Fig. 4 Aldehyde and heterocycle containing substrates for AO. The position which is oxidized is marked by a *black dot*

(Mendel and Schwarz 2011). The biochemical basis of the pathology of SO deficiency is unclear and merits further investigation. Fatal brain damage may be due to the accumulation of a toxic metabolite, possibly SO_3^{2-}, which is a strong nucleophile that can react with a wide variety of cell components. It has been reported that sulfite reacts with protein disulfides to form sulfonated cysteine derivatives, and since the integrity of disulfide bonds is crucial to the tertiary structure and thus protein function, the disruption of protein structure by sulfitolysis may result in altered cellular activities leading to biochemical lesions (Feng et al. 2007). Alternatively, a deficiency in the reaction product (sulfate, SO_4^{2-}) may disturb normal fetal and neonatal development of the brain. In addition, the nature of the lesion in human sulfite oxidase deficiency (with the central nervous system (CNS) being disproportionately affected) suggests that the principal problem is likely to be lipid peroxidation in the brain. Specifically, the cell membranes of the CNS myelin sheath are unique in possessing high concentrations of sulfatides and related lipids, which is likely the root cause of the sensitivity of the CNS to SO deficiency. It was also shown that sulfite-mediated oxidative stress is accompanied by a depletion of intracellular ATP, which is possibly due to the inhibition of glutamate dehydrogenase by high sulfite concentrations in mitochondria (Mendel and Schwarz 2011).

mARC

More recently, investigation of the aerobic reduction of amidoxime structures led to the discovery of a so far unknown molybdenum-containing enzyme system (Hille et al. 2011). It was named "mitochondrial amidoxime reducing component" (mARC), because initially N-reduction of amidoxime structures was studied with this enzyme purified from mammalian liver mitochondria. After recombinant expression of human mARC, it became clear that besides SO, XOR, and AO, another molybdenum-containing enzyme exists. The human genome harbors two mARC genes, referred to as hmARC1 and hmARC2, which are organized in a tandem arrangement on chromosome 1 (designated as MOSC1 and MOSC2 in the databases). These two enzymes are localized in or are associated with mitochondria. The two enzymes form the catalytical part of a three-component enzyme system, consisting of mARC, heme/cytochrome b_5, and

NADH/FAD-dependent cytochrome b_5 reductase. The catalytic core itself is a monomer with a molecular mass of 33 kDa. Both hmARC1 and hmARC2 catalyze the N-reduction of a variety of N-hydroxylated substrates such as N-hydroxy-cytosine, albeit with different specificities (Hille et al. 2011). However, although many N-hydroxylated compounds have been found to serve as substrates for native and recombinant mARC proteins, the physiological substrates, and thus the physiological function, of mARC proteins are as yet unknown. Based on the finding that hmARC1 and hmARC2 likewise catalyze the reduction of N-hydroxylated base analogs with high efficiency (N-hydroxy-cytosine), it appears likely that one of the physiological functions of human mARC proteins could be to prevent accumulation of such mutagenic substances in the cell (Hille et al. 2011). In addition, hmARC1 and hmARC2 have recently been suggested to act as regulators for the L-arginine-dependent biosynthesis of NO by catalyzing the controlled elimination of the NO precursor N4-hydroxy-L-arginine. However, because immunolocalization of mARC proteins in mouse cells has shown that at least mARC2 localizes exclusively to the mitochondria, the putative reduction of N4-hydroxy-L-arginine is likely to be limited to this organelle. A role for mammalian mARC proteins in the detoxification of N-hydroxylated substrates appears likely, given the relatively high abundance of these proteins in typical detoxification organs such as liver and kidney (Hille et al. 2011).

Uptake of Molybdate and Synthesis of the Catalytic Molybdenum Center in the Eukaryotic Cell

The biological active form of molybdenum is the oxyanion molybdate. Molybdate enters the cell via specific uptake systems, which have been studied in detail in bacteria. Here, highly specific uptake systems exist, which belong to the family of ATP-binding cassette transporters (ABC Transporter) and are able to bind molybdate specifically under low concentrations (Hagen 2011). In higher eukaryotes, the uptake of molybdate still remains obscure. Recently, first molybdate-transporting proteins have been identified in algae and plants. Two proteins (MOT1 and MOT2) belonging to the large sulfate carrier superfamily with nine putative membrane-spanning domains were shown to transport molybdate with high affinity across cellular membranes (Hagen 2011). Unexpectedly, none of them were found to reside in the plasma membrane surrounding the cell, but they were localized to the endomembrane system or to the mitochondrial envelope. Thus, further studies have to clarify the role of these MOT-transporters in the cell and how molybdate enters through the cytoplasmic membrane (Hagen 2011).

Once having entered the cell, molybdenum has to be attached to its cofactor scaffold, thereby converting it to Moco and gaining biological activity (Leimkühler et al. 2011). Moco is a tricyclic pyranopterin with a unique ene-dithiolate group and a terminal phosphate (Fig. 1). What is the task of the pterin moiety of Moco? Obviously, the pterin functions in binding and coordinating the catalytic metal correctly within the active center of a given Mo-enzyme where it can have different ligands. It also could control of the redox behavior of the molybdenum atom or participate in the electron transfer to or from molybdenum via the delocalized electrons within the pterin. Thus, the role of the pterin still remains obscure (Leimkühler et al. 2011).

In all higher organisms including humans, Moco is synthesized by a conserved biosynthetic pathway that can be divided into three steps, according to the biosynthetic intermediates (Fig. 5): the synthesis of cyclic pyranopterin monophosphate (cPMP), conversion of cPMP into MPT by introduction of two sulfur atoms, and insertion of molybdate to form Moco. Six proteins were identified catalyzing Moco biosynthesis in humans. Genes and the encoded proteins were named in humans as MOCS (molybdenum cofactor synthesis), but different names were given in plants, fungi and, bacteria (see Table 1) (Mendel and Schwarz 2011).

The biosynthesis of Moco starts from 5′-GTP, which results in the formation of cPMP, the first stable intermediate of Moco biosynthesis (Fig. 5). cPMP is an oxygen sensitive 6-alkyl pterin with a cyclic phosphate group at the C2' and C4' atoms (Leimkühler et al. 2011). The *MOCS1* locus encodes two proteins, MOCS1A and MOCS1B, which are involved in the conversion of 5′-GTP into cPMP. MOCS1A belongs to the superfamily of S-Adenosyl Methionine (SAM)-dependent Radical Enzymes. Members of this family catalyze the formation of protein and/or substrate radicals by reductive cleavage of SAM by a [4Fe–4S] cluster. MOCS1A is a protein containing two

Molybdenum and Ions in Living Systems, Fig. 5 The biosynthesis of Moco in humans. Shown is a scheme of the bio synthetic pathway for Moco biosynthesis in humans. The names of the proteins involved in the reactions are colored in *blue*, and the identified and characterized molybdoenzymes of each human molybdoenzyme family are shown in *blue*

oxygen-sensitive Fe–S clusters each coordinated by only three cysteine residues. The N-terminal [4Fe–4S] cluster, present in all radical SAM proteins, binds SAM and carries out the reductive cleavage of SAM to generate the 5′-deoxyadenosyl radical, which subsequently initiates the transformation of 5′-GTP bound through the C-terminal [4Fe–4S] cluster. The role of MOCS1B in humans is yet unknown, but it is believed that it participates in pyrophosphate release upon the rearrangement reaction (Mendel and Schwarz 2011).

The next step involves the conversion of cPMP to MPT in which two sulfur atoms are incorporated in the C1′ and C2′ positions of cPMP (Leimkühler et al. 2011) (Fig. 5). This reaction is catalyzed by MPT synthase, a protein consisting of two small (~10 kDa) and two large subunits (~21 kDa), encoded by *MOCS2A* and *MOCS2B*, respectively. It was shown that MPT synthase carries the sulfur in form of a thiocarboxylate at the C-terminal glycine of MOCS2A. The first sulfur is added at the C2′ position of cPMP, resulting in a hemisulfurated cPMP intermediate, before the sulfur at the C1′ position is added by a second sulfurated MOCS2A subunit. During the reaction, cPMP remains bound to one MOCS2B subunit (Mendel and Schwarz 2011).

The regeneration of sulfur at the C-terminal glycine of MOCS2A resembles the first step of the ubiquitin-dependent protein degradation. It was shown that the E1-like protein MOCS3 activates the C-terminus of MOCS2A by addition of an acyl-adenylate. In the second reaction, the activated MOCS2A acyl-adenylate is converted to a thiocarboxylate by action

Molybdenum and Ions in Living Systems, Table 1 Nomenclature of the proteins involved in Moco biosynthesis in bacteria and eukaryotes

Step	Bacteria *E. coli*	Fungi *A. nidulans*	Plants *A. thaliana*	Humans *H. sapiens*
1.	MoaA	CnxA	Cnx2	MOCS1A
	MoaC	CnxC	Cnx3	MOCS1B
2.	MoaD	CnxG	Cnx7	MOCS2A
	MoaE	CnxH	Cnx6	MOCS2B
	MoeB	CnxF	Cnx5	MOCS3
3.	MogA	CnxE	Cnx1	Geph.
	MoaB			
4.	MoeA	CnxE	Cnx1	Geph.
	MobA	–	–	–
	MobB			
		HxB	ABA3	HMCS

of a persulfide formed at the C-terminal rhodanese-like domain of MOCS3 (Mendel and Schwarz 2011).

After synthesis of the dithiolene moiety in MPT, the chemical backbone is built for binding and coordination of the molybdenum atom. In humans, the gephyrin gene encodes a 2-domain protein with an N-terminal domain that forms an MPT-AMP intermediate and a C-terminal domain that inserts the molybdate anion and splits the AMP from the activated intermediate. Thus, active Moco is formed (Mendel and Schwarz 2011) (Fig. 5). Apart from the molybdenum cofactor biosynthesis, gephyrin is also involved in synaptic anchoring of inhibitory ligand-gated ion channels. Alternative splicing has been proposed to contribute to gephyrin's functional diversity within the cell.

The completed Moco can be directly inserted into sulfite oxidase or mARC. For XDH and AO, Moco is further modified by an exchange of the equatorial oxygen ligand by sulfur, forming the sulfurated or mono-oxo form of Moco. This reaction is carried out by a Moco sulfurase (HMCS) in humans, a two-domain protein with an N-terminal L-cysteine desulfurase domain, and a C-terminal Moco binding (MOSC) domain (Hille et al. 2011) (Fig. 5). The Moco likely gets sulfurated while bound to the HMCS protein and is then inserted into XDH and AO.

Toxicity and Deficiency of Molybdenum

Under normal dietary conditions, the molybdenum content of tissues is quite low (0.1–1.0 μg/g wet weight). It appears that in normal dietary status, all tissue molybdenum is quantitatively associated with MPT, and the highest concentrations are found in liver, kidney, adrenal gland, and bone (Coughlan 1983). Usually, dietary deficiency of molybdenum does not occur in humans. In contrast, the effect of excessive molybdenum intake on animals is more severe, and the symptoms vary among species. Ruminants such as cows and sheep show extreme sensitivity to dietary molybdenum levels starting from 2 to 30 ppm. The symptoms resemble those of copper deficiency, and treatment with supplemental copper usually reverses them. It had been suggested that the underlying biochemical events leading to molybdenum toxicosis take place within the gastrointestinal tract where the interactions among dietary copper, molybdenum, and sulfur are maximal and likely insoluble molybdenum–sulfur–copper complexes are formed (Rajagopalan 1988).

The consequences when no Moco is formed in humans are devastating (Mendel and Schwarz 2011). Human Moco deficiency resides in a loss of activity of all Mo-dependent enzymes: SO, XDH, AO, and mARC. Moco deficiency is a hereditary recessive disorder that affects neonates soon after birth. While patients are born normally, they develop intractable seizures within days after birth and require intensive care and anticonvulsant treatment. Symptoms develop shortly after birth, when the metabolism starts to operate and toxic metabolites are building up within the body (Mendel and Schwarz 2011). In Moco deficiency, the major cause of the disease resides in the loss of SO, which effectively removes sulfite from the human body. Thus, the symptoms are similar to SO deficiency which results in the buildup of sulfite, ultimately causing severe neurological damage, disordered autonomic function, exaggerated startle reactions, dysmorphic facial features, alterations in muscle tone, progressive cerebral palsy, microencephaly, seizures, and death (Feng et al. 2007). Patients that survive the acute initial phase of seizures survive only few years with basically no neuronal development. Patients are unable to make any coordinated movements, need to be tube fed, and show no signs of communication with their environment. The incidence of Moco deficiency is below 1:100,000; however, many missed, nondiagnosed, and nonreported cases are suspected, and therefore, prevalence determination is limited (Mendel and Schwarz 2011). Due to the lack of any causative treatment, death in early childhood has been the usual outcome. While isolated sulfite oxidase deficiency cannot be treated to date, first attempts to treat Moco deficiency were successful. In 2008, a baby was diagnosed with a mutation in the *MOCS1*-locus shortly after birth. It was decided to start a treatment with purified cPMP on day 36 of life. As starting dose, 80 μg cPMP per kg body weight was administered. Within days after treatment was started, urinary markers of sulfite oxidase and XDH deficiency (S-sulfocysteine as well as xanthine and uric acid) returned to almost normal readings and stayed constant. Clinically, the patient became more alert a few days after the treatment started, the convulsions and twitching disappeared within the first 2 weeks as documented by an electroencephalogram showing the return of rhythmic elements and markedly reduced epileptiform discharges (Mendel and Schwarz 2011). This therapy so far seems promising; however, the further development of the child after long-term treatment has to be monitored.

Cross-References

▶ Copper, Biological Functions
▶ Cytochrome c Oxidase, CuA Center
▶ Iron, Physical and Chemical Properties
▶ Iron-Sulfur Cluster Proteins, Nitrogenases
▶ Reactive Oxygen Species
▶ Selenocysteine

References

Coughlan MP (1983) The role of molybdenum in human biology. J Inherit Metab Dis 6(Suppl 1):70–77
Feng C, Tollin G, Enemark JH (2007) Sulfite oxidizing enzymes. Biochim Biophys Acta 1774:527–539
Garattini E, Mendel R, Romao MJ, Wright R, Terao M (2003) Mammalian molybdo-flavoenzymes, an expanding family of proteins: structure, genetics, regulation, function and pathophysiology. Biochem J 372:15–32
Hagen WR (2011) Cellular uptake of molybdenum and tungsten. Coord Chem Rev 255:1117–1128
Harrison R (2004) Physiological roles of xanthine oxidoreductase. Drug Metab Rev 36:363–375
Hille R (1996) The mononuclear molybdenum enzymes. Chem Rev 96:2757–2816
Hille R, Nishino T, Bittner F (2011) Molybdenum enzymes in higher organisms. Coord Chem Rev 255:1179–1205
Leimkühler S, Wuebbens M, Rajagopalan KV (2011) The history of the discovery of the molybdenum cofactor and novel aspects of its biosynthesis in bacteria. Coord Chem Rev 255:1129–1144
Mendel RR, Schwarz G (2011) Molybdenum cofactor biosynthesis in plants and humans. Coord Chem Rev 255:1145–1158
Rajagopalan KV (1988) Molybdenum: an essential trace element in human nutrition. Annu Rev Nutr 8:401–427
Schwarz G, Mendel RR, Ribbe MW (2009) Molybdenum cofactors, enzymes and pathways. Nature 460:839–847

Molybdenum and Ions in Prokaryotes

Anne Walburger and Axel Magalon
Laboratoire de Chimie Bactérienne (UPR9043),
Institut de Microbiologie de la Méditerranée,
CNRS & Aix-Marseille Université, Marseille, France

Synonyms

Mo cofactor; Mo-enzyme; Molybdate; Molybdenum cofactor; Molybdoenzyme

Definition

In the periodic table, ▶ molybdenum (Mo) is a chemical element with the atomic number 42 classified as a transition metal and which is indispensable to most life forms. The utility of Mo in biology is based on its coordination that allows the metal to catalytically cycle through the IV (fully reduced, 2 electrons), V (half oxidized, 1 electron), and VI (fully oxidized, 0 electron) oxidation states. Coordination of Mo is ensured by an organic cofactor, the so-called ▶ molybdenum cofactor (Moco), common to all ▶ molybdoenzymes with the exception of

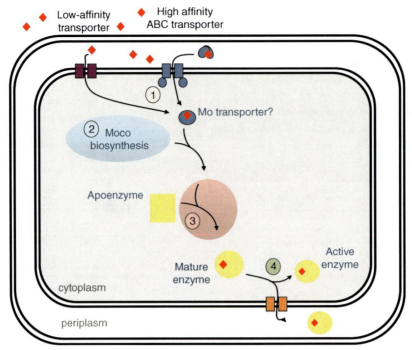

Molybdenum and Ions in Prokaryotes, Fig. 1 The fate of molybdenum in a prokaryotic cell. Molybdate is taken up into the cytoplasm through either high-affinity ABC transporters or low-affinity ones (step 1) where it might be scavenged or stored by a Mo transporter which would be made up by a complex of proteins involved in Moco biosynthesis. Molybdate is then complexed to the molybdopterin to form the molybdenum cofactor (Moco) (step 2). Moco is inserted into the resident apoenzymes in coordination with others metal cofactors yielding mature molybdoenzymes (step 3). The mature molybdoenzyme may be further localized in the periplasm through translocation of the inner membrane by the Tat machinery dedicated to folded proteins (step 4). Proper molybdenum homeostasis is ensured through different levels of genetic regulation at each four steps

nitrogenase. Moco is incorporated into enzymes and forms part of the active site that executes key transformations in the metabolism of nitrogen, sulfur, and carbon compounds. More than 100 molybdenum-containing enzymes are now known in prokaryotes and eukaryotes (Zhang and Gladyshev 2008). Molybdenum in eukaryotes is described in ▶ Molybdenum in Biology and ▶ Molybdenum and Its Ions in Living System. Instead of Mo, ▶ tungsten (W) can be found in the cofactor in bacteria and in ▶ Archaea, making it the heaviest element with a biological function. Both cofactors Moco and Wco have comparable properties but are not equally incorporated into metalloproteins. It is described in ▶ Tungsten Cofactors, Binding Proteins, and Transporters in Biological Systems and in ▶ Tungsten in Biological Systems. The fate of molybdenum in prokaryotes is summarized in Fig. 1.

Uptake of Molybdenum

Molybdenum is bioavailable in the form of its oxyanion ▶ molybdate (MoO_4^{2-}). Notably, it has been reported that the combined deficiency of oxygen and of Mo in the ancient deep ocean may have delayed the evolution of eukaryotic life on Earth for nearly 2 billion years as eukaryotes cannot fix nitrogen and must acquire it from prokaryotes. Molybdate is not very abundant in the environment (~1 ppm in the Earth's crust); thus, high-affinity transporters have been developed by prokaryotes to scavenge it in the presence of competing anions. Due to its similar structural characteristics to sulfate (SO_4^{2-}), thiosulfate ($S_2SO_3^{2-}$), selenate (SeO_4^{2-}), and selenite ($SeSO_3^{2-}$), it may also be taken up through low-efficiency anion transport system. In bacteria, the first identified and widely distributed Mo transporter in prokaryotes was the high-affinity ModABC transport system which has been widely characterized at the genetic, biochemical, and structural level. ModA binds the molybdate (and tungstate to a lesser extent) in the periplasm and delivers it to ModB, the membrane integral channel, while ModC energizes the molybdate transport through ATP hydrolysis. The ModABC system is widespread in prokaryotes; besides two additional classes of Mo/W ABC transport systems were identified: WtpABC (dual specificity for Mo/W) and TupABC (W-specific) (reviewed by Hagen 2011). The molybdate/tungstate binding proteins of these transport systems differ not only in terms of specificity and oxoanion coordination chemistry but also at the primary sequence level. Moreover, the occurrence of more than one gene encoding for molybdate/tungstate binding proteins is quite common. The implication is that, by employing more than one of these transport systems, an organism could possibly fine-tune its molybdate and tungstate uptake capabilities. Finally, to distinguish Mo from W, exquisitely discriminating systems have evolved at different levels such as metal uptake, metal insertion into the cofactor, and dedicated chaperone-assisted cofactor incorporation into molybdoenzymes.

Biosynthesis of Molybdenum Cofactors

Once molybdate is taken up into the cell, it is complexed to a specific cofactor to be catalytically active. With the exception of the bacterial nitrogenase (not reviewed in this section), the core structure of the molybdenum cofactor is a pterin-based cofactor (▶ molybdopterin or MPT) with a C6-substituted pyrano ring, a terminal phosphate, and a unique dithiolate group binding Mo (Fig. 2). The metal can be attached to one or two pterin moieties substituted, in some cases, with nucleotides (guanine, cytosine, thymine, and inosine) and with additional terminal oxygen and sulfur ligands. A common feature is the extreme oxygen sensitivity of the molybdenum cofactor (Moco) and its lability outside the molybdoenzyme. Although widespread in all kingdoms, Moco is synthesized by a conserved biosynthetic pathway divided in four steps; a fifth one being present only in prokaryotes and which consists in a final modification by a nucleotide (Fig. 2). Using a combination of biochemical, genetic, and structural approaches, Moco biosynthesis in *Escherichia coli* has been extensively studied for decades. These studies allowed the identification of at least 17 genes involved in Moco biosynthesis and clustered in six different *mo* loci, the corresponding proteins being highly conserved in other organisms.

Biosynthesis initiates with the conversion of GTP into cPMP (cyclic pyranopterin monophosphate) catalyzed by MoaA and MoaC by a yet not fully chemically understood reaction. The subsequent step consists in the incorporation of two sulfur atoms into cPMP by the MPT synthase, a heterotetrameric complex of two small and two large subunits, MoaD and MoaE, respectively. To allow sulfur transfer to the small subunit of MPT synthase, *E. coli* MoeB must catalyze the prior adenylylation of the C-terminal glycine residue of MoaD in a process which is notably similar to the action of the ubiquitin-activating enzyme. MoeB serves, as such, as the MPT synthase sulfurase, the protein responsible for regenerating the thiocarboxylate group at the C-terminus of MoaD. AMP-activated MoaD becomes sulfurated by sulfide transfer through the formation of a thiocarboxylate group on the terminal glycine residue. In *E. coli*, the pyridoxal-phosphate-dependent L-cysteine desulfurase IscS is the primary physiological sulfur-donating enzyme for the generation of the thiocarboxylate of MPT synthase. After the formation of the sulfurated MPT synthase, introduction of the dithiolene moiety to cPMP yielding MPT completes the formation of the chemical backbone necessary for binding and coordination of the molybdenum atom in Moco. On completion of MPT synthesis, the metal is transferred by a multistep reaction involving the MogA and MoeA proteins. MogA adenylates MPT in a Mg^{2+}- and ATP-dependent way and forms MPT-AMP which is subsequently transferred to MoeA. In the presence of Mg^{2+} and molybdate, bound MPT-AMP is hydrolyzed, and molybdenum is transferred to the MPT dithiolate, resulting in Moco release. This is the form of cofactor exclusively found in eukaryotes but also in prokaryotic members of the sulfite oxidase family. It is worth noting that Moco can be further modified by the addition of mononucleotides such as GMP, CMP, IMP, or AMP to the phosphate group of the MPT. In *E. coli*, MocA catalyzes the addition of CMP to the phosphate group of Moco, while MobA catalyzes the addition of GMP to the Moco in a MobB-facilitated process.

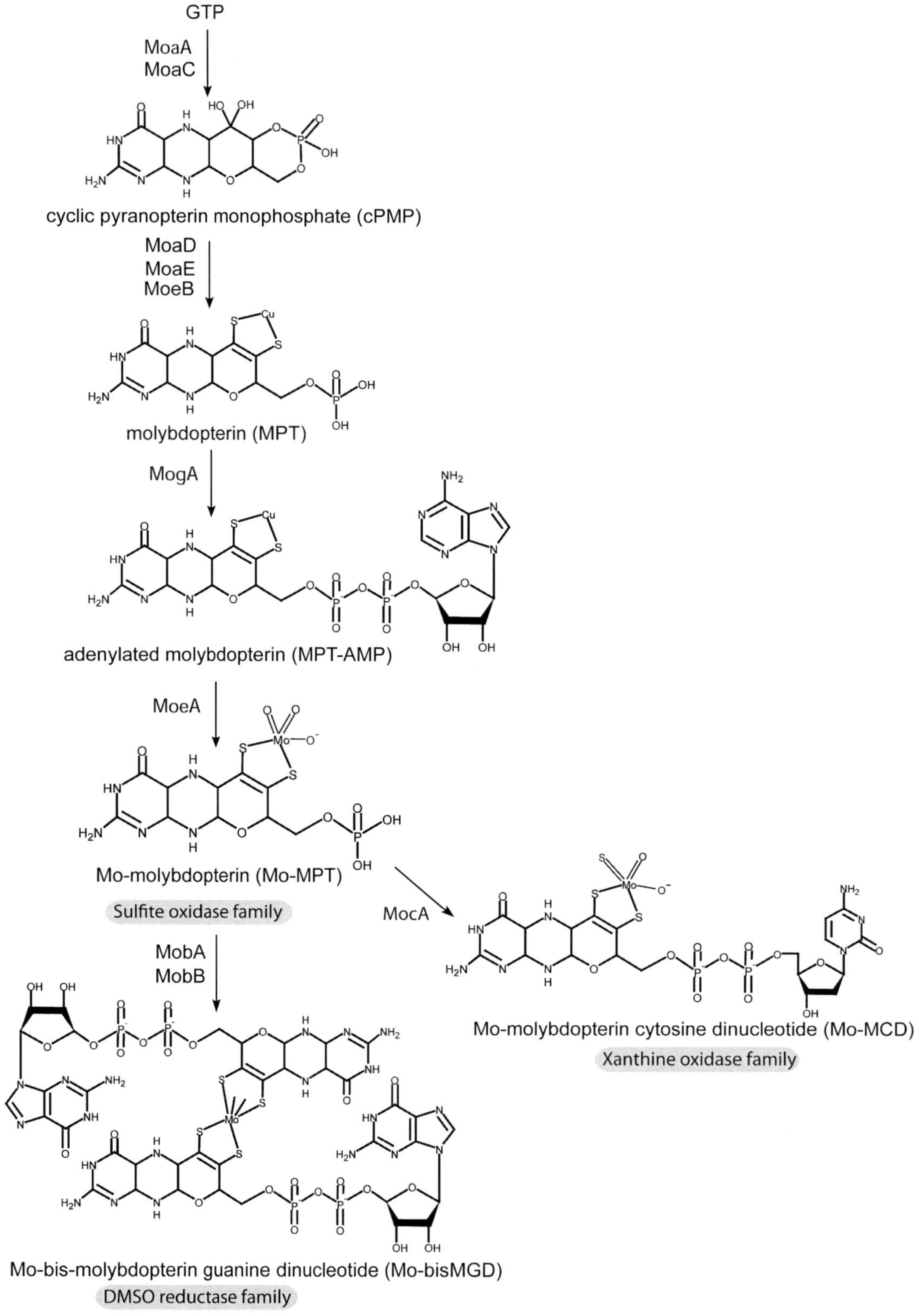

Molybdenum and Ions in Prokaryotes, Fig. 2 (continued)

Furthermore, two of the MPT-GMP moieties are ligated to the molybdenum atom via the dithiolene group of MPT, forming bisMGD (bis-molybdopterin guanine dinucleotide) through a remaining enigmatic step (Fig. 2). Finally, the detection of multiple interactions between biosynthetic proteins from final stages of Mo-bisMGD biosynthesis, i.e., MogA, MoeA, MobA, and MobB proteins, provides not only strong support to the concerted action of these proteins but also to the intricate relationship between metal incorporation and nucleotide addition steps. Formation of such complexes would ensure both the fast and protected transfer of reactive and oxygen-sensitive intermediates within the reaction sequence from MPT to Mo-bisMGD (i.e., Mo-bis-molybdopterin guanine dinucleotide), the major form of cofactor found in *E. coli* molybdoenzymes. Moco biosynthesis is reviewed in (Magalon and Mendel 2008; Schwarz et al. 2009; Leimkuhler et al. 2011) and in ▶ Molybdenum Cofactor, Biosynthesis and Distribution.

Structural and Functional Diversity of Molybdoenzymes

The prokaryotic molybdoenzymes fall into different families based on the type of cofactor bound (Figs. 2 and 3) (Rothery et al. 2008; Dobbeck 2011; Magalon et al. 2011). The sulfite oxidase family contains a single molybdopterin cofactor and a conserved cysteine as the protein-Mo ligand and is exemplified by the *E. coli* YedY protein. The xanthine oxidase family, which includes the *Rhodobacter capsulatus* xanthine dehydrogenase, contains a Mo atom coordinated by a single molybdopterin, lacks a protein-Mo ligand, but contains additional sulfur in the coordination sphere. The cofactor found in the xanthine oxidase family may be further substituted by cytosine yielding a Mo-MCD cofactor. Finally, most prokaryotic enzymes fall in the dimethyl sulfoxide reductase family harboring a Mo atom coordinated by two nucleotide-substituted molybdopterin. The Mo atom is further coordinated by a cysteine, selenocysteine, serine, or aspartate protein Mo-ligand.

The sulfite oxidase family consists not only of prokaryotic members but also of sulfite oxidase of higher organisms and assimilatory nitrate reductase of plants. Very few sulfite oxidases from prokaryotes were characterized to date despite abundance as revealed by genome data analysis. They have been shown to be involved in very different metabolic processes such as energy generation from sulfur compounds, host colonization, sulfite detoxification, and organosulfonate degradation (Kappler 2011). Members of xanthine oxidase include xanthine dehydrogenases involved in purine catabolism and aldehyde oxidases which oxidize numerous aromatic and nonaromatic heterocycles and aldehydes to yield their respective carboxylic acids. Finally, enzymes of the dimethyl sulfoxide reductase family include a broad variety of architecture ranging from single subunit to heterotrimeric complexes which can even be membrane-associated and of cellular location with members present in all compartments. As such, prominent examples of single subunit systems harboring Mo-bisMGD as the sole prosthetic group are the periplasmic trimethylamine *N*-oxide reductase (TorA), the periplasmic dimethyl sulfoxide reductase (DorA), and the cytoplasmic biotin sulfoxide reductase (BisC) only found in some prokaryotes. At the next level of complexity, the Mo-bis-MGD-containing subunit can include a [4Fe-4S] cluster and pair up with a heme-containing electron-transfer subunit as in the periplasmic nitrate reductase NapAB found in many eubacteria. At the most complex level, these molybdo-subunits can be part of multisubunit proteins either periplasmic or membrane-associated of which a large number have been extensively characterized at the structural, functional, and biophysical level. The most prominent examples of these systems are the membrane-associated ▶ nitrate reductase (NarGHI) and ▶ formate dehydrogenase (FdnGHI) or the periplasmic polysulfide dehydrogenase (PsrABC).

Molybdenum and Ions in Prokaryotes, Fig. 2 Biosynthesis of various forms of molybdenum cofactor which define distinct families of prokaryotic molybdoenzymes. The biosynthesis of Mo-MPT occurs in 4 steps common to all organisms. Enzymes which harbor the Mo-MPT as cofactor belong to the sulfite oxidase family. During a fifth step, a nucleotide is added to the Mo-MPT. MocA adds CMP, and enzymes that harbor Mo-MCD as cofactor belong to the xanthine oxidase family. MobA adds GMP together with a second molecule of Mo-MGD yielding Mo-bisMGD. Enzymes that harbor Mo-bisMGD belong to the DMSO reductase family

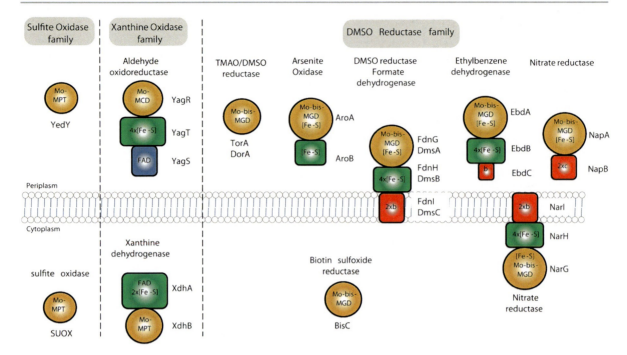

Molybdenum and Ions in Prokaryotes, Fig. 3 Structural and functional diversity among molybdoenzymes. Molybdoenzymes are divided into three families according to the type of Moco present in their catalytic subunits: the sulfite oxidase family harbors Mo-MPT, the xanthine oxidase family harbors sulfurated Mo-MPT with or without the addition of a nucleotide, and the DMSO reductase family harbors Mo-bisMGD. Molybdoenzymes can be localized into the cytoplasm, anchored to the membrane, or localized into the periplasm. They can be found as monomeric single subunit or as multimeric complexes. Oligomerization of the individual molybdoenzymes is not represented in this figure. Subunits harboring the Moco are colored in *yellow*, the ones harboring [Fe-S] in *green*, the ones harboring hemes in *red*, and the ones harboring FAD only in *blue*. All represented molybdoenzymes are found in *E. coli* unless the cytoplasmic sulfite oxidase (SUOX) from *Deinococcus radiodurans*, the monomeric DMSO reductase (DorA) from *R. capsulatus*, the arsenite oxidase (AroAB) from *Ralstonia sp. 22*, and the ethylbenzene dehydrogenase (EbdABC) from *Azoarcus-like strain EbN1*

The quinol/nitrate oxidoreductase of the anaerobic respiratory chain of most prokaryotes, an archetypal member of these elaborated and complex members of the dimethyl sulfoxide reductase family, is described here (Fig. 4) (Blasco et al. 2001).

NarGHI is a nonexported membrane-bound complex composed of three subunits that bind eight redox centers: (1) a catalytic subunit (NarG, 140 kDa) containing a Mo-bisMGD cofactor and a proximal [4Fe-4S] cluster (FS0) with unique properties, (2) an electron transfer subunit (NarH, 60 kDa) carrying one [3Fe-4S] cluster (FS4) and three [4Fe-4S] clusters (FS1 to FS3), (3) and a quinol-oxidizing membrane-bound subunit (NarI, 22 kDa) containing two *b*-type hemes (b_D and b_P). Quinol binding occurs at a specific Q_D site of NarI in close proximity to heme b_D and at which semiquinone intermediates are stabilized. The two electrons issued from this oxidative reaction are transferred to the other metal centers of the heterotrimeric complex facing the cytoplasm, while the two protons are released in the periplasm, thus contributing to the generation of a transmembrane proton gradient. Most interesting is that a second hydrophobic cleft of the NarI subunit is occupied by a cardiolipin molecule, anionic phospholipid. Binding of the lipid molecule is responsible for fine-tuning of the Q_D site and heme b_D center geometries allowing quinol binding, stabilization of a semiquinone intermediate, and thus generation of an active state of the enzymatic complex. At the opposite end of the complex, nitrate is reduced to nitrite through a two-electron oxidative reaction occurring at the Mo center. The electron transfer relay extends over a distance of approximately 98 Å with inter-center distances (edge to edge) between redox components of 5.4–11.2 Å.

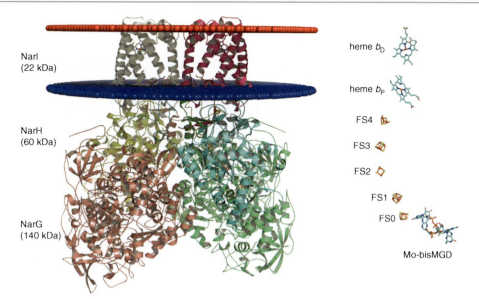

Molybdenum and Ions in Prokaryotes, Fig. 4 Structure of the quinol/nitrate oxidoreductase A of *Escherichia coli*. The homodimer is represented on the left with elements of secondary structure shown as cartoon. NarG is colored in *pink* and in *green*, NarH in *yellow* and in *cyan*, and NarI in *gray* and *fuchsia*. The cytoplasmic membrane is represented as two *ellipses*, one colored in *red* at the interface with the periplasm and the other one colored in *blue* at the interface with the cytoplasm. Metal centers are shown on the left. The Mo-bisMGD is buried in NarG close to the [Fe-S] cluster (FS0). NarH harbors 4 [Fe-S] clusters, FS1, FS2, FS3, and FS4. NarI harbors two *b*-type hemes, b_P (P as proximal) and b_D (D as distal). Pdb ID: 1q16

In some cases, functioning of these enzymes of the dimethyl sulfoxide reductase family is part of an anaerobic respiratory process (e.g., denitrification or anaerobic respiration of arsenic and selenium oxyanions). Together with the formate dehydrogenase (FdnGHI), the nitrate reductase (NarGHI) is part of the "redox loop mechanism" that allows the generation of a proton gradient. Formate oxidation is coupled to quinol (QH_2) generation subsequently oxidized by NarGHI. Since many respiratory molybdoenzymes are not membrane bound, the energy could theoretically be lost. Through transitory associations with transmembrane subunits which exhibit quinol/quinone oxidoreductase activity, they participate to proton motive force (pmf) formation or to quinone/quinol recycling, thus ensuring a continuous supply of oxidized quinone for the quinone-reducing electron input components.

Biogenesis of Molybdoenzymes

As Moco is labile and oxygen-sensitive once liberated from proteins, it was assumed that Moco does not occur in a "free state"; rather, Moco should be transferred immediately after biosynthesis to the apoenzyme or that it could be bound to a carrier protein that protects and stores Moco until further use. Moreover, X-ray crystallographic studies of all known molybdoenzymes revealed that the Moco is not located at the surface of the protein, but it is buried deeply within the enzyme and, in some cases, in close proximity to [Fe-S] clusters. This observation suggests that Moco needs to be incorporated before the completion of folding and oligomerization of enzyme subunits.

Within the prokaryotic cell, successful synthesis and assembly of molybdoenzymes is an intricate process that requires several steps such as the synthesis of the different subunits in the cytoplasm, their assembly, the incorporation of various types of metal or organic cofactors, and the anchoring of the complex to the membrane. In the case of periplasmic or outer membrane molybdoenzymes, the assembly and ▸ metal cofactor incorporation steps take place in the cytoplasm prior to translocation across the inner membrane via the twin-arginine translocation (Tat) apparatus, a machinery dedicated to the export of folded proteins. Importantly, in addition to general chaperones, system-specific chaperones often assist formation of active molybdoenzymes. Altogether,

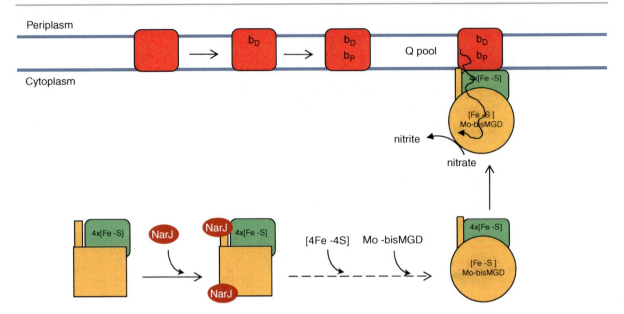

Molybdenum and Ions in Prokaryotes, Fig. 5 Biogenesis model of the nitrate reductase A (NarGHI) from *Escherichia coli*. Upon synthesis and assembly of a NarGH complex harboring the four [Fe-S] in NarH (likely inserted cotranslationally), NarJ binds NarG on two different sites. Recognition and binding of the N-terminus of NarG (sharing similarities to twin-arginine signal peptide of exported molybdoenzymes) prevents premature membrane anchoring of the NarGH complex until all metal insertion and folding events are completed. NarJ binding to a second undefined site of NarG allows sequential insertion of the [Fe-S] cluster and of the Mo-bisMGD. Once maturated, the holoNarGH complex reaches its final localization by binding NarI. Two *b*-type hemes are inserted sequentially into NarI prior interaction with the NarGH complex

these system-specific and general chaperones may function to stabilize the substrates against misfolding and proteolysis, such that a certain level of structure is acquired before Moco insertion can proceed, as well as to help escort folded substrates to the Tat translocon while preventing early engagement. The gene encoding the system-specific chaperone is usually organized in operon with the genes encoding the structural subunits of the molybdoenzyme, thus facilitating its isolation. However, there are several cases of molybdoenzymes for which no specific chaperone has been yet identified. For instance, enzymes from the sulfite oxidase family have no specific chaperone associated. In these cases, it is not understood how these enzymes achieve their biogenesis (Magalon et al. 2011).

Biogenesis of the heterotrimeric and membrane-bound nitrate reductase A from *E. coli* has been one of the first molybdoenzyme whose maturation pathway has been extensively studied (Fig. 5). Initial biochemical and genetic studies indicated that the NarJ protein encoded by the *narGHJI* operon plays an essential role in nitrate reductase activity promoting correct assembly of the enzyme complex without being part of the final structure. In the absence of NarJ, a global defect in metal incorporation into NarGHI is observed. In addition to both metal cofactors of the catalytic subunit NarG (i.e., FeS and Mo-bisMGD), the proximal heme b_P is absent due to loss of coordination between maturation of the NarI and NarGH components. Finally, the absence of NarJ did not appear to affect the stability or the cellular distribution of the apoenzyme which remains largely associated to the cytoplasmic membrane. In fact, it has been demonstrated that NarJ orchestrates metal cofactor insertion, subunit assembly, and membrane-anchoring steps. To achieve this multifunctional character, NarJ interacts with two distinct sites of the NarG catalytic subunit to perform distinct functions. NarJ binding on the N-terminus of NarG represents part of a chaperone-mediated quality control process preventing membrane anchoring of the soluble and cytoplasmic NarGH complex before all maturation and folding events have been completed. In particular, NarJ ensures complete maturation of the

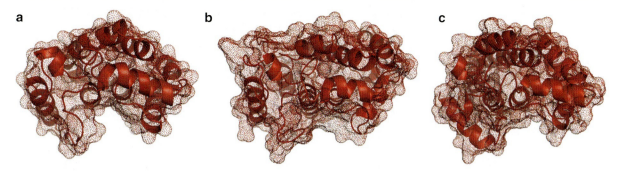

Molybdenum and Ions in Prokaryotes, Fig. 6 Structure of system-specific chaperones with an all-alpha-helical fold. (**a**) NarJ from *Archaeoglobus fulgidus* (PDB ID code 2o9x), (**b**) DmsD from *Escherichia coli* (PDB ID code 3EFP), (**c**) TorD monomer from *Shewanella massilia* (PDB ID code 1n1c)

b-type cytochrome NarI by proper timing for membrane anchoring of the cytoplasmic NarGH complex. A second yet undefined NarJ-binding site within the catalytic subunit NarG is responsible for sequential insertion of the FeS cluster (FS0) followed by Mo-bisMGD. The exact function of NarJ in this process is unclear. Nevertheless, NarJ is an indispensable component of the Moco insertional process in authorizing the interaction of the apoenzyme with a complex made up of several cofactor biosynthetic proteins in charge of Moco delivery.

The multifunctional character of this system-specific chaperone raises the question, at a structural level, how this can be achieved. Indeed, the X-ray structure of NarJ-like protein from *Archaeoglobus fulgidus* indicates an all-helical fold (Protein Data Bank (PDB) ID code 2o9x), a feature shared by other bacterial system-specific chaperones such as DmsD (PDB ID codes: 1s9u, 3efp, 3cw0) or TorD (PDB ID code 1n1c), thus forming a new family of chaperones (Pfam PF02613) (Fig. 6). It is tempting to speculate that the ability of NarJ to recognize and interact with distinct sites of a metalloprotein partner or with several metalloprotein partners is based on a structural flexibility. Initial nuclear magnetic resonance (NMR) experiments support this notion by the observation of one or several flexible regions in the protein as well as a global conformational change upon binding the N-terminus of NarG. Such a structural flexibility of the chaperone appears to be a common feature of several members of this new family of chaperones as deduced from the observation of several disordered regions.

An additional level of complexity is attained with the periplasmic location of numerous molybdoenzymes. Since acquisition of the metal centers and in particular of Moco is strictly cytoplasmic events, assembly and folding of such molybdoenzymes proceeds before their translocation through the Tat translocon. Coordination of the folding and assembly events with the translocation of the molybdoenzyme is achieved through the action of system-specific chaperones which specifically interact with the Tat signal peptide located at the N-terminus of the catalytic subunit. A paradigmatic example of such coordination can be illustrated with the chaperone TorD dedicated to the maturation of the periplasmic and monomeric trimethylamine *N*-oxide reductase (TorA) harboring Mo-bisMGD as sole prosthetic group. The TorA-specific chaperone, TorD, interacts with the N-terminus end of the apoenzyme to prevent its export into the periplasm; meanwhile, TorD promotes Moco insertion into the apoenzyme through binding to a second site. A similar situation is encountered with the heterotrimeric and periplasmically facing dimethyl sulfoxide reductase complex (DmsABC) whose maturation strictly relies on the action of the DmsD chaperone.

A number of periplasmic or periplasmically oriented multimeric molybdoenzymes of the dimethyl sulfoxide reductase family have been genetically or biochemically characterized in different bacteria or Archaea. Considering their relatedness to the NarGHI or DmsABC complexes and the existence of an additional gene encoding for a NarJ/TorD/DmsD-like chaperone protein in the corresponding operons, it is tempting to speculate that folding and assembly of these molybdoenzymes will follow the same trend. Exceptions are the periplasmic and multimeric arsenite oxidase (Aox), polysulfide (Psr), and arsenate reductase (Aro) enzymes which do not possess in their respective operons an additional gene encoding for a system-specific chaperone. However, the presence of a Tat signal peptide and of a Fe-S cluster together with the Mo-bisMGD in the catalytic subunit supports the hypothesis that their maturation pathway will need assistance by an as yet unidentified system-specific chaperone which may be located elsewhere in the genome.

Aside from the functions classically associated with the above-mentioned system-specific chaperones, it appears that maturation of the xanthine oxidase family strictly depends on the presence of a sulfur atom on the Mo coordination sphere, a reaction performed by the associated chaperone. The best-studied system is the case of the chaperone XdhC, a hypothetical member of the MOSC (i.e., *Mo*co *s*ulfurase *C*-terminal domain) protein superfamily (see ▶ Molybdenum-enzymes, MOSC Family), in the biogenesis of the *R. capsulatus* xanthine dehydrogenase (XdhAB). This enzyme consists of a cytoplasmic heterotetrameric complex that catalyzes the hydroxylation of hypoxanthine and xanthine, the last two steps in purine degradation. The XdhA subunit contains two [2Fe-2 S] clusters in addition to FAD, while the XdhB subunit binds Mo-MPT. Functional synthesis of the *R. capsulatus* XdhAB complex requires the presence of XdhC, encoded by the *xdhABC* operon, which entails binding of Moco and its insertion into the XdhB subunit. In that sense, XdhC differs considerably from the above-mentioned system-specific chaperones for Mo-bisMGD-containing enzymes as it does not proofread its metalloenzyme substrate. XdhC specifically promotes the sulfuration of Mo-MPT by interaction with a cysteine desulfurase, which transfers the sulfur to Moco bound to XdhC. XdhC protects the sulfurated form of Moco from oxidation before its transfer into apoXdhAB. Importantly, to prevent all available Mo-MPT in the cell from being converted to Mo-bisMGD and to guarantee a Mo-MPT supply for XdhAB, XdhC interacts with MoeA and MobA proteins involved in the final stages of Moco synthesis. Whereas interaction with MoeA allows Mo-MPT transfer to XdhC, its interaction with MobA prevents Mo-bisMGD formation.

Regulation of Genes Related to Molybdenum Metabolism

Molybdenum homeostasis can be controlled at several levels into the prokaryotic cell such as molybdate uptake, molybdenum cofactor biosynthesis, and its utilization by molybdoenzymes, thus ensuring proper utilization of the metal in response to the energetic demand or to environmental conditions. At first, molybdate uptake can be downregulated by a transcriptional repressor under conditions of high molybdate availability. Upon binding of molybdate, the transcriptional regulator ModE undergoes conformational changes allowing its binding on the operator region of the *modABC* operon, thus preventing further binding of the RNA polymerase. Interestingly, ModE is also responsible for upregulation of the expression of the *moaABCDE* operon involved in Moco biosynthesis and of the expression of some molybdoenzymes.

An additional level of regulation is exerted at the translational level through the action of a riboswitch. Riboswitches are structured RNA domains that selectively bind metabolites or metal ions and function as gene control elements. Riboswitches are commonly found in the 5′ untranslated regions of mRNAs. A highly conserved RNA motif has thus been identified upstream of genes encoding molybdate transporters, molybdenum cofactor biosynthesis enzymes, and proteins that utilize Moco as a cofactor. In *E. coli*, this riboswitch controls the *modABC* operon in response to cofactor production by binding Moco and represses the translation of all genes under its control. Such a Moco-sensing riboswitch would participate to the Mo homeostasis by ensuring that, if there are sufficient levels of Moco, not only the molybdate transport but also the Moco biosynthetic proteins and the molybdoenzymes are not produced. Interestingly, this highly conserved RNA is able to discriminate between molybdenum- and tungsten-containing cofactors.

Finally, most of prokaryotic molybdoenzymes being expressed under anaerobic conditions, the next level of regulation is made through the fumarate nitrate regulation (FNR) transcription factor, a global regulatory protein whose activity is regulated by O_2 and which ensures a tight control of the aerobic/anaerobic transition. Additional control levels of the expression of individual molybdoenzymes are obtained through the participation of two-component systems such as the NarXL or NarPQ which provide response to nitrate and nitrite or TorSR toward trimethylamine *N*-oxide. Overall, Mo homeostasis is maintained through a complex array of overlapping transcription and translational factors.

Cross-References

- ▶ Molybdenum and Ions in Living Systems
- ▶ Molybdenum Cofactor, Biosynthesis and Distribution
- ▶ Molybdenum in Biological Systems
- ▶ Molybdenum-enzymes, MOSC Family
- ▶ Tungsten Cofactors, Binding Proteins, and Transporters in Biological Systems
- ▶ Tungsten in Biological Systems

References

Blasco F, Guigliarelli B et al (2001) The coordination and function of the redox centers of the membrane-bound nitrate reductases. Cell Mol Life Sci 58(2):179–193

Dobbeck H (2011) Structural aspects of mononuclear Mo/W enzymes. Coord Chem Rev 255(9–10):1104–1116

Hagen W (2011) Cellular uptake of molybdenum and tungsten. Coord Chem Rev 255(9–10):1117–1128

Kappler U (2011) Bacterial sulfite-oxidizing enzymes. Biochim Biophys Acta 1807(1):1–10

Leimkuhler S, Wuebbens M et al (2011) The history of discovery of the molybdenum cofactor and novel aspects of its biosynthesis in bacteria. Coord Chem Rev 255(9–10):1129–1144

Magalon A, Mendel R (2008) Biosynthesis and insertion of the molybdenum cofactor. In: Böck A, Curtiss R III, Kaper JB et al (eds) EcoSal—*Escherichia coli* and *Salmonella*: cellular and molecular biology. ASM Press, Washington, DC

Magalon A, Fedor J et al (2011) Molybdenum enzymes in bacteria and their maturation. Coord Chem Rev 255(9–10):1159–1178

Rothery RA, Workun GJ et al (2008) The prokaryotic complex iron-sulfur molybdoenzyme family. Biochim Biophys Acta 1778(9):1897–1929

Schwarz G, Mendel RR et al (2009) Molybdenum cofactors, enzymes and pathways. Nature 460(7257):839–847

Zhang Y, Gladyshev VN (2008) Molybdoproteomes and evolution of molybdenum utilization. J Mol Biol 379(4):881–899

Molybdenum Cofactor

- ▶ Molybdenum and Ions in Prokaryotes
- ▶ Molybdenum Cofactor, Biosynthesis and Distribution

Molybdenum Cofactor, Biosynthesis and Distribution

Ralf R. Mendel
Department of Plant Biology, Braunschweig University of Technology, Braunschweig, Germany

Synonyms

Molybdenum cofactor; Molybdopterin

Definition

The molybdenum cofactor is a unique tricyclic pterin scaffold, named molybdopterin, which coordinates molybdenum via a dithiolene group. The task of the pterin is to bind and position the catalytic metal molybdenum correctly within the active center of a given molybdenum enzyme, and depending on the enzyme class, molybdenum can have different ligands.

The transition element molybdenum (Mo) is an essential micronutrient for animals, plants, and most microorganisms where it forms part of the active centers of metalloenzymes (Mendel and Schwarz 2011; Schwarz and Mendel 2006). Mo seems to be biologically inactive until it becomes complexed by a special pterin to form the molybdenum cofactor (Moco), thus gaining biological activity (Schwarz et al. 2009). More than 50 Mo-containing enzymes are known, most of them of bacterial origin, while only a handful Mo enzymes were found among eukaryotes. There is another type of Mo-containing cofactor which is found only once in nature, namely, in bacterial nitrogenase, forming the so-called FeMo-cofactor that consists of two partial cubanes ($MoFe_3S_3$ and Fe_4S_3) which are joined by three bridging sulfurs (Hu and Ribbe 2011). Nitrogenase is required for biological nitrogen fixation, which is an essential step in the nitrogen cycle in the biosphere. In contrast to nitrogenase, all other Mo-containing enzymes characterized to this end contain the pterin-type cofactor. The biosynthesis of Moco involves the complex interaction of six proteins and is a multistep process, which also requires reducing equivalents, iron, ATP, and copper. Among higher organisms, most of this knowledge derives from studies in plants and humans which may be surprising – but the yeast *Saccharomyces cerevisiae* as model organism plays no role in Mo research as it belongs to those organisms that do not contain Mo enzymes while many other yeasts such as *Pichia pastoris* do need Moco (Zhang et al. 2011).

Molybdenum Uptake

What route does Mo take from its entry into the cell, via formation and modification of Moco up to its insertion into apo-metalloenzymes? Organisms take up Mo in the form of its oxyanion molybdate. It requires specific uptake systems to scavenge molybdate in the presence of competing anions (Hagen 2011). These Mo uptake systems were studied in detail in bacteria, while in higher organisms only recently, first molybdate-transporting proteins have been identified in algae and plants where two proteins (MOT1 and MOT2) belonging to the large sulfate carrier superfamily were shown to transport molybdate with high affinity across cellular membranes. Most remarkably, these transporters were located in the interior of the cell, while the Mo transporter in the plasma membrane is still lacking. Obviously, a whole system of molybdate transporters is existing – at least among plants. To this end, nothing is known about the molecular uptake mechanism of molybdate in mammals, but it may be that in addition to a possible high-affinity system, molybdate could also nonspecifically enter the mammalian cell through the sulfate and/or phosphate uptake system.

The Molybdenum Cofactor

Once having entered the cell, Mo has to be attached to its cofactor scaffold, thereby converting to Moco. This cofactor is a tricyclic pteridin that coordinates the metal via a dithiolene group (Fig. 1, right upper part). Because of the unique nature of the pterin in Moco, the metal-free form of the cofactor is called molybdopterin or metal-containing pterin (MPT; Fig. 1), and the latter reflects the fact that not only Mo but also tungsten can be coordinated by this pterin scaffold (Schwarz et al. 2009). The task of the pterin is to bind and position the catalytic metal Mo correctly within the active center of a given Mo enzyme, and depending on the class of Mo enzyme, Mo can have different ligands. Another possible role of the pterin moiety could be control of the redox behavior of the Mo atom. In addition, the pterin might also participate in the electron transfer to or from Mo via the delocalized electrons within the pterin. Moco is not located on the surface of the protein; rather, it is buried deeply within the interior of the enzyme and a tunnel-like structure makes it accessible to the appropriate substrates (Hille et al. 2011). During its lifetime, the Mo enzyme does not liberate Moco. In vitro, however, one can remove Moco from the holoenzyme whereafter Moco loses Mo and undergoes rapid and irreversible loss of function due to oxidation. To this end, there are no indications for a Moco recycling mechanism in the cell. In bacteria, the MPT-based core moiety of Moco can be further modified to form bis-MPT equivalents or dinucleotide variants of Moco (Leimkühler et al. 2011).

Molybdenum Cofactor Biosynthesis

A mutational block of Moco biosynthesis leads to the loss of essential metabolic functions because all enzymes depending on Moco lose their activity, which ultimately causes death of the organism. In humans, plants, and all other eukaryotes studied so far, a conserved biosynthetic pathway that can be divided into four steps, according to the biosynthetic intermediates cPMP, MPT, adenylated MPT, and Moco (Fig. 1), synthesizes Moco. Always six gene products (= proteins) catalyzing Moco biosynthesis have been identified (named CNX) in plants and MOCS in humans (Mendel and Schwarz 2011; Schwarz et al. 2009). The eukaryotic Moco biosynthesis genes are homologous to their counterparts in bacteria (Leimkühler et al. 2011).

Step 1: Guanosine 5′-triphosphate (GTP) is transformed into cyclic pyranopterin monophosphate (cPMP), also known as precursor Z (Fig. 1). In comparison to Moco and MPT, cPMP is the most stable intermediate with an estimated half-life of several hours at a low pH. By combining labeling studies with ^{1}H-NMR analysis, it has been confirmed that during the conversion of GTP into cPMP, each carbon atom of the ribose and the ring atoms of the guanine are incorporated into cPMP. cPMP is still sulfur-free but has already the tricyclic pyranopterin structure similar to the mature cofactor. Step 1 of Moco biosynthesis seems to be highly conserved between

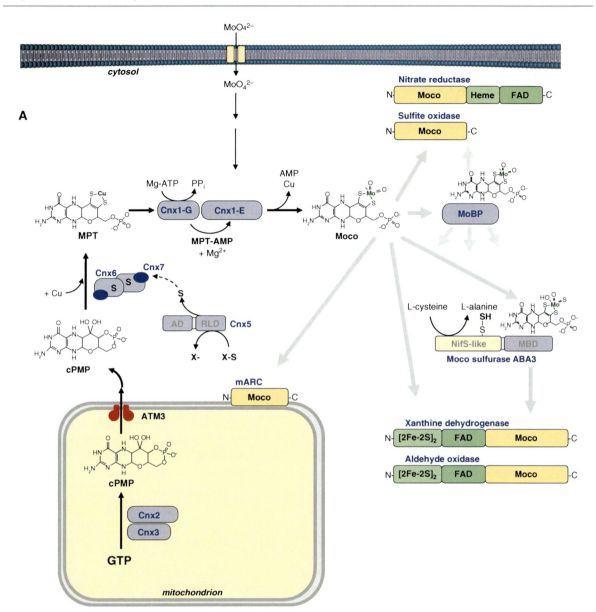

Molybdenum Cofactor, Biosynthesis and Distribution, Fig. 1 Organization of biosynthesis, distribution, and maturation of Moco in higher organisms (here shown for plants). The basic steps of Moco biosynthesis are shown starting from the conversion of GTP to cPMP in the mitochondria, and all subsequent steps proceed in the cytosol. Moco biosynthesis enzymes (named Cnx) and Moco-binding proteins (MoBPs) are shown in *blue*. MPT synthase, consisting of Cnx6 and Cnx7, is sulfurated by Cnx5, with the primary sulfur donor (X-S) mobilized by the rhodanese-like domain of Cnx5 (RLD) being unknown. The adenylation domain of Cnx5 (AD) is required for adenylation and activation of the small MPT synthase subunit Cnx7. It is assumed that copper (Cu) is inserted directly after dithiolene formation. The individual reactions of Cnx1 and its products (Moco, pyrophosphate PP_i, AMP, copper) are indicated. Moco can be either bound to a Moco-binding protein (MoBP); to the five Mo-enzymes nitrate reductase, sulfite oxidase, xanthine dehydrogenase, aldehyde oxidase, and mARC; or to the Moco-binding C-terminal domain (MBD) of the Moco sulfurase ABA3. The ABA3 C-terminus is the sulfuration platform for Moco. The NifS-like domain of ABA3 generates a protein-bound persulfide, which is transferred to Moco bound to its C-terminus which in turn exchanges nonsulfurated for sulfurated Moco. The domain structure of the Mo-enzyme monomers is given in the figure. It is obvious that the eukaryotic Mo enzymes are evolutionary related to each other: xanthine dehydrogenase and aldehyde oxidase form a pair, sulfite oxidase forms a pair with nitrate reductase, and mARC arose from ABA3-MBD (Modified after Mendel and Schwarz 2011)

higher organisms and bacteria because the plant and human genes are able to replace the function of their homologs in bacteria. Step 1 is catalyzed by two proteins (Cnx2 and Cnx3) that – in plants and humans – were found to localize to the matrix fraction of mitochondria (Mendel and Schwarz 2011). As cPMP synthesis takes place in mitochondria while all subsequent steps of Moco biosynthesis have been demonstrated to be localized in the cytosol, export of cPMP from the mitochondria into the cytosol is required to allow further processing to Moco. This task is fulfilled by the transporter protein Atm3, which is localized in the inner membrane of mitochondria and belongs to the family of ATP-binding cassette (ABC) transporters. Surprisingly, Atm3 functions also in exporting a precursor compound essential for Fe-S cluster synthesis in the cytosol.

Step 2: In the second stage, sulfur is transferred to cPMP in order to generate MPT. This reaction is catalyzed by the enzyme MPT synthase, a heterotetrameric complex of two small (Cnx7) and two large (Cnx6) subunits that stoichiometrically converts cPMP into MPT (Fig. 1). Each of the small subunits within the tetrameric enzyme complex carries a single sulfur atom as thiocarboxylate at the highly conserved C-terminus. The two sulfur atoms are not simultaneously transferred to cPMP; rather, the sulfurs become sequentially inserted into cPMP. After MPT synthase has transferred the two sulfur atoms to cPMP, it has to be resulfurated in order to regenerate the enzyme for the next reaction cycle of cPMP conversion. This separate resulfuration is catalyzed by the enzyme MPT-synthase sulfurase (Cnx5), which is a two-domain protein (Fig. 1) that activates Cnx7 by adenylation (carried out by the Cnx5 N-terminal domain) followed by a sulfur transfer reaction (carried out by the Cnx5 C-terminal domain which has a rhodanese function). The identity of the donor for the reactive mobile sulfur is yet unknown, but a redundant function of different persulfide-generating systems like cysteine desulfurases is possible.

Step 3: In the next step, MPT has to be activated to become receptive for Mo insertion. Therefore, Mo has to be transferred to MPT in order to form Moco, thus linking molybdate uptake to the MPT pathway. This mechanism has been first uncovered in plants where the protein Cnx1 was found to catalyze a complex sequence of reactions (Schwarz and Mendel 2006). In bacteria, this step is catalyzed by two proteins (Leimkühler et al. 2011), while during evolution to higher organisms, these two proteins were fused to a two-domain protein. The C-terminal Cnx1 domain (= Cnx1-G) adenylates MPT and forms the intermediate MPT-AMP that remains bound to Cnx1-G. The crystal structure of the Cnx1-G revealed a copper bound to the MPT dithiolate sulfurs, whose nature was confirmed by anomalous scattering of the metal (Kuper et al. 2004). Up to now the function of copper is unknown, but it might play a role in sulfur transfer to cPMP, in protecting the MPT dithiolate from oxidation, and/or presenting a suitable leaving group for Mo insertion. The origin of copper is also unclear, but it is reasonable to assume that it is transferred in vivo via the action of cytoplasmic chaperones thereafter binding to the dithiolate group of MPT just after the latter has been formed, i.e., at the end of step 2 of Moco biosynthesis.

Step 4: In the final step of Moco biosynthesis, MPT-AMP has to be converted into mature Moco: MPT-AMP is transferred to the N-terminal domain of Cnx1 (= Cnx1-E), thereby building a product-substrate channel. Cnx1-E cleaves the adenylate, releases copper, and inserts Mo, thus yielding active Moco. In vitro studies with Cnx1-G-bound MPT-AMP revealed an inhibition of Moco synthesis in the presence of 1 μM $CuCl_2$, providing a link between Mo and copper metabolism (Mendel and Schwarz 2011). Remarkably, the human homolog of Cnx1 is the protein gephyrin which fulfills a dual role: It functions not only as Mo insertase, but it serves also as anchor protein for inhibitory neuroreceptors in the postsynaptic membrane of neurons (Mendel and Schwarz 2011).

Distribution of Moco: Storage and Transfer

In higher organisms, the steps beyond Moco biosynthesis are less well known and, therefore, research focuses on Moco transport, allocation, and insertion into apo-Mo-enzymes (Mendel and Schwarz 2011). Here one has to keep the cellular context in mind as Moco biosynthesis occurs to be microcompartmentalized in a multiprotein-biosynthesis complex localized in the cytosol of the cell. After completion of biosynthesis, Moco either has to be allocated and inserted into the apo-Mo-enzymes, or Moco has to become bound to a carrier protein that protects and stores Moco until further use, thus providing a way to buffer supply and demand of Moco. In bacteria, a complex of proteins synthesizing the last steps of

Moco biosynthesis donates the mature cofactor to apoenzymes assisted by enzyme-specific chaperones (Leimkühler et al. 2011). Nearly each bacterial Mo enzyme has a private chaperone available (Magalon et al. 2011); however, in higher organisms, no Mo-enzyme-specific chaperones have been found yet. Moco is highly unstable once liberated from proteins, and it loses the Mo atom and undergoes rapid and irreversible loss of function due to oxidation. Therefore, it was assumed that Moco does not occur free but permanently protein bound in the cell. A cellular Moco distribution system should meet two demands: (1) It should bind Moco subsequent to its synthesis, and (2) it should maintain a directed flow of Moco from the Moco donor Cnx1 to the Moco-dependent enzymes. The availability of sufficient amounts of Moco is essential for the cell to meet its changing demand of Mo enzymes.

A Moco-storage/carrier protein has been described for the green alga *Chlamydomonas reinhardtii* (a homotetramer that can hold four Moco molecules) donating Moco to the Mo-enzyme nitrate reductase. In the higher plant *Arabidopsis thaliana*, recently, a novel protein family was identified (Mendel and Schwarz 2011) consisting of eight members (Fig. 1) that all can bind Moco and are therefore named Moco-binding proteins (MoBPs). They seem to be involved in the cellular distribution of Moco because in the cytoplasm of a living cell, they were found to undergo protein-protein interaction both with the "Moco-donor" protein Cnx1 and with the "Moco-user" protein nitrate reductase. Obviously, land plants with differentiated organs need more MoBPs than a unicellular and motile alga. It is, however, still open whether these proteins represent the default-pathway for Moco-allocation to its users or whether it is only a buffer, and the default-way goes directly from CNX1 to the appropriate apo-Mo-enzymes.

Insertion of Moco into Mo Enzymes and Final Maturation

Among eukaryotes, nothing is known how FAD, iron-sulfur centers, and heme are bound to the Mo enzymes. For Moco, however, the first crystallographic analyses of Mo enzymes made evident that the cofactor is deeply buried within the holoenzyme so that Moco could only have been incorporated prior to or during completion of folding and dimerization of the apoprotein monomers. As there are three classes of Mo enzymes known in higher organisms (Hille et al. 2011) also differences in the insertion of Moco might be considered. For insertion of Moco into the target apoenzymes either (still unknown) chaperone proteins would be needed or the Moco-carrier/binding proteins could become involved at the stage that has been shown for the *Arabidopsis* MoBPs. Once being introduced into the apoprotein, Mo is coordinated by additional ligands present in the active center of the respective Mo enzyme. Moco forms an important part of the active center, but it represents not the only constituent of the center because the additional ligands supplied by the apoprotein are equally important for the specificity of the reaction. Like every other protein, also Mo enzymes are finally degraded after having fulfilled their tasks. One can assume that Moco is degraded as well because free Moco released from proteins is extremely labile and sensitive to oxidation.

Enzymes of the xanthine oxidoreductase family require a final step of maturation during or after insertion of Moco, i.e., the addition of a terminal sulfido group to the Mo center in order to gain enzymatic activity (Hille et al. 2011). This sulfur ligand is added by a separate enzymatic reaction catalyzed by the Moco sulfurase ABA3 (Fig. 1). ABA3 is a homodimeric two-domain protein with its N-terminal domain sharing structural and functional homologies to bacterial cysteine desulfurases. In a pyridoxal phosphate-dependent manner, the N-terminal domain of ABA3 decomposes L-cysteine to yield alanine and elemental sulfur, the latter being bound as a persulfide to a highly conserved cysteine residue of ABA3 (Fig. 1). The C-terminal domain of ABA3 shares a significant degree of similarity to the third class of Mo enzymes in eukaryotes (= mARC proteins) and was shown to bind sulfurated Moco, which receives the terminal sulfur via intramolecular persulfide-transfer from the N-terminal domain. It is likely that subsequent to Moco-sulfuration, ABA3 exchanges nonsulfurated for sulfurated Moco, thus activating its target Mo enzyme.

Cross-References

▶ Molybdenum and Ions in Living Systems
▶ Molybdenum and Ions in Prokaryotes
▶ Molybdenum in Biological Systems

References

Hagen WR (2011) Cellular uptake of molybdenum and tungsten. Coord Chem Rev 255:1117–1128

Hille R, Nishino T, Bittner F (2011) Molybdenum enzymes in higher organisms. Coord Chem Rev 255:1179–1205

Hu Y, Ribbe MW (2011) Biosynthesis of nitrogenase FeMoco. Coord Chem Rev 255:1218–1230

Kuper J, Llamas A, Hecht HJ et al (2004) Structure of molybdopterin-bound Cnx1G domain links molybdenum and copper metabolism. Nature 430:803–806

Leimkühler S, Wuebbens MM, Rajagopalan KV (2011) The history of the discovery of the molybdenum cofactor and novel aspects of its biosynthesis in bacteria. Coord Chem Rev 255:1129–1143

Magalon A, Fedor JG, Walburger A et al (2011) Molybdenum enzymes in bacteria and their maturation. Coord Chem Rev 255:1159–1178

Mendel RR, Schwarz G (2011) Molybdenum cofactor biosynthesis in plants and humans. Coord Chem Rev 255:1145–1158

Schwarz G, Mendel RR (2006) Molybdenum cofactor biosynthesis and molybdenum enzymes. Annu Rev Plant Biol 57:623–647

Schwarz G, Mendel RR, Ribbe MW (2009) Molybdenum cofactors, enzymes and pathways. Nature 460:839–847

Zhang Y, Rump S, Gladyshev VN (2011) Comparative genomics and evolution of molybdenum utilization. Coord Chem Rev 255:1206–1217

Molybdenum Cofactor–Containing Enzymes

▶ Molybdenum in Biological Systems

Molybdenum in Biological Systems

Russ Hille
Department of Biochemistry, University of California, Riverside, CA, USA

Synonyms

Moco-containing enzymes; Molybdenum cofactor–containing enzymes; Molybdenum-containing enzymes

Definition

Molybdenum-containing enzymes are those metalloproteins possessing the second-row transition metal molybdenum in their active sites. All are oxidoreductases with the metal present as Mo(VI) in the oxidized form.

Background

Molybdenum: Physical and Chemical Properties

Molybdenum is a Group 6 element, and the only second-row transition metal that is widely distributed in biology. It is found in variety of minerals, but its most prevalent form in the environment is the oxidized $Mo^{VI}O_4^{2-}$ molybdate ion. Molybdate is very water soluble, and due to its solubility molybdenum is the most abundant transition metal in seawater; as a consequence, molybdate is highly bioavailable. In addition, because of its occurrence as molybdate, molybdenum is unusual among transition metals in that its uptake, transport, and utilization in biological systems is as an anion rather than a cation.

Molybdenum: Low Molecular Mass Compounds

Beginning in the mid-1980s, several classes of mononuclear model compounds have been developed that utilize bulky ligands to prevent the otherwise facile formation of μ-oxo dimers (Berg and Holm 1985; Roberts et al. 1988). Systems consisting of $LMo^{VI}O_2$ cores were the first to be developed, and these have proven to be excellent mimics of the active sites of one of the three major categories of molybdenum-containing enzymes (see below). These models are themselves robust catalysts, efficiently reacting with suitable oxygen atom acceptors (e.g., phosphines) to form an $LMo^{IV}O$ species which then can react with an oxygen atom donor (such as dimethyl sulfoxide, DMSO) to return to the parent $LMo^{VI}O_2$ species (Berg and Holm 1985). Other models with an $L_2Mo^{VI}O(OR)$ make-up, where L represents a bidentate enedithiolate ligand such as $^-S(CN)C=C(CN)S^-$ and OR is an alkoxide ligand, have been developed that mimic features of the active sites of a second major family of molybdenum enzymes (Lim and Holm 2001). Very recently, $LMo^{VI}OS$ models have been synthesized that provide mimics of the final major category of molybdenum enzymes (Doonan et al. 2008). These efforts have demonstrated that all three types of enzyme active site represent intrinsically stable chemical species whose inherent chemical reactivity is put to use in the enzyme-catalyzed reaction. A common aspect in the reaction mechanisms of the vast majority of molybdenum-containing enzymes is the labilization of one of the oxo ligands of the metal, either a Mo=O or Mo-OH, which permits oxygen atom transfer in the course of the biological reaction. The model chemistry of molybdenum as it pertains to biological systems has recently been thoroughly reviewed (Enemark et al. 2004).

Molybdenum and Its Ions in Biological Systems

Cellular uptake of molybdate by prokaryotes is via a broadly distributed ATP-binding cassette (ABC) transporter system that consists of three components: a small extracellular binding protein (ModA in *E. coli*), a membrane-spanning component (ModB), and an intracellular ATP-binding component (ModC) (see, e.g., (Schwarz and Mendel 2006) for a review). The ModABC transporter was one of the first such systems to be characterized crystallographically, demonstrating that the membrane-integral component alternates between configurations with its molybdate binding site exposed to either the exterior or interior of the cell in a manner that is driven by ATP hydrolysis (Hollenstein et al. 2007). Molybdenum cofactor transporters have also been identified in *Chlamydomonas* (Ataya et al. 2003; Fischer et al. 2006). Molybdate is also taken up by the phosphate transporter in many organisms, an observation that reflects the similarity of the two polyanions.

There is evidence that some organisms may possess molybdenum storage proteins (Ataya et al. 2003; Fischer et al. 2006), although in most cases molybdate is rapidly and specifically incorporated into enzymes requiring the metal, and the intracellular concentration of free molybdate is otherwise thought to be quite low.

There are several intracellular routes for molybdate utilization once taken up by the cell. In nitrogen-fixing bacteria such as *Klebsiella pneumoniae* and *Azotobacter vinelandii*, molybdate is incorporated into a preassembled Fe/S cluster (either Fe_6S_9 or Fe_7S_9) on the NifB gene product by NifEN and NifH prior to insertion of the now mature M cluster into the enzyme nitrogenase (Fig. 1) (Hu et al. 2008). For other intracellular destinations, molybdate is first incorporated into a pyranopterin cofactor synthesized by the biosynthetic pathway, as shown in Fig. 2 (Schwarz and Mendel 2006). As shown, the insertion reaction itself involves the displacement of copper from an enedithiolate side chain of the cofactor with concomitant hydrolysis of an AMP adduct of the cofactor (Kuper et al. 2004; Llamas et al. 2004). In bacteria, the molybdenum-containing cofactor can be further modified by its elaboration as the dinucleotide of guanosine, cytosine, or adenosine prior to incorporation into specific apoenzyme targets. For a great many molybdenum-containing enzymes, particularly in prokaryotes, a second equivalent of the pyranopterin ligand is incorporated into the cofactor prior to insertion in a process that is not well understood at present. For other enzymes, such as the sulfite dehydrogenase from the bacterium *Starkeya novella*, or in eukaryotic enzymes such as sulfite oxidase and nitrate reductase, the cofactor is incorporated into the apoprotein without further chemical modification. Still for other enzymes, including bacterial xanthine dehydrogenases, eukaryotic xanthine oxidoreductases, and aldehyde oxidases from both these sources, an acid-labile sulfur must be incorporated into the molybdenum coordination sphere as a catalytically essential Mo=S group prior to insertion into the apoprotein (Massey and Edmondson 1970). In *R. capsulatus*, sulfur incorporation is catalyzed by the XdhC gene product (part of an *xdhABC* operon whose first two genes encode the two subunits of the organism's xanthine dehydrogenase); XdhC also is required for insertion of the now sulfurated cofactor into the apoprotein (Neumann et al. 2007; Schumann et al. 2008).

Molybdenum-Containing Enzymes

The vast majority of molybdenum in the cell is found in the active sites of enzymes. Apart from the unique $MoFe_7S_{11}$ cluster in the active site of nitrogenase, as mentioned above, molybdenum enzymes can be grouped into three families based on the structure of their active sites, as shown in Fig. 1 (Hille 1996). Enzymes within each family also share significant amino acid sequence homologies and fundamental structural motifs. The first family consists of xanthine oxidoreductase and related enzymes that typically catalyze the oxidative hydroxylation of an activated carbon center (in a wide range of aromatic heterocycles, or aldehydes that become oxidized to carboxylic acids). These enzymes possess an $LMo^{VI}OS(OH)$ active site with the Mo=O group occupying the apical position in a square pyramidal coordination geometry in the oxidized state. Nicotinate dehydrogenase is an exception to this generalization, with an equatorial

Molybdenum in Biological Systems, Fig. 1 *Metal centers in molybdenum-containing enzymes. Top*, the M and P clusters of nitrogenase. In the former case, the central ion designated ? has recently been shown likely to be a carbon atom (Lancaster et al. 2011). *Bottom*, consensus active site structures for members of the xanthine oxidoreductase, sulfite oxidase, and DMSO reductase families of mononuclear enzymes. In these structures, the enedithiolate ligand of the pyranopterin cofactor is coordinated to the metal as shown

Mo=Se rather than Mo=S (Wagener et al. 2009). Another interesting variation on the theme of this family, both from the standpoint of structure and the reaction catalyzed, is the CO dehydrogenase from aerobic bacteria such as *Oligotropha carboxidovorans*, which catalyzes the oxidation of CO to CO_2. The active site of this enzyme is a unique binuclear Mo/Cu center with a bridging μ-sulfido ligand spanning the two metals (Dobbek et al. 2002). The molybdenum coordination sphere otherwise resembles that of other members of this family, with the bridging sulfido ligand occupying the position of the Mo = S seen in other enzymes, and a Mo = O in place of the equatorial Mo-OH of other family members.

The second family of mononuclear molybdenum enzymes includes sulfite oxidase and sulfite dehydrogenase, as well as the assimilatory nitrate reductases of fungi and higher plants. These enzymes have an $LMo^{VI}O_2(S\text{-}cys)$ active site in the oxidized state, again in a square pyramidal coordination geometry, with a cysteine ligand from the polypeptide completing the molybdenum coordination sphere. The third family consists of bacterial enzymes that have the general structure $L_2Mo^{VI}OX$, with two equivalents of the bidentate pterin ligand, a Mo = O and a sixth ligand X contributed by the polypeptide, all in a trigonal prismatic coordination sphere. X may be a serine (as in DMSO reductase, Schindelin et al. 1996), a cysteine (the periplasmic NAP dissimilatory reductases from *Desulfovibrio desulfuricans* (Dias et al. 1999) or *E. coli* (Jepson et al. 2007)), selenocysteine (formate dehydrogenase H from *E. coli*) (Boyington et al. 1997), aspartate (the NarGHI nitrate reductase from *E. coli*) (Jornakka et al. 2004), or even hydroxide (arsenite oxidase from *Alcaligenes faecalis*) (Ellis et al. 2001). The disposition of the two pterin ligands in these enzymes, with only a very small dihedral angle between the two enedithiolates of the pterin ligands, is considerably different than is observed in model bis-enedithiolate compounds, where the coordination geometry is octahedral and the dihedral angle between the two pterins is approximately 90° (Lim and Holm 2001).

Both these latter families generally catalyze straightforward oxygen atom transfer reactions that

Molybdenum in Biological Systems, Fig. 2 The biosynthetic pathway for the pyranopterin cofactor common to all mononuclear molybdenum enzymes

involve donation to a substrate lone pair of electrons (or the reverse) (Pietsch and Hall 1996), although enzymes such as polysulfide reductase catalyze even simpler oxidation-reduction reactions. In the case of sulfite oxidase and similar enzymes, the enzyme cycles between $LMo^{VI}O_2(S\text{-}cys)$ and $LMo^{IV}O(OH)(S\text{-}cys)$ states, transferring its equatorial Mo=O group to substrate in the course of the reaction. The equatorial Mo=O is regenerated by hydroxide from solvent, which displaces product from the molybdenum coordination sphere and deprotonates upon reoxidation of the metal. The reaction goes in the reverse direction in the case of the assimilatory nitrate reductases, a fact that has been justified on thermodynamic grounds (Donahue and Holm 1993). For DMSO reductase and related enzymes, the reaction cycles between $L_2Mo^{VI}OX$ and $L_2Mo^{IV}X$ states, with the reduced form having a square-pyramidal coordination geometry that contrasts with the trigonal prismatic geometry seen with the oxidized enzyme (Schindelin et al. 1996). By contrast, enzymes of the xanthine oxidoreductase family catalyze more complicated reactions that typically entail cleavage of a C-H bond. In the case of xanthine oxidoreductase itself, the reaction has been shown to involve base-assisted proton abstraction from the equatorial Mo-OH group (by a universally conserved active site glutamate residue), followed by nucleophilic attack on the carbon center to be hydroxylated and hydride transfer to the $Mo^{VI}=S$. This chemistry yields an $LMo^{IV}O(SH)(OR)$ intermediate, with the molybdenum formally reduced to the IV state and product coordinated to the metal by the catalytically introduced hydroxyl group. The catalytic cycle is completed by oxidation of the molybdenum center and displacement of product by hydroxide from solvent. Depending on the reaction conditions, a paramagnetic $Mo^VOS(OR)$ species may be generated that elicits the well-studied "very rapid" EPR signal (Hille 2002, 2005).

With few exceptions (e.g., the DMSO reductase from *Rhodobacter* species and the higher plant sulfite oxidases), molybdenum-containing enzymes possess redox-active centers in addition to the molybdenum center. Nitrogenase, for example, has the Fe_8S_7 P cluster (Kim and Rees 1992) that is capable of accepting multiple reducing equivalents (Fig. 1). Most members of the xanthine oxidase family of mononuclear enzymes possess two [2Fe-2S] clusters of the spinach ferredoxin variety and flavin adenine dinucleotide (FAD) (Enroth et al. 2000) (although the well-studied aldehyde oxidoreductase from *Desulfovibrio gigas* lacks the FAD (Huber et al. 1996)). Vertebrate sulfite oxidases have a b-type cytochrome in addition the molybdenum (Kisker et al. 1997), and the plant nitrate reductases possess a cytochrome and FAD as well (Campbell 2001). Members of the DMSO family have by far the most diverse make-up (Hille 1996), with many possessing a [3Fe-4S] or [4Fe-4S] cluster in the same polypeptide as the molybdenum center. In addition, a considerable number possess one or more additional subunits with [4Fe-4S] clusters and/or hemes; many of these last enzymes are peripheral or integral membrane proteins that draw on or contribute to the reducing equivalents of the membrane-integral quinone pool. Invariably in these more complex enzymes, the molybdenum center participates directly in only one half of the full catalytic cycle (usually the reductive) and simply obtaining reducing equivalents obtained for reduction of substrate from other redox-active centers of the enzyme.

Tissue Distribution of Molybdenum Enzymes and Their Role in Human Health and Disease

The human genome encodes four molybdenum-containing enzymes: xanthine oxidoreductase, sulfite oxidase, aldehyde oxidase and a recently discovered protein that is a component of the mitochondrial amidoxime-reducing system designated mARC (Havemayer et al. 2006). As part of the purine degradative pathway, xanthine oxidoreductase is present at detectable levels in almost all tissues but is particularly prevalent in the liver, kidney, lung, and intestinal mucosa. Aldehyde oxidase is found predominantly in the liver, while sulfite oxidase levels are highest in the liver, lung, and heart.

There are two fundamental types of genetic lesion that result in loss of molybdenum enzyme activity: (1) mutations in the structural genes encoding the enzymes that result in structural instability or loss of activity and (2) mutations in one or another of the enzymes involved in the biosynthetic pathway for the pterin cofactor that result in loss of the ability to synthesize the cofactor. The latter are the most serious since all four of the molybdenum-containing enzymes lose activity; individuals born with a homozygous defect rapidly develop profound difficulties principally associated with dysfunction of the central nervous system and usually die within the first 2 months of life (Johnson and Duran 2001).

Another pleiotropic mutation that may occur is in the proteins responsible for insertion of the catalytically essential Mo=S group of the molybdenum centers of xanthine oxidoreductase and aldehyde oxidase. Sulfite oxidase and mARC, which do not require this sulfur, are not affected. As with mutations in the structural genes for xanthine oxidoreductase and aldehyde oxidase, these mutations are frequently asymptomatic, although individuals do present with xanthinuria, i.e., elevated serum concentrations of xanthine (reflecting the loss of xanthine oxidoreductase activity, specifically). Symptoms may include xanthine stones in the urinary tract (particularly the kidney), and very infrequently individuals manifest a predisposition toward duodenal ulcers. Mutations in the structural gene for sulfite oxidase that lead to loss of activity yield symptoms similar to those observed when pterin cofactor biosynthesis is compromised. A number of point mutations have been identified clinically and surprisingly, from a consideration of the protein's known structure, most of these do not map to the active site molybdenum center of the enzyme but rather to the dimerization domain of the protein, suggesting that protein stability is compromised (Kisker et al. 1997). Given that the principal tissues affected by loss of sulfite oxidase activity are in the central nervous system, where membranes of, e.g., the myelin sheath have high concentrations of sulfatides and related sulfur-containing compounds, it seems likely that the symptoms reflect a dysfunction in lipid metabolism rather than protein metabolism (involving cysteine and methionine degradation).

Two additional aspects of xanthine oxidoreductase specifically are of clinical relevance. The first of these has to do with the fact that while the enzyme is normally present in tissues as a dehydrogenase that utilizes NAD^+ as the acceptor of reducing equivalents that enter the enzyme in the course of the reductive half-reaction, under certain pathophysiological conditions (e.g., ischemia-reperfusion injury), a conformational change occurs in the protein's FAD domain, induced by limited proteolysis or disulfide bond formation, that results in utilization of O_2 rather than NAD^+ with resultant generation of superoxide ion and peroxide upon reoxygenation of the affected tissue (Enroth et al. 2000). These reactive oxygen species have been implicated in the collateral tissue damage associated with stroke and cardiac infarction (see Hille and Nishino 1995 for a discussion).

Inhibitors of xanthine oxidoreductase are also clinically important for two specific reasons. First, they are effective in the treatment of gout and gouty arthritis by reducing the high plasma uric acid levels that are the root cause of the symptoms. Drugs such as Zyloprim (allopurinol) (Elion et al. 1966; Massey et al. 1970) and the newly FDA-approved Uloric (febuxostat) (Komoriya et al. 2004; Takano et al. 2005; Becker et al. 2005) have proven highly effective in the treatment of these conditions. Second, given that xanthine oxidoreductase frequently acts on and inactivates a variety of heterocyclic drugs – including a range of chemotherapeutic agents – its inhibition in tandem drug therapies permits use of significantly lower doses of the primary drug. Indeed, the 1985 Nobel Prize for Physiology or Medicine was awarded to Gertrude Elion and George Hitchings for their use of allopurinol in a tandem drug therapy with the first-generation chemotherapeutic agent 6-mercaptopurine.

Cross-References

▶ CO-dehydrogenase/Acetyl-CoA Synthase
▶ Molybdenum and Ions in Living Systems
▶ Molybdenum and Ions in Prokaryotes
▶ Molybdenum-enzymes, MOSC Family
▶ Tungsten in Biological Systems

References

Ataya FS, Witte CP, Galván A, Igeño MI, Fernández E (2003) *Mcp1* encodes the molybdenum cofactor carrier protein in *Chlamydomonas reinhardtii* and participates in protection, binding, and storage functions of the cofactor. J Biol Chem 278:10885–10890

Becker MA, Schumacher HR, Wortmann RL, MacDonald PD, Eustace D, Palo WA, Vernillet L, Joseph-Bridge N (2005) Febuxostat compared with allopurinol in patients with hyperuricemia and gout. N Engl J Med 353:2450–2461

Berg JM, Holm RH (1985) A model for the active sites of oxo-transfer molybdoenzymes: reactivity, kinetics and catalysis. J Am Chem Soc 107:925–932

Boyington JC, Gladyshev VN, Khangulaov SV, Stadtman TC, Sun PD (1997) Crystal structure of formate dehydrogenase H: catalysis involving Mo, molybdopterin, selenocysteine and an Fe_4S_4 cluster. Science 275:1305–1308

Campbell WH (2001) Structure and function of eukaryotic NAD (P)H:nitrate reductase. Cell Mol Life Sci 58:194–204

Dias JM, Than ME, Humm A, Huber R, Bourenkov GP, Bartunik HD, Bursakov S, Calvete J, Caldeira J, Carneiro C, Moura JJG, Moura I, Romao MJ (1999) Crystal structure of the first dissimilatory nitrate reductase as 1.9 Å solved by MAD methods. Structure 7:65–79

Dobbek H, Gremer L, Kiefersauer R, Huber R, Meyer O (2002) Catalysis at a binuclear [CuSMoO(OH)] cluster in a CO dehydrogenase resolved at 1.1-Å resolution. Proc Natl Acad Sci USA 99:15971–15976

Doonan CJ, Rubie ND, Peariso K, Harris HH, Knottenbelt SZ, George GN, Young CG, Kirk ML (2008) Electronic structure description of the *cis*-MoOS unit in models for molybdenum hydroxylases. J Am Chem Soc 130:55–65

Elion GB, Kovensky A, Hitchings GH, Metz E, Rundles RW (1966) Metabolic studies of allopurinol as an inhibitor of xanthine oxidase. Biochem Pharmacol 15:863–880

Ellis P, Conrads T, Hille R, Kuhn P (2001) Crystal structure of the 100 kDa arsenite oxidase from *Alcaligenes faecalis* in two crystal forms at 1.64 and 2.03 Å. Structure 9:125–132

Enemark JH, Cooney JJA, Wang J-J, Holm RH (2004) Synthetic analogues and reaction systems relevant to the molybdenum and tungsten oxotreansferases. Chem Rev 104:1175–1200

Enroth C, Eger BT, Okamoto K, Nishino T, Nishno T, Pai EF (2000) Crystal structures of bovine milk xanthine dehydrogenase and xanthine oxidase: structure-based mechanism of conversion. Proc Natl Acad Sci USA 97:10723–10728

Fischer K, Llamas A, Tejada-Jiminez M, Schrader N, Kuper J, Ataya FS, Galván A, Mendel RR, Fernandez E, Schwarz G (2006) Function and structure of the molybdenum cofactor carrier protein from *Chlamydomonas reinhardtii*. J Biol Chem 281:30186–30194

Havemeyer A, Bittner F, Wollers S, Mendel R, Kunze T, Clement B (2006) Identification of the missing component in the mitochondrial benzamidoxime prodrug-converting system as a novel molybdenum enzyme. J Biol Chem 281:34796–34802

Hille R (1996) The mononuclear molybdenum enzymes. Chem Rev 96:2757–2816

Hille R (2002) Molybdenum and Tungsten in biology. Trends Biochem Sci 27:360–367

Hille R (2005) Molybdenum-containing hydroxylases. Arch Biochem Biophys 433:107–116

Hille R, Nishino T (1995) Xanthine oxidase and xanthine dehydrogenase. FASEB J 9:995–1003

Hollenstein K, Frei DC, Locher KP (2007) Structure of an ABC transporter in complex with its binding protein. Nature 466:213–216

Holm R, Donahue JP (1993) A thermodynamic scale for oxygen atom transfer reactions. Polyhedron 12:571–589

Hu Y, Fay AW, Lee C, Yoshizawa J, Ribbe MW (2008) Assembly of the MoFe protein. Biochemistry 47:3971–3981

Huber R, Hof P, Duarte RO, Moura JJG, Moura I, LeGall J, Hille R, Archer M, Romão M (1996) A structure-based catalytic mechanism for the xanthine oxidase family of molybdenum enzymes. Proc Natl Acad Sci USA 93:8846–8851

Jepson BJN, Mohan S, Clarke TA, Ggates AJ, Cole JA, Butler CS, Butt JN, Hemmings AM, Richardson DJ (2007) Spectropotentiometric and structural analysis of the periplasmic nitrate reductase from *Escherichia coli*. J Biol Chem 282:6425–6437

Johnson JL, Duran M (2001) Molybdenum cofactor deficiency and isolated sulfite oxidase deficiency. In: Scriver RC, Beaudet AL, Sly WS, Valle D, Childs B, Vogelstein B (eds)

The metabolic basis of inherited disease, vol 1, 6th edn. McGraw-Hill, New York, pp 1463–1475

Jornakka M, Richardson D, Byrne B, Iwata S (2004) Architecture of NarGH reveals a structural classification of Mo-*bis*BGD enzymes. Structure 12:95–104

Kim J, Rees DC (1992) Crystallographic structure and functional implications of the nitrogenase molybdenum-iron protein from *Azotobacter vinelandii*. Nature 360:553–560

Kisker C, Schindelin H, Pacheco A, Wehbi WA, Garrett RM, Rajagopalan KV, Enemark JH, Rees, DC (1997) Molecular basis of sulfite oxidase deficiency from the structure of sulfite oxidase. Cell 91:973–983

Komoriya K, Hoshidea S, Takeda K, Kobayashia H, Kuboa J, Tsuchimotoa M, Nakachib T, Yamanakac H, Kamatanic N (2004) Pharmacokinetics and pharmacodynamics of Febuxostat (TMX-67), a non-purine selective inhibitor of xanthine oxidase/xanthine dehydrogenase(NPSIXO) in patients with gout and/or hyperuricemia. Nucleos Nucleot Nucl Acids 23:1119–1122

Kuper J, Llamas A, Hecht H-J, Mendel RR, Schwarz G (2004) Structure of the molybdopterin-bound Cnx1G domain links molybdenum and copper metabolism. Nature 430:803–806

Lancaster KM, Roemelt M, Ettenhuber P, HU YL, Ribbe MW, Neese F, Bergmann U, DeBeer S (2011) X-ray emission spectroscopy evidences a central carbon in the nitrogenase iron-molybdenum cofactor. Science 334:974–977

Lim BS, Holm RH (2001) Bis(dithiolene)molybdenum analogues relevant to the DMSO reductase enzyme family: synthesis, structures and oxygen atom transfer reaction and kinetics. J Am Chem Soc 123:1920–1930

Llamas A, Mendel RR, Schwarz G (2004) Synthesis of adenylylated molybdopterin. J Biol Chem 279:55241–55246

Massey V, Edmondson DE (1970) On the mechanism of inactivation of xanthine oxidase by cyanide. J Biol Chem 245:6595–6598

Massey V, Komai H, Palmer G, Elion GB (1970) On the mechanism of inhibition of xanthine oxidase by allopurinol and other pyrazolo[3,4-*d*]pyrimidines. J Biol Chem 245: 2837–2844

Neumann M, Stöcklein W, Walburger A, Magalon A, Leimkühler S (2007) Identification of a *Rhodobacter capsulatus* L-cysteine desulfurase that sulfurates the molybdenum cofactor when bound to XdhC and before its insertion into xanthine dehydrogenase. Biochemistry 46:9586–9595

Pietsch MA, Hall MB (1996) Theoretical studies on models for the oxo-transfer reaction of dioxomolybdenum enzymes. Inorg Chem 35:1273–1278

Roberts SA, Young CG, Kipkee CA, Cleland WE Jr, Yamanouchi K, Carducci MD, Enemark JH (1988) Dioxomolybdenum[VI] complexes of the hydrotris (3,5-dimethyl-1-pyrazolyl)borate ligand. Synthesis and oxygen atom transfer reactions. Inorg Chem 29:3650–3656

Schindelin H, Kisker C, Hilton J, Rajagopalan KV, Rees DC (1996) Crystal structure of DMSO reductase: redox-linked changes in molybdopterin coordination. Science 272: 1615–1621

Schumann SM, Möller N, Anker SD, Lendzian F, Hildebrandt P, Leimkühler S (2008) The mechanism of assembly and cofactor insertion into *Rhodobacter capsulatus* xanthine dehydrogenase. J Biol Chem 283:16602–16611

Schwarz G, Mendel RR (2006) Molybdenum cofactor biosynthesis and molybdenum enzymes. Annu Rev Plant Biol 57:623–647

Takano Y, Hase-Aoki K, Horiuchi H, Zhao L, Kasahara Y, Kondo S, Becker MA (2005) Selectivity of febuxostat, a novel non-purine inhibitor of xanthine oxidase/xanthine dehydrogenase. Life Sci 76:1835–1847

Wagener N, Pierik A, Ibdah A, Hille R, Dobbek H (2009) The Mo-Se active site of nicotinate dehydrogenase. Proc Natl Acad Sci 106:11055–11060

Molybdenum, Physical and Chemical Properties

Fathi Habashi
Department of Mining, Metallurgical, and Materials Engineering, Laval University, Quebec City, Canada

Molybdenum is a lustrous silver-white solid with typically metallic properties. It is a transition metal; hence it is less reactive than the typical metals and more reactive than the less typical metals. It is a refractory metal used in lamp and lighting industries, in electronic and semiconductor industries, in high-temperature and vacuum furnace construction, in glass and ceramic industries, and in nuclear technology. Molybdenum-containing catalysts are used for a broad range of reactions. Molybdenum disulfide is an important solid lubricant, being used to reduce wear, friction, and sustain lubrication under boundary sliding conditions. Molybdenum is an essential trace element for enzymes which fix nitrogen in leguminous crops. Molybdenum compounds in general are of a low order of toxicity.

Molybdenum is extracted as a by-product of copper production from porphyry ores and such ores are the major source of rhenium which is associated with molybdenite, MoO_2. Molybdenite in quartz veins does not contain rhenium.

Physical Properties

Atomic number	42
Atomic weight	95.94
Relative abundance in Earth's crust, %	2.3×10^{-4}
Density at 20°C, g/cm^3	10.22

(continued)

Crystal structure	Body centered cubic
Atomic radius, pm	139
Melting point, °C	2,617
Boiling point, °C	4,612
Latent heat of fusion at melting point, kJ/mol	35.6
Mean specific heat (0–100°C), J kg^{-1} K^{-1}	251
Thermal conductivity (0–100°C), W m^{-1} K^{-1}	137
Electrical resistivity at 20°C, μΩ·cm	5.7
Temperature coefficient (0–100°C), K^{-1}	4.35×10^{-3}
Young's modulus, GPa	324.8
Rigidity modulus, GPa	125.6
Bulk modulus, GPa	261.2
Poisson's ratio	0.293
Linear coefficient of thermal expansion (0–100°C), K^{-1}	5.1×10^{-6}
Standard electrode potential, V	−0.200

Chemical Properties

Molybdenum is produced by the reduction of high purity MoO_3 with hydrogen. The metal retains its luster almost indefinitely in air. On prolonged heating in air below 600°C, it becomes covered with its trioxide; at 600°C the oxide sublimes and rapid oxidation occurs. Molybdenum burns in oxygen at 500–600°C. It is attacked by fluorine when cold and by chlorine and bromine when hot. Dilute acids and concentrated hydrochloric acid have very little effect on the metal. Moderately concentrated nitric acid dissolves it but concentrated nitric acid soon passivates the surface and reaction ceases.

Molybdenum is dissolved by a mixture of concentrated nitric and concentrated hydrofluoric acids. It is practically unaffected by alkaline solutions and very nearly so by fused alkali-metal hydroxides. However, fused oxidizing salts such as sodium peroxide, sodium or potassium nitrate or perchlorate dissolve the metal rapidly. It reacts on heating with carbon, boron, nitrogen, and silicon and forms many alloys.

As a transition metal, molybdenum may have a valency of +2, +3, +4, +5, or +6. It also forms many complexes and colored compounds. Its aqueous chemistry is complicated. It is dominated by oxo-species which are prone to dimerize or polymerize. Typical compounds of molybdenum(VI) are: the trioxide, MoO_3, sodium molybdate, Na_2MoO_4, ammonium heptamolybdate, $(NH_4)_6Mo_7O_{24} \cdot 4H_2O$, and $(NH_4)_6Mo_7O_{24} \cdot 2H_2O$. Well-characterized oxides are molybdenum trioxide, MoO_3, and molybdenum dioxide, MoO_2. Molybdenum blues, so called because of their blue color, are mixed oxide hydroxides of molybdenum (VI) and molybdenum (V). They are amorphous solids that are soluble in water and alcohols. A typical species is $(Mo_3^{6+}Mo_3^{5+}O_{18}H)^-$.

Molybdenite, MoS_2, is the major molybdenum mineral but molybdenum trisulfide, MoS_3, is precipitated when hydrogen sulfide is passed through an acidified molybdate solution. It is like the sulfides of tin, arsenic, and antimony, dissolves in yellow ammonium sulfide giving a tetrathio salt, MoS_4^{2-}. When heated in nitrogen at ca. 350°C, molybdenum trisulfide decomposes to amorphous molybdenum disulfide. Heated in air, it begins to oxidize to molybdenum trioxide at 200°C.

References

Sebenik R et al (1997) Molybdenum. In: Habashi F (ed) Handbook of extractive metallurgy. Wiley-VCH, Weinheim, pp 1361–1402

Molybdenum-Containing Enzymes

▶ Molybdenum in Biological Systems

Molybdenum-enzymes, MOSC Family

Florian Bittner
Department of Plant Biology, Braunschweig
University of Technology, Braunschweig, Germany

Synonyms

Abscisic acid deficient 3 (ABA3): Low osmotic stress 5 (LOS5), Sirtinol resistant 3 (SIR3), Freezing sensitive 1 (FRS1), Altered chloroplast import 2 (ACI2); Molybdenum cofactor sulfurase C-terminal domain-containing 1/2 (MOSC-1/MOSC-2): mitochondrial amidoxime-reducing component 1/2 (mARC-1/mARC-2), Candidate diabetes associated kidney gene 7 (CDK7)

Definition

Molybdenum cofactor sulfurase C-terminal (MOSC) domain proteins form a small protein family in eukaryotes and prokaryotes with its members being characterized by the requirement of molybdenum cofactor and the presence of a strictly conserved cysteine residue that appears to be essential for activity of these enzymes. While the physiological function of the name-giving member of this protein family, molybdenum cofactor sulfurase, is to activate molybdenum enzymes of the xanthine oxidase family by introducing L-cysteine-derived sulfur into their molybdenum center, the physiological role of all other MOSC domain proteins is as yet largely unknown.

History of Molybdenum Cofactor (Moco) Sulfurases, the Name-Giving Members of the MOSC Domain Family

Molybdenum enzymes can generally be divided into three subfamilies: the dimethylsulfoxide reductase family, which is exclusive to prokaryotes, and the sulfite oxidase and the xanthine oxidase families that both have representatives in eukaryotes as well as in prokaryotes (Hille 1996). As a major feature, enzymes of the xanthine oxidase family such as the most prominent members aldehyde oxidase and xanthine oxidoreductase contain a specific sulfur atom that can be abstracted from the enzyme by cyanide treatment in the form of thiocyanate, which leads to inactivation of the respective enzyme (Hille et al. 2011). Yet, cyanide-inactivated forms of aldehyde oxidase and xanthine oxidoreductase can be reactivated in vitro by sulfide treatment under strongly reducing conditions, and this demonstrated that the cyanolizable sulfur is a ligand of the molybdenum metal in the Moco of aldehyde oxidase and xanthine oxidoreductase. However, how this sulfur is incorporated into the Moco of xanthine oxidase family enzymes in vivo remained unknown until the detailed characterization of the *maroon-like* (*ma-l*) mutant from *Drosophila melanogaster*. Due to the reduced ability to form the red eye pigment drosopterin, these mutant flies are characterized by a maroon-like eye color caused by the loss of xanthine oxidoreductase activity. In addition to xanthine oxidoreductase, the activities of aldehyde oxidase and another xanthine oxidase family enzyme, pyridoxal oxidase, were absent in *ma-l* mutants, which were ascribed to the requirement of a terminal sulfur ligand of the Moco of these proteins. Since sulfite oxidase was unaffected in *ma-l* flies, it was concluded this enzyme belongs to a different family of molybdenum enzymes that does not require a terminal sulfur ligand. By use of the *Neurospora crassa* mutant *nitrate nonutilizer-1* (*nit-1*), which carries a mutational block in Moco biosynthesis but contains the apoform of the molybdenum enzyme nitrate reductase, extracts of *ma-l* flies have been shown to contain wild-type levels of Moco as they were capable of fully reconstituting the *N. crassa* apo-nitrate reductase. Based on this experiment, the possibility of the *ma-l* mutation affecting the biosynthetical pathway of Moco was excluded and instead, the mutation was postulated to affect a posttranslational sulfuration that is essential for the activity of xanthine oxidoreductase, aldehyde oxidase, and pyridoxal oxidase, but not of sulfite oxidase. Since the lost molybdenum enzyme activities were reconstituted in extracts of *ma-l* flies by sulfide/dithionite treatment, it was confirmed that the *ma-l* mutation indeed affected the incorporation of sulfur into the Moco of these enzymes, thereby demonstrating that the activation of xanthine oxidase family enzymes is genetically controlled.

A phenotype similar to *ma-l* mutant flies is exhibited by the *Aspergillus nidulans* mutant *hypoxanthine nonutilizer gene B* (*hxB*), which is unable to grow on hypoxanthine or nicotinate as sole nitrogen source due to the simultaneous loss of xanthine oxidoreductase and nicotinate hydroxylase activities. In contrast to these enzymes, nitrate reductase is not affected in *hxB* mutants as its molybdenum center, like that of sulfite oxidase, does not require a terminal sulfur ligand. Thus, the mutation in *A. nidulans hxB* mutants is highly similar to the mutation in the fruit fly *ma-l* mutants as it affects the enzymes of the xanthine oxidase family, whereas the enzymes of the sulfite oxidase family (nitrate reductase and sulfite oxidase) are unaffected. According to their proposed functions, the proteins encoded by the *ma-l* locus in fruit fly and the *hxB* locus in *A. nidulans* were designated as Moco sulfurases. Cloning and computational analysis of the *ma-l* and *hxB* cDNAs revealed that eukaryotic Moco sulfurases consist of two domains, an NH_2-terminal domain with similarities to bacterial cysteine desulfurases and a C-terminal domain without significant sequence similarities to functionally described proteins. The function of the C-terminal domain therefore remained unknown and was assumed to be limited to the recognition of the Moco sulfurase target enzymes. Interestingly, Moco sulfurase enzymes consisting of these two domains appeared to occur exclusively in eukaryotes but not in prokaryotes.

Deficiencies ascribed to a loss of function of the Moco sulfurase gene have been described later also in cattle, humans, silkworm, tomato, tobacco, and *Arabidopsis thaliana*. As known from the fruit fly *ma-l* mutants and the *A. nidulans hxB* mutants, the main features of these mutants are reduced or abolished activities of xanthine oxidase family enzymes and unaffected activities of sulfite oxidase family enzymes. Although the biochemical phenotype is basically indistinguishable, the pathology is significantly different in animals and plants. Moco sulfurase deficiency in animals, in particular mammals, is referred to as xanthinuria type II, an autosomal-recessive disorder mainly caused by the loss of xanthine oxidoreductase activity. Affected individuals are characterized by reduced levels of uric acid in serum and urine with concomitant accumulation of xanthine and hypoxanthine. While in humans, xanthinuria type II only rarely leads to xanthine stones in the kidney, urinary tract infection, and acute renal failure, it has been observed to result in early death in other mammals such as cattle, most likely due to a lower solubility of accumulating xanthine and hypoxanthine. Up to now, no symptoms have been reported as yet that can be ascribed to the loss of aldehyde oxidase activity and, thus, the symptoms of xanthinuria type II resemble that of xanthinuria type I, which is caused by a mutation in the xanthine oxidoreductase structural gene. In contrast to animals, Moco sulfurase deficiency in plants is primarily caused by impaired aldehyde oxidase activity and is characterized by aberrant stomatal control, excessive water loss, a wilty phenotype, reduced seed dormancy, and reduced tolerance to a number of abiotic stresses such as drought stress, cold stress, and salinity (Bittner and Mendel 2010). All symptoms are caused by reduced levels of the plant hormone abscisic acid, which is produced by a specific aldehyde oxidase through the oxidation of abscisic aldehyde.

At the present time, detailed biochemical information on Moco sulfurases is available only for the ABA3 protein from *A. thaliana*. As typical for eukaryotic Moco sulfurases, each ABA3 monomer of approximately 92 kDa can be subdivided into two domains: the NH_2-terminal domain of approximately 56 kDa shares significant similarities to the bacterial cysteine desulfurases SufS, NifS, and IscS, which provide sulfur for a number of cellular processes such as iron-sulfur cluster biogenesis, thiamin synthesis, lipoic acid synthesis, and thio modification of tRNAs (Kessler 2006). As typical for all hitherto described cysteine desulfurases, the NH_2-terminal domain of ABA3 (also referred to as ABA3-NifS) possesses a pyridoxal phosphate cofactor, which is crucial for the L-cysteine desulfurase activity of the protein, as well as a conserved cysteine residue to which the abstracted sulfur is bound in a form of persulfide. Recent experiments suggest that the persulfide is only transiently bound to this conserved cysteine and subsequently is subject of intramolecular transfer from the NH_2-terminal domain to the 35 kDa C-terminal domain, which has been found to bind Moco in its sulfur-free as well as in its sulfurated form (Hille et al. 2011). Based on these observations, distinct functions can be ascribed to the two domains of ABA3 (Fig. 1): the NH_2-terminal NifS-like domain catalyzes the decomposition of L-cysteine, thereby converting the sulfur into a form that can be stored and transferred within the Moco sulfurase protein. The C-terminal domain acts as an acceptor of the NH_2-terminal domain-derived sulfur, binds the desulfo form of Moco, and inserts the sulfur into the ABA3-bound Moco. However, the mechanism by which ABA3 finally activates its target enzymes xanthine oxidoreductase and aldehyde oxidase is presently unclear. Two possibilities are discussed, the first involving the transfer only of the terminal sulfur that is bound to the Moco of the C-terminal domain, the second involving the transfer of the entire sulfurated Moco from ABA3 to xanthine oxidoreductase and aldehyde oxidase. In the first case, the C-terminal domain would act as a scaffold for the assembly of a molybdenum-sulfur center, from which the sulfur is released and passed to the desulfo Moco of xanthine oxidoreductase and aldehyde oxidase. Subsequently, the ABA3-bound Moco would again be in the desulfo form and could accept a new sulfur from the NH_2-terminal domain. In the second case, the reaction catalyzed by ABA3 would be an exchange reaction in which desulfo Moco of xanthine oxidoreductase and aldehyde oxidase is replaced by sulfurated Moco from the C-terminal domain. The desulfo Moco that has been extracted from the target enzymes and that has subsequently been bound to the

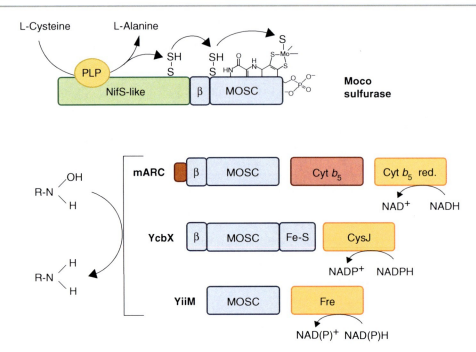

Molybdenum-enzymes, MOSC Family, Fig. 1 Domain structures of known MOSC domain proteins and their respective redox partners. The name-giving Moco sulfurase consists of the NH$_2$-terminal NifS-like cysteine desulfurase domain that abstracts sulfur from L-cysteine in a pyridoxal phosphate (PLP)-dependent manner. The sulfur is transferred within the Moco sulfurase protein via a chain of cysteine residues and is finally bound to the C-terminally bound Moco. Eukaryotic mARC proteins with an NH$_2$-terminal organellar targeting peptide, *E. coli* YcbX and YiiM as well as their respective redox partners with their basic structures are shown (β β-barrel subdomain, Fe–S iron-sulfur cluster, Cyt b_5 heme/cytochrome b_5, Cyt b_5 red FAD/NADH cytochrome b_5 reductase)

Moco sulfurase ABA3 could be regenerated with sulfur provided by the NH$_2$-terminal domain. The observation that the affected proteins in Moco sulfurase-deficient mutants have wild-type levels of Moco and exclusively lack the terminal sulfur ligand neither favors nor precludes any of the two alternatives. Rather, besides the more apparent first alternative, also the second alternative must be seriously considered since on the one hand, ABA3-bound Moco can simply be removed from the protein without disturbing the integrity of the protein, and on the other hand, Moco is bound by cofactor-free ABA3 with high affinity when provided exogenously. This suggests that the Moco is bound to the surface of the C-terminal domain of ABA3 rather than in a deeply buried binding pocket as is found in the "true" enzymes of the xanthine oxidase family. These properties may allow ABA3 to transfer its bound sulfurated Moco to xanthine oxidoreductase and aldehyde oxidase and to take over the desulfo Moco from these enzymes subsequently.

Other MOSC Domain Proteins in Eukaryotes

Besides the name-giving Moco sulfurases, other Moco sulfurase C-terminal (MOSC) domain proteins are encoded by the genomes of most eukaryotes (Hille et al. 2011), except those that do not have molybdenum metabolism including several unicellular eukaryotes like parasites and most yeasts. A basic feature of eukaryotic MOSC domain proteins is that it shares a certain degree of sequence similarity with the C-terminal domain of eukaryotic Moco sulfurases, albeit the MOSC classification originally refers to the presence of a conserved cysteine residue rather than to sequence similarities or the ability to bind Moco (Anantharaman and Aravind 2002). First evidence for these MOSC domain proteins representing a new subfamily of molybdenum enzymes came from the purification of a MOSC-2 protein from the outer membrane of pig liver mitochondria (Hille et al. 2011; Havemeyer et al. 2011). While its physiological function is as yet unknown, it has been demonstrated to be the central part of a three-component system that catalyzes the activation of N-hydroxylated prodrugs and subsrates such as amidoximes, N-hydroxy-sulfonamides, and N-hydroxy-guanidines. As such it acts in concert with heme-containing cytochrome b_5 and FAD-containing cytochrome b_5 reductase, which deliver electrons derived from NADH that go from the FAD and likely through the heme to the Moco of the MOSC protein that provides the substrate-binding site (Fig. 1). Interestingly, this electron transport chain resembles that of eukaryotic nitrate reductase, despite the fact that it consists of separate proteins while nitrate reductase combines all cofactors in a single polypeptide chain. Based on its subcellular localization in mitochondria, its ability to reduce amidoximes (i.e., N-hydroxylated amidines), and the fact that it represents part of three-component system, the newly identified MOSC protein was designated *mitochondrial amidoxime-reducing component*, *mARC*.

After cloning of the two MOSC genes that are encoded by the human genome and establishment of a recombinant human system that contains all three components of the prodrug-converting system, several N-hydroxylated compounds were tested as putative substrates of physiological relevance. For instance, the N-hydroxylated base analogue N-hydroxy-cytosine is reduced to cytosine with high efficiency by human MOSC/mARC proteins, providing a mechanism for detoxification of these base analogues which otherwise would be misincorporated into DNA and cause accumulation of mutations. The ability of human mARC proteins to reduce the nitric oxide precursor N^4-hydroxy-L-arginine in vitro suggests another physiological role of mARC proteins. Nitric oxide is a physiological mediator with diverse functions in maintenance of vascular homeostasis, neuronal signaling, and inhibition of tumor cell growth. However, uncontrolled production of nitric oxide can lead to cellular damage and a number of serious conditions. Tight control of nitric oxide synthesis by nitric oxide synthases, which catalyze the oxidation of L-arginine to L-citrulline and nitric oxide via the intermediate N^4-hydroxy-L-arginine, is therefore crucial for the cell. In this respect, MOSC/mARC proteins may have an important function in the regulation of intracellular nitric oxide levels by controlled elimination of the nitric oxide precursor N^4-hydroxy-L-arginine to prevent excess formation of nitric oxide.

As a structural feature, eukaryotic MOSC/mARC proteins not only harbor the Moco-binding MOSC domain but also a NH_2-terminal β-barrel subdomain that has been discussed to be involved in substrate interaction as well as membrane association of these proteins. Moreover, most MOSC/mARC proteins, except Moco sulfurases and several MOSC/mARC proteins from plants, carry another extension on their extreme NH-terminal end that is likely to be required for mitochondrial import. Consistent with this, mouse MOSC-1/mARC-1 and MOSC-2/mARC-2 have been shown to be most abundant in liver and kidney mitochondria. Although the enzyme from pig liver has originally been isolated from the outer mitochondrial membrane, a submitochondrial localization of mouse MOSC/mARC proteins in the inner membrane, as well as a localization in peroxisomal membranes, has been demonstrated. MOSC/mARC proteins represent the simplest eukaryotic molybdenum enzymes in that they have the lowest molecular weight (average of 35 kDa) and consist of only a single distinct MOSC domain that binds Moco as the only prosthetic group. The present knowledge on MOSC/mARC proteins suggests that they are not active as stand-alone proteins, but rather act along with other redox-active proteins such as heme/cytochrome b_5 and the FAD-binding NADH-cytochrome b_5 reductase. Yet, the existence of other physiological redox partners must be considered since certain MOSC domain proteins may be located in other compartments than mitochondria.

Another important aspect concerns the structure of the molybdenum center in eukaryotic MOSC/mARC proteins and probably all proteins of the MOSC domain superfamily. In contrast to the molybdenum center of xanthine oxidase family enzymes that besides an apical oxo ligand and an equatorial hydroxyl ligand bind the aforementioned terminal sulfur ligand, the molybdenum center of human MOSC/mARC proteins is not coordinating a terminal sulfur. This has been demonstrated on the one hand by cyanide treatment which neither released sulfur in the form of thiocyanate nor significantly affected the activities of the human MOSC/mARC proteins. Consistent with this, coincubation of apo-MOSC/mARC proteins with a cofactor-loaded Moco carrier protein that binds the cofactor in the sulfur-free form reconstituted the activity of the MOSC/mARC proteins. Both experiments confirmed that human MOSC/mARC proteins do not require a sulfurated Moco in xanthine oxidase family enzymes.

On the other hand, partially reduced human MOSC/mARC proteins developed EPR signals characteristic of the Mo(V) (d^1) state. Both spectra closely resembled the so-called low-pH EPR signal seen with sulfite oxidase family enzymes, which do not have a terminal sulfur ligand but a cysteine-sulfur ligand derived from the protein. While initial studies with all possible cysteine-to-serine variants suggested that cysteine is not involved in coordination of the molybdenum center in human MOSC/mARC proteins, the reinvestigation of these variants undoubtedly has demonstrated that mutation of the MOSC-typical conserved cysteine residue abrogates enzymatic activity and perturbs the EPR spectrum to a significant degree. It was thus concluded that MOSC/mARC proteins indeed provide a cysteine sulfur to the molybdenum center, thereby allowing classification of MOSC/mARC proteins as new members of the sulfite oxidase family.

MOSC Domain Proteins in Prokaryotes

While in eukaryotes all MOSC domain proteins appear to fuse either to a β-barrel subdomain or to an additional NH_2-terminal cysteine desulfurase domain as in case of Moco sulfurases, prokaryotic MOSC domain proteins appear to be more diverse (Anantharaman and Aravind 2002). In *Escherichia coli*, two MOSC domain proteins were described with the YiiM protein representing the simplest case that exclusively consists of a MOSC domain (which, apart from the conserved cysteine, is not significantly similar to the C-terminal domain of Moco sulfurases) and the YcbX protein representing a more complex case with an NH_2-terminal β-barrel subdomain and a C-terminal iron-sulfur cluster-binding domain. Even more complex MOSC domain proteins are found in other prokaryotic genomes such as that of *Vibrio cholerae* that encodes a protein consisting of a β-barrel subdomain, a MOSC domain, a FAD/NADH-binding domain, and a C-terminal iron-sulfur cluster-binding domain (NH_2- to C-terminal end). While the latter protein is likely to function independent from other redox-active proteins, YiiM and YcbX have been shown to rely on specific redox proteins such as the FMN/FAD protein CysJ and the NAD(P)H-flavin oxidoreductase Fre, respectively. As observed for eukaryotic MOSC domain proteins, a preference for *N*-hydroxylated substrates such as the base analogue 6-*N*-hydroxylaminopurine has been demonstrated for the *E. coli* MOSC domain proteins as well (Hille et al. 2011). Both YcbX and YiiM appear to be crucial for the detoxification of *N*-hydroxylated base analogues as mutagenesis of the *ycbX* and *yiiM* genes rendered the bacteria sensitive to exogenously applied 6-*N*-hydroxylaminopurine, with rapid accumulation of mutations and cell death. While it is likely that this activity is of physiological relevance, it remains to be investigated whether or not it represents the only physiological function of YcbX and YiiM. Very recent studies on YcbX and YiiM including EPR spectroscopy and mutagenesis of the conserved MOSC cysteine residue indicated that both proteins share the same type of Moco as seen in their eukaryotic MOSC/mARC counterparts, thereby classifying these enzymes likewise as members of the sulfite oxidase family of molybdenum enzymes.

Cross-References

▶ Molybdenum Cofactor, Biosynthesis and Distribution

References

Anantharaman V, Aravind L (2002) MOSC domains: ancient, predicted sulfur-carrier domains, present in diverse metal-sulfur cluster biosynthesis proteins including molybdenum cofactor sulfurases. FEMS Microbiol Rev 207:55–61

Bittner F, Mendel RR (2010) Cell biology of molybdenum. In: Hell R, Mendel RR (eds) Cell biology of metals and nutrients, Plant cell monographs. Springer, Heidelberg

Havemeyer A, Lang J, Clement B (2011) The fourth mammalian molybdenum enzyme mARC: current state of research. Drug Metab Rev 43(4):524–539

Hille R (1996) The mononuclear molybdenum enzymes. Chem Rev 96:2757–2816

Hille R, Nishino T, Bittner F (2011) Molybdenum enzymes in higher organisms. Coord Chem Rev 255:1179–1205

Kessler D (2006) Enzymatic activation of sulfur for incorporation into biomolecules in prokaryotes. FEMS Microbiol Rev 30:825–840

Molybdoenzyme

▶ Molybdenum and Ions in Living Systems
▶ Molybdenum and Ions in Prokaryotes

Molybdopterin

▶ Molybdenum Cofactor, Biosynthesis and Distribution
▶ Molybdenum and Ions in Living Systems

Monocation/Monovalent Metal Cation

▶ Calcium Ion Selectivity in Biological Systems

Monocopper Blue Proteins

Kyle M. Lancaster[1] and Harry B. Gray[2]
[1]Department of Chemistry and Chemical Biology, Cornell University, Ithaca, NY, USA
[2]Beckman Institute, California Institute of Technology, Pasadena, CA, USA

Synonyms

Blue copper proteins; Type 1 copper proteins

Definition

Monocopper blue proteins (MBPs) are small (< 200 residues) proteins with metal-binding domains accommodating a single type 1 or blue Cu ion. Type 1 Cu arises principally due to a highly covalent, Cu-S interaction between the metal and an endogenous Cys thiolate sidechain. In the oxidized state (Cu^{II}), this interaction gives rise to an intense ligand-to-metal charge-transfer (LMCT) band that gives these proteins their eponymous blue color. Additionally, the Cu-S(Cys) covalency acts in concert with polypeptide-mediated structural restraints to purpose the MBPs for efficient electron transfer (ET). MBPs modulate $Cu^{II/I}$ reduction potentials, reduce nuclear reorganization energy, and enhance donor-acceptor electronic coupling. MBPs are divided into the cupredoxin and phytocyanin families. Cupredoxins are constituents of biological ET chains in bacteria, archaea, and plants. Phytocyanins are plant proteins involved in signal transduction and organismal defense/tissue repair. MBPs have been central to the development of bioinorganic chemistry and molecular biology, as well as to the study of biological ET reactions.

Introduction. The first monocopper blue proteins (MBPs) were discovered in the 1950s by biochemists investigating bacterial respiration. This discovery was facilitated by the proteins' intense blue color (Fig. 1). Because they share similar roles with Fe-containing ferredoxins, these proteins have been generally termed "cupredoxins." The structural and functional diversity of the cupredoxin superfamily has been the subject of a detailed review (Dennison 2005). Cupredoxins are ubiquitous in bacteria and archaea and are widely distributed in plants. They are soluble proteins generally comprising 90 – 150 amino acids in their mature forms. Cupredoxins include the azurin, pseudoazurin, plastocyanin, auracyanin, amicyanin, rusticyanin, halocyanin, and sulfocyanin families. Cupredoxins whose roles have been elucidated are electron transfer (ET) proteins that participate in biochemical processes including photosynthesis, denitrification, and cellular respiration. However, the physiological functions of many cupredoxins remain unknown.

Phytocyanins, which are unique to plant proteomes, represent a distinct class of MBPs from cupredoxins (Hart et al. 2011). Known families of phytocyanins are plantacyanins, mavicyanins, stellacyanins, and uclacyanins. While plantacyanins are single-domain proteins, stellacyanins and uclacyanins are chimeric proteins whose type 1 Cu domains are fused to C-terminal domains resembling glycosylated arabinogalactan proteins (AGP). These AGP domains are typically involved in cellular signaling, and can also participate in repair of tissue damage. Additionally, the AGP domains contain C-terminal hydrophobic signaling tags for modification by glycosylphosphatidylinositol (GPI) for cell-surface anchoring. As such, phytocyanins are implicated in plant signal transduction and defense.

MBP genes, with the exception of those for plastocyanins, encode a $\sim$ 20 amino acid signal peptide at the N-terminus that directs protein localization. Bacterial and archaeal cupredoxins are compartmentalized to the periplasmic space between the plasma membrane and cell wall. Phytocyanins are secreted into the extracellular space where the C-terminal GPI modifications anchor them to the cell surface. Plant plastocyanins

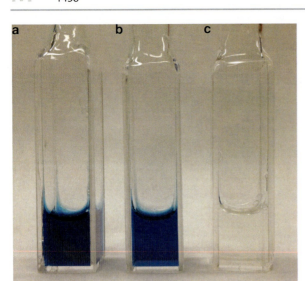

Monocopper Blue Proteins, Fig. 1 *Pseudomonas aeruginosa* azurin in aqueous buffer. (**a**) 2 mM Cu^{II} azurin, (**b**) 1 mM Cu^{II} azurin, (**c**) 1 mM Cu^{I} azurin (Photograph courtesy of Jeffrey J. Warren)

are localized to chloroplasts where they serve as electron carriers in the photosynthetic ET chain.

Molecular Structure. As of January 2012, the structures of over 100 naturally occurring and engineered MBPs have been deposited in the RCSB Protein Data Bank (http://www.rcsb.org). MBP tertiary structure consists of a series of ß-strands organized into a "Greek-key" ß-sandwich or ß-barrel motif (Hart et al. 2011). This tertiary structure is built around a genetically conserved core of hydrophobic residues that define the protein folding mechanism as well as confer high thermodynamic stability. Individual families are distinguished by variant intramolecular interactions between ß-strands, the orientation and number of these ß-strands, and the nature of the polypeptide loops connecting the ß-strands (Fig. 2). Typically, MBP ß-sandwiches consist of 8 ß-strands. However, auracyanin and amicyanin feature 9 ß-strands, while rusticyanin has 13. Additional secondary structure elements may be present that further distinguish MBP families.

The metal-binding sites in MBPs are situated at the extreme ends of the proteins; by convention they are said to be at the "north" end (Fig. 2). MBP metal-binding sites universally comprise two His N-donors and a CysS-donor in a pseudotrigonal arrangement (Fig. 3) (Gray et al. 2000). While the Cu-N(His) distances (1.9 – 2.2 Å) are unremarkable, the Cu-S (Cys) distances are short (2.1 – 2.3 Å). The short Cu-S(Cys) interactions are a hallmark of type 1 copper. The inner-sphere ligands are often complemented by a weakly interacting axial Met thioether at 2.8 – 3.5 Å. Stellacyanins and umecyanins substitute a Gln amide in place of this thioether, with Cu-O(Gln) distances on the order of 2.4 Å. Azurin's metal-binding site contains a backbone carbonyl that contributes a fifth ligand opposite the methionine thioether. The Cu-O(Gly) distance varies from 2.4 to 2.6 Å depending on azurin variant. Tomato plantacyanin has a Val in the axial position and as such only has the minimal N-N-S trigonal coordination sphere.

MBP inner-sphere ligands are situated on the loops connecting ß-strands. The C-terminal His, Cys, and the axial ligand (where present) are found together on one loop with a $CysX_{2-4}HisX_{2-4}$(Met/Gln) consensus sequence. The remaining His and axial carbonyl are present on another loop approximately 40 residues N-terminal with respect to the Cys-His-(Axial) loop. The lengths of these loops are highly variable across MBPs, and they profoundly influence spectroscopic properties, reduction potentials, and ET transfer reactivity (Battistuzzi et al. 2009). Additionally, loops are exchangeable between MBPs resulting in chimeric proteins whose properties are characteristic of the loop donor.

The Cys thiolate in MBPs participates in an outer-sphere hydrogen bonding network (Fig. 3) (Gray et al. 2000; Lancaster 2012). This network has been implicated in inducing the "entatic" or "rack-bound" state that in part tunes blue copper for efficient ET reactivity. The hydrogen bonds enforce a ~ 0° dihedral angle on the Cys sidechain that is conserved across MBPs. Additionally, sterics have been demonstrated to constrain equatorial His residues in some MBPs. In cases where both Cu^{II} and Cu^{I} forms of an MBP have been structurally characterized, the active sites are nearly superimposable, with the exception of amicyanin where one His is unbound in the Cu^{I} form.

Crystallographic data have shown that the active sites of apo-MBPs are minimally perturbed from the holo-forms. However, apoprotein solution studies reveal disorder in the loops bearing the inner-sphere ligands. This suggests that the "entatic" or "rack-induced" states of MBPs may not be preformed but rather are instead produced upon metal coordination. Once bound, metals confer thermodynamic stability to

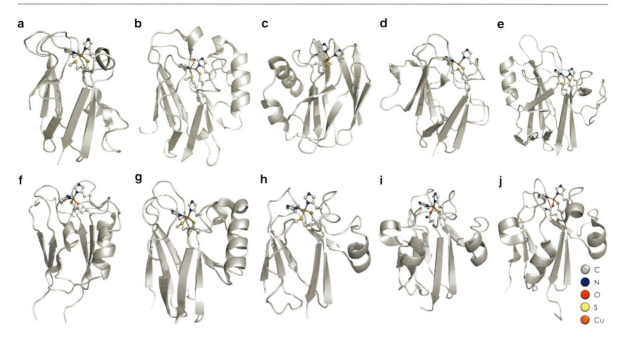

Monocopper Blue Proteins, Fig. 2 X-ray crystal structures reveal considerable structural diversity among MBPs despite their shared ß-sandwich tertiary structure. (**a**) Poplar plastocyanin (PDBID: 1PNC), (**b**) *Pseudomonas aeruginosa* azurin (4AZU), (**c**) *Alcaligenes faecalis* pseudoazurin, (**d**) *Paracoccus denitrificans* amicyanin (3L45), (**e**) *Thiobacillus ferooxidans* rusticyanin (1RCY), (**f**) Cucumber stellacyanin (1JER), (**g**) *Chloroflexus aurantiacus* auracyanin A (2AAN), (**h**) Spinach plantacyanin (1F56), (**i**) Horseradish umecyanin (1X9R), (**j**) Zucchini mavicyanin (1WS7)

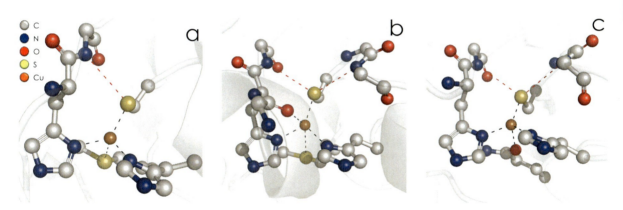

Monocopper Blue Proteins, Fig. 3 MBP active sites feature a conserved pseudotrigonal N_2S inner coordination sphere. The S-donating Cys is constrained by hydrogen bonds from backbone amides. Families of MBP are distinguished by variable complements of weak axial ligands. Plastocyanins (**a**, 1PNC) feature an axial thioether S-donor, azurins (**b**, 4AZU) have an additional axial interaction from a carbonyl O-donor, and stellacyanins (**c**, 1JER) substitute an amide O-donor for the commonly encountered thioether. Inner-sphere interactions are indicated by *black dashed lines*; outer-sphere interactions are indicated by *red dashed lines*

MBPs and also facilitate protein folding (Wittung-Stafshede 2004).

Electronic Structure. The electronic structure of type 1 Cu has been the subject of extensive study and has been recently reviewed (Solomon and Hadt 2011). MBPs are spectroscopically distinguished in their oxidized forms by an intense ligand-to-metal charge-transfer (LMCT) band near 16,000 cm^{-1} that dominates their electronic absorption spectra (Fig. 4a). MBPs also display very small $^{63,65}Cu$ hyperfine couplings in their EPR spectra relative to aqueous Cu^{II} (Fig. 4b). Type 1 $^{63,65}Cu$ hyperfine couplings range

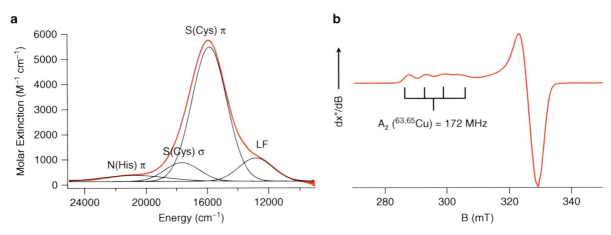

Monocopper Blue Proteins, Fig. 4 (a) Electronic absorption spectrum of *Pseudomonas aeruginosa* azurin. Individual excitations are indicated. (b) X-band (9.75 GHz) EPR spectrum of *Pseudomonas aeruginosa* azurin. The small ^{63,65}Cu hyperfine coupling characteristic of type 1 Cu is highlighted

from 60 to 285 MHz. By comparison, "normal" or type 2 ^{63,65}Cu hyperfine couplings as observed in aqueous Cu^{II} are $\geq$ 450 MHz. Spectroscopic, biochemical, and theoretical methods have established that these two features of type 1 Cu are attributable to a high degree of covalency between the equatorial Cys thiolate and Cu. The singly occupied Cu 3d orbital (conventionally the Cu $3d_{x2-y2}$) is oriented for maximal overlap with S π-orbitals. This affords the physical mechanism underlying the intensity of the 16,000 cm^{-1} absorption band, which is formally assigned as S π →Cu $3d_{x2-y2}$ LMCT. The unpaired electron in an oxidized blue copper site equally occupies the Cu 3d and S π orbitals. This delocalization attenuates the spin-dipolar contribution to the Cu hyperfine coupling, resulting in the small values observed for type 1 Cu proteins.

In addition to the dominant S π → Cu $3d_{x2-y2}$ LMCT, type 1 Cu absorption spectra feature several less intense LMCT bands. Very weak LMCT bands are observed at energies $\geq$ 20,000 cm^{-1} that have been assigned to transitions from Met and His molecular orbitals to Cu $3d_{x2-y2}$. Additionally, there is an S(Cys)σ → Cu $3d_{x2-y2}$ band occurring at 19,000 – 22,000 cm^{-1}. The relative intensities of the S(Cys) π and S(Cys) σ LMCT bands have been understood in terms of a progression of Cu electronic structure referred to as "coupled distortion" (Solomon and Hadt 2011). This progression arises due to perturbations of the Cu-S π-covalency, resulting in the perturbation of blue type 1 Cu sites to green "type 1.5" and red type 2 Cu. As axial ligand interactions strengthen, the $3d_{x2-y2}$ orbital rotates, favoring overlap with the S σ orbital. As a result, the absorptivity of the higher energy S(Cys) σ LMCT band increases relative to the S(π) transition effecting progression in color from blue to green. In the extreme case of a strongly interacting axial cysteine as encountered in the cupredoxin-like protein nitrosocyanin, the dominant LMCT is at sufficiently high energy to give rise to red Cu. Cu-S covalency is also attenuated as the magnitude of this coupled distortion increases, resulting in increasing ^{63,65}Cu hyperfine values. Green Cu proteins have hyperfine values near 300 MHz, intermediate between type 1 and type 2 Cu, while red Cu hyperfine couplings are near 450 MHz and are characteristic of type 2 copper.

The type 1 Cu ligand field (LF) is similar to that of tetrahedral Cu^{II}. LF excitations have been identified and assigned on the basis of low-temperature electronic absorption and magnetic circular dichroism spectroscopies. The highest LF excitations, $\sim$14,000 – 12,000 cm^{-1}, are $3d_{x2-y2}$ → $3d_{xz/yz}$. The $3d_{xy}$ to $3d_{x2-y2}$ excitation follows at $\sim$10,000 cm^{-1}. The lowest-lying LF excitation is from Cu d_{x2-y2} → $3d_{z2}$ and occurs at $\sim$5,000 cm^{-1}. This excitation results in EPR g_z values larger than would be expected in cases of high metal-ligand covalency. Additionally, it provides a mechanism for accelerated electron spin relaxation, conferring uncommonly narrow NMR line widths to paramagnetically shifted resonances corresponding to inner-sphere ligands in the spectra of oxidized type 1 Cu proteins.

Electron Transfer Reactivity. MBPs have figured extensively in the study of biological ET. Early studies concerned intermolecular reactions between the cupredoxins azurin, plastocyanin, and stellacyanin with cytochromes as well as redox-active coordination complexes. MBP electron self-exchange rates have been determined by stopped-flow kinetics, freeze-quench EPR measurements, and NMR relaxation experiments. MBPs have also been used to study intramolecular ET between both endogenous and covalently appended redox agents to Cu. In the latter case, the cupredoxin azurin has been engineered into scaffolds for measurements of multistep intramolecular ET kinetics.

ET reactivity may be parameterized with reduction potentials ($E°$), donor-acceptor coupling (κ), and reorganization energy (λ) (Marcus and Sutin 1985). The aqueous $E°$ of the $Cu^{II/I}$ couple is +0.15 V versus normal hydrogen electrode (NHE). However, the $Cu^{II/I}$ $E°'$ values in naturally occurring and engineered MBPs can range from +0.12 V to nearly +1.0 V versus NHE. $E°'$ values are tuned by many factors, including variation of axial ligand complement, metal-binding site hydrophobicity, protein surface electrostatics, and outer-sphere hydrogen bonding networks involving the inner-sphere ligands bound to Cu (Lancaster 2012). *Rhus vernicifera* stellacyanin, with its strongly interacting axial Gln, has among the lowest MBP $E°'$ values at 0.18 mV versus NHE. *Pseudomonas aeruginosa* azurin, with its thioether and carbonyl axial ligands, has an $E°'$ of 0.30 V versus NHE. Poplar plastocyanin, featuring only a Met axial ligand, has $E°'$ of 0.38 V versus NHE. *Thiobacillus ferooxidans* rusticyanin also features a single Met axial ligand, but its Cu binding site is deeply buried in the protein's hydrophobic interior, resulting in a 0.68 V $E°'$.

κ is optimized for MBP ET reactivity by both the type 1 Cu electronic structure as well as the surrounding polypeptide (Gray and Winkler 2010; Solomon and Hadt 2011). The anisotropic covalency between Cu and the Cys thiolate strongly couples the metal ion to the polypeptide backbone. This interaction heavily weights intramolecular ET pathways involving the Cys. Hydrogen bonding from the "rack" has also been implicated in defining ET routes through MBPs. The "rack" enforced, $\sim 0°$ dihedral angle between the S and C_α of the equatorial Cys is thought to maximize superexchange mediated by these sidechains. Additionally, the hydrogen bonds to S provide an additional route for ET between Cu and the rest of the protein.

Many MBPs feature a patch of hydrophobic residues on the protein surface surrounding one of the equatorial His sidechains. These patches facilitate docking with intermolecular ET partners; this in turn enhances κ for ET routes involving equatorial His sidechains.

MBPs exert profound control over λ values associated with $Cu^{II/I}$ ET reactions. Cu coordination complexes are intrinsically slow ET agents due in large part to large λ values (~ 2 eV). Outer-sphere coordination networks in combination with the strong Cu-S bonds in MBPs constrain Cu in geometries that are equally suitable for the coordination of Cu^{II} and Cu^{I} (Gray et al. 2000; Lancaster 2012). Where crystallographically characterized, the Cu^{II} and Cu^{I} forms of MBPs are almost structurally indistinguishable (amicyanin being a notable exception). ET experiments with the rack-constrained, hard-ligand "type zero copper" variant of azurin have demonstrated that the rack alone lowers $\lambda \sim$ twofold relative to Cu coordination complexes. The highly covalent Cu-S interaction further reduces λ. Resonance Raman spectroscopic analyses estimate the inner-sphere λ of type 1 Cu at ~ 0.2 eV, and investigations of activation less intramolecular ET using Ru-labeled azurin fixes λ (total) at ~ 0.8 eV for MBPs.

Engineered Monocopper Blue Proteins. The primary amino acid sequence for the cupredoxin azurin from *Pseudomonas aeruginosa* was among the earliest reported for any protein. Azurin was also the first MBP to be recombinantly expressed in *E. coli*, making it an early platform for site-directed mutagenesis. Hosts of azurin variants have been produced with modifications to metal complement, inner-sphere ligands, structural residues, and outer-sphere coordination networks (Lu 2011; Lancaster 2012). Concerted effects of inner- and outer-sphere tuning have permitted the generation of azurin variants with $E°'$ values spanning a 0.6 V range. Loop-directed mutagenesis has enabled the construction of an azurin variant incorporating a mixed-valent, binuclear Cu_A center. The Cys112Asp/Met121Leu azurin variant gives rise to a "type zero" copper center, an efficient ET site comprised entirely of "hard" ligands (O/N-donors). Additionally, azurin has been used to study the effects of nonstandard amino acids on type 1 Cu electronic structure and physical properties. Installation of a strongly donating axial homocysteine thiolate converts the type 1 Cu center of azurin to "red copper." Efforts have also been made to convert orthologous proteins into MBPs.

Conclusion. The thorough and cross-disciplinary study of MBPs has yielded valuable insights into biological coordination chemistry, biophysics, and bioenergetics. Among many emergent applications, efforts are underway to purpose MBPs for cancer therapy as well as to develop devices for data storage based on MBP redox state. Despite thorough elucidation of the electronic structure and reactivity of MBPs, fundamental questions regarding their physiological roles remain. Coupled to their extreme versatility as design and experimental platforms, this void in understanding ensures that MBPs will remain subjects of active inquiry.

Cross-References

▶ Copper, Biological Functions
▶ Copper-Binding Proteins
▶ Plastocyanin

References

Battistuzzi G, Borsari M et al (2009) Active site loop dictates the thermodynamics of reduction and ligand protonation in cupredoxins. Biochim Biophys Acta 1794:995–1000
Dennison C (2005) Investigating the structure and function of cupredoxins. Coord Chem Rev 249:3025–3054
Gray HB, Winkler JR (2010) Electron flow through metalloproteins. Biochim Biophys Acta 1797:1563–1572
Gray HB, Malmström BG, Williams RJP (2000) Copper coordination in blue proteins. J Biol Inorg Chem 5:551–559
Hart PJ, Nersissian AM, DeBeer George S (2011) Copper proteins with type 1 sites. In: Encyclopedia of inorganic and bioinorganic chemistry. Wiley, Chichester
Lancaster KM (2012) Biological outer-sphere coordination. Struct Bond 142:119–153
Lu Y (2011) Metalloprotein design & engineering. In: Encyclopedia of inorganic and bioinorganic chemistry. Wiley, Chichester
Marcus RA, Sutin N (1985) Electron transfers in chemistry and biology. Biochim Biophys Acta 811:265–322
Solomon EI, Hadt RG (2011) Recent advances in understanding blue copper proteins. Coord Chem Rev 255:774–789
Wittung-Stafshede P (2004) Role of cofactors in folding of the blue-copper protein azurin. Inorg Chem 43:7926–7933

Monodispersity

▶ Gold Nanoparticles, Biosynthesis

Monomethylarsonous Acid Methyltransferase

▶ Arsenic Methyltransferases

Monooxygenase

▶ Iron Proteins, Mononuclear (non-heme) Iron Oxygenases

Monovalent Cations

▶ Potassium, Physical and Chemical Properties
▶ Sodium, Physical and Chemical Properties

Monovalent Cations in Tryptophan Synthase Catalysis and Substrate Channeling Regulation

Michael F. Dunn
Department of Biochemistry, University of California at Riverside, Riverside, CA, USA

Synonyms

Activate: stimulate; Allosteric: other site; Catalytic pathway: reaction pathway; Conformation: fold; Coordination geometry: bonding geometry; Dimeric unit: subunit dimer; Enzyme: catalyst; Inhibit: hinder; Inner coordination sphere: coordination bond; Intermediate: metabolite; Mimic: imitate; Site: cleft; Substrate: reactant

Definition

Monovalent cations such as Na^+ and K^+ play important roles in proteins and enzymes by providing structural stabilization and/or modulation, and by functioning as effectors of biological function. It is now known that monovalent cations influence the activities of many enzymes, and this modulation of activity plays a significant role in the regulation of biological processes. This entry presents the bienzyme complex tryptophan synthase as a paradigm for exploring the mechanisms by which monovalent cations regulate enzyme catalysis.

Background

There are a large number of enzymes now known to be activated and/or inhibited by Na^+ or K^+, See Woehl and Dunn (1995a, b), Peracchi et al. (1995); (chapters in this issue). In most of these enzyme systems, nonphysiological monovalent cations (MVCs) can be substituted into the MVC site. Such substitutions almost inevitably result in enzymes with modified catalytic properties. Hence, substitutions with Li^+, Cs^+, Rb^+, or MVC mimics such as NH_4^+ and guanidinium ion (Woehl and Dunn 1995a, b; Peracchi et al. 1995; Fan et al. 2000) can provide site-specific probes of MVC function and mechanism. The structural characterization of MVC sites in enzymes is relatively recent. Because the x-ray-scattering factors of Na^+, K^+ H_2O, and H_3O^+ are similar (10–18 electrons), structures with well-defined MVC sites in enzymes only began to appear in the mid-1990s (Woehl and Dunn 1995a). The majority of the MVC-activated enzymes for which x-ray structures are available show the MVC site located near but not within bonding distance of either the bound substrate or the active site catalytic groups. *Thus, the MVCs which bind to these sites are classified as cofactors which modulate enzyme function through allosteric ("other site") interactions*, Woehl and Dunn (1995a).

Tryptophan Synthase

The tryptophan synthase $\alpha_2\beta_2$ bienzyme complex is a MVC-activated pyridoxal phosphate (PLP)-requiring enzyme that catalyzes the last two steps in the synthesis of L-tryptophan (L-Trp) (Fig. 1). These transformations are designated as the α- and β-reactions. In the α-reaction, 3-indole-D-glycerol 3′-phosphate (IGP) is cleaved to indole and D-glyceraldehyde 3-phosphate (G3P) (Fig. 1a, [1]). During the β-reaction (Fig. 1a, [2], and Fig. 1b), indole produced at the α-site is transferred from the α-site into the β-site via a 25 Å-long tunnel. In the β-site, L-serine (L-Ser) reacts with PLP through a series of transient intermediates to give the quasi-stable α-aminoacrylate intermediate, E(A-A) (Fig. 1b, Stage I). Indole then reacts with the E(A-A) intermediate to ultimately give L-Trp (Stage II). The intramolecular transfer of indole from the α-site to the β-site was the first known example of substrate channeling, and the first channeling system to be characterized at near atomic resolution (see Hyde et al. in Dunn et al. 2008). Substrate channeling in tryptophan synthase is regulated by a set of allosteric interactions that synchronize the α- and β-reactions and prevent the escape of indole from the α-β dimeric unit, Dunn et al. (2008). These allosteric interactions switch the α- and β-subunits between open conformations of low activity and closed conformations of high activity. The formation of the closed E(A-A) activates the α-site by $\sim$30-fold (Fig. 1b, Stage I), conversion of the closed $E(Q)_3$ to the open $E(Aex_2)$ deactivates the α-site (Stage II). The binding of IGP to the α-site activates the conversion of $E(Aex_1)$ to E(A-A) by $\sim$10-fold. *This switching between open (inactive) and closed (active) conformations serves to synchronize the α- and β-reactions and prevent escape of indole*.

Structure of the Monovalent Cation-Binding Site

The $\alpha_2\beta_2$ bienzyme complex was discovered by Peracchi et al. (1995) to be activated by MVCs, and this was independently shown by Woehl and Dunn (1995b). These two initial publications and subsequent work demonstrated that at 25°C, Li^+, Na^+, K^+ NH_4^+, Cs^+ Rb^+, and guanidinium ion all activate both the $\alpha\beta$- and β-reactions and MVC binding plays an essential role in mediating communication between the α- and β-subunits within each $\alpha\beta$-dimeric unit (Rhee et al. (1996); Pan et al. (1997); Woehl and Dunn (1999); Fan et al. (2000); Weber-Ban et al. (2001); Dierkers et al. (2009)).

The structures with different MVCs bound to the MVC site have been determined by single crystal x-ray diffraction (Rhee et al. 1996) (Fig. 2). The MVC site substituted with Na^+, K^+, or Cs^+ is formed by a loop structure in each β-subunit and located about 8 Å from the β-active site. The specific bonding interactions between the MVC and the MVC site depend on the ionic radius and charge density of the MVC (Rhee et al. 1996), see Fig. 2. In these structures, the formal positive charge on the MVC is satisfied by coordination of three to six carbonyl oxygens (depending on the MVC) of the surrounding polypeptide residues (βVal231, βGly232, βGly268, βLeu304, βPhe306, and βSer308) and by

Monovalent Cations in Tryptophan Synthase Catalysis and Substrate Channeling Regulation, Fig. 1 (a) The α-reaction (1) and the β-reaction (2) catalyzed by the tryptophan synthase bienzyme complex. (b) Stages I and II of the β-reaction. L-Ser reacts with the internal aldimine, E(Ain) in Stage I to give the α-aminoacrylate intermediate, E(A-A). Indole, formed in the α-reaction, reacts with E(A-A) in Stage II to give L-Trp (Reprinted with permission from Dierkers et al. (2009). Copyright (2009) American Chemical Society)

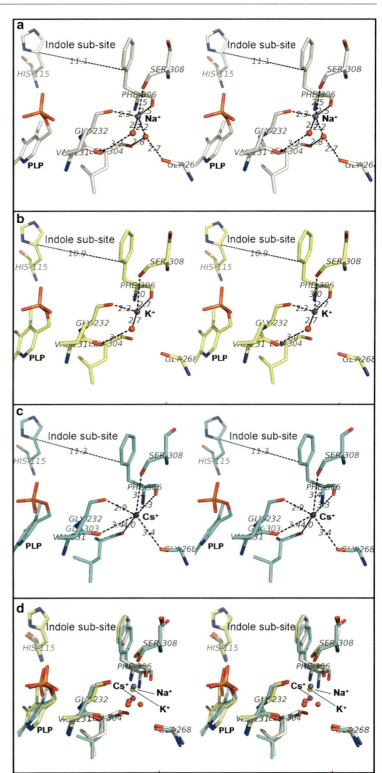

Monovalent Cations in Tryptophan Synthase Catalysis and Substrate Channeling Regulation, Fig. 2 (a)–(c) Stereoscopic drawings comparing the monovalent cation sites of the Na^+, K^+, and Cs^+ complexes of the internal adimine, E(Ain), form of the enzyme. Color schemes: C is *gray*, *yellow* or *cyan* respectively in (a), (b), and (c). CPK colors are used for all other atoms. *Dashed lines* show interatomic distances. (d) Stereo diagram of the aligned monovalent cation sites for the Na^+, K^+, and Cs^+ sites. The Na^+ and K^+ structures are closely similar. Cs^+ is displaced by ~2.1 Å from the position occupied by Na^+, and side chain and main chain atoms of Phe306 in the Cs^+ structure are displaced relative to the corresponding atoms in the Na^+ and K^+ structures. Na^+ is *blue*, K^+ is *yellow*, and Cs^+ is *magenta* (Reprinted with permission from Dierkers et al. (2009). Copyright (2009) American Chemical Society)

coordination of water. In the Na⁺ complex, Na⁺ is 5-coordinate with bonds to only three of the six carbonyls (βGly232, βPhe306, and βSer308) and two water molecules. In the K⁺ 4-coordinate complex, only a single water molecule is coordinated, while Cs⁺ is 6-coordinate with an inner coordination sphere consisting of all six carbonyls but no waters (Fig. 2). Notice that, the ionic radius (see Woehl and Dunn 1995a) of Na⁺ (1.02 Å) is too small to fill the protein cavity and therefore Na⁺ recruits two waters into its coordination sphere (Fig. 2a). The ionic radius of K⁺ (1.38 Å) nearly fits, and only a single water is needed to complete the coordination sphere (Fig. 2b), while the ionic radius of Cs⁺ (1.70 Å) fills the cavity (Fig. 2c) (Rhee et al. 1996; Dierkers et al. 2009). In this complex, Cs⁺ is shifted ~1.1 Å from the position occupied by Na⁺ (Fig. 2d) (Rhee et al. 1996). Figure redrawn from Dierkers et al. (2009).

Structural Perturbations by Monovalent Cations

The different structures of the Na⁺, K⁺, and Cs⁺ inner coordination complexes perturb the surrounding protein structure (Fig. 2d) (Rhee et al. 1996) with the consequence that the β-subunit indole subsite is deformed, particularly with regard to the position of the βPhe306 side chain (Dierkers et al. 2009). *This deformation of the indole subsite exerts strong effects on catalysis in the β-reaction, and alters communication between the α- and β-sites* (Dierkers et al. 2009). The removal of all MVCs renders the β-site essentially inactive (Fig. 3). This loss of activity is completely restored by the binding of MVCs (Woehl and Dunn 1995b; Dierkers et al. 2009). The binding of MVCs alters the equilibrium distribution of intermediates in the β-reaction (Fig. 1b, Fig. 3). The E(A-A) intermediate is stabilized by the binding of Cs⁺ or NH₄⁺ and by the MVC-free enzyme. The external aldimine, E(Aex₁) (Fig. 1), is stabilized by Na⁺. Quinonoidal intermediates and quinonoid analogues are stabilized by Cs⁺ and NH₄⁺, while the Na⁺ and MVC-free enzymes give only trace amounts of these quinonoid species (Fig. 3). Figure 3 compares the UV/Vis absorption spectra of the species formed in the reaction of the L-Trp analogue, L-histidine (L-His). These spectra show that L-His reacts with enzyme-bound PLP to give the L-His external aldimine and quinonoid species, and the distribution of these species is remarkably dependent on the particular MVC bound to the β-subunit. Only small amounts of the L-His quinonoid

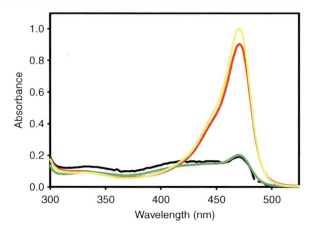

Monovalent Cations in Tryptophan Synthase Catalysis and Substrate Channeling Regulation, Fig. 3 UV/Vis absorption spectra showing monovalent cation effects on the reaction of the L-tryptophan analogue, L-histidine (L-His), with the β-sites of tryptophan synthase (Dierkers et al. 2009). In this example, the α-site is occupied by a tight binding IGP analogue. L-His reacts with E(Ain) to give an equilibrating mixture of the L-His quinonoid and the L-His external aldimine forms of the enzyme. Only trace amounts of the L-His quinonoid species (λ_{max} 466 nm) are formed for the MVC-free (*green*) and Na⁺ (*black*) forms of the enzyme. In contrast to this behavior, the Cs⁺ (*red*) and NH₄⁺ (*yellow*) forms of the enzyme give the L-His quinonoid species almost exclusively (Reprinted with permission from Dierkers et al. (2009). Copyright (2009) American Chemical Society)

species are formed with the MVC-free and Na⁺ forms of the enzyme. Essentially, all of the β-sites are converted to the L-His quinonoid when the MVC site is occupied by either Cs⁺ or NH₄⁺ (Dierkers et al. 2009).

The long dimension of the indole subsite is defined by the side chains of βHis115 and βPhe306 (Figs. 2 and 4). The diameter of the cavity defined by these two residues is decreased in the Na⁺ complex by about 2 Å in comparison to the Cs⁺ structure (Fig. 4). Comparison of the structures of the Na⁺ and Cs⁺ forms of the enzyme shows Cs⁺ favors β-subunit structures poised for the switch to the closed conformation with a preformed subsite for indole (Dierkers et al. 2009). The Na⁺ complex favors the open conformation of the β-subunit. The indole subsites of the open Na⁺ complexes are partially collapsed, distorted, and too small to accommodate indole (Dierkers et al. 2009). *The work of* Woehl and Dunn (1999) *and* Dierkers et al. (2009) *show that the relative energies of intermediates along the pathway of the β-reaction are different for each of the MVCs investigated.* Furthermore, the

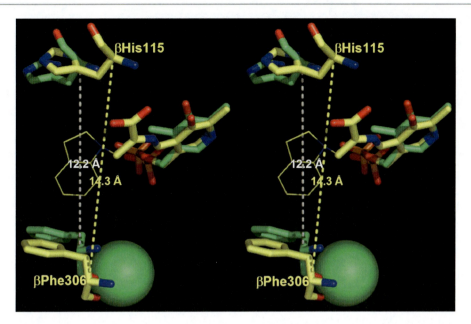

Monovalent Cations in Tryptophan Synthase Catalysis and Substrate Channeling Regulation, Fig. 4 The indole analogue, indoline, reacts with the α-aminoacrylate intermediate, E(A-A) (Fig. 1), to form a quasi-stable quinonoid species, E(Q)$_{indoline}$. A stereoscopic drawing comparing x-ray structures of the indole subsites of the open Na$^+$ E(Ain) complex with the closed Cs$^+$ E(Q)$_{indoline}$ complex is shown. Coloring scheme: the C and Cs$^+$ atoms are *yellow* in the Cs$^+$ E(Q)$_{indoline}$ complex, the C and Na$^+$ atoms of the Na$^+$ E(Ain) complex are *green*. All other atoms are shown with CPK coloring. The indoline ring of the Cs$^+$ E(Q)$_{indoline}$ complex is shown in wireframe. The *dashed lines* measure the distances between the Cα atoms of βHis115 and βPhe306 in the two complexes. Notice that the distance between the side chains of βHis115 and βPhe306 is too short in the Na$^+$ E(Ain) complex to accommodate the indoline ring (Reprinted with permission from Dierkers et al. (2009). Copyright (2009) American Chemical Society)

energetics of the switching between open and closed conformation states of the α- and β-subunits are strongly perturbed. Detailed analyses of the kinetics and thermodynamics of ligand binding establish that, in tryptophan synthase the bound, monovalent cation exerts strong effects on two steps in the catalytic pathway (Woehl and Dunn 1999). For example, in comparison to the MVC-free enzyme, the affinity for L-Ser is increased about fivefold, and the rate of formation of the (Na$^+$)E(Aex$_1$) intermediate is stimulated about threefold; while in Stage II of the β-reaction, the MVC-free E(A-A) is essentially inactive (Dierkers et al. 2009), while the binding of Na$^+$, K$^+$, or Cs$^+$ strongly activates the MVC-free E(A-A).

Summary

The x-ray structure data base for the MVC site of the tryptophan synthase complex shows that, owing to the difference in ionic radius (and hence charge density) of the MVC, the complexes of Na$^+$, K$^+$, and Cs$^+$ exhibit significantly different coordination geometries within the same MVC site. These structural differences propagate out from the MVC site, causing a perturbation of the indole subsite of the β-subunit that exerts strong effects on the catalytic properties of the β-site.

The overall action of MVC effector binding is to introduce a fine-tuning of catalysis by modulating both the activation energies and the relative ground state energies of select steps in the β-reaction. To achieve this modulation, there is a preferential (i.e., tighter) binding interaction of the metal ion to one of two preexisting protein conformations (Woehl and Dunn 1995a, 1999). The preferential interaction with certain intermediates then makes possible changes in the relative positions of transition-state and ground-state energies which reflect the preferential interaction. The result is a different set of rate constants for the metal-bound and metal-free enzyme species that is selective for specific steps in the mechanism (Woehl and Dunn 1995a, 1999). *The remarkable dependence* of quinonoid formation on the binding of an activating metal ion in tryptophan synthase strongly suggests that metal ion-binding stabilizes the binding of indole to the indole subsite so that the indole ring is correctly aligned for the nucleophilic attack on the Cβ of the E(A-A) intermediate. Consequently, the MVC cofactor plays an important role in the stereoelectronic control of the formation of this C–C bond in tryptophan synthase catalysis.

Cross-References

▶ K^+-Dependent Na^+/Ca^{2+} Exchanger
▶ Potassium in Health and Disease
▶ Potassium, Physical and Chemical Properties
▶ Potassium-Binding Site Types in Proteins
▶ Sodium, Physical and Chemical Properties
▶ Sodium-Binding Site Types in Proteins
▶ Sodium-Coupled Secondary Transporters, Structure and Function
▶ Sodium/Glucose Co-transporters, Structure and Function
▶ Sodium-Hydrogen Exchangers, Structure and Function in Human Health and Disease

References

Dierkers AT, Niks D et al (2009) Tryptophan synthase: structure and function of the monovalent cation site. Biochemistry 48:10997–11010
Dunn MF, Niks D et al (2008) Tryptophan synthase: the workings of a channeling nanomachine. Trends Biochem Sci 33:254–264
Fan YX, McPhie P et al (2000) Regulation of tryptophan synthase by temperature, monovalent cations, and an allosteric ligand. Evidence from Arrhenius plots, absorption spectra, and primary kinetic isotope effects. Biochemistry 39:4692–4703
Pan P, Woehl E et al (1997) Protein architecture, dynamics, and allostery in tryptophan synthase channeling. Trends Biochem Sci 22:22–27
Peracchi A, Mozzarelli A et al (1995) Monovalent cations affect dynamic and functional properties of the tryptophan synthase $\alpha_2\beta_2$ complex. Biochemistry 34:9459–9465
Rhee S, Parris KD et al (1996) Exchange of K^+ or Cs^+ for Na^+ induces local and long-range changes in the three-dimensional structure of the tryptophan synthase $\alpha_2\beta_2$ complex. Biochemistry 35:4211–4221
Weber-Ban E, Hur O et al (2001) Investigation of allosteric linkages in the regulation of tryptophan synthase: the roles of salt bridges and monovalent cations probed by site-directed mutation, optical spectroscopy, and kinetics. Biochemistry 40:3497–3511
Woehl EU, Dunn MF (1995a) The roles of Na^+ and K^+ in pyridoxal phosphate enzyme catalysis. Coord Chem Rev 144:147–197
Woehl EU, Dunn MF (1995b) Monovalent metal ions play an essential role in catalysis and intersubunit communication in the tryptophan synthase bienzyme complex. Biochemistry 34:9466–9476
Woehl EU, Dunn MF (1999) Mechanisms of monovalent cation action in enzyme catalysis: the tryptophan synthase α-, β- and $\alpha\beta$-reactions. Biochemistry 38:7131–7141

Motor Neuron Disease

▶ Copper-Zinc Superoxide Dismutase and Lou Gehrig's Disease

Multiminicore Disease (MmD)

▶ Selenium and Muscle Function

Multiple Labeling

▶ Palladium, Colloidal Nanoparticles in Electron Microscopy

Muscle Calcium-Binding Protein

▶ Parvalbumin

Myopathy with Congenital Fiber Type Disproportion (CFTD)

▶ Selenium and Muscle Function

Myosin Light Chain

▶ Calcium-Binding Proteins, Overview

Myosin Light Chain (MLC)

▶ Calcium Sparklets and Waves

Myosin Light Chains

▶ Calcium in Biological Systems

Myotonic Dystrophies

▶ Selenium and Muscle Function

N

Na(+)/K(+)-Exchanging ATPase

▶ Sodium/Potassium-ATPase Structure and Function, Overview

Na,K-Activated ATPase

▶ Sodium/Potassium-ATPase Structure and Function, Overview

Na/K-ATPase, Na/K Pump, Sodium- and Potassium-Dependent Adenosine Triphosphatase

▶ Rubidium in Biological Systems and Medicine

NA: Nicotianamine

▶ Iron Proteins, Plant Iron Transporters

Na$^+$/Ca^{2+}-K$^+$ Exchanger

▶ Potassium Dependent Sodium/Calcium Exchangers

Na$^+$/H$^+$ Exchanger

▶ Sodium-Hydrogen Exchangers, Structure and Function in Human Health and Disease

Na$^+$/K$^+$ Pump

▶ Sodium/Potassium-ATPase Structure and Function, Overview

Na$^+$/Sugar Symporter

▶ Sodium/Glucose Co-transporters, Structure and Function

Na$^+$-Dependent Glucose Transporter

▶ Sodium/Glucose Co-transporters, Structure and Function

Na$^+$-Dependent Proteins

▶ Sodium-Binding Site Types in Proteins

V.N. Uversky et al. (eds.), *Encyclopedia of Metalloproteins*, DOI 10.1007/978-1-4614-1533-6,
© Springer Science+Business Media New York 2013

Nanobio Conjugates

▶ Nanosilver, Next-Generation Antithrombotic Agent

Nanocrystalline Silver

▶ Silver, Pharmacological and Toxicological Profile as Antimicrobial Agent in Medical Devices

Nanoparticles

▶ Gold Nanoparticle Platform for Protein-Protein Interactions and Drug Discovery

Nanosilver, Next-Generation Antithrombotic Agent

Siddhartha Shrivastava
Rensselaer Nanotechnology Center, Rensselaer Polytechnic Institute, Troy, NY, USA
Center for Biotechnology and Interdisciplinary Studies, Rensselaer Polytechnic Institute, Troy, NY, USA

Synonyms

Antithrombotic; Blood platelets; Cell signaling; Fibrin; Integrins; Nanobio conjugates; Silver nanoparticles (AgNPs)

Definition

Thrombolytic therapy in acute stroke has reduced ischemia; however, it is also associated with increased incidence of intracerebral hemorrhage and expanding stroke. Platelets and fibrin are the major components of thrombi and are an obvious target for majority of antithrombotic therapies. Silver nanoparticles have shown to have innate antiplatelet properties with nanosilver. It can effectively prevent platelet activation in response to physiological agonists, under both in vitro as well as ex vivo conditions, and immobilize and stabilize proteins. Furthermore, nanosilver can also significantly retard fibrin polymerization kinetics both in pure and plasma-incorporated systems and hence can impede thrombus formation. Together with its inherent antiplatelet and antibacterial properties, capacity to inhibit fibrin polymerization can open up possibilities of employing nanosilver as future antithrombotic agent.

The Current Approach

Thrombotic disorders have emerged as a serious threat to society. The conventional management of thrombotic and cardiovascular disorders is based on the use of heparin, oral anticoagulants, and aspirin. Despite progress in the sciences, these drugs still remain a challenge and mystery. The development of low molecular weight heparins (LMWHS) and the synthesis of heparinomimetics (compounds with heparin-like properties) represent a refined use of heparin. Additional drugs will continue to develop. However, none of these drugs will ever match the polypharmacology (the treatment of diseases by modulating more than one target) of heparin. Aspirin still remains the leading drug in the management of thrombotic and cardiovascular disorders. The newer antiplatelet drugs such as adenosine diphosphate receptor inhibitors, GPIIb/IIIa inhibitors, and other specific inhibitors have limited effects and have been tested in patients who have already been treated with aspirin. Warfarin provides a convenient and affordable approach in the long-term outpatient management of thrombotic disorders. The optimized use of these drugs still remains the approach of choice to manage thrombotic disorders. The new anticoagulant targets, such as tissue factor, individual clotting factors, recombinant forms of serpins (antithrombin, heparin cofactor II, and tissue factor pathway inhibitors), recombinant activated protein C, thrombomodulin, and site-specific serine protease inhibitors complexes, have also been developed. There is a major thrust on the development of orally bioavailable anti-Xa and anti-IIa agents, which are slated to replace oral anticoagulants. Both the anti-factor Xa and anti-IIa agents have been developed for oral use and have provided impressive clinical results. However, safety concerns related to liver enzyme elevations and thrombosis rebound have been reported with their use. For these reasons, the US Food and Drug Administration did not approve the orally active antithrombin agent Ximelagatran for several indications. The synthetic pentasaccharide (Fondaparinux) has undergone clinical development. Unexpectedly, Fondaparinux also produced major bleeding problems at minimal dosages. Fondaparinux exhibits only one of the multiple pharmacologic effects of heparins. Thus, its therapeutic index will be proportionately narrower. The focus has now shifted to regulating and maintaining platelets, the cellular component of thrombus, in an inactive state. Antiplatelet drugs like inhibitors of P2Y12 receptor, integrin $\alpha_{IIb}\beta_3$, cyclooxygenase, and phosphodiesterase are also widely used in therapy but with serious limitations. Phosphatidylinositol-3-kinase inhibitors have been proposed as potential antithrombotic therapy in view of the role of this enzyme in platelet activation and plug stability; however, these too have major limitations for use as drugs. In general, the severity of acute coronary syndrome is regulated by both magnitude and stability of the thrombus. Platelets and fibrin are the major components of thrombi. Since fibrin is available in large concentration at lesion sites and in all types of thrombi, it is also an obvious target for the majority of antithrombotic therapies. The current thrombolytic therapy in acute stroke has reduced ischemia; however, it is also associated with increased incidence of intracerebral hemorrhage and expanding stroke. Therefore, target of thrombolytic therapy has repositioned to reducing magnitude of hemorrhagic complications while maintaining therapeutic efficacy.

Why Nanosilver

Considering the complicities of above limitations, there is a need for a drug molecule or agent which can interact with both, cellular components of the thrombus, i.e., platelets and with fibrin, the protein component of thrombi. Nanomaterials have shown potential not only due to their nanometer scale size that enables them to enter even the smallest compartment within the cell but they can also be synthesized and assembled in various shapes, sizes, and architectures, allowing them to achieve the required properties. Additionally, they can also be decorated with functional groups, receptors, signal peptides, and other desired molecules to provide them with target specificity and to further enhance their pharmacological properties. To attain such customizable properties in a single molecule has always been a challenge for the traditional medicines.

Silver nanoparticles, of size range 10–20 nm, are established to penetrate small capillaries following systemic administration and are eliminated from the system by the liver and kidneys. These properties opened up the possibility of delivering these nanomedicines to specific issues with proper control. Furthermore, its demonstrated antimicrobial potential and ability to conjugate and modify the protein structure and function made it the candidate of choice for testing its antithrombotic potential. But toxicity of silver, as with any other nanopreparations, remains a concern. To tackle the issue, researchers have successfully synthesized biocompatible nanoparticles of silver using glucose as reducing agent (Shrivastava et al. 2007). Although their detailed toxicity profiling is yet to be done, initial studies using a mouse model indicated it did not exhibit any undesirable side effects after its intravenous administration. Mice survived and reproduced normally after the dose treatment.

These biocompatible nanoparticles of silver were then analyzed for their effect on platelets (Shrivastava et al. 2009). Administered intravenously, nanosilver protected mice platelets from being aggregated in response to platelet agonists. The results were reproducible in platelets obtained from human blood and activated by different agonists (thrombin, collagen, and ADP). Particles were also equally effective in precluding the aggregation of cells obtained from diabetic patients. These effects were innate to silver and not exhibited by gold nanoparticles of comparable sizes.

Researchers attribute this inhibition to the conformational modulation of platelet surface integrins $\alpha_{IIb}\beta_3$ (a kind of receptor on platelet surface responsible for its ligation with fibrin), thus preventing the latter's interaction with fibrinogen. Detailed characterization studies revealed that the observed regulation of integrins was the outcome of the altered activity of many intracellular enzymes and proteins due to their conjugation with internalized nanosilver particles. These results which are the consequence of well-known property of silver nanoparticles to immobilize and stabilize proteins promote them as a potential antiplatelet agent.

The antithrombotic potential of biocompatible silver nanoparticles is further strengthened by analyzing their interaction with fibrin(ogen); the protein content of the clot (Shrivastava et al. 2011). Nanosilver was found to significantly retard the fibrin polymerization kinetics both in pure and plasma-incorporated systems and hence to impede thrombus formation. Spectroscopic and electron microscopic analysis revealed interaction of nanosilver with fibrinogen with identifiable conformational changes in protein secondary structure. Resistance to thrombin-induced clot formation in presence of nanosilver could be attributable to inhibition of enzymatic activity of thrombin due to its conjugation with nanoparticles, resulting in reduced proteolysis of fibrinogen, or prevention of fibrin polymerization, or both. Although from studies the modulation of thrombin activity could not be ruled out, the conformational changes observed in fibrin(ogen) in presence of silver nanoparticles and inhibition of atroxin (a snake venom known to polymerize fibrinogen independent of thrombin action)-induced polymerization favor the alternative possibility. Much higher concentration of nanosilver was required to retard the fibrin polymerization in natural plasma. This is due to high propensity of nanosilver to conjugate with plasma proteins like albumin and globulins, which are present in abundance.

The ability of nanosilver to retard fibrin polymerization, in association with its antiplatelet property, projects silver nanoparticles as possible antithrombotic agent after appraisal of issues like specificity, silver toxicity, and bleeding risk. Although from initial studies, bleeding tendency was not observed following intravenous administration of nanosilver in mice, it is premature to predict whether silver would be more specific than the current drugs (warfarin and heparin) and has less bleeding risk. These nanoparticles can be used effectively with coronary stents, where the antibacterial property of nanosilver would complement its antithrombotic propensity.

Silver nanoparticles have potential to form bioconjugates with novel physicochemical properties and hence are of considerable importance to cater unmet biomedical goals but, it would be too early and untimely to predict that nanosilver conjugates will be more specific and less toxic than current available remedies. To ascertain this, detailed characterization of the nano-bio interaction is a prerequisite. The initial cellular events that take place at the nano-bio interface mimic to a certain extent the natural adhesive interaction of cells with the extracellular matrix (ECM). In fact, the living cells cannot interact directly with foreign materials, but they readily attach to the adsorbed layer of soluble matrix proteins that adsorb on the bare nanomaterial upon contact with physiological fluids in vivo or culture medium in vitro, or to the layer of immobilized proteins on the composite. Hence, the concentration, distribution, mobility, and orientation of the adsorbed protein layer on a surface play a fundamental role in the biofunctionality of a synthetic material and are important factors in understanding the biological response of a substrate. Cells recognize these proteins via integrins - a family of cell surface receptors – that provide transmembrane links between the extracellular matrix (ECM) and the cellular cytoskeleton. Abnormalities in the cell-ECM integrin-mediated interactions are associated with pathologic situations that include tumor formation. Besides, integrin-mediated adhesion involves not only the receptor-ligand but also post-ligation interactions with multiple binding partners. Thus, the cell-nanomaterial interaction is a complex multistep process starting with adsorption of proteins, followed by cell adhesion and spreading, and late events, related to cell growth, differentiation, matrix deposition, and cell functioning. Measurement and quantification of these parameters starts with the detailed understanding of the interaction between proteins and nanomaterials as each individual protein can exhibit different characteristics on the same nanosurface. Hence, like all nascent medicines and medical devices, critical appraisal of benefits and side effects of nanosilver with intense cytotoxicity studies will be required to guarantee its safety as a therapeutic agent.

Cross-References

▶ Metallothioneins and Silver
▶ Silver as Disinfectant
▶ Silver-Induced Conformational Changes of Polypeptides

References

Shrivastava S, Berra T, Roy A et al (2007) Characterization of enhanced antibacterial effects of novel silver nanoparticles. Nanotechnology 18:225103–225112

Shrivastava S, Berra T, Singh SK et al (2009) Characterization of antiplatelet properties of silver nanoparticles. ACS Nano 3(6):1357–1364

Shrivastava S, Berra T, Mukhopadhyay A et al (2011) Negative regulation of fibrin polymerization and clot formation by nanoparticles of silver. Colloids Surf B Biointerfaces 82(1):241–246

Nanostructures

▶ Silicon Nanowires

Nanowired Cerebrolysin and Superior Neuroprotective Effects in Silver Neurotoxicity

▶ Silver, Neurotoxicity

NCKX

▶ Potassium Dependent Sodium/Calcium Exchangers

NCS-1

▶ Calcium, Neuronal Sensor Proteins

Nedaplatin

▶ Platinum Anticancer Drugs

Neodymium

Takashiro Akitsu
Department of Chemistry, Tokyo University of Science, Shinjuku-ku, Tokyo, Japan

Definition

A lanthanoid element, the third element (cerium group) of the f-elements block, with the symbol Nd, atomic number 60, and atomic weight 144.24. Electron configuration [Xe]$4f^6 6s^2$. Neodymium is composed of stable (^{142}Nd, 27.2%; ^{143}Nd, 12.2%; ^{145}Nd, 8.3%; ^{146}Nd, 17.2%; ^{148}Nd, 5.7%) and two radioactive (^{144}Nd, 12.2%; ^{150}Nd, 5.6%) isotopes. Discovered by C. A. von Welsbach Mosander in 1885. Neodymium exhibits oxidation states II, III, and IV; atomic radii 181 pm, covalent radii 202 pm; redox potential (acidic solution) Nd^{3+}/Nd -2.431 V; Nd^{3+}/Nd^{2+} -2.6 V; electronegativity (Pauling) 1.14. Ground electronic state of Nd^{3+} is $^4I_{9/2}$ with S = 3/2, L = 6, J = 9/2 with $\lambda = 290$ cm^{-1}. Most stable technogenic radionuclide ^{150}Nd (half-life 6.7 $\times 10^{18}$ years). The most common compounds: Nd$_2$O$_3$, NdCl$_3$, and NdS. Biologically, neodymium is of low to moderate toxicity, can cause a fatty liver by intravenous infusion (Atkins et al. 2006; Cotton et al. 1999; Huheey et al. 1997; Oki et al. 1998; Rayner-Canham and Overton 2006).

Cross-References

▶ Lanthanide Ions as Luminescent Probes
▶ Lanthanide Metalloproteins
▶ Lanthanides and Cancer
▶ Lanthanides in Biological Labeling, Imaging, and Therapy
▶ Lanthanides in Nucleic Acid Analysis
▶ Lanthanides, Physical and Chemical Characteristics

References

Atkins P, Overton T, Rourke J, Weller M, Armstrong F (2006) Shriver and Atkins inorganic chemistry, 4th edn. Oxford University Press, Oxford/New York

Cotton FA, Wilkinson G, Murillo CA, Bochmann M (1999) Advanced inorganic chemistry, 6th edn. Wiley-Interscience, New York

Huheey JE, Keiter EA, Keiter RL (1997) Inorganic chemistry: principles of structure and reactivity, 4th edn. Prentice Hall, New York

Oki M, Osawa T, Tanaka M, Chihara H (1998) Encyclopedic dictionary of chemistry. Tokyo Kagaku Dojin, Tokyo

Rayner-Canham G, Overton T (2006) Descriptive inorganic chemistry, 4th edn. W. H. Freeman, New York

Neptunium, Physical and Chemical Properties

Fathi Habashi
Department of Mining, Metallurgical, and Materials Engineering, Laval University, Quebec City, Canada

Neptunium is silvery in appearance, chemically reactive, and is found in three allotropes:
- α-neptunium, orthorhombic, density 20.45 g/cm^3
- β-neptunium above 280°C, tetragonal, density 19.36 g/cm^3 at 313°C
- γ-neptunium above 577°C, cubic, density 18 g/cm^3 at 600°C

Neptunium has the largest liquid range of any element, 3,363°C, between the melting point and boiling point. It is a radioactive metal that is a by-product of nuclear reactors and plutonium production

$$^{238}_{92}U - ^{1}_{0}n \rightarrow ^{239}_{92}U \xrightarrow[23\,min]{\beta^-} ^{239}_{93}Np \xrightarrow[2.355\,d]{\beta^-} ^{239}_{94}Pu$$

^{237}Np is irradiated with neutrons to create ^{238}Pu, an alpha emitter for radioisotope thermal generators for spacecraft and military applications. ^{237}Np will capture a neutron to form ^{238}Np and beta decay with a half-life of 2 days to ^{238}Pu.

$$^{237}_{93}Np + ^{1}_{0}n \rightarrow ^{238}_{93}Np \xrightarrow[2.117\,d]{\beta^-} ^{238}_{94}Pu$$

Nineteen neptunium radioisotopes have been characterized, with the most stable being ^{237}Np with a half-life of 2.14 million years, ^{236}Np with a half-life of 154,000 years, and ^{235}Np with a half-life of 396.1 days. All of the remaining isotopes have half-lives that are less than 4.5 days, and the majority of these have half-lives that are less than 50 min. This element also has four meta states, with the most stable being ^{236m}Np ($t_{1/2}$ 22.5 h). Neptunium has a unique decay chain ending with thallium and not lead as in the other series (Fig. 1). ^{237}Np is fissionable.

Physical Properties

Atomic number	93
Atomic weight	237
Relative abundance in Earth's crust, %	0
Density, g/cm^3	20.45
Atomic radius, pm	155
Melting point, °C	637
Boiling point, °C	4,000
Heat of fusion, kJ·mol^{-1}	3.20
Heat of vaporization, kJ·mol^{-1}	336
Molar heat capacity, J·mol^{-1}·K^{-1}	29.46

Chemical Properties

^{237}Np is produced through the reduction of ^{237}NpF$_3$ with barium or lithium vapor at around 1,200°C

$$2\,NpF_3 + 3\,Ba \rightarrow 2\,Np + 3\,BaF_2$$

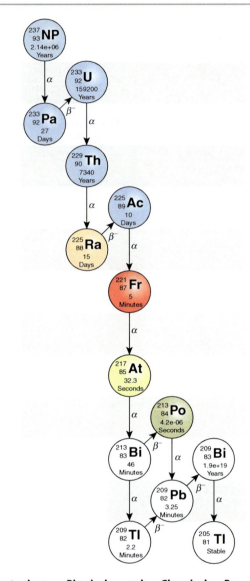

Neptunium, Physical and Chemical Properties, Fig. 1 Neptunium decay chain

It is most often extracted from spent nuclear fuel rods as a by-product in plutonium production. Neptunium has four ionic oxidation states in solution: Np^{3+} (pale purple), Np^{4+} (yellow green), NpO$_2^+$ (green blue), and NpO$_2^{2+}$ (pale pink). Neptunium (III) hydroxide is insoluble in water and does not dissolve in excess alkali. It is susceptible to oxidation in contact to air forming neptunium (IV). Neptunium forms tri- and tetrahalides, e.g., NpF$_3$, NpF$_4$, NpCl$_4$, NpBr$_3$, NpI$_3$, and oxides of various compositions, e.g., Np$_3$O$_8$ and NpO$_2$. Neptunium (VI) fluoride, NpF$_6$, is volatile like uranium hexafluoride.

References

Asperey LB, Penneman RA (1967) The chemistry of the actinides. Chem & Eng News, 31 July, pp 75–91

Habashi F (2003) Metals from ores. An introduction to extractive metallurgy. Métallurgie Extractive Québec, Québec City 2003. Distributed by Laval University Bookstore, www.zone.ul.ca

http://en.wikipedia.org/wiki/Neptunium

Seaborg GT (1966) Progress beyond plutonium. Chem & Eng News, 20 June, pp 76–88

Nervous Cell's Resistance

▶ Lithium, Neuroprotective Effect

Neuronal Calcium

▶ Calcium in Nervous System

Neuropathology

▶ Silver Impregnation Methods in Diagnostics

Neuroplasticity

▶ Lithium, Neuroprotective Effect

Neurotrophic

▶ Lithium, Neuroprotective Effect

Nickel

▶ Nickel Ions in Biological Systems

Nickel Biochemistry

▶ Nickel in Bacteria and Archaea

Nickel Carcinogenesis

Kazimierz S. Kasprzak[1] and Wojciech Bal[2]
[1]Chemical Biology Laboratory, Frederick National Laboratory for Cancer Research, Frederick, MD, USA
[2]Institute of Biochemistry and Biophysics, Polish Academy of Sciences, Warsaw, Poland

Synonyms

Nickel carcinogenicity; Nickel tumorigenicity; Nickel-induced cancer; Nickel-induced tumors

Definition

Nickel compounds are capable of inducing malignant tumors in humans and animals.

Human Epidemiology

The development of respiratory tract cancers in a Welsh nickel refinery workers was first reported in 1933. By 1949, nasal and paranasal sinuses and lung cancers were proclaimed in Great Britain as occupational diseases among nickel workers. To date, the carcinogenicity of nickel derivatives has been confirmed by wider epidemiology studies in humans and experimental bioassays in animals (IARC 1990; Kasprzak and Salnikow 2007).

Carcinogenic Nickel Compounds

The original respiratory tract malignancies were associated with chronic exposure of workers to nickel-containing dusts and fumes from roasting and smelting processes. Although it was believed for years that only water-insoluble nickel components of the fumes and dusts (e.g., nickel sulfides and oxides) were to blame, more recent studies indicate that aerosols of water-soluble nickel compounds are carcinogenic to humans as well. Tobacco smoking was found to be only a weak to moderate confounder (Grimsrud et al. 2003).

Besides occupational exposures, nickel present in endoprostheses, bone-fixing plates and screws, and other implanted medical devices made of metal alloys has been suspected, but not proven, to be the cause of sporadic local tumors.

The International Agency for Research on Cancer (IARC) concluded that nickel sulfate and the combinations of nickel sulfides and oxides encountered in the nickel refining industry are carcinogenic to humans (group 1) and that metallic nickel is possibly carcinogenic to humans (group 2B).

There is no published evidence of cancer risk from general environmental and dietary nickel exposures (Kasprzak and Salnikow 2007).

Histopathologic Types of Nickel Cancer

The sinonasal cancers found in nickel workers were squamous cell carcinomas (48%), anaplastic and undifferentiated carcinomas (39%), adenocarcinomas (6%), transitional cell carcinomas (3%), and other malignant tumors (4%). Lung tumors included squamous cell carcinomas (67%); anaplastic, small cell, and oat cell carcinomas (15%); adenocarcinomas (8%); large cell carcinomas (3%); and other malignant tumors (7%) (Sunderman et al. 1989).

Carcinogenesis in Experimental Animals

Carcinogenic Nickel Compounds

Since 1943, when the first inhalatory experiment on nickel was performed in mice, experiments on laboratory animals revealed carcinogenicity of nickel compounds with low or no aqueous solubility (e.g., $Ni(OH)_2$, Ni_3S_2, NiS, NiO) following inhalation or parenteral administration. Carcinogenesis of soluble nickel compounds was studied less extensively, yielding positive results only in some experiments. Therefore, insoluble nickel compounds are generally considered as stronger carcinogens than soluble compounds most likely because of longer retention at the target tissue (IARC 1990; Kasprzak et al. 2003; Kasprzak and Salnikow 2007).

Species and Strain Susceptibility

No absolute species specificity has been observed in nickel carcinogenesis. However, rats are more susceptible than mice, hamsters, or rabbits, and substantial variations in susceptibilities occur within rat and mouse strains. The reasons for these differences may include different abilities of phagocytes to ingest and metabolize nickel-containing particles and differences in the capacity of antioxidant systems among animals of different species and strains (IARC 1990; Kasprzak and Salnikow 2007).

Exposure Routes

Nickel compounds induce tumors at virtually all application sites, though with different efficacy. In rats, various organs appeared to be susceptible to nickel via local administration, but intraocular and intramuscular injections yielded the highest tumor incidences (Sunderman 1984). The routes of administration that produced local tumors in animals include inhalation and intramuscular, subcutaneous, intrarenal, intraperitoneal, intraocular, intracranial, and intra-articular injections.

Inhalation

A review of inhalation carcinogenesis by selected nickel compounds has been published by Oller et al. (1997). Pulmonary tumors were induced in rats inhaling water-insoluble $Ni(CO)_4$ vapor, or Ni_3S_2, NiO, and feinstein (refinery dust–containing Ni_3S_2, NiO, and metallic nickel). In contrast, however, soluble nickel sulfate, $NiSO_4$, was found not to be carcinogenic. Mice were more resistant to nickel carcinogenesis by the inhalatory route than rats.

A single intratracheal instillation of Ni_3S_2 resulted in the development of only one tumor in 26 rats likely because of fast clearance of the deposited particles. But Ni_3S_2 readily induced carcinomas and sarcomas in tracheal transplants, which could not eliminate the carcinogen. Thus, prolonged exposure to a nickel carcinogen is critical for tumor development (Kasprzak and Salnikow 2007).

Intramuscular and Subcutaneous Injections

The intramuscular and subcutaneous injections of water-insoluble Ni_3S_2 or NiO resulted in the development of local sarcomas in rats and mice of various strains and in other animals. The yield of these tumors (predominantly rhabdomyosarcomas and fibrosarcomas) was often 100% in 1 year, and they readily metastasized to the lung and other organs (Kasprzak et al. 2003; Kasprzak and Salnikow 2007).

Intraperitoneal Injection

Multiple intraperitoneal injections of water-soluble nickel acetate increased lung tumor incidence in strain

A mice. A single intraperitoneal injection of nickel acetate followed by dietary treatment with a tumor promoter, sodium barbital, resulted in the development of renal cortical adenomas in male F344 rats. Likewise, nickel acetate injected intraperitoneally to pregnant F344 rats produced tumors in the offspring: pituitary tumors developed without barbital and renal tumors developed only with barbital coadministration. These results indicate that nickel acetate is a complete carcinogen for fetal rat pituitary gland and a potent initiator of carcinogenesis in fetal rat kidney (Kasprzak et al. 2003; Kasprzak and Salnikow 2007).

Intrarenal Injection
Kidney tumors developed in rats following intrarenal injections of Ni_3S_2. Histologically, most of these tumors resembled the sarcomatous variant of the renal mesenchymal tumor. The Ni_3S_2 injections also caused erythrocytosis due to induction of erythropoietin, a part of the hypoxia-mimicking effect of nickel that may be involved in the mechanisms of carcinogenesis (Salnikow and Kasprzak 2007).

Intratesticular Injection
Malignant testicular tumors developed in rats after intratesticular injection of Ni_3S_2. The tumors were classified as fibrosarcomas, malignant fibrous histiocytomas, and rhabdomyosarcomas. This histologic spectrum of tumors indicated that Ni_3S_2 caused malignant transformation of undifferentiated pluripotent mesenchymal cells (Kasprzak et al. 2003; Kasprzak and Salnikow 2007).

Intraocular Injection
Intraocular injections of Ni_3S_2 to rats resulted in the development of local malignant tumors. Interestingly, Ni_3S_2 also induced ocular tumors in an evolutionarily distant species, the Japanese newt *Cynops pyrrhogaster* (Kasprzak and Salnikow 2007).

Other Routes of Exposure
Malignant sarcomas developed in rats at the site of intracranial and intra-articular injections of water-insoluble Ni_3S_2. In contrast, no tumors developed following treatment of hamsters with Ni_3S_2 by multiple applications into the cheek pouch, oral cavity, or digestive tract. Likewise, no malignant tumors occurred in rats after injections of Ni_3S_2 into the submaxillary gland or into the liver. And there is no evidence of cancer induction by soluble nickel administered in drinking water (Kasprzak and Salnikow 2007).

Interactions with Other Carcinogens
Coadministration of nickel compounds with organic carcinogens produced a significant synergistic effect. A single intramuscular injection of Ni_3S_2 with 3,4-benzo[a]pyrene to rats induced more sarcomas than injection of the carcinogens alone. A similar effect was observed after intratracheal instillation of the same compounds to rats. Most of the muscle sarcomas caused by nickel alone or in combination with 3,4-benzo[a]pyrene were rhabdomyosarcomas typical for nickel carcinogenesis in rats, whereas exposure to 3,4-benzo[a]pyrene alone produced fibrosarcomas (Kasprzak et al. 2003; Kasprzak and Salnikow 2007).

Interactions with Essential Metals
The essential metals manganese, magnesium, and zinc, but not calcium, coadministered intramuscularly with Ni_3S_2 to rats, reduced local tumor incidence in a dose-dependent manner. Magnesium was the strongest, and zinc was the weakest inhibitor. However, separate administration of the essential metals did not produce this effect. Magnesium also strongly inhibited renal carcinogenesis by Ni_3S_2 in the rat (Kasprzak et al. 2003; Kasprzak and Salnikow 2007).

Co-injection of Ni_3S_2 with iron, either metallic powder (Fe°) or Fe(III) sulfate, inhibited Ni_3S_2 carcinogenicity in the rat muscle. However, when co-injected with Ni_3S_2 into the rat kidney, Fe° did not affect the final yield of tumors but significantly accelerated their development. Neither Fe° nor Fe(III) produced muscle or kidney tumors by themselves. In strain A mice, addition of magnesium or calcium acetates lowered the incidence of lung adenomas promoted by nickel acetate (IARC 1990; Kasprzak et al. 2003; Kasprzak and Salnikow 2007).

Malignant Transformation of Cultured Cells

Soluble and insoluble nickel compounds are capable of transforming cultured human and rodent cells to a malignant phenotype that includes altered morphology, immortalization, anchorage independence, and development of tumors upon injection to athymic mice. Rodent cells are transformed more easily than human cells. At equitoxic doses, insoluble Ni_3S_2, NiO,

Ni_2O_3, and soluble Ni(II) acetate showed equal transforming effects in BHK-21 cells. The ability of nickel to immortalize cultured cells was found to be higher than that of other strong carcinogens, such as benzo[a]pyrene diol epoxide, N-methyl-N-nitrosourea, and γ- or X-rays. In addition to the cell-transforming ability, typical for tumor-initiating agents and complete carcinogens, nickel also has the properties of tumor promoters, i.e., agents that facilitate the development of tumors initiated by other carcinogens (Kasprzak et al. 2003; Kasprzak and Salnikow 2007).

Mechanistic Considerations

Biological Perspectives

Genotoxic Effects
Exposure of cells to nickel compounds, both soluble and insoluble, results in structural damage to chromosomes. In cultured rodent cells, the damage is detectable microscopically as micronuclei formation, regional decondensation, deletions, and sister chromatid exchange (SCE). In human lymphocytes, nickel sulfate increased SCEs frequency, while Ni_3S_2 enhanced micronuclei formation. Other types of genotoxic damage by nickel included mutations in cultured human lung cancer cells, insertion mutation in rat kidney epithelial cells, and deletion mutations in Chinese hamster ovary cells. The G → T transversion mutation, indicative of oxidative DNA damage, was found in the K-ras gene in renal tumors induced by Ni_3S_2. The same type of point mutation in the p53 gene was associated with nickel-induced human lung tumors (Kasprzak et al. 2003; Kasprzak and Salnikow 2007).

Despite numerous reports of DNA and chromatin damage in nickel-treated cells, the mutagenic potential of this metal has been usually described as low. This notion is based on the often conflicting results of mutagenesis assays in bacteria, fruit fly, some mammalian cells, and whole animals. Thus, nickel was not mutagenic to *E. coli* and *S. typhimurium* but did mutate *Corynebacterium* sp. 887 (*hom*) bacteria. However, nickel acted as a potent co-mutagen with alkylating agents in some *E. coli* and *S. typhimurium* strains. Treatment with Ni_3S_2 of nasal mucosa and lung cells isolated from wild-type mice resulted in DNA fragmentation that might be mutagenic. In contrast, when Ni_3S_2 was applied to respiratory tract cells of transgenic mice and rats, the mutation frequency of the lacZ and lacI reporter genes was not increased. Also, no increase in ouabain-resistant or 6-thioguanine-resistant colony formation was found in human fibroblasts even at concentrations of Ni_3S_2 that increased 200-fold the frequency of anchorage independence of these cells. No 6-thioguanine-resistant colonies were formed in V79 cells exposed to NiS or NiO. So, the mutagenicity of nickel can be observed only in selected experimental models (Kasprzak et al. 2003; Kasprzak and Salnikow 2007; Bal et al. 2011; Kasprzak 2011).

Epigenetic Effects
Alterations in Gene Expression Epigenetic toxicity of nickel that may assist in malignant cell transformation manifests itself through deregulation of gene expression. The latter may be tested experimentally at the genome level and the levels of signaling, transcription, and post-transcriptional modifications. At the genome level, analysis of nickel-exposed cells and animals revealed changes in the expression of genes controlling inflammation, stress, cell proliferation, and cell death. Those rather nonspecific effects have been accompanied by a more typical effect of nickel: upregulation of genes responding to hypoxia, especially those involved in glucose transport and glycolysis. This response is regulated through activation of the HIF-1 transcription factor. Over the time of a prolonged exposure to nickel, which is essential for carcinogenesis, the hypoxia-like reaction may facilitate selection of cells into a glycolytic phenotype, common among cancer cells. It may also assist in selection of cells that are resistant to apoptosis, another cancer hallmark. In addition, nickel suppression of the inflammatory/immune responses, observed in animals, impairs recognition and destruction of malignant cells. Altogether, the hypoxia-mimicking and immunosuppressive effects of nickel have the potential of promoting tumor development (Salnikow and Kasprzak 2007).

Since lung is the major target for nickel carcinogenesis in humans, changes in gene expression in lung cells exposed to nickel have been studied extensively. These changes included increased expression of some metal-sequestering protein genes and hypoxia-inducible genes, as well as enhancement of extracellular matrix repair, cell proliferation, and oxidative stress. At the same time, the expression of surfactant-associated proteins was decreased (Salnikow and Kasprzak 2007). In particular, upregulated were the calcium-dependent NDRG1/Cap43 protein, and proteins regulating iron functions, lactotransferrin, ferritin, heme oxygenase, and erythropoietin. This may lead to disturbances in calcium signaling (see Disruption of Calcium Homeostasis) and iron homeostasis. The latter will assist in ROS generation and oxidative stress with all its genotoxic and epigenetic consequences (see Mediation of Oxidative Damage).

The epigenetic effects of nickel also include activation of K-ras oncogene and inhibition of tumor suppressor genes Rb, p16, p53, FHIT, Zac-1, and Gas-1. Transcription factors TGF-β, AP-1, ATF-1, NF-κB, NF-AT, and especially HIF-1 are activated by nickel. The activation of HIF-1 is the result of loss of proline hydroxylation in its component HIF-1α. This, in turn, follows nickel-induced depletion of ascorbate (see "Effect on Ascorbic Acid, Glutathione, and Other Antioxidant Systems"). The HIF-1 transcription factor upregulates expression of numerous genes involved in glucose transport and glycolysis, and other genes, including those coding for carbonic anhydrase IX, ceruloplasmin, erythropoietin, inducible nitric oxide synthase; NDRG-1/Cap43; vascular endothelial growth factor (VEGF); and many others. The downregulation of the Zac-1 and Gas-1 genes, mentioned above, is also HIF-1 dependent (Salnikow and Kasprzak 2007).

Inhibition of DNA Repair Enzymes Inhibition of any element of the DNA repair systems has the potential of assisting in cell mutation and carcinogenesis. Nickel was found to inhibit base and nucleotide excision repair. Such inhibition enhanced DNA damage inflicted by UV radiation; ROS; benzo[a]pyrene-7,8-diol 9,10-epoxide; and alkylating agents. The proteins targeted by nickel in DNA repair machinery include the XPA zinc-finger protein and O^6-methylguanine-DNA methyltransferase, but not Fpg-glycosylase. Effects of nickel and other metals were also tested on the bacterial and human deoxyribo- and ribonucleotide phosphohydrolases, MutT and MTH1, which prevent potentially mutagenic utilization of damaged nucleoside triphosphates for DNA and RNA synthesis. Nickel inhibited both enzymes. In addition, through the depletion of ascorbate, nickel has the potential of inhibiting alkyl DNA dioxygenases ABH1 and ABH2 (Kasprzak and Salnikow 2007).

DNA Methylation and Histone Modifications Nickel enhances methylation of certain cytosine residues in DNA and affects acetylation, methylation, and other modifications of histones in cultured cells. These effects alter chromatin structure and gene expression and may thus assist in cell transformation. DNA hypermethylation has a gene-silencing effect, and this was found in the promoter region of the p16 tumor suppressor gene of nickel-induced murine tumors. The acetylation of all four core histones has been reported to be downregulated by nickel through inhibition of specific histone acetyltransferases. This effect was strongest in histone H2B and weakest in histone H3. The decrease in histone acetylation is expected to inhibit gene expression.

Analysis of core histones extracted from nuclei of cells exposed to nickel revealed changes in their monoubiquitination status and presence of certain unique structural modifications, such as truncation and deamidation. The truncation was found in histones H2A and H2B: histone H2A lost its C-terminal tail's ending – SHHKAKGK by the direct Ni(II) action, whereas histone H2B was truncated on both termini in two identical, KAVTK, repeats, typical for the action of calpains. Besides the truncation, histone H2B also had its Gln-22 residue deamidated and Met-59 and Met-62 residues oxidized to sulfoxides, a signature of oxidative stress. Since the enzymatic activity of calpains depends on calcium, the truncation of H2B may be indicative of disruption by nickel of calcium homeostasis (Salnikow and Kasprzak 2007; Bal et al. 2011).

Disruption of Calcium Homeostasis Nickel blocks calcium channels and disturbs intracellular calcium homeostasis that results in the experimentally observed rapid proliferation of nickel-transformed cells in low-calcium media. Since cytoplasmic Ca(II) pulses regulate expression of genes associated with cell growth, differentiation, and apoptosis, one can presume that derangement of such pulses is pathogenic and may lead to malignant transformation. These data prompted broader investigations of nickel/calcium interactions in gene expression signaling reviewed in more detail by Salnikow and Kasprzak (2007).

Chemical Perspectives
Nickel Interactions with Biomolecules
Pathogenic effects of nickel such as cell transformation and cancer are the ultimate results of the primary chemical interactions of nickel with biomolecules and essential metal ions. These interactions include

binding of nickel cations by various cell and tissue components and the enhancement of ROS generation by the bound metal. Ingested particles of water-insoluble nickel compounds (e.g., Ni_3S_2) may generate ROS in the process of their oxidative solubilization in cell and tissue fluids. Hence, nickel may disturb functions of biomolecules due to changes in their conformational and molecular integrity that follow its binding and/or damage caused by ROS. The general rules of nickel binding to proteins and other molecules are presented elsewhere in this encyclopedia. Below, only these molecules, which seem to be relevant to carcinogenesis, are considered.

Chromatin: DNA, Histones, and Protamines Binding of nickel to DNA is relatively weak. Therefore, in nuclear and sperm chromatin, the major nickel binders are the proteins, especially histones and protamines. Briefly, strong nickel-binding motifs are present in protamine P2 (RTHGQSHYRR N-terminal motif) and in core histones H3 (-CAIH-) and H2A (-TESHHKAKGK C-terminal motif). Importantly, nickel coordinated in -CAIH-, RTHGQSHYR-, and -SHHKAKGK complexes can mediate oxidative damage to other molecules (see "Mediation of Oxidative Damage"). The Ni(II)-SHHKAKGK complex is derived from the Ni(II)-TESHHKAKGK complex in histone H2A owing to the nickel-facilitated hydrolysis of the E-S bond. Since the C-terminal tail of H2A is involved in maintaining chromatin structure, its truncation may affect gene expression (Salnikow and Kasprzak 2007).

Regulatory Proteins The DAN protein has a nickel-binding motif – PHSHAHPHP. Since DAN has tumor-suppressive activity, its possible interaction with nickel may assist in carcinogenesis. Similarly, cullin-2, a component of the E3 ligase complex that ubiquitinates HIF-1α for proteosomal destruction, binds nickel at several sites. The binding prevents the ubiquitination and may thus assist in triggering the hypoxic gene response by nickel. Lipovitellin-2β and importin-α appear to be proteins whose binding of nickel also may deregulate gene expression. Nickel can also interact with the iron regulatory protein-1 (IRP-1), a central regulator of iron homeostasis. Replacement by nickel of one iron ion in the 4Fe-4 S cluster of IRP-1 inhibits its enzymatic activity; this also may contribute to the nickel-induced hypoxic response (Kasprzak and Salnikow 2007; Bal et al. 2011).

Mediation of Oxidative Damage
The involvement of reactive oxygen species (ROS) in nickel toxicity and carcinogenesis was reviewed recently by Kasprzak (2011). ROS may be generated by both soluble and insoluble nickel compounds. This is possible when Ni(II) is complexed by certain natural ligands like peptides and proteins, especially these forming square planar nickel complexes. Reactions of such complexes with O_2 or H_2O_2 yield the hydroxyl radical ˙OH and other radicals. The oxidation of water-insoluble nickel sulfides may involve both nickel and sulfur and lead to generation of not only Ni(II)- but also sulfur-derived ROS and other reactive intermediates (e.g., the sulfite anion). This enables the sulfides to produce a wider spectrum of oxidative damage than other nickel compounds and may be responsible for their high carcinogenic activity.

Lipid Peroxidation The relevance of lipid peroxidation to carcinogenesis can be related to reactive intermediates, including ROS, and specifically to promutagenic nature of the DNA adducts with some products of this peroxidation, e.g., 4-hydroxynonenal. The formation of lipid peroxides is the result of oxidation of polyenic fatty acid residues of phospholipids. Lipid peroxides decompose to epoxy-fatty acids, alkanes, alkenes, alkanals, alkenals, hydroxyalkenals, and aldehydes, many of them being able to attack and further damage other molecules. Nickel, like some other transition metals, was found to promote lipid peroxidation in both in vitro and in vivo experiments (Kasprzak and Salnikow 2007; Kasprzak 2011).

Protein Oxidation Protein oxidation is mechanistically involved in a wide variety of pathologic effects, including cancer development, and nickel may catalyze it. The major targets are the side chains of Cys, His, Arg, Lys, and Pro. Radical intermediates arising in protein oxidation may lead to protein fragmentation and cross-linking, e.g., with DNA. Because of the site-specific nature of nickel binding, protein damage may also be site specific. For example, Ni(II) bound to the RTH motif in the 15-mer peptide RTHGQSHYRRRHCSR-amide, modeling the N-terminal sequence of human protamine P2, catalyzes oxidation by H_2O_2 of not only Arg-1 and His-3 but also Tyr-8. The reason for this is a strong structuring effect of nickel on the peptide ligand that brings Tyr-8 close to the metal center. The bound nickel also shifts all the positive Arg side chains to one side of the molecule that increases binding of the peptide with DNA.

DNA Damage In nuclear chromatin, nickel is likely to associate predominantly with histones. The protein-bound nickel may catalyze ROS generation and thus induce oxidative damage to all chromatin components, including DNA. Generally, oxidative effects found in DNA of animals and cultured cells treated with nickel are cross-links of various types, strand scission, depurination, and base modifications.

The most common effect of nickel in chromatin is DNA-protein cross-linking. This effect can be induced in two ways: through the formation of mixed-ligand nickel complexes or through generation of strong covalent bonds directly between DNA and the proteins; nickel was found to generate both. Intra- and interstrand cross-linking in DNA is also possible, but in case of nickel, the formation of such cross-links could only be deduced from the double CC → TT mutations typical for intrastrand cross-links developing in DNA exposed to nickel and hydrogen peroxide. The presence of cross-links in chromatin may lead to morphologic aberrations of chromosomes. Such aberrations were observed in lymphocytes of workers exposed to nickel compounds. Another common ROS effect is DNA cleavage (strand scission). In vitro, nickel was found to promote DNA cleavage by H_2O_2 predominantly at the cytosine, thymine, and guanine residues (Kasprzak and Salnikow 2007; Bal et al. 2011).

The spectrum of damage produced by ROS in DNA bases includes over 20 DNA base products, including 8-oxoguanine (8-oxo-dG), the most popular indicator of oxidative DNA damage. Under air, nickel assisted in DNA base oxidation in isolated chromatin, but not in pure DNA, thus indicating participation of chromatin proteins in producing this effect.

Following in vivo exposure to soluble nickel, elevated amounts of at least one damaged DNA base were found in organs of rats and mice. In the lungs of rats, both insoluble Ni_3S_2 and NiO and soluble nickel sulfate, instilled intratracheally, increased pulmonary 8-oxo-dG levels. But in cultured HeLa cells, only Ni_3S_2 was active. This difference seems to indicate different mechanisms of the damage in vitro and in vivo by the same compounds, perhaps owing to contribution of ROS generated by inflammatory cells in the rat lung. In mice treated with soluble nickel, renal 8-oxo-dG levels were increased only in the BALB/c strain, which has low glutathione and glutathione-peroxidase levels compared with other strains.

DNA damage resulting from exposure to nickel also includes depurination. Nickel produced apurinic sites in DNA and mediated in vitro hydrolysis of 2′-deoxyguanosine. The depurination occurred concurrently with DNA strand scission, and both effects could be the result of ROS attack on DNA's sugar moiety. Modified sugars constitute alkali-labile sites that have been found in DNA from nickel-treated cells. ROS are eventually also responsible for the generation of DNA adducts with the major end products of lipid peroxidation, 4-hydroxynonenal and malondialdehyde, and products of protein oxidation.

Effect on Ascorbic Acid, Glutathione, and Other Antioxidant Systems The oxidative damage that follows nickel treatment may be aggravated by depletion by nickel of major cellular antioxidants ascorbate and glutathione or inhibition of antioxidant enzymes. Nickel is capable of depleting intracellular ascorbate due to catalytic oxidation and hydrolysis of both ascorbic and dehydroascorbic acid. The underlying chemistry includes ROS generation. Nickel also inhibits ascorbic acid transporters SVCT1 and SVCT2. The depletion of ascorbate leads to the inhibition of cellular hydroxylases that, in turn, may result in activation of the HIF-1-dependent hypoxic response, impairment of assembly of proteins with collagen-like domains, and/or inactivation of ascorbate-dependent DNA and histone demethylases and DNA dealkylases. These effects may thus be mechanistically involved in carcinogenesis through enhancement of DNA damage by endogenous oxidants and prevention of DNA repair.

In cultured cells, nickel depletes yet another natural antioxidant, glutathione (GSH). Ni(II) sensitizes GSH to oxidation by air only in alkaline, but not in neutral solutions. Therefore, a direct impact of nickel on GSH oxidation may be limited, and the depletion of cellular GSH by nickel, found experimentally, would be a secondary effect owed to scavenging by GSH of ROS generated by other nickel-dependent mechanisms. The importance of GSH in the defense against ROS has been indicated by a marked increase of GSH production in cells resistant to nickel and hydrogen peroxide. GSH levels are also higher in muscles of C57BL mice, which are less sensitive to nickel carcinogenesis than muscles of C3H mice having lower GSH levels.

Nickel may also affect the enzymatic components of cellular antioxidant defense. In vitro, nickel inhibited superoxide dismutase (SOD) and catalase (CAT). However, in animals, nickel effects on these enzymes and on glutathione peroxidase (GSH-Px) and glutathione reductase (GSSG-R) appeared to be ambivalent, depending on the animal, target cells, nickel concentration, and regimen of exposure. Nonetheless, inhibition of these enzymes by nickel observed in animals under certain conditions might enhance oxidative damage.

Conclusion

Nickel is an occupational respiratory tract carcinogen in humans and a multi-tissue carcinogen in experimental animals. It also is a potent cell-transforming agent in vitro.

Nickel carcinogenicity is associated with its potential to damage cellular components through both direct and indirect attack on proteins, nucleic acids, and smaller molecules. The type and extent of the damage depend on the intracellular dose of nickel ions that, in turn, is a function of physicochemical properties of particular nickel derivatives, their ability to enter the cell and/or to dissolve within the cell.

The observed wide spectrum of the genotoxic and epigenetic effects of nickel that may lead to cancer is the ultimate result of the primary chemical interactions of nickel, including its binding to certain cellular ligands and redox activity of the resulting complexes. The ligands are amino acids, peptides, proteins, and other molecules, but not DNA. Nickel binding may be damaging to the function of a bioligand by itself because of conformational alterations or nickel-assisted hydrolysis of certain peptide bonds. However, the major damage, identified thus far in nickel-exposed cells and animals, appears to be inflicted by ROS that are generated in reactions of nickel complexes with oxygen and its metabolic derivatives (e.g., $^{\bullet}O_2^-$, H_2O_2, lipid peroxides). In humans and animals, ROS may also originate from inflammatory cells activated by nickel, especially by the insoluble particulate compounds. ROS are capable of inflicting oxidative damage to both the nickel-bound and neighboring biomolecules such as lipids, proteins, and nucleic acids. The strongest association of oxidative damage with nickel carcinogenesis stems from the promutagenic properties of many DNA base products generated by ROS and found in nickel-exposed cells in vitro and in vivo. Other DNA lesions like strand breaks and apurinic sites, which have also been found, may lead to mutations too (Kasprzak and Salnikow 2007; Kasprzak 2011). Therefore, the formation of all these lesions in DNA may be considered as a tumor-initiating effect, which can be further strengthened by the capacity of nickel to inhibit DNA repair.

Physiologically generated ROS serve as messengers in gene expression signaling. Therefore, it seems probable that the increase in redox reactions driven by intracellular nickel may disturb orderly generation of the messenger ROS and affect redox-regulated proteins, such as NF-κB, AP-1, p53, K-ras, Bcl-2, HIF-1, and others. This may disturb orderly progression of the cell cycle. Also, binding of nickel to histones may sensitize them to oxidation and hydrolytic damage and lead to changes in chromatin structure. Effects of this type along with oxidative damage to other proteins constitute epigenetic effects, which through alterations in gene expression, may assist in the promotion and progression of tumors.

The most profound epigenetic effects of nickel are those caused by activation of HIF-1 transcription factor. HIF-1 is involved in upregulation of numerous genes that maintain high glycolytic rate typical of cancer. Nickel exposure is thus capable of promoting selection of cells that acquire a phenotype similar to that of cancer cells. The mechanisms of HIF-1 activation involve depletion by nickel of cellular ascorbate with the resulting inactivation of prolyl hydroxylases. Proline hydroxylation is also crucial for proper assembly of collagen, extracellular matrix, and other proteins having collagen-like domains, such as the lung surfactant. Hence, the depletion of ascorbate can be especially deleterious for lung epithelial cells, the major target for nickel carcinogenesis in humans.

Future Directions

The investigations of nickel cancer in humans require further epidemiologic studies of possible threat posed by inhalation of nickel-containing welding fumes and wearing endoprostheses, bone-fixing plates and screws, and other internal medical devices (e.g., pacemakers and stents) made of nickel alloys. More efforts should also be focused upon epidemiology of environmental nickel exposures and co-exposures with other carcinogens (Bal et al. 2011). Animal models for such studies should include species that, like humans, depend on dietary ascorbate (e.g., guinea pigs and knock-out mice and rats).

The understanding of the molecular mechanisms of nickel carcinogenesis greatly benefits from research on the chemical basis of nickel interactions with cellular and tissue ligands and essential metals. Therefore, there is an urgent need for more investigations in this field. They will allow for identification of more nickel-binding biomolecules and better prediction of pathogenic effects of the binding (especially protein hydrolysis and ROS generation) and eventually lead to faster development of preventative measures against nickel cancer in humans. One such measure would include ascorbic acid supplementation to counteract the inhibition by nickel of HIF-1α prolyl hydroxylase and other hydroxylases, particularly those regulating DNA and histone functions. Here, further efforts are necessary to test the efficacy of ascorbate against nickel toxicity and carcinogenicity.

Acknowledgment This entry is dedicated to the memory of Dr. F. William Sunderman, Jr. (1931–2011) whose pioneering investigations of nickel toxicology and carcinogenesis have been inspirational to so many followers.

Cross-References

▶ Calcium in Health and Disease
▶ Iron Homeostasis in Health and Disease
▶ Magnesium in Biological Systems
▶ Magnesium in Health and Disease
▶ Metals and the Periodic Table
▶ Nickel-Binding Sites in Proteins
▶ Nickel Carcinogenesis
▶ Nickel Hyposensitization
▶ Nickel Superoxide Dismutase
▶ Nickel Transporters
▶ Nickel, Physical and Chemical Properties
▶ Nickel Ions in Biological Systems
▶ Urease
▶ Zinc and Zinc Ions in Biological Systems
▶ Zinc Finger Folds and Functions
▶ Zinc Homeostasis, Whole Body

References

Bal W, Protas AM, Kasprzak KS (2011) Genotoxicity of metal ions: chemical insights. Met Ions Life Sci 8:319–373

Grimsrud TK, Berge SR, Martinsen JI et al (2003) Lung cancer incidence among Norwegian nickel-refinery workers 1953–2000. J Environ Monit 5:190–197

IARC (1990) IARC monographs on the evaluation of carcinogenic risks to humans: chromium, nickel and welding, vol 49. International Agency for Research on Cancer, WHO, Lyon, pp 257–445

Kasprzak KS (2011) Role of oxidative damage in metal-induced carcinogenesis. In: Banfalvi K (ed) Cellular effects of heavy metals. Springer, New York, pp 237–259

Kasprzak KS, Salnikow K (2007) Nickel toxicity and carcinogenesis. In: Sigel A, Sigel H, Sigel RKO (eds) Metal ions in life sciences: nickel and its surprising impact in nature, vol 2. Wiley, Chichester, pp 619–660

Kasprzak KS, Sunderman FW Jr, Salnikow K (2003) Nickel carcinogenesis. Mutat Res 533:67–97

Oller AR, Costa M, Oberdorster G (1997) Carcinogenicity assessment of selected nickel compounds. Toxicol Appl Pharmacol 143:152–166

Salnikow K, Kasprzak KS (2007) Nickel-dependent gene expression. In: Sigel A, Sigel H, Sigel RKO (eds) Metal ions in life sciences: nickel and its surprising impact in nature, vol 2. Wiley, Chichester, pp 581–618

Sunderman FW Jr (1984) Carcinogenicity of nickel compounds in animals. In: Sunderman FW Jr (ed) Nickel in the human environment. IARC, Lyon, pp 127–142

Sunderman FW Jr, Morgan LG, Andersen A et al (1989) Histopathology of sinonasal and lung cancers in nickel refinery workers. Ann Clin Lab Sci 19:44–50

Nickel Carcinogenicity

▶ Nickel Carcinogenesis

Nickel Cofactor

▶ Nickel Ions in Biological Systems

Nickel Desensitization

▶ Nickel Hyposensitization

Nickel Enzymes

▶ Nickel Ions in Biological Systems

Nickel Hyposensitization

Mario Di Gioacchino
Occupational Medicine and Allergy, Head of Allergy and Immunotoxicology Unit (Ce.S.I.), G. d'Annunzio University, Via dei Vestini, Chieti, Italy

Synonyms

Nickel desensitization; Nickel vaccination

Definition

Hyposensitization refers to a specific therapeutic procedure that allows an allergic (i.e., sensitized) subject to become tolerant toward the sensitizing antigen, with reduction or absence of symptoms. In the case of Nickel hyposensitization, the procedure allows a part of sensitized patients affected by Allergic Contact Dermatitis (ACD) to safely make contact with Nickel containing objects, and those with Systemic Nickel Allergy Syndrome (SNAS) to freely eat Nickel containing foods. The complete disappearance of symptoms is observed in 50% of these patients, while the rest shows a variable degree of tolerance or, in a minority of subjects, no effects at all. No side effects are observed during and after the treatment. The hyposensitization treatment exerts its clinical effects through a modulation of the allergic immune reaction specific for Nickel.

Animal Experiments

Experimental studies in animal models demonstrate the possibility to induce immunological tolerance to complete antigens and haptens, including Nickel, by repeated oral administrations, through a mechanism involving suppressor and T regulatory lymphocytes. Moreover, oral feeding with an antigen can prevent a subsequent sensitization. In fact, in guinea pigs, the repeated oral administration of vegetable proteins could induce a state of antigen-specific tolerance in the animals (Wells and Osborne 1911) and a preventive treatment by repeated oral administration of chloro-2,4-dinitrobenzene renders guinea pigs unable to become subsequently sensitized (Chase 1946). Animals treated by oral route with Nickel and Chromium powder or metallic salts incorporated into pelleted foods fail to react to subsequent immunization whereas control animals not pretreated become clearly hypersensitive (Vreeburg et al. 1984). The induction of tolerance is in part related to the metal speciation, as hexavalent chromium ($K_2Cr_2O_2$) induces a full tolerance, while trivalent chromium ($CrCl_3$) induces a lower degree of tolerance and metallic chromium is ineffective. This preventive effect seems to be a consequence of an isotype-specific immunoregulation in gut-associated lymphoid tissues, part of which could be mediated by IgG-specific suppressor T lymphocytes and IgA-specific helper T lymphocytes, as shown after albumin repeated feeding in mice (Richman et al. 1981).

Oral administration of Nickel (both as $NiSO_4$ and $NiCl_2$) not only can prevent a subsequent sensitization for long time but, when administered to mice already sensitized to Nickel, is able to induce a long-term persistent desensitization. This effect is mediated by antigen-presenting cells (APCs) and by subsets of T lymphocytes (CD4-8+ T cells and T regulatory cells). In fact, Roelofs-Haarhuis and colleagues (2003) demonstrated that T suppressor cells and tolerogenic APCs induced by Nickel oral administration participate to a positive feedback loop that can enhance and maintain tolerance when activated by the antigen in the presence of a danger signal. Under these conditions, APCs and T suppressor effector cells infectiously spread the tolerance to naive T cells and APCs of a naive recipient.

Effects of Nickel Hyposensitization in Humans

The oral administration of capsules containing Nickel to ACD patients seems to be followed in humans by a reduction of symptoms associated to metal exposure. The first attempt was made by Sjövall et al. (1987), who, having observed that patients with Nickel ACD reported an improvement of their hand eczema and metal sensitivity after a positive oral provocation test with Nickel salts, administered orally capsules containing 5 mg of $NiSO_4$ to Nickel-sensitized patients for 6 weeks; this treatment led to the reduction of the degree of contact allergy, measured as an increase in the lower dose of $NiSO_4$ able to induce positive patch test reactions before and after treatment. No effects were observed with a dose of 0.5 mg. Many other authors applied protocols for Nickel hyposensitization with positive results in the majority of patients; however, those studies were conducted with a limited number of subjects and for a limited period of treatment. Instead, when applied subcutaneously, the hyposensitization protocol failed to produce the beneficial effects in Nickel-sensitized patients.

In 1995, the attention of clinicians was focused on the subset of patients with Nickel ACD that showed symptoms after ingestion of Nickel containing foods, mainly vegetables (▶ Nickel in Plants). A multicenter Italian study confirmed the efficiency of Nickel hyposensitization in this group of patients. The oral administration of very low doses of Nickel sulfate tablets (0.1 ng daily for the first year, then every other day for the second and third years) to 51 patients led to the disappearance of symptoms in 29 of the 30 patients who completed the treatment course (Panzani et al. 1995). This subset of patients, all affected by "typical" ACD, develop cutaneous manifestations (eczematous lesions, angioedema, and urticaria) localized at different sites than those generated by the direct contact with Nickel and suffer from gastrointestinal symptoms (meteorism, diarrhea, colic, etc.) after the ingestion of Nickel containing food (▶ Nickel in Plants). Such symptoms disappear or diminish after a low Nickel diet and recur or intensify after an oral Nickel challenge; besides, the severity of symptoms and the number of symptomatic patients are related to the Nickel dose orally administered. These findings lead to consider the cutaneous lesions as a consequence of the ingestion of Nickel.

This condition is variably named "Systemic Contact Dermatitis" or "Systemic Nickel Allergy Syndrome" (SNAS). The second definition seems to represent better the clinical picture and the pathogenesis of the disease. This issue was discussed in September 2005 in the congress "Allergy Vaccination" in Cyprus whose proceedings were published in the International Journal of Immunopathology and Pharmacology, volume18, supplement 4, 2005. SNAS affect almost 20% of patients with Nickel ACD, as shown by a wide clinical survey (1,086 patients, F: M = 1,023:63) performed between 1987 and 2003. These patients are not affected by alteration of Nickel metabolism (Andreassi et al. 1998). Moreover, SNAS and ACD originate from partially different pathogenic processes. In fact, other than cytokines of the Th1 pattern (INFγ and TNFα), SNAS appear to be mediated also by interleukin-4, interleukin-13, and interleukin-5, Th2 cytokine typical of the type I immune reactions and not usually active in the type IV immune responses, which characterizes ACD. Particularly, IL-5 is an immunological trait that distinguishes SNAS from ACD patients and healthy controls (Boscolo et al. 1999). In SNAS patients, but not in ACD, oral challenge is associated with infiltration of lymphocytes (mostly CD4+ T cells) in the duodenal epithelium, and decrease of CD8+ T cells, as consequence of cell apoptosis (Di Gioacchino et al. 2000).

Following these studies, a research group from the Catholic University of Rome made a large clinical trial, involving 214 patients, all affected by SNAS (characterized by positive patch test to Nickel, cutaneous and/or gastrointestinal symptoms after ingestion of Nickel containing food, clinical benefit from a Nickel free diet and positive oral Nickel challenge). A group of 136 patients were treated for 12 months with a very low dose of Nickel (up to 0.2 ng per day) while following a Nickel free diet. The control group (78 patients) just consumed Nickel free food for the same period. After 1 year, patients gradually resumed Nickel containing food: the majority of the Nickel-treated patients (94 out of 136, 69%) showed a clear clinical improvement in their condition, even complete remission in 47% of them, that could reintroduce Nickel containing food with no sign of disease; instead, only a minority of the patients of the control group (17.9%) could reintroduce dietary Nickel without showing symptoms.

The Nickel doses used for the hyposensitization treatment in the various studies were quite different ranging from 1 ng to 5 mg and not justified by dose finding studies, until 2010 when it was established that SNAS patients tolerate 1.5 μg Nickel/week without side effects, whereas ACD patients can receive very higher doses, up to 50 mg.

At present Nickel hyposensitization in SNAS is administered at the cumulative dose of 1.5 μg/week and proves to be effective in reducing symptoms and drug consumption. Besides, the treatment induces significant modulation of the immune system. The clinical benefits are maintained at least for 1 year, the longest period of follow-up that has been evaluated in controlled trials so far (Minelli et al. 2010). This study was designed as an open clinical trial to evaluate several pharmacological, clinical, and immunological features: the tolerability of increasing oral doses of Nickel, the safety and clinical efficacy of a long period treatment (1 year) with the highest Ni tolerated dose, and the potential for oral Nickel hyposensitization to modify some immunological

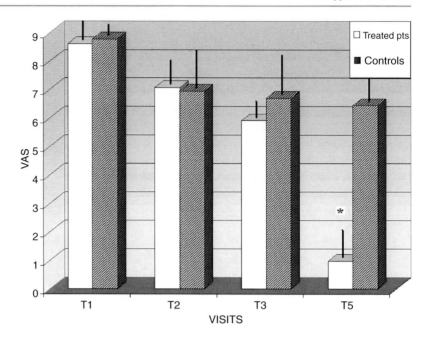

Nickel Hyposensitization, Fig. 1 Visual analogical scale (VAS) of NiOH-treated patients and controls at each of the four visits. At T5, the values of VAS of treated patients were significantly lower than those observed in the control group. The higher values express the most severe clinical pictures. Values are expressed as mean±SD. * = $p < 0.001$ respectively T1, T2, and T3 (Chi Square) (From Minelli et al. (2010). Int J Immunopathol Pharmacol)

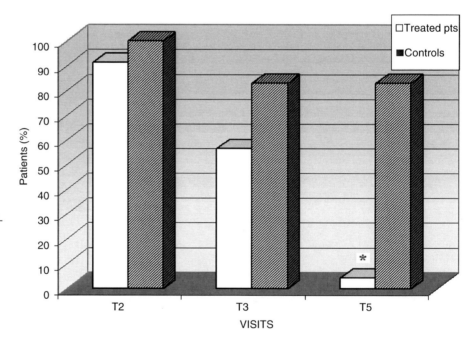

Nickel Hyposensitization, Fig. 2 Percentages of NiOH-treated patients and controls which necessitated rescue medications at the various stages of the treatment. *=$p<0.001$ respectively T2 and T3 (Chi Square) (From Minelli et al. (2010). Int J Immunopathol Pharmacol)

traits of SNAS. Also in this study two groups of patients were compared: one treated by low Nickel diet and the oral administration of Nickel, and the other (control group) treated by Nickel low diet alone. SNAS clinical severity and drug consumption showed a significant reduction in the group of Nickel-treated patients compared to the control group (Figs. 1 and 2). In agreement with that finding, 90% of the treated patients and none of the control group tolerated the Nickel-rich food reintroduction.

Oral Nickel hyposensitization, with high doses of metal, has been proposed also for ACD patients; however, the clinical trials set up so far involved a limited number of ACD patients and the treatment was administered for a short period of time. The most recent clinical trial (Bonamonte et al. 2011) studied 28 Nickel

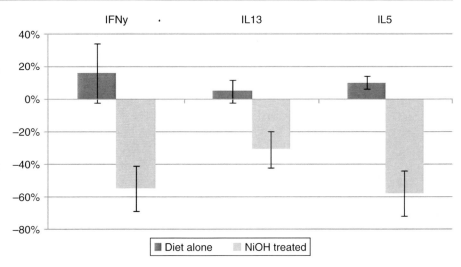

Nickel Hyposensitization, Fig. 3 Pre/post NiOH treatment changes (%) in in vitro Ni induced IFNγ, IL5 and IL13 release by PBMCs of patients treated by NiOH, or by diet alone. NiOH treated patients showed a significant decline in cytokine release with respect to the values detected before treatment (From Minelli et al. (2010). Int J Immunopathol Pharmacol)

ACD patients who received a daily dose of 50 μg of $NiSO_4$ in cellulose capsules for 3 months. In 26 patients, that completed the study, oral hyposensitization ameliorated clinical manifestations, despite continued Nickel exposures; moreover, the threshold of skin responsiveness to nickel increased and the T lymphocyte responsiveness to the metal in vitro decreased. During the 1-year follow-up, 50% of the patients experienced relapses of the clinical manifestations at the sites of topical exposure to Nickel, likely as consequence of the short period of treatment.

Mechanisms of Nickel Hyposensitization

The effectiveness of Nickel hyposensitization is related to the immune modulation that this treatment induces. After 3 months, T lymphocytes from ACD-treated patients are less responsive when exposed to the metal in vitro, in terms of cell proliferation and cytokine release, than T cells before treatment (Bonamonte et al. 2011). Similarly a significant decrease in the INF-γ, IL-13, and IL-5 Nickel-driven release by peripheral blood mononuclear cells (Fig. 3) is induced by hyposensitization in SNAS patients after 1-year treatment (Minelli et al. 2010). Of particular importance is that lymphocytes from hyposensitized SNAS patients produce a significant higher amount of IL-10 with respect to baseline level (Minelli et al. 2010); noteworthy, an increase in IL-10 producing lymphocytes, CD25+ T regulatory cells, is characteristic of a successful hyposensitization with vaccines in respiratory allergy. Moreover CD25+ Treg cells are able to control the activation of both naive and effector Nickel specific T cells in healthy subjects (Cavani and De Luca 2010). The finding of increased IL-10 secretion by cultured PBMC from Nickel hyposensitized SNAS patients suggests that the described treatment is able to reverse the impaired development of efficient regulatory T cells that is the cause of the clonal expansion of effector T lymphocytes in allergic pathologies.

Conclusions

Nickel hyposensitization is effective in patients suffering from Nickel allergy. However, reliable scientific data are available only for patients affected by Systemic Nickel Allergy Syndrome. The majority of such patients can safely consume Nickel containing foods after 1-year treatment. The clinical experiences in ACD patients, although positive and encouraging, are scarce in term of number of patient treated and length of hyposensitization course that is followed, after a relatively short period of time, by a relapse of cutaneous symptoms.

In any case Nickel hyposensitization is able to modulate immune responses to Nickel restoring a state of tolerance that seems to be mediated by T regulatory lymphocytes.

A double blind placebo controlled trial in SNAS patients started in the beginning of 2011, ending July 2012, to definitively establish the clinical usefulness of this treatment.

Cross-References

▶ Nickel in Plants

References

Andreassi M, Di Gioacchino M, Sabbioni E, Pietra R, Masci S, Amerio P, Bavazzano P, Boscolo P (1998) Serum and urine nickel in nickel-sensitized women: effects of oral challenge with the metal. Contact Dermatitis 38:5–8

Bonamonte D, Cristaudo A, Nasorri F, Carbone T, De Pità O, Angelini G, Cavani A (2011) Efficacy of oral hyposensitization in allergic contact dermatitis caused by nickel. Contact Dermatitis. doi:10.1111/j.1600-0536.2011.01940.x

Boscolo P, Di Gioacchino M, Sabbioni E, Benvenuti F, Conti P, Reale M, Bavazzano P, Giuliano G (1999) Expression of lymphocyte subpopulations, cytokine serum levels, and blood and urinary trace elements in asymptomatic atopic men exposed to an urban environment. Int Arch Occup Environ Health 72:26–32

Cavani A, De Luca A (2010) Allergic contact dermatitis: novel mechanisms and therapeutic perspectives. Curr Drug Metab 11(3):228–233

Chase MW (1946) Inhibition of experimental drug allergy by prior feeding of the sensitizing agent. Proc Soc Exp Biol Med 61:257–259

Di Gioacchino M, Boscolo P, Cavallucci E, Verna N, Di Stefano F, Di Sciascio M, Masci S, Andreassi M, Sabbioni E, Angelucci D, Conti P (2000) Lymphocyte subset changes in blood and gastrointestinal mucosa after oral nickel challenge in nickel-sensitized women. Contact Dermatitis 43:206–211

Minelli M, Schiavino D, Musca F, Bruno ME, Falagiani P, Mistrello G, Riva G, Braga M, Turi MC, Di Rienzo V, Petrarca C, Schiavone C, Di Gioacchino M (2010) Oral hyposensitization to nickel induces clinical improvement and a decrease in TH1 and TH2 cytokines in patients with systemic nickel allergy syndrome. Int J Immunopathol Pharmacol 23:193–201

Panzani RC, Schiavino D, Nucera E, Pellegrino S, Fais G, Schinco G, Patriarca G (1995) Oral hyposensitization to nickel allergy: preliminary clinical results. Int Arch Allergy Immunol 107:251–254

Richman LK, Graeff AS, Yarchoan R, Strober W (1981) Simultaneous induction of antigen-specific IgA helper T cells and IgG suppressor T cells in the murine Peyer's patch after protein feeding. J Immunol 126:2079–2083

Roelofs-Haarhuis K, Wu X, Nowak M, Fang M, Artik S, Gleichmann E (2003) Infectious nickel tolerance: a reciprocal interplay of tolerogenic APCs and T suppressor cells that is driven by immunization. J Immunol 171:2863–2872

Sjövall P, Christensen OB, Möller H (1987) Oral hyposensitization in nickel allergy. J Am Acad Dermatol 17:774–778

Vreeburg KJ, de Groot K, von Blomberg M, Scheper RJ (1984) Induction of immunological tolerance by oral administration of nickel and chromium. J Dent Res 63:124–128

Wells HG, Osborne TB (1911) The biological reactions of the vegetable proteins. J Infect Dis 8:66–124

Nickel in Bacteria and Archaea

R. Gary Sawers
Institute for Microbiology, Martin-Luther University Halle-Wittenberg, Halle (Saale), Germany

Synonyms

Nickel biochemistry; Nickel metabolism

Definition

Nickel is an important trace element for many prokaryotic microorganisms, also referred to as microbes, in the domains of bacteria and archaea. The Ni^{2+} cation is generally the metal species that is specifically transported by microorganisms. Nickel is associated with enzymes as a component of mononuclear, dinuclear, or multinuclear prosthetic groups, and these are essential for catalytic function. Seven microbial enzymes that contain nickel have been characterized in detail. At least one further enzyme recently characterized and found in microbes is a nickel metalloenzyme. There are also further proteins specifically involved in nickel metabolism: These are either required for transport of nickel or its insertion into the prosthetic groups of nickel enzymes.

Nickel as an Essential Trace Element in Microbes

Although nickel is a transition metal that is comparatively abundant in the lithosphere, it is less abundant in aqueous systems (<10 nM), where it exists as the soluble Ni^{2+} cation. The concentration of Ni^{2+} was, however, higher in the oceans early in evolution, prior to the generation of molecular oxygen (Nitschke and Russell 2009; Thauer et al. 2010). Due to the low concentrations of the metal in freshwater and marine environments, its importance in biological systems was overlooked for a long time until evidence began to accumulate in the 1970s that nickel was important for optimal growth of certain microorganisms. Nickel was unequivocally demonstrated to be essential for growth of the methanogenic archaeon *Methanobacterium thermoautotrophicum* in 1979 (Thauer et al. 2010). This finding then led to numerous further studies that in the meantime have identified nickel to be an *essential trace element* for a wide variety of bacteria and archaea. Nickel is required in the *prosthetic groups* of a total of eight enzymes found in either bacteria or archaea, including [NiFe]-hydrogenases, CO dehydrogenase (CODH), acetyl-CoA synthase/decarbonylase, methyl-coenzyme M reductase (MCR), superoxide dismutase (Ni-SOD), glyoxylase I, urease, and acireductone dioxygenase (Ragsdale 2009). No microbe has been identified to date that synthesizes all of these enzymes, although the [NiFe]-hydrogenases are widespread.

A recent *phylogenomic analysis* of all sequenced bacterial and archaeal genomes (Zhang and Gladyshev 2010) revealed that the distribution of nickel-containing enzymes extends to almost all *phyla*. Exceptions where no genes encoding characterized nickel transport systems or metalloenzymes are present include phyla representing intracellular parasites or endosymbionts, such as Spirochetes, Chlamydia, and Rickettsia, which is in accord with the reduction in genome size of these bacteria due to their highly specialized lifestyles. The deeply branching thermophilic Thermotogae phylum also lacks a characterized nickel metalloproteome. Surprisingly, the gammaproteobacterial order *Xanthomonadales*, many of whose species are plant pathogens, also lacks known nickel enzymes.

Among the archaea nickel, metalloenzymes appear to be widespread. Of the archaeal genomes whose sequences have completed to date, only that of *Nanoarchaeum equitans* lacks the coding capacity for nickel enzymes. While *Nanoarchaeum equitans* has a parasitic lifestyle and has an extremely limited metabolic capacity, the majority of other *Euryarchaeota* and *Crenarchaeota* are reliant on hydrogen as an electron source, and thus [NiFe]-hydrogenases are abundant among these archaea. Moreover, as the *methanogens* fix CO_2 by the *Wood-Ljungdahl pathway*, they also contain genes encoding CODH and acetyl-CoA synthase/decarbonylase.

Nickel Enzymes

Nickel in the active site of enzymes can be mononuclear, dinuclear, or can form part of a heteronuclear cofactor, often together with iron, as in the [NiFe]-hydrogenases. Generally, the biologically relevant

Nickel in Bacteria and Archaea, Table 1 Reactions catalyzed by nickel metalloenzymes

Enzyme	Reaction
Urease	$H_2N–CO–NH_2 + H_2O \rightarrow 2NH_3 + CO_2$
Hydrogenase	$H_2 \leftrightarrow 2H^+ + 2e^-$
CO Dehydrogenase	$CO + H_2O \leftrightarrow CO_2 + 2H^+ + 2e^-$
Acetyl-CoA synthase	$CO + CH_3\text{-THF} + CoASH \leftrightarrow H_3C–CO–SCoA + THF$
Methyl-CoM reductase	$H_3C–CoM + HS–CoB \leftrightarrow CoM–S–S–CoB + CH_4$
Superoxide dismutase	$2O_2^- + 2H^+ \leftrightarrow H_2O_2 + O_2$
Glyoxalase I	$H_3C–CO–CHO$ (methylglyoxal) $+ GSH \rightarrow$ S-lactoyl-GSH

oxidation states of nickel are Ni^+, Ni^{2+}, and Ni^{3+}, and depending on the means by which nickel is ligated to the protein, it usually functions either as a redox catalyst, for example, as in the case of hydrogenase or CO-dehydrogenase where nickel is liganded by cysteinyl sulfurs, or as a *Lewis acid*, as in urease where N and O ligands are used. MCR in the methanogenic archaea is unusual in that it has a prosthetic group in which a porphinoid coordinates nickel to the enzyme.

The redox-driven reactions catalyzed by nickel metalloenzymes are usually reversible hydrogenation reactions ($2 H^+ + 2e^-$ transfers) involving simple inorganic substrates such as H_2, CO, CO_2, carbonyls, or O_2^- (see Table 1). Some of these reactions are considered to have been important in the evolution of life on this planet, especially in the synthesis of acetate and formate from CO or CO_2 via hydrogenation (Huber and Wächtershäuser 1997). These reactions were possibly driven by nickel- and iron sulfide-based minerals found deep in the ocean at *hydrothermal vents*. The retention of Ni- and Fe-based catalysts in the active sites of [NiFe]-hydrogenases and acetyl CoA synthase/decarbonylase throughout evolution is perhaps testament to the importance of these reactions for primordial *chemolithoautotrophic metabolism* (Huber and Wächtershäuser 1997; Nitschke and Russell 2009).

Distribution of Nickel Enzymes Within Microbes

As mentioned above, not all of the known nickel enzymes are found in a single microorganism, and

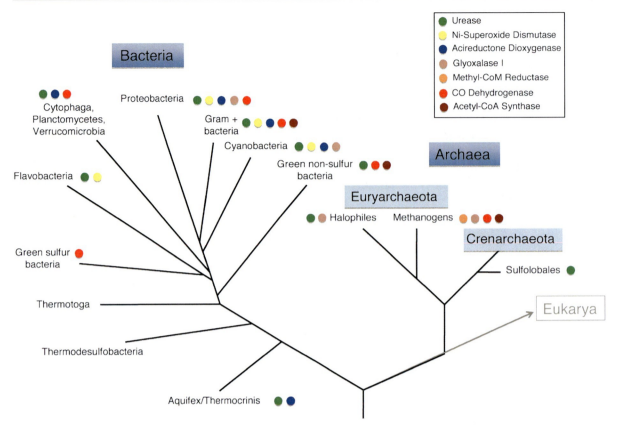

Nickel in Bacteria and Archaea, Fig. 1 Distribution of characterized nickel-containing enzymes in the domains of bacteria and archaea. This phylogenetic tree is drawn schematically and is not meant to indicate precise evolutionary distances. The color key in the upper right portion of the figure indicates the enzymes represented. The occurrence of the respective enzymes in the depicted phyla is indicated by the colored circle to the right or left of the phylum name. Note that due to their wide distribution in all of the phyla, except *Thermodesulfobacteria*, the [NiFe]-hydrogenases are not shown. Please see text for details

their distribution within microbes generally reflects the functional requirement in the habitat(s) in which the respective host is found. Essentially, in urease, glyoxalase I, and acireductone dioxygenase, nickel has a role as a Lewis acid, while in the remaining enzymes, it performs a redox function (Ragsdale 2009). The next sections will give a brief overview of the nickel enzymes employing Lewis acid–type reactions followed by the redox-based catalysts and of the bacteria and archaea in which these enzymes are found and those from which they have been best characterized.

Urease

Urease was the first nickel enzyme to be identified and characterized. It catalyzes the hydrolysis of the urea to ammonia and carbamate, with the latter spontaneously decomposing to ammonia and bicarbonate (Kaluarachchi et al. 2010; Kuchar and Hausinger 2004; Ragsdale 2009). Urease is an important enzyme in the *global nitrogen cycle* and is found in a broad range of bacterial phyla. Initially characterized in the enterobacterial genera *Proteus* and *Klebsiella,* it has now been identified as a key virulence factor in *Helicobacter* species, which colonize the gastric mucosa of humans often causing gastritis, ulcers, and even gastric cancer (Maier et al. 2007). Urease in *Helicobacter* species can account for up to 10% of the total protein, and the ammonia produced presumably helps offset the acidic pH in the immediate environment of the bacterium.

Figure 1 presents a schematic distribution of the key nickel metalloenzymes in the main bacterial and archaeal phyla. Although widespread among the proteobacteria, Gram-positives and cyanobacteria, urease is less frequently found among the more deeply branching phyla of bacteria. It is present in a few *Flavobacterium* and *Cytophaga* species, which are usually found in aquatic habitats, and apart from sporadic instances of urease being found in two green non-sulfur species, *Deinococcus radiodurans* and in *Thermocrinis albus*, which is a bacterium close to the root of the branch point where bacteria and archaea diverged, it is essentially absent from other bacterial phyla.

Despite its importance in the nitrogen cycle, urease is not abundant among the archaea and has only been identified in the genomes of certain extreme *halophiles*, such as *Natronomonas pharaonis* and *Haloarcula marismortui*, and in two genera of the order *Sulfolobales* (Fig. 1).

Glyoxalase I

Toxic methylglyoxal is derived from dihydroxyacetone phosphate, and it is proposed to be important in certain bacteria helping to control carbon flux. Methylglyoxal is converted by nickel-containing glyoxalase I to lactoylglutathione, which spontaneously hydrolyzes to lactate. The enzyme is restricted to proteobacteria, such as *Escherichia coli*, and some cyanobacteria. Glyoxalase I is not at all widely distributed among bacteria, and orthologs have been found so far only in halophilic and some methanogenic Euryarchaeota.

Acireductone Dioxygenase (ACR)

Like the two previously described enzymes, ACR (1,2-dihydroxy-3-keto-5-methylthiopentene dioxygenase) is restricted to certain groups of bacteria and is not found at all in archaea. The enzyme has a role in the methionine salvage pathway where it generates 3-methylthiopropionate, formate, and carbon monoxide if nickel is present in the enzyme (Ragsdale 2009). ACR is limited to one species in the *Aquificales* and 2 in the *Planctomycetales* orders; otherwise, it is absent from the deeply branching bacteria. It is predominantly found in proteobacteria and some Gram-positive aerobes, such as *Nocardia* and *Frankia*. ACR, a member of the *cupin*-2 *superfamily*, has also been identified in the genomes of certain cyanobacterial genera, principally *Synechococcus* and *Prochlorococcus* species. Because it has an absolute requirement for molecular oxygen, it is found only in aerobes or facultative aerobes.

Nickel-Superoxide Dismutase (Ni-SOD)

Superoxide is a *reactive oxygen species* generated as a by-product of respiration with oxygen, and as such, SOD enzymes evolved only after the advent of *oxygenic photosynthesis* approximately 2.5 billion years ago. Superoxide damages *iron-sulfur clusters* in enzymes and consequently is detrimental to many metabolic processes. Ni-SOD, like all SOD enzymes, catalyzes the dismutation of superoxide to oxygen and hydrogen peroxide. Homotetrameric Ni-SOD was only identified comparatively recently in the high-GC Gram-positive bacterium *Streptomyces seoulensis* and has subsequently been identified in the genomes of a number of marine bacteria. It is not found in archaea, and apart from certain cyanobacterial genera, such as *Trichodesmium*, *Synechococcus*, and *Prochlorococcus*, it is mainly found in the Gram-positive Actinobacteria and deep-sea γ-proteobacteria, such as *Moritella* and *Alteromonas* (Fig. 1).

Methyl-Coenzyme M Reductase (MCR)

In contrast to all other nickel enzymes described so far, MCR is found exclusively in methanogenic archaea (Thauer et al. 2010). The methanogens belong to the phylum Euryarchaeota, and they are obligate anaerobes. MCR catalyzes the final step in *methanogenesis* in which the methyl group bound to coenzyme M is reduced to gaseous CH_4. Recent studies have revealed that the reverse reaction, also performed by MCR in methanogens found in communities with *sulfate-reducing bacteria*, initiates the anaerobic oxidation of methane in marine sediments, thus preventing release of this potent *greenhouse gas* to the atmosphere.

Carbon Monoxide Dehydrogenase (CODH)

CODH catalyzes the reversible reduction of CO_2 to CO. In aerobes, CODH is a *molybdoenzyme*, while in anaerobes, the enzyme has a nickel-containing prosthetic group. Ni-CODH is found either on its own or in complex with acetyl-CoA synthase (ACS) (Kuchar and Hausinger 2004; Ragsdale 2009). The enzyme is prevalent in the SO_4^{2-}-reducing deltaproteobacteria and is also found in *Campylobacter jejuni*, an epsilon proteobacterium. Ni-CODH is found in the low-GC Gram-positive acetogens and in isolated species belonging to both the green sulfur and the green non-sulfur bacteria as well as planctomycetes (Fig. 1).

The archaeal species that have Ni-CODH are Euryarchaeota and generally methanogens. The exceptions are the non-methanogenic genera *Thermococcus*, *Archaeoglobus,* and *Ferroglobus*, which also have the enzyme.

Acetyl-CoA Synthase (ACS)

All microorganisms that have nickel-containing ACS also have a Ni-CODH, and together they form a protein complex that catalyzes the reversible Wood-Ljungdahl reductive acetyl-CoA pathway. Microorganisms can use the pathway either as a means of fixing CO_2 or to cleave acetate, for example, in the acetate-utilizing methanogens, such as *Methanosarcina barkeri*.

In general, the *homoacetogenic bacteria* belonging to the low-GC Firmicutes and the same Euryarchaeota that have CODH synthesize an ACS. As with CODH, ACS is also found in some *Dehalococcoides* species, which belong to the deeply branching green non-sulfur bacteria.

[NiFe]-Hydrogenases

Hydrogenases catalyze the reversible conversion of H_2 to protons and electrons and are the most widespread nickel metalloenzymes within the microbes. Like CODH/ACS, [NiFe]-hydrogenases are very old enzymes (Huber and Wächtershäuser 1997; Nitschke and Russell 2009), and this is reflected in their distribution throughout the bacteria and archaea. [NiFe]-hydrogenases have been particularly well characterized in *E. coli*, and two types are found: the hydrogen-oxidizing enzymes typified by hydrogenase 1 (Hyd-1) and the hydrogen-evolving enzymes to which Hyd-3 belongs. The latter class also includes those that form the *energy-converting hydrogenases* (Ech) found in many strict anaerobes and archaea. These enzymes share amino acid sequence similarity with components of NADH-dehydrogenase and are indeed likely to represent the progenitors of complex I (Nitschke and Russell 2009; Thauer et al. 2010).

Orthologs of Hyd-1 and Hyd-3 are distributed throughout the archaea and bacteria, and it is probably easier to indicate where they are not found rather than discuss where they occur. Apart from the intracellular endosymbionts and parasites described earlier, essentially, the only phylum in which [NiFe]-hydrogenase is not found is Thermodesulfobacteria. It was also initially thought that the cyanobacteria lacked [NiFe]-hydrogenases; however, examples can be found in both single-celled cyanobacteria such as *Gloeothece* and *Synechococcus*, as well as filamentous genera like *Nostoc*, *Anabaena*, and *Lyngbya*.

Other Nickel-Dependent Enzymes

The nickel enzymes described above represent those which have been identified and characterized to date. However, adopting solely bioinformatic and comparative genomic approaches based on information gleaned from the coordination geometry of nickel in the prosthetic groups of characterized nickel metalloenzymes has severe limitations when trying to identify new nickel enzymes. A recent study that used a combined more traditional protein purification approach coupled with highly sensitive mass spectrometric techniques (Cvetkovic et al. 2010) unraveled the metalloproteome of the hyperthermophilic archaeon *Pyrococcus furiosus*. The study revealed new nickel enzymes with largely unknown functions. Future use of this approach with other microorganisms might prove useful in identifying new nickel-containing enzymes.

Accessory Proteins Involved in Nickel Metabolism

In order to supply nickel in the requisite amounts for nickel-dependent enzymes, specific Ni^{2+} transport systems, accessory proteins for nickel insertion into enzymes, proteins for nickel storage, and general nickel ion homeostasis are required. The use of genetically tractable model organisms has greatly aided the elucidation of these various systems (Eitinger et al. 2005; Kaluarachchi et al. 2010; Maier et al. 2007). Subsequent to identification of the respective genes, it was then possible through comparative genomics to examine the distribution of these accessory proteins throughout microorganisms in general.

Nickel Transport

Due to its comparatively low abundance in aqueous systems, nickel acquisition requires a specific nickel-uptake system. Most bacteria that have nickel enzymes have a specific *ATP-binding cassette (ABC) transport* system encoded by the *nik* operon, which is dedicated to the high affinity and energy-driven uptake of nickel ions into the cell (Eitinger et al. 2005). Affinities are in the low nanomolar range, in accordance with nickel concentrations in the ocean. Particular components of the Nik system are also present in archaea. For example, the transmembrane NikC (permease), NikD, and NikE (both ATP-binding proteins) components are found in the methanogens and numerous other archaeal genera; however, the NikA Ni^{2+}-binding periplasmic component is absent, as is the NikB permease component.

The Nik secondary transport system is supplemented in some bacteria by permeases, such as HupE and UreH, the latter of which is prevalent in cyanobacteria, halophilic archaea, and the Gram-positive bacteria, which also have ureases. UreH's function as a Ni^{2+} transporter was identified through complementation studies of an *E. coli nik* mutant. A permease-like protein, SodT, encoded by a gene adjacent to the gene encoding Ni-SOD is found in the genome of certain *Streptomyces* species, and this has 8 transmembrane helices like UreH and is suggested to be a further *secondary transporter* specific for Ni^{2+} (Eitinger et al. 2005).

Finally, NixA has been shown to be a high-affinity Ni^{2+} permease (with affinity in the low nM range) identified initially in *Helicobacter* species (Maier et al. 2007). This protein is also prevalent in proteobacteria, Gram-positives, as well as in the archaeal genus *Sulfolobus*.

Nickel Efflux

Nickel homeostasis is important because, when present in excess Ni^{2+} needs to be detoxified. Although it is unclear whether there are efflux systems that are exclusively specific for nickel, the so-called RND-type trans-envelope efflux systems demonstrate specificity for particular groups of transition metals. For example, *Helicobacter* species have a CznABC efflux pump that functions to export cadmium, zinc, and nickel (Maier et al. 2007). These protein complexes function as *energy-driven efflux pumps* and are comparatively widespread among the bacteria; however, no Czn homologs are found in the archaea.

Nickel Accessory Proteins

A dedicated system of proteins for the incorporation of Ni^{2+} into [NiFe]-hydrogenases has been particularly well characterized in *E. coli, Ralstonia eutropha,* and *Helicobacter pylori* (Kaluarachchi et al. 2010). What is clear from these studies is that introduction of Ni^{2+} into the active site occurs after the cyanide- and CO-liganded iron atom is introduced. Four proteins play key roles in the insertion of Ni^{2+} and completion of active site biosynthesis: the GTPase HypB, the Zn^{2+}- and Ni^{2+}-binding protein HypA (or its homolog HybF), the metallochaperone SlyD, and finally, a hydrogenase-specific protease, which removes specifically a C-terminal peptide from the hydrogenase large subunit subsequent to nickel insertion. Notably, HypA, HypB, and SlyD are required to enhance the efficiency of nickel insertion, and if high nickel concentrations are supplied in the growth medium, this obviates the requirement for these three proteins. However, because in their natural habitats most bacteria are not exposed to high Ni^{2+} concentrations, these proteins have been conserved throughout evolution. All microorganisms that synthesize [NiFe]-hydrogenases have the HypA and HypB proteins. Notably, the signal for the protease to activate its proteolytic function appears to be the presence of the nickel ion itself in the protein.

A dedicated complex of proteins is also involved in active site generation of urease. Notably, studies with mutants in *H. pylori* have demonstrated that HypA, HypB, and SlyD also have a function in Ni^{2+} insertion into urease. The UreD, E, F, and G proteins are encoded in most genomes of bacteria that make urease. Like HypB, UreG is a GTPase, but its precise function is unclear. UreE delivers nickel, and in many, but not all, bacteria, it has a histidine-rich C-terminal domain that can bind 5–6 Ni^{2+} ions. In those microorganisms in which UreE lacks a His-rich domain, this domain is often found instead to be present on UreG. UreE is absent in the archaea that have urease, while UreD and UreF are present. Notably, as well as being found in the halophilic archaea and *Sulfolobales*, which have urease, a UreG ortholog is also found in the methanogenic archaea, which apparently do not have urease. What the function of this protein might be is unclear.

The model organism that has been used to study nickel insertion into CODH is *Rhodospirillum rubrum*, and this has identified three proteins CooC, T, and J that are required for maturation of CODH. CooC has a nucleotide-binding domain and belongs to the Ras-like GTPase superfamily, while CooJ has a His-rich C-terminal domain with 16 histidines that can complex up to 4 Ni^{2+} ions. Thus, each maturation process of a nickel enzyme seems to have in common GTPase-like and a Ni^{2+}-binding/storage component. CooT is a small protein of 65 amino acids, and its function in nickel insertion is unclear, and it seems to be restricted to only a few bacterial species; a similar limited distribution is also observed for CooJ. CooC, on the other hand, is present in all bacteria and archaea that synthesize a CODH.

A specific nickel insertion system for ACS has not been categorically demonstrated. Whether a degree of overlap between the nickel insertion systems of hydrogenases, or CODH, exists remains to be established.

Nickel Sequestration and Storage

A number of other cellular proteins have been shown to bind nickel, although these studies have so far been restricted to model organisms (for reviews see Kaluarachchi et al. 2010; Maier et al. 2007). The *metallochaperone* SlyD has been suggested to be a storage protein for nickel. Indeed, the chaperone GroES has also been shown to bind 2 Ni^{2+}/monomer. The heat-shock protein HspA and the small His-rich Hpn Ni-binding proteins identified in *H. pylori* are proposed to be involved in nickel homeostasis and detoxification, respectively. Hpn is exclusive to *H. pylori*, while HspA is more widespread and is found in a large number of bacterial phyla, as well as in several methanogens.

The protein networks identified so far in bacteria that are required to ensure nickel insertion and homeostasis suggests that there may be further nickel-binding proteins and chaperones that are specialized for yet to be discovered nickel metalloproteins.

Cross-References

- Acireductone Dioxygenase
- CO-Dehydrogenase/Acetyl-CoA Synthase
- Methyl Coenzyme M Reductase
- [NiFe]-Hydrogenases
- Nickel Ions in Biological Systems
- Nickel Superoxide Dismutase
- Nickel Transporters
- Nickel in Plants
- Nickel, Physical and Chemical Properties
- Nickel-Binding Sites in Proteins
- NikR, Nickel-Dependent Transcription Factor
- Urease

References

Cvetkovic A, Menon AL, Thorgersen MP, Scott JW, Poole FL, Jenney FE Jr, Lancaster WA, Praissman JL, Shanmukh S, Vaccaro BJ, Trauger SA, Kalisiak E, Apon JV, Siuzdak G, Yannone SM, Tainer JA, Adams MW (2010) Microbial metalloproteomes are largely uncharacterized. Nature 466:779–782

Eitinger T, Suhr J, Moore L, Smith JAC (2005) Secondary transporters for nickel and cobalt ions: themes and variations. Biometals 18:399–405

Huber C, Wächtershäuser G (1997) Activated acetic acid by carbon fixation on (Fe, Ni)S under primordial conditions. Science 276:245–247

Kaluarachchi H, Chung KCC, Zamble DB (2010) Microbial nickel proteins. Nat Prod Rep 27:681–694

Kuchar J, Hausinger RP (2004) Biosynthesis of metal sites. Chem Rev 104:509–525

Maier RJ, Benoit SL, Seshadri S (2007) Nickel-binding and accessory proteins facilitating Ni-enzyme maturation in *Helicobacter pylori*. Biometals 20:655–664

Nitschke W, Russell MJ (2009) Hydrothermal focusing of chemical and chemiosmotic energy, supported by delivery of catalytic Fe, Ni, Mo/W, Co, S and Se, forced life to emerge. J Mol Evolut 69:481–496

Ragsdale SW (2009) Nickel-based enzyme systems. J Biol Chem 284:18571–18575

Thauer RK, Kaster AK, Goenrich M, Schick M, Hiromoto T, Shima S (2010) Hydrogenases from methanogenic archaea, nickel, a novel cofactor, and H_2 storage. Annu Rev Biochem 79:507–536

Zhang Y, Gladyshev VN (2010) General trends in trace element utilization revealed by comparative genomic analyses of Co, Cu, Mo, Ni, and Se. J Biol Chem 285:3393–3405

Nickel in Plants

Paulo Mazzafera[1], Tiago Tezotto[2] and Joe C. Polacco[3]
[1]Departamento de Biologia Vegetal, Universidade Estadual de Campinas/Instituto de Biologia, Cidade Universitária, Campinas, SP, Brazil
[2]Departamento de Produção Vegetal, Universidade de São Paulo/Escola Superior de Agricultura Luiz de Queiroz, Piracicaba, SP, Brazil
[3]Department of Biochemistry/Interdisciplinary Plant Group, University of Missouri, Columbia, MO, USA

Synonyms

Essential nutrient for plants; Urease activation

Definitions

Nickel has been suggested to be an essential micronutrient in plant metabolism because it is part of the active site of the enzyme urease, which catalyzes the hydrolysis of urea to ammonia and bicarbonate. Urease is an abundant protein in seeds of some plant species and is also found, as a separate isoform where studied, in lower levels in vegetative tissues of many, if not all, plants. A deficiency of urease activity leads to accumulation of urea and subsequent formation of necrotic lesions in leaves. The dependence of urease activity on Ni indicates, at the least, that

Ni is a conditionally essential micronutrient, under conditions in which urea is the major N source, either applied as a fertilizer or liberated during mobilization of N stored mainly as arginine. Ni can be remobilized among plant organs, which may be a desirable characteristic under a demand for urease be activity, which is particularly the case of plants fertilized with urea.

Ni as an Essential and Sometimes an Undesired Element

Nickel has been suggested to be an essential micronutrient in plant metabolism, based on its essential role in the catalytic metallocenter of plant ureases. Urease catalyzes the following hydrolysis of urea:

$$(NH_2)_2C = O + 3H_2O \rightarrow 2NH_4^+ + HCO_3^- + OH^-$$

Urease (EC 3.5.1.5, urea amidohydrolase) is the only known Ni-containing enzyme in plants, unlike the eubacteria and archaea, which collectively produce seven know Ni enzymes. Ni deficiency in plants leads to accumulation of urea and subsequent formation of necrotic lesions in leaves (Eskew et al. 1983). Supplementation of Ni to some plant species seems to have a marked effect on plant growth. However, only a few plants failed to complete their life cycle in the absence of Ni, one of the rules to classify a mineral as essential for plants (Gerendás et al. 1999). In fact, compliance with this rule is also inadequate for other elements (Gerendás et al. 1999). On the other hand, a number of reports described the damage caused by Ni when present in plant tissues above normal limits. Many, if not most reports did not explore the functional mechanisms of this Ni toxicity. Ni toxicity in plants has been documented in soils with a naturally high content of this metal – serpentine soils – and as a consequence of application of sewage sludge as an alternative fertilizer. The most obvious primary mechanism related to Ni toxicity is the displacement of divalent cations (Ca, Mg, Mn, Fe, Cu, and Zn) with chemical similarities to Ni, from key metabolic enzymes or compounds (Yusuf et al. 2011).

Metabolic Origin of Urea

Two major progenitors have been invoked for urea generated endogenously in plants: arginine and ureides.

Arginine: This amino acid is hydrolyzed to ornithine and urea by the action of arginase (L-Arg amidinohydrolase, EC 3.5.3.1). Several reports support the idea that there is a large flux of N moving from the guanido group of arginine to ammonia by way of urea hydrolysis, particularly in germinating seeds in which arginine may represent almost 20% of total seed N, half of which is potentially mobilizable via urea.

Ureides: The route of purine breakdown is especially significant in soybean and other tropical legumes since the ureides allantoin and allantoate (purine breakdown products) represent the major form of fixed N exported via the xylem to aerial portions of the plant. In microorganisms and algae the two "ureido" groups of allantoate (and hence, all ureide N) are usually hydrolytically released as urea, thus making it tempting to suggest the same for plants. However, a number of reports using urease-negative soybean and a combination of genomic, genetic, and chemical approaches in soybean and *Arabidopsis* indicated that urea is not released during ureide conversion to ammonia in these plants (Werner et al. 2009).

Urea as a Fertilizer

Urea is provided to crop plants by foliar spray or directly applied to the soil. Spraying has limited use because it provides only part of the total N demand. Because of the high percentage of N in the molecule (46%), the cost per Kg N in urea is lower than that for other N-containing fertilizers, an explanation why urea comprises at least half of all N applied to crops worldwide (http://faostat.fao.org).

It is accepted that in the field urea is mainly converted to ammonium and nitrate by both soil and phylloplane microbes, but once directly taken up by the plant it must be converted to ammonia before its N can be assimilated into amino acids. However, it is still unclear how much urea is directly absorbed by plants although the existence of an active urea transporter and facilitated passive urea transport in the root membranes (reviewed in Witte 2011) has been demonstrated. So, if urea is taken up by plants they must efficiently hydrolyze it, since plants fertilized on urea or nitrate present the same internal concentration of urea in their tissues (Witte 2011).

A number of reports showed that urea-supported plants eventually showed N-deficiency, though it is not clear if plants were supplied with supplemental Ni. Urea-supported *Arabidopsis* grew better when Ni was also provided (Witte 2011). Assimilation of N in urea without its undergoing hydrolysis has been proposed, but at least in those plants where urease activity has been experimentally blocked, no other assimilatory pathway is evident (Witte 2011).

Ni as a Fertilizer

It is obvious that the clearest agronomic response to Ni fertilization is related to N supplementation as urea. However, a neglected aspect in defining a role for Ni in plants has been the doses used in several studies, which in general are over the normal concentrations for a micronutrient in plants. In other words, there is a clear lack of studies using low rates of Ni.

The plant nickel requirement is the lowest of any essential element but soil supply seems to be adequate as long as soil conditions are favorable, i.e., conducive to availability. Additionally, the amount of Ni absorbed by plants will also depend upon the accessibility of Ni to the roots and the Ni uptake process in the root cell membrane.

Ni has been already included in fertilization of some crops but without any accurate research on rate and application formula. In Brazil, Ni has been used as a fertilizer in soybean fields dedicated to seed production in order to improve seed quality (germination and vigor). The rate used is 0.25 Kg ha^{-1} NiSO$_4$.6H$_2$O and it is applied at the beginning of the flowering period. Ni has also been sprayed in different crops at doses varying from 0.25 to 0.5 Kg Ni ha^{-1} (potato, coffee, citrus, *Phaseolus*, maize, soybean, and wheat – http://www.agronomianet.com.br/SAIS/SULFATO%20DE%20NIQUEL.xls).

Ni in a High N-Demanding Crop: Coffee

The direct relationship between urease and N metabolism was evidenced by several studies where Ni-deficient plants showed altered amino acid profiles (reviewed in Gerendás et al. 1999; Yusuf et al. 2011). However, the subject has not been fully explored when urea is the main source of N for plant growth, particularly in high N-demanding species.

Coffee is a high N-demanding crop and normal fertilization rates may vary from 150 to 450 kg N ha^{-1}. However, depending on the climate in which coffee is cultivated in Brazil different fertilization management regimes are adopted and N application rates may reach 800 kg N ha^{-1} (Cerrado – Brazilian savannah – of the Bahia State, provided as fertirrigation), returning a bean yield of up to 2,400 Kg ha^{-1}. The Brazilian mean productivity is about 1,300 Kg ha^{-1} (http://www.carvalhaes.com.br/safra/anexos/conab2010-2.pdf).

Ni seems to have an important relationship with N metabolism in coffee. Adult coffee trees in the field respond to high rates of N-urea fertilization with increased foliar contents of total N and total soluble protein, and nitrate reductase and glutamine synthetase activities (Reis et al. 2009). In these N-fertilized trees, even without Ni supplementation, Ni content increased in the leaves and particularly in the fruits indicating remobilization. Ni has relative high phloem mobility in plant tissues (Yusuf et al. 2011). In a similar trial, coffee trees cultivated in two distinct climatic regions in Brazil exhibited increased urease activity, amino acids and total N as fertilization with urea was increased (Neto 2009). In both trials (Neto 2009; Reis et al. 2009) urea-N induced high coffee bean yields and no toxic symptoms were recorded even at very high fertilization rates.

In another experiment, where the aim was to evaluate the toxic effect of Ni and not N fertilization, Ni (as NiSO$_4$) was supplied to coffee trees at levels that might easily kill other plants (35, 105, and 210 g Ni plant^{-1}). No sign of toxicity was observed (Tezotto et al. 2012). Control and treated trees received N from a N-P-K formulation (20-00-20). Ni did not affect coffee bean production or bean characteristics (size and weight) over 2 years. As the Ni supply increased there was a steady increase of the metal in leaves, stems, and beans, with leaves having the highest content and fruits the lowest. In addition, the same coffee trees showed a threefold increase of leaf protein and more leaves were found per branch. The leaves showed delayed senescence, and the fruits delayed maturation. Maturation and senescence delay might be related to inhibition of ethylene biosynthesis due to the substitution of Ni^{2+} for Fe^{2+} in ACC oxidase and the formation of an inactive enzyme-metal complex. Recent results of greenhouse experiments confirmed this field trial (T. Tezotto, P. Mazzafera, and J.L. Favarin, unpublished).

Considering that Ni can be redistributed among coffee tree organs (Reis et al. 2009), accumulation of

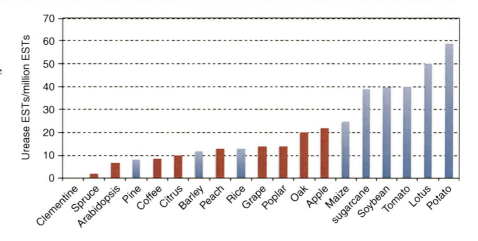

Nickel in Plants, Fig. 1 Estimates of the number of urease-related ESTs per million ESTs for several species listed in Gene Index Project

Ni in the stems is an interesting aspect to be explored in this crop (Tezotto et al. 2012). Transport of Ni in plants is related to its chemical properties by which it binds organic acids, amino acids, and amines, in addition to the existence of specific membrane transporters (Yusuf et al. 2011).

Ni and Urease in Woody Plants

As discussed above, coffee trees may have the capacity to store Ni in the stem, which could be then remobilized to actively growing tissues, such as beans. This is one of the few examples of Ni movement and urease in woody plants.

Most of what it is known about plant urease comes from studies with soybean and a few other annual herbaceous plants (*Arabidopsis*, watermelon, pigeonpea, etc.). Unfortunately, the literature is sparse regarding urease in woody plants. Summarizing three publications between 1950 and 1960, Shim et al. (1973) reported that urease activity was detected in fresh extracts of young leaves of peach, apricot, and cherry, but no activity was found in grape, pear, and apple. However, urease activity was found in acetone powders of young apple leaves (Shim et al. 1973). Urease activity was subsequently confirmed (Shim et al. 1973) in 1-year and 9-year-old apple trees (*Malus pumila* Retd.) and highest activity was found in leaves, roots, and bark. Urease activity in the leaves and bark increased with urea application.

In Pecan (*Carya illinoinensis*), Ni deficiency caused changes in different N-containing compounds related to ureides and amino acids as well as in organic acids. Ni deficiency led to a reduction of xylem sap-associated urease activity (Bai et al. 2006). In leaves of the same species Ni deficiency disrupted ureide catabolism, resulting in accumulation of xanthine, allantoic acid, ureidoglycolate, and citrulline, but reduction of urea concentration, and urease activity (Bai et al. 2006). The authors also observed accumulation of glycine, valine, isoleucine, tyrosine, tryptophan, arginine, and total free amino acids, and lower concentrations of histidine and glutamic acid under Ni deficiency. Allantoin concentration was not changed.

In a search for information on urease in plants "urease" keyword search was carried out among 18 species in the Gene Index Project (http://compbio.dfci.harvard.edu/tgi/plant.html), and a few EST sequences were retrieved. The selection of the nonwoody species was random. The same search in a coffee EST project – CafEST – returned only two sequences (http://bioinfo04.ibi.unicamp.br). These positive "hits" may be overestimates because some of the ESTs showed high identity with microbial ureases and may therefore be the consequence of contamination of the plant tissues used to produce the cDNA libraries. These data were normalized for the total number of ESTs for each species and are shown in Fig. 1.

Although some annual crops also have few EST sequences in the Gene Index database and the tissues used to produce the cDNA libraries varied from plant to plant in each sequencing project, it is noteworthy that woody species tended to have lower urease sequences, and therefore, urease expression. A detailed analysis of all cDNA libraries and reanalysis of the EST sequences, excluding possible contaminants, might result in more solid information on urease expression in trees.

How Do Trees Respond to Urea and Ni Fertilization?

Most of what is known about Ni distribution in plant tissues is related to Ni toxicity studies in herbaceous hyper-accumulating species. Pulford et al. (2002) were able to classify 20 willow varieties growing on sludge-amended soil into two groups, according to their ability to partition heavy metals (Ni, Cu, Cd, Zn) in the wood and bark. Regarding Ni, one group had relatively low Ni in the bark, and the second had relatively high Ni in this tissue. Ni content in the wood was similar in the two groups. Cu followed exactly the same distribution pattern observed for Ni, and trees of the second (high Ni) group performed poorly in terms of survival and biomass production. The authors discuss the possible relationship between low performance and Ni and Cu toxicity. The interpretation that Ni might have been redistributed to young growing tissues in the second group, which, theoretically, are active sinks, is suggested, thus causing metabolic disturbance.

Ni also accumulates in the bark of other tree species and the proportion between bark and wood seems to determine the metal concentration in the total biomass. Zn, Cu, Ni, and Cd in stems of *Salix viminalis* increase with height because as the stem narrows toward the shoot apex, the proportion of bark increases (Sander and Ericsson 1998). But increases in Ni, from the lowest to the highest stem position, were the greatest followed by Cu, Zn, and Cd. It is worthy of mention that the trees used in this study were not part of a metal toxicity study.

It is known from experiments with herbaceous plants that foliar application of urea leads to a large accumulation of urea and severe necrosis if Ni is not available for activation of urease. However, it is speculated that in woody plants accumulation of Ni in the stem (bark) may attenuate such effects. In crops demanding high N, such as coffee, this strategy allied to remobilization capacity would be crucial for the efficiency of N assimilation when urea is used as N source. Searching the CafEST project several nucleotide sequences identifying protein membrane transporters related to Ni uptake in bacteria were found. While several might be due to microbial contaminant, cDNAs, at least one was already proven to be specifically involved in such a role in *Arabidopsis thaliana* (AtDUR3 – At5g45380, reviewed in Witte 2011).

Pulse-chase-like experiments, where Ni is offered for a short interval to a plant and urease activity/expression and metal distribution are determined later during different developmental stages, with and without urea supply, might provide clues about the importance Ni remobilization on plant N economy. The possibility that this may also be true for herbaceous plants as well is not excluded.

Cross-References

▶ [NiFe]-Hydrogenases
▶ Nickel in Bacteria and Archaea
▶ Nickel Ions in Biological Systems
▶ Nickel Superoxide Dismutase
▶ Nickel Transporters
▶ Nickel, Physical and Chemical Properties
▶ Nickel-Binding Sites in Proteins
▶ Urease

References

Bai C, Wood CC, Reilly BW (2006) Nickel deficiency disrupts metabolism of ureides, amino acids, and organic acids of young pecan foliage. Plant Physiol 140:433–443

Eskew DL, Welch RM, Cary EE (1983) Nickel: an essential micronutrient for legumes and possibly all higher plants. Science 222:621–623

Gerendás J, Polacco JC, Freyermuth SK, Sattelmacher B (1999) Significance of nickel for plant growth and metabolism. J Plant Nutr Soil Sci 162:241–256

Neto AP (2009) Metabolism of nitrogen and concentration of nutrients in coffee plants irrigated with different rates of nitrogen (in Portuguese), University of São Paulo, Piracicaba, p 94

Pulford ID, Riddell-Black D, Stewart C (2002) Heavy metal uptake by willow clones from sewage sludge-treated soil: the potential for phytoremediation. Int J Phytorem 4:59–72

Reis AR, Favarin JL, Gallo LA, Malavolta E, Moraes MF, Lavres Junior J (2009) Nitrate reductase and glutamine synthetase activity in coffee leaves during fruit development. Rev Bras Ciên Solo 33:315–324

Sander ML, Ericsson T (1998) Vertical distributions of plant nutrients and heavy metals in *Salix viminalis* stems and their implications for sampling. Biomass Bioenerg 14:57–66

Shim K-K, Splittstoesser WE, Titus JS (1973) Changes in urease activity in apple trees as related to urea applications. Physiol Plant 28:327–331

Tezotto T, Favarin JL, Azevedo RA, Alleoni LRF, Mazzafera P (2012) Coffee is highly tolerant to cadmium, nickel and zinc: plant and soil nutritional status, metal distribution and bean yield. Field Crops Res 125:25–34

Werner AK, Romeis T, Witte C-P (2009) Ureide catabolism in *Arabidopsis thaliana* and *Escherichia coli*. Nat Chem Biol 6:19–21

Witte CP (2011) Urea metabolism in plants. Plant Sci 180:431–438

Yusuf M, Fariduddin Q, Hayat S, Ahmad A (2011) Nickel: an overview of uptake, essentiality and toxicity in plants. Bull Environ Contam Toxicol 86:1–17

Nickel Ions in Biological Systems

Stéphane L. Benoit and Robert J. Maier
Department of Microbiology, The University of Georgia, Athens, GA, USA

Synonyms

Nickel; Nickel cofactor; Nickel enzymes; Nickel maturation

Definition

Contribution of nickel ions to the catalytic activity of several enzymes essential for plants, fungi, eubacteria, or archaebacteria.

Introduction

Nickel (chemical symbol: Ni, Z = 28, atomic weight = 58.69) is closely related to cobalt and belongs to group 10 of the periodic table. Nickel is usually found in the 0 or +2 oxidation state, but oxidation states +1, +3, and even +4 have been demonstrated. However, these latter ion species are usually not stable in aqueous solutions. Nickel is ubiquitous but not distributed equally; for instance, it is a natural constituent of plant tissues and normal concentrations range from 0.05 to 10 μg per g of dry matter, but up to 1 mg nickel per g can be found in some "hyperaccumulators" plant species (Küpper and Kroneck 2007) (also see ▶ Nickel in Plants); by contrast, nickel concentrations in human serum are in the nanomolar range. The latter is of major significance to human health as some pathogens require nickel. Although nickel's essentiality in higher organisms is questionable (for instance, there are no known nickel-dependent enzymes in humans), nickel deficiency appears to cause health problems. For instance, rats grown under nickel-deficient conditions show no abnormality, but first-generation pups continued in nickel-deficient conditions show significant abnormalities. These include a malnourished appearance (rough coat and uneven hair development) and ultrastructural changes of the liver. The abnormalities were more severe in second- and third-generation nickel-deficient animals. These results, along with studies on nickel-deficient chickens, suggest possible roles for nickel in higher organisms in membrane stability or in lipid metabolism (Denkhaus and Salnikow 2002).

Nickel can be found associated with nucleotides and their constituents (sugar, base, phosphate) (Sigel and Sigel 2007), but its biological role is essentially defined by its use as a cofactor of enzymes found in all phyla of life (plants, fungi, eubacteria, and archaebacteria). These "nickel-enzymes" (i.e., nickel is an essential cofactor for their catalytic activity) catalyze eight distinct biological activities, a modest number when compared to other metalloenzymes such as iron- or copper-containing enzymes. However, the importance of nickel-enzymes is highlighted by the fact that seven of these eight enzymes (glyoxalase I being the exception) use or produce gases that play significant roles in the carbon, nitrogen, and oxygen cycles (Ragsdale 2009).

In this chapter, the current knowledge on the role of nickel ions in biological systems is summarized, with a special emphasis on the eight nickel-enzymes known so far as well as the accessory proteins involved in the nickel delivery steps required for nickel-enzyme maturation. Some nickel activation pathways have been extensively studied, and it is now known that in some cases, they even share common accessory proteins (e.g., hydrogenase and urease). By contrast, the mechanisms by which other nickel-enzymes (glyoxalase I or acireductone dioxygenase) obtain the nickel cofactor remain to be characterized. The structure and metallocenter assembly of each protein, as well as pre-catalytic events such as nickel transport, nickel-sensing, and nickel-dependent regulation will be covered in depth by other contributors of the encyclopedia.

Acireductone Dioxygenase

Acireductone dioxygenase (ARD; EC 1.13.11.54) is the only known nickel-containing oxygenase

(Pochapsky et al. 2007). A member of the cupin family, it is an enzyme of the methionine salvage pathway first characterized in *Klebsiella pneumoniae*. Interestingly, the binding of either Ni^{2+} or Fe^{2+} to the same catalytic site leads to a change of protein conformation and generates distinct products: using 1,2-dihydroxy-3-oxo-5-(methylthio)pent-1-ene (acireductone) and O_2, Ni^{2+}-bound ARD generates methylthiopropionate, formate, and CO, while Fe^{2+}-bound ARD produces methylthioketobutyrate and formate.

CO Dehydrogenase/Acetyl-CoA Synthase

CO Dehydrogenase (CODH; EC 1.2.99.2) catalyzes the reversible oxidation of CO to CO_2 (Lindahl and Graham 2007). CODH allows anaerobic microorganisms to use CO as their sole carbon source and therefore plays a significant role in the global carbon cycle, with an estimated 10^8 t of CO removed from the atmosphere every year (Ragsdale 2009). CODH can be found as a monofunctional enzyme with ten iron atoms and one nickel per monomer or as a bifunctional CODH/Acetyl-CoA synthase (ACS; EC 2.3.1.169) containing 14 irons and 3 nickel (Ragsdale 2009). As revealed by crystal structure, CODHs possess five metal clusters, including the catalytic nickel-containing "C-cluster." Upon interaction, CODH and ACS form an $\alpha 2\beta 2$ heterotetrameric complex, whose activity is central to the Wood-Ljungdahl pathway. The CO generated in the CODH in the C-cluster is channeled to the catalytic A-cluster of ACS (through a 70-Å-long molecular tunnel) and reacts with CoASH and CH3-CFeSP to generate acetyl-CoA and CFeSP.

Most of the current knowledge on CODH biosynthesis comes from studies on *Rhodospirillum rubrum*. Two proteins, CooC and CooJ, have been shown to be required for CODH activity in the absence of supplemental nickel. CooC appears to be an NTPase, and CooJ is a putative Ni chaperone with a histidine-rich C-terminus domain. The role of a third protein (encoded on the *cooCTJ* operon) with limited homology to HypC is not yet defined.

Many organisms containing ACS and CODH possess a gene called *acsF* that encodes for a putative NTPase similar to CooC. While purified AcsF from *Moorella thermoacetica* showed weak ATP hydrolysis activity, its contribution to the ACS Ni-maturation is still not clear.

Glyoxalase I

Glyoxalase I (GlxI; EC 4.4.1.5; also called glyoxylase I) and II catalyze the detoxification of methylglyoxal into lactate; glyoxalase I converts methylglyoxal in the presence of glutathione into *S*-D-lactoylglutathione that is further metabolized into D-lactate by glyoxalase II (Sukdeo et al. 2007). Methylglyoxal, a by-product of several metabolic pathways, is highly toxic due to its involvement in the formation of advanced glycation end products. For a long time, all GlxI enzymes were believed to be zinc-enzymes, like the human GlxI. However, it was later shown that some GlxI enzymes are more active with nickel, like the bacterial GlxI or the glyoxalase enzyme of protozoan trypanosomatids.

Methyl-Coenzyme M Reductase

Methyl-coenzyme M reductase (MCR; EC 2.8.4.1) is the terminal enzyme of methane biosynthesis in methanogenic archaebacteria, therefore contributing to the global carbon cycle. (Jaun and Thauer 2007). The enzyme catalyzes the conversion of methyl-coenzyme M (methyl-2-thioethanesulfonate) and coenzyme B (N7-mercaptoheptanoylthreonine phosphate; CoBSH) into methane (CH_4) and an heterodisulfide CoB-S-S-CoM. Analysis of the crystal structure of the inactive Ni(II)-MCR revealed that the protein is organized as a $(\alpha\beta\gamma)_2$ heterotrimeric complex and that Ni is associated as a F_{430}-cofactor (Ni hydrocorphin) in the α-subunit. Only the Ni(I) form of the MCR enzyme can initiate catalysis. In addition, MCR homologs have been discovered in microorganisms carrying out the opposite process, namely, the anaerobic oxidation of methane (AOM) with sulfate. Experiments with purified MCR from *Methanothermobacter marburgensis* have recently revealed that the enzyme can convert methane into methyl-coenzyme M (Scheller et al. 2010).

Nickel Superoxide Dismutase

Superoxide dismutases (SOD; EC 1.15.1.1) are ubiquitous enzymes that protect cells from oxidative damage, by catalyzing the dismutation of superoxides (O_2^-) into oxygen (O_2) and hydrogen peroxide (H_2O_2). Superoxides, which are by-products of oxygen respiration or produced by macrophages or neutrophils, are harmful to Fe-S clusters and combine with nitric oxide to generate another toxic product, peroxynitrite. Based on their metal cofactors, several classes of SOD have been defined: The Cu-Zn SODs are mostly found in eukaryotes; the Fe-SODs, Mn-SODs, or cambialistic Fe/Mn (either Fe or Mn can be used) SODs are broadly distributed among prokaryotes, protists, or mitochondria; while the most recently discovered Ni-SODs are almost exclusively prokaryotic (*Streptomyces* genus and cyanobacteria). Ni-SODs are homo-hexamers built from right-handed 4-helix bundles. According to the crystal structure, each contains an amino-terminal "Ni-hook" with the motif His-Cys-X-X-Pro-Cys-Gly-X-Tyr that coordinates Ni(III). The biosynthesis, the structure, and the role of Ni-SOD are detailed in the entries ▶ Nickel Superoxide Dismutase and ▶ Nickel-Binding Sites in Proteins.

NiFe-Hydrogenase

Hydrogenases (EC 1.12.*x*.*x*) catalyze the interconversion of molecular hydrogen (H_2) and its component protons and electrons. They are found in bacteria and eukaryotes (hydrogenosomes of protozoa and chloroplasts of green algae). H_2-"evolving" hydrogenases facilitate disposal of excess cellular reducing equivalents (formate or a reduced electron carrier such as ferredoxin can serve as the reductant), while H_2-"uptaking" hydrogenases couple H_2 oxidation to the reduction of various terminal electrons acceptors (e.g., O_2, NO_3^-, SO_4^{2-}, CO_2, or fumarate) (Vignais and Billoud 2007). Three classes of hydrogenases have been defined: the NiFe-hydrogenases, the FeFe-hydrogenases, and the iron-sulfur cluster-free Fe-hydrogenases. NiFe-hydrogenases have heterodimeric structures; the large subunit contains the NiFe catalytic site and the small subunit coordinates one to three Fe-S centers. Ni is bound to the cysteines of two CxxC motifs. NiFe-hydrogenases are widely distributed in archaea and bacteria. In *Ralstonia eutropha*, one of the three NiFe-hydrogenases has been shown to be an H_2 sensor mediating transcriptional regulation. Some pathogenic bacteria (e.g., *Salmonella enterica* serovar Typhimurium and *Helicobacter* species *H. pylori* and *H. hepaticus*) can use high-energy reductant H_2 as an energy source to aid respiratory growth and to augment virulence during host colonization.

The most informative system for identifying the roles of the NiFe-hydrogenase accessory proteins has been *Escherichia coli*, where the HypABCDEF accessory proteins and the HybFG orthologues play roles in maturation of either some of or all three NiFe-hydrogenases. Based on the current model for the *E. coli* hydrogenase 3, HypC, HypD, HypE, and HypF protein complexes incorporate Fe, and HypA and HypB subsequently deliver Ni. Indeed, disruption of *hypA* or *hypB* genes leads to deficiency in hydrogenase activity that can be partially restored upon addition of nickel to the growth medium, indicating that both gene products are involved in the nickel delivery to apo-hydrogenase. HypA serves as a nickel chaperone, and HypB, a GTPase, is a regulator that thermodynamically controls the donation of the metal to the hydrogenase apoprotein or release of the nickel-free chaperone. Unlike *E. coli* HypB, some HypB homologs (e.g., *Bradyrhizobium japonicum*) have histidine-rich amino-terminal domains that appear to play a role in nickel storage. The *B. japonicum* HypB binds 18 nickel atoms per dimer and appears to be also involved in hydrogenase regulation. Other proteins involved in NiFe-hydrogenase maturation in bacteria include SlyD and HspA. SlyD, a member of the FK506-binding protein family of peptidylprolyl isomerases (PPIases), has a histidine-rich C-terminal extension and has been shown to interact with HypB in *E. coli* and *H. pylori*. Binding of Ni results in inhibition of the PPIase activity. In *H. pylori*, the HspA protein is also needed for full NiFe-hydrogenase maturation. *H. pylori* HspA is a homolog of bacterial heat shock protein GroES, with a unique histidine-rich, nickel-binding carboxy-terminal domain.

Urease

Urease (EC 3.5.1.5) was the first enzyme to be isolated (from jack beans, in 1926) and crystallized and the first shown to contain nickel (Dixon et al. 1975). The active site of urease contains two nickel atoms bound to histidines and an aspartate and bridged by a carboxylated lysine and a solvent molecule. The enzyme catalyzes the cleavage of urea into ammonia and carbamate, which spontaneously hydrolyzes into bicarbonate and a second molecule of ammonia. Urease therefore contributes significantly to the global nitrogen cycle. It is found in bacteria, fungi, algae, and higher plants (but not in animals). Ureases play different roles depending upon the organism: for instance, they aid plants to assimilate urea-supplied nitrogen, while ureases are considered a virulence factor for some pathogenic bacteria such as the gastric colonizer *Helicobacter pylori*, which relies on urease to neutralize the gastric pH. All ureases are nickel-containing enzymes that share significant homology; however, they differ in their quaternary structure. Plant and fungal ureases are homo-oligomers of 90 kDa subunits, while bacterial ureases are heterodimeric or trimeric proteins. Still, nickel is key to their catalytic activity, thus their function.

Most of the current knowledge on urease maturation derives from studies on *Klebsiella aerogenes* (Carter et al. 2009). Four accessory proteins are usually required for the production of active urease: UreD (also called UreH in some bacteria), UreE, UreF, and UreG. However, some urease-containing bacteria lack any identifiable homologs of urease maturation factors: an example is *Bacillus subtilis* (Carter et al. 2009). Prokaryotic and eukaryotic urease activation complexes generally seem to be conserved despite limited protein sequence conservation for UreF and UreD. The nickel activation of urease relies on interactions between the four accessory proteins and the urease apoenzyme to form successive complexes in the following order: UreD/UreDF/UreDFG/UreDFGE. Among the four proteins, UreE and UreG have been well characterized. UreE is the main Ni-chaperone that delivers Ni^{2+} to the urease catalytic site. Some UreE (e.g., *K. aerogenes*) have histidine-rich sequences conferring Ni-sequestering ability, while UreG is a GTPase, whose exact role is still undefined. There is a connection between NiFe-hydrogenase and urease maturation in *Helicobacter* species: hydrogenase maturation proteins HypA and HypB are also needed for full urease maturation in *Helicobacter* species (Maier et al. 2007). However, these proteins are not required for urease maturation in other microorganisms (e.g., *Rhizobium leguminosarum*). In addition, species like *H. pylori* possess two small histidine-rich nickel-binding proteins, called Hpn and Hpn-like, that have been shown to play a role in nickel-urease maturation. They are hypothesized to store nickel reserves (Maier et al. 2007).

Concluding Comments

Nickel ions play an important role in biological systems, because they serve as cofactors of key enzymes that are vital to global cycles. The eight Ni-enzymes described above have been well studied in recent years, and the understanding of the process by which nickel is incorporated into the catalytic center of these enzymes has advanced considerably. However, some of the nickel activation pathways are still largely unknown (e.g., for GlxI, ARD, or SOD). Besides the eight proteins described herein, there have been reports of a few additional Ni-utilizing enzymes that include QueD, a *Streptomyces* sp. FLA monocupin dioxygenase quercetinase and AraM, a *B. subtilis* glycerol-1 phosphate dehydrogenase (Li and Zamble 2009). However, many more Ni-enzymes are likely still uncharacterized, as suggested by a recent study in *Pyrococcus furiosus* (Cvetkovic et al. 2010), leading to the expectations that the knowledge of the roles of nickel in biology will be significantly expanded.

Cross-References

▶ Acireductone Dioxygenase
▶ CO-Dehydrogenase/Acetyl-CoA Synthase
▶ Metals and the Periodic Table
▶ Methyl Coenzyme M Reductase
▶ [NiFe]-Hydrogenases
▶ Nickel in Bacteria and Archaea
▶ Nickel in Plants
▶ Nickel Superoxide Dismutase
▶ Nickel Transporters
▶ Nickel, Physical and Chemical Properties
▶ Nickel-Binding Proteins, Overview
▶ Nickel-Binding Sites in Proteins
▶ Urease

References

Carter EL, Flugga N, Boer JL et al (2009) Interplay of metal ions and urease. Metallomics 1(3):207–221

Cvetkovic A, Menon AL, Thorgersen MP et al (2010) Microbial metalloproteomes are largely uncharacterized. Nature 466(7307):779–782

Denkhaus E, Salnikow K (2002) Nickel essentiality, toxicity, and carcinogenicity. Crit Rev Oncol Hematol 42(1):35–56

Dixon NE, Gazzola TC, Blakeley RL et al (1975) Letter: Jack bean urease (EC 3.5.1.5). A metalloenzyme. A simple biological role for nickel? J Am Chem Soc 97(14):4131–4133

Jaun B, Thauer RK (2007) Methyl-coenzyme M reductase and its nickel corphin coenzyme F_{430} in methanogenic archaea. In: Sigel A, Sigel H, Sigel R (eds) Nickel and its surprising impact in nature. Wiley, West Sussex

Küpper H, Kroneck P (2007) Nickel in the environment and its role in the metabolism of plants and cyanobacteria. In: Sigel

A, Sigel H, Sigel R (eds) Nickel and its surprising impact in nature. Wiley, West Sussex
Li Y, Zamble DB (2009) Nickel homeostasis and nickel regulation: an overview. Chem Rev 109(10):4617–4643
Lindahl P, Graham D (2007) Acetyl-coenzyme A synthase and nickel-containing carbon monoxide dehydrogenases. In: Sigel A, Sigel H, Sigel R (eds) Nickel and its surprising impact in nature. Wiley, West Sussex
Maier RJ, Benoit SL, Seshadri S (2007) Nickel-binding and accessory proteins facilitating Ni-enzyme maturation in *Helicobacter pylori*. Biometals 20(3–4):655–664
Pochapsky T, Ju T, Dang M et al (2007) Nickel in acireductone dioxygenase. In: Sigel A, Sigel H, Sigel R (eds) Nickel and its surprising impact in nature. Wiley, West Sussex
Ragsdale S (2009) Nickel-based enzyme systems. J Biol Chem 284(28):18571–18575
Scheller S, Goenrich M, Boecher R et al (2010) The key nickel enzyme of methanogenesis catalyses the anaerobic oxidation of methane. Nature 465(7298):606–608
Sigel RK, Sigel H (2007) Complex formation of nickel (II) and related metal ions with sugar residues, nucleobases, phosphates, nucleotides and nucleic acids. In: Sigel A, Sigel H, Sigel R (eds) Nickel and its surprising impact in nature. Wiley, West Sussex
Sukdeo N, Daub E, Honek J (2007) Biochemistry of the nickel-dependent glyoxalase I enzymes. In: Sigel A, Sigel H, Sigel R (eds) Nickel and its surprising impact in nature. Wiley, West Sussex
Vignais PM, Billoud B (2007) Occurrence, classification, and biological function of hydrogenases: an overview. Chem Rev 107(10):4206–4272

Nickel Maturation

▶ Nickel Ions in Biological Systems

Nickel Metabolism

▶ Nickel in Bacteria and Archaea

Nickel Metalloregulatory Protein

▶ NikR, Nickel-Dependent Transcription Factor

Nickel Sensor

▶ NikR, Nickel-Dependent Transcription Factor

Nickel Superoxide Dismutase

Kelly C. Ryan and Michael J. Maroney
Department of Chemistry, Lederle Graduate Research Center, University of Massachusetts at Amherst, Amherst, MA, USA

Synonyms

Nickel-containing superoxide dismutase; NiSOD; sodN

Definition

Nickel superoxide dismutase is a nickel-containing enzyme that catalyzes the disproportionation of two equivalents of superoxide to molecular oxygen and hydrogen peroxide.

Introduction

Superoxide Dismutases

Reactive oxygen species (ROS) including superoxide ($O_2^{\bullet-}$) are a by-product of aerobic metabolism and if not removed may cause significant cellular damage. In multicellular organisms, superoxide and ROSs have been linked to age-related degenerative processes, cellular death, and many diseases including cardiovascular disease, neurodegenerative diseases (fALS, Parkinson's, and Alzheimer's), and cancer (Bafana et al. 2011). Superoxide dismutases (SODs) are metalloenzymes that catalyze the disproportionation of two equivalents of superoxide to molecular oxygen and hydrogen peroxide at or near the diffusion limit (1, 2, 3), and are thus the first line of defense against oxidative stress (Bafana et al. 2011).

$$M^{(n+1)} + O_2^{\bullet-} \rightarrow M^{n+} + O_2 \quad (1)$$

$$M^{n+} + O_2^{\bullet-} + 2H^+ \rightarrow M^{(n+1)} + H_2O_2 \quad (2)$$

$$2O_2^{\bullet-} + 2H^+ \rightarrow O_2 + H_2O_2 \quad (3)$$

All known SODs utilize a redox metal center that cycles between oxidized and reduced states that differ

Nickel Superoxide Dismutase, Table 1 Ni, Mn, and Fe(III/II) and Cu(II/I) reduction potentials of model complexes and SODs

Complex	Coordination	$E_{1/2}$ versus NHE (V)	References
Ni_{aq}	NiO_6	+2.26	Bryngelson and Maroney (2007)
$Ni(bipy)_3$	NiN_6	+1.72	
$Ni([13]aneN_4)$	NiN_4	+1.12	
$Ni[14]aneN_4)$	NiN_4	+0.92	
$Ni[H_{-3}(gly)_4]$	NiN_4	+0.79	
Ni-1	NiS_2N_2	+0.30	
$[Ni(ema)](Et_4N)_2$	NiS_2N_2	−0.30	
$O_2 + e^- \leftrightarrow O_2^{\bullet-}$		−0.16	Miller (2008)
$O_2^{\bullet-} + e^- + 2H^+ \leftrightarrow H_2O_2$		+0.89	
NiSOD (wild-type)	NiS_2N_3	+0.29	Herbst et al. (2009)
Y9F	NiS_2N_3	+0.30	
D3A	NiS_2N_3	+0.31	
Fe_{aq}	FeO_6	+0.77	Miller (2008)
FeSOD	FeO_3N_2	+0.22	
Mn_{aq}	MnO_6	+1.51	
MnSOD	MnO_3N_2	+0.30	
Cu_{aq}	CuO_6	+0.34	Konecny et al. (1999)
CuZnSOD	$CuON_4$	+0.28–0.40	

by one electron, $M^{n+/(n+1)+}$, during catalysis – the so-called ping-pong mechanism. There are three classes of SODs based on sequence homology: (1) CuZnSOD, (2) FeSOD and MnSOD, and (3) NiSOD. The first two classes of SODs have been well characterized (Abreu and Cabelli 2010; Bafana et al. 2011). More recently, the third class of SOD with nickel in the active site has been characterized from *Streptomyces* sp. (Bryngelson and Maroney 2007). In order for a metalloenzyme to be an active SOD, the reduction potential of the metal cofactor must lie between the two half reactions, optimal midpoint +0.36 V versus NHE (1, 2), and (Table 1). Even though Cu_{aq} can satisfy this requirement, the protein is utilized to sequester the metal cofactor and superoxide substrate from cellular environment. In Fe and MnSODs, although the active sites of these enzymes are strikingly similar, the protein must tune the redox properties of each metal by more than 0.4 V compared to Fe_{aq} and Mn_{aq}, to achieve SOD activity (Abreu and Cabelli 2010).

Utilizing Nickel

One feature that distinguishes NiSODs from the CuZnSOD, FeSOD, and MnSOD proteins are the ligands employed to bind their respective metal cofactors. The Cu, Fe, and Mn enzymes utilize exclusively N/O-donors including three His, one Asp, and H_2O/OH^- (Abreu and Cabelli 2010). NiSOD is the only known SOD to utilize S-donors in the active site of the enzyme. The NiSOD protein binds one Ni cofactor via the N-terminal amine, the backbone amide of Cys2, the side chain imidazole group of His1, and the side chain thiolates of Cys2 and Cys6 (Fig. 1). Regardless of the active site's structural differences, the roles played by the rest of the protein include promoting metal selectivity, adjusting the redox potential of the metal center into an appropriate range to facilitate catalysis (1, 2, 3), and controlling substrate access to, and product release from, the metal sites.

Sulfur-donor ligands play a critical role in the redox biochemistry of nickel. Without these electron-rich donors, the Ni(III/II) couple would lie at potentials that are above those that can be accessed by biological oxidants. For example, the redox potential of Ni_{aq} is estimated to be +2.26 V based on calculations (Zilbermann et al. 2005). In biology, thiolate (cysteinate) or sulfide ligation is closely associated with nickel centers that play a redox role. This includes the active sites of acetylocoenzyme A synthase (ACS), hydrogenases (H_2ases), carbon monoxide dehydrogenase (COdH), methylcoenzyme M reductase (MCR), and NiSOD (Maroney 1999). Among these enzymes, NiSOD is unique in that nickel is the only cofactor present in the active site; ACS contains two Ni-cofactors and an FeS-cluster, H_2ase contains an Fe cofactor with CO and CN^- ligands, COdH contains a NiFes-cluster, and MCR contains a reduced tetrapyrrole.

Molecular Biology

Phylogeny

NiSOD is primarily a bacterial enzyme; however, the gene encoding NiSOD, *sodN*, has also been observed in eukaryotic green algae (Dupont et al. 2008). Phylogenetic trees, constructed by Dupont and coworkers (Dupont et al. 2008), contain *sodN* sequences from model organism genomes and the Sargasso Sea

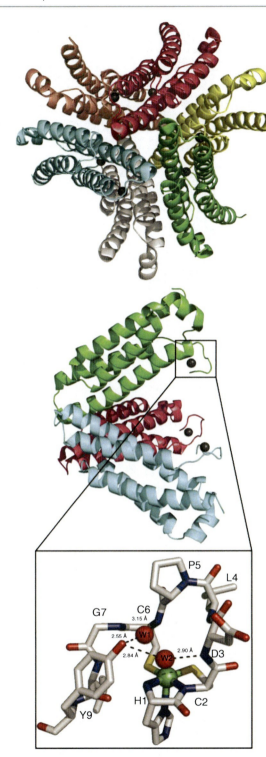

metagenome. Eighty-five NiSOD sequences included in the analysis produce trees containing four major clusters: I-actinomycetes, II-cyanobacteria, IIIa-gammaproteobacteria and IIIb-bacteroidetes, and IV-planctomycetes, cyanobacteria and deltaproteobacteria (Dupont et al. 2008). The data support a view that NiSOD is an evolutionary adaptation that emerged in response to the decreased availability of iron in marine environments that was a consequence of the evolution of oxygenic photosynthesis (Dupont et al. 2008). Other organisms may have acquired NiSOD via horizontal gene transfers (Dupont et al. 2008). NiSOD was first discovered in *Streptomyces* species that may also contain genes encoding an FeZnSOD (*sodF*) (Bryngelson and Maroney 2007), an association that is not ubiquitous. Selective expression of the two SODs depends on the metal availability in the cell. When Ni is present in the growth medium, *sodN* is upregulated and *sodF* is downregulated. Upon Ni depletion, NiSOD is not expressed, but *sodF* is upregulated to compensate for the loss of NiSOD (Bryngelson and Maroney 2007).

The *sodN* gene contains an N-terminal pro-sequence located directly upstream from the sequence encoding the Ni-binding site. The gene encoding the peptidase responsible for peptide cleavage (*sodX*) is found adjacent to the *sodN* gene in the operon (Dupont et al. 2008). The pro-sequence, although conserved in organisms within a species, varies in length and sequence in *sodN* genes from different species (Dupont et al. 2008). However, the residues involved in Ni binding (His1, Cys2, and Cys6) are invariant across all species (Fig. 2). In addition, several other residues in the Ni-binding motif, dubbed the "Ni-hook," (HCDXPCGXY) are strongly conserved. Asp3 is involved in H-bonding to Cys2 and Tyr9, Pro5 is proposed to help stabilize the turn in the Ni-hook domain (Barondeau et al. 2004), and Tyr9 is a residue implicated in a substrate/product gating mechanism in

Nickel Superoxide Dismutase, Fig. 1 Crystal structures of oxidized NiSOD (1T6U) from *S. coelicolor*: Looking down the threefold axis of the hexamer formed by interdigitation of two trimers (*top*), a single trimeric motif (*middle*), Ni-hook motif (*bottom*), with residues 1–9 and showing the active site water molecules and the associated H-bonding network

Nickel Superoxide Dismutase, Fig. 1 (continued)

```
S. Coelicolor            1   --------MLS-----RLFAPKVTVSABCDLPCGVYDPAQARIEAESVKAVQEKMAGNDDP--------------HFQTRATVIKEQRAELAKHH   68
S. avermitilis           1   --------MLS-----RLFAPKVKVSABCDLPCGVYDPAQARIEAESVKAVQEKMAGNDDP--------------HFQARATVIKEQRAELAKHH   68
S. sp.                   1   --------MLS-----RLFAPKVKVSABCDLPCGVYDPAQARIEAESVKAVQEKYQANDDA--------------DYRTRAILIKEQRAELAKHH   68
S. seoulensis            1   --------MLS-----RLFAPKVKVSABCDLPCGVYDPAQARIEAESVKAIQEKMAANDDL--------------HFQIRATVIKEQRAELAKHH   68
Actinosynnema mirum      1   --------MRLLSRIRELHRAIPEATABCDLPCGVYDPAQARIEAESIKAVQEKYQANEDP--------------EFRQRAVLIKEQRSELVKHH   73
Micromonospora sp.       1   --------MRL-----PRILVPSVTVSABCDLPCGVYDPAQARIEAESVKAIAEKYQANTDP--------------EFRTRALIIKEQRAELKHH   69
Janibacter sp.           1   --------MRL----P-ALFSIIDVSABCDLPCGVYDPAQARIEAESIKAIIAKVADNDDP--------------DFRTRAVLIKEQRSQLVKEH   68
Nocardioides sp.         1   --------MFV-----R-LFAPTLEVSABCDLPCGVYDPAQARIEAESIKAIIAKVADNDDP--------------DFRTRAILIKEQRSELVKHH   68
Frankia sp.              1   --------MSL-----LHPKTQVSABCDLPCGVYDPAQARIEAQSVKAIQEKYQGNEDP--------------VFRARALSIKEERADLVKHH   66
Synechococcus sp.        1   ------MLSTVISALKALFPAPEVQABCDGPCGVYDPSSARIAAEAVVSMTKKILDLEHP--EGGDTQAMASYQNTMSRFIAIKEEQAQITKDE   86
Trichodesmium erythraeum 1   MGISLKIMLNTIATQIRKWFSAPEVSABCDGPCGVYDPASARIYAEAVLSMTKKILDLDPK--AGDHKTA-----NTLSRYIAIKEEQAQRTKED   88
Crocosphaera watsonii    1   ------MLNRIAHQIKQFFPAPEVSABCDGPCGVYDPASSRVYAEAVLSMTKKIADLEAKVPSPDDKAATVAYHNTLSRYIAVKEEQAQRTKEE   88
Synechococcus sp.        1   ------MLRSALSTLACALPAPVAEABCDGPCGVYDPASARVAAEAVLSMTKKLKAMEAP--ASGDAAALAAYNNTFGRYVAIKEEEAQRTKKE   86
Marinomanas sp.          1   -------MMBSVIKTLDKVIGFKTAQABCDIPCKIYDPSAAQLAALSCVRLMDLIKEVEEK--------ASLGVADFAQISRLVSEKESACAEVKEA   82
Marinomanas sp.          1   -------MLHAITKKLVGNTRVATVSABCDIPCKIYDPATAQIAVLTMIRLLDLINELDAK--------ESLSIKDQAQLIRLVNEKEAHGLKVKEE   82

S. Coelicolor            69  VSVLWSDYFKPPHFEKYPELBQLVNDTLKALSAA--KGSKDPATGQKALDYIAQIDKIFWETK--------KA-----------------131
S. avermitilis           69  VSVLWSDYFKPPHFEKYPELBQLVNDTLKALSAA--KASTDPATGQKALDYIAQIDKIFWETK--------KA-----------------131
S. sp.                   69  VSVLWSDYFKPPHFEKYPELBQLVNDTLKALSAA--KGSNDPATGQKALDLIAQIDKIFWETK--------KA-----------------131
S. seoulensis            69  LDVLWSDYFKPPHFESYPELHTLVNEAVKALSAA--KASTDPATGQKALDYIAQIDKIFWETK--------KA-----------------131
Actinosynnema mirum      74  LWVLWTDYFKPAPHFEKYPQLHDLFNRATKAAGASGTKGNMDPATGQQLLDLIAEIDKIFWETK--------QA-----------------138
Micromonospora sp.       70  LWVLWTDYFKPAHFEKYPNLHTLFNEATKLAGASGAKGSVDPARADELLQKIDEISKIFWETK--------QA-----------------134
Janibacter sp.           69  LWVLWTDYFKPPHFEKFPNLHSLINEATKLAGASGTKGEFDAAKADELLAKIDEIDAIFWETK--------KA-----------------133
Nocardioides sp.         69  LWVLWTDYFKPPHFEKYPQLHTLVNEATKLAGAAGTKGELDAGKADELLAKIDQIAEIFWETK--------KA-----------------133
Frankia sp.              67  LWVLWTDYFKPNHLEKYPELHDLFWQATKAAGAAGAKGSVDPADGQKLIDKIDEIAKVFWETK--------AAA----------------132
Synechococcus sp.        87  LLILWTDYFKPVHLEKYPDLHDTFWKAAKLCSAC--KVEVSTQHCQELMDAVQKIHNIFWETKSRDVSWYTAS----------------157
Trichodesmium erythraeum 89  LLILWTDYFKPVHLEKYPDLHDTFWKAAKLCSAC--KVEVSLEHATELLAAVEKIHNMFWATKERDVTWYKAS---------------159
Crocosphaera watsonii    89  LLILWTDYFKPVHLEAYPDLHDMFWKAAKLCSSC--KVEVNLAHANQLLMEVEKIHNIFWESKGRKVEWVKAS---------------159
Synechococcus sp.        86  LLILWTDYFKPDHLATFPDLHDTFWKAAKLCSAC--KVNIDQAKAEELMAAVEKIHGMFWQSKGRSDAWVTAS--------------157
Marinomanas sp.          83  VRIIWGDYFKAPQFEQVPNVHDLVHSIMLQASKC--KQGIDRVEGEKLVTLVNNFAEAFWFTKD--VLTYRAPCPYPPMLDVIYPDLKIKNRA   171
Marinomanas sp.          83  VRVIWGDYFKQPQFDQVPHVHELVHQIMLQASKA--KQGVSRDDALGLLTLVNEFAEAFWLTKG--VATYKAIAPYLPSEVVVYPKLDAI---   168
```

Nickel Superoxide Dismutase, Fig. 2 Sequence alignments for NiSOD, Ni-hook motif outlined in box, from various organisms using BLAST and COBALT tools (Gasteiger et al. 2003)

the active site (Herbst et al. 2009). Several residues downstream from the Ni-hook motif are also strongly conserved including Lys27, Arg39, Lys44, Glu45, Lys52, Phe63, Ser86, Lys89, Phe111 (Dupont et al. 2008). These residues serve as oligomeric contacts (Arg39, Glu45, Lys52, Ser86, and Lys89), support for the folded tertiary structure (Phe63, Phe111, and Lys89), and potentially in electrostatic guidance of the superoxide substrate (Lys27). Most of the weakly conserved residues in sodN are located on the solvent surface of the protein away from the active site (Dupont et al. 2008).

Expression, Processing, and Metallocenter Assembly

Several strategies for production of active WT-NiSOD, as well as mutant constructs of NiSOD have been reported (Barondeau et al. 2004; Bryngelson et al. 2004; Wuerges et al. 2004). Native NiSOD can be isolated directly from Streptomyces sp., where N-terminal processing and nickel incorporation occur in vivo. Addition of nickel to the growth culture increases the level of sodN transcript and maturation of the expressed protein more than ninefold within an hour of addition (Bryngelson and Maroney 2007). Methods have also been developed for the expression of recombinant wild-type and mutants of NiSOD in Streptomyces lividans TK24 ΔsodN, and Streptomyces coelicolor A3 (Bryngelson and Maroney 2007). However, the native processing with sodX is less well studied, and heterologous expression in Escherichia coli with in vitro or in vivo processing have been used to produce protein(s) used in structural and mechanistic studies (Choudhury et al. 1999; Barondeau et al. 2004; Bryngelson et al. 2004; Bryngelson and Maroney 2007). The in vitro processing method includes insertion of sodNΔ, lacking the N-terminal leader sequence, into the pET-30Xa-LIC vector, resulting in an N-terminal His-tagged fusion protein with a factor Xa protease cleavage site, IEGR, directly adjacent to the His1 residue. The expressed fusion protein is purified by immobilized metal-affinity chromatography and results in an inactive enzyme that does not bind nickel. Protease cleavage between the Arg and His residue results in the properly processed 117 amino acid apo-NiSOD. Following reduction of an intramolecular disulfide between Cys2 and Cys6 with DTT, nickel incorporation and fully recombinant and active NiSOD are produced (Bryngelson et al. 2004). The in vivo production of NiSOD in E. coli is accomplished in two differing methods. A similar sodNΔ is inserted into the pET26b vector where the N-terminal leader sequence is replaced by the E. coli pelB signal sequence. The addition of $NiCl_2$ at induction and the presence of the pelB peptidase in *E. coli* produce an active NiSOD enzyme in vivo (Barondeau et al. 2004). An alternate in vivo production of NiSOD is achieved by isolation of *sodN* and *sodX* (the upstream peptidase) from *Prochlorococcus marinus* and insertion into a streptomycin resistance-conferring derivative of plasmid pCH675AF to produce p*sodNX* allowing for translational coupling. *E. coli* SG12041 p*sodNX* transformants produce active NiSOD and have been shown to restore oxygen and superoxide tolerance in *sodA/BΔ E. coli* (Bryngelson and Maroney 2007).

Structural Biology

X-Ray Absorption Spectroscopy

The first information available regarding the active site structure of NiSOD was provided by Ni K-edge XAS studies of native *Streptomyces seoulensis* NiSOD (Bryngelson and Maroney 2007). The studies were performed on frozen solutions of as-isolated and dithionite-reduced NiSOD, and the XANES analysis indicated a 5-coordinate square pyramidal to 4-coordinate square planar change in geometry upon reduction of the enzyme. The EXAFS analysis indicated mixed-donor atom ligands, including S-donors, in the Ni site, consistent with N_3S_2- and N_2S_2-donor ligand sets in the as-isolated and reduced NiSOD samples, respectively (Bryngelson and Maroney 2007). As previously described in section "Utilizing Nickel," S-donors are utilized in Ni enzymes to lower the potential of the Ni(III/II) couple so that it is accessible to biological redox agents. Since there are only three sulfur-containing residues in NiSOD (Cys2, Cys6, and Met28) and a single point mutation (M28L) indicated residue 28 is not a metal binding ligand, XAS studies correctly identified an N-terminal nickel-binding site including the side chains of the two Cys residues (Bryngelson et al. 2004). The identities of the three N-donors were later identified as the N-terminal amine, the His1 imidazole and the Cys2 backbone amidate by high-resolution crystallography, as discussed in section "Crystal Structures."

Crystal Structures

High-resolution crystal structures are available for NiSOD from *S. seoulensis* and *S. coelicolor* (wild-type and mutant forms) (Barondeau et al. 2004; Wuerges et al. 2004; Herbst et al. 2009). The wild-type NiSOD crystal structures of NiSOD from these two organisms are highly similar, (not surprisingly considering the 90% identity in their amino acid sequences), and show that the protein is a homohexamer composed of monomers that are 4-helix bundles that place the nickel sites in a nearly octahedral arrangement with Ni–Ni distances of ~25 Å (Fig. 1). This structure is distinct from other SODs (CuZnSODs are homodimeric or monomeric β-barrel proteins; Fe/MnSODs are typically dimers or tetramers and their structures contain α-helical domains with some β-barrel structure). Although the mechanistic advantage, if any, of the hexameric structure has not been established, the hexameric quaternary structure has been confirmed in solution by ultracentrifugation, ESI-MS, and size-exclusion chromatography (Wuerges et al. 2004; Ryan et al. 2010). Size-exclusion chromatography studies demonstrated that hexamer formation does not depend on N-terminal processing of the His-tag leader sequence, reduction of the intermolecular disulfide bond between Cys2 and Cys6, or Ni incorporation (Ryan et al. 2010). The hexamer is approximately spherical with an exterior diameter of 60 Å and a large interior void with a 20-Å diameter that is filled with water and co-crystallized ions. The hexamer can be viewed as a dimer of trimers, where three subunits form a trimer arranged like the legs of a tripod supported by salt bridges between Arg39 and Glu45. The trimers interdigitate to form the hexamer, which is stabilized by hydrophobic ($Trp59_A$-$Trp59_F$), salt bridge (Lys52-Asp61, Lys52-Asp4, and Lys89-Asp3), and H-bonding ($Ser56_A$-$Ser56_F$, $His75_A$-$Gln76_F$, $Gln76_A$-$His75_F$) interactions (Barondeau et al. 2004). An additional hexamer contact includes a H-bonding interaction between Glu17 and His1, the role of which will be discussed further in section "Mechanism." The NiSOD monomers are stabilized by extensive hydrophobic packing, salt bridges, and H-bonding, and the exterior surface of the monomer contains polar and charged side chains.

The crystal structures also reveal details regarding the structure of the active site of NiSOD and the residues utilized in metal binding. As determined by XAS studies, the two S-donors are derived from Cys2 and Cys6 and the N-donor lost upon reduction of the nickel site is confirmed to be the imidazole side chain of His1. The structures show that the apical histidine imidazole ligand is disordered, the result of a mixture of Ni(III)/His_{on} and Ni(II)/His_{off} sites in the hexameric protein. In addition, the crystal structures identify the remaining N-donors as the backbone amidate of Cys2 and the N-terminal amine.

Thus, all five ligands are derived from three residues in the N-terminus of the protein; His1, Cys2, and Cys6 (Barondeau et al. 2004; Wuerges et al. 2004). In addition, there are two non-liganded water molecules in the active site that are involved with H-bonding to Asp3, Cys6, and Tyr9 (Barondeau et al. 2004) (Fig. 1). In the apo-NiSOD crystal structure, all residues prior to Val8 are disordered, but upon nickel binding they form a well-ordered Ni-hook domain that is facilitated by a trans to cis isomerization of Pro5 that allows for Cys6 binding (Barondeau et al. 2004).

Second coordination-sphere residues that influence the structure of the active site of NiSOD include Tyr9 and Asp3. There are four tyrosine residues in NiSOD, two of which are conserved (Tyr9 and 62) (Fig. 2). Single mutants, D3A-, Y9F- and Y62F-NiSOD, and one double mutant, Y9F/Y62F-NiSOD, were constructed, and two of these mutants, Y9F-NiSOD and D3A-NiSOD, were crystallized and structurally characterized using XRD. The structure of the Y9F mutant yielded a structure of the enzyme with chloride bound – the first crystal structure of an anion complex of NiSOD. The nature of the anion complex was confirmed by Br^- substitution and comparison of anomalous difference electron density maps. The position of the anion in the active site was predicted by earlier work (Barondeau et al. 2004) and involves interactions with the amide N atoms of residues 3, 6, and 7. These interactions position the anion close to, but not within, bonding distance of the Ni center (Ni–Cl = 3.54 Å), consistent with an outer sphere redox mechanism and with the notion that residues in the second coordination sphere inhibit anion access to the Ni center. The D3A-NiSOD mutant has a structure where the largest perturbation in the Ni site is the movement of the Y9 to position the OH group 1-Å closer to the Ni center (Herbst et al. 2009). The repositioning of Y9 in D3A-NiSOD suggests that this mutant may favor the formation of a peroxo complex by stabilizing it with a H-bonding interaction with Y9, a suggestion that is supported by the extreme peroxide sensitivity of catalysis by this mutant (Herbst et al. 2009).

Mechanism

Spectroscopy

Several spectroscopic techniques have been used to study the electronic and structural properties of the nickel site in NiSOD and its redox chemistry, including XAS, EPR (ENDOR and ESEEM), UV/Vis, MCD, and resonance Raman. Ni K-edge XAS studies have been performed on various mutants of NiSOD including first coordination-sphere residues involved in nickel binding and second sphere residues that are involved in substrate/product gating. First coordination-sphere mutants include H1A, H1Q, C2S, C6S, and C2/6S (Bryngelson et al. 2004; Ryan et al. 2010). XANES analysis of the two His1 mutants indicate a 4-coordinate planar Ni environment similar to that of dithionite-reduced native NiSOD (Bryngelson et al. 2004). The Cys-to-Ser single mutations were designed to investigate if one S-donor would be sufficient to produce a redox-active nickel site (Ryan et al. 2010). The XAS data indicate that the two single-Cys mutant proteins contain five-coordinate pyramidal nickel sites that lack any S-donor ligand, regardless of which cysteine residue is mutated. In fact, the two single-Cys mutants have nickel site structures that are essentially identical to that determined for the C2/6S double mutant that lacks both S-donor ligands. Features arising from atoms in the second and third coordination-spheres are consistent with the presence of an imidazole ligand, presumably from His1 (Ryan et al. 2010). Coupled with results from MCD, the results are consistent with the formation of high-spin, $S = 1$ Ni(II) center in the Cys mutants, indicating that both Cys residues are required to form the native low-spin site (Johnson et al. 2010).

The low sulfur content of NiSOD also allowed for XAS studies using the S K-edge to investigate the molecular and electronic structure of the active site (Bryngelson and Maroney 2007). XAS spectra of oxidized NiSOD contain two pre-edge features that confirm the presence of thiolate ligands are bound to a Ni(III) center. Photoreduction of the sample in the x-ray beam showed that the Ni(II) center in the photoproduct retained thiolate ligation. However, the spectrum of H_2O_2-reduced NiSOD contains only a broad edge lacking any pre-edge features, a spectrum that is similar to that obtained from thioether complexes and consistent with protonation of the cysteine ligands. This result suggests that the S-donors are possible sites of protonation during turnover (2) (Bryngelson and Maroney 2007).

The resting (oxidized) state of the enzyme exhibits a rhombic EPR spectrum with g-values = 2.306, 2.232, and 2.016 (Choudhury et al. 1999; Wuerges et al. 2004; Bryngelson and Maroney 2007). This is consistent with a low-spin d^7 Ni(III) ion in the active site, and upon incorporation of ^{61}Ni (I = 3/2) additional hyperfine splitting is observed. Integration of the signal intensity indicates that in WT-NiSOD, only half of the nickel sites are in the III oxidation state (Herbst et al. 2009), a result that is parallels crystallographic results that show a mixture of His$_{on}$ (Ni(III)) and His$_{off}$ sites in the hexamer (*vide supra*) (Barondeau et al. 2004; Wuerges et al. 2004). Large N-hyperfine coupling is observed for the g = 2.016 signal ($A_z \sim 30$ G), due to interaction between the singly occupied Ni d_z^2 orbital and the ^{14}N (I = 1) nucleus of the apical His1 imidazole ligand (Bryngelson and Maroney 2007). Protein expression in ^{15}N (I = ½) enriched media resulted in an EPR spectrum with a two-line hyperfine pattern for the g = 2.016 signal (Wuerges et al. 2004). EPR spectra from mutant NiSODs show two trends: (1) signals resemble wild-type NiSOD (with spin quantification confirming mixtures of oxidation states) or (2) EPR silent species. The EPR signal from oxidized NiSOD is lost upon reduction of the enzyme with dithionite, consistent with a Ni(III/II) redox couple utilized in SOD catalysis. In addition, ENDOR and ESEEM studies of NiSOD provided further details regarding the interaction of the apical His1 imidazole ligand with the Ni(III) center. Field-dependent Q-band ^{14}N CW ENDOR was used to determine nuclear hyperfine and quadrupole coupling tensors involving the donor δ-N atom of the apical His1 imidazole ligand. ^{14}N ESEEM was used to probe the weaker hyperfine coupling associated with the unbound ε-N of His1. The analysis suggests that the His1 H-bonding network with Glu17 in the reduced form of NiSOD is weakened in the oxidized state of the enzyme, consistent with a His$_{on/off}$ catalytic mechanism (Lee et al. 2010).

The UV/Vis spectra of the oxidized wild-type and dithionite reduced NiSOD differ significantly. Oxidized NiSOD exhibit two absorption maxima (which are distinct from other SODs (Bryngelson and Maroney 2007)). Oxidized NiSOD exhibits an absorption maximum at ~378 nm that has a relatively high extinction coefficient ($\epsilon \sim 6,800$ M^{-1} cm^{-1}) and has been assigned to a S → Ni(III) LMCT transition. CD and MCD spectra allow for resolution of ten transitions in the UV/Vis spectrum of oxidized NiSOD. Variable-temperature variable-field MCD studies reveal a temperature-dependent C-term that is consistent with a paramagnetic S = 1/2 d^7 Ni(III) nickel center in oxidized NiSOD. Upon reduction of NiSOD, these bands are much weaker indicating a dramatic change in electronic structure to one that is diamagnetic, where the UV–vis is dominated by ligand field transitions, and closely resembles synthetic planar Ni(II) complexes with N_2S_2 coordination (Bryngelson and Maroney 2007).

Resonance Raman spectra exhibit three protein-derived vibrational features at 359, 365, and 391 cm^{-1}. The intense peaks at 349 and 365 cm^{-1} are assigned to the Ni(III)–S stretching modes involving the Cys2 and Cys6 residues, which is supported by the fact that these peaks are enhanced by excitation wavelengths at 413 and 407 nm, which are near the LMCT S → Ni(III) electronic transition. When a lower-energy excitation wavelength, 488 nm, is used these features are not observed. The higher-energy vibrational feature at 391 cm^{-1} was assigned to a combination of the two Ni-S stretching and bending of the Cys2 chelate backbone or the stretching mode of the Cys2 backbone amide which could be enhanced by the LMCT excitation (Bryngelson and Maroney 2007).

Kinetics

Studies of NiSOD using pulse-radiolytic generation of superoxide confirm the enzyme operates via the "ping–pong" mechanism described for other SODs, at a rate near the diffusion limit, and exhibits a pH dependence similar to MnSOD (Fig. 3) (Bryngelson and Maroney 2007; Abreu and Cabelli 2010). Although no sequence homology exists between NiSOD and MnSOD, and NiSOD lacks the aqua/hydroxo ligand utilized in MnSOD, structural homology involving the positions of Y9 and Y62 with an analogous Tyr residue (Y34) and F66 in MnSOD exists (Herbst et al. 2009). The kinetic studies of the Tyr9 mutation show the onset of substrate saturation effects that support the proposed role of Tyr9 in controlling access to the active site (Herbst et al. 2009). Together these results suggest that NiSOD and MnSOD represent examples of convergent evolution (Herbst et al. 2009).

Studies of additional mutants involving the His1 residue (H1Q and H1A), confirm the important functional role of His1. Both mutant forms of NiSOD are isolated in the reduced form of the enzyme and, although the enzyme may still be active, the rate of catalysis is at least two orders of magnitude slower than wild-type NiSOD (Table 2), presumably because the Ni(III) form of the enzyme is more difficult

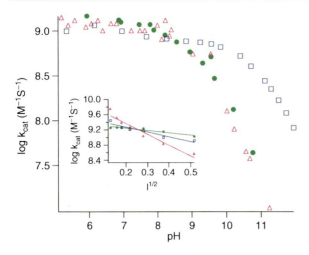

Nickel Superoxide Dismutase, Fig. 3 The dependence of k_{cat} on pH and ionic strength (*inset*) for SODs: NiSOD closed circles, MnSOD open triangles, and CuZnSOD open squares (Figure adapted from Choudhury et al. 1999)

Nickel Superoxide Dismutase, Table 2 NiSOD kinetic properties and relative activity

NiSOD sample	Activity	References
NiSOD native	1.30[a]	Choudhury et al. (1999)
NiSOD wild-type	0.71[a], 100[b]	Wuerges et al. (2004), Herbst et al. (2009)
H1A, C, D, K, N, Q, R, W, or Y	<5[b]	Wuerges et al. (2004)
D3A	0.21[a], 23[b]	Wuerges et al. (2004), Herbst et al. (2009)
Y9F	0.32[a], 78[b]	Wuerges et al. (2004), Herbst et al. (2009)
Y62F	0.64[a]	Herbst et al. (2009)
Y9/62 F	0.24[a]	Herbst et al. (2009)

[a]Activity measured by pulse radiolytic generation of superoxide $k_{calc} \times 10^9$ M^{-1} s^{-1} at pH 7.5
[b]Activity reported as % activity in cell extracts using a standard cytochrome c assay

to obtain or is less stable (Barondeau et al. 2004; Wuerges et al. 2004).

The catalytic rate constant of SODs as a function of ionic strength has been determined for several types of this enzyme (Fig. 3 inset). The k_{cat} of MnSOD exhibits the strongest effect and CuZnSOD has a slight effect respect to changes of solution ionic strength. In CuZnSOD, ionic strength dependence is attributed to lysine electrostatic steering of superoxide to the active site (Bryngelson and Maroney 2007; Abreu and Cabelli 2010). As compared to CuZnSOD, NiSOD shows very little ionic strength dependence (Bryngelson and Maroney 2007; Abreu and Cabelli 2010), indicating there is virtually no electrostatic steering involved in the mechanism.

Model Studies

Computational Models

The spectroscopic and structural studies contributed to the development of detailed theoretical models of the NiSOD active site that provided additional mechanistic insights. Initial structures of *Streptomyces* sp. NiSOD active-sites were derived from crystallographically determined coordinates (Barondeau et al. 2004; Wuerges et al. 2004), and constraints were imposed for typical backbone atoms and to position second sphere residues (Glu17 and Arg47). A hybrid-density functional method (B3LYP) combined with the triple-ζ basis set (TZV) was used for the calculations of spin restricted models in the case of Ni(II) and spin-unrestricted models for Ni(III) (Bryngelson and Maroney 2007). Geometry-optimized active site models, referred to as ox^{1-3} and *red* by Fiedler et al. (Bryngelson and Maroney 2007) reproduce the equatorial ligand distances within 0.05 Å as compared to the crystallographic data. However, predictions of the axial Ni–N_{His1} distance in the ox^1 model, which includes only the ligand residues, was inconsistent with the XRD results. In the ox^2 and ox^3 models, H-bonding interactions of His1 with Glu17 and Arg47 were included. The ox^2 model includes both second-sphere residue interactions and ox^3 only includes the interaction between His1 and Glu17. Although the addition of these interactions lengthens the Ni–N_{ax} distance to values that are closer to the crystallographic distance, neither model properly predicts the distance observed in the crystal structures of NiSOD (Ni–N_{ax} = 2.07 (ox^1), 2.14 (ox^2), 2.16 (ox^3), 2.35–2.63 Å (XRD)) (Bryngelson and Maroney 2007). The difference in the apical imidazole ligand distance could be due to the lack of steric/electronic forces not included in the truncated active-site model(s) or to the mixture of oxidized/reduced His1 positions in the crystal structures that could produce a larger Ni–N_{ax} distance.

To probe the electronic structure of the reduced active site, the *red* model, which contains Ni(II) and lacks the apical imidazole ligand, was examined. The change in coordination geometry at the nickel center to a planar NiN_2S_2 complex strongly stabilizes the Ni d_{z^2}-based MO and destabilizes the Ni π-orbitals, leading to a highly covalent Ni(II) site with extensive mixing of the Ni and ligand-based orbitals. Additionally, an analogous Ni(III) ox-His$_{off}$ model was studied, and the redox-active MO was primarily Ni d-orbital based (70%). Together these results indicate that the active-site has evolved to favor metal-based rather than ligand based redox properties in NiSOD (Bryngelson and Maroney 2007).

Computational models have also been used to investigate the mechanistic roles of the Cys residues (Pelmenschikov and Siegbahn 2006; Prabhakar et al. 2006; Bryngelson and Maroney 2007). These studies indicate that the S-donor residues Cys2 and Cys6 are crucial in tuning the redox properties of the metal cofactor and serve as a possible source of protons for the oxidative half-reaction. From calculations on a Ni(II)SOD active-site model, it was proposed that the highly covalent character of the Ni-S$_{Cys}$ bonds derives from σ-bonding interactions (Bryngelson and Maroney 2007). The net nonbonding π-interactions or filled/filled interactions promote Ni(II)-based oxidation over S-based oxidation and assist in redox tuning of the metal center. The calculations indicate that protonation of the Cys side chains in the reduced state of the enzyme has little effect on the Ni-S bond distances (Bryngelson and Maroney 2007). This prediction is supported by structural studies of synthetic model compounds that show only small changes in M-S distances upon protonation and alkylation of model compounds (Bryngelson and Maroney 2007). One investigation of the solvent-accessible side of the equatorial plane in NiSOD suggested the side chain of Cys2 has more electron density than Cys6 (Pelmenschikov and Siegbahn 2006). This finding implicates the S-donor of Cys2 as a protonation site for the oxidative half reaction However, in another study the H-bonding network involving Cys6, Tyr9, and a water molecule was suggested to be the source of protons, indicating that the protonation of Cys6 leads to a thermodynamically favorable proton transfer in the formation of H$_2$O$_2$ in NiSOD (Prabhakar et al. 2006). The theoretical predictions that the Cys ligands can be protonated in the reduced enzyme are supported by S K-edge XAS studies on enzyme reduced with hydrogen peroxide (*vide supra*) (Bryngelson and Maroney 2007).

Synthetic Model Complexes

In addition to computational studies modeling the active site of NiSOD, several synthetic model complexes have been designed to investigate the role of the different types of N-donor ligands (amine, amide, and imidazole), the S-donors in the active-site, and how the S-donors are protected from oxidation by $O_2^{\bullet-}$ or the O_2/H_2O_2 products during catalysis.

Active-site models were designed with N_2S_2-, N_2S-, N_3S_2-, and N_3S-donor ligands, where different types of N-donors were involved (Gennari et al. 2010; Mathrubootham et al. 2010). Several spectroscopic, biochemical, and computational methods were utilized to compare the physical, electronic, and functional properties of these model systems with those of NiSOD. The results from studies of synthetic models support the conclusion from computation models that the presence of the mixed amine/amidate N-donors in NiSOD acts to promote metal-centered oxidation (Mathrubootham et al. 2010).

Cysteine derived S-donors are utilized in nickel-redox biochemistry (Maroney 1999), because they stabilize the Ni(III/II) redox couple and facilitate electron transfer. However, the coordination of thiolates in Ni(II) planar centers are also known to promote sulfur oxidation to form S-oxygenates (Mathrubootham et al. 2010). In NiSOD, it has been suggested that protonation of the metal bound cysteine thiolates and the electronic preference for metal centered redox (Bryngelson and Maroney 2007) help to protect the cysteinate ligands from oxidation. To address the electronic effects of H-bonding to the Cys ligands in the active site, model complexes containing planar N_2S_2-donor ligand sets where one S is modified, [Ni(nmp)(SR)]$^{-1}$ where R = C_6H_4-p-Cl, ^{t}Bu, o-benzoylaminobenzenethiol, N-(2-mercaptoethyl)benamide, and N-acetyl-L-cysteine methyl ester, were studied by UV–vis, cyclic voltammetry, electronic structure calculations, and were assayed for SOD activity (Gale et al. 2010). The studies determined that as the donor strength of the functionalized thiolate decreases (as in the aryl-S ligands) the oxidation potential of the Ni(II)N$_2$S$_2$ complex increases, and that aryl-S ligands also provide some protection from thiolate oxidation. These findings suggest that decreased electron density resulting from H-bonding at Cys6 plays a significant role in modulating the redox potential of the Ni center, and in protecting the reduced enzyme from oxidative modification (Gale et al. 2010).

Peptide Studies

The fact that all five of the ligands in the active site are derived from the first six amino acids in the protein, and that the "nickel-hook" motif is defined by the first several amino acids, makes model studies using synthetic peptides very accessible (Shearer et al. 2009). Peptides including the first twelve amino acids, [Ni(SOD^{M1})], where the N-terminus is unmodified or acylated, and the His1 ε-nitrogen is functionalized in a number of ways (unmodified, N-Me, N-DNP, and N-Tos) have been studied. Both M1-series peptides with unmodified imidazole ligands bind Ni in a 1:1 ratio, but M1-Ac has a lower affinity for Ni presumably due to the decreased covalent and increased ionic character of the amide bonds. The Ni redox potentials of the two Ni-peptides are both higher than for SODs: Ni(II)SOD^{M1}: 0.70(2) V, Ni(II)SOD^{M1}-AC: 0.49(3) V, and their activities measured by the xanthine/xanthine oxidase assay with NBT formazan are one to two orders of magnitude lower than SODs (Neupane and Shearer 2006). Neither of these M1 peptides could be isolated in the Ni(III) form. The M1-peptides with modified imidazole ligands were constructed to simulate the electron density changes expected to result from the H-bonding network that exists in the enzyme between His1 and Glu17/Arg47. The $E_{1/2}$ of the four Ni-peptides are all higher than that for SODs (~0.091 V vs Ag/Ag$^+$); Ni(II)SOD^{M1}-Im-H: 0.70(2) V, Ni(II)SOD^{M1}-Im-Me: 0.282(4) V, Ni(II)SOD^{M1}-Im-DNP 0.470(10) V, and Ni(II)SOD^{M1}-Im-Tos 0.598(5) V. The Lewis basicity of the functionalized imidazole donor is reflected in the SOD activity and in the electronic properties of the nickel center in the peptides. Decreased Lewis basicity (e.g., Ni(SOD^{M1}-Im-Tos) decreases the electron density at the Ni(III) center and improves the SOD activity (Shearer et al. 2009). This study supports the hypothesis that the subtle electronic changes due to the H-bonding network involving the His1 donor are directly linked to optimizing the redox chemistry in the enzyme.

Another set of peptides were constructed to examine the role of His1 and test the hypothesis that it is bound during catalysis and unbound to protect the sulfurs from oxidative damage when the enzyme is at rest. The M2-series peptides contained the first 7 amino acids in the NiSOD sequence, [Ni(SOD^{M2})], and modifications were made at residue 1 to include His, Ala, or Asp. All three M2 peptides are monomers, form 1:1 complexes with Ni, and have K_d values that are similar to those of the M1-series peptides. In contrast to Ni(II)SOD^{M2}A1/D1 mutant peptides, Ni(II)SOD^{M2}H1 CV studies suggest the His1 residue is bound during catalysis and acts to stabilize the Ni(III) oxidation state. Mutation of the His1 residue to Ala results in an anodic peak in the cyclic voltammogram that is ~250 mV more positive than observed in Ni(II) SOD^{M2}H1 (Neupane et al. 2007). Although the redox properties of all three metallopeptides fall within the potential range defined by the O$_2^{•-}$ oxidation/reduction couples, Ni(II)SOD^{M2}H1: 0.52(1) V, Ni(II) SOD^{M2}A1: 0.67(1) V, Ni(II)SOD^{M2}D1: 0.48(1) V), Ni(II)SOD^{M2}A1/D1 display significantly reduced SOD activity (over an order of magnitude lower), compared with Ni(II)SOD^{M2}H1. Although removing the apical imidazole ligand reduces the SOD activity of the peptide model, it does not completely abolish activity. Nonetheless, it is clear that the axial His1 ligand optimizes the active site for catalysis.

Conclusions

In many respects, NiSOD resembles other SODs, which catalyze O$_2^{•-}$ disproportionation near the diffusion limit over a broad pH range. However, NiSOD takes a different tack to arrive at novel solutions to the design features required for SOD catalysis. Instead of using a metal that supports redox chemistry in aqueous solution, nickel, with no aqueous redox chemistry, was selected. To acquire a redox potential that is optimum for catalyzing O$_2^{•-}$ disproportionation, cysteinate ligands not found in other SODs are employed, despite their sensitivity to oxidation. The remaining ligands are designed to help stabilize Ni(III) and prevent S-centered redox processes. The protein also employs a mechanism involving a tyrosine residue to act as a gate-keeper for substrate binding and product release that is reminiscent of that operating in MnSODs. Thus, similar functions have evolved using a variety of structures tailored to each metal cofactor.

Cross-References

▶ Acireductone Dioxygenase
▶ Catalases as NAD(P)H-Dependent Tellurite Reductases
▶ CO-Dehydrogenase/Acetyl-CoA Synthase
▶ Copper-Zinc Superoxide Dismutase and Lou Gehrig's Disease
▶ Methyl Coenzyme M Reductase

- ▶ [NiFe]-Hydrogenases
- ▶ Nickel in Bacteria and Archaea
- ▶ Nickel Ions in Biological Systems
- ▶ Nickel, Physical and Chemical Properties
- ▶ Nickel-Binding Proteins, Overview
- ▶ Nickel-Binding Sites in Proteins
- ▶ Urease

References

Abreu IA, Cabelli DE (2010) Superoxide dismutases-a review of the metal-associated mechanistic variations. Biochim Biophys Acta 1804(2):263–274

Bafana A, Dutt S et al (2011) The basic and applied aspects of superoxide dismutase. J Mol Catal B: Enzym 68(2):129–138

Barondeau DP, Kassmann CJ et al. (2004) Nickel Superoxide Dismutase Structure and Mechanism. Biochemistry 43(25):8038–8047

Bryngelson PA, Arobo SE et al (2004) Expression, reconstitution, and mutation of recombinant Streptomyces coelicolor NiSOD. J Am Chem Soc 126(2):460–461

Bryngelson PA, Maroney MJ (2007) Nickel superoxide dismutase. Met Ions Life Sci 2:417–443

Choudhury SB, Lee J-W et al (1999) Examination of the Nickel Site Structure and Reaction Mechanism in Streptomyces seoulensis Superoxide Dismutase. Biochemistry 38(12):3744–3752

Dupont CL, Neupane K et al (2008) Diversity, function and evolution of genes coding for putative Ni-containing superoxide dismutases. Environ Microbiol 10(7):1831–1843

Gale EM, Narendrapurapu BS et al (2010) Exploring the effects of H-bonding in synthetic analogues of nickel superoxide dismutase (Ni-SOD): experimental and theoretical implications for protection of the Ni-SCys bond. Inorg Chem 49(15):7080–7096

Gasteiger E, Gattiker A et al (2003) ExPASy: the proteomics server for in-depth protein knowledge and analysis. Nucleic Acids Res 31:3784–3788

Gennari M, Orio M et al (2010) Reversible apical coordination of imidazole between the Ni(III) and Ni(II) oxidation states of a dithiolate complex: a process related to the Ni superoxide dismutase. Inorg Chem 49(14):6399–6401

Herbst RW, Guce A et al (2009) Role of conserved tyrosine residues in NiSOD catalysis: a case of convergent evolution. Biochemistry 48(15):3354–3369

Johnson OE, Ryan KC et al (2010) Spectroscopic and computational investigation of three Cys-to-Ser mutants of nickel superoxide dismutase: insight into the roles played by the Cys2 and Cys6 active-site residues. J Biol Inorg Chem 15(5):777–793

Konecny R, Li J et al (1999) CuZn Superoxide Dismutase Geometry Optimization, Energetics, and Redox Potential Calculations by Density Functional and Electrostatic Methods. Inorg Chem 38(5):940–950

Lee HI, Lee JW et al (2010) ENDOR and ESEEM investigation of the Ni-containing superoxide dismutase. J Biol Inorg Chem 15(2):175–182

Maroney MJ (1999) Structure/function relationships in nickel metallobiochemistry. Curr Opin Chem Biol 3(2):188–199

Mathrubootham V, Thomas J et al. (2010) Bisamidate and mixed amine/amidate NiN2S2 complexes as models for nickel-containing acetyl coenzyme A synthase and superoxide dismutase: an experimental and computational study. Inorg Chem 49(12):5393–5406

Miller AF (2008) Redox tuning over almost 1 V in a structurally conserved active site: lessons from Fe-containing superoxide dismutase. Acc Chem Res 41(4):501–510

Neupane KP, Shearer J (2006) The influence of amine/amide versus bisamide coordination in nickel superoxide dismutase. Inorg Chem 45(26):10552–10566

Neupane KP, Gearty K et al (2007) Probing variable axial ligation in nickel superoxide dismutase utilizing metallopeptide-based models: insight into the superoxide disproportionation mechanism. J Am Chem Soc 129(47):14605–14618

Pelmenschikov V, Siegbahn PEM (2006) Nickel superoxide dismutase reaction mechanism studied by hybrid density functional methods. J Am Chem Soc 128(23):7466–7475

Prabhakar R, Morokuma K et al (2006) A DFT study of the mechanism of Ni superoxide dismutase (NiSOD): role of the active site cysteine-6 residue in the oxidative half-reaction. J Comput Chem 27(12):1438–1445

Ryan KC, Johnson OE et al (2010) Nickel superoxide dismutase: structural and functional roles of Cys2 and Cys6. J Biol Inorg Chem 15(5):795–807

Shearer J, Neupane KP et al (2009) Metallopeptide based mimics with substituted histidines approximate a key hydrogen bonding network in the metalloenzyme nickel superoxide dismutase. Inorg Chem 48(22):10560–10571

Wuerges J, Lee JW et al (2004) Crystal structure of nickel-containing superoxide dismutase reveals another type of active site. Proc Natl Acad Sci USA 101(23):8569–8574

Zilbermann I, Maimon E et al (2005) Redox chemistry of nickel complexes in aqueous solutions. Chem Rev 105(6):2609–2625

Nickel Transporters

Thomas Eitinger
Institut für Biologie/Mikrobiologie,
Humboldt-Universität zu Berlin, Berlin, Germany

Synonyms

Metal transporters; Transition-metal uptake

Definition

Import systems that transport the transition metal ion across cell membranes in order to provide it for incorporation into nickel-containing enzymes.

Background

Nickel is a trace nutrient for prokaryotes, certain fungi, algae, and plants. It represents an essential cofactor of a group of unrelated metalloenzymes catalyzing central reactions in energy and nitrogen metabolism, in detoxification processes, and in a side reaction of the methionine salvage pathway (▶ Nickel in Bacteria and Archaea, ▶ Nickel-Binding Sites in Proteins, ▶ Nickel Superoxide Dismutase, ▶ Acireductone Dioxygenase, ▶ [NiFe]-Hydrogenases, ▶ Urease, ▶ CO-Dehydrogenase/Acetyl-CoA Synthase). The focus of this entry is on import systems that transport Ni^{2+} ions with high affinity and specificity under physiologically relevant conditions into microbial cells. Most of the transporters discussed here have closely related counterparts among the ▶ Cobalt Transporters that function in the uptake of Co^{2+} ions. From a bioenergetic point of view, nickel importers can be classified into primary and secondary active transport systems (Eitinger et al. 2005; Rodionov et al. 2006) (Table 1). The latter include the long-known nickel/cobalt transporter (NiCoT) family and the weakly related systems UreH and HupE/UreJ. Primary active nickel importers contain subunits with an ATP-binding cassette (ABC) and are driven by ATP hydrolysis. NikABCDE is a canonical ABC transporter with restricted distribution. The more recently identified NikMNQO and NikKLMQO systems are members of the energy-coupling factor (ECF) family of micronutrient importers. These systems are widespread in bacteria and archaea and represent a novel type of ABC transporter (Rodionov et al. 2009; Eitinger et al. 2011). Another recent observation is energy-dependent transport of nickel ions across the outer membrane of Gram-negative bacteria, and the available data on this process is included in this entry.

NiCoT

High-affinity Ni^{2+}-uptake systems are known since the early 1990s. The first system reported was a one-component transporter in a β-proteobacterium that is required for the activity of nickel-dependent enzymes such as NiFe hydrogenase and urease. This transporter is an integral membrane protein with eight transmembrane domains (TMD) and is encoded adjacent to a hydrogenase operon (reviewed in Eitinger 2001). Later, homologs were identified in many Gram-negative and Gram-positive bacteria, in thermoacidophilic archaea, and in ascomycetous and basidiomycetous fungi (Eitinger et al. 2005; Zhang et al. 2009). The distribution among all kingdoms of life characterizes the NiCoTs as an ancient protein family. Analyses of NiCoT sequences predict two four-helix segments formed by the N-terminal and C-terminal parts of the proteins (Fig. 1). The two halves are connected by a large and highly charged cytoplasmic loop. A couple of conserved sequence motifs mainly located within the TMDs have been found to be essential for transport activity. Of special importance is TMD II harboring the +HAXDADH (+, R or K; X, V, F, or L) motif which serves as the signature sequence for NiCoTs. This segment interacts spatially with a histidine- or asparagine-containing region in TMD I together forming a central part of the selectivity filter that controls velocity and ion selectivity of the transport process (Degen and Eitinger 2002). Site-directed mutagenesis was applied to investigate the consequences of amino-acid replacements introduced into a cobalt- and a nickel-preferring NiCoT. These experiments showed that an increase of transport velocity results in a decreased specificity. The data indicated that preference of a NiCoT for either Co^{2+} or Ni^{2+} ion cannot be predicted solely based on the amino-acid sequence level.

Functional genomics proved a powerful method to predict the substrate preference of NiCoTs. The analyses uncovered that colocalization and/or coregulation of NiCoT genes with genes implicated in nickel or cobalt metabolism is a reliable indicator of the NiCoT's preference for Ni^{2+} or Co^{2+} (Rodionov et al. 2006) (▶ NikR, Nickel-Dependent Transcription Factor).

Nickel Transporters, Table 1 Classification and distribution of nickel-uptake systems

System	Mechanistic type	Occurrence
NikABCDE	Primary, canonical ABC-type	Prokaryotes
NikMNQO NikKLMQO	Primary, ECF-type	Prokaryotes
NiCoT	Secondary	Prokaryotes, fungi
UreH	Secondary	Prokaryotes, (plants?)
HupE/UreJ	Secondary	Prokaryotes
TBDT	TonB-dependent outer-membrane transporter	Gram-negative bacteria

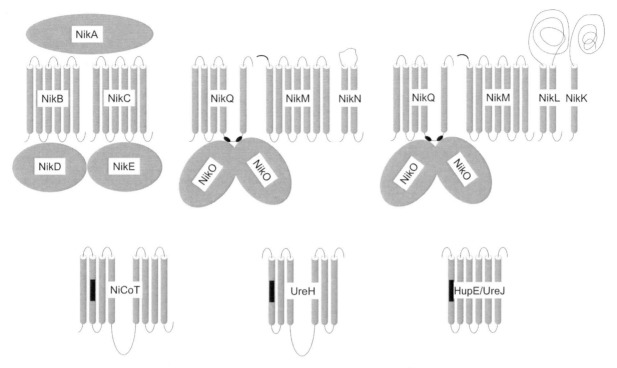

Nickel Transporters, Fig. 1 Characteristics of microbial nickel-uptake systems. See the text for details. Transmembrane helices are indicated as cylinders. The contact sites of the T unit NikQ with the ABC ATPases of the ECF-type Ni^{2+} transporters is depicted. The strongly conserved N-terminus of NikM proteins is highlighted. Nik(MN) fusions present in many NikMNQO-type Ni^{2+} transporters are indicated by a dashed line. Black bars indicate the signature sequence in TMD II of NiCoTs and related sequences in TMD I of UreH and HupE/UreJ. Outer-membrane nickel transporters of Gram-negative bacteria are not shown

UreH

Many *ureH* genes are located in bacterial hydrogenase and urease gene clusters (Rodionov et al. 2006; Zhang et al. 2009). UreH proteins are characterized by a six-transmembrane-domain architecture with the N- and C-terminus facing the exterior of the cells. TMD I contains a conserved sequence with KHALEPDH as the consensus that strongly resembles the signature sequence of NiCoTs. The amino-acid sequence predicts two three-helix bundles that are separated by a hydrophilic loop, rich in histidine residues. Experimental evidence confirms that UreH functions as a high-affinity nickel transporter (Eitinger et al. 2005).

A subtype (called SodT) of the UreH family is found in marine cyanobacteria and is encoded adjacent to the genes for ► Nickel Superoxide Dismutase and its cognate maturation endopeptidase. Within the characteristic sequence motif in TMD I [HVV(S/G)GADHL], the two histidine residues are separated by six amino-acid residues rather than five amino-acid residues as in the case of standard UreH and NiCoT proteins.

Relatives of the SodT proteins are present in plants, but their function remains enigmatic. The plant SodT-like proteins are produced with an N-terminal bipartite thylakoid transit peptide suggesting a localization within the thylakoid membranes of chloroplasts. Indeed, the SodT homolog At2g16800 of *Arabidopsis thaliana* clearly localizes to the chloroplast (Eitinger et al. 2005). Recent analyses of the homologs of rice (*Oryza sativa*) and the tobacco relative *Nicotiana benthamiana* confirmed their localization in thylakoids. These studies assigned a role in normal chloroplast development to SodT and uncovered leaf variegation and necrosis ("zebra necrosis") due to the absence of SodT. Urease activity – the only known nickel-dependent enzyme activity in plants (► Nickel in Plants) – was not altered in mutant leaves, and thus, a link of SodT to nickel metabolism is not obvious (Li et al. 2010).

HupE/UreJ

As indicated by the designation HupE/UreJ, those proteins represent a family whose members – in many cases – are encoded within ▶ NiFe Hydrogenase (*hup*, "hydrogen uptake") or ▶ Urease (*ure*) gene clusters. HupE/UreJ are integral membrane proteins with six TMDs in the mature state. Many HupE/UreJ proteins are produced as precursors with a predicted N-terminal signal peptide which is cleaved off at a conserved site releasing an N-terminal histidine residue of the processed proteins. TMD I of mature HupE/UreJ proteins contain a HPXXGXDH motif which resembles the signature sequence of NiCoTs and the related signatures in UreH and SodT proteins (Eitinger et al. 2005). A role in Ni^{2+} uptake is conceivable (and has been experimentally confirmed in a few cases) for the large number of HupE/UreJ homologs encoded in hydrogenase and urease clusters. Another set of *hupE* genes, mainly found in cyanobacteria, is genomically unlinked to nickel metabolism. Bioinformatic analyses predicted a role of those HupE proteins as transporters for Co^{2+} ion in the metabolic context of biosynthesis of the cobalt-containing coenzyme B_{12} (Rodionov et al. 2006).

NikABCDE

NikABCDE belongs to the PepT dipeptide and oligopeptide transporter family of canonical ABC importers and was originally identified as a high-affinity nickel-uptake system in *E. coli* (reviewed in Eitinger et al. 2011). Homologs with an experimentally confirmed or predicted function in nickel uptake were mainly found in proteobacteria, a few firmicutes species, and certain archaea. Based on amino-acid sequence comparisons, it is difficult to divide members of the PepT family into transporters with a function in Ni^{2+} or peptide uptake. Thus, predictions of NikABCDE systems as nickel transporters took into account the genomic colocalization with nickel metabolism and the presence of binding sites for the nickel-dependent repressor NikR upstream of *nikABCDE* genes (Rodionov et al. 2006; Zhang et al. 2009). NikABCDE systems are composed of pairs of transmembrane proteins (NikB, NikC), pairs of ABC ATPases (NikD, NikE), and the extracytoplasmic solute-binding protein NikA. Several crystal structures of NikA have been reported, but the molecular basis of selective binding of Ni^{2+} ion has not yet been unraveled. The structural data suggest that metal ions are not bound in a free form but rather in complex with a yet unidentified organic cofactor. To give an even more complex picture, NikA was shown to bind heme at a site remote from the nickel-binding site (reviewed by Eitinger et al. 2011).

NikMNQO/NikKLMQO

ECF transporters, a novel type of ABC importers for micronutrients in prokaryotes, were originally identified during functional genomic analysis of nickel and cobalt metabolism (Rodionov et al. 2006). These systems consist of an energy-coupling factor, composed of a conserved transmembrane protein (T component) and pairs of ABC ATPases (A components). Substrate specificity is conveyed through an S component which is made up of a single integral membrane protein (in most cases), or multiple membrane proteins as in the case of transporters for Ni^{2+} ions or Co^{2+} ions (Eitinger et al. 2011). In contrast to canonical ABC importers, all ECF systems are devoid of extracytoplasmic soluble solute-binding proteins. The list of substrates of ECF transporters includes many vitamins and intermediates of salvage pathways in addition to Ni^{2+} and Co^{2+} ions.

The basic module of ECF-type Ni^{2+} transporters consists of the T unit NikQ, the A unit NikO, and the partial S unit NikM. Analyses of related ECF-type Co^{2+} transporters have revealed that the basic CbiM-CbiQ-CbiO module is inactive. Activity is dependent on the presence of CbiN, a small integral membrane protein with two TMDs and an extracytoplasmic loop in between (Siche et al. 2010). Based on this and on a couple of additional observations (reviewed in Eitinger et al. 2011), it was concluded that S units of Ni^{2+} und Co^{2+} transporters need components in addition to the M proteins. Whereas CbiN represents this additional component in the case of the Co^{2+} transporters, variations occur among the Ni^{2+} transporters, and NikN or NikK plus NikL represent the additional part(s) of the S unit in most cases. NikN displays a similar topology as CbiN, but the two proteins fall into different branches in a phylogenetic tree (Rodionov et al. 2006; Zhang et al. 2009). Unlike in the case of the cobalt transporters which contain CbiM and CbiN as two peptide chains, NikM and NikN are fused to a single peptide chain in many cases. NikK and NikL contain one and two TMDs, respectively, and an extended extracytoplasmic domain (Fig. 1). NikM proteins contain the strongly conserved sequence MHIPD at the N terminus which is likely involved in metal recognition. As indicated in Fig. 1, NikQ proteins – like all T units of ECF transporters (Eitinger et al. 2011) – harbor two three-amino-acid motifs with Ala-Arg-Gly as the consensus in a cytoplasmic loop. These regions are candidate contact sites for interactions with the ABC ATPases.

TonB-Dependent Outer-Membrane Transporters (TBDT)

Until recently, active transport of solutes across the outer membrane of Gram-negative bacteria was only known for Fe^{3+} ions in complex with organic siderophores and for coenzyme B_{12}-related cobalamins. Recently, however, functional genomics and biochemical analyses identified a large number of outer-membrane transporters for alternate solutes (Schauer et al. 2008). Outer-membrane transporters consist of a C-terminal β-barrel domain and an N-terminal plug domain, and they bind their substrates with high affinity. They depend on ExbB-ExbD-TonB, a complex of proteins located in the cytoplasmic membrane. TonB contains an extensive periplasmic segment that interacts with the outer-membrane receptors.

Nickel-regulated genes for outer-membrane transporters and nickel-regulated *exbBD-tonB* clusters were identified in proteobacteria (Rodionov et al. 2006). These data and the outcome of experimental analyses with *Helicobacter pylori* (reviewed in Schauer et al. 2008) support the notion that Ni^{2+} ions cross the outer membrane of many Gram-negative bacteria by a TonB-dependent active transport mechanism.

Cross-References

- Acireductone Dioxygenase
- CO-Dehydrogenase/Acetyl-CoA Synthase
- Cobalt Transporters
- [NiFe]-Hydrogenases
- Nickel in Bacteria and Archaea
- Nickel in Plants
- Nickel Superoxide Dismutase
- Nickel-Binding Sites in Proteins
- NikR, Nickel-Dependent Transcription Factor
- Urease

References

Degen O, Eitinger T (2002) Substrate specificity of nickel/cobalt permeases: insights from mutants altered in transmembrane domains I and II. J Bacteriol 184:3569–3577

Eitinger T (2001) Microbial nickel transport. In: Winkelmann G (ed) Microbial transport systems. Wiley-VCH, Weinheim

Eitinger T, Suhr J, Moore L, Smith JAC (2005) Secondary transporters for nickel and cobalt ions: theme and variations. Biometals 18:399–405

Eitinger T, Rodionov DA, Grote M, Schneider E (2011) Canonical and ECF-type ATP-binding cassette importers in prokaryotes: diversity in modular organization and cellular functions. FEMS Microbiol Rev 35:3–67

Li J, Pandeya D, Nath K et al (2010) ZEBRA-NECROSIS, a thylakoid-bound protein, is critical for the photoprotection of developing chloroplasts during early leaf development. Plant J 62:713–725

Rodionov DA, Hebbeln P, Gelfand MS, Eitinger T (2006) Comparative and functional genomic analysis of prokaryotic nickel and cobalt uptake transporters: evidence for a novel group of ATP-binding cassette transporters. J Bacteriol 188:317–327

Rodionov DA, Hebbeln P, Eudes A et al (2009) A novel class of modular transporters for vitamins in prokaryotes. J Bacteriol 191:42–51

Schauer K, Rodionov DA, de Reuse H (2008) New substrates for TonB-dependent transport: do we only see the "tip of the iceberg"? Trends Biochem Sci 33:330–338

Siche S, Neubauer O, Hebbeln P, Eitinger T (2010) A bipartite S unit of an ECF-type cobalt transporter. Res Microbiol 161:824–829

Zhang Y, Rodionov DA, Gelfand MS, Gladyshev VN (2009) Comparative genomic analyses of nickel, cobalt and vitamin B_{12} utilization. BMC Genomics 10:78

Nickel Tumorigenicity

- Nickel Carcinogenesis

Nickel Vaccination

- Nickel Hyposensitization

Nickel, Physical and Chemical Properties

Fathi Habashi
Department of Mining, Metallurgical, and Materials Engineering, Laval University, Quebec City, Canada

Nickel is a hard silver white metal, which forms cubic crystals. It is malleable, ductile, and has superior strength and corrosion resistance. The metal is a fair conductor of heat and electricity and exhibits magnetic properties below 345°C.

Physical Properties

Atomic number	28
Atomic weight	58.69
Melting point, °C	1,455
Boiling point, °C	2,730
Relative density, 25°C	8.9
Volume increase on melting, %	4.5
Heat of fusion at m.p, J/g	302
Heat of sublimation at 25°C, J/g	7,317
Heat of vaporization (T_{crit}), J/g	6,375
Thermal conductivity $(0 - 100°C)$, W m^{-1}K^{-1}	88.5
Heat capacity $(0 - 100°C)$, J g^{-1}K^{-1}	0.452
Electrical resistivity at 20°C, $\mu\Omega$/cm	6.9
Temperature coefficient of electrical resistivity $(0 - 100°C)$, K^{-1}	6.8×10^{-3}
Thermal expansion coefficient $(0 - 100°C)$, K^{-1}	13.3×10^{-6}
Modulus of elasticity, GPa	199.5
Brinell hardness	85

Chemical Properties

Nickel is a transition metal having the electronic configuration 2, 8, 16, 2. It dissolves in ammoniacal solutions in presence of oxygen forming ammine complex:

$$Ni + 1/2 O_2 + 2NH_3 + H_2O \rightarrow [Ni(NH_3)_2]^{2+} + 2OH^-$$

Nickel powder is precipitated from nickel sulfate solution by hydrogen at high temperature and pressure:

$$Ni^{2+} + H_2 \rightarrow Ni + 2H^+$$

Nickel sulfide undergoes oxidation in neutral medium at ambient conditions to NiSO$_4$:

$$NiS + 2O_{2(aq)} \rightarrow NiSO_4$$

In acidic medium, however, elemental sulfur forms:

$$NiS + 1/2 O_{2(aq)} + 2H^+ \rightarrow Ni^{2+} + S + H_2O$$

Nickel sulfide is oxidized at high temperature to NiO and sulfur dioxide:

$$NiS + 1 1/2 O_2 \rightarrow NiO + SO_2$$

Molten nickel sulfide, Ni$_3$S$_2$, undergoes conversion reaction when oxygen is blown through:

$$Ni_3S_2 + 2O_2 \rightarrow 3Ni + 2SO_2$$

This is a basic reaction in the metallurgy of nickel. Carbon monoxide reacts with nickel powder at 60°C to form the volatile nickel tetracarbonyl, Ni(CO)$_4$. This reaction is reversible, with the carbonyl compound decomposing to carbon monoxide and nickel at higher temperatures (180°C):

$$Ni + 4CO \leftrightarrows Ni(CO)_4$$

This reaction is the basis for the refining of nickel. No other metal forms similar carbonyl compounds under such mild conditions at atmospheric pressure. Nickel carbonyl is highly toxic.

References

Habashi F (1999) Textbook of hydrometallurgy. Métallurgie Extractive Québec, Québec, distributed by Laval University Bookstore

Habashi F (2002) Textbook of pyrometallurgy. Métallurgie Extractive Québec, Québec City, distributed by Laval University Bookstore

Kerfoot DGE (1997) Chapter 12. In: Habashi F (ed) Handbook of extractive metallurgy. Wiley-VCH, Heidelberg

Nickel-Binding Proteins, Overview

Anne Volbeda and Juan C. Fontecilla-Camps
Metalloproteins; Institut de Biologie Structurale J.P. Ebel; CEA; CNRS; Université J. Fourier, Grenoble, France

Definition

Ni-binding proteins may be separated into four classes: enzymes that use the special chemical properties of nickel to catalyze a small number of biologically and environmentally important redox and nonredox reactions, chaperones that function in the maturation of Ni-containing enzyme active sites, transporters taking care of nickel-trafficking, and nickel-responsive DNA-binding proteins that function as transcriptional regulators.

Introduction

Nickel is a 3-D transition metal with a very rich coordination chemistry (Fig. 1), as exemplified by the different ligand environments that are observed in current nickel-containing proteins (see ▶ Nickel-Binding Sites in Proteins). In its most common Ni(II) oxidation state, eight 3d electrons must be distributed over five 3d orbitals, giving either a paramagnetic high-spin or a diamagnetic low-spin species. High-spin ($S = 1$) Ni(II) has a larger metal radius and typically displays longer ligand distances than low-spin ($S = 0$) Ni(II). As predicted by ligand field theory, $S = 0$ compounds are observed in square planar coordination environments, which involve four ligands, and trigonal bipyramidal ones involving five ligands. In this case, the highest energy 3d orbital of the nickel, which is the one that overlaps most extensively with the ligand positions, is unoccupied. An octahedral environment with six ligands normally results in an $S = 1$ compound: two single electrons with parallel spins occupy the two highest 3d orbitals that have similar energies, providing a significant stabilization. In a tetrahedral environment of four ligands, three 3d orbitals are high in energy, which likewise leads to an $S = 1$ state as the most stable one. Other Ni(II) coordination environments like square pyramidal (square planar with a fifth apical ligand) and tetragonal (similar to octahedral with long bonds to the two apical ligands) can give rise to both $S = 0$ and $S = 1$ states, depending on the strength of the bonds, which in turn depend on the ligand type.

Nickel is also a redox active metal: oxidation of Ni(II) gives Ni(III), which with seven 3d electrons gives either an $S = 3/2$ or an $S = 1/2$ state, whereas reduction gives Ni(I) with $S = 1/2$. These paramagnetic states can easily be detected with EPR (electron paramagnetic resonance) spectroscopy, unless there is a strong antiferromagnetic coupling with a nearby metal ion in an odd-electron state. The redox potentials that are required to obtain such states with model compounds are not often within physiological range, but they can be shifted to biologically more relevant values by interactions with a protein environment. As discussed further below, this is observed in the group of nickel-containing enzymes that catalyze redox reactions. Ni(III) compounds are most often octahedral, whereas Ni(I) ions, which have a larger radius and require therefore more space, display mostly a tetrahedral coordination. Very uncommon are compounds with Ni(IV) or Ni(0). Due to the reducing effect of X-ray-generated photoelectrons, the presence of Ni(0) in some crystal structures of Ni-containing proteins obtained from X-ray diffraction data cannot be excluded, but such a very reduced state may not be biologically relevant.

The versatile chemical properties of nickel allow it to catalyze quite different types of reactions when bound to a protein (Table 1). The protein environment may distort coordination geometries and/or stabilize high-energy structures, thus leading to a decrease of the free-energy difference of the rate-limiting transition state relative to the starting state of the reaction process. This results in large rate enhancements. However, compared to other transition metals like iron, manganese, copper, and zinc, nickel is used by only a small number of enzymes. This may be due to a combination of its relatively low bioavailability and its redox activity, which leads to facile reactions with molecular oxygen (O_2). A lot of nickel is present in the earth core, but the concentration of dissolved nickel ions in seawater is reported to be, on average, only about 0.00015 micrograms per liter (9 nM; and only ≈0.5 nM in the human body, but more Ni may be

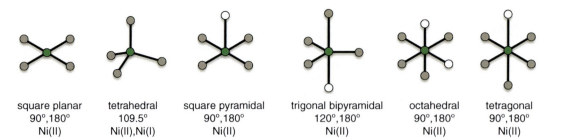

Nickel-Binding Proteins, Overview, Fig. 1 Typical nickel coordination environments (Ni in *green*, ligands in *gray* and *white*)

Nickel-Binding Proteins, Overview, Table 1 Reactions catalyzed by Ni-containing enzymes

Enzyme	Reaction	
Hydrogenase[a]	$H_2 \Leftrightarrow 2H^+ + 2e^-$	(1)
CO dehydrogenase[a]	$CO_2 + 2e^- + 2H^+ \Leftrightarrow CO + H_2O$	(2)
Acetyl coenzyme A synthase[b]	$CH_3{}^- - [Co^{3+}] + SCoA^- + CO \Leftrightarrow CH_3(CO) - SCoA + [Co^+]$	(3)
Methyl coenzyme M reductase[c]	$CH_3-SCoM + HS-CoB \Leftrightarrow CH_4 + CoMS-SCoB$	(4)
Superoxide dismutase[a]	$2O_2{}^{\bullet-} + 2H^+ \rightarrow H_2O_2 + O_2$	(5)
Urease	$H_2N - CO - NH_2 + H_2O \rightarrow H_2NCOOH + NH_3$	(6)
Acireductone dioxygenase	$H_3CSC_2H_4-C_3H_3O_3 + O_2 \rightarrow H_3CSC_2H_4-COOH + HCOOH + CO$	(7)

[a]Reactions (1), (2), and (5) are also catalyzed by unrelated non-Ni-containing metalloenzyme(s)
[b][Co^{3+}] and [Co^+] denote a methyl-carrying cobalt-containing protein, and CoA is coenzyme A
[c]CoM and CoB are coenzymes M and B, respectively

present in chelated forms). According to geological and geochemical evidence, Ni availability may have been considerably higher when the Earth was young and the first life forms were just developing. At that time, reducing conditions prevailed and as a consequence, O_2 concentrations were very low. In addition, meteor impacts and volcanic ejections of magma, which provide the major sources of nickel, were much more frequent than today. From the function of the few currently known nickel-containing enzymes, nickel appears to be particularly well suited to catalyze reactions involving small gaseous compounds (Table 1), including molecular hydrogen (H_2), carbon monoxide (CO), carbon dioxide (CO_2), methane (CH_4), and ammonia (NH_3). H_2 and CO, which may have been produced in large quantities by volcanic outbursts, are carriers of chemical energy. Before the biological invention of photosynthesis, developing life-forms may have depended on these diatomic molecules as a source of reducing power. In addition, it does not seem unlikely that CO and CO_2 were the main sources of carbon for the first life forms, and NH_3 could have been a nitrogen source.

The use of nickel by specific enzymes is possible thanks to special nickel-binding proteins that ensure a highly controlled management of its cellular distribution and storage. Some of these proteins are involved in the active transport of nickel between the environment and the cytosol. This is required both when there is too little and, conversely, when there is too much nickel present, as the latter may have toxic effects. Specific nickel-storing proteins decrease the free nickel concentration. Nickel chaperones are needed for the maturation of Ni-dependent enzymes. In addition, nickel-sensing proteins are used to modulate the expression of genes coding for proteins involved in Ni-dependent processes. The various types of nickel-binding proteins and enzymes will be discussed in the following sections, with cross-references to more detailed chapters in the encyclopedia.

Nickel-Binding Proteins, Overview, Fig. 2 Cα-chain traces of five Ni-containing redox enzymes. H$_2$ase is a heterodimer, CODH a homodimer, MCR a dimer of heterotrimers, and SOD a hexamer. Metal sites are shown as *spheres*. Color code: Ni *green*, Fe *red-brown*, S *yellow*, and Mg *blue*

Nickel-Dependent Redox Enzymes

Five nickel-containing redox enzymes have been, and still are extensively investigated: (a) [NiFe]-hydrogenase (NiFe-H$_2$ase), (b) carbon monoxide dehydrogenase (Ni-CODH), (c) acetyl coenzyme A synthase (ACS), (d) methyl coenzyme M reductase (MCR), and (e) superoxide dismutase (Ni-SOD). The catalyzed reactions are shown in Table 1. The enzymes that catalyze reactions 1–4 (Table 1) normally require strict anaerobic conditions to function (Fontecilla-Camps et al. 2009). They may have been among the first enzymes to have evolved, possibly around 3.8 billion years ago, as they are only found in "primitive" prokaryotic organisms, including autotrophic bacteria and archaea, and reactions 1–4 are involved in primordial chemistry. An important common feature of NiFe-H$_2$ase, Ni-CODH and MCR is the location of the Ni-binding active site deeply buried in the protein interior (Fig. 2). This may be due to the redox-active role of nickel, and it may prevent unwanted side reactions with other redox-active molecules that are abundant in the intracellular environment. The ACS active site is much more exposed and easily inactivated by the exchange of a labile Ni ion with other metal ions like Zn^{2+} and Cu^+. ACS is classified as a redox enzyme because the overall three-substrate reaction includes two redox steps: oxidative addition of a methyl cation and reductive elimination of acetyl CoA (in addition to CO-binding, formation of an acetyl intermediate and its transfer to CoA). ACS can organize the arrangement of its three approximately globular polypeptide domains in three different ways (Volbeda et al. 2009). These different conformations might be necessary for the recognition of the three very different substrates (see reaction 3 in Table 1).

Ni-SOD probably evolved after the invention of the complex photosynthetic machinery, producing O_2 from water, which led to the great oxidation event of the earth atmosphere. According to geological evidence, this happened about 2.4 billion years ago. An O_2 molecule readily takes up a low-potential electron, for example, from a reduced metal ion like Ni(II) (in the so-called Fenton reaction), to form the extremely reactive superoxide radical. SOD enzymes neutralize this very toxic species (reaction 5). The nickel site in Ni-SOD is located at the surface, but apparently it is not very reactive toward other putative redox partners in the cell. Its accessibility may facilitate a rapid detoxification function.

Further comparison of the structures of the five Ni-dependent redox enzymes reveals several similarities. One of these is the presence of long, to a great extent hydrophobic, tunnels in NiFe-H$_2$ase, MCR, Ni-CODH, and ACS. These tunnels are ideally suited to guide the diffusion of hydrophobic reaction substrates and products, including gases, between the enzyme surface and the active site. Ni-CODH and ACS form a stable complex that in one conformation shows the tunnel to extend over a distance of about 70 Å from the Ni-CODH to the ACS active site, allowing the immediate consumption (by reaction 3) of CO that is produced by the otherwise endergonic reaction 2. The tunnel-containing domains in Ni-CODH and ACS have similar polypeptide folds, suggesting that they are evolutionary related. Furthermore, both NiFe-H$_2$ase and Ni-CODH contain three iron-sulfur clusters that are suitably positioned to provide a rapid pathway for electron transfer between the active site and the surface, where enzyme specific redox partners may transiently bind.

Nickel-Binding Proteins, Overview, Fig. 3 Cα-chain traces and Ni sites of hexameric urease and monomeric ARD

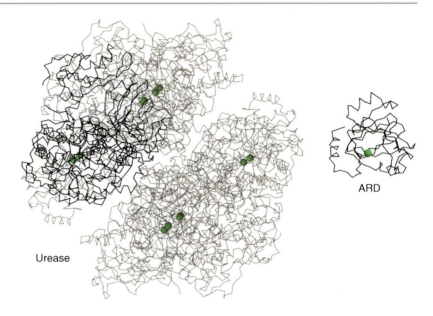

There are also common features in nickel binding. The nickel ions of NiFe-H$_2$ase and Ni-CODH, as well as one of the two Ni ions of ACS, are each involved in three approximately coplanar Ni-S bonds and one additional apical Ni-S bond. In these three enzymes, two *cis* sites (i.e., perpendicular to each other) appear available for substrate binding that, if occupied, would complete an octahedral coordination (Fig. 1). So far, it has not been possible to determine with certainty which of the latter sites is used for catalytic turnover. In addition, each of these three Ni ions is in the immediate vicinity of at least one Fe ion. Close to the first Ni, ACS has a second Ni ion very similar to its counterpart in Ni-SOD, with bonds to two Cys-S atoms and two successive main chain N atoms in a square planar arrangement. The superoxide substrate of Ni-SOD most likely binds at an apical position, trans to an imidazole ligand. MCR has a unique porphyrin-related, highly reduced tetrapyrrole cofactor called corphin or F$_{430}$, in which the Ni binds to four N atoms in a distorted square planar coordination, with an additional apical bond to a glutamine amide O atom. The trans apical position is most likely used for the binding of a methyl reaction intermediate, which is very different from the methyl binding in ACS that has been proposed to take place in the plane of three S ligands.

Typical redox states seem to vary between Ni(II) and Ni(III) in NiFe-H$_2$ase and Ni-SOD, and between Ni(II) and Ni(I) in Ni-CODH and ACS, but the catalytic involvement of some of these states is still being debated. In MCR, there is evidence for the involvement of all three redox states, that is, Ni(I), Ni(II), and Ni(III), in the catalytic mechanism. According to recent studies (Amara et al. 2011), a Ni-bound hydride intermediate may be involved in the catalytic mechanism of Ni-CODH. This provides an additional similarity with NiFe-H$_2$ase, where a hydride intermediate results from the heterolytic cleavage of H$_2$. For more details on the function of the various redox enzymes, their structures, nickel binding sites, and special substrates like coenzymes A, B, and M, see ▶ [NiFe]-Hydrogenases, ▶ CO-Dehydrogenase/Acetyl-CoA Synthase, ▶ Nickel Superoxide Dismutase, and ▶ Methyl Coenzyme M Reductase.

Nonredox Nickel-Dependent Enzymes

Two well-known Ni-binding nonredox enzymes are urease and acireductone dioxygenase (Ni-ARD). Urease is a particularly efficient enzyme, catalyzing the hydrolysis of urea 14 orders of magnitude more rapidly than it would proceed in its absence. The produced carbamic acid (reaction 6 in Table 1) spontaneously breaks down to carbon dioxide and ammonia. A dinuclear nickel site is bound at the bottom of a deep cleft, next to the C-terminal region of a parallel 8-stranded β-barrel (Fig. 3). The two Ni ions are in a distorted octahedral and distorted square-pyramidal coordination environment, respectively; this includes two histidine ligands, a bridging carbamated lysine, a bridging water or hydroxide, a terminally bound water, and, for the octahedral nickel, an additional aspartate. For catalysis, the Lewis acid properties of nickel are used. Other Ni-specific factors, like its uniquely versatile coordination chemistry, may also play a role because there are no enzymes with other metals known that can compete with urease for the hydrolysis of urea. In contrast to the redox enzymes discussed above, urease is not restricted to prokaryotes since it is also found in all plants. For more information see ▶ Urease.

ARD is a relatively small enzyme (Fig. 3) that can catalyze two different reactions of the methionine salvage pathway, depending on the metal that is bound to its active site. In the presence of Ni, the acireductone ($H_3CSC_2H_4$-$C_3H_3O_3$) substrate is oxidized to 3-(methylthio)propanoic acid, formic acid, and CO (reaction 7 in Table 1), whereas with Fe, no CO is formed and the products are 4-(methylthio)-2-oxobutanoic and formic acid. In the Ni enzyme, the metal is in a distorted octahedral environment involving three orthogonal histidine ligands, one glutamate, and two water molecules that are most likely exchangeable with substrate O atoms. ARD with Ni has so far only been characterized in a bacterium (*Klebsiella pneumonia*), where it may also bind Fe. The Fe-binding version of the enzyme seems to be much more widespread and is also found in mammals. For more details on this remarkable enzyme, see, ▶ Acireductone Dioxygenase.

Several other Ni-dependent nonredox enzymes have been reported. These include peptide deformylase, quercetinase, and glyoxalase I. However, each of them may also function with other divalent cations capable of performing Lewis acid chemistry. For certain enzymes like glyoxalase I, the Ni-bound form is the most active one in prokaryotes, whereas in eukaryotes, Zn is preferred. This might reflect adaptations that were due to changing conditions in the relative bioavailability of these metals during evolution.

Nickel Chaperones

The proper assemblage of nickel in most enzyme active sites involves the assistance of special maturation proteins called nickel metallochaperones (Li and Zamble 2009) or, more briefly, nickel chaperones. From a functional viewpoint, the Ni sites of these proteins have to be relatively labile because they must transfer the metal from a Ni source, which may itself be a protein or some other intracellular Ni chelator, to the enzyme. At present, the molecular details of Ni transfer, requiring intermolecular ligand exchange, are far from being understood. The involved protein-protein complexes are necessarily dynamic and therefore difficult to characterize. Because nickel is weakly bound, details of the binding interactions are also difficult to obtain. Nevertheless, significant progress has been made with the identification and further analysis of nickel-binding properties of enzyme-specific nickel chaperones. Here, these proteins are discussed only summarily, emphasizing the crystallographically obtained information.

Insertion of nickel in the NiFe-H_2ase active site requires the Ni carriers HypA and SlyD as well as the GTP-dependent Ni insertase HypB. In certain organisms, maturation of NiFe-H_2ase may be interconnected with that of urease. For example, in the gastric pathogen *Helicobacter pylori*, HypA has been shown to interact with the Ni carrier UreE that is needed for the maturation of urease. The latter further requires three other nickel chaperones, UreF, UreG, and UreH, UreG being a GTP-dependent Ni insertase. Like HypB, it is related to two ATP-dependent nickel chaperones of Ni-CODH and ACS, CooC and AcsF. Other cytosolic Ni carriers that may be involved in enzyme maturation are HspA, Hpn, and so-called Hpn-like, which have been characterized in *H. pylori*. These proteins may in addition function in intracellular nickel sequestration and detoxification. In this respect, it is also interesting to mention *Streptococcus suis* Dpr, a dodecameric metal storage protein belonging to the ferritin superfamily, which is capable of binding nickel inside a big internal cavity. Crystal structures have been reported for HypA, HypB, SlyD, UreE, CooC, and Dpr, but only the SlyD, UreE, and Dpr structures contain significant amounts of bound nickel (Fig. 4). In the case of *Thermus thermophilus* SlyD, the nickel is octahedrally bound by six histidines, but three of these are located in a His-tag used for purification purposes, and one of the latter comes from a different SlyD molecule. In addition, no hydrogenase has been characterized in *T. thermophilus*. Therefore, the functional significance of the observed nickel site may be questioned. Interestingly, in the case of UreE, four

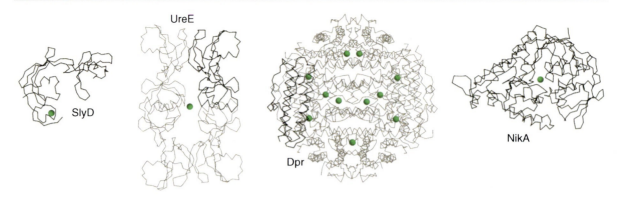

Nickel-Binding Proteins, Overview, Fig. 4 Cα-chain traces and Ni sites of monomeric SlyD and NikA, tetrameric UreE, and dodecameric Dpr

different protein molecules bind to Ni(II), each contributing one histidine ligand. The observed interactions between UreE monomers might mimic those in complexes with other nickel-transferring proteins. As the metal lies on a pseudofourfold symmetry axis, it has a square planar coordination base. This favors the binding of Ni(II) relative to other metals that could also be present in the cytosolic environment, like Cu(I) and Zn(II), which prefer a tetrahedral coordination.

Nickel Transporters

Nickel transport between the cell and its environment involves mostly membrane proteins and, in gram-negative bacteria having an outer (OM) and a cytoplasmic membrane (CM), additional periplasmic proteins that may be either soluble or membrane-attached. When the nickel concentration in the environment is high enough, passive diffusion of Ni ions or small Ni chelates may be enough to satisfy the cellular demands for the production of Ni-dependent enzymes. Nonspecific OM porins (trimeric β-barrel proteins) and much more specific CM permeases as NiCoT (nickel-cobalt transporters) allow such passive transport. The latter are found in gram-negative and gram-positive bacteria, as well as in archaea and fungi. However, in more common situations, there is too little nickel present in the cell, whereas sometimes there is too much of it, which may cause toxic effects. This explains the evolution of complex energy-consuming active Ni uptake and Ni efflux systems (Li and Zamble 2009; Ma et al. 2009).

Active nickel import is performed, for example, by ATP-driven ABC (ATP-binding cassette) transporters, such as the NikABCDE system in *Escherichia coli*. NikA is a periplasmic Ni-binding protein; NikB and NikC form a heterodimeric pore through the cytoplasmic membrane (CM), and NikD and NikE constitute the cytosolic ATPase subunits. Crystallographic studies have shown that NikA (Fig. 4) actually binds a small, not yet fully characterized Ni-chelate, which may be a nickelophore. The presence of specific metallophores may be an important factor to allow metal specificity. In *H. pylori*, a Ni-specific outer membrane (OM) receptor, called FrpB4, has been found to function as a TonB-dependent acid-induced transporter of Ni(II), probably as a nickelophore, in combination with a periplasmic- and IM-spanning TonB/ExbB/ExbD complex. The periplasmic TonB protein component reversibly interacts with the N-terminal plug region (the TON box) of the OM-spanning β-barrel of FrpB4 and somehow uses the proton gradient across the CM through the membrane-bound ExbB and ExbD subunits as an energy source to expel nickel to the cytosol. Active nickel efflux systems include the so-called Cnr (cobalt-nickel resistance) proteins. These form a CnrABC complex where CnrA spans the CM, CnrB the periplasm, and CnrC is bound at the inner side of the OM.

Although some structures of related transport systems are known, structural information on nickel transporters is quite rare due to the difficulties that are involved in the crystallization of membrane protein complexes, and many molecular details of active nickel transport remain to be determined. For more information see ▶ Nickel Transporters.

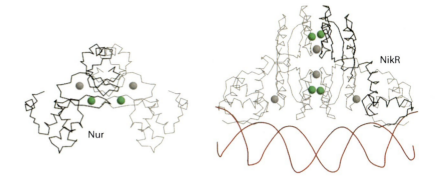

Nickel-Binding Proteins, Overview, Fig. 5 Cα-chain traces and metal sites of dimeric Nur and tetrameric NikR (Bound DNA is shown as a *red* double helix)

Nickel-Sensing Transcriptional Regulators

In order to keep an optimal concentration of nickel inside the cell, uptake and efflux processes must be properly balanced. This homeostasis is obtained thanks to the presence of specific nickel-sensing systems that function as transcriptional regulators. For nickel, several single-component transcriptional regulators have been characterized (Li and Zamble 2009; Ma et al. 2009). They are structurally similar to other transcriptional regulatory proteins, containing a DNA- and a metal- or small-molecule-binding domain. Some are single-gene regulators, whereas others regulate the expression of a large number of genes. The DNA-binding conformation is allosterically regulated by the presence or absence of bound nickel. The Ni-bound form generally acts as a gene activator for the expression of Ni-storage and Ni-efflux proteins and as a gene repressor for Ni-uptake proteins, Ni-dependent enzymes, and their chaperones, whereas the Ni-free form produces the opposite effect. A few examples will be very briefly described.

Some transcriptional regulators have been characterized that control the efflux of nickel and cobalt. These include NmtR and KmtR in *Mycobacterium tuberculosis*, which belong to the ArsR family of homodimeric transcriptional regulators, and RcnR in *E. coli*, which may also function as a homodimer but belongs to a different family. These proteins function as gene repressors in the metal-free form. SOD expression in *Streptomyces coelicolor* is regulated by Nur, which belongs to the Fur family. In the Ni-bound form, Nur activates the expression of Ni-SOD and represses the expression of the unrelated Fe-SOD, whereas in the absence of Ni, it operates in the opposite direction. The crystal structure of Nur has been reported (Fig. 5), showing it is a homodimer with two metal binding sites per monomer, one of which is occupied by Ni and the other by either Ni or Zn. The first metal site lies close to the subunit interface, whereas the second one is located at the interface between the DNA-binding and dimerization domains. Further studies will be needed to understand how this sensor works. Structurally, the best characterized nickel-sensing transcriptional regulator, NikR, controls the expression of the Nik operon and, in *H. pylori*, of several other genes, including those involved in TonB-dependent active nickel import. It functions with a homotetrameric core of C-terminal Ni-binding domains that are flanked by two dimer-forming DNA-binding N-terminal domains (Fig. 5). There is one high-affinity square-planar Ni(II) coordination site per molecule. The nickel ligands consist of a thiolate and two imidazoles from one molecule and a third imidazole from another one. It has been found that the same site has higher affinity for Cu(II), but not for Cu(I). Because under the anaerobic conditions of the cytosol, Cu(II) is converted to Cu(I), the metal site becomes nickel specific. Additional low-affinity nickel-binding sites appear to be important for the allosteric regulation of the conformation of NikR. For more details see ▶ NikR, Nickel-Dependent Transcription Factor.

Concluding Remarks

The structures of the Ni-proteins discussed in the text correspond to the following deposition codes in the protein data bank, www.pdb.org (Berman et al. 2000): 1YQW, 1OAO, 1HBN, 1T6U, 3LA4, 1VR3, 1F9Z, 3E3U, 3A43, 2HF8, 3CGM, 3NY0, 3CXN, 3KJI, 2XJO, 3DP8, 3EYY and 2HZV. More references may be found there. Nickel-containing enzymes like CODH and MCR have a significant influence on the global carbon cycle. The former keeps atmospheric CO levels low and the latter is the exclusive producer of CH_4. The use of urea as a major soil fertilizer is only possible thanks to its conversion by urease to ammonia. Ni-H_2ases are central to the energy metabolism of many bacteria and archaea. Urease and Ni-H_2ase are also medically important, as both enzymes are virulence factors for certain human pathogens. Continued fundamental studies of Ni-using enzymes, chaperones, transporters, and transcriptional regulators are necessary to understand more fully why and how living cells utilize nickel instead of other metals and how this use is regulated.

Cross-References

- ▶ Acireductone Dioxygenase
- ▶ CO-Dehydrogenase/Acetyl-CoA Synthase
- ▶ Methyl Coenzyme M Reductase
- ▶ [NiFe]-Hydrogenases
- ▶ Nickel Superoxide Dismutase
- ▶ Nickel Transporters
- ▶ Nickel-Binding Sites in Proteins
- ▶ NikR, Nickel-Dependent Transcription Factor
- ▶ Urease

References

Amara P, Mouesca JM, Volbeda A, Fontecilla-Camps JC (2011) Carbon monoxide dehydrogenase reaction mechanism: a likely case of abnormal CO_2 insertion to a Ni-H$^-$ bond. Inorg Chem 50:1868–1878

Berman HM, Westbrook J, Feng Z, Gilliland G, Bhat TN, Weissig H, Shindyalov IN, Bourne PE (2000) The protein data bank. Nucleic Acids Research 28:235–242

Fontecilla-Camps JC, Amara P, Cavazza C, Nicolet Y, Volbeda A (2009) Structure-function relationships of anaerobic gas-processing metalloenzymes. Nature 460:814–822

Li Y, Zamble DB (2009) Nickel homeostasis and nickel regulation: an overview. Chem Rev 109:4617–4643

Ma Z, Jacobsen FE, Giedroc DP (2009) Metal transporters and metal sensors: how coordination chemistry controls bacterial metal homeostasis. Chem Rev 109:4644–4681

Volbeda A, Darnault C, Tan X, Lindahl PA, Fontecilla-Camps JC (2009) Novel domain arrangement in the crystal structure of a truncated acetyl-CoA synthase from *Moorella thermoacetica*. Biochemistry 48:7916–7926

Nickel-Binding Sites in Proteins

Jodi L. Boer[1] and Robert P. Hausinger[1,2]
[1]Department of Biochemistry and Molecular Biology, Michigan State University, East Lansing, MI, USA
[2]Department of Microbiology and Molecular Genetics, 6193 Biomedical and Physical Sciences, Michigan State University, East Lansing, MI, USA

Synonyms

Metallocenters of nickel-containing proteins

Definition

Protein Ni-binding sites refer to specific amino acids, backbone atoms, or cofactors that cooperate to bind Ni ions or larger Ni metallocenters to polypeptides. These sites may have catalytic activity, be involved in Ni transport or homeostasis, represent adventitious Ni binding to a site meant for another metal, or be engineered into a protein to facilitate purification.

Introduction

Ni is essential to many life forms, but excess levels are toxic and carcinogenic; thus, cells have devised mechanisms to sense Ni concentrations and maintain homeostatic control of this metal ion (Hausinger 1993; Sigel et al. 2007). Ni-binding sites occur in Ni-dependent enzymes and a wide variety of proteins that transport or deliver Ni, regulate transcription of genes involved in Ni metabolism, or adventitiously bind the metal ion.

The most common oxidation state for Ni in biology is Ni^{2+}, although redox-active sites are present in some Ni enzymes leading to Ni^{+1} or Ni^{+3} states. The metal ion is most often found in six-coordinate octahedral configuration; however, planar or tetrahedral four-coordinate geometries and various five-coordinate ligand environments are known. The imidazole group of His side chains is the most common amino acid ligand of Ni, with other amino acids coordinating the metal via the sulfur atoms of Cys or Met and the carboxylate groups of Glu and Asp residues

(Sigel et al. 2007). Backbone amides and amino-terminal amine groups bind the metal ion in some proteins, and Ni can be part of a larger metallocluster or incorporated into a tetrapyrrole bound to some proteins.

This essay focuses on features of Ni-binding sites in proteins with well-defined structures but also mentions a few less well-characterized examples where these sites are partially characterized by spectroscopic and mutagenic approaches.

Ni-Binding Sites in Enzymes

Figure 1 illustrates the active sites with known structures of Ni-containing enzymes. All are crystal structures except for that of acireductone dioxygenase, which is derived from NMR studies. Enzymes utilize the Ni to perform a wide variety of reactions, from hydrolysis to redox chemistry. A key aspect of Ni binding at enzyme active sites is that the metal ion must retain at least one open coordination site in order to bind the substrate. This open site may be occupied by water in the resting enzyme. Ni ligands in these enzymes vary widely, typically involving amino acid side chains but also sometimes utilizing backbone atoms and nonprotein cofactors (Hausinger 1993; Mulrooney and Hausinger 2003; Sigel et al. 2007; Li and Zamble 2009; Ragsdale 2009).

Glyoxalase I binds the hemithioacetal derived from addition of glutathione and methylglyoxal and then uses Ni to catalyze its conversion to S-D-lactoylglutathione. The substrate likely displaces one or both water ligand(s) when coordinating to the Ni, and the oxidation state remains Ni^{2+} throughout the reaction. This protein illustrates how multiple peptide chains, in this case two, can cooperate to form a single Ni-binding site.

Acireductone dioxygenase uses the Ni at its active site to catalyze the cleavage of 1,2-dihydroxy-3-keto-5-methylthiopentane (acireductone) by reaction with oxygen, producing methylthiopropionate, formic acid, and carbon monoxide. Ni likely remains in the 2+ oxidation state during the reaction, acting as a Lewis acid. This enzyme alternatively binds Fe^{2+}, resulting in the distinct products formic acid plus the α-keto acid precursor of Met, which is used as part of a salvage pathway.

Ni-superoxide dismutase catalyzes the disproportionation of two molecules of superoxide to form oxygen and hydrogen peroxide, identical in chemistry to that of Cu/Zn-, Mn-, and Fe-superoxide dismutases, which are unrelated in sequence. In this case, Ni cycles between Ni^{2+} in a square planar configuration and Ni^{3+} in a square pyramidal coordination. Unique to the Ni superoxide dismutase, the protein backbone forms part of the coordination sphere of the metal, using both the N-terminal amine and amide nitrogen ligands. His, a fifth ligand in the oxidized form of the enzyme, swings away from the metal ion in the reduced form.

Methyl coenzyme M reductase catalyzes the final step in methane formation in methanogenic archaea. The enzyme reacts methyl coenzyme M (methyl-S-thioethanesulfonate) with coenzyme B (N-7-mercaptoheptanoylthreonine phosphate), producing methane and disulfide-linked CoM-CoB. The active site contains an F430 cofactor, which is a Ni-containing tetrapyrrole related to, but extensively modified from, sirohemes and corrinoids. A Gln side chain coordinates at an axial ligand position, and substrates react at the other axial site. A very similar Ni-tetrapyrrole is found in an enzyme related in sequence to methyl coenzyme M reductase, but in that case, it participates in anaerobic methane oxidation. The Ni cofactors in these enzymes undergo redox chemistry that includes Ni^{+1} along with Ni^{+2} and/or Ni^{+3} states.

Urease catalyzes the hydrolysis of urea to ammonia and carbamate, which then dissociates into another molecule of ammonia and bicarbonate. The urease active site contains a dinuclear Ni center, with the two metals bridged by a Lys carbamate. The metallocluster is thought to both bind the urea and activate a water molecule for nucleophilic attack.

[NiFe] hydrogenases catalyze the reversible reduction of protons to H_2. The active site contains both Ni and Fe, with the Ni coordinated by four Cys residues (or in some cases, three Cys and one Se-Cys), while Fe is coordinated by two of the same Cys residues and several diatomic molecules identified as cyanide or carbon monoxide. There is a bridging ligand as well, with its identity depending on the state of the enzyme. Several other iron-sulfur clusters typically are present to serve as a conduit for electrons to electron carrier proteins. The mechanism is still being investigated, although spectroscopic evidence shows the Ni cycles between +2 and +3 oxidation states.

Carbon monoxide dehydrogenase (CODH) catalyzes the reversible oxidation of CO to CO_2. The protein contains a catalytic C-cluster with variations

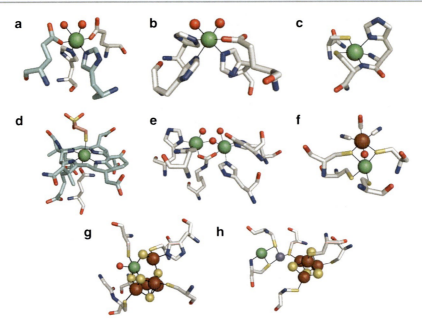

Nickel-Binding Sites in Proteins, Fig. 1 Active sites of Ni-containing enzymes. (**a**) Ni-glyoxalase I (1F9Z, *Escherichia coli*) catalyzes a critical step in the cellular conversion of methylglyoxal to lactate. The Ni (*green sphere*) bridges two subunits (carbons colored by subunit) and is bound by two His and two Glu, with two coordinated waters (*red spheres*). (**b**) Acireductone dioxygenase (1ZRR, NMR structure, *Klebsiella oxytoca*) binds Ni using three His and one Glu, leaving two open coordination sites filled by water and catalyzes a reaction between acireductone and O_2 to form carbon monoxide, formic acid, and methylthiopropionate (the Fe(II)-bound enzyme generates different products). (**c**) Ni-superoxide dismutase (1T6U, *Streptomyces coelicolor*) catalyzes the disproportionation of two molecules of superoxide into H_2O_2 and O_2. In the oxidized form (shown), the Ni is coordinated by two Cys, the amino-terminal amine, a backbone amide, and an axial His. In the reduced form, the His swings away, leaving a square-planar geometry (not shown). (**d**) Methyl coenzyme M reductase (1MRO, *Methanothermobacter margburgensis*) catalyzes the reduction of methyl coenzyme M with coenzyme B, producing methane and a mixed disulfide. The structure shown contains the Ni-tetrapyrrole, termed coenzyme F430, with axial coordination by Gln and coenzyme M. (**e**) Urease (1FWJ, *Klebsiella aerogenes*) catalyzes the hydrolysis of urea to NH_3 and carbamate, which further decomposes to NH_3 and H_2CO_3. Each Ni is coordinated by two His, terminal waters, and, in one case, by Asp, and the metals are bridged by a lysine carbamate and a hydroxyl group. (**f**) [NiFe]-hydrogenase (1YRQ, *Desulfovibrio fructosovorans*) catalyzes the reversible formation of H_2 from protons and electrons. The Ni is coordinated by four Cys residues, two of which also coordinate Fe (that has a carbon monoxide and two cyanide ligands), plus a bridging ligand whose identity and presence depend on the state of the enzyme. (**g**) Carbon monoxide dehydrogenase (CODH, 1SU7, *Carboxydothermus hydrogenoformans*) catalyzes the interconversion of CO and CO_2. The Ni is part of a [Ni-4Fe-5S] metallocluster, where the composition varies for proteins from different sources. (**h**) Acetyl CoA synthase (ACS, 2Z8Y, *Morella thermocetica*) catalyzes the biosynthesis of acetyl CoA from two C-1 units at the metallocluster shown. One Ni is bound in square-planar geometry to two backbone amides and two Cys, while a second Ni (shown here replaced by Cu in *blue*) is bound to the same two Cys as well as another that is linked to a [4Fe-4S] cluster. A third Ni (not shown) at a distant site in this protein is in a cluster that functions like and closely resembles that in CODH

on a [Ni-4Fe-5S] cluster, depending on the source organism. Ni is the likely site of CO binding, and one of the Fe atoms activates a hydroxide to attack the CO to form the product. The two electrons generated by this oxidative reaction are transferred through other iron-sulfur clusters in the protein to reduce partner electron carrier proteins.

Acetyl coenzyme A (CoA) synthase (ACS) often is found in a large complex that includes CODH. In this protein complex, CO_2 is reduced by the metallocenter of CODH to form CO, and at a separate active site, this toxic intermediate becomes linked to a methyl group, provided by a corrinoid/iron-sulfur protein, and to CoA, thus forming acetyl CoA. The CO-producing and CO-consuming sites are linked by an ~70 Å long molecular tunnel through the protein complex. In addition to the Ni cluster in the CODH, the acetyl CoA synthase active site contains two Ni atoms as well as a [4Fe-4S] cluster. One relatively stable Ni is bound via two backbone amides and two Cys residues.

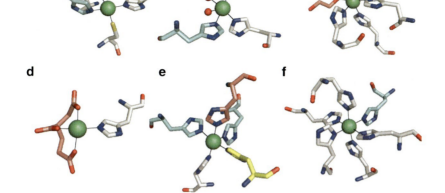

Nickel-Binding Sites in Proteins, Fig. 2 Ni-binding sites in Ni-sensor or Ni-delivery proteins. (**a**) *E. coli* NikR (2HZV) is a tetrameric Ni-responsive repressor with four high-affinity binding sites, each coordinated in square-planar geometry by three His and one Cys between subunit pairs. (**b**) NikR from *Helicobacter pylori* (3LGH) binds Ni in the same manner as the *E. coli* protein in some subunits and forms a distinct five- or six-coordinate site bound to three His in other subunits. (**c**) Nur (3EYY, *Streptomyces coelicolor*) is a Ni-responsive repressor that is structurally distinct from NikR. The Ni is coordinated by three His, with malonate and ethylene glycol occupying additional sites in the crystal structure shown. (**d**) NikA (3DP8, *E. coli*) is a periplasmic protein that binds Ni to a single His, additionally using an organic metallophore that was modeled as butane-1,2,4-tricarboxylate. (**e**) UreE (3NY0, *Helicobacter pylori*) is a metallochaperone involved in urease maturation. The structure shown depicts Ni bound nonsymmetrically to five His derived from four different subunits. (**f**) SlyD (3CGM, *Thermus thermophilus*) is a peptidyl-prolyl isomerase with six His coordinating the metal. The *E. coli* protein has been implicated in Ni metabolism during hydrogenase maturation

The second Ni, easily replaced by Cu or Zn, is coordinated by the same two Cys along with a third Cys that also serves as a ligand to the iron-sulfur cluster. An analogous protein complex found in acetate-degrading methanogens carries out nearly the reverse reaction. It splits acetyl CoA into a CO, CoA, and methyl group bound to CoM; in this case, oxidation of CO to CO_2 is used to provide electrons for conversion of methyl CoM into methane. Several mechanisms have been proposed for the C-C and C-S bond formation/cleavage events, with variations in the order of substrate binding and the oxidation states of the Ni sites.

Ni-Binding Sites in Noncatalytic Proteins

Noncatalytic Ni-binding proteins contain metallocenters (Fig. 2) that function in Ni sensing, regulation, transport, and delivery (Li and Zamble 2009). These proteins use the same amino acids as already noted for the Ni enzymes, but an open coordination site is not required. As with some enzymes, nonprotein ligands may be used.

Ni Sensors and Regulators

Ni sensors and regulators play important roles in Ni homeostasis by allowing organisms to maintain constant cellular Ni concentrations through modulation of the transcription of genes encoding Ni uptake or efflux pumps. From a structural perspective, the best studied example of such a protein is NikR. This protein is a repressor of the *nikABCDE* gene cluster encoding a Ni uptake system in *Escherichia coli* and other bacteria. NikR is a tetramer consisting of two DNA-binding domains and four metal-binding domains. The NikR tetramer binds four Ni ions, each coordinated to a Cys and three His (one from a different chain) in a square-planar geometry. NikR from *Helicobacter pylori* has additional five- or six-coordinate Ni-binding sites and is a global regulator rather than the more specific *E. coli* homolog (Li and Zamble 2009).

Nur is a Ni-dependent regulator that controls expression of Ni- and Fe- superoxide dismutases. It uses three His residues to bind one face of Ni, while the remaining octahedral coordination sites can bind other ligands; in the crystal structure, those sites are occupied by malonate and ethylene glycol from the crystallization buffer (Li and Zamble 2009).

The Ni sensor RcnR regulates the Ni efflux transporter RcnA. Although no crystal structure is available, site-directed mutagenesis and X-ray absorption spectroscopic results indicate that Ni is coordinated by using two His, one Cys, the amino terminus, a backbone nitrogen, and another unknown ligand (Li and Zamble 2009). Members of the ArsR/SmtB family also regulate Ni-related genes, although the structural details of the metal-binding sites are not known (Ma et al. 2009).

Ni Uptake and Efflux Systems

Much less is known about the structure of Ni-binding sites in proteins that import or export this metal ion. The NikABCDE transport system belongs to the ATP-binding cassette transporters. NikA is a periplasmic protein that binds Ni using a single His residue and an unidentified organic complex, modeled in the crystal structure as butane-1,2,4-tricarboxylate. NikB and NikC are transmembrane proteins, and NikD and NikE are nucleotide-binding proteins. The amino acid ligands used to transiently coordinate Ni as it is taken into the cell remain undefined. In contrast, some potential Ni-binding ligands are proposed for the NiCoT family proteins (transporting Ni, Co, or both, depending on the protein) on the basis of mutagenesis studies. Nevertheless, the mechanism for discriminating between these metals and the structural details of the metal-binding sites are largely unknown (Mulrooney and Hausinger 2003; Li and Zamble 2009).

In addition to importing Ni, cells must export this metal ion when in excess. Very little is known about how the Ni efflux proteins bind the metal. For example, *E. coli* uses RcnA as a nickel exporter; however, no details about its metal-binding site are known.

Metal Delivery

Some Ni-containing enzymes require intracellular chaperones to deliver the Ni to the active site (Mulrooney and Hausinger 2003). For example, ureases generally require four accessory proteins for enzyme activation with UreE serving as a metallochaperone that delivers the Ni. This protein binds Ni using His residues which are sometimes localized in His-rich C-termini of the dimeric protein. Mutagenesis studies have shown that the His-coordinated Ni-binding site at the interface of the UreE dimer (or in the UreE tetramer formed at elevated protein concentrations) is needed for Ni transfer to the urease apoprotein. The UreG and UreD urease accessory proteins also have been shown to bind Ni with the metal ligands still undefined, and urease activation may involve a series of Ni transfers (Kaluarachchi et al. 2010).

[NiFe] hydrogenases also require accessory proteins for their assembly. HypA binds one molecule of Zn, which is likely structural, as well as one Ni atom at a second site, the ligands for which are unknown. HypB also binds Ni and Zn and has different numbers of metal-binding sites depending on the organism. Some HypB sequences also have a His-rich tail that can bind multiple Ni atoms. A His-rich metal-binding domain of *E. coli* SlyD is known to play a role in maturation of its cognate hydrogenase, and the Ni-bound structure is available for SlyD from a thermophilic microorganism; however, not every SlyD homolog possesses this domain (Kaluarachchi et al. 2010).

CODH also has accessory proteins necessary for enzyme activation. A CooC-like protein from *Carboxydothermus hydrogenoformans* has been shown to bind Ni, but the homolog in *Rhodospirillum rubrum* did not exhibit detectable Ni binding. CooJ contains a multi-His metal-binding motif at its C-terminus and binds four Ni per monomer, but the structural details of those binding sites are unknown (Kaluarachchi et al. 2010).

Ni-binding sites also are important in proteins that function in extracellular metal delivery within multicellular organisms. Although a structure is not available, extensive studies have shown how serum albumin binds Ni at its amino terminus, including the use of the amino terminal amine, two backbone amides, an Asp, and a His (Hausinger 1993).

Ni-Substituted Proteins

Many proteins bind Ni in place of the native metal (Fig. 3). The first row transition elements Fe, Ni, Cu, and Zn all have similar atomic radii, and proteins use the same amino acids to coordinate them. How proteins discriminate between different metals is an open question in the biochemistry of metals. In peptide

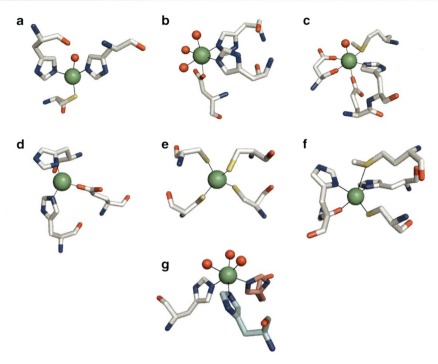

Nickel-Binding Sites in Proteins, Fig. 3 Ni-substituted protein metallocenters. (**a**) Peptide deformylase (1BS7, *E. coli*) is an Fe^{2+} enzyme but can also bind Ni^{2+} using two His and a Cys to form a site capable of catalyzing the removal of the formyl group from the N-terminus of polypeptide chains. (**b**) Lysyl hydroxylase JMJD6 (3K2O, *Homo sapiens*), an α-ketoglutarate-dependent dioxygenase, is inhibited by Ni replacing Fe^{2+} at its active site consisting of a two-His-one-carboxylate motif. (**c** and **d**) DtxR (2TDX, *Corynebacterium diphtheriae*) regulates genes that encode proteins important to iron uptake and storage. In vivo, it is specific for iron, but nickel and cobalt bind at two sites (one involving one His, one Met, one Glu, and the side chain and amide of an Asp; the second is composed of two His and one Glu) in vitro and promote DNA binding. (**e**) Rubredoxin (1R0J, *Clostridium pasteurianum*) contains Fe coordinated by four Cys natively, but the Ni^{2+}-substituted protein has been crystallized. (**f**) Azurin (1NZR, *Pseudomonas aeruginosa*) is a Cu-containing electron carrier that was crystallized with Ni bound to two His (including one His amide), one Cys, and one Met. (**g**) Ornithine transcarbamylase (2W37, *Lactobacillus hilgardii*) binds metals, including Ni, using three His side chains at its threefold symmetry axis that is distant from the active site

deformylase, Ni^{2+} substitutes for the physiological, but oxygen-labile, Fe^{2+} and still allows the enzyme to turn over, albeit with reduced kinetics. In other cases, Ni substitution for Fe results in a nonfunctional enzyme, as in the lysyl hydroxylase JMJD6 and other α-ketoglutarate dependent dioxygenases. Indeed, substitution of this metal into other members of this enzyme family leads to the hypoxia-mimicking effects of Ni^{2+} (Chen and Costa 2009). DtxR, an Fe^{2+}-sensitive regulator, also binds Ni^{2+} which promotes its binding to the appropriate sites on DNA in vitro. Due to its relatively stable Ni^{2+} redox state, substitution of this metal into Fe or Cu proteins such as rubredoxin or azurin eliminates their electron transfer activities. Finally, there are examples of adventitious Ni-binding sites that are unrelated to protein function, such as the site in ornithine transcarbamylase that connects three subunits. Noncatalytic metal-binding sites can participate in protein stabilization.

Ni-Binding to His-Tagged Proteins

A common purification strategy is to create a fusion between the protein of interest and a poly-His tag. The His-rich region serves as an affinity tag for binding to immobilized-metal ion chromatography beads containing attached Ni. This tag, though useful for purification, can cause protein aggregation due to the His residues from multiple proteins coalescing to

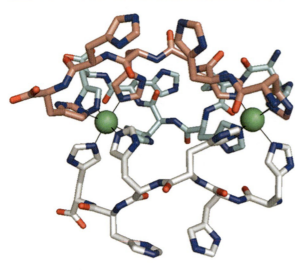

Nickel-Binding Sites in Proteins, Fig. 4 His-tag used for affinity purification. A commonly used protein purification procedure is to construct an expression system where the gene of interest is fused to a sequence encoding a poly-His tag, allowing for isolation by immobilized-metal ion chromatography. The structure shown (1Q3I, a sodium, potassium-ATPase) illustrates the mode by which the His region of three tags can aggregate when binding Ni, with each metal ion coordinated by six His chelate nickel. Figure 4 shows the poly-His tail from three subunits of a tagged protein coming together to bind Ni.

Cross-References

- CO-Dehydrogenase/Acetyl-CoA Synthase
- Methyl Coenzyme M Reductase
- [NiFe]-Hydrogenases
- Nickel Ions in Biological Systems
- Nickel Superoxide Dismutase
- Nickel Transporters
- Nickel-Binding Proteins, Overview
- NikR, Nickel-Dependent Transcription Factor
- Urease

References

Chen H, Costa M (2009) Iron- and 2-oxoglutarate-dependent dioxygenases: an emerging group of molecular targets for nickel toxicity and carcinogenicity. Biometals 22:191–196

Hausinger RP (1993) Biochemistry of nickel. Plenum, New York

Kaluarachchi H, Chan Chung KC, Zamble DB (2010) Microbial nickel proteins. Nat Prod Rep 27:681–694

Li Y, Zamble DB (2009) Nickel homeostasis and nickel regulation: an overview. Chem Rev 109:4617–4643

Ma Z, Jacobsen FE, Giedroc DP (2009) Coordination chemistry of bacterial metal transport and sensing. Chem Rev 109:4644–4681

Mulrooney SB, Hausinger RP (2003) Nickel uptake and utilization by microorganisms. FEMS Microbiol Rev 27:239–261

Ragsdale SW (2009) Nickel-based enzyme systems. J Biol Chem 284:18571–18575

Sigel A, Sigel H, Sigel RKO (eds) (2007) Nickel and its surprising impact in nature. Wiley, Chichester

Nickel-Containing Superoxide Dismutase

▶ Nickel Superoxide Dismutase

Nickel-Dependent Transcription Factor

▶ NikR, Nickel-Dependent Transcription Factor

Nickel-Induced Cancer

▶ Nickel Carcinogenesis

Nickel-Induced Tumors

▶ Nickel Carcinogenesis

Nicotinic Acid Hydroxylase

William Self
Molecular Biology & Microbiology, Burnett School of Biomedical Sciences, University of Central Florida, Orlando, FL, USA

Eubacterium barkeri (originally *Clostridium barkeri*) was isolated as a strict anaerobe capable of using nicotinic acid as a primary carbon and nitrogen source from the Potomac River mud by Earl Stadtman's group in the early 1970s (Stadtman et al. 1972). In an effort to understand the catabolic breakdown of nicotinic acid better, the enzyme catalyst nicotinic acid hydroxylase (NAH) was isolated from this model organism and studied biochemically. NAH was studied for nearly 10 years before it was discovered that it contained both molybdenum and selenium (Dilworth 1982, 1983). This enzyme catalyzes the hydroxylation of nicotinic acid to 6-hydroxynicotinic acid with the concomitant reduction of pyridine nucleotide (Gladyshev et al. 1994, 1996). Like other members of the selenium-dependent molybdenum hydroxylase family, selenium is found as a dissociable cofactor and is believed to be a ligand of the Mo cofactor (Gladyshev et al. 1996; Wagener et al. 2009). Indeed, this is the best studied member of this small class of enzymes, and both electron paramagnetic resonance and crystallographic studies support the presence of selenium as a direct ligand of the Mo cofactor. As such, this selenium atom is analogous to the cyanolyzable sulfur described for the non-selenium-dependent member of this family of enzymes (Coughlan et al. 1980).

The physiological role of this enzyme has been the subject of more recent investigation (Alhapel et al. 2006). A series of enzymes were identified after cloning of a 23 kb region of the genome of *E. barkeri* that are capable of converting nicotinic acid to pyruvate and propionate. A similar gene cluster was also identified in sequenced genomes from a variety of organisms including *Bradyrhizobium* and *Burkholderia* sp., indicating that this strategy to metabolize nicotinic acid is widely used in microbes. Since nicotinic acid indeed is a constituent of all living things – this pathway has an important role in the turnover of nutrients in microbes in a variety of environmental settings. Thus, also this type of selenium cofactor (labile cofactor) has only been identified in a small number of enzymes; the role these enzymes play in metabolism may be underappreciated.

References

Alhapel A, Darley DJ, Wagener N, Eckel E, Elsner N, Pierik AJ (2006) Molecular and functional analysis of nicotinate catabolism in *Eubacterium barkeri*. Proc Natl Acad Sci USA 103:12341–12346

Coughlan MP, Johnson JL, Rajagopalan KV (1980) Mechanisms of inactivation of molybdoenzymes by cyanide. J Biol Chem 255:2694–2699

Dilworth GL (1982) Properties of the selenium-containing moiety of nicotinic acid hydroxylase from *Clostridium barkeri*. Arch Biochem Biophys 219:30–38

Dilworth GL (1983) Occurrence of molybdenum in the nicotinic acid hydroxylase from *Clostridium barkeri*. Arch Biochem Biophys 221:565–569

Gladyshev VN, Khangulov SV, Stadtman TC (1994) Nicotinic acid hydroxylase from *Clostridium barkeri*: electron paramagnetic resonance studies show that selenium is coordinated with molybdenum in the catalytically active selenium-dependent enzyme. Proc Natl Acad Sci USA 91: 232–236

Gladyshev VN, Khangulov SV, Stadtman TC (1996) Properties of the selenium- and molybdenum-containing nicotinic acid hydroxylase from *Clostridium barkeri*. Biochemistry 35:212–223

Stadtman ER, Stadtman TC, Pastan I, Smith LD (1972) *Clostridium barkeri* sp. n. J Bacteriol 110:758–760

Wagener N, Pierik AJ, Ibdah A, Hille R, Dobbek H (2009) The Mo-Se active site of nicotinate dehydrogenase. Proc Natl Acad Sci USA 106:11055–11060

[NiFe]-Hydrogenases

Juan C. Fontecilla-Camps and Anne Volbeda
Metalloproteins; Institut de Biologie Structurale
J.P. Ebel; CEA; CNRS; Université J. Fourier,
Grenoble, France

Definition

[NiFe]-hydrogenases catalyze the heterolytic cleavage of the most simple of chemical compounds, molecular hydrogen, following the reactions $H_2 \leftrightarrow H^- + H^+$ and $H^- \leftrightarrow 2e^- + H^+$. Under sufficiently reducing conditions, they are also able to catalyze the production of H_2 from two protons and two electrons.

Introduction

Many microorganisms can use hydrogen as a source of reducing power or generate it through the reduction of protons to get rid of low-potential electrons, or both. The corresponding reaction, $H_2 \leftrightarrow 2H^+ + 2e^-$, is mediated by metalloenzymes called hydrogenases. The use of catalytic transition metal centers is explained by the significant increase in the acidity of molecular hydrogen when bound to them. So far, three phylogenetically unrelated classes of hydrogenases have been identified: "true" [NiFe]- and [FeFe]-hydrogenases which have hydrogen as their only substrate or product, and a third class where hydrogen uptake is coupled to methenyltetrahydromethanopterin reduction (Vignais and Billoud 2007). The respective crystal structures in combination with spectroscopic evidence have shown that the active sites contain low-spin iron coordinated to CO, indicating that a $Fe(CO)_x$ unit is the common natural solution to hydrogen catalysis (Fontecilla-Camps et al. 2007; Pandelia et al. 2010).

Out of the six structures of periplasmic [NiFe]-hydrogenases that have been determined so far, five correspond to enzymes from sulfate-reducing bacteria, and one is from a purple sulfur bacterium. They all are heterodimers and share the topology first described for the *Desulfovibrio gigas* enzyme (pdb code 1FRV). The active sites of [NiFe]-hydrogenases are buried in the protein and, consequently, there is a need for electron and proton transfer between the catalytic center and the molecular surface. The same applies to molecular hydrogen that has to access the active site or escape from it. Motivated by the unusual structure of the active site of these enzymes, studies aimed at elucidating their assembly process have been carried out in recent years, and both relevant structural and functional data are now available. The relative sensibility of hydrogenases to molecular oxygen is also a major current subject of study. Besides its basic scientific interest, this aspect is central to the possible use of these enzymes in biofuel cells or the large-scale production of "biohydrogen." The points mentioned above are addressed here from a structural perspective.

[NiFe]-Hydrogenase Architecture

The first atomic model of [NiFe]-hydrogenase was published in 1995 (Volbeda et al. 1995). The small subunit, which contains three FeS clusters, is composed of two structural domains called I_S and II_S (Fig. 1). I_S has a flavodoxin-like topology, and its proximal (relative to the active site) [Fe$_4$S$_4$] cluster binds at approximately the location that occupies the phosphate group of the FMN nucleotide in flavodoxin. The remaining mesial [Fe$_3$S$_4$] and distal [Fe$_4$S$_4$] clusters bind domain II_S which lacks extensive secondary structure elements. The surface-exposed His185 is a ligand to one of the irons of the distal cluster. All the remaining protein

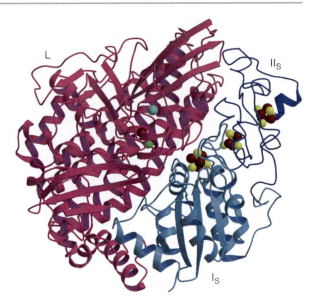

[NiFe]-Hydrogenases, Fig. 1 Structure of *D. gigas* [NiFe]-hydrogenase. *Arrows* depict β-strands, *ribbons* α-helices, and *spheres* metal sites with color codes: Ni *green*, Fe *red-brown*, Mg *cyan*, S *yellow*

ligands to the FeS clusters are cysteine thiolates. Some [NiFe]-hydrogenases, such as NAD(P)$^+$-reducing and energy-converting (Ech) enzymes, lack II_S.

The hydrogenase large (L) subunit contains the active site (Fig. 1). In the initial 2.85 Å resolution electron density map, there were two strong peaks at this site. The Ni site was assigned to the peak bound to the thiolate of Cys530, based on EXAFS analyses of homologous [NiFeSe]-hydrogenases that showed that the selenocysteine corresponding to Cys530 is a ligand to Ni. The second strong electron density peak was tentatively assigned to iron.

Crystallographic evidence for an active iron site in the *D. gigas* enzyme was obtained using 3 Å resolution anomalous scattering X-ray data collected above (1.733 Å) and below (1.750 Å) the iron absorption edge. A significant peak for the second active site metal ion was only observed with data collected at 1.733 Å (Volbeda et al. 1996). A similar analysis was performed later by Higuchi et al. using *D. vulgaris* Miyazaki [NiFe]-hydrogenase crystals. In that study, the previous assignment of the nickel ion position was also confirmed crystallographically.

At the end of the large subunit chain tracing, it was noticed that there was no electron density for the 15 C-terminal residues predicted by the gene sequence. This was unexpected because the electron density disappeared beyond His536, which was deeply buried in the hydrogenase structure. This puzzle was solved by a report stating that maturation of the large subunit of *D. gigas* [NiFe]-hydrogenase requires the proteolytic cleavage of a 15-residue C-terminal peptide. Higher-resolution structures of the *D. vulgaris* Miyazaki (pdb code 1H2A) and *D. gigas* enzymes (pdb code 2FRV) showed a Mg^{2+} ion bound to the C-terminal histidine.

Structures of [NiFe]-hydrogenases from *D. fructosovorans*, *D. desulfuricans*, *Allochromatium vinosum* and the [NiFeSe] enzymes from *Desulfomicrobium baculatum* and *D. vulgaris* Hildenborough were subsequently reported (pdb codes 1FRF, 1E3D, 3MYR, 1CC1, and 2WPN). The [NiFeSe] enzymes show some differences with respect to the [NiFe]-hydrogenases: (1) the mesial $[Fe_3S_4]$ cluster (Fig. 1) is replaced by $[Fe_4S_4]$; (2) a transition metal that, based on an inductively coupled plasma analysis, corresponds to a 14th iron ion of the enzyme replaces the Mg^{2+} ion; (3) near the active site, there is a bound H_2S molecule or Cl^- ion; (4) a selenocysteine replaces the terminal *Cys530* Ni ligand (*D. gigas* numbering, in italics). Other differences are mostly located at the enzyme surface.

A 2.54 Å resolution analysis of *D. gigas* [NiFe]-hydrogenase indicated that the active site Fe ion was connected to three elongated electron density peaks. It was known from FTIR spectroscopic studies using *A. vinosum* [NiFe]-hydrogenase that the enzyme contained species giving rise to three intrinsic high-frequency redox-sensitive bands. Using ^{13}C and ^{15}N labeling, these bands were assigned to two CN^- and one CO. Because the same FTIR bands are present in the *D. gigas* enzyme, they were assigned to the iron diatomic ligands, named L1, L2, and L3, modeled within the elongated electron density peaks mentioned above. L3, which occupies a hydrophobic cavity, was assigned to CO, and L1 and L2 that hydrogen-bond protein atoms (Fig. 2) were modeled as CN^-.

The active site contains two *cis* sites available for substrate binding: a bridging site between Fe and Ni, called E2, and a Ni-terminal one called E1 (Fig. 2). Most oxidized [NiFe]-hydrogenase samples display a ready state that is easy to activate and an unready state that requires a long time for reductive activation. In the initial structural analyses of oxidized [NiFe]-hydrogenase, E2 was modeled as containing either an oxo ligand or a sulfide ion. Electrochemical studies on

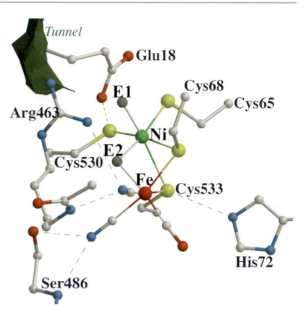

[NiFe]-Hydrogenases, Fig. 2 The nickel–iron active site

air-exposed [NiFe]-hydrogenase adsorbed onto a graphite electrode indicated that the unready state results from less reducing conditions than the ready state, suggesting it contains only partially reduced oxygen species. Indeed, there is crystallographic evidence for peroxide binding to E2 in the unready states (Fontecilla-Camps et al. 2007 and references therein). Conversely, the structures of two reductively activated hydrogenases did not display observable electron density for either E1 or E2. Because hydrons are not usually detected in protein crystals at medium resolution, this observation does not rule out the possibility of hydride binding to the active site after reductive treatment. In fact, it is very likely that a hydron species binds to E1 as the competitive inhibitor CO does.

Enzyme Maturation

The hydrogenase maturation process can be divided into (1) apoenzyme synthesis, (2) nickel and iron transport and storage, (3) CN/CO ligand synthesis and partial active site assembly, and (4) nickel insertion, generally followed by a C-terminal amino acid stretch proteolysis and burial of the nascent C-terminal region in the large subunit core (Böck et al. 2006). In addition, the general-purpose ISC (*i*ron–*s*ulfur *c*luster) proteins mediate

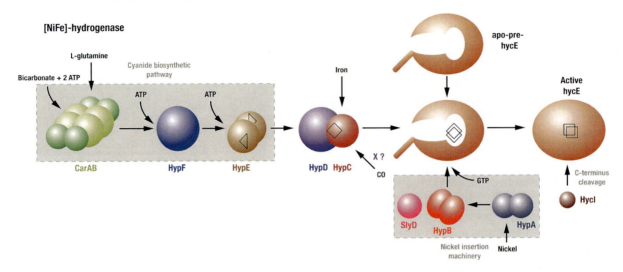

[NiFe]-Hydrogenases, Fig. 3 Maturation of [NiFe]-hydrogenase (see text)

FeS cluster assembly. Some microorganisms use a cytoplasmic hydrogenase-like regulatory H_2-sensor protein to detect hydrogen. A signaling pathway that results in the expression of [NiFe]-hydrogenase structural genes is further composed of a kinase and a DNA-binding protein. The second step that involves iron and nickel transport is highly regulated because these metal ions can be toxic. Steps 3 and 4 are more complex because they require the synthesis of the active site ligands and its assembly including both Fe and, after proteolysis, Ni insertion. Using *E. coli* hydrogenase 3 as an example, this process is carried out by six gene products from the *hyp* operon and a nickel-dependent protease (Fig. 3). CN^- synthesis in *E. coli* depends on HypE and HypF. The latter, with a mass of 82 kDa, is homologous at its N-terminal region to eukaryotic acylphosphatases and to O-carbamoyltransferases at its C-terminal stretch. The three-dimensional structure of the N-terminal domain of HypF (pdb code 1GXT) has confirmed its structural similarity to acylphosphatases. HypF first hydrolyzes carbamoyl phosphate (CP), which itself is produced by the enzyme CarAB from L-glutamine and bicarbonate; subsequently it transfers the resulting carbamoyl group to the C-terminal Cys of HypE. There is experimental evidence for the formation of a HypF/HypE complex.

The 35 kDa monomeric HypE (pdb code 2RB9) has similarities with aminoimidazole ribonucleotide synthase (PurM) and selenophosphate synthase (SelD). ATP-dependent PurM dehydrates aminoimidazole ribonucleotide, and its ATP-binding motif is also found in HypE. CN^- synthesis involves transfer of carbamoyl from HypF to the HypE C-terminal cysteine where it is dehydrated to form a HypE–SCN complex. Initially, CO was also thought to be synthesized from CP. However, this hypothesis has been ruled out by recent experiments using $^{13}CO_2$ (CO_2 is a CP precursor). Because only CN^- was labeled in this experiment, CO must be synthesized via a different route. Roseboom et al. have proposed that CO is synthesized from either acetate or one of its precursors.

The 9.6 kDa HypC (pdb code 2Z1C) interacts very strongly with both the precursor of the hydrogenase large subunit HycE and with HypD (pdb code 2Z1D). The latter is a 41.4 kDa protein that coordinates a $[Fe_4S_4]$ cluster. The HypCD complex is postulated to get the CN^- ligands from HypE–SCN and subsequently transfer them to the hydrogenase nascent active site Fe ion, although the mechanism is not known. Another unsolved point is the origin of the Fe ion. The HypC–preHycE complex, which is stable under reducing conditions, dissociates in the presence of alkylating agents. This suggests that HypC forms a covalent complex with preHycE through its C-terminal cysteine. It is thought that iron inserts to hydrogenase as a $FeCO(CN)_2$ unit, preceding nickel insertion. In fact, the HypC–preHycE complex dissociates in the presence of nickel.

The GTPase HypB is implied in nickel transport/insertion. However, in a $HypB^-$ mutant, the enzymatic activity could be restored by adding nickel to the growth medium, implying that HypB is not essential for nickel insertion. When nickel is replaced by zinc in vitro, hydrogenase maturation does not take place because the required protease activity is nickel-specific. Two structures of HypB are known (pdb codes 2HF9 and 2WSM). Some HypBs have a histidine-rich stretch at their N terminus. If this region is deleted, these proteins bind Ni with a 1:1 stoichiometry indicating that the histidine-rich stretch may just serve as a reservoir for nickel ions. Downstream the histidine-rich sequence, *E. coli* HypB has a CxxCGC motif that corresponds to a high-affinity Ni-binding site, which is not essential for nickel insertion into hydrogenase.

HypA (pdb code 3A43) interacts with HypB and binds one nickel ion per molecule. HypA may be the component that ultimately provides nickel to preHycE, but this has not been clearly established. HypB could provoke the conformational change needed for nickel delivery by HypA. Finally, mutation of the proline cis/trans isomerase SlyD (pdb code 3LUO) that interacts with HypB reduces hydrogenase activity; this effect is abolished by added nickel.

The last step in the maturation of [NiFe]-hydrogenases generally involves the proteolysis of a large subunit C-terminal extension. The protease is extremely specific, and each hydrogenase in a given organism will be processed by its corresponding isoform. The structure of the 17.5 kDa endopeptidase HybD (pdb code 1CFZ) consists of a twisted five-stranded β-sheet sandwiched by three α-helices on one side and two on the other and contains an adventitious cadmium ion. The as-purified HybD and related HycI (pdb code 2E85) do not contain metal, but upon in vitro nickel addition, they can cleave their respective substrates. Based on this observation and on the fact that HybD is insensitive to standard protease inhibitors, it has been concluded that the enzyme recognizes the nickel ion bound to preHycE. It has also been postulated that the observed cadmium ion occupies the binding site of the physiologically relevant nickel ion. Extensive site-directed mutagenesis at endopeptidase sites has indicated that most of the mutations do not completely block proteolysis. Thus, it is tempting to conclude that it is largely the active site-bound nickel ion, which determines the properties of the cleavage site. Some [NiFe]-hydrogenases lack a large subunit C-terminal extension and, consequently, do not require proteolytic processing to mature. Typical examples are cytoplasmic H_2 sensors and Ech hydrogenases.

Electron Transfer

Soluble periplasmic [NiFe]-hydrogenases are hydrogen-uptake enzymes that can generate a proton motive force across the cytoplasmic membrane used to synthetize ATP. Hydrogen oxidation also produces electrons that are transferred to the cytoplasmic side of the membrane. Membrane-bound periplasmic hydrogenases are complexed to cytochrome b, which directly transfers electrons to a quinone. The membrane-bound energy-converting (Ech) [NiFe]-hydrogenases use ferredoxins as redox partners. In addition, the Ech enzymes translocate protons from the cytoplasm to its exterior.

In [NiFe]-enzymes from *Desulfovibrio* sp., electrons are transferred from the active site to the redox partner via a proximal [Fe_4S_4], a mesial [Fe_3S_4], and a distal [Fe_4S_4] cluster (Fig. 1). The center-to-center 12 Å distance between consecutive redox centers in this pathway is typical for electron transfer in redox proteins. Less obvious is the location of the [Fe_3S_4] cluster between the two [Fe_4S_4] centers because it has a much more positive redox potential than that of hydrogen uptake. This, in turn, means that the [Fe_3S_4] cluster should be predominantly reduced during catalysis. Because the proximal and distal [Fe_4S_4] clusters are separated by about 20 Å, it is difficult to envision direct electron transfer between them. The *D. fructosovorans* [NiFe]-hydrogenase [Fe_3S_4] cluster was converted to [Fe_4S_4] in order to clarify its role in electron transfer. Although more air-sensitive, the mutated enzyme was as active as the native one, indicating that internal electron transfer is not the rate limiting step. Consequently, some other process, such as substrate access to the active site, hydrogen heterolytic cleavage, proton transfer to the solution, or external electron transfer, must be limiting. Studies using different electron acceptors favor the latter option.

A specific redox partner recognition site in [NiFe]-hydrogenases is constituted by a series of acidic residues that form a crown around the distal [Fe_4S_4] His ligand in *Desulfovibrio* sp. enzymes. This crown is likely to interact with positively charged regions surrounding an exposed heme on soluble c-type cytochromes. The lifetime of the electron transfer complex is probably short, given its rather high K_m value (7.4 μM for cytochrome c_3 binding to [NiFe]-hydrogenase in *D. gigas*).

Although automated rigid docking approaches have limited applications, they may be improved when combined with NMR spectroscopic data. This approach has been used to study the complex formed between *D. vulgaris* Miyazaki F cythchrome c_3 and periplasmic [NiFe]-hydrogenase (Yahata et al. 2006). The available cytochrome NMR and hydrogenase X-ray structures (pdb codes 1IT1 and 1H2A) were used in this study. Chemical shift perturbation analyses using oxidized [NiFe]-hydrogenase confirmed that it interacts with the region around heme IV, which has the highest redox potential at pH 7.0. The docking model with the shortest distance between heme IV and the distal [Fe$_4$S$_4$] cluster suggested an electron transfer pathway involving four conserved acidic residues from the enzyme and four lysines from the cytochrome. Conversely, electron transfer from reduced c_3 to hydrogenase, which in vitro can reduce protons, involves a contact interface close to heme III, which has the lowest redox potential. Thus, depending on the direction of electron transfer, different cytochrome hemes may be involved.

Proton Transfer

In *Dm. baculatum* [NiFeSe]-hydrogenase, the unusual SeC is a terminal Ni ligand (like the homologous *Cys530* in the *D. gigas* enzyme). The 2.15 Å resolution structure of an active, reduced form of [NiFeSe]-hydrogenase indicated that the Se atom forms a hydrogen bond with the carboxylate of the invariant *Glu18* (Fig. 2). These residues and the Ni ion display higher temperature (B) factors than surrounding atoms, which may result from a mixture of protonation states for these species in the crystal. This suggests that *Cys530* and *Glu18* are components of the proton transfer pathway. Replacement of *Glu18* by Gln in *D. fructosovorans* hydrogenase kept its Ni-related EPR spectroscopic properties but resulted in loss of catalytic activity and H/D isotope exchange (Dementin et al. 2004). As shown by active *para*-H$_2$/*ortho*-H$_2$ conversion, the mutated enzyme was still able to heterolytically cleave H$_2$, but proton/deuteron transfer between solvent and the active site was impaired. As already predicted from the crystallographic results, *Glu18* is essential for proton transfer.

Defining proton transfer pathways is difficult because, as one moves away from the active site, many seem possible. The active site and the proximal [Fe$_4$S$_4$] cluster are connected by hydrogen bonds involving the small subunit *Desulfovibrio* invariant *Thr18*, *Glu16*, and *Glu73*. This connection suggests that reduction of the proximal cluster provides a driving force for proton transfer from the active site. Indeed, the potential of several redox couples in hydrogenases changes as a function of pH indicating that reduction is coupled to protonation. Although *Glu16* is replaced by Gly in *Dm. baculatum* [NiFeSe]-hydrogenase, a water molecule substitutes its carboxylate group, so this proton transfer pathway may be conserved.

In another possible path conserved in all the crystal structures, the Mg^{2+} ion is connected to *Glu18* through four water molecules, the C-terminal carboxylate, an additional water molecule, and *Glu46*. This pathway could be used for proton transfer to the redox partner, cytochrome c_3, because its lowest-energy docking solution involves a contact close to the Mg^{2+} site. Electron transfer should take place in a different transient complex having a contact interface close to the distal [4Fe–4S] cluster. The occurrence of both proton and electron transfer would agree with the redox-Bohr effect observed for electron exchange between hydrogenase and cytochrome c_3 in *D. vulgaris* (Hildenborough). Although a Fe ion replaces Mg^{2+} in the *Dm. baculatum* [NiFeSe]-hydrogenase structure, the coordination sphere is the same. In some hydrogenases, the C-terminal histidine is replaced by arginine, which could replace Mg^{2+} and form a salt bridge with *Glu46*.

Still another pathway has been proposed for the oxidized *D. desulfuricans* [NiFe]-hydrogenase (pdb code 1E3D). It involves the approach of the protonated *Glu18* toward a water molecule, which interacts with the conserved *Asp528*. Otherwise, the side chain of *Cys530* could rotate and donate a proton to *Asp528* via *Arg463*. The crystal structures of *D. desulfuricans* and *D. fructosovorans* [NiFe]-hydrogenases show that *Cys530* can position its Sγ at hydrogen bonding distance from *Arg463*.

Oxygen Sensitivity

The oxygen tolerance of microorganisms varies depending on their normal habitat. Consequently, they can have hydrogenases with different levels of resistance to inactivation by O$_2$. Understanding the structural basis of high-level oxygen tolerance is

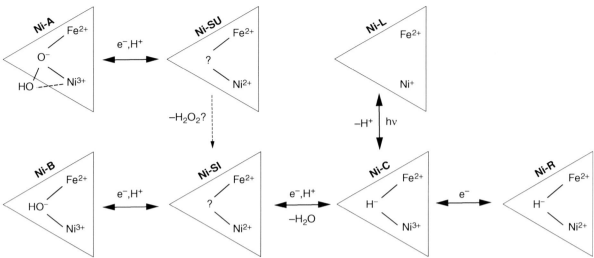

[NiFe]-Hydrogenases, Fig. 4 Redox states of the Ni-Fe site, going to increasingly reducing conditions from *left* to *right*

a major endeavor in contemporary technologically oriented hydrogenase research.

Oxidized inactive states of the Ni–Fe site. The active sites of all the oxidized unready hydrogenase structures contain several species. It is generally accepted that the ready Ni–B form corresponds to Ni(III) with a bridging hydroxo ligand (Fig. 4). The structures of the unready Ni–A and Ni–SU states are less well defined having in principle both the peroxo ligand and cysteine sulfenates as possible oxygen-generated components (the electron in reaction 1 is likely to come from the oxidation of the $[Fe_3S_4]^0$ cluster):

$$O_2 + Ni(II) + e^- + H^+ \rightarrow Ni(III)-OOH^- \quad (1)$$

$$Ni(III)-OOH^- + Cys-S^- \rightarrow Ni(III)-OH^- + Cys-SO^- \quad (2)$$

The bridging site is occupied by a (hydr)oxo ligand in a significant fraction of the unready hydrogenase molecules as indicated by the fact that the peroxo oxygen atom proximal to the Fe ion refines to higher occupancy than the distal one in the structure. Both $H_2^{17}O$ and $^{17}O_2$ modify the EPR spectra of the oxidized states of hydrogenases. One possible explanation for this observation is the oxidation of Ni(II)–OH$^-$/Cys–SO$^-$ (Ni–SU) to yield Ni(III)–OOH$^-$/Cys–S$^-$(Ni–A), that is, the reverse of reaction 2. However, the equilibrium of reaction 2 should be displaced to the right, as S–O bonds are stronger than O–O bonds. Accordingly, in biochemical reactions involving CysS–OH intermediates, these often progress by forming the stronger CysS–SCys bond. The (hydro) peroxo ligand that generates the Ni–A signal should correspond to a stable species because the activation energy required to displace the reaction to the right in (2) is too high. For similar reasons, it is difficult to postulate that Ni–SU corresponds to Ni(II)–OH$^-$/Cys–SO. An alternative model for Ni–SU is Ni(II)–OOH$^-$/Cys–S$^-$. In this case, the Cys–SO species would correspond to dead enzyme. Note that anaerobically purified hydrogenases are significantly more active than their reactivated, aerobically purified counterparts. Sulfenic acid formation (reaction 2) might also explain the progressive and irreversible hydrogenase inactivation noted in electrochemical analyses. Cys–SO should stabilize Ni (II) because sulfenate is a less good σ donor than thiolate.

If the Ni–A and Ni–SU states correspond to Ni (III)–OOH$^-$/Cys–S$^-$ and Ni(II)–OOH$^-$/Cys–S$^-$, respectively, how can the peroxo ligand contain oxygen from either $H_2^{17}O$ or $^{17}O_2$? The following reaction, analogous to the reported exchange of peroxide with water in microperoxidase-8, may explain this conundrum:

$$\begin{aligned} Ni(II)-OOH^- - Fe(II) + e^- + H^+ &\rightarrow \\ \rightarrow Ni(III)-O^{2-}-Fe(II) + H_2O &\rightarrow \\ \rightarrow Ni(II)-OOH^- - Fe(II) + e^- + H^+ & \end{aligned} \quad (3)$$

The bridging ligand will be labeled if the (hydro) peroxo-containing species generated by reaction 1 is produced in the presence of $^{17}O_2$. Conversely, in the presence of $H_2^{17}O$, the labeled oxygen atom will exchange according to (3), the electron source being again the reduced $[Fe_3S_4]$ cluster. Reaction 3 also explains the lack of $H_2^{17}O$ exchange in the Ni–A form because it does not have the electron required to form $Ni(III)–O^{2-}–Fe(II)$. As shown by the structures of unready hydrogenase, the two peroxo oxygen atoms lie close to Ni, explaining previous $^{17}O_2$ EPR and $H_2^{17}O$ ENDOR spectroscopic results. The possibility of O exchange between H_2O and a peroxo ligand has already been evoked by Lamle et al. using an electrochemical approach.

Hydrophobic tunnels in [NiFe]-hydrogenases. [NiFe(Se)]-hydrogenases contain specific hydrophobic tunnels that facilitate access of H_2 to the active site. In order to determine the possible role of the tunnels in catalysis, low-resolution X-ray diffraction data were collected from a *D. fructosovorans* hydrogenase crystal exposed to high Xe pressure (Montet et al. 1997). From the electron density map calculated with these data, it was possible to identify ten Xe sites, all located inside the hydrophobic tunnels (Fig. 5). In the most relevant experiment, the Xe sites were used as starting points for a molecular dynamics simulation with hydrogen molecules. In several trajectories, H_2 approached the active site with minimal gas-Fe and gas-Ni distances of 1.8 Å and 1.4 Å, respectively. The average of 80 trajectories indicated that statistically, the gas only occupied the observed cavities.

Teixeira and coworkers carried out a similar theoretical study using the three-dimensional model of the *D. gigas* [NiFe]-hydrogenase. The two studies concerning gas diffusion reached similar conclusions and underscored the importance of the hydrophobic tunnels for both (correct) substrate access to the active site and hydrogen storage inside [NiFe]-hydrogenases.

A possible reason for the *Dm. baculatum* [NiFeSe]-hydrogenase high oxygen tolerance is the presence of Se in the active site, which may allow H_2 passage while obstructing access of the bulkier O_2 through the larger van der Waals radius of selenium compared to sulfur. In addition, as noted before, the nearby tunnel residue *Val110* is substituted by a leucine, which also leads to a decreased tunnel diameter.

Hydrogen sensors related to [NiFe]-hydrogenases. In the Knallgas bacterium *Ralstonia eutropha*,

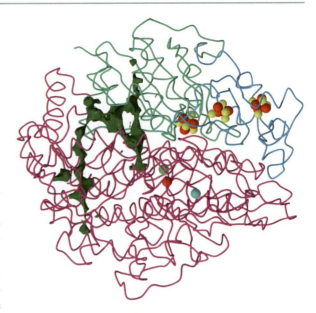

[NiFe]-Hydrogenases, Fig. 5 Hydrophobic tunnel network (in *green*) in *D. fructosovorans* [NiFe]-hydrogenase. Atom color codes as in Fig. 1

H_2-dependent transcription is directed by a signal transduction apparatus consisting of the response regulator HoxA and its cognate His protein kinase HoxJ. A third component, the regulatory hydrogenase (RH) senses H_2 and is also found in *Rhodobacter capsulatus*. As [NiFe]-hydrogenases, the sensors are able to catalyze H_2 uptake, H_2 evolution, and H–D exchange. The oxidized Ni–A and Ni–B forms are inaccessible in these proteins, which are insensitive to both O_2 and CO. RH is only found in the more reduced active Ni–SI and Ni–C states. In addition, in RHs, there is no coupling between the active site Ni and the proximal cluster, an interaction typical of standard [NiFe]-hydrogenases.

One plausible reason for the resistance to inhibition by oxygen in RHs would be limited access of O_2 to the Ni–Fe site. In [NiFe]-hydrogenases, the gas tunnel connecting the protein surface to the active site disappears near the nickel ion (Fig. 5), where highly conserved valine and leucine residues are found. In RH and the H_2-sensing HupUV, these two amino acids are respectively replaced by the more massive Ile and Phe. The RH large subunit Ile62 and Phe110 have been mutated to Val and Leu, respectively. When purified aerobically, the Ile62Val, Phe110Leu, and Ile62Val/Phe110Leu mutants displayed greatly reduced H_2 uptake activities. Like [NiFe]-hydrogenases, the mutants were reductively activated with dithionite $+H_2$. A similar result was obtained for the *R. capsulatus* sensor. These studies indicate that the hydrophobic tunnel can serve as a filter, modulating gas access to the active site as a function of its size.

Oxygen-insensitive [NiFe]-hydrogenases from Ralstonia eutropha. The Knallgas bacterium *Ralstonia eutropha* has a membrane-bound [NiFe]-hydrogenase connected to the respiratory chain by a b-type cytochrome (MBH) and a cytoplasmic soluble one (SH), which reduces NAD^+ using H_2. Both enzymes are very resistant to oxygen and CO and do not show the corresponding Ni-based EPR signals of other [NiFe]-hydrogenases. MBH resembles *Desulfovibrio* sp. oxygen-sensitive hydrogenases, so the basis for its resistance is not well understood. SH contains the heterodimeric complexes HoxFU and HoxHY, corresponding to the FMN-binding diaphorase and the hydrogenase, respectively. Both HoxFU and HoxHY bind FMN. Friedrich et al. have carried out an extensive mutagenesis program concerning this enzyme.

Substrate Binding and Catalysis

Because hydrons are not detectable by X-ray crystallography at the resolutions observed so far, hydrogenase crystal structures have only provided indirect evidence of substrate binding. A hydride is likely to be bound at E2 in crystal structures of reduced, active enzyme (Fig. 2) because it would complete a square pyramidal Ni coordination. Many different stable Ni–Fe site intermediates have been observed by EPR and FTIR spectroscopy (Fig. 4). Crystallographic evidence for bound oxygen species in the active site of unready (Ni–A and Ni–SU) and ready enzyme states (Ni–B) has been discussed above.

Redox reactions with the FeS clusters and proton transfer reactions generate Ni–Fe states that are at thermodynamic equilibrium so that getting hydrogenase crystals in a single homogeneous state is difficult. The Ni–C state should be two electrons more reduced than the Ni–B state, according to redox titrations of *D. gigas* [NiFe]-hydrogenase. These two paramagnetic states have been assigned to Ni^{3+} species, suggesting that the two additional electrons in the Ni–C form reside in a hydride. The Ni–C/Ni–R and one $[Fe_4S_4]^{2+}/[Fe_4S_4]^+$ redox couple are at redox equilibrium with H_2. Consequently, the enzyme in the Ni–C state should be able to react with hydrogen, indicating that both E1 and E2 can bind hydrons. As mentioned above, the putative hydride may be bound at the bridging E2 site leaving E1 open for substrate binding. Accordingly, it was observed that CO binds terminally to the Ni at E1 site in the crystal structure of *D. vulgaris* Miyazaki F hydrogenase (pdb code 1UBK).

HYSCORE and ENDOR spectroscopies have provided direct evidence for a bound hydride in the Ni–C state of the regulatory [NiFe]-hydrogenase from *R. eutropha*. In addition, a single-crystal EPR analysis of the same form of *D. vulgaris* Miyazaki F hydrogenase showed that the hydride binds at the Ni–Fe bridging E2 position. This result has been confirmed by more recent ENDOR/HYSCORE evidence using the same enzyme (Pandelia et al. 2010).

Concluding Remarks

More references on the proteins corresponding to the pdb codes cited above can be found in the protein data bank, www.pdb.org (Berman et al. 2000). Although the three-dimensional structures cannot provide a complete picture of hydrogen biocatalysis, without the spatial models, especially of the active sites, the interpretation of EPR, XAS, FTIR, and other spectroscopic data would be severely limited. The same applies to the study of active site cluster assembly and the electrochemical characterization of hydrogenases. Besides allowing the formulation of catalytic mechanisms, the active site structures are being used as an inspiration for the synthesis of biomimetic models. The greenhouse effects derived from fossil fuel utilization represent a threat to future generations. For this reason, the production of hydrogen by oxygen-evolving photosynthetic organisms has become a plausible solution to global energy requirements. Thus, it has become urgent to either design or discover oxygen-resistant hydrogenases. Preliminary action in the right direction is represented by the site-directed mutagenesis of amino acid residues lining the hydrophobic tunnels that lead to the buried active site of hydrogenases, as studied by Rousset et al. A few designed mutants of *D. fructosovorans* hydrogenase have been found to be significantly more oxygen-resistant than the wild type. Internal tunnels can discriminate between oxygen and hydrogen molecules based on gas size and the nature of tunnel residues. In this respect, the crystal structures are fundamental.

References

Berman HM, Westbrook J, Feng Z, Gilliland G, Bhat TN, Weissig H, Shindyalov IN, Bourne PE (2000) The Protein Data Bank. Nucleic Acids Research 28:235–242

Böck A, King PW, Blokesch M, Posewitz MC (2006) Maturation of hydrogenases. Adv Microb Physiol 51:1–71

Dementin S, Burlat B, De Lacey AL, Pardo A, Adryanczyk-Perrier G, Guigliarelli B, Fernandez VM, Rousset M (2004) A glutamate is the essential proton transfer gate during the catalytic cycle of [NiFe] hydrogenase. J Biol Chem 279:10508–10513

Fontecilla-Camps JC, Volbeda A, Cavazza C, Nicolet Y (2007) Structure/function relationships of [NiFe]- and [FeFe]-hydrogenases. Chem Rev 107:4273–4303

Montet Y, Amara P, Volbeda A, Vernède X, Hatchikian EC, Field MJ, Frey M, Fontecilla-Camps JC (1997) Gas access to the active site of Ni-Fe hydrogenases probed by X-ray crystallography and molecular dynamics. Nat Struct Biol 4:523–526

Pandelia ME, Ogata H, Lubitz W (2010) Intermediates in the catalytic cycle of [NiFe] hydrogenase: functional spectroscopy of the active site. ChemPhysChem 11:1127–1140

Vignais MV, Billoud B (2007) Occurrence, classification and biological function of hydrogenases: an overview. Chem Rev 107:4206–4272

Volbeda A, Charon MH, Piras C, Hatchikian EC, Frey M, Fontecilla-Camps JC (1995) Crystal structure of the nickel-iron hydrogenase from *Desulfovibrio gigas*. Nature 373:580–587

Volbeda A, Garcin E, Piras C, De Lacey AL, Fernandez VM, Hatchikian EC, Frey M, Fontecilla-Camps JC (1996) Structure of the [NiFe] hydrogenase active site: evidence for biologically uncommon Fe ligands. J Am Chem Soc 118:12989–12996

Yahata N, Saitoh T, Takayama Y, Ozawa K, Ogata H, Higuchi Y, Akutsu H (2006) Redox interaction of cytochrome c_3 with [NiFe] hydrogenase from *Desulfovibrio vulgaris* Miyazaki F. Biochemistry 45:1653–1662

NikR, Nickel-Dependent Transcription Factor

Caroline Fauquant and Isabelle Michaud-Soret
iRTSV/LCBM UMR 5249 CEA-CNRS-UJF, CEA/Grenoble, Bât K, Université Grenoble, Grenoble, France

Synonyms

Nickel metalloregulatory protein; Nickel sensor; Nickel-dependent transcription factor

Definition

NikR, a nickel sensor, regulates the transcriptional expression of numerous genes involved in nickel homeostasis by acting as a repressor or an activator according to the regulated gene and the microorganism concerned.

An Overview of the Subject

Nickel is required for several microbial enzymes activity such as urease or hydrogenase, but is potentially toxic (▶ Nickel Ions in Biological Systems, ▶ Nickel-Binding Proteins, Overview). Therefore, nickel homeostasis must be tightly controlled. NikR proteins are the major actors in the regulation of nickel homeostasis in bacteria (▶ Nickel in Bacteria and Archaea) at the level of gene expression and does not have any equivalent in eukaryotes (Dosanjh and Michel 2006 and references therein). The first member of the NikR family described was NikR from *Escherichia coli* (EcNikR) (De Pina et al. 1999). EcNikR only regulates the gene expression of nickel import proteins (▶ Nickel Transporters).

Later on, another NikR (HpNikR) was found to play a critical role in *Helicobacter pylori* since the survival of this human gastric pathogen relies on the production of active urease and hydrogenase. HpNikR is pleiotropic and controls the expression (activation or repression) of a large set of genes involved in different stresses (Danielli and Scarlato 2010; Ernst et al. 2007).

At the moment, functional and structural information have been mainly described on the three NikRs from *E. coli*, *H. pylori*, and *Pyrococcus horikoshii* (Chivers and Tahirov 2005; Dian et al. 2006; Schreiter et al. 2006). Here, an overview on NikR proteins is provided.

NikR and the Gene Expression Regulation in Response to Nickel

NikR is a transcriptional factor that represses or activates the expression of specific genes in response to nickel intracellular concentration. Functional information has been mainly obtained by gene expression analysis in *E. coli* and in *H. pylori*, molecular biology and protein/DNA interaction studies followed by footprint, gel shift assays, and fluorescence anisotropy experiments. A summary of the current knowledge derived from these data is presented in Fig. 1 and described below.

In *E. coli* (Fig. 1a), nickel is required for the activity of enzymes such as [NiFe] hydrogenases (noted HydNiFe), which reversibly catalyze H₂ oxidation especially under anaerobic growth (▶ [NiFe]-Hydrogenases).

The nickel (free or bound to a nickelophore) probably crosses the outer membrane via porines (passive mechanism) while nickel goes through the inner membrane via the inner transporter NikABCDE, an ATP-binding cassette-type transporter allowing Ni(II) specific import and encoded by the *nikABCDE* operon. This operon encodes five proteins: NikA a periplasmic nickel-binding protein; NikB and NikC integral inner membrane proteins; and NikD and NikE membrane-associated proteins with intrinsic ATPase activity. The expression of the *nikABCDE* operon is highly regulated in a nickel-dependent way as well as in two nickel-independent ways. When intracellular nickel levels reach a threshold, nickel-bound NikR binds to the *nik* promoter region to repress the transcription of *nikABCDE*. The *nikABCDE* operon is also negatively regulated by the NarLX two-component system complex in response to nitrate and positively regulated by the oxygen sensor Fumarate Nitrate Regulator (FNR) under anaerobic conditions. Nickel ions imported by this Nik transporter would be essentially dedicated to NiFe-hydrogenases in anaerobic growth conditions. Nickel chaperones such as HypA and SlyD/HypB are involved in their maturation. A second and unknown Ni(II) ions transporter has been proposed to import nickel to the other Ni(II)-proteins. Nickel overload is also prevented by systems like the nickel/cobalt efflux pump RcnA encoded by the *rcnAB* operon. Its expression is regulated by the transcription factor RcnR. In the absence of an *excess* of Ni(II) (or Co(II)), RcnR represses the transcription of *rcnAB*.

In *H. pylori* (Fig. 1b), a human pathogen classified as a class 1 carcinogen, nickel is needed to colonize the acidic gastric niche persistently as cofactor for essential metalloproteins involved in acid acclimation, respiration, and detoxification such as urease and hydrogenase. Urease catalyzes the hydrolysis of urea into ammonia and bicarbonate, two buffering compounds that allow the bacterium to maintain its intracellular neutrality (▶ Urease). This enzyme, composed of the UreA and UreB subunits, is very abundant (up to 10% of total soluble cellular proteins) and requires 24 nickel ions per active enzymatic complex. *H. pylori* synthesizes another Ni(II)-containing enzyme that is important for colonization: the [NiFe] hydrogenase composed of HydA and HydB. Ni(II) acquisition from the external environment relies on porines or on FrpB4, an outer membrane (OM) transporter energized by the TonB machinery that is also involved in iron uptake. Then, nickel is transported through the cytoplasmic membrane via the NixA permease. In the cytoplasm, Ni(II) ions can be bound by nickel chaperones such as UreE, HypA/B, and HspA needed for the urease and hydrogenase maturation, respectively. *H. pylori* also exhibits unique proteins involved in nickel storage and/or nickel delivery including three proteins that bind Ni(II) with high affinity: Hpn, Hpn2, and HspA. Hpn is a Histidine-rich protein that represents up to 2% of the total proteins and Hpn2 (or Hpn-like) is a Histidine/Glutamine-rich protein. Both proteins play an important role in Ni(II) detoxification and storage. Another abundant protein, HspA, is a GroES co-chaperone homologue that through its unique histidine- and cysteine-rich C-terminal extension sequesters nickel and acts as a specialized nickel chaperone for hydrogenase maturation. Finally, nickel efflux is possible through a metal pump CznABC that is able to export nickel, zinc, and cadmium across the cytoplasmic membrane.

In response to an increase in intracellular nickel concentration, HpNikR regulator binds Ni(II) and then regulates, directly or indirectly, not only the expression of the genes coding for the proteins involved in nickel uptake, utilization, and storage, mentioned above, but also genes whose products are involved in iron homeostasis and in acid stress response. It is remarkable in that, depending on its target gene it acts in response to nickel either as an activator or a repressor (Danielli and Scarlato 2010).

Ni(II)-HpNikR is able to induce the transcription of genes encoding enzyme using nickel as cofactor (*ureAB*), and nickel storage proteins (*hpn, hpn2*...) as indicated in Fig. 1b. In the same way, Ni(II)-HpNikR leads upon specific DNA binding also to the repression of genes transcription encoding enzyme using nickel as cofactor (*hydABC*), proteins involved in nickel import (*nixA, fecA3, frpB4*), the energetic machinery (*exbB/D tonB*), *nikR* itself, and the ferric uptake regulator (*fur*), as indicated in Fig. 1b. Only the efflux of nickel via the

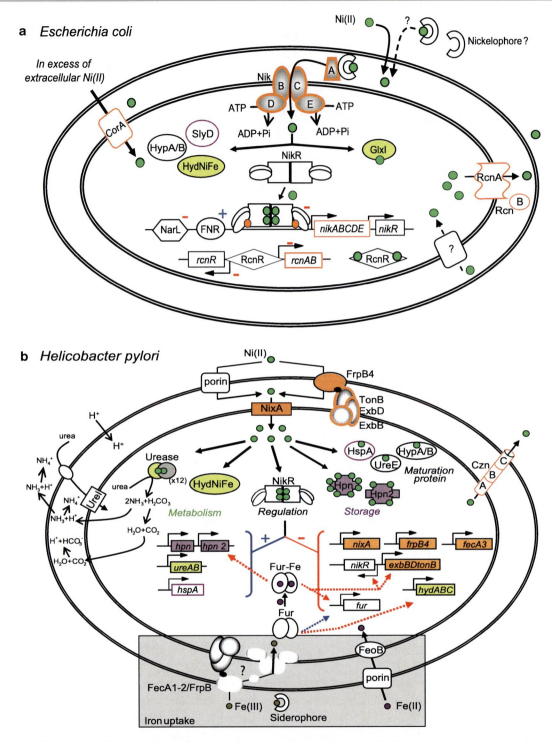

NikR, Nickel-Dependent Transcription Factor, Fig. 1 Schematic representation of the NikR regulons in (**a**) *E. coli* and (**b**) *H. pylori*. Whereas EcNikR controls only the expression of one operon, HpNikR is a pleiotropic metalloregulator which controls by activation or repression the expression of a tenth of genes or operons in response to nickel binding. Gene activation is in *blue* (+); gene repression is in *red* (−). There is an overlapping between NikR (nickel dependent), Fur (iron dependent), and the ArsRS two-components (acid sensitive; not shown here for more clarity) regulons. In *gray box* putative iron uptake is shortly schematized

CznABC transporter is not regulated by HpNikR. Nickel homeostasis is also regulated by Fur, which controls *nikR* expression, and by the ArsRS two-component system in response to acidity. In *H. pylori*, there is an overlap between the nickel and iron metabolisms. First, transport of both ions relies on the TonB machinery. Second, several genes are co-regulated by NikR and Fur, the expression of the latter regulator being itself controlled by NikR. Furthermore, both genes are autoregulated and have their gene expression which varies in a pH-dependent manner (Danielli and Scarlato 2010). The role of these overlapping regulations in *H. pylori* may exist in order to respond to several metallic stresses (starvation or excess) (Bahlawane et al. 2010).

EcNikR and HpNikR DNA-Binding Sequences

Whereas EcNikR represses the single *nikABCDE* operon by binding to a perfect inverted repeat sequence GTATGA-n16-TCATAC, HpNikR binds to promoters from multiple genes that contain poorly conserved inverted repeats (TATWATT-n11-AATWATA with W = adenine or thymine). A pseudo-consensus motif TRWYA-n15-TRWYA, with Y = thymine or cytosine, R = adenine or guanine, and W = adenine or thymine, has recently been proposed from comparative analysis of *H. pylori* and *Helicobacter mustalae* NikR operators. Genes regulated by HpNikR can be classified into two groups in view of the dissociation constants obtained from experimental determination (footprinting, gel shift assays, or fluorescence anisotropy experiments) in presence of nickel ions: (1) high affinity promoters: PureA/PnixA/PfecA3/PfrpB4 with Kds in the 10 nM range; (2) low affinity promoters: PnikR/PexbB/Pfur with Kds in the micromolar range (Li and Zamble 2009). Metal binding to a secondary metal-binding site (X sites which will be described below) in presence of manganese ions excess was shown to increase the DNA/protein affinity to some of this low affinity class of promoters from micromolar to 10 nM range (Bahlawane et al. 2010). This supplementary regulation level may allow a tuning of the HpNikR regulon as a function of the metal excess (Ni(II) or other transition metals dications). Recently, De Reuse and colleagues demonstrated a hierarchical regulation of the NikR-mediated nickel response, which may be partially explained by this supplementary regulation level.

NikR Structural Properties

NikR proteins (*E. coli*, *H. pylori*, and *P. horikoshii* NikR structures) are homotetramers or more exactly dimer of dimers, composed of two main domains: a tetramerization domain (TD) with a ßαßßαß fold structurally homologous to the ACT domain ("Aspartate kinase-Chorismate mutase-TyrA" regulatory small-molecule-binding domain) flanked by two dimeric Ribbon-Helix-Helix (RHH) DNA-binding domains (DBD) at the N-terminus (Dosanjh and Michel 2006; Schreiter et al. 2006) (Figs. 2, 3). NikR is the only known metal-regulated member of the RHH family of prokaryotic DNA-binding proteins. In these proteins the N-terminal sequences of two monomers intertwine to produce an antiparallel ß-sheet that binds in the major groove with specific DNA-protein contacts. Alignments of available sequence (BLAST of the NCBI database) reveal numerous homologues of NikR from a variety of archea and eubacteria species. Figure 2 presents the alignment of the best characterized NikR together with their secondary structures. A unique feature of HpNikR is its N-terminal extension which was unresolved in the structure (Dian et al. 2006) but has been shown to play an important role in the response of HpNikR to specific promoters in function of pH and metal excess (Li and Zamble 2009).

NikRs were observed by X-ray crystallography in different conformations (Fig. 3): an open conformation in which the DBDs are linearly placed on each side of the TD (holo-EcNikR and holo-PhNikR) and a closed *trans*-conformation where the DBDs are placed on opposite sides (holo-HpNikR). Whereas the Ec and PhNikR structures are quite symmetrical at the level of their tetramerization interfaces, HpNikR shows an asymmetry which leads to modification of the surface properties of the tetramer interface with one locked side (solvent accessibility reduced) and one unlocked side (see Figs. 3d, 4). Whereas modeling and dynamic studies as well as NMR study suggested that several conformations of apo- and holo- HpNikR exist and interconvert in solution (Musiani et al. 2010), SAXS (Small-Angle X-ray Scattering) data suggest mainly a *cis*-conformation (Bahlawane et al. 2010). Modeling and dynamic studies of EcNikR proposed an open conformation in solution.

However, the conformation suited for NikR proteins to bind their operator was established by the crystal structure of holo-EcNikR/DNA complex

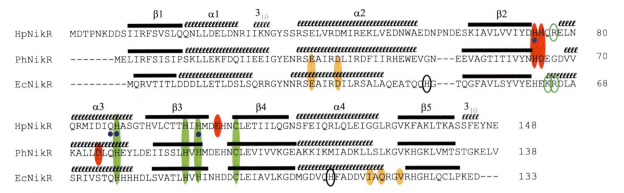

NikR, Nickel-Dependent Transcription Factor, Fig. 2 Sequence alignment of NikR proteins from *H. pylori* (O25896), *P. horikoshii* (O58316), and *E. coli* (P0A6Z6) with their corresponding secondary structures. Residues involved in the formation of metal-binding sites are indicated as follows: high affinity site in *green*, intermediate site in *blue*, external site in *red*, and an interdomain site in *orange*. Residues *encircled in green* are involved in polar interaction with DNA backbone phosphates in the holo-EcNikR-DNA complex and potentially in holo-HpNikR-DNA complex. Residues *encircled in black* indicate residues involved in Ni(II) binding in a low affinity site. Protein references are from UniProt

(Schreiter et al. 2006) to be a closed *cis*-conformation where the two DBDs are located on the same side and are bound to the two halves of the operator sequence.

DNA-binding activation of NikR depends on the incorporation of metal ions at specific binding sites: a high affinity metal-binding site (HA), an interdomain site (a potassium site in case of EcNikR), and several secondary nickel-binding sites. The description of these sites will be done in the following part.

The most important modification observed in the crystal structure of holo-NikRs in the presence of DNA (resolved structure or models) compared to the holo-NikRs alone (resolved structure) is the rotation of the DBD to reorient their antiparallel ß-strands toward the DNA. This movement allows each DBD to occupy the DNA major groove of the operator inverted repeat half-site. Moreover, no difference could be observed between the TD interfaces of holo- and DNA-bound EcNikR. In this crystal structure, Holo-EcNikR-DNA is in a closed *cis*-conformation.

Three classes of interaction have been observed between EcNikR and the DNA: (1) Specific interactions where hydrogen bonds are made by the side chains of Arg-3 and Thr-5 from the ß-sheets of the DBD and the nucleotide bases in the DNA major groove, (2) nonspecific polar interactions between the DBD and the phosphate backbone of DNA, (3) nonspecific interactions between the TD (two EcNikR subunits) and the DNA phosphates (Fig. 3 insert 2).

The second important modification in Holo EcNikR-DNA crystal structure is the presence of a potassium-binding site "an interdomain site" at the new interface created between the DBD and the TD (see following part, and Fig. 3 insert 2). Coordination of the metal ion in this site could allow locking these domains into the DNA-bound conformation.

Metal-Binding Sites Properties

High affinity site – NikRs are able to bind Ni(II) in a high affinity site (HA) located at the tetramerization interface. In solution, all NikR proteins purified so far may bind Ni(II) to the four affinity sites. However, the asymmetry of the HpNikR interface probably causes a two by two nickel binding to the HA sites in solution and the filling of only two HA sites in the crystal structure of holo-HpNikR obtained at low pH. Recently, Ciurli and colleagues have resolved the crystal structure of HpNikR at pH 7.3 showing that four nickel ions are bound then in the four high affinity sites.

The architecture of the HA sites is conserved among all characterized NikR proteins (Ec, Hp, and Ph) (Li and Zamble 2009). There, nickel ions are coordinated in a square planar geometry by a cysteine and two histidines from one subunit and one histidine from the adjacent subunit (Fig. 3 insert 1) (▶ Nickel-Binding Sites in Proteins; ▶ Nickel, Physical and Chemical Properties). When HA sites are occupied, a conserved

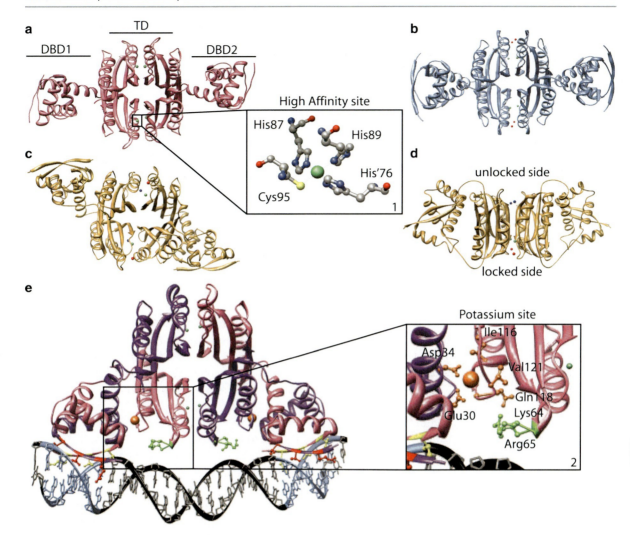

NikR, Nickel-Dependent Transcription Factor, Fig. 3 NikRs structures in distinct conformations resolved by X-ray. (**a**) Holo-EcNikR in an open conformation (pdb code: 2HZA, resolved at 2.1 Å), *insert 1*: Ni(II) in the high-affinity site (HA) is coordinated by four residues (His87, His89, Cys95 from one subunit, and His'76 from an adjacent subunit) in a square planar geometry. (**b**) Holo-PhNikR in an open conformation (pdb codes: 2BJ1, resolved at 3.0 Å) (Chivers and Tahirov 2005). Ni(II) ions are located in the high affinity and external sites. (**c**) and (**d**) Holo-HpNikR with three distinct sites (HA, intermediary and external sites) in a closed *trans*-conformation, two views of the structure are shown to consider the asymmetrical organization of the TD (locked and unlocked side) (pdb code: 2CAD, resolved at 2.3 Å) (Dian et al. 2006). (**e**) Holo-EcNikR-DNA in a closed *cis*-conformation (pdb code: 2HZV, resolved at 3.1 Å) (Schreiter et al. 2006). Each subunit of a dimer is indicated in *pink* and *purple*. In *red* and *yellow* are indicated residues involved in hydrogen bonds with the DNA major groove; *insert 2*: K$^+$ in the interdomain site is coordinated by five residues in *orange* and residues (Lys64, Arg65) in the loop leading into helix α3 involved in polar interactions with DNA backbone phosphates are in *green*

network of hydrogen bonds is formed at the tetramerization interface that connects nickel ions two by two. Mutagenesis studies demonstrated that HA site residues were essential to the DNA-binding activity of NikR. In vitro, other metals are able to bind to these sites allowing protein/DNA interactions in a specific manner. The metal-binding affinities to the HA sites follow the Irving-Williams series ($Mn^{2+} < Co^{2+} < Ni^{2+} < Cu^{2+} > Zn^{2+}$) suggesting that they are not preorganized for nickel binding. This metal binding to this site has been extensively studied using various spectroscopic methods such as UV-visible spectroscopy and X-ray absorption spectroscopy. The ligand to metal charge transfer band (LMTC) from

S → Ni(II) around 300 nm is a spectroscopic signature which allows to follow the Ni(II) binding and to determine directly or indirectly (using EGTA competition assays) the dissociation constants for Ni(II)-NikR. Two range of dissociation constants have been obtained for both EcNikR and HpNikR according to the technique used for the measurements: picomolar range (competition assays) and sub-micromolar range (direct UV-visible spectroscopic measurements as well as filter binding and isothermal calorimetric experiments). These differences, which caused discrepancy between research groups, are not yet fully understood but may come from different protein preparations, questionable EGTA-Ni dissociation constants values, or kinetic question due to the conformational change of the protein upon metal binding. Indeed, the accessibility of the high affinity site is probably different between the apo- and the holoprotein (buried or not environment, $k_{on} \neq k_{off}$).

The potassium site – The structure determination of holo-EcNikR-DNA complex allows the discovery of two potassium ions per EcNikR tetramer at the hinge of the DBD and TD (Schreiter et al. 2006). These ions are coordinated by two strictly conserved amino acid side chains, Glu-30 and Asp-34, from the DBD, and by the three backbone carbonyl oxygens of Ile-116, Gln-118, and Val-121, from a loop of the TD (Fig. 3 insert 2). Mutagenesis of the conserved ligands and DNA-binding assays demonstrate that this metal-binding site is necessary and the presence or not of potassium ions influences EcNikR binding to its DNA target.

In PhNikR, nickel ions have been found in the same type of site as this potassium site in EcNikR but was not present in nickel soaked HpNikR, even if the corresponding residues were conserved. Their mutation did not affect the DNA-binding properties of HpNikR.

Other metal binding sites – Secondary nickel binding sites have been described in NikR proteins but appear more species-specific as described in Fig. 4.

Several low affinity sites have been described in EcNikR and HpNikR in solution and in the structures. In HpNikR, the Ni(II)-soaked protein contains only two HA sites filled, two nickel ions are found in an intermediary site, and two last in an external site (involving His-74/His-75) localized at the tetramerization interface inside and at the surface, respectively.

Ni(II) in the intermediary site is in an octahedral geometry and coordinated by His-74, His-101 from one monomer, and His'-88 and a water-mediated interaction of Gln'-87 from the adjacent monomer.

Ni(II) in the external site is in an octahedral geometry and coordinated by the side chains of His-74, His-75, the carbonyl group of Glu-104 from one monomer, and a citrate molecule to complete the coordination sphere. The occupancy of the external site triggers small conformation changes in the β2α3 loop containing Gln-76 and Arg-77. The equivalent residues of EcNikR (Lys-64 and Arg-65) were found to interact directly with the DNA backbone in the EcNikR/DNA complex. It was recently proposed, based on mutant studies and structural data, that the metallation of this external secondary site triggers the disruption of a salt bridge between the aspartate D85 and the arginine R77. This could release the Arg-77 required for HpNikR DNA binding to low-affinity promoters. This external metal-binding site does not seem to be specific for nickel ions binding. It was proposed that this site could also accept other metal ions with octahedral geometry which can fit with manganese ions as well. This site is proposed to allow a tuning of the NikR gene expression regulation in response to metal ions excess. In PhNikR structure, nickel ions were also found in an external secondary site in a distorted trigonal pyramid geometry coordinated by one conserved residue His-64 (conserved in NikRs) and other not conserved residues.

Six secondary low affinity metal-binding sites have been seen in the EcNikR structures after soaking with nickel ions. Two residues His-110 and His-48 involved in two of these sites are also suggested to be responsible for the massive aggregation specific to EcNikR observed in solution upon nickel excess addition.

The N-terminal extension found in HpNikR has also been proposed to be involved in metal ion binding or low pH activation of apo-HpNikR triggering a binding to PureA and an increase of the *ureA* expression level (Li and Zamble 2009).

Mechanism of Action of NikR Proteins

NikR/DNA interactions were proposed to rely on protein activation by Ni(II) binding to high-affinity (HA), a potassium site or secondary external sites (Bahlawane et al. 2010; Li and Zamble 2009; Zambelli et al. 2008). Structural and functional studies reveal

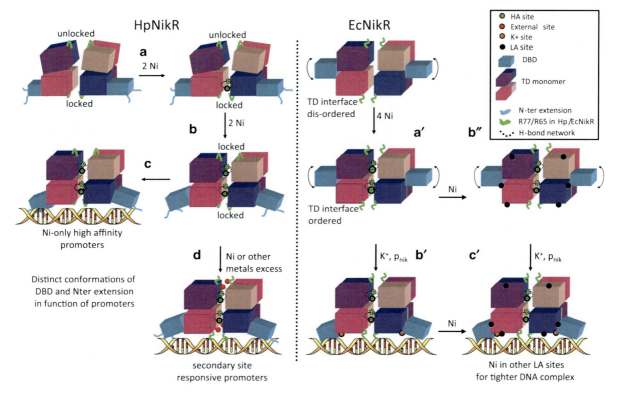

NikR, Nickel-Dependent Transcription Factor, Fig. 4 Schematic representation of the regulation mechanisms of the NikR proteins in *H. pylori* (*left*) and *E. coli* (*right*). EcNikR and HpNikR use similar allosteric regulation of their ACT domains by nickel entry at the HA sites (**a**, **b**, and **a'**). These changes are secured by the formation of the hydrogen-bound network. In both proteins this H-bond network is essential to the activation for DNA binding. Although this mechanism is likely to be common for NikR proteins, the conformation of all NikR proteins may not be the same in solution in apo form. This activation step would be sufficient for HpNikR to bind to high affinity promoter regions (**c**). In a second step, species-specific mechanisms exist that are adapted to the nickel requirement by each organism. For HpNikR, the X site plays a role in modifying the position of DNA-interacting residues (such as Q76 and R77) from the β2-β3 loops (**d**) (Bahlawane et al. 2010). In EcNikR, this site seems unnecessary since these residues (K64, R65) can readily interact with DNA without nickel-induced structural changes (Schreiter et al. 2006). Instead, EcNikR final conformation seems to require an interdomain site located between the TD and DBD domains that can bind potassium ion and lock the DBDs in a closed *cis*-conformation (**b'**). Nickel binding in a third class of sites, LA sites, would improve its DNA affinity (**b''**, **c'**). HpNikR mechanism might be completed by its unique N-terminal extension of 9 amino acids required for specific DNA binding. This N-terminal extension has recently been shown to be involved in the pH-responsive DNA-binding activity of HpNikR. Moreover, it adopts, as the DBD, distinct conformation in function of promoter, as recently proposed by Chivers' group

features from which a model for HpNikR activation was proposed with: (1) HA sites and a hydrogen bond network are required for DNA binding and (2) metallation of a unique secondary external site (X) modulates HpNikR DNA binding to low affinity promoters by disruption of a salt bridge (Fig. 4).

EcNikR structural and functional studies reveal features from which a model for activation was proposed with: (1) HA sites and a hydrogen bond network are required for DNA binding, (2) final conformation seems to require an interdomain site located between the TD and DBD domains that can bind potassium ion and lock the DBDs in a closed *cis*-conformation, and (3) nickel binding in a third class of sites, low affinity sites, different of the secondary external site (X) found in HpNikR, would improve its DNA affinity (Schreiter et al. 2006; Li and Zamble 2009).

The HpNikR model could explain how it modulates the expression of various genes as a function of the metal ions concentrations. This regulation mechanism, specific to *H. pylori*, may allow for a gradual response to large variability in metal concentrations encountered in the gastric environment as a function of diet and/or of metal exposure. The mechanism also reflects possible differences of nickel regulation requirements between *H. pylori* and *E. coli*. HpNikR not only inhibits the import of nickel when excess is reached, but can also increase the expression level of urease and other nickel-utilizing proteins. The evolution of the secondary external X site in HpNikR, located in the vicinity of DNA-binding residues, might provide an additional metal-based regulation mechanism. This mechanism might not be conserved in EcNikR and would explain a stringent and direct repression of nickel import in *E. coli*.

In conclusion, in vitro characterizations of the properties of the NikR proteins, including extensive structure-function analysis and the identification of metal-binding sites, have been reported. In vivo previous studies have mostly focused on analysis of individual genes and more information are now needed on expression hierarchy of the positive or negative HpNikR targets in response to nickel or other metal ions. Global studies would allow deciphering the intricate regulation and interferences between iron and nickel homeostasis control through metalloregulators such as NikR.

Acknowledgments Christelle Bahlawane is greatly acknowledged for discussions. Because of the restricted number of cited references allowed we were not able to cite all the important references and we would like to apologize for this and thank all the researchers involved in this NikR field.

Cross-References

▶ [NiFe]-Hydrogenases
▶ Nickel in Bacteria and Archaea
▶ Nickel Ions in Biological Systems
▶ Nickel Transporters
▶ Nickel, Physical and Chemical Properties
▶ Nickel-Binding Proteins, Overview
▶ Nickel-Binding Sites in Proteins
▶ Urease

References

Bahlawane C, Dian C, Muller C, Round A, Fauquant C, Schauer K, de Reuse H, Terradot L, Michaud-Soret I (2010) Structural and mechanistic insights into *Helicobacter pylori* NikR activation. Nucleic Acids Res 38:3106–3118
Chivers PT, Tahirov TH (2005) Structure of *Pyrococcus horikoshii* NikR: nickel sensing and implications for the regulation of DNA recognition. J Mol Biol 348:597–607
Danielli A, Scarlato V (2010) Regulatory circuits in *Helicobacter pylori*: network motifs and regulators involved in metal-dependent responses. FEMS Microbiol Rev 34:738–752
De Pina K, Desjardin V, Mandrand-Berthelot MA, Giordano G, Wu LF (1999) Isolation and characterization of the *nikR* gene encoding a nickel-responsive regulator in *Escherichia coli*. J Bacteriol 181:670–674
Dian C, Schauer K, Kapp U, McSweeney SM, Labigne A, Terradot L (2006) Structural basis of the nickel response in *Helicobacter pylori*: crystal structures of HpNikR in Apo and nickel-bound states. J Mol Biol 361:715–730
Dosanjh NS, Michel SL (2006) Microbial nickel metalloregulation: NikRs for nickel ions. Curr Opin Chem Biol 10:123–130
Ernst FD, van Vliet AHM, Kist M, Kusters JG, Bereswill S (2007) The role of nickel in environmental adaptation of the gastric pathogen *Helicobacter pylori*. In: Sigel A, Sigel H, Sigel RKO (eds) Nickel and its surprising impact in nature, Vol. 2. John Wiley and Sons, Ltd, Chichester. doi:10.1002/9780470028131
Li Y, Zamble DB (2009) Nickel homeostasis and nickel regulation: an overview. Chem Rev 109:4617–4643
Musiani F, Branimir B, Magistrato A, Zambelli B, Turano P, Losasso V, Micheletti C, Ciurli S, Carloni P (2010) Computational study of the DNA-binding protein *helicobacter pylori* NikR: the role of Ni^{2+}. J Chem Theory Comput 6:3503–3515
Schreiter ER, Wang SC, Zamble DB, Drennan CL (2006) NikR-operator complex structure and the mechanism of repressor activation by metal ions. Proc Natl Acad Sci USA 103:13676–13681
Zambelli B, Danielli A, Romagnoli S, Neyroz P, Ciurli S, Scarlato V (2008) High-affinity Ni^{2+} binding selectively promotes binding of *Helicobacter pylori* NikR to its target urease promoter. J Mol Biol 383:1129–1143

Niobium, Physical and Chemical Properties

Fathi Habashi
Department of Mining, Metallurgical, and Materials Engineering, Laval University, Quebec City, Canada

Physical Properties

Niobium is a refractory metal with a high melting point. It occurs mainly in association of tantalum minerals. Both niobium and tantalum are transition metals, i.e., they are less reactive than the typical metals but more reactive than the less typical metals. The transition metals are characterized by having the outermost electron shell containing two electrons and the next

Niobium, Physical and Chemical Properties, Fig. 1 The similarity of atomic radii of Zr-Hf, Nb-Ta, Mo-W, etc., due to the lanthanide contraction

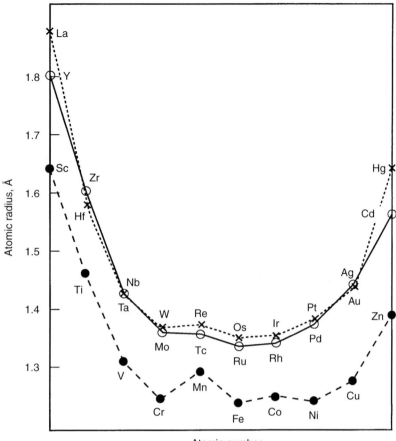

inner shell an increasing number of electrons. Although niobium [and tantalum] has two electrons in the outermost shell and it would have been expected that to have valency of 2, yet its main valency is 5.

Atomic number	41
Atomic weight	92.91
Density at 20°C, g cm^{-3}	8.57
Relative abundance in Earth's crust, %	2.4×10^{-3}
Crystal structure	Body-centered cubic ($a = 3.3 \times 10^{-10}$ m)
Melting point, °C	$2{,}468 \pm 10$
Boiling point, °C	4,927°C
Linear coefficient of thermal expansion, K^{-1}	6.892×10^{-6}
Specific heat, kJ kg^{-1} K^{-1}	0.26
Latent heat of fusion, kJ kg^{-1}	290
Latent heat of vaporization, kJ kg^{-1}	7,490
Thermal conductivity at 0°C, W cm^{-1} K^{-1}	0.533
Electrical resistivity at 0°C, μΩ · cm	15.22
Temperature coefficient 0–600°C, K^{-1}	0.0396

(*continued*)

Electrochemical equivalent, mg C^{-1}	0.19256
Standard electrode potential, $E°$ Nb/Nb^{+5}, V	−0.96
Magnetic susceptibility at 25°C	2.28×10^{-6}
Superconductivity T_C, K	9.13
Spectral emissivity at 650×10^{-10} m (at 2,003 K)	2.28×10^{-6}
Ionization potential, eV	6.67
Work function, eV	4.01

The mechanical properties, like those of most refractory metals, are influenced by the purity of the metal, the production method, and the mechanical treatment. Even small amounts of interstitial impurities increase the hardness and strength but reduce the ductility.

Chemical Properties

Niobium and tantalum are closely associated because of the lanthanide contraction (Fig. 1). Niobium is resistant to most organic and inorganic acids, with the exception of HF, at temperatures up to 100°C. Concentrated sulfuric acid above 150°C causes embrittlement. The resistance toward alkaline solutions is lower. Because niobium has a marked tendency to form oxides, hydrides, nitrides, and carbides, its use in air is limited to temperatures up to 200°C.

Niobium pentoxide Nb_2O_5, *m.p.* 265°C, *b.p.* 1,495°C, is a colorless powder that can only be dissolved by fusion with acidic or alkaline fluxes such as NaOH or $KHSO_4$, or in hydrofluoric acid. It is prepared by hydrolyzing solutions of alkali-metal niobates or niobium pentachloride, or by precipitation from hydrofluoric acid solutions with alkali-metal hydroxides or ammonia. The oxide hydrate is filtered, washed, and calcined at 800–1,100°C. The temperature and treatment time determine which of the various crystalline modifications is formed. Nearly all the phase changes are irreversible.

Niobium pentachloride, $NbCl_5$, *m.p.* 209.5°C, *b.p.* 249°C, is produced by chlorination of ferroniobium, niobium metal, or niobium scrap. It forms strongly hygroscopic yellow crystals that react with water to form $NbOCl_3$ or $Nb_2O_5 \cdot xH_2O$. Niobium trichloride, $NbCl_3$, disproportionates between 900°C and 1,000°C. Niobium oxychloride, $NbOCl_3$, is a colorless, crystalline compound that sublimes at 400°C and partly decomposes into niobium pentoxide and niobium pentachloride on further heating.

References

Eckert J, Starck HC (1997) Niobium and niobium compounds. In: Habashi F (ed) Handbook of extractive metallurgy. Wiley-VCH, Weinheim, pp 1403–1416

Habashi F (2003) Metals from ores. An introduction to extractive metallurgy. Métallurgie extractive Québec, Quebec City, Canada. Distributed by Laval University Bookstore, www.zone.ul.ca

NiSOD

▶ Nickel Superoxide Dismutase

Nitrile Hydratase

▶ Cobalt Proteins, Overview
▶ Cobalt-containing Enzymes
▶ Nitrile Hydratase and Related Enzyme

Nitrile Hydratase and Related Enzyme

Masafumi Odaka[1] and Michihiko Kobayashi[2]
[1]Department of Biotechnology and Life Science, Graduate School of Technology, Tokyo University of Agriculture and Technology, Koganei, Tokyo, Japan
[2]Graduate School of Life and Environmental Sciences, Institute of Applied Biochemistry, The University of Tsukuba, Tsukuba, Ibaraki, Japan

Synonyms

Cysteine sulfenic acid; Cysteine sulfinic acid; Hydration; Metallochaperone; Nitrile hydratase; Non-corrin cobalt; Self-subunit swapping

Definition

Nitrile hydratase and thiocyanate hydrolase which belongs to the same protein family are a non-corrin Co^{3+} or nonheme Fe^{3+} enzyme, the former being used for industrial production of acrylamide and nicotinamide. Structural studies revealed the unique structure of their metallocenter involving two oxidized cysteine ligands. Various studies including time-resolved X-ray crystallography indicated the cysteine sulfenic acid ligand is involved in the catalysis. A novel maturation mechanism for Co-type nitrile hydratase with its activator protein, named "self-subunit swapping," is proposed.

Introduction

Nitrile hydratase (NHase; EC 4.2.1.84) catalyzes the hydration of various aliphatic and aromatic nitriles to the corresponding amides. NHase is well known for its use for the industrial production of acrylamide. Currently, more than 30 kt of acrylamide is produced by the bioprocess involving this enzyme. Basic analyses of NHase and development of its use as a biocatalyst have been extensively performed by Yamada, Asano, Nagasawa, Kobayashi, and their colleagues. Also, NHase has been used for the industrial production of nicotinamide in China (for detailed reviews, please see Kobayashi et al. 1992; Kobayashi and Shimizu 1998). NHase is unique in the unusual structure of its catalytic cobalt or iron center (Endo et al. 2001; Kovacs 2004; Yano et al. 2008). In this entry, the focus is on the studies regarding the structures and reaction mechanism of NHase and thiocyanate hydrolase, which belongs to the same protein family.

Structural Studies on NHase

NHase is composed of two distinct subunits, α and β, with similar molecular weights of 22–24 k (Kobayashi et al. 1992). The amino acid sequences of both subunits are conserved among all known Co- and Fe-type NHases. Most NHases exist as a hetero-tetramer, but *Rhodococcus rhodochrous* J1 had been shown to have two NHase gene clusters, each of which is induced by a distinct inducing agent. The produced enzymes are termed J1 L-NHase and J1 H-NHase, respectively. J1 L-NHase has the normal hetero-tetramer structure, while J1 H-NHase was shown to have the $(\alpha\beta)_{10-11}$ structure (Kobayashi et al. 1992). J1 H-NHase is very stable and has been used as an industrial biocatalyst for acrylamide production.

Incipient studies on the structure of the metallocenter have focused on the iron-type NHases from *Rhodococcus erythropolis* sp. R312, N774, and N771 (ReNHase) (Endo et al. 2001). These NHases must be identical because all of them have the same amino acid sequences. Electron spin resonance (ESR) spectroscopic studies revealed that Fe-type NHase was the first mononuclear low-spin nonheme Fe^{3+} enzyme (Endo et al. 2001). Later, Co-type NHase was shown to have a low-spin non-corrin Co^{3+} center by ESR and X-ray absorption spectroscopy (Kovacs 2004). Endo and his colleagues studied the photoreactivity of ReNHase: In vivo and in vitro, it is gradually inactivated in the dark, but the activity is immediately recovered on light illumination. By measuring Fourier transform infrared (FTIR) spectra before and after photoreaction, it was shown that a nitric oxide (NO) molecule bound to the iron in the inactive state was released on light illumination to activate the enzyme (Endo et al. 2001).

The first determined structure was that of the active state of ReNHase (Endo et al. 2001). Each subunit has a long N-terminal loop that wraps around the core region of the other subunit to form a tight $\alpha\beta$ heterodimer. Two heterodimers interact to yield a $(\alpha\beta)_2$ hetero-tetramer with an interface of 1,560 $Å^2$. The catalytic Fe center is located at the interface of the α and β subunits in a heterodimer. Apparently, no allosteric interaction is observed between two $\alpha\beta$ heterodimers. All proteinous ligand atoms are provided by the metal-binding motif, Cys_1-Thr_1/Ser_1-Leu-Cys_2-Ser_2-Cys_3, in the α subunit. The Fe ion is coordinated to two main chain amide nitrogen atoms from Ser_2 and Cys_3 and three sulfur atoms from Cys_1, Cys_2, and Cys_3. The sixth ligand cannot be observed because of the resolution limit. Subsequently, the high-resolution structure of the ReNHase in the nitrosylated state revealed that Cys_2 and Cys_3 are posttranslationally oxidized to cysteine sulfinic acid (Cys-SO_2H) and cysteine sulfenic acid (Cys-SOH), respectively. Both modifications were confirmed by mass spectrometry (Endo et al. 2001). FTIR and X-ray absorption spectroscopic studies combined with density functional calculation demonstrated the former existed as Cys-SO_2^-, but the protonation state of the latter has not been unequivocally because the studies revealed the contradictory ionization states (Yamanaka et al. 2010). Thus, the latter residue is described as Cys-SO(H). The structure of the Co-type NHase from *Pseudonocardia thermophila* JCM 3095 (PtNHase) is highly conserved except for the unconserved loop region between the N-terminal helix and C-terminal sheet domains of the β subunit (Miyanaga et al. 2001). Figure 1 shows the structure around the Co center of *Pt*NHase, in which all protein ligands including two posttranslationally oxidized Cys residues are completely conserved within the Fe-type one. The sixth ligand site is occupied by a water molecule. The structure of the NHase metallocenter is very unusual. First, the coordination of the amidate nitrogens from the peptide

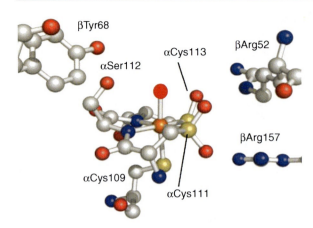

Nitrile Hydratase and Related Enzyme, Fig. 1 Structure around the Co center of NHase from *Pseudonocardia thermophila* JCM 3095. *Gray, red, blue, yellow,* and *orange spheres* represent carbon, oxygen, nitrogen, sulfur, and cobalt atoms, respectively

backbone is very rare, having been found only in a few proteins. Second, three Cys ligands exist in three different oxidation states. In fact, this unusual metallocenter structure has only been reported for NHases and a related enzyme, thiocyanate hydrolase (discussed later). The functions of these unique features will be discussed below.

Reaction Mechanism of NHase

Based on the crystal structure, three reaction mechanisms had been proposed (Kobayashi and Shimizu 1998; Kovacs 2004; Yano et al. 2008). (a) A substrate trapped in the pocket is attacked by the coordinated hydroxide ion, which has been activated by the metal. (b) The coordinated hydroxide ion activates a solvent molecule to attack the substrate positioned in the pocket. (c) The substrate is replaced by a solvent molecule and coordinated to the metal ion. Subsequently, the solvent molecule activated by the base located in the pocket attacks the carbon atom of the nitrile group. In all cases, the metal ion functions as a Lewis acid. To understand the catalytic mechanism, various model complexes mimicking the reaction center of cobalt- as well as iron-type NHases have been synthesized and characterized (for detailed reviews, Mascharak 2002; Kovacs 2004; Yano et al. 2008). Marscharak and his colleagues synthesized three Fe^{3+}-type complexes with similar mixed nitrogen and sulfur ligand fields: $[Fe^{III}(PyPepS)_2]^-$ (**1**), $[Fe^{III}(PyPepSO_2)_2]^-$ (**2**), and $[Fe^{III}(PyAS)]^+$ (**3**) (Fig. 2). They studied the effect of the oxygenation of sulfur ligands as well as the amide ligands by cyclic voltammetry. By comparing the redox potentials of the Fe ion in **1** and **2**, they concluded that the electron donation of a sulfur ligand was weakened by its oxygenation. In contrast, the redox half potential ($E_{1/2}$) corresponding to Fe^{3+}/Fe^{2+} of **3** is about 1.0 V larger than that of **1**, indicating that the amide coordination provided a strong ligand field at the Fe^{3+} center. The stabilization of the Co^{3+} oxidation state by amide ligands was also suggested by the Co^{3+} complex mimicking the equatorial plane of NHase, $[Co(N_2S_2)]^-$. Chottard and his colleagues showed that a bis(sulfenato-S)Co^{3+} complex (**4**) catalyzed the hydration of acetonitrile to acetamide but that the corresponding bis(sulfinato-S)Co^{3+} complex (**5**) did not (Fig. 3). Kovacs and her coworkers have studied the Co-type NHase model complexes in the low-spin trivalent state and the ligand exchange reaction. They concluded that the trans-thiolate sulfur played an important role in promoting the ligand exchange at the sixth site. Thus, in NHase, the Lewis acidity of the metal ion is likely to be fine-tuned by the unique ligand environment including two Cys oxidations, two amide coordinations, and trans-thiolate sulfur. Very recently, Kovacs and her coworkers succeeded in obtaining Co-containing complexes mimicking the NHase amide-bound (**6**) and iminol-bound (**7**) intermediates (Fig. 4) by performing the hydration of acetonitrile-coordinated Co complexes (Swartz et al. 2011), which strongly suggest the reaction scheme (c).

Odaka and coworkers have studied the roles of two oxidized cysteine ligands using a Fe-type enzyme, ReNHase (Endo et al. 2001; Hashimoto et al. 2008). ReNHase reconstituted anaerobically from unmodified subunits with a Fe^{2+} ion showed no catalytic activity but acquired it when the corresponding cysteine ligands were oxidized under air. They found 2-cyano-2-propyl hydroperoxide specifically oxidized αCys114-SO(H) to Cys-SO$_2$H, to inactivate completely the enzyme. These findings indicate that the oxidized Cys ligands, especially αCys114-SO(H), are essential for the catalytic activity. Recently, ReNHase was shown to hydrolyze aliphatic small isonitrile into the corresponding amine and carbon monoxide (CO) (Hashimoto et al. 2008). The k_{cat} for the hydrolysis of *tert*-butylisonitrile (tBuNC) was 10^5

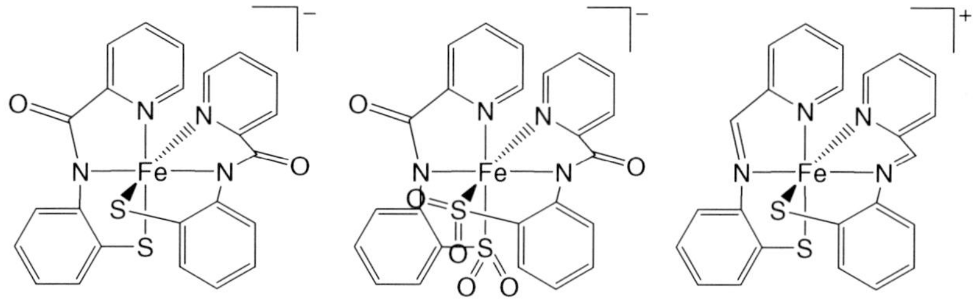

Nitrile Hydratase and Related Enzyme, Fig. 2 Structures of [Fe^{III}(PyPepS)$_2$]$^-$ (**1**), [Fe^{III}(PyPepSO$_2$)$_2$]$^-$ (**2**), and [Fe^{III}(PyAS)]$^+$ (**3**)

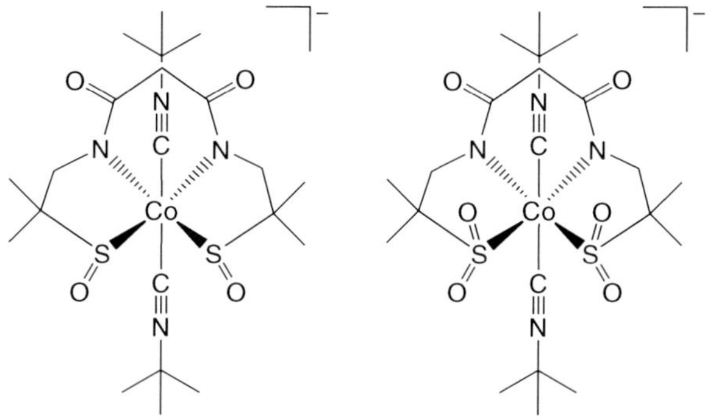

Nitrile Hydratase and Related Enzyme, Fig. 3 Structure of [Co^{III}(L-N$_2$SOSO)(tBuNC)$_2$]$^-$ (**4**) and [Co^{III}(L-N$_2$SO$_2$SO$_2$)(tBuNC)$_2$]$^-$ (**5**)

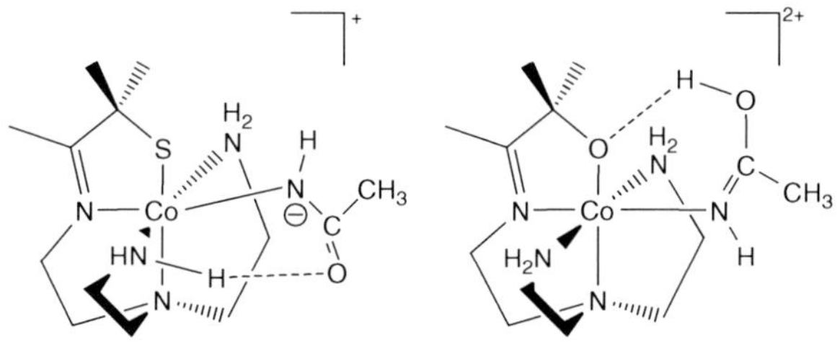

Nitrile Hydratase and Related Enzyme, Fig. 4 Structure of [Co^{III}(S^{Me2}N$_4$(tren))(NHC(O)CH$_3$)]$^+$ (**6**) and [Co^{III}(O^{Me2}N$_4$(tren))(NHC(OH)CH$_3$)]$^{2+}$ (**7**)

times lower than that for the hydration of methacrylonitrile, indicating that tBuNC was likely to be a substrate suitable for time-resolved studies. Taking advantage of the photoreactivity of ReNHase, Hashimoto et al. performed time-resolved X-ray crystallography of the catalytic reaction of ReNHase using *tert*-butylisonitrile (tBuNC). Crystals of the nitrosylated inactive ReNHase were soaked in a solution containing tBuNC in the dark. The catalytic reaction was initiated by the photo-induced denitrosylation in crystals, and the reaction was terminated by flush cooling with N$_2$ gas at 95 K. Figure 5 shows snapshots of the structure around the catalytic center before tBuNC soaking (a) and at 0-, 18-, 120-, and 440-min illumination (b–e, respectively). Before illumination (Fig. 5b), an NO molecule is bound to the Fe ion and a tBuNC molecule is located in the substrate-binding pocket with its isonitrile group facing opposite to the NO molecule. Probably, the *tert*-butyl group is too bulky to be located at another position in the nitrosylated state. After illumination, the electron density of NO and tBuNC are less clear, suggesting the movement of both molecules (Fig. 5c). Then, the electron density of the NO molecule disappeared and the isonitrile group of tBuNC bound to the sixth coordination site (Fig. 5d). Finally, the shape of the electron density of tBuNC is significantly altered (Fig. 5e). The electron density of the *tert*-butyl group is about 1.0 Å away from the Fe ion, and an extra electron density is observed near the carbon atom of the isonitrile group and the oxygen atom of the sulfenate group of αCys114-SO(H). Because one of the products is *tert*-butylamine (tBuNH$_2$), the structure at 440-min reaction was recalculated based on the model containing a tBuNH$_2$ molecule in the reaction cavity (Fig. 5f). tBuNH$_2$ was well fitted to the 2Fo-Fc electron density (gray mesh), and two corresponding positive electron densities (green mesh) were observed near the Fe ion and the sulfenate oxygen of αCys114-SO(H) in the Fo-Fc electron density map. These electron densities could be assigned to a carbon atom of the isonitrile of tBuNC and a solvent water molecule (H$_2$Oa), which might form another product, a CO molecule. Based upon these results, Hashimoto et al. proposed a catalytic mechanism in which the substrate coordinated to the metal ion was attacked nucleophilically by a water molecule activated by the oxygen atom of the

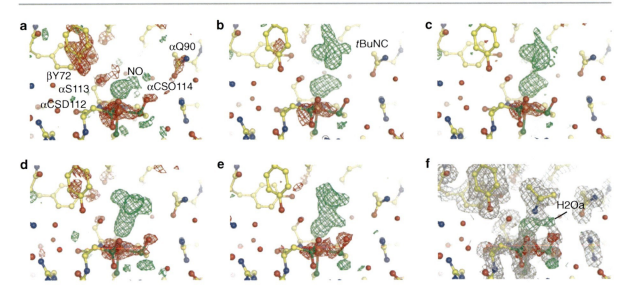

Nitrile Hydratase and Related Enzyme, Fig. 5 Time-resolved X-ray crystallographic study of the hydrolysis of *tert*-butylisonitrile (tBuNC) catalyzed by Fe-type NHase. (**a**) Structure around the Fe center of the nitrosylated inactive ReNHase without tBuNC and (**b–e**) with tBuNC after illumination for 18 (**c**), 120 (**d**), and 440 min (**e**). (**f**) Fo-Fc electron density (3.0 σ contour, *green*, and −3.0 σ contour, *red*) and 2Fo-Fc electron density (1.0 σ contour, *gray*) superimposed on the refined structure at 440 min. tBuNH$_2$ was included in the calculation. αCSD112 and αCSO114 indicate αCys112-SO$_2^-$ and αCys114-SO(H), respectively. *Yellow, blue, red, green*, and *brown spheres* represent carbon, nitrogen, oxygen, sulfur, and iron atoms, respectively

sulfenyl group of αCys114-SO(H) (Fig. 6a). Theoretical calculation studies on the first shell of ReNHase revealed the possibility that the sulfenyl oxygen could act as a catalytic base when the substrate was coordinated to the metal ion. This reaction mechanism is also supported by the recent intermediate structure demonstrated by Kovacs and coworkers (Fig. 4: **6** and **7**) (Swartz et al. 2011). In the present mechanism, αCys114 should be deprotonated as Cys-SO$^-$, because the sulfenate group activates a water molecule to attack the substrate. Further studies including neutron crystallography to clarify the protonation state of the αCys-SO(H) ligand will be required.

Several amino acid residues important for the catalysis have been revealed by site-directed mutagenesis and kinetic studies. Two conserved Arg residues, βArg56 and βArg141, in ReNHase, which were hydrogen-bonded to two oxidized Cys residues, were essential, because the activity decreased to 0.2% of the wild-type level when one of them was replaced by Lys (Endo et al. 2001). Miyanaga et al. (2004) replaced βTyr68 of the Co-type NHase from *Pseudonocardia thermophila* JCM 3095 (PtNHase) with Phe, because the residue was hydrogen-bonded to one of the ligand residues, αSer112. The resulting mutant, βY68F, exhibited a 10 times higher K_m and a 100 times smaller k_{cat}, showing βTyr68 is responsible for the catalysis. Holz and his coworkers have studied the pH and temperature dependence of the kinetic parameters of PtNHase very carefully (Mitra and Holz 2007). Based on the results and the crystal structure, they proposed the novel reaction mechanism in which βTyr68 forms a catalytic triad with the Ser ligand, αSer112, and an adjacent residue, βTrp72, and functions as a catalytic base (Fig. 6b). However, this mechanism should be revised, because the αS113A mutant ReNHase, from which the hydroxyl group of αSer113 (corresponding to αSer112 of PtNHase) was removed, exhibited about 23% of the k_{cat}/K_m value of the wild type (Yamanaka et al. 2010). Recently, Brodkin et al. studied the participation of amino acid residues involved in the second and third shells around the metallocenter of Co-type NHase from *Pseudomonas putida* NRRL-18668 (PpNHase) (Brodkin et al. 2011). Among the residues selected, four conserved ones, αAsp164, βGlu56, βHis147, and βHis71, have significant effects on the catalytic efficiency. Interestingly, βGlu56 and βHis147 form a hydrogen bond network with βArg149 and αCys115-SO$_2$H (corresponding to βArg141 and αCys112-SO$_2$H of ReNHase, respectively),

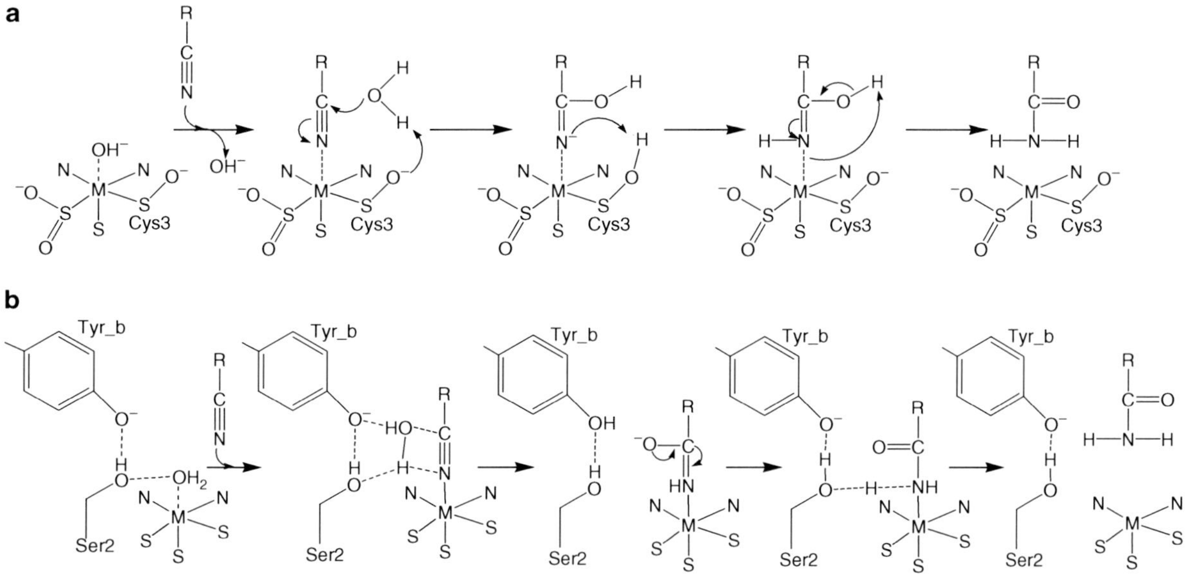

Nitrile Hydratase and Related Enzyme, Fig. 6 Proposed catalytic mechanisms of NHase. (**a**) A water molecule activated by the sulfenate group of the Cys-SO(H) ligand nucleophilically attacks the nitrile carbon (Hashimoto et al. 2008). (**b**) The Ser ligand ionizes the phenolate oxygen of the strictly conserved Tyr (Tyr_b), which activates a water molecule (Mitra and Holz 2007)

and αAsp164 interacts with αCys112 (non-oxidized Cys ligand). These findings are consistent with the reaction mechanism in Fig. 6a.

A Novel NHase Family Enzyme, Thiocyanate Hydrolase

Thiocyanate (SCN^-) is one of the major pollutants in the wastewater from a coal gasification factory. Thiocyanate hydrolase (SCNase) is isolated from *Thiobacillus thioparus* THI 115, a chemolithoautotrophic eubacterium metabolizing SCN^- to sulfate as the sole energy source. SCNase is the first enzyme exhibiting microbial thiocyanate degradation and catalyzes the hydrolysis of SCN^- to produce carbonyl sulfide and ammonia ($SCN^- + 2H_2O \rightarrow COS + NH_3 + OH^-$). SCNase is composed of three kinds of subunits, α, β, and γ, that form a $(\alpha\beta\gamma)_4$ hetero-dodecameric structure. SCNase shows high amino acid sequence similarities with NHase. The SCNase γ subunit corresponds to the NHase α subunit, while the SCNase β and α subunits do to the N-terminal and C-terminal halves of the NHase β subunit. UV-vis absorption and ESR spectra indicated that SCNase has a low-spin Co^{3+} center in a similar coordination sphere to in the case of NHase (Katayama et al. 2006). The crystal structure of SCNase has revealed that four αβγ heterotrimers interact at the N-terminal extension of the β subunit to form the $(\alpha\beta\gamma)_4$ structure and that the structure of the αβγ heterotrimer is very similar to that of the NHase αβ heterodimer including the Co^{3+} center (Arakawa et al. 2007). All protein ligands are conserved in NHase including two oxidized Cys residues, $\gamma Cys131-SO_2^-$ and γCys133-SO(H), respectively. These results indicate that SCNase belongs to the same protein family with NHase. Also, SCNase is the second protein that has two oxidized Cys ligands in its Co^{3+} active center. SCNase with $\gamma Cys131-SO_2^-$ and unmodified γCys133 was activated during storage at $-80\ °C$ concomitant with the oxidation of γCys133 to Cys-SO(H). In contrast, SCNase lost its catalytic activity after further oxidation of γCys133 to $Cys-SO_2^-$. These results indicate that, as in the case of NHase, the oxidation state of γCys133-SO(H) is essential for the catalytic activity of SCNase (Arakawa et al. 2009).

The substrate-binding pocket of SCNase is small and highly hydrophilic. Its surface is positively charged, because seven Arg residues surround the Co^{3+} center. On the other hand, the pocket of NHase consists of a much larger number of residues, and thus, the pocket is larger than that of SCNase. In addition, there is an aromatic cluster at the top of the pocket that makes it rather hydrophobic. One can assume that the small and positively charged pocket of SCNase is suitable for the small and anionic substrate, SCN^-, while the broader and hydrophobic pocket of NHase is preferable for nitriles. Thus, the difference in the substrate pocket rationally explains the differences in the substrate preference between SCNase and NHase (Arakawa et al. 2007).

Unique Maturation System of NHase

Both Co- and Fe-type NHases require an activator protein encoded downstream of their structural genes for their functional expression. Without coexpression of the activator protein, a Co-type NHase is expressed as the apoprotein, while a Fe-type one is as inclusion bodies. Interestingly, the Co-type NHase activator is completely different from the Fe-type one, suggesting that each activator protein assists the maturation of the corresponding NHase in a different manner. The Fe-type NHase activators are a 44–47 kDa protein and have the metal-binding motif $C_{73}XCC_{76}$ and the putative GTP-binding motif, Walker A motif $G_{13}XXXXGKT_{20}$, and the base recognition motif $S_{179}KXD_{182}$ (amino acid numbers are based on the NHase activator of *Rhodococcus erythropolis* N771). The Fe-type NHase activators may incorporate a Fe ion and deliver it to NHase in a GTP-dependent manner. The replacement of one of three Cys residues in the CXCC motif results in the loss of the functional expression of Fe-type NHase. On the other hand, the Co-type NHase activator is very unique, because it has no known metal-binding motif and is rather small, its molecular mass being 10–17 kDa. In addition, it exhibits indisputable amino acid sequence similarity to no protein reported except the NHase β subunit.

Recently, Kobayashi and his colleagues discovered the activator-dependent maturation mechanism of Co-type NHase (Fig. 7) (Zhou et al. 2008; 2009; 2010). They found the activator protein of J1 L-NHase (termed e) formed a complex with an α subunit, $αe_2$, this stoichiometry being confirmed by size exclusion chromatography and sedimentation equilibrium. The $αe_2$ complex incorporates one Co^{2+} ion, and the Co-bound $αe_2$ complex (holo-$αe_2$) possesses the $Cys-SO_2^-$ modification. Although the Cys-SO(H)

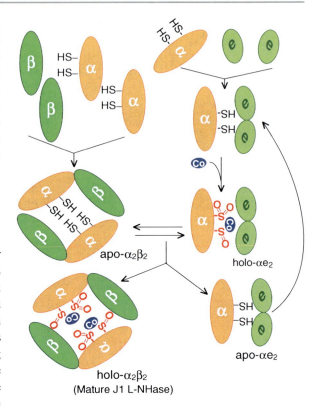

Nitrile Hydratase and Related Enzyme, Fig. 7 Self-subunit swapping maturation of J1 L-NHase. The α and β subunits of J1 L-NHase, its activator (e), and the Co ion are shown in *orange, green, light green*, and *blue*, respectively. Two oxidized Cys ligands are shown in *red*

modification was not confirmed because of its chemical instability during preparation, the posttranslational oxidation of the Cys ligands was shown to occur before the assembly of the α and β subunits. When holo-$αe_2$ (+Co, +Cys oxidation(s)) is incubated with the apo form of J1 L-NHase (apo-$α_2β_2$) (−Co, −Cys oxidation), the exchange of the α subunits between holo-$αe_2$ and apo-$α_2β_2$ occurs to generate apo-$αe_2$ (−Co, −Cys oxidation) and holo-$α_2β_2$ (+Co, +Cys oxidation), respectively (Fig. 7). This mechanism is unique in the following points. (a) The activator protein does not deliver a metal ion to the apoprotein but enables the α subunit to incorporate Co ion. (b) Not the metal cofactor but the Co-containing α subunit with Cys modification(s) is transported to the apoenzyme. Kobayashi and his coworkers named this unprecedented maturation mechanism "self-subunit swapping" (Zhou et al. 2008). The mechanism has been confirmed for J1 H-NHase too (Zhou et al. 2010). However, α subunit swapping did not occur between the apoprotein of J1 L-NHase and the α-activator complex of J1 H-NHase, and vice versa, revealing that self-subunit swapping is an enzyme-specific reaction (Zhou et al. 2010).

Acetonitrile hydratase (ANHase) from *Rhodococcus jostii* RHA1 is a novel type of NHase (Okamoto et al. 2010). ANHase consists of α and β subunits like other NHases, but both subunits are much larger than those of other NHases and exhibit no significant sequence similarity with any proteins in the databases. In addition to one Co ion, ANHase contains two Cu and one Zn ion per αβ heterodimer. Co seems to be catalytically essential, because ANHase expressed was inactive in the Co-depleted growth media but not in either Cu- or Zn-depleted media. In ANHase, an activator protein, AnhE, was encoded between the structural genes of the α and β subunits and has been shown to assist the delivery of a Co ion to its α subunit during the maturation. The purified AnhE exists in a monomer-dimer equilibrium, and the dimer form is stabilized by divalent metal ions including Co^{2+}, Cu^{2+}, Zn^{2+}, and Ni^{2+}. This combination with the results of isothermal titration calorimetry and UV-vis absorption spectroscopic studies suggests AnhE provides a Co ion to ANHase. Although the detailed mechanism has not been elucidated, AnhE might be a new type of Co-type NHase activator.

Due to the page limit, all of the related papers cannot be referenced. Please check on PubMed using the corresponding key words.

Cross-References

▶ Cobalt Proteins, Overview

References

Arakawa T, Kawano Y, Kataoka S et al (2007) Structure of thiocyanate hydrolase: A new nitrile hydratase family protein with a novel five-coordinate cobalt(III) center. J Mol Biol 366:1497–1509

Arakawa T, Kawano Y, Katayama Y, Nakayama H, Dohmae N, Yohda M, Odaka M (2009) Structural basis for catalytic activation of thiocyanate hydrolase involving metal-ligated cysteine modification. J Am Chem Soc 131:14838–14843

Brodkin HR, Novak WRP, Milne AC, D'Aquino JA, Karabacak NM, Goldberg IG, Agar JN, Payne MS, Petsko GA, Ondrechen MJ, Ringe D (2011) Evidence of the participation of remote residues in the catalytic activity of Co-type nitrile hydratase from *Pseudomonas putida*. Biochemistry 50:4923–4935

Endo I, Nojiri M, Tsujimura M et al (2001) Fe-type nitrile hydratase. J Inorg Biochem 83:247–253

Hashimoto K, Suzuki H, Taniguchi K et al (2008) Catalytic mechanism of nitrile hydratase proposed by time-resolved X-ray crystallography using a novel substrate, tert-butylisonitrile. J Biol Chem 283:36617–36623

Katayama Y, Hashimoto K, Nakayama H et al (2006) Thiocyanate hydrolase is a cobalt-containing metalloenzyme with a cysteine-sulfinic acid ligand. J Am Chem Soc 128:728–729

Kobayashi M, Shimizu S (1998) Metalloenzyme nitrile hydratase: Structure, regulation and application to Biotechnology. Nat Biotechnol 16:733–736

Kobayashi M, Nagasawa T, Yamada H (1992) Enzymatic synthesis of acrylamide: a success story not yet over. Trends Biotechnol 10:402–408

Kovacs JA (2004) Synthetic analogues of cysteinate-ligated non-heme iron and non-corrinoid cobalt enzymes. Chem Rev 104:825–848

Mascharak PK (2002) Structural and functional model of nitrile hydratase. Coord Chem Rev 225:201–214

Mitra S, Holz RC (2007) Unraveling the catalytic mechanism of nitrile hydratases. J Biol Chem 282:7397–7404

Miyanaga A, Fushinobu S, Ito K et al (2001) Crystal structure of cobalt-containing nitrile hydratase. Biochem Biophys Res Commun 288:1169–1174

Miyanaga A, Fushinobu S, Ito K, Shoun H, Wakagi T (2004) Mutational and structural analysis of cobalt-containing nitrile hydratase on substrate and metal binding. Eur J Biochem 271:429–438

Okamoto S, Petegem FV, Patrauchan MA, Eltis LD (2010) AnhE, a metallochaperone involved in the maturation of a cobalt-dependent nitrile hydratase. J Biol Chem 285:25126–25133

Swartz RD, Coggins MK, Kaminsky W, Kovacs JA (2011) Nitrile hydration by thiolate- and alkoxide-ligated Co-NHase analogues. Isolation of Co(III)-amidate and Co(III)-iminol intermediates. J Am Chem Soc 133:3954–3963

Yamanaka Y, Hashimoto K, Ohtaki A et al (2010) Kinetic and structural studies on roles of the serine ligand and a strictly conserved tyrosine residue in nitrile hydratase. J Biol Inorg Chem 15:655–665

Yano T, Ozawa T, Masuda H (2008) Structural and functional model systems for analysis of the active center of nitrile hydratase. Chem Lett 37:672–677

Zhou Z, Hashimoto Y, Shiraki K, Kobayashi M (2008) Discovery of posttranslational maturation by self-subunit swapping. Proc Natl Acad Sci, USA 105:14849–14854

Zhou Z, Hashimoto Y, Kobayashi M (2009) Self-subunit swapping chaperone needed for the maturation of multimeric metalloenzyme nitrile hydratase by a subunit exchange mechanism also carries out the oxidation of the metal ligand cysteine residues and insertion of cobalt. J Biol Chem 284:14930–14938

Zhou Z, Hashimoto Y, Cui T, Washizawa Y, Mino H, Kobayashi M (2010) Unique biogenesis of high-molecular-mass multimeric metalloenzyme nitrile hydratase: intermediates, and a proposed mechanism for self-subunit swapping maturation. Biochemistry 49:9638–9648

Nitrite Reductase

Pedro Tavares and Alice S. Pereira
Departamento de Química, Faculdade de
Ciências e Tecnologia, Requimte, Centro de Química
Fina e Biotecnologia, Universidade Nova de Lisboa,
Caparica, Portugal

Synonyms

Copper nitrite reductase; Copper-containing nitrite reductase; Cu-containing respiratory nitrite reductase

Definition

Copper-containing nitrite reductase (CuNiR, EC 1.7.2.1, formerly EC 1.7.99.3) is an enzyme that catalyzes the one-electron reduction of nitrite (NO_2^-) to nitric oxide (NO) and water. This enzyme is found in the periplasm of some bacteria, as well as in fungus, and is involved in the dissimilatory pathway of nitrogen metabolism. This enzyme contains multiple copper centers and should not be confused with other nitrite reductases, in particular the heme-containing cytochrome cd_1 that is also found in denitrifiers (and with which shares the same ENZYME entry).

Biological Function and Occurrence

The dissimilatory pathway of nitrogen metabolism comprises four steps (Zumft 1997): Nitrate reductase catalyzes the reduction of nitrate to nitrite and water. In a second step, nitrite is reduced to nitric oxide and water by a nitrite reductase. The third step is accomplished by nitric oxide reductase that will catalyze the reduction of nitric oxide to nitrous oxide and water. Finally, nitrous oxide reductase will catalyze the reduction of nitrous oxide to dinitrogen. Two types of nitrite reductases can catalyze the second step. Although the most studied denitrifying bacteria contain heme nitrite reductases (cytochrome cd_1), copper-containing nitrite reductases are present in a greater number of genera (both Gram-negative and Gram-positive eubacteria as well as Archae) (Rinaldo and Cutruzzolà 2007). There is no rational reason for the existence of the two different types of enzymes and, in bacteria, the copper-containing and heme enzymes were never found to coexist. Under conditions favoring high nitric oxide concentrations, CuNiR might also function as a nitric oxide reductase.

Structure and Cofactors

CuNiR are the product of *nirK* gene (depending on the organisms the designation can also be *nirU* or *aniA*) that contains a leader sequence (corresponding to the first 18–38 amino acids), which is in agreement with the periplasmatic enzyme localization. It is possible to find 11 CuNiR protein sequences in UniProtKB/Swiss-Prot (http://www.uniprot.org/) that are manually annotated. The mature gene product has a molecular mass between 34 and 38 kDa, corresponding to 330–370 amino acids length. The primary structure has 60–80% positional identity between different species. With a single documented exception (found in *Hyphomicrobium denitrificans* A3151, see below) CuNiRs have a trimeric quaternary structure, as first shown for the nitrite reductase of *Alcaligenes cycloclastes* IAM 1013 (Suzuki et al. 1999). Since then approximately 100 3D structures were resolved for different experimental conditions (e.g., in variable pH, substoichiometric copper loading, nitrite soaked and a complex with a redox partner). The trimer is composed by three equal subunits packed tightly together in a way that approximately 30% of the molecular surface per monomer is buried, forming a central channel of 5–6 Å (Fig. 1).

Each monomer can be divided into two plastocyanin-like domains with Greek-key ß barrel topology (Fig. 2). In the monomer N-terminal domain, one copper ion is coordinated by two histidine and one cysteine residues in a trigonal planar structure, and by one methionine residue in an axial position. Such center is classified as a class I, type-1 copper center, similar to the ones found in plastocyanin or pseudoazurin (and in other cupredoxins). This copper center is responsible for mediating electron transfer between the physiological redox partners and the active site.

The protein active site is located in the monomer interface, at a distance of 12.5–13 Å from the type-1 copper center. The active site is a type-2 copper center, with the copper ion being coordinated by two histidine residues from one monomer and another from an adjacent monomer, as well as a water molecule (Fig. 3).

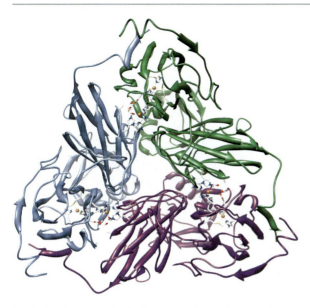

Nitrite Reductase, Fig. 1 Representation of the trimeric quaternary structure of *Alcaligenes faecalis* CuNiR (oxidized nitrite soaked, PDB ID: 1AS6). The three equivalent monomers are colored *blue*, *green*, and *purple* (Figure prepared with UCSF Chimera http://www.cgl.ucsf.edu/chimera/)

Nitrite Reductase, Fig. 2 Structure representation of an *Alcaligenes faecalis* CuNiR monomer (oxidized nitrite soaked, PDB ID: 1AS6). The two plastocyanin-like domains are colored *blue* and *pink* (Figure prepared with UCSF Chimera http://www.cgl.ucsf.edu/chimera/)

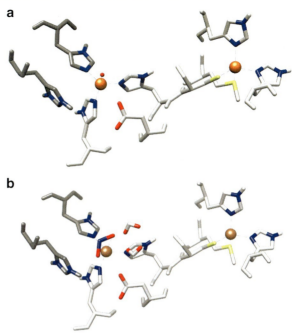

Nitrite Reductase, Fig. 3 Detailed view of copper center coordination. Structures from *Alcaligenes faecalis* CuNiR oxidized (**a**, PDB ID: 1AS7) and oxidized nitrite soaked (**b**, PDB code: 1AS6) were used. Copper atoms are colored *orange* and amino acid residues from different monomers are colored *light* and *dark gray*. Ligands of type-1 copper center are His95, Cys136, His145, and Met150, and of type-2 copper center are His100, His135, and His306. The copper centers are at a 12.6 Å distance. Non-ligand residues Asp98 and His255 that are involved in a H-bonded network in the vicinity of the active site are also depicted (Figure prepared with UCSF Chimera http://www.cgl.ucsf.edu/chimera/)

One of these histidines is contiguous to the cysteine residue that acts as ligand to the type-1 copper center, thus providing a bridging mode between the two copper centers (similar to what was observed in other multicopper oxidases such as laccase, ascorbate oxidase, or ceruloplasmin).

Two conserved residues, an aspartate and a histidine, are positioned close to the center and are proposed to be involved in the catalytic mechanism. The aspartate residue binds to the copper-bound water as well as to second water molecule that bridges to the non-ligand histidine residue. In accordance with the fact that the type-2 copper site is the active site, in crystallographic structures obtained by soaking crystals in nitrite, it was observed that the nitrite molecule will replace the copper-bound water. In this nitrite-bound form, an oxygen of the nitrite molecule, instead of water, is bridging between the non-ligand aspartate and histidine residues. These and other hydrogen bonding interactions maintain the integrity and orientation of type-2 copper ligands that are fundamental for enzyme activity (Murphy et al. 1997).

Nitrite Reductase, Fig. 4 Representation of the hexameric quaternary structure of *Hyphomicrobium denitrificans* CuNiR (PDB ID: 2DV6). For each trimeric unity, monomers are colored *blue*, *green*, and *purple*. The overall structure can be described as a trigonal prism-shaped hexamer: (**a**) top view; (**b**) side view (Figure prepared with UCSF Chimera http://www.cgl.ucsf.edu/chimera/)

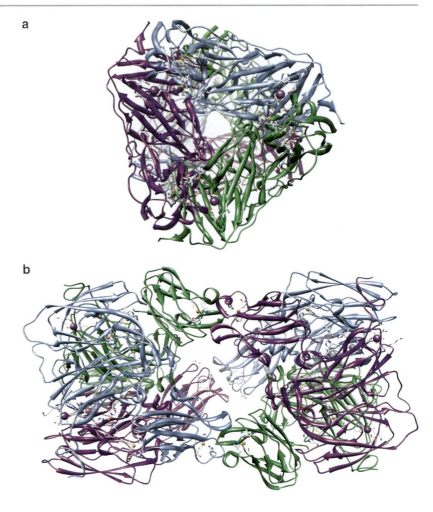

The type-2 copper center is located at the bottom of a 12–13 Å deep cavity, which has one face that is hydrophobic in nature and that is thought to be the way out for the product of the reduction reaction, which is nitric oxide. The stoichiometry of type-2 copper sites is preparation dependent, which initially caused some confusion in the determination of copper ions per enzyme ratio. Nevertheless, preparations with very small amounts of type-2 copper centers (less than 0.5 copper ions per enzyme) were useful to understand some structural features. For example, it is known that in the absence of this center the enzyme retains the trimeric structure with small structural changes. Not surprisingly, the trimer is stabilized by hydrogen bonding, and to a less extent via water molecules and salt bridges.

So far only one exception to the above described structure was found. The CuNiR isolated from *Hyphomicrobium denitrificans* A3151 (a methylotrophic denitrifying bacterium) shows a trigonal prism-shaped hexameric molecule (Nojiri et al. 2007). Two tightly associated trimers, giving rise to one threefold and three twofold axes, form the homohexamer (Fig. 4). Each monomer is constituted by 447 amino acids organized in two regions: N-terminal (residues 24–131) and C-terminal (residues 163–445). Between the two regions exists a 30 amino acid length linker sequence (residues 132–162). The monomer can also be defined by the existence of three Greek-key β barrel domains. Domain I (located in the N-terminal region) contains one type-1 copper center with a distorted trigonal bipyramidal geometry resembling the copper site of azurins. The copper ion is coordinated by two histidines and one cysteine residues positioned in a trigonal plane, with one methionine and one glutamate in axial positions. The monomer contains two more copper ions. Domain II harbors another type-1 copper center, in this case a class I center coordinated by two histidine, one cysteine, and one methionine residues. This second type-1 copper center is at a distance of 24.1 Å from the other type-1 copper center located in domain I. A third copper ion (the putative active site) is coordinated by three histidine residues, two from domain II and the third one from domain III of different monomers, much like type-2 copper centers of other characterized CuNiRs. The coordination is completed by one solvent molecule. The type-2 center is 12.6 Å away from the closest type-1 copper center. Unsurprisingly, hydrogen bonding plays an important role in the active site, with the involvement of a histidine and aspartate residues. The unique hexameric structure observed is formed by interactions between: (1) N-terminal region (domain I) of one monomer and C-terminal region (domains II and III) of an adjacent monomer; and (2) N-terminal regions of different monomer, much alike in other cupredoxin dimers. The type-1 copper in this region is instrumental for dimerization of the trimer.

Biochemical Properties

On searching the early literature of CuNiRs an apparent heterogeneity in molecular masses, catalytic activities, and spectroscopic properties could be found (see below). However, it was readily established that copper was required for activity (Zumft 1997; Rinaldo and Cutruzzolà 2007). The purification procedure can be somewhat complicated (including buffer extraction of acetone dried cells, ammonium sulfate precipitation, and several chromatographic steps) which probably justifies the reported range of 1.56–4.6 copper ions per enzyme. The specific activity value of nitrite reduction also varies from 1 to 380 $\mu mol.min^{-1}.mg^{-1}$ and the determined K_m values range from 30 to 740 μM. The optimum pH value is usually in the range of 6.5–7.4. CuNiR can accept electrons from different redox partners such as azurin, pseudoazurin, or cytochromes c, c_2, c_{552} and c_{553}. Small molecules can be used as an electron source in enzymatic assays. In this case, the use of sodium dithionite/methyl viologen, dithionite/benzyl viologen, and ascorbate/phenazine methosulfate were reported. From the reported inhibiters which include diatomic molecules (cyanide, carbon monoxide), diethyl dithiocarbamate has been extensively used since it has the capacity of inhibiting CuNiR activity but not the heme-containing cytochrome cd_1 nitrite reductases.

Several intracellular recombinant expression systems were reported for *A. cycloclastes*, *A. xylosoxidans*, *R. sphaeroides* and *A. faecalis* S-6. For the latter, yields of 35–45 mg/L of culture were obtained. In some cases His-tag systems were used to facilitate purification. Also, it was possible to develop expression systems including leader sequences designed to deliver the gene product into the periplasm of *E. coli* (as was the case of *A. xylosoxidans* and *A. faecalis* S-6).

Spectroscopic Properties

According to spectroscopic properties of type-1 copper centers, CuNiRs can be divided into two well-defined groups: blue and green enzymes.

Blue CuNiRs show an intense absorption band (ε between 2,000 and 5,000 M^{-1} cm^{-1}) centered at approximately 590 nm caused by an allowed ligand-to-metal, S(Cys) → Cu(II), charge-transfer transition. Due to the lack of sulfur ligation, no significant absorption lines in the visible region are observed for the type-2 copper center. An axial EPR signal is observed, with $g_\| > g_\perp > 2$ and small hyperfine splitting (3–7 mT). The type-2 copper center also contributes to the EPR spectrum, displaying an axial signal with larger hyperfine coupling constants (12–20 mT). Attribution of EPR signals is facilitated by comparison of native and type-2 copper-free CuNiR forms. It was also observed that resolved low magnetic field EPR lines attributed to type-2 copper site shift in position in the presence of anions such as nitrite, nitrate, or azide, as expected for an active site.

"Green CuNiRs" are green or bluish-green enzymes that in addition to an absorption band around 600 nm also show absorption near 460 nm. Resonance Raman studies show that excitation within either bands leads to similar spectrum, pointing to the likelihood for both electronic transition to have S(Cys) → Cu(II) charge-transfer character. The type-1 copper center EPR signal is known to be rhombic, with larger hyperfine constants. The possibility that these differences were due to a more tetragonal distortion in the "green" variant, when compared to the blue copper center present in plastocyanin was initially put forward. Structural data show that the bond length is similar between blue and "green" type-1 copper variants, but a difference in the orientation of the methionine ligand was detected. Replacement of methionine by threonine leads to an optical blue center (and a raise in redox potential of ca. 100 mV), but no change was observed in hystidine or cysteine hyperfine couplings or in g values and Cu nuclear hyperfine coupling, as judged by Q-Band ENDOR (Veselov et al. 1998). Using ENDOR in both native and type-2 copper depleted enzyme samples in conjunction with ^{15}N and ^{14}N nitrite it was possible to provide additional evidence that the substrate binds to type-2 copper via its oxygen atoms and by displacing one proton, probably on a water molecule bound to the copper ion. It was also possible to show, using infrared spectroscopy, that the vibrational mode of carbon monoxide is similar to that of CO bound to hemocyanin. Once more, the type-2 copper depleted enzyme was used to show that this binding is specific of the active site.

It should be noted that for the hexameric *Hyphomicrobium denitrificans* A3151 CuNiR a blue type-1 copper site is located in domain I and a "green" type-1 copper site is located in domain II.

Catalytic Mechanism

Nitrite reduction accomplished by this class of enzymes can be expressed by the following redox reaction:

$$NO_2^- + e^- + 2H^+ \rightarrow N_2O + H_2O$$

A mechanism for this monoelectronic nitrite reduction can be drawn from kinetic, spectroscopic, and crystallographic data obtained for native and mutant enzymes. Four mechanistic steps can explain the overall reaction (Fig. 5).

In the resting state of the enzyme water is bound to type-2 copper ion. This water ligand is hydrogen bonded to a carboxylate group of a nearby aspartate residue. This carboxylate group is hydrogen bonded to a second water molecule that is in turn hydrogen bonded to a histidine residue, giving rise to a H-bonding network. In the first catalytic step, the copper-bound water is displaced by the nitrite ion with the possible release of a hydroxide. Nitrite binds to the oxidized copper via its oxygen atoms, being also hydrogen bonded to the carboxylate group by one of these atoms. Nitrite induces an increase in the formal redox potential of type-2 copper ion, thus facilitating the intramolecular electron transfer from a type-1 copper center. The aspartate and histidine residues that establish the H-bonding network control the intramolecular electron transfer, and the reaction proceeds through the reduction of the type-2 copper ion and nitrite protonation. This enables the N-O bond cleavage and the formation of a copper nitrosyl intermediate. At this point we must consider the possible binding modes of the nitrosyl intermediate. One possibility is that, after bond cleavage, both nitric oxide and hydroxide formed are binding an oxidized copper ion with the sequential release of nitric oxide and the protonation of hydroxide (possibly by a locally ordered solvent molecule) to restore the resting state enzyme (Fig. 5, alternative A). However, the structure obtained for reduced CuNiR crystals in NO-saturated solution revealed an unprecedented side-on binding mode (Antonyuk et al. 2005; Tocheva et al. 2004). If this structure is representative of the nitrosyl intermediate, a more plausible description of this mechanistic step is that protonation and intramolecular electron transfer trigger a rearrangement of the bound nitrite leading to bond cleavage and water release with the formation of a nitrosyl intermediate in which the nitrogen and oxygen atoms are nearly equidistant from the type-2 copper ion (Fig. 5, alternative B). The resting state of the enzyme is obtained after nitric oxide being displaced by a water molecule and escaping by the hydrophobic side of the active site. Nevertheless, an end-on nitric oxide binding mode is also observed under limiting exposure with nitric oxide gas. Geometry-optimized DFT calculations indicate that both binding modes are best described by a Cu$^+$NO$^\bullet$ electronic configuration and that the end-on structure is more stable by 7 kcal/mol (Ghosh et al. 2009). It was also found that the end-on binding mode corresponds to a global energy minimum, while the side-on mode corresponds to a local minimum.

Also, the aspartate residue that is involved in the H-bonding network is playing a key role in determining the nitrosyl intermediate configuration, since it is most probable that in the side-on configuration the aspartate residue is hydrogen bonded to the oxygen atom of nitric oxide, perpendicularly to the plane Cu-N-O, while for the end-on case this residue assumes a parallel orientation. It is also observed that stronger inhibitors, for example azide, are able to establish a hydrogen bond to the same aspartate residue.

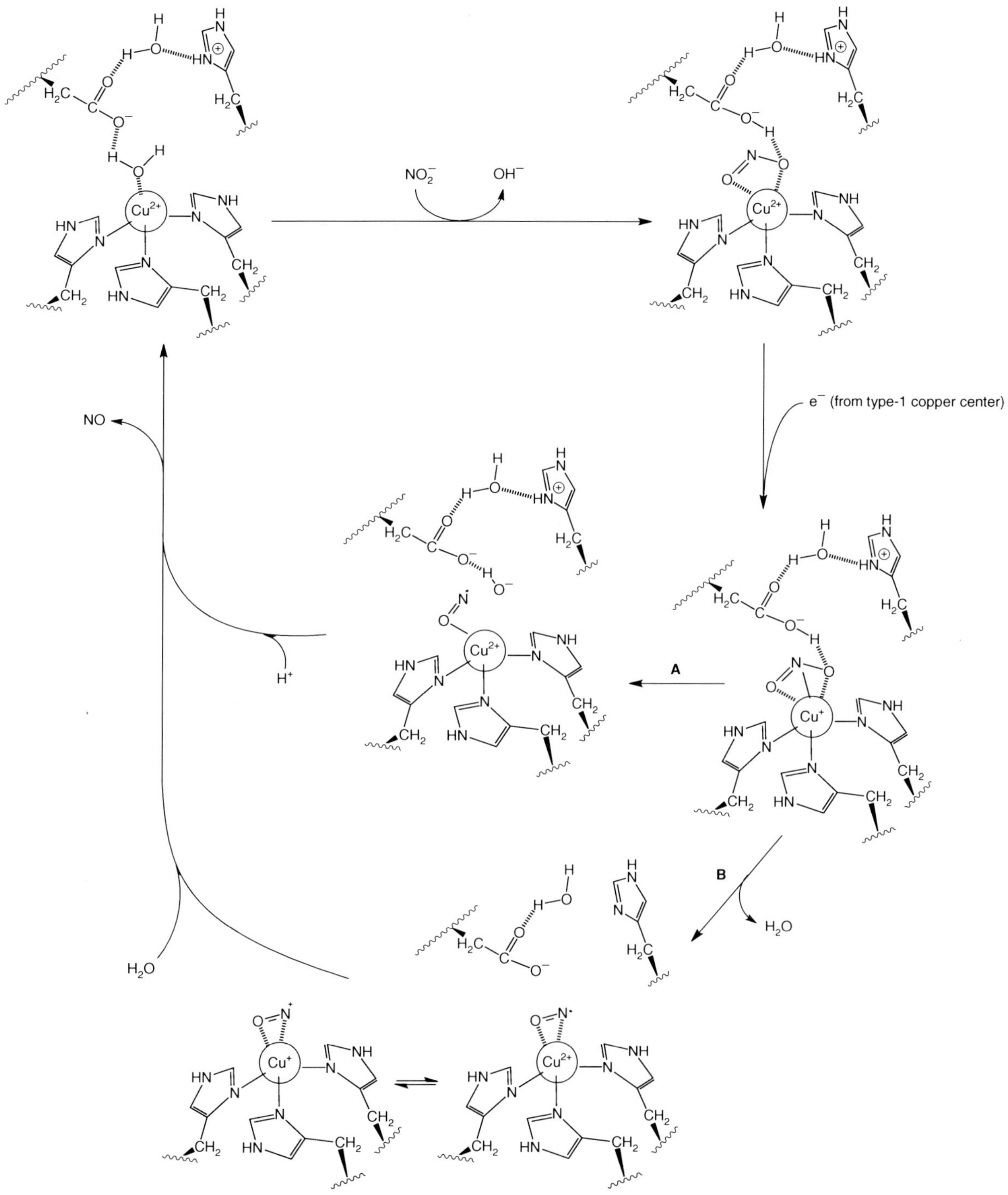

Nitrite Reductase, Fig. 5 Proposed mechanism for the monoelectronic nitrite reduction by CuNiRs. (**a**) and (**b**) are alternative mechanistic steps depending on the formation of an end-on (**a**) or side-on (**b**) nitrosyl intermediates

As described above, type-2 copper center reduction plays an important role in the reaction mechanism. Recently, a high-resolution crystal structure of the electron transfer complex between a CuNiR, from *Alcaligenes xylosoxidans*, and its redox partner, cytochrome c_{551}, was obtained (Fig. 6) (Nojiri et al. 2009).

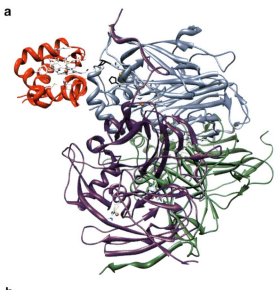

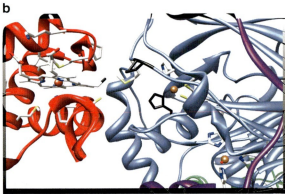

Nitrite Reductase, Fig. 6 (a) Structure representation of the electron transfer complex between *Alcaligenes xylosoxidans* CuNiR and cytochrome c_{551} (PDB ID: 2ZON). CuNiR equivalent monomers are colored *blue*, *green*, and *purple*, and cytochrome c_{551} is colored *red*. Heme CBC carbon atom, CuNiR Pro88 and His139 are colored *black* (postulated electron transfer pathway). (b) Detailed view of the docking interface and cofactors. The heme iron is at 10.5 Å of distance from the type-1 copper ion (Figure prepared with UCSF Chimera http://www.cgl.ucsf.edu/chimera/)

methyl carbon, and a proline and a histidine ligand from CuNiR. Dynamic of a complex formation between *A. faecalis* CuNiR and pseudoazurin was studied by nuclear magnetic resonance spectroscopy and isothermal titration calorimetry. The nature of the interaction depends on the redox state of pseudoazurin. In the reduced state, the binding of pseudoazurin to CuNiR exhibits a slow mode with high affinity as well as a weak fast mode, while in the oxidized state a single fast binding process is sufficient to explain the obtained data.

It was shown that the hydrophobic electron-transfer path results by desolvation in the docking region due to close contact between the electron donor protein and CuNiR. Contact between hydrophobic patches located near heme and type-1 copper center puts these cofactors only 10.5 Å apart, which can justify the observed second-order electron-transfer rate constant of $(4.8 \pm 0.2) \times 10^6$ M^{-1} s^{-1}. A proposed heme iron to type-1 copper pathway involves the exposed CBC

Cross-References

▶ Ascorbate Oxidase
▶ Copper-Binding Proteins
▶ Laccases
▶ Monocopper Blue Proteins
▶ Nitrous Oxide Reductase
▶ Plastocyanin

References

Antonyuk SV, Strange RW, Sawers G, Eady RR, Hasnain SS (2005) Atomic resolution structures of resting-state, substrate- and product-complexed Cu-nitrite reductase provide insight into catalytic mechanism. Proc Natl Acad Sci USA 102(34):12041–12046

Ghosh S, Dey A, Sun Y, Scholes CP, Solomon EI (2009) Spectroscopic and computational studies of nitrite reductase: proton induced electron transfer and backbonding contributions to reactivity. J Am Chem Soc 131(1):277–288

Murphy ME, Turley S, Adman ET (1997) Structure of nitrite bound to copper-containing nitrite reductase from Alcaligenes faecalis. Mechanistic implications. J Biochem 272(45):28455–28460

Nojiri M, Xie Y, Inoue T, Yamamoto T, Matsumura H, Kataoka K, Deligeer K et al (2007) Structure and function of a hexameric copper-containing nitrite reductase. Proc Natl Acad Sci USA 104(11):4315–4320

Nojiri M, Koteishi H, Nakagami T, Kobayashi K, Inoue T, Yamaguchi K, Suzuki S (2009) Structural basis of interprotein electron transfer for nitrite reduction in denitrification. Nature 462(7269):117–120

Rinaldo S, Cutruzzolà F (2007) Nitrite reductases in denitrification. In: Bothe H, Ferguson SJ, Newton WE (eds) Biology of the nitrogen cycle, 1st edn. Elsevier, Amsterdam

Suzuki S, Kataoka K, Yamaguchi K, Inoue T, Kai Y (1999) Structure-function relationships of copper-containing nitrite reductases. Coord Chem Rev 192:245–265

Tocheva E, Rosell F, Mauk A, Murphy M (2004) Side-on copper-nitrosyl coordination by nitrite reductase. Science 304(5672):867–870

Veselov A, Olesen K, Sienkiewicz A, Shapleigh J, Scholes C (1998) Electronic structural information from Q-band ENDOR on the type 1 and type 2 copper liganding environment in wild-type and mutant forms of copper-containing nitrite reductase. Biochemistry 37(17):6095–6105

Zumft W (1997) Cell biology and molecular basis of denitrification. Microbiol Mol Biol Rev 61(4):533–616

Nitrogen Fixation

▶ Iron-Sulfur Cluster Proteins, Nitrogenases

Nitrogen Oxide(s) Scavengers

▶ Heme Proteins, the Globins

Nitrous Oxide Reductase

Simone Dell'Acqua[1,2], Sofia R. Pauleta[1], Isabel Moura[1] and José J. G. Moura[1]
[1]REQUIMTE/CQFB, Departamento de Química, Faculdade de Ciências e Tecnologia, Universidade Nova de Lisboa, Caparica, Portugal
[2]Dipartimento di Chimica, Università di Pavia, Pavia, Italy

Synonyms

Copper enzyme; Copper-sulfur biological center; CuA center; CuZ center; Last step of the denitrification pathway; Nitrous oxide reduction

Definition of the Subject

The concentration of nitrous oxide, a potent greenhouse gas, has increased in the last decades. Nitrous oxide reductase is the last enzyme of the denitrification cycle and is the only known enzyme to catalyze in vivo the reduction of nitrous oxide to molecular nitrogen. This enzyme was first isolated in 1972, though most of its spectroscopic and kinetic properties were only determined in the late 80s and 90s, and up to now many of this aspects are still under study. This enzyme presents two copper centers, the electron transferring center, a mixed-valent binuclear CuA center, similar to the one found in cytochrome c oxidase, and CuZ center, the catalytic center. The X-ray structure of nitrous oxide reductase revealed that CuZ center is a unique copper center in biology, a tetranuclear μ_4-sulfide-bridged copper center.

Nitrous Oxide Reductase

The last step of the denitrification pathway is the reduction of N_2O to N_2 ($N_2O + 2H^+ + 2e^- \rightarrow N_2 + H_2O$), which is catalyzed by the copper containing enzyme nitrous oxide reductase (N_2OR). This enzyme has an important role on the activation of N_2O, by lowering the activation barrier of that reaction. In spite of being thermodynamically favorable, since the $\Delta G^{\circ\prime}$ is negative (−339.5 kJ/mol), this reaction is kinetically inert.

Nitrous oxide reductase, as being part of the denitrification pathway, contributes to the ATP synthesis in the absence of oxygen. In addition, it also plays an important environmental role by contributing to the decrease of the amount of N_2O, a green house gas.

In particular, the fascinating structure of the tetranuclear copper-sulfur center, CuZ, which constitutes the catalytic site of nitrous oxide reductase, makes this enzyme unique and a subject of numerous studies in the field of bioinorganic chemistry.

Structural Studies of N_2O Reductase

Nitrous oxide reductase was isolated from the first time in 1982 from *Pseudomonas stutzeri* (Zumft and Matsubara 1982) and over the last years it has been isolated and characterized from many other organisms (Dell'Acqua et al. 2011a and references therein). However, its structure has only been determined more recently by X-ray crystallography. Three structures have been reported so far (from *Marinobacter hydrocarbonoclasticus* [previously known as *Pseudomonas nautica*], at 2.4 Å resolution (Brown et al. 2000b), *Paracoccus denitrificans*, 1.6 Å (Brown et al. 2000a), and *Achromobacter cycloclastes*, 1.7 Å (Paraskevopoulos et al. 2006)), revealing a unique feature: the catalytic site, CuZ center, is a completely new copper center, since it is a tetranuclear copper cluster

Nitrous Oxide Reductase, Fig. 1 *Panel A:* Representation of *Marinobacter hydrocarbonoclasticus* nitrous oxide reductase functional dimer. On the left, the dimer is colored according to subunit, with one monomer colored *grey* and the other *purple*, CuA and CuZ centers are colored in *dark blue* and *light blue*, respectively. On the right, the surface of the dimer is colored according to subunit evidencing the functional dimer. *Panel B:* CuA and CuZ centers have their ligands colored according to element and the copper ions in CuZ center are numbered I, II, III and IV. The Figure was created with Chimera using 1QNI

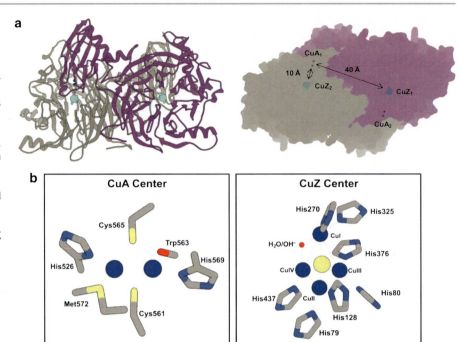

with a μ_4-sulfide bridge and in which all the copper ions, except one are coordinated by the side chains of two histidine residues (Fig. 1b). In addition, the structure shows that this enzyme is a functional homodimer (*vide infra*), containing an additional copper center, CuA, which functions as the electron transfer center (Fig. 1a).

Each monomer of nitrous oxide reductase has two domains. The N-terminal domain has a β-propeller fold with seven blades, and contains the catalytic tetranuclear copper center, CuZ. The C-terminal domain has a cupredoxin fold and harbors the binuclear copper center, CuA. The two copper centers in the same monomer are 40 Å apart but only at 10 Å in between the two different monomers, a distance that enables an efficient electron transfer. Thus, nitrous oxide reductase is the typical example of a functional homodimer, since the electron transfer from CuA to CuZ center can only occur between the centers located in two different monomers (Fig. 1a).

The structure of *Wolinella succinogenes* N$_2$OR has not been determined yet. However, its primary sequence reveals that this enzyme presents an additional domain at the C-terminal, which contains a *c*-type heme (Teraguchi and Hollocher 1989; Zhang et al. 1992). It is a unique feature only found in a few host-associated organisms from the Epsilonproteobacteria, as *Wolinella*, *Campylobacter* and *Sulfurimonas* genera, and also in bacteria from the *Denitrovibrio* genus.

CuA center, the electron transfer center, is very similar to the one present in cytochrome *c* oxidase (Fig. 1b), a binuclear copper center with two cysteines side chains bridging the two Cu ions (Savelieff and Lu 2010). Besides these two ligands, the copper ions are coordinated by the side-chains of two histidines, one methionine and a carbonyl from a tryptophan residue.

CuZ center, the catalytic center, has four copper ions and a sulfur that bridges all the copper ions (Fig. 1b). As mentioned, the copper ions, numbered CuI, CuII and CuIII are coordinated by two histidine side-chains, while Cu IV is only coordinated by one histidine side-chain (Fig. 1b). In the three structures of nitrous oxide reductase, another molecule is present in between CuI and CuIV ions, which has been modeled as being a hydroxyl or a water molecule. This position has been proposed to be the substrate binding site (*vide infra*).

Moreover, the X-ray structure of N$_2$OR inhibited with iodine has also been determined, showing the inhibitor molecule bound between CuI and CuIV (Paraskevopoulos et al. 2006), in the same position where the extra density was observed and modeled as a OH$^-$/H$_2$O molecule (Brown et al. 2000a).

Nitrous Oxide Reductase, Fig. 2 The UV-visible absorption (**a**) and EPR (**b**) spectra of different forms of N_2OR. (i) *purple* form; (ii) *pink* form; (iii) *blue* form. The EPR spectra were recorded at 9.66 GHz (X-band), at 30 K. A schematic representation of each of the nitrous oxide reductase forms is shown in (**c**). Copper ions are *blue* colored in the (II) oxidation state or *light grey* in the (I) oxidation state. In the CuZ non-active forms, the $Cu_I(II)$ ion(s) is(are) represented as *blue square(s)* to indicate that these forms are inactive. In either CuZ or CuA centers, the unpaired electrons are delocalized within the cluster through the sulfur(s) atom(s), represented as *yellow spheres*, that in the case of CuA center are two thiolates, from two cysteine side-chain, and in CuZ center a sulfide

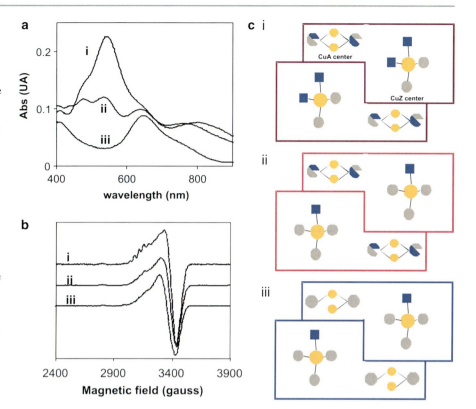

Spectroscopic Properties of N_2O Reductase

The enzyme can be isolated in different oxidation states (*purple*, *pink* and *blue*) by changing the procedure followed during the purification (in the presence or in the absence of oxygen) (Zumft and Kroneck 2007).

In the most oxidized state, nitrous oxide reductase is in the *purple form*, with both CuA and CuZ centers contributing to absorption bands observed in its visible spectrum (Fig. 2a, Ci). CuA center is in a mixed valence state, $[Cu^{1.5+}\text{-}Cu^{1.5+}]$, with the electron being shared between the two copper ions and the absorption bands at 480, 540 and ~800 nm are attributed to this center ((Savelieff and Lu 2010) and references therein). CuZ center is in the state $[2Cu^{2+}\text{-}2Cu^+]$, and contributes to the visible spectrum with absorption bands at 550 and 640 nm (Fig. 2a) (Farrar et al. 1998). It is noteworthy to mention that when CuA center is reduced, $[Cu^+\text{-}Cu^+]$, the solution of N_2OR has still a purple color.

Other redox forms of nitrous oxide reductase that can be isolated have either a pink (pink form) or a blue color (blue form) (Zumft 1997). These forms, contrary to the purple form, have been mainly obtained when the enzyme is purified in the presence of oxygen. In the *blue form*, CuZ center is in the $[1Cu^{2+}\text{-}3Cu^+]$ state, while CuA center is reduced (with both copper ions as Cu^+) and thus, does not contribute to the visible spectrum, which presents a single absorption band at 640 nm (Fig. 2a, Ciii). Finally, the *pink form* of nitrous oxide reductase has CuZ center as $[1Cu^{2+}\text{-}3Cu^+]$ but CuA center is oxidized, $[Cu^{1.5+}\text{-}Cu^{1.5+}]$ (Zumft 1997) (Fig. 2a, Cii).

The EPR spectrum of the purple form only presents the contribution of CuA center, since CuZ center is in a diamagnetic state, $S = 0$ $[2Cu^{2+}\text{-}2Cu^+]$ (Farrar et al. 1991). CuA center exhibits an axial spectrum with $g_\parallel = 2.18$ and $g_\perp \sim 2.03$ and seven lines of hyperfine splitting with $A_\parallel = 38$ G, due to the interaction of the electron spin with the nuclei of the Cu ions that have a nuclear spin of $I = 3/2$ (Antholine et al. 1992). Therefore, the unpaired electron is completely delocalized between the two copper ions, resulting in a $S = 1/2$ spin state (Fig. 2b).

The blue form of nitrous oxide reductase has also been extensively investigated using several spectroscopic techniques, namely EPR, resonance Raman, MCD and EXAFS spectroscopies. In this form of the enzyme, CuA center is reduced, diamagnetic, and thus CuZ center is the only observable paramagnetic center.

MCD spectroscopy showed that N_2OR in the blue form is in a $S = 1/2$ spin state, while the EXAFS analysis revealed that CuZ center is in the $[1Cu^{2+}\text{-}3Cu^+]$ state (Chen et al. 2004).

The distribution of the unpaired electron spin (of a single Cu^{2+}) of CuZ center was determined by EPR spectroscopy. The X-band EPR spectrum of the blue form shows some hyperfine structure in the $g_\parallel$ region (Fig. 2b), but the analysis of the hyperfine coupling is not clear due to the broad signal.

In the case of *Marinobacter hydrocarbonoclasticus* N_2OR, the Q-band EPR spectrum of the blue form was used to map the system's g values (Chen et al. 2002). This spectrum has the features of an almost axial spectrum with $g_\parallel = 2.16$ and $g_\perp = 2.04$. These values indicate that the single spin resides predominantly in a Cu $d_{x2\text{-}y2}$ orbital (Chen et al. 2002). Moreover, the hyperfine pattern observed in the X-band EPR spectrum can only be explained considering that the spin density is distributed between two copper ions. Since the ratio of the hyperfine coupling constants is approximately 5:2, this is the ratio of the spin densities on the two copper ions of CuZ center (Chen et al. 2002). The crystal structure, spin state and copper oxidation states were used in a spin unrestricted DFT calculation to obtain a description of the ground state wave function of the CuZ center. The calculated wave function is partially delocalized between the four coppers ions of the CuZ center, with CuI being the dominantly oxidized copper (42%), CuII is the second copper, which has a significant spin density (16%), along with the μ_4-bridging sulfide (14%), while CuIII and CuIV present 8% and 3% of the spin density, respectively (Chen et al. 2002).

A similar study, performed on the blue form of *Paracoccus pantotrophus* N_2OR using the coordinates of *Paracoccus denitrificans* CuZ center, proposes that the spin density is generally delocalized over the four coppers ions of CuZ center (CuI 20.1%, CuII 9.5%, CuIII 4.8% and CuIV 9.2%) and that there is also spin density distributed over the μ_4-sulfide ion and the oxygen ligand that is bound in between CuI and CuIV (Oganesyan et al. 2004).

Activation of Nitrous Oxide Reductase

As mentioned, nitrous oxide reductase can be isolated in different redox states. In common these forms of the enzyme have a low catalytic activity towards the reduction of nitrous oxide, with maximum activity below 10 U/mg (U/mg = $\mu molN_2O/min \times mgN_2OR$) (Table 1), and thus it has been proposed that nitrous oxide reductase requires activation for maximum activity (Ghosh et al. 2003).

However, the mechanism of activation is still unknown. In the early studies of this enzyme, it was shown that CO and prolonged alkaline treatment was efficient to activate the enzyme (Coyle et al. 1985; Riester et al. 1989), while in more recent studies a prolonged incubation of the enzyme with reduced dyes, such as methyl viologen and benzyl viologen, was demonstrated to activate the enzyme (Ghosh et al. 2003; Chan et al. 2004). A main difference in the conclusion from these studies is that in early studies, it was proposed that the catalytically active form of the enzyme was the fully oxidized purple form, while in the latest studies it has been argued that the catalytically active form is the one with CuZ center in the fully reduced state.

Indeed, it was observed that incubation of the enzyme with reduced methyl viologen, increases its activity, while the characteristic EPR signal of N_2OR with CuZ center in the blue form decreases, until the enzyme is in the fully reduced state with CuZ center as $[4Cu^+]$ (Fig. 3). Therefore, it has been proposed that nitrous oxide reductase requires activation for maximum activity, and that the catalytically active form of the enzyme is the fully reduced one, with CuA center as $[Cu^+\text{-}Cu^+]$ and CuZ center as $[4Cu^+]$ (Ghosh et al. 2003; Chan et al. 2004).

The analysis of the data presented in Table 1 together with the pH dependence activity studies performed in *Achromobacter cycloclastes* N_2OR, opens the question as whether the mechanism of activation involves the deprotonation/protonation of a ligand or a transfer of two hydrides, since the proton concentration has multiple effects on both the activation and the catalytic activity (Fujita and Dooley 2007). Recently, spectroscopic, computation and kinetic studies on *Achromobacter cycloclastes* and *Marinobacter hydrocarbonoclasticus* N_2OR have shown that the effect of pH on the rate of reduction of CuZ presents a pKa of 9.0 (Ghosh et al. 2007).

Nitrous Oxide Reductase, Table 1 Maximum activity of nitrous oxide reductase in the different redox forms

Organism	Redox form (purification condition)	Treatment	Electron donor	Activity (U/mg)[a]	References
Pseudomonas stutzeri	Purple (anaerobic)	As-isolated	BV	4	Coyle et al. (1985)
		Activated alkaline	BV	60	
	Pink (anaerobic)	As-isolated	BV	2	
		Activated alkaline	BV	11	
Achromobacter cycloclastes	Pink (aerobic)	As-isolated	MV	86	Hulse and Averill (1990)
	Blue (anaerobic)	As-isolated	MV	7	Fujita et al. (2007)
	Fully reduced	Activated MV	MV	124	Chan et al. (2004), Fujita et al. (2007)
Marinobacter hydrocarbonoclasticus	Blue (aerobic)	As-isolated	MV	0.01	Dell'Acqua et al. (2008)
	Fully reduced	Activated MV	MV	160	
Pseudomonas aeruginosa	Purple (anaerobic)	As-isolated	BV	0.5	SooHoo and Hollocher (1991)
		Activated alkaline	BV	27	
Paracoccus pantotrophus	Purple (anaerobic)	As-isolated	MV	3	Rasmussen et al. (2002)
	Pink (aerobic)	As-isolated	MV	9	
Paracoccus denitrificans	Purple (anaerobic)	Activated BV	BV	122	Snyder and Hollocher (1987)
Alcaligenes xyloxidans	Purple (anaerobic)	As-isolated	MV	6	Ferretti et al. (1999)

[a]U/mg = µmol N_2O/min × mg N_2OR. Legend: BV is reduced benzyl viologen and MV is reduced methyl viologen

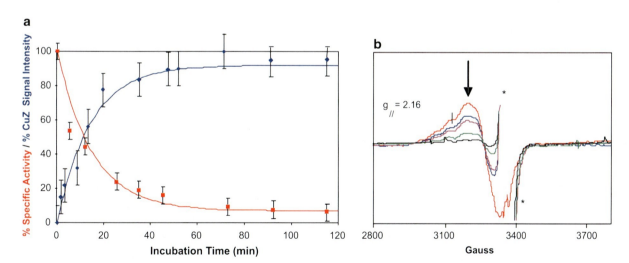

Nitrous Oxide Reductase, Fig. 3 Panel A: *Marinobacter hydrocarbonoclasticus* nitrous oxide reductase activity (*blue line*) and CuZ EPR signal intensity (*red line*) versus incubation time with reduced methyl viologen. Panel B: EPR spectra of nitrous oxide reductase used in the activity assays, showing the CuZ EPR signal of the enzyme after 0 min (*red*, 100% CuZ as [$1Cu^{2+}$-$3Cu^+$]), 5 min (*blue*), 12 min (*pink*), 35 min (*green*), and 115 min (*black*) of incubation with reduced methyl viologen. The EPR spectra were measured at 77 K, 9.319 GHz. (*) Methyl viologen radical signal (Figure adapted from Ghosh et al. 2003)

The observed effect has been proposed to be the protonation equilibrium of a neighboring lysine residue (K412 and K397 for *Achromobacter cycloclastes* and *Marinobacter hydrocarbonoclasticus* N_2OR, respectively) (Ghosh et al. 2007), which would raise the E^o of CuZ center and provide a proton to lower the barrier for both the N–O cleavage and reduction of the CuZ center to the fully [4 Cu^{1+}] state during turnover.

Binding of the Substrate to Nitrous Oxide Reductase

CuZ center presents an empty coordination site between CuI and CuIV. Attending to the fact that CuIV is coordinated by only one histidine side-chain and the extra electron density observed in the X-ray structure of all N_2ORs, has been modeled as a OH^-/H_2O molecule bridging those two copper ions (Brown et al. 2000a, b), it has been proposed that this site is the N_2O binding site (Brown et al. 2000b).

The N_2O molecule could bind to CuZ center in different modes, however the lowest energy structure for the binding mode of N_2O to CuZ center was determined by DFT calculations to be the μ-1,3-N_2O bridging mode when CuZ center is in the fully reduced state [$4Cu^{1+}$] (Ghosh et al. 2003; Chen et al. 2004). Nevertheless, when CuZ center is in the redox state [$1Cu^{+2}$-$3Cu^{1+}$], N_2O is proposed to bind CuZ center through the N-atom to CuI in an end-on mode (Ghosh et al. 2003).

In fact, N_2O binding in a bent μ–1,3 bridging mode to the fully reduced CuZ center is most efficient due to the strong back – bonding from CuI and CuIV ions. The CuZ to N_2O charge transfer and strong interaction of the μ-1,3-N_2O ligand with the two redox active CuI and CuIV ions provide a low activation energy pathway for the N-O cleavage process, which involves two electrons. The N-O bond is weakened, which facilitates the cleavage through simultaneous transfer of two electrons from CuZ center. After N-O cleavage, the proposed catalytic cycle consists of a sequence of alternating protonation/one electron reduction steps, which returns the CuZ center to the fully reduced state (Gorelsky et al. 2006).

The created electron holes on CuI and CuIV by the two-electron reduction of N_2O can be efficiently delocalized through the bridging sulfide exchange pathways, contributing to the thermodynamics and kinetics of this process, by lowering the geometric and electronic reorganization energies between fully reduced and oxidized CuZ states.

Intermediate Catalytically Active Species, Mechanism of Reduction, Catalysis and Inactivation of the Enzyme

The reaction between pre-activated N_2OR and N_2O has been investigated in order to gain information on the catalytic mechanism and in particular to identify intermediate species during turnover. In particular, the reaction between fully reduced N_2OR and a stoichiometric amount of N_2O, monitored by UV-visible spectroscopy allowed the identification of an intermediate in the turnover cycle of the enzyme, named $CuZ°$ (Dell'Acqua et al. 2010). This intermediate species of nitrous oxide reductase has CuA center in the oxidized state (with absorption bands at 480 and 540 nm), CuZ center in the [$1Cu^{+2}$-$3Cu^{1+}$] state, with an absorption band at 680 nm, and its catalytic activity is identical to the enzyme in the fully reduced state.

The life-time of this intermediate species is very small (decay with $k = 0.3$ min^{-1}), and it was observed that it decays to an enzyme form, in which CuZ center has an absorption band at 640 nm (similar to the blue form), and contrary to $CuZ°$, very low catalytic activity.

The comparison of the EPR spectrum of the blue form of N_2OR with that of $CuZ°$, lead to the proposal that the redox state of CuZ center is identical, [$1Cu^{+2}$-$3Cu^{1+}$], and that subtle but important local structural rearrangements must occur during this inactivation process. Most likely, these changes must be the reverse of the ones that occur during N_2OR activation.

The identification of the species $CuZ°$ as an intermediate species in the catalytic cycle of N_2OR, lead to the proposal of the mechanism of reduction, catalysis and inactivation of the enzyme that is presented in Fig. 4. In this mechanism the non-active CuZ state characterized by the 640 nm band does not participate in the catalytic cycle, since it has a slow reduction (activation process) which leads to the fully reduced state. The fast turnover cycle ($k_{cat} = 320$ s^{-1}, calculated for *Marinobacter hydrocarbonoclasticus* N_2OR with methyl viologen as electron donor (Dell'Acqua et al. 2008)) implies that re-reduction of the N_2O-oxidized copper center must be rapid in the

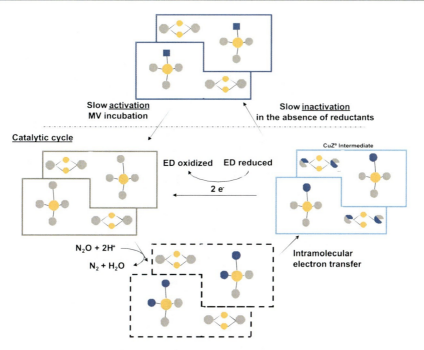

Nitrous Oxide Reductase, Fig. 4 Schematic diagram of the proposed mechanism of reduction, catalysis and inactivation of the catalytic center CuZ. Copper ions are *blue* colored in the (II) oxidation state or *light grey* in the (I) oxidation state. In the CuZ non-active form, the $Cu_I(II)$ ion is represented as a *blue square* to indicate that this form is inactive. In either CuZ or CuA centers, the unpaired electrons are delocalized within the cluster through the sulfur(s) atom(s), represented as *yellow spheres*, that in the case of CuA center are two thiolates, from two cysteine side-chain, and in CuZ center a sulfide (Figure adapted from (Dell'Acqua et al. 2011a). The dashed intermediate species has not yet been observed nor isolated

enzymatic turnover and thus excludes the involvement of the non-active CuZ state. This is also in agreement with the fact that all the as-purified forms of N_2OR present very low catalytic activity (Table 1), unless subjected to a prolonged activation process.

As a simple proton transfer would be faster than the timescale of conversion of $CuZ°$ in the redox-equivalent, inactive form, it is likely that this would involve some minor structural changes in CuZ center to meet the geometrical requirement needed to optimize the formation of the bridged ligand at the CuZ cluster after N_2O release.

Identification and Interaction with the Physiological Electron Donor: The Electron Transfer Route

Small electron-transfer proteins, either *c*-type cytochromes or type 1 copper proteins, have been used as electron donors to N_2OR in order to mimic the physiological conditions in in vitro assays. Indeed, it was identified that a periplasmic *c*-type cytochrome is the electron donor of N_2OR isolated from *Rhodobacter capsulatus*, *Rhodobacter sphaeroides f. sp. denitrificans*, *Wolinella succinogenes*, *Marinobacter hydrocarbonoclasticus* and *Paracoccus pantotrophus* (Itoh et al. 1989; Richardson et al. 1991; Berks et al. 1993; Zhang and Hollocher 1993; Dell'Acqua et al. 2008). In the case of *Paracoccus pantotrophus* N_2OR, the enzyme can also accept electrons from a periplasmic type 1 copper protein, pseudoazurin (Berks et al. 1993), which is also the electron donor to *Achromobacter cycloclastes* N_2OR (Fujita et al. 2009).

Artificial electron donors, such as the mitochondrial cytochrome *c* isolated from horse heart or bovine heart have been shown to be competent electron donors to *Paracoccus pantotrophus* and *Achromobacter cycloclastes* N_2OR, respectively (Rasmussen et al. 2005; Fujita et al. 2007).

Furthermore, direct electron transfer experiments showed that the small electron donor proteins can directly donate electrons to CuA center but cannot

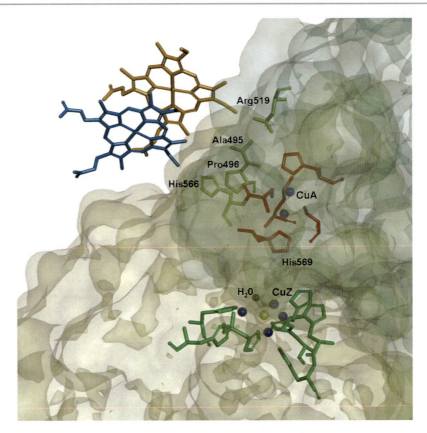

Nitrous Oxide Reductase, Fig. 5 Electron transfer pathway from *Marinobacter hydrocarbonoclasticus* cytochrome c_{552} to *Marinobacter hydrocarbonoclasticus* nitrous oxide reductase. Two possible orientations of cytochrome c_{552} are shown (for clarity only the heme group of cytochrome c_{552} is displayed). The conserved residues involved in the electron transfer from the heme group to CuA center (Asp519, Ala495, Pro496, His566) are represented in *yellow*. The proposed electron transfer pathway from CuA center to CuZ center involves His569, which is a CuA ligand. After, the electron is transferred to the oxygen of a water molecule or hydroxyl group bounded in between CuI and CuIV of CuZ center. The ligands of CuA center (His526, Cys561, Cys565, His569, Met572) are colored red, while the ligands of CuZ center (His79, His80, His128, His270, His325, His376, His437) are represented in green. The copper ions of CuA and CuZ centers are represented as blue spheres, while the sulfide in CuZ center is represented as a yellow sphere

per si reduce CuZ center (Rasmussen et al. 2005; Dell'Acqua et al. 2008). On the contrary, small reduced dyes (methyl viologen or benzyl viologen) are able to reduce both CuA and CuZ centers, at different rates: while CuA center is rapidly reduced, CuZ center is slowly reduced (Zumft and Kroneck 2007). Thus, these small dyes are either able to directly interact with CuZ center or donate electron through a different pathway.

In the case of *Marinobacter hydrocarbonoclasticus* N$_2$OR, a comparative study between the physiological (cytochrome c_{552}) and artificial electron donor (methyl viologen) has been performed, evidencing that the rate limiting step in the reaction is the intermolecular electron transfer in the first case, whereas the electron donation is faster than the substrate reduction when methyl viologen is used as electron donor. This difference is also translated into the different pH behavior that is observed on the activity when these electron donors are used in the assay. A pK_a of 6.6 was determined for methyl viologen with high activity being maintained at high pH values. However, a pK_a of 8.3 was determined for cytochrome c_{552}, and the activity decreases at high pH values.

Since there is no residue in the interface to attribute this pKa, the attention must be turned to the reduction process of CuZ center. This pK_a could be associated with a water ligand in the Cu^{2+}-H_2O-Cu^+ state and the deprotonation of this state will lower the redox potential making it difficult to CuA reduce CuZ center. This problem would not exist when methyl viologen is used as electron source, considering that this small reductant can deliver electrons directly to CuZ center without the need to go through CuA center.

In the model structures for the electron transfer complexes of *Marinobacter hydrocarbonoclasticus*, *Paracoccus denitrificans* and *Achromobacter cycloclastes* N_2OR with their physiological electron donors, the small donor proteins (*Marinobacter hydrocarbonoclasticus* cytochrome c_{552}, and both pseudoazurin and cytochrome c_{550} from *Paracoccus denitrificans* and pseudoazurin from *Achromobacter cycloclastes*) bind at N_2OR surface near CuA center in accordance with the direct electron transfer experiments (Berks et al. 1993; Mattila and Haltia 2005; Dell'Acqua et al. 2011b). In these molecular docking studies, the exposed heme methyl of small cytochrome c is positioned at 5Å of a conserved patch in N_2OR surface. This conserved patch is composed by Ala495-Pro496, Asp519 and His566 (residues numbered according to *Marinobacter hydrocarbonoclasticus* N_2OR primary sequence), in which the carboxyl group of Asp519 has a hydrogen bond to His526, the terminal ligand of CuA center (Fig. 5) (Mattila and Haltia 2005; Dell'Acqua et al. 2008).

The electron transfer pathway from CuA center to CuZ center is proposed to go through the imidazole ring of His569 (numbered according to *Marinobacter hydrocarbonoclasticus* N_2OR primary sequence) to the water molecule modeled in between CuI and CuIV of CuZ center (Mattila and Haltia 2005). This electron transfer route is an analogue of the one proposed for the electron transfer from CuA center to the heme a in cytochrome c oxidase (Wang et al. 2002).

Future Perspectives

Since the discovery of the presence of the two unusual multicopper centers, CuA and CuZ, nitrous oxide reductase has been extensively investigated.

Ten years after the determination of the first crystal structure of N_2OR, many progresses have occurred specially regarding the unique CuZ center, namely the possibility of a structural basis for the interpretation of spectroscopic data supported by theoretical calculations; the identification of the different oxidation states; the proposal of electron transfer routes and a better understanding of the catalytic cycle; as well as, the characterization of active species involved in the enzyme turnover.

In particular, the identification of the fully reduced state and CuZ° as highly reactive forms of N_2OR, when compared to any of the as-isolated forms of the enzyme, led to the proposal of a mechanism for the enzyme activation and catalytic cycle. However, the structural differences between this CuZ° intermediate and CuZ in the [$1Cu^{2+}$-$3Cu^+$] state, and also the structure of the catalytically competent CuZ intermediate in the [$2Cu^{2+}$-$2Cu^+$] state still remains to be identified and fully characterized. Moreover, the structural changes that must occur in the enzyme, either around CuZ center and/or CuA center during the activation of the enzyme have also been clarified.

3D structural elucidations of the different redox forms of the enzyme, complemented with other spectroscopic tools is envisaged with most interest, as well as, a better definition of the extra electron density identified in between CuI and CuIV ions of CuZ center. In fact, during the revision of this manuscript the structure of *Pseudomonas stutzeri* nitrous oxide reductase with both copper centers in the oxidized state was reported (Pomowski et al. 2011). In this structure the CuZ center presents an additional sulfur atom bridging CuI and CuIV, where the extra electron density was observed in the other structures. This structure opens new basis for the interpretation of the structural and electronic properties of N_2ORs in different forms, which recover a crucial aspect in the study of this fascinating enzyme.

Regarding the electron transfer route and the activation of the enzyme, the extensive characterization of the interaction of N_2OR with its physiological partners will be crucial, since the results obtained so far with the strong reductants (methyl and benzyl viologen), argue that the electron route promoted is not physiologically relevant.

Finally, N_2OR plays a central role in the detoxification of the green-house gas, nitrous oxide. Therefore, the understanding of the catalytic cycle and activation mechanism of nitrous oxide reductase will open, for sure, a door for new ideas to improve or implement novel biotechnological applications of this enzyme, such as biosensors and removal of nitrogen compounds from wastewater.

Acknowledgements We would like to thank Fundação para a Ciência e Tecnologia, MCTES for financial support (project PTDC/QUI/64638/2006 and PTDC/QUI-BIQ/116481/2010 to I.M.).

Cross-References

▶ Copper, Biological Functions
▶ Copper, Physical and Chemical Properties
▶ Copper-Binding Proteins
▶ Cytochrome c Oxidase, CuA Center

References

Antholine WE, Kastrau DH, Steffens GC et al (1992) A comparative EPR investigation of the multicopper proteins nitrous-oxide reductase and cytochrome c oxidase. Eur J Biochem 209:875–881

Berks BC, Baratta D, Richardson J et al (1993) Purification and characterization of a nitrous oxide reductase from *Thiosphaera pantotropha*. Implications for the mechanism of aerobic nitrous oxide reduction. Eur J Biochem 212:467–476

Brown K, Djinovic-Carugo K, Haltia T et al (2000a) Revisiting the catalytic CuZ cluster of nitrous oxide (N_2O) reductase. Evidence of a bridging inorganic sulfur. J Biol Chem 275:41133–41136

Brown K, Tegoni M, Prudencio M et al (2000b) A novel type of catalytic copper cluster in nitrous oxide reductase. Nat Struct Biol 7:191–195

Chan JM, Bollinger JA, Grewell CL et al (2004) Reductively activated nitrous oxide reductase reacts directly with substrate. J Am Chem Soc 126:3030–3031

Chen P, DeBeer GS, Cabrito I et al (2002) Electronic structure description of the μ_4-sulfide bridged tetranuclear CuZ center in N_2O reductase. J Am Chem Soc 124:744–745

Chen P, Gorelsky SI, Ghosh S et al (2004) N_2O reduction by the mu4-sulfide-bridged tetranuclear CuZ cluster active site. Angew Chem Int Ed Engl 43:4132–4140

Coyle CL, Zumft WG, Kroneck PM et al (1985) Nitrous oxide reductase from denitrifying *Pseudomonas perfectomarina*. Purification and properties of a novel multicopper enzyme. Eur J Biochem 153:459–467

Dell'Acqua S, Pauleta SR, Monzani E et al (2008) Electron transfer complex between nitrous oxide reductase and cytochrome c_{552} from *Pseudomonas nautica*: kinetic, nuclear magnetic resonance, and docking studies. Biochemistry 47:10852–10862

Dell'Acqua S, Pauleta SR, de Sousa PMP et al (2010) A new CuZ active form in the catalytic reduction of N2O by nitrous oxide reductase from *Pseudomonas nautica*. J Biol Inorg Chem 15:967–976

Dell'Acqua S, Pauleta SR, Moura I et al (2011a) The tetranuclear copper active site of nitrous oxide reductase: the CuZ center. J Biol Inorg Chem 16:183–194

Dell'Acqua S, Moura I, Moura JJG et al (2011b) The electron transfer complex between nitrous oxide reductase and its electron donors. J Biol Inorg Chem 16:1241–1254

Farrar JA, Thomson AJ, Cheesman MR et al (1991) A model of the copper centres of nitrous oxide reductase (*Pseudomonas stutzeri*). Evidence from optical, EPR and MCD spectroscopy. FEBS Lett 294:11–15

Farrar JA, Zumft WG, Thomson AJ (1998) CuA and CuZ are variants of the electron transfer center in nitrous oxide reductase. Proc Nalt Acad Sci USA 95:9891–9896

Ferretti S, Grossmann JG et al (1999) Biochemical characterization and solution structure of nitrous oxide reductase from Alcaligenes xylosoxidans (NCIMB 11015). Eur J Biochem 259(3):651–659

Fujita K, Dooley DM (2007) Insights into the mechanism of N_2O reduction by reductively activated N_2O reductase from kinetics and spectroscopic studies of pH effects. Inorg Chem 46:613–615

Fujita K, Chan JM, Bollinger JA et al (2007) Anaerobic purification, characterization and preliminary mechanistic study of recombinant nitrous oxide reductase from *Achromobacter cycloclastes*. J Inorg Biochem 101:1836–1844

Fujita K, Ijima F, Obara Y et al (2009) Direct electron transfer from pseudoazurin to nitrous oxide reductase in catalytic N_2O reduction. J Biol Inorg Chem 14(Suppl 1):S11–S20

Ghosh S, Gorelsky SI, Chen P et al (2003) Activation of N_2O reduction by the fully reduced μ_4-sulfide bridged tetranuclear CuZ cluster in nitrous oxide reductase. J Am Chem Soc 125:15708–15709

Ghosh S, Gorelsky SI, George SD et al (2007) Spectroscopic, computational, and kinetic studies of the μ_4-sulfide-bridged tetranuclear CuZ cluster in N_2O reductase: pH effect on the edge ligand and its contribution to reactivity. J Am Chem Soc 129:3955–3965

Gorelsky SI, Ghosh S, Solomon EI (2006) Mechanism of N_2O reduction by the μ_4-S tetranuclear CuZ cluster of nitrous oxide reductase. J Am Chem Soc 128:278–290

Hulse CL, Averill BA (1990) Isolation of a high specific activity pink, monomeric nitrous oxide reductase from Achromobacter cycloclastes. Biochem Biophys Res Commun 166(2):729–735

Itoh M, Matsuura K, Satoh T (1989) Involvement of cytochrome bc_1 complex in the electron transfer pathway for N_2O reduction in a photodenitrifier. Rhodobacter sphaeroides f. s. denitrificans. FEBS Lett 251:104–108

Mattila K, Haltia T (2005) How does nitrous oxide reductase interact with its electron donors? – a docking study. Proteins 59:708–722

Oganesyan VS, Rasmussen T, Fairhurst S et al (2004) Characterisation of [Cu4S], the catalytic site in nitrous oxide reductase, by EPR spectroscopy. Dalton Trans 7:996–1002

Paraskevopoulos K, Antonyuk SV, Sawers RG et al (2006) Insight into catalysis of nitrous oxide reductase from high-resolution structures of resting and inhibitor-bound enzyme from *Achromobacter cycloclastes*. J Mol Biol 362:55–65

Pomowski A, Zumft W, Kroneck P et al (2011) N_2O binding at a [4Cu:2S] copper-sulphur cluster in nitrous oxide reductase. Nature 477:234–237

Rasmussen T, Berks BC et al (2002) Multiple forms of the catalytic centre, CuZ, in the enzyme nitrous oxide reductase from Paracoccus pantotrophus. Biochem J 364(Pt 3):807–815

Rasmussen T, Brittain T, Berks BC et al (2005) Formation of a cytochrome c-nitrous oxide reductase complex is obligatory for N_2O reduction by *Paracoccus pantotrophus*. Dalton Trans 21:3501–3506

Richardson DJ, Bell LC, McEwan AG et al (1991) Cytochrome c_2 is essential for electron transfer to nitrous oxide reductase from physiological substrates in *Rhodobacter capsulatus* and can act as an electron donor to the reductase in vitro. Correlation with photoinhibition studies. Eur J Biochem 199:677–683

Riester J, Zumft WG, Kroneck PM (1989) Nitrous oxide reductase from *Pseudomonas stutzeri*. Redox properties and spectroscopic characterization of different forms of the multicopper enzyme. Eur J Biochem 178:751–762

Savelieff MG, Lu Y (2010) Cu-A centers and their biosynthetic models in azurin. J Biol Inorg Chem 15:461–483

Snyder SW, Hollocher TC (1987) Purification and some characteristics of nitrous oxide reductase from Paracoccus denitrificans. J Biol Chem 262(14):6515–6525

SooHoo CK, Hollocher TC (1991) Purification and characterization of nitrous oxide reductase from Pseudomonas aeruginosa strain P2. J Biol Chem 266(4):2203–2209

Teraguchi S, Hollocher TC (1989) Purification and some characteristics of a cytochrome *c*-containing nitrous oxide reductase from *Wolinella succinogenes*. J Biol Chem 264:1972–1979

Wang K, Geren L, Zhen Y et al (2002) Mutants of the CuA site in cytochrome *c* oxidase of *Rhodobacter sphaeroides*: II. Rapid kinetic analysis of electron transfer. Biochemistry 41:2298–2304

Zhang CS, Hollocher TC (1993) The reaction of reduced cytochromes *c* with nitrous oxide reductase ot *Wolinella succinogenes*. Biochim Biophys Acta 1142:253–261

Zhang C, Jones AM, Hollocher TC (1992) An apparently allosteric effect involving N_2O with the nitrous oxide reductase from *Wolinella succinogenes*. Biochem Biophys Res Commun 187:135–139

Zumft WG (1997) Cell biology and molecular basis of denitrification. Microbiol Mol Biol Rev 61:533–616

Zumft WG, Kroneck PM (2007) Respiratory transformation of nitrous oxide (N_2O) to dinitrogen by Bacteria and Archaea. Adv Microb Physiol 52:107–227

Zumft WG, Matsubara T (1982) A novel kind of multi-copper protein as terminal oxidoreductase of nitrous oxide respiration in *Pseudomonas perfectomarinus*. FEBS Lett 148:107–112

Nitrous Oxide Reduction

▶ Nitrous Oxide Reductase

NMR spectroscopy of Gallium in Biological Systems

João Paulo André
Centro de Química, Universidade do Minho, Braga, Portugal

Synonyms

Group 13 elements NMR in biology; Quadrupolar NMR in biology

Definition

Even though no biological role is known for gallium, it is a useful element in NMR structural studies of biologically relevant systems such as metalloproteins and radiopharmaceutical agents with application in clinical diagnosis. Gallium has two isotopes suitable for NMR spectroscopy, ^{69}Ga and ^{71}Ga, both being quadrupolar nuclei (nuclear spin $I > ½$) of moderate frequency. (Relevant magnetic properties are summarized in Table 1.) In spite of the lower natural abundance, ^{71}Ga is usually the most favorable isotope for direct NMR observations, due to higher receptivity and narrower line width than ^{69}Ga. The chemical shifts and line widths of the NMR signals originated by gallium may give structural information on the coordination environment of the trivalent metal ion.

^{27}Al ($I = 5/2$), ^{115}In ($I = 9/2$), and ^{205}Tl ($I = 1/2$), the group companion elements of gallium, are also adequate for NMR spectroscopy (André and Mäcke 2003).

Principles of Quadrupolar NMR

Nuclei with nuclear spin $I > ½$ are called quadrupolar given that they have a nonzero nuclear quadrupole moment Q (a nonspherical charge distribution within the nucleus). The main relaxation pathway for quadrupolar nuclei is the quadrupolar relaxation mechanism. For isotropic molecular movements in liquids, this mechanism is characterized by an electric interaction between the quadrupole moment and fluctuating electrical field gradients present at the site of the nucleus. This interaction is modulated by the molecular tumbling.

The decays of the longitudinal and transverse magnetizations of a quadrupolar nucleus consist of the sum of I (if I is integer) or $I + 1/2$ (if I is half-integer) exponentials corresponding to single quantum transitions between the $2I + 1$ allowed nuclear Zeeman energy levels. For example, in the case of gallium ($I = 3/2$), the observed magnetization is due to two components: I, the central transition ($m_I = ½ \rightarrow -½$) and II, the outer transitions ($m_I = 3/2 \rightarrow ½$ and $m_I = -½ \rightarrow -3/2$). The relaxation of the individual components is highly dependent on the motion and frequency of the nucleus under study. For half-integer quadrupolar nuclei, three situations of molecular motion can be considered (Aramini and Vogel 1998; Drakenberg et al. 1997).

NMR spectroscopy of Gallium in Biological Systems, Table 1 NMR properties of gallium

	^{69}Ga	^{71}Ga
Nuclear spin, I	3/2	3/2
Isotopic abundance	60.4	39.6
NMR frequency[a]	24.00	30.50
Quadrupole moment Q (10^{-28} m^2)	0.178	0.112
Relative receptivity[b,c] R_x	0.042, 237	0.056, 319
Relative line width[d] W_x	5.94	2.35
Relative peak height[e] H_x	3.4	11.6

Adapted from Delpuech (1983)
[a]In MHz for an induction of 2.348 T (^{1}H at 100 MHz)
[b]In relation to ^{1}H or ^{13}C (first and second values, respectively)
[c]Computed as the ratio R of the receptivities $\alpha_x \gamma_x I_x(I_x + 1)$ at constant field of the mentioned isotope and of the reference nucleus
[d]Computed as the ratio W_x of the values taken by the function $(2I + 3)Q^2/I^2(2I - 1)$ for the mentioned isotope and for ^{27}Al nuclei
[e]Computed as 100 times the ratio of the values taken by the function R_x/W_x for the mentioned isotope and for ^{27}Al nuclei

Rapid Isotropic Motion (Extreme Narrowing Conditions)

This situation occurs when the correlation τ_c time of the molecule to which the metal ion is bound is short (small molecules tumbling fast in solution) in comparison to the inverse Larmor frequency ($\omega_0 \tau_c \ll 1$). In such case, the longitudinal and transverse magnetizations will follow a single exponential decay with the same time constant. The general expression for the T_1 and T_2 relaxation times for any $I > 1/2$ nucleus is given by:

$$\frac{1}{T_1} = \frac{1}{T_2} = \pi \Delta \nu_{1/2} = \frac{3\pi^2}{10} \frac{(2I+3)}{I^2(I^2-1)} \chi^2 \tau_c \left(1 + \frac{\eta^2}{3}\right) \quad (1)$$

The parameter χ, quadrupolar coupling constant, represents the magnitude of the interaction between the nuclear quadrupole moment (Q) and the electric field gradient at the nucleus, with q_{zz} as its biggest component.

$$\chi = \frac{e^2 Q q_{zz}}{h} \quad (2)$$

e is the electron charge and h the Planck's constant. For complexed metal ions, the higher the symmetry of the coordination environment, the smaller the electric field gradient at the nucleus due to the ligands, and consequently the lower the value of χ.

η is the asymmetry parameter of the electric field gradient at the position of the nucleus; its value varies between 0 and 1 and it gives the deviation of the electric field from axial symmetry, which mainly depends on the lack of spherical symmetry of the p electron density. Cubic, tetrahedral, octahedral, or spherical symmetry have normally a zero field gradient ($q = 0$), which gives rise to sharp signals. Asymmetry in the ligand field produces an increase in the NMR line width (Delpuech 1983; Akitt 1987).

Intermediate Isotropic Motion (Near-Extreme Narrowing)

In this limit (important for low-frequency nuclei bound to relatively small proteins), $\omega_0 \tau_c$ is no longer small compared to unity and the relaxation cannot be described exactly as monoexponential. In practice, the relaxation will appear exponential up to $\omega_0 \tau_c \approx 1.5$, even though T_1 and T_2 are different and field dependent. The following equations apply to half-integer quadrupolar nuclei when $\omega_0 \tau_c \leq 1.5$ (Aramini and Vogel 1998):

$$\frac{1}{T_1} = \frac{3\pi^2}{100} \chi^2 \frac{2I+3}{I^2(I^2-1)} \\ \times \left[\frac{2\tau_c}{1+(\omega_0 \tau_c)^2} + \frac{8\tau_c}{1+4(\omega_0 \tau_c)^2} \right] \quad (3)$$

$$\frac{1}{T_2} = \frac{3\pi^2}{100} \chi^2 \frac{2I+3}{I^2(I^2-1)} \\ \times \left[3\tau_c + \frac{5\tau_c}{1+(\omega_0 \tau_c)^2} + \frac{2\tau_c}{1+4(\omega_0 \tau_c)^2} \right] \quad (4)$$

Slow Isotropic Motion (Nonextreme Narrowing Conditions)

This is the case when $\omega_0 \tau_c \gg 1$. Under such conditions, the relaxation of a half-integer nucleus is not exponential anymore and, as a result, the shape of the NMR signal is not Lorentzian. The concept of single relaxation does not apply and multiple time constants are necessary to describe the magnetization decay (Drakenberg et al. 1997). For half-integer quadrupolar nuclei in this situation (moderate to high frequency nuclei bound to large proteins), the central transition

($m_I = 1/2 \to -1/2$) can originate a relatively narrow signal, according to (5), while the peaks due to all outer components are broadened beyond detection.

$$\Delta v_{1/2}(m_I = 1/2 \to -1/2) = k\left(\frac{\chi^2}{v_0^2 \tau_c}\right)$$
$$I = \frac{3}{2}, \quad k = 2.0 \times 10^{-2}$$
$$I = \frac{5}{2}, \quad k = 7.2 \times 10^{-3} \quad (5)$$
$$I = \frac{7}{2}, \quad k = 4.5 \times 10^{-3}$$

Therefore, the linewidth of this component decreases with increasing $\omega_0\tau_c$. Additionally, under these conditions, one should theoretically observe only 40, 25.7, or 19.0% of the signal for $I = 3/2$, 5/2, and 7/2 nuclei, respectively (Aramini and Vogel 1998).

Another important feature of the signal due to the central transition in this slow motion limit is the field dependence of its chemical shift. Although this shift is always smaller than the width of the broader components in the resonance, it will be significant compared to the central transition signal (Drakenberg et al. 1997).

Another noteworthy aspect is the fact that the intensity of the central transition signal of a half-integer quadrupolar nuclei in this motional situation is affected by the pulse angle (Drakenberg et al. 1997). An effective pulse length, t_p, of 90° for this component is much shorter than for the same nucleus under extreme narrowing conditions, according to the equation:

$$t_p(m_I = 1/2 \to -1/2) = \frac{t_p}{I + \frac{1}{2}} \quad (6)$$

This is valid when the radiofrequency pulse strength is much less than the quadrupole coupling constant (Aramini and Vogel 1998).

Coordination Aspects of Gallium

The +3 oxidation state of gallium is the most stable in aqueous solution. In the pH range of 3–7, Ga^{3+} can hydrolyze to insoluble trihydroxide if its concentration exceeds nanomolar level. Nevertheless, this precipitation can be avoided in the presence of stabilizing agents. At physiological pH, the solubility of gallium is high due to the almost exclusive formation of $[Ga(OH)_4]^-$.

The coordination chemistry of Ga^{3+} is quite similar to that of the high spin Fe^{3+} ion. To this, contribute the same charge of both ions, similar ionic radii (62 pm for Ga^{3+} and 65 pm for Fe^{3+}), and the same major coordination number of six (Ga^{3+} chelates sometimes are four and five coordinated).

Ga^{3+} is classified as a hard Lewis acid, forming thermodynamically stable complexes with ligands that are hard Lewis bases. Thus, ligands with oxygen and/or nitrogen donor atoms (like carboxylate, phosphonate, phenolate, hydroxamate, and amine groups) constitute good chelating agents for this ion. Gallium(III) is suitable for complexation with polydentate ligands, both cyclic and open chain structures. The majority of ligands designed for Ga^{3+} are hexadentate although several chelates have been reported which are stable in vivo and have coordination numbers of four and five (André and Mäcke 2003).

Macrocyclic chelators, in particular triaza ligands, are very adequate for Ga^{3+} chelation due to their high conformational and size selectivities, allowing a good fit of the relatively small cation in the macrocyclic cavity. Triaza macrocyclic ligands with different types of pendant arms (carboxylates, alkylphosphinates, methylenephosphonates) have proved to be suitable ligands regarding the stable complexes they form with the gallium ion. The complexes formed are usually octahedral or pseudo-octahedral, showing a C_3 symmetry axis, with the three nitrogen atoms occupying one facial plane and the oxygen atoms from the pendant arms occupying the other (N_3O_3 systems). For this reason, the complexes are highly symmetrical at the coordination center and the metal ion gives origin to sharp gallium NMR signals (André and Mäcke 2003 and references therein).

Protein Studies

Proteins are often too big for complete structural determination in solution by current multidimensional NMR techniques. (^1H NMR signals are intrinsically broad due to slow molecular tumbling.) NMR spectroscopy of group 13 elements, particularly of $^{71/69}Ga$, has been used to investigate directly the metals in their specific binding sites in transferrins and to reveal subtle inter-site differences. Human serum transferrin is the protein that transports Fe^{3+} ions and it is a member of a small group of monomeric non-heme proteins (MW *circa* 76–81 kDa), which includes lactoferrin, ovotransferrin, and melanotransferrin.

It has two binding sites for ferric ions (these are found in six-coordinate, distorted octahedral coordination geometry) which are identified as C-terminal and N-terminal sites. Two tyrosines, one histidine, and one aspartic acid constitute four ligating groups to the metal ion. It requires a synergistic anion for the formation of stable metal complexes (in vivo the CO_3^{2-} as a bidentate ion serves this purpose by coordinating directly to the metal in the fifth and sixth coordination positions). Since serum transferrin is normally only about 30% saturated with iron, it retains a relatively high capacity for binding other metal ions.

Vogel and Aramini demonstrated the feasibility of using NMR quadrupolar metal nuclei to probe the metal ion binding sites in large proteins based on the detection of the central transition ($m_I = 1/2 \rightarrow m_I = -1/2$) of a half-integer quadrupolar nucleus ($I = n/2, n = 3, 5, 7$), which is facilitated by increasing nuclear resonance frequency and protein size ($\omega_0 \tau_c \gg 1$). These authors showed that important information about the metal ion binding site, namely the symmetry of the site (i.e., χ, the quadrupole coupling constant) and the motion of the bound metal ion (i.e., τ_c, the rotational correlation time), may be extracted from the magnetic field's dependence of the chemical shift and the line width of the signal due to the bound metal ion (Aramini and Vogel 1993; Germann et al. 1994; Aramini et al. 1994).

Vogel and Aramini investigated the binding of Al^{3+} to ovotransferrin and its half molecules in the presence of ^{13}C-enriched carbonate and oxalate using ^{27}Al and ^{13}C spectroscopy (Aramini and Vogel 1993). They pointed out that the detection of the central transition of quadrupolar nuclei bound to large proteins depends considerably on a number of factors: (i) the strength of the external magnetic field (ω_0); (ii) the dimensions and motion of the macromolecule (τ_c, temperature, solution viscosity), (iii) the intrinsic quadrupole moment of the specific nucleus (Q), and (iv) the nature of the electric field gradient at the metal ion binding site.

$^{71/69}Ga$ NMR studies with ovotransferrin (oTf) have shown that the metal ion interacts preferentially with the N-site of the intact protein, as previously found for the Al^{3+} binding to oTf when carbonate serves as the synergistic anion (Aramini and Vogel 1993). In the presence of oxalate, oTf exhibits no site preference for Ga^{3+}. The isotropic chemical shifts of the oTf-bound $^{71/69}Ga$ NMR signals fall well within the range of Ga^{3+} bound to six oxygen-containing ligands (+40 to +80 ppm) (Germann et al. 1994; Aramini et al. 1994). Using the observed chemical shifts and line width of the protein-bound ^{71}Ga signals at two magnetic fields, the quadrupole coupling constants (χ) for the ^{71}Ga nucleus as well as the rotational correlation time (τ_c) of the bound metal were calculated.

Complexes with Relevance for Nuclear Medicine

^{67}Ga (γ, $t_{1/2}$ 3.25 days), ^{68}Ga (β^+, $t_{1/2}$ 68 min) are radionuclides that find a wide scope of applications in diagnostic radiopharmaceuticals due to their emitting properties and their suitable half-lives. In particular, ^{68}Ga is nowadays a very much-sought-after radionuclide for positron emission tomography (PET).

Metal radionuclides have to be chelated with suitable ligands that form kinetically and thermodynamically stable complexes in vivo or, otherwise the radiometal can be donated to endogenous high-affinity binding sites such as those located on the serum transferrin which displays two iron-binding sites with high affinity for this metal ion (Maecke and André 2007).

^{69}Ga NMR spectroscopy has shown that the complex Ga(NODASA) has a remarkable stability with respect to acid-catalyzed dissociation, similarly to what Parker et al. have found for Ga(NOTA) (André and Mäcke 2003 and references therein). The complexes in the solid state are approximately octahedral, and ^1H NMR studies have suggested that these structures are maintained in solution. The pair of C_3 symmetric "facial" N_3 and O_3 donors lead to a minimal electric field gradient at the metallic center in the "x-y" plane and consequently to the observation of narrow ^{69}Ga resonances (Table 2).

The coordination chemistry of gallium with α-aminoalkylphosphinic acid ligands based on triazacyclononane (L1, L2, and L3 in Table 2) was investigated by Parker et al. and compared to that of the analogous α-aminocarboxylate ligands (André and Mäcke 2003 and references therein). The ^{71}Ga NMR signals of the phosphinate complexes appear at lower frequencies (between +130 and +140 ppm, Table 2) than the signal of Ga^{3+} in a tris(amino)tris(carboxylate) environment. Moreover, it was found that increasing the size of the alkyl group on the phosphorus atom increased the ^{71}Ga line width (L3 in Table 2). As in each of these cases, the electric field gradient about the

NMR spectroscopy of Gallium in Biological Systems, Table 2 NMR chemical shifts and line widths of some complexes of ^{71}Ga(III)

Ligand	δ (ppm)	$v_{1/2}$ (Hz)
H$_2$O (1:6)	0	53
NOTA	+171	210 320 (^{69}Ga)
NOTAC6[a]	+165.5	528.5
NOTAC8[a]	+165.8	621.8
NODASA	+165 (^{69}Ga)	1,000
NOTP	+110	434
NOTMA	+149	
L1	+132	560
L2	+139	200
L3	+130	1,220
TTHA-(BuA)$_2$	+134	35,000
H$_3$ppma (1:2)	−62.3	50
TAMS	+34	3,400
TACS	+18	1,000
TAPS	+57	1,230

The data presented are from several authors in various publications and, if not otherwise stated, are summarized in André and Mäcke (2003)
[a]de Sá et al. (2010)
NOTA: 1,4,7-triazacyclononane-1,4,7-triacetic acid
NOTAC6: 1,4,7-triazacyclononane-1-hexanoic acid-4,7-diacetic acid
NOTAC8: 1,4,7-triazacyclononane-1-octanoic acid-4,7-diacetic acid
NODASA: 1,4,7-triazacyclononane-1-succinic acid-4,7-diacetic acid
NOTP: 1,4,7-triazacyclononane-1,4,7-tris-(methylenephosphonic acid)
L1: 1,4,7-triazacyclononane-1,4,7-triyltrimethylenetris (phenylphosphinic acid)
L2: 1,4,7-triazacyclononane-1,4,7-triyltrimethylenetris (methylphosphinic acid)
L3: 1,4,7-triazacyclononane-1,4,7-triyltrimethylenetris (benzylphosphinic acid)
TTHA-(BuA)$_2$: triethylenetetramine-N,N′,N″,N‴-tetraacetic acid-N,N‴-bis(butylamide)
H$_3$ppma: tris(4-phenylphosphinato)-3-methyl-3-azabutyl)amine
TAMS: 1,1,1-tris(((2-hydroxy-5-sulfobenzyl)amino)methyl) ethane
TACS: cis,cis-1,3,5-tris((2-hydroxy-5-sulfobenzyl)amino) cyclohexane
TAPS: 1,2,3-tris((2-hydroxy-5-sulfobenzyl)amino)propane

quadrupolar nuclei is very similar, the line width change was attributed to a change in molecular tumbling (τ_c).

When the oxygen donors were from phenolates, in tripodal aminophenolate ligand complexes (TAMS, TACS, and TAPS), Caravan et al. have found that the chemical shift range of ^{71}Ga moves to even lower frequencies (+18 to +57 ppm) but still lies in the range that is expected for octahedral Ga^{3+} complexes. (André and Mäcke and references therein). Orvig et al. demonstrated the formation of highly symmetrical bicapped bisligand complexes of gallium (and of aluminum and indium) with the N$_4$O$_3$ tripodal phosphonic ligand H$_3$ppma, which give very narrow signals in the ^{27}Al NMR spectra (and in the ^{71}Ga and ^{115}In spectra) as a result of a S_6 symmetry (Table 2) (André and Mäcke 2003 and references therein).

Cross-References

▶ Gallium Uptake and Transport by Transferrin
▶ Gallium(III) Complexes, Inhibition of Proteasome Activity
▶ Gallium, Therapeutic Effects

References

Akitt JW (1987) Aluminum, gallium, indium and thallium. In: Mason J (ed) Multinuclear NMR. Plenum, New York, pp 259–292
André JP, Mäcke H (2003) NMR spectroscopy of group 13 metal ions: biologically relevant aspects. J Inorg Biochem 97:315–323
Aramini JM, Vogel HJ (1993) Aluminum-27 and carbon-13 NMR studies of aluminum(3+) binding to ovotransferrin and its half-molecules. J Am Chem Soc 115:245–252
Aramini JM, Vogel HJ (1998) Quadrupolar metal ion NMR studies of metalloproteins. Biochem Cell Biol 76:210–222
Aramini JM, McIntyre DD, Vogel HJ (1994) Gallium(3+) binding to ovotransferrin and its half-molecules: a multinuclear NMR study. J Am Chem Soc 116:11506–11511
de Sá A, Prata MIM, Geraldes CFGC, André JP (2010) Triaza-based amphiphilic chelators: synthetic route, in vitro characterization and in vivo studies of their Ga(III) and Al(III) chelates. J Inorg Biochem 104:1051–1062
Delpuech JJ (1983) NMR of newly accessible nuclei, vol 2. Academic Press, London
Drakenberg T, Johansson C, Forsén S (1997) Metal NMR for the study of metalloproteins. In: Reid DG (ed) Protein NMR techniques, vol 60, Methods in molecular biology. Humana Press, Totowa
Germann MW, Aramini JM, Vogel HJ (1994) Quadrupolar metal ion NMR study of ovotransferrin at 17.6 T. J Am Chem Soc 116:6971–6972
Maecke HR, André JP (2007) PET chemistry: the driving force in molecular imaging. 62nd Ernst-Schering-Research-Foundation on PET Chemistry. In: Schubiger PA, Lehmann L and Friebe M (eds) Ernst schering research foundation workshop. Ernst Schering Res Fdn. Berlin, Germany, 62:215–242

NMR Structure Determination of Protein-Ligand Complexes using Lanthanides

Michael Assfalg
Department of Biotechnology, University of Verona, Verona, Italy

Synonyms

Complex, adduct; Ligand, Inhibitor; Structure, spatial arrangement

Definition

Computing spatial coordinates of protein–ligand supramolecular complexes or adducts from structural restraints determined from perturbations of the NMR spectroscopy signals of nuclei located in the proximity of paramagnetic lanthanide ions.

Traditionally, NMR-based protein structure determination has relied heavily on the nuclear Overhauser effect (NOE), an interaction between two nuclei, usually two hydrogen atoms, that arises if they are sufficiently close in space (usually less than ∼6–7 Å). Hundreds or thousands of such interactions are typically measured for a well-ordered protein and converted into distance bounds during the structure determination process. NOEs between protein and ligand can likewise be used to define the structure of a protein complex. The assignment of intra- and intermolecular NOE peaks can be a very laborious and difficult task and, therefore, the use of more readily attainable alternative or complementary structural restraints has gained significant interest. In this respect, paramagnetic metal ions offer outstanding opportunities for NMR studies of protein–ligand complexes, providing structural information accessible with relative ease. Among the paramagnetic metal ions, lanthanide ions constitute an invaluable tool, offering a variety of paramagnetic features. The lanthanide ion bound to a molecular scaffold constitutes a powerful spectroscopic probe, generating perturbations of the NMR signals of nuclei present even at long distances from the metal (up to ∼40 Å) and becoming the center of a reference frame where the spatial coordinates of the affected nuclei can be accurately determined. Applications are not only restricted to metalloproteins since, even in the absence of metal binding sites, paramagnetic ion labeling strategies are available.

Paramagnetic-derived structural restraints can be used for the analysis of protein–ligand complex structure determination in the same way as they are employed for the calculation of the tertiary structure of a single macromolecule. However, in the former case, additional knowledge about the binding equilibrium is also required. Indeed, the measured paramagnetic effect on the molecular component which does not incorporate the metal ion corresponds to the effect of the bound state, scaled by the bound fraction of the considered species: $PE^{measured} = PE^{bound} \times f^{bound}$, where $f^{bound} = [bound]/[bound + free]$. f^{bound} can be calculated at any concentration if the value of the dissociation constant (K_d) is known.

A further aspect that may differ in the analysis of protein–ligand complexes as compared to single molecule problems relates to the structure calculation method itself. The most rigorous approach to the calculation of the structure of a complex requires sufficient experimental data to define both the structures of the individual components and their arrangement relative to each other. However, such a full dataset might not be easily obtainable or there could be an interest in the description of the sole ligand binding site, which is often the case in drug discovery programs. In these situations, faster approaches are better suited. Data-driven docking procedures can be used to bring together the two components of a complex if their structures in isolation are known (van Dijk et al. 2005). The attainable resolution will depend on the nature, number, and quality of the NMR-derived structural restraints. In their simplest implementation, docking algorithms comprise a rigid-body approach in which the two molecules are brought together using a target energy function that includes a term describing the available NMR data. More accurate solutions can be determined if side chains around the binding site are allowed to move under a specified force field in order to best accommodate the ligand. The most widely used protein structure calculation software, including CYANA and XPLOR-NIH, as well as the docking program HADDOCK, have already been integrated by modules, allowing the incorporation of most paramagnetic-based restraints.

In the following, the paramagnetic effects most relevant for protein–ligand complex structure determination are briefly described, followed by a summary of the possibilities to incorporate lanthanide ions in these systems. A large variety of structure calculation strategies including paramagnetic restraints are possible, thereby preventing the outline of a unique protocol. Three application examples based on rather distinct experimental approaches are further described, possibly providing a broad overview of the versatility of lanthanide-based structural calculations.

Paramagnetic Effects as Restraints for NMR Structure Calculation

Paramagnetism originates from unpaired electrons and is conveniently described in terms of a magnetic susceptibility tensor χ (often decomposed in an isotropic component χ_{iso} and an anisotropic part $\Delta\chi$). As the magnetic moment of electrons is almost three orders of magnitude larger than that of protons, the presence of paramagnetic substances causes significant perturbations in NMR spectra (Bertini 2001). The effects are evident at great distances from the paramagnetic center and, therefore, are of particular interest for the study of biological systems. The most prominent effects caused by lanthanide ions in the NMR spectra of biomolecules are pseudocontact shifts (PCSs), paramagnetic relaxation enhancements (PRE), and residual dipolar couplings (RDCs).

Pseudocontact Shifts

The presence of metal ions with a nonzero $\Delta\chi$ tensor causes changes in chemical shifts of nearby nuclei, called paramagnetic shifts. The latter result from both through-bond (contact shifts) and through-space (pseudocontact shifts) interactions. For stably bound lanthanide ions, contact shifts can often be neglected because the unpaired electrons reside in the inner f-orbitals that are shielded from the ligand field and because they do not readily delocalize into coordinating ligand molecules due to the ionic character of the interaction. PCSs are independent of the isotropic component of the χ tensor and are generally described in terms of the axial and rhombic components of the $\Delta\chi$ anisotropy tensor:

$$\Delta\delta^{PCS} = \frac{1}{12\pi r^3} \times \left[\Delta\chi_{ax}(3cos^2\theta - 1) + \frac{3}{2}\Delta\chi_{rh}sin^2\theta cos2\varphi \right]$$

where $\Delta\delta^{PCS}$ is the difference in chemical shifts measured between diamagnetic and paramagnetic samples and r (the distance between the metal ion and the nuclear spin), θ and φ are the polar coordinates describing the position of the nuclear spin with respect to the principal axes of the $\Delta\chi$ tensor. The above interaction, therefore, depends on the orientation of the molecule in solution and does not average to zero as the molecule tumbles.

PCS data possess both long-range distance and angular dependence and, therefore, can be exploited as a tool in NMR structural studies. PCSs values are determined from chemical shift changes extracted typically from ^1H–^{15}N- and/or ^1H–^{13}C-HSQC experiments (but also from ^1H and ^{13}C one-dimensional spectra), recorded for both the paramagnetic (e.g., Dy^{3+}) and diamagnetic (e.g., La^{3+}) states of a sample. PCSs give access to a coordinate system that is tied to a specific molecule and centered about the metal ion and that can be used as a reference frame to position other nuclear spins belonging to a molecular partner by their PCSs.

Paramagnetic Relaxation Enhancements

Paramagnetic centers cause line broadening in the NMR spectrum due to enhancements of the nuclear transverse relaxation rates R_2. In addition, a paramagnetic-induced increase in the longitudinal relaxation rate R_1 also occurs. The relaxation enhancements are essentially due to dipolar interaction mechanisms termed Solomon and Curie-spin mechanism. The former is predominant when electronic lifetimes are long, as in the case of Gd^{3+}, while the latter mechanism becomes more important in case of paramagnetic lanthanides other than Gd^{3+}, characterized by short electronic lifetimes, and particularly at high magnetic fields. PREs are proportional to the nuclear spin gyromagnetic ratio γ_I and to the isotropic component of the χ tensor (χ_{iso}), which in turn depends on the magnitude of the electron gyromagnetic ratio γ_S. Both Solomon and Curie mechanisms show a metal-nucleus distance dependency of r^{-6}, although additional mechanisms such as cross correlation between chemical shift anisotropy and dipolar shift anisotropy may sometimes cause more complicated dependencies. Paramagnetic centers with predominantly Solomon relaxation mechanisms are insensitive to cross correlation effects, making Gd^{3+} the lanthanide ion of choice for distance measurements by PRE.

Transverse relaxation enhancement data can be derived in several different ways, for example, by measuring the decrease in the intensity of signals in a ^1H–^{15}N HSQC spectrum of the protein after introduction of the paramagnetic moiety, taking advantage of the following relationship:

$$PRE(R_2) = \frac{1}{t} ln\left(\frac{I^{para}}{I^{dia}}\right)$$

where I^{para} and I^{dia} are the peak intensities of resonances in the presence of a paramagnetic and diamagnetic ion, respectively, and t is the total time during the periods of the HSQC pulse sequence in which the amide proton magnetization is in the transverse plane and undergoing paramagnetic relaxation.

Similarly, proton longitudinal relaxation times can be measured using a modified HSQC pulse sequence incorporating a classical inversion recovery module (180°-τ-90°-) and measuring peak intensities at varying time delays. Again, paramagnetic contributions $PRE(R_1)$ are evaluated as the difference in R_1 measured for the paramagnetic and diamagnetic samples.

Incorporation of the derived distances into a structure calculation requires some care, given that many of the paramagnetic tags are relatively mobile. The observed PRE effects will be a time average of all the conformations sampled by the tag, and so rather than treating the paramagnetic center as a fixed point in structure calculations, it can be treated as an ensemble average. Alternatively, the data can be implemented as loose/ambiguous restraints that can help guide the structure.

Residual Dipolar Couplings

The dipolar interaction between nuclei gives rise not only to relaxation effects such as the NOE but also to a splitting of NMR signals that is dependent both on the proximity of the two nuclei and on the orientation of their internuclear vector relative to the applied magnetic field. The approximately isotropic tumbling of a protein in solution results in these dipolar couplings being averaged to zero. Only under certain specific conditions, dipolar interactions are not totally canceled out and small dipolar splittings are observed in the solution NMR spectra (of the order of up to a few tens of Hertz).

Molecular alignment can be obtained by use of liquid crystal media or when the molecule itself has sufficiently high magnetic susceptibility anisotropy. All lanthanide ions with nonzero χ-tensor anisotropy generate weak molecular alignment with the external magnetic field and, consequently, RDCs. The alignment tensor A originates and is proportional to the $\Delta\chi$ tensor, with the same axes directions:

$$A = \frac{B_0^2}{15\mu_0 kT}\Delta\chi.$$

where B_0 is the magnetic field strength, μ_0 the induction constant, k the Boltzmann constant, and T the temperature.

The measurement of RDCs provides information about how bond vectors between observed nuclei are aligned relative to the external magnetic field, which in turn yields the orientation of these bond vectors relative to each other. By independently measuring the RDCs of the two components of a protein–ligand complex, the orientations of the two molecules relative to each other can be calculated. In practice, a tensor describing the interaction of the internuclear vectors with the external magnetic field must be calculated; measured RDC values are then compared with values that are back-calculated from the angle between each internuclear vector and the alignment tensor. Finally, the orientation of each component is optimized with respect to a single alignment tensor. These data provide powerful long-range geometric constraints, particularly in cases where the interaction is weak.

Lanthanides in Proteins and Ligands

Lanthanides form stable trivalent ions without any known essential role in biology. Although these metal ions are not found naturally bound to proteins or small ligands, they can be incorporated artificially in molecular frameworks (Scheme 1). The latter possibility has opened the way to an increasing number of NMR applications in structural biology and drug discovery. The great interest toward lanthanides resides in the strong paramagnetism associated with most trivalent

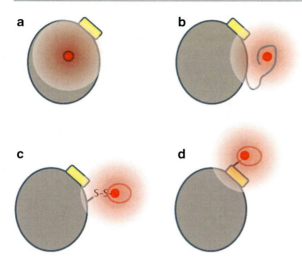

NMR Structure Determination of Protein-Ligand Complexes using Lanthanides, Scheme 1 Approaches to incorporate paramagnetic lanthanides in protein–ligand complexes. (**a**) Replacement of physiological metal ions with lanthanide ions within the protein binding site. (**b**) Protein fusion with lanthanide-binding peptides. (**c**) Chemical protein tagging with metal chelators. (**d**) Site-specific chemical ligand tagging with metal complexes. The protein is represented in *gray*, the ligand in *yellow*, the lanthanide ion in *red*, the fusion peptide in *light blue*, and the metal chelator in *orange*. The paramagnetic effect of the metal ion is depicted as an isotropic influence *shaded region* for simplicity

cations, which makes them very attractive as spectroscopic probes. Particularly, they have a recognized capability of providing long-range restraints to be used for molecular structure calculations (Pintacuda et al. 2007).

The paramagnetism of lanthanides arises from unpaired electrons in the f-orbitals of their trivalent ions, Ln^{3+}. All lanthanide ions are paramagnetic, except La^{3+} and Lu^{3+}. The chemical properties of Ln^{3+} ions are quite similar, and their size is comparable, ranging from 1.00 Å (Lu^{3+}) to 1.17 Å (La^{3+}). As a consequence, the binding affinities for proteins and ligands do not differ to a large extent, and it is possible to exchange ion at the best convenience without altering molecular structures. For example, it becomes possible to use diamagnetic lanthanide ions as reference probes, allowing the measurement of paramagnetic shifts simply as the difference in chemical shifts observed in the presence of a paramagnetic or diamagnetic lanthanide. In addition to La^{3+} and Lu^{3+}, also Y^{3+}, a chemically related nonlanthanide ion with the same radius as Dy^{3+}, is often used as diamagnetic reference.

Gd^{3+} is the only paramagnetic lanthanide ion with an isotropic environment of the unpaired electrons, resulting in a long electronic relaxation time that gives rise to strong relaxation enhancements in NMR spectra without significant changes in chemical shifts. In contrast, the electronic relaxation times of all other paramagnetic Ln^{3+} ions are very short and magnetic anisotropy is strong due to nonisotropic populations of f-orbitals, resulting in large PCSs and reduced relaxation enhancements. Placed in the same chemical environment, different Ln^{3+} ions display different magnitudes of the χ tensor and its associated anisotropy. Therefore, lanthanides with small $\Delta\chi$ tensors can be exploited to obtain structural information close to the metal ion, while highly paramagnetic ions can be used to observe long-distance effects.

Many of the early applications of lanthanide-containing biomolecules were concerning metalloproteins, since an appropriate Ln^{3+} ion can be incorporated through replacement of an original ion in the metal binding site. For example, Ca^{2+} ions of calcium-binding proteins can be effectively substituted by lanthanide ions (Bertini et al. 2001). However, more general applicability has been achieved by the introduction of lanthanide-binding tags (Otting 2010). A variety of metal ion tags has been described. Metal binding peptide motives can be attached to the N- or C-termini of proteins as fusion peptides. Alternatively, lanthanide-binding peptide (LBP) tags or chemical compounds can be site-specifically attached to proteins exploiting formation of disulfide bonds.

Applications of lanthanide-incorporating low-molecular-weight ligands have been essentially restricted to Gd compounds. Due to their outstanding relaxation enhancing properties, Gd^{3+} complexes are the most frequently used paramagnetic contrast agents in clinical MR imaging (Caravan 2006). Since the free metal ions are poorly tolerated, they must be coordinated by a strongly binding ligand that occupies most of the available coordination sites. Recently, the search for new-generation contrast agents has been strongly directed to providing them with high tissue and/or organ specificity. To this purpose, drugs have been synthesized according to a general scheme which envisions three essential components: the recognition synthon, a spacer, and the Gd^{3+} chelate. Recognition synthons such fatty or bile acid moieties render the complexes specific binders of lipid carrier proteins such as albumin or fatty acid binding proteins.

Because the relativity of Gd agents is increased when binding to slow-moving macromolecules, there is a growing interest in the elucidation of three-dimensional structures of protein/gadolinium-contrast-agent adducts in order to improve molecular design.

In principle, lanthanide-incorporating ligands could find widespread use in complex structure calculations as an alternative way to introduce paramagnetic probes in protein–ligand systems.

Example Applications

Lanthanide-Binding Proteins and Transferred Paramagnetic Shifts

Many protein targets, and consequently protein–ligand complexes, have large molecular weights and are, therefore, accessible with difficulty by solution NMR spectroscopy. Indeed, the long rotational correlation time of these systems determines fast nuclear spin relaxation that rapidly cancels the observable magnetization during experiment. This observation has prompted Otting and coworkers to develop a method for protein–ligand structure determination relying on transferred PCS measurements (John et al. 2006). The proposed approach requires the presence of a lanthanide ion bound to the target protein near the ligand binding site and rapid exchange between the free and bound forms of the low-molecular-weight binder. The latter requirement ensures that information about the ligand bound state is transferred to the averaged NMR signal. In the presence of a large excess of ligand, NMR parameters such as PCSs can be measured from NMR spectra of the same quality as that of the free ligand molecule, without suffering from the limitations due to the large molecular weight of the complex. A structural model of the protein obtained by X-ray crystallography or NMR is further necessary to allow the determination of the complete three-dimensional structure of the complex. The method was verified with the ternary 30-kDa complex between the lanthanide-labeled N-terminal domain of the ε exonuclease subunit from the E. coli DNA polymerase III (ε186), the subunit θ, and thymidine. ε186 displays high affinity for single lanthanide ions in the active site. Thymidine binds weakly to the protein, with a dissociation constant of about 7 mM, and exchanging fast between free and bound forms on the NMR time scale.

The magnetic susceptibility tensor parameters of a number of lanthanide ions (Dy^{3+}, Tb^{3+}, Er^{3+}) with respect to the protein were determined using amide PCS of ε186 from ^1H–^{15}N HSQC and HNCO spectra. These parameters agree with large PCS for the entire active site and line broadening beyond detection for all proton nuclei within a distance of about 15 Å from the metal. Simple ^1H and ^{13}C NMR spectra of thymidine were used to evaluate PCSs induced on binding to a lanthanide-incorporating target protein. In titration experiments, concentration-dependent relaxation enhancements and PCSs were observed (Fig. 1), allowing the determination of bound ligand fraction f^{bound} through a fitting procedure. The latter information is essential to extract the paramagnetic effects experienced by thymidine in the bound state.

In order to obtain precise PCSs measurements, it was chosen to analyze natural isotopic abundance ^{13}C NMR spectra. Indeed, the lower gyromagnetic ratio of this nucleus compared to that of the proton renders it less susceptible to paramagnetic line broadening. Thus, it becomes possible to observe signals of nuclei at shorter distance from the paramagnetic center and to detect even small peak shifts in the absence of significant Δv_{para} that would compromise spectral resolution. Figure 2 displays the comparison between the ^{13}C NMR spectrum recorded for ε186/θ/Dy^{3+} and ε186/θ/La^{3+}. The differential line broadening indicated by different peak intensities for the paramagnetic sample constitutes an indication of the distance of the ^{13}C nuclei from Dy^{3+}. $\Delta\delta_{para}^{bound}$ values, determined from the observed $\Delta\delta_{para}$ and the calculated f^{bound}, were obtained for thymidine in the presence of ε186/θ/Dy^{3+}, ε186/θ/Tb^{3+}, and ε186/θ/Er^{3+}. The diamagnetic shifts due to thymine binding were subtracted using ε186/θ/La^{3+} as a reference. Contributions from contact shifts and anisotropic chemical shifts were considered negligible, allowing to equate $\Delta\delta_{para}^{bound}$ with $\Delta\delta^{PCS}$.

The structure of ε186/θ/thymidine was calculated using the software XPLOR-NIH, with the modifications used to incorporate refinement against paramagnetic observables. The experimental restraints included only PCSs, and the available susceptibility tensor parameters were introduced. Four independent calculations were run by including the PCSs data corresponding to one of the three paramagnetic ions or including all datasets simultaneously. During the docking procedure, the backbone atom coordinates

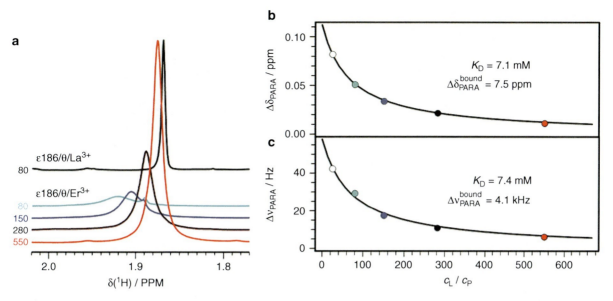

NMR Structure Determination of Protein-Ligand Complexes using Lanthanides, Fig. 1 (a) ^{1}H NMR resonance of the methyl group of thymidine in the presence of diamagnetic ε186/θ/La^{3+} and paramagnetic ε186/θ/Er^{3+} at increasing thymidine concentrations. The spectra are labeled with the ligand/protein ratio. The weak peak at 1.89 ppm originates from the protein. (b) Paramagnetic shift and (c) paramagnetic line broadening (= line width at half-height minus line width observed in the diamagnetic spectrum) of the methyl ^{1}H NMR resonance plotted versus the ligand/protein ratio. The *solid line* represents the best fitting curve (Reproduced with permission from John et al. (2006). Copyright 2006 American Chemical Society)

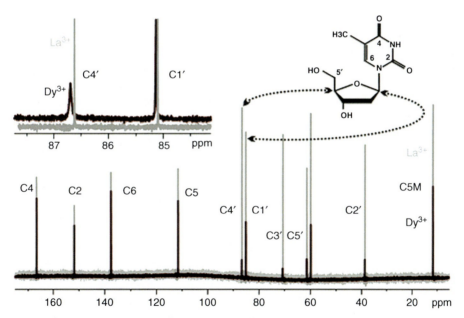

NMR Structure Determination of Protein-Ligand Complexes using Lanthanides, Fig. 2 Natural abundance ^{13}C NMR spectra of 50 mM thymidine in the presence of 90 μM ε186/θ/Dy^{3+} (*red*) and ε186/θ/La^{3+} (*gray*), respectively. The diamagnetic spectrum (128 scans) is scaled eightfold with respect to the paramagnetic spectrum (1,024 scans). The signal at 60 ppm originates from Tris buffer. The resonances of C4′ and C1′ of thymidine are expanded in the insert, highlighting their different paramagnetic line broadenings and shifts (Reproduced with permission from John et al. (2006). Copyright 2006 American Chemical Society)

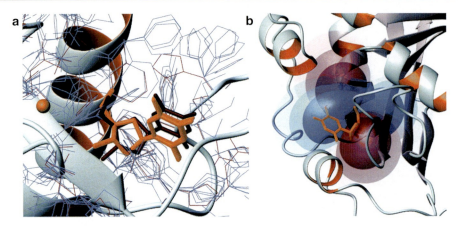

NMR Structure Determination of Protein-Ligand Complexes using Lanthanides, Fig. 3 (a) Superposition of the seven best structures of the ε186/thymidine complex calculated using data from all three lanthanides Dy^{3+}, Tb^{3+}, and Er^{3+}. The protein backbone is represented as a *gray* and *yellow* ribbon. Protein side chains and thymidine molecules from the calculated structures are shown in *blue* and *yellow*, respectively. Protein side chains and the TMP molecule in the crystal structure of ε186 are overlaid in *red*. (b) Best structure superimposed with the $\Delta\chi$ tensor of Er^{3+}. The tensor is represented by *blue* and *red* lobes for positive and negative PCS, respectively, contoured at PCS isosurfaces of ±5, ±2.5, and ±1.25 ppm (Reproduced with permission from John et al. (2006). Copyright 2006 American Chemical Society)

were kept fixed, while those of the side chains were allowed to be flexible. Similar solutions were found from all calculations, with the ligand positioned in an analogous location and orientation as the TMP molecule in the crystal structure of ε186 (Fig. 3a). In Fig. 3b, the best structure of the selected cluster is superimposed onto the $\Delta\chi$ tensor of Er^{3+}, represented by isosurfaces of positive and negative PCS.

The quality of the determined complex structure was evaluated as high on the basis of the good correlation between experimental and back-calculated PCSs, thereby demonstrating the feasibility of the proposed approach. One advantage of the use of transferred PCS over that of transferred NOE and transferred cross correlated relaxation is that paramagnetic effects are much larger and present the advantage of being scalable by the use of a range of different lanthanide ions.

Lanthanide-Tagged Proteins: Combined Use of Pseudocontact Shifts and Residual Dipolar Couplings

Prestegard and coworkers have proposed the use of a combination of orientationally sensitive measurements (RDCs) and long-range distance- and orientation-dependent data (PCSs) as an approach for NMR structure determination of protein–ligand complexes (Zhuang et al. 2008). They demonstrated the feasibility of the method investigating the Galectin-3/lactose complex. Galectin-3 is a member of the galectin family of lectins defined by a conserved ~14-kDa carbohydrate recognition domain showing affinity for β-galactosides. Lactose is a small soluble analog of these ligands displaying weak binding affinity to galectin-3 (K_d = 0.2 mM) and undergoing fast exchange on and off the protein binding site.

In this study, a fusion construct was designed to add a short polypeptide (YIDTNNDGWYEGDELLA) to the C-terminus of the protein. The peptide was previously shown to present a high affinity for lanthanide ions. Tagged galectin-3 samples were prepared, incorporating the diamagnetic metal ion Lu^{3+} and the paramagnetic metal ion Dy^{3+}. The described protocol for the complex structure determination included the separate measurement of RDCs and PCSs for the protein and the ligand (in the bound form), followed by docking of the lactose into the protein binding site driven by the obtained structural restraints. As expected, the presence of Lu^{3+} produced chemical shift changes of protein amide ^1H–^{15}N resonances, as well as peak intensity attenuations; however, the signals from about 35 amino acid residues could be safely assigned and analyzed.

RDCs of the protein were obtained by comparison of the cross peak position difference between ^1H–^{15}N HSQC and TROSY spectra of the paramagnetic

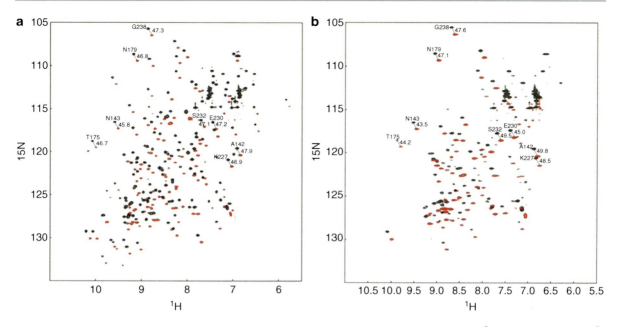

NMR Structure Determination of Protein-Ligand Complexes using Lanthanides, Fig. 4 ^{1}H–^{15}N HSQC and TROSY overlays for 0.3 mM galectin-3–LBT at a ^{1}H frequency of 600 MHz with (**a**) 0.3 mM Lu^{3+} and (**b**) 0.3 mM Dy^{3+} (Reproduced with permission from Zhuang et al. (2008). Copyright 2008 The Protein Society)

sample, with respect to the difference observed for the same set of spectra collected for the diamagnetic sample (Fig. 4). The cross peaks of the fully decoupled HSQC spectrum appear displaced from the undecoupled TROSY cross peaks by one half the sum of scalar and dipolar couplings in both dimensions. The difference in the offsets between the two samples corresponds to one half the RDC coupling at each site. The RDC data were used to determine the order tensor elements and to back-calculate PCSs, as in the absence of mobility, both sets of data relate to the same susceptibility tensor. It was therefore possible to define the position of the metal center in best agreement with the experimental data. The good correlation between experimental and calculated values also supported the modeling of the tag as rigid with respect to the protein.

RDCs and PCSs of the ligand were measured on a sample containing galectin-3 and lactose at a molar ratio of 1:5 in order to reach appropriate signal-to-noise and corrected by the known bound ligand fraction. The data were obtained from ^{1}H–^{13}C HSQC spectra taken with natural abundance material and measuring the splitting difference in the ^{13}C dimension between the paramagnetic and diamagnetic samples. Using available crystal structure coordinates from a similar system, the order tensor elements for the ligand were derived. The obtained values were similar to those found for the protein. It is also expected that the actual alignment frame orientations are shared between galectin-3 and lactose, thereby allowing the determination of the relative orientation of protein and ligand once the molecular coordinates are transformed to their principal alignment frames. It has to be mentioned that four solutions are equally possible due to the insensitivity of RDCs to inversion of axes, and independent observations are required to select the correct orientation. In the present case, reference was made to the crystal structure to solve the ambiguity.

The complex structure calculation was performed by docking the ligand into the protein, considered as separate entities within the XPLOR-NIH software, including both orientational and translational constraints. The principal order parameters for the protein were used together with the experimentally determined RDCs and PCSs for both galectin-3 and lactose. The ten lowest energy structures showed a realistic positioning of the ligand within the binding site, although with a significant deviation from that of N-acetyllactosamine in the crystal structure (heavy atoms RMSD = 4.3 Å). The calculation was repeated by including also a single intermolecular NOE, and the agreement to the crystal structure improved

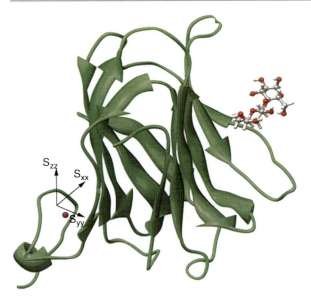

NMR Structure Determination of Protein-Ligand Complexes using Lanthanides, Fig. 5 Final ligand–protein complex structure determined by XPLOR-NIH using ligand and protein RDCs, PCSs, and a single ligand–protein NOE (Reproduced with permission from Zhuang et al. (2008). Copyright 2008 The Protein Society)

NMR Structure Determination of Protein-Ligand Complexes using Lanthanides, Scheme 2 Compound 1: a lipid-conjugated lanthanide ion (Ln^{3+}) chelating molecule

significantly (RMSD = 1.91 Å). The final ligand–protein complex structure is depicted in Fig. 5.

The described study demonstrates that the combined PCSs- and RDCs-based approach can be an alternative to classical NOE-based structure calculation methods. Given the long distance separating the paramagnetic ion from the ligand, rather small PCS values were measured, which resulted in a relatively low precision of distance constraints and forced the inclusion of one intermolecular NOE to attain a better positioning of lactose. It can be expected that the use of different lanthanide ions will provide the possibility to add further orientational and distance constraints.

Lanthanide-Incorporating Ligands

Lipid-conjugated gadolinium complexes constitute a new class of compounds with significant potential for the development of potent and specific MRI contrast agents. Indeed, the lipophilic moiety may be designed to specifically interact with lipid binding sites in proteins such as the abundantly expressed extracellular albumin and the intracellular fatty acid binding proteins. Binding of the contrast agents to these macromolecules increases the compound's relaxivity, allowing for enhanced sensitivity of the MR measurement, and can offer new routes for drug distribution and excretion. Bile acid-based gadolinium chelates have shown specificity for the hepatic tissue, being able to enter the cells by means of a protein-mediated active transport mechanism. Compound 1 (Scheme 2) and several related compounds have been shown to interact with a bile acid binding protein (BABP)($K_d \sim 5 \times 10^{-6}$ M, Assfalg et al. 2007), a member of the fatty acid binding protein family considered the putative bile acid carrier in the frame of the enterohepatic circulation. The NMR structure of the protein–ligand adduct has been determined in order to provide the basis for an improved drug design (Tomaselli et al. 2008).

NMR experiments were performed on a ^{15}N- or a ^{15}N,^{13}C- labeled protein sample in the presence of the Gd^{3+} containing compound 1 (Gd-1), its diamagnetic Y^{3+} analog (Y-1), or a combination of both ligands. Y-1 alone was used to run experiments for resonances assignment of the protein and the ligand in the complex, as well as to extract intermolecular distance information from conventional edited/filtered heteronuclear NOESY experiments. A total number of 14 intermolecular NOEs were found, involving mostly the three methyl groups of the sterol scaffold. The majority of the bile acid proton signals fall in the so-called diamagnetic hump of the spectrum and are severely overlapped. Their assignment is possible only in part once the first structural models have been calculated.

Gd^{3+} determines cancelation of signals from nuclei that are close in space and significant relaxation rate enhancements even at long distance. However, no chemical shift effects are induced, allowing straightforward resonance assignments. Due to the small size

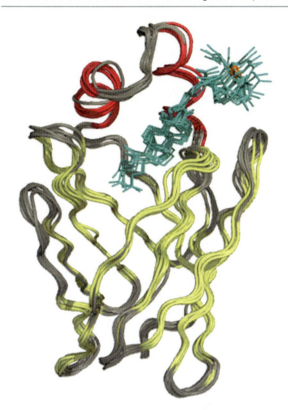

NMR Structure Determination of Protein-Ligand Complexes using Lanthanides, Fig. 6 Lowest-energy solution structures calculated for the complex BABP/Gd-1 (PDB: 2k62). The protein is shown in *ribbon* representation, color coded according to secondary structure (*yellow*: sheet, *magenta*: helix). The ligand is represented in *cyan sticks*, while the metal ion is indicated by an *orange sphere* (Reproduced with permission from Tomaselli et al. (2008). Copyright 2008 American Chemical Society)

of bile acid binding proteins (~14 kDa), saturation of the ligand binding site with a gadolinium-containing compound would destroy most protein signals. While it is possible to add nonsaturating amounts of the paramagnetic ligand, a complete structure determination requires that all protein molecules are bound to the ligand in order to be able to describe the holo state. In the considered study, a diamagnetic dilution approach has been followed, where only a small percentage of the saturating concentration of ligand involved Gd-1, the remaining part being provided by Y-1. In this way, almost all protein amide resonances remained observable, and paramagnetic enhancements of longitudinal relaxation rates could be measured with reasonable signal-to-noise ratio.

R_1 data were measured using an inversion recovery scheme coupled to a ^{15}N-editing sequence. PRE(R_1) was calculated as the difference in R_1 observed between a paramagnetic and a diamagnetic protein–ligand adduct. PRE(R_1) values up to 2.7 s^{-1} were determined and converted into distances according to the relation PRE(R_1) = $K_{dip} \times r^{-6}$. The proportionality constant K_{dip} was estimated based on independently acquired knowledge about the relevant correlation times of the system.

The calculation of the tertiary structure of the BABP/Gd-1 adduct was performed with HADDOCK with the inclusion of the intermolecular NOE- and PRE-derived distance restraints together with the whole set of protein intramolecular NOEs. This procedure was adopted to ensure highest precision of the solution by simultaneously exploiting the great capability of the docking algorithm to search in the intermolecular conformational space. Distance restraints derived from relaxation measurements consisted in upper and lower bounds set to ±2 Å of the calculated distances. The structure calculation led to a final ensemble of seven structures (Fig. 6) displaying backbone and global pairwise RMSD values for protein atom coordinates of 0.66 ± 0.11 and 1.42 ± 0.21 Å, respectively, and an RMSD referred to the ligand coordinates of 0.88 ± 0.3 Å, indicative of a very high precision.

Cross-References

▶ Lanthanide Metalloproteins
▶ Lanthanides in Biological Labeling, Imaging, and Therapy
▶ Lanthanides, Physical and Chemical Characteristics
▶ Lanthanum, Physical and Chemical Properties

References

Assfalg M et al (2007) NMR structural studies of the supramolecular adducts between a liver cytosolic bile acid binding protein and gadolinium(III)-chelates bearing bile acids residues: molecular determinants of the binding of a hepatospecific magnetic resonance imaging contrast agent. J Med Chem 50(22):5257–5268

Bertini I (2001) Solution NMR of paramagnetic molecules: applications to metallobiomolecules and models. Elsevier, Amsterdam

Bertini I et al (2001) Magnetic susceptibility tensor anisotropies for a lanthanide ion series in a fixed protein matrix. J Am Chem Soc 123(18):4181–4188

Caravan P (2006) Strategies for increasing the sensitivity of gadolinium based MRI contrast agents. Chem Soc Rev 35(6):512–523

John M et al (2006) Structure determination of protein–ligand complexes by transferred paramagnetic shifts. J Am Chem Soc 128(39):12910–12916

Otting G (2010) Protein NMR using paramagnetic ions. Annu Rev Biophys 39:387–405

Pintacuda G et al (2007) NMR structure determination of protein–ligand complexes by lanthanide labeling. Acc Chem Res 40(3):206–212

Tomaselli S et al (2008) Solution structure of the supramolecular adduct between a liver cytosolic bile acid binding protein and a bile acid-based gadolinium(III)-chelate, a potential hepatospecific magnetic resonance imaging contrast agent. J Med Chem 51(21):6782–6792

van Dijk ADJ, Boelens R, Bonvin AMJJ (2005) Data-driven docking for the study of biomolecular complexes. FEBS J 272(2):293–312

Zhuang T, Lee H, Imperiali B, Prestegard JH (2008) Structure determination of a Galectin-3-carbohydrate complex using paramagnetism-based NMR constraints. Protein Sci 17:1220–1231

Non-catalytic Zinc

▶ Zinc Structural Site in Alcohol Dehydrogenases

Non-corrin Cobalt

▶ Cobalt Proteins, Overview
▶ Nitrile Hydratase and Related Enzyme

Non-physiological/Non-biogenic/"Alien" Metal Cation

▶ Calcium Ion Selectivity in Biological Systems

Novel Therapeutics

▶ Tellurium in Nature

NPs

▶ Gold Nanoparticle Platform for Protein-Protein Interactions and Drug Discovery

NRAMP: Natural Resistance-Associated Macrophage Protein

▶ Iron Proteins, Plant Iron Transporters

Nuclear Factor of Activated T-Cell C3 (NFATc3)

▶ Calcium Sparklets and Waves

Nutritional Muscular Dystrophy

▶ Selenium and Muscle Function

Nutritionally Essential Metal

▶ Copper, Biological Functions

O

Obstruction

► Cadmium Exposure, Cellular and Molecular Adaptations

Oncomodulin

► Parvalbumin

Opposition

► Cadmium Exposure, Cellular and Molecular Adaptations

ORF, Open Reading Frame

► Zinc Aminopeptidases, Aminopeptidase from Vibrio Proteolyticus (Aeromonas proteolytica) as Prototypical Enzyme

Organic and Inorganic Arsenic

► Arsenic in Nature

Organic Mercury

► Mercury Toxicity

Organogermanium Compounds

► Germanium-Containing Compounds, Current Knowledge and Applications

Organometallic Complex/Protein Composite

► Palladium, Coordination of Organometallic Complexes in Apoferritin

Organometallic Inhibitors of Protein Kinases

► Platinum(II) Complexes, Inhibition of Kinases

Organometalloenzyme

► Palladium, Coordination of Organometallic Complexes in Apoferritin

V.N. Uversky et al. (eds.), *Encyclopedia of Metalloproteins*, DOI 10.1007/978-1-4614-1533-6,

Organometalloprotein

▶ Palladium, Coordination of Organometallic Complexes in Apoferritin

Organotin

▶ Tin Complexes, Antitumor Activity
▶ Tin, Toxicity

Origin of Life

▶ Iron-Sulfur Cluster Proteins, Ferredoxins

Osmium Complexes with Azole Heterocycles as Potential Antitumor Drugs

Iryna N. Stepanenko, Gabriel E. Büchel, Bernhard K. Keppler and Vladimir B. Arion
Institute of Inorganic Chemistry, University of Vienna, Vienna, Austria

Synonyms

Antitumor drug – anticancer drug; Complex – coordination or organometallic compound

Definition

1. Antimetastatic activity – activity against spreading of cancer cells via blood or lymph vessels
2. Antiproliferative activity – activity against the reproduction of similar (here cancer) cells
3. Cytotoxicity – producing a toxic effect that destroys living (here cancer) cells
4. In vitro (literally: within glass, lat.) – a test that is done in a laboratory vessel or generally out of a living organism
5. In vivo (literally: within the living, lat.) – a test that is done within a whole living organism

Introduction

Inorganic compounds have been known as therapeutic agents for 5,000 years (Orvig and Abrams 1999). However, the history of metal-based anticancer therapy has begun only with the discovery of the antitumor activity of cisplatin, cis-[Pt(NH$_3$)$_2$Cl$_2$], in 1965 (Rosenberg et al. 1965). Now this growing field of research includes both coordination and organometallic compounds containing gallium, gold, titanium, iron, and ruthenium (Hartinger and Dyson 2009; Jakupec et al. 2008; Peacock and Sadler 2008; van Rijt and Sadler 2009).

The first biological applications of osmium complexes were reported in the 1950s. Dwyer et al. (1952) studied the toxicity of tris-2,2'-bipyridine osmium(II) complex to mice and its inhibitory action on cholinesterase as a target enzyme (Koch et al. 1956), while Taylor et al. (1975) investigated the curariform paralysis of rats and mice by osmium(II) complexes bearing modified terpyridine and 2,2'-bipyridine ligands. Osmium tetroxide with reputation of highly toxic inorganic compound is clinically used for synovectomy in arthritic patients, while osmium carbohydrate derivatives belong to potential antiarthritic agents (Bessant et al. 2003; Goldstein et al. 2005; Hinckley et al. 1983). To the best of our knowledge the antitumor activity of osmium complexes was first tested by Craciunescu et al. on [Os^{IV}X$_6$]$^{2-}$ salts (where X = Cl, Br) with piperazine, antipyrine, thiazole, and sulphonamide derivatives (Craciunescu 1977; Craciunescu et al. 1991; Doadrio et al. 1977a, b, 1980) and neutral Os^{III}L$_3$ and Os^{IV}L$_2$Cl$_2$ complexes with dithiocarbamate or xanthate derivatives (Craciunescu et al. 1992) in the late 1970s – early 1990s. A new phase in anticancer osmium history started in the first decade of the twenty-first century when in search of a new alternative to known ruthenium anticancer therapeutics the heavier congener was chosen as an alternative metal for inorganic drug design.

Investigation of osmium compounds is of particular interest not only because it complements the growing family of ruthenium compounds which exhibit antitumor activity but also because of well-established differences between the metals. These are reflected in the preparation approaches to their coordination and organometallic compounds, the preference for higher oxidation states, the stronger π back-donation from lower oxidation states, and the much stronger spin-orbit coupling of the heavier congener, as well as the marked differences in metal-ligand exchange kinetics, which are essential for their medicinal applicability as anticancer drugs (Singh et al. 2006; Stepanenko et al. 2007; Peacock and Sadler 2008). The advantage of using osmium analogues with a cytotoxicity similar to that of their ruthenium congener lies in the higher inertness of osmium species under conditions relevant for drug formulation (Büchel et al. 2009; Peacock et al. 2007). For these reasons osmium is another metal that deserves attention in the development of effective inorganic antitumor drugs.

A family of azole complexes with osmium in different oxidation states (II, III, IV, VI) (Büchel et al. 2009, 2011a, b, 2012; Cebriàn-Losantos et al. 2007; Stepanenko et al. 2007, 2008; Chiorescu et al. 2007; Ni et al. 2012) and osmium(VI) complexes with Schiff-base ligands (Ni et al. 2011) have been reported recently, along with a large number of organoosmium (II) arene compounds (Bruijnincx and Sadler 2009; Casini et al. 2010; Dorcier et al. 2006; Filak et al. 2010; Fu et al. 2011; Hanif et al. 2010a, b; Schmid et al. 2007a, b), which possess varied antiproliferative activity, from complexes displaying activity in nanomolar concentrations to those being inactive at much higher concentrations in both in vitro and in vivo assays. The interest in osmium azole complexes is fuelled by the fact that two ruthenium(III) complexes, namely, (H_2im)[trans-RuCl$_4$(dmso)(Him)] (NAMI-A, Him = imidazole) and (H_2ind)[trans-RuCl$_4$(Hind)$_2$] (KP1019, Hind = indazole) (Fig. 1), are now under clinical trials (Clarke 2003; Groessl et al. 2011; Heffeter et al. 2010).

For platinum-based chemotherapeutics the covalent binding to the nucleophilic nitrogen atoms of DNA is the widely accepted biological event which is responsible for the induction of apoptosis in tumor cells. Nevertheless, the emerged on this basis approach for the design of new metal-based anticancer drugs that bind preferably to DNA (classical anticancer drugs) underwent critical discussion recently and new target proteins have been suggested for cancer therapy (Ang

Osmium Complexes with Azole Heterocycles as Potential Antitumor Drugs, Fig. 1 Ruthenium(III) complexes in clinical trials: (H_2im)[trans-RuCl$_4$(dmso)(Him)] (NAMI-A, Him = imidazole) (left) and (H_2ind)[trans-RuCl$_4$(Hind)$_2$] (KP1019, Hind = indazole) (right)

and Dyson 2006). As a result nonclassical anticancer drug candidates are now under active investigation, e.g., antimetastatic coordination (NAMI-A) and organoruthenium compounds [Ru(η^6-arene)(pta)Cl$_2$], where pta = 1,3,5-triaza-7-phosphatricyclo[3.3.1.1]-decane (Casini et al. 2010; Dyson and Sava 2006; Ang and Dyson 2006).

Unfortunately, the mechanism of biological activity of known organoosmium(II) compounds has not yet been established at least at the molecular level. Compared to analogous ruthenium compounds their slower hydrolysis or diminished hydrolytic lability, as well as the absence of correlation between DNA binding/hydrolysis and cytotoxicity, suggest that organoosmium(II) compounds have a different mechanism of action and DNA may not necessarily be the target (Bruijnincx and Sadler 2009; Fu et al. 2010; van Rijt et al. 2010).

This entry covers a series of eight different types of azole-based osmium complexes (Fig. 2), namely, trans,cis,cis-[Os^{II}Cl$_2$(dmso)$_2$(Hazole)$_2$] (**A**), cis,fac-[Os^{II}Cl$_2$(dmso)$_3$(Hazole)] (**B**), (H_2azole)[trans-Os^{III}Cl$_4$(dmso)(Hazole)] (**C**), mer-[Os^{III}Cl$_3$(Hazole)$_3$] (**D**), trans-[Os^{III}Cl$_2$(Hazole)$_4$]Cl (**E**), cis-[Os^{III}Cl$_2$(Hazole)$_4$]Cl (**F**), (Cation)[Os^{IV}Cl$_5$(Hazole)] (**G**), and trans-[Os^{IV}Cl$_4$(Hazole)$_2$] (**H**). These osmium complexes will be compared with analogous ruthenium azole compounds from the point of view of their synthetic feasibility with emphasis on the

Osmium Complexes with Azole Heterocycles as Potential Antitumor Drugs, Fig. 2 Types (A–H) of osmium complexes and their Hazole (L) ligands (**1–5**)

differences between these two metals in substitution chemistry, as well as structure-property relationships, which can be useful for future drug design. These osmium complexes, stable under biologically relevant conditions, are considered as potential nonclassical anticancer drug candidates, whose mechanisms of action remain to be elucidated.

Synthesis of Osmium Complexes with Azole Heterocycles

Osmium(II) and osmium(III)–Dmso/Hazole Complexes

Trans-[Os^{II}Cl$_2$(dmso)$_4$], *cis*-[Os^{II}Cl$_2$(dmso)$_4$], and [(dmso)$_2$H][Os^{III}Cl$_4$(dmso)$_2$] were used as starting compounds for the synthesis of osmium–dmso/Hazole complexes. Prepared by reduction of Os^{VIII}O$_4$ in the presence of N$_2$H$_4$·2HCl and concentrated HCl (Brauer 1981) or by a cation exchange reaction of (NH$_4$)$_2$[Os^{IV}Cl$_6$] (McDonagh et al. 1997), the H$_2$[Os^{IV}Cl$_6$] was further reduced to Os(III) and Os(II) by SnCl$_2$·2H$_2$O in the presence of dimethyl sulfoxide at 85–90 °C with the formation of [(dmso)$_2$H][Os^{III}Cl$_4$(dmso)$_2$] (Cebrián-Losantos et al. 2007) and *trans*-[Os^{II}Cl$_2$(dmso)$_4$] (Alessio et al. 2003; McDonagh et al. 1997). The last compound was converted into *cis*-[Os^{II}Cl$_2$(dmso)$_4$] in dimethyl sulfoxide at 150 °C (Alessio et al. 2003). Whereas both [(dmso)$_2$H][Os^{III}Cl$_4$(dmso)$_2$] and *trans*-[Os^{II}Cl$_2$(dmso)$_4$] contain *S*-coordinated dmso ligands, *cis*-[Os^{II}Cl$_2$(dmso)$_4$] exists as a mixture of *cis,fac*-[Os^{II}Cl$_2$(dms*o*)(dms*o*)$_3$], containing three facially *S*-coordinated dmso ligands and one dmso ligand bound to osmium via oxygen, and thermodynamically more stable *cis*-[Os^{II}Cl$_2$(dmso)$_4$], isomers (Alessio et al. 2003).

Compounds [(dmso)$_2$H][*trans*-M^{III}Cl$_4$(dmso)$_2$], *cis,fac*-[M^{II}Cl$_2$(dms*o*)(dms*o*)$_3$], and *trans*-[M^{II}Cl$_2$(dmso)$_4$] have also been reported for ruthenium (Alessio 2004). However *cis*-[M^{II}Cl$_2$(dmso)$_4$], with all dmso ligands coordinated to metal via sulfur, is known only for osmium(II). Osmium(III) NAMI-A analogues have been synthesized much later than the related ruthenium-azole-dmso compounds because of the lack of suitable starting materials. Although evidence

Osmium Complexes with Azole Heterocycles as Potential Antitumor Drugs, Fig. 3 Synthetic routes to complexes **A**, **B**, and **C**. For labeling used see Fig. 2

for the formation of [trans-Os^{III}Cl$_4$(dmso)$_2$]$^-$ in solution was reported (Komozin 2000), [(dmso)$_2$H][Os^{III}Cl$_4$(dmso)$_2$] was isolated and characterized as a solid only quite recently (Cebrià n-Losantos et al. 2007).

The treatment of trans-[OsCl$_2$(dmso)$_4$] with azole ligands in a 1:2 molar ratio in dry boiling ethanol afforded complexes trans,cis,cis-[OsCl$_2$(dmso)$_2$(Hazole)$_2$] (**A**: L = **1, 2, 4, 5**) with 80–88% yields. The reaction starting from cis-[OsCl$_2$(dmso)$_4$] under similar conditions resulted in formation of complexes cis,fac-[OsCl$_2$(dmso)$_3$(Hazole)] (**B**: L = **1, 2, 4, 5**) with 46–85% yields (Fig. 3) (Stepanenko et al. 2007). Osmium(III) NAMI-A analogues (H$_2$azole) [Os^{III}Cl$_4$(dmso)(Hazole)] (**C**: L = **1–5**) were synthesized by reaction of [(dmso)$_2$H][trans-Os^{III}Cl$_4$(dmso)$_2$] with excess azole ligand on heating in dry acetone, ethanol, or methanol in 50–80% yields (Fig. 3) (Cebrià n-Losantos et al. 2007). The formation of complexes of types **A**, **B**, and **C** was explained by a remarkable destabilizing trans influence of the two dmso ligands. Complexes **A** were converted into complexes **B** by heating in dimethyl sulfoxide at 125°C in 70–75% yields (Fig. 3) and two pathways for such transformation were proposed and discussed (Stepanenko et al. 2007).

The ruthenium analogues of type **B** were prepared starting from cis,fac-[Ru^{II}Cl$_2$(dmso)(dmso)$_3$] and azole ligand in molar ratio 1:1 by replacement of the more labile O-bonded dmso under mild conditions (Henn et al. 1991; Alessio 2004; Taqui Khan et al. 1992). Excess azole ligand in boiling solvents leads to the substitution of two dmso ligands with formation of [Ru^{II}Cl$_2$(dmso)$_2$(Hazole)$_2$]. The latter is identified often as a mixture trans,cis,cis- (type **A**) and cis,cis,cis- isomers (Alessio 2004). The trans,trans,trans-[Ru^{II}Cl$_2$(dmso)$_2$(Hazole)$_2$] isomer is also well-documented (Reisner et al. 2005a). In the case of osmium, because of its increased kinetic inertness compared to ruthenium, only compounds **A** and **B** were isolated (Stepanenko et al. 2007). [Ru^{III}Cl$_4$(dmso)(Hazole)]$^-$ compounds (type **C**) were obtained under milder conditions than the osmium analogues as

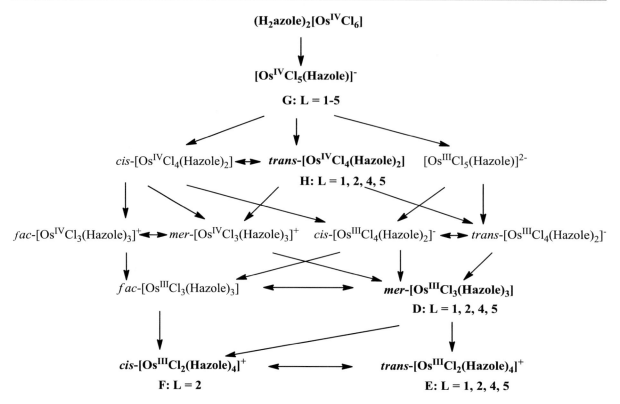

Osmium Complexes with Azole Heterocycles as Potential Antitumor Drugs, Fig. 4 Synthetic routes to complexes D, E, F, G, and H. For labeling used see Fig. 2

expected due to the higher reactivity of ruthenium species (Alessio et al. 1993; Reisner et al. 2004).

Osmium(III) and osmium(IV)–Hazole Complexes

The synthetic route to osmium–Hazole compounds is shown in Fig. 4 (complexes drawn in bold style were isolated as solids) and includes the concurrent formation of *mono*-, *bis*-, *tris*-, *tetra*-substituted osmium(IV) and osmium(III) complexes.

The precursors $(H_2azole)_2[Os^{IV}Cl_6]$ were obtained by reaction of $[(dmso)_2H]_2[Os^{IV}Cl_6]$ with azole heterocycles in a 1:3 molar ratio in dry ethanol at room temperature in 79–97% yields (Büchel et al. 2009; Rudnitskaya et al. 1994; Stepanenko et al. 2008). $[Os^{IV}Cl_6]^{2-}$ is well-documented with varying organic cations (e.g., protonated phenanthroline, thiazole, quinoline, bipyridine, piperazine, piperidine, acridine), and the interest in their use as potent therapeutics is a thing of the past (Craciunescu et al. 1991; Doadrio et al. 1977a).

Replacement of the chloride ligands in $[OsCl_6]^{2-}$ by an azole heterocycle in boiling isoamyl alcohol (110 °C for **D**: L = 1, 2, 4; **E2**; **F2**) or hexanol-1 (**D5**; **E**: L = 1, 4, 5) in the presence of free azole in molar ratio 1:1 or 1:2 for **D**, 1:4 for **E**, **F** leads to *mer*-$[Os^{III}Cl_3(Hazole)_3]$ and *trans*-$[Os^{III}Cl_2(Hazole)_4]Cl$ in 36–83% (**D**: L = 1, 4, 5; **E**: L = 1, 2, 5), and 5–10% yields (**D2**, **E4**) (Stepanenko et al. 2008; Chiorescu et al. 2007). Osmium(III) compounds were obtained in alcohol in the absence of any reducing agent. The *mer*-$[Os^{III}Cl_3(Hazole)_3]$ (**D**) is an intermediate in the synthesis of *trans*-$[Os^{III}Cl_2(Hazole)_4]Cl$ (**E**) and its yield is significantly increased when the amount of azole heterocycle used relative to the amount of $(H_2azole)_2[Os^{IV}Cl_6]$ is 1:1 and the reaction is carried out under milder conditions. Nevertheless this synthesis is accompanied by the concurrent formation of *tris*- and *tetra*-substituted products which can be separated chromatographically. The fourth azole heterocycle coordinates to osmium upon substitution of the chlorido ligand *trans* to the azole ligand in *mer*-$[OsCl_3(Hazole)_3]$. The substitution of one of the two mutually *trans*-arranged chlorido ligands is less favored (<4% yield for **F2**) and was observed only for the strongest net electron-donor ligand imidazole (Fig. 4) (Stepanenko et al. 2008).

The complexes trans-[Ru^{III}Cl$_2$(Hazole)$_4$]Cl were synthesized starting from [Ru^{III}Cl$_3$(EtSPh)$_3$], (H$_2$azole)[Ru^{III}Cl$_4$(Hazole)$_2$], and mer-[Ru^{III}Cl$_3$(Hazole)$_3$] (where Hazole = Hpz, Him, Htrz, Hbzim) (Reisner et al. 2005b). Trans-[Ru^{II}Cl$_2$(Hind)$_4$] obtained by reaction of (H$_2$ind)[Ru^{III}Cl$_4$(Hind)$_2$] with indazole can be oxidized with H$_2$O$_2$ in the presence of HCl and indazole to produce trans-[Ru^{III}Cl$_2$(Hind)$_4$]Cl (Jakupec et al. 2005). The mer-[Ru^{III}Cl$_3$(Hazole)$_3$] is an intermediate in the synthesis of trans-[Ru^{III}Cl$_2$(Hazole)$_4$]Cl. Formation of cis-[Ru^{III}Cl$_2$(Hazole)$_4$]Cl or isomerization of trans-[Ru^{III}Cl$_2$(Hazole)$_4$]Cl into the cis-isomer was not observed. The ruthenium compounds are kinetically more labile toward ligand exchange or isomerization than the analogous osmium species. The minor yield of cis-[Os^{III}Cl$_2$(Him)$_4$]Cl along with predominant formation of the trans-isomer suggests that the transient cis-isomers in the case of ruthenium were not discovered because of their rapid isomerization into trans-species.

To avoid deeper (tris-, tetra-) substitution in [Os^{IV}Cl$_6$]$^{2-}$ the mono- (**G**) and bis- (**H**) substituted osmium complexes were obtained starting from (H$_2$azole)$_2$[Os^{IV}Cl$_6$] (Fig. 4). The reactions performed with suspensions of the starting compounds in hexanol-1 at 160 °C generate trans-[Os^{IV}Cl$_4$(Hazole)$_2$] (**H**: L = **1, 2, 5**) in 44–82% yields (Büchel et al. 2011a). The products **H4** with 1H- and 2H-indazole (vide infra) was obtained by heating (H$_2$ind)$_2$[Os^{IV}Cl$_6$] in the solid state at 150 °C (Büchel et al. 2011a, b). The main product obtained in 19% yield was trans-[Os^{IV}Cl$_4$(2H-ind)$_2$] (**H4 (2H)**), whereas trans-[Os^{IV}Cl$_4$(1H-ind)$_2$] (**H4 (1H)**) was isolated as a side product. The imidazole (**H2**) and pyrazole (**H1**) products can be obtained in alcohols at lower temperatures (85–130 °C). To quench the reaction after the first substitution step in [OsCl$_6$]$^{2-}$ a large organic cation (tetrabutylammonium, n-Bu$_4$N$^+$) facilitating the precipitation of the desired [Os^{IV}Cl$_5$(Hazole)]$^-$ (**G**) complexes was employed. The (n-Bu$_4$N)[Os^{IV}Cl$_5$(Hazole)] salts (**G**: L = **1, 3, 4 (1H), 5**) generated in boiling ethanol in 24–79% yields can be further converted into sodium and azolium salts, to improve aqueous solubility, in 91–98 and 52–78% yields, respectively (**G**: L = **1, 3, 4 (1H)**) (Büchel et al. 2009). The indazolium salts **G4 (1H)** and **G4 (2H)** can be synthesized directly from (H$_2$ind)$_2$[Os^{IV}Cl$_6$] in boiling ethanol and separated by crystallization in 22 and 27% yields, respectively (Büchel et al. 2012). The imidazole analogue of (n-Bu$_4$N)[Os^{IV}Cl$_5$(Hazole)] was synthesized in isoamyl alcohol at 100 °C in minor yield, whereas in boiling ethanol {(Bu$_4$N)$_2$[Os^{IV}Cl$_6$]}$_2$·[Os^{IV}Cl$_4$(Him)$_2$] is formed (Büchel et al. 2009). Thus, the synthesis of mono- (**G**) and bis- (**H**) substituted osmium complexes was performed according to the following scheme, known as the Anderson rearrangement reaction (Davies et al. 1995):

$$(H_2azole)_2[Os^{IV}Cl_6] \xrightarrow{-HCl} (H_2azole)$$
$$\times [Os^{IV}Cl_5(Hazole)] \xrightarrow{-HCl} [Os^{IV}Cl_4(Hazole)_2].$$

The mono- and bis-azole substituted ruthenium complexes were reported only with a central ion in a 3+ formal oxidation state, while the osmium in analogous compounds is stabilized in a higher oxidation state (4+). To the best of our knowledge, ruthenium (IV) complexes with coordinated azole ligands have not yet been reported. Among ruthenium(IV) complexes only [Ru^{IV}(H$_2$L)Cl$_2$] (where H$_4$L = 1,2-cyclohexanediaminetetraacetic acid) was tested for its antiproliferative activity in vitro and in vivo (Vilaplana et al. 2004). The milestone compounds in the field of antitumor active azole-based ruthenium complexes, i.e., [Ru^{III}Cl$_5$(Hazole)]$^{2-}$ and trans-[Ru^{III}Cl$_4$(Hazole)$_2$]$^-$ (KP1019 analogues), can be easily prepared from RuCl$_3$ and excess azole heterocycles in concentrated or diluted HCl solutions (Keppler et al. 1987a, b, 1989; Reisner et al. 2004, 2005b; Lipponer et al. 1996). Note that on using 1,2,4-triazole, the cis-[Ru^{III}Cl$_4$(Htrz)$_2$]$^-$ complex was also reported, whereas for bis-substituted osmium(IV) compounds only trans-isomers are known (Büchel et al. 2011a; Arion et al. 2003). Attempts to reduce trans-[Os^{IV}Cl$_4$(Hazole)$_2$] to osmium(III) complexes both electrochemically and chemically are underway in our laboratory.

Coordination Modes of Azoles in Osmium Complexes

The studied azole heterocycles (Fig. 2) belong to families of 1,2- (Hpz, Hind), 1,3- (Him, Hbzim) or 1,2,4- (Htrz) azoles and can be classified as mono- (Hpz, Him, Htrz) or bi- (Hind, Hbzim) cyclic heterocycles.

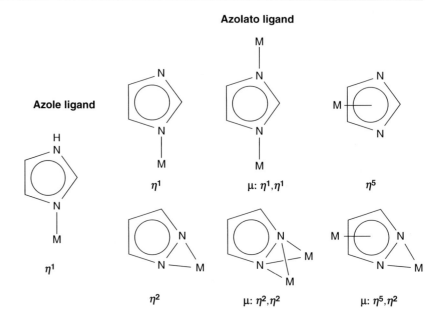

Osmium Complexes with Azole Heterocycles as Potential Antitumor Drugs, Fig. 5 Examples of coordination modes for imidazole, imidazolato, and pyrazolato ligands

They are referred to as 1*H*-azoles. Usually azoles with two nitrogen atoms act as monodentate (η^1) ligands toward metal ions (Fig. 5). In deprotonated form the azolato ion can adopt several coordination modes including η^1, η^2, η^5 non-bridging and $\mu:\eta^1,\eta^1$, $\mu:\eta^2,\eta^2$, $\mu:\eta^5,\eta^2$ bridging modes (Fig. 5) (Cortes-Llamas et al. 2006; El-Kadri et al. 2005, 2006; Nief 2001; Perera et al. 2000; Plass et al. 2002; Yu et al. 2005). The coordination chemistry of 1,2,4-triazole/1,2,4-triazolato ligands varies more due to the increased number of nitrogen atoms and tautomerism (Fuhrmann et al. 1997; Haasnoot 2000; Jeffrey et al. 1983). The bridging coordination modes such as 1,2-$\mu:\eta^1,\eta^1$, 1,4-$\mu:\eta^1,\eta^1$, and 1,2,4-$\mu:\eta^1,\eta^1,\eta^1$ are the most common for 1,2,4-triazole/1,2,4-triazolato ligands, whereas η^1 coordination is rare (Haasnoot 2000).

The osmium complexes reported herein contain azole heterocycles acting as monodentate (η^1) ligands. 1,2,4-Triazole crystallizes in the solid state exclusively as the 1*H*-tautomer. Upon coordination to osmium(III) in **C3** and osmium(IV) in **G3** the triazole ligand adopts the 4*H*-tautomeric form and coordinates via N2 (Fig. 6) (Büchel et al. 2009; Cebriàn-Losantos et al. 2007; Fuhrmann et al. 1997; Jeffrey et al. 1983). The N2 atom in triazole is less basic than N4, therefore such behavior is less expected. This mode of triazole coordination was also documented for ruthenium triazole based complexes of type **D** and *trans/cis*-[Ru^{III}Cl$_4$(Htrz)$_2$]$^-$ (Reisner et al. 2005b;

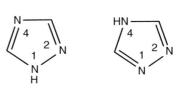

Osmium Complexes with Azole Heterocycles as Potential Antitumor Drugs, Fig. 6 1*H*- (*left*) and 4*H*- (*right*) 1,2,4-triazole tautomers

Arion et al. 2003). In contrast, the 1*H*-tautomeric form coordinated to ruthenium(III) via N4 was discovered in (H$_2$trz)[*trans*-Ru^{III}Cl$_4$(dmso)(Htrz)] (type **C**) or via N2 in *trans*-[Ru^{III}Cl$_2$(Htrz)$_4$]Cl (type **E**) (Reisner et al. 2004; 2005b).

The indazole heterocycle is usually referred to as 1*H*-indazole although two other possible tautomers exist (Fig. 7). The dominant tautomer in the gaseous phase as well as in aqueous solution is 1*H*-indazole (Schmidt et al. 2008; Stadlbauer 2002). In addition, X-ray diffraction studies confirm this preference in the solid state (Benetollo et al. 1990; Nan'ya et al. 1987).

As expected, the ligation of indazole in (**A4**, **B4**, **C4**, **D4**, **E4**, **G4** (**1*H***) and **H4** (**1*H***)) takes place via the atom N2 (as 1*H*-ind). This is confirmed by well-documented crystallographic studies on ruthenium and other metal indazole complexes (Jakupec et al. 2005; Pieper et al. 2001; Reisner et al. 2004). Therefore, it is quite surprising that the Anderson

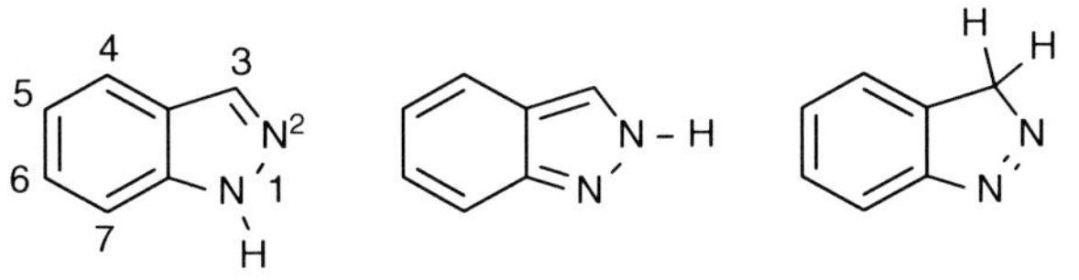

Osmium Complexes with Azole Heterocycles as Potential Antitumor Drugs, Fig. 7 The 1H- (*left*), 2H- (*middle*) and 3H-tautomers (*right*) of indazole

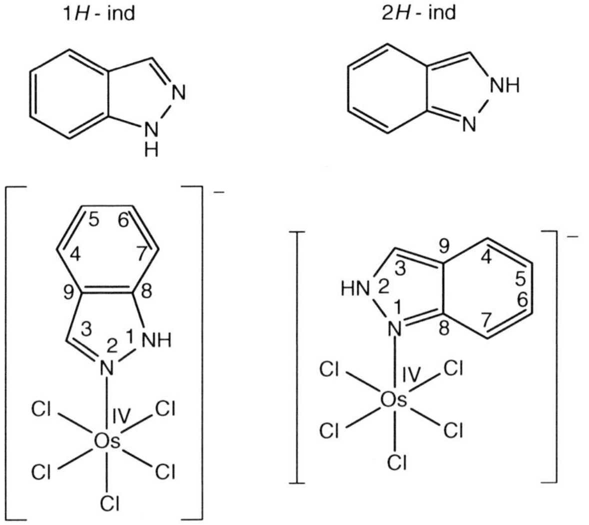

Osmium Complexes with Azole Heterocycles as Potential Antitumor Drugs, Fig. 8 Coordination of indazole in $[Os^{IV}Cl_5(1H\text{-ind})]^-$ (*left*) and $[Os^{IV}Cl_5(2H\text{-ind})]^-$ (*right*)

rearrangement reaction of $(H_2ind)_2[Os^{IV}Cl_6]$, both in the presence of tetrabutylammonium chloride and in its absence, resulted in formation of $[Os^{IV}Cl_5(2H\text{-ind})]^-$ along with $[Os^{IV}Cl_5(1H\text{-ind})]^-$. In the first complex the 2H-form of indazole is adopted and bound to osmium(IV) via the nitrogen atom N1 (Fig. 8) (Büchel et al. 2012). This mode of coordination is unprecedented in the coordination chemistry of indazole. The indazolium salts (**G4 (1H), G4 (2H)**) can be easily separated by fractional crystallization from water/acetone (Büchel et al. 2012).

The second example of the ortho-quinoid indazole tautomer stabilization upon coordination to metal is trans-$[Os^{IV}Cl_4(2H\text{-ind})_2]$ (**H4 (2H)**), which was isolated as a blue solid in low yield (19%) (Büchel et al. 2011a). Trans-$[Os^{IV}Cl_4(1H\text{-ind})_2]$ (**H4 (1H)**) was obtained as red crystals but only as a minor product (Büchel et al. 2011b). The remaining theoretically possible isomer trans-$[Os^{IV}Cl_4(1H\text{-ind})(2H\text{-ind})]$ has not been isolated yet. Elucidation of the reasons for stabilization of 1H-ind and 2H-ind tautomeric forms in **G4**, **H4** is one of the remaining tasks in this intriguing chemistry.

The binding modes and the tautomeric forms adopted by azole heterocycles were established by X-ray diffraction and NMR spectroscopy.

X-Ray Crystallographic Evidence for Azole Coordination Modes

Most of the osmium complexes were characterized by X-ray diffraction, which confirmed the coordination mode of the 1H-azoles (Hpz, Him, Hind, Hbzim) (Büchel et al. 2009, 2011b, 2012; Cebrián-Losantos et al. 2007; Chiorescu et al. 2007; Stepanenko et al. 2007, 2008). In the case of indazole the stabilization of a 2H-tautomer due to the coordination to a metal center was also observed in the **G (2H)** and **H (2H)** complexes (Büchel et al. 2011a, 2012), whereas in the 1,2,4-triazole systems (**C, G**) the 4H-tautomer coordinated via N2 was found (Büchel et al. 2009; Cebrián-Losantos et al. 2007). Here the focus will be on the intriguing 4H-trz and 2H-ind binding modes.

The 1,2,4-triazole as a monodentate ligand prefers coordination via N4 to first or second row transition metal ions. The 1,2,4-triazole coordination via N2 was first reported for trans/cis-$[Ru^{III}Cl_4(Htrz)_2]^-$ complexes in 2003 (Arion et al. 2003). Therefore, the structures of $(H_2trz)[Os^{III}Cl_4(dmso)(4H\text{-trz})]$ (**C3**) (Cebrián-Losantos et al. 2007) and $(Bu_4N)[Os^{IV}Cl_5(4H\text{-trz})] \cdot C_2H_5OH$ (**G3**) (Büchel et al. 2009) are novel examples of a stabilization via complexation of a tautomer which is the minor one in the free state (Fig. 9). The established mode of 1,2,4-triazole coordination is corroborated by participation of all nitrogen atoms in strong intermolecular hydrogen bonding interactions.

The structures of $[Os^{IV}Cl_5(2H\text{-ind})]^-$ (in **G (2H)**), $[Os^{IV}Cl_5(1H\text{-ind})]^-$ (in **G (1H)**), $[Os^{IV}Cl_4(2H\text{-ind})_2]$ (**H (2H)**), and $[Os^{IV}Cl_4(1H\text{-ind})_2]$ (**H (1H)**) are shown in Fig. 10, where the indazole is stabilized in quinoid (2H-ind) and benzenoid (1H-ind) tautomeric forms, respectively. Indazole acts mainly as a monodentate neutral ligand (1H-ind) in metal complexes binding to metal ions via N2. In a few cases, it was found to be deprotonated, acting as a bridging ($\mu:\eta^1,\eta^1$) or rarely monodentate (η^1, via N1 or N2) indazolato ligand

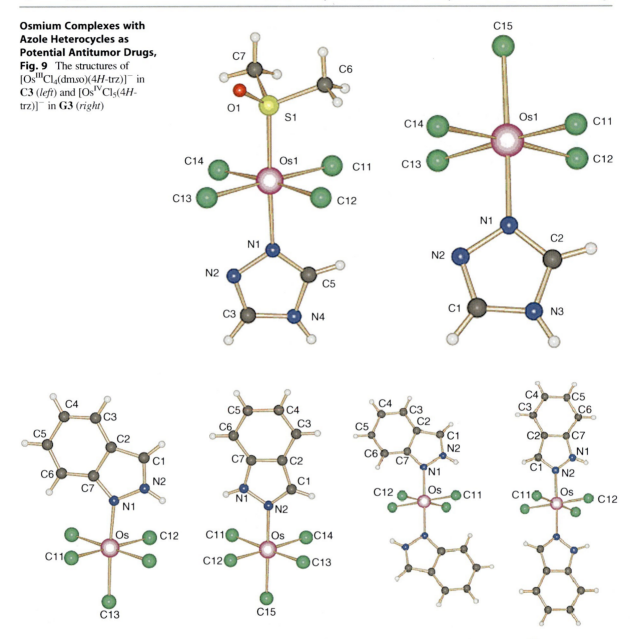

Osmium Complexes with Azole Heterocycles as Potential Antitumor Drugs, Fig. 9 The structures of $[Os^{III}Cl_4(dmso)(4H\text{-}trz)]^-$ in **C3** (*left*) and $[Os^{IV}Cl_5(4H\text{-}trz)]^-$ in **G3** (*right*)

Osmium Complexes with Azole Heterocycles as Potential Antitumor Drugs, Fig. 10 The structures of $[Os^{IV}Cl_5(2H\text{-}ind)]^-$ in **G (2H)**, $[Os^{IV}Cl_5(1H\text{-}ind)]^-$ in **G (1H)**, $[Os^{IV}Cl_4(2H\text{-}ind)_2]$ in **H (2H)** and $[Os^{IV}Cl_4(1H\text{-}ind)_2]$ in **H (1H)** (from *left* to *right*)

(Cortes-Llamas et al. 2006; De la Cruz et al. 2002; Espinet et al. 2000; Fackler et al. 1997; Rendle et al. 1975; Schuecker et al. 2008).

NMR Evidence for a Ligand Coordination Mode in Solution

One- and two-dimensional NMR spectroscopy was applied to diamagnetic (d^6) osmium(II) (**A**, **B**), paramagnetic (d^5) osmium(III) (**C**, **D**, **E**, **F**), and (d^4) osmium(IV) (**G**, **H**) complexes. The proton resonance signals of osmium(III) complexes are broad and strongly shifted to negative values and can be used as fingerprints for compound identification. The proton resonances of osmium(IV) complexes, especially $[Os^{IV}Cl_5(Hazole)]^-$ (**G**), are sharp, well resolved (preserve their multiplicity), shifted slightly (except for the

Osmium Complexes with Azole Heterocycles as Potential Antitumor Drugs, Table 1 ^{1}H NMR resonances of coordinated indazole in **A4, B4, C4, D4, E4, G4, H4** complexes compared to free indazole (the numbering scheme for NMR assignment is given in Fig. 7)

Comp.	Metal	Solvent	^{1}H NMR resonances, ppm
Hind (1H)[a]		dmso-d_6	13.1 (s, H$_1$), 8.1 (s, H$_3$), 7.78 (d, H$_4$), 7.58 (d, H$_7$), 7.36 (t, H$_6$), 7.13 (t, H$_5$)
Hind (1H)		CDCl$_3$	10.01 (s, H$_1$), 8.09 (s, H$_3$), 7.77 (d, H$_4$), 7.51 (d, H$_7$), 7.4 (t, H$_6$), 7.18 (t, H$_5$)
(H$_2$ind)$^{+}$[b]		dmso-d_6	8.08 (s, H$_3$), 7.75 (d, H$_4$), 7.53 (d, H$_7$), 7.34 (t, H$_6$), 7.11 (t, H$_5$)
A4 (1H)	Os(II)	CDCl$_3$	12.74 (s, H$_1$), 8.68 (s, H$_3$), 7.65 (d, H$_4$), 7.47 (d, H$_7$), 7.4 (t, H$_6$), 7.16 (t, H$_5$)
B4 (1H)	Os(II)	CDCl$_3$	14.43 (s, H$_1$), 9.2 (s, H$_3$), 7.73 (d, H$_4$), 7.51 (d, H$_7$), 7.43 (t, H$_6$), 7.19 (t, H$_5$)
C4 (1H)	Os(III)	dmso-d_6	4.29, 3.31, 2.8, 1.8, −9.3, −12.6
D4 (1H)	Os(III)	dmso-d_6	12.13, 8.62, 4.51, 4.06, 3.96, 3.3, 2.43, 2.08, −0.66, −9.49, −21.19, −24.03
E4 (1H)	Os(III)	dmso-d_6	7.39, 6.28, 5.83, 4.69, 2.77, −24
G4 (1H)	Os(IV)	dmso-d_6	17.76 (s, H$_1$), 10.85 (d, H$_7$), 8.23 (t, H$_5$), 5.9 (d, H$_4$), 3.06 (t, H$_6$), −4.54 (s, H$_3$)
G4 (2H)	Os(IV)	dmso-d_6	14.25 (s, H$_2$), 6.66 (t, H$_5$), 4.52 (d, H$_7$), 2.81 (d, H$_4$), −0.43 (t, H$_6$),−14.54 (s, H$_3$)
H4 (1H)	Os(IV)	dmso-d_6	11.78, 8.45, 5.06, −4.01, −8.68
H4 (2H)	Os(IV)	dmso-d_6	13.68, 5.18, 0.58, −3.28, −4.44

[a]Reference (AIST 2012)
[b](H$_2$ind)$^+$ in **C4, G4 (1H), G4 (2H)**

Osmium Complexes with Azole Heterocycles as Potential Antitumor Drugs, Table 2 ^{13}C NMR resonances of coordinated indazole in **A4, B4, G4** complexes compared to free indazole (the numbering for NMR assignment is given in Fig. 8)

Comp.	Metal	Solvent	^{13}C NMR resonances, ppm
Hind (1H)[a]		dmso-d_6	139.86 (C$_8$), 133.36 (C$_3$), 125.78 (C$_6$), 122.80 (C$_9$), 120.40 (C$_4$), 120.09 (C$_5$), 109.99 (C$_7$)
Hind (1H)[a]		CDCl$_3$	140.20 (C$_8$), 134.76 (C$_3$), 126.84 (C$_6$), 123.24 (C$_9$), 121.00 (C$_4$), 120.91 (C$_5$), 109.82 (C$_7$)
(H$_2$ind)$^{+}$[b]		dmso-d_6	140.32 (C$_8$), 133.82 (C$_3$), 126.41 (C$_6$), 123.22 (C$_9$), 120.98 (C$_4$), 120.67 (C$_5$), 110.56 (C$_7$)
A4 (1H)	Os(II)	CDCl$_3$	139.9 (C$_8$), 135.5 (C$_3$), 128.9 (C$_6$), 122.6 (C$_9$), 122.4 (C$_5$), 121.3 (C$_4$), 110.4 (C$_7$)
B4 (1H)	Os(II)	CDCl$_3$	140.4 (C$_8$), 137.2 (C$_3$), 129.2 (C$_6$), 122.8 (C$_9$), 122.6 (C$_5$), 121.5 (C$_4$), 110.9 (C$_7$)
G4 (1H)	Os(IV)	dmso-d_6	200.66 (C$_3$), 173.67 (C$_8$), 163.74 (C$_6$), 139.58 (C$_4$), 106.16 (C$_5$), 81.88 (C$_7$), 75.94 (C$_9$)
G4 (2H)	Os(IV)	dmso-d_6	299.7 (C$_3$), 184.29 (C$_8$), 177.15 (C$_6$), 157.09 (C$_4$), 104.60 (C$_5$), 99.06 (C$_7$), 58.55 (C$_9$)

[a]Reference (AIST 2012)
[b](H$_2$ind)$^+$ in **G4 (2H)**

protons which are closer to the metal center) compared to those observed in diamagnetic compounds and can be easily assigned by 2D NMR spectroscopy. The indazole and pyrazole complexes with two (Hind) and one (Hpz) ring azole ligands, correspondingly, demonstrate this behavior clearly (Tables 1–3).

Indazole, both in free and coordinated states, shows two singlets (NH (H$_1$ or H$_2$) and H$_3$), two doublets (H$_4$ and H$_7$), and two triplets (H$_5$ and H$_6$) in ^{1}H NMR spectra. Osmium complexes with one (**B4, C4, G4**), two (**A4, H4**), and four (**E4**) coordinated indazole ligands are characterized by one set of indazole signals (five CH and one NH resonances, sometimes the last one is undetectable, as in **H4**), whereas the complex with three *mer*-coordinated indazole ligands (**D4**) shows two sets of signals for indazole molecules, which were not assigned because of the paramagnetism of osmium(III). One of these sets belongs presumably to the two *trans*-arranged indazole ligands. The coordination of indazole to osmium(II) via N2 (1H-ind) can be established mainly by the nearest proton resonances H$_1$ and H$_3$ which are downfield shifted in **A4, B4** compared to those in metal-free indazole. Due to coordination to the osmium(IV) in **G4** the order in which proton signals appear changes dramatically. The NH protons closest to the osmium (IV) in **G4 (1H)** and **G4 (2H)** give downfield shifted resonances, whereas H$_3$ peaks are shifted to negative values and differ from each other significantly by ca. 10 ppm. The different coordination mode of indazole

Osmium Complexes with Azole Heterocycles as Potential Antitumor Drugs, Table 3 ^{1}H NMR resonances of coordinated pyrazole in **A1, B1, C1, D1, G1, H1** complexes compared to free pyrazole

Comp.	Metal	Solvent	^{1}H and ^{13}C NMR resonances, ppm	Numbering scheme for NMR assignment
Hpz (^{13}C^a)		CDCl$_3$	10.26 (s, H$_1$), 7.63 (s, 2H, H$_3$ + H$_5$), 6.37 (s, H$_4$) 133.59 (C$_3$ + C$_5$), 104.81 (C$_4$)	
(H$_2$pz)$^{+b}$		dmso-d$_6$	7.71 (s, 2H, H$_3$ + H$_5$), 6.33 (s, H$_4$) 134.25 (C$_3$ + C$_5$), 106.16 (C$_4$)	
A1	Os(II)	CDCl$_3$	13.05 (s, H$_1$), 7.83 (t, H$_3$), 7.65 (t, H$_5$), 6.40 (q, H$_4$) 139.1 (C$_3$), 129.8 (C$_5$), 106.4 (C$_4$)	
B1	Os(II)	CDCl$_3$	14.44 (s, H$_1$), 8.60 (t, H$_3$), 7.66 (t, H$_5$), 6.40 (q, H$_4$) 140.7 (C$_3$), 130.7 (C$_5$), 107.3 (C$_4$)	
C1	Os(III)	dmso-d$_6$	−2.88, −5.59, −9.4, −14.7	
D1	Os(III)	dmso-d$_6$	10.51, 3.30, 0.26, −2.45, −5.26, −6.06, −13.05, −21.63	
G1	Os(IV)	dmso-d$_6$	15.71 (s, H$_1$), 6.39 (s, 1H), −2.44 (s, 1H), −2.71 (s, 1H) 191.99 {−2.44}, 181.33 {−2.71}, 71.18 {6.39}	
H1	Os(IV)	dmso-d$_6$	19.21 (s, H$_1$), 6.47 (s, H$_4$), −2.97 (s, H$_5$), −5.59 (d, H$_3$) 218.41 (C$_3$),181.14 (C$_5$), 73.52 (C$_4$)	

^{a}Reference (AIST 2012)
^{b}(H$_2$pz)$^+$ in **C1, G1**

has a strong effect on the position of NH (124.7 ppm in [Os^{IV}Cl$_5$(1H-ind)]$^-$ and 85.9 ppm in [Os^{IV}Cl$_5$(2H-ind)]$^-$) and C$_3$ signals (200.66 ppm in [Os^{IV}Cl$_5$(1H-ind)]$^-$ and 299.7 ppm in [Os^{IV}Cl$_5$(2H-ind)]$^-$). Thus, the 2D NMR spectroscopy of **G4** complexes can be used as a diagnostic method to distinguish between the coordination modes of indazole tautomers bound to osmium(IV). Carbon resonances of coordinated indazole in **A4, B4, G4** complexes are summarized in Table 2.

Pyrazole is characterized by three (H$_3$ and H$_5$ resonances are overlapped) or four proton resonances in the free state and in osmium complexes, respectively (Table 3). As for indazole complexes, the osmium (III) complex with three *mer*-coordinated pyrazole ligands (**D1**) displays two sets of pyrazole ligand resonances, whereas complexes with one (**B1, C1, G1**) or two (**A1, H1**) pyrazole heterocycles show one set of signals. Upon coordination via N2 in osmium(II) complexes (**A1, B1**) the proton signals of H$_1$ and H$_3$, nearest to the coordinated atom, are downfield shifted, whereas two other protons (H$_4$ and H$_5$) preserve their positions (Table 3). The full assignment of proton resonances in paramagnetic osmium complexes was done only for osmium(IV) complex **H1**. The comparison of the ^{1}H and ^{13}C NMR spectra of **G1** and **H1** makes the assignment of protons in position 4 unambiguous: H$_4$ and C$_4$ resonances in **G1** and **H1** are almost at the same values (6.39 or 6.47 ppm for H$_4$; 71.18 and 73.52 ppm for C$_4$). The other two signals with negative chemical shift values belong to H$_3$ and H$_5$ protons and the most upfield shifted resonance in **H1** originates from the H$_3$ proton nearest to the place of coordination.

Electrochemical Properties of Osmium Complexes

The electrochemical properties of osmium complexes studied by cyclic voltammetry in organic solvents (**A, B, D, E, G, H**) or aqueous phosphate buffer (**C, E**) at platinum (**A, B, D**) or carbon-disk (**C, E, G, H**) working electrodes are summarized in Tables 4 and 5.

The Os$^{II/III}$ response (undetected for **D2, D5** in the range of potentials measured) was found to be a reversible one-electron process for all other complexes studied. The Os$^{III/IV}$ response was not observed for complexes **A** and **B**. For complexes **C, D** and **E** redox wave Os$^{III/IV}$ is reversible, while for **G** and **H** it is irreversible. ΣE_L was calculated for all complexes but **G4 (2H)** and **H4 (2H)** by using the ligand parameters (E_L in V) collected in the literature (E_L (Cl) = −0.24, E_L (s-dmso) = 0.57,

Osmium Complexes with Azole Heterocycles as Potential Antitumor Drugs, Table 4 Cyclic voltammetric data for complexes **A, B, D, E, G, H** in organic solvents[a]

Complex	Solvent	$E_{1/2}$ (Os$^{II/III}$)[b]	$E_{1/2}$ (Os$^{III/IV}$)	ΣE_L	References
A1	CH$_3$CN	0.89	–	1.06	(Stepanenko et al. 2007)
A2	DMF	0.71	–	0.84	
A4	CH$_3$CN	0.94	–	1.18	
A5	DMF	0.80	–	0.86	
B1	CH$_3$CN	1.57	–	1.43	
B2	DMF	1.40	–	1.32	
B4	CH$_3$CN	1.66	–	1.49	
B5	DMF	1.51	–	1.33	
D1	CH$_3$CN	−0.65	1.02	−0.12	(Chiorescu et al. 2007)
D2	CH$_3$CN	–	0.5	−0.45	
D4	CH$_3$CN	−0.38	1.14	0.06	
D5	CH$_3$CN	–	0.66	−0.42	
E1	CH$_3$CN	−0.20	1.35	0.32	(Stepanenko et al. 2008)
E4	CH$_3$CN	0.1	1.55	0.56	
E5	CH$_3$CN	−0.40	1.10	−0.08	
G1 (n-Bu$_4$N$^+$)	DMSO	−1.66	0.03	−1.0	(Büchel et al. 2009)
G4 (n-Bu$_4$N$^+$, **1H**)	DMSO	−1.39	0.12	−0.94	
G4 (H$_2$ind$^+$, **1H**)	DMSO	−1.33	0.13	−0.94	(Büchel et al. 2012)
G4 (H$_2$ind$^+$, **2H**)	DMSO	−1.43	0.03	–	
G5 (n-Bu$_4$N$^+$)	DMSO	−1.78	−0.03	−1.1	(Büchel et al. 2009)
H1	DMSO	−1.22	0.50	−0.56	(Büchel et al. 2011a)
H2	DMSO	−1.55	0.31	−0.78	
H4 (**2H**)	DMSO	−1.17	0.42	–	
H5	DMSO	−1.41	0.37	−0.76	

[a]Supporting electrolyte (Bu$_4$N)[BF$_4$], internal standard [Fe(η^5-C$_5$H$_5$)$_2$], potentials are quoted relative to NHE;
[b]E_p (Os$^{II/III}$) for complexes **G** and **H**

Osmium Complexes with Azole Heterocycles as Potential Antitumor Drugs, Table 5 Cyclic voltammetric data for complexes **C, E** in aqueous phosphate buffer (pH 7)[a]

Complex	$E_{1/2}$ (Os$^{II/III}$)	$E_{1/2}$ (Os$^{III/IV}$)	ΣE_L	References
C1	0.16	1.40	−0.19	(Cebrià n-Losantos et al. 2007)
C2	0.12	1.33	−0.3	
C4	0.17	1.41	−0.13	
C5	0.12	1.27	−0.29	
E2	−1.17	0.36	−0.12	(Stepanenko et al. 2008)
E5	−1.21	0.31	−0.08	

[a]Internal standard methyl viologen, potentials are quoted relative to NHE

E_L (Hpz) = 0.20, E_L (Him) = 0.09, E_L (Hind) = 0.26, E_L (Hbzim) = 0.10) (Reisner et al. 2008).

Slopes and intercepts for Os$^{II/III}$ and Os$^{III/IV}$ species with different net charges measured in organic solvents and aqueous phosphate buffer were established by plotting $E_{1/2}$ against ΣE_L (Table 6, Figs. 11–13). Note that the 1−, 0, 1+, 2+ (in organic solvents) or 1+, 2+ (in water) Os$^{II/III}$ species reported by Lever demonstrate that the complex net charge has no marked effect on the osmium-centered redox potentials measured in aqueous or organic solvents (Table 6) (Lever 1990). In our experiments all Os$^{II/III}$ (and Os$^{III/IV}$, correspondingly) charged species lie on the same line in organic solvents (Figs. 11, 12),

Osmium Complexes with Azole Heterocycles as Potential Antitumor Drugs, Table 6 Slope and intercept data for $Os^{II/III}$ and $Os^{III/IV}$ species with different net charges in organic solvents and aqueous solutions

Complex	Number of data points	Os^n/Os^{n+1}	Solvent	Complex net charges	S_M	I_M	R-value
(Lever 1990)	18	Os^{II}/Os^{III}	Water	1+/2+, 2+/3+	1.61	−1.30	0.99
	80		Organic	1−/0, 0/1+, 1+/2+, 2+/3+	1.01	−0.40	0.98
A, B, D, E, H, G	20	Os^{II}/Os^{III}	Organic	3−/2−, 2−/1−, 1−/0, 0/1+	1.29	−0.39	0.99
D, E, H, G	14	Os^{III}/Os^{IV}	Organic	2−/1−, 1−/0, 0/1+, 1+/2+	0.99	1.06	0.99

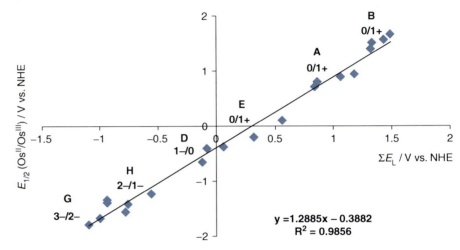

Osmium Complexes with Azole Heterocycles as Potential Antitumor Drugs, Fig. 11 Plot of $Os^{II/III}$ potentials ($E_{1/2}$ for **A**, **B**, **D**, **E**; E_p for **G** and **H**) against ΣE_L in organic solvents

whereas in aqueous solution the slope (S_M) and intercept (I_M) of (1)

$$E = S_M \cdot \Sigma E_L + I_M \quad (1)$$

are highly dependent on the complex net charge (Fig. 13). Calculating S_M and I_M values for $Os^{III/IV}$ species with an octahedral environment in organic solvents involving 2−/1−, 1−/0, 0/1+, 1+/2+ net charged complexes results in:

$$E = 0.99 \cdot \Sigma E_L + 1.06.$$

Compounds involving 3−/2−, 2−/1−, 1−/0, 0/1+ net charged $Os^{II/III}$ centers are characterized by a slope (1.29) and intercept (−0.39) which are close to the S_M and I_M values (1.01; −0.40 correspondingly) reported by Lever (Table 6) (Lever 1990).

The effect of the complex net charge on the redox potential in aqueous solution was also reported for analogous ruthenium azole compounds (Reisner et al. 2008). Unfortunately, due to the small number of experimental potentials measured for osmium complexes in water the estimation of the S_M and I_M values is questionable.

The application of Lever's equation (1) has allowed the estimation of the yet unknown E_L ligand parameter for the 2H-ind tautomer as an average of two values measured in **G4 (2H)** and **H4 (2H)**: E_L(2H-ind) = 0.16 V versus NHE (Table 7; only $Os^{III/IV}$ response was taken due to the possible change of indazole coordination mode on deeper reduction). It is lower than those of 1H-ind, indicating that 2H-ind is a stronger net electron-donor than the 1H-tautomer.

Antiproliferative Activity of Osmium Complexes

The Os(II) complexes (**A**, **B**) which are stable toward hydrolysis in aqueous solution show only marginal cytotoxicity in 41M (ovarian carcinoma), SK-BR-3 (mammary carcinoma), and SW480 (colon carcinoma) cell lines with IC$_{50}$ values >1,000 μM (Egger et al. 2008).

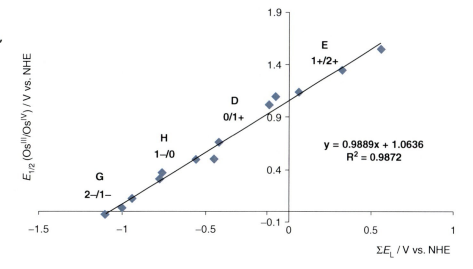

Osmium Complexes with Azole Heterocycles as Potential Antitumor Drugs, Fig. 12 Plot of $Os^{III/IV}$ potentials ($E_{1/2}$) against ΣE_L in organic solvents

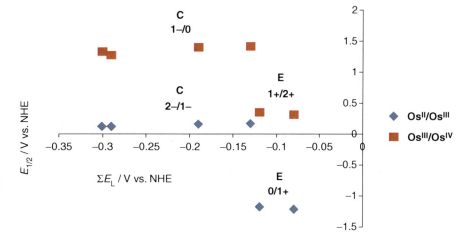

Osmium Complexes with Azole Heterocycles as Potential Antitumor Drugs, Fig. 13 Plot of Os^{II}/Os^{III} and Os^{III}/Os^{IV} potentials ($E_{1/2}$) against ΣE_L in aqueous phosphate buffer

Unlike the Os(III)-NAMI-A analogues (C), which are also resistant to hydrolysis, show reasonable antiproliferative activity in vitro with IC_{50} values in the 10^{-5} to 10^{-4} M concentration range in the two human cell lines HT-29 (colon carcinoma) and SK-BR-3 (mammary carcinoma) (Cebriàn-Losantos et al. 2007; Egger et al. 2008). They are more cytotoxic than the hydrolytically more labile Ru(III)-NAMI-A complexes. This effect is most pronounced in the case of the indazole complex (C4), which by one order of magnitude is a more potent antiproliferative agent than its ruthenium congener (HT-29 cells, 21 ± 1 μM) (Cebriàn-Losantos et al. 2007). The antimetastatic activity of Os(III)-NAMI-A analogues is yet to be investigated. The most cytotoxic compounds are *tetra*-substituted osmium(III) complexes (E) with IC_{50} values in the 10^{-6} to 10^{-5} M concentration range in the three human cell lines CH1 (ovarian carcinoma), A549 (non-small cell lung carcinoma) and SW480 (colon carcinoma) (Stepanenko et al. 2008). The pyrazole complex (E1) showed the highest activity (with IC_{50} values from 1.0 to 3.2 μM) in three cell lines (Stepanenko et al. 2008). Among osmium(IV) complexes, which are also hydrolytically stable, the *bis*-substituted osmium complexes (H) are more potent antiproliferative agents than the corresponding *mono*-substituted azolium complexes (G, H_2azole^+), at least for the pyrazole and indazole

Osmium Complexes with Azole Heterocycles as Potential Antitumor Drugs, Table 7 E_L ligand parameter for 2H-ind tautomer

Os^n/Os^{n+1}	$E = S_M \cdot \Sigma E_L + I_M$	E_L (1H-ind or 2H-ind)		
		G4 (H$_2$ind$^+$, **1H**)	G4 (H$_2$ind$^+$, **2H**)	H4 (**2H**)
Os^{III}/Os^{IV}	$E = 0.99 \cdot \Sigma E_L + 1.06$	0.26	0.16	0.16

Osmium Complexes with Azole Heterocycles as Potential Antitumor Drugs, Table 8 Comparison of $E_{1/2}$ (E_p) potentials (in organic solvents) and cytotoxicity in vitro of pyrazole and indazole osmium compounds

Complex	$E_{1/2}$ (Os^{II}/Os^{III})	$E_{1/2}$ (Os^{III}/Os^{IV})	IC$_{50}$, µM			References
			CH1	A549	SW480	
G1 (n-Bu$_4$N$^+$)	−1.66	0.03	–	–	–	
G1 (Na$^+$)	–	–	26 ± 6	77 ± 20	18 ± 3	(Büchel et al. 2009)
G1 (H$_2$pz$^+$)	–	–	142 ± 20	229 ± 27	105 ± 3	
H1	−1.22	0.50	115 ± 14	>160	120 ± 5	(Büchel et al. 2011a)
E1	−0.20	1.35	1.0 ± 0.2	3.2 ± 0.8	2.3 ± 0.2	(Stepanenko et al. 2008)
G4 (n-Bu$_4$N$^+$, **1H**)	−1.39	0.12	–	–	–	(Büchel et al. 2009)
G4 (Na$^+$, **1H**)	–	–	73 ± 24	247 ± 29	114 ± 5	
G4 (H$_2$ind$^+$, **1H**)	−1.33	0.13	98 ± 23	224 ± 18	110 ± 6	(Büchel et al. 2012)
G4 (H$_2$ind$^+$, **2H**)	−1.43	0.03	92 ± 20	113 ± 17	100 ± 15	
H4 (**2H**)	−1.17	0.42	53 ± 4	181 ± 11	41 ± 14	(Büchel et al. 2011a)

series (Büchel et al. 2009, 2011a). The complex containing 2H-indazole (**G4 (H$_2$ind$^+$, 2H)**) might be advantageous over 1H-indazole-containing analogue (**G4 (H$_2$ind$^+$, 1H)**) with regard to inhibition of tumor cell growth (Büchel et al. 2009, 2012). The replacement of the azolium cation by sodium in pyrazole compounds (**G1**) results in a 3.0- to 5.8-fold enhancement of cytotoxicity (IC$_{50}$ values in the 10^{-5} M range), while the difference is reduced for indazole complexes (**G4 (1H)**) in the CH1 cell line and completely abolished for indazole complexes (**G4 (1H)**) in A549 and SW480 cell lines and triazole complexes (**G3**) in all three cell lines (Büchel et al. 2009, 2012).

Recently reported correlations of redox potentials and antiproliferative activity for a homologous series of indazole-based {[RuCl$_{(6-n)}$(Hind)$_n$]$^{(3-n)-}$ (n = 0–4)}, NAMI-A type and KP1019 related ruthenium(III) complexes, namely, that an increased reduction potential results in an enhanced antiproliferative activity in the same cell line (Reisner et al. 2008), was inspected for the homologous pyrazole and indazole osmium(III, IV) series (Table 8).

The lack of a clear-cut correlation between antiproliferative activity and reduction potentials within the two series of osmium complexes indicates that the reduction potential cannot play a major role in their biological activity.

Conclusions

The performed synthetic work resulted in the preparation of 44 new osmium(II, III, IV) complexes with azole heterocycles demonstrating some difference in the substitution chemistry at ruthenium and at osmium centers. The successful synthesis of the desired osmium complexes cannot be performed in a routine way simply by following the protocols developed for ruthenium species. The osmium complexes display different kinetics to that of their ruthenium analogues, resembling the differences in the substitution chemistry of Pt(II) and Pd(II).

The electrochemical properties of osmium complexes in organic solvents and in aqueous solution have been described. The linear regression $E_{1/2}$ against ΣE_L enabled the estimate of the slope (S_M) and intercept (I_M) values for Os$^{III/IV}$ process in organic solvents, as follows: $E = 0.99 \Sigma E_L + 1.06$, which allows for the prediction of the redox potentials for other osmium complexes with the same redox couples. It should be also noted, that the ligand parameter (E_L) for 2H-indazole was determined for the first time.

Among reported osmium complexes some of them, e.g., Os(III)-NAMI-A analogues (**C**) and *tetra*-substituted osmium(III) complexes (**E**), can be selected as drug candidates for further preclinical development. The established structure-property relationships, when the replacement of the chlorido ligands by azole heterocycles enhances the antiproliferative activity (as in case **E**), or the stabilization of certain tautomers results in an increased antiproliferative activity (**G4** (H$_2$ind$^+$, 1H-ind) and **G4** (H$_2$ind$^+$, 2H-ind)), provide a rational approach for purposeful drug design. The synthesis of ruthenium 2H-indazole complexes, and evaluation of their cytotoxicity are worthy of attention in order to obtain more effective antitumor drugs.

The observed hydrolytic stability of osmium complexes and their reasonable antiproliferative activity in vitro indicates that hydrolysis might not be a prerequisite for the antitumor activity. Moreover the reduction potential cannot be the main factor for contributing to the antiproliferative activity. The mechanism of action of osmium complexes remains an open question and the influence of other conditions (e.g., cellular uptake, interactions with different molecular targets) should be investigated.

Cross-References

▶ Hexavalent Chromium and Cancer
▶ Lanthanides and Cancer
▶ Magnesium
▶ Osmium, Physical and Chemical Properties
▶ Polonium and Cancer
▶ Selenium-Binding Protein 1 and Cancer
▶ Tin Complexes, Antitumor Activity

References

AIST (2012) RIO-DB Spectral Database for Organic Compounds SDBC
Alessio E (2004) Synthesis and reactivity of Ru-, Os-, Rh-, and Ir-halide-sulfoxide complexes. Chem Rev 104:4203–4242
Alessio E, Balducci G, Lutman A et al (1993) Synthesis and characterization of two new classes of ruthenium(III)-sulfoxide complexes with nitrogen donor ligands (L): Na[trans-RuCl$_4$(R$_2$SO-S)(L)] and mer,cis-RuCl$_3$(R$_2$SO-S)(R$_2$SO-O)(L). The crystal structure of Na[trans-RuCl$_4$(DMSO-S)(NH$_3$)]·2DMSO, Na[trans-RuCl$_4$(DMSO-S)(Im)]·H$_2$O·Me$_2$CO (Im = imidazole) and mer,cis-RuCl$_3$(DMSO-S)(DMSO-O)(NH$_3$). Inorg Chim Acta 203:205–217
Alessio E, Serli B, Zangrando E et al (2003) Geometrical and linkage isomers of [OsCl$_2$(dmso)$_4$] – the complete picture. Eur J Inorg Chem 3160–3166
Ang WH, Dyson PJ (2006) Classical and non-classical ruthenium-based anticancer drugs: towards targeted chemotherapy. Eur J Inorg Chem 4003–4018
Arion VB, Reisner E, Fremuth M et al (2003) Synthesis, X-ray diffraction structures, spectroscopic properties, and in vitro antitumor activity of isomeric (1H-1,2,4-Triazole)Ru(III) complexes. Inorg Chem 42:6024–6031
Benetollo F, Del Pra A, Baiocchi L (1990) Molecular conformation of benzydamine hydrochloride. Farmaco 45:1361–1367
Bessant R, Steuer A, Rigby S et al (2003) Osmic acid revisited: factors that predict a favorable response. Rheumatology 42:1036–1043
Brauer G (1981) Handbuch der präparativen anorganischen Chemie, III, 1742–1744. Ferdinand Enke Verlag, Stuttgart
Bruijnincx PCA, Sadler PJ (2009) Controlling platinum, ruthenium, and osmium reactivity for anticancer drug design. Adv Inorg Chem 61:1–62
Büchel GE, Stepanenko IN, Hejl M et al (2012) Osmium(IV) complexes with 1H- and 2H-indazoles: Tautomer identity versus spectroscopic properties and antiproliferative activity. J Inorg Biochem DOI: 10.1016/j.jinorgbio.2012.04.001
Büchel GE, Stepanenko IN, Hejl M et al (2009) [Os^{IV}Cl$_5$(Hazole)]$^-$ complexes: synthesis, structure, spectroscopic properties, and antiproliferative activity. Inorg Chem 48:10737–10747
Büchel GE, Stepanenko IN, Hejl M et al (2011a) En route to osmium analogues of KP1019: synthesis, structure, spectroscopic properties and antiproliferative activity of trans-[Os^{IV}Cl$_4$(Hazole)$_2$]. Inorg Chem 50:7690–7697
Büchel GE, Stepanenko IN, Hejl M et al (2011b) 5th EuCheMS conference on NITROGEN LIGANDS in coordination chemistry, metal-organic chemistry, bioinorganic chemistry, materials & catalysis, Granada, 4–8 Sept 2011, p. 158
Casini A, Hartinger CG, Nazarov AA et al (2010) Organometallic antitumor agents with alternative modes of action. Topics in Organometallic Chemistry 32:57–80
Cebrián-Losantos B, Krokhin AA, Stepanenko IN et al (2007) Osmium NAMI-A analogs: synthesis, structural and spectroscopic characterization, and antiproliferative properties. Inorg Chem 46:5023–5033
Chiorescu I, Stepanenko IN, Arion VB et al (2007) Theoretical study on OsCl$_3$(azole)$_3$: isomerism, redox properties, electronic and vibrational spectra. J Biol Inorg Chem 12 (Suppl. 1): 226
Clarke MJ (2003) Ruthenium metallopharmaceuticals. Coord Chem Rev 236:209–233
Cortes-Llamas SA, Hernandez-Perez JM, Ho M et al (2006) Indazolato derivatives of boron, aluminum, and gallium: characterization and solvent-dependent regioisomeric structures through p-p interactions in the solid state. Organometallics 25:588–595
Craciunescu DG (1977) Molecular interactions between the complex anions of transition metals and nitrogenous organic

bases of biological importance. An Real Acad Farm 43:265–292

Craciunescu DG, Molina C, Parrondo-Iglesias E et al (1991) Study on the pharmacological (trypanocidal, antitumor and anti-AIDS) effects of the new ionic ruthenium(III) and osmium(IV) complexes. An Real Acad Farm 57:221–240

Craciunescu DG, Molina C, Parrondo-Iglesias E et al (1992) Dual pharmacological activities (trypanocidal and antitumor) displayed by the new neutral osmium(III) and osmium(IV) complexes and by the new anionic ruthenium (III) complexes. An Real Acad Farm 58:207–231

Davies JA, Hockensmith CM, Kukushkin V Yu et al (1995) Synthetic coordination chemistry – principles and practice. World Scientific, Singapore, pp 392–396

De la Cruz R, Espinet P, Gallego AM et al (2002) Structural and dynamic studies in solution of anionic dinuclear azolato-bridged palladium(II) complexes. J Organomet Chem 663:108–117

Doadrio A, Craciunescu D, Ghirvu C et al (1977a) On the preparation and antitumoral assay of some piperazine and antipiryne complex salts with hexachloroosmate(IV), hexachloroplatinate(IV), tetrachloroplatinate(II), tetrachloropalladate(II), and tetrachloroaurate(III) anions. An Quim 73:1220–1223

Doadrio A, Craciunescu D, Ghirvu G (1977b) New piperazinium complex salts with antitumor action. An Quim 73:1042–1046

Doadrio A, Craciunescu D, Ghirvu C (1980) Structure-antitumor activity relationship for the platinum(2+), palladium(2+), palladium(4+), and osmium(4+) complexes and complex salts with sulfonamide derivatives. An Real Acad Farm 46:153–166

Dorcier A, Ang WH, Bolano S et al (2006) In vitro evaluation of rhodium and osmium RAPTA analogues: the case for organometallic anticancer drugs not based on ruthenium. Organometallics 25:4090–4096

Dwyer FP, Gyarfas EC, Rogers WP et al (1952) Biological activity of complex ions. Nature 170:190–191

Dyson PJ, Sava G (2006) Metal-based antitumor drugs in the post genomic era. Dalton Trans 16:1929–1933

Egger A, Cebrián-Losantos B, Stepanenko IN et al (2008) Hydrolysis and cytotoxic properties of osmium(II)/(III)-DMSO-azole complexes. Chem Biodivers 5:1588–1593

El-Kaderi H, Heeg MJ, Winter CH (2005) Synthesis, structure, and ligand redistribution equilibria of mixed ligand complexes of the heavier group 2 elements containing pyrazolato and p-diketiminato ligands. Eur J Inorg Chem 2081–2088

El-Kadri OM, Heeg MJ, Winter CH (2006) Synthesis, structural characterization, and properties of chromium(III) complexes containing amidinato ligands and η^2-pyrazolato, η^2-1,2,4-triazolato, or η^1-tetrazolato ligands. Dalton Trans 4506–4513

Espinet P, Gallego AM, Martinez-Ilarduya JM et al (2000) Dinuclear azolato-bridged complexes of the nickel group metals with haloaryl ligands: a reinvestigation of their behavior in solution. Inorg Chem 39:975–979

Fackler JP Jr, Staples RJ, Raptis RG (1997) Crystal structure of cis-bis(triphenylphosphine)platinum(II)-chloro-N-indazolate, PtCl [P(C_6H_5)_3]_2(C_7H_5N_2). Z Kristallogr 212:157–158

Filak LK, Muehlgassner G, Jakupec MA et al (2010) Organometallic indolo[3,2-c]quinolines versus indolo[3,2-d] benzazepines: synthesis, structural and spectroscopic characterization, and biological efficacy. J Biol Inorg Chem 15:903–918

Fu Y, Habtemariam A, Pizarro AM et al (2010) Organometallic osmium arene complexes with potent cancer cell cytotoxicity. J Med Chem 53:8192–8196

Fu Y, Habtemariam A, Basri AMBH et al (2011) Structure-activity relationships for organometallic osmium arene phenylazopyridine complexes with potent anticancer activity. Dalton Trans 40:10553–10562

Fuhrmann P, Koritsanszky T, Luger P (1997) Crystal structure of ((triphenylphosphoranylidene)methyl)triphenylphosphonium octacarbonyl-diiron, [(C_6H_5)_3PCHP(C_6H_5)_3]_2[Fe_2(CO)_8]. Z Kristallogr 212:213–220

Goldstein S, Czapski G, Heller A (2005) Osmium tetroxide, used in the treatment of arthritic joints, is a fast mimic of superoxide dismutase. Free Radical Bio Med 38:839–845

Groessl M, Zava O, Dyson PJ (2011) Cellular uptake and subcellular distribution of ruthenium-based metallodrugs under clinical investigation versus cisplatin. Metallomics 3:591–599

Haasnoot JG (2000) Mononuclear, oligonuclear and polynuclear metal coordination compounds with 1,2,4-triazole derivatives as ligands. Coord Chem Rev 200–202:131–185

Hanif M, Nazarov AA, Hartinger CG et al (2010a) Osmium(II)-versus ruthenium(II)-arene carbohydrate-based anticancer compounds: similarities and differences. Dalton Trans 39:7345–7352

Hanif M, Henke H, Meier SM et al (2010b) Is the reactivity of M (II)-arene complexes of 3-hydroxy-2(1H)-pyridones to biomolecules the anticancer activity determining parameter? Inorg Chem 49:7953–7963

Hartinger CG, Dyson PJ (2009) Bioorganometallic chemistry-from teaching paradigms to medicinal applications. Chem Soc Rev 38:391–401

Heffeter P, Böck K, Atil B et al (2010) Intracellular protein binding patterns of the anticancer ruthenium drugs KP1019 and KP1339. J Biol Inorg Chem 15:737–748

Henn M, Alessio E, Mestroni G et al (1991) Ruthenium(II)-dimethyl sulfoxide complexes with nitrogen ligands: synthesis, characterization and solution chemistry. The crystal structures of cis, fac-RuCl_2(DMSO)_3(NH_3) and trans, cis, cis-RuCl_2(DMSO)_2(NH_3)_2·H_2O. Inorg Chim Acta 187:39–50

Hinckley CC, Bemiller JN, Strack LE et al (1983) Platinum, gold, and other metal chemotherapeutic agents (ACS symposium series) 209:421–437, Lippard SJ Ed, American Chemical Society

Jakupec MA, Reisner E, Eichinger A et al (2005) Redox-active antineoplastic ruthenium complexes with indazole: correlation of in vitro potency and reduction potential. J Med Chem 48:2831–2837

Jakupec MA, Galanski M, Arion VB et al (2008) Antitumor metal compounds: more than theme and variations. Dalton Trans 2:183–194

Jeffrey GA, Ruble JR, Yates JH (1983) Neutron diffraction at 15 and 120 K and ab initio molecular-orbital studies of the molecular structure of 1,2,4-triazole. Acta Cryst B39:388–394

Keppler BK, Henn M, Juhl UM et al (1989) New ruthenium complexes for the treatment of cancer, in Progress in clinical biochemistry and medicine, vol 10, pp 41−69, Springer, Berlin

Keppler BK, Rupp W, Juhl UM et al (1987a) Synthesis, molecular structure, and tumor-inhibiting properties of imidazolium trans-bis(imidazole)tetrachlororuthenate(III) and its methyl-substituted derivatives. Inorg Chem 26:4366–4370

Keppler BK, Wehe D, Endres H et al (1987b) Synthesis, antitumor activity, and X-ray structure of bis(imidazolium) (imidazole)pentachlororuthenate(III), $(ImH)_2(RuImCl_5)$. Inorg Chem 26:844–846

Koch JH, Gyarfas EC, Dwyer FP (1956) Iron storage and transport in iron-depleted rats with notes on combined iron and copper deficiency. Aust J Biol Sci 9:371–381

Komozin PN (2000) EPR of substituted halo Ru(III) and Os(III) complexes. Zh Neorg Khim 45:662–674

Lever ABP (1990) Electrochemical parametrization of metal complex redox potentials, using the ruthenium(III)/ruthenium(II) couple to generate a ligand electrochemical series. Inorg Chem 29:1271–1285

Lipponer K-G, Vogel E, Keppler BK (1996) Synthesis, characterization and solution chemistry of indazolium trans-tetrachlorobis(indazole)ruthenate(III), a new anticancer ruthenium complex. IR, UV, NMR, HPLC, investigations and antitumor activity. Crystal structures of 1-methylindazolium trans-tetrachlorobis-(1-methylindazole)ruthenate(III) and its hydrolysis product trans-monoaquatrichlorobis (1-methylindazole)ruthenate(III). Met Based Drugs 3:243–260

McDonagh AM, Humphrey MG, Hockless DC (1997) Selective preparation of cis- or trans-dichlorobis{(R, R)-1,2-phenylenebis(methylphenylphosphine-P)}osmium(II) from dimethylsulfoxide complex precursors. Tetrahedron Asymmetr 8:3579–3583

Nan'ya S, Katsuraya K, Maekawa E et al (1987) Synthesis of 2H-indazole-4,7-dione derivatives from 3-phenylsydnone and p-toluquinone. The crystal and molecular structure of 6-bromo-5-methyl-2-phenyl-2H-indazole-4,7-dione. J Heterocycl Chem 24:971–975

Ni W-X, Man W-L, Cheung MT-W et al (2011) Osmium(VI) complexes as a new class of potential anti-cancer agents. Chem Commun 47:2140–2142

Ni W-X, Man W-L, Ho M et al (2012) Osmium(VI) nitride complexes bearing azole heterocycles: A new class of antitumor agents. Chem Sci 3:1582–1588

Nief F (2001) Heterocyclopentadienyl complexes of group-3 metals. Eur J Inorg Chem 891–904

Orvig C, Abrams MJ (1999) Medicinal inorganic chemistry: introduction. Chem Rev 99:2201–2203

Peacock AFA, Habtemariam A, Moggach SA et al (2007) Chloro half-sandwich osmium(II) complexes: influence of chelated N, N-ligands on hydrolysis, guanine binding, and cytotoxicity. Inorg Chem 46:4049–4059

Peacock AFA, Sadler PJ (2008) Medicinal organometallic chemistry: designing metal arene complexes as anticancer agents. Chem Asian J 3:1890–1899

Perera JR, Heeg MJ, Winter CH (2000) η^5-Imidazolato complexes of ruthenium. Organometallics 19:5263–5265

Pieper T, Sommer M, Galanski M et al (2001) $[RuCl_3ind_3]$ and $[RuCl_2ind_4]$: two new ruthenium complexes derived from the tumor-inhibiting RuIII compound HInd (OC-6-11)-$[RuCl_4ind_2]$ (ind = indazole). Z Anorg Allg Chem 627:261–265

Plass W, Pohlmann A, Subramanian PS et al (2002) Magnetic interactions through imidazolate-bridges: synthesis, spectroscopy, crystal structure and magnetic properties of μ-imidazolato-bridged copper(II) complexes. Z Anorg Allg Chem 628:1377–1384

Reisner E, Arion VB, Guedes da Silva MFC et al (2004) Tuning of redox potentials for the design of ruthenium anticancer drugs – an electrochemical study of [trans-$RuCl_4L$ (DMSO)]$^-$ and [trans-$RuCl_4L_2$]$^-$ Complexes, where L = Imidazole, 1,2,4-Triazole, Indazole. Inorg Chem 43:7083–7093

Reisner E, Arion VB, Rufinska A et al (2005a) Isomeric $[RuCl_2(DMSO)_2(indazole)_2]$ complexes: ruthenium(II)-mediated coupling reaction of acetonitrile with 1H-indazole. Dalton Trans 14:2355–2364

Reisner E, Arion VB, Eichinger A et al (2005b) Tuning of redox properties for the design of ruthenium anticancer drugs: part 2. Syntheses, crystal structures, and electrochemistry of potentially antitumor $[Ru^{III/II}Cl_{6-n}(Azole)_n]z$ (n = 3, 4, 6) complexes. Inorg Chem 44:6704–6716

Reisner E, Arion VB, Keppler BK et al (2008) Electron-transfer activated metal-based anticancer drugs. Inorg Chim Acta 361:1569–1583

Rendle DF, Storr A, Trotter J (1975) Crystal and molecular structures of pyrazolyl-, 3-methylpyrazolyl-, and indazolylgallium dimethyl dimers. Can J Chem 53:2930–2943

Rosenberg B, VanCamp L, Krigas T (1965) Inhibition of cell division in Escherichia coli by electrolysis products from a platinum electrode. Nature 205:698–699

Rudnitskaya OV, Buslaeva TM, Lyalina NN (1994) Osmium (IV) dimethyl sulfoxide complexes. Zh Neorg Khim 39:922–924

Schmid WF, John RO, Arion VB et al (2007a) Highly antiproliferative ruthenium(II) and osmium(II) arene complexes with paullone-derived ligands. Organometallics 26:6643–6652

Schmid WF, John RO, Muehlgassner G et al (2007b) Metal-based paullones as putative CDK inhibitors for antitumor chemotherapy. J Med Chem 50:6343–6355

Schmidt A, Beutler A, Snovydovych B (2008) Recent advances in the chemistry of indazoles. Eur J Org Chem 4073–4095

Schuecker R, John RO, Jakupec MA et al (2008) Water-soluble mixed-ligand ruthenium(II) and osmium(II) arene complexes with high antiproliferative activity. Organometallics 27:6587–6595

Singh P, Sarkar B, Sieger M et al (2006) The metal − NO interaction in the redox systems $[Cl_5Os(NO)]^{n-}$, n = 1–3, and cis-$[(bpy)_2ClOs(NO)]^{2+/+}$: calculations, structural, electrochemical, and spectroscopic results. Inorg Chem 45:4602–4609

Stadlbauer W (2002) Product class 2: 1H- and 2H-indazoles. Sci Syn 12:227–235

Stepanenko IN, Cebrián-Losantos B, Arion VB et al (2007) The complexes $[OsCl_2(azole)_2(dmso)_2]$ and $[OsCl_2(azole)(dmso)_3]$: synthesis, structure, spectroscopic properties and catalytic hydration of chloronitriles. Eur J Inorg Chem 400–411

Stepanenko IN, Krokhin AA, John RO et al (2008) Synthesis, structure, spectroscopic properties, and antiproliferative activity in vitro of novel osmium(III) complexes with azole heterocycles. Inorg Chem 47:7338–7347

Taqui Khan MM, Khan NM, Kureshy RI et al (1992) Synthesis, characterization and X-ray crystal structure studies of ruthenium(II) pyrazole complexes and their interaction with small molecules. Polyhedron 11:431–441

Taylor DB, Callahan KP, Shaikh I (1975) Synthesis of a bifunctional coordination complex of osmium with curariform activity. J Med Chem 18:1088–1094

van Rijt SH, Mukherjee A, Pizarro AM et al (2010) Cytotoxicity, hydrophobicity, uptake, and distribution of osmium(II) anticancer complexes in ovarian cancer cells. J Med Chem 53:840–849

van Rijt SH, Sadler PJ (2009) Current applications and future potential for bioinorganic chemistry in the development of anticancer drugs. Drug Discov Today 14:1089–1097

Vilaplana RA, Castineiras A, Gonzalez-Vilchez F (2004) Synthesis, structure, properties and biological behavior of the complex $[Ru^{IV}(H_2L)Cl_2] \cdot 2H_2O$ (H_4L = 1,2-cyclohexanediaminetetraacetic acid). Bioinorg Chem Appl 2:275–292

Yu Z, Knox JE, Korolev AV et al (2005) Synthesis, characterization, and hydrolysis products of $(\eta^2\text{-tBu}_2pz)AlH(\mu:\eta^1, \eta^1\text{-tBu}_2pz)_2AlH_2$ – structural characterization of a complex containing η^1-, η^2-, and $\mu:\eta^1$, η^1-pyrazolato ligands and a complex containing a terminal hydroxo ligand. Eur J Inorg Chem 330–337

Osmium, Physical and Chemical Properties

Fathi Habashi
Department of Mining, Metallurgical, and Materials Engineering, Laval University, Quebec City, Canada

Osmium is a member of the group of six platinum metals which occur together in nature and is located in the center of the Periodic Table in the Iron–Cobalt–Nickel group:

Fe	Co	Ni
Ru	Rh	Pd
Osmium	Ir	Pa

Physical Properties

Osmium is a transition metal like the other members of the group. They are less reactive than the typical metals but more reactive than the less typical metals. The transition metals are characterized by having the outermost electron shell containing two electrons and the next inner shell an increasing number of electrons. Because of the small energy differences between the valence shells, a number of oxidation states occur.

Atomic number	76
Atomic weight	190.2
Relative abundance in Earth's crust, %	1×10^{-6}
Abundance of major natural isotopes	189 (16.1%)
	190 (26.4%)
	192 (41.0%)
Crystal structure	Hexagonal closed pack
Lattice constants at 20°C	
a, nm	0.27341
c, nm	0.43197
Atomic radius, nm	0.134
Melting point, °C	3,045
Boiling point, °C	5,025
Specific heat at 25°C c_p, J g^{-1} K^{-1}	0.13
Thermal conductivity λ, W m^{-1} K^{-1}	87
Density at 20°C, g/cm^3	22.61[a]
Brinell hardness number	250
Young's modulus E, N/mm^2	559 170
Specific electrical resistance at 0°C, $\mu\Omega$ cm	8.12
Temperature coefficient of electrical resistance (0–100°C), K^{-1}	0.0042

[a]Osmium is the densest metal known

Chemical Properties

Although solid osmium is unaffected by air at room temperature, the powder will give off osmium tetroxide, OsO_4, a highly toxic vapor with a characteristic odor, hence the metal's name (the Greek word *osme*, a smell or odor). It is the most important compound of osmium available as a pale yellow solid. Osmium remains in the residue when dissolving platinum ores in aqua regia. It is used to add hardness to alloys. It is also used for fountain pen tips, instrument pivots, and electrical contacts. Its oxidation states are 8, 6, 4, 3, 2, 0, and −2.

References

Habashi F (2003) Metals from ores. An introduction to extractive metallurgy. Métallurgie extractive Québec, Quebec City, Canada. Distributed by Laval University Bookstore, www.zone.ul.ca

Renner H (1997) Platinum group metals. In: Habashi F (ed) Handbook of extractive metallurgy. Wiley-VCH, Weinheim, pp 1269–1360

Overcome Cisplatin Resistance

▶ Gold Complexes as Prospective Metal-Based Anticancer Drugs

Oxaliplatin

▶ Platinum Anticancer Drugs

Oxaliplatin, Clinical Use in Cancer Patients

Alexander Stein
Hubertus Wald Tumor Center, University Cancer Center Hamburg (UCCH), University Hospital Hamburg-Eppendorf (UKE), Hamburg, Germany

Synonyms

trans-L-dach (1R, 2R-diaminocyclohexane) oxalatoplatinum (L-OHP)

Definition of the Subject

Oxaliplatin is a highly active anticancer drug licensed for treatment of colorectal cancer (CRC) with a strong impact on improved prognosis of this disease. Based on chemical structure, pharmacodynamics and pharmacokinetics clinical development and current clinical use of oxaliplatin in CRC will be discussed, with particular regard to the distinct disease settings (locally advanced, R0 resectable liver metastases, palliative), combination with other agents and prevention of toxicity. Furthermore, use in other tumor types especially gastric and gastroesophageal junction cancer will be reviewed.

Introduction

Oxaliplatin is a platinum (II) analogue similar to carboplatin and cisplatin. The drug was discovered in 1976 at Nagoya City University by Yoshinori Kidani, and subsequently developed for treatment of patients with colorectal cancer (CRC) and other digestive tract cancers. Oxaliplatin was granted US patent in 1979 and subsequently in-licensed by debiopharm. In 1994 the drug was licensed to sanofi-aventis. Two years later, oxaliplatin gained approval in France and later on in Europe for colorectal cancer (CRC). Currently, there are several drugs licensed for the treatment of CRC. For postoperative chemotherapy after resection of localized CRC, fluoropyrimidines (FP) and oxaliplatin are currently the only available compounds with proven efficacy. For advanced (metastatic) CRC, oxaliplatin and irinotecan are the most important cytotoxic compounds, both combined with FP and with or without monoclonal antibodies (moAb), e.g., bevacizumab against vascular endothelial growth factor (VEGF) or cetuximab/panitumumab against epidermal growth factor receptor (EGFR), which at least in CRC only display activity in case of Kirsten rat sarcoma viral oncogene (KRAS) wildtype, a downstream signal in the EGFR pathway (Cunningham et al. 2010).

Chemistry

Oxaliplatin, or *trans*-L-dach (1R, 2R-diaminocyclohexane) oxalatoplatinum (L-OHP), differs from cisplatin and carboplatin in its possession of a bulky diaminocyclohexane (DACH) moiety and the presence of an oxalate "leaving group" (Raymond et al. 1998). In plasma oxaliplatin rapidly undergoes nonenzymatic transformation into intermediates (Ehrsson et al. 2002). The antitumor activity of the drug seems to be limited to free platinum species present in the ultrafilterable fraction of plasma.

Pharmacodynamics

There are several mechanisms of action described for oxaliplatin, although induction of deoxyribonucleic acid (DNA) lesions seems to be the main cytotoxic effect. Others are arrest and inhibition of nucleic acid synthesis as well as immunologic mechanisms. Oxaliplatin induces different types of DNA cross-links (intra- and inter-strand as well as protein) with

DNA intra-strand cross-links predominantly triggering DNA lesion. DNA synthesis is blocked via direct inhibition of thymidylate synthase, similar to the antimetabolite effect of FP, without being additive if both drugs are combined. Synthesis of messenger ribonucleic acid (RNA) is blocked by platinum-DNA adducts which either bind to transcription factors or inhibit RNA polymerase. Furthermore, oxaliplatin seems to induce immunogenic signals on the surface of cancer cells before apoptosis, triggering interferon gamma production and interaction with toll-like receptor-4 on dendritic cells, resulting in immunogenic death of cancer cells (Alcindor and Beauger 2011).

Pharmacokinetics and Metabolism

Pharmacokinetic data are available for the ultrafilterable fraction containing all active and inactive unbound platinum species. Elimination of unbound platinum from the ultrafiltrate is triphasic, and consists of two short phases with half-lives of 0.28–0.43 (t1/2 [alpha]) and 16.3–16.8 (t1/2 [beta]) hours, depending on the administered dose (85 or 130 mg/m^2). The long terminal phase has a half-life (t1/2 [gamma]) of 273–391 h. Steady-state ultrafiltrate drug concentrations are attained during the first cycle of treatment with oxaliplatin, and inter- and intrapatient variabilities in ultrafilterable platinum exposure are low. The pharmacokinetic disposition of oxaliplatin is similar after hepatic arterial or intravenous infusion. Oxaliplatin binds to plasma proteins, mainly albumin, and erythrocytes. After 2 h infusion, about 15% of the drug remains in the blood, whereas the majority is cleared from plasma by covalent binding to tissue and renal elimination. Elimination of oxaliplatin is predominately via urinary excretion (53.8 +/− 9.1%), whereas fecal excretion accounts for only 2.1 +/− 1.9% of the administered dose 5 days after administration (Graham et al. 2000). Renal clearance correlates with the glomerular filtration rate. In a small sample of patients, mild (creatinine clearance, 40–59 ml/min) or moderate (creatinine clearance, 20–39 ml/min) renal impairment did not alter maximal platinum concentration, and although platinum exposure increased with increasing renal dysfunction, incidence or severity of toxicities did not. Clearance of platinum does not seem to be influenced by hepatic impairment. Even in patients with severe hepatic dysfunction (bilirubin >3.0 mg/dl), or after liver transplantation single agent oxaliplatin with 130 mg/m^2 every 3 weeks was well tolerated as demonstrated in a dose escalation study (Alcindor and Beauger 2011).

Combination of Oxaliplatin with Other Cytotoxic Agents

Oxaliplatin demonstrated only modest activity as single agent, with an overall response rate of 10–20% in single arm phase II [50, 51] and 1.3% in a second line randomized phase III trial [52], and is thus combined with other chemotherapeutic agents, usually with fluoropyrimidines. Synergistic activity of oxaliplatin and FP has been shown in vitro and in vivo. However, the influence of oxaliplatin on pharmacokinetics of 5-fluorouracil (5FU), at least if modulated with folinic acid, seems to be negligible. Oxaliplatin and FP may be combined with moAbs for treatment of metastatic CRC, whereas combinations with moAbs have failed in adjuvant treatment of stage II/III disease.

There is a reasonable preclinical rationale for synergistic activity of oxaliplatin and EGFR moAbs, although sequence of administration might be important as demonstrated in an in vitro study with an antagonistic effect for anti-EGFR treatment followed by chemotherapy compared to a synergistic antiproliferative effect for the vice versa schedule. Current clinical data from phase III trials have raised concerns about this combination, although the delivery of the FP, either orally or via bolus or continuous infusion, is likely to be responsible for the varying results of adding an EGFR moAb to FP and oxaliplatin. Whereas infusional 5FU schedules (e.g., FOLFOX regimen) have shown significant benefit in terms of response and PFS for the addition of the EGFR moAb cetuximab, neither bolus 5FU (Nordic FLOX) nor oral capecitabine (XELOX) regimens displayed any survival benefit for the combination in KRAS wild type. Similar disadvantageous results were recently reported in gastric cancer for the addition of the EGFR moAb panitumumab to oxaliplatin, epirubicin, and the oral FP capecitabine. Unfortunately, neither preclinical nor clinical data are currently available, which might explain these findings. Furthermore, there are no randomized data about EGFR moAbs combined with single agent FP available, allowing further insight into the efficacy or potential interactions of this combination without oxaliplatin. In two small single arm phase II trials with overall about 100 patients, all treated with oral FP, either capecitabine or S1, combined with cetuximab, retrospective analysis of KRAS status revealed an RR of 24.1% in KRAS mutant compared to 36.8–48.3% in KRAS wild type, indicating that there might be some benefit of the combination. In KRAS mutant CRC patients, the addition of an EGFR moAb to chemotherapy with 5FU/oxaliplatin (FOLFOX regimen) has shown decreased disease-free survival (DFS) and OS compared to chemotherapy alone. Although the interaction of KRAS mutation and effects of the combination treatment are not fully elucidated yet, recent in vitro analysis in KRAS mutant cell lines indicated some interesting explanations. The antagonistic effects of treatment with oxaliplatin and EGFR inhibitors, although at least preclinically this effect was only noted for the tyrosine kinase (TK) inhibitor gefitinib and not the EGFR moAB panitumumab, might be induced by phosphorylation of protein kinase B (AKT) (downstream TK of EGFR pathway). Thus, oxaliplatin might somehow induce downstream pathway activation despite receptor inhibition, leading to the mentioned effects.

Furthermore, in rectal cancer combining EGFR moAbs with FP-based chemoradiation has shown disappointing results in terms of response at the rectal primary tumor. Based on the available data, molecular interactions between oxaliplatin, FP, radiation, and anti-EGFR moAbs remain largely unclear and further research to elucidate the underlying mechanism is urgently warranted (Grothey and Lenz 2012).

Clinical Development of Oxaliplatin

Phase I Studies

Starting 1984 several phase I trials were performed evaluating different dosages and administration schedules (Raymond et al. 1998). After using adequate antiemetic prophylaxis, neurotoxicity was the major dose-limiting toxicity. Mean tolerable dose was found to be either 130–135 mg/m^2 every 3 weeks administered as 2 h infusion or 125–150 mg/m^2 given constantly or chronomodulated over 5 days every 3 weeks.

Phase II Studies

Oxaliplatin with 130 mg/m^2 administered over 2 h every 3 weeks or chronomodulated over 5 days was evaluated in several phase II trials in either previously untreated or 5FU refractory colorectal cancer patients. Whereas single agent oxaliplatin showed objective response rates of 20–24% in previously untreated patients, thus being in the range of single agent 5FU or irinotecan, the activity in pretreated CRC patients was significantly lower with an RR of 10–11%. In the next step oxaliplatin short or continuous infusion were combined with 5FU and leucovorin (LV). High response rates of up to 66% were shown with chronomodulated oxaliplatin and 5FU schedules in previously untreated patients. A variety of different FOLFOX schedules as well as chronomodulated regimens were established, showing meaningful activity in previously treated patients.

Phase III Studies in Colorectal Cancer

Stage II/III Colorectal Cancer

Three large randomized phase III trials investigated the role of adjuvant treatment with oxaliplatin combined with fluoropyrimidines (either 5FU or capecitabine) after complete resection of stage II/III (UICC) CRC (MOSAIC, NSABP C-07 and NO 16968/XELOXA) showing a clear DFS and OS benefit for oxaliplatin-based regimens in stage III (Table 1). Similar and consistent results could be documented in the control groups of further three phase III trials combining oxaliplatin-based regimens with either bevacizumab or cetuximab (AVANT, NSABP C-08, NO147) with a similar DFS rate at 3 years. Furthermore, a current analyses of four NSABP trials (C-04–C-08) revealed a 2–3% increase in 5-year overall survival by the addition of oxaliplatin to 5FU in stage II independent of clinical risk factors. However, adding oxaliplatin to 5FU/LV significantly increases grade 3/4 (according to the common toxicity criteria from the national cancer institutes – NCI CTC AE) adverse events, e.g., neutropenia, febrile neutropenia, diarrhea, and especially neurosensory toxicity, whereas mortality and rate of secondary malignancies were similar. Current standard of care is adjuvant treatment for 6 months with FP and oxaliplatin after resection of a stage III colon cancer. In patients with stage II disease with increased individual risk as determined by clinical estimation, similar treatment or single agent FP might be considered.

Oxaliplatin, Clinical Use in Cancer Patients, Table 1 Outcome of adjuvant (after complete surgical resection) oxaliplatin-based chemotherapy in localized colon cancer (stage II/III according to UICC)

Trial	Regimen	Pat.	Stage	DFS rate	HR (95% CI)	p-value	OS rate	HR (95% CI)	p-value
MOSAIC	LV5FU2 vs. FOLFOX4	2,246		After 5 year 67.4 vs. 73.3%	0.8 (0.68–0.93)	0.003	After 6 year 76 vs 78.5%	0.84 (0.71–1.00)	0.046
		40%	II	79.9 vs. 83.7%	0.84 (0.62–1.14)	0.258	86.8 vs 86.9%	1.00 (0.70–1.41)	0.986
		60%	III	58.9 vs. 66.4%	0.78 (0.65–0.93)	0.005	68.7 vs 72.9%	0.80 (0.65–0.97)	0.023
NSABP C-07	5FU/LV vs. FLOX	6,789		After 5 year 64.2 vs. 69.4%	0.81 (0.70–0.93)	0.002	After 6 year 73.5 vs 77.7%	0.85 (0.72–1.01)	0.06
		29%	II						
		71%	III						
NO 16968 (XELOXA)	5FU/LV vs. XELOX	1,886	III	After 5 year 59.8 vs. 66.1%	0.80 (0.69–0.93)	0.0045	After 7 year 67 vs. 73%	0.84 (0.71–1.00)	0.047

LV5FU2 de Gramont infusional fluorouracil, *FOLFOX4* LV5FU2 with oxaliplatin, *5FU* 5-fluorouracil, *LV* leucovorin, *FLOX* bolus 5FU/LV (Roswell park) with oxaliplatin, *XELOX* capecitabine with oxaliplatin, *pat* patients, *DFS* disease-free survival, *OS* overall survival, *HR* hazard ratio, *CI* confidence interval, *ref* reference

The use of oxaliplatin in the perioperative treatment for rectal cancer was investigated in several phase III trials (ACCORD 12/0405-Prodige 2, STAR-01, NSABP R-04, CAO/ARO/AIO-04, PETACC 6). Whereas toxicity was increased by the addition of oxaliplatin to preoperative FP-based chemoradiation in all trials, significantly higher rate of pathological complete responses could only be shown in the German CAO/ARO/AIO-04 trial. Data are summarized in Table 2. However, survival data need to be awaited before final conclusion can be drawn, although oxaliplatin-based postoperative chemotherapy for up to 6 months in total is/was scheduled only in the CAO/ARO/AIO-04 and the still ongoing PETACC 6 trial.

Clearly R0 Resectable Colorectal Liver Metastases

Surgical resection of colorectal liver metastases (CLM) is a potentially curative treatment, with reported 5-year OS rates ranging from 20% to 40%. Current management strategies are using either peri- or postoperative systemic treatment. Perioperative chemotherapy with FOLFOX4 (3 months pre- and 3 months postoperatively) was investigated in the EORTC 40983 trial, randomizing 364 patients, compared to resection alone, demonstrating superior PFS rate at 3 years for perioperative treatment in patients undergoing resection (33.2 vs. 42.4%, HR 0.73, 95% CI 0.55–0.97; p = 0.025). Recent OS update revealed a nonsignificant trend toward an increased 5-year OS rate of 4.1% (52.4 vs. 48.3%, p = 0.34). However, the trial was not powered for OS benefit, which was a secondary endpoint. This approach represents a current standard for management of resectable CLM. Early results of oral FP capecitabine and oxaliplatin (XELOX regimen) and bevacizumab administered either preoperatively or postoperatively in comparison to XELOX showed promising results.

Metastatic Colorectal Cancer

Adding oxaliplatin to 5FU/LV significantly increased objective response rates (RR) (16/22.3% vs. 53/50.7%) and PFS (6.1/6.2 vs. 8.7/9.0 months) in two phase III trials in previously untreated patients, respectively. However, the FOLFOX4 regimen adding 85 mg/m^2 oxaliplatin delivered over 2 h every 2 weeks to the de Gramont regimen (LV5FU2) evolved as the standard 5FU and oxaliplatin regimen according to the results of EFC 2962 trial. In the EFC 4584 trial superior RR (9.9% vs. 0% vs. 1.3%) and time to progression (TTP) (4.6 vs. 2.7 vs. 1.6 months) for the combination of FOLFOX4 regimen compared to 5FU/LV or oxaliplatin alone in second line setting, after failure of bolus 5FU/LV and irinotecan (IFL), could be shown. Although RR and PFS or time to progression (TTP) were significantly increased with the addition of oxaliplatin, no OS benefit could be demonstrated in these two trials. The North Central Cancer Treatment Group (NCCTG) – Intergroup study was able to demonstrate a survival benefit for oxaliplatin combination,

Oxaliplatin, Clinical Use in Cancer Patients, Table 2 Role of oxaliplatin combined with neoadjuvant fluoropyrimidine-based chemoradiation for localized rectal cancer

TRial	Regimen	RT dose/fraction (Gy)	Stage	Pat.	Grade 3/4 AE	p-val	pCR rate	p-val	crm+	p-val
STAR 1	5FU (225 mg/m^2/day) vs. 5FUOx (60 mg/m^2 weekly × 6)	50.4/1.8	T3/4 and/or N+	747	8 vs. 24%	<0.001	16 vs. 16%	0.904	7 vs. 4%	0.24
ACCORD 12/0405-Prodige 2	Cap (45 Gy/800 mg/m^2/bid) vs CapOx (50 Gy/Cap + Ox 50 mg/m^2 weekly × 5)	45/1.8 50/2	≥T3 or ≥T2 lower rectum	598	11 vs. 25%	<0.001	13.9 vs. 19.2%	0.09	19.3 vs. 9.9%	0.02
NSABP R04	5FU (225 mg/m^2/day) vs. Cap (825 mg/m^2/bid)	46 + boost	T3/4 and/or N+	1,437	na		18.8 vs. 22.2%	0.12	na	
	5FU/Cap vs. 5FU or Cap + Ox (50 mg/m^2 weekly × 5)			1,166	only GI AE 6.6 vs 15.4%	<0.001	19.1 vs. 20.9%	0.46	na	
CAO/ARO/AIO 2004	5FU (1,000 mg/m^2 d1-5 + 29–33) vs. 5FU (250 mg/m^2 d1-14 + 22–35) + Ox (50 mg/m^2 d1,8,22,29)[a]	50.4/1.8	T3/4 and/or N+	1,265	22 vs. 23%	na	12.8 vs. 16.5%	0.045	6 vs. 5%	na

5FU 5-fluorouracil, *Ox* oxaliplatin, *Cap* capecitabine, *pat* patients, *AE* adverse event, *pCR* complete pathological response, *crm+* involved circumferential margins, *GI* gastrointestinal ref: reference, *na* not available
[a] Both regimens followed postoperative with 4 months of 5FU or mFOLFOX6 respectively

comparing FOLFOX to IFL or a combination of oxaliplatin and irinotecan (IROX) in previously untreated patients. RR, TTP, and OS were significantly increased with FOLFOX compared to IFL (45% vs. 31%, 8.7 vs. 6.9 months, and 19.5 vs. 15.0 months). The IROX regimen showed a similar RR (35%) and TTP (6.5 months) and a slightly better OS (17.4 months) than IFL. Furthermore, the FOLFOX regimen had significantly lower rates of severe nausea, vomiting, diarrhea, febrile neutropenia, and dehydration. After failure of first-line FP, the IROX regimen showed significantly increased RR, TTP and OS compared to irinotecan alone. However, the current standard approach uses a sequence of both drugs each combined with FP with or without moAb. Therefore, IROX regimen might thus be considered for special treatment situations, e.g., FP intolerance or complete dihydropyrimidine dehydrogenase deficiency (Simpson et al. 2003).

In the early development of oxaliplatin, different schedules (infusion for 2–6 h or continuously over 4–5 days) were developed in several phase III trials comparing chronomodulated with constant schedules for oxaliplatin and 5FU and the addition of oxaliplatin to chronomodulated or constant 5FU. However, randomized comparison between FOLFOX delivered in 2 days and chronomodulated regimen of 5FU and oxaliplatin (chronoFLO) delivered in 4 days showed similar RR, PFS, and OS. Thus, the shorter and easier to administer FOLFOX is currently the preferred option.

The next generation of clinical trials investigated the substitution of 5FU and leucovorin (LV) by capecitabine (XELOX). A pooled analysis of several randomized phase II and III trials demonstrated similar PFS and OS, but lower RR (Odds ratio = 0.85; 95% confidence interval (CI): 0.74–0.97; p = 0.02) for XELOX compared to FOLFOX.

After establishing the different regimens with FP and oxaliplatin, further potential combination partners were evaluated. Compared to 5FU/LV and irinotecan (FOLFIRI), the three drug combination of 5FU/LV, irinotecan, and oxaliplatin (FOLFOXIRI) improved RR by independent review (34 vs. 60%; p < 0.0001), secondary R0 resection rate of metastases (6 vs. 15%; p = 0.033), PFS, (6.9 vs. 9.8 months; hazard ratio [HR] 0.63; p = 0.0006), and OS (16.7 vs. 22.6 months; HR 0.70; p = 0.032). The efficacy of FOLFOXIRI was further underlined by the results of a Greek phase III trial with a similar design, yet different and less dosage of irinotecan and oxaliplatin, showing a trend toward increased RR, PFS, and OS.

Following the advances of targeted therapies, anti-VEGF and anti-EGFR-moAbs were combined with oxaliplatin-based regimen. The addition of bevacizumab to FOLFOX4 or XELOX significantly prolonged PFS from 8 to 9.4 months (HR 0.83, 97.5% CI 0.72–0.95; p = 0.0023). Independently reviewed response rates did not differ between bevacizumab and placebo containing regimens (38 vs. 38%). There was a trend toward increased OS with 19.9 vs. 21.3 months (HR 0.89, 97.5% CI 0.76–1.03; p = 0.077). Adding bevacizumab to FOLFOX4 significantly increased response rate (8.3 vs. 22.7%, p < 0.0001), PFS (4.7 vs. 7.3 months, HR 0.61; p < 0.0001), and OS (10.8 vs. 12.9 months, HR 0.75; p = 0.0011) after failure of 5FU and irinotecan.

Combining oxaliplatin with anti-EGFR antibodies should be done with caution in regard of the above-mentioned issues on FP administration and scheduling. Currently, randomized data has shown superior efficacy in terms of RR and PFS for the combination of FOLFOX with cetuximab or panitumumab in previously untreated, KRAS wild type patients. The COIN trial compared FOLFOX (34%) or XELOX (66%) with or without cetuximab and intermittent chemotherapy in a three-arm trial in KRAS wild type first-line population. Although RR increased with the addition of cetuximab (57 vs. 64%; p = 0.049), PFS (8.6 vs. 8.6 months, HR 0.96, 95% CI 0.82–1.12; p = 0.60) and OS (17.9 vs. 17.0 months, HR 1.04, 95% CI 0.87–1.23; p = 0.67) were similar. Subgroup analyses suggested that adding cetuximab to XELOX did not result in increased survival (HR 1.02, 95% CI 0.82–1.26), whereas FOLFOX and cetuximab seem to be beneficial compared to FOLFOX alone (HR 0.72, 95% CI 0.53–0.98).

Phase III Studies in Gastric and Gastroesophageal Junction (GEJ) Cancer

Although not licensed for gastric cancer, oxaliplatin has been evaluated compared to cisplatin in advanced gastric and GEJ cancer and compared to observation alone after curative D2 gastrectomy in combination with capecitabine in several randomized phase III trials (Van Cutsem et al. 2011). Regarding the equivalence of both platins, non-inferiority was proven in the 2×2 factorial REAL 2 study. In this trial four epirubicin (E) containing regimens were evaluated comparing oxaliplatin (O) and cisplatin (C) as well as 5-FU (F) and capecitabine (X) (ECF vs. ECX vs. EOF vs. EOX), demonstrating the best efficacy for the EOX regimen. Although formally negative, as superiority could not be demonstrated, a German phase III trial displayed a nonsignificant trend toward better PFS (5.8 vs. 3.9 months, p = 0.077), OS (10.7 vs. 8.8 months), and RR (34.8 vs. 24.5%, p = 0.012) in favor of the oxaliplatin containing 5-FU-based regimen (FLO). Whereas cisplatin was associated with higher incidences of grade 3/4 (NCI CTC v2.0) neutropenia, alopecia, renal toxicity, and thromboembolism, use of oxaliplatin led to higher incidences of grade 3/4 diarrhea and neuropathy. After curative D2 gastrectomy, the combination of capecitabine and oxaliplatin (XELOX) was superior compared to observation alone in terms of DFS (HR 0.56, 95% CI 0.44–0.72, p < 0.0001) in the recently published CLASSIC trial.

Oxaliplatin in Other Tumor Types

Oxaliplatin combinations have furthermore been evaluated in a broad variety of tumor types including lung cancer, esophageal, biliary, and pancreatic cancer.

Use of Oxaliplatin in Specific Patient Groups

Pooled analysis including 3,742 patients (614 patients ≥70 years) from four randomized trials in adjuvant and metastatic CRC comparing FOLFOX to either 5FU/LV or IFL demonstrated similar relative benefit in terms of response rate and survival and no increase in adverse events or 60-day mortality for the elderly patients selected in this trials. There are conflicting data about the use of oxaliplatin in CRC patients older than 70 years with some analyses demonstrating a detriment (ACCENT database including MOSAIC and NSABP C-07) compared to another (NSABP C-08, XELOXA, X-ACT, and AVANT) with a reduced but still statistical significant DFS and OS benefit. However, it is yet unclear what might serve as the most likely explanation for less benefit of elderly patients: Currently, reduced oxaliplatin administration (less cycles) and increased toxicity in elderly patients are under discussion, as well as death of second cancers. The FOCUS 2 trial compared upfront single agent FP to combination of FP and oxaliplatin (with a 20% dose reduction of both drugs) in patients, which were not eligible for standard palliative first-line combination for mCRC as judged by the investigator. The trial recruited 459 patients with median age of 74 years (range 35–87) and 29% with ECOG performance status 2. For the upfront oxaliplatin-based combination, a trend for better PFS (HR 0.84, 95%CI 0.69–1.01), with only modest increase in grade 3/4 toxicities (32 vs. 38%, p = 0.17), was demonstrated. Furthermore, the patient-centered endpoint (overall treatment utility) favored the use of oxaliplatin.

Currently, the feasibility of a standard FP and oxaliplatin regimen (FOLFOX4) was shown in a retrospective analysis with 24 HIV-positive patients with concomitant antiretroviral therapy.

Safety and Tolerability

Toxicity Profile

Oxaliplatin induces moderate myelotoxic effects presumably by affecting bone marrow progenitor cells. Furthermore, DNA adducts can be found in leukocytes. Whereas grade 3/4 neutropenia is common, occurring in about 40% of patients, febrile neutropenia is rare. Thrombocytopenia and anemia is usually mild, although hemolysis and secondary immune thrombocytopenia can be caused by hypersensitivity reactions. Mild to moderate gastrointestinal adverse events are mainly nausea, vomiting, and diarrhea, which are caused by nonspecific toxic effects on mucosa. Neurotoxicity, especially sensory, is the major side effect limiting treatment continuation. There are two distinct patterns of peripheral sensory neuropathy (PSN). Whereas acute PSN is characterized by paresthesia, dysesthesia, or allodynia affecting extremities and the oropharyngolaryngeal area either during or shortly after oxaliplatin administration, often triggered by exposure to cold, chronic PSN is a cumulative side effect occurring after repeated oxaliplatin administration and essentially involves distal extremities. Incidence of grade 3/4 neurotoxicity is up to 20%, with 11.9/2.8/0.7% of grade 1/2/3 PSN remaining 4 years after treatment, respectively. Acute PSN seems to be induced by affection of voltage-gated sodium channels, whereas chronic PSN is associated with atrophy and mitochondrial dysfunction in the dorsal root ganglia cells by accumulation of platinum compounds.

There have been reports from post-marketing surveillance of prolonged prothrombin time and INR occasionally associated with hemorrhage in patients on anticoagulants, veno-occlusive disease of liver also known as sinusoidal obstruction syndrome, and pulmonary fibrosis, as well as other interstitial lung diseases (Alcindor and Beauger 2011).

Prevention of Neurologic Toxicity

Clinical Strategies for Prevention of Cumulative Neurotoxicity

Triggered by oxaliplatin-induced neurotoxicity, different treatment strategies were investigated to reduce cumulative dose of oxaliplatin, especially in CRC. In both French OPTIMOX (Opt) strategies, FOLFOX for 3 months was administered followed by either maintenance with 5FU/LV (Opt 1) or complete stop (Opt 2) followed by scheduled restart of FOLFOX after 6 months (Opt 1) or in case of disease progression whenever it occurs (Opt 2). Complete stop of chemotherapy seemed to be inferior compared to maintenance in terms of duration of disease control and PFS. The CONcePT trial used a different strategy with an alternating schedule of 8 weeks mFOLFOX7 and bevacizumab followed by 8 weeks 5FU/LV and bevacizumab compared to continuous mFOLFOX7 and bevacizumab. Although the trial was prematurely terminated, time to treatment failure (5.6 vs. 4.2 months) and PFS (12.0 vs. 7.3 months) seemed to favour intermittent (and maintenance) compared to continuous treatment. More recently, the already mentioned COIN trial compared an Opt 2 like strategy with continuous treatment in 1,630 patients. Non-inferiority of both strategies could not be proven because upper limits of both CIs for the HR were greater than the predefined non-inferiority margins. However, although complete treatment discontinuation might be an option for patients with less aggressive disease, e.g., normal platelet count at baseline, no liver metastases, less than two metastatic sites, normal LDH, no symptoms, maintenance treatment seem to be the preferred strategy. Re-induction of treatment with oxaliplatin or FOLFOX in case of progressive disease showed an RR of 20% (after FP maintenance) up to 32% (after complete stop), further underlining the feasibility of this approach.

Prophylactic Measures

In a recently published Cochrane review, none of the potential chemoprotective agents (acetylcysteine, amifostine, calcium and magnesium – CaMg, gluthatione, Org 2766, oxycarbazepine, diethyldithiocarbamate, or vitamine E) prevent or limit the neurotoxicity (Albers et al. 2011). However, a more recent study, which was not part of the review (NCCTG N04C7) using intravenous CaMg before and after oxaliplatin infusion in patients treated with FOLFOX in the adjuvant treatment setting, showed neuroprotective properties. With regard to the safety issue and assumed inferior oxaliplatin efficacy in patients receiving CaMg, preliminary results of the CONcePT study in the palliative setting demonstrated no evidence for inferior antitumor activity when CaMg was given, which is also in accordance with the results of the NEUROXA study. However, with remaining incertitude about the risk/benefit ratio, CaMg might only be used with caution.

Cross-References

▶ Platinum Anticancer Drugs
▶ Platinum-Containing Anticancer Drugs and Proteins, Interaction
▶ Platinum-Resistant Cancer
▶ Platinum(II) Complexes, Inhibition of Kinases
▶ Proteomic Basis of Cisplatin Resistance
▶ Zinc, Metallated DNA-Protein Crosslinks as Finger Conformation and Reactivity Probes

References

Albers JW, Chaudhry V et al (2011) Interventions for preventing neuropathy caused by cisplatin and related compounds. Cochrane Database Syst Rev 2:CD005228
Alcindor T, Beauger N (2011) Oxaliplatin: a review in the era of molecularly targeted therapy. Curr Oncol 18(1):18–25
Cunningham D, Atkin W et al (2010) Colorectal cancer. Lancet 375(9719):1030–1047
Ehrsson H, Wallin I et al (2002) Pharmacokinetics of oxaliplatin in humans. Med Oncol 19(4):261–265
Graham MA, Lockwood GF et al (2000) Clinical pharmacokinetics of oxaliplatin: a critical review. Clin Cancer Res 6(4):1205–1218
Grothey A, Lenz HJ (2012) Explaining the unexplainable: EGFR antibodies in colorectal cancer. J Clin Oncol 30(15):1735–1737
Raymond E, Chaney SG et al (1998) Oxaliplatin: a review of preclinical and clinical studies. Ann Oncol 9(10):1053–1071
Simpson D, Dunn C et al (2003) Oxaliplatin: a review of its use in combination therapy for advanced metastatic colorectal cancer. Drugs 63(19):2127–2156
Van Cutsem E, Dicato M et al (2011) The diagnosis and management of gastric cancer: expert discussion and recommendations from the 12th ESMO/World Congress on Gastrointestinal Cancer, Barcelona, 2010. Ann Oncol 22(Suppl 5):v1–v9

Oxidative Challenge

▶ Cadmium and Oxidative Stress

Oxidative Stress

▶ Arsenic, Free Radical and Oxidative Stress
▶ Magnesium and Inflammation
▶ Tellurium and Oxidative Stress

Oxidative Stress Response

▶ Cadmium and Stress Response

Oxygen Free Radicals, Reactive Oxygen Species

▶ Magnesium in Health and Disease

Oxygen Oxidoreductase

▶ Ascorbate Oxidase

Oxygenase

▶ Iron Proteins, Mononuclear (non-heme) Iron Oxygenases

P

p101

▶ Phosphatidylinositol 3-Kinases

p110α

▶ Phosphatidylinositol 3-Kinases

p110β

▶ Phosphatidylinositol 3-Kinases

p110γ

▶ Phosphatidylinositol 3-Kinases

p110δ

▶ Phosphatidylinositol 3-Kinases

p150

▶ Phosphatidylinositol 3-Kinases

p50α

▶ Phosphatidylinositol 3-Kinases

p55α

▶ Phosphatidylinositol 3-Kinases

p55γ

▶ Phosphatidylinositol 3-Kinases

p84

▶ Phosphatidylinositol 3-Kinases

p85α

▶ Phosphatidylinositol 3-Kinases

p85β

▶ Phosphatidylinositol 3-Kinases

V.N. Uversky et al. (eds.), *Encyclopedia of Metalloproteins*, DOI 10.1007/978-1-4614-1533-6,
© Springer Science+Business Media New York 2013

p87

▶ Phosphatidylinositol 3-Kinases

PABA-Peptide Hydrolase

▶ Zinc Meprins

Palladium Allergy

▶ Palladium, Biological Effects

Palladium Complex-induced Release of Cyt c from Biological Membrane

Sanaz Emami[1], Hedayatollah Ghourchian[1] and Adeleh Divsalar[2]
[1]Department of Biophysics, Institute of Biochemistry and Biophysics (IBB), University of Tehran, Tehran, Iran
[2]Department of Biological Sciences, Tarbiat Moallem University, Tehran, Iran

Synonyms

Cytochrome c (Cyt c): small heme protein

Definition

▶ Cyt c is one of the important metalloproteins that is bound to the outer surface of the inner mitochondrial membrane. It is a highly conserved and a peripheral protein (12 kDa) and consists of 104 amino acids. Its redox center consists of a single iron ion complexed by four nitrogen atoms of the porphyrin ring in which two residues of histidine and methionine (His-18 and Met-80) serve as axial ligands of the protein. Because of its heme group, Cyt c can shuttle electrons between respiratory chain complexes III (Cyt reductase) and IV (Cyt c oxidase). Without Cyt c, electron flow would be interrupted with two potentially lethal consequences – loss of ATP synthesis and overproduction of the radical superoxide anion – as a result of incomplete oxidation. At physiological pH, Cyt c is mostly protonated and therefore electrostatically binds to acidic phospholipids, which are abundantly present in mitochondrial inner membrane. Under normal physiological conditions, phospholipid cardiolipin (CL) anchors Cyt c to the mitochondrial inner membrane. The acyl chains of CL are inserted into a hydrophobic channel in Cyt c, while another acyl chain from CL extends into the phospholipids bilayer. These electrostatic and hydrophobic interactions between CL and Cyt c must be breached in order for Cyt c to leave mitochondria. Cyt c is also important mainly due to its ability to leave the mitochondrial membrane and functions as a death messenger in programmed cell death or apoptosis. During apoptosis, Cyt c is released from the outer membrane of mitochondria. In the other words, Cyt c is released from the outer surface of the inner membrane and passed across the intermembrane space through the outer membrane to cytosol. This allows it to interact with apoptosis protease activating factor-1 in the cytosol. This, in turn, activates pro-caspase-9. Caspase-9 then triggers other caspases. It is known that apoptosis contributes significantly to the cytotoxic effects of many traditional anticancer drugs (Kaufmann and Earnshaw 2000). The cytotoxicity of metal complexes is due to the apoptotic and pro-apoptotic activity. Primitive events in apoptosis are known to occur in mitochondria and endoplasmic reticulum. In this case, Cyt c serves as an important target for anticancer drugs in which it is a crucial factor in apoptosis. Previous studies have demonstrated that a wide range of anticancer agents, including chemotherapeutic agents, hormones, and various biologicals, induce apoptosis in malignant cells in vitro (Arends and Wyllie 1991). Many such anticancer drugs actually kill target cells by inducing apoptosis. The application of metal complexes as therapeutic drugs was first mentioned 5,000 years ago (Orvig and Abrams 1999). The major classes of metal-based anticancer drugs include platinum (II), palladium (II), gold (I) and (III), metalloporphyrins, ruthenium (II) and (III), bismuth (III), and copper (II) compounds. The use of metallopharmaceuticals such as platinum drugs (e.g., cisplatin) for cancer treatment illustrates the utility of metal complexes as therapeutic agents (Shimizu and Rosenberg 1973). Platinum (II) complexes, including cisplatin and carboplatin, are widely used in treating several human diseases such as ovarian, testicular, lung, urinary bladder, head, and neck cancers. However, cisplatin and its analogues produce several side effects including injury to renal tubular epithelial cells. This manifests as either acute renal failure or chronic syndrome characterized by renal electrolyte wasting (Divsalar et al. 2011). Because palladium chemistry is similar to that of platinum, it was speculated that BOPN might also exhibit antitumor activities with lower side effects than those caused by platinum (II) complexes. Attempts have been made to synthesize Pd(II) complexes with such activities. Palladium complexes are expected to have lower kidney toxicity than cisplatin due to the lowered probability of replacing the tightly bound chelate ligands of Pd(II) with sulfhydryl groups of proteins in kidney tubules. It has been reported (Genova et al. 2004) that Pd(II) complexes also have significant cytotoxicity against certain human tumor cell lines. It has been generally accepted that the ultimate target for platinum and palladium anticancer agents is likely to be DNA. But the reaction mechanism of cisplatin by which it induces cell death is still to be established. However, cisplatin inhibits the biosynthesis of DNA, and evidences for its in vivo activity at the transcription level had been reported (Mansouri-Torshizi et al. 2008). Platinum group metal complexes therefore offer potential as antitumor agents. BOPN has already been introduced as an example of palladium complexes which has the potential to be used as one of the promising drugs in cancer therapy in the future. Contemporary research focuses on the way that this complex would induce its effects in initiating apoptosis (Emami et al. 2011).

Release of Cyt c from the Membrane

In this entry, the function of Pd complexes as anticancer drugs in releasing Cyt c from the biological membrane has been discussed. For this purpose, an artificial monolayer containing a mixture of 11-mercaptoundecanoic acid (MUA) and 11-mercapto-1-undecanol (MU) was assembled on a gold electrode and used as a model membrane to adsorb Cyt c. Due to the existence of both hydrophobic and electrostatic interactions between Cyt c and the assembled membrane, this model membrane is considered to be a rough analogue of a biological membrane. MUA is a long-chain thiol, which can form a self-assembled monolayer on gold, giving a negatively charged surface, which can electrostatically attract the positively charged lysine residues near the heme cavity of Cyt c. MU acts as spacer between carboxylic acid groups and provides a stable monolayer by reducing the charge density on the electrode surface. For determining the release of Cyt c from the model membrane, the modified gold electrode was inserted in the electrochemical cell and then the BOPN was gradually added to the solution. As mentioned earlier, for electrochemistry investigations, the cleaned gold electrode was first modified with a mixture of MUA and MU. Afterward, Cyt c was adsorbed on the assembled membrane. The initial binding interaction contains a purely electrostatic association between the negatively charged surface of the membrane and the positively charged Cyt c molecules. This interaction leads to the hydrophobic interaction of Cyt c with the membrane and subsequently causes the partial or complete penetration of Cyt c into the membrane interior (Salamon and Tollin 1996). One of the critical points of the interaction of Cyt c with the membrane is electrostatic and hydrophobic interaction. By these means, not only the electrostatic interaction of Cyt c is important for the electron transferring processes, but also the hydrophobic interaction which leads to insertion of Cyt c into the membrane is vital and must occur to allow Cyt c to transfer electrons through the membrane from the aqueous phase to the electrode surface (Salamon and Tollin 1996). In these measurements, a stable and quasi-reversible cyclic voltammogram (CV) was observed for Cyt c immobilized on the modified electrode. The results confirm that the CV response was reduced in the presence of different concentrations of BOPN. On the other hand, the number of BOPN molecules bound per imbedded Cyt c was calculated to be about 0.5. This indicates that each BOPN binds to two Cyt c molecules while the association constant was determined as $1.4 \times 10^5 \, M^{-1}$. Also the kinetic binding constant of BOPN to Cyt c was obtained by addition of BOPN to the protein solution and recording the current at every 1 min intervals. Thus, binding of BOPN to Cyt c was confirmed based on the values of association constant ($1.4 \times 10^5 \, M^{-1}$) and the rate of binding constant ($0.21 \, min^{-1}$). These results revealed the strong binding between Cyt c protein and BOPN. On the other hand, the CV height which was decreased in the presence of different concentrations of BOPN revealed that BOPN can breach the electrostatic and hydrophobic interaction of Cyt c and membrane. This result was confirmed by the UV–vis spectrum in which the samples were transferred into a cuvette and the amount of released Cyt c was estimated by UV–vis spectrophotometery. Then the cell solution was sampled before and after each addition of BOPN. The UV–vis spectrum revealed that by increasing the concentration of BOPN, a progressive change in the absorption UV–vis spectrum is observed at around 280, which is a peak attributed to the aromatic amino acids, and 408 nm peak from the Soret band. The result proved the release of Cyt c from the self-assembled monolayer due to BOPN addition. In order to reveal the specificity of BOPN as an anticancer on conformational change and release of Cyt c, the effects of Ba^{2+} and Zn^{2+} as two non-anticancer metal ions were investigated, but neither change in conformation nor in release of Cyt c was observed.

Effects of BOPN on the Conformation of Cyt c

To study the effect of BOPN on conformational changes of Cyt c, both the intrinsic and extrinsic fluorescence emission spectra of Cyt c in the absence and presence of different concentration of BOPN were investigated. Fluorescence is commonly used as a method of analysis in which the molecules of the analyte are excited by irradiation at a certain wavelength and their emitted radiation of a different wavelength is measured. The tryptophan fluorescence is a sensitive measure of the overall conformation of the protein. Cyt c has special behavior when its fluorescence emission is measured. The fluorescence spectrum of the single tryptophan residue at position 59 of native Cyt c at pH 7.0 is almost completely quenched by the heme group and represents only 2.0% of free tryptophan fluorescence in aqueous solution. Increasing the concentration of BOPN brought the emission intensity to a maximum. This result indicates that BOPN binds to Cyt c and inhibits the intrinsic fluorescence quenching. This led us to the conclusion that BOPN affects the dissociation of the Met-80 and His-18 coordination and opens up the heme crevice resulting in a substantial increase of Trp-59 fluorescence. Thus, the results confirm that by adding BOPN to the cell solution, the fluorescence emission appeared and its intensity was increased by increasing the concentration. Due to the limitation of Cyt c surface binding sites, the emission intensity reaches a plateau at higher concentrations of BOPN. It was concluded that at higher BOPN concentrations, the fluorescence quenching could be inhibited completely due to conformational changes in Cyt c and its heme, prosthetic group. In order to identify the surface hydrophobic sites on proteins, neutral Nile red was used as a polarity-sensitive probe. Many reports in the literature confirm that the fluorescence of the probe increases upon binding to hydrophobic sites on the protein (Cardamone and Puri 1992). Increasing in emission intensity of Nile red also confirms that BOPN can change the Cyt c conformation through changing the accessible polarity site of Cyt c.

Effects of BOPN on the Structure of Cyt c

Secondary structure can be determined by circular dichroism (CD) spectroscopy in the "far-UV" spectral region (190–250 nm). At these wavelengths, the chromophore is the peptide bond, and the signal arises when it is located in a regular, folded environment. Alpha-helix, beta-sheet, and random coil structures each give rise to a characteristic shape and magnitude of the CD spectrum. Thus, far-CD spectra were used for determination of the alteration in secondary structure of a protein in the presence of BOPN. The peptide bond is the principle absorbing group, and studies in this region can give information on the secondary structure. The CD spectrum of Cyt c shows the typical features of the α-helical protein structure at 222 and 208 nm that are predominantly associated with α-helical amide transitions. In order to consider the effect of BOPN on the conformational changes of Cyt c, far-CD spectra were recorded. The results revealed that the ordered helical structure of Cyt c gradually disappears with increasing concentration of BOPN and the protein adopts a random-like structure. CD deconvolution program was used for measuring the percentage of helix, antiparallel, parallel, beta-turn, and random coil structure. Thus, the BOPN causes a conformational change from α-helical protein structure to random coil polypeptide with less organized tertiary structure.

Discussion

In this entry, the effects of BOPN on Cyt c have been discussed as the first important factor in apoptosis. Cyt c has one function in mitochondria and an entirely different function in the cytosol. As mentioned, Cyt c is at the very heart of life as a carrier of electrons in the mitochondria, and it is a key opening the door to cell death. It was found that the function of BOPN as an anticancer agent was linked to binding sites on Cyt c and inducing the conformational change of Cyt c. Strong binding between Cyt c and BOPN results in release of Cyt c from the mitochondrial membrane (Emami et al. 2011). On the other hand, the single tryptophan residue in Cyt c serves as a built-in fluorescent label. Thus, the fluorescence spectrum of the single tryptophan residue of native Cyt c is almost completely quenched by the heme group and represents only small amount of free tryptophan fluorescence in aqueous solution. As a result, Cyt c itself has no significant emission intensity. As noted, BOPN inhibits the intrinsic fluorescence quenching of Cyt c. Therefore, BOPN leads the dissociation of the Met-80 coordination and opens up the heme crevice resulting in a substantial increase of tryptophan fluorescence. It induces the conformational change in Cyt c and alters the α-helical structure to random coil. The electrochemical experiments showed that the anticancer BOPN can also breach the electrostatic and hydrophobic interactions between Cyt c and biological membrane so that Cyt c is released from the biological membrane. Release of Cyt c unleashes the killer caspases; this leads to proteolytic destruction of the cancer cells. Thus, BOPN functions as an anticancer drug and triggers apoptosis by inducing release of Cyt c from mitochondria.

Conclusion

This work highlights the effects of BOPN as an anticancer compound on the Cyt c protein. It provides an insight into how metallopharmaceutical drugs such as BOPN induce an apoptosis pathway in a cell. Cytosolic Cyt c is necessary for the initiation of the apoptotic program. BOPN can breach the interaction between Cyt c and the inner mitochondrial membrane and change its structure. Thus, it has been demonstrated how BOPN might provide a promising approach in future cancer therapy and therefore provide useful information to design better metal anticancer complexes with lower side effects in the future.

Cross-References

- ▶ Gold Complexes as Prospective Metal-Based Anticancer Drugs
- ▶ Gold(III) Complexes, Cytotoxic Effects
- ▶ Palladium, Biological Effects
- ▶ Palladium, Coordination of Organometallic Complexes in Apoferritin
- ▶ Platinum Anticancer Drugs

References

Arends MJ, Wyllie AH (1991) Apoptosis: mechanisms and roles in pathology. Int Rev Exp Pathol 32:223–254

Cardamone M, Puri NK (1992) Spectrofluorimetric assessment of the surface hydrophobicity of proteins. Biochem J 282:589–593

Divsalar A, Saboury AA, Ahadi L, Zemanatiyar E, Mansouri-Torshizi H, Ajloo D, Sarma RH (2011) Biological evaluation and interaction of a newly designed anti-cancer Pd(II) complex and human serum albumin. J Biomol Struct Dyn 29(2):283–296

Emami S, Ghourchiana H, Divsalar A (2011) Release of Cyt c from the model membrane due to conformational change induced by anticancer palladium complex. Int J Biol Macromol 48:243–248

Genova P, Varadinova T, Matsanz AI, Marinova D, Souza P (2004) Toxic effects of bis(thiosemicarbazone) compounds and its palladium(II) complexes on herpes simplex virus growth. Toxicol Appl Pharmacol 197:107–112

Kaufmann SH, Earnshaw WC (2000) Induction of apoptosis by cancer chemotherapy. Exp Cell Res 256:42–49

Mansouri-Torshizi H, Moghaddam M, Divsalar A, Saboury AA (2008) 2,2-Bipyridine-butyldithio-carbamato-platinum(II) and palladium(II) complexes: Synthesis, characterization, cytotoxicity and rich DNA-binding studies. Bioorg Med Chem 16:9616–9625

Orvig C, Abrams MJ (1999) Medicinal inorganic chemistry: introduction. Chem Rev 99:2201–2204

Salamon Z, Tollin G (1996) Surface plasmon resonance studies of complex formation between cytochrome c and bovine cytochrome c oxidase incorporated into a supported planar lipid bilayer. I. Binding of cytochrome c to cardiolipin/phosphatidylcholine membranes in absence of oxidase. Biophys J 71:848–857

Shimizu M, Rosenberg B (1973) A similar action to UV-irradiation and a preferential inhibition of DNA synthesis in *E coli* by antitumor platinum compounds. J Antibiot 26:243–245

Palladium Cytotoxicity

▶ Palladium, Biological Effects

Palladium Hypersensitivity

▶ Palladium, Biological Effects

Palladium Toxicity

▶ Palladium, Biological Effects

Palladium, Biological Effects

John Wataha
Department of Restorative Dentistry, University of Washington HSC D779A, School of Dentistry, Seattle, WA, USA

Synonyms

Palladium allergy; Palladium cytotoxicity; Palladium hypersensitivity; Palladium toxicity

Definitions

Palladium (Pd), an element with atomic number 46, occurs in row (period) five and group ten of the periodic table as a transition metal. Palladium is a member of the platinum-group metals (ruthenium, rhodium, palladium, osmium, iridium, and platinum, Fig. 1), which share chemical properties and tend to occur together in the Earth's crust. Palladium has no known natural biological role. Named after the asteroid "Pallas," palladium was discovered by William Wollaston in 1803 (Wataha and Hanks 1996). Wollaston dissolved crude platinum ore into aqua regia, precipitated out the platinum, and then added cyanide to precipitate palladium cyanide. Palladium was purified by heating the cyanide compound and extracting the metal.

Palladium Chemistry, Physical Properties, and Metallurgy

As a transition metal, palladium has multiple oxidation states (+1, +2, +4, +6); although, only the +2 and +4 states are common. Furthermore, it has an electronic structure in which all of its outer d-shell electrons are paired ($3d^{10}$), accounting for its relative resistance to corrosion and chemical reactivity. Because of this limited chemical reactivity, palladium is considered a noble metal, along with gold and the other platinum-group metals (Fig. 1). The outer s-shell orbitals of palladium are empty ($5s^0$); this permits alternate electronic configurations for coordination bonding (Fig. 1).

Palladium reacts with chlorine and fluorine at high temperatures to form PdF_2 and $PdCl_2$, both of which are somewhat water soluble, more so if they are hydrated (e.g., $PdCl_2.6H_2O$). Palladium also reacts with nitric, sulfuric, acetic, and to some extent with hydrochloric acids, and will form sulfides, phosphides, and acetates (Wataha and Hanks 1996). PdO is not water soluble. Palladium(II) oxide is dark black, whereas palladium(II) chloride is deep red and palladium(II) acetate is an orange-rust color. Palladium forms coordination compounds with electrophilic ligands such as cyanide.

Palladium is the least dense of the platinum-group metals (12.02 g/cc, 25°C) and also the lowest melting (1,555°C) (Wataha and Shor 2010). However, it is considerably higher in density and melting point than other elements with which it is often alloyed for biomedical applications (e.g., gallium, silver, and copper). Palladium is moderately hard and stiff, with a Vicker's hardness of 461 MPa and modulus of 121 GPa (Wataha and Shor 2010). In dentistry, where palladium is alloyed with other elements for intraoral prostheses, these values are intermediate among dental alloys. Palladium is a silvery-white metal, and has an often exploited property of hydrogen absorption.

Metallurgically, palladium exists in a face-centered cubic lattice, and is miscible with copper, silver, and gold; all of these share this type of lattice structure. Alloys of palladium with these metals are common in dentistry and other biomedical applications. Although palladium has more limited solubility with gallium or platinum, the latter are sometimes added to palladium-containing biomedical alloys to improve strength or modulus. Recently, nickel-chromium and

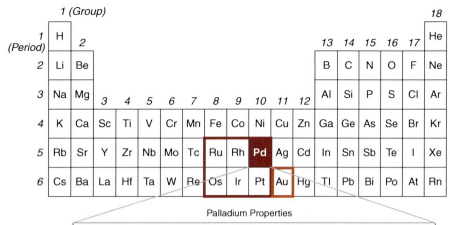

Palladium, Biological Effects, Fig. 1 Palladium (symbol Pd) occurs in the periodic table as a period 5, group 10 metal. It is a member of the platinum-group metals (*red box*) and a noble metal (*red* and *orange boxes* together). Palladium's outer shell electron configuration minimizes reactivity with other elements, yet allows coordination with organic ligands. The properties of palladium greatly influence its biological properties and uses in biomedical applications

cobalt-chromium alloys with added palladium (about 25 wt.%) have been proposed for use in dentistry. Corrosion of dental alloys in intraoral environments is one significant, potential exposure route of humans to palladium.

Abundance, Uses, and Human Exposure

Palladium occurs at about 0.015 mg/kg (ppm) in the Earth's crust with other platinum-group metals, which is by far the highest concentration among the platinum-group metals, but far below other common metals (e.g., copper is 60 mg/kg) (Ravindra et al. 2004; Wataha and Shor 2010). For this reason, palladium is considered a precious metal and is traded as such as a commodity. The rare abundance makes palladium costs fluctuate significantly on commodity markets in response to changes in supply or use, and these fluctuations have had significant effects on its use in humans, particularly in dentistry. Since 1990, the price of palladium has fluctuated between approximately USD125 and USD1100/oz. The most productive mines for palladium are in Russia (Siberia), South Africa, and Canada (Ontario); about 40% of the world's supply comes from Siberia and another 40% from South Africa (Kielhorn et al. 2002). Other smaller deposits occur in the United States (Montana), Australia, and South America.

Palladium has an increasing number of uses, and the total demand for this metal has increased dramatically from about 100 metric tons/year in 1988 to about 300 metric tons/year in 2001, and still more today (Kielhorn et al. 2002). The primary cause of this increased demand has been the use of palladium in automobile emission control catalytic converters. Because small amounts of particulate palladium are emitted into the atmosphere from cars with catalytic converters (4-108 ng/km driven (Kielhorn et al. 2002)), this use is a significant source of human exposure, but not directly (see below). Palladium also is a critical component of diodes, transistors, integrated circuits, capacitors, and various semiconductors, and is a major component of jewelry and dental alloys. The catalytic properties of palladium also are exploited in the synthesis of several industrially important organic monomers such as vinyl acetate or terephthalic acid. Finally, several uses of palladium exploit its ability to filter or absorb hydrogen; for example, dental impression materials have employed particulate palladium to absorb hydrogen gas released during setting of the impression that would otherwise ruin the models used to fabricate dental prostheses.

The uses of palladium determine its human exposure and therefore its potential biological effects in humans. The greatest source of potential exposure is to particulate in the air created by release of palladium from automobile catalytic converters. As would be expected, the amount of palladium exposure from air and dusts varies significantly by location, but is likely about 0.3 ng/day on average. By contrast, concentrations of palladium in soil range from 0 to 47 µg/kg, and some studies have reported palladium in grasses (0.3–1.3 μg/kg), pigeons (100–800 μg/kg dry weight), and in seaweeds (0.25 μg/kg) (Kielhorn et al. 2002). These latter values suggest that palladium has significant mobility in the environment in spite of its low chemical reactivity and no known normal biological role (Ravindra et al. 2004).

Humans ingest relatively small amounts of palladium through diet, probably less than 2 μg/day. Relatively little (<0.03 μg/person/day) comes through drinking water, and even less from the air (Kielhorn et al. 2002). Thus, the primary effect of particulate palladium emitted from automobile catalysts appears to be as a source of environmental load that ultimately ends up in foods. There is no known nutritional role for palladium in humans.

Individuals who have dental prostheses containing palladium may be exposed to palladium via corrosion of the metal to palladium(II) ions in the saliva. Exact exposures have not been accurately determined, and are influenced by the type of alloy, oral forces (e.g., acidic diets, tooth brushing) specific to the individual, and amount of alloy surface exposed. Crude estimates suggest that a range of 1.5–15 μg/day in such individuals is possible (Kielhorn et al. 2002).

Occupational exposures to palladium in mines, refineries, and technical areas are another potential route of exposure (e.g., dental technicians who fabricate dental prostheses). Miners appear to have the lowest exposures (<0.003 μg Pd/m^3 of air). Refinery exposures are higher, ranging from 0.001 to 0.36 μg/m^3, and the highest levels have been reported in dental labs, in the range of 3.5–9.8 μg/m^3. The above values are estimates and robust data are not available (Kielhorn et al. 2002).

Fears about Biological Effects of Palladium

The use of palladium-containing dental alloys increased dramatically when the United States deregulated the price of gold in the 1970s. By the late 1980s, palladium was a common constituent of many dental alloys because gold prices had increased 20-fold. In the early 1990s, several articles appeared in the non-peer-reviewed German literature or popular press, claiming that palladium alloys were releasing milligrams of palladium into patients and causing or contributing to a wide variety of illnesses (Wataha and Shor 2010). A public hysteria ensued, and for several years, alloys without palladium were widely sought by patients, then dentists. As a consequence, several "palladium-free" alloys were developed and still exist today. With the exception of hypersensitivity, none of the original claims of biological harm attributed to palladium were ever substantiated, and controlled studies, primarily in vitro, showed that palladium corrosion from these alloys was much lower than originally claimed (Wataha and Hanks 1996). Yet residual fears remain about the biological effects of palladium in some countries, especially in Europe.

Toxicity in Animals and Humans

Distribution, Elimination, and Systemic Toxicity

Virtually all information on the toxicity of palladium is from animal models, primarily rats, rabbits, and mice. Human studies are unknown. It is generally accepted that palladium ions (mostly Pd(II)) are the mediators of nearly all biological effects of palladium.

In animals, the absorption, distribution, and fate of palladium ions (Pd(II)) are dependent on the route and dose of administration. More than 99% of $PdCl_2$ administered to rats in a single oral dose is eliminated in 3 days, yet if administered intratracheally, 40 days are required to eliminate 95% of the ions. If administered intravenously, only 80% of the palladium is eliminated after 40 days (Fig. 2). Palladium administered intravenously is primarily distributed to the spleen, lung, and bone and is eliminated through the urine and feces. Orally administered palladium is primarily eliminated through the feces. In these single dose studies, palladium ions did not cross the placenta but was present in milk (Wataha and Hanks 1996). A recent report found that when 10–250 ng/mL potassium hexachloro-palladate was administered orally to rats daily for up to 90 days, administered doses below 250 ng/mL were eliminated rapidly through the feces, but 250 ng/mL also was eliminated through the urine. Palladium did not accumulate in liver, lung, spleen, or bone; most absorbed metal was in the kidneys, up to 124 ng/g (dry weight) after 250 ng/mL for 90 days.

The systemic toxicity of palladium ions also is dependent on the route of administration and the solubility of the administered compound; compounds with higher solubility, such as $PdCl_2.6H_2O$, are more toxic (Fig. 2). The lethal dose (50% frequency, abbreviated as LD50)

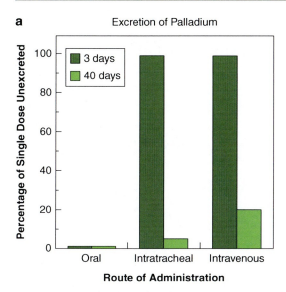
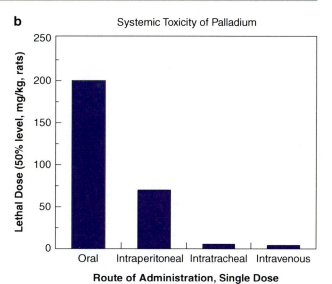

Palladium, Biological Effects, Fig. 2 (**a**) The excretion of palladium in rats depends on it route of administration. Single oral doses are rapidly and completely excreted via the feces. Intratracheal or intravenous doses are retained much longer and may remain in the body for more than 40 days. In these latter cases, some of the palladium will be excreted via the kidneys and urine. (**b**) Single doses of palladium can be lethal in rats, but the lethal dose depends on the route of administration. Intratracheal and intravenous doses are 20-fold more potent than oral doses; intraperitoneal doses are intermediate in potency

of PdCl$_2$ in rats has been reported as 200 mg/kg after oral administration, 70 mg/kg via intraperitoneal administration, 6 mg/kg from intratracheal administration, and 5 mg/kg from intravenous administration (Wataha and Hanks 1996; Kielhorn et al. 2002) (Fig. 2). In mice, LD50 doses are 87 mg/kg for the intratracheal route and 1,000 mg/kg for the oral route. Toxicity in rabbits occurs in the milligram to kilogram range of concentrations. Acute signs of toxicity for palladium (various animal models) are reduced eating and drinking, emaciation, ataxia and "tiptoe gait," convulsions, and cardiovascular problems. Changes in kidney, liver, lung, and intestinal histology have all been observed at higher doses of administration, even via the oral route. Histological changes in the airway tissues also have been reported from chronic intratracheal administration. Palladium has been reported to alter the biochemical function of enzymes in the liver and kidneys.

Mutagenicity, Teratogenicity, and Carcinogenicity

The mutagenicity of compounds is assessed primarily by their ability to cause genetic changes in bacteria that can be quickly and easily measured, despite uncertain correlation of these results with higher organisms. Using bacterial systems, palladium compounds appear to have no mutagenic effects. However, palladium ions interact with DNA and inhibit DNA synthesis in vitro and in vivo (Kielhorn et al. 2002). There are inadequate data to determine the teratogenicity of palladium ions or their effect on reproduction or development.

Hypersensitivity (Allergy)

Incidence of Hypersensitivity

There is little doubt that the predominant biological effect of palladium in humans is its ability to induce sensitization (allergy). Hypersensitivity to palladium is most commonly encountered from jewelry, or in dentistry where palladium-containing alloys are used for dental prostheses. Like all other biological effects of palladium, hypersensitivity is mediated through the intraoral release of palladium ions, most commonly palladium(II).

The incidence of hypersensitivity to palladium is not well defined despite extensive investigation over 40 years (Fauschou et al. 2011); much has been ascertained through case reports. Palladium causes allergic contact dermatitis or allergic contact granuloma in sensitized individuals (Thyssen and Menné 2010). Other common symptoms include localized stomatitis or oral lichen planus adjacent to a dental restoration containing palladium. Swelling of the lips and cheeks, dizziness, asthma, or chronic urticaria, have been reported less frequently (Kielhorn et al. 2002). The role of the dental prosthesis has been supported by resolution of the symptoms when an alternate, metal-free restoration replaces the original restoration. The role of palladium in many of these cases is uncertain.

Studies with patch tests, where $PdCl_2$ is applied to intact skin, or prick tests, where $PdCl_2$ is applied to a small laceration of the skin, have been most used to establish the incidence of palladium hypersensitivity. Based on these studies, the incidence is fairly broad, from 2% to 36%, but with most studies in the 2–3% range (Wataha and Hanks 1996). However, these studies nearly all used subjects predisposed to allergy or with signs of hyperimmune status (e.g., patient had dermatitis). Larger scale studies that have been done on school-age children suggest an incidence palladium allergy of 7–8% biased to females, but as with smaller scale studies, the population tested was selected from patients with dermatitis. A comprehensive literature evaluation has recently estimated the prevalence of palladium allergy at 7.8% in patients with dermatitis and 7.4% in dental patients with oral disease; these estimates are subject to the same population biases mentioned above (Thyssen and Menné 2010). Based on a few retrospective studies, the prevalence of hypersensitivity to palladium appears to have increased from 1996 to 2008 (from 0.1% to 1.8%); the influence of nickel allergy was not determined (see discussion below). The problem of biased sampling blurs the true incidence of palladium allergy in the general population. Because of this situation, the incidence of palladium hypersensitivity in the general population is not known with any certainty; rigorously controlled, randomized studies are not available.

The release of palladium from dental alloys is generally low (Wataha and Hanks 1996), and the incidence of allergic reactions from dental appliances containing palladium is therefore not common. A number of studies report that when patients are exposed to pure palladium metal, allergic reactions are infrequent (0–7%), even when the patient was known to be hypersensitive to palladium on the basis of a $PdCl_2$ patch test.

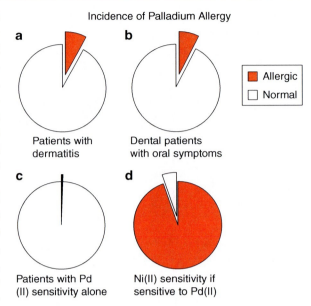

Palladium, Biological Effects, Fig. 3 The incidence of allergy to palladium(II) ions correlates with allergy to nickel and other conditions. (**a**) The incidence of Pd(II) allergy in people with dermatitis is about 7–8%, and is similar in dental patients who report oral allergic symptoms of some sort (**b**). Allergy to Pd(II) alone, with no other metal allergy, is apparently rare, probably << 1% (**c**). Individuals who are Pd(II) sensitive nearly always also are allergic to Ni(II) (**d**)

Cross-Reactivity with Nickel

There is little doubt that nickel sensitivity plays some role in how frequently patients react to palladium, although a true cross-sensitivity has not been proven. The majority (93–100%) of patients who exhibit positive patch tests to palladium are hypersensitive to nickel (Wataha and Hanks 1996). Exclusive sensitivity to palladium is relatively rare. One study has estimated that the so-called mono-sensitization (Pd(II) allergy without accompanying allergy to other metals) was about 0.2–0.5% in patients with dermatitis or oral symptoms secondary to dental restoration (Thyssen and Menné 2010) (Fig. 3).

Occupational Exposure Effects

Individuals working in refineries, dental laboratories, and in occupations where palladium catalysts are used are most likely to experience adverse effects from palladium exposure, and effects are nearly always correlated with hypersensitivity (Kielhorn et al. 2002). A common presumption, which remains unproven for palladium, is that occupational exposure to palladium causes hypersensitivity. Refinery workers where the platinum-group metals are processed most often exhibit allergy-related diseases in respiratory passages, skin, and eyes. Dental technicians appear to be affected primarily by dusts that are generated during polishing of alloys and other materials. Higher rates of lung cancer and pneumoconiosis have been reported in this group. However, it is important to note that the role of palladium in contributing to these problems has not been established, and that other potentially provocative dusts (acrylics, other metals, ceramics, and asbestos) might contribute to these effects. Interestingly, workers in the automobile catalyst industry have an incidence of palladium hypersensitivity that is within those of other groups.

Cellular and Molecular Effects

Enzyme Inhibition and Cytotoxicity

Although in vivo studies of the toxicity of palladium date to the 1930s, studies of cellular effects and cytotoxicity occurred later (1969), beginning with the effects of palladium ions on cellular enzymes such as catalase, α- chymotrypsin, lysozyme, peroxidase, ribonuclease, and trypsin (Wataha and Hanks 1996). These initial studies were followed by studies on other enzymes. Nearly all work in this area tested the effects of Pd(II), primarily as the chloride salt, but also as the sulfate and nitrate. Pd(II) inhibited enzyme activities at concentrations ranging from 20 to 500 μmol/L. Inhibition was attributed to binding of Pd(II) to the enzymes, leading to altered peptide conformation and destruction of catalytic activity.

Beginning in 1986, Pd(II) (almost exclusively as the chloride salt) was investigated as an inhibitor of cellular function; these studies were done mostly in mouse cells, but also in human cells (Wataha and Hanks 1996). To some extent, these studies paralleled the introduction of automobile catalytic converters that contained palladium. Cellular function inhibition (usually to 50% normal values) occurs at 100–400 μmol/L; DNA synthesis is inhibited at 100–300 μmol/L, protein synthesis at 340 μmol/L; mitochondrial respiration at 360 μmol/L, and total cellular protein content at 370 μmol/L (Wataha et al. 1991). Later studies using mouse and human monocytes reported cytotoxicities at 280–405 μmol/L, respectively. Nearly all of the above reports used exposure intervals of 24–72 h.

Several cytotoxicity investigations on Pd(II) were done in the context of the potential toxicity of dental alloys because of the "palladium" scare in dentistry in the early 1990s (see previous discussion). The fear was that palladium ions released from dental alloys would cause oral or systemic toxicity. These studies compared Pd(II) cytotoxicity with other constituents of dental alloys, and provide an opportunity to compare the cytotoxicity of palladium with other metals in which equivalent conditions were used. In mouse fibroblasts, Pd(II) is relatively low in toxicity (300 μmol/L) among transition metals where toxicities range from 1.1 μmol/L (Cd(II)) to >435 μmol/L (In (III)) (Fig. 4); the rank order of cytotoxicity varies somewhat depending on which parameter of cell function was measured (Wataha et al. 1991).

Pd(II) exerts its cellular suppression across a relatively narrow concentration range relative to other metals, suggesting that it causes cellular suppression via relatively specific mechanisms. In mouse macrophages and human monocytes, Pd(II) also is among the least cytotoxic metals. Why Pd(II) is relatively low in cytotoxicity is not well understood, although some investigators feel that cytotoxicity is limited by the relatively low rate of cellular uptake of Pd(II) (0.11 fmol/cell/μM/h) versus other metal ions (range was 0.11–45.3 fmol/cell/μM/h) (Wataha et al. 1993). Other investigators believe that the ability of Pd(II) to bind cellular macromolecules is relatively limited compared to metals such as Cd(II) or Hg(II). The accuracy of these proposed mechanisms remains uncertain. Also unclear are explanations of why Pd(II) cytotoxicity, like other metal ions, is dependent on the cell type.

Inflammatory Cell Signaling Effects

As part of an assessment of how dental restorations might influence oral health, the effects of palladium on inflammatory cell signaling have been investigated; this work has been done exclusively in vitro because in vivo studies are exceptionally difficult to fund and design. The rationale for these studies is that dental restorations (containing and potentially releasing palladium) are in intimate long-term contact with oral gingival tissues and harbor biofilms that often cause inflammation. Two questions have been posed: (1) Can Pd(II) trigger similar inflammatory pathways?, (2) Can Pd(II) amplify or suppress the inflammatory pathways triggered by the biofilms? Studies have focused on

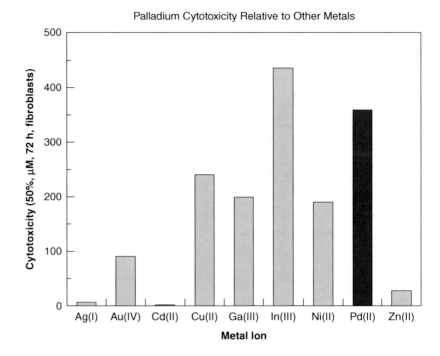

Palladium, Biological Effects, Fig. 4 When toxicity is tested in cell culture, palladium(II) is less toxic than other metal ions. The above data were collected after exposures to metal ions for 72 h in mouse fibroblasts, measuring cellular mitochondrial activity

monocytes, which mediate chronic inflammation in dental diseases, and on the NFκB inflammatory pathway, which plays a major role in regulating the secretion of cytokines that cause gingival inflammation.

Whereas metal ions such as Au(III) and Ni(II) cause marked (three to five fold) increases in the secretion of interleukin 6 (IL6) from activated human monocytic cells at sublethal concentrations, Pd(II) has little effect. Pd(II) is less effective than these other ions in altering tumor necrosis factor-alpha (TNFα) and interleukin 1-beta(IL1β) secretion. Furthermore, Pd(II) does not trigger secretion of these cytokines from un-activated monocytes. These results suggest a relatively minor role for palladium corrosion products from dental alloys in any alteration of inflammatory processes in oral tissues. Pd(II) alone has no significant effect on activation of the NFκB pathway components (p65, IκBα) in un-activated monocytes. However, in activated monocytes, Pd(II) increases nuclear translocation of p65 by 1.5-fold at sublethal concentrations of 75–200 μmol/L; similar effects are observed for Au (III) (75–200 μmol/L) and Hg(II) (10–10 μmol/L) but not Ni(II) (10–90 μmol/L). On the other hand, Pd(II) does not alter the levels or breakdown of IκBα in activated cells, suggesting that this arm of the NFκB pathway is unaffected. At 200 μmol/L, Pd(II) increases levels of reactive oxygen species (ROS) 3.5-fold over controls in activated monocytes. ROS has been investigated because of their known role in activation of the NFκB pathway.

Bacterial Effects

Palladium ions have been assessed for their ability to suppress the growth of oral bacteria, specifically *Aggregatibacter actinomycetemcomitans*, *Actinomyces naeslundii*, *Fusobacterium nucleatum*, *Porphyromonas gingivalis*, and *Prevotella intermedia* (Chung et al. 2011). These bacteria play roles in periodontal disease or caries. Palladium (II) shows a modest ability to suppress the growth of these bacteria by >50% of controls at 1,500 μmol/L or greater. Pd(II) was less potent than Au(III) or Pt(IV).

Therapeutic Uses

Palladium has been proposed as a chemotherapeutic for a variety of diseases and conditions, none successfully. In the 1970s, palladium chloride was tested to treat tuberculosis at a large daily dosage of 18 mg, but was not effective. Other early studies proposed to use palladium as a germicide, and still others attempted to use palladium(II) hydroxide to treat obesity. Obesity studies used injections of 5–7 mg/day of a colloidal $Pd(OH)_2$ suspension for 36 months. Patients lost an average of 19 kg of weight, but suffered necrosis at the injection sites.

Palladium ions and palladium-organic compounds have been explored as anti-neoplastic agents based on the success of platinum compounds, specifically cisplatin. Many compounds have been explored, but none are in common use clinically. Pd(II)-titanate compounds have been explored for properties of growth suppression of cancer cells in vitro. The concept was that the insoluble titanates could be sequestered at the site of a tumor and deliver palladium ions to suppress cell growth. However, these complexes are less effective in reducing the growth of fibroblasts or oral cancer cells than other metal-titanate complexes such as cisplatin, Hg(II), Au(III), or Pt(IV). Titanate-metal complexes also have been explored as bactericides, but Pd(II) is not as effective in inhibiting bacterial growth as other metal ions.

Cross-References

▶ Chromium and Allergic Reponses
▶ Gold Complexes as Prospective Metal-Based Anticancer Drugs
▶ Metals and the Periodic Table
▶ Nickel Carcinogenesis
▶ Nickel-Binding Sites in Proteins
▶ Palladium Complex-induced Release of Cyt c from Biological Membrane
▶ Palladium, Colloidal Nanoparticles in Electron Microscopy
▶ Palladium, Coordination of Organometallic Complexes in Apoferritin
▶ Palladium, Physical and Chemical Properties
▶ Platinum Anticancer Drugs
▶ Platinum, Physical and Chemical Properties

References

Chung WO, Wataha JC, Hobbs DT, An J, Wong JJ, Park CH, Dogan S, Elvington MC, Rutherford RB (2011) Peroxotitanate- and monosodium metal-titanate compounds as inhibitors of bacterial growth. J Biomed Mater Res A 97A:348–354

Fauschou A, Menné T, Johansen JD, Thyssen JP (2011) Metal allergen of the 21st century- a review on exposure, epidemiology and clinical manifestations of palladium allergy. Contact Dermatitis 64:185–195

Kielhorn J, Melber C, Keller D, Mangelsdorf I (2002) Palladium- a review of exposure and effects to human health. Int J Hyg Environ Health 205:417–432

Ravindra K, Bencs L, Van Grieken R (2004) Platinum group elements in the environment and their health risk. Sci Total Environ 318:1–43

Thyssen JP, Menné T (2010) Metal allergy- a review on exposures, penetration, genetics, prevalence, and clinical implications. Chem Res Toxicol 23:309–318

Wataha JC, Hanks CT (1996) Biological effects of palladium and risk of using palladium in dental casting alloys. J Oral Rehabil 23:309–320

Wataha JC, Hanks CT, Craig RG (1991) The in vitro effects of metal cations on eukaryotic cell metabolism. J Biomed Mater Res 25:1133–1149

Wataha JC, Hanks CT, Craig RG (1993) Uptake of metal cations by fibroblasts in vitro. J Biomed Mater Res 27:227–232

Wataha JC, Shor K (2010) Palladium alloys for biomedical devices. Expert Rev Med Devices 7:489–501

Palladium, Colloidal Nanoparticles in Electron Microscopy

Marie Vancová
Institute of Parasitology, Biology Centre of the Academy of Sciences of the Czech Republic and University of South Bohemia, České Budějovice, Czech Republic

Synonyms

Analytical and high resolution electron microscopy; Immunolabeling; Metal nanoconjugates; Multiple labeling

Definitions

The colloidal palladium solution contains dispersed particles visible using only electron microscopy. Internal structure of nanoparticles can be studied in high-resolution electron microscopes: field emission scanning electron microscope (FESEM) with resolution around 1 nm and higher; field emission transmission electron microscope (FETEM), or scanning transmission electron microscopy (STEM) with resolution less than 1 nm. Analytical electron microscopic techniques allow distinguishing elements present in the specimens. Recently, palladium nanoparticles of different sizes and shapes have been used for ultrastructural localizations of antigens.

Introduction

Nanoparticles composed of different metals (e.g., colloidal palladium, silver, platinum) or bimetallic nanoparticles have been recently used as electron-dense markers in several electron microscopic localization studies. Gold nanoparticles have been used predominantly for this purpose for many decades in the so-called immunogold labeling technique. For labeling studies, only nanoparticles in the size range of 5–15 nm can be used because they display both high labeling efficiency and satisfactory spatial resolution. These nanoparticles are conjugated to biomolecules that bind to exposed epitopes and thus enable precise localization of structures under investigation. Nanoparticles composed of different metals extend a number of suitable markers for simultaneous labeling of multiple antigens (co-localization of antigens). Colloidal palladium conjugates of uniform sizes but having various shapes (e.g., spherical, umbonate, or faceted) are distinguished by transmission electron microscopy or scanning electron microscopy (Fig. 1, Meyer et al. 2005). Metal nano-conjugates of different elemental composition, but of same sizes and shapes, may be differentiated using only high resolution analytical electron microscopy.

High-Resolution Analysis of Palladium Nanoparticles

The first high-resolution analysis of spherical palladium nanoparticles of diameters ranging from 1.4 nm to 5 nm was performed by Heinemann and Poppa in 1985. They observed the extension of the lattice parameter with reduction of the particle size. However, other observations describe a decrease of the lattice parameters with a decrease in the palladium particle size (Lamber et al. 1995). One of the reasons for such variation might be an incorporation of contaminating

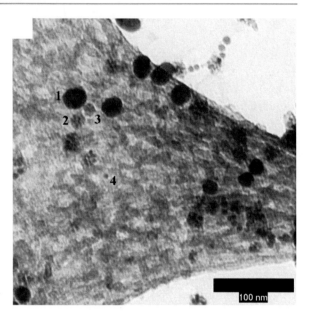

Palladium, Colloidal Nanoparticles in Electron Microscopy, Fig. 1 TEM micrograph showing multiple antibody labeling of a human platelet. 18-nm Au particle (*1*), umbonate 18-nm Pd particle (*2*), hexagonal 15-nm Pd particle (*3*), and 5-nm Au particle (*4*) (From Meyer et al. 2005, used with permission of Microsc Microanal (Cambridge University Press))

atoms, for example, oxygen, carbon, or hydrogen, into the palladium crystal lattice. The lattice parameter is dependent on the shape of nanoparticle as well.

The periodicity in the lattice (positions of atoms) can be visualized at high resolution using different techniques: electron diffraction (fast Fourier transforms calculated from an image captured using CCD camera, Fig. 2), selected area electron diffraction, surface electron energy-loss fine-structure spectroscopy as well as X-ray diffraction (XRD). High resolution TEM and corresponding fast Fourier transforms were used to characterize the shapes of palladium nanoparticles. Palladium particles adopt face-centered cubic (fcc) cubo-octahedral (Fig. 2), icosahedral, decahedral, and twinned basic shapes (José-Yacamán et al. 2001).

Palladium nanoparticles of size of 3–5 nm observed using field emission TEM working at accelerating voltage 200 kV are stable under electron beam irradiation (current density about 10 $Å/cm^2$) for the first 10–30 s, then structural fluctuation appears. The dissipation of palladium nanoparticles was described during intensive electron beam irradiation (10^3 $Å/cm^2$), probably due to diffusion (Fig. 3) (Tanaka et al. 2002).

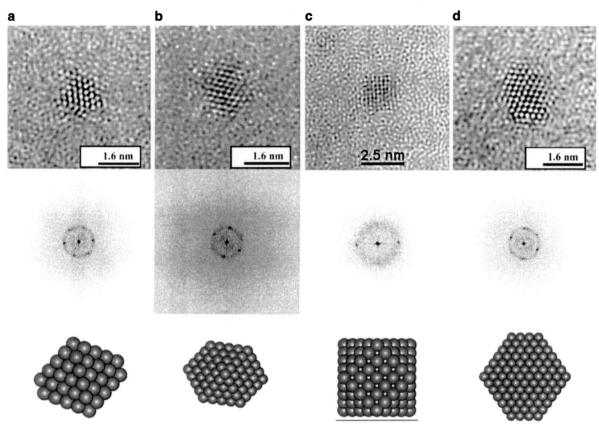

Palladium, Colloidal Nanoparticles in Electron Microscopy, Fig. 2 TEM micrographs of Pd particles with FCC structure in a (110) orientation (**a**, **b**), a (100) orientation (**c**), and a distorted (110) orientation (**d**). The corresponding FFT and models showing the orientation (From José-Yacamán et al. 2001, used with permission of J Mol Catal A: Chemical (Elsevier))

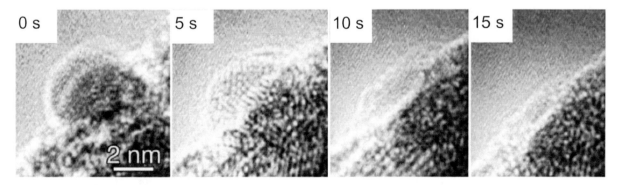

Palladium, Colloidal Nanoparticles in Electron Microscopy, Fig. 3 Sequential images of palladium nanoparticle during strong beam irradiation (From Tanaka et al. (2002), used with permission of Micron (Elsevier))

Visualization of Palladium Nanoparticles in FESEM

Principle of FESEM: A focused beam of electrons (produced by a field emission gun) is scanned over a specimen, interacts with its atoms, and generates various signals. As the electron beam excites an electron in the sample, secondary electrons are emitted from the surface and therefore enable visualization of the surface topography. Backscattered electrons are beam electrons reflected back from the specimen. The production of backscattered electrons depends on the atomic number of elements in the specimen. Therefore, metal nanoparticles imaged using a detector of backscattered electrons appear brighter than the background containing lower atomic number elements present in the biological specimens (Fig. 4).

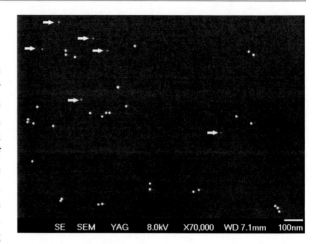

Palladium, Colloidal Nanoparticles in Electron Microscopy, Fig. 4 Gold and palladium (*arrows*) nanoparticles (both 10 nm) visualized using backscattered electrons in FE-SEM

Visualization of Palladium Nanoparticles in STEM

The principle of STEM is similar to SEM, but in addition, transmission modes of imaging are available. Transmitted electrons leaving the sample at low angles to the optical axis are detected using a bright field detector. Beam electrons scattered at high angles by their interactions close to the atom nuclei in the sample are detected using either an annular dark field or high angle annular dark field (HAADF) detector. The intensity of the scattering is dependent on the atomic number of sample elements (Z contrast image). As seen in Fig. 5a, HAADF-STEM is able to visualize distinct parts of bimetallic gold-palladium nanoparticles as confirmed with elemental composition analysis (Fig. 5b) at very high resolution (Mayoral et al. 2012). New generation of aberration-corrected HAADF-STEM achieves resolution of 0.8 Å.

How to Distinguish Between Nanoparticles Composed of Different Elements?

The elements present in the specimens can be distinguished using analytical electron microscopic techniques that use as a signal either X-rays or transmitted electrons that have lost a measurable amount of energy due to inelatic scattering (energy loss electrons). An *energy-dispersive X-ray* (EDX) detector can be used to image and distinguish elements present in the sample as well as to compare their X-ray intensities (qualitative and quantitative analyses). The energy of the emitted X-rays is characteristic of the elemental composition of the sample. In the SEM, X-rays are generated after the interaction of sample with a fine electron beam throughout the whole interaction volume (primary electrons penetrate inside the sample and slow down; the signal is generated from the whole volume). X-ray resolution is influenced by the used accelerating voltages and the atomic number of the element (different size of the interaction volume). Higher spatial resolution of EDX analysis offers the STEM, but only for thin specimens. EDX spectroscopy (also energy-dispersive spectroscopy, EDS) is a more sensitive analytical technique for heavy elements than for light elements in contrast to *electron energy-loss spectrometry* (EELS). EELS is a more efficient mapping technique in which the images are formed with transmitted, inelastically scattered electrons. By combining STEM with EELS, information about distribution of elements in the sample is available at atomic resolution.

Energy-filtered transmission electron microscopy (EFTEM) is the next analytical technique that allows accurate chemical mapping with spatial resolution about 1 nm. Similar to EELS, the images are formed with transmitted, inelastically scattered electrons that go through an energy filter where only electrons with defined energy are separated. Using the EFTEM microscope equipped with a cryostage, Bovin et al. (2000)

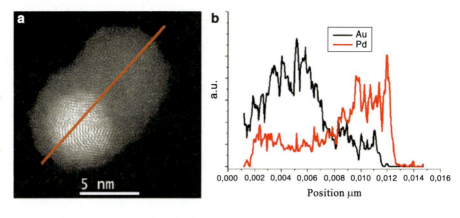

**Palladium, Colloidal Nanoparticles in Electron Microscopy,
Fig. 5** Aberration-corrected HAADF-STEM micrograph of one Au-Pd nanoparticle (**a**) and elemental composition analysis (**b**). The *orange line* shows the distribution of elements (EDS) (From Mayoral et al. 2012, used with permission of Micron (Elsevier))

observed the aggregation of palladium nanoparticles as a function of pH and ionic strength by low dose techniques. In this case, a frozen solution containing palladium nanoparticles was imaged. However, the direct observation of inorganic nanoparticles in liquids can be performed only using specially designed liquid cells. The cell is loaded into the specimen holder of the TEM. This new instrumentation allows direct observation of details of dynamical diffusion or motions of particles in thin liquid layers in real time (de Jongea et al. 2009).

Preparation Techniques for Observation of Palladium Nanoparticles in EM

Correct sample preparation is a crucial step for successful EM observation. These techniques differ according to types of electron microscopes.

The *thin-film method* can be used for direct visualization of distributed palladium nanoparticles in all types of electron microscopes (FESEM, TEM, and STEM). Firstly, a TEM grid is covered either by a carbon or plastic film (for example Formvar film) coated with carbon. Glow discharge can be used to charge the support film to enhance its hydrophilic properties. Next, a drop of colloidal solution is placed on the coated grid. The excess solution is removed by filter paper to make a thin and uniform monolayer of nanoparticles. The grid with the absorbed material is either dried in vacuum or at room temperature.

The thin-film method is, for example, used commonly for the determination of precise particle size and size distribution. A precise measurement of the particle size is possible only in correctly calibrated high-resolution electron microscopes. Micrographs with thousands of particles are captured at the same magnifications with a high-resolution CCD camera. Counting and measuring of particles can be performed using image analysis software (e.g., freeware program Image J; http://rsbweb.nih.gov/ij/) and statistical evaluation of results.

Hydrated samples can be visualized in the electron microscopes equipped with a cryostage (*cryo EM*). The grid with the thin layer of absorbed material is immersed (plunge freezing) in a cryogenic liquid (e.g., liquid ethane, liquid nitrogen or slushed liquid nitrogen). The grid with nanoparticles embedded in a thin layer of vitreous ice is transferred directly into the electron microscope. Sublimation can be used to remove surfacial layer of the ice.

Furthermore, either ultrathin resin *sections* or cryosections of the material containing nanoparticles or sections labeled with palladium conjugates can be analyzed using the above-mentioned EM techniques. For analytical techniques such as EFTEM, the use of additional staining agents, such as osmium tetroxide, uranyl acetate, and lead citrate, should be avoided. Thickness of sections/samples may influence the background signal in both EELS and EFTEM (signal-to-noise ratio) analyses. Therefore, critical point drying of thawed cryosections (cut at a thickness of 100 nm) after labeling has been used instead of methylcellulose embedding to avoid increased thickness of sections (Bleher et al. 2008).

Conjugation of Palladium Nanoparticles to Biomolecules

The main prerequisite of utilization of palladium nanoparticles in biological applications is their

Palladium, Colloidal Nanoparticles in Electron Microscopy, Fig. 6 Formation of amide bonds between carboxylated Pd nanoparticles (capped with dihydrolipoic acid, DHLA) and proteins containing amine groups using 1-ethyl-3-(3-dimethylaminopropyl)-carbodiimide (EDC) (From Vancová et al. 2011, used with permission of Microsc Microanal (Cambridge University Press))

conjugation to selected molecules that mediate binding to structures under study.

The palladium colloids used for conjugation reactions have been generated by the chemical reduction of metal salts with either sodium citrate according to Turkevich and Kim (1970), or sodium ascorbate according to Brintzinger (Meyer and Albrecht 2003). Synthesized nanoparticles are stabilized due to the ion layer forming at the particle surface (electrostatic stabilization). Surfaces of particles are hydrophilic and thus suitable for subsequent conjugation reactions.

In contrast to palladium nanoparticles, gold nanoparticles have been used for immuno-detection of antigens in electron microscopy applications for decades. Gold as well as palladium nanoparticles can be attached *non-covalently* to proteins or nucleic acids by hydrophobic interactions. In order to reduce the electrostatic attractions between the positively charged groups of protein molecules and the negatively charged surface of gold particles and to let the hydrophobic interactions to prevail, it is necessary to adjust the pH of the colloid solution close to the isoelectric point of the protein intended for conjugation. Stabilized and unstabilized gold nanoconjugates can be easily recognized by change in the color of the solution after addition of salt. The red color (stabilized gold colloid) can be change to the purple-black color of the solution containing aggregated gold nanoparticles. This effect is used in a coagulation test that enables one to balance/adjust important conditions (pH, ionic strength, protein concentration) during the non-covalent conjugation. The fine-tuning of these conditions increases the stability of the prepared conjugate. The color of water-soluble nano-sized palladium colloids is light yellow in contrast to the gray color of the solution containing coagulated palladium nanoparticles. However, the color may vary and depends on particle concentration and size, as well as on the use of stabilizers. The fine-tuning of the conjugation parameters is hindered by very small difference between colors of stabilized and unstabilized palladium colloid solution. Therefore, all parameters have to be calculated empirically and the biological activity of such palladium conjugates should be tested. However, several successful localization studies have already been performed, for example, labeling of ultrathin cryosections of skeletal muscle tissue (Bleher et al. 2008) or labeling of human platelets (Meyer et al. 2005).

Biomolecules attached to palladium nanoparticles by *covalent conjugation* have been used as electron-dense markers in two studies. In both reports, the first step of the conjugation starts with surface modification of palladium nanoparticles using thiolic compounds (either dihydrolipoic acid or 11-mercaptoundecanoic acid). These stabilizing agents contain carboxylic acid groups in addition to thiol groups, localized on the other end of the molecule. Carboxylic groups can be coupled either to DNA (Wang et al. 2012) or to amine groups of proteins where react to form an amide bond linkage (Fig. 6, Vancová et al. 2011).

Conclusion

Palladium nanoparticles are the subject of substantial research predominantly due to their excellent catalytic properties with applications as hydrogen sensors, hydrogen storage, fuel cell catalysts, etc. In future, these properties might be used in biology (intracellular catalysts) and potentially in medicine. Electron microscopy is an essential tool for structural characterization and experimental examination of properties of palladium nanoparticles. Moreover, palladium nanoparticles are used as alternative markers for ultrastructural localization of antigens next to the traditional and widely used gold nanoparticles. Labeling densities obtained by both metal conjugates are comparable (Vancová et al. 2011).

Cross-References

▶ Biomineralization of Gold Nanoparticles from Gold Complexes in Cupriavidus Metallidurans CH34
▶ Colloidal Silver Nanoparticles and Bovine Serum Albumin
▶ Gold Nanomaterials as Prospective Metal-based Delivery Systems for Cancer Treatment
▶ Gold Nanoparticle Platform for Protein-Protein Interactions and Drug Discovery
▶ Gold Nanoparticles and Proteins, Interaction
▶ Palladium, Biological Effects

References

Bleher R, Kandela I, Meyer DA, Albrecht RM (2008) Immuno-EM using colloidal metal nanoparticles and electron spectroscopic imaging for co-localization at high spatial resolution. J Microsc 230(Pt 3):388–395
Bovin JO, Huber T, Balmes O, Malm JO, Karlsson G (2000) A new view on chemistry of solids in solution–cryo energy-filtered transmission electron microscopy (cryo-EFTEM) imaging of aggregating palladium colloids in vitreous ice. Chemistry 6:129–132
de Jongea N, Peckys DB, Kremers GJ, Pistona DW (2009) Electron microscopy of whole cells in liquid with nanometer resolution. PNAS 106:2159–2164
Heinemann K, Poppa H (1985) In-situ TEM evidence of lattice expansion of very small supported palladium particles. Surf Sci 156:265–274
José-Yacamán M, Marin-Almazo M, Ascencio JA (2001) High resolution TEM studies on palladium nanoparticles. J Mol Catal A Chem 173:61–74
Lamber R, Wetjen S, Jaeger NI (1995) Size dependence of the lattice parameter of small palladium particles. Phys Rev B 51:10968–10971
Mayoral A, Deepak FL, Esparza R, Casillas G, Magen C, Perez-Tijerina E, José-Jacamán M (2012) On the structure of bimetallis noble metal nanoparticles as revealed by aberration corrected scanning transmission electron microscopy (STEM). Micron 43:557–564
Meyer DA, Albrecht RM (2003) Sodium ascorbate method for the synthesis of colloidal palladium particles of different sizes. Microsc Microanal 9:1190–1191
Meyer DA, Oliver JA, Albrecht RM (2005) A method for the quadruple labeling of platelet surface epitopes for transmission electron microscopy. Microsc Microanal 11:142–143
Navaladian S, Viswanathan B, Varadarajan TK, Viswanath RP (2009) A rapid synthesis of priented palladium nanoparticles by UV irradiation. Nanoscale Res Lett 4:181–186
Tanaka M, Takeguchi M, Furuya K (2002) Behavior of metal nanoparticles in the electron beam. Micron 33:441–446
Turkevich J, Kim G (1970) Palladium: preparation and catalytic properties of particles of uniform size. Science 169:873–879
Vancová M, Šlouf M, Langhans J, Pavlová E, Nebesářová J (2011) Application of colloidal palladium nanoparticles for labeling in electron microscopy. Microsc Microanal 17:810–816
Wang Z, Li H, Zhen S, He N (2012) Preparation of carboxyl group-modified palladium nanoparticles in an aqueous solution and their conjugation with DNA. Nanoscale 4:3536–3542

Palladium, Coordination of Organometallic Complexes in Apoferritin

Takafumi Ueno and Satoshi Abe
Department of Biomolecular Engineering, Graduate School of Bioscience and Biotechnology, Tokyo Institute of Technology, Yokohama, Japan

Synonyms

Artificial metalloenzyme; Artificial metalloprotein; Organometallic complex/protein composite; Organometalloenzyme; Organometalloprotein

Definition

Ferritin (Fr) is an iron-storage protein comprising 24 subunits that assemble to form a hollow cage-like structure with 8-nm and 12-nm inner and outer diameters, respectively (Fig. 1). The intact cage is stable over a wide pH range (2–11) and a temperature of approximately 80°C. Another important property of this protein cage is the perforation of the protein shell by small channels located at the junctions of three subunits. These channels are required for the transport of several metal ions, small metal complexes, and other organic molecules. These unique properties were initially mainly used for the preparation of various different metal nanoparticles, such as pure metals, metal oxides, and quantum dots, as manganese and uranyl oxides could be synthesized in the Fr cage. Coordination of organometallic Pd complexes

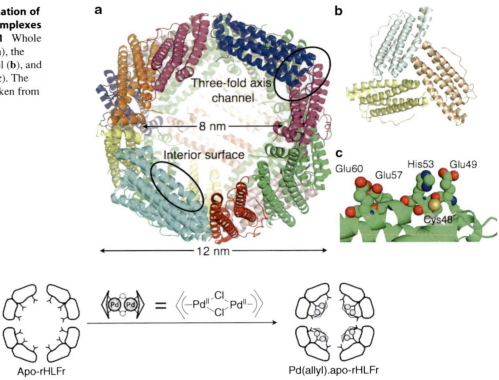

Palladium, Coordination of Organometallic Complexes in Apoferritin, Fig. 1 Whole structure of apo-Fr (a), the threefold axis channel (b), and the interior surface (c). The apo-Fr structure is taken from PDB ID 1AEW

Palladium, Coordination of Organometallic Complexes in Apoferritin, Fig. 2 Schematic drawing of preparation of **Pd(allyl)·apo-rHLFr** (Abe et al. 2008) (Reproduced with permission from American Chemical Society)

(Pd(allyl), allyl = $\eta^3-C_3H_5$) in apoferritin (apo-Fr) was observed in 2008, using high-resolution X-ray crystal structure analysis at 1.7 Å resolution by Ueno and his co-workers (Figs. 2 and 3). This is the first crystal structure of a single-protein cage containing numerous organometallic compounds. This composite can also serve as a catalytic nanoreactor for the Suzuki-Miyaura coupling reaction. The crystal structure shows that the binding sites for Pd (allyl) are divided into two regions, the threefold axis channel and the accumulation center, as observed for **Pd^{2+}·apo-Fr** (Fig. 3). Pd (allyl) complexes maintain the same dinuclear structure at each of the binding regions. In the threefold axis channel, Cys126 binds to two Pd(allyl) complexes as a bridging ligand. His114 and a water molecule coordinate with different Pd complexes, thereby retaining the dinuclear structure. At the accumulation center, the same unique structure is formed by Cys48 and His49, but with Glu45 involved instead of a water molecule. Sequence mutants automatically complete the unique coordination layout of the numerous Pd(allyl) dinuclear complexes by further appending or removing Cys and His residues. Moreover, the properties of Fr are applicable for the coordination of other organometallic compounds. When Rh(norbornadiene) complexes were fixed in the Fr cage, the composite was observed to catalyze the polymerization reaction of phenylacetylene. Ru(p-cymene) moiety coordinated with His49, Glu53, and His173 at a position adjacent to the accumulation center by retaining a mononuclear geometry similar to that of the reported Ru complex. The crystal structures of dimethylaminomethyl-ferrocene/apoferritin composites were determined, and a new anomalous peak was observed at the twofold symmetry axis cavity of the apo-rHLFr molecule. Different types of metal complexes were individually arranged at the interior. It is difficult to construct specific accumulation patterns by using synthetic multidentate ligands alone. Thus, it is believed that the technology would lead to the development of many different types of metal complexes within the confined interior of protein cages for the manufacture of useful materials, such as biocatalysts, biosensors, and metal-containing drugs.

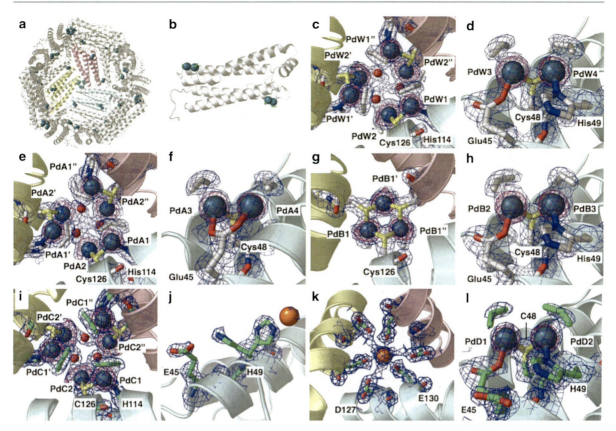

Palladium, Coordination of Organometallic Complexes in Apoferritin, Fig. 3 Crystal structures of Pd(allyl)·apo-rHLFrs. Whole (a) and monomer (b) structures of Pd(allyl)·apo-rHLFr; Threefold axis channels and accumulation centers of Pd(allyl)·apo-rHLFr (c, d); Pd(allyl)·apo-H49A-rHLFr (e, f); Pd(allyl)·apo-H114A-rHLFr (g, h); Pd(allyl)·apo-C48A-rHLFr (i, j); and Pd(allyl)·apo-C126A-rHLFr (k, l), respectively (Abe et al. 2008, 2010). Palladium, cadmium, and oxygen of water are shown as *spherical* models colored with *blue*, *orange*, and *red*, respectively. The selected $2|F_o| - |F_c|$ electron density maps are shown at 1.0σ (Reproduced with permission from American Chemical Society)

Introduction

Organometallic compounds have attracted considerable attention for applications in producing metal-containing drugs and bio-imaging reagents, and for many other biological functions (Hillard and Jaouen 2011). The variation in their coordination geometries and organic ligand structures enables the compounds to bind to specific sites of proteins. Several hybrids consisting of proteins and organometallic complexes can be utilized as catalysts that are active in aqueous solution. This approach to the development of organometalloenzymes is one of the fast-growing fields in both organometallic and bioinorganic chemistry. Although various types of organometallic complexes have been conjugated to proteins, via covalent or non-covalent conjugation techniques, there are a few composites that have demonstrated a high level of performance because of the low stability and the limited applicability to proteins and metal complexes. One of the most successful methods for preparing organometallic compounds/protein composites has been the use of avidin/biotin technology, after Whitesides and co-workers reported an avidin/biotin-Rh complex enzyme that could promote asymmetric olefin hydrogenations. However, new methods are required for the preparation of robust composites with high reactivity and precise coordination design. Thus, organometallic-ferritin composites were employed as candidates for preparing new organometalloenzymes.

Ferritin (Fr) is an iron-storage protein comprising 24 subunits that assemble to form a hollow cage-like structure with 8-nm and 12-nm inner and outer diameter, respectively (Fig. 1) (Theil 1987). Iron atom is accumulated in the form of Fe^{2+} ion, then oxidized, and finally stored as a cluster of ferric oxyhydroxide within the interior cage formed by the protein subunits. The intact cage is stable over a wide pH range (2–11) and a temperature of approximately 80°C. An important property of this protein cage is the perforation of the protein shell by small channels named "threefold axis channels" that are located at the junctions of three subunits (Fig. 1b). These channels are required for the transport of several metal ions and small metal complexes that coordinate to the amino acid residues exposed on the interior surface (Fig. 1c). Thus, there have been many reports on the preparation of various types of nanoparticles, such as pure metals, metal oxides, and quantum dots, by using the basic functions of ferritin. Ueno and Watanabe et al. developed a size-selective hydrogenation biocatalyst by encapsulation of a Pd nanoparticle in apo-Fr (Ueno et al. 2004, 2009). Moreover, they succeeded in improving the catalytic reactivity by in situ reduction to form core/shell bimetallic Au/Pd nanoparticles in apo-Fr (Suzuki et al. 2009). After these initial successful demonstrations of biocatalyst design, they reported the catalytic reactions and the crystal structures of organometallic complexes/apo-Fr composites (Abe et al. 2008, 2009). In this entry, the authors have described about the coordination structures and catalytic properties of apo-Fr and the mutants containing organometallic palladium complexes.

Preparation and Catalytic Reaction of Organometallic Palladium Complexes in Ferritin

Water-soluble ligands and polymer-bound catalysts have been previously developed for use in aqueous-phase organometallic catalysis (Cornils and Herrmann 2004). Fr is a candidate that satisfies many of the requirements for a successful catalyst because the metal complexes that accumulate in the cage are expected to promote the catalytic reactions in aqueous solution under ordinary conditions. The number and reactivity of the complexes will be controlled by the coordination environment in the cage. In order to test the suitability of the protein cage for organometallic catalysis, a Pd(allyl) (allyl = $\eta^3-C_3H_5$) complex was employed as a catalyst accumulated in apo-Fr to construct a biocatalyst for the Suzuki-Miyaura coupling reactions, which is one of the most important carbon-carbon cross-coupling reactions catalyzed by organometallic complexes.

Recombinant L-chain apo-Fr from horse liver (Apo-rHLFr) containing Pd(allyl) complexes (**Pd(allyl)·apo-rHLFr**) was constructed by the treatment of apo-rHFr with 100 equivalents of [Pd(allyl)Cl]$_2$ in 50 mM Tris/HCl (pH 8.0) (Fig. 2). The crystal structure shows that the binding sites of Pd(allyl) are divided into two regions, the threefold axis channel and the accumulation center, as observed for **Pd^{2+}·apo-Fr** (Fig. 3a, b) (Ueno et al. 2009). Pd(allyl) complexes maintain the same dinuclear structures at each of the binding regions. In the threefold axis channel, Cys126 binds to two Pd(allyl) complexes as a bridging ligand (Fig. 3c). His114 and a water molecule coordinate with different Pd complexes for retaining the dinuclear structure. At the accumulation center, the same unique structure is formed by Cys48 and His49, but with the Glu45 involved instead of a water molecule (Fig. 3d). To alter the coordination structure of Pd(allyl) complexes in the apo-rHLFr cage, four mutants, in which His49, His114, Cys48, or Cys126 was replaced with alanine (Ala), and their composites with the Pd complexes were prepared. The crystal structure of **Pd(allyl)·apo-H49A-rHLFr** shows that the number of Pd atoms in the cage is 96, which is identical to the value of **Pd(allyl)·apo-rHLFr**. The coordination structures of the dinuclear Pd complex at the accumulation site were accurately conserved because each of the Oε1 and Oε2 atoms of the carboxylate moiety of Glu45 is separately bound to two Pd atoms as a bidentate ligand (Fig. 3f). These results suggest that the conformational changes of His49 and Glu45 contribute to the stability of the square planar thiolato-bridged dinuclear Pd(allyl) complexes. On the other hand, **Pd(allyl)·apo-H114A-rHLFr** has thiolato-bridged trinuclear Pd complexes at the threefold axis channels, due to the deletion of His114 (Fig. 3g). The PdB1-PdB1′ distance (3.24 Å) indicates that the complexes have no direct bonding interaction among the Pd atoms. The coordination geometry of each Pd atom is a square-planar structure with an allyl ligand and two Sγ atoms at the threefold axis channel with a typical six-membered ring structure. The crystal structure of **Pd(allyl)·apo-C48A-rHLFr** shows that there are two conformers of His49 with no Pd(allyl) complexes at the accumulation center (Fig. 3j). This indicates that Cys48 is a crucial residue for the formation of a dinuclear Pd(allyl) complex and that the side chain of His49 is highly flexible. The imidazole ring of His49 is expected to adopt several conformations that enable it to adapt to various metal coordination geometries, because the conformation of His49 is dramatically altered by the coordination of the metal ions when metal ions, such as Pd^{2+} and Au^{3+}, are bound to the accumulation center (Ueno et al. 2009; Suzuki et al. 2009). The crystal structure of **Pd(allyl)•apo-C126A-rHLFr** shows that there are no Pd(allyl) complexes bound to the threefold axis channel as a result of the Cys to Ala mutation (Fig. 3k). The quantity of Pd atoms in **Pd(allyl)•apo-C126A-rHLFr** (37 ± 4) determined by quantitative analysis is slightly lesser than the number observed in the crystal structure (48). However, **Pd(allyl)•apo-C48A-rHLFr** contains 62.5 ± 1.5 Pd complexes, which is greater than the observed quantity of Pd(allyl) complexes in the crystal structure (48). This variation in the number of complexes observed for the Cys-deleted mutants suggests that Cys126 is a critical residue for the formation of Pd(allyl) dinuclear complexes and that this residue plays a role in facilitating the uptake of the Pd complexes at the threefold axis channel.

The experiments using Cys and His substitutions show that (1) Cys126 accelerates the incorporation of metal complexes by inducing a conformational change at the threefold axis channel, and (2) Cys48 is essential for the formation of dinuclear Pd(allyl) complexes at the accumulation center. The quantity of Pd(allyl) complexes bound to the interior surface of apo-rHLFr depends on the number of Cys residues at the appropriate sites. Apo-rHLFr is, thus, able to bind 96 Pd(allyl) complexes because apo-Fr has a total of 48 Cys residues, which are indicated to be necessary for maintaining the dinuclear Pd(allyl) structures. His and Glu residues, which have flexible side-chain structures, in the vicinity of the Cys residues are required for the stabilization of the dinuclear $[Pd(allyl)]_2(Cys)$ structures. This suggests that the introduction of additional Cys residues within the apo-rHLFr cage would promote an increase in the inner capacity of the cage to accommodate greater quantities of Pd(allyl) complexes. The coordination structures of Pd(allyl) complexes could be controlled by cooperatively altering the positions of His and Glu residues on the interior surface of apo-rHLFr.

In order to evaluate the catalytic activities of these composites, a Suzuki-Miyaura coupling reaction of 4-iodoaniline and phenylboronic acid to yield 4-phenylaniline was carried out. The turnover frequencies (TOF = [product (mol)] per **Pd(allyl) •apo-rHLFr** per hour) of the coupling reactions were determined using ^{1}H-NMR based on the consumption of 4-iodoaniline and the formation of the product. The activity of **Pd(allyl)•apo-rHLFr** (TOF = 3500) was almost identical to that of **Pd(allyl)•apo-H49A-rHLAFr** (3400). This is due to structural conservation of dinuclear $[Pd(allyl)]_2$ with bridging ligation of Asp45, which can occur even after the deletion of His49. The activity of **Pd(allyl)•apo-C48A/H49A-rHLFr** (1900) was approximately half that of **Pd(allyl)•apo-rHLFr**, due to lack of Pd(allyl)-binding sites at the accumulation center. These results suggest that each Pd dinuclear complex at both binding areas has similar activity in the catalytic reaction. On the other hand, **Pd(allyl)•apo-H114A-rHLFr** (900) and **Pd(allyl)•apo-C126A-rHLFr** (830) showed approximately an activity that is fourfold lower than that of **Pd(allyl)•apo-rHLFr** (3500). This decrease in the activities of **Pd(allyl)•apo-H114A-rHLFr** suggests that the trinuclear Pd cluster at the threefold axis channel interferes with the penetration of substrates or that the geometry of Pd complexes at the threefold axis channel is different from **Pd(allyl)•apo-rHLFr.** In the Pd(allyl)-binding sites of **Pd(allyl)•apo-C126A-rHLFr**, there are no Pd complexes at the threefold axis channel; however, the Pd(allyl) dinuclear center is completely maintained at the accumulation center. The entry of the phenylboronic acid into the apo-rHLFr cage might be inhibited at the channel by electrostatic repulsion due to the six negatively charged carboxylic acids of Glu and Asp residues exposed on the surface of the channel. Thus, the catalytic activities of the **Pd(allyl)·apo-rHLFrs** are expected to be improved by increasing the number of Pd(allyl) complexes at suitable positions in apo-rHLFr via the introduction of Cys residues.

Coordination Design of Organometallic Palladium Complexes in Ferritin

The crystal structure of **Pd(allyl)•apo-rHLFr** reveals that one subunit of apo-rHLFr has two binding domains for Pd(allyl) complexes: one located on the interior surface of the apo-rHLFr cage, centered at Cys48 in the accumulation center, and the other at

Cys126 in the threefold axis channel (Fig. 3). It has been found that replacing either Cys or His residues in these two binding sites with Ala dramatically changes the metal-binding structures within the cages of the **Pd(allyl)•apo-rHLFr** mutants (Abe et al. 2008, 2010). Thus, to rearrange the coordination structures of Pd(allyl) complexes in the apo-rHLFr cage by appending or removing Cys and His residues, two factors should primarily be considered: (1) Cys48 is essential for the formation of the dinuclear Pd(allyl) complexes at the accumulation center and (2) His49, which has the flexible side chain in the vicinity of the Cys residue, is necessary to control the diverse directions of the dinuclear Pd(allyl) moieties in the cage. On the basis of these hypotheses, new Pd(allyl)•apo-rHLFr composites were prepared using the rationally designed mutants apo-E45C/C48A, E45C/R52H, and E45C/H49A/R52C-rHLFrs in which Pd complexes can be bound with specifically confined coordination structures that are different from those in **Pd(allyl)•apo-rHLFr** (Wang et al. 2011).

Comparison of the accumulation centers of **Pd(allyl)•apo-rHLFr** and **Pd(allyl)•apo-E45C/C48A-rHLFr** shows that the substitution of Cys48 with Ala deletes the [Pd(allyl)]$_2$ moiety bound to the native Cys of apo-rHLFr (Fig. 4). Replacement of Glu45 with Cys results in a new dinuclear center of PdE1-PdE2 (Fig. 4b). The reasons for choosing the residue sequence at a position 45 as the new location for the Cys are that this position, on the interior surface of the apo-rHLFr cage, avoids potential interference from adjacent subunits and also that there are appropriate distances between this position and both His49 and His173, so that these two His residues, His49 and His173, act as supporting ligands for the PdE1-PdE2 pair (Fig. 4b). The crystal structure indicates that the positions of PdE1 and PdE2 are relatively fixed. These mutations are regarded as operations that relocate Cys48 from the interior surface of apo-rHLFr to a new position, where a Cys residue works together with neighboring His residues to produce a new binding site for a dinuclear Pd complex.

In the second mutant apo-E45C/R52H-rHLFr, Cys45 is added while conserving the native Cys48. Thus, the PdF1-PdF2 pair coordinated with Cys45, His49, and His173 has a structure analogous to PdE1-PdE2 in **Pd(allyl)•apo-E45C/C48A-Fr**, which indicates a certain flexibility of the residues in this binding site (Fig. 4c). Another pair, PdF3-PdF4, is stabilized by the thiol group of Cys48, the imidazole group of the mutated His52, and one water molecule. The relatively large shift between PdW3 and PdF3 (1.88 Å) demonstrates that a different location of the coordinating His will cause conformational alteration of Pd atoms. Coordination with the amino acid residues, water molecules, and allyl ligands produces a typical square planar structure for all the Pd atoms. Apo-E45C/H49A/R52H-rHLFr is derived from apo-E45C/R52H-rHLFr; however, a comparison of the crystal structures of **Pd(allyl)•apo-E45C/R52H-rHLFr** and **Pd(allyl)•apo-E45C/H49A/R52H-rHLFr** showed no significant structural difference around Cys48 (Fig. 4d). The shifts of PdF3-PdG3 and PdF4-PdG4 are just 0.21 and 0.27 Å, respectively. However, this mutation remarkably alters the geometry of the dinuclear Pd complex bound to Cys45. The PdG1-PdG2 pair is oriented perpendicularly to PdF1-PdF2, with coordination of His173 and one molecule of H$_2$O to PdG1 and PdG2, respectively (Fig. 4c, d). This observation proves the significance of His in modulating the coordination structure of Pd(allyl) with a Cys residue. **Pd(allyl)•apo-E45C/H49A/R52H-rHLFr** contains 144 Pd atoms, in accordance with ICP/BCA data (154 ± 3 Pd atoms per Fr), supposing very rigid coordination structures of the four Pd(allyl) moieties.

The catalytic reactivities of the **Pd(allyl)•apo-rHLFr** composites for the Suzuki-Miyaura coupling reaction of 4-iodoaniline and phenylboronic acid were assessed. The TOF value for **Pd(allyl)•apo-E45C/C48A-rHLFr** (4200) is slightly higher than that of **Pd(allyl)•apo-rHLFr** (3500). The TOF values of **Pd(allyl)•apo-E45C/R52H-rHLFr** and **Pd(allyl)•apo-E45C/H49A/R52H-rHLFr** (4300 and 4200, respectively) are only slightly higher than that of **Pd(allyl)•apo-rHLFr**, although the number of Pd complexes in these composites are 1.5-fold more than those in **Pd(allyl)•apo-E45C/C48A-rHLFr**. Our previous work shows the following: (1) the substrates penetrate the Fr cage through the threefold axis channels before they react with the Pd complexes on the interior surface of apo-rHLFr and (2) the replacement of His with Glu or a water molecule for coordinating with the Pd(allyl) moiety has little effect on the catalytic reactivity (Abe et al. 2008). Thus, the process of the penetration is thought to be the rate-limiting step. Future work concentrating on further improving the catalytic activity of

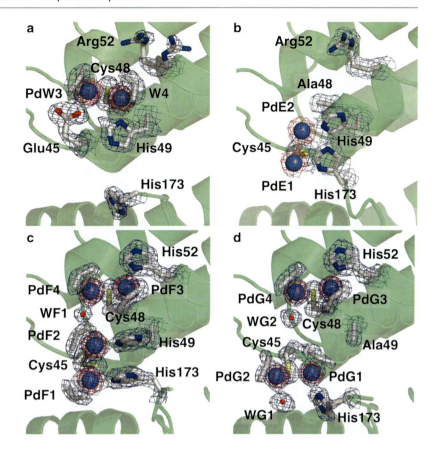

Palladium, Coordination of Organometallic Complexes in Apoferritin, Fig. 4 Crystal structures of the accumulation centers of Pd(allyl)·apo-rHLFr (**a**), Pd(allyl)·apo-E45C/C48A-rHLFr (**b**), Pd(allyl)·apo-E45C/R52H-rHLFr (**c**), and Pd(allyl)·apo-E45C/H49A/R52H-rHLFr (**d**) (Wang et al. 2011). The Pd atoms are indicated as *sphere* models. The selected $2|F_o| - |F_c|$ electron density maps at 1.0σ are shown (Reproduced with permission from The Royal Society of Chemistry)

apo-rHLFr should involve the manipulation of the threefold axis channels for the diffusion process as well as the coordination structure of the Pd(allyl).

Summary and Future Directions

It has been demonstrated that a variety of coordination structures of multinuclear Pd complexes, including dinuclear and trinuclear complexes, are constructed by protein engineering of the interior surface of apo-rHLFr. These results suggest that multinuclear metal complexes with various coordination structures could be prepared by the deletion or introduction of key residues, such as His and Cys, at appropriate positions on the interior surface of Fr. To design such structures, two factors should be primarily considered: (1) Cys residues are essential components for capturing two Pd complexes for the formation of the dinuclear structures at the accumulation center and (2) His residues, which have flexible side chains, are needed in the vicinity of the Cys residues in order to control the diverse direction of the dinuclear Pd(allyl) moieties at the center. Sequence mutants automatically complete the unique coordination layout of numerous Pd(allyl) dinuclear complexes by the formation of further appendage or removal of Cys and His residues. Moreover, the properties of Fr are also applicable for other organometallic compounds. A Ru(*p*-cymene) moiety could coordinate with His49, Glu53, and His173 at a position adjacent to the accumulation center, thus retaining a mononuclear geometry similar to that of the reported Ru complex (Takezawa et al. 2011). The crystal structures of dimethylaminomethylferrocene/apoferritin composites were determined; they showed the existence of a new anomalous peak at the twofold symmetry axis cavity of the apo-rHLFr molecule (Niemeyer et al. 2008). Different types of metal complexes are individually arranged in different ways on the interior surface. It is difficult to construct specific accumulation patterns by using synthetic multidentate ligands alone. Thus, it is believed that the technology would lead to the development of many different types of metal complexes at the confined interior of protein cages, for the manufacture of useful materials, such as biocatalysts, biosensors, and metal-containing drugs.

Cross-References

▶ Apoferritin, Activation by Gold, Silver, and Platinum Nanoparticles
▶ Iron Proteins, Ferritin
▶ Palladium Complex-induced Release of Cyt c from Biological Membrane
▶ Palladium, Physical and Chemical Properties
▶ Palladium-catalysed Allylic Nucleophilic Substitution Reactions, Artificial Metalloenzymes
▶ Palladium-Mediated Site-Selective Suzuki-Miyaura Protein Modification
▶ Radium, Physical and Chemical Properties

References

Abe S, Niemeyer J et al (2008) Control of the coordination structure of organometallic palladium complexes in an apo-ferritin cage. J Am Chem Soc 130:10512–10514
Abe S, Hirata K et al (2009) Polymerization of phenylacetylene by rhodium complexes within a discrete space of apo-ferritin. J Am Chem Soc 131:6958–6960
Abe S, Hikage T et al (2010) Mechanism of accumulation and incorporation of organometallic Pd complexes into the protein nanocage of apo-ferritin. Inorg Chem 49:6967–6073
Cornils B, Herrmann WA (2004) Aqueous-phase organometallic catalysis. Wiley-VCH, Weinheim
Hillard EA, Jaouen G (2011) Bioorganometallics: future trends in drug discovery, analytical chemistry, and catalyst. Organometallics 30:20–27
Niemeyer J, Abe S et al (2008) Noncovalent insertion of ferrocenes into the protein shell of apo-ferritin. Chem Commun 44:6519–6521
Suzuki M, Abe M et al (2009) Preparation and catalytic reaction of Au/Pd bimetallic nanoparticles in apo-ferritin. Chem Commun 45:4871–4873
Takezawa Y, Bockmann P et al (2011) Incorporation of organometallic Ru complexes into apo-ferritin cage. Dalton Trans 40:2190–2195
Theil EC (1987) Ferritin: structures, gene regulation, and cellular function in animals, plants, and microorganisms. Annu Rev Biochem 56:289–315
Ueno T, Suzuki M et al (2004) Size-selective olefin hydrogenation by a Pd Nano cluster provided in an apo-ferritin cage. Angew Chem Int Ed 43:2527–2530
Ueno T, Abe M et al (2009) Process of accumulation of metal ions on the interior surface of apo-ferritin: crystal structures of a series of apo-ferritins containing variable quantities of Pd(II) ions. J Am Chem Soc 131:5094–5100
Wang ZY, Takezawa Y et al (2011) Definite coordination arrangement of organometallic palladium complexes accumulated on the desired interior surface of apo-ferritin. Chem Commun 47:170–172

Palladium, Physical and Chemical Properties

Fathi Habashi
Department of Mining, Metallurgical, and Materials Engineering, Laval University, Quebec City, Canada

Palladium is a member of the group of six platinum metals which occur together in nature and is located in the center of the Periodic Table in the Iron – Cobalt – Nickel group.

Fe	Co	Ni
Ru	Rh	*Palladium*
Os	Ir	Pt

Physical Properties

Atomic number	46
Atomic weight	106.42
Relative abundance, %	1×10^{-6}
Density at 20°C, g/cm^3	12.02
Major natural isotopes	105 (22.3%)
	106 (27.3%)
	108 (26.5%)
Crystal structure	Face-centered cubic
Atomic radius, nm	0.138
Melting point, °C	1,554
Boiling point, °C	2,940
Specific heat at 25°C, C_p, J g^{-1} K^{-1}	0.23
Thermal conductivity λ, W m^{-1} K^{-1}	75
Brinell hardness	52
Tensile strength σ_B, N/mm^2	196.2
Specific electrical resistance at 0°C, $\mu\Omega$ cm	9.92
Temperature coefficient of electrical resistance (0–100°C), K^{-1}	0.0038
Thermoelectric voltage versus Pt at 100°C E, V	−0.57

Chemical Properties

Because of the small energy differences between the valence shells, a number of oxidation states occur: 0, +2, +3, +4. The divalent state is the most common. Palladium dissolves in HCl–Cl$_2$ to form tetrachloropalladate (II) ion, [PdCl$_4$]$^{2-}$:

$$Pd + 2HCl + Cl_2 \rightarrow H_2[PdCl_4]$$

Evaporation of this solution in hydrochloric acid yields PdCl$_2$. The addition of ammonia to a solution of H$_2$[PdCl$_4$] causes the formation of tetraamminepalladium (II) chloride, [Pd(NH$_3$)$_4$]Cl$_2$

$$H_2[PdCl_4] + 6NH_3 \rightarrow [Pd(NH_3)_4]Cl_2 + 2NH_4Cl$$

which on acidification yields the sparingly soluble light-yellow *trans*-diamminedichloropalladium (II), [PdCl$_2$(NH$_3$)$_2$]:

$$[Pd(NH_3)_4]Cl_2 + 2H^+ \rightarrow [PdCl_2(NH_3)_2] + 2NH_4^+$$

Dichlorodiamminepalladium (II), [Pd(NH$_3$)$_2$Cl$_2$], decomposes at 900°C in an inert atmosphere to yield palladium sponge:

$$3[Pd(NH_3)_2Cl_2] \rightarrow 3Pd + 4NH_4Cl + 2HCl + N_2$$

Impure ammonium hexachloropalladate(IV), (NH$_4$)$_2$[PdCl$_6$], can be purified by dissolving in ammonia then acidification:

$$3(NH_4)_2[PdCl_6] + 20NH_3 \rightarrow 3[Pd(NH_3)_4]Cl_2 + 12NH_4Cl + N_2$$

$$[Pd(NH_3)_4]Cl_2 + 2HCl \rightarrow [PdCl_2(NH_3)_2] + 2NH_4Cl$$

Pure ammonium hexachloropalladate, (NH$_4$)$_2$[PdCl$_6$], is precipitated by NH$_4$Cl from aqua regia leach solution of palladium as a red powder. The powder is decomposed at 900°C to produce palladium sponge:

$$(NH_4)_2[PdCl_6] \rightarrow Pd + 2NH_4Cl_2 + 2Cl_2$$

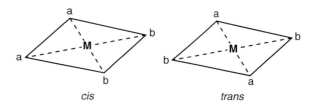

Palladium, Physical and Chemical Properties, Fig. 1 Cis and trans-isomers of palladium xxx

Isomerism

Crystals having the same number and kind of atoms (formula) but different structures may form on crystallization; these are called isomers and the phenomenon is called isomerism. In the case of geometrical isomerism the isomers have the same structural formula but the atoms have a different distribution in space. When the like groups are close together, they are called the *cis* form, and when far apart, they are called the *trans* form (from Latin *cis* = on the same side, and *trans* = across). Geometrical isomerism is possible in square planer and octahedral complexes but not in tetrahedral arrangement because in the latter case all four coordinating groups are equidistant from one another. For example: *Square planer complexes*. When NH$_3$ is added to a solution containing PtCl$_4^{2-}$ ions and the solution is evaporated, the crystals obtained differ in physical properties from those obtained when HCl is added to a solution containing [Pt(NH$_3$)$_4$]$^{2+}$ ions and the solution similarly evaporated:

$$PtCl_4^{2-} + 2NH_3 \rightarrow [Pt(NH_3)_2Cl_2] + 2Cl^-$$

$$[Pt(NH_3)_4]^{2+} + 4HCl \rightarrow [Pt(NH_3)_2Cl_2] + 2NH_4Cl + 2H^+$$

Both crystals obtained have the same chemical composition; the difference is attributed to different geometrical orientation. In the first case, the *cis*-isomer is formed while in the second, it is the *trans* as shown in Fig. 1. It can be seen that the four coordinating groups are equidistant from the central atom, but they are not equidistant from one another.

References

Habashi F (1999) A textbook of hydrometallurgy, 2nd edn. Métallurgie extractive Québec, Québec City. Distributed by Laval University Bookstore, www.zone.ul.ca

Habashi F (2003) Metals from ores. An introduction to extractive metallurgy. Métallurgie extractive Québec, Quebec City, Canada. Distributed by Laval University Bookstore, www.zone.ul.ca

Remy H (1956) Comprehensive inorganic chemistry, vol 2. Elsevier, Amsterdam

Renner H (1997) Platinum group metals. In: Habashi F (ed) Handbook of extractive metallurgy. Wiley-VCH, Weinheim, pp 1269–1360

Palladium-catalysed Allylic Nucleophilic Substitution Reactions, Artificial Metalloenzymes

Wouter Laan and Paul C. J. Kamer
School of Chemistry, University of St Andrews, St Andrews, UK

Synonyms

Asymmetric allylic alkylation/amination; Hybrid catalysts; Synthetic metalloenzymes

Definitions

Artificial metalloenzymes refer to catalysts resulting from combining a catalytically active organometallic compound with a macromolecular host, which may be a protein or DNA.

Palladium-catalyzed allylic nucleophilic substitution refers to a catalytic reaction in which a leaving group in the allylic position of an allyl compound is displaced by a nucleophile via formation of an intermediate palladium complex with the three allyl carbons coordinated as a π-complex.

Catalysis: Transition Metal Complexes and Enzymes

Catalysis makes chemical conversions faster and more selective. This allows reactions to occur with higher efficiency under milder conditions, thereby reducing the waste production and energy consumption associated with synthetic processes. Evidently, the catalytic proficiency of enzymes for the fast and highly controlled construction of chemical bonds under mild conditions is crucial for Life. In addition, catalysis has a great impact on the quality of our lives by playing a key role in the synthesis of bulk chemicals like fuels, plastics, and detergents, and fine chemicals such as pharmaceuticals, agrochemicals, and fragrances. In fact, it is estimated that currently about 80% of all industrial chemical synthesis routes contain at least one catalytic step using either an enzyme or a synthetic catalyst.

Due to their reactivity, metals play a key role in catalysis. It is estimated that about 40% of all enzymes are metalloenzymes while organometallic compounds, particularly transition-metal complexes, dominate the realm of synthetic catalysts.

Transition metals like palladium, rhodium, ruthenium, etc., catalyze a wide array of chemical reactions. While the performance of transition-metal catalysts depends largely on the nature of the metal, it can be significantly influenced by organic ligands coordinated to the metal center. Via steric and electronic effects, this can lead to not only a dramatic change in activity, but also the product distribution of the reaction can be greatly altered by the ligand. These effects allow for controlling the rate and selectivity of a reaction by changing the structure of the ligand.

The broad reaction scope and the ability to tune catalyst performance for a given reaction have made transition-metal catalysis one of the most important research areas in modern chemistry. This has led to the development of a vast array of ligand structures. Various classes of ligands with different donor-atoms have been explored, including phosphorus-donor ligands (e.g., phosphines, phosphites) nitrogen-donor ligands, and carbon-donor ligands (e.g., carbenes).

Three of the most important types of selectivity in a chemical reaction are:
- *Enantioselectivity*: The degree to which one of the two possible enantiomers (the so-called R or S enantiomer) of a chiral product is preferentially produced in a chemical reaction. This is quantitatively expressed by the enantiomeric excess (e.e.), which can range from 0 (equal amounts of both enantiomers produced) to 100 (exclusive formation of one enantiomer).

- *Regioselectivity*: The degree to which the product of reaction at one site dominates over reaction at other sites.
- *Chemoselectivity*: The degree to which a chemical reagent preferentially reacts with one of two or more different functional groups. Chemoselectivity is high when reaction occurs with only a limited number of different functional groups.

Many biologically active compounds, such as pharmaceuticals, agrochemicals, flavors, and nutrients, are chiral. As the different enantiomers of a chiral compound can have dramatically different effects in a biological environment, more than 50% of today's top-selling drugs are single enantiomers. Therefore, the development of highly efficient and enantioselective catalysts for asymmetric synthesis is one of the most active areas in catalysis research.

The catalytic activity and selectivity of transition-metal catalysts is almost exclusively controlled by the coordinating ligands, the so-called first coordination sphere. In contrast, enzymes invoke high control over the second coordination sphere to control their activity and selectivity. The second coordination sphere refers to the manifold of non-covalent interactions provided by the protein scaffold, such as hydrogen bonding, hydrophobic, and electrostatic interactions. Studies of synthetic mimics of enzyme active sites have indicated that second-sphere substrate interactions are crucial for the catalytic proficiency of enzymes. This enables enzymes to employ highly efficient substrate recognition and orientation and stabilization of reactive intermediates, affording very high reaction rates (the turnover frequencies for biocatalysts are usually in the range of $10–10,000$ s^{-1}) and excellent enantio-, regio-, and chemoselectivities. Transition-metal catalysts are typically much slower (turnover frequencies of $1–10$ s^{-1} or less), and also the selectivities of enzymatic catalysis are seldom equaled by chemical catalysts.

Because of their outstanding catalytic performance under mild conditions, enzymes are very attractive catalysts for synthetic processes. For many reactions, the application of enzymes as biocatalysts is highly effective, and enzymatic processes are increasingly implemented in industrial production processes. Also, the application of protein engineering methods such as directed evolution makes it increasingly facile to tailor the properties of enzymes, allowing for expansion of the substrate-scope, tuning the selectivity, or increasing the stability under operating conditions. However, the reaction scope of enzymes is limited, that is, for many important chemical conversions, such as the allylic substitution reaction, natural enzymes are not available.

Catalytic Allylic Substitution

Catalytic allylic substitution reactions are metal-mediated reactions between allylic electrophiles and carbon or heteroatom nucleophiles, affording the products of formal S_N2 or S_N2' substitutions (Fig. 1). Commonly used allylic electrophiles are allyl chlorides, acetates, carbonates, or other types of ester derived from allylic alcohols. The nucleophile is usually a so-called soft nucleophile such as the anion of a $\beta-$dicarbonyl compound, or an amine or the anion of an imide. Reactions involving carbon nucleophiles are often called *allylic alkylations*, whereas reactions with amine nucleophiles are usually referred to as *allylic aminations*. The allylic substitution reaction is compatible with many functional groups including hydroxy, amino, ester, and amide groups. The remaining olefinic bond in the reaction products can be functionalized further by other catalytic transformations like Heck olefination. Therefore, the products are valuable intermediates in the synthesis of natural products and pharmaceuticals. The exceptionally broad scope and the utility of its products make the catalyzed allylic substitution reaction a synthetically very useful reaction.

The first transition-metal-mediated allylic substitution reaction was discovered in 1965 by Tsuji, then working at Toray Industries (Tsuji et al. 1965), who showed that palladium coordinated to an allyl compound facilitates the reaction of the allyl with carbon nucleophiles, generating products containing a new carbon-carbon bond. Since then, extensive research has been carried out in the field, and even though a variety of transition-metals has been found to catalyze this reaction, palladium has received the greatest attention.

Using a chiral phosphine-ligand, Trost and coworkers were the first to develop an enantioselective allylic substitution reaction. They also developed some of the most useful catalysts for enantioselective allylic alkylation and extensively demonstrated the utility of

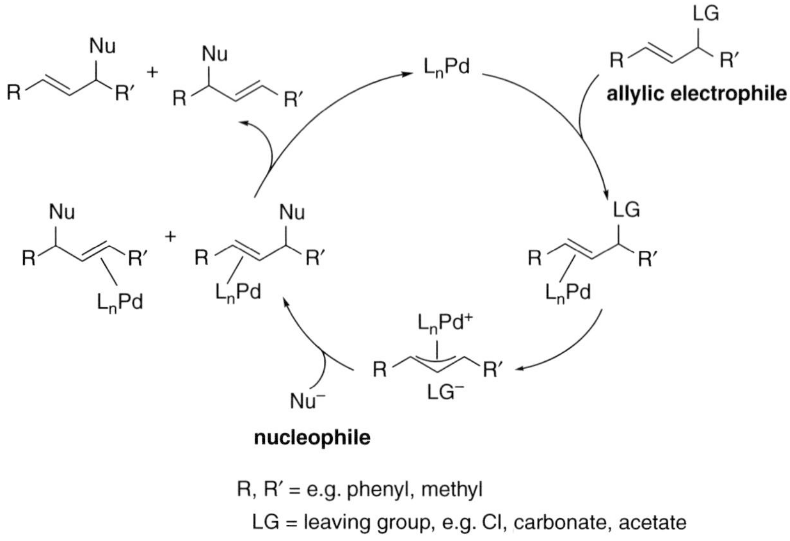

Palladium-catalysed Allylic Nucleophilic Substitution Reactions, Artificial Metalloenzymes,
Fig. 1 Catalytic cycle of palladium-catalyzed allylic substitution

R, R' = e.g. phenyl, methyl
LG = leaving group, e.g. Cl, carbonate, acetate

these reactions for natural product synthesis. Because of the important contributions of the two researchers to the field, palladium-catalyzed allylic substitution reactions have become known as "Tsuji–Trost" reactions.

Artificial Metalloenzymes

Whereas the fields of transition-metal and biocatalysis have for a very long time developed independently, the desire to merge the best features of both worlds has recently spawned a new field integrating both disciplines: artificial metalloenzymes. Artificial metalloenzymes are hybrid catalysts composed of a catalytically active transition-metal complex anchored to a biomolecular host, which can be a protein or DNA. The motivation underpinning the development of such hybrid systems is to create catalysts that combine the broad catalytic scope of transition-metal catalysis with the high activity and selectivity under mild reaction conditions characteristic of enzymes. Embedding transition-metal catalysts within biomacromolecular scaffolds is envisioned to allow for harnessing the chirality and second coordination sphere of the host to provide enzyme-like control over transition-metal-catalyzed reactions (Deuss et al. 2011).

In 1978, Wilson and Whitesides reported the research that laid the foundation for the now thriving field of artificial metalloenzymes (Wilson and Whitesides 1978). In their visionary work, they introduced a well-established organometallic catalyst in a protein host and applied the resulting hybrid system in an archetypical transition-metal catalysis reaction, the asymmetric hydrogenation of olefins. Avidin is a remarkably robust tetrameric protein, which binds the vitamin biotin or its chemically modified derivatives with such high affinity that association is considered to be irreversible ($K_a \sim 10^{15}$ M^{-1}). Biotin contains a valeric acid side chain, which provides a convenient handle for the synthesis of biotin-conjugates. Wilson and Whitesides exploited this to embed a phosphine-based rhodium catalyst in a protein environment. A biotinylated phosphine-ligand was synthesized and subsequently used to create a biotinylated rhodium–diphosphine complex. Simple mixing of this catalyst with a solution of avidin afforded a supramolecularly assembled artificial metalloenzyme, which was applied in the asymmetric hydrogenation of N-acetamidoacrylic acid, providing the product N-acetylalanine in 41% e.e., with quantitative conversion (Fig. 2).

A number of other proteins have been used for the development of artificial metalloenzymes via supramolecular anchoring of organometallic catalysts, such as myoglobin and BSA. Also the direct coordination of a transition-metal catalysts to protein-residues (dative-anchoring) and the site-specific covalent modification of a reactive residue (typically cysteine) have been employed to develop protein-based artificial metalloenzymes active in a variety of chemical

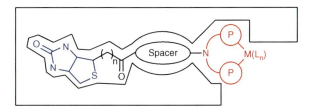

Palladium-catalysed Allylic Nucleophilic Substitution Reactions, Artificial Metalloenzymes, Fig. 2 Schematic representation of the biotin-avidin approach for artificial metalloenzyme development. M = transition metal (e.g., Rh, Pd); L = additional ligands (e.g., allyl)

conversions not accessible using conventional enzymes (Deuss et al. 2011).

Because artificial metalloenzymes are composites of a biomolecular scaffold and a synthetic catalytic moiety, the performance of these hybrid systems can be modulated by changing the structure of the scaffold or the structure of the introduced catalyst. The body of work of Ward and coworkers provides a particularly illustrative example of how this can be applied to optimize the performance of artificial metalloenzymes. Inspired by the report of Wilson and Whitesides, Ward and coworkers started to substantially expand the avidin–biotin platform for artificial metalloenzyme development. They included the related protein streptavidin, and firmly demonstrated that the performance of the hybrids can be optimized by synergistic tuning of both the structure of the protein and the organometallic catalyst, a strategy for which the term "chemogenetic" optimization was coined. Various biotinylated ligands were synthesized with different linker structures (e.g., amino-acid spacers) and different phosphine-ligands. Each spacer places the metal moiety in a distinct chiral environment, which may have a significant effect on both the activity and the selectivity. The screening of various combinations of protein-mutants and ligand structures yielded hydrogenation-catalysts with enantioselectivities up to 96%. In some cases, the artificial metalloenzymes displayed a higher activity than the corresponding rhodium–ligand complex alone and also substrate-specific systems were found. By anchoring different types of catalysts, the approach was also successfully extended to other types of reactions, including transfer-hydrogenation, sulfoxidation, and olefin metathesis (Ward 2011).

Directed evolution is a very powerful tool for the optimization of the properties of proteins. This approach uses repeating cycles of random mutagenesis followed by screening for the best hits in a Darwinian fashion, and is increasingly applied to improve enzyme performance and protein properties. The group of Reetz has successfully applied this approach to optimize the structure of the protein scaffold in artificial metalloenzymes based on the rhodium diphosphine–modified biotin-streptavidin system. Although successful, the procedure was reported to be labor intensive (Reetz et al. 2006).

Ward et al. have implemented a strategy whereby evolution of the enzyme is achieved by chemogenetic optimization guided by rational design, an approach which they have named "designed evolution." In this process, the mutagenesis in every optimization cycle is guided by design choices based on structural information. This significantly reduces the number of possible mutants employed in the procedure, while the optimization process is still very effective (Ward 2011).

DNA is another scaffold that has been employed for artificial metalloenzyme development. The sugar–phosphate backbone provides a source of chirality, and oligonucleotides have been found to exhibit molecular-recognition properties that rival those of proteins and can even function as catalysts in their own right (DNA- and RNAzymes). Moreover, the well-known base-pairing rules allow the rational engineering of the secondary and tertiary structure of DNA. These features make DNA well suited for engineering selective hybrid catalysts. A prominent approach to DNA-based artificial metalloenzymes has been elaborated by Roelfes and coworkers. Using copper catalysts which bind to the double-stranded DNA helix by intercalation and/or groove binding, in combination with commercially available salmon testes and calf thymus DNA or synthetic DNA, selective catalysts for a number of synthetically important chemical conversions, such as the asymmetric Diels–Alder, Michael addition, and Friedel–Crafts alkylation reactions, have been developed (Megens and Roelfes 2011).

Artificial Metalloenzymes Active in Palladium-Catalyzed Allylic Substitution

Ward and coworkers demonstrated in 2008 that the reaction repertoire of the biotin–avidin artificial metalloenzyme platform could be extended to palladium-catalyzed asymmetric allylic alkylation. Initial evaluation of various ligand–protein combinations in the substitution reaction of 1,3-diphenylpropenyl

Palladium-catalysed Allylic Nucleophilic Substitution Reactions, Artificial Metalloenzymes,
Fig. 3 Examples of allylic substitution reactions catalyzed by palladium-containing artificial metalloenzymes

acetate and dimethyl malonate (Fig. 3) afforded little conversion to the alkylation product but resulted predominantly in the hydrolysis of the substrate. However, including the surfactant didodecyldimethylammonium bromide (DMB; 40 equivalents relative to the protein) gave rise to a significant increase in the yield. With a combination of chemical and genetic optimization of the artificial metalloenzymes, the product of the reaction was obtained with up to 95% conversion and enantiomeric excesses ranging from 90% R to 82% S (Pierron et al. 2008). The nature of the amino-acid spacer linking the diphosphine-ligand to the biotin moiety was found to be critical. The highest selectivity was obtained with an o-aminobenzoate spacer (Fig. 4 left) combined with the S112A streptavidin mutant. The enantioselectivity of the reaction could be completely reversed to 82% of the S product by changing to a R-proline spacer (Fig. 4 right) in combination with the S112G-V47G double mutant. It was also found that the presence of the host protein leads to an acceleration of the alkylation reaction compared to the protein-free biotinylated catalysts.

Recently, the group of Kamer reported the modification of the native cysteine of the photoactive yellow protein (PYP) using 1,1′-carbonyldiimidazole (CDI)-activated carboxylic acid derivatives of mono- and diphosphines, resulting in the covalent anchoring of the ligands via a thioester bond in high yields and selectivity. The similar anchoring of palladium–phosphine complexes afforded metalloenzymes (Fig. 5) active as catalysts for the allylic amination reaction between 1,3-diphenylpropenyl acetate and benzylamine (Laan et al. 2010). In aqueous buffered

Palladium-catalysed Allylic Nucleophilic Substitution Reactions, Artificial Metalloenzymes, Fig. 4 Ligand structures leading to high R (*left*) and S (*right*) selectivity in allylic substitution reactions catalyzed by palladium-containing artificial metalloenzymes based on the avidin-biotin system

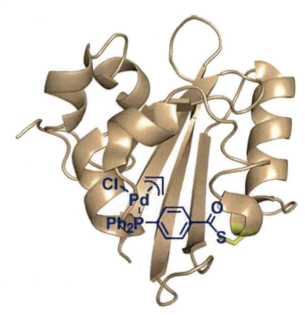

Palladium-catalysed Allylic Nucleophilic Substitution Reactions, Artificial Metalloenzymes, Fig. 5 Schematic representation of PYP covalently modified with a palladium-phosphine complex by a thioester bond

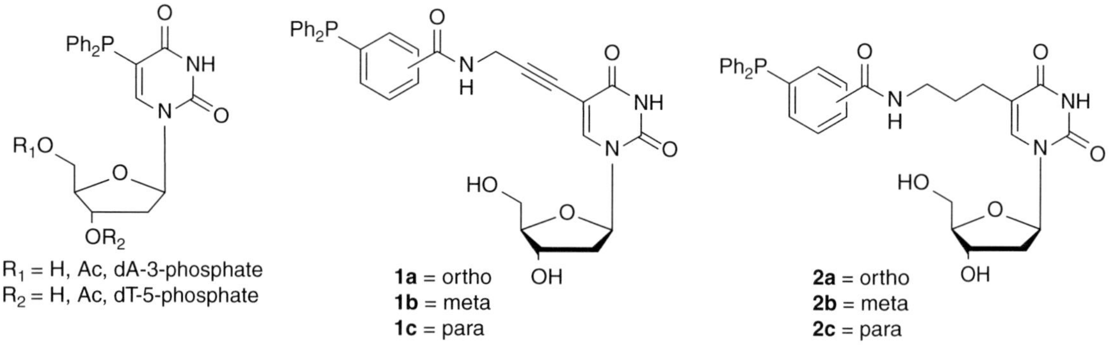

Palladium-catalysed Allylic Nucleophilic Substitution Reactions, Artificial Metalloenzymes, Fig. 6 Oligonucleotide based ligands for palladium catalyzed allylic substitution reactions

solution, irreproducible conversions were observed, likely stemming from substrate insolubility and/or inaccessibility of the transition-metal catalyst. To counter these irreproducibility issues in the catalysis reactions, 50% DMF was included in the reaction mixtures. Under these conditions, good conversions were obtained, but no enantioselectivity was observed. This could result from the (partial) denaturation of the protein due to the high concentration of organic solvent.

Kamer and coworkers also reported an approach for the synthesis of phosphine-modified nucleotides and their application in palladium-catalyzed allylic substitution (Ropartz et al. 2007). The palladium-catalyzed coupling of diphenylphosphine to the commercially available 5-iodo-2′-deoxyuridine (IdU) led to the functionalized nucleotide 5-diphenylphosphino-2′-deoxyuridine (dppdU, Fig. 6) in high yield. Functionalization of IdU-containing trinucleotides anchored on a solid-support was also achieved, although higher palladium loadings were required. The phosphine-containing nucleotides were used as ligands for palladium-catalyzed asymmetric allylic amination reactions. Good enantiomeric excesses were obtained (up to 82% for the S enantiomer) with the mononucleotide ligand dppdU. The functionalized trinucleotides also showed good conversion, albeit at the expense of the enantioselectivity.

An alternative approach to prepare phosphine-modified mono- and oligonucleotides has also been reported (Nuzzolo et al. 2010). Coupling of a linker-compound containing a chemically masked amine to IdU, followed by deprotection of the amine, provided an amino-modified deoxyuridine. Reduction of the triple bond in the linker to a single bond by hydrogenation afforded a more flexible and reactive amino-group on the deoxyuridine. Subsequently, carboxylic acid containing (di)phosphines was activated and reacted with the functionalized bases, affording seven different phosphine-modified deoxyuridines, with yields being significantly higher when the hydrogenated amino-modified deoxyuridine was employed. Application of the same approach to oligonucleotides (15–16 mers) containing the hydrogenated amino-modified deoxyuridine afforded several phosphine-modified oligonucleotides, which were purified by ethanol precipitation. A preliminary catalytic study was performed with two types of ligands, **1** and **2** (Fig. 6). Allylic alkylation of 1,3-diphenylallyl acetate with dimethyl malonate and allylic amination with benzylamine as nucleophile resulted in 100% conversion using ligands **1b**, **1c**, **2b**, and **2c** while ligands **1a** and **2a** gave less than 10% conversion. Unfortunately, no enantioselectivity was obtained.

DNA-Based Metalloenzymes Active in Iridium-Catalyzed Allylic Substitution

The group of Jäschke synthesized several diene-modified oligonucleotides of 19 bases by modification of an introduced 4-triazolyldeoxyuridine. These modified oligonucleotides were tested in the iridium-catalyzed allylic amination of phenyl allyl acetate. Several complementary DNA and RNA strands were used including some that form loops in either the complementary or the ligand-containing strand. The e.e. of both the substrate and the product was recorded as the conversion was found to be around 45% for all systems. The reactions gave e.e. values up to 27%, and interestingly, it appeared that the configuration of the major enantiomer was dependent on the complementary strand used (Fournier et al. 2009). The influence of the complementary strands on the catalyst performance holds great promise for combinatorial screening and catalyst optimization as the sequence of the complementary strand can easily be varied using automated DNA synthesis.

Combining transition-metal catalysts with biomacromolecules provides access to artificial enzymes with unnatural catalytic activities. This is a challenging but exciting and fast developing area in catalysis research. The powerful molecular recognition of biopolymers such as proteins and DNA can be combined with traditional catalyst design and ligand optimization of transition-metal catalysis. Potent biomolecular techniques such as site-directed mutagenesis and directed evolution can boost the optimization process for these hybrid catalysts even further.

Undoubtedly, the diversity of the available approaches will provide effective and efficient systems for challenging late transition-metal-catalyzed reactions. The discussed successes in palladium-catalyzed allylic substitution illustrate the great potential of these hybrid catalysts.

Cross-References

▶ Artificial Selenoproteins
▶ Palladium, Coordination of Organometallic Complexes in Apoferritin
▶ Palladium, Physical and Chemical Properties

References

Deuss PJ, den Heeten R, Laan W, Kamer PCJ (2011) Bioinspired catalyst design and artificial metalloenzymes. Chem Eur J 17:4680–4698
Fournier P, Fiammengo R, Jaeschke A (2009) Allylic amination by a DNA-diene-iridium(I) hybrid catalyst. Angew Chem Int Ed 48:4426–4429
Laan W, Munoz BK, den Heeten R, Kamer PCJ (2010) Artificial metalloenzymes through cysteine-selective conjugation of phosphines to photoactive yellow protein. Chembiochem 11:1236–1239
Megens RP, Roelfes G (2011) Asymmetric catalysis with helical polymers. Chem Eur J 17:8514–8523
Nuzzolo M, Grabulosa A, Slawin AMZ, Meeuwenoord NJ, van der Marel GA, Kamer PCJ (2010) Functionalization of mono- and oligonucleotides with phosphane ligands by amide bond formation. Eur J Org Chem 17:3229–3236
Pierron J, Malan C, Creus M, Gradinaru J, Hafner I, Ivanova A, Sardo A, Ward TR (2008) Artificial metalloenzymes for asymmetric allylic alkylation on the basis of the biotin-avidin technology. Angew Chem Int Ed 47:701–705
Reetz MT, Peyralans JJP, Maichele A, Fu Y, Maywald M (2006) Directed evolution of hybrid enzymes: evolving enantioselectivity of an achiral Rh-complex anchored to a protein. Chem Commun 41:4318–4320
Ropartz L, Meeuwenoord NJ, van der Marel GA, van Leeuwen PWNM, Slawin AMZ, Kamer PCJ (2007) Phosphine containing oligonucleotides for the development of metallodeoxyribozymes. Chem Commun 15:1556–1558
Tsuji J, Takahashi H, Morikawa M (1965) Organic syntheses by means of noble metal compounds XVII. Reaction of π-allylpalladium chloride with nucleophiles. Tetrahedron Lett 6:4387–4388
Ward TR (2011) Artificial metalloenzymes based on the biotin-avidin technology: enantioselective catalysis and beyond. Acc Chem Res 44:47–57
Wilson ME, Whitesides GM (1978) Conversion of a Protein to a Homogeneous Asymmetric Hydrogenation Catalyst by Site-Specific Modification with a Diphosphinerhodium(I) Moiety. J A Chem Soc 100:306–307

Palladium-Mediated Site-Selective Suzuki-Miyaura Protein Modification

Christopher D. Spicer[1,2] and Benjamin G. Davis[1]
[1]Chemistry Research Laboratory, Department of Chemistry, University of Oxford, Oxford, UK
[2]St. Hilda's College, University of Oxford, Oxford, UK

Synonyms

Suzuki biology; Suzuki-Miyaura couplings at unnatural amino acids

Definition

In organic chemistry, the Suzuki-Miyaura reaction refers to the cross coupling of an aryl or vinyl halide with an aryl/vinyl boronic acid. The reacton is catalyzed by a transition metal complex, usually palladium (0), although reactions with nickel (0) are also known. Reactions proceed with high specificity, in high yields, to generate non-labile carbon-carbon bonds. Recently, the reaction has been applied to the chemical post-translational modification of proteins and peptides. The unnatural amino acid *p*-iodophenylalanine has been most widely utilized, due to the facile incorporation of such residues into proteins by amber-stop codon suppression.

Post-translational Modification of Proteins

The post-translational modification (PTM) of proteins is common in both eukaryotic and prokaryotic organisms, greatly increasing the range of protein function and structure available to cells. In order to study, manipulate, and utilize such modified proteins, it is important to be able to access homogenously modified constructs. Isolation from natural sources is hindered by complex purification procedures, low yields, and the often rapid reversibility of modification. There is therefore a need for chemical tools to access such proteins in both an in vivo and an in vitro setting. Yet limitations continue to remain due to the relatively small number of synthetic methodologies available for site-selective chemical modification of amino acids. Such reactions have traditionally relied upon reactions at nucleophilic substituents, particularly cysteine or lysine residues (Chalker et al. 2009a).

Recent advances in both biotechnology and synthetic chemistry have opened up the possibility of using metal-mediated reactions for protein modification. One such reaction is the palladium-mediated Suzuki-Miyaura cross coupling between aryl or vinyl halides and aryl/vinyl boronic acids. This has found widespread utility in organic chemistry over the last 30 years. Herein the key developments that have led to this reaction becoming an attractive tool for PTM and synthetic biology are described.

The characteristics of the Suzuki-Miyaura coupling offer a number of potential benefits over traditional methods for PTM which have led to its study. Firstly, the reactive groups associated with it (aryl halides and boronic acids) are very rarely found in nature, thus rendering such reactions bio-orthogonal (i.e., they do not interfere with native biochemical processes). This allows a high degree of selectivity, whilst the reaction also possesses high functional group tolerance. Secondly, the polar nature of boronic acids may allow the installation of hydrophobic modifications, previously difficult to access in aqueous systems due to their low water solubility. Finally, the reaction generates an inherently stable and irreversible carbon-carbon bond, with the resultant biaryl having numerous advantageous characteristics such as an intrinsic fluorescence.

In order to assess the merits and potential impact of the Suzuki-Miyaura cross coupling, it is useful to consider the two main challenges faced when developing new chemical methods for protein modification. Firstly, any modification should be site-selective and high yielding in order to access homogeneous constructs without the need for significant purification. Secondly, the chosen reaction conditions must not disturb protein tertiary and quaternary structure, that is, near neutral pH, at or below physiological temperature, with water as the sole reaction media. Each shall be addressed in turn with respect to the Suzuki-Miyaura cross coupling.

Site-Selective Introduction of Aryl Halides into Proteins

Much work has been undertaken to extend past the 20 proteinogenic amino acids commonly used by all living organisms, that is, those commonly used by cells to generate the primary structure of a protein. As such, a number of methods now exist for the incorporation of unnatural amino acids into proteins. These include the use of auxotrophic prokaryotic strains, breaking the degeneracy of the genetic code and codon reassignment. One such technique, "amber-stop codon suppression," relies on the reassignment of the TAG amber-stop codon to incorporate the desired unnatural amino acid (De Graaf et al. 2009). Practically, this is achieved by repeated cycles of mutagenesis to generate an orthogonal aminoacyl-tRNA synthetase/suppressor tRNA pair (aaRS/tRNA$_{CUA}$) specific for the TAG codon, that is, one that is acylated with the desired unnatural amino acid by the host cell machinery, but is not acylated by any native host synthetases (See Fig. 1). This is often achieved via transformation of the host organism with an aaRS/tRNA$_{CUA}$ pair from another unrelated species. The resultant system results in a competition between incorporation of the unnatural amino acid and protein truncation in response to the amber-stop codon.

One of the most successful systems relies upon the reassignment of the tyrosine aaRS/tRNA$_{CUA}$ pair from the archaebacteria *Methanocaldococcus jannaschii* (*M. jann*). Optimization of this system over the last 20 years has allowed for the incorporation of a range of tyrosine analogues at levels approaching those seen for native amino acids, in *E. coli* (Young et al. 2010). One such analogue is the unnatural aryl iodide amino acid L-*p*-iodophenylalanine (*p*IPhe, see Fig. 1). This residue

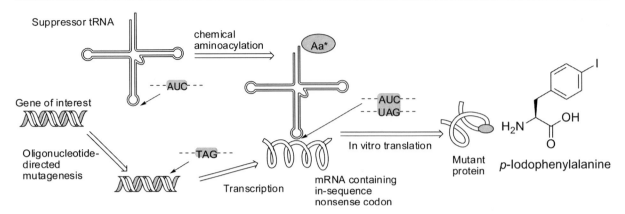

Palladium-Mediated Site-Selective Suzuki-Miyaura Protein Modification, Fig. 1 Amber-stop codon suppression and the structure of the unnatural amino acid pIPhe (Adapted with permission from A. J. de Graaf, M. Kooijman, W. E. Hennink and E. Mastrobattista, *Bioconjugate Chem.*, 2009, 20, 1281–1295. Copyright 2009 American Chemical Society)

uses iodine to mimic the hydroxyl group of tyrosine in an expanded aryl pocket in the tRNA synthetase. Such an amino acid may act as the aryl halide substrate for transition metal-mediated cross couplings. An alternative system utilizes a mutated aaRS/tRNA$_{CUA}$ pair for the amino acid pyrrolysine (an archaebacteria-specific amino acid very rarely seen in other domains). In particular, the archaebacteria *Methanosarcina mazei* pair has been mutated to allow the incorporation of both acetylated lysine derivatives and structurally unrelated tyrosine derivatives, including pIPhe (Wang et al. 2011).

An alternative strategy is to install the iodide residue via chemical transformation. Specifically, the mono- or di-iodination of tyrosine residues can be achieved through the use of Barluenga's reagent (IPy$_2$BF$_4$) (Espuña et al. 2006). The reaction has a high specificity toward tyrosine; however, the high natural abundance of such residues can be seen to limit the potential applications. The presence of multiple potential iodination sites often makes the reaction difficult to control and leads to heterogeneous mixtures.

Development of Palladium Catalysts for Protein Modification

While the related, palladium-mediated Sonogashira and Heck couplings had previously been demonstrated on short peptides containing pIPhe, in low yields and mixed organic solvents, it was not until 2005 that the Suzuki-Miyaura coupling was demonstrated on such a system by Ojida et al. (2005). pIPhe was incorporated into a 34 amino acid long WW domain peptide of the Pin1 protein via solid phase peptide synthesis. The use of sodium tetrachloropalladate (Na$_2$PdCl$_4$) in a 1:1 mixture of glycerol and water led to high conversions at 40°C for the cross coupling to a series of boronic acids. While the reaction was limited to a short synthetic peptide, this work demonstrated the plausibility of using such couplings in protein modification. In particular, the orthogonality of the coupling toward proteinogenic amino acids was shown. This work was followed by the first reported palladium-mediated reactions on longer protein sequences by Kodama et al. (See Fig. 2). In two separate reports, they described Heck and Sonogashira couplings at pIPhe incorporated into a Ras protein, albeit in low yields (2% for Heck and 25% for Sonogashira) (Kodama et al. 2006, 2007).

In spite of this promising precedence, early attempts to undertake Suzuki-Miyaura cross couplings were largely unsuccessful. Partial reaction was demonstrated by Brustad et al. who incorporated the unnatural aryl boronic acid residue p-boronophenylalanine (pBPhe) via amber-stop codon suppression, into the protein T4 lysozyme (Brustad et al. 2008). Suzuki-Miyaura coupling in the presence of tris(dibenzylideneacetone)dipalladium (Pd$_2$(dba)$_3$) and a fluorescent aryl iodide proceeded in 30% conversion (See Fig. 2). This however required denaturing conditions, elevated temperature (70°C), and long reaction times (12 h), which are not viable for most protein substrates.

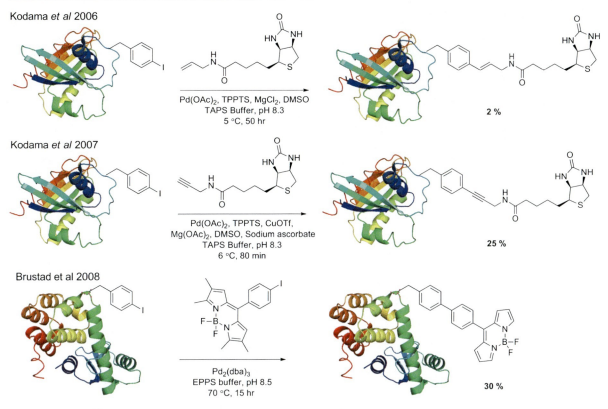

Palladium-Mediated Site-Selective Suzuki-Miyaura Protein Modification, Fig. 2 First examples of palladium-mediated Heck, Sonogashira, and Suzuki couplings on protein substrates

This led to a call for the development of new palladium catalysts capable of catalyzing the reaction at 37°C in aqueous media.

In organic synthesis, common surfactants such as Triton X-100 have often been used in order to solubilize hydrophobic reagents, thus promoting aqueous cross couplings. However, such conditions are not applicable to the chemical modification of proteins. A viable alternative is the formation of a water-soluble palladium complex, via the use of hydrophilic ligands. Water-soluble phosphine ligands such as TPPTS and Amphos have previously been used for small molecule coupling in mixed organic/aqueous systems (Polshettiwar et al. 2010). Such ligands, however, are often highly expensive and prone to oxidation, requiring the use of extensively degassed solvents and inert atmosphere.

Recently, Chalker et al. screened a range of water-soluble ligands in the model cross coupling of a glucose-based boronic acid to a protected *p*IPhe derivative (Chalker et al. 2009b). This led to the discovery of the bis-anion of 2-amino-4,6-dihydroxypyrimidine (ADHP) as a cheap ligand for the formation of a water- and air-stable, phosphine-free palladium (II) precatalyst. The catalyst was shown to catalyze Suzuki-Miyaura cross coupling at 37°C at pH 8, with water (or buffered solution) as the only solvent, on both protected amino acids and short peptides. As a model protein system, subtilisin *Bacillus lentus* (SBL) was functionalized with an aryl iodide moiety via nucleophilic substitution at cysteine 156 (See Fig. 3). A range of boronic acids could be coupled to this "tagged" protein with >95% conversion in as little as 30 min. The palladium (II) precatalyst (50 equiv.) was reduced to the active palladium (0) species via the sacrificial homo-coupling of a small amount of boronic acid, present in a large excess (500 equiv.). The catalyst was also shown to be applicable to organic synthesis, with a wide range of boronic acids and aryl halides being coupled with high efficiency in a microwave-promoted reaction. This ADHP catalyst has recently been adopted in the pharmaceutical industry for small molecule synthesis.

Palladium-Mediated Site-Selective Suzuki-Miyaura Protein Modification,

Fig. 3 Suzuki-Miyaura coupling of a range of boronic acids to a cysteine-linked aryl halide on SBL, using the ADHP catalyst system (Reprinted with permission from J. M. Chalker, C. S. C. Wood and B. G. Davis, *J. Am. Chem. Soc.*, 2009, 131, 16346–16347. Copyright 2009 American Chemical Society)

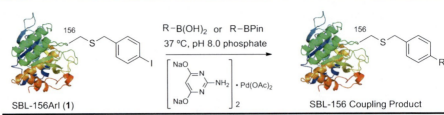

Entry	R-B(OH)$_2$ / R-BPin	Time	Mass Calculated	Mass Observed	Conversion
1	Ph-B(OH)$_2$	30 min	27035	27035	> 95%
2	Me-C$_6$H$_4$-B(OH)$_2$	30 min	27049	27049	> 95%
3	F-C$_6$H$_4$-B(OH)$_2$	30 min	27053	27053	> 95%
4	NC-C$_6$H$_4$-B(OH)$_2$	30 min	27060	27060	> 95%
5	MeO$_2$S-C$_6$H$_4$-B(OH)$_2$	30 min	27113	27113	> 95%
6	furan-3-B(OH)$_2$	30 min	27025	27025	> 95%
7	HO-C$_6$H$_4$-B(OH)$_2$	30 min	27051	27052	> 95%
8	alkenyl-B(OH)$_2$	30 min	27041	27041	> 95%
9	glycoside-vinyl-BPin	60 min	27177	27176	> 95%

Applications of Pyrimidine-Palladium Complexes

Spicer and Davis subsequently applied this ADHP catalyst to the first efficient Suzuki-Miyaura coupling at a genetically incorporated unnatural amino acid (Spicer and Davis 2011). Amber-stop codon suppression was used to install *p*IPhe site selectively into a model maltose-binding protein (MBP). A decreased reaction rate at the amino acid, compared to a cysteine-linked aryl halide, was attributed to an increase in steric crowding. However, the reaction still reached completion within 2 h when furan-3-boronic acid was utilized as a coupling partner (See Fig. 4). During the course of this work, a potent small molecule palladium scavenger, 3-mercaptopropionic acid, was identified. The use of this scavenger, post-coupling, disrupted nonspecific binding of palladium to the protein surface which had previously resulted in a loss of protein signal during mass spectrometry. In addition to allowing facile reaction monitoring, such scavenging also enabled protein purification by dialysis or size-exclusion chromatography. Wang et al. also subsequently utilized this Suzuki-Miyaura coupling strategy to confirm the incorporation of *p*IPhe into the Z-domain protein via a pyrrolysine aaRS/tRNA$_{CUA}$ (Wang et al. 2011). Coupling was undertaken with a fluorescent dansyl-based boronic acid and the ADHP catalyst designed by Chalker et al. After 5 h, a clear fluorescent band could be observed upon UV irradiation of an SDS-PAGE gel, corresponding to the labeled protein.

Recently, Spicer et al. demonstrated the use of Suzuki-Miyaura couplings for the cell surface labeling of *E. coli* (Spicer et al. 2012). An engineered plasmid coding for the *E. coli* cell membrane protein outer membrane protein C (OmpC) was transformed into an OmpC knockout *E. coli* strain, JW-2203-1.

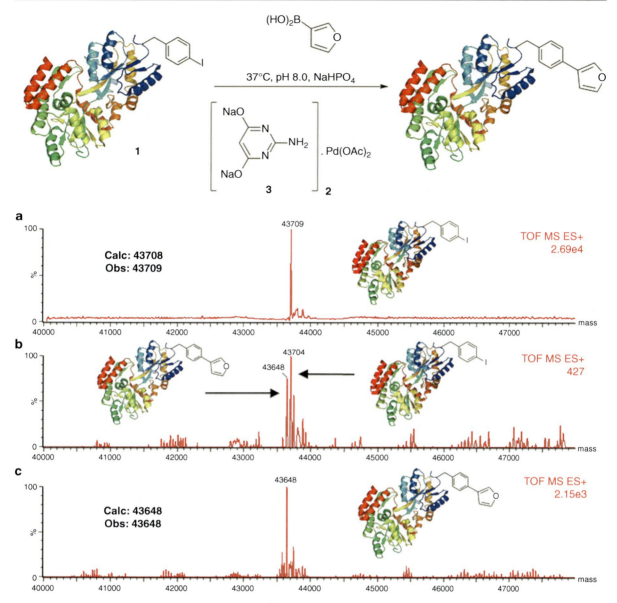

Palladium-Mediated Site-Selective Suzuki-Miyaura Protein Modification, Fig. 4 Suzuki-Miyaura coupling at a genetically encoded pIPhe residue on MBP, using the ADHP catalyst system. (**a**) $t = 0$ h, (**b**) $t = 1$ h, (**c**) $t = 2$ h (Reprinted with permission from C. D. Spicer and B. G. Davis, *Chem. Comm.*, 2011, 47, 1698–1700. Reproduced by permission of The Royal Society of Chemistry)

Induction of protein expression and amber-stop codon suppression led to cell surface "tagging" of the cells with aryl halide pIPhe residues at position Y232. A novel fluorescein-based fluorescent boronic acid was used as a reporter for cross coupling, with labeled cells being visualized by fluorescence microscopy (See Fig. 5). Importantly in the absence of palladium, no labeling was observed, ruling out the possibility of boronic acid coordination to diols or sugars present on the cell surface. Similar reactions in the absence of boronic acid or aryl halide also failed to result in any labeling, supporting a Suzuki-Miyaura mechanism. The specifically labeled protein could be observed by SDS-PAGE gel of the cellular membrane fraction, with labeling confirmed to be taking place at OmpC. The toxicity of the catalyst was also evaluated via flow

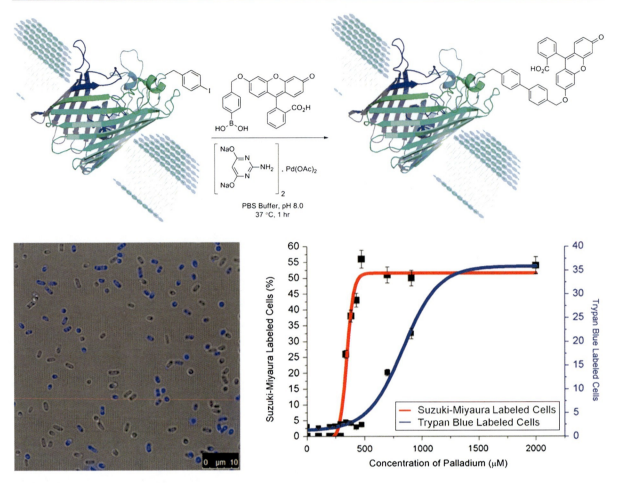

Palladium-Mediated Site-Selective Suzuki-Miyaura Protein Modification, Fig. 5 Cell surface labeling of *E. coli* protein OmpC with a fluorescein boronic acid. The reaction showed a "switch-like" dose response and negligible toxicity (Adapted with permission from C. D. Spicer, T. Triemer and B. G. Davis, *J. Am. Chem. Soc.*, 2012, 134, 800–803. Copyright 2012 American Chemical Society)

cytometry using trypan blue as a viability stain. Toxicity was shown to be negligible at the catalyst loadings required for efficient labeling to take place. Importantly, a "switch-like" dose response was observed for this cellular chemistry, highlighting the strong potential for the use of chemical control in synthetic biology.

Future Directions

While recent work has demonstrated that the Suzuki-Miyaura cross coupling is a potentially useful addition to the "toolbox" for post-translational modification of proteins, a number of key challenges remain. Currently, couplings are mostly limited to applications in prokaryotic systems or derived proteins. However, recent developments in the field of amber-stop codon suppression are opening the door to incorporation of unnatural amino acids in eukaryotic cells and even in whole organisms at low levels. This could potentially lead to in vivo applications of the Suzuki-Miyaura coupling in eukaryotes. Another challenge is the ability to undertake intracellular chemistry. Reactions are currently limited to cell surface or ex vivo settings. As such, the development of cell-permeable boronic acids and catalysts would be highly advantageous.

The ability of the Suzuki-Miyaura coupling to efficiently install non-labile carbon-carbon bonds in a "switch-like" manner, with modifications previously inaccessible via traditional methods, renders it unique among the tools for protein modification. Many applications can be envisaged, and with much research now being undertaken to explore the scope and limitations of the reaction, it is likely that it will find increasing favor in the area of synthetic biology in the coming years.

Cross-References

- ▶ Boron: Physical and Chemical Properties
- ▶ Metals and the Periodic Table
- ▶ Palladium, Biological Effects
- ▶ Palladium, Physical and Chemical Properties
- ▶ Palladium-catalysed Allylic Nucleophilic Substitution Reactions, Artificial Metalloenzymes

References

Brustad E, Bushey ML, Lee JW, Groff D, Liu W, Schultz PG (2008) A genetically encoded boronate-containing amino acid. Angew Chem Int Ed 47:8220–8223

Chalker JM, Bernardes GJL, Lin YA, Davis BG (2009a) Chemical modification of proteins at cysteine: opportunities in chemistry and biology. Chem Asian J 4:630–640

Chalker JM, Wood CSC, Davis BG (2009b) A convenient catalyst for aqueous and protein Suzuki-Miyaura cross-coupling. J Am Chem Soc 131:16346–16347

De Graaf AJ, Kooijman M, Hennink WE, Mastrobattista E (2009) Nonnatural amino acids for site-specific protein conjugation. Bioconjug Chem 20:1281–1295

Espuña G, Andreu D, Barluenga J, Pérez X, Planas A, Arsequell G, Valencia G (2006) Iodination of proteins by IPy_2BF_4, a new tool in protein chemistry. Biochemistry 45:5957–5963

Kodama K, Fukuzawa S, Nakayama H, Kigawa T, Sakamoto K, Yabuki T, Matsuda N, Shirouzu M, Takio K, Tachibana K, Yokoyama S (2006) Regioselective carbon–carbon bond formation in proteins with palladium catalysis. Chembiochem 7:134–139

Kodama K, Fukuzawa S, Nakayama H, Sakamoto K, Kigawa T, Yabuki T, Matsuda N, Shirouzu M, Takio K, Yokoyama S, Tachibana K (2007) Site-specific functionalization of proteins by organopalladium reactions. Chembiochem 8:232–238

Ojida A, Tsutsumi H, Kasagi N, Hamachi I (2005) Suzuki coupling for protein modification. Tetrahedron Lett 46:3301–3305

Polshettiwar V, Decottignies A, Len C, Fihri A (2010) Suzuki–Miyaura cross-coupling reactions in aqueous media: green and sustainable syntheses of biaryls. ChemSusChem 3:502–522

Spicer CD, Davis BG (2011) Palladium-mediated site-selective Suzuki-Miyaura protein modification at genetically encoded aryl halides. Chem Commun 47:1698–1700

Spicer CD, Triemer T, Davis BG (2012) Palladium-mediated cell-surface labeling. J Am Chem Soc 134:800–803

Wang YS, Russell WK, Wang Z, Wan W, Dodd LE, Pai PJ, Russell DH, Liu WR (2011) The de novo engineering of pyrrolysyl-tRNA synthetase for genetic incorporation of l-phenylalanine and its derivatives. Mol Biosyst 7:714–717

Young TS, Ahmad IY, Yin JA, Schultz PG (2010) An enhanced system for unnatural amino acid mutagenesis in *E. coli*. J Mol Biol 395:361–374

Panchromium

▶ Vanadium Metal and Compounds, Properties, Interactions, and Applications

Participation of Zinc in Blood Coagulation Mechanisms

▶ Zinc in Hemostasis

Particulate Matter

▶ Vanadium Pentoxide Effects on Lungs

Particulate Methane Monooxygenase

Stephen M. Smith and Amy C. Rosenzweig
Departments of Molecular Biosciences and of Chemistry, Northwestern University, Evanston, IL, USA

Synonyms

Biological methane oxidation catalyst; Membrane-bound methane monooxygenase; pMMO

Definition

Particulate methane monooxygenase, typically abbreviated pMMO, is a multisubunit, integral membrane enzyme that oxidizes methane to methanol in methanotrophic bacteria.

Methanotrophic Bacteria

Methanotrophs are gram-negative bacteria that use methane as their sole source of carbon and energy. Methanotrophs have attracted increasing attention from the scientific community for their potential use in bioremediation applications, including degradation of carcinogenic halogenated groundwater pollutants and environmental removal of methane, a potent greenhouse gas, generated by both natural and anthropogenic sources (Semrau et al. 2010). Methanotrophs have been traditionally classified into two types based on membrane architecture, growth temperatures, phospholipid composition, and carbon assimilation pathways (Semrau et al. 2010). Type I methanotrophs, including the genera *Methylomicrobium*, *Methylobacter*, and *Methylomonas*, have bundled membranes, grow at 30°C, contain 14 and 16 carbon phospholipids, and utilize the ribulose monophosphate pathway for carbon assimilation. Type II methanotrophs, including the genera *Methylosinus* and *Methylocystis*, have paired membrane architectures, also typically grow at 30°C, contain 18 carbon phospholipids, and utilize the serine pathway for multicarbon compound generation. Type X methanotrophs, characterized by the well-studied genus *Methylococcus*, are a subset of the type-I methanotrophs, but grow at higher temperatures (>40°C) and contain low levels of serine pathway enzymes. All of these methanotrophs are classified within the phylum *Proteobacteria*, as either *Gammaproteobacteria* (type I and type X) or *Alphaproteobacteria* (type II). Several recently identified species, including methanotrophs grouped within the phylum *Verrucomicrobia*, defy this classification scheme (Semrau et al. 2010).

Methane Monooxygenases

The first step of methanotroph metabolism, the conversion of methane to methanol, is catalyzed by enzymes called methane monooxygenases (MMOs) (Hakemian and Rosenzweig 2007). Methanol is then converted to formaldehyde by methanol dehydrogenase (MDH) at which point it is either used for carbon assimilation or further processed to formate and then carbon dioxide by formaldehyde and formate dehydrogenases. The methane oxidation reaction is of particular interest because methane is the most inert hydrocarbon (C–H bond dissociation energy 104 kcal/mol). Abundant natural gas deposits, composed primarily of methane, are underutilized as chemical feedstocks and alternative energy sources because current industrial catalysts require extreme temperatures, are costly, and produce waste (Balasubramanian et al. 2010). The MMO enzymes can oxidize methane at ambient temperature and pressure and thus could serve as inspiration for the development of new cost-effective and environmentally friendly catalysts (Himes and Karlin 2009).

MMOs are multisubunit metalloenzyme complexes that exist in two distinct forms. The well-studied soluble methane monooxygenase (sMMO) contains a nonheme, carboxylate-bridged diiron active site that can oxidize a wide range of hydrocarbon substrates, including alkanes, alkenes, and cyclic and aromatic compounds. Much is known about the structure and mechanism of sMMO (Hakemian and Rosenzweig 2007; Semrau et al. 2010), but this enzyme is only expressed by a few strains of methanotrophic bacteria under conditions of copper stress (Murrell et al. 2000). The most abundant biological methane oxidation catalyst, particulate methane monooxygenase (pMMO), is a membrane-bound enzyme and is expressed by all but one genus of methanotrophs at high copper-to-biomass ratios (Hakemian and Rosenzweig 2007; Murrell et al. 2000). The hydrocarbon substrate range for pMMO is more limited, including alkanes and alkenes up to five carbon atoms in length. In the strains of methanotrophic bacteria that express both sMMO and pMMO, differential expression is controlled by copper availability, but the mechanism of this "copper switch" remains a mystery (Murrell et al. 2000). In contrast to sMMO, details of the pMMO active site and mechanism are not clear and have been controversial (Chan and Yu 2008; Rosenzweig 2008).

Architecture of pMMO

The pMMO enzyme is composed of three polypeptides: pmoB or α (~47 kDa), pmoA or β (~24 kDa), and pmoC or γ (~22 kDa). Early biochemical studies of purified pMMO from *Methylococcus capsulatus* (Bath) suggested that the polypeptide arrangement was either an $\alpha\beta\gamma$ monomer or an $\alpha_2\beta_2\gamma_2$ dimer (Lieberman and Rosenzweig 2004). The exact

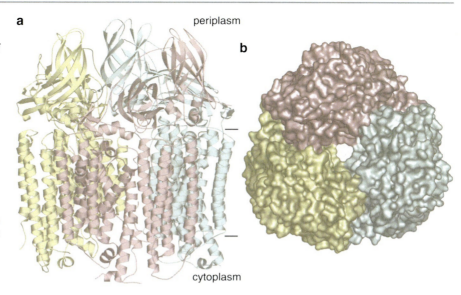

Particulate Methane Monooxygenase, Fig. 1 Overall architecture of pMMO (PDB accession code 1YEW). (**a**) Cartoon diagram of the pMMO trimer with the three αβγ protomers shown in *pink*, *yellow*, and *cyan*. The two *black* lines indicate the approximate location of the membrane, and the periplasmic and cytoplasmic sides of the membrane are indicated. (**b**) Surface representation of the pMMO trimer viewed perpendicular to the membrane normal from the cytoplasmic side

oligomerization state of pMMO was not determined until 2005 when the 2.8-Å resolution crystal structure of pMMO from *M. capsulatus* (Bath) was solved (Lieberman and Rosenzweig 2005). In 2008, the crystal structure of pMMO from *M. trichosporium* OB3b was solved to 3.9-Å resolution, and the $\alpha_3\beta_3\gamma_3$ oligomerization state was also observed (Semrau et al. 2010). The pMMO trimer forms a cylinder-like structure approximately 105-Å long and 90 Å in diameter with six soluble β-barrel domains and 42-43 transmembrane α-helices (Fig. 1a). A central cavity extends through the trimer, approximately 11 Å in diameter in the soluble region, widening to 22 Å in the membrane region (Fig. 1b) (Lieberman and Rosenzweig 2005). The functional significance of this trimeric architecture is not known. The pmoB subunit consists of an N-terminal cupredoxin-like domain tethered to the membrane by two transmembrane helices followed by a long loop region connecting to a C-terminal cupredoxin-like domain. The pmoA and pmoC subunits contribute seven and five transmembrane helices, respectively.

The pMMO structure has also been examined by electron microscopy (Balasubramanian and Rosenzweig 2007; Hakemian and Rosenzweig 2007). Cryo electron microscopy (cryo EM) studies of *M. capsulatus* (Bath) pMMO resulted in a 23-Å resolution electron density map into which the crystal structure could be modeled (Hakemian and Rosenzweig 2007). This finding is significant because the cryo EM preparation retains catalytic activity whereas the detergent necessary for pMMO crystallization, undecyl-β-D-maltoside, results in low activity or inactivated enzyme. The fact that these two structures are consistent indicates that the crystal structure represents a functional form of pMMO. A bulge of electron density in the membrane-spanning region of the cryo EM structure is believed to be a "belt" of dodecyl-β-D-maltoside molecules used for solubilization. In addition, holes or voids of electron density in the area between the soluble and membrane-spanning regions might represent a substrate entry path. A second cryo EM study suggests that MDH interacts with the soluble cupredoxin-like domains of the pmoB subunit (Balasubramanian and Rosenzweig 2007). An interaction with MDH is exciting, but further analysis is required to establish the nature of this putative complex.

Metal Content and Spectroscopy

All catalytically active pMMO preparations contain copper (Lieberman and Rosenzweig 2004; Semrau et al. 2010). However, the reported number of copper ions varies, often substantially, and has been quite contentious within the field. Iron is also present in some active pMMO samples. For purified *M. capsulatus* (Bath) pMMO, copper stoichiometries are in the ranges of 12–20 (Chan and coworkers), 8–10 (DiSpirito and coworkers), 2–3 (Rosenzweig and coworkers), and 2 (Dalton and coworkers) per 100 kDa αβγ protomer (Hakemian and Rosenzweig 2007; Lieberman and Rosenzweig 2004; Semrau et al. 2010). Of the 8–10 copper ions measured by DiSpirito and coworkers, only two are thought to be inherent to the protein. The remaining 6–8 copper ions are associated with methanobactin, a copper chelator produced by methanotrophs (Semrau et al. 2010). Taking this into account, the reported copper contents of purified pMMO from *M. capsulatus* (Bath) fall into two groups: low copper (2–3) and high copper (12–20). Although the iron content of purified pMMO from *M. capsulatus* (Bath) is less variable, ranging from 0 to 2 iron ions per 100 kDa αβγ protomer, its origin as either a contaminant or an actual pMMO cofactor is controversial (Hakemian and Rosenzweig 2007; Semrau et al. 2010). Purified *M. trichosporium* OB3b pMMO typically contains two copper ions per 100 kDa αβγ protomer with only trace amounts of iron (Hakemian and Rosenzweig 2007). Electron paramagnetic resonance (EPR) and X-ray absorption (XAS) spectroscopies have been used extensively to probe the pMMO metal centers (Rosenzweig 2008). All preparations of pMMO from *M. capsulatus* (Bath) and *M. trichosporium* OB3b exhibit a type-2 Cu^{II} EPR signal. XAS data indicate the presence of both Cu^{II} and Cu^{I}, and extended X-ray absorption fine structure (EXAFS) data are best fit with a first coordination shell composed of oxygen and/or nitrogen ligands and a second shell Cu–Cu interaction at a short distance of 2.5–2.6 Å.

Metal Centers in the pMMO Crystal Structure

In the *M. capsulatus* (Bath) pMMO structure, each αβγ protomer contains three metal binding sites: a mononuclear and a dinuclear copper site, both within the soluble cupredoxin domains of pmoB, and a site within the pmoC subunit occupied by zinc (Fig. 2) (Balasubramanian and Rosenzweig 2007; Lieberman and Rosenzweig 2005; Rosenzweig 2008). Three highly conserved histidine residues, His 33, His 137, and His 139, coordinate the dinuclear copper center. Residue His 33 is the N-terminal residue of pmoB and is modeled as bidentate to one of the copper ions with both the amino-terminal nitrogen and the side-chain δ nitrogen within coordinating distance. The two copper ions are separated by ~2.6 Å, a distance supported by both the crystallographic analysis and the EXAFS data. This dinuclear copper center is apparently also present in the crystal structure of *M. trichosporium* OB3b pMMO (Rosenzweig 2008). The ligands to the mononuclear copper center are His 48 and His 72, with Gln 404 perhaps interacting with this site through a coordinated solvent molecule (Lieberman and Rosenzweig 2005). Residue His 48 is not conserved and is often replaced with asparagine, as is the case for *M. trichosporium* OB3b pMMO (Hakemian and Rosenzweig 2007). It is therefore not surprising that this location is devoid of metal ions in the *M. trichosporium* OB3b pMMO structure (Rosenzweig 2008). Detailed characterization of both copper centers, including geometric parameters and the possible presence of exogenous ligands, will require significantly higher resolution crystallographic data, however.

The site occupied by zinc is the only metal binding site observed within the transmembrane regions of pMMO. The zinc ion is coordinated by pmoC residues Asp 156, His 160, and His 173, all strictly conserved in known methanotrophs. Residue Glu 195 from pmoA may also be present in the coordination sphere, but the electron density assigned to this residue is poorly defined. This site is not believed to contain a zinc ion in vivo, but instead results from the use of 200 mM zinc acetate for crystallization. Purified pMMO contains less than 0.2 Zn ions per monomer (Lieberman and Rosenzweig 2005). In the *M. trichosporium* OB3b pMMO structure, which was determined without the addition of exogenous zinc, there is evidence for copper at this location (Rosenzweig 2008). Although the ligands to this crystallographically observed metal site are reminiscent of the sMMO diiron active site, there is no crystallographic evidence for iron in the pMMO structure (Lieberman and Rosenzweig 2005). It remains unclear whether metal binding at this site is physiologically or functionally relevant. Adjacent to this site lies a patch of hydrophilic residues, including Glu 154 from pmoC and His 38, Met 42, Asp 47, Asp 49, and Glu 100 from pmoA. These residues are highly conserved and are notable because they generate a hydrophilic region within the transmembrane domain.

Active Site Models

Beyond the general belief that the pMMO is a metalloenzyme, there has been no consensus

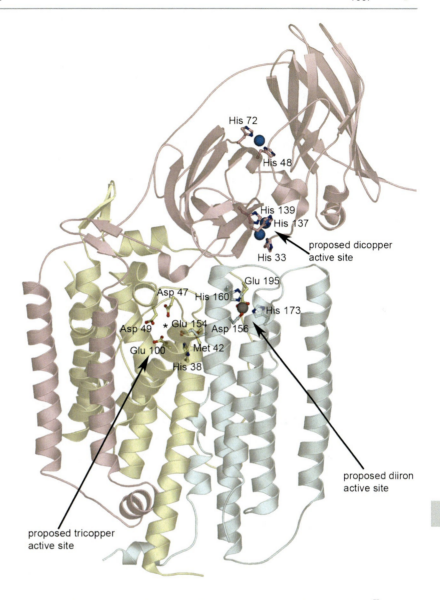

Particulate Methane Monooxygenase, Fig. 2 One αβγ protomer of pMMO (PDB accession code 1YEW). The individual subunits are shown in *pink* (pmoB), *yellow* (pmoA), and *cyan* (pmoC). The copper ions in the soluble regions of pmoB are shown as *blue* spheres. The intramembrane zinc ion is shown as a *gray* sphere. Ligands to the copper and zinc sites are shown as sticks and labeled. Residues comprising the intramembrane hydrophilic patch are also shown as sticks, and the location of the proposed tricopper center is indicated with an *asterisk*

regarding the nature and location of the pMMO active site. Three distinct models have been put forth on the basis of biochemical, spectroscopic, and crystallographic data. In one model, Chan and coworkers have proposed that 5–7 tricopper clusters (>15 copper ions total) function in either catalysis (C-clusters) or electron transfer (E-clusters) (Chan and Yu 2008). The C-clusters are suggested to activate dioxygen and facilitate the hydroxylation of methane, and the E-clusters are proposed to shuttle reducing equivalents from NADH to the C-clusters. Evidence for these tricopper clusters derives from an isotropic EPR signal at a g-value of approximately 2.1, which is interpreted as a ferromagnetically coupled $Cu^{II}Cu^{II}Cu^{II}$ cluster and can be distinguished from the type-2 Cu^{II} signal using redox potentiometry and EPR. Chan and coworkers modeled one of these tricopper clusters into the hydrophilic patch of conserved residues within the pmoA subunit (Fig. 2). In this computational model, the three copper ions are bridged by an oxygen atom with Glu 154 (from pmoC) and His 38 coordinating Cu1, Asp 47 and Met 42 coordinating Cu2, and Asp 49 and Glu 100 coordinating Cu3 (Chan and Yu 2008; Rosenzweig 2008). The remainder of the >15 copper ions are proposed to bind cooperatively with micromolar affinity to the C-terminal cupredoxin domain of pmoB. This domain was termed the "copper sponge" and suggested to be stabilized by binding of at least 10 Cu^I ions. Several aspects of the Chan model are not consistent with the pMMO literature. In particular, the EPR signal attributed to the tricopper center has not been detected by other researchers, and the crystal structures show no bound metal ions either in the intramembrane hydrophilic patch or in the C-terminal cupredoxin domain of pmoB.

The second model, proposed by DiSpirito and coworkers, implicates iron in pMMO activity. Preparations from these researchers typically contain two copper ions and two iron ions per pMMO protomer (Lieberman and Rosenzweig 2004; Semrau et al. 2010). Most important to this model is the finding that purified pMMO from *M. capsulatus* (Bath) exhibits a Mössbauer spectroscopic signal with parameters similar to those attributed to the sMMO diferric active site (Rosenzweig 2008; Semrau et al. 2010). The concentration of the species giving rise to this spectrum accounts for approximately 10% of the total iron in the purified sample, but appears to correlate with enzyme activity since purified pMMO has, at best, only about 10% of whole cell activity. These data suggest that the active site of pMMO is a diiron center, which is proposed to reside at the site occupied by zinc in the *M. capsulatus* (Bath) pMMO crystal structure (Fig. 2). In this model, the copper present in purified pMMO functions as an electron donor (Semrau et al. 2010). The putative diiron active site has been controversial for several reasons. First, the Mössbauer signal could result from contaminating sMMO or other diiron-containing proteins such as hemerythrin, although control experiments suggest these proteins are not present. Second, Chan and coworkers have reported pMMO activity for purified preparations containing no detectable iron (Lieberman and Rosenzweig 2004). Third, only zinc and copper have been shown to bind at the proposed location of the diiron active site (Rosenzweig 2008).

The third active site model for pMMO involves methane oxidation at the dicopper center in the pmoB subunit (Balasubramanian and Rosenzweig 2007; Balasubramanian et al. 2010). Upon removal of metal ions from *M. capsulatus* (Bath) membrane-bound pMMO by cyanide treatment, all enzyme activity is abolished, and addition of 2–3 equivalents of copper per 100 kDa pMMO protomer can restore 90% of methane oxidation activity and 70% of propylene epoxidation activity. EXAFS data indicate that these reconstituted pMMO membranes contain a copper cluster similar to that detected in purified pMMO. Addition of iron does not restore or improve pMMO activity. Taken together, these data indicate that copper, not iron, is present in the active site and that more than 2–3 copper ions are not needed for activity, and in fact, inhibit activity (Balasubramanian et al. 2010).

In another set of experiments, soluble fragments of the *M. capsulatus* (Bath) pmoB subunit (called spmoB) lacking the two transmembrane helices, but maintaining the ligands to the mononuclear and dinuclear copper centers, were expressed in *E. coli*. Copper-loaded spmoB samples bind ~3 Cu as predicted by the *M. capsulatus* (Bath) crystal structure. Most importantly, copper-loaded spmoB contains a copper cluster detected by EXAFS and exhibits methane oxidation and propylene epoxidation activities. Mutations in spmoB that disrupt the mononuclear copper center decrease enzyme activity whereas alterations of the dinuclear copper center ligands abolish activity (Balasubramanian et al. 2010). These data are therefore most compatible with the active site being located at the crystallographic dicopper center. This hypothesis is also consistent with recent work identifying an oxodicopper^{II} species as the active site in a Cu-ZSM-5 zeolite that oxidizes methane (Himes and Karlin 2009). The dicopper active site model is not without issues, however. The ligands to the proposed dicopper center (His 33, His 137, His 139 in *M. capsulatus* (Bath) pMMO), while strictly conserved among pMMOs from *Proteobacteria*, are not present in the putative pMMO enzymes from *Verrucomicrobia* (Semrau et al. 2010). Also, the protocol to remove metals from membrane-bound pMMO is more selective for copper, leaving open the possibility that the small amount of the remaining metal associated with pMMO might somehow contribute to catalytic activity. Finally, the spmoB protein, while able to oxidize methane and propylene, has significantly less activity than the native or copper-reconstituted pMMO (Balasubramanian et al. 2010).

Future Directions

As detailed above, the proposed pMMO active site models attempt to explain biochemical and spectroscopic data in the context of the overall structure of pMMO (Rosenzweig 2008). The varying results from different researchers highlight the difficulty working with integral membrane metalloenzymes and have precluded strong consensus within the field. Future work should focus on high-resolution structural characterization of the pMMO metal centers, development of a mutagenesis system for native pMMO, generation of more robust catalytic fragments, and identification of the methane and methanol binding sites. These pursuits will solidify the identification of the active site and set the stage for probing the catalytic mechanism in detail.

Cross-References

▶ Ascorbate Oxidase
▶ Biological Copper Transport
▶ Catechol Oxidase and Tyrosinase
▶ Copper-Binding Proteins
▶ Copper, Mononuclear Monooxygenases
▶ Copper, Physical and Chemical Properties
▶ Cytochrome c Oxidase, CuA Center
▶ Formate Dehydrogenase
▶ Heme Proteins, Cytochrome c Oxidase
▶ Laccases
▶ Nitrous Oxide Reductase
▶ Ribonucleotide Reductase

References

Balasubramanian R, Rosenzweig AC (2007) Structural and mechanistic insights into methane oxidation by particulate methane monooxygenase. Acc Chem Res 40:573–580
Balasubramanian R, Smith SM et al (2010) Oxidation of methane by a biological dicopper centre. Nature 465:115–119
Chan SI, Yu SSF (2008) Controlled oxidation of hydrocarbons by the membrane-bound methane monooxygenase: the case for a tricopper cluster. Acc Chem Res 41:969–979
Hakemian AS, Rosenzweig AC (2007) The biochemistry of methane oxidation. Annu Rev Biochem 76:223–241
Himes RA, Karlin KD (2009) Copper-dioxygen complex mediated C-H bond oxygenation: relevance for particulate methane monooxygenase (pMMO). Curr Opin Chem Biol 13:119–131
Lieberman RL, Rosenzweig AC (2004) Biological methane oxidation: regulation, biochemistry, and active site structure of particulate methane monooxygenase. Crit Rev Biochem Mol Biol 39:147–164
Lieberman RL, Rosenzweig AC (2005) Crystal structure of a membrane-bound metalloenzyme that catalyses the biological oxidation of methane. Nature 434:177–182
Murrell JC, McDonald IR et al (2000) Regulation of expression of methane monooxygenases by copper ions. Trends Microbiol 8:221–225
Rosenzweig AC (2008) The metal centres of particulate methane monooxygenase. Biochem Soc Trans 36:1134–1137
Semrau JD, DiSpirito AA et al (2010) Methanotrophs and copper. FEMS Microbiol Lett 34:496–531

Parvalbumin

Michael T. Henzl
Department of Biochemistry, University of Missouri, Columbia, MO, USA

Synonyms

Avian thymic hormone; Muscle calcium-binding protein; Oncomodulin

Definition

Parvalbumins are small calcium-binding proteins, members of the EF-hand protein family, that are expressed in vertebrates. Typically, their two divalent ion-binding sites exhibit high affinity for Ca^{2+} and physiologically relevant affinity for Mg^{2+}. Particularly abundant in certain skeletal muscle fibers and neurons, parvalbumins function primarily as mobile cytosolic Ca^{2+} buffers. They influence the duration of intracellular Ca^{2+} signals and play key roles in muscle relaxation following contraction and neuronal recovery following excitation. Select isoforms are believed to have additional specialized functions.

Introduction

Parvalbumins are small, vertebrate-specific EF-hand proteins (Berchtold 1996). Henrotte isolated and crystallized the protein from carp skeletal muscle extracts in 1952. Although they were observed soon thereafter in other fish and in amphibians, their presence in reptiles, birds, and mammals was not reported until 1974. The name "parvalbumin" ("parvus" = small; "albumin" = soluble protein) was suggested by Pechère, based on diminutive size and high solubility. Ca^{2+}-binding activity was reported in 1971, Mg^{2+}-binding activity in 1977.

In 1982, Heizmann detected parvalbumin in specific rat neurons. Abundance in muscle and neurons, coupled with Ca^{2+}-binding capability, suggested a role as a relaxation factor. Subsequent detection in other tissues has fostered the more general view of parvalbumin as a Ca^{2+} buffer.

Parvalbumins are prolate ellipsoids. The crystal structure of carp parvalbumin revealed the structural element responsible for the Ca^{2+}-binding activity – the "EF hand" – and provided a blueprint for interpreting other EF-hand protein sequences. The molecule includes two tightly associated structural domains – a 70-residue metal ion-binding module harboring two EF-hand motifs and a 40-residue N-terminal domain.

Involvement in muscle and neuronal function has made parvalbumin the focus of myriad physiological studies. It is also a popular subject for physical studies because it offers a tractable model system for studying protein behavior.

Genetics

In common with other EF-hand proteins, parvalbumins are believed to have evolved from a primordial 40-residue precursor consisting of a metal ion-binding loop and flanking helices (Kretsinger 1980). Gene duplication events ultimately produced an ancestral four-domain protein, the putative forerunner of calmodulin and troponin C. Deletion of domain 1 from that ancestral protein produced the parvalbumin precursor, which subsequently suffered deletions in domain 2 that abolished Ca^{2+}-binding activity.

Sublineages. The parvalbumin family includes α and β sublineages – distinguished by isoelectric point, sequence length, and several lineage-specific sequence characteristics. pI values are typically <5.0 for β-parvalbumins and >5.0 for α isoforms. However, the frog α isoform has an isoelectric point of 4.88.

The α C-terminal helices are generally one residue longer than β C-terminal helices.

Whereas α isoforms typically contain 109 residues, β-parvalbumins contain just 108. However, pike α- (pI 5.0) parvalbumin contains 108 residues, rather than 109, the result of a single-residue deletion in the N-terminal region. And cod parvalbumin, a β isoform on the basis of pI (4.75) and sequence, includes 113 residues.

Whereas β-parvalbumins harbor cysteine at consensus position 18, α isoforms harbor phenylalanine or valine. Residue 85 is leucine in the α lineage and phenylalanine (with rare exception) in the β lineage.

Parvalbumin isoform number decreases as one ascends the evolutionary ladder. Fish and amphibians express up to seven, commonly between two and five. Reptiles and birds express three – one α and two βs. Mammals express just two – one α and one β. Generally identified by species and isoelectric point, several isoforms have acquired names. The avian pI 4.3 β isoform is called avian thymic hormone (ATH); the avian pI 4.6 β isoform is called parvalbumin 3 (PV3); and the mammalian β isoform is called "oncomodulin."

Chromosomal locations. The mouse α-parvalbumin gene resides on chromosome 15 (5 E); the rat α gene is on chromosome 7 (7q34); and the human gene resides on chromosome 22 (22q12-q13.1). The mouse β-parvalbumin gene resides on chromosome 5 (5 G1-5 G3), and the rat β gene is located on chromosome 12 (12p11). In human, two β loci have been identified on chromosome 7, at 7p22.1 and 7q21.2. The putative gene products are denoted oncomodulin (OCM) and oncomodulin 2 (OCM2). The translated coding sequences exhibit just two differences. Whereas OCM harbors arginine and asparagine at residues 19 and 41, respectively, OCM2 harbors glutamine and serine. One could argue that OCM and OCM2 constitute distinct isoforms and that mammals express three parvalbumins. However, both gene products share the sequence eccentricities that distinguish the mammalian β-parvalbumin from other parvalbumins and are likely functionally equivalent.

Gene structure. All mammalian parvalbumin genes include five exons, with the coding sequence contained in exons II through IV (Pauls et al. 1996). The gene for the rat β-parvalbumin is noteworthy for insertion of a viral LTR immediately upstream from the transcription start site. This promoter element may explain why the protein is expressed in tumors and transformed cell lines from rat, but not other species.

Regulation. Parvalbumin expression is regulated by estrogen via the ERβ receptor. In a murine ERβ knockout, parvalbumin levels are increased in males and decreased in females. ERβ colocalizes with parvalbumin-positive neurons in the cortex, amygdala, basal forebrain, and hippocampus.

Electrical activity and light can influence parvalbumin expression. Chronic low-frequency stimulation of fast-twitch muscles or denervation significantly lowers parvalbumin levels, implying that parvalbumin expression is under the control of fast-type motor neuron activity. Circadian oscillations in PV expression have been observed in AII amacrine neurons from rat retina, as well as in the ciliated ependymal layer of the third ventricle at the level of the suprachiasmatic nuclei. In the sensory cells of the mammalian auditory organ, or organ of Corti (OC), expression of the α and β isoforms is developmentally regulated and is believed to be influenced by efferent innervation.

Structure

Primary structure. Sequence length ranges from 106 to 113 residues, commonly 108 or 109. With few exceptions, parvalbumins are devoid of tryptophan, and many lack tyrosine. In the absence of tryptophan, the UV absorbance spectra display the characteristic vibronic transitions of phenylalanine, with features at 267 (sh), 262, 258, 255, and 250 nm. Parvalbumins are also characterized by low cysteine, methionine, proline, histidine, and arginine contents.

The parvalbumin sequence includes 24 invariant residues. Ten are associated with the hydrophobic core, nine with the metal ion-binding loops. Of the latter, eight are aspartyl or glutamyl residues, involved in metal ion coordination, whereas G98 is required for steric reasons. Three of the remaining invariant residues are solvent accessible. Finally, R75 and E81 form a structurally important hydrogen-bonded network.

The EF-hand motifs, 30–31 residues in length, consist of a central metal ion-binding loop and flanking amphipathic helices. Scheme 1 displays the numbering scheme suggested by Kretsinger. In the binding loop, the coordinating groups are positioned at the approximate vertices of an octahedron. Proceeding from N to C, the ligating oxygen atoms appear in the order x, y, z, $-y$, $-x$, and $-z$. Within the helical elements, nonpolar side chains occupy the third,

Parvalbumin, Scheme 1 EF-hand numbering scheme

sixth, and tenth positions. Note that the exiting, or C-terminal, helix begins within the binding loop.

Secondary structure. The far-UV circular dichroism (CD) spectra display pronounced minima at 222 and 208 nm, characteristic of helical proteins. At 25°C and physiological ionic strength, removal of divalent ions has little impact on the CD spectrum, implying retention of helical structure. β-structure is absent, excepting a short antiparallel segment linking the two metal ion-binding loops. Extended loops link the six helical segments, and the N-terminus also adopts a nonregular secondary structure.

Tertiary structure. The parvalbumin structure includes six helical segments (A-F), organized into two domains – the N-terminal AB domain and the metal ion-binding domain (CD-EF domain). The AB domain, ≈40 residues, includes the A and B helices and an intervening loop. The metal ion-binding domain harbors two EF-hand motifs, named the CD and EF sites after the helical elements flanking the binding loop. The CD and EF binding loops span residues 51–62 and 90–101, respectively. They are related by a twofold symmetry axis. In Ca^{2+} sensor proteins – notably calmodulin – Ca^{2+} binding provokes solvent exposure of apolar surface, which can associate with biological targets. In parvalbumin, the corresponding residues are occluded by the AB domain, preventing calmodulin-like interactions.

The PDB includes over 30 parvalbumin entries. The majority are Ca^{2+}-bound structures, solved by X-ray crystallography. However, the carp (pI 4.25) parvalbumin structure has been determined with Cd^{2+} present in the CD and EF binding loops (PDB 1CDP), and the pike β (pI 4.10)-parvalbumin has likewise been crystallized with Mn^{2+} bound at both sites (2PAL). Regardless of lineage or metal ion-binding properties, the Ca^{2+}-bound structures are remarkably similar.

Ca^{2+}-bound parvalbumin. The CD and EF sites exhibit the Ca^{2+} coordination geometry emblematic of the EF-hand family. The ligating groups are arrayed in a pseudooctahedral manner fashion about

the bound Ca^{2+}. Although coordination is monodentate at the $+x$, $+y$, $+z$, $-y$, and $-x$ positions, the invariant glutamyl side chain at $-z$ is a bidentate ligand. Thus, the Ca^{2+} coordination symmetry is pentagonal bipyramidal. As in all classic EF-hand motifs, the $-y$ ligand is a main-chain carbonyl. Sequence variations are confined to the $+x$, $+y$, $+z$, and $-x$ positions.

In the CD loop, aspartyl groups reside at $+x$ and $+y$; glutamate resides at $-x$ (except in the mammalian β isoform, which employs aspartate); and a seryl hydroxyl is the $+z$ ligand. The glutamate at $-x$ is unique to the parvalbumins. In other EF-hand proteins, including rat β, water serves as the proximal ligand at $-x$. Incorporation of serine at $+z$ is likewise restricted to the parvalbumin family. This sequence characteristic is also nearly invariant – the cod β isoform being a rare exception, in which glutamate replaces serine.

In the EF loop, aspartyl residues contribute the $+x$, $+y$, and $+z$ ligands. For steric reasons – a consequence of the glutamate at $-x$ in the CD site – glycine invariably resides at the $-x$ position in the EF site, and a water molecule serves as the actual ligand. Interestingly, although site 1 in calmodulin shares the same ligand array as the EF site in parvalbumin, its affinity for Ca^{2+} is more than three orders of magnitude lower.

Very high-resolution structures have been reported for the rat α and pike β (pI 4.10) isoforms (PDB codes 1RWY and 2PVB, respectively). Despite their different lineages, they exhibit virtually identical average Ca^{2+}-ligand bond lengths (Table 1). The average bond distances and ligand-Ca^{2+}-ligand bond angles (Table 2) are nearly identical in the two sites.

Although 40 residues apart in the primary structure, the CD and EF loops are in close proximity, the bound Ca^{2+} ions separated by just 12 Å. In common with all paired EF-hand motifs, the CD and EF loops are hydrogen-bonded in the bound state, although not necessarily in the apoprotein. The main-chain carbonyl and amide nitrogen from residue 18 in each motif (usually residues 58 and 97) form a short fragment of antiparallel β structure.

Mg^{2+}-bound parvalbumin. There are presently no structures in the PDB for an isoform with Mg^{2+} bound at both the CD and EF sites. However, pike β (pI 4.10)-parvalbumin has been crystallized with Ca^{2+} in the CD site and Mg^{2+} in the EF site (PDB 4PAL). Replacement of Ca^{2+} with Mg^{2+} contracts the EF-site coordination sphere, the average metal-oxygen distance shrinking by 0.25 Å. Significantly, E101 coordinates Mg^{2+} in a monodentate manner, rather than bidentate, producing octahedral coordination geometry.

Divalent ion-free parvalbumin. The similarity of the Ca^{2+}-bound parvalbumins prompted speculation that differences in divalent ion affinity might be reflected in the apoprotein structures. Isoforms that undergo significant structural rearrangement upon metal ion binding, it was suggested, should exhibit reduced affinity, because the energetic cost of the conformational change would be subtracted from the intrinsic binding energy. The solution structures of the Ca^{2+}-free rat α (high-affinity) and rat β (low-affinity) are consistent with this hypothesis.

Parvalbumin, Table 1 Representative Ca^{2+}-ligand bond distances

	CD site		EF site	
Position	Coordinating atom	Distance (Å)	Coordinating atom	Distance (Å)
$+x$	Asp-51-OD1	2.27	Asp-90-OD1	2.32
$+y$	Asp-53-OD1	2.32	Asp-92-OD1	2.35
$+z$	Ser-55-OG	2.53	Asp-94-OD1	2.34
$-y$	Phe-57-O	2.33	Lys-96-O	2.36
$-x$	Glu-59-OE1	2.36	HOH-O	2.40
$-z$	Glu-62-OE1	2.42	Glu-101-OE1	2.44
$-z$	Glu-62-OE2	2.52	Glu-101-OE2	2.53
	Average	2.39	Average	2.39

Parvalbumin, Table 2 Ligand-Ca^{2+}-ligand angles[a]

Ligands	Angle (°)
CD site	
D51-OD1 ($+x$) – Ca^{2+} – E59-OE1 ($-x$)	166.1
D53-OD1 ($+y$) – Ca^{2+} – E62-OE2 ($-z$)	74.0
S55-OG ($+z$) – Ca^{2+} – D53-OD1 ($+y$)	79.1
F57-O ($-y$) – Ca^{2+} – S55-OG ($+z$)	75.6
E62-OE1 ($-z$) – Ca^{2+} – F57-O ($-y$)	80.2
Average equatorial angle	77.2
EF site	
D90-OD1 ($+x$) – Ca^{2+} – HOH-O ($-x$)	161.3
D92-OD1 ($+y$) – Ca^{2+} – E101-OE2 ($-z$)	75.1
D94-OD1 ($+z$) – Ca^{2+} – D92-OD1 ($+y$)	80.9
K96-O ($-y$) – Ca^{2+} – D94-OD1 ($+z$)	79.5
E101-OE1 ($-z$) – Ca^{2+} – K96-O ($-y$)	81.9
Average equatorial angle	79.4

[a]Axial ligands reside at $\pm x$, equatorial at $\pm y$ and $\pm z$

largely solvent accessible, F49 moves into the core. In the Ca^{2+}-free protein, F49 and I50 associate noncovalently with L85 (Fig. 1b). Rearrangement of the core is accompanied by straightening of the D helix, reorientation of the C helix, and remodeling of the AB/CD-EF interface. Whereas the antiparallel β segment joining the CD and EF binding loops remains intact in the Ca^{2+}-free α isoform, the hydrogen bonds are broken in rat β.

Evidently, however, not all Ca^{2+}-linked structural changes reduce binding affinity. In ATH, Ca^{2+} removal causes a major reorientation of the B helix (Fig. 1c), resulting in exposure of a large hydrophobic surface. Nevertheless, ATH retains high affinity for Ca^{2+}. The critical factor may be rearrangement of the hydrophobic core. In contrast to rat β-parvalbumin, Ca^{2+} binding has almost no impact on the ATH internal packing (Schuermann et al. 2010).

Physical Properties

The physical properties of parvalbumin have been studied by many investigators over the years. The structure and physical properties have been reviewed in detail by Permyakov and Kretsinger (2011).

Conformational Stability

Parvalbumin conformational stability is strongly dependent on ligation state. The melting temperatures of Ca^{2+}-bound parvalbumins, determined by differential scanning calorimetry (DSC), can exceed 100°C. Carp (pI 4.25) parvalbumin denatures at 90°C in 0.1 mM Ca^{2+} at pH 7.0, and the rat α and β isoforms denature at 100.2°C and 91.4°C, respectively, in the presence of 1.0 mM Ca^{2+} at pH 8.5. Interestingly, pike α-parvalbumin exhibits two transitions – at 90.4°C and 123.8°C. The first transition is believed to reflect uncoupling of the AB and CD-EF domains and unfolding of the former; the second corresponds to unfolding of the metal ion-binding domain.

Extrapolating DSC data for Ca^{2+}-bound carp (pI 4.25) parvalbumin to 25°C yield an estimated stability of 14.6 ± 1.0 kcal mol^{-1}. Parvalbumin is believed to be a major cause of fish-associated food allergy. The stability of the Ca^{2+}-bound protein undoubtedly contributes to its allergenic potency, rendering it resistant to thermal denaturation and proteolytic digestion.

Parvalbumin, Fig. 1 Stereoviews of superimposed structures of Ca^{2+}-free and Ca^{2+}-bound parvalbumins. (**a**) rat α-parvalbumin in the Ca^{2+}-free (silver, PDB 2JWW) and Ca^{2+}-bound (cyan, PDB 1RWY), and avian thymic hormone. (**b**) Rat β-parvalbumin in the Ca^{2+}-free (silver, PDB 2NLN) and Ca^{2+}-bound (green, PDB 1RRO) forms. Upon Ca^{2+} removal, F66 and F70 withdraw from the protein interior and, conversely, F49 enters the core. F49, F66, and F70 have been colored orange in the Ca^{2+}-free structure for emphasis. (**c**) Avian thymic hormone in the Ca^{2+}-free (silver, PDB 2KYC) and Ca^{2+}-bound (magenta, PDB 2KYF) states. With the exception of helix B, the structures are very similar

The rat α-PV structure is similar in the Ca^{2+}-bound and Ca^{2+}-free states, with significant differences confined to loop regions of the molecule (Fig. 1a). By contrast, Ca^{2+}-free and Ca^{2+}-loaded rat β-parvalbumin (oncomodulin) differs substantially. Ca^{2+} removal triggers reorganization of the hydrophobic core. Whereas F66 and F70 vacate the interior and become

Parvalbumins are less stable in the Ca^{2+}-free state, and because monovalent ions can occupy the EF-hand motifs (vide infra), the apparent stability is sensitive to the identity and concentration of the major solvent cation. For example, at low ionic strength (1.0 mM NaP_i, 1.0 mM EDTA), rat β-parvalbumin unfolds at 41.1°C. Addition of either Na^+ or K^+ significantly stabilizes the protein. Thus, in 0.20 M NaCl, 0.005 M NaP_i, 0.010 M EDTA, pH 7.4, the observed melting temperature (T_m) increases to 52.9°C. If the NaCl is replaced by KCl, the apparent T_m is comparable, albeit somewhat lower, at 50.4°C.

Divalent ion-free rat α-parvalbumin is even less stable than rat β, unfolding at 37.7°C in 2.0 mM NaP_i, 5.0 mM EDTA. Na^+ stabilizes the protein. For example, the T_m increases to 46.0°C in 0.20 M NaCl, 0.010 M EDTA, pH 7.4. However, K^+ has little influence on stability – for example, the T_m is just 35.5°C in 0.20 M KCl, 0.005 M KP_i, 0.01 M EDTA. Although both rat isoforms remain folded at low ionic strength, the pike pI 4.2 and pI 5.0 isoforms are disordered.

Urea denaturation data for Ca^{2+}-free rat β-parvalbumin in buffered saline at 25°C yield a conformational stability of 4.0 ± 0.1 kcal mol^{-1}. CPV3 exhibits a corresponding stability of 4.8 ± 0.1 kcal mol^{-1}.

Mg^{2+}-bound parvalbumins display intermediate stability. For example, in 20 mM Mg^{2+}, rat β-parvalbumin unfolds at 68.5°C, and pike α-parvalbumin unfolds at 77.2°C in 1.0 mM Mg^{2+}.

Hydrodynamic Properties

Parvalbumins are monomeric, even at millimolar concentrations. In saline at 20°C, rat α-parvalbumin displays an apparent sedimentation coefficient of 1.41 ± 0.01 in both the Ca^{2+}-bound (1.0 mM Ca^{2+}) and Ca^{2+}-free (5.0 mM EDTA) forms. The apparent S value for the Ca^{2+}-bound rat β isoform is slightly larger, 1.60 ± 0.01 S, and decreases perceptibly, to 1.56 S, in the absence of Ca^{2+}. Ca^{2+}-bound ATH sediments with an S value of 1.45 ± 0.01 S at 20°C. The Ca^{2+}-free protein exhibits a somewhat lower S value, 1.35 ± 0.01 S, a reflection perhaps of the aforementioned helical reorientation that accompanies Ca^{2+} removal.

The rotational correlation times (τ_c), measured at 20°C, for Ca^{2+}-free and Ca^{2+}-bound CPV3, are 6.92 (0.04) and 6.78 (0.02) ns, respectively. The Ca^{2+}-free forms of ATH and rat β-parvalbumin exhibit τ_c values of 6.68 (0.04) and 6.96 ns at the same temperature. The corresponding value for the rat α isoform is 6.49 (0.03) at 25°C.

Backbone Dynamics

NMR ^{15}N relaxation studies suggest that the parvalbumin structures are rigid in both Ca^{2+}-free and Ca^{2+}-bound states. The order parameter for the N–H bond, an indication of the mobility of the polypeptide chain, can assume values between 0 (least ordered) and 1.0 (most ordered). CPV3 displays an average order parameter of 0.95 in both Ca^{2+}-free and Ca^{2+}-bound states at 20°C. Under the same conditions, Ca^{2+}-free ATH displays the identical value. The rat α and β isoforms both exhibit average order parameters of 0.92 at 20°C and 25°C, respectively. The Ca^{2+}-bound proteins exhibit corresponding values of 0.85 and 0.84 at 32°C and 37°C, respectively.

Metal Ion Affinities

The parvalbumin CD and EF sites are typically "Ca^{2+}/Mg^{2+}" sites –with association constants for Ca^{2+} and Mg^{2+} in excess of $10^7 M^{-1}$ and $10^4 M^{-1}$, respectively. The affinity for Mg^{2+} is physiologically relevant. Assuming an intracellular Mg^{2+} concentration of ≈1 mM, the parvalbumin sites will be largely occupied by Mg^{2+} at "resting-state" cytosolic Ca^{2+} concentrations ($\leq 10^{-7}$ M). Because Mg^{2+} dissociation is slow (vide infra), Ca^{2+} binding is limited by Mg^{2+} release.

Although widely perceived as interchangeable cytosolic Ca^{2+} buffers, specific parvalbumins exhibit distinctive ion-binding properties (Table 3). The mammalian β-parvalbumin (oncomodulin), the avian β (pI 4.6) isoform, CPV3, and whiting β (pI 4.44)-parvalbumin are notable examples. The physical basis for the disparate behavior is unclear, but structural features outside the binding loop are evidently significant. The rat β isoform exhibits atypical residues at six positions within the CD site (49, 50, 57, 58, 59, and 60). However, replacement with the consensus residues does not restore binding affinity unless L85 is concomitantly replaced by phenylalanine (Henzl et al. 2008). This finding relates perhaps to the aforementioned interaction in the Ca^{2+}-free protein involving F49, I50, and L85, which is abolished upon Ca^{2+} binding.

Parvalbumin, Table 3 Representative parvalbumin binding constants[a]

Protein	$k_{1,Ca}$	$k_{2,Ca}$	$k_{1,Mg}$	$k_{2,Mg}$	Method	Conditions
Rat α	1.3×10^8	1.2×10^8	2.0×10^4	1.6×10^4	ITC	0.15 M Na$^+$, pH 7.4, 25°C
	1.4×10^8	1.2×10^8	1.0×10^4	7.4×10^3	ITC	0.15 M K$^+$, pH 7.4, 5°C
	1.9×10^9	1.0×10^9	1.7×10^5	4.8×10^4	ITC	0.15 M K$^+$, pH 7.4, 5°C
ATH	2.4×10^8	1.0×10^8	2.2×10^4	1.2×10^4	ITC	0.15 M Na$^+$, pH 7.4, 25°C
Rabbit α	1.5×10^8	1.5×10^8	6.2×10^4	6.2×10^4	Flow dialysis	0.15 M K$^+$, pH 7.6, 22°C
Frog pI 4.88	1.3×10^8	1.3×10^8	4.8×10^4	4.8×10^4	Flow dialysis	0.15 M K$^+$, pH 7.6, 22°C
Frog pI 4.50	4.5×10^8	4.5×10^8	3.7×10^4	3.4×10^4	Flow dialysis	0.15 M K$^+$, pH 7.6, 22°C
Rat β	2.3×10^7	1.5×10^6	9.4×10^3	160	ITC	0.15 M Na$^+$, pH 7.4, 25°C
	3.1×10^7	3.8×10^6	1.4×10^4	280	ITC	0.15 M K$^+$, pH 7.4, 25°C
CPV3	4.5×10^7	2.4×10^7	5.0e4	2.2×10^4	ITC	0.15 M Na$^+$, pH 7.4, 25°C
Whiting pI 4.44	5×10^8	6×10^6	n.d.	n.d.	Fluorescence	0.05 M Tris, pH 7.5, 20°C

[a]Microscopic binding constants in units of M^{-1}. Uncertainties are on the order of ±0.3, except for whiting (±1)

Due to the coulombic nature of the protein-ion interaction, binding parameters are sensitive to pH and ionic strength. The ionic radius of Na$^+$ is comparable to Ca^{2+}, and the monovalent cation competes with the divalent ion for the EF-hand motifs. By contrast, the larger K$^+$ ion does not generally bind to vacant EF-hand motifs. For example, whereas rat α-parvalbumin binds one equivalent of Na$^+$ with an estimated affinity of 630 M^{-1}, it does not bind K$^+$. As a consequence, the average binding constant, measured at pH 7.4, increases from 1.2×10^8 M^{-1} to 1.4×10^9 M^{-1} when 0.15 M NaCl is replaced by 0.15 M KCl. Interestingly, the rat β isoform – which binds nearly two equivalents of Na$^+$ – also binds a single K$^+$ ion. Resultingly, the binding constants are less sensitive to the identity of the solvent cation.

In binding studies performed to date, either the two Ca^{2+}-binding sites behave equivalently, or the first Ca^{2+} is bound with higher affinity than the second. Given that the CD and EF binding loops are linked by hydrogen bonds, the absence of positive cooperativity is counterintuitive. However, the diagnosis of cooperativity is only unambiguous in proteins containing identical binding sites. When two sites differ significantly in affinity, binding can appear noncooperative, even when there is a favorable interaction between them. There is circumstantial evidence for cooperative interactions in parvalbumin. For example, the Ca^{2+} affinities of both sites in rat α-parvalbumin increase nearly tenfold when K$^+$ replaces Na$^+$ as the major solvent cation, although just one site binds Na$^+$.

Ca^{2+} and Mg^{2+} Binding Kinetics

The rate constants for Ca^{2+} dissociation from frog parvalbumin IVb (pI 4.75) were measured by stopped flow spectrophotometry – monitoring the rise in Tb^{3+} luminescence following rapid mixing of Ca^{2+}-loaded parvalbumin and Tb^{3+}. The data were well accommodated by a single exponential. The rate constants measured at 0°C, 10°C, and 20°C were 0.19 (0.01), 0.46 (0.02), and 1.03 (0.03) s^{-1}. When corresponding Mg^{2+} dissociation data were modeled with a single exponential, the apparent rate constants at the three temperatures were 0.93 (0.02), 1.76 (0.07), and 3.42 (0.14). However, better agreement was obtained for the Mg^{2+} data with a double exponential model, suggesting that the kinetics of Mg^{2+} release differ for the CD and EF sites. The resulting pairs of rate constants were 1.22/0.46 s^{-1}, 3.36/1.04 s^{-1}, and 9.41/1.75 s^{-1}. Although faster than Ca^{2+} release, Mg^{2+} dissociation is nevertheless slow, requiring on the order of 100 ms^{-1}. Because the CD and EF sites will be primarily occupied by Mg^{2+} at "resting-state" cytosolic Ca^{2+} concentrations, the rate of Ca^{2+} binding is determined by the slow kinetics of Mg^{2+} release.

Lanthanide Ion Binding

Lanthanide ions substitute readily for Ca^{2+} in EF-hand motifs – typically binding 2–3 orders of magnitude more tightly than Ca^{2+} – and have convenient spectroscopic properties. Because the ionic radius decreases regularly across the series, the ions can be used to probe the size and flexibility of an EF-hand motif. Whereas the EF site of carp (pI 4.25) parvalbumin is flexible – exhibiting virtually identical affinities for large (e.g., Ce^{3+}), intermediate (e.g., Gd^{3+}), and small (e.g., Lu^{3+}) lanthanides – the CD site displays a distinct preference for the larger ions.

Impact of the AB Domain

The AB domain is critical for stability. The isolated CD-EF domains of rat α- and β-parvalbumins are disordered, and their Ca^{2+} affinities are greatly reduced. In buffered saline at pH 7.4 and 25°C, the α CD-EF domain exhibits positively cooperative Ca^{2+} binding, with macroscopic stepwise association constants of 3.7 (0.1) × 10^3 and 8.6 (0.1) × 10^4 M^{-1} – a 10.4 kcal mol^{-1} reduction in standard binding free energy, relative to the intact protein. The apparent Ca^{2+} association constants for the rat β CD-EF domain, measured by equilibrium dialysis, are 4.2 (0.1) × 10^3 and 6.1 (0.1) × 10^3 M^{-1} – an 8.3 kcal mol^{-1} reduction in binding free energy, relative to the intact rat β isoform. The isolated AB and CD-EF domains associate in the presence of Ca^{2+}.

Parvalbumin Function

Parvalbumin is most abundant in skeletal muscle and the CNS. In rat, α-parvalbumin has also been detected in a variety of other tissues – notably kidney, adipose tissue, bone, teeth, reproductive tissues (prostate, testes, seminal vesicles, ovaries), and the inner hair cells (IHCs) of the auditory organ (organ of Corti). Parvalbumin function in muscle and nerve is described below. Although it is presumed to function as a Ca^{2+} buffer in other tissue settings, the details remain to be elucidated.

Muscle

Parvalbumin concentrations can exceed 1 mM in skeletal muscle. Restricted to fast-twitch fibers in mammals, it has been detected in slow-, intermediate-, and fast-twitch fibers in fish (Arif 2009). Expression levels in fish muscle are highest in fast-twitch fibers and lowest in slow-twitch. Within a given fiber type, longitudinal variations in muscle relaxation rate are observed, with parallel alterations in parvalbumin content. In sheepshead, for example, the relaxation rate is significantly higher in anterior slow-twitch muscle than posterior slow-twitch muscle, and parvalbumin levels are correspondingly higher.

Parvalbumin functions as a soluble relaxation factor in muscle (Gillis 1985). Because the Ca^{2+}-binding rate in vivo is limited by Mg^{2+} dissociation, parvalbumin is classified as a "slow" Ca^{2+} buffer. In contrast to calbindin, a "fast" Ca^{2+} buffer, parvalbumin has no effect on the rapid rise in intracellular Ca^{2+} concentration ($[Ca^{2+}]_i$) at the beginning of a Ca^{2+} transient. Rather, its impact is restricted to the subsequent $[Ca^{2+}]_i$ decay. Immediately following the abrupt increase in cytosolic Ca^{2+} at the onset of contraction, the CD and EF sites are occupied by Mg^{2+} and, therefore, unable to bind Ca^{2+}. Thus, Ca^{2+} binds to the regulatory sites in troponin C (TnC), triggering the conformational change in troponin I that facilitates actin-myosin interaction. Subsequently, following Mg^{2+} dissociation, the parvalbumin sites compete Ca^{2+} away from TnC, until it is subsequently resequestered by the sarcoplasmic reticulum (SR). The synthetic Ca^{2+} chelator EDTA mimics the impact of parvalbumin on muscle relaxation.

Consistent with the proposed role, injection of parvalbumin cDNA into rat slow-twitch muscle increases the relaxation rate, without altering contraction kinetics. Conversely, ablation of the murine parvalbumin gene lowers relaxation rates in fast-twitch muscle fibers from $PV^{-/-}$ animals. Thus, parvalbumin elevates the initial $[Ca^{2+}]_i$ decay rate. The difference between wild-type and $PV^{-/-}$ mice should disappear after long stimulation or during long tetanic stimulations because the parvalbumin sites should eventually be saturated with Ca^{2+}. Interestingly, the faster relaxation rate persists for wild-type muscles under these conditions, suggesting that parvalbumin may facilitate Ca^{2+} sequestration by the SR.

Although not expressed in cardiac muscle, parvalbumin improves diastolic function in failing heart muscle. Compromised diastolic function is believed to reflect a diminished rate of Ca^{2+} sequestration following contraction. In rat, adenoviral-mediated in vivo delivery of the parvalbumin gene produced cardiac expression of the protein at levels comparable to those seen in fast-twitch skeletal muscle and significantly improved the heart muscle relaxation rate.

Neurons

Parvalbumin is expressed in a subset of GABAergic interneurons in the CNS. In adult vertebrates, GABA is primarily an inhibitory neurotransmitter – that is, binding to receptors on the postsynaptic neuron decreases the probability of a subsequent action potential in that neuron. In the cortex, parvalbumin is present in approximately 5% of the GABAergic interneurons. The protein is also expressed in several interneuron types in the hippocampus, and it is expressed in Purkinje cells and molecular layer interneurons in the cerebellum. Local parvalbumin concentrations in axon terminals can approach 1 mM. Although most parvalbumin interneurons are fast spiking, not all fast-spiking neurons express parvalbumin.

Ca^{2+} signals in parvalbumin-positive neurons exhibit biexponential decay – the "signature" of parvalbumin expression (Schwaller 2009). The initial decay of $[Ca^{2+}]_i$ is more rapid than in control neurons, but the decrease in $[Ca^{2+}]_i$ is actually slower at later times. Both kinetic components have functional significance.

The impact at short times is revealed by repeated electrical stimulation of a parvalbumin-positive cerebellar interneuron synaptically connected to a Purkinje cell. In wild-type mice, following the second stimulus, the Purkinje cell exhibits a decreased electrical response. This result, termed depression, presumably reflects diminished GABA release from the interneuron. However, in $PV^{-/-}$ animals, the second stimulus produces an increased response by the Purkinje cell. This result, termed facilitation, presumably reflects the higher residual $[Ca^{2+}]_i$ in the presynaptic terminals when the second stimulus is applied.

This difference between wild-type and $PV^{-/-}$ animals is greatest at short interspike intervals (30 ms) when the protein has the largest impact on $[Ca^{2+}]_i$ decay. Thus, the ability of parvalbumin to prevent facilitation is maximal at a frequency within the range of gamma oscillations (30–80 Hz). Gamma oscillations are associated with several brain activities, including memory storage and retrieval. In one study, the maximal amplitude of gamma oscillations recorded in area CA3 was higher in $PV^{-/-}$ mice than in control mice, possibly due to increased facilitation of GABA release.

At long times following stimulation, parvalbumin binding sites are largely occupied by Ca^{2+}, and the protein acts transiently as a Ca^{2+} source. This activity, reflected in the slow $[Ca^{2+}]_i$ decay component, leads to strong, delayed GABA release at interneuron-interneuron synapses subsequent to the action potential–stimulated burst of neurotransmitter release. The impact of parvalbumin throughout the entire Ca^{2+} signal suggests that the protein may modulate synaptic plasticity – that is, the strength of the connection between neurons.

Parvalbumin-positive neurons are believed to stabilize neuronal networks. Although $PV^{-/-}$ mice do not display spontaneous epileptic activity, chemically induced seizures are more severe than in wild-type mice. $PV^{-/-}$ mice also display mild impairment of motor coordination and motor learning. There is intense interest in parvalbumin as a marker for various CNS pathologies – including Alzheimer's, Huntington's, Parkinson's, epilepsy, schizophrenia, bipolar disorder, and depression.

Kidney

In rat, α-parvalbumin is expressed in the distal convoluted tubule and proximal collecting duct. In mouse, the protein is detected in the early distal convoluted tubule, which plays an essential role in NaCl reabsorption. It colocalizes with NCC, the thiazide-sensitive Na+/Cl − cotransporter. In $PV^{-/-}$ kidney, expression of NCC is downregulated. Conversely, overexpression in murine DCT cells in vitro induces endogenous NCC expression. These observations suggest that parvalbumin may be a Ca^{2+}-dependent modulator of NCC levels. Colocalization is also observed between parvalbumin and TRPM6, the channel involved in renal Mg^{2+} reabsorption.

Mammalian β-Parvalbumin

Detected in rat tumor tissue in 1979 (MacManus and Whitfield 1983), a normal expression site for the mammalian β isoform went unrecognized for 16 years. Then, in 1995, the protein was detected in the guinea pig OC. Subsequent work revealed that expression is restricted to the outer hair cell (OHC) population, where levels can reach 0.5 mM. The function of the protein in the OHCs, believed to amplify acoustic signals, remains unclear. The alternative name for the protein, "oncomodulin," reflects its discovery in rat tumor tissue and an apparent calmodulin-like capacity to stimulate cAMP phosphodiesterase.

Avian β-Parvalbumins

In birds, as in mammals, the α isoform is expressed in skeletal muscle and neurons. Both β isoforms are expressed in the thymic cortex. In addition, avian thymic hormone is detected in chick retina, choroid, sclera, and extraocular muscle. Expression increases in the choroid during recovery from induced myopia. ATH may participate in contraction/relaxation of extravascular smooth muscle in the choroidal stroma or modulate synaptic transmission in choroidal neurons.

The sequences of CPV3 and rat β-parvalbumin are identical at 74 of 108 positions, implying a common precursor. Interestingly, just as oncomodulin is expressed in the OHCs of the OC, CPV3 is expressed in the sensory cells – that is, hair cells – of the avian auditory organ (basilar papilla).

Extracellular Functions of PV

Several parvalbumins exhibit extracellular functions. The avian β-parvalbumins are endocrine factors in the avian immune system. In fact, ATH was purified from chicken thymus on the basis of its capacity to stimulate T cell differentiation. It circulates in the blood on a 5-day cycle and stimulates development of cell-mediated immunity. Receptors for CPV3 have likewise been detected in peripheral lymphoid tissue.

The mammalian β isoform (oncomodulin) is also expressed by macrophages. Upon activation, they secrete the protein into the extracellular medium, where it displays potent nerve growth factor activity. Oncomodulin binds to retinal ganglion cells in a cAMP-dependent manner. Following optic nerve injury, treatment with oncomodulin, along with a nonhydrolyzable cAMP analog and zymosan, stimulates far more extensive axonal outgrowth from the site of injury than other known nerve growth factors.

Cross-References

▶ Calcium in Biological Systems
▶ Calcium in Health and Disease
▶ Calcium in Heart Function and Diseases
▶ Calcium in Nervous System
▶ Calcium-Binding Protein Site Types
▶ Calcium-Binding Proteins
▶ Calcium-Binding Proteins, Overview
▶ EF-Hand Proteins
▶ Lanthanide Ions as Luminescent Probes
▶ Lanthanide Metalloproteins
▶ Magnesium Binding Sites in Proteins

References

Arif SH (2009) A Ca^{2+}-binding protein with numerous roles and uses: parvalbumin in molecular biology and physiology. Bioessays 31:410–421

Berchtold MW (1996) Parvalbumin. In: Celio MR, Pauls T, Schwaller B (eds) Guidebook to the calcium-binding proteins. Oxford University Press, New York

Gillis JM (1985) Relaxation of vertebrate skeletal muscle. A synthesis of the biochemical and physiological approaches. Biochim Biophys Acta 811:97–145

Henzl MT, Davis ME, Tan A (2008) Leucine-85 is an important determinant of divalent ion affinity in Rat β-Parvalbumin (oncomodulin). Biochemistry 47:13635–13646

Kretsinger RH (1980) Structure and evolution of calcium-modulated proteins. CRC Crit Rev Biochem 8:119–174

MacManus JP, Whitfield JF (1983) Oncomodulin: a calcium-binding protein from hepatoma. Calcium Cell Funct 4:411–440

Pauls TL, Cox JA, Berchtold MW (1996) The Ca^{2+}-binding proteins parvalbumin and oncomodulin and their genes: New structural and functional findings. Biochim Biophys Acta 1306:39–54

Permyakov EA, Kretsinger RH (2011) Calcium binding proteins. Wiley, Hoboken, pp 221–235

Schwaller B (2009) The continuing disappearance of "pure" Ca^{2+} buffers. Cell Mol Life Sci 66:275–300

Schuermann JP, Tan A, Tanner JJ, Henzl MT (2010) Structure of avian thymic hormone, a high-affinity avian beta-parvalbumin, in the Ca^{2+}-free and Ca^{2+}-bound states. J Mol Biol 397:991–1002

Patch Test

▶ Chromium and Allergic Reponses

Pb in Plants

▶ Lead in Plants

Pb(II)-RNA Coordination

▶ Lead and RNA

Pb^{2+} Excretion/Efflux

▶ Lead Transporters

Pb^{2+} Resistance

▶ Lead Transporters

Pb^{2+} Uptake

▶ Lead Transporters

p-Diphenol Oxidase

▶ Laccases

PEF

▶ Penta-EF-Hand Calcium-Binding Proteins

Penta-EF-Hand Calcium-Binding Proteins

Masatoshi Maki
Department of Applied Molecular Biosciences,
Graduate School of Bioagricultural Sciences,
Nagoya University, Nagoya, Japan

Synonyms

Five-EF-hand calcium-binding proteins; PEF

Definition

Proteins that possess a domain of five serially repetitive EF-hand-like motifs.

General Features

Eukaryotes have a group of proteins that possess a domain of five serially repeated helix-loop-helix Ca^{2+}-binding motifs (EF-hands). The domain of approximately 170 amino acid residues encompassing the five EF-hand motifs is named a penta-EF-hand (abbreviated PEF) domain (Maki et al. 2002). Some EF-hands in the PEF domain lack the ability to bind Ca^{2+} due to substitutions of amino acids essential for metal coordination. Generally, a pair of juxtaposed EF-hands within a protein molecule forms a stable subdomain or so-called lobe in the case of calmodulin that has four EF-hands (N-lobe and C-lobe). X-ray crystal analyses of several PEF domains in different types of mammalian proteins have revealed the following common structures. (1) Eight α-helices instead of ten α-helices are present as schematically shown for human ALG-2 in Fig. 1a. The helices α4 and α7 are shared by neighboring EF-hands (α4, EF2 and EF3; α7, EF4 and EF5). (2) An EF2-EF3-connecting helix (α4) is shorter by seven residues than that of calmodulin (Fig. 1b). (3) Dimerization of the PEF molecules occurs by pairing EF5s of the same PEF (for homodimerization) or a closely related PEF (for heterodimerization) (Fig. 2). An antiparallel β-sheet structure is formed between the calcium-binding loops of the two EF5s (Figs. 1c, 2c); even though this β-sheet is absent in the Ca^{2+}-free form of ALG-2 (Fig. 1a, b). (4) While calpain catalytic subunits have cysteine proteinase domains and C2-like domains, other PEF proteins have various lengths of preceding Gly/Pro-rich sequences containing some hydrophobic residues (human PEF proteins: peflin, 113 amino acids (aa); calpain small subunit 1, 83 aa; grancalcin, 43 aa; sorcin 29 aa; ALG-2, 23 aa). This N-terminal tail is not evolutionarily conserved (*Drosophila* ALG-2, 8 aa; *Caenorhabditis elegans* ALG-2, 6 aa). Plant and lower eukaryotic PEF proteins also have N-terminal tails of biased amino acid compositions.

Evolution of PEF Genes

Homologs of PEF genes have been found in lower animals, plants, fungi, and protists. Neither sorcin nor PEF-containing calpains have been found in fungi, suggesting divergence of the ALG-2/sorcin/calpain clan in the animal kingdom (Maki et al. 2002). Moreover, phylogenetic analyses have indicated a slightly closer relation of the ALG-2/peflin subfamily to the sorcin/grancalcin subfamily than to the calpain subfamily among all human PEF proteins as shown in Fig. 3.

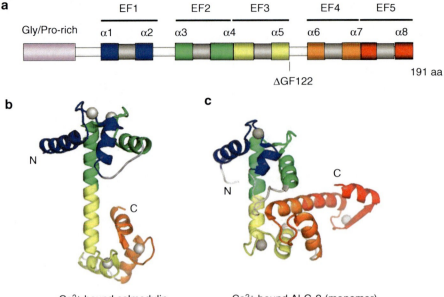

Penta-EF-Hand Calcium-Binding Proteins, Fig. 1 The structure of ALG-2. (**a**) Schematic presentation of the ALG-2 structure. Human ALG-2 has 191 amino acids (aa) and five EF-hands that consist of eight α-helices. An alternatively spliced shorter isoform has a deletion of two residues (121-GlyPhe-122, designated ΔGF122) in the loop connecting EF3 and EF4. X-ray crystal 3D-structures of (**b**) calmodulin and (**c**) ALG-2. Ribbon presentations of Ca^{2+}-bound forms of calmodulin (*left panel*, PDB ID: 1CLL) and the monomer of ALG-2 (*right panel*, PDB ID: 2ZN9). *Gray spheres*, calcium atoms. Structural figures were generated with PyMol (DeLano Scientific LLC, Palo Alto, CA, USA)

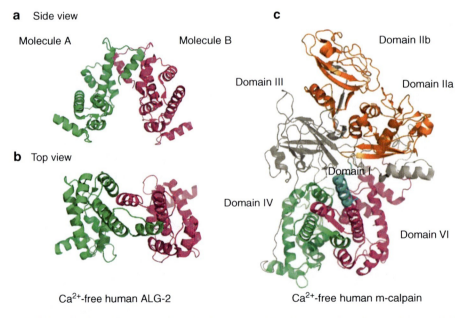

Penta-EF-Hand Calcium-Binding Proteins, Fig. 2 3D-structures of the ALG-2 dimer and *m*-calpain. (**a**, **b**) Ribbon presentations of Ca^{2+}-free forms of the ALG-2 dimer (PDB ID: 2ZND) shown as *side view* (**a**) and *top view* (**b**). Molecule A (*green*), Molecule B (*magenta*). (**c**) Structure of the human *m*-calpain (calpain 2) in Ca^{2+}-free form (PDB ID:1KFU). The PEF domains of the large subunit and the small subunit are represented by *green* and *magenta* ribbons, respectively. Domain II (protease domain, subdivided into IIa and IIb), *orange*; domain III, *dark gray*; N-terminal anchor helix, *cyan*; linker region in domain I, *light gray*

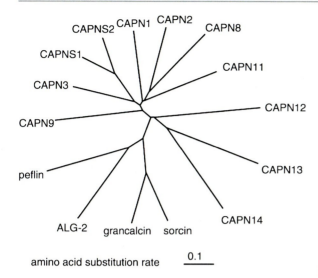

Penta-EF-Hand Calcium-Binding Proteins, Fig. 3 Phylogenetic tree of human PEF proteins. Amino acid sequences of human PEF proteins were aligned by using the Clustal X program, and a bootstrap file was created. The phylogenetic tree was drawn with the tree-viewing program TreeView in radial presentation. Scale bar, amino acid substitution rate

```
                        Ca²⁺-binding loop
                        x  y  z -y -x  -z
ALG-2       FLWNVFQRVDKDRSGVISDTELQQALSNG
peflin      EAYSWFQSVDSDHSGYISMKELKQALVNC
CAPNS1      QFRRLFAQLAGD-DMEVSATELMNILNKV
CAPN1       NFKALFRQLAGE-DMEISVKELRTILNRI
sorcin      PLYGYFAAVAGQ-DGQIDADELQRCLTQS
grancalcin  SVYTYFSAVAGQ-DGEVDAEELQRCLTQS
calmodulin  EFKEAFSLFDKDGDGTITTKELGTVMRSL
```

Penta-EF-Hand Calcium-Binding Proteins, Fig. 4 Alignment of amino acid sequences of EF1s. Amino acid sequences of EF1s from human PEF proteins (ALG-2, peflin, CAPNS1, CAPN1, sorcin, and grancalcin) and calmodulin are aligned. Residues conserved in four or more sequences are highlighted with a *light blue* background. Positions of calcium-coordinating oxygen atoms are marked with x, y, z, -y, -x, and -z. Generally, oxygen atoms at -x and -y come from water and peptide carbonyl oxygen, respectively

PEF proteins, however, can be classified into two groups based on the primary structures of EF1s: Group I (ALG-2, peflin) and Group II (calpain subfamily, sorcin and grancalcin) (Fig. 4). Crystal structures of the PEF domains of rat and pig calpain small subunits revealed that EF1s of Group II proteins have a substitution at the x position from Asp to Ala and lack one residue in the Ca^{2+}-binding loop. Calcium-coordinating oxygen atoms come from the carbonyl oxygen atom of Ala and water at the x and y positions, respectively. A closer evolutionary relation between the sorcin and calpain subfamilies than between sorcin and ALG-2 is supported by the fact that positions of introns are relatively well conserved between sorcin and calpains, but the conservation is limited between ALG-2 and calpains.

ALG-2 Subfamily

ALG-2 was originally identified as one of the factors associated with cell death of mouse T cell hybridoma induced by stimulants such as T-cell receptor cross-linking, Fas, and glucocorticoid, and it was so named "apoptosis-linked gene 2" (Vito et al. 1996). No apparent phenotypes, however, were observed in *ALG-2* gene knockout mice, even in the immune system, raising arguments regarding the direct roles of ALG-2 in pro-apoptotic functions (Jang et al. 2002). Later studies using cultured cells show conflicting results regarding apoptosis. ALG-2 was reported to play roles as a stimulating cofactor during endoplasmic reticulum (ER)-stress-induced apoptosis by treatment of immortalized Apaf-1$^{-/-}$ mouse embryonic fibroblasts with thapsigargin, an ER-calcium-ATPase inhibitor (Rao et al. 2004). On the other hand, knockdown of ALG-2 induced accumulation of HeLa cells in the G_2/M cell cycle phase and an increase in the number of early apoptotic and dead cells (Høj et al. 2009). Despite the fact that a function in cell death has not been clearly shown, the gene name *PDCD6* has been adopted from "programmed cell death 6." *PDCD6*, a chromosomal locus at 5p15.33, is a potential prognostic marker for malignant progression in pulmonary adenocarcinoma with bronchioloalveolar features, and patients with high expression levels of *PDCD6* have low survival rates. In contrast, low expression level of *PDCD6* is correlated with poor overall survival of gastric cancer patients. The cellular apoptotic program is partly driven by cell cycle-controlling factors such as p53 tumor suppressor, and dysregulation of the delicate balance between proliferation and cell death can lead to malignant progression of tumors.

ALG-2 associates with a number of cellular proteins in a Ca^{2+}-dependent manner, such as ALIX, annexin A7, annexin A11, TSG101, phospholipid scramblase 3 (PLSCR3), Sec31A, Scotin, RBM22, ASK1, Raf-1, and mucolipin-1. Except for the last three proteins, most ALG-2-interacting proteins have

Pro-rich regions. Sequence comparison and mutational analyses of ALG-2-binding sites in ALIX, PLSCR3, and Sec31A have revealed the presence of at least two types of ALG-2-binding motifs in the Pro-rich regions. Motif I, PPYPX(2)PGYP (P, Pro; Y, Tyr; X, variable residue), is represented in ALIX and PLSCR3, and Motif II, PXPGFX(7)PXAXG, is represented in Sec31A and PLSCR3. Both motifs are found in the Pro-rich region of PLSCR3 and their ALG-2 binding capacities have been experimentally demonstrated.

Fluorescence photometric analyses have revealed that binding of Ca^{2+} to ALG-2 induces conformational changes to exposed hydrophobic surfaces. Analyses of the crystal structures of Ca^{2+}-free and -bound forms of ALG-2 and the Zn^{2+}-bond form of the complex with an ALIX peptide (Protein data bank, PDB, ID codes: 2ZND, 2ZN9 and 2ZNE, respectively), have demonstrated that binding of Ca^{2+} to ALG-2 enables the side chain of Arg125, present in the loop connecting EF3 and EF4, to move sufficiently to make a primary hydrophobic pocket (Pocket 1), created by the EF3-EF4-connecting loop and EF5, accessible to the critical PPYP (ProProTyrPro) motif (Ca^{2+}/EF3-driven arginine switch model) (Suzuki et al. 2008). The bottom of the hydrophobic pocket is formed by Tyr180 from another molecule of dimerized ALG-2, and substitution of Tyr180 with Ala abolishes the binding and dimerization. The second hydrophobic pocket (Pocket 2), created by residues in EF2, EF3, and Phe122, accommodates the last YP of Motif I in the ALIX peptide. The naturally occurring alternatively spliced shorter isoform lacks two residues, 121-GlyPhe-122, and does not bind ALIX but binds Sec31A and PLSCR3 that contain Motif II (Shibata et al. 2008). In the crystal structure of the Ca^{2+}-bound isoform (PDB ID code 3AAJ), a wall created by the main chain of 121-GlyPheGly-123 facing the two pockets is reconfigured. The side chain of Arg123 (corresponding to Arg125 in the major isoform) is placed to partially block Pocket 1. It is conceivable that Motif I and Motif II bind ALG-2 at different surfaces of ALG-2, but this remains to be clarified. Each monomer of the ALG-2 homodimer has one ALIX-binding site, indicating the presence of two binding sites in the dimer. ALG-2 has the capacity to bridge two different proteins, for example, ALIX and TSG101, and functions as a Ca^{2+}-dependent adaptor protein (Okumura et al. 2009).

Peflin, a 30-kDa protein, is the closest paralog of ALG-2 in mammals (Fig. 4). It has the longest N-terminal tail (111 aa) in the PEF family except for calpain catalytic subunits. The N-terminal tail contains nine repeats of [AP]PGGPYGGP (Ala or Pro in brackets). Peflin forms a heterodimer with ALG-2 via their EF5s and is present in the cytoplasm, whereas ALG-2 occurs both in the cytoplasm and nucleus. ALG-2 does not have a nuclear localization signal, and thus it may translocate to the nucleus either in a passive manner or by binding to its nuclear localizing interacting proteins. Peflin may inhibit nuclear translocation of ALG-2. Addition of Ca^{2+} to the cell lysate causes translocation of peflin to the membrane fraction together with ALG-2. Interacting proteins other than ALG-2 have not been validated yet. The peflin gene seems to have evolved more rapidly than the ALG-2 gene. It is found in *Drosophila* but not in *Caenorhabditis elegans*.

Calpain Subfamily

Calpains are non-lysosomal intracellular cysteine proteinases. Calpain 1 (μ-calpain) and calpain 2 (m-calpain) are ubiquitously expressed in mammalian tissues and are activated by micromolar and millimolar levels of Ca^{2+}, respectively, in vitro protease assays. As shown in Table 1, in the human genome, 14 calpain catalytic subunit genes (*CAPN1, CAPN2, CAPN3, CAPN5, CAPN6, CAPN7, CAPN8, CAPN9, CAPN10, CAPN11, CAPN12, CAPN13, CAPN14,* and *CAPN15*) and two calpain small subunit genes (*CAPNS1* or designated *CAPN4* and *CAPNS2*) have been identified. Typical calpains contain a short N-terminal heterogeneous domain (domain I), a cysteine protease domain (domains IIa and IIb), a C2-like domain (domain III), and a PEF domain (domain IV). The PEF domain is missing in calpains 5, 6, 7, 10, and 15, and is replaced with an unrelated domain.

The catalytic triad cysteine residues consist of Cys in domain IIa and His and Asn in domain IIb, but Cys is substituted with Lys in calpain 6. Both calpains 1 and 2 are heterodimers (a unique 80-kDa catalytic subunit and common 30-kDa regulatory subunit). The small

Penta-EF-Hand Calcium-Binding Proteins, Table 1 Human proteins containing PEF domains

Groups	Protein names	Alternative names	Gene symbols	Amino acid residues	UniProtKB/Swiss-Prot annotations			Calcium-binding EF-hand No.	
					Accession codes	Annotated EF-hands	EF-hand No. in PEF	Experimental	Potential
I	*ALG-2 subfamily*								
	ALG-2	PDCD6	*PDCD6*	191	O75340	5	1,2,3,4,5	1,3,5	2
	Peflin		*PEF1*	284	Q9UBV8	5	1,2,3,4,5		1,3
II	*Calpain subfamily*								
	Calpain 1	μCL, μ-calpain	*CAPN1*	714	P07384	4	1,2,3,5	2,3	
	Calpain 2	mCL, m-calpain	*CAPN2*	700	P17655	3	2,3,5	2,3	
	Calpain 3	p94	*CAPN3*	821	P20807	4	1,2,3,5		2,3
	Calpain 8	nCL-2	*CAPN8*	703	A6NHC0	3	2,3,5		2,3
	Calpain 9	nCL-4	*CAPN9*	690	O14815	3	1,2,3		2,3
	Calpain 11		*CAPN11*	739	Q9UMQ6	2	2,3		2,3
	Calpain 12		*CAPN12*	719	Q6ZSI9	1	3		3
	Calpain 13		*CAPN13*	669	Q6MZZ7	2	2,5		
	Calpain 14		*CAPN14*	684	A8MX76	3	2,3,5		2
	Calpain small subunit 1	CSS1, 30K	*CAPNS1*	268	P04632	5	1,2,3,4,5	1,2,3 (s)	
	Calpain small subunit 2	CSS2, 30K-2	*CAPNS2*	248	Q96L46	4	2,3,4,5		2,3
	Sorcin subfamily								
	Sorcin	CP-22	*SRI*	198	P30626	4	1,2,3,4		2,3
	Grancalcin		*GCA*	217	P28676	4	1,2,3,4	1,3	4

(s), by similarity to the experimentally identified calcium-binding sites in pig and rat small subunit proteins

subunit has a Gly-rich N-terminal tail (domain V) followed by a PEF domain (domain VI). The two carboxy-terminal PEF domains (large IV and small VI) form the major heterodimer interface that holds the two subunits together through the pairing of their EF5s. Domain I interacts with domain VI and functions as an anchor helix, which is released and autolyzed upon Ca^{2+}-dependent activation (Fig. 2c). Calpastatin, an endogenous inhibitory protein, has four units of inhibitory domains that bind to conventional calpains in the presence of Ca^{2+}. Each inhibitory domain consists of subdomains A, B, and C, among which subdomains A and C bind to the Ca^{2+}-bound PEF domains of the large and small subunits, respectively. Although the subdomain B, an inhibitory core, passes through the active site cleft, it escapes cleavage by looping out and around the active site cysteine residue (Hanna et al. 2008).

Knocking out genes encoding calpain 1 (*Capn1*) and calpain 2 (*Capn2*) in mice results in abnormal platelet function and in embryonic lethality, respectively. Since the regulatory subunit of calpain 1 (small subunit, gene product of *CAPNS1*) is identical to that of calpain 2, deficiency of the small subunit is embryonically lethal in mice. Calpain 3 is specifically expressed in skeletal muscle, and loss-of-function mutations are associated with limb-girdle muscular dystrophy type 2A (LGMD2A). Calpain 8 and calpain 9, both of which are specifically expressed in the gastrointestinal tract and mutually form a heterodimer, are involved in gastric mucosal defense against external stresses. More information about calpains is available at http://calpain.org/. The *Drosophila melanogaster* genome encodes three typical calpains (UniProtKB accession codes: Calpain-A, Q11002; Calpain-B, Q9VT65; Calpain-C, Q9VXH6) as well as one atypical calpain (Calpain-D, P27398). Similar to mammalian calpains, the PEF domains of *Drosophila* calpains also lack one residue in the potential Ca^{2+}-binding loop of EF1.

Sorcin Subfamily

Sorcin, a 22 kDa protein, was first discovered as a gene amplified together with the P-glycoprotein gene in multidrug-resistant cancer cells. Sorcin is highly expressed in human myelogenous leukemia K562 cells and is purified as a homodimer. It is expressed in a wide variety of cells and involved in Ca^{2+}-homeostasis by modulating Ca^{2+} channel activities of the cardiac ryanodine receptor (RYR2), the pore-forming subunit of the voltage-dependent L-type Ca^{2+}-channel, and the cardiac Na$^+$-Ca^{2+} exchanger (Zamparelli et al. 2010). Sorcin also associates with annexin A7 (also designated annexin VII or synexin) in a Ca^{2+}-dependent fashion and inhibits annexin-mediated aggregation of chromaffin granules in the adrenal medulla. The N-terminal regulatory domain of annexin A7 is rich in Gly, Tyr, and Pro, and a "GYP motif" is thought to serve as a binding site for sorcin. Heterodimerization with its closest paralog, grancalcin, has also been reported.

Grancalcin is the closest paralog of sorcin. Grancalcin protein, 28 kDa in size, was isolated from the cytosol of unstimulated neutrophils using antiserum that was raised against membranes from neutrophils activated with phorbol 12-myristate 13-acetate (PMA). Grancalcin displays a Ca^{2+}-dependent translocation to the granules and plasma membrane of neutrophils, suggesting roles in granule-membrane fusion and degranulation of neutrophils. It associates with leukocyte plastin (L-plastin) in the absence of Ca^{2+}, and the grancalcin-L-plastin complex dissociates in the presence of Ca^{2+}. Orthologs of sorcin and grancalcin are found in vertebrates but not in *Drosophila* or *Caenorhabiditis elegans*.

Non-mammalian PEF Proteins

Some non-mammalian PEF proteins are erroneously annotated in protein sequence databases such as UnitProtKB (http://www.uniprot.org/help/uniprotkb). The sequence of *Schistosoma japonicum* sorcin (UniProtKB accession code: Q94743) actually corresponds to ALG-2 as evidenced by a high conservation of the EF1 sequence. The sequence of *Zea mays* grancalcin (UniProtKB accession code: B6SNX9) is more similar to the sequence of Group I PEF proteins in the EF1 sequence, indicating that naming this protein grancalcin is not appropriate. It is notable that Group II PEF proteins are missing from nematodes, yeast, plants, and lower eukaryotes. While the calpain large (catalytic) subunit is found in *Drosophila*, the calpain small (regulatory) subunit and sorcin as well as grancalcin are absent in the fruit fly. Since peflin is not found in nematodes, the ancestral *ALG-2* gene might be a prototype of PEF proteins in the animal–fungi kingdom. PEF genes seem to have a complex evolutionary history and exhibit gene duplications and probably gene losses in some taxa. For example, the fission yeast *Schizosaccharomyces pombe* lacks readily recognizable PEF sequences in the currently available genome database.

A unique PEF protein in *Saccharomyces cerevisiae*, named Pef1p, binds calcium and zinc and homodimerizes in vitro and in vivo like mammalian homologs. Disruption of PEF1 induces bi-budded cell morphology. GFP-fused Pef1p accumulates at the neck between the dividing mother and daughter cells. The cellular slime mould *Dictyostelium discoideum* possesses two PEF genes (*alg2-A* and *alg2-B*), which are nonessential, and *alg-2A* and *alg-2B* null strains display a normal developmental phenotype. However, in contrast to the developmentally arrested *noxA*$^-$, *noxB*$^-$, and *p22phox*$^-$ strains, the double knockout strains *alg-2B*$^-$/*noxA*$^-$ and *alg-2B*$^-$/*noxB*$^-$, as well as *alg-2B*$^-$/*p22*$^{phox-}$, show a restored developmental phenotype, suggesting involvement of *alg-2B* in the NADPH oxidase system.

Conclusions

Functions of PEF proteins remain unclear except for the clear regulatory function of the calpain small subunit. Unlike calmodulin, PEF proteins do not exhibit a significant change from the closed conformational state in the absence of Ca^{2+} to the open conformational state in the presence Ca^{2+}. Nevertheless, some PEF proteins show Ca^{2+}-dependent interactions with target proteins. Recent findings of ALG-2 as a Ca^{2+}-dependent adaptor protein shed new light on the physiological functions of PEF proteins and indicate the importance of dimerization for PEF proteins to exert their functions.

Cross-References

▸ Annexins
▸ C2 Domain Proteins
▸ Calcium-Binding Protein Site Types
▸ Calcium-Binding Proteins, Overview
▸ Calmodulin
▸ EF-Hand Proteins
▸ Zinc-Binding Sites in Proteins

References

Hanna RA, Campbell RL, Davies PL (2008) Calcium-bound structure of calpain and its mechanism of inhibition by calpastatin. Nature 456:409–412

Høj BR, la Cour JM, Mollerup J et al (2009) ALG-2 knockdown in HeLa cells results in G2/M cell cycle phase accumulation and cell death. Biochem Biophys Res Commun 378:145–148

Jang IK, Hu R, Lacanà E et al (2002) Apoptosis-linked gene 2-deficient mice exhibit normal T-cell development and function. Mol Cell Biol 22:4094–4100

Maki M, Kitaura Y, Satoh H et al (2002) Structures, functions and molecular evolution of the penta-EF-hand Ca^{2+}-binding proteins. Biochim Biophys Acta 1600:51–60

Okumura M, Ichioka F, Kobayashi R et al (2009) Penta-EF-hand protein ALG-2 functions as a Ca^{2+}-dependent adaptor that bridges Alix and TSG101. Biochem Biophys Res Commun 386:237–241

Rao RV, Poksay KS, Castro-Obregon S et al (2004) Molecular components of a cell death pathway activated by endoplasmic reticulum stress. J Biol Chem 279:177–187

Shibata H, Suzuki H, Kakiuchi T et al (2008) Identification of Alix-type and non-Alix-type ALG-2-binding sites in human phospholipid scramblase 3: Differential binding to an alternatively spliced isoform and amino acid-substituted mutants. J Biol Chem 283:9623–9632

Suzuki H, Kawasaki M, Inuzuka T et al (2008) Structural basis for Ca^{2+}-dependent formation of ALG-2/Alix peptide complex: Ca^{2+}/EF3-driven arginine switch mechanism. Structure 16:1562–1573

Vito P, Lacanà E, D'Adamio L (1996) Interfering with apoptosis: Ca^{2+}-binding protein ALG-2 and Alzheimer's disease gene ALG-3. Science 271:521–525

Zamparelli C, Macquaide N, Colotti G et al (2010) Activation of the cardiac Na^+–Ca^{2+} exchanger by sorcin via the interaction of the respective Ca^{2+}-binding domains. J Mol Cell Cardiol 49:132–141

Pentavalent Antimonial Therapy Against Leishmaniases, Molecular Bases

▸ Antimony-Based Therapy of Leishmaniases, Molecular and Cellular Rationale

Peptide Toxins

▸ Potassium Channels and Toxins, Interactions

Permeation

▸ Potassium Channel Diversity, Regulation of Potassium Flux across Pores
▸ Potassium Channels, Structure and Function

Peroxidases

▸ Heme Proteins, Heme Peroxidases

PEZ: Phenolics Efflux Zero

▸ Iron Proteins, Plant Iron Transporters

pH Regulation

▸ Sodium-Hydrogen Exchangers, Structure and Function in Human Health and Disease

Phagocytes

▸ Magnesium and Inflammation

Phosphatidylinositol 3-Kinases

John E. Burke and Roger L. Williams
Medical Research Council, Laboratory of Molecular Biology, Cambridge, UK

Synonyms

p101; p110α; p110β; p110γ; p110δ; p150; p50α; p55α; p55γ; p84; p85α; p85β; p87; PI3K; PI3K-C2α; PI3K-C2β; PI3K-C2γ; PIK3C2A; PIK3C2B; PIK3C2G; PIK3C3; PIK3CA; PIK3CB; PIK3CD; PIK3CG; PIK3R1; PIK3R2; PIK3R3; PIK3R4; PIK3R5; PIK3R6; vps15; vps34

Definition

Phosphatidylinositol 3-kinases (PI3Ks) are a family of metalloenzymes that phosphorylate the 3′ hydroxyl group of membrane-bound inositol phospholipids. These enzymes play important signaling roles because phospholipids that they produce serve as recruitment signals for lipid-binding proteins. The recruitment of downstream lipid-binding proteins to membranes controls various cellular processes including autophagy, cell cycle progression, cellular proliferation, growth, and survival. The PI3K family of enzymes consists of multiple catalytic and regulatory isoforms with both specialized and overlapping functions. Specific members of the PI3K family are frequently misregulated in human diseases including cancer and diabetes.

Introduction

Phosphatidylinositol 3-kinases (PI3Ks) catalyze the transfer of a phosphate group from ATP to the 3′ hydroxyl group of inositol phospholipids. They are primarily cytosolic enzymes and must be recruited to a membrane surface to carry out their lipid kinase activity. The 3′-phosphorylated phosphoinositide products of PI3Ks represent less than 1% of the total amount of phosphorylated phosphoinositides in a cell (Rommel et al. 2010a). This very small proportion of PI3K lipid products controls the recruitment of many lipid-binding proteins that play key roles in effecting numerous important cellular processes, and therefore, signaling through the PI3K family must be tightly regulated. The misregulation of the production of phosphoinositides has been demonstrated in many human diseases, and various mutations in PI3Ks, predominantly in class IA PI3Kα, are identified in human cancers. Inhibition of specific PI3K family members is an important area of drug development, and several inhibitors of PI3K activity are currently in clinical trials (Courtney et al. 2010).

The PI3K family of enzymes encompasses three different classes defined by structural and biochemical properties. All classes of the PI3K family share a conserved catalytic core composed of a C2 domain, a helical domain, and a C-terminal catalytic domain (Fig. 1) (Vanhaesebroeck et al. 2010). The C2 domain of PI3Ks binds to acidic phospholipids, but does not require the presence of Ca^{2+} for lipid binding, in contrast to some other C2 domains found in membrane-associated enzymes such as phospholipase A2. However, all PI3Ks require the presence of a metal cation (Mg^{2+}, Mn^{2+}, or Ca^{2+}, depending on the PI3K isoform) coordinated to ATP to catalyze the phosphate transfer reaction from ATP to lipid substrate. The catalytic domain responsible for lipid kinase activity of PI3Ks has a two-lobed structure that is shared with protein kinases (Fig. 2). The ATP complexed with metal cations is sandwiched in the cleft between the two lobes. There are several conserved elements in the catalytic domain that are required for lipid kinase activity, including the activation loop responsible for substrate binding, the catalytic loop necessary for phosphate transfer, and the C-terminal helix essential for lipid interactions and maintaining a closed inactive state in the absence of lipid substrate (Miller et al. 2010; Williams et al. 2009). The various family members differ from each other in their preference for the head group of the inositol lipid substrate, activation by the type of metal cation, as well as the association with various regulatory subunits. We review herein our current understanding of the structure, mechanism, and regulation of each of the three classes of PI3K, as well as the role of the entire family of PI3Ks in a range of human disease states.

Phosphatidylinositol 3-Kinases

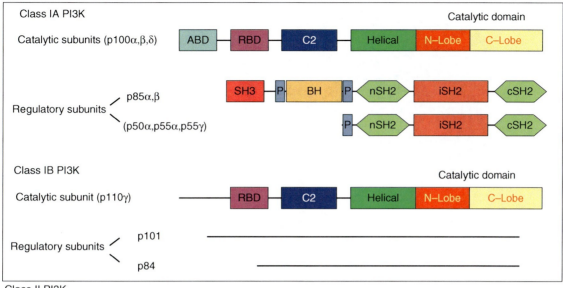

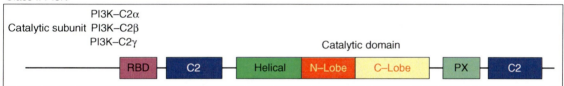

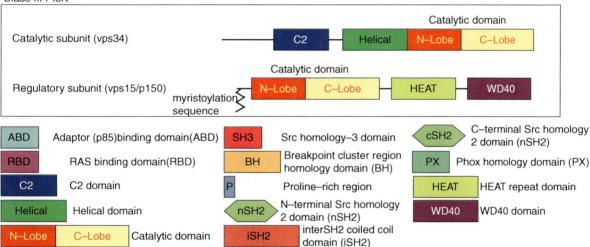

Phosphatidylinositol 3-Kinases, Fig. 1 Domain architecture of the three classes of PI3Ks with the different domain abbreviations described in the legend

Class I PI3K

The class I PI3Ks were the first to be identified and are the most well characterized of the PI3K family. Class I PI3K activity has been demonstrated to play key roles in tumorigenesis, growth, and survival. These PI3Ks can phosphorylate PtdIns, PtdIns4P, and PtdIns(4,5)P$_2$ phosphoinositides in vitro, and this activity is

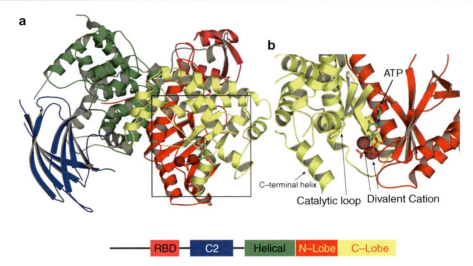

Phosphatidylinositol 3-Kinases, Fig. 2 Structure of the catalytic subunit of class IB p110γ in complex with ATP and two divalent cations (pdb:1E8X) highlighting conserved features of the catalytic core of PI3Ks. (**a**) The domain architecture of the RBD, C2, helical, and catalytic domains is colored on the structure according to the schematic. (**b**) View of the ATP-binding site located between the N-lobe and C-lobe of the catalytic domain. The metal-binding sites as well as important lipid kinase features are labeled. The RBD, C2, and helical domains have been removed for clarity

dependent on the presence of Mg^{2+}; however, in vivo, the main substrate is $PtdIns(4,5)P_2$, and the main product is $PtdIns(3,4,5)P_3$ (Rommel et al. 2010a). One of the best characterized functions of the class I PI3Ks is to recruit the Ser/Thr kinase protein kinase B (PKB), otherwise known as Akt, to the lipid membrane via binding of the pleckstrin homology (PH) domain of PKB to $PtdIns(3,4,5)P_3$. At the lipid membrane, PKB gets phosphorylated, and this leads to stimulation of its protein kinase activity, which subsequently activates pathways important in both growth and survival. The class I PI3K isoforms are activated downstream of numerous important signaling pathways, including receptor tyrosine kinases (RTKs) and G protein–coupled receptors (GPCRs). A major difference among the class I PI3K isoforms is their relative ability to be activated by these different upstream signaling pathways. Class I PI3K is split into two distinct subgroups, class IA, and class IB, based on interactions with regulatory subunits.

The class IA subgroup exists as an obligate heterodimer between a p110 catalytic subunit, of which there are three different isoforms (p110α, p110β, and p110δ), and a p85-type regulatory subunit, of which there are five isoforms (p85α, p85β, p55γ, p55α, and p50α) (Fig. 1) (Vanhaesebroeck et al. 2010). Both the catalytic and the regulatory subunits have a multidomain organization, which enables fine-tuning of the PI3K catalytic activity. The kinase domain resides at the C-terminus of the catalytic subunit in all PI3Ks. In class IA catalytic subunits, the kinase domain is preceded by an N-terminal adaptor-binding domain (ABD) that binds with high affinity to the regulatory subunit, followed by a RAS-binding domain (RBD), C2, and the helical domains. The full-length p85 regulatory subunits consist of an array of signaling modules: src-homology 3 (SH3) domain that binds to PXXP motifs in interacting partners; a BCR homology (BH) domain that can bind to Cdc42 and Rac1; and two src-homology 2 (SH2) domains, an N-terminal SH2 domain (nSH2), and a C-terminal SH2 domain (cSH2) separated by a long coiled-coil region termed the inter-SH2 domain (iSH2). The p55 and p50 subunits lack the SH3 and BCR domains. The p85 regulatory subunit is known to play three distinct functions in its interaction with the class IA PI3Ks: stabilization of the catalytic subunit, inhibition of the lipid kinase activity, and activation by RTKs mediated by SH2 domains binding phosphorylated phosphotyrosine residues present in RTKs and their adaptor proteins (Williams et al. 2009).

Several crystal structures of the class IA PI3K isoforms in complex with different constructs of the regulatory subunit p85 have been solved and have shed light on some of the determinants of inhibition. These structures have been particularly useful in determining the mechanism of activation of cancer-linked mutations that lie at the interface of the catalytic and regulatory subunits (Rommel et al. 2010b). There are many cancer-linked mutations in both p110α and p85α, and discussion of all of them is beyond the scope of this review. Specific cancer-linked mutations are highlighted, and an excellent review of all cancer-linked mutations can be found in (Rommel et al. 2010b). The ABD domain of the catalytic subunit has a large hydrophobic interface with the iSH2 coiled-coil domain of the regulatory subunit. This contact is necessary and sufficient for the class IA p110 catalytic subunit to bind to the p85 regulatory subunit. The iSH2 domain also makes extensive contacts with both the C2 and kinase domains of the catalytic subunit. Cancer-linked mutations in both p110α and p85α have been discovered that map to the interface between the iSH2 and the C2 domains (N345K in p110α, and D560K and D565K in p85α). These mutations activate PI3K in vitro and increase its transforming abilities in vivo (Rommel et al. 2010b). The most potently activating mutation of p110α is located in the C-lobe of the kinase domain (H1047R). This mutation is shown to activate the enzyme by altering how the enzyme interacts with a membrane surface (Rommel et al. 2010b).

The role of the nSH2 and cSH2 domains of p85 in regulating PI3K activity has been partly elucidated by structures of the p110α isoform in complex with the nSH2 domain and the p110β isoform in complex with the cSH2 (Rommel et al. 2010b; Zhang et al. 2011). The presence of a p85 construct containing only the nSH2 and iSH2 domains is the minimal regulatory construct needed to fully inhibit the p110α isoform. Binding of tyrosine-phosphorylated receptors or their adaptor proteins by the nSH2 domain break the interaction between the nSH2 and helical domain activating lipid kinase activity (Rommel et al. 2010a). The nSH2 domain forms a contact with the helical domain that is mediated by E545 and K379 in p110α and p85α, respectively. A charge reversal mutation of either of these residues (E545K in p110α or K379E in p85α) mimics this phosphopeptide activation. Many cancer-linked helical domain mutations identified in p110α probably function by breaking the nSH2-helical domain contact (E542K, E545K, E546Q). The p110β isoform requires the presence of both the nSH2 and cSH2 domains for full inhibition of lipid kinase activity. The cSH2 domain of p85 forms a contact with the C-lobe of the kinase domain of p110β. A mutation of a key contact residue in the cSH2 domain (Y685A) completely abolishes cSH2 inhibition (Zhang et al. 2011). This contact between the cSH2 and the catalytic domain appears to be incompatible with p110α due to several subtle structural differences between p110α and p110β that may sterically prevent this interaction. This differential isoform regulation by the cSH2 may partially explain the presence of activating cancer-linked mutations in p110α, but not the other isoforms.

The class IB subgroup is composed of a single catalytic subunit, p110γ, which can form a complex with two regulatory subunits, p101 or p84/p87. The presence of the regulatory subunit switches the substrate preference from PtdIns to $PtdIns(4,5)P_2$ (Williams et al. 2009). The p110γ is activated downstream of GPCRs. This activation is potentiated by the presence of the p101 or p84–p87 regulatory subunit. The class IB PI3K contains the same conserved catalytic core found in all PI3Ks (C2, helical, and kinase domains) as well as a RBD. The N-terminal region of p110γ is likely to have a similar fold as the class IA ABD domain; however, due to sequence differences, this p110γ region does not bind to p85-type regulatory subunits, and instead is necessary (although not sufficient) for binding to the p84–p87 or p101 regulatory subunits. The structure of the free catalytic subunit of p110γ (lacking the N-terminal domain) was the first of the PI3K isoforms to be determined, and numerous structures including complexes with the small GTPase RAS, as well as with ATP and multiple PI3K inhibitors, have shed light on the mode of regulation and inhibition of this enzyme (Fig. 2) (Williams et al. 2009). However, there is no structural information on the regulatory subunit, or how it interacts with the catalytic subunit.

The class I PI3K isoforms are differentially activated downstream of diverse signaling complexes. All class IA isoforms can be activated downstream of RTKs through the SH2 domains of p85 regulatory subunits binding to phosphorylated tyrosines of RTKs and adaptor proteins. Interestingly, there is evidence of selective p85-mediated recruitment of specific p110 isoforms to specific activated receptors (Vanhaesebroeck et al. 2010). Class IB p110γ is not activated by RTKs due to lack of SH2 domains in the p84–p87 or p101 regulatory subunits. The p110β and p110γ isoforms are able to be activated downstream of GPCRs, by Gβγ heterodimers from heterotrimeric

G proteins, while p110α and p110δ are not. The mechanism of Gβγ activation of p110β and p110γ is currently unknown. All class I PI3Ks contain a RAS-binding domain (RBD), and both p110α and p110γ have been shown to be activated directly by the small GTPase RAS. The role of RAS in activating p110δ and p110β is still ambiguous, with the possibility of different RAS family members being responsible for activation (Vanhaesebroeck et al. 2010).

There are also slight differences among the class I catalytic isoforms in how they respond to the presence of Mn^{2+} versus Mg^{2+} as the metal cofactor. The lipid kinase activity is maximal in the presence of Mg^{2+}, but there is significant protein kinase activity in the presence of Mn^{2+}. The p110α and p110β have *protein kinase* activity towards p85 in the presence of Mn^{2+}, while p110β, p110δ, and p110γ autophosphorylate a specific serine residue near the C-terminus of the kinase domain in the presence of Mn^{2+}. The autophosphorylation of p110β and p110δ causes a decrease in lipid kinase activity, but has no effect on p110γ (Foukas and Shepherd 2004).

Role of Class I PI3Ks in Disease

The class I PI3 kinases have long been associated with cancer due to the transforming capabilities of viral oncogenes associated with PI3K lipid kinase activity, and this link has been underscored by the discovery of hot-spot somatic mutations in both p110α and p85α in tumors of various tissues. Overexpression of p85α mutants together with either p110α, p110β, or p110δ leads to cell transformation. However, it is not clear whether endogenous p85α mutants activate only the endogenous p110α isoform or if they also function through other class IA isoforms. The main hot-spot mutations E542K, E545K, H1047L, and H1047R in p110α described above represent over 80% of all cancer-linked mutations discovered in p110α. The incidence of these mutations is greater than 25% in some forms of cancer (Rommel et al. 2010b). The antagonist of the PI3K pathway, lipid phosphatase PTEN (phosphatase and tensin homolog deleted on chromosome 10), is a tumor suppressor that dephosphorylates the 3' position of PtdIns $(3,4,5)P_3$. It is also frequently mutated or deleted in cancer highlighting the importance of PtdIns $(3,4,5)P_3$ in cancer development. The prevalence of these activating mutations has made inhibition of p110α, a major target for cancer treatment. Many pan-PI3K inhibitors as well as p110α-specific inhibitors are currently in clinical trials for various forms of cancer (Courtney et al. 2010).

A small number of p110γ mutations have been identified in tumors, but these have not yet been characterized, so it is not known whether they are gain-of-function mutations. Differential expression of the p110δ and p110γ isoforms has been detected in several forms of leukemia. Overexpression of wild-type p110β, p110δ, and p110γ isoforms induces oncogenic transformations of cultured cells, while overexpression of wild-type p110α does not (Rommel et al. 2010b). The p110β isoform also appears to play a key role in the oncogenesis of cancers that have a loss of the PTEN gene. The depletion of p110β but not p110α in these cancers decreased tumorigenesis and cellular proliferation (Courtney et al. 2010). In summary, all class I PI3K isoforms can be oncogenic, and this can be caused by either overexpression or specific activating mutations. Determining the various roles of all of the class I isoforms of PI3K in cancer provides an exciting framework for the development of new cancer therapies.

The p110γ and p110δ isoforms of PI3K are highly expressed in immune cells, and this led to the initial hypothesis that they would be important in immune responses such as allergy and arthritis. The use of both knockout and "knockin" (expressing lipid kinase dead enzyme) mouse models for p110γ and p110δ have shown protection from various inflammatory diseases in these mice. A full examination of these mouse models and their implications are reviewed in (Ghigo et al. 2010). The generation of specific p110γ and p110δ-specific inhibitors might be key tools in targeting and treating chronic inflammatory diseases.

Class II PI3K

The class II PI3Ks were the last of the classes of PI3Ks to be discovered based on sequence homology with class I and class III. In mammals, three different isoforms have been identified (PI3K-C2α, PI3K-C2β, and PI3K-C2γ). The class II PI3Ks play a role in insulin secretion, clathrin-mediated exocytosis, and smooth muscle cell contraction (Falasca and Maffucci 2007; Vanhaesebroeck et al. 2010). Besides the RBD domain and the catalytic core, they contain an extra C2 domain and a phox homology (PX) domain at the C-terminus, as well as an N-terminal extension that is unique to each isoform (Fig. 1). The class II PI3Ks have the highest lipid kinase activity for PtdIns, but can also weakly phosphorylate PtdIns4P. In the PI3K-C2α isoform, clathrin binding to the N-terminal extension activates lipid kinase activity, and even though PtdIns3P remains the major product, there is an increase in the relative levels of PtdIns $(3,4)P_2$ produced, along with a very small amount of PtdIns $(3,4,5)P_3$. However, most current evidence suggests that, in vivo, the preferred substrate is PtdIns (Falasca and Maffucci 2007).

The class II PI3Ks are unique from the other classes in that they have similar lipid kinase activity in the presence of Ca^{2+} and Mg^{2+} and show no lipid kinase activity in the presence of Mn^{2+}. The Ca^{2+} activation is due to the binding of Ca^{2+}-ATP in the active site, with no requirement for Ca^{2+}-mediated lipid binding by either of the two C2 domains present. No regulatory partners for class II PI3K have been identified yet, and it has been proposed that its lipid kinase activity is regulated by the presence of the unique N-terminal and C-terminal extensions. The structure of the PX domain of class II PI3K showed that this domain binds specifically to PtdIns$(4,5)P_2$; however, the regulatory role of this interaction in cells is not fully understood (Vanhaesebroeck et al. 2010). The N-terminal extensions contain many proline-rich regions that could act as important targets for protein-protein interactions mediated by SH3 domains. The N-terminal extension of PI3K-C2α and PI3K-C2β binds to clathrin and activates lipid kinase activity, where the N-terminal extension of PI3K-C2β can interact with growth factor receptor-bound protein (Grb2), mediating localization to activated tyrosine receptors. The small GTPase TC10 activates PI3K-C2α in cells, but a direct link between PI3K and TC10 has not been demonstrated. There is a possibility that this interaction is mediated by the RAS-binding domain. However, it is known that PI3K-C2β neither binds to nor is activated by RAS (Vanhaesebroeck et al. 2010). Further structural information will be critical in fully understanding the regulation of class II PI3K.

Class III PI3K

The class III PI3 kinase was originally identified as a protein that was important in the endosomal sorting pathway in yeast, which led to the naming of this enzyme as vacuolar protein sorting protein 34 (Vps34) (Backer 2008). It was later identified as a PI3 kinase due to the discovery of p110α and its sequence similarity to Vps34. This enzyme has a well-defined role in autophagy, membrane trafficking, and cell signaling. Vps34 shares similar catalytic core domain architecture with the class I PI3Ks, but it does not contain either ABD or RBD domains. This enzyme is also unique from the other classes in that it has its maximal lipid kinase activity in the presence of Mn^{2+}, and it can phosphorylate only PtdIns to generate PtdIns3P. The structure of the helical and kinase domains of the Vps34 catalytic subunit has been recently solved and showed similarities with class I PI3K structures. Interestingly, this structure suggests that for full activation of lipid kinase activity, the helical C-terminal tail located at the very C-terminus of the kinase domain adopts an open conformation essential for interaction with the lipid membranes (Miller et al. 2010).

The class III PI3K is associated with a regulatory subunit known as Vps15 (also known as p150 in mammals). The presence of Vps15 is known to activate Vps34 activity in cells; however, the exact mechanism of this activation is unknown. The Vps15 regulatory subunit is composed of a putative serine/threonine protein kinase domain, followed by a series of HEAT repeats, and a C-terminal WD40 domain (Fig. 1). The presence of the myristoylation sequence of Vps15 has led to the proposal that this subunit activates Vps34 through targeting the catalytic subunit to the membrane surface. The exact structural elements regulating binding of Vps15 to Vps34 are unknown; however, it has been shown that the C-terminal 57 residues in Vps34 are necessary and sufficient for Vps15 binding (Backer 2008).

In yeast and mammalian cells, both Vps34 and Vps15 are predominantly found in large multiprotein complexes, and the presence of protein-binding partners governs the role in controlling different intracellular processes such as autophagy and vesicle-mediated intracellular traffic. In yeast, Vps34 and Vps15 are found in three predominant complexes: one, consisting of Vps34 and Vps15 along with Atg6/Vps30 and Vps38, is critical in endosomal trafficking; another, with Vps34 and Vps15 along with Atg6/Vps30 and Atg14, is critical in controlling autophagy; while the third one, composed of Vps34 and Vps15 along with guanine nucleotide-binding protein 1α (Gpa1), is critical in pheromone response. In mammalian cells, Vps34 interacts with the mammalian orthologues of Atg6/Vps30, Atg14, and Vps38 known as Beclin-1, Atg14/Barkor, and UVRAG (UV irradiation resistance-associated gene), respectively. These complexes also contain many other associated proteins, including Bif-1 (Bax-interacting factor 1), BECN1-regulated autophagy protein 1 (AMBRA1), and Rubicon. These complexes are required for autophagy and vesicular trafficking in mammalian cells. Specific cellular functions of Vps34 may be modulated either by different regulatory interactions or by cellular localization dependent on the binding of diverse protein partners. It has also been shown that Vps34 plays a role in the regulation of the mammalian target of rapamycin (mTOR) by nutrients. The exact mechanism of this regulation is currently unknown.

Role of Class II and Class III PI3Ks in Human Disease

The role of class II and class III PI3Ks in human disease has remained enigmatic, with very few studies identifying misregulation in these enzymes correlated with disease. However, both of these classes play critical roles in cellular pathways that are linked to human diseases such as autophagy, endocytosis, and exocytosis. So far, there have been initial reports about mutations in class III PI3K that are associated with schizophrenia, but the exact mechanism of this association is unclear (Backer 2008). There are also preliminary reports that the mRNA level for class II PI3K-C2α are diminished in type 2 diabetic individuals compared to nondiabetic individuals. Continued study into the cellular and physiological roles of both class II and class III PI3K is required to establish their role in human disease. The development of specific class II and class III inhibitors will be a key tool in understanding their physiological roles.

Conclusions

An incredible wealth of knowledge exists for the structure, regulation, and function of the various classes of PI3Ks. However, numerous questions remain, particularly the exact role of each isoform in cellular signaling pathways. Future studies are needed to identify all of the PI3K-interacting partners required for lipid kinase regulation and localization, as well as for determining cross talk between both upstream and downstream members of the pathway. Continued structural, biochemical, biological, and clinical research will be key in solving fundamental questions about the regulation of PI3K and its role in various human diseases.

Cross-References

▶ C2 Domain Proteins

References

Backer JM (2008) The regulation and function of Class III PI3Ks: novel roles for Vps34. Biochem J 410:1–17

Courtney KD, Corcoran RB, Engelman JA (2010) The PI3K pathway as drug target in human cancer. J Clin Oncol 28:1075–1083

Falasca M, Maffucci T (2007) Role of class II phosphoinositide 3-kinase in cell signalling. Biochem Soc Trans 35:211–214

Foukas LC, Shepherd PR (2004) Phosphoinositide 3-kinase: the protein kinase that time forgot. Biochem Soc Trans 32: 330–331

Ghigo A, Damilano F, Braccini L, Hirsch E (2010) PI3K inhibition in inflammation: toward tailored therapies for specific diseases. Bioessays 32:185–196

Miller S, Tavshanjian B, Oleksy A, Perisic O, Houseman BT, Shokat KM, Williams RL (2010) Shaping development of autophagy inhibitors with the structure of the lipid kinase Vps34. Science 327:1638–1642

Rommel C, Vanhaesbroeck B, Vogt PK (eds) (2010a) Phosphoinositide 3-kinase in health and disease volume I. Current topics in microbiology and immunology. Springer, Berlin/Heidelberg

Rommel C, Vanhaesbroeck B, Vogt PK (eds) (2010b) Phosphoinositide 3-kinase in health and disease volume II. Current topics in microbiology and immunology. Springer, Berlin/Heidelberg

Vanhaesebroeck B, Guillermet-Guibert J, Graupera M, Bilanges B (2010) The emerging mechanisms of isoform-specific PI3K signalling. Nat Rev Mol Cell Biol 11:329–341

Williams R, Berndt A, Miller S, Hon WC, Zhang X (2009) Form and flexibility in phosphoinositide 3-kinases. Biochem Soc Trans 37:615–626

Zhang X, Vadas O, Perisic O, Anderson KE, Clark J, Hawkins PT, Stephens LR, Williams R (2011) Structure of lipid kinase p110β/p85β elucidates an unusual SH2-domain mediated inhibitory mechanism. Mol Cell 41(5):567–578

Photoautotroph Metazoan

▶ Thallium, Distribution in Plants

Photosynthetic Organisms

▶ Magnesium in Plants

Photothermal Therapy

▶ Gold Nanomaterials as Prospective Metal-based Delivery Systems for Cancer Treatment

Physiological/Biogenic Metal Cation

▶ Calcium Ion Selectivity in Biological Systems

PI3K

▶ Phosphatidylinositol 3-Kinases

PI3K-C2α

▶ Phosphatidylinositol 3-Kinases

PI3K-C2β

▶ Phosphatidylinositol 3-Kinases

PI3K-C2γ

▶ Phosphatidylinositol 3-Kinases

PIK3C2A

▶ Phosphatidylinositol 3-Kinases

PIK3C2B

▶ Phosphatidylinositol 3-Kinases

PIK3C2G

▶ Phosphatidylinositol 3-Kinases

PIK3C3

▶ Phosphatidylinositol 3-Kinases

PIK3CA

▶ Phosphatidylinositol 3-Kinases

PIK3CB

▶ Phosphatidylinositol 3-Kinases

PIK3CD

▶ Phosphatidylinositol 3-Kinases

PIK3CG

▶ Phosphatidylinositol 3-Kinases

PIK3R1

▶ Phosphatidylinositol 3-Kinases

PIK3R2

▶ Phosphatidylinositol 3-Kinases

PIK3R3

▶ Phosphatidylinositol 3-Kinases

PIK3R4

▶ Phosphatidylinositol 3-Kinases

PIK3R5

▶ Phosphatidylinositol 3-Kinases

PIK3R6

▶ Phosphatidylinositol 3-Kinases

Pitchblende

▶ Polonium and Cancer

Plant-Mediated Bioremediation

▶ Lead and Phytoremediation

Plants: Plantae

▶ Thallium, Distribution in Plants

Plasma Membrane Ca^{2+}-ATPase (PMCA)

▶ Calcium ATPases

Plastocyanin

Ivano Bertini[1,2] and Roberta Pierattelli[3]
[1]Magnetic Resonance Center (CERM) – University of Florence, Sesto Fiorentino, Italy
[2]Department of Chemistry, University of Florence, Sesto Fiorentino, Italy
[3]CERM and Department of Chemistry "Ugo Schiff", University of Florence, Sesto Fiorentino, Italy

Synonyms

Blue copper protein; Cupredoxin; Type-1 copper protein

Definition

Plastocyanins are small proteins ($\simeq$ 10 kDa) belonging to a larger class of metalloproteins, named blue copper proteins, which contain Cu in the active site. Plastocyanins play an essential role in oxygenic photosynthesis in higher plant, cyanobacteria, and algae, where they transfer electrons between photosystems II and I receiving an electron from cytochrome f (Cyt f), a subunit of the membrane-bound cytochrome b_6f complex, and donating the electron to the P700+ photoreaction center of photosystem I.

Structure and Function

The name plastocyanin was introduced by Katoh and Takamiya in the early 1960s when isolating this protein from spinach chloroplasts (Katoh 2003). Plastocyanin from poplar was the first blue copper protein to be structurally characterized, and its structure was determined by the Freeman group in 1978 (Colman et al. 1978). Since then, structures of plastocyanins from several plants, green algae, and cyanobacteria have been solved either by X-ray diffraction or by nuclear magnetic resonance (NMR) spectroscopy (www.pdb.org). Several of the structures have been determined with the copper ion in both the +1 and +2 oxidation states and at several pH values.

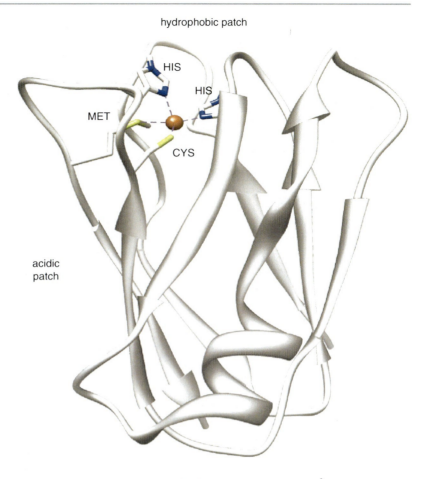

Plastocyanin, Fig 1 Ribbon drawing of the cyanobacterium *Synechocystis sp.* PCC 6803 Cu(II) plastocyanin solution structure showing the secondary structure elements of the protein (PDB ID 1J5D, Bertini et al. 2001a). The metal ion is represented as a sphere of arbitrary dimension. The picture was rendered with Chimera 1.5.3

They present a β-barrel fold defined by two β-sheets composed by eight strands following a key Greek motif (Fig. 1). Generally, there is a kink in the second strand such as its first half is part of the first sheet while the second half is part of the second sheet. Strand 5 is variable in length and generally linked to a small portion of a helix. The helical content is variable in proteins from different sources.

Although the low primary sequences homology of plastocyanins from plants, algae, and cyanobacteria (www.expasy.org) the structure of the copper-binding site is generally conserved (Gray et al. 2000). It is located in the so-called northern region of the molecule, following the naming convention by Freeman. The copper ion sits in a depression close to the protein surface, without being solvent exposed, thus preventing the adventitious binding of external ligands and the occurrence of side reactions. The copper binding geometry can be described as a flattened tetrahedron in which the three equatorial ligands (a S_γ-Cys and two N_δ-His) are strongly bound ($r_{Cu-X} < 2.2$ Å) while a Met residue constitute the apex. The Cu(II)-S_δ-Met distance is variable in plastocyanins from different sources but generally around 2.8–2.9 Å. A conserved Phe residue makes contact with the Met and may help in fixing the position of the axial ligand. A hydrogen bond network in the metal site is largely conserved. In particular, the peptide NH of an Asn residue next to the second His ligand is hydrogen bonded to the bound S_γ-Cys atom in most plastocyanins. This hydrogen bond is relevant for maintaining the geometry of the metal site.

Mobility studies performed by NMR on both Cu(I) and Cu(II) plastocyanins pointed out that plastocyanin is a rigid molecule in both oxidation states, in the submillisecond time scale, consistent with its β-barrel structure. On the other hand, in the milliseconds and longer time scales, they show significant protein flexibility on the loops connecting the β-strand region that contains the Cys and the two His copper ligands (Bertini et al. 2001a, b).

Plastocyanins display reduction potentials ranging from 340 to 370 mV that decrease as the pH is lowered (Sykes 1991). In the interaction with inorganic redox partners, the protein is almost inactive at pH 4.0. In this condition, the ring of one of the His copper ligands (the one more solvent exposed) rotates around the C_β-C_γ bond and flips away from the copper. Such rearrangement makes the copper coordination trigonal planar with an S_γ-Cys, an N_δ-His, and an S_δ-Met as ligands. The copper moves toward the bound methionine, with a shortening of the Cu-S bond to 2.5 Å. It has been proposed that this change of geometry, stabilizing the Cu(I) form of the protein rendering it inactive, is used in vivo as a cellular switch to turn off electron transfer if needed.

All plastocyanins have a hydrophobic surface surrounding the exposed histidine of the copper-binding site, the so-called hydrophobic patch. In plant plastocyanins, acidic residues are located on either side of the highly conserved tyrosine identifying the so-called acidic patch. Algal plastocyanins, and those from vascular plants in the family *Apiaceae*, contain similar acidic residues but are shaped differently from those of plant plastocyanins. In cyanobacteria, the distribution of charged residues on the surface is different from eukaryotic plastocyanins, and variations among different bacterial species are large. Although the acidic patches are not conserved in bacteria, the hydrophobic patch is always present. These hydrophobic and acidic patches are believed to be the recognition sites for the other proteins involved in electron transfer. A series of mutagenesis experiments as well as experiments involving chemically modified proteins revealed that the interaction between plastocyanin and its electron donor Cyt f is highly electrostatic, suggesting that the conserved Tyr in the acidic patch is the electron entry site (Kannt et al. 1996). However, NMR analysis of the complex of spinach plastocyanin and the soluble domain of turnip cytochrome f revealed that the complex has a single orientation involving both the hydrophobic and the acidic patches. In the latter case, the exposed His residue has been identified as the electron entry site (Ubbink et al. 1998).

Site-directed mutagenesis studies were also helpful for providing details about the interaction with P700+. It appears that the main interaction site of plastocyanin with photosystem I is on the hydrophobic patch even if the acidic patch, when present, is also important in controlling the binding and that the electron transfer occurs through the more exposed histidine ligand (Haehnel et al. 1994).

The electronic structure of plastocyanins' copper site has been studied by a wide range of spectroscopic and computational methods (Solomon et al. 2004; Solomon and Hadt 2011). Copper(I) possesses a d^{10} closed shell, being diamagnetic. As a consequence, the reduced ion is spectroscopically silent, and its studies have been limited to X-ray absorption and photoelectron spectroscopies. Instead, Cu(II) is a d^9 metal ion with one unpaired electron and consequently is paramagnetic. The unusual coordination features of this metal center in plastocyanins induce unique spectroscopic properties relative to normal Cu(II) complexes. The most striking characteristic of these proteins is indeed the blue color due to an intense absorption band ($\varepsilon \simeq 5,000$ M^{-1} cm^{-1}) in the 600 nm spectral region ascribed to πCys$\rightarrow$Cu(II) ligand-to-metal charge transfer transition. Its intensity is attributed to the orientation adopted by the $d_{x^2-y^2}$ orbital which favors overlap with a π orbital of the S_γ-Cys and gives rise to a strongly covalent Cu(II)-Cys bond. A second band at *ca.* 450 nm has been assigned as a pseudo-σCys$\rightarrow$Cu(II) ligand-to-metal charge transfer transition based on resonance Raman experiments. The relative intensity of these bands is a distinctive feature of each blue copper protein (Malkin and Malmström 1970).

Electron paramagnetic resonance spectra of plastocyanins exhibit characteristic small $A_{//}$ values (63 × 10^{-4} cm^{-1}, compared to 164 × 10^{-4} cm^{-1} in $CuCl_4^{2-}$). This is attributable to the strong covalency of the Cu(II)-thiolate bond that reduces the electron spin density on the copper ion, as also supported by theoretical calculations. NMR spectroscopy confirmed the existence of a large electron delocalization onto the coordinated Cys ligand, and it has been used to elucidate the electronic properties of the metal center (Bertini et al. 2000).

The magnetic moment associated with the unpaired electron present in Cu(II) plastocyanin induces a non-negligible effect on the magnetic properties of nearby nuclei through magnetic coupling. This coupling between electron and nuclear spins (hyperfine coupling) affects both the chemical shifts and the relaxation rates of the nuclei in the proximity of the paramagnetic center. The extent of the latter effect determines the detectability of the NMR lines of these nuclei, since NMR signal linewidths are

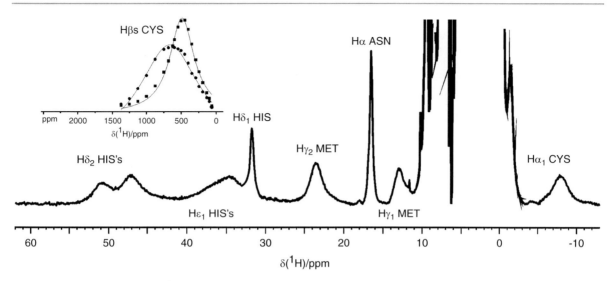

Plastocyanin, Fig. 2 800 MHz ^{1}H NMR spectra of oxidized spinach plastocyanin recorded in H$_2$O solution. In the insets, the far-downfield regions containing signals not observable in direct detection are shown. The positions and the linewidths of these signals in the oxidized species were obtained using saturation transfer experiments over the far-downfield region by measuring the intensity of the exchange connectivity with the corresponding signal in the reduced species and performing a Lorentzian fit to the data points (Bertini et al. 2001a)

proportional to the nuclear transverse relaxation rates R$_2$ (Banci et al. 2002). The low molecular weight of plastocyanin makes it possible to detect a large number of hyperfine-shifted signals with high sensitivity and resolution (Fig. 2), even at very high magnetic fields, since Curie relaxation is negligible. Once located, the hyperfine-shifted signals may be assigned by 1D or 2D saturation transfer experiments in a sample containing a mixture of the oxidized and reduced forms of the protein. From the assignment of the hyperfine-shifted signals of Cu(II), plastocyanin information on the electron delocalization onto the metal ligands can be obtained calculating the contact and pseudocontact contributions to the hyperfine shifts. In most cases, since the magnetic anisotropy of the Cu(II) ion is low, the observed shifts can be approximated to the contact contribution, which can be used as an initial criterion to compare the electron spin density among proteins (Bertini et al. 2000).

Cross-References

▶ Copper-Binding Proteins
▶ Copper, Biological Functions
▶ Monocopper Blue Proteins

References

Banci L, Pierattelli R, Vila AJ (2002) NMR studies on copper protein. Adv Prot Chem 60:397–449

Bertini I, Fernández CO, Karlsson BG, Leckner J, Luchinat C, Malmström BG, Nersissian AM, Pierattelli R, Shipp E, Valentine JS, Vila AJ (2000) Structural information through NMR hyperfine shifts in blue copper proteins. J Am Chem Soc 122:3701–3707

Bertini I, Ciurli S, Dikiy A, Fernández CO, Luchinat C, Safarov N, Shumilin S, Vila AJ (2001a) The first solution structure of an oxidized paramagnetic copper (II) protein: the case of plastocyanin from the cyanobacterium *Synechocystis* PCC6803. J Am Chem Soc 123:2405–2413

Bertini I, Bryant DA, Ciurli S, Dikiy A, Fernández CO, Luchinat C, Safarov N, Vila AJ, Zhao J (2001b) Backbone dynamics of plastocyanin in both oxidation states. Solution structure of the reduced form and comparison with the oxidized state. J Biol Chem 276:47217–47226

Colman PM, Freeman HC, Guss JM, Murata M, Norris VA, Ramshaw JAM, Venkatappa MP (1978) X-Ray crystal-structure analysis of plastocyanin at 2.7 Å resolution. Nature 272:319–324

Gray HB, Malmström BG, Williams RJ (2000) Copper coordination in blue proteins. J Biol Inorg Chem 5:551–559

Haehnel W, Jansen T, Gause K, Klösgen RB, Stahl B, Michl D, Huvermann B, Karas M, Herrmann RG (1994) Electron transfer from plastocyanin to photosystem I. EMBO J 13:1028–1038

Kannt A, Young S, Bendall DS (1996) The role of acidic residues of plastocyanin in its interaction with cytochrome f. Biochim Biophys Acta 1277:115–126

Katoh S (2003) Early research on the role of plastocyanin in photosynthesis. Photosynth Res 76:255–261

Malkin R, Malmström BG (1970) The state and function of copper in biological systems. Adv Enzymol Relat Areas Mol Biol 33:177–244

Solomon EI, Hadt RG (2011) Recent advances in understanding blue copper proteins. Coord Chem Rev 255:774–789

Solomon EI, Szilagyi RK, George SD, Basumallick L (2004) Electronic structures of metal sites in proteins and models: contributions to function in blue copper proteins. Chem Rev 104:419–458

Sykes AG (1991) Plastocyanin and the blue copper proteins. Struct Bonding 75:175–224

Ubbink M, Ejdebäck M, Karlsson BG, Bendall DS (1998) The structure of the complex of plastocyanin and cytochrome f, determined by paramagnetic NMR and restrained rigid-body molecular dynamics. Structure 6:323–335

Platinum (IV) Complexes, Inhibition of Porcine Pancreatic Phospholipase A2

Tina Kamčeva[1,2] and Marijana Petković[1]
[1]Laboratory of Physical Chemistry, Vinča Institute of Nuclear Sciences, University of Belgrade, Belgrade, Serbia
[2]Laboratory of Clinical Biochemistry, Section of Clinical Pharmacology, Haukeland University Hospital, Bergen, Norway

Synonyms

Binding of platinum to proteins; Enzyme involved in phospholipid metabolism; Metal-protein interactions; Secretory phospholipase A_2 group IB

Definition

Phospholipases A_2 are a group of enzymes, which catalyze hydrolysis of ester bond at the sn-2 position of phospholipids, generating important second messenger molecules. Secretory type of this enzyme has been shown to play a role in metastatic potential of a tumor, as it has been found that this enzyme is overexpressed in tumors with higher metastatic potential. The effect of platinum complexes on the activity of this group of enzymes has not been investigated in detail. The mechanism of the interaction between Pt(IV) complexes, potential novel antitumor drugs, and secretory type of PLA_2 is important, because this enzyme is considered as target enzyme for antitumor therapy.

Phospholipases

Phospholipases are large family of enzymes which all have the same function, to hydrolyze phospholipids into fatty acids and other lipophilic compounds.

First enzyme belonging to this group was discovered in 1986, when it was isolated from human pancreatic juice. The subsequent isolation of enzymes with phospholipase activity from various mammalian exocrine glands and snake venoms revealed that all these enzymes are relatively small (14 kDa), have highly conserved structure, and that they are very stable enzymes containing high number of cross-linked cysteine disulfide bridges.

At first, it was shown that phospholipases have just digestive role, but with the discovery of new types of secretory phospholipases in mammalian neutrophils and intracellular phospholipases in human platelets and neutrophils, more physiological functions of these enzymes were acknowledged. Phospholipases have a major role in homeostasis of cellular membrane and phospholipid turnover, host defense, digestion and uptake of fat and cholesterol, and as a "toxic" component of the venoms of snakes and insects. Products of phospholipase activity are precursors for the generation of various bioactive signaling compounds, which are important for inter- and intracellular signal transduction (Aloulou et al. 2012).

On the basis of the ester bond that is cleaved within a phospholipid molecule, phospholipases are classified in four major groups: A, B, C, or D. Phospholipases A_1 (PLA_1) and A_2 (PLA_2) cleave the acyl ester bond at either sn-1 or sn-2 position (Fig. 1), respectively. As a result, phospholipase A_2 releases arachidonic acid and lysophospholipid.

Phospholipase B hydrolyzes ester bonds at both sn-1 and sn-2 positions and it generally acts on lysolecithin. Phospholipases C catalyze hydrolysis of glycerophosphate bond (Fig. 1), releasing diacylglycerol and a phosphate-containing head group; its most important function is in cell signaling. Phospholipases D hydrolyze the P–O bond after phosphate group (Fig. 1), releasing phosphatidic acid and an alcohol. All these enzymes are further classified into subgroups according to their structural/or functional properties. For example, group of enzymes named phospholipases A_2 contains 14 subgroups (I-XIV) of enzymes identified to date and categorized into five main categories as given in the Table 1. These are:

Platinum (IV) Complexes, Inhibition of Porcine Pancreatic Phospholipase A2, Fig. 1 Classification of phospholipases on PLA_1, PLA_2, PLC, and PLD, according to site of their action on phospholipid

secreted phospholipases (sPLAs), intracellular or cytosolic Ca^{2+}-dependent phospholipases (cPLAs), Ca^{2+}-independent phospholipases iPLAs, platelet-activating factor acetylhydrolase phospholipases A_2 (PAF-AHs), and lysosomal phospholipases A_2 (Six et al. 2000).

The diversity of the phospholipase function from digestion of nutrients and remodeling of cell membrane to the formation of bioactive molecules suggests that these enzymes are critical to life. They represent a complex cellular mechanism in regulating turnover and rearrangement of membrane phospholipids, which is very important in a wide range of intracellular and extracellular functions. That is why this enzyme is more and more popular as a biological therapeutic target in development of new drugs for prevention and treatment of various diseases (Aloulou et al. 2012).

Because secreted phospholipase A_2 from bovine pancreas is the earliest discovered and the most investigated enzyme in this superfamily, it is most often used as a model in experiments. Even though the molecular mechanism of secreted phospholipase A_2 action is not well understood, many studies have examined the role of this enzyme in eicosanoid release and, consequently, the potential effect of enzyme specific inhibitors in reducing inflammatory process in animal models. Since secretory phospholipase A_2 has been implicated in chronic inflammatory diseases, such as atherosclerosis and asthma, it is a relevant target for anti-inflammatory drugs. On the other hand, this enzyme has been investigated in development and growth of human solid tumors and tumor cell proliferation, as a molecular target in chemotherapy and tumor sensitization to radiation therapy, as well as in the new strategies for phospholipase-activated targeted drug delivery via liposomes.

Because of this, it is not surprising that the interaction of phospholipase A_2 and chemotherapeutic agents is of interest in biochemistry nowadays.

Metal Complexes as Therapeutic Agents

Drugs based on transition metal complexes, such as platinum or ruthenium, are of particular interest because of several reasons (▶ Platinum Anticancer Drugs).

First of all, inorganic complexes with metal center and variety of ligands have a certain advantage over organic compounds, since metal can exist in more oxidation states. This feature is important due to the fact that when pharmaceutical compound enters the body, it reacts with a number of oxidative and reductive species where metal compound can change its oxidation state and continue its function, not being inactivated or trapped. Beside this, different oxidation states of metals determine their reactivity and stability and that particular property can be used for design of new drugs based on transition metals. For example, ruthenium exists in oxidation states II, III, and IV and it is known that one with the oxidation state III is more inert than other two. The idea of application of ruthenium, in this example, as a chemotherapeutic drug is to administer complex based on ruthenium in the oxidation state III so that the reaction of the complex with biomolecules in healthy, oxygen-rich tissues is preserved. Since the tumor tissue has low level of oxygen, it represents the reductive environment and ruthenium in metal complex changes its oxidation state from III to II. This form of ruthenium complex is more active and the change of oxidation state acts as an activator for the drug. In the case of platinum complexes, Pt (IV) complexes are more inert than Pt (II) complexes and they are supposed to undergo reduction in the cancer cells (Dabrowiak 2009).

Another reason why transition metal complexes are relevant compounds in chemotherapy is the reaction of ligand exchange, which has been exploited in therapy to target cells. Namely, metal complexes can have various ligands, bound to the central ion through weak chemical bonds, called coordinate covalent bonds and they can change their structure in different environment, releasing and/or exchanging weakly bound ligands. For example, the most used chemotherapeutic agent based on platinum, *cis*-platinum, has four monodentate ligands, two amino groups and two chloride ions. The concentration of chloride ions in blood is about 100 mM, which is sufficiently high to preserve

Platinum (IV) Complexes, Inhibition of Porcine Pancreatic Phospholipase A2, Table 1 Classification of phospholipases A_2 on 14 subgroups (I–XIV) and five categories and their characteristics

Group			Initial/common source	Name	Size/kDa
I	A		Cobra/krait venom	$sPLA_2$	13–15
	B		Mammal pancreas		
II	A		Human synovial fluid/platelets		
			Rattlesnake/viper venom		
	B		Gaboon viper venom		
	C		Rat/murine testis		15
	D		Human/murine pancreas/spleen		14–15
	E		Human/murine brain/hearth/uterus		
	F		Murine testis/embryo		16–17
III			Bee/lizard/scorpion/human		15–18
IV	A		Human U937 cells/platelets	$cPLA_2$	85
			Raw 264.7 cells/rat kidney		
	B		Human pancreas/liver/hearth/brain	$cPLA_2$	114
	C		Human hearth/skeletal muscle	$cPLA_2$	61
V			Mammal hearth/lung/macrophages	$sPLA_2$	14
VI	A-1		P388D1 macrophages, CHO cells	$iPLA_2$-A	84–85
	A-2		Human lymphocytes, testis	$iPLA_2$-B	88–90
	B		Human hearth/skeletal muscle	$iPLA_2$ or $iPLA_2$-2	88
VII	A		Human/murine/porcine/bovine plasma	PAF-AH	45
	B		Human/bovine liver/kidney	PAF-AH(II)	40
VIII	A		Human brain	PAF-AHIb $_1$	26
	B		Human brain	PAF-AHIb $_2$	26
IX			Snail venom	$sPLA_2$	14
X			Human spleen/thymus/leukocytes		
XI	A		Green rice shoots	PLA_2-I	12.4
	B		Green rice shoots	PLA_2-II	12.9
XII			Mammal hearth/kidney/skin/muscle	$sPLA_2$	18.7
XIII			Parvovirus		<10
XIV			Symbiotic fungus/streptomyces		13–19

the *cis*-platin in the stable, nonreactive state. Nevertheless, when it enters the cell through diffusion or active transport (▶ Biological Copper Transport), it reaches the chloride-ion poor environment (around 3–20 mM in the cytosol) which results in the complex hydrolysis (release of both chloride ions and binding of two water molecules) and activation; only in this state, *cis*-platin can bind DNA strands (Dabrowiak 2009).

In many studies, it has been proved that platinum complexes bind N-donor atoms on guanine residues on DNA strands, causing distortion of DNA (▶ DNA-Platinum Complexes, Novel Enzymatic Properties). This, in turn, prevents transcription and replication and provokes cell apoptosis. On the other hand, platinum complexes can react with a number of other compounds in the cell, such as peptides and proteins, containing N-, S-, or O-donors (▶ Platinum-Containing Anticancer Drugs and Proteins, Interaction). All proteins are different combinations of 20 common amino acids of which seven have donor atoms in their side chains and are potential targets for metallo-drugs from the standpoint of coordination chemistry. These are aspartic and glutamic acid, histidine, lysine, methionine, cysteine, and tyrosine. Large proteins usually have several disulfide bridges or disulfide bonds in their structure, formed between two cysteines in amino acid sequence, that enables the protein folding and obtaining tertiary or quaternary structure. Indirectly, this enables the protein to fulfill specific function. Cysteine residues react with heavy/transition metal ions, such as platinum, which can deform and inactivate the protein (▶ Terpyridine Platinum(II) Complexes as Cysteine Protease Inhibitors) (▶ Platinum- and Ruthenium-Based Anticancer Compounds, Inhibition of Glutathione Transferase P1-1) (▶ Platinum(II) Complexes, Inhibition of Kinases). It is known that platinum binds to albumin, insulin, metallothionein, cytochrome C, ubiquitin, myoglobin, Na, K-ATPase, and glutathione S-transferase. The interaction of metallo-drugs with proteins has been considered as one mechanism by which tumor cells acquire resistance to therapy (▶ Platinum-Resistant Cancer).

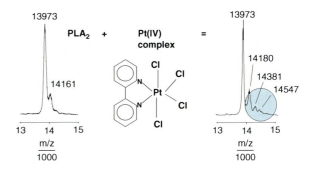

Platinum (IV) Complexes, Inhibition of Porcine Pancreatic Phospholipase A2, Fig. 2 Example of binding of platinum (IV) complex for the pancreatic phospholipase A_2 detected by MALDI-TOF mass spectrometry

Interaction of Pt(IV) Complex with Phospholipase A_2: Binding and Inhibition

Porcine pancreatic phospholipase A_2 belongs to group IB of secretory phospholipases A_2. All subgroups of secretory phospholipases A_2 share common features such as low molecular weight, a catalytic histidine-aspartic (His-Asp) acid dyad, Ca^{2+} bound in the active site, and conserved disulfide bonds. This subgroup of the enzyme requires higher concentrations of Ca^{2+} ions (millimolar range) for their activation (▶ Calcium-Binding Proteins, Overview). Subgroup IB of secreted phospholipases A_2 has seven disulfide bonds, and it is supposed that those are the potential sites for interaction with platinum compounds (▶ Platinum-Containing Anticancer Drugs and Proteins, Interaction).

It is important to highlight that there are several possible outcomes of the interaction of platinum complex with the phospholipase A_2. The platinum compound can bind to the enzyme or can undergo weak hydrophobic interaction without binding. If the platinum complex binds the enzyme, the bond formed can affect the activity of the enzyme and cause inhibition, or it can have no effect on the enzyme activity. This depends of the site where platinum is bound. Hence, investigating the interaction between platinum complexes and the enzyme considers detecting of the potential binding of the complex for the phospholipase and investigating whether eventual binding causes changes in enzyme's activity.

In order to determine binding of metal complexes, among them, those that contain Pt(IV) as the metal center, for the enzyme, several techniques can be used, such as NMR (nuclear magnetic resonance); "soft" ionization mass spectrometric methods: MALDI (matrix-assisted laser desorption and ionization), ESI (electron spray ionization) or FAB (fast atom bombardment) mass spectrometry; chromatographic methods, etc.

Regardless of the method, the main idea for determining whether the complex is bound or not is to compare appropriate spectra or chromatogram of pure enzyme and enzyme previously incubated with metal complex. Eventual changes can be then analyzed and attributed to the corresponding structural changes. Nevertheless, it is important to be aware that not every change in the spectrum confirms binding of two compounds. One should also keep in mind during analysis that no observable change in the spectrum obtained by one method does not eliminate the possibility of binding of metal complex to the enzyme. It is important therefore to use different scientific techniques and approaches in order to detect different types of interaction. MALDI-TOF MS (matrix-assisted laser desorption and ionization) together with other "soft" ionization methods is one of the most promising technique for determination of the interaction between metal complexes and phospholipase A_2. By the assistance of matrices, indirect ionization, and desorption of the biological sample, avoiding unnecessary fragmentation became possible (Schiller et al. 2007). In the Fig. 2, the MALDI-TOF spectra of the pure enzyme and the enzyme incubated with platinum (IV) complex are presented.

The binding of the complex for the enzyme is demonstrated simply by detecting additional signals on higher m/z values than those characteristic for the phospholipase A_2. Additionally, the m/z difference should correspond to the mass of complex or its fragment.

Inhibition

Like it was said earlier in the text, the binding of tested compound for the enzyme does not automatically mean that the compound is inhibitor of the enzyme and the enzyme activity should not be furthermore investigated (▶ Terpyridine Platinum(II) Complexes as Cysteine Protease Inhibitors). Investigation of the effect of potential inhibitor on the enzyme activity is usually done by comparing the amount of a substrate consumed (in this case a phospholipid) or of the product formed (corresponding lysophospholipid) in the reaction catalyzed by an enzyme in the presence of this substance with values obtained without it. In the case of measuring activity of different types of phospholipases, the main obstacle is that enzyme is soluble in water, but its substrate as well as products are hydrophobic and do not dissolve in the same polar solvent. It is well known that lipids in water spontaneously aggregate in micelles, ball-shaped structures with polar heads on the border and hydrophobic tails in the center. Because of this, the most used technique for determining enzyme activity, UV/VIS (ultraviolet/visible) spectroscopy, is not the method of choice. Using detergents or fluorescent labeling of phospholipids prior using other methods complicates the procedure and does not resemble the conditions in the cell membranes. Here, the possibilities of previously described mass spectrometric method in determining phospholipase activity in the presence of platinum (IV) compound are presented. This method has several advantages over other methods, such as simple and fast preparation of samples, spectra acquisition and analysis. It is possible not only to detect substrate and product of the reaction and the metal complex in the same spectra, but also to monitor the changes in their amounts with the time. The quantification of species by mass spectrometric methods requests the usage of internal standards, which are added to all samples in the same concentration. If the goal is to observe the decrease of the substrate (phospholipid) with the time of reaction of hydrolysis by phospholipase A_2, internal standard should be another phospholipid, with similar structure and slightly different mass. It is important for internal standard to have as much as possible similar ionization and desorption properties, in order to undergo the same processes and to the same extent as substrate under the same conditions. Likewise, if the goal is to monitor and to quantify increase of the product, lysophospholipid, appropriate lysophospholipid with slightly different mass is used. The principle of monitoring the effect of the activity of phospholipase A_2 is illustrated in the Fig. 3.

To determine whether platinum (IV) complex inhibits the enzyme, it is necessary to calculate and compare kinetic parameters from several spectra in the absence and the presence of the metal complex under the same conditions. Knowing that internal standard is always added in the same amount, it is possible to calculate amounts of other species in the sample from the proportion of known standard concentration and appropriate mass peak intensities. Hence, it is possible to determine kinetic parameters, such as Km and Vmax, (constants for Michaelis-Menten's kinetics) and compare their values with and without metal complex added in the reaction sample. In this case, it is possible simply to compare the amount of product obtained by hydrolysis of the same amount of phospholipid by phospholipase A_2 with the amount of the same product when the platinum (IV) complex is added. When platinum (IV) complex is added in the sample, it binds for the phospholipases and affects its activity, because less lysophospholipid is produced compared to the case of incubation without the transition metal complex (Kamceva et al. 2011) (Radisavljevic et al. 2012).

Nevertheless, knowing kinetic parameters gives valuable information for the type of inhibition and consequently the mechanism of interaction of platinum (IV) complex with the enzyme. According to mechanism by which an enzyme binds the inhibitor and/or substrate, inhibitors may exhibit competitive, noncompetitive, or uncompetitive inhibition. Competitive inhibition is the mechanism where the inhibitor binds to the active site on the enzyme before the substrate, thus preventing binding of substrate. Noncompetitive inhibitors bind to the enzyme equally irrespectively on the interaction with the substrate; if one of these states (binding to the pure enzyme or to the enzyme-substrate complex) is more favored, this type of inhibition is called mixed inhibition. In the case of

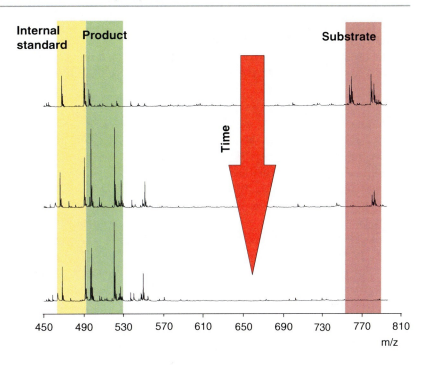

Platinum (IV) Complexes, Inhibition of Porcine Pancreatic Phospholipase A2, Fig. 3 Determining the activity of phospholipase A_2 by quantifying the amount of substrate (phospholipid) or product (lysophospholipid) with the time of reaction by MALDI-TOF mass spectrometry

inhibition of pancreatic phospholipase A_2 by platinum (IV) complex, the type of inhibition is uncompetitive or anticompetitive. Uncompetitive inhibition means that the inhibitor interacts with the enzyme already complexed with the substrate (phospholipid) and prevents generation of a product (lysophospholipid and free fatty acid). On a molecular level, this type of inhibition suggests that platinum (IV) complex binds to the phospholipase A_2 in the presence of substrate, phospholipids. When substrate binds to the active site, the enzyme changes its conformation and sites for potential binding of platinum complex become available.

Even though the inhibition of phospholipase A_2 by platinum and ruthenium complexes as well as the mechanism of this inhibition is explained, it is not yet known if the inhibition by metal complexes will have therapeutic effect and whether this type of the enzyme can be considered as target molecule for cancer therapy or not.

Cross-References

▶ Calcium-Binding Proteins, Overview
▶ Platinum- and Ruthenium-Based Anticancer Compounds, Inhibition of Glutathione Transferase P1-1
▶ Platinum Anticancer Drugs
▶ Platinum(II) Complexes, Inhibition of Kinases
▶ Platinum-Containing Anticancer Drugs and Proteins, Interaction
▶ Terpyridine Platinum(II) Complexes as Cysteine Protease Inhibitors

References

Aloulou A, Ali YB, Bezzine S, Garqouri Y, Gelb MH (2012) Phospholipases: an overview. Methods Mol Biol 861:63–85

Dabrowiak JC (2009) Metals in medicine. Willey, Chichester

Kamceva T, Flemmig J, Damnjanovic B, Arnhold J, Mijatovic A, Petkovic M (2011) Inhibitory effect of platinum and ruthenium bipyridyl complexes on porcine pancreatic phospholipase A_2. Metallomics 10:1056–1063

Radisavljevic M, Kamceva T, Bugarcic ZD, Petkovic M (2012) Inhibitory effect of cisplatin and [Pt(dach)Cl2] on the activity of phospholipase A2. J Enzyme Inhib Med Chem. doi:10.3109/14756366.2012.666539. Available online via DIALOG. http://informahealthcare.com/doi/pdf/10.3109/14756366.2012.666539

Schiller J, Suss R, Fuchs B, Muller M, Zschornig O, Arnold K (2007) MALDI-TOF MS in lipidomics. Front Biosci 12:2568–2579

Six DA, Dennis EA (2000) The expanding superfamily of phospholipase A_2 enzymes: classification and characterization. Biochim Biophys Acta 1488:1–19

Platinum- and Ruthenium-Based Anticancer Compounds, Inhibition of Glutathione Transferase P1-1

Lorien J. Parker[1,2], Paul J. Dyson[3], Mario Lo Bello[4] and Michael W. Parker[1,2]
[1]Biota Structural Biology Laboratory, St. Vincent's Institute of Medical Research, Fitzroy, VIC, Australia
[2]Department of Biochemistry and Molecular Biology, Bio21 Molecular Science and Biotechnology Institute, The University of Melbourne, Parkville, VIC, Australia
[3]Institut des Sciences et Ingénierie Chimiques, Ecole Polytechnique Fédérale de Lausanne (EPFL) SB ISIC-Direction, Lausanne, Switzerland
[4]Department of Biology, University of Rome "Tor Vergata", Rome, Italy

Synonyms

Detoxification of anticancer metallodrugs; Metallocompound protein binding

Definition

Glutathione S-transferases (GSTs) are intracellular enzymes that can bind small molecule toxins leading to their eventual removal from the cell. A major class of anticancer drugs are metallocompounds, particularly those that contain platinum (widely used in the clinic) and ruthenium (in clinical trials). GSTs can bind and react with certain metallocompounds and metalloinhibitors of GSTs can be designed.

What are Glutathione S-Transferases?

The glutathione S-transferases (GSTs; EC 2.5.1.18) are a family of detoxification proteins. The main role of these enzymes is to catalyze the conjugation reaction of the tripeptide glutathione (GSH, γ-Glu-Cys-Gly) with a number of toxic compounds found or introduced into the cell. Subsequently, in the next step of detoxification, these GSH-conjugates are pumped out of the cell (Mannervik and Danielson 1988).

The conjugation reaction occurs through the nucleophilic addition of the thiol of GSH to an electrophilic center of a toxic nonpolar compound (1):

$$RX + GSH \rightarrow GSR + XH \quad (1)$$

where R may be an alkyl or aryl group and X may be a wide variety of organic and inorganic ions. GSTs have also been shown to be involved in the detoxification of antibiotics, insecticides, herbicides, and environmental pollutants. GSTs catalyze many other types of reactions, all involving the co-substrate GSH, including Michael additions to α,β-unsaturated ketones, nucleophilic aromatic substitutions, and epoxide ring-opening reactions. GST-catalyzed addition reactions are usually studied using ethacrynic acid (EA) while aromatic substitution reactions are studied most often using the model substrate 1-chloro-2,4-dinitrobenzene (Mannervik and Danielson 1988).

In addition to their catalytic role in detoxification, GSTs possess many other functions including structural roles such as S-crystallins, peroxidase and isomerase activity, dehalogenation reactions, transcription factors, prostaglandin synthesis, cell protection from radiation damage, a target for transglutaminase in neural cells committed to apoptosis, isomerization of maleylacetate, nitric oxide (NO) transport, thiol transferase activity, dehydroascorbate reduction, inhibition of the proapoptotic action of Bax, modulation of calcium channels and chloride channel function (Oakley 2005).

GSTs are ubiquitously expressed in aerobic organisms and can be grouped into three main families: the cytosolic, mitochondrial, and microsomal enzymes (the latter called MAPEGS). At least seven cytosolic classes of GSTs have been identified in mammals including Alpha, Mu, Pi, Sigma, Theta, Zeta, and Omega. Kappa class GSTs are found in the mitochondrial matrix. In addition to these enzymes, there are other closely related enzymes that do not have the typical GST enzymatic function but nevertheless adopt the same three-dimensional folds as the cytosolic GSTs (e.g., chloride intracellular channels or CLICs and elongation factor 1B).

The expression patterns of GSTs vary considerably depending on the tissue and cell types. For example, the liver contains high levels of the Alpha class enzyme but very little Pi, whereas the Pi class enzyme dominates in the intestine. GSTP1-1 is generally more

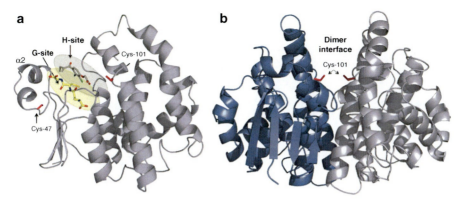

Platinum- and Ruthenium-Based Anticancer Compounds, Inhibition of Glutathione Transferase P1-1, Fig. 1 Three-dimensional fold of GST P1-1. The figure shows the location of the two most reactive cysteines, Cys-47 and Cys-101 (*red bonds*) and the location of the G-site, including helix α2 that forms one of the "walls" of the active site, and the hydrophobic substrate-binding site or H-site. (**a**) Monomeric fold. (**b**) Dimeric organization

highly expressed in proliferating cells than in fully differentiated ones. Consequently, the role of this enzyme in a number of different diseases, in particular cancers, has been intensively explored.

What Do Glutathione Transferases Look Like?

Generally, cytosolic GSTs exist as either homo- or heterodimers with subunits of molecular weight of approximately 25 kDa. The structure of human Pi class GST (hGSTP1-1) is shown in Fig. 1 (Reinemer et al. 1992). The N-terminal domain of the polypeptide, comprising approximately one third of the protein, consists of a β-sheet of four strands with three flanking α-helices. This topology belongs to the thioredoxin superfamily fold. The loop connecting helix α2 to beta-strand β3, referred to as the *cis*-Pro loop, contains a characteristic *cis*-proline residue that is strictly conserved throughout the GST superfamily. The N-terminal domain also contains the glutathione-binding site (G-site; Fig. 1a). The all α-helical C-terminal domain comprises two thirds of the protein. In hGSTP1-1, there are five α-helices. This domain contributes residues involved in binding hydrophobic surfaces of small molecular weight toxins. The hydrophobic substrate-binding site, or H-site (Fig. 1a), is highly variable and is responsible for the catalytic diversity that is characteristic of the GST family.

The GSTP1-1 dimer has a markedly V-shaped subunit interface (Fig. 1b) that is rather open to solvent but which is predominantly hydrophobic, particularly in the subunit contact regions. The subunits, related by a twofold axis, interact to form the dimer structure illustrated in Fig. 1b, utilize a novel "lock and key" mechanism (Reinemer et al. 1992). The "key" residue is an aromatic residue from the loop proceeding strand β3 that is Tyr-49 in the hGSTP1-1. The "lock" is formed by hydrophobic residues from helices α4 and α5 of the neighboring subunit. In the Pi class enzyme, the dimer interface is fairly large and creates a V-shaped crevice that is solvent accessible as can be seen in Fig. 1b.

None of the cysteine residues of hGSTP1-1 (four per subunit) are located in the active site and their nonessential character has been demonstrated by site-directed mutagenesis. Of the four, the most reactive to chemical reagents are Cys-47 and Cys-101. Crystallographic analysis reveals that Cys-47 is located at the C-terminal end of helix α2, the most flexible helix in GSTP1-1, with its thiol group partly buried in a small hydrophobic pocket defined by residues Trp-38, Leu-43, Leu-52, Tyr-63, and the aliphatic portion of Lys-54 (Reinemer et al. 1992; Fig. 1a). When GSH is bound to GSTP1-1, helix α2, which forms one wall of the G-site, is restricted in its movement by interactions with GSH. This results in the burial of the Cys-47 residue from solvent making it inaccessible for modification. Although Cys-47 is not directly involved in either substrate binding or catalysis, chemical modification of its side chain (in the absence of GSH) inactivates the enzyme reversibly, possibly by blocking a structural transition during catalysis or by inducing

Platinum- and Ruthenium-Based Anticancer Compounds, Inhibition of Glutathione Transferase P1-1, Fig. 2 Chemical structures of (**a**) cisplatin, (**b**) ethacrynic acid, (**c**) ethacraplatin, (EACPT) and (**d**) etharapta

a conformational change that is unfavorable for GSH binding. Either way the structural integrity of this cysteine residue seems to be crucial in maintaining an active conformation of the enzyme. Under physiological conditions, where GSH is present at high concentrations, Cys-101 is the only reactive cysteine residue in the protein. Cys-101 is located at the very bottom of the dimer interface crevice (Fig. 1b).

GST and Binding of Nonsubstrate Compounds

GSTs were originally identified as proteins that bind large lipophilic molecules (> 400 Da), and thus, it was hypothesized that their function was to store and/or transport such molecules in the cell (Litwack et al. 1971). The so-called ligandins could bind to a range of physiological molecules including hematin, vitamin E, bilirubin, steroids, thyroid hormones, fatty acids, and a range of drugs. Early work suggested these molecules bound at the GST dimer interface and some nonmammalian GST structures did reveal that lipophilic molecules bind there (McTigue et al. 1995; Ji et al. 1996). However, it appears most, if not all, such ligands bind in the H-site of human GSTP1-1 (Oakley et al. 1999).

GST Involvement in Drug Resistance

Overexpression of GSTP1-1 has been observed in cancer tissues of the lung, colon, ovary, testis, bladder, mouth, and kidney (Hayes et al. 2005). Overexpression of this isozyme has been associated with transformation to malignancy and the development of drug resistance to cancer treatment. For example, GSTP1-1 knockout mice are much more susceptible to the cytotoxic effects of anticancer drugs than their matched normal controls (Henderson et al. 1998). The literature is replete with correlative studies demonstrating elevated hGSTP1-1 levels in a variety of drug-resistant cell lines. A number of GST inhibitors have shown promising results in the reversal of resistance and prevention of tumor development, but many suffer from a lack of specificity and hence exhibit undesirable side effects in preclinical and clinical trials (Hayes et al. 2005).

The GST Inhibitor Ethacrynic Acid

Ethacrynic acid (EA; Fig. 2), a well-known GSTP1-1 inhibitor, has been investigated as a potential therapeutic molecule. It reached phase II clinical trials but was subsequently withdrawn due to severe dose-limiting toxicities (Hayes et al. 2005). The structure of GSTP1-1 in complex with EA showed it bound to the active site of the enzyme in both productive and nonproductive modes in the absence of GSH (i.e., binding in the active site with orientations conducive to attack by the co-substrate GSH or not) and formed a conjugate with GSH when it was present.

GSTs as a Target for Drug Design

The results from the EA studies prompted research into the design of GSTP1-1 inhibitors that could prevent both the GST-catalyzed conjugation of anticancer drugs and their noncatalytic sequestration by the enzyme. Such an inhibitor would prevent GST from "mopping up" the supply of drug in the cell, resulting in an increase in available drug for tumor-attacking activity and avoiding the development of GST-based resistance that is caused by increased expression of the enzyme in cancer cells.

Metal-Based Anticancer Drugs

The metal-based anticancer drug cisplatin (*cis*-diamminedichloroplatinum(II); Fig. 2a) and related platinum-based drugs carboplatin and oxaliplatin are used in about half of all tumor therapies, particularly sarcomas, some carcinomas (e.g., small cell lung cancer, and ovarian cancer), lymphomas, and germ cell tumors. These drugs exert their cytotoxicity through the formation of interstrand and intrastrand cross-links with DNA, which, in turn, disrupts transcription, translation, and repair mechanisms. The targeted cancer cells subsequently undergo programmed cell death or apoptosis. The use of these drugs is often limited due to severe side effects and reoccurrence of the disease due to intrinsic and acquired resistance.

The development of the next generation of platinum drugs has focused on octahedral platinum (IV) complexes due to their relative chemical inertness and generally better oral bioavailability due to the higher coordination sphere compared to platinum(II) centers, allowing for the introduction of ligands that can be tailored to optimize pharmacokinetic properties.

An alternative approach to improve the druggable properties of platinum-based drugs has been to search for other metal centers with more suitable properties. Compounds with ruthenium centers possess many advantages over platinum-based compounds including a broad range of synthetic chemistry possibilities, a range of oxidation states (II to IV), and a markedly lower toxicity compared to similar platinum-based drugs. Good examples are KP1019, a ruthenium-based compound with no significant toxic side effects, which shows promising activity against colorectal tumors, and NAMI-A, a ruthenium drug candidate with selectivity for solid tumors, that has successfully completed human phase I clinical trials (Ang et al. 2009).

GSTs and Metallodrug Design

GST overexpression is strongly correlated with the development of resistance in cisplatin treatment and inhibition of GSTs can reverse this resistance, leading to the hypothesis that GSTs may be directly involved in this resistance process (Hayes et al. 2005). With this in mind, novel metallocompounds have recently been synthesized that comprise platinum(IV) centers with the GST inhibitor, EA, as ligands (Fig. 2). The aim of these studies have been to develop prodrugs which target overexpressed GST enzymes in cancer cells, leading to enzymatic inhibition concomitant with cleavage and release of the metallodrug from the enzyme, thereby overcoming resistance and leading to cell death.

Ethacraplatin

Ethacraplatin is a trans-platinum(IV) carboxylate complex with two axial ethacrynic acid moieties (EACPT; Fig. 2c). EACPT decreases GST activity in vitro by over 90% and in cell-based assays lowers the activity of cellular GST to less than 30%. Interestingly, EACPT significantly inhibits GST activity more than would be expected by EA alone. EACPT also inhibits the cell viability of a variety of carcinoma cell lines (Parker et al. 2011).

The three-dimensional atomic structure of GSTP1-1 complexed to EACPT, as determined by X-ray crystallography, reveals that the platinum center binds at the dimer interface and the EA ligands are cleaved and translocated to the active site, thus inhibiting the enzyme (Parker et al. 2011). XANES studies demonstrate that the platinum center is reduced from +IV to +II in the process (Parker et al. 2011). The presence of a metal-binding site at the dimer interface had not been previously documented and hence was a surprising feature of the crystal structure.

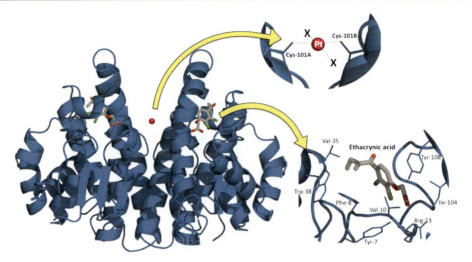

Platinum- and Ruthenium-Based Anticancer Compounds, Inhibition of Glutathione Transferase P1-1, Fig. 3 Crystal structure of the GSTP1-1 EACPT complex (Parker et al. 2011). EACPT is shown in stick fashion, the platinum center by a *pink sphere* and the enzyme as *blue ribbon*. The intact compound is thought to bind at the dimer interface in a relatively symmetric manner. The compound is then cleaved and the EA moieties transverse to the substrate-binding site while the platinum center remains bound to the dimer interface via Cys-101 residues. The identity of the two of the metal ligands was not clear from the electron density maps and hence are notated as "X"

This work led to the hypothesis that GSTP1-1 involvement in the resistance of cancer cells to metallodrugs might be partly due to sequestering of such drugs at the dimer interface in addition to detoxification through GSH-catalyzed reactions at the active site (Fig. 3).

The crystal structure of the GST-EACPT complex showed the platinum center was bound at the dimer interface by binding to the Cys-101 residues from each monomer. This binding may contribute to the surprising increase in inhibitory activity of EACPT compared to EA alone.

Etharapta

A promising class of ruthenium anticancer drugs are ruthenium(II) arene PTA complexes (RAPTA; Fig. 2d) which are water soluble and highly selective for cancer cells and display both anti-angiogenic and anti-metastatic effects in vivo. RAPTA-based compounds with EA ligands have been synthesized, so-called etharapta compounds, as certain metastatic tumors contain high levels of GSTs. The resulting compounds exhibit dual activity of GST inhibition and cancer cell cytotoxicity (Ang et al. 2009). Like ethacraplatin, the complex was shown to be a more effective GST inhibitor than EA alone, suggesting that the metal center also is involved in inhibition of the enzyme.

Structures of GSTP1-1 etharapta complexes derived from crystals of GSTP1-1 incubated with etharapta for varying incubation times reveal that the metal complex binds at the dimer interface first with the metal center bound through Cys-101 residues of opposing monomers and then over time, the metal center is released from the protein while the EA ligands traverse to the active sites where they bind and inhibit the enzymatic activity (Fig. 4).

GSTP1-1 Dimer Interface is a Metal-Binding Site

Early studies of GSTs suggested the dimer interface might bind large hydrophobic molecules although no solid experimental evidence has ever been presented that this is the case in the human GST pi class enzyme (see above). Instead the crystallographic studies with platinum- and ruthenium-based compounds have revealed the dimer interface, by virtue of closely spaced cysteine residues from neighboring monomers, is a potent metal-binding site. This observation has led to the suggestion that one mode of GST-based resistance might involve sequestering metallodrugs at the

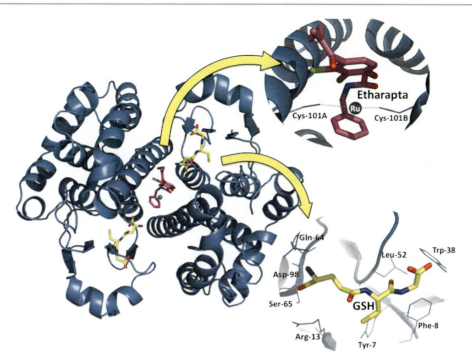

Platinum- and Ruthenium-Based Anticancer Compounds, Inhibition of Glutathione Transferase P1-1, Fig. 4 Crystal structure of the GSTP1-1 etharapta complex (Ang et al. 2009). Etharapta is shown in stick fashion, the ruthenium center by a *green sphere*, and GSH in *yellow bonds*. GSTP1-1 is shown as *blue ribbon*. The intact compound is thought to bind at the dimer interface initially. The compound is then cleaved and the EA moieties transverse to the substrate-binding site while the ruthenium center remains bound to the dimer interface via Cys-101 residues. There is evidence to suggest that at some point, the activated metal center is released leading to cell apoptosis. Cellular studies show a two-phase apoptosis process in which the first wave of apoptosis may be attributed to initial inhibition of GSTP1-1 by the etharapta complex and the second wave to elimination of an activated ruthenium drug from the enzyme that interferes with other biomolecular targets

subunit interface. An interesting result from these studies is that binding at the dimer interface appears to enhance the activity of known GST inhibitors such as EA that bind at the active site. Perhaps surprisingly, the metallocompounds are activated at the dimer interface by some unknown mechanism and the GST inhibitory ligands can then translocate to the enzyme active sites. Taken together, these results suggest GST inhibitors that block both the dimer interface and active site may prove potent anticancer adjuvants in future drug discovery efforts.

Cross-References

▶ Platinum Anticancer Drugs
▶ Platinum-Containing Anticancer Drugs and Proteins, Interaction
▶ Platinum-Resistant Cancer

References

Ang WH, Parker LJ, De Luca A, Juillerat-Jeanneret L, Morton CJ, Lo Bello M, Parker MW, Dyson PJ (2009) Rational design of an organometallic glutathione S-transferase inhibitor and its interactions with GST P1-1. Angew Chem Int Ed 48:3854–3857

Hayes JD, Flanagan JU, Jowsey IR (2005) Glutathione transferases. Annu Rev Pharmacol Toxicol 45:51–88

Henderson CJ, Smith AG, Ure J, Brown K, Bacon EJ, Wolf CR (1998) Increased skin tumorigenesis in mice lacking pi class glutathione S-transferase. Proc Natl Acad Sci USA 95:5275–5280

Ji X, von Rosenvinge EC, Johnson WW, Armstrong RN, Gilland GL (1996) Location of a potential transport binding site in a sigma class glutathione transferase by X-ray crystallography. Proc Natl Acad Sci USA 93:8208–8213

Litwack G, Ketterer B, Arias IM (1971) Ligandin: a hepatic protein which binds steroids, bilirubin, carcinogens and a number of exogenous organic anions. Nature 234:466–467

Mannervik B, Danielson UH (1988) Glutathione transferases – structure and catalytic activity. CRC Crit Rev Biochem 23:283–337

McTigue MA, Williams DR, Tainer JA (1995) Crystal structures of a schistosomal drug and vaccine target: glutathione S-transferase from *Schistosoma japonica* and its complex with the leading antischistosomal drug praziquantel. J Mol Biol 246:21–27

Oakley AJ (2005) Glutathione transferases: new functions. Curr Opin Struct Biol 15:716–723

Oakley AJ, Lo Bello M, Nuccetelli M, Mazzetti AP, Parker MW (1999) The ligandin (non-substrate) binding site of human Pi class glutathione transferase is located in the electrophile binding site (H-site). J Mol Biol 291:913–926

Parker LJ, Italiano LC, Morton CJ, Hancock NC, Ascher DB, Aitken JB, Harris HH, Campomanes P, Rothlisberger U, De Luca A, Lo Bello M, Ang WH, Dyson PJ, Parker MW (2011) Studies of glutathione transferase P1-1 bound to a platinum(IV)-based anticancer compound reveal the molecular basis of its activation. Chem Eur J 17:7806–7816

Reinemer P, Dirr HW, Ladenstein R, Huber R, Lo Bello M, Federici G, Parker MW (1992) Three-dimensional structure of class pi glutathione S-transferase from human placenta in complex with S-hexylglutathione at 2.8 Å resolution. J Mol Biol 227:214–226

Platinum Anticancer Drugs

Rabbab Oun[1] and Nial J. Wheate[2]
[1]Strathclyde Institute of Pharmacy and Biomedical Sciences, University of Strathclyde, Glasgow, UK
[2]Faculty of Pharmacy, The University of Sydney, Sydney, NSW, Australia

Synonyms

Carboplatin; Cisplatin; Heptaplatin; Lobaplatin; Nedaplatin; Oxaliplatin

Definition

A platinum-based anticancer drug is any agent that contains one or more platinum atoms in the oxidation state of II or IV, contains mono- or multidentate non-labile am(m)ine carrier ligands and labile chlorido or bidentate carboxylate ligands, and which acts as a prodrug; it undergoes aquation to yield a reactive complex capable of forming coordinate bonds with DNA bases. Its binding to DNA prevents replication and transcription which causes cell death through apoptosis.

Introduction

There are currently six platinum drugs with marketing approval in various regions throughout the world: cisplatin, carboplatin, oxaliplatin, nedaplatin, lobaplatin, and heptaplatin (Fig. 1). Platinum-based drugs are used for the treatment of human and animal tumors and are currently indicated for bladder, testicular, ovarian, non-small cell lung, small cell lung, and colorectal cancers as well as melanomas, lymphomas, and myelomas (Wheate et al. 2010).

History, Development, and New Drugs in Clinical Trial

While cisplatin was first synthesized in the late nineteenth century by Michele Peyrone (Kauffman et al. 2010), its anticancer properties were accidently discovered by Barnett Rosenberg in the 1960s while studying the effects of electric fields on the growth of *Escherichia coli* bacteria (Rosenberg et al. 1965). Cisplatin was approved for use by the US Food and Drug Administration in the 1970s. Carboplatin was developed as a less-toxic derivative of cisplatin and is used to treat the same cancers as cisplatin (approved late 1980s), although it is used as the platinum drug of choice in the treatment of ovarian carcinomas. Oxaliplatin gained marketing approval in the 1990s in Europe and in the last decade in the USA and is currently only approved for the treatment of colorectal cancers. There are a number of platinum-based drugs that are currently in various stages of clinical trials, including satraplatin, picoplatin, Prolindac, Lipoplatin, and Lipoxal. Satraplatin is a mononuclear drug based on platinum (IV) which requires in vivo reduction to platinum (II) before it becomes active. Picoplatin is a sterically hindered drug with reduced protein and glutathione binding. The remaining three drugs are polymer-based agents containing the active components of cisplatin (Lipoplatin) or oxaliplatin (Prolindac and Lipoxal).

Mechanism of Action

Platinum agents act as prodrugs and must undergo ligand substitution reactions with water to become active (Fig. 2). The severity of drug toxicity is largely related to the aquation rate of the drug; the slower its aquation, the less toxic it is. In the blood stream where the chloride concentration is high, approximately 100 mM, the drugs remain intact. Upon entering

Platinum Anticancer Drugs, Fig. 1 The six platinum anticancer drugs with marketing approval for the treatment of human and animal tumors

Platinum Anticancer Drugs, Fig. 2 A simplified reaction scheme of the aquation of cisplatin to form its reactive species which then goes on to form coordination bonds with guanosine and adenosine bases in DNA

a cell, whether healthy or cancerous, the chloride concentration is significantly lower, as low as 4 mM, which causes the drugs to undergo aquation. The resultant complexes, containing one coordinated water ligand, then undergo a second aquation step to yield complex now containing only coordinated am(m)ine and water ligands.

The target of platinum drugs is nuclear DNA. As an electrophile, platinum most readily binds to the N7 atom of guanosine bases and to a lesser extent, the N7 atom of adenosine bases. Binding can also occur at the adenosine N1 and the guanosine O6. Each platinum drug will bind two bases either through nucleosides on the same strand (intrastrand binding) or through individual bases, on different strands (interstrand binding). The most readily formed DNA adduct is the 1,2-GpG intrastrand adduct. Binding of platinum via this mode typically causes the DNA helix to unwind and bend, preventing DNA transcription and replication. This DNA damage subsequently initiates apoptosis.

Toxicity

Systemic toxicity of all platinum anticancer drugs is an ongoing issue with their application and can be a major cause of dose reduction for patients during subsequent dose cycles. The dose limiting toxicity (DLT) is different for most of the drugs. For cisplatin, the DLT is nephrotoxicity, which can be managed through the use of pre- and post-hyperhydration and diuresis. Other side effects include: nausea (controllable through the coadministration of antiemetics), ototoxicity, peripheral neurotoxicity and neuropathy, and myelosuppression (which may require the administration of antibiotics). The DLT of carboplatin is myelosuppression, although up to 30% of patients may also develop a hypersensitivity reaction to the drug. The DLT of oxaliplatin is neurotoxicity; for nedaplatin, it is myelosuppression; for lobaplatin, it is thrombocytopenia; and for heptaplatin, it is nephrotoxicity and/or intra-abdominal bleeding.

Formulation

Platinum drugs are formulated as powders for reconstitution or as prepared aqueous solutions. Cisplatin supplied as a lyophilized powder also contains sodium chloride (to prevent premature aquation of the drug upon dissolution) and mannitol (to manage nephrotoxicity). Cisplatin dilutions are prepared in 5% dextrose solutions. Carboplatin powders contain no other active pharmaceutical ingredients or excipients and are reconstituted in either 5% dextrose or 0.9% saline. Oxaliplatin is only supplied as a powder for reconstitution due to its instability once dissolved (shelf life as little as 6 h). It is only reconstituted in 5% dextrose and never in saline, as the chloride ions may displace the oxalate ligand, making the drug significantly more toxic. Formulation information on nedaplatin, lobaplatin, and heptaplatin is not readily available.

Protein Interactions and Binding

Platinum in its II + and IV + oxidation states is a soft electrophile and binds rapidly to soft nucleophiles. The most abundant nucleophile relevant to platinum drugs is the amino acid cysteine which contains a thiol group. Platinum drugs will also bind to the thioether group of methionine residues in peptides and proteins. There are several important proteins that are known to bind platinum within blood serum, and peptides/proteins located either on the surface of the cells or within cells.

Glutathione is a tripeptide consisting of glycine and cysteine residues connected to a glutamate residue through the latter's γ-carboxylate group. It is found in millimolar concentrations inside cells and is utilized as an antioxidant. Glutathione is found in two forms: the reduced form and the oxidized form, two glutathione peptides connected through a disulfide bond. Glutathione binding to platinum results in the degradation and deactivation of the drugs through the displacement of the drugs' ligands (Reedijk 1999). The initial binding step usually involves the displacement of one of the labile drug ligands (chloride or water) by coordination to the cysteine residue's sulfur atom. Ring closure can occur through the displacement of a second ligand on the drug by the cysteine's amine group (Fig. 3). Reaction with a second reduced glutathione molecule results in the displacement of the next two drug ligands, yielding either a *cis*- or *trans*-glutathione-chelated complex. While glutathione is able to rapidly displace mono-dentate-binding ligands, bi- and tridentate ligands, such as the biscarboxylate of carboplatin or the oxalate ligand of oxaliplatin, it has been shown to be more resistant to displacement by the tripeptide. In addition, it was thought that glutathione was unable to displace amine-based multidentate ligands, such as the 1R,2R-diaminohexane ligand of oxaliplatin, although this has now been shown to occur and may also play a role in the action/inaction of some platinum complexes (Kemp et al. 2008).

Serum Proteins. Human blood serum is a mixture of proteins, salts, and cells all of which can interact with platinum drugs affecting both their structure and transport (Timerbaev et al. 2006). The most abundant protein in the blood is serum albumin, a 66 kDa protein involved in small molecule transport, metabolism, and osmotic blood pressure regulation. It is a single-chain protein containing 17 disulfide bridges and one free cysteine residue (Cys34) located in a binding pocket. This site has a strong affinity for soft metals and as such, rapidly and irreversibly binds to all platinum-containing drugs. In fact, it is estimated that up to 90% of platinum is albumin bound within only a few minutes post administration. Binding of platinum drugs by serum albumin can have three different effects. Protein binding can remove the drug from systemic circulation, significantly decreasing the dose that is available to tumors. Binding by the cysteine residue has a similarly destructive effect on the drugs' structure that glutathione has, resulting in the displacement of both the chloride and am(m)ine ligands. Serum albumin may also have a positive effect, and it is thought that it may play a role in the reduction of platinum(IV)-based drugs, like satraplatin, to their platinum(II) forms. To a lesser extent, platinum drugs can also bind other serum proteins such as transferrin, an iron-binding blood plasma glycoprotein, and gamma globulins, which include immunoglobulins.

Copper Transporters. For many years, it was thought that uncharged platinum drugs entered cancerous cells through a process of passive diffusion. While this is still a possible route of entry for these drugs, it is now known that their uptake is much more complex and also involves active transport mechanisms.

Platinum Anticancer Drugs, Fig. 3 A simplified reaction of cisplatin with two tripeptide glutathione molecules, showing how the drug can be degraded with the eventual loss of its chloride and ammine ligands

A major means by which platinum drugs can be drawn into cells is through copper transporters (Wang et al. 2011). The most important protein in this class is human copper transporter 1 (hCTR1); an amino acid protein located within the cell membrane as a homotrimer. The N-terminus of the protein is located in the extracellular environment with the C-terminus located within the protein's core, pointing towards the cytosol. Like all platinum-binding proteins, hCTR1 is sulfur rich, although in this case, the important residues are methionine rather than cysteine. Each hCTR1 contains two methionine-rich motifs giving a hCTR1 channel up to 24 methionine residues which can bind to platinum drugs. Copper transporters are also associated with the efflux of platinum drugs out of cells. Two P-type ATPases are thought to be involved including ATP7A and ATP7B.

Organic Cation Transporters. Organic cation transporter proteins move monoamine neurotransmitter chemicals, some xenobiotics, and some clinically used drugs into and out of cells. The interaction of platinum drugs with these proteins has only recently been discovered. The organs most affected by platinum drugs, e.g., the kidneys, have been found to express a large number of OCT proteins on their cell surfaces, which may help explain why the platinum drugs are toxic to such organs (Burger et al. 2011).

Sodium, Calcium, and Potassium Channels. The use of platinum drugs is associated with peripheral neurotoxicity and neuropathy, which manifests as numbness, tingling, hyperexcitability, and impairment of deep tendon reflexes. Signaling and communication between neurons involved in muscle contraction is dependent on very small changes in sodium, potassium, and calcium concentrations inside the cell (Kiernan 2007). The toxicity associated with platinum drugs therefore arises from the binding of the drugs to the ion channels within the membranes of the neurons, affecting the influx and efflux of the cations. However, a specific mechanism by how the platinum drugs bind and interfere with these channels is not yet known.

High-Mobility Group Proteins. Platinum interactions are not exclusively with proteins while the drugs are in systemic circulation, diffusing in a cell, or during intracellular transport but can also occur after the drugs have bound to DNA. High-mobility group (HMG) proteins are involved in DNA replication and transcription, and more importantly, DNA repair. In this respect, the protein HMG1 is known to be involved in the recognition and repair of cisplatin adducts on

DNA (Ohndorf et al. 1999). The protein recognizes both 1,2-GpG and 1,2-ApG intrastrand cross-links formed by cisplatin and carboplatin through binding to the drug adduct from the DNA's minor groove. Binding of platinum to the DNA kinks the helix, forming a notch into which the protein intercalates its 37-phenylalanine residue, thus marking the site for repair.

Cross-References

▶ Platinum (IV) Complexes, Inhibition of Porcine Pancreatic Phospholipase A2
▶ Platinum- and Ruthenium-Based Anticancer Compounds, Inhibition of Glutathione Transferase P1-1
▶ Platinum Complexes and Methionine Motif in Copper Transport Proteins, Interaction
▶ Platinum Complexes and Methionine Motif in Copper Transport Proteins, Interaction
▶ Platinum Interaction with Copper Proteins
▶ Platinum(II) Complexes, Inhibition of Kinases
▶ Platinum-Containing Anticancer Drugs and Proteins, Interaction
▶ Platinum-Resistant Cancer
▶ Terpyridine Platinum(II) Complexes as Cysteine Protease Inhibitors
▶ Zinc, Metallated DNA-Protein Crosslinks as Finger Conformation and Reactivity Probes

References

Kauffman GB, Pentimalli R, Doldi S, Hall MD (2010) Michele Peyrone (1813–1883). Discoverer of cisplatin. Platinum Metals Rev 54:250–256
Kemp S, Wheate NJ, Pisani MP, Aldrich-Wright JR (2008) Degradation of fully coordinated platinum(II)-based DNA intercalators by reduced L-glutathione. J Med Chem 51:2787–2794
Kiernan MC (2007) The pain with platinum: oxaliplatin and neuropathy. Eur J Cancer 43:2631–2633
Ohndorf UM, Rould MA, He Q, Pabo CO, Lippard SJ (1999) Basis for recognition of cisplatin-modified DNA by high-mobility-group proteins. Nature 399:708–712
Reedijk J (1999) Why does cisplatin reach guanine-N7 with competing S-donor ligands available in the cell? Chem Rev 99:2499–2510
Rosenberg B, Van Camp L, Krigas T (1965) Inhibition of cell division in *Escherichia coli* by electrolysis products from a platinum electrode. Nature 205:698–699
Timerbaev AR, Hartinger CG, Aleksenko SS, Keppler BK (2006) Interactions of antitumor metallodrugs with serum proteins: advances in characterization using modern analytical methodology. Chem Rev 106:2224–2248
Wang X, Du X, Li H, Chan DS-B, Sun H (2011) The effect of the extracellular domain of human copper transporter (hCTR1) on cisplatin activation. Angew Chem Int Ed 50:2706–2711
Wheate NJ, Walker S, Craig GE, Oun R (2010) The status of platinum anticancer drugs in the clinic and in clinical trials. Dalton Trans 39:8113–8127

Platinum Complexes and Methionine Motif in Copper Transport Proteins, Interaction

Fabio Arnesano
Department of Chemistry, University of Bari "Aldo Moro", Bari, Italy

Synonyms

Copper transport proteins; Platinum anticancer drugs

Definition

Passive diffusion across the cellular membrane and active/facilitated transport can both play a role in the cellular uptake of platinum-based antitumor drugs that are widely used in the treatment of several forms of cancer. The major copper influx transporter Ctr1, located in the plasma membrane, can bind platinum complexes through methionine-rich motifs in the amino-terminal extracellular domain and is also involved in platinum-drug uptake.

Description

Platinum-based anticancer drugs such as cisplatin, carboplatin, and oxaliplatin (Fig. 1) are among the most potent chemotherapeutic agents for the treatment of various malignant tumors; however, severe side effects as well as drug resistance have limited their clinical applications. One possible mechanism of tumor cell resistance is the impaired intracellular accumulation of platinum drugs, which is partly attributed to reduced drug uptake.

Platinum Complexes and Methionine Motif in Copper Transport Proteins, Interaction, Fig. 1 Molecular structures of Pt(II) complexes. Cisplatin, carboplatin, and oxaliplatin are currently used in the clinics; transplatin, the inactive isomer of cisplatin, is also reported

Early studies noted that cellular uptake of cisplatin was linear, concentration dependent, and nonsaturable, all hallmarks of passive drug entry into the cell. Barnett Rosenberg, the discoverer of cisplatin, stated that, "*the drug is passively transported across the cellular membrane—no active transport is necessary,*" and this paradigm has ruled most drug design approaches (Hall et al. 2008).

Cisplatin is a simple neutral inorganic compound having square planar geometry and containing a Pt(II) center bonded to two non-labile ammine ligands (called *carrier ligands*) and two labile chlorido ligands (*leaving groups*) in a *cis*-configuration. Transplatin has the same composition, but the ligands are in a *trans*-arrangement. Whereas cisplatin is active against a wide variety of tumors, transplatin is inactive and toxic. As with all chemotherapeutic drugs, cisplatin also has drawbacks, including intrinsic or acquired resistance and toxicity, which have both a negative influence on the cure rate. Some of the severe safety issues have been brought to a solution by the introduction of *carboplatin*, which is equally effective, but with a more acceptable side-effect profile. Carboplatin contains two ammines as carrier ligands and 1,1-cyclobutanedicarboxylato as leaving group. The efforts to circumvent mechanisms of resistance and thereby broaden the clinical utility of this class of agents also resulted in the discovery of *oxaliplatin* (active in patients with colorectal cancer). Oxaliplatin contains R,R-1,2-diaminocyclohexane as carrier ligand and an oxalato as leaving group (Arnesano and Natile 2009) (Fig. 1).

Cisplatin is administered as a formulation that includes 154 mM NaCl. One of the main assumptions is that the neutral, intact drug is the species that traverses the lipid bilayer by *passive diffusion*. Outside the cell, aquation/hydrolysis of neutral cisplatin is suppressed (but not prevented) owing to the high chloride levels (~100 mM) present in blood plasma and equivalent media, compared with the low (4 mM) chloride concentration in the cytoplasm.

In the first *aquation* step, a key process in the activation of cisplatin, a chloride leaving group is replaced by water (H_2O) giving a singly positively charged aquachloro species that can then be hydrolyzed to a hydroxo ligand (OH^-) yielding the chlorohydroxo species *cis*-[PtCl(OH)(NH$_3$)$_2$], that is neutral and could also potentially diffuse across the lipid bilayer. The proportion of aquachloro to chlorohydroxo species is dependent on the pH of the solution. The aquated species is more reactive (labile) and can subsequently bind to the N7 nitrogen of guanine or adenine on nuclear DNA, the generally accepted cytotoxic target of cisplatin, but could also undergo rapid reaction with other biomolecular targets and deactivation (Fig. 2).

At high cisplatin concentrations other nonsaturable systems, such as fluid-phase *endocytosis*, may also contribute to cisplatin uptake. The latter process allows compounds to enter the cells via membrane-lined invaginations (Hall et al. 2008).

More recently, active/facilitated transport pathways have been identified. Among systems involved in facilitated influx of platinum drugs, *organic cation transporters* have been shown to play an essential role in the cellular entry of oxaliplatin. Moreover, *copper transporter 1 (Ctr1)*, a major copper influx transporter, has also been found to facilitate the uptake of cisplatin and other platinum drugs in yeast, mammalian, and human cells. Other proteins involved in copper homeostasis, such as the copper efflux transporters *ATPases 7A and 7B*, have been discovered to mediate platinum-drug efflux (Safaei and Howell 2005; Hall et al. 2008; Gupta and Lutsenko 2009) (Fig. 2).

A connection between copper and platinum trafficking was suggested by observations of bidirectional *cross-resistance* of copper and platinum drugs in a variety of cell lines: cells selected for resistance to high levels of copper were found to be resistant to platinum drugs and vice versa. Moreover, yeast cells

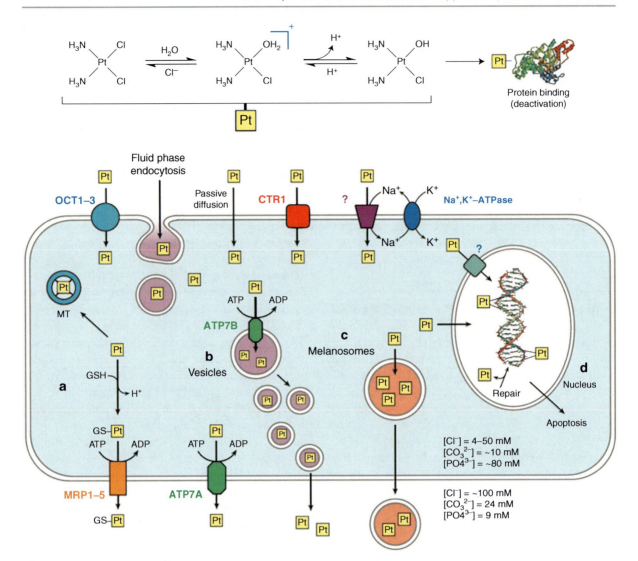

Platinum Complexes and Methionine Motif in Copper Transport Proteins, Interaction, Fig. 2 Schematic representation of the mechanism affecting and controlling the cellular accumulation of platinum chemotherapeutics exemplified by cisplatin. In the extracellular environment, cisplatin can be aquated or react with biomolecules, resulting in a complex speciation profile. These species may enter the cell or cross-react with extracellular proteins such as serum albumin, reducing the bioavailable pool of drug. Neutral platinum drugs can enter the cell by passive diffusion across the lipid bilayer or carrier-mediated import proteins like organic cation transporters (OCT) or copper transporter 1 (CTR1). Inside the cell platinum drugs can be deactivated by binding to thiol-rich metallothioneins (MT) or chelated by glutathione (GSH) and effluxed from the cell via the GS-X pumps (MRP1-5). Platinum drugs can also be ensnared in subcellular organelles such as vesicles via ATP7B influx, or melanosomes in melanoma cells, followed by exocytosis to expel platinum from the cells (Reproduced from Hall et al. 2008)

in which the CTR1 gene was deleted showed increased resistance to cisplatin, that correlated with reduced uptake of cisplatin and reduced formation of the critical platinum–DNA adducts compared with control cells. In contrast, studies of a variety of human cell lines in which Ctr1 was overexpressed revealed that simply increasing the level of Ctr1 did not always correlate with an increase in cell death. This latter observation was inconsistent with a mechanism based on direct transport through Ctr1; and it raised the possibility that changes in human Ctr1 levels may induce additional pathways that regulate the removal/detoxification of the drug (Safaei and Howell 2005; Hall et al. 2008; Gupta and Lutsenko 2009).

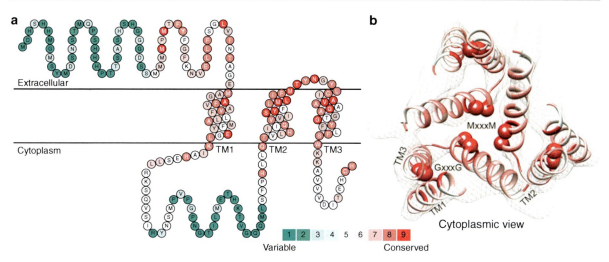

Platinum Complexes and Methionine Motif in Copper Transport Proteins, Interaction, Fig. 3 Evolutionary conservation profile (shown by the *color bar*) of hCtr1 full sequence and TM model. (**a**) Predicted topology of hCtr1. (**b**) Intracellular view of hCtr1 structural model (TM domain), fitted onto the cryoelectron microscopy map. Residues of the MxxxM and GxxxG motifs are shown as *spheres*, and the three TM helices are marked on one subunit (Reproduced from Schushan et al. 2010)

Human Ctr1 (hCtr1) is a member of a highly conserved family of copper transporters. Despite a large difference in molecular size and a lack of extensive sequence conservation, Ctr proteins share a global overall design that subdivides each transporter into three regions: (1) an extracellular N-terminal domain, (2) a membrane-embedded domain composed of three transmembrane (TM) helices, and (3) an intracellular C-terminal domain of variable length. The Ctr signature motif MxxxM-X_{12}-GxxxG correlates a strictly conserved MxxxM-motif at the extracellular end of the second transmembrane helix (TM2) with an almost entirely conserved GxxxG-motif in the center of the third transmembrane helix (TM3) (Fig. 3).

The structure of hCtr1, solved by electron crystallography to an in plane resolution of 7 Å, is a channel-like trimer which creates a pore that stretches across the membrane bilayer at the interface between the subunits. TM2 is the key element lining the pore and contains functionally important methionine residues of the MxxxM-motif that participate in Cu(I)-coordination during transport across the membrane. Toward the extracellular exit the pore measures 8 Å across, toward the intracellular end the pore widens into a large aqueous cavity, measuring 22 Å across (De Feo et al. 2009). Aligned with and sealing both ends of the pore, the extracellular N-terminal domain and the intracellular C-terminal domain of hCtr1 provide additional metal binding sites. The unique feature of the N-terminal domain of Ctr1 is the presence of methionine (Met)-rich regions arranged as MxxM or MxM motifs containing 3–5 methionine residues per *Mets* motif. Yeast Ctr1 (yCtr1) has 8 Mets motifs, with a total of 30 methionine residues in a 140-residue extracellular domain, whereas hCtr1 has two Mets motifs in a 65-residue extracellular region. The Met-rich motifs in the N-terminal extracellular domain play a key role in transporting copper ions as well as platinum-based antitumor drugs (Fig. 3).

An octapeptide (MTGMKGMS, called *Mets7*) based on the sequence of the seventh Met-rich motif of the N-terminal domain of yCtr1 (MSGMSGM), with slight modifications to make it more amenable to water solution investigations, was used to explore the interaction with several structurally diverse platinum drugs and analogues (Fig. 1). Electrospray ionization mass spectrometry (ESI-MS) showed that Mets7 binds to platinum in 1:1 stoichiometry, cisplatin loses both ammine ligands and finally coordinates to Mets7 in the *naked* form, where all the original ligands are replaced by Mets7 residues (Fig. 4). When the reaction was extended to transplatin, a clinically ineffective trans-Pt(II) complex, no release of ammine ligands was observed (Arnesano et al. 2007).

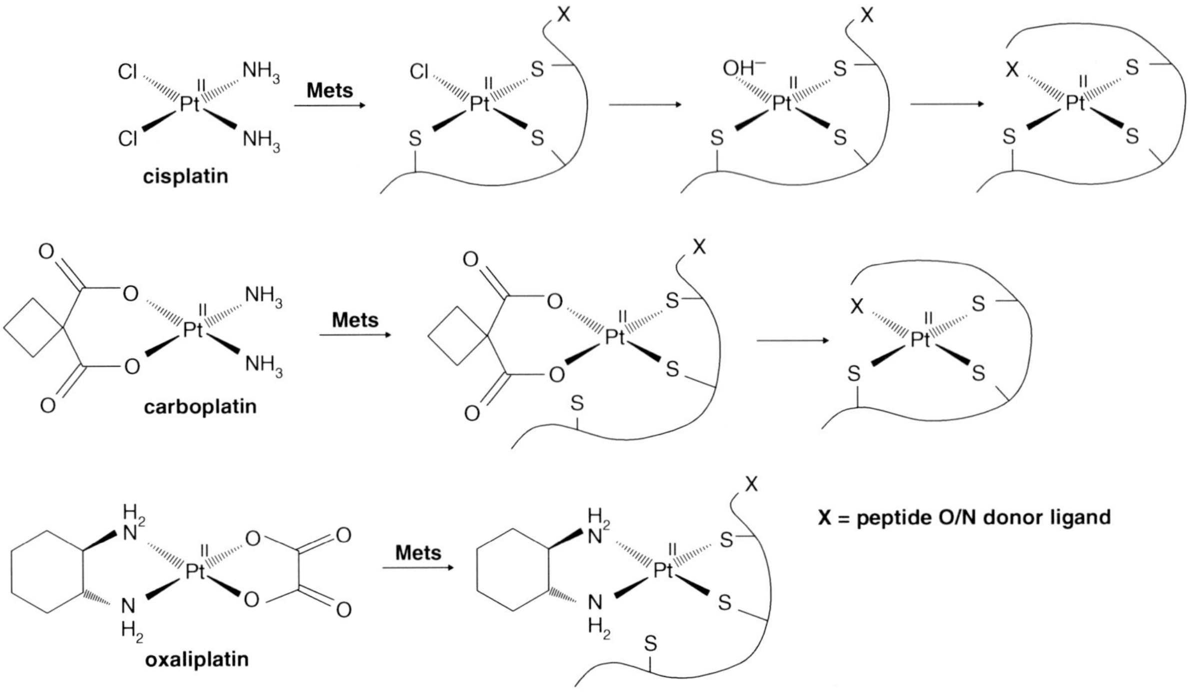

Platinum Complexes and Methionine Motif in Copper Transport Proteins, Interaction, Fig. 4 The reaction pathways between Met-rich peptides corresponding to the N-terminal domain of Ctr1 and clinically relevant Pt(II) complexes

Similar results were obtained using peptides corresponding to the two Met-rich motifs of hCtr1, the first one encompassing residues 7–14 (MGMSYMDS) and the second one spanning residues 40–47 (MMMMPMTF), and were also confirmed by using a 20mer (MDHSHHMGMSYMDSNSTMQP, called hCtr1-N20) and a 55mer-peptide (MDHSHH-MGMSYMDSNSTMQPSHHHPTTSASHSHGGGD-SSMMMMPMTFYFGFKNVE, called hCtr1-N55) corresponding to amino acid sequences of the N-terminal domain of hCtr1 (residues 1-20 and 1-55, respectively) (Crider et al. 2010; Wu et al. 2009; Wang et al. 2011).

Using nuclear magnetic resonance (NMR) spectroscopy one could conclude that methionine residues are involved in platinum binding to Mets7 as well as to the longer construct hCtr1-N55. ^{195}Pt NMR and X-ray absorption spectroscopy (XAS) data confirmed that platinum is bound to sulfur atoms of methionines and that the adducts of cisplatin and transplatin with Mets7 have S_3X (X = O, N, or Cl at high chloride concentration) and S_2N_2 coordination environments, respectively. Additional ligands are a chloride or a hydroxyl oxygen or a deprotonated amide nitrogen in the case of cisplatin and the two ammine ligands in the case of transplatin. Circular dichroism (CD) indicated that Mets7 tends to adopt a random-coil structure in aqueous solution. Upon interaction with cisplatin the octapeptide undergoes a large conformational change and a structural model of the platinum adduct obtained by quantum mechanics/molecular mechanics (QM/MM) calculations showed that Mets7 forms two β-turns, in accordance with experimental data (Arnesano et al. 2007; Nguyen et al. 2012) (Fig. 5).

Molecular dynamics (MD) simulations confirmed that hCtr1-N20 folds into a conformation which is neither a α-helix nor a β-sheet structure. Analysis over the MD trajectory also revealed that all of the methionine residues of hCtr1-N20 are exposed and accessible to solvents (Wu et al. 2009). Moreover, an all-atom model of hCtr1 trimer, using the low-resolution structure as a starting point and ab initio modeling of the extracellular domains, showed that the interfaces between N-terminal domains of adjoining monomers are supported exclusively by hydrophobic interactions. Platinum binding to the Met-rich

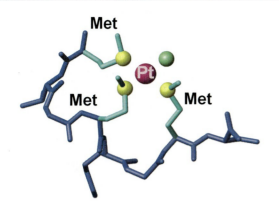

Platinum Complexes and Methionine Motif in Copper Transport Proteins, Interaction, Fig. 5 Structural model of the adduct between cisplatin and Mets7. Sulfur atoms of methionines are shown as *yellow spheres*, the chloride ion as a *green sphere* and the platinum in *magenta*

motifs was also found to stabilize trimeric hCtr1 at the plasma membrane, possibly owing to cisplatin cross-linking of adjacent monomers (Fig. 6).

The observation that NH_3 is released rapidly upon the initial binding of cisplatin is due to the *trans-labilizing effect* of the sulfur ligands, which induces fast release of ammine ligands in a *trans* position. A similar effect dictates the reactions between cisplatin and transplatin with Zn7 metallothionein. Both complexes bind to cysteines of Zn7 metallothionein, whereas the formation of the adducts is different: the two ammines are released in the case of cisplatin, but both of them remain bound in the case of transplatin. These data illustrated again the determining role of *cis/trans* geometry in the interactions with proteins or peptides and revealed the impact of nonleaving or leaving groups on the reactivity of Pt(II) complexes (Arnesano and Natile 2009).

To address this latter issue, carboplatin and oxaliplatin were included in the studies (Crider et al. 2010; Wu et al. 2009).

For the reactions with Met-rich motifs, carboplatin and cisplatin share similar reaction patterns, with platinum undergoing complete loss of all original ligands and solely bound to the peptides. Nevertheless, the binding rate of carboplatin is much slower than that of cisplatin. In the case of cisplatin, removal of all original ligands from the platinum center occurs within a few hours of reaction, whereas loss of ligands from carboplatin happens only after 48 h of reaction. Carboplatin initially can lose both ammine ligands and still keep the cyclobutanedicarboxylate group, but finally coordinates to Met-rich motifs as a naked platinum ion. The *chelating effect* of cyclobutanedicarboxylate ligand, which provides two donor atoms to platinum thus forming a six-membered chelate ring, may account for this reaction pathway (Fig. 3).

In contrast, in the reaction between Met-rich motifs and oxaliplatin, the methionines are able to efficiently displace the oxalato group, whereas the chelating diaminocyclohexane ligand remains coordinated to platinum throughout the whole time course owing to its significant stability resulting from a five-membered chelate ring and its steric hindrance (Fig. 3). A further Pt(diaminocyclohexane) unit can then bind to the peptides thereby forming a second adduct. Unlike the reactions with cisplatin and carboplatin, where the free Met-rich peptides were nearly completely recovered following treatment with a *chelator*, the Pt(diaminocyclohexane) units bound to methionines are not easily removed by competitive chelation. Oxaliplatin, on the other hand, exhibits a binding rate comparable to that of cisplatin (Crider et al. 2010).

Since the binding step is a substitution process in which either chloride or oxygen is replaced by the thioether sulfur of methionines, the binding rate is largely dependent on the *leaving group*, which is believed to be hydrolyzed once the platinum drug enters the cell. The chelating effect and steric hindrance of the cyclobutanedicarboxylate bulky moiety in carboplatin result in a much slower hydrolysis rate and binding of carboplatin to Met-rich motifs.

It is still an open question how platinum adducts with Ctr1 are activated and delivered to the intracellular targets. The varied reactivity of platinum drugs toward Met-rich motifs suggests that the transporter role of Ctr1 for various Pt(II) agents may be different thus resulting in a different spectrum of anticancer activity.

The affinity of Pt(II) for the sulfur of methionine is similar to that of Cu(I), and it has been hypothesized that it may pass through the trimer via a series of *transchelation* reactions in a manner similar to copper (Wang et al. 2011). However, other evidence has shown that Ctr1-facilitated cellular entry of platinum drugs may have a distinct mechanism compared to the uptake of copper ion.

While extracellular Met-rich motifs of Ctr1 are essential for platinum-drug uptake, they are not strictly

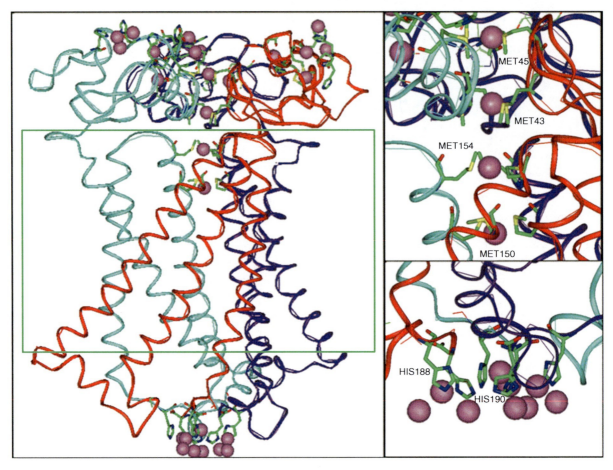

Platinum Complexes and Methionine Motif in Copper Transport Proteins, Interaction, Fig. 6 All-atom model of hCtr1 trimer and predicted copper-binding sites (magenta spheres). Sites are found at each of the four stacked *methionine-triads* (displayed in the *upper right panel*), at the lower end of the vestibule on the cytoplasmic side of the protein (*lower right panel*), and at various Met-rich sites in the N-terminal domain. The predicted TM region is delimited by the green rectangle (Reproduced from Tsigelny et al. 2012)

required for copper uptake when the metal ion is abundant, although they enhance transport when copper is scarce. Quite possibly these N-terminal motifs serve to funnel copper toward the transmembrane pore. Key to this function are the $M^{40}MMMPM^{45}$ motif located proximal to the start of TM1 of hCtr1 and methionines M^{150} and M^{154} of the conserved MxxxM-motif in TM2 (Abada and Howell 2010).

When the protein is assembled as a trimer, four methionine-triads at residues 43, 45, 150, and 154 of each subunit of hCtr1 align into four stacked rings which line the channel and could contribute directly to copper conduction along the central pore by providing closely spaced three-coordinate binding sites for copper (Fig. 6, *upper right panel*).

Deletion of the $M^{40}MMMPM^{45}$ motif or of the entire first 45 amino acids impairs cisplatin uptake by hCtr1, where as conversion of the methionines to isoleucines in the MxxxM-motif of TM2, while impeding copper transport, actually increases cisplatin uptake. Furthermore, conversion of methionines to alanines in both extracellular Mets motifs of hCTR1 (A^7GASYA^{12} and $A^{40}AAAPA^{45}$) prevents the cisplatin-induced stabilization of Ctr1 trimers, presumably due to loss of binding. By contrast, copper still binds to the protein and induces *endocytosis* (Gupta and Lutsenko 2009).

Differences in the chemistry of copper and platinum may translate into differences in metal-induced protein conformational changes and mechanisms of metal transport by Ctr1. Apart from the fact that both Cu(I) and Pt(II) prefer *soft*, sulfur-rich coordination environments, their coordination chemistry is otherwise different. Whereas Cu(I) can adopt linear, trigonal, or tetrahedral coordination geometries, Pt(II) is square planar. In addition, the ligands on Pt(II) are less likely to be *kinetically labile*, although inner-sphere ligand exchange reactions can occur, with their likelihood depending on the *trans* effect of the ligands (Crider et al. 2010).

The speciation of the platinum bound to *single-methionine* peptides is different from that observed for the parent reaction of cisplatin with Met-rich peptides in that ammine ligands are initially retained on platinum. On the other hand, cisplatin binding to Mets motifs of Ctr1 (containing at least three methionines) involves the removal of all the ligands but, like the multimerization of Ctr1, this is a process that occurs slowly relative to the kinetics of transport of cisplatin. Since the uptake of cisplatin in vivo may only take minutes, it is unlikely that the *intact* anticancer drug molecule may migrate efficiently by a methionine-based sulfur–sulfur exchange through the pore formed by hCtr1.

Although the trimeric hCtr1 pore is quite expandable and modeling studies of the transmembrane helices predict that these structures are dynamic (Figs. 3 and 6), Ctr1 seems to be ideally suited for transporting kinetically labile, *monovalent* cations. Isolated Met-rich motifs of Ctr1 bind Cu(I) selectively over Cu(II); additionally, Ctr1 is unable to transport Zn(II) or other *divalent* metal ions but readily takes in Ag(I). Given the different charges and radii of Pt(II) and Cu(I) ions, it is also unlikely for platinum to be transported in the *naked* form by Ctr1. Moreover, since the ammine ligands are present in the DNA adducts formed by clinically important Pt(II) complexes, the potential replacement of the pendant ligands on platinum drugs if they pass through Ctr1 may have important consequences on the mechanism of action (Gupta and Lutsenko 2009; Crider et al. 2010; Abada and Howell 2010).

There remains the possibility that Ctr1 delivers cisplatin into the cell by chelating it via the N-terminal methionines followed by endocytosis or that Ctr1 regulates a distinct platinum-drug influx mechanism. This different mechanism could be copper-dependent, and the observed decrease in cisplatin uptake in Ctr1 mutants could be due to a derangement in copper metabolism (Arnesano and Natile 2009; Gupta and Lutsenko 2009; Abada and Howell 2010).

In the first hypothesis, cisplatin can be taken into cells as a *passenger* during the normal course of copper-dependent *endocytosis* of Ctr1 (Fig. 7, *panel A*). This has been interpreted as a mechanism that helps limit the toxicity of high extracellular levels of copper. However, low concentrations of cisplatin (2 μM) were also shown to induce *internalization, ubiquitination*, and *proteasomal degradation* of hCtr1 by a process that involves *macropinocytosis*. Proteasome inhibitors such as *bortezomib* were found to prevent cisplatin-induced downregulation of Ctr1 thus increasing cisplatin uptake by 1.6–2.4-fold in a concentration-dependent manner. The endocytic process, which requires the copper chaperone Atox1, is *rapid* and provides an explanation for why the greatest effect of Ctr1 on cisplatin transport is observed during the early phase of cisplatin uptake (Abada and Howell 2010).

Moreover, downregulation of Ctr1 by cisplatin is functionally significant as it reduces the subsequent uptake of copper. In this scenario, the involvement of Ctr1 in the uptake of cisplatin could be indirect: copper deficiency due to Ctr1 downregulation may result in metabolic changes that facilitate entry, retention, and/or efficacy of the platinum drug by *alternative routes*.

If the mechanism of cisplatin uptake involves endocytosis of drug bound to Ctr1 and/or degradation of the cisplatin–Ctr1 complex, then additional steps are needed to complete the process: cisplatin needs to be released from the endocytic vesicles/lysosomes into the cytosol or delivered to the nucleus, mitochondria, and microsomal compartments, all of which are known to accumulate platinum after drug exposure (Arnesano and Natile 2009; Gupta and Lutsenko 2009; Abada and Howell 2010) (Fig. 7, *panel B*).

Late endosome and lysosome/vacuole compartments of mammalian and yeast cells contain a Ctr1 homologue, named *Ctr2*. Human Ctr2 has significant structural similarity to hCtr1 despite a low amino acid sequence homology. It also contains an MxM motif at the N-terminus as well as a conserved MxxxM-motif in TM2. Ctr2 functions to release *intracellular* copper stores; however, there is also evidence in mammalian systems that Ctr2 can function as a copper influx transporter in case of *extracellular* copper excess (Abada and Howell 2010).

Ctr2 has a large effect on the accumulation of both cisplatin and carboplatin. However, instead of limiting

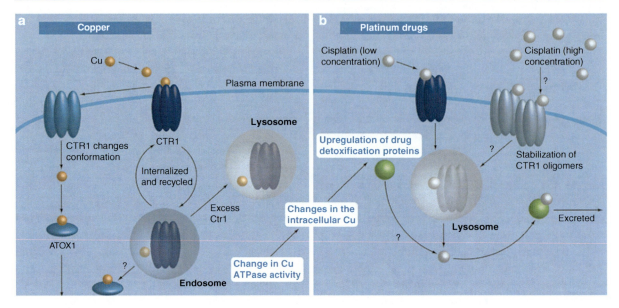

Platinum Complexes and Methionine Motif in Copper Transport Proteins, Interaction, Fig. 7 Proposed associations between mammalian copper metabolism and the transport of cisplatin. (**a**) Copper enters the cell through Ctr1, upon release copper is accepted by the metallochaperone Atox1. High levels of copper induce downregulation (endocytosis and degradation) of Ctr1. (**b**) Cisplatin entry into the cell depends on the presence of Ctr1. The mechanism of uptake may involve stabilization of Ctr1 oligomers and endocytosis (Adapted from Gupta and Lutsenko 2009)

import, knock down of CTR2 increased the uptake of both drugs by a factor of two- to threefold, an effect opposite to that of knocking out CTR1. Likewise, copper *chelators* were able to decrease the expression level of Ctr2, thus increasing platinum-drug uptake. This increase in uptake translated to an increase in formation of platinum–DNA adducts and cytotoxicity. Indeed, pretreatment with the copper chelator *tetrathiomolybdate* was shown to improve cisplatin efficacy in an in vivo mouse cervical cancer model. In contrast, exposure to copper resulted in an increase in CTR2 mRNA levels and protein half-life (Abada and Howell 2010).

The decreased Ctr2 half-life observed following *copper starvation* is at least in part due to increased proteasomal degradation as bortezomib can block this effect. Again this post-transcriptional regulation is dependent on the copper chaperone Atox1. Atox1 binds both Cu(I) and cisplatin, using two cysteine residues in a CxxC motif, and may function as a copper/platinum sensor involved in a switch between up- and downregulation of Ctr1 and Ctr2 (Abada and Howell 2010).

The interaction of clinically relevant Pt(II) complexes with Met-rich motifs, therefore, represents the initial step of a series of molecular events leading to platinum-drug uptake, intracellular distribution, and cytotoxicity. All these processes are intricately connected to copper metabolism and the success of chemotherapy is also dependent on a delicate balance of the effects on the homeostasis of this essential metal ion.

Cross-References

▶ Biological Copper Transport
▶ Copper, Physical and Chemical Properties
▶ Copper-Binding Proteins
▶ Oxaliplatin
▶ Platinum Anticancer Drugs
▶ Platinum Interaction with Copper Proteins
▶ Platinum, Physical and Chemical Properties
▶ Platinum-Containing Anticancer Drugs and Proteins, Interaction
▶ Platinum-Resistant Cancer
▶ Proteomic Basis of Cisplatin Resistance

References

Abada P, Howell SB (2010) Regulation of cisplatin cytotoxicity by Cu influx transporters. Met Based Drugs 2010:317581

Arnesano F, Natile G (2009) Mechanistic insight into the cellular uptake and processing of cisplatin 30 years after its approval by FDA. Coord Chem Rev 253:2070–2081

Arnesano F, Scintilla S, Natile G (2007) Interaction between platinum complexes and a methionine motif found in copper transport proteins. Angew Chem Int Ed Engl 46:9062–9064

Crider SE, Holbrook RJ, Franz KJ (2010) Coordination of platinum therapeutic agents to Met-rich motifs of human copper transport protein 1. Metallomics 2:74–83

De Feo CJ, Aller SG, Siluvai GS et al (2009) Three-dimensional structure of the human copper transporter hCTR1. Proc Natl Acad Sci USA 106:4237–4242

Gupta A, Lutsenko S (2009) Human copper transporters: mechanism, role in human diseases and therapeutic potential. Future Med Chem 1:1125–1142

Hall MD, Okabe M, Shen DW et al (2008) The role of cellular accumulation in determining sensitivity to platinum-based chemotherapy. Annu Rev Pharmacol Toxicol 48:495–535

Nguyen TH, Arnesano F, Scintilla S et al (2012) Structural determinants of cisplatin and transplatin binding to the Met-rich motif of Ctr1: a computational spectroscopy approach. J Chem Theory Comput 8:2912–2920

Safaei R, Howell SB (2005) Copper transporters regulate the cellular pharmacology and sensitivity to Pt drugs. Crit Rev Oncol Hematol 53:13–23

Schushan M, Barkan Y, Haliloglu T et al (2010) C(alpha)-trace model of the transmembrane domain of human copper transporter 1, motion and functional implications. Proc Natl Acad Sci USA 107:10908–10913

Tsigelny IF, Sharikov Y, Greenberg JP et al (2012) An all-atom model of the structure of human copper transporter 1. Cell Biochem Biophys 63:223–234

Wang X, Du X, Li H et al (2011) The effect of the extracellular domain of human copper transporter (hCTR1) on cisplatin activation. Angew Chem Int Ed Eng l50:2706–2711

Wu Z, Liu Q, Liang X et al (2009) Reactivity of platinum-based antitumor drugs towards a Met- and His-rich 20mer peptide corresponding to the N-terminal domain of human copper transporter 1. J Biol Inorg Chem 14:1313–1323

Platinum Interaction with Copper Proteins

Maria E. Palm-Espling and Pernilla Wittung-Stafshede
Department of Chemistry, Chemical Biological Center, Umeå University, Umeå, Sweden

Synonyms

Binding of platinum to metalloproteins; Cisplatin uptake and cellular pharmacology; Copper transporting proteins and resistance of platinum anticancer drugs; Metal-protein interactions

Definition

Platinum originating from platinum anticancer drugs have been shown to bind copper transporting proteins in human cells. This is suggested to be one of the mechanisms by which cells evolve resistance toward these drugs. Copper and platinum are similar in some aspects and both easily form bonds with sulfur. The sulfur containing copper binding site is therefore a likely target for platinum interaction.

Platinum Anticancer Drugs and Copper Proteins

Platinum anticancer drugs are used worldwide to treat different types of cancer, for example, testicular, colon, head, and lung cancer (▶ Platinum Anticancer Drugs). The first platinum anticancer drug to be used in the clinics was cisplatin, a chemically simple molecule whose anticancer effect was discovered in the early 1960s. Since then, newer generations of platinum anticancer drugs have been developed in order to mimic side effects and alter activity. Still, one problem remains, the cancer cells develop resistance against the platinum anticancer drugs during treatment (▶ Platinum-Resistant Cancer). The resistance mechanism is multifactorial and several causes have been proposed, one of them involves the cells own copper transporting machinery.

Platinum Anticancer Drugs

Cisplatin (Fig. 1) was the first metal-based agent to be used worldwide in the clinics for treatment of cancer. It is effective against lung, ovarian, bladder, testicular, head, and neck cancer. The substance had been synthesized and reported by Michele Peyrone in 1845. More than hundred years later, in the early 1960s, the cytotoxic effect of cisplatin was discovered by Professor Barnett Rosenberg at Michigan State University. He and his group were studying the effects of electric fields on the growth of cells with help of platinum mesh electrodes and *Escherichia coli* bacteria. They found that when applying the electric current, the number of growing cells was reduced. The bacteria also adapted long filaments, indicating that the bacterial cells were growing but not dividing. To their surprise, they found that the electric current itself had no effect on cell division, but the current was causing

Platinum Interaction with Copper Proteins, Fig. 1 The three platinum anticancer substances used in the clinics today; cisplatin, carboplatin, and oxaliplatin

a chemical reaction dissolving platinum from the electrodes into the medium. By synthesizing several platinum complexes for testing on both bacterial cells and later tumor-bearing mice, they found that *cis*-diamminedichloroplatinum(II), cisplatin, exhibited antitumor activity. Even though there was resistance and doubts against using a toxic metal as a drug, Rosenberg with help of Michigan State University and the Research Corporation patented his new discovery. Cisplatin was approved for use by the United States Food and Drug Administration (FDA) in 1978 and become instantly successful. Cisplatin still saves and prolongs life of cancer patients and has an annual sale of over 2 billion US dollars. The discovery of cisplatin opened new possibilities in the field of medicinal research. Earlier, identifying new agents was focused on small organic molecules and natural compounds. But almost overnight, cisplatin expanded the drug discovery landscape to include metal-based molecules which has resulted in a number of new platinum-based antitumor agents in the clinics and other metal-based drugs in clinical trials.

When cisplatin enters the cell, it becomes aquated, meaning that the chloride ligands are exchanged to water. This is due to the lower concentration of NaCl in the cytoplasm compared to the blood plasma. The aquated molecule can cross-link DNA through a covalent coordinate bond to the nitrogen atom on guanine and also to a lesser extent adenine. This DNA damage prevents replication and transcription, which leads to apoptosis and cell death. Since platinum and platinum derivatives are prone to bind to sulfur-containing molecules like cysteine and methionine, proteins rich in these amino acids are suggested to be involved in cisplatin side reactions and resistance (▶ Terpyridine Platinum(II) Complexes as Cysteine Protease Inhibitors).

Carboplatin (*cis*-diammine(1,1-cyclobutanedicarboxylato)platinum(II), Fig. 1) is the second-generation platinum anticancer drug and has less side effects than cisplatin. It was approved by the United States Food and Drug Administration (FDA) in 1989. The main difference between carboplatin and cisplatin is the six-membered chelate ring that makes the drug less chemically reactive and prone to hydrolyze and thereby reduces some of the unwanted side reactions.

Oxaliplatin (1,2-diaminocyclohexane oxalate platinum(II), Fig. 1) is the third-generation platinum-based anticancer drug and was approved by the United States Food and Drug Administration (FDA) in 2002. It has a more complex structure than previous platinum-based anticancer drugs and is also effective against some tumors that have become resistant to cisplatin and carboplatin. Oxaliplatin is especially effective against colon cancer. The tumor specificity is most likely mediated by oxaliplatin uptake of the organic cation transporter one and two (OCT1/2) (Dabrowiak 2009).

A major problem with platinum-based anticancer therapy is that cancer cell exposure to the drugs causes resistance. This can be noticed in the clinics when tumors stop responding and thereby stop shrinking after the given dose drug. Increasing the dose will increase overall toxicity but not gaining any enhanced antitumor effect, and the treatment is forced to stop. The mechanism behind cisplatin resistance is an important question to solve, and much research has been performed on the topic (▶ Platinum-Containing Anticancer Drugs and Proteins, Interaction). It seems like the resistance of cisplatin is multifactorial and both intrinsic and acquired. From a molecular point of view, the resistance can be caused mainly in three ways. First, the cell can block the influx of the drug. Second, if the drug does enter, the cell can efflux it back outside or block it from reaching the target. Finally, if the drug reaches the target, the DNA, it can be removed by a repair system. Many pathways in the cell have been discussed when it comes to cisplatin resistance. The pathway controlling the copper homeostasis in the cell become interesting when cisplatin-resistant cancer cells showed an overexpression of the copper transporting protein ATP7B (Dabrowiak 2009).

Copper Transporting Proteins

Copper is an essential trace metal for the human body and an important cofactor for several proteins in our cells. The reactions mediated by some of these proteins involve interconversion of copper between two oxidation states, Cu^{II} and Cu^{I}. Because of its ability to easily oxidize, it is highly toxic if left free in the cytoplasm. For this reason, there is a highly conserved sophisticated system within the cell that transports the copper ions (▶ Copper-Binding Proteins). A common characteristic of these proteins is that they contain methionine- or cysteine-rich motifs building up copper binding sites. Extracellular Cu^{II} is believed to be converted to Cu^{I} by enzymes on the cell surface, and copper enters the cell as Cu^{I} by the copper transporter Ctr1. After entry, the copper is passed on to chaperones that transport the copper to specific enzymatic complexes. Copper chaperone Atox1 transports copper to the ATPases ATP7A/B in the trans-Golgi network for use in the secretory pathway. If the intracellular copper concentration gets too high, ATP7A/B can translocate and export copper out of the cell (Fig. 2) (Gupta and Lutsenko 2009).

Ctr1 is the main copper uptake transporter in human cells. The protein is a stable homotrimer positioned in the cell membrane. The trimer forms a central pore that can function as a channel. Each monomer is composed of a 28-kDa protein chain with an extracellular N-terminus, three transmembrane helices, an intracellular loop between the first and second helix, and an intracellular C-terminus (Fig. 3a). Studies of crystallized Ctr1 with electron microscopy showed a series of rings of methionines, histidines, and cysteines lingering the inside of the pore to facilitate copper transport. The last ring on the intracellular side of the channel, at the C-terminus, is composed of cysteines. The relative binding affinities are such that this part could transfer copper to Atox1, although this has not yet been demonstrated. The expression of Ctr1 is regulated at both transcriptional and translational levels in human cells. Elevated copper concentrations are known to trigger endocytosis of Ctr1, and at a transcriptional level, Ctr1 is downregulated by intracellular copper. Later studies have uncovered a second potential copper transporter, Ctr2. It is not located in the plasma membrane as Ctr1

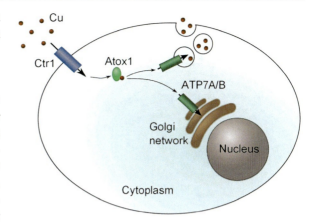

Platinum Interaction with Copper Proteins, Fig. 2 The proteins used for copper transport to the secretory pathway in a human cell

but mostly in the membranes of intracellular compartments. The role of Ctr2 in copper transport and homeostasis is still unclear (Howell et al. 2010).

Atox1 is a copper chaperone that transports copper from Ctr1 to ATP7A/B. It is a 7.5-kDa protein with a $\beta\alpha\beta\beta\alpha\beta$ ferredoxin-like fold. The protein can bind one Cu^{I} ion in a conserved metal binding site (MXCXXC) positioned in a surface exposed loop (Fig. 3b) (Anastassopoulou et al. 2004).

ATP7A and ATP7B are two homologous copper ATPases in human cells. They use the energy of ATP hydrolysis to transport Cu^{I} from the cytosol across cellular membranes. ATP7A/B are 160–170-kDa membrane proteins consisting of several domains. The transmembrane part consists of eight helices forming a pore for copper transport. ATP binding and hydrolysis take place in the cytosol at the domains called N- and P-domains. Another domain called the A-domain is also in the cytosol and is required for conformational changes during ATP hydrolysis. In the N-terminal part are six metal-binding domains (Fig. 3c). The structures of the metal-binding domains are highly similar to Atox1 and each domain can bind one Cu^{I} in a MXCXXC motif. The sequence homology of ATP7A and ATP7B is high (60%); however, their functions are not identical. The proteins are mostly located in separate tissues; for example, ATP7A is abundant in the intestine (pumping copper into the bloodstream), whereas ATP7B can be found in the liver (pumping excess copper out of the body via the bile). When intracellular copper concentration becomes too high, ATP7A/B exports the copper out

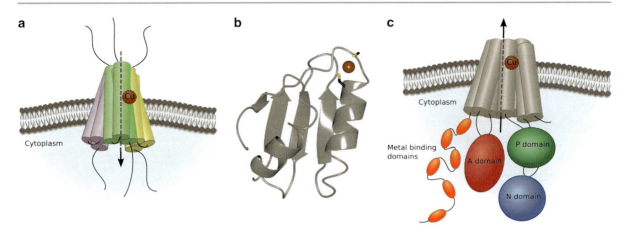

Platinum Interaction with Copper Proteins, Fig. 3 (a) Schematic picture of the copper uptake transporter Ctr1. (b) Structure of the copper chaperone Atox1 (1TL4, Anastassopoulou et al. 2004). (c) Schematic picture of the copper ATPases ATP7A/B

of the cell. This event is most likely indirect since ATP7A/B can pump copper into intracellular vesicles that later fuse with the cell membrane and exports copper via exocytosis. The proteins can also coexist in some tissues and then compensate for the loss of function of the other. In addition to their copper export function, ATP7A/B is required for delivery of copper cofactors to various copper-containing proteins. The copper proteins in the human body acquire their metal from the copper ATPases in the trans-Golgi network within the secretory pathway (Fig. 2). Under basal conditions, ATP7A/B is located at the trans-Golgi network, where they receive copper from Atox1 that transfer the copper ion to one of the metal-binding domains. Initial copper binding results in a conformational change that allows for copper transfer and additional copper binding to ATP7A/B. Copper can be loaded to empty metal-binding domains by Atox1 via a transient heterocomplex where copper is shared by both proteins. However, Atox1 may also deliver copper directly into the transmembrane pore, as was shown in the case of the yeast homolog. Whether copper delivered to the metal-binding domains goes into the transmembrane pore via intraprotein transfer or act as regulators is not yet fully established. Inside the trans-Golgi network, the copper is incorporated into newly synthesized proteins and enzymes that require copper for their function (Gupta and Lutsenko 2009).

Ctr1 Interactions with Platinum Anticancer Drugs

Several studies have shown the importance of Ctr1 when it comes to cisplatin drug uptake and cytotoxicity, even though the exact mechanism is not known. The absence of Ctr1 renders yeast and mammalian cells resistant to copper, cisplatin, carboplatin, and oxaliplatin, whereas overexpression of Ctr1 is found to sensitize cells to the toxic effects of these agents. It was also shown in yeast that copper and cisplatin compete and reduce one another's uptake. These results strongly suggest Ctr1 as a key transporter of cisplatin into the cell (Gupta and Lutsenko 2009). Ctr1 was also shown to regulate the cellular accumulation of cisplatin in several cell studies. An in vivo study in mouse showed that knockout of Ctr1 completely eliminates responsiveness of tumors to cisplatin and an in vivo study in rat showed that cisplatin, carboplatin, and oxaliplatin were effective only in cells with high levels of Ctr1. During the acquisition of cisplatin resistance, the expression of Ctr1 is not altered; however, there seems to be a failure to glycosylate the transporter. That the protein is not glycoslylated makes it a target for proteases, which cuts the N-terminal domain and thereby inactivates the transport function of Ctr1. When exposing human cancer cells to cisplatin, Ctr1 is degraded and both cisplatin and copper uptake is reduced. This downregulation involves endocytosis where Ctr1 first have been marked by ubiquination. From a study with Atox1-/- cells, the downregulation and ubiquination of Ctr1 was shown to be dependent on Atox1. When it comes to transport mechanism, electron microscopic analysis of Ctr1 suggests the pore to be 8 Å in diameter, sufficient for Cu^+ but too small for cisplatin or the other platinum anticancer drugs (Howell et al. 2010). The transporter contains three monomers and is flexible in size, and

FRET (fluorescence resonance energy transfer) measurements have shown that copper and cisplatin have different effects on Ctr1 conformation (Gupta and Lutsenko 2009). Both Cu^+ and Pt^{2+} are soft Lewis acids (▶ Platinum, Physical and Chemical Properties) that form weak bonds with methionine. Ctr1 has rings of methionines lingering the inside of the pore that can coordinate the metals. The specificity of Cu^+ for the channel is due to its acidity, making it bind methionine with affinity high enough to pass through but not so high that the pore is plugged. Since Pt^{2+} has a similar affinity as Cu^+ for methionines, it may pass through as well, but the exact mechanism and how many drug ligands that go through remains unclear (Howell et al. 2010). Binding of cisplatin, carboplatin, and oxaliplatin to a methionine-rich peptide corresponding to the N-terminal domain of Ctr1 has been well documented (▶ Platinum Complexes and Methionine Motif in Copper Transport Proteins, Interaction). In these studies, binding rate and extent of platinum binding to each peptide are dependent on the characteristics of the drug ligands around the platinum ion. Most original drug ligands are also lost in the binding process, raising the question if platinum enters the cell as the free metal ion (Crider et al. 2010). Many evidences are pointing at Ctr1 as a main entry of cisplatin into the cells. If degradation of Ctr1 also is the main determinant of platinum resistance needs to be studied further.

Atox1 Interactions with Platinum Anticancer Drugs

Atox1 has in the latest years become interesting in platinum anticancer drug resistance after the findings that Ctr1 and ATP7A/B could mediate cisplatin resistance. Despite its own part in cisplatin resistance, it also shares a high structural similarity to the metal-binding domains on ATP7A/B. Results from studying cisplatin-Atox1 interactions can also be discussed in the terms of the metal binding domains of the copper ATPases. In vivo studies have shown that Atox1 is required for the Ctr1-mediated uptake of cisplatin into the cell and that downregulation or defunctionalization of Atox1 can contribute to cisplatin resistance (Howell et al. 2010). At a more mechanistic level, in vitro using crystallography cisplatin was shown to bind to the copper binding site of Atox1 (Boal and Rosenzweig 2009). Two crystal structures were solved, one monomer where cisplatin has lost all its original ligands and is coordinated by the two cysteines (Cys12 and Cys15) in the copper binding site and one reducing agent originating from the buffer used in the experiment. The second structure is an Atox1 dimer bridged by cisplatin that is bounded by one cysteine each from the copper binding sites (Cys15). In the dimer, cisplatin has retained its amine ligands. Overall, in these structures, platinum seems to mimic copper by fitting into the copper binding site. Looking at the cisplatin-Atox1 interactions in solution, as opposed to "frozen" states as in crystallography, gives a more complex picture. Using CD spectroscopy Atox1-cisplatin complexes could be further demonstrated in solution and also monitored over time (Palm et al. 2011). The time dependence experiments showed that the complexes are not stable but the protein unfolded and aggregated over time, in hours to days. This relates to the in vivo studies where defunctionalizing of Atox1 could contribute to cisplatin resistance. At lower cisplatin concentrations, the complex is stable and might in vivo follow the copper route and reach the copper ATPases. Atox1-cisplatin binding was also demonstrated with NMR (nuclear magnetic resonance spectroscopy), both in vitro and when Atox1 was overexpressed in *E. coli* cells (Arnesano et al. 2011). Cisplatin was shown, as in previous cases, to bind in the copper binding site between the cysteines by losing its chloride ligands. With time also the amine ligands left and protein dimers started to form. Carboplatin, whose bioactive form is similar to cisplatin, and oxaliplatin were recently shown to bind Atox1 in vitro. These initial experiments with Atox1 and cisplatin suggest a role in cell resistance for Atox1 and will hopefully be accompanied by further studies.

ATP7A/B Interactions with Platinum Anticancer Drugs

The first suggestion that copper ATPases are involved in resistance of platinum anticancer drugs came after the observation that cells resistant to cisplatin had elevated levels of ATP7B. Of the two human copper ATPases, ATP7B seems to be more involved in the platinum drug resistance than ATP7A. In a mouse model of ovary cancer, a downregulation of ATP7B increased the efficacy of cisplatin treatment and reduced the growth of tumors by 40%. This first mouse model targeting ATP7B illustrates the therapeutic potential of modulating ATP7B function or expression (Gupta and Lutsenko 2009). ATP7B has been shown to be involved in cisplatin efflux from the cell by accumulating cisplatin into vesicles to be exported out of the cell. The exact mechanism is not clear; however, since direct transport of platinum by ATP7B was shown to occur very slow and at a nonphysiological pH of ~3–4. An alternative explanation could be that the transport is indirectly mediated by ATP7B. Experiments showing noncompetition between copper and cisplatin transport of ATP7B support that hypothesis. There is though little doubt that cisplatin binds to ATP7B and that this also stimulates catalytic phosphorylation. A number of experiments have shown cisplatin binding to the metal-binding domains of ATP7B, both in vitro and in vivo, and the evidences for specific cisplatin binding to the copper binding sites of the metal-binding domains in ATP7B are strong (Dmitriev 2011; Gupta and Lutsenko 2009; Safaei et al. 2012). In both bacterial and mammalian cell cultures, overexpression of the metal-binding domains alone has induced resistance toward cisplatin, raising the question if the protein sequesters the drug and thereby makes it unavailable for binding to its target, the DNA. Even though these proteins are low abundance, they each have at least six potential platinum binding sites, significantly increasing the total availability. The metal binding sites may not be the final destination of the drug but instead work as high-affinity kinetic traps that hand the platinum over to more abundant possible platinum-binding proteins, like glutathione or metallothionein (Dmitriev 2011). The copper binding site (CXXC) in the metal-binding domains at ATP7B was found to be crucial for cell resistance and accumulation of cisplatin in a recent cell study. Interestingly, a difference in trafficking pattern was also seen when comparing wild-type ATP7B to a mutant with serines instead of cysteines in the copper binding sites of the metal-binding domains. While wild-type protein after exposure to cisplatin becomes dispersed to vesicles throughout the cell, the mutant was unable to relocalize from the trans-Golgi network (Safaei et al. 2012). One point that should not be forgotten is that binding of platinum drug to copper proteins also alters copper homeostasis in the cell. Copper overload induce a wide-ranging detoxification response. Copper accumulation due to ATP7B inactivation was shown to greatly upregulate expression of metallothioneins, alter DNA-repair machinery, and alter the redox balance in the cell – changes that all can contribute to an altered effect of platinum anticancer drugs. However, studies altering also copper concentrations indicate that cisplatin resistance appears to correlate better with the amount of ATP7B in the cells than with intracellular copper concentration or copper resistance. This supports the conclusion that direct effects of drug binding to ATP7B are causing resistance rather than indirect effects of altered copper homeostasis, but further studies are needed. To summarize, there are three possible ways for ATP7A/B to increase cell resistance to platinum anticancer drugs. First, the proteins can directly or indirectly increase efflux of platinum derivatives out of the cell. Second, they could sequester the drugs by binding to the many metal binding sites. Third, binding of drug to these proteins may alter copper homeostasis and cell environment which can directly or indirectly induce cell resistance (Dmitriev 2011).

Outlook

Solving the resistance mechanism of the platinum anticancer drugs is a very difficult task because of the drug's high reactivity and the complexity of human cells. The finding of some key resistance mechanisms could enable development of new drugs aiming at prolonging the anticancer effect. The copper transporting system has been found to be important in terms of cisplatin sensitivity. Copper deficiency has been found in platinum-resistant cell lines, as would be the result of Ctr1 degradation and ATP7B overexpression. These results are highly suggestive of a link between copper transport and platinum resistance. Copper transporting proteins as Ctr1, ATP7A/B, and Atox1 have all been shown to bind cisplatin and affect its behavior in cells. Copper ions are never found as free ions in a human cell; instead, they are protein bound from Ctr1 transferred to chaperones and further on to cellular targets. Is this also the case for platinum? Exactly how cisplatin is utilizing the copper transport pathways is difficult to say and further studies are needed. A better understanding of transcriptional and metabolic changes associated with changes in activity and behavior of the copper transporting proteins is also needed. In recent studies, the proteasome inhibitor bortezomib was found to prevent cisplatin-induced downregulation of Ctr1. Combinations of platinum anticancer drug, bortezomib, and in some cases also radiotherapy are now being tested in phase I trials (Howell et al. 2010).

Cross-References

- ▶ Copper-Binding Proteins
- ▶ Platinum Anticancer Drugs
- ▶ Platinum Complexes and Methionine Motif in Copper Transport Proteins, Interaction
- ▶ Platinum, Physical and Chemical Properties
- ▶ Platinum-Containing Anticancer Drugs and Proteins, Interaction
- ▶ Platinum-Resistant Cancer
- ▶ Terpyridine Platinum(II) Complexes as Cysteine Protease Inhibitors

References

Anastassopoulou I, Banci L, Bertini I et al (2004) Solution structure of the apo and copper(I)-loaded human metallochaperone HAH1. Biochem 43:13046–13053

Arnesano F, Banci L, Bertini I et al (2011) Probing the interaction of cisplatin with the human copper chaperone Atox1 by solution and in-cell NMR spectroscopy. J Am Chem Soc 133:18361–18369

Boal AK, Rosenzweig AC (2009) Crystal structure of cisplatin bound to a human copper chaperone. J Am Chem Soc 131:14196–14197

Crider SE, Holbrook RJ, Franz KJ (2010) Coordination of platinum therapeutic agents to met-rich motifs of human copper transport protein1. Metallomics 2:74–83

Dabrowiak JC (2009) Metals in medicine. Wiley, West Sussex

Dmitriev OY (2011) Mechanism of tumor resistance to cisplatin mediated by the copper transporter ATP7B. Biochem Cell Biol 89:138–147

Gupta A, Lutsenko S (2009) Human copper transporters: mechanism, role in human diseases and therapeutic potential. Future Med Chem 1:1125–1142

Howell SB, Safaei R, Larson CA et al (2010) Copper transporters and the cellular pharmacology of the platinum-containing cancer drugs. Mol Pharmacol 77:887–894

Palm ME, Weise CF, Lundin C et al (2011) Cisplatin binds human copper chaperone Atox1 and promotes unfolding *in vitro*. Proc Natl Acad Sci USA 108:6951–6956

Safaei R, Adams PL, Maktabi MH et al (2012) The CXXC motifs in the metal binding domains are required for ATP7B to mediate resistance to cisplatin. J Inorg Biochem 110:8–17

Platinum(II) Complexes, Inhibition of Kinases

David J. Mann[1] and Ramon Vilar[2]
[1]Division of Molecular Biosciences, Department of Life Sciences, Imperial College London, South Kensington, London, UK
[2]Department of Chemistry, Imperial College London, South Kensington, London, UK

Synonyms

Enzyme inhibition; Organometallic inhibitors of protein kinases; Platinum-containing bioactive kinase inhibitors

Definition

Protein kinases are enzymes that catalyze the transfer of the terminal phosphate from ATP to a substrate protein. Many kinases are important regulators of cellular processes, and inhibition of those kinases can have substantial therapeutic benefits. Platinum-based compounds have proven effective as protein kinase inhibitors allowing compounds of specific three-dimensional configuration to be investigated.

Protein Phosphorylation and Protein Kinases

Regulation of cellular functions has to be accurately orchestrated to coordinate physiological processes and maintain viability. Rather than generate new proteins in response to every changing situation, cells often respond to changes in their environment through tuning the activity of existing proteins as required. This is generally achieved by posttranslational modification of existing proteins with covalent attachment of phosphate groups to amino acid side chains (typically in mammals, those with a hydroxyl group: serine, threonine, and tyrosine) being the most prevalent. These posttranslational modifications can affect protein function by, for example, influencing binding partners or changing protein conformation. This addition of phosphate to a protein is executed by members of the protein kinase family. These enzymes catalyze the transfer of the terminal phosphate of adenosine triphosphate (ATP) to the target protein.

In humans, the protein kinase family has more than 500 different members, each of which can target a variety of substrates for phosphorylation and thus regulate the activity of a large part of the functions and responses of a cell. Given the importance of protein kinases in such regulation, it is not surprising that many diseases are associated with aberrant protein phosphorylation due to overactive protein kinases. Such diseases include many forms of cancer, neurological disease, and also many infectious diseases in which the pathogen alters the regulation of key protein kinases to allow productive infection.

Protein Kinase Inhibitors

Given this widespread backdrop of key protein kinases regulating cellular processes, it is not surprising that many of these protein kinases make excellent drug targets for therapeutic intervention. There is a wide range of small molecule inhibitors of protein kinases available. The majority of these inhibitors target the ATP binding site of the kinase as this forms a structurally conserved, deep pocket into which ligands can stably bind with high affinity, thereby preventing access for ATP and hence blocking phosphotransfer activity. However, this also presents a problem: because the ATP binding pocket is structurally conserved among all members of the family, it is often difficult to generate protein kinase inhibitors that are selective for a particular protein kinase, and this problem has to be overcome in order to produce therapeutically useful compounds.

Traditionally, enzyme inhibitors have been based on purely organic compounds. While this has been a successful strategy and has yielded a wide range of inhibitors – and indeed the pharmaceutical industry has relied on this type of compound for the majority of existing drugs – it is becoming more and more apparent that new strategies are needed to generate compounds with a wider range of features to target particular enzymes more selectively. Metal complexes have emerged as an attractive approach in this regard. In this type of compound, the metal center plays a structural role by organizing different organic ligands in a precise geometrical configuration to confer the desired selectivity to the compound for a given target. This approach has several advantages as compared to the use of purely organic compounds. Firstly, metal centers can display a wide range of geometries (linear, bent, trigonal planar, tetrahedral, square planar, trigonal bipyramidal, octahedral), and by combining them with a relatively small number of ligands, it is possible to easily generate a large library of related compounds which in turn can be very useful to establish structure-activity relationships for the complexes toward a given target. In addition to their attractive structural features, metal centers can also provide the compounds with unique functionality such as magnetism, luminescence, or catalytic activity making them very attractive as tools to understand biological processes.

Metal complexes have been successfully targeted to a wide range of different proteins (Meggers (2009) for a recent review); herein, we will focus on the use of platinum complexes as kinases inhibitors.

Platinum complexes are generally inert, can be easily designed to be air and water stable, and, depending on the oxidation state of platinum and the coordinated ligands, can yield different geometries, the most common being square planar (with Pt^{II}) and octahedral (with Pt^{IV}). These features make platinum complexes attractive as scaffolds for the development of kinase inhibitors. In addition, the interest in this area has also been partly fuelled by the fact that several successful anticancer drugs are based on platinum complexes (cisplatin and its analogues). While the main mode of action of these drugs is proposed to be via interactions with DNA, there is an ongoing interest in establishing whether these complexes also have other targets such as enzymes.

Figure 1 summarizes a selection of platinum complexes that have been reported as inhibitors of protein kinases. Low-affinity kinase inhibitors have been reported (Child et al. 2011) with relatively simple ligands organized around a platinum(II) center. For example, planar platinum(II) complexes **1–5** were investigated as inhibitors of two protein kinases that are intimately involved with cellular proliferation and cancer: the cell division regulator cyclin-dependent kinase 2 (cdk2) and the mitogen-activated protein kinase (MAPK, also known as extracellular signal-regulated kinase (ERK)). Complexes **2** and **3**, which

Platinum(II) Complexes, Inhibition of Kinases, Fig. 1 Chemical structures of the well-established kinase inhibitor staurosporine and selected examples of platinum(II) complexes which have shown to have inhibitory activity against protein kinases

contain the amino-substituted phenanthroline, were shown to be the best inhibitors in the series tested with micromolar activity toward both the kinases tested. It is important to point out that the metal-free (uncoordinated) ligands in these examples did not display inhibitory activity toward the kinases tested highlighting the essential role played by the platinum (II) center as the organizing scaffold to yield compounds of the right shape.

Ground-breaking studies from Meggers et al. have shown that far greater inhibitory potency can be achieved using more complex ligands around an organizing platinum(II) center (Williams et al. 2007). These structures of these complexes were inspired by the natural product staurosporine (Fig. 1), an indolocarbazole alkaloid originally isolated from the bacterium *Streptomyces staurospeus* that is a well-established and highly potent inhibitor that targets the ATP binding site of many protein kinases. Complex **6** was identified from a small library of 28 complexes prepared by simply changing the substituents on the coordinated pyridines. Complex **6** was shown to be a nanomolar inhibitor of glycogen synthase kinase 3α (GSK3α, a protein kinase involved in cell proliferative control and also implicated in the progression of Alzheimer's disease). The crystal structure of this complex demonstrated that the three different ligands were organized around the platinum(II) center to yield a compound that mimicked the overall three-dimensional shape of staurosporine, and this study nicely illustrates the binding affinities that can be achieved by judicious choice of ligands.

While the above examples have used platinum(II) to provide a defined geometry to potentially inhibitory complexes, others have used the metal center as a convenient means of linking together distinct functional units to generate pleiotropic compounds (see Fig. 2). The multikinase inhibitor L^1 was coordinated to a platinum(II) complex (**7**) with the aim of linking it to a carrier protein (lysozyme) that is known to accumulate selectively in the proximal tubule cells of the kidney, a cell type that forms the basis of some renal cancers. Thus, the authors linked a general kinase inhibitor with a targeting agent via the platinum(II) center (**8**). Testing of this agent demonstrated that **8** could cause a substantial decrease in the level of tyrosine phosphoproteins in kidney cells whereas the unconjugated individual components were less effective (Harmsen et al. 2011). Interestingly, complex **8** showed inhibited the production of a range of tyrosine phosphorylated peptides phosphorylated by several distinct protein kinases, indicating that the arrangement of groups in this compound was active against multiple targets as desired. Ligand docking calculations with the protein tyrosine kinase c-KIT showed that L^1 likely binds to the ATP pocket with the pyridine sticking out allowing coordination to platinum(II) as well as leaving enough space for the targeting protein. Direct testing of these compounds against c-KIT and another tyrosine kinase (platelet-derived growth factor receptor β, PDGFRβ) demonstrated that **7** was more effective than L^1 against c-KIT and as effective an inhibitor of PDGFRβ. This example highlights the ability of platinum to not only act as a linker between the multikinase inhibitor and the drug

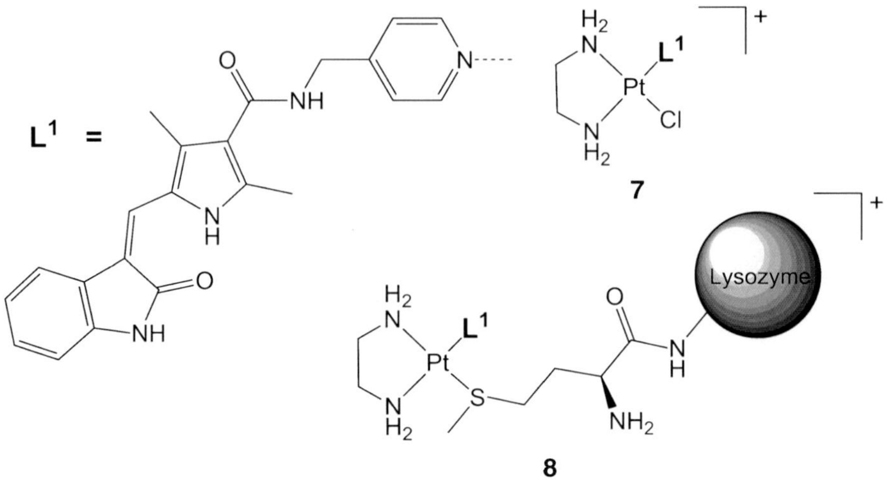

Platinum(II) Complexes, Inhibition of Kinases,

Fig. 2 Chemical structures of a multikinase inhibitor (L^1) and its conjugates to platinum (II) (**7**) and the targeting protein lysozyme (**8**) via coordination to the metal center

carrier but also to potentially enhance the inhibitory activity of the parent compound. It is important to note, however, that protein kinase inhibitors that target multiple enzymes are currently less favored as therapeutic agents than those that are highly selective for a single protein kinase.

Conclusions

From the examples currently available in the literature, it is clear that a platinum(II) core is a highly flexible way of bringing specific ligands into particular geometric configurations and that highly potent protein kinase inhibitors can be developed around such a metal core. However, to date, all studies have simply taken one or two protein kinases and demonstrated inhibition without attention to selectivity. Given that the majority of protein kinases share a structurally conserved ATP binding site and that most protein kinase inhibitors reported to date target this site, off-target activity is a substantial problem. The examples of platinum(II)-based protein kinase inhibitors described above have been inspired by known active compounds; while this has proved a relatively effective strategy, exploration of novel ligands may provide enhanced potency and more importantly enhanced selectivity. The ability of metals such as platinum to hold ligands in defined geometries may allow rapid exploitation of chemical space to identify selective protein kinase inhibitors with substantial ligand efficiency. There is a long history of the use of platinum as an anticancer agent making further exploitation through development of novel agents more acceptable for therapeutic applications. Taken together, these observations should encourage further exploration of the potential of platinum(II) complexes as enzyme inhibitors.

Cross-References

▶ Phosphatidylinositol 3-Kinases
▶ Platinum (IV) Complexes, Inhibition of Porcine Pancreatic Phospholipase A2
▶ Platinum- and Ruthenium-Based Anticancer Compounds, Inhibition of Glutathione Transferase P1-1
▶ Platinum Anticancer Drugs
▶ Platinum-Resistant Cancer
▶ Terpyridine Platinum(II) Complexes as Cysteine Protease Inhibitors

References

Child ES, Georgiades SN, Rose KN, Stafford VS, Patel CBK, Steinke JHG, Mann DJ, Vilar R (2011) Inhibtion of mitogen-activated protein kinase (MAPK) and cyclin-dependent kinase 2 (cdk2) by platinum(II) phenanthroline complexes. J Chem Biol 4:159–165

Harmsen S, Dolman MEM, Nemes Z, Lacombe M, Szokol B, Pato J, Keri G, Orfi L, Storm G, Hennink WE, Kok RJ (2011) Development of a cell-selective and intrinsically active multikinase inhibitor bioconjugate. Bioconjug Chem 22:540–545

Meggers E (2009) Targeting proteins with metal complexes. Chem Commun 2009:1001–1010

Williams DS, Carroll PJ, Meggers E (2007) Platinum complex as a nanomolar protein kinase inhibitor. Inorg Chem 46:2944–2946

Platinum(II), Terpyridine Complexes, Inhibition of Cysteine Proteases

Yan-Chung Lo[1,2] and Andrew H.-J. Wang[2]
[1]The Genomics Research Center, Academia Sinica, Taipei, Taiwan
[2]Institute of Biological Chemistry, Academia Sinica, Taipei, Taiwan

Synonyms

Cysteine protease activity inhibited by terpyridine platinum(II); Terpyridine platinum(II) complexes as cysteine protease inhibitors

Definition

Many proteins contain metal ions and are characterized as metalloproteins, such as zinc finger, manganese superoxide dismutase, and metalloprotease. The metal ion binds with proteins through coordination bonds or covalent bonds which stabilize protein structure or provide biological function. An enzyme that conducts proteolysis activity is named a protease. Proteases include serine proteases, cysteine proteases, threonine proteases, metalloprotease, and others. Intriguingly, terpyridine platinum(II) [TP-Pt(II)] complexes can coordinate with histidine residue of serine protease such as chymotrypsin or form covalent bond with cysteine residue of cysteine protease to form a metalloprotein, thereby inhibiting the protease function.

The Metal Ion–Containing Proteins

Metalloproteins constitute a large number of biologically active proteins which are responsible for several important functions. Several metals, such as Au, Pt, Zn, etc., have been widely investigated for their biological functions. For example, the Zn^{2+} ion–containing proteins (such as zinc finger, histone deacetylase, or pyroglutamyl cyclase) play important roles in transcriptional regulation and posttranslational modification. Cisplatin (cis-diamminedichloroplatinum(II)), a platinum(II) based complex, is the first metal-based anticancer drug. The gold complexes (like Auranofin) could inhibit human thioredoxin reductase 1 (hTrxR 1) and glutathione reductase (GR) activities. Thus, metalloproteins find myriad of applications in different biological processes. But how these transition metal ions affect biological processes are not yet fully understood. This entry will elucidate the interaction between platinum(II) complexes and cysteine proteases, and to use this knowledge in building a concept of drug design.

Platinum (II) Complexes Are Widely Studied in Both DNA and Protein Functions

Cisplatin, carboplatin, and oxaliplatin, belonging to the platinum(II) complexes, are commonly used for the treatment of cancer (in Fig. 1). The molecular mechanism of their effect is via their binding to the N-7 positions of two adjacent guanine bases (like alkylating agent) to disrupt DNA function, leading to cell death (Wang and Lippard 2005). Although cisplatin binds to thiol groups, but it is easily excreted out by urine, other platinum compounds (like TP-Pt(II)) have been explored as chemotherapeutics against cancer.

TP-Pt(II) complexes, one of aromatic-amine platinum(II) complexes (Fig. 1), have been synthesized and studied for more than three decades and is well reviewed recently (Cummings 2009). The fourth ligand of platinum(II) atom in all the complexes is thiolate or chloride ion so that the complex bears only a single positive charge. When the substitution ligand is chloride, the TP-Pt(II) chloride complex can interact with DNA by coordinating with guanine N-7. While measuring the DNA-binding ability of TP-Pt(II) complexes, the binding affinity toward calf thymus DNA increased with increasing aromatic planar surface extension and decreased by steric effect of stacking surface of side chain. When the substitution ligand is thiolate, like compound VI (Fig. 1) [(HET)-TP-Pt(II)], it intercalates into deoxyCpG site from major groove (Wang et al. 1978).

TP-Pt(II) complexes can interact not only with DNA but also protein to affect their functions.

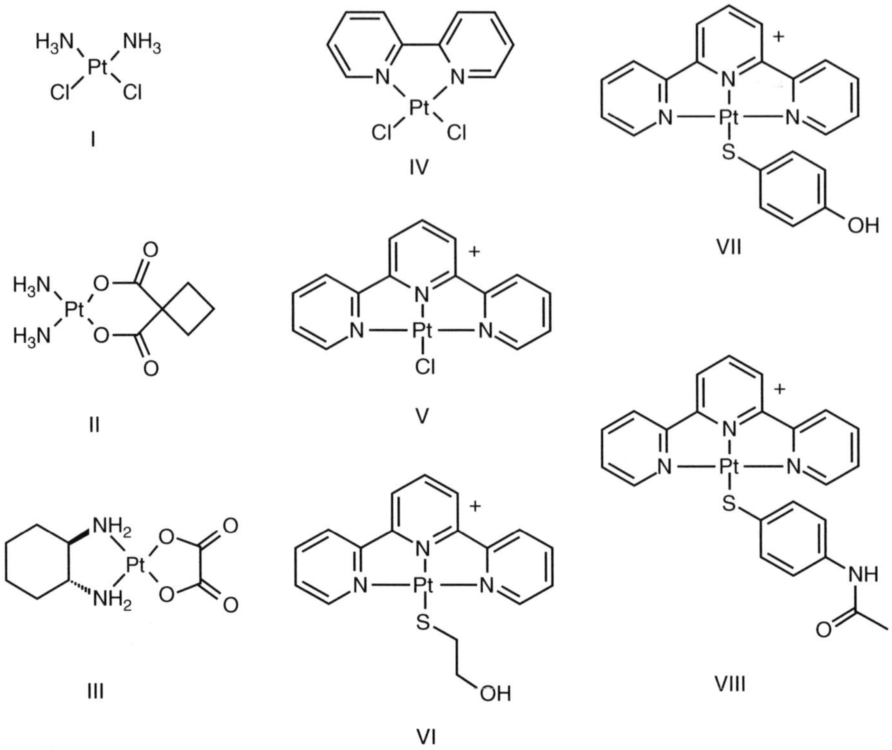

Platinum(II), Terpyridine Complexes, Inhibition of Cysteine Proteases,
Fig. 1 The platinum(II) complexes. *I* Cisplatin; *II* Carboplatin; *III* Oxaliplatin; *IV* Bipyridine platinum(II) dichloride; *V* Terpyridine platinum(II) chloride; *VI* Hydroxylethanethiolate terpyridine platinum(II); *VII* 4-Hydroxylthiophenolate terpyridine platinum(II); *VIII* N-Acetyl-4-aminothiophenolate terpyridine platinum(II)

For example, TP-Pt(II) complexes are irreversible inhibitors of parasite trypanothione reductase (TryR) and hTrxR 1, which play important roles in parasites (*Leishmania donovani*, *Trypanosoma cruzi*, and *Trypanosoma brucei*) and cancer cells (HeLa, gilioma, and head-and-neck squamous carcinomas), respectively. This inhibitory effect may be the reason for their cellular cytotoxicity. The interactions between platinum (II) complexes (e.g., cisplatin, TP-Pt(II), etc.) and serine/cysteine proteases (e.g., chymotrypsin, papain, cathepsin B, caspases, etc.) have been investigated. TP-Pt(II) chloride was proposed to label the active-site Cys25 of papain, based on a kinetic assay, and tagged on active-site histidine residue of chymotrypsin as shown by mass spectrometer. These platinum(II) complexes have common properties which easily interact with cysteine, methionine, and histidine residues of proteins. Although the effect of the platinum(II) complexes on DNA has been well studied, direct observation of how platinum(II) complexes affect protein function is rare.

To sum up, TP-Pt(II) complexes can inhibit TryR, TrxR 1, and serine/cysteine protease activities (Lo et al. 2009, 2011). Since cysteine proteases play important roles in a number of biological processes, it will be of interest to investigate how platinum(II) complexes interact with them.

Bioisosterism of Cysteine Proteases

Bioisosterism is a strategy of medicinal chemistry for the rational design of new drugs or of modified compounds (Thornber 1979). Bioisosteres are substituents or groups that have chemical or physical similarities which imparts broadly similar biological properties to a chemical compound. For example, the structures of TP-Pt(II) complexes (Fig. 1, compounds VI and VII) are similar to those of topoisomerase inhibitors, N-[4-(acridin-9-ylamino)-3-methoxyphenyl]methanesulfonamide (*m*-AMSA) and N-[2-(dimethylamino)ethyl]acridine-4-caboxamide (DACA) (McFadyen et al. 1985; Lo et al. 2009). In addition, TP-Pt(II) complexes effectively inhibit not only hTrxR 1 but also mammalian topoisomerases. Owing to the similar structure and function, compounds VI and VII and *m*-AMSA and DACA are bioisosteres and, thus, supposedly play the role of topoisomerase inhibitor.

Cysteine proteases play important roles in a number of biological processes and can be classified into three groups according to their structural folds (Fig. 2,

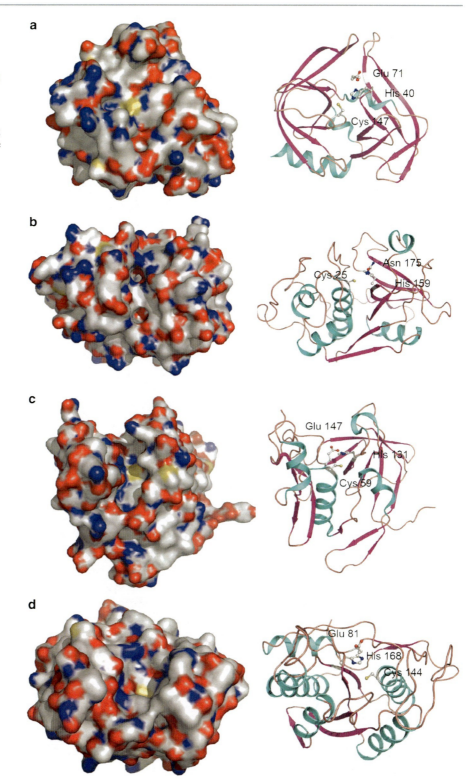

Platinum(II), Terpyridine Complexes, Inhibition of Cysteine Proteases, Fig. 2 The diagram of cysteine proteases presented surface and ribbon. (**a**) Chymotrypsin-like fold (CVB3 3C^{pro}; PDB: 2ZTZ). (**b**) Papain-like fold (papain; PDB: 2CIO). (**c**) Papain-like fold (GspA; PDB: 3A2Z). (**d**) Others (PGP I; PDB: 3RNZ)

Platinum(II), Terpyridine Complexes, Inhibition of Cysteine Proteases, Table 1 The catalytic triad and structure fold for cysteine proteases

Protease	Fold	Active site
CVB3 3C^{pro}	Chymotrypsin like	Cys147, His40, Glu71
Papaya Papain	Papain	Cys25, His159, Asn175
Cathepsin B	Papain like	Cys29, His199, Asn219
GspA	Papain like	Cys59, His131, Glu147
PGP I	Others	Cys144, His168, Glu81

Table 1) (Tong 2002), viz. cysteine proteases with chymotrypsin-like fold, e.g., the 3C-protease (3C^{pro}) from human coxsackievirus strain B3; cysteine proteases with papain-like fold, e.g., the N-terminal domain of glutathionylspermidine synthetase/amidase (GspA) from *Escherichia coli*; and all other cysteine proteases with less conserved structural folds, e.g., pyroglutamyl peptidase I (PGP I) from *Bacillus amyloliquefaciens* were grouped separately. Though cysteine proteases have different structural folds, they can be viewed as bioisosteres owing to their common catalytic triad or diad.

The active site of hTrxR 1 is composed of Cys59, Cys64, His472, Glu477, Cys497, SeC498. His472 and Glu477 act as acid–base catalysts, facilitating the electron and proton transfers. TP-Pt(II) complexes have been found to bind to the essential pair of C497-SeC498 in the active site. Further, the catalytic triad of the cysteine protease (e.g., papain, cathepsin B, etc.) has acid–base catalytic property, and contains active-site cysteine, along with a histidine and a glutamate or asparagine. Hence cysteine proteases and hTrxR 1 possess similar active-site residues (cysteine or selenocysteine, histidine, and a glutamate or asparagine), they are bioisosteres. Thus, the compounds (like TP-Pt(II) complexes) interacting with hTrxR 1 are likely to exhibit inhibitor activity to its bioisoestere, i.e., cysteine protease.

Direct Evidence of TP-Pt(II) Interaction with Cysteine Residue of Cysteine Proteases

Several studies have indicated that platinum(II) complexes may interact with many cysteine proteases. To address whether the platinum(II) complexes (Fig. 1, compound V, VII, and VIII) can directly interact with cysteine proteases and inhibit their functions, several biophysical methods were applied.

The primary method for examining the interaction between TP-Pt(II) complexes and cysteine proteases is enzyme inhibition assay. The inhibition assay can screen effective inhibitors rapidly and offer a clue to the interactions between TP-Pt(II) complexes and cysteine proteases. Notably, the active-site environments of cysteine proteases (like, 3C^{pro}, GspA, and PGP I) are more hydrophobic; therefore, substitution of platinum complexes with aromatic ring may increase inhibitory activity against the cysteine proteases. This concept has been demonstrated previously (Lo, et al. 2011) which showed that the inhibitory effects of platinum(II) complexes on PGP I depends on the substitution of platinum complex with aromatic rings and their inhibitory effects are in decreasing order as follows: polypyridine platinum (II) complexes > cisplatin > carboplatin.

Furthermore, to address the detailed features of the interaction of TP-Pt(II)-cysteine proteases, X-ray crystallographic and MALDI-TOF-TOF spectrometric analyses were applied. The color of TP-Pt(II) chloride solution is red. Without TP-Pt(II), the crystal of PGP I is colorless (Fig. 3a). When TP-Pt(II) interacts with cysteine protease, the color of crystal becomes pink (Fig. 3b). This is a simple way to observe the formation of TP-Pt-protein adducts and this interaction can further be confirmed by detailed X-ray crystallographic analysis. For example, as shown in Fig. 3c, the electron density map shows clearly the covalent bond formation between platinum(II) atom of TP-Pt(II) and sulfur atom of active-site cysteine residue of PGP I, proving that TP-Pt(II) directly interacts with cysteine residue of cysteine proteases.

However, X-ray crystallography has its own limitations, e.g., growing crystal of a protein in its active form. In those cases of failing getting crystals, mass-spectrometry-based analytical tools are more useful. In this approach, it is advisable to first identify the peptide sequence associated with TP-Pt(II) by digesting TP-Pt(II)-protein adducts (like TP-Pt(II)-PGP I adduct, Fig. 4a) with trypsin and analyzing them by MALDI-TOF. This approach was used by Lo et al. (2011), where they analyzed TP-Pt(II)-GspA and TP-Pt(II)-3C^{Pro} adducts by MALDI-TOF-TOF to label the TP-Pt(II) complex on the associated residue. However, such mass spectrometric method may not always work since the trypsin-digested peptides might not be eluted thoroughly. For example, although Lo et al. could detect active site for TP-Pt(II)-

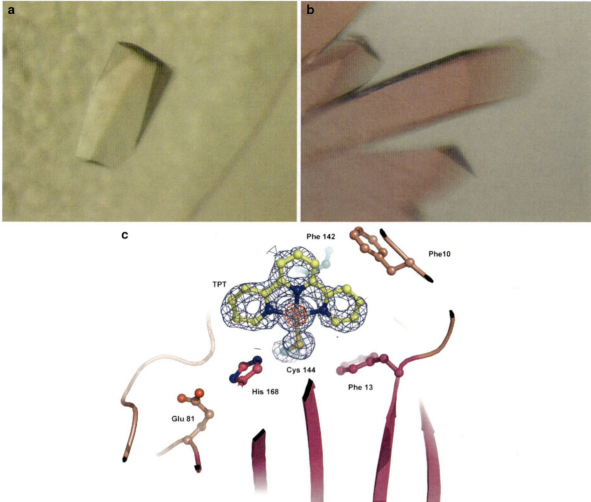

Platinum(II), Terpyridine Complexes, Inhibition of Cysteine Proteases, Fig. 3 (**a**) The crystal of PGP I is colorless. (**b**) The color of TP-Pt(II)-PGP I adduct is *pink*. (**c**) The electron density map of platinum-sulfur bond in the TP-Pt(II)-PGP I cocrystal

GspA, TP-Pt(II)-3C^{Pro}, he could neither detect it in the trypsin-digested peptide in TP-Pt(II)-PGP I adducts (Fig. 4b, lower panel), nor in the peptide containing active-site cysteine of PGP I (Fig. 4b, upper panel), probably owing to the difficult elution of the active-site-containing peptide from ZipTip.

It is worthy to note that not all cysteine enzymes can interact with TP-Pt(II) complexes specifically via the active site, probably due to the active-site shape that is not suitable to accommodate the TP-Pt(II) complex. The cocrystal structure of TP-Pt(II) and SsPTP (protein tyrosine phosphatase from *Sulfolobus solfataricus*) reveals that TP-Pt(II) binds at the entrance of the active site, stacking on the tryptophan side chain, but not coordinating to the active-site cysteine.

The Ligand Replacement Effect of TP-Pt(II) Complexes

The pK_a value of cysteine residue varies depending on the surrounding environment of the active site; thus making cysteine a critical residue to define the biological properties of a protein. For example, the pK_a value of the active-site cysteine of GspA is ∼3.05; and of Cys 34 residue of human albumin is ∼5.0, while the normal pK_a value is 8.5. The binding preference of TP-Pt(II) depends on the availability of low pK_a cysteine residue at the active site of protein, which is more

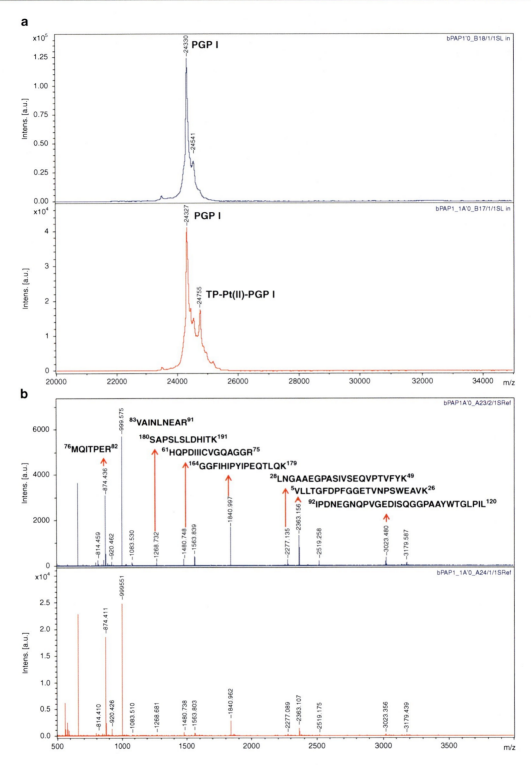

Platinum(II), Terpyridine Complexes, Inhibition of Cysteine Proteases, Fig. 4 (a) The total molecule weight mass spectrum of PGP I tagged with TP-Pt(II) or not. (b) Trypsin digestion of PGP I was pre-incubated with or without TP-Pt(II) complexes before trypsin digestion and the following MS analysis. The peptide containing active site, whose M/Z is 3161.5, could not be found here reactive for metal complexes (e.g., Pt) than any other amino-acid residue. For example, TP-Pt(II) complexes bind to the active-site histidine residue of serine proteases, whereas in cysteine proteases, it bind to the cysteine residue, instead of histidine, in the active site. The sulfhydryl group of the low pK_a cysteine residue of proteins will form a reactive nucleophilic thiolate which attaches the platinum atom of TP-Pt(II) complexes and replaces their displaceable ligand. A sulfhydryl group is a stronger nucleophile than an imidazolium cation in the active site of papain-like fold protein. Further, the ligand exchange also occurs in other metal complexes (like Ru and Au) (Fricker 2010).

Thus, toward building a concept of drug design, one first needs to identify the amino-acid residue interacting with the inhibitor, followed by due consideration given to the ligand replacement based on the charge of the ligand. Thereby, the inhibitor design can be refined to be more effective. This methodology is explained, with an example, in the following part of this entry.

Using TP-Pt(II)-GspA as a Model for Drug Design

A number of innovative, rational drug design approaches have been invented, such as high-throughput screening, combinatorial chemistry, ligand-based drug design, and fragment-based drug design. Ideally, the inhibitors of the target protein can be identified by random screening and the pharmacophore of the drug then can be generalized by the common features of different inhibitors. If this target protein has a known ligand which interacts with it, the ligand could be modified as an inhibitor. The fragment-based drug design is to link several fragments which act at different sites of the target protein to make a more potent inhibitor. TP-Pt(II) complex can potentially bind on the cysteine proteases; therefore, it can be considered as an ideal pharmacophore of cysteine proteases inhibitor.

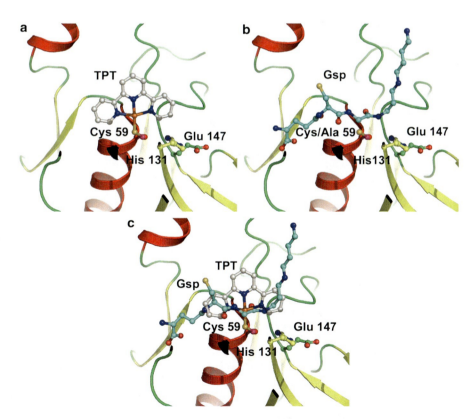

Platinum(II), Terpyridine Complexes, Inhibition of Cysteine Proteases, Fig. 5 (continued)

d

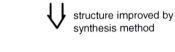

Platinum(II), Terpyridine Complexes, Inhibition of Cysteine Proteases, Fig. 5 (a) The TP-Pt(II)-GspA adduct was modeled in silico and presented by PyMOL. (b) The Gsp-GspA interaction was presented by PyMOL (PDB: 3A2Y). (c) The superposition of TP-Pt(II)-GspA and Gsp-GspA was presented by PyMOL. (d) The scheme of design of GspA inhibitors

In trypanosome, there are many cysteine proteases which could be potential chemotherapeutic targets, e.g., papain-like fold containing cysteine proteases viz. *Trypanosoma brucei* cathepsin B-like enzyme (*Tb*Cat), Rhodesain, and trypanothione synthetase/amidase (TSA). TP-Pt(II) complexes have been linked, as an inhibitor, with GspA, which contains a papain-like cysteine protease domain. This domain is the highly conserved structure in the N terminus of TSA. Therefore, TP-Pt(II) could also be an inhibitor against TSA. To make TP-Pt(II) compound specifically targeting TSA, the following approach is proposed. From the result of TP-Pt(II) tagging on the active-site cysteine residue of GspA, TP-Pt(II)-GspA adduct is modeled *in silico* (presented by PyMOL in Fig. 5a). The Gsp (glutathionylspermidine)-GspA crystal complex (PDB: 3A2Y) was retrieved (presented by PyMOL in Fig. 5b) (Pai et al. 2011). Gsp is the substrate of GspA. The superposition of TP-Pt(II)-GspA and Gsp-GspA structures revealed that TP-Pt(II) complex overlaps with cysteine and glycine of Gsp, as shown in Fig. 5c. Therefore, TP-Pt(II) could also be modeled as an inhibitor of GspA. To this effect, a hypothesis is proposed for drug design (Fig. 5d). Further, based on two ligands, TP-Pt(II) complex and Gsp, another inhibitor model I, model II, can be designed (Fig. 5d). In inhibitor model II [5-(4-(2-formylphenyl)-1H-imidazol-1-yl)-5-oxopentanamide], the aldehyde group is substituted to platinum atom (at No. 4) coupled with a decreases in aromatic ring and ring size. The aldehyde group also forms pseudo ring via hydrogen bond with nitrogen atom of imidazole ring (Fig. 5d, shown as purple color dashed line). Using bioisosterism concept, the imidazole could also be substituted to its bioisoestere triazole (inhibitor model III).

To assess the reliability of this approach, the GspA/TSA inhibitors were searched from the published literature. This approach retrieved a compound, designated as DDU 86439 [(N-(3-(dimethylamino)propyl)-2-(3-(3-fluorophenyl)-1H-indazol-1-yl)acetamide)]. This compound was identified in a high-throughput drug screening, thus confirming to the concept of building of drug design.

The aforementioned case study indicates that TP-Pt(II) chloride complex is a potent cysteine proteases inhibitor and it can serve as a pharmacophore for further drug design. Furthermore, polypyridine complexes are chelating compounds and can coordinate with a variety of metal ions (like Au^{3+}, Pd^{2+}, Zn^{2+}, and Ru^{2+}). These metals have different coordination numbers and these metal complexes exhibit a high structural diversity owing to their geometries. In conclusion, TP-Pt(II) complex can not only be a pharmacophore for design of cysteine proteases inhibitor, but other metal complexes can also be a pharmacophore for design of other protein inhibitors.

References

Cummings SD (2009) Platinum complexes of terpyridine: interaction and reactivity with biomolecules. Coord Chem Rev 253:1495–1516

Fricker SP (2010) Cysteine proteases as targets for metal-based drugs. Metallomics 2:366–377

Lo YC, Ko TP, Su WC, Su TL, Wang AH (2009) Terpyridine-platinum(II) complexes are effective inhibitors of mammalian topoisomerases and human thioredoxin reductase 1. J Inorg Biochem 103:1082–1092

Lo YC, Su WC, Ko TP, Wnag NC, Wang AH (2011) Terpyridine platinum(II) complexes inhibit cysteine proteases by binding to active-site cysteine. J Biomol Struct Dyn 29:267–282

McFadyen WD, Wakelin LPG, Roos IAG, Leopold VA (1985) Activity of platinum(II) intercalating agents against murine leukemia L1210. J Med Chem 28:1113–1116

Pai CH, Wu HJ, Lin CH, Wang AH (2011) Structure and mechanism of *Escherichia coli* glutathionylspermidine amidase belonging to the family of cysteine; histidine-dependent amidohydrolases/peptidases. Protein Sci 20:557–566

Thornber CW (1979) Isosterism and molecular modification in drug design. Chem Soc Rev 8:563–580

Tong L (2002) Viral proteases. Chem Rev 102:4609–4626

Wang D, Lippard SJ (2005) Cellular processing of platinum anticancer drugs. Nat Rev Drug Discov 5:307–320

Wang AH, Nathans J, van der Marel G, van Boom JH, Rich A (1978) Molecular structure of a double helical DNA fragment intercalator complex between deoxy CpG and a terpyridine platinum compound. Nature 276:471–474

Platinum, Physical and Chemical Properties

Fathi Habashi
Department of Mining, Metallurgical, and Materials Engineering, Laval University, Quebec City, Canada

Platinum is a member of the group of six platinum metals which occur together in nature and is located in the center of the Periodic Table in the Iron–Cobalt–Nickel group:

Fe	Co	Ni
Ru	Rh	Pd
Os	Ir	*Platinum*

Physical Properties

Platinum is a transition metal like the other members of the group. They are less reactive than the typical metals but more reactive than the less typical metals.

The transition metals are characterized by having the outermost electron shell containing two electrons and the next inner shell an increasing number of electrons. Because of the small energy differences between the valence shells, a number of oxidation states occur.

Atomic number	78
Atomic weight	195.08
Relative abundance, %	5×10^{-7}
Density at 20 °C, g cm^{-3}	21.45
Major natural isotopes	194 (32.9 %)
	195 (33.8 %)
	196 (25.3 %)
Crystal structure	Face centered cubic
Atomic radius, nm	0.139
Melting point, °C	1,772
Boiling point, °C	4,170
Specific heat at 25 °C, C_p, J g^{-1} K^{-1}	0.13
Thermal conductivity λ, W m^{-1} K^{-1}	73
Brinell hardness	50
Tensile strength σ_B, N mm^{-2}	137.3
Specific electrical resistance at 0 °C, $\mu\Omega$ cm	9.85
Temperature coefficient of electrical resistance (0–100 °C), K^{-1}	0.0039

Chemical Properties

Because of the small energy differences between the valence shells, a number of oxidation states occur: 0, +2, +4, and +5. The tetravalent state is the most common. Platinum dissolves in concentrated HCl saturated with chlorine to form hexachloroplatinate ion, $[PtCl_6]^{2-}$, from which hexachloroplatinic (IV) acid, $H_2[PtCl_6]\cdot 6H_2O$, can be crystallized as reddish brown crystals (m.p. 150 °C). Ammonium hexachloroplatinate (IV), $(NH_4)_2[PtCl_6]$, is precipitated as a yellow powder from solution of hexachloroplatinic acid by adding ammonia:

$$H_2[PtCl_6] + NH_3 \rightarrow (NH_4)_2[PtCl_6]$$

The powder is decomposed at 800 °C to produce platinum sponge:

$$3(NH_4)_2[PtCl_6] \rightarrow 3Pt + 2NH_4Cl + 16HCl + 2N_2$$

Platinum (IV) chloride, $PtCl_4$, is produced by dehydration of $H_2PtCl_6\cdot 6H_2O$ at 300 °C in a stream of chlorine. It is a red-brown, crystalline, hygroscopic powder. Above 380 °C, $PtCl_4$ liberates chlorine and forms platinum (II) chloride, $PtCl_2$, which is stable between 435 °C and 580 °C. Above 580 °C, further decomposition occurs to yield metallic platinum. Platinum (IV) chloride is soluble in water. Platinum (II) chloride can also be obtained from a solution of $H_2[PtCl_4]$ by evaporation under vacuum. Chlorides of platinum can be produced by reacting chlorine with finely divided platinum.

Platinum has been in such demand for jewelry, for laboratory crucibles, dishes, and boats, and for electrodes for electrogravimetry. The corrosion and erosion resistance of platinum metals toward molten glass is not matched by any other material. All platinum group metals have strong catalytic activity, especially for hydrogenation reactions. Platinum metals and their alloys are widely used in thermocouples.

References

Habashi F (2003) Metals from ores. An introduction to extractive metallurgy. Métallurgie extractive Québec, Quebec City, Canada. Distributed by Laval University Bookstore, www.zone.ul.ca

Renner H (1997) Platinum group metals. In: Habashi F (ed) Handbook of extractive metallurgy. Wiley-VCH, Weinheim, pp 1269–1360

Platinum-Containing Anticancer Drugs and Proteins, Interaction

Radu Silaghi-Dumitrescu and Cristina Bischin
Department of Chemistry and Chemical Engineering, Babes-Bolyai University, Cluj-Napoca, Romania

Synonyms

Binding of platinum to metalloproteins; Cellular pharmacology of platinum-based drugs; Metal-protein interactions; The resistance mechanism of the platinum anticancer drugs

Definition

Platinum-based anticancer drugs can bind to a range of proteins, especially at sulfur atoms, affecting their

Platinum-Containing Anticancer Drugs and Proteins, Interaction, Fig. 1 Structures of three common platinum-containing anticancer drugs (Bischin et al. 2011)

conformation and functions. One possible mechanism of resistance and toxicity induced by these drugs may be explained according to this interaction, relevant to the therapeutic effect.

One of the most important drugs employed in treating various types of cancers is cisplatin. Unlike most other drugs, cisplatin is a simple inorganic compound, [$H_6Cl_2N_2Pt$] (cf. Fig. 1) (Che and Siu 2010). After administration, one of the chloride ligands at the cisplatin platinum center is displaced by water. Subsequent displacement of this water allows the platinum to coordinate to a nitrogen atom in DNA. Thereafter, cross-linking of two DNA bases may occur via displacement of the other chloride ligand from platinum. This binding then triggers DNA repair mechanisms, which in turn may activate apoptosis (Che and Siu 2010; Cossa et al. 2009). Oxaliplatin and carboplatin are two other widely used platinum derivatives, featuring organic ligands at the metal (cf. Fig. 1); they show lower toxicity than cisplatin, allowing higher dosage during treatment. The limiting toxicity of oxaliplatin is peripheral sensory neuropathy, also seen with cisplatin. The targets and mechanisms of action for these drugs are different; for instance, oxaliplatin can be used with the patients who have developed resistance to cisplatin (▶ Oxaliplatin). Carboplatin is better retained in the body, so that its effect is longer lasting (Che and Siu 2010; Bischin et al. 2011, ▶ Platinum Anticancer Drugs).

Side effects such as nausea, progressive peripheral sensory neuropathy, fatigue, vomiting, alopecia, hematological suppression, renal damage still fuel interest into the mechanisms of interaction of cisplatin and related compounds with a wider range of biomolecules, not only DNA (Cossa et al. 2009). Cisplatin-derived platinum were indeed found to bind to a range of proteins, as demonstrated by elemental analyses, chromatography, and mass spectrometry. It was in fact estimated that more than 95% of the cisplatin that has entered a cell is bound to proteins and peptides, rather than to DNA (Bischin et al. 2011).

Structurally Characterized Platinum-Protein Complexes

Numerous crystal structures are available with platinum bound to proteins. However, not all cases involve cisplatin and its congeners: the presence of platinum may at times be fortuitous; nevertheless, even in those cases, hints for a possible physiological relevance are given by these binding events. In all of these structures, the metal preferentially binds to sulfur-containing aminoacids (methionine, cysteine) and nitrogen-containing ones (especially histidine). Table 1 gives a review of these interactions.

Aminoacids

According to the HSAB theory (hard-soft acid–base), Pt(II) is a soft center due to its large radius and relatively small charge (e.g., as opposed to Pt(IV)). The ideal ligands for such a metal center are either the soft ones (such as the sulfur in methionine and cysteine) or the borderline ones (such as the nitrogen in the histidine imidazole ring) (Holm et al. 1996, ▶ Platinum, Physical and Chemical Properties). Indeed, cisplatin has high affinity for thiol groups such as those in cysteine and in its derivatives/(poly)peptides glutathione and metallothionein. Furthermore, cisplatin raises the level of thiol groups in cells, which is at least in part responsible for the resistance developed against this drug (Platinum-resistant cancer); thiol-blocking reagents favor cisplatin-DNA binding in vivo. Pt-methionine adducts were detected in the urine of cisplatin-treated patients, pointing out to this amino acid as an additional target of platinum. Using nucleopeptides containing both a sulfur-based ligand and a nitrogen base, it was shown that the reaction with platinum complexes leads to platinum coordination to the sulfur atom, but that this sulfur is subsequently substituted by the nitrogen from the nucleobase;

Platinum-Containing Anticancer Drugs and Proteins, Interaction, Table 1 Experimentally determined structure for platinum-bound proteins (Adapted after Bischin et al. 2011)

Protein name	Amino acid interacting with Pt
Neisseria gonorrhoeae pilin	His
Transposon Tn5	Cys
Human DNA repair endonuclease HAP1	Lys Met
Thermococcus litoralis trehalose/maltose-binding protein	Met
Glycoprotein Ib (GPIb)	Thr
Bacillus subtilis trp repressor binding protein	Gln, His, Lys, Met, Thr
Clostridium thermocellum cellulosomal xylanase Z	Met
Mycobacterium tuberculosis shikimate kinase	Met
Escherichia coli adaptor protein ClpS	Met
Truncated human EGFR ectodomain bound to TGFalpha	Met
Pyrococcus furiosus Pfu-1218608	Met
Human complement regulator CD55	His, Met
Pitrilysin	Arg, His
Escherichia coli OmpX membrane protein	Met
Thermus thermophilus ClpB chaperone	Arg, Gln, Lys, Met, Thr, Tyr
N-terminal domain of human platelet receptor glycoprotein Ib-alpha	His
Resistin	Met
Escherichia coli response regulator NarL	Met, Glu
Mor protein of bacteriophage Mu	His, Met
Endoplasmic reticulum protein Rp19	Cys, Lys
Bacillus subtilis YfhH hypothetical protein	Met
Talin rod spanning residues 482-789	Ser, Arg, Lys
Rice alpha-galactosidase complexed with D-galactose	Met
Arabidopsis thaliana Acyl-CoA oxidase 1	Met, Glu, His, Asn
Bacillus subtilis putative fructokinase	Met
Yersinia enterocolitica type III secretion chaperone SycT	His
Epstein-Barr virus oncogene BARF1	Met, His, Glu
Tapes japonica lysozyme	Met
Hen egg white lysozyme	His
DCN-1 protein	Met
Pyrococcus horikoshii OT3 PH0725	Met, Glu, His
PhoP regulator	His
Bacillus subtilis FapR protein	Met
S445C mutant of *Pseudomonas sp.* MIS38 lipase	Cys
Human thioredoxin reductase 1	Cys
Human copper chaperone	Cys
PacI-DNA enzyme complex	Met

chelate situations were also found with a Pt-(S, N) coordination pattern stable toward Pt-S dissociation (Bischin et al. 2011).

Glutathione (GSH)

The cysteine-containing tripeptide glutathione reaches concentrations as high as 10 mM. Cisplatin forms several types of adducts with cisplatin, all involving Pt-S bonds. Administration of glutathione together with cisplatin was found to increase the efficiency of anticancer treatment; the proposed explanations for this effect include a reduced toxicity due to platinum ligation by glutathione, a more general antioxidant effect of the peptide, or formation of more active platinum species under the influence of glutathione. In this connection, a decrease in glutathione concentration at the kidney accompanies the nephrotoxic side effects of this drug. Furthermore, cisplatin resistance mechanisms involve a rise in glutathione concentration and elimination of Pt-GSH adducts from the cell (Bischin et al. 2011).

Metallothionein (MT)

Despite its relatively low molecular weight, metallothionein features 20 cysteines and 9 methionines, which allow it to bind at least up to seven metals at a time and function in metal homeostasis – with regulatory and antioxidant roles also cited, including in the context of apoptosis. The naturally bound Zn in MT is easily displaced by platinum, as proved by high-performance liquid chromatography (HPLC), atomic absorption spectroscopy, and ultraviolet (UV) absorption spectroscopy. MT increases are part of the resistance response triggered by cisplatin (Bischin et al. 2011).

Thioredoxin (TRX) and Thioredoxin Reductase

Thioredoxin is an antioxidant small protein which uses electrons provided by thioredoxin reductase from NADPH. In certain types of tumors increased TRX levels were found to correlate with cisplatin resistance. Cisplatin and oxaliplatin inhibit TRX directly (Witte et al. 2005). A few other Pt-containing compounds were found to bind to the selenocysteine active site in thioredioxin reductase, acting as inhibitors (Che and Siu 2010; Bischin et al. 2011).

Serum Albumin

Known to bind and transport a wide range of metabolites and exogenous substances, and being the most abundant protein in blood serum (~ 0.6 mM), featuring a free thiol as well as 17 disulfide bridges, albumin is expected to bind platinum compounds efficiently. It was indeed shown that ~ 24 h after cisplatin administration, up to 98% of the total platinum can be bound to blood proteins – and especially to albumin. Interestingly, this was found not to limit the cytotoxicity of the drug; however, it did limit its urinary excretion and it was in fact found that administration of albumin together with cisplatin diminishes the nephrotoxicity. Cisplatin causes distinct variations in protein conformation including considerable decrease of the helical structure and a change in binding affinity toward other molecules (Ivanov et al. 1998).

Evidence for platinum binding to serum albumin has come from capillary isoelectric focusing (CIEF) with whole column imaging detection (WCID) fluorescence spectroscopy, FT-IR, circular dichroism (CD), inductively coupled plasma mass spectrometry, (ICPMS) capillary electrophoresis (CE), and NMR-spectroscopy (Ivanov et al. 1998). Met298 is the main target for cisplatin in albumin, with Met87, Met446 and Cys34 also shown to play a role. Cisplatin can induce inter-protein cross-linking of albumin via sulfur-platinum interactions (Ivanov et al. 1998). It is important to note that each platinum-containing drug reacts in different manners with albumin; thus, oxaliplatin is slightly more reactive than cisplatin toward albumin, while nedaplatin and lobaplatin bind only very weakly to serum proteins (Bischin et al. 2011).

Other Antioxidants

Cisplatin induces acute as well as chronic decreases in several major plasma antioxidants. This may be due to oxidative stress-induced consumption of antioxidants as well as to renal loss of water-soluble low molecular weight antioxidants due to hyperfiltration in combination with a specific cisplatin-related renal tubular defect. This leads to increased oxidative damage to normal tissues such as renal tubular cells. The degree to which antioxidant-rich nutrients may protect against cisplatin-induced oxidative damage DNA retain the antitumor efficacy is still a subject of study (Bischin et al. 2010).

The administration of controlled amounts of antioxidants such as vitamin E, ascorbate, selenium, and carotenoids, before or after treatment with cisplatin, was reported to protect against nephrotoxicity in human and animals, without compromising the antitumor activity. Aminoguanidine prevents tubular damage and perivascular inflammation observed in kidneys as a result of cisplatin administration. Furthermore, the levels of malondialdehyde were reduced, and the changes in liver glutathione and in serum urea levels normally caused by cisplatin were alleviated by aminoguanidine (Bischin et al. 2010).

Resveratrol, a natural molecule with antioxidant, antifungal, anti-inflammatory, antiplatelet, and anticancer action exerts a powerful antioxidant effect on generation of reactive oxygen species and lipid peroxidation in blood platelets induced by platinum compounds. This beneficial effect was observed by the production of thiobarbituric acid reactive substances (TBARS), the level of conjugated diene, and the generation of superoxide anion radicals and other reactive oxygen species. Moreover, indicators of renal injury, such as increased serum creatinine levels, urinary volume, and urinary protein caused by the administration of cisplatin, were also significantly reduced with resveratrol (Bischin et al. 2010).

Carnosine, a biological dipeptide predominating in long-lived tissues such as skeletal muscles and brain, protects against cisplatin-induced nephrotoxicity in mice. The effects were evaluated by plasma creatinine, urea, malondialdehyde, nitrate, superoxide dismutase, and catalase activities (Bischin et al. 2010).

Apocynin, which is used as a specific NADPH oxidase inhibitor, was shown to ameliorate the renal histological damage and the increase in blood urea nitrogen, serum creatinine, and urinary excretion of total protein, N-acetyl-β-D-glucosamidase and glutathione-S/transferase induced by cisplatin (Bischin et al. 2010).

Lycopene, a naturally occurring carotenoid in tomatoes has attracted considerable attention as a potential chemopreventive agent in rats. It is an efficient antioxidant and has a singlet oxygen and free radical scavenging capacity and has a protective effect against cisplatin-induced nephrotoxicity and oxidative stress (Bischin et al. 2010).

Transferrin (TF)

The serum iron transporter transferrin is found in blood at relatively large concentrations (~35 μM) and is also able to bind cisplatin. Nuclear magnetic resonance (NMR) measurements have identified methionines 256 and 499 as the main residues that serve as ligands to the platinum, with Cys 34, aspartates, and tyrosines also involved. Nanoelectrospray ionization quadrupole time-of-flight mass spectrometry (nano ESI-QTOF-MS) and size-exclusion high-performance liquid chromatography/inductively coupled plasma mass spectrometry (HPLC/ICPMS) were also employed for characterization of Pt-transferrin interactions. TF was proposed as a carrier for cisplatin, as a means to allow controlled release of the drug within the organism (Bischin et al. 2011).

Ubiquitin

Ubiquitin's role in key cellular processes such as cell cycle control, DNA repair, and protein degradation make it a sensitive target for drugs. The N-terminal methionine and His68 are the two main targets for cisplatin and related compounds in ubiquitin according to ESI-MS and NMR data. This binding then interferes with proteolysis and ubiquitination processes (Bischin et al. 2011).

Metalloproteinase

Since matrix metalloproteinases are important in events such as tissue remodeling and cancer cell invasion and metastasis, their interaction with anticancer drugs is also of interest. A series of platinum complexes were shown to inhibit matrix metalloproteinase (MMP-3) by binding to the enzyme as shown by ESI-MS, CD, and NMR-spectroscopy (Che and Siu 2010). A decrease of the MMP-2 level in cisplatin treatment was noted, and proposed to be involved in the therapeutic effect (Bischin et al. 2011).

Hemoglobin (Hb) and Myoglobin

Binding of of cisplatin, oxaliplatin, and carboplatin to hemoglobin was demonstrated with nanoelectrospray ionization quadrupole time-of-flight mass spectrometry (nanoESI-QTOF-MS) and size-exclusion high-performance liquid chromatography/inductively coupled plasma mass spectrometry (HPLC/ICPMS). Cisplatin–Hb complexes were formed at clinically relevant concentrations of cisplatin and Hb, with differences found between cisplatin, oxaliplatin, and carboplatin in terms of adduct formation not only with pure hemoglobin but also with blood. Hb cysteine and proline residues are involved in these interactions; it was also noted that platinum binding can induce heme release from Hb, which not only affects oxygen-transporting capacity but is also a source of oxidative stress. Hb's high concentrations in blood and the key role that the Fe(II) role plays in Hb, make this protein particularly sensitive to changes in redox status. In fact, under stress conditions such as physical effort or certain pathological conditions Hb engages in reactions with oxidative stress agents – primarily peroxide – yielding free radicals and highly oxidizing states at the iron (ferryl, Compound II) (▶ Heme Proteins, the Globins). As a result, Hb is a sensitive target for the stress imposed by cisplatin in patients. In relation to this, it was shown that incubation of whole blood with cisplatin led to irreversible sequestration of a significant part of the platinum into the erythrocytes within 2 h (Bischin et al. 2011).

According to mass spectrometry data, myoglobin (Mb) can also bind cisplatin – with up to two Pt per Mb and using as ligands His116, Ser117, Lys118, and His 119 (Bischin et al. 2011).

Cytochrome c

Cytochrome *c* serves as key component of the electron transport chain and is also implicated, most likely via redox reactions involving peroxides, in apoptosis; it may thus be expected to be a sensitive target for exogenous compounds such as cisplatin (Kagan et al. 2005). Based on data from electrospray ionization mass spectrometry (ESI-MS), tandem mass spectrometry (MS/MS), Fourier transform mass spectrometry(FT-MS), and NMR, cisplatin and carboplatin were found to bind to cytochrome *c*, with the iron ligand Met65 and with Met80 as important ligands for platinum. Also found to bind platinum were Cys14, Cys17, His18, His26, and His33. These binding events were not found to limit the cytotoxic effects of the metal (Bischin et al. 2011).

Copper-Binding Proteins

The copper transporter 1 (CTR1), normally responsible for the uptake of Cu from extracellular space, was shown to play an important role in the accumulation of platinum-based drugs. Cisplatin interacts with CTR1 both at the ^{40}MXXM^{45} tetratpeptide and at a site that produces the conformational changes that trigger degradation. It was proposed that CTR1 may mediate the transport of cisplatin through a pore it forms in the plasma membrane, as well as via endocytosis (Larson et al. 2010). Two other Cu transporters involved in cisplatin efflux are ATP7A and ATP7B (Bischin et al. 2011; Boal and Rosenzweig 2009, ▶ Platinum Complexes and Methionine Motif in Copper Transport Proteins, Interaction, ▶ Platinum Interaction with Copper Proteins).

Membrane Proteins

As an integral membrane multidrug resistance-associated protein belonging to the ATP-binding cassette superfamily, MRP1 confers resistance to several drugs, including cisplatin. This is due mislocalization of MRP1 as well as to modifications in the level of glycosylation of the protein, even when the level of gene expression remains the same.

With other membrane proteins such as folate-binding protein (FBP), an increase of the protein concentration in the cytoplasm was reported, as well as effects on endocytosis and membrane vesicle recycling (Bischin et al. 2011).

Cytoskeletal Platinum-Binding Proteins

Platinum-containing drugs induce defects in the cytoskeletal system. Actin and filamin are known to bind carboplatin and to be downregulated in cisplatin-resistant cells; also downregulated were β-tubulin, keratin, and dynamin 2. With actin, filamin, and dynamin important in endocytosis, it was proposed that platinum-induced defects of the cytoskeletal system cause a reduced uptake of cisplatin and carboplatin (Bischin et al. 2011).

DNA-Processing Enzymes

Topoisomerase II is inhibited by Pt(II)-(2,2':6'2'-terpiridine) via mechanisms involving interaction with DNA as well as direct binding of platinum to the enzyme, including interaction with a thiol group (Che and Siu 2010; Bischin et al. 2011).

Cisplatin and related compounds are thus known to bind to several classes of proteins, affecting aspects as diverse as structural, antioxidant, electron transfer, small molecule or ion transport, or DNA processing. In several cases, detailed structural analyses of these binding phenomena are available. These interactions should be considered among the mechanisms whereby platinum-containing drugs induce toxic side effects in humans – if not, as is the case with other metal-containing drugs, as also relevant to the therapeutic effect.

Cross-References

▶ Heme Proteins, the Globins
▶ Platinum Complexes and Methionine Motif in Copper Transport Proteins, Interaction
▶ Oxaliplatin
▶ Platinum Anticancer Drugs
▶ Platinum Interaction with Copper Proteins
▶ Platinum-Resistant Cancer
▶ Platinum, Physical and Chemical Properties

References

Bischin C, Taciuc V, Silaghi-Dumitrescu R (2010) Effects of antioxidants in cisplatin toxicology. In: Radu Silaghi-Dumitrescu GG (ed) Metal elements in environment, medicine and biology. Eurobit Publishing House, Timisoara, pp 265–270

Bischin C, Lupan A, Taciuc V et al (2011) Interactions between proteins and platinum-containing anti-cancer drugs. Mini Rev Med Chem 11:214–224

Boal AK, Rosenzweig AC (2009) Crystal structures of cisplatin bound to a human copper chaperone. J Am Chem Soc 131:14196–14197

Che C-M, Siu FM (2010) Metal complexes in medicine with a focus on enzyme inhibition. Curr Opin Chem Biol 14:255–261

Cossa G, Gatti L, Zunino F et al (2009) Strategies to improve the efficacy of platinum compounds. Curr Med Chem 16:2355–2365

Holm RH, Kennepohl P, Solomon EI (1996) Structural and functional aspects of metal sites in biology. Chem Rev 96:2239–2314

Ivanov AI, Christodoulou J, Parkinson JA et al (1998) Cisplatin binding sites on human albumin. J Biol Chem 273:14721–14730

Kagan VE, Tyurin VA, Jiang J et al (2005) Cytochrome c acts as a cardiolipin oxygenase required for release of proapoptotic factors. Nat Chem Biol 1:223–232

Larson AC, Adams PL, Jandial DD et al (2010) The role of the N-terminus of mammalian copper transporter 1 in the cellular accumulation of cisplatin. Biochem Pharmacol 80:448–454

Witte AB, Anestål K, Jerremalm E et al (2005) Inhibition of thioredoxin reductase but not glutathion reductase by the major classes of alkylating and platinum-containing anticancer compounds. Free Radic Biol Med 39:696–703

Platinum-Containing Bioactive Kinase Inhibitors

▶ Platinum(II) Complexes, Inhibition of Kinases

Platinum-Resistant Cancer

Niels Eckstein and Bodo Haas
Federal Institute for Drugs and Medical Devices (BfArM), Bonn, Germany

Synonyms

Chemoresistance to platinum-based drugs

Definition

Cisplatin, carboplatin, and oxaliplatin are alkylating anticancer drugs used in chemotherapy regimens for the treatment of various tumor entities (e.g., testicular, ovarian, and lung cancer). Exposure of platinum-based drugs to the tumor over extended periods of time can lead to resistance formation. Various molecular and cellular mechanisms of action contribute to a resistant phenotype. Chemotherapy resistance finally forms a major hurdle in the clinical management of the disease.

Introduction

Cancer is one of the most deadly diseases worldwide with more than 1.5 Mio new cases annually in the USA alone. US cancer statistics are annually published by Siegel and Jemal in the CA Cancer Journal for Clinicians (Siegel et al. 2012). Despite tremendous efforts dedicated to conquering malignant diseases, the age-adjusted mortality rate has remained almost unchanged for most cancers over the past five decades.

Chemotherapy is a central strategy for the systemic treatment of cancer. Unfortunately, cancer cells have the propensity to become resistant to therapy, which is the major limitation of current therapeutic concepts. Cancer patients are treated by repeated cycles of chemotherapy and the clinical course of some cancers is often characterized by series of relapses. Relapses usually occur as therapy-refractory tumors, which represent the main problem in the clinical management of cancer. Clinical data show that the manifestation of chemotherapy resistance is a time-consuming complex biological process.

Cisplatin was introduced into the market as the first *platinum-based anticancer drug* in the 1970s. Since that time, it is an important cytostatic chemotherapy drug. Some tumor entities even lost their fright after introduction of cisplatin into the therapeutic armamentarium. For instance, cure rates of up to 95% are reported for testicular cancer depending on the tumor-stage at the time of diagnosis.

As depicted above (Fig. 1), currently, three platinum complexes are approved for the use in humans in the EU and the USA: cisplatin, carboplatin, and oxaliplatin. The pioneering discovery of the pharmacologic effects of cisplatin – initially known as Peyronnes salt – was a random discovery by the American physician A. Rosenberg. A comprehensive overview of known mechanisms of resistance including the "cancer stem cell theory" is given in B. A. Teichers compendium "Cancer Drug Resistance" (Teicher 2006).

Platinum-Resistant Cancer, Fig. 1 Structure formulas of approved platinum-containing anticancer drugs

Cisplatin   Carboplatin   Oxaliplatin

Mechanism of Action of Platinum-Containing Anticancer Drugs

Cisplatin is inactive when orally administered and, thus, the prodrug cisplatin must be toxicated endogenously. The active principle formed inside the cell is the electrophile aquo-complex (Fig. 2). High extracellular chloride concentrations (~100 mM) prevent extracellular formation of the active complex. Upon entering the cell, in a low chloride environment (~2-30 mM), the aquo-complex is formed. The active principle is preferentially built as a shift in the reaction balance in accordance to the principle of Le Chatelíer. The mechanism of action of the aquated complex is covalent cross-linking of DNA nitrogen nucleophiles. This molecular mechanism is very similar to the one of bifunctional N-LOST derivatives, another group of alkylating anticancer drugs. The cisplatin bisaquo-complex prefers an electrophilic reaction with N-7 nitrogen atoms of adenine and guanine. 1,2 or 1,3 intra-strand cross-links are preferentially built (to an extent of about 90%). Affected are genomic and mitochondrial DNA molecules (Wang and Lippard 2005).

Carboplatin mechanistically acts similar to cisplatin. However, a slower pharmacokinetic profile and a different spectrum of side effects has been reported (Stewart 2007). The mechanism of action of oxaliplatin substantially differs from cis- and carboplatin, which might be explained by the lipophilic cyclohexane residue (so-called DACH-residue, di-amino-cyclo-hexane) (Eckstein 2011).

The binding to cellular macromolecules takes place with decreasing prevalence of RNA > DNA > protein. The first-line indications of platinum compounds are testicular cancer, ovarian cancer, small cell and non-small cell lung cancer, esophageal cancer, osteosarcoma, cervical and endometrial cancer, and cancers of the head and neck and the bladder. Platinum compounds have a broad range of side effects. Problematic

Platinum-Resistant Cancer, Fig. 2 Activation of the prodrug cisplatin. Due to its high extracellular and low intracellular concentration, chloride as a chemical separation group is prevalently substituted by the H_2O-nucleophile inside the cell

are nephro- and ototoxicity, but the therapy-limiting side effect is its extraordinary high potential to cause nausea and emesis. Thus, cisplatin usually is administered together with ondansetron or another potent anti-emetogens with a 5-HT_3 antagonizing mechanism of action. The dose-limiting factor of cisplatin, however, is nephrotoxicity. Augmenting nephrotoxicity becomes evident in an increase in serum urea, creatinine, and serum uric acid. Nephrotoxic disorders are clinically manifest with atypical kinetics: they appear after 2 days or 2 weeks. In particular, high-dose or dose-dense therapies carry the risk of acute renal failure. Of particular importance is also neurotoxicity. Peripheral neurotoxicity is associated with hearing loss, visual disturbances, and seizures, the latter can be enhanced by nephrotoxic electrolyte disturbances (hypomagnesemia). Peripheral neuropathy can manifest itself in a declining perception of the hands and feet, known as hand-foot syndrome. Carboplatin has a diminished nephro- and ototoxicity, but can cause bone marrow depression, while oxaliplatins most characteristic side effect is a dose-dependent neurotoxicity.

Drug Resistance

The events in resistant cells are complex and not fully elucidated yet. Multicausal mechanisms of resistance

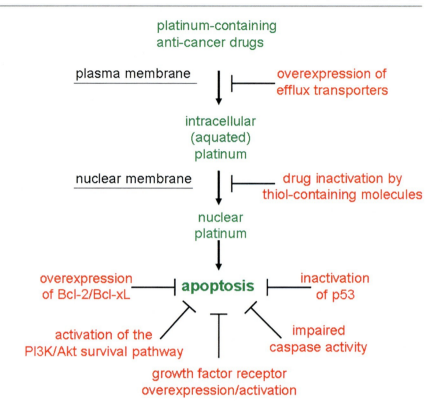

Platinum-Resistant Cancer, Fig. 3 Mechanisms of resistance to platinum-based anticancer drugs; *green*: mechanism of action; *red*: potential mechanisms of resistance

complicate the treatment of refractory tumors. Various molecular mechanisms are involved (Fig. 3).

Among others, *general mechanisms of drug resistance* are:
- P-glycoprotein mediated diminished drug accumulation
- Elevated drug inactivation
- Diminished DNA repair and elevated DNA damage tolerance
- Enhanced expression of anti-apoptotic genes
- Inactivation of the p53 pathway

However, this knowledge did not lead to resounding clinical strategies to overcome resistance and many of these mechanisms are not fully understood.

Specific principles of platinum resistance are:
- Reduced uptake or increased efflux of platinum-based compounds by transporters
- Cellular compartmentation
- Detoxification of bioactive platinum aquo-complexes by sulfur-containing molecules
- Increased activity of DNA repair
- All mechanisms are summarized and reviewed in Stewart (2007), Wang and Lippard (2005)

Cis- and carboplatin generate mutual cross-resistant cells. In contrast, oxaliplatin resistant cells are often not cross-resistant to cisplatin, pointing to a different mechanism of action. Cisplatin resistance occurs intrinsically (e.g., colon carcinomas) or acquired (e.g., ovarian carcinomas). Some tumor specimens show no tendency to acquire resistance at all (e.g., testicular cancer). Regarding transport processes, reduced accumulation of platinum compounds in the cytosol can be caused by reduced uptake, increased efflux, or cellular compartmentation. Several ATP-binding cassette transport proteins are involved like MRP2 and MRP6, Ctr1 and Ctr2, or ATP7A and ATP7B.

Unexpectedly, the degree of reduced intracellular cisplatin accumulation is often not proportional to the level of resistance. Thus, several mechanisms of cisplatin resistance emerge simultaneously. For instance, glutathione, thioredoxin, and metallothioneins are cysteine-rich peptides, capable of detoxicating the highly reactive aquated cisplatin-complex. Cisplatin resistance in ovarian cancer is directly proportional to increased intracellular glutathione. However, increased glutathione levels are reversible but resistance is not. Other mechanisms are acquired imbalance of apoptotic pathways, and overexpression of receptor tyrosine kinases: ERB B1-4, IGF-1R, VEGFR 1-3, and PDGF receptor family members. ERB B2 (HER-2), for instance, activates PI3K/AKT signaling resulting in prevention of apoptosis. Experiments with recombinant expression of ERB B2 confirmed this mechanism (Eckstein 2011).

Apoptosis Triggered by Platinum-Containing Anticancer Drugs

Benign cells in a multicellular organism are balanced at equilibrium of proliferation and cell death. The Greek-derived word apoptosis (απόπτωσις) literally means "autumnally falling leaves," describing a subject to be doomed. It is often referred to as programmed cell death. Recently, other mechanisms of programmed cell death have been identified, like autophagy, paraptosis, and mitotic catastrophe. To this end, apoptosis more accurately is defined as cell death induced by caspases, a specialized family of aspartate-specific cysteine proteases that carry out this form of cellular suicide. Two apoptosis pathways are divided, the intrinsic and the extrinsic apoptotic signaling pathway. The intrinsic apoptotic signaling cascade accounts for the most common cause of cell death induced by cytotoxic drugs like cisplatin. Accordingly, it is triggered among others by DNA damage. This pathway is balanced by pro- and anti-apoptotic members of the Bcl-2 protein family. Bcl-2 is a 26 kDa anti-apoptotic protein, which was first discovered in B-cell lymphoma cells after a 14/21 translocation. This put the *bcl-2* gene under the control of the immunoglobulin-promoter and led to overexpression, thereby triggering the cells resistant to apoptosis. The tumor-suppressor protein p53 is a pivotal point for the activation of the intrinsic apoptotic pathway: p53 responds to diverse cellular stresses by arresting cell cycle progression through expression of p53 target genes such as the mitotic inhibitors p27 and p21. After unrepairable DNA damage (e.g., by platinum-containing cytostatic drugs), p53 triggers cell death via the expression of apoptotic genes (*puma*, *noxa*, etc.) and by inhibiting the expression of anti-apoptotic genes (reviewed in Ashkenazi and Herbst 2008; Eckstein 2011).

Cross-References

▶ Hexavalent Chromium and Cancer
▶ Lanthanides and Cancer
▶ Polonium and Cancer
▶ Selenium-Binding Protein 1 and Cancer

References

Ashkenazi A, Herbst RS (2008) To kill a tumor cell: the potential of proapoptotic receptor agonists. J Clin Invest 118:1979–1990

Eckstein N (2011) Platinum resistance in breast and ovarian cancer cell lines. J Exp Clin Cancer Res 30:91

Siegel R, Naishadham D, Jemal A (2012) Cancer statistics, 2012. CA Cancer J Clin 62:10–29

Stewart DJ (2007) Mechanisms of resistance to cisplatin and carboplatin. Crit Rev Oncol Hematol 63:12–31

Teicher BA (2006) Cancer drug resistance. Humana Press, Totowa

Wang D, Lippard SJ (2005) Cellular processing of platinum anticancer drugs. Nat Rev Drug Discov 4:307–320

Plutonium, Physical and Chemical Properties

Fathi Habashi
Department of Mining, Metallurgical, and Materials Engineering, Laval University, Quebec City, Canada

Plutonium is a silvery lustrous metal that exhibits five phase transitions between ambient temperature and its melting point. Because of the different densities of each phase, large changes in expansion are observed during phase transitions which complicates its use for military purposes and as a fuel in nuclear power stations. It is produced in nuclear reactors when uranium 238 absorbs neutrons forming uranium 239 which is a beta emitter with a short half-life; its daughter neptunium 239 also emits an electron to form plutonium 239.

$$^{238}_{92}U + ^{1}_{0}n \rightarrow ^{239}_{92}U \rightarrow ^{239}_{93}Np + e^-$$
$$\downarrow$$
$$^{239}_{94}Pu + e^-$$

Plutonium 239 (half-life 2.4×10^4 years) is an alpha emitter; it undergoes fission with the emission of several neutrons and can maintain a chain reaction. Thus, a nuclear reactor using uranium as a fuel not only produces energy but also produces another nuclear fuel. Under certain conditions, it is possible to generate fissionable material at a rate equal to or greater than the rate of consumption of the uranium. Such a reactor is known as a *breeder reactor*. Fifteen isotopes of plutonium have been synthesized; long-lived isotopes. About 400 t of plutonium have been produced in nuclear power stations excluding the former Eastern-bloc countries. The ^{239}Pu stockpiled in weapons is estimated to exceed 100 t. About 10 t of plutonium are distributed on earth as a result of military activities (bomb tests, satellite accidents, bombers, rockets, etc.) and releases from nuclear facilities.

Physical Properties

Atomic number	94
Atomic weight	244.06
Relative abundance in Earth's crust,%	Trace
Density, g/cm^3	15.92–19.86
Melting point, °C	640 ± 2
Thermal conductivity α-Pu at ambient temperature, W cm^{-1} K^{-1}	0.084–0.041

Chemical Properties

Plutonium metal readily reacts with oxygen and moisture at ambient temperature, especially in bulk quantities due to self-heating. The surface turns from grayish to dark blue and finally green oxide layers. In aqueous sodium chloride solution, it is converted to Pu(O)H. Plutonium dissolves rapidly in hydrochloric or hydrobromic acid but slowly attacked by hydrofluoric acid. It undergoes passivation in nitric acid but dissolves in the presence of fluoride ions. Sulfuric acid converts plutonium slowly to its sulfate.

Five oxidation states exist: Pu^{3+}, Pu^{4+}, PuO_2^+, PuO_2^{2+}, and PuO_5^{3-} of which up to four can be present simultaneously in significant quantities. The ions possess characteristic absorption bands that are used in spectrophotometric analysis. The PuO_2^+ cation disproportionates readily at low acidity. The Pu^{4+} cation forms complexes with anions of inorganic acids. The nitrate complexes $Pu(NO_3)_x^{(4-x)+}$ (with $x = 1$ to 4) are employed for the separation of plutonium by solvent extraction and anion exchange. With organic compounds, various complexes have been observed. Ethylenediaminetetraacetic acid forms complexes with plutonium cations of all valences and is often used to decontaminate surfaces or to wash out ingested plutonium.

Plutonium from spent nuclear fuel is recovered by extraction with tri-*n*-butyl phosphate. Nonaqueous recovery by electrorefining from molten alkali metal and alkaline earth metal chlorides is an alternative for metallic fuels. Plutonium metal is prepared from its oxide by reduction with calcium in molten $CaCl_2$ or $CaCl_2$–CaF_2. Plutonium forms hard, black, metallic hydrides of the type PuH_{2+x} and PuH_3. They are inert to cold water but react slowly at 90°C. Four carbides of Pu exist: Pu_3C_2, PuC_{1-x}, Pu_2C_3, and PuC_2. They decompose at high temperature, are oxidized when heated, and dissolve in oxidizing acids with the formation of CO_2 and carboxylic acids. The plutonium oxygen phase diagram exhibits the existence of several oxides.

Plutonium hydroxides and hydrated oxides are obtained by precipitation from aqueous solution. With excess alkali-metal hydroxides or NH_4OH, Pu^{3+} precipitates as blue or pale purple $Pu(OH)_3 \cdot xH_2O$, Pu^{4+} as olive-green $Pu(OH)_4 \cdot xH_2O$, and PuO_2^{2+} as $PuO_2(OH)_2 \cdot H_2O$. Addition of H_2O_2 to solutions of Pu^{4+} in dilute acid leads to the formation of plutonium peroxide.

Several plutonium sulfides exist, including PuS, Pu_3S_4, Pu_2S_3, PuS_2, and various nonstoichiometric sulfides. All are black powders except PuS, which is brown. The following oxysulfides are known: PuOS, $Pu_4O_4S_3$, Pu_2O_2S, and $Pu_2O_2S_3$. Three selenides have been identified: PuSe, Pu_2Se_3, and $PuSe_{2-x}$. In the Pu–Te system, the following compounds have been observed: PuTe, $PuTe_2$, and two modifications of Pu_2Te_3. Plutonium trifluoride and tetrafluoride hydrates are obtained by precipitation from aqueous solution. They undergo dehydration when heated in HF atmosphere to give PuF_3 and PuF_4. Plutonium hexafluoride, PuF_6, is volatile (mp 52°C, bp 62.15°C).

Because of its high radiotoxicity, plutonium is handled exclusively in airtight glove boxes which are under reduced pressure to prevent leakage of aerosols or volatile compounds in case of a containment rupture.

References

Habashi F (2003) Metals from ores. An introduction to extractive metallurgy. Métallurgie Extractive Québec, Québec City. Distributed by Laval University Bookstore, www.zone.ul.ca

Kock L (1997) Plutonium. In: Habashi F (ed) Handbook of extractive metallurgy. Wiley, Weinheim, pp 1685–1694

Seaborg GT (1966) Progress beyond plutonium. Chem Eng News 44:76–88

$PM_{2.5}$

▶ Vanadium Pentoxide Effects on Lungs

pMMO

▶ Particulate Methane Monooxygenase

Po-210. ^{210}Po

▶ Polonium in Terrestrial Environment

Polonium

▶ Polonium in Terrestrial Environment

Polonium and Cancer

Vincenzo Zagà[1] and Charilaos Lygidakis[2]
[1]Department of Territorial Pneumotisiology, Italian Society of Tobaccology (SITAB), AUSL of Bologna, Bologna, Italy
[2]Regional Health Service of Emilia Romagna, AUSL of Bologna, Bologna, Italy

Synonyms

Alpha rays; Alpha-radioactivity; Chronic obstructive pulmonary disease; Fertilizers; Lead; Lung cancer; Pitchblende; Polyphosphates; Radioactivity; Radium; Radon; Smoking; Tobacco; Tobacco filters; Tobacco leaves; Tobacco plants; Tobacco smoke

Definition of the Subject

The alpha-radioactive polonium-210 (Po-210) is one of the most powerful carcinogenic agents of tobacco smoke and responsible for the histotype shift of lung cancer from squamous cell type to adenocarcinoma. Po-210 may be found in the tobacco plants due to meteorological events and environmental dust, as well as due to the fertilizers, which are rich in polyphosphates containing radium (Ra-226) and its decay products: lead-210 (Pb-210) and Po-210. As a result, Pb-210 and Po-210 reach the bronchopulmonary apparatus by smoking a cigarette; their carcinogenic activity will manifest especially in patients with compromised mucociliary clearance. The radiological risk from Po-210 in a smoker of 20 cigarettes per day for a year is equivalent to the one deriving from 300 chest X-rays, with an autonomous oncogenic capability of four lung cancers per 10,000 smokers.

The History of Polonium

Polonium-210 (Po-210) was the first radioactive element discovered by Marie Sklodowska-Curie and Pierre Curie in 1898. As a matter of fact, Henri Becquerel presented the first paper of Pierre and Marie Curie on a "substance nouvelle radio-active," which was found in pitchblende, in the *Académie des Sciences* on 18 July 1898. The announcement of the discovery of a new element introduced a new word: *radio-actif*. The *trait d'union*, however, would resist against the first publications on radioactivity.

The symbol Po, written by Pierre Curie, was first seen in some laboratory notes in 13 July 1898, during an experiment of sulfur sublimation. On the second last page of a notebook, the term "polonium" appeared just 2 days before the presentation in the *Académie des Sciences*. For their work on radioactivity, the Curies were awarded the 1903 Nobel Prize in Physics, while Marie Curie also received the Nobel Prize in Chemistry in 1911; tragically, Pierre had been killed in an accident on 20 April 1906.

Polonium, also called "radium F," was discovered by the Curies while they were investigating the cause of pitchblende's persistent radioactivity, even after the removal of uranium and radium. Their work was remarkable, considering the means available in the late nineteenth century and the fact that the element can be found in uranium ores at about 0.1 mg per ton.

Chemistry

Polonium-210 (Po-210) is a radioactive element found in the Mendeleev's periodic table with atomic number 84 and atomic weight 210; it is the last radioactive nuclide from the uranium (U-238) family. It results from the decay of natural radon (Rn-222), a gas universally distributed in the atmosphere. By α-emission, Rn-222 produces the transient Po-218 that, after a sequence of alpha and beta emissions, finally decays to Po-210. Polonium has more than 30 radioisotopes, but Po-210 is the most dangerous of all. This alpha-emitting isotope has a half-life of 138.38 days, an effective biological half-time of 46 days, while its precursor, lead-210 (Pb-210), has a half-life of 22 years (International Agency for Research on Cancer (IARC) 2001). It is a high-energy α-particle emitter (5.3 MeV), but it can also emit gamma photons (803 keV) with an emission probability of nearly 1×10^{-5}; it decays to stable Pb-206.

Polonium is present in nature in pitchblende containing rocks and constitutes $2.1 \times 10^{-4}\%$ of the Earth's crust. The major resources of pitchblende are located in Canada, the USA, Congo, and South Africa. Nonetheless, it is so rare to find it that it is more convenient to create it in the lab, by bombarding Bi-209 with neutrons. It has a melting point of 254°C and a boiling 1 of 962°C. Moreover, polonium presents a simple cubic structure at room temperature, which seems to be unstable regarding the interatomic interaction. The crystal structure of polonium was first determined by Beamer and Maxwell in 1946. Two structures have been identified: the alpha (simple cubic structure) and the beta (rhombohedral structure).

Main Use

Due to its high alpha radiation emissions and the difficult extraction process, Po-210 use is rather limited. The main applications are:
1. As a resource of neutrons when it is mixed with beryllium. There is a high risk of exposure to polonium during production of these sources (International Agency for Research on Cancer (IARC) 2001).
2. As an energy resource for satellites and other space devices.
3. In antistatic devices of some precision instruments and in brushes that eliminate dust gathered on photographic film. For the latter use, polonium is sealed and shielded to minimize radiation risk.
4. In devices that eliminate static charges in textile mills, though less dangerous and easier to find beta-ray sources are now used.

Toxicity

Polonium is a highly toxic, pure alpha emitter, with elevated specific activity (1,017 Bq kg^{-1}), and is dangerous to handle even in milligram amounts. Alpha rays, which are formed by helium-4 (He-4) nucleus, are highly ionizing and, therefore, are particularly harmful for living tissues. Alpha rays are the least penetrating type of radiation, and they manage to travel only a few centimeters in air. They can be easily stopped by obstacles, such as a sheet of paper, and they can penetrate living tissues by only a few microns. In fact, since they lose all of their energy after a short distance, they can be dangerous for tissues only when substances emitting alpha particles enter the organism by respiration or ingestion.

In addition, alpha rays are highly ionizing and, therefore, are particularly harmful for living tissues. One milligram of polonium can emit as many alpha particles as 5 g of radium. The impact on humans can be devastating, as it can cause considerable damage by causing cell death; promoting a massive, progressive, and rapid necrosis; and not allowing the organism enough time to replace the quantity of dead cells.

The maximum allowable body burden for ingested polonium is 1,100 Bq, which is equivalent to a particle weighing only 6.6×10^{-6} μg. The considerable energy of its high linear energy transfer alpha radiation (5.305 MeV) turns Po-210 highly radiotoxic by internal contamination, a fact that became notorious with the murder of the former Soviet spy, Alexander Litvinenko in 2006.

Contamination Ways

Po-210 can contaminate the human body through natural ways, such as water, air, plants, and various foods.

Spencer et al. report that 77.3% of the daily Po-210 intake of an adult male comes from food, 4.7% from water, and 0.6% from air (Spencer et al. 1977). Nonetheless, inhaling cigarette smoke can supply much more Po-210 (17.4%) than water and air combined (Spencer et al. 1977). Fifty to ninety percent of the ingested Po-210 will promptly leave the body in feces, but the remaining fraction enters the blood circulation.

From Earth to Tobacco

Multinational tobacco manufacturers have been aware of the alpha-radioactivity presence in tobacco smoke since the 1960s (Zagà et al. 2011). As a matter of fact, in 1959, a Canadian health official first suggested to Philip Morris executives to reduce the level of tobacco radioactivity. Turner and Radley were among the first that talked about the radioactivity of tobacco in 1960, after having observed the presence of naturally occurring alpha-ray activity of foods.

Po-210 and its precursor, lead 210 (Pb-210), which are present in tobacco plants, may be absorbed in three main associated ways (Zagà et al. 2011):

1. Directly from terrain that contains uranium, through the plant's roots.
2. Coating on leaves as a result of meteorological events, rain, snow, and environmental dust. In fact, Radon-222, a product of U-238 decay, is a noble and volatile gas that can partially escape from terrain into the atmosphere and create Pb-210 and Po-210. These are absorbed by atmospheric dust, creating the Aitken particles that consequently lie on the leaves. The numerous trichomes of tobacco plants resemble filamentous pores and are metal accumulators, particularly of Pb-210 and Po-210. The quantity of the latter will then increase, as there is further Pb-210 decay and this may be the principal way Po-210 enters tobacco plants. Experiments have confirmed that radon and lead decay products are highly concentrated in the trichomes of leaves (Zagà et al. 2011). Additionally, accumulation mechanisms of Pb-210 on trichomes of tobacco have been widely discussed and studied by others. Conversely, Turner and Radley suggested that the additional alpha activity of Pb-210 and Po-210 in the lungs from the radium content in cigarettes is less than 1% of the atmospheric radon inhaled by both smokers and nonsmokers (Turner and Radley 1960), while Singh and Nilekani have identified the importance of the fertilizers employed (Singh and Nilekani 1976).
3. Calcium polyphosphate fertilizers are enriched with radium, which is chemically similar to calcium, and derive from soil that contains pitchblende and apatite. Interestingly, according to several studies, Indian cigarettes, which are made of scarcely fertilized tobacco, are 6–15 times less radioactive compared to the American ones, which derive from intensively fertilized plants (Singh and Nilekani 1976).

From Tobacco to Lungs

By lighting a cigarette, tobacco burns, reaching 800–900°C when inhaling, and smoke is created, which is composed of a corpuscular (5%) and a gas phase (95%). Po-210 and Pb-210 are adsorbed in the insoluble particles of the corpuscular phase. The latter is present in a high quantity and is a weak alpha ($<1 \times 10^{-5}$), gamma, beta, and X emitter, as well as a precursor of the former. All these inhaled particles are deposited in the bronchopulmonary apparatus and particularly in segmental bronchi bifurcations, due to ciliary action. According to measurements by Cohen et al., radium and thorium are also present in cigarettes; however, 99% of the radioactivity comes from Po-210 (Cohen et al. 1980), which remains in the bronchopulmonary apparatus after inhalation.

All these particles have a different fate based on the efficacy of the mucociliary clearance. This mechanical purification is reduced gradually in smokers with chronic obstructive pulmonary disease (COPD), resulting in the accumulation of insoluble Pb-210 particles, which decay to Po-210 over time.

It has become a common clinical observation that smokers and ex-smokers with moderate to severe COPD have a higher incidence of lung cancer. The more severe COPD becomes, the greater the risk of radioactive load accumulation is. Polonium radiation in the bronchial epithelium depends on the particle concentration of these areas and on the time of their permanence. Since the half-life of polonium is 138.38 days and of lead 22 years, which decays afterward into polonium, there is a significant cancer risk due to chronic exposure to low levels of insoluble alpha-emitting particles; these are responsible for high radiation doses in small tissue areas particularly in the bifurcations (hot spots). This process is facilitated by the abovementioned impaired mucociliary clearance of smokers. In fact, metaplastic lesions may be present in the ciliated epithelium of all heavy smokers. Po-210 of the insoluble particles becomes even more penetrative because of zones with damaged or scarcely ciliated epithelium, where mucous mainly stagnates.

Radioactive particles reach various organs and tissues through pulmonary and systemic circulation and cause mutations of the genetic cellular structure, deviations of the standard cellular characteristics, accelerated aging, and quicker death due to a wide range of diseases. As a matter of fact, Po-210 levels in smokers are significantly higher in blood, urine, liver, kidney, heart, and psoas muscle (Radford and Hunt 1964; Zagà et al. 2011). Little et al. estimated that Po-210 concentration in blood is 63.64 mBq/kg of blood in smokers and 28.12 mBq/kg of blood in nonsmokers (Little and McGandy 1966).

Po-210 Quantity in Tobacco Smoke

Po-210 alpha radioactivity in tobacco smoke depends on several variables: the geographic region of tobacco growth, the storage time and modality, the existence of a filter, its length and composition, and the way of smoking (Zagà et al. 2011). Common filters, found in the cigarettes of commerce, are able to reduce Po-210 activity on average by 4.6% (Skwarzec et al. 2001), but resin filters may reduce lung exposure to alpha radiation even more. Various authors reported that a percentage of the Po-210 contained in cigarettes may be found in mainstream smoke, ranging from 3.7% to 58% (Khater 2004). Approximately 50% of cigarette's Po-210 is transferred with the smoke, 35% remains in the stub, and 15% is found in the ash.

From the 1960s to the dawn of the new millennium, tobacco radioactivity and the associated biological risk have been noted to be high. Such an observation was also made by the Complex Unit of the Institute of Chemical, Radiochemical, and Metallurgic Sciences of University of Bologna (SMETEC); the Italian National Agency for New Technologies, Energy, and Sustainable Economic Development (ENEA); and the Italian Society of Tobaccology (SITAB) (Zagà et al. 2011). It has been confirmed that a cigarette with tobacco of western origin emits 75 mBq of alpha radioactivity from Po-210, distributed in mainstream (5 mBq, 6.7%), sidestream (1.2 mBq, 1.6%), and ash (68.8 mBq, 91.7%).

At the end of the 1990s, several studies suggested that there was a reduction in the alpha radioactivity of tobacco smoke compared to previously known data. This was also observed in a recent study on the content of Pb-210 and Po-210 of the ten cigarette brands with the highest consumption in Italy in 2010 (Zagà et al. 2011). In this research, there was no significant difference in the content of Pb-210 and Po-210: the mean concentration of Pb-210 was 14.6 ± 2.7 mBq/cigarette, while that of Po-210 was 15.8 ± 2.2 mBq/cigarette.

Po-210 and Second-Hand Smoking

Passive smokers are exposed to the same components as active smokers, including radioactive elements. Po-210 in second-hand smoke is 50–70% the quantity found in active smoke. Moreover, passive smokers are exposed to environmental pollution from radon as well as from Po-210 of cigarette smoke, both of which increase lung cancer risk. There is evidence that Po-210 and Rn-222, alongside other carcinogenic substances, play a significant role in lung cancer development due to second-hand smoke (Brennan et al. 2004).

Po-210 and Narghilé Smoke

Po-210 is also present in narghilé smoke (Zagà et al. 2011). An international multidisciplinary team (from Egypt, Arabia, and France), coordinated by Khater, has recently published a pioneering study on narghilé (shisha, hookah) tobacco radioactivity. Before this research, only very few data were available on this issue.

The research was based on the measurement of some natural radionuclide activity and the estimation of the internal radiation dose due to narghilé tobamel (moassel) smoking. Tobamel is a fashionable flavored tobacco-molasses mixture (with added glycerol) currently used in narghilé. However, there are other forms such as jurak, similar to tobamel, but unflavored, containing minced fruits and no glycerol. The results of the study revealed a wide range of radioactivity concentrations (in Bq/kg dry weight): U-238 = 55 Bq (19–93), Th-234 = 11 Bq

(3–23), Ra-226 = 3 Bq (1.2–8), Pb-210 = 14 Bq (3–29), Po-210 = 13 Bq (7–32), Th-232 = 7 Bq (4–10), and K-40 = 719 Bq (437–1,044). The researchers concluded that the average concentrations of natural radionuclides in moassel tobacco pastes were comparable to their concentration in Greek cigarettes and tobacco leaves, and lower than that of Brazilian tobacco leaves. Finally, it has also been observed that the radioactivity concentration in tobacco products basically depends on the tobacco content itself, not on other ingredients such as molasses, glycerol, or fruits.

Po-210 Carcinogenicity in Tobacco Smoke

Even though 85 to 90 out of a 100 lung cancers are caused by tobacco smoke, less than 20% of smokers get lung cancer (International Agency for Research on Cancer (IARC). 1986). Many factors may influence individual susceptibility to lung cancer in smokers. Polonium is among them, albeit it is still less considered or even ignored as a carcinogenic substance (Muggli et al. 2008). When associated to other mutagenic and carcinogenic nonradioactive substances, which are inhaled with tobacco smoke (such as aromatic hydrocarbons, cadmium, N-nitrosamine), it seems to constitute the principal etiological factor for lung cancer (Martell 1983). Martell studied workers from uranium mines and suggested that given their continuous exposure to low doses of Po-210, that could be the cause of lung and other types of cancer.

Long-term tissue exposure to alpha radiation can induce cancer either by itself or in association with other nonradioactive carcinogenic substances. Polonium-210 emits alpha particles, which have a penetration limit of about 40 µm or less in animal tissue but a very high damaging effect. Since the late 1990s, IARC has identified Po-210 as a carcinogenic element for laboratory animals, classifying it among the group 1 agents (International Agency for Research on Cancer (IARC) 2004). DNA chromosome damage by exposure to alpha radiation is 100 times greater than the one caused by other types of radiation (Marmorstein 1986). Little et al. estimated that the radiation dose of the bronchial epithelium of bifurcations in the inferior lobes of people smoking for 25 years would be 200 rem (Little and Radford 1967). This can be explained by the local accumulation of Pb-210 insoluble particles. According to Martell, the cumulative dose of alpha radiation in bronchial bifurcations of smokers that die of lung cancer is approximately 1600 rem (80 rad). This dose is sufficient to induce a malignant transformation caused by alpha-particle interaction with basal cells (Martell and Sweder 1981).

Black et al. reported that Po-210 radiation dose in heavy smokers was up to 82.5 mrad per day (Black and Bretthauer 1968). Radford et al. estimated that the radiation dose for a person smoking two packs of cigarettes a day may be up to 30.1 rad a year or 752.5 rad over a 25-year period (Radford and Hunt 1964). Such a radiation exposure dose rate was about 150 times higher than the approximately 5 rem per 25 years received from natural background radiation sources.

Mechanism of Action

Ionizing radiation, including Po-210, could silence the tumor suppressor gene p16(INK4a) by promoter methylation. Inactivation of this gene was found in lung cancers of both smokers and radiation-exposed nonsmoker workers. It seems that such an inactivation plays a major role in carcinogenesis.

Biological Harm

But what is the level of biological damage caused by tobacco smoke Po-210? Estimating the damage is a very difficult and complicated task. Using the 1990 ENEA data on the average time of Po-210 presence in lungs, which is 53 days, the data of the BEIR IV Committee on lung cancer risk after exposure to radon and its decay products (Pb-210, Po-210) (National Research Council (U.S.). Committee on the Biological Effects of Ionizing Radiations., United States. Environmental Protection Agency. et al. 1988), and the data of the International Commission on Radiological Protection (ICRP), which are based on the survivors of the bomb A of Hiroshima, it is possible to estimate the lung cancer risk, which is 4×10^{-4} year^{-1} (four cases per 10,000 smokers per year, which corresponds to nearly 5,000 cases for the 11.1 million Italian smokers). This estimate does not take into account the promoter role of Po-210 (cocarcinogen) in the bronchopulmonary cancer and the overall carcinogenic activity of all substances.

To render the biological harm deriving from Po-210 in smoke more comprehensible, it has been compared to the damage caused by radiation in conventional chest X-rays. Since the dose of a modern chest radiograph is 0.03 rem, a smoker of 30 cigarettes per day receives a radiation dose equivalent to approximately 250-300 chest X-rays per year (Marmorstein 1986; Evans 1993). However, the alpha radioactivity alone does not cause the steep rise of the carcinogenic risk; instead, it is the combined and multiplicative action of each carcinogenic and cocarcinogenic component responsible for such consequence.

A Histotype Shift

There is evidence that in the last 40 years a histotype change of lung cancers has been noticed, shifting from squamous cell carcinoma to adenocarcinoma, in which the bronchoalveolar (BAC) subtype is also included. The abovementioned shift was observed in the early 1970s and has been noted ever since (Devesa et al. 2005). Some of the factors that have induced such a shift are:

1. The utilization of different varieties of tobacco in the US cigarette blends. This change reduced benzopyrenes in smoke, but produced an increase of nearly 50 % in nicotine-derived nitrosaminoketone (NNK) in the last quarter of the twentieth century (Boyle 2004).
2. The introduction of low-tar, low-nicotine, filtered cigarettes since the mid-1950s. Some studies suggested that there was a reduction in the risk for lung cancer by 2.5 times in smokers of filtered cigarettes. It has also been suggested, however, that the common cellulose acetate filters have contributed to the aforementioned histotype change.
3. The massive introduction of polyphosphate fertilizers in tobacco cultivations, contributing alpha radiation (from Pb-210 and Po-210) and TSNAs significantly (Hoffmann and Hoffmann 1997), especially in Western cigarettes rather than in the ones from poor agricultural areas like India (Singh and Nilekani 1976). As a matter of fact, the prevalent histotype of lung cancer has not changed since the 1960s as several studies (Behera and Balamugesh 2004) showed; Viswanathan et al., for instance, reported 50.5% squamous cell carcinomas versus 28.4% adenocarcinomas in 1962.

Ways of Reducing the Radioactive Load of Tobacco Smoke

Regulating and reducing this harmful radiation, which comes from fertilizers, could help reduce lung cancer incidence. Tobacco radiation could be reduced by applying various solutions, which may also work combined with (Zagà et al. 2011):

1. Use of alternative polyphosphate sources, such as organic fertilizers from animals.
2. Use of ammonium phosphate as a fertilizer, instead of calcium phosphate.
3. Different storage methods; Po-210 radioactivity of tobacco may rise over time while in storage.
4. Genetic modifications of tobacco plants with significant reduction of trichomes concentration on the leaves, on which Pb-210 and Po-210 accumulate.
5. Resin filters may decrease lung exposure to alpha radiation. Conversely, common filters reduce Po-210 activity, on average, by 4.6% (Skwarzec et al. 2001).
6. Removal of radioactive lead and polonium from tobacco could also be made by washing the leaf with water containing hydrogen peroxide and/or a wetting agent.
7. Perhaps the simplest and most applicable solutions would be the quantitative decrease in polyphosphate use in tobacco cultivations and the regulation of the maximum acceptable level of alpha radiation of cigarettes, which should also be clearly indicated on the packet (Muggli et al. 2008).

The Deceit of Big Tobacco

While multinational tobacco manufacturers have been aware of the alpha-radioactivity presence in tobacco smoke since the 1960s, they have covered it up strategically.

Nowadays, the most comprehensive resource on this matter is the Legacy Tobacco Documents Library, a vast databank, which was established by the University of California. Thirteen million classified documents from the major US tobacco manufacturers have been released for a total of 80 million pages. Among these documents, one can find the lawsuit of the State of Minnesota against Philip Morris Incorporated et al., in which there are 481 confidential documents and memorandums on the alpha radioactivity from Po-210 in tobacco smoke.

The archives reveal that Philip Morris has been aware of the lead and polonium existence in cigarettes since the 1960s, as was also discussed in various articles (Muggli et al. 2008). In these internal documents, it can be seen that there was a clear interest in polonium's radioactivity and the induction of bronchogenic carcinomas in laboratory animals and presumably in humans. In fact, there was a recommendation to avoid any public attention to the problem for fear of *"waking a sleeping giant"* (Muggli et al. 2008).

Several declassified documents and notes have revealed that the manufacturers chose to neglect their own researchers regarding the ways of removal of lead and polonium, mainly due to the financial and commercial burden. In 1980, one confidential memorandum revealed that the issue was mainly caused by calcium phosphate fertilizers employed in tobacco cultivations. Moreover, cigarette manufacturers knew about studies regarding the possibility of decreasing tobacco and smoke radioactivity by using ammonium phosphate instead of calcium phosphate as fertilizer.

However, it is not to be excluded that the reduction in alpha radioactivity, as observed since the end of the nineties, is due to an intervention of the tobacco manufacturers. So far, the majority of public opinion still ignores the presence of polonium radioactivity in tobacco smoke and the serious public health threat that it represents. Yet, from a communicative and motivational point of view, it could become a great opportunity for prevention and smoking cessation. For now, it seems that something has changed in the media and scientific world since the widely covered Litvinenko case and some other studies (Muggli et al. 2008).

Conclusions

Polonium-210 represents one of the principal causes of lung cancer and its shift from squamous cell carcinoma to adenocarcinoma. Provided that it is true that tobacco manufacturers have been aware of the presence of Po-210 in smoke since the early 1960s and concealed its existence intentionally in various ways, it is likely that the medical and scientific sector is guilty of having ignored it.

It is necessary that the medical and scientific world becomes aware and conscious of this problem, creating systematic educational programs of tobaccology in the university curricula of the Medical Sciences courses.

Likewise, governments should force manufacturers to introduce cigarettes with low Po-210 concentration and place a clear indication about this on the packet in order to reduce smokers' risk.

Finally, since people fear everything that is radioactive, perhaps it would be useful to create an adequate information campaign so as to enable and accelerate smokers' motivational pathways and increase the efficacy of antismoking programs.

Cross-References

▸ Lead in Plants
▸ Lead, Physical and Chemical Properties
▸ Polonium in Terrestrial Environment
▸ Radium, Physical and Chemical Properties

References

Behera D, Balamugesh T (2004) Lung cancer in India. Indian J Chest Dis Allied Sci 46(4):269–281

Black SC, Bretthauer EW (1968) Polonium-210 in tobacco. Radiol Health Data Rep 9(3):145–152

Boyle P (2004) Tobacco and public health: science and policy. Oxford University Press, Oxford/New York

Brennan P, Buffler PA et al (2004) Secondhand smoke exposure in adulthood and risk of lung cancer among never smokers: a pooled analysis of two large studies. Int J Cancer 109(1):125–131

Cohen BS, Eisenbud M et al (1980) Alpha radioactivity in cigarette smoke. Radiat Res 83(1):190–196

Devesa SS, Bray F et al (2005) International lung cancer trends by histologic type: male:female differences diminishing and adenocarcinoma rates rising. Int J Cancer 117(2):294–299

Evans GD (1993) Cigarette smoke = radiation hazard. Pediatrics 92(3):464–465

Hoffmann D, Hoffmann I (1997) The changing cigarette, 1950–1995. J Toxicol Environ Health 50(4):307–364

International Agency for Research on Cancer (IARC) (1986) Tobacco smoking. IARC monographs programme on the evaluation of the carcinogenic risk of chemicals to humans, vol 38. IARC, Lyon, pp 127–135

International Agency for Research on Cancer (IARC) (2001) Ionizing radiation, part 2: some internally deposited radionuclides. IARC, Lyon

International Agency for Research on Cancer (IARC) (2004) Tobacco smoke and involuntary smoking. IARC monographs programme on the evaluation of the carcinogenic risk of chemicals to humans, vol 83. IARC, Lyon, pp 80–83

Khater AE (2004) Polonium-210 budget in cigarettes. J Environ Radioact 71(1):33–41

Little JB, McGandy RB (1966) Measurement of polonium-210 in human blood. Nature 211(5051):842–843

Little JB, Radford EP Jr (1967) Polonium-210 in bronchial epithelium of cigarette smokers. Science 155(762):606–607

Marmorstein J (1986) Lung cancer: is the increasing incidence due to radioactive polonium in cigarettes? South Med J 79(2):145–150

Martell EA (1983) alpha-Radiation dose at bronchial bifurcations of smokers from indoor exposure to radon progeny. Proc Natl Acad Sci USA 80(5):1285–1289

Martell EA, Sweder KS (1981) The roles of polonium isotopes in the etiology of lung cancer in cigarette smokers and uranium miners. In: International conference on radiation hazards in mining. Colorado School of Mines, Golden, CO

Muggli ME, Ebbert JO et al (2008) Waking a sleeping giant: the tobacco industry's response to the polonium-210 issue. Am J Public Health 98(9):1643–1650

National Research Council (U.S.). Committee on the Biological Effects of Ionizing Radiations, United States. Environmental Protection Agency et al (1988) Health risks of radon and other internally deposited alpha-emitters. National Academy Press, Washington, DC

Radford EP Jr, Hunt VR (1964) Polonium-210: a volatile radioelement in cigarettes. Science 143:247–249

Singh DR, Nilekani SR (1976) Measurement of polonium activity in Indian Tobacco. Health Phys 31(4):393–394

Skwarzec B, Ulatowski J et al (2001) Inhalation of 210Po and 210Pb from cigarette smoking in Poland. J Environ Radioact 57(3):221–230

Spencer H, Holtzman RB et al (1977) Metabolic balances of 210Pb and 210Po at natural levels. Radiat Res 69(1):166–184

Turner RC, Radley JM (1960) Naturally occurring alpha activity of cigarette tobacco. Lancet 1:1197

Zagà V, Lygidakis C et al (2011) Polonium and lung cancer. J Oncol 2011:860103

Polonium in Terrestrial Environment

Bertil R. R. Persson
Department of Medical Radiation Physics, Lund University, Lund, Sweden

Synonyms

Po-210. ^{210}Po; Polonium; Polonium-210; Radioactive element; α-Radiation emitter

Definitions

The element polonium has atomic number 84 and is located in the fifth period of the periodic system: O-S-Se-Te-Po. Its chemical properties have some similarities to its closest neighbors which are tellurium (Te) above and bismuth (Bi) to the left. Pierre and Marie Curie used the similarity to Bi in the separation process they used (July 1898) in their discovery of polonium. The most widely available isotope of polonium in the terrestrial environment is ^{210}Po which is the final radioactive decay product in the natural uranium-238 decay series. Polonium-210 decays with a half-life of 138.4 days emitting α-particles (100% 5.29 MeV; 0.00122% 4.52 MeV) and single gamma rays (0.803 MeV) with very low abundance (1.23×10^{-5}). Other long-lived polonium isotopes are the artificially produced ^{208}Po (2.898 y) and ^{209}Po (102 y) which both decay through α-decays 5.11 and 4.88 MeV, respectively, and are used as yield determinants in the radiochemical analysis of ^{210}Po.

Origin of ^{210}Po in the Terrestrial Environment

The presence of ^{210}Po in the ground can be traced to the decay of ^{238}U.

$$^{238}U > ^{234}Th > ^{234}Pa > ^{234}U > ^{230}Th > ^{226}Ra > ^{222}Rn(3.82 days).$$

After the first five decays, Rn-222 is formed which is a noble gas diffusing out from the ground into the atmosphere where it decays to the following short-lived products which attach to airborne small particles:

$$^{222}Rn(3.82 days) > ^{218}Po(RaA 3.10 min)$$
$$> ^{214}Pb(RaB 26.8 min) > ^{214}Bi(RaC 19.9 min)$$
$$> ^{214}Po(RaC' 164.3 \mu s).$$

The decay products following ^{214}Po are longer lived:

$$^{214}Po(RaC' 164.3 \mu s) > ^{210}Pb(RaD 22.20 y)$$
$$> ^{210}Bi(RaE 5.01 d) > ^{210}Po(RaF 138.4 d) > ^{206}Po(stable).$$

The concentration of those long-lived products in air increases with height and reaches a maximum in the stratosphere.

Polonium in Terrestrial Environment, Table 1 The distribution of activity concentrations of ^{210}Po in soil from different countries Bq kg^{-1} dry weight

Country	^{210}Po Min Bq/kg	^{210}Po Max Bq/kg	^{210}Po Average Bq/kg	SE	References
Russia	30	50	40	5	Dowdall and O'Dea (1999)
South India	17	228	123	53	Hasanen (1977)
USA	8	128	68	30	Myrick et al. (1983)
Germany	33	207	120	44	Schuttelkopf (1982)
Brazil	32	70	51	10	Santos et al. (1990)
Turkey	13	135	74	31	Akyil et al. (2002)
Turkey	10	870	330	160	Akyil et al. (2008)
Syria	1,2	110	56	27	Al-Masri et al. (2008)
Brazil U-mine	30	92	59	29	Santos et al. (1993)
Brazil control	3	38	14	11	Santos et al. (1993)
Turkey	26	41	37	6	Ekdal et al. (2006)
India, Kaiga (0–5 cm)	17	228	83	52	Karunakara et al. (2000)
India, Kaiga (5–10 cm)	7	142	44	33	Karunakara et al. (2000)
India, Kaiga (10–25 cm)	4	67	23	16	Karunakara et al. (2000)
Turkey fertilized	37	57	45	8	Bolca et al. (2007)
Turkey unfertilized	33	41	38	5	Bolca et al. (2007)
India, Kalpakkam			44		Iyengar et al. (1980)
India, Mysore	8	37	22	15	Nagaiah et al. (1995)
India, Ullai	1	14	8	6	Radhakrishna et al. (1993)
India, coastal Karnataka	4	45	24	21	Siddappa et al. (1994)
USA, Colorado, Iowa	29	59	44	15	USAEC (1980)
USA, Wyoming, Control	2	9	5	3	Ibrahim and Whicker (1987)
USA, Wyoming, U-Mine	21	81	51	30	Ibrahim and Whicker (1987)
Finland, Olkiluoto (0–10 cm)	40	43	41	2	Brown et al. (2011)
Norway, Dovrefjell (0–6 cm)	11	78	44	34	Brown et al. (2011)
World average	1	870	60	13	This work (SE)

Levels of ^{210}Po in the Terrestrial Environment

Levels of ^{210}Po in Soil

Soil consists of particles of different minerals as well as organic matter in various stages of degradation. ^{210}Po in soils originates either as a product from the radioactive decay of radionuclides of ^{238}U series present in the soil (supported) or the result of the deposition of radon decay products from the atmosphere (unsupported). Airborne particles with attached ^{210}Pb and ^{210}Po are deposited on the earth's surface through fallout, which results in accumulation of the final long-lived ^{210}Pb (22.3 y) in plants and the top layer of soil, where it decays to ^{210}Bi (5 d) > ^{210}Po (140d) and finally to stable ^{206}Pb.

The levels of ^{210}Pb and ^{210}Po contained in the top layer of soil can be correlated with the amount of atmospheric precipitation. But the ingrowth from ^{238}U series present in the soil supported that ^{210}Pb is the main source of ^{210}Po in soil and established an equilibrium with a ratio close to 1. Due to the different amounts of clay and organic colloids in various soils, the ^{210}Po content varies with soil type (Parfenov 1974).

The activity concentrations of ^{210}Po in soils from various locations in the world are given in Table 1 and displayed in Fig. 1. The world average of the activity concentrations of ^{210}Po in soil is 60 ± 13 (SE) Bq kg^{-1}

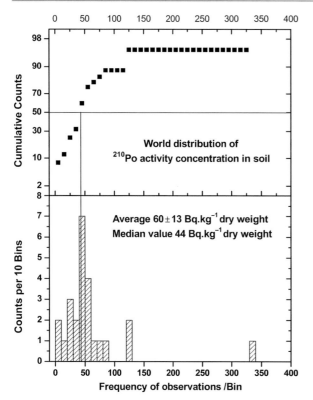

Polonium in Terrestrial Environment, Fig. 1 The distribution of the activity concentrations of ^{210}Po in soils from various locations

dry weight and the median value is 44 (50%) Bq kg^{-1} dry weight.

Groundwater Concentrations of ^{210}Po

The average activity concentrations of ^{210}Po in water sources are given in Table 2. ^{210}Po concentrations in household water from private drilled wells have been observed to be quite high (maximum 6,500 but median 48–107 mBq l^{-1}). In water from dug wells, the concentrations of ^{210}Po are lower (maximum 120 but median 5 mBq l^{-1}). But from public water supplies, the concentrations of ^{210}Po are usually very low (median 3–5 mBq l^{-1}).

Polonium Soil-to-Plant Transfer Factor

Uptake of radionuclides from soil to plant is characterized using a transfer factor (TF), which is defined as the ratio of radionuclide activity concentration per unit mass concentrations (Bq kg^{-1} dry wt.) of plant (C_{plant}) and soil (C_{soil}), respectively: TF = C_{plant}/C_{soil}.

Average values of transfer factors of polonium estimated (mBq kg^{-1} dry wt. plant/Bq kg^{-1} dry wt. soil) for various crop groups, crop compartments, and crop/soil combinations are displayed in Table 3. The upper part of Table 3 shows the current established values (IAEA 1994). The lower data are compiled from an extensive compilation of recently published data on transfer factors (Vandenhove et al. 2009).

The TF for a given type of plant and for a given radionuclide can vary considerably depending on several factors such as the physical and chemical properties of the soil, environmental conditions, and chemical form of the radionuclide in soil. The overall average in Table 3 including and excluding deposition that is shown by the two lowest beams indicates that about 7–8% of ^{210}Po present in the soil is transferred to plants, while the transfer factors for non-leafy plants, maize, and cereals are extremely low.

The generally accepted worldwide average of the transfer factor for ^{210}Po in vegetables and fruit is 1 (mBq kg^{-1} dry wt. plant/Bq kg^{-1} dry wt. soil) and for grain 2 (mBq kg^{-1} dry wt. plant/Bq kg^{-1} dry wt. soil) with corresponding values for ^{210}Pb 0.1 and 5 (Staven et al. 2003). In soils with high content of ^{226}Ra and its daughters ^{210}Pb and ^{210}Po, however, the transfer factors can be much higher (Al-Masri et al. 2008; Al-Masri et al. 2010).

Atmospheric ^{210}Po Deposition Transfer to Plants

Mosses and lichens are slow-growing perennials that have high interception potentials for aerosols in precipitation and, therefore, contain significantly higher ^{210}Po and ^{210}Pb concentrations than vascular plants and fungi. The median activity concentrations are in mosses 2,000 Bq kg^{-1} dry weight, lichens 200 Bq kg^{-1} dry weight, and leafy plants 2–20 Bq kg^{-1} dry weight.

Peat bogs are characterized of being covered by primitive plants that grow from the top, while the dead bottom develops to peat. Peat is a heterogeneous mixture of partially humified remains of several groups of plants together with some inorganic material. Estimates of the ^{210}Po activity concentrations in mosses, lichens, and peat at different locations are given in Table 4.

Polonium in Terrestrial Environment, Table 2 Estimates of the ^{210}Po activity concentrations in groundwater and drinking water at different locations around the world (Vesterbacka et al. 2005)

Type of plant	Location	^{210}Po Min mBq l^{-1}	^{210}Po Max mBq l^{-1}	^{210}Po Average mBq l^{-1}	References
Reference value	Worldwide			5	UNSCEAR (2000)
Recommendation	EU			100	WHO (2004)
Surface water	Finland	1.6	2.0	1.9	Vesterbacka (2007)
Lake water	Finland	1.0	6.5		Vesterbacka (2007)
Drilled wells	Finland			48	Vesterbacka (2007)
Water works	Finland			3	Vesterbacka (2007)
Groundwater	Brazil			3	Bonotto and Bueno (2008)
Mineral water	Italy	<0.04	21	1.8	Desideri et al. (2007)
Groundwater	California USA	0.25	555	<26	Ruberu et al. (2007)
Drilled wells	N. Nevada, USA	0.4	6,500	107	Seiler (2011)
Mineral water	Italy	0.12	11.3	3	Jia and Torri (2007)
Mineral water	Austria	0.4	6.1	1.9	Wallner et al. (2008)

Polonium in Terrestrial Environment, Table 3 Average polonium soil-to-plant transfer factors (mBq kg^{-1} dry wt. plant/Bq kg^{-1} dry wt. soil) for crop groups, crop compartments, and crop/soil combinations (IAEA 1994; Vandenhove et al. 2009)

Plant group	Min	Max	Average TF*1000	Rel. SD
Wheat grain-grain			2.30	
Potato			7.00	
Vegetables			1.20	
Grasses			0.90	
Cereals-grain	0.224	0.26	0.24	0.11
Maize-grain	0.018	0.466	0.24	1.31
Rice-grain			17	
Leafy vegetables			19	0.91
Non-leafy vegetables	0.016	0.37	0.19	1.30
Legumes-pods	0.06	1.02	0.48	0.96
Root crops-roots	0.24	49	12	1.38
Root crops-shoots	58	97	77	0.35
Tubers	0.143	34	8.0	1.44
Natural pastures	22	1,020	259	1.25
All cereals	0.018	16.8	3.6	2.09
Pastures/grasses	18	1,020	259	1.25
Fodder	0.016	97	25	1.40
All excluding deposition	0.016	1,020	56	2.86
All including deposition	0.016	1,020	74	2.16

^{210}Po and ^{210}Pb in Leaves of Tobacco Plant

High activity concentrations of ^{210}Po and ^{210}Pb are found in leaves of the tobacco plant and its products partly due to the contribution of ^{210}Po from atmospheric fallout (Berger et al. 1965). But ^{210}Po and ^{210}Pb are also absorbed by the roots of the plant (Marsden 1964; Tso et al. 1966).

Since the tobacco leaves are processed for a long time, the unsupported ^{210}Po fallout present in fresh tobacco leaves decays to a great deal before use. Elderly samples of tobacco harvested at different years (1938–2010) in various countries have activity concentrations of ^{210}Pb in the range of 9–21 Bq kg^{-1} with an average of 13 ± 3 Bq kg^{-1} and of ^{210}Po in the range of 6–45 Bq kg^{-1} with an average of 18 ± 12 Bq kg^{-1}. The average ^{210}Po activity content per cigarette is estimated to 15 ± 5 mBq. But the activity of ^{210}Po in smoke inhaled from a cigarette is ranging from 2 to 20 mBq due to various burning temperature.

^{210}Po and ^{210}Pb in Vegetables

The activity concentrations of ^{210}Po in various types of vegetables mBq kg^{-1} fresh weight are given in Table 5. In leafy vegetables, the minimum activity concentration of ^{210}Po was 4 mBq kg^{-1} and corresponding maximum 9,360 mBq kg^{-1}. The estimated average of ^{210}Po in leafy vegetables is 320 ± 190 mBq kg^{-1} which is about 3 times the UNSCEAR reference value of 100 mBq kg^{-1}.

In root and fruit vegetables, the minimum activity concentration of ^{210}Po was 1 mBq kg^{-1} and

Polonium in Terrestrial Environment, Table 4 Estimates of the ^{210}Po activity concentrations in mosses, lichens, and peat at different locations

Type of plant	Location	^{210}Po Min Bq kg^{-1} dry wt.	^{210}Po Max Bq kg^{-1} dry wt.	^{210}Po Average Bq kg^{-1} dry wt.	SE	References
Mosses						
Polytrichum	N. Sweden	300	960	630	330	Holm et al. (1981)
Sphagnum	N. Sweden	185	700	443	258	Holm et al. (1981)
Alectoria	N. Sweden	570	640	605	35	Holm et al. (1981)
Pterobryopsis tumida	S. India			2,724	13	Karunakara et al. (2000)
Grimmia pulvinata	W. Turkey	1,228	1,228	1,228		Ugur et al. (2003)
Lycopodium cernuum	Syria			1,322		Al-Masri et al. (2005)
Funaria hygrometrica	Syria			2,392		Al-Masri et al. (2005)
Lichens						
Cladonia alpestris	Ctr. Sweden			250	30	Persson (1970 #359), Persson (1974 #351)
Cladonia arbuscula	Vågå Norway			140	27	Skuterud et al. (2005)
Cladonia arbuscula	E Namdal Norway			141	11	Skuterud et al. (2005)
Cladonia arbuscula	Dovrefjell Norway			138		Brown et al. (2011)
Cladonia stellaris	Dovrefjell Norway			30		Brown et al. (2011)
Peat						
Sphagnum	N. Sweden			192	37	Malmer and Holm (1984)
Sphagnum	S. Sweden			439	117	Malmer and Holm (1984)

corresponding maximum 430 mBq kg^{-1}. The estimated average of ^{210}Po in leafy vegetables is 81 ± 18 mBq kg^{-1} which is about twice the UNSCEAR reference value of 40 mBq kg^{-1}.

^{210}Po and ^{210}Pb in Cereals and Grain Products

The activity concentrations of ^{210}Pb and ^{210}Po in various types of cereals and grain products mBq kg^{-1} fresh weight are given in Table 6. For ^{210}Pb, the minimum was 15 mBq kg^{-1} and maximum 1,940 mBq kg^{-1} and average of all reported values was 137 ± 35 mBq kg^{-1} which is more than twice the UNSCEAR's reference value of 50 mBq kg^{-1}. For ^{210}Po, the minimum was 30 mBq kg^{-1} and maximum 10,200 mBq kg^{-1} and average of all reported values was 239 ± 77mBq kg^{-1} which is about 4 times the UNSCEAR's reference value of 60 mBq kg^{-1}. The dietary intake of cereals and grain products is about 140 kg y^{-1} which is the 2nd highest of all food items (UNSCEAR 2000).

^{210}Po and ^{210}Pb in Food Chains Grass-Cattle-Milk-Meat

The activity concentrations of ^{210}Pb and ^{210}Po in various types of milk and meat products mBq kg^{-1} fresh weight are given in Table 7.

For ^{210}Pb in milk products, the minimum was 5 mBq kg^{-1} and maximum 88 mBq kg^{-1} and average of all reported values was 81 ± 19 mBq kg^{-1} which is twice the UNSCEAR's reference value of 15 mBq kg^{-1}. For ^{210}Po in milk products, the minimum was 2 mBq kg^{-1} and maximum 80 mBq kg^{-1} and average of all reported values was 59 ± 13 mBq kg^{-1}, which is four times the UNSCEAR's reference value of 15 mBq kg^{-1}.

For ^{210}Pb in meat, the minimum was 15 mBq kg^{-1} and maximum 140 mBq kg^{-1} and average of all

Polonium in Terrestrial Environment, Table 5 Activity concentrations of ^{210}Pb and ^{210}Po in various types of vegetables mBq kg^{-1} fresh weight

Type of vegetable Country	^{210}Po Min mBq kg^{-1}	^{210}Po Max mBq kg^{-1}	^{210}Po Average mBq kg^{-1}	SD	References
Leafy vegetables					
Worldwide	4	430	94	±52	UNSCEAR (2000)
Egypt	45	1,800	1,270	±370	Din (2011)
Korea	19	9,360	216	±176	Lee et al. (2009)
India, Kalpakkam	26	653	215	±112	Kannan et al. (2001)
England and Wales	4	140	60	±20	McDonald et al. (1999)
Poland	31	82	55	±22	PietrzakFlis et al. (1997)
Average			320	±190	
Reference value			100		UNSCEAR (2000)
Root and fruit vegetables					
Worldwide	12	210	75	±27	UNSCEAR (2000)
Egypt	10	230	150	±70	Din (2011)
Korea	153	167	160	±7	Lee et al. (2009)
India, Kalpakkam	10	180	44	±43	Kannan et al. (2001)
England and Wales	1	13	5	±4	McDonald et al. (1999)
Poland	28	49	37	±11	PietrzakFlis et al. (1997)
Korea	40	80	60	±28	Lee et al. (2009)
England and Wales	16	430	121	±78	McDonald et al. (1999)
Poland	28	213	80	±75	PietrzakFlis et al. (1997)
Average			81	±18	
Reference value			40		UNSCEAR (2000)
Other vegetables					
Egypt	140	460	270	±50	Din (2011)
India, Kalpakkam	10	108	31	±21	Kannan et al. (2001)
Average			150	±120	

reported values was 32 ±13 mBq kg^{-1} which is the same as the UNSCEAR's reference value of 80 mBq kg^{-1}. For ^{210}Po in meat, the minimum was 21 mBq kg^{-1} and maximum 120 mBq kg^{-1} and average of all reported values was 70 ± 39 mBq kg^{-1}, which does not differ significantly from the UNSCEAR's reference value of 60 mBq kg^{-1}. The dietary intake of milk and meat products is 170 kg y^{-1}, the highest of all food items of terrestrial origin (UNSCEAR 2000).

^{210}Po and ^{210}Pb in Food Chains Lichen-Reindeer-Man or Lichen-Caribou-Man

Lichens are slow-growing perennials that have high interception potentials for aerosols in precipitation and, therefore, contain significantly higher ^{210}Pb and ^{210}Po concentrations than vascular plants. The ^{210}Po/^{210}Pb activity ratio in lichens is typically equal to 1 as ^{210}Po approaches secular equilibrium with ^{210}Pb. The activity concentrations in lichens of *Cladonia family* which is grazed by reindeer and caribou vary between 110 and 430 Bq kg^{-1} dry weight with an average of 243 ± 11 Bq kg^{-1} dry weight (Persson 1970a; Persson 1972; Persson and Holm 2011; Persson et al. 1974).

The activity concentrations of ^{210}Pb in reindeer meat vary between 200 and 1,200 mBq kg^{-1} fresh weight with an average of 500 ± 100 mBq kg^{-1} fresh weight. The activity concentrations of ^{210}Po in reindeer meat vary between 1,700 and 13,300 mBq kg^{-1} fresh weight with an average of 7,800 ± 1,300 mBq kg^{-1} fresh weight. This maintains about 10 times higher ^{210}Po concentration in soft tissues of residences regularly consuming caribou or reindeer meet than the 80 mBq kg fresh weight estimated for the general population (Hill 1966; Persson 1970b).

Polonium in Terrestrial Environment, Table 6 Activity concentrations of ^{210}Pb and ^{210}Po in various types of cereals and grain products Bq kg^{-1} fresh weight

Sample types	^{210}Pb Min mBq kg^{-1}	^{210}Pb Max mBq kg^{-1}	^{210}Pb Average mBq kg^{-1}	^{210}Po Min mBq kg^{-1}	^{210}Po Max mBq kg^{-1}	^{210}Po Average mBq kg^{-1}	References
Cereals						300	Cunha et al. (2001)
Macaroni				150	250	200	Cunha et al. (2001)
Cereals				40	400	220	Parfenov (1974)
Wheat	78	118	97				Santos et al. (2002)
Pasta	14	206	38				
Grains			38			42	Wu et al. (2008)
Flour	200	480	400	220	670	400	BfS (2003)
Cereal grains	200	1,940	300	40	10,200	1,400	BfS (2003)
Flour			400	100	600	400	Globel et al. (1966)
Bread			90	40	200	90	Globel et al. (1966)
Wheat				30	45	37	Kannan et al. (2001)
Wheat				14	15	15	Khandekar (1977)
Wheat flour	74	150	112	75	109	92	PietrzakFlis et al. (1997)
Rye flour	110	200	155	117	155	136	PietrzakFlis et al. (1997)
Cereals	22	110	66	27	31	29	Smithbriggs et al. (1986)
Whole-grain products			81				Morse and Welford (1971)
Flour			48				Morse and Welford (1971)
Macaroni			34				Morse and Welford (1971)
Bakery products			67				Morse and Welford (1971)
Cereals				70	400	208	Din (2011)
Rice				137	140	140	Lee et al. (2009)
Barley				110	150	150	Lee et al. (2009)
Soybean				200	210	210	Lee et al. (2009)
Average	15	1,940	137 ± 35	30	10,200	239 ± 77	

In bone of reindeer, the concentrations of ^{210}Pb vary between 111 and 481 Bq kg^{-1} fresh weight with an average of 282 ± 48 Bq kg^{-1} fresh weight (Holtzman 1966; Kauranen and Miettinen 1969; Persson 1972; Persson 1970b). The ^{210}Pb in bone of caribou varies between 160 and 870 Bq kg^{-1} fresh weight with an average of 450 ± 80 Bq kg^{-1} fresh weight (Beasley and Palmer 1966; Blanchar and Moore 1970; Holtzman 1966; Macdonald et al. 1996).

^{210}Po in Wildlife

The activity concentration of ^{210}Po in some species of terrestrial wildlife in Norway at Dovrefjell, as well as in amphibians and reptiles from Olkiluoto, Finland, are given in Table 8.

The activity concentrations of ^{210}Po in earthworms were similar to those in small mammals in the range of 21–122 Bq kg^{-1} dry weight. In the pectoral muscle of willow grouse, the activity concentration of ^{210}Po was in the range of 1.9–4.7 Bq kg^{-1} dry weight (Brown et al. 2011).

Dietary Intake of ^{210}Po and ^{210}Pb in General Food

^{210}Po and ^{210}Pb ingested with foodstuffs is considered to contribute significantly to the internal radiation dose to man. The mean annual intakes of ^{210}Po and ^{210}Pb in the adult diet are about 58 Bq and 30 Bq, respectively, causing effective doses of 0.070 mSv y^{-1} from ^{210}Po and 0.021 mSv y^{-1} from ^{210}Pb (UNSCEAR 2000).

Polonium in Terrestrial Environment, Table 7 Activity concentrations of ^{210}Pb and ^{210}Po in various milk and meat products Bq kg^{-1} fresh weight

Country	^{210}Pb Min mBq kg^{-1}	^{210}Pb Max mBq kg^{-1}	^{210}Pb Average mBq kg^{-1}	SD	^{210}Po Min mBq kg^{-1}	^{210}Po Max mBq kg^{-1}	^{210}Po Average mBq kg^{-1}	SD	References
Milk products									
Worldwide	5	88	25	10	2	80	23	10	UNSCEAR (2000)
Syria			22.5				194		Al-Masri et al. (2004)
India, Kalpakkam					8	12	10	1	Kannan et al. 2001)
Poland	18	29	24	6	16	23	20	4	Pietrzak-Flis et al. (1997)
Slovenia	43	60	54	6	30	65	48	16	Štrok and Smodiš 2010)
UK			35	1					Smithbriggs and Bradley (1984)
Average			32	13			59	77	This work
Reference value			15				15		UNSCEAR (2000)
Meat products									
Worldwide	15	140	67	17	37	120	81	13	UNSCEAR (2000)
India, Kalpakkam					21	100	28	6	Kannan et al. (2001)
Poland	98	106	102	15	99	102	101	15	Pietrzak-Flis et al. (1997)
UK			74	1					Smithbriggs and Bradley (1984)
Average			81	19			70	38	This work
Reference value			80				60		UNSCEAR (2000)

Polonium in Terrestrial Environment, Table 8 Summary of the activity concentrations of ^{210}Po studied in some species of terrestrial wildlife in Norway at Dovrefjell and in amphibians and reptiles from Olkiluoto, Finland, in 2007 (Brown et al. 2011)

Animal species	^{210}Po Bq kg^{-1} Dry wt. Min	^{210}Po Bq kg^{-1} Dry wt. Max	^{210}Po Bq kg^{-1} Dry wt. Average	SD	^{210}Po Bq kg^{-1} Fresh wt. Average
Red earthworm – *Lumbricus rubellus*	28	69	49	20	8.3
Grey worm – *Aporrectodea caliginosa*	21	123	57	36	9.7
Common shrew – *Sorex araneus*	20	84	37	17	11.2
Bank vole – *Myodes glareolus*	40	86	65	17	19.5
Willow grouse – *Lagopus lagopus*	1.9	4.7	3.3	1.4	1
Viper – *Vipera berus*	21	22	21	0.5	5.3
Frog – *Rana temporaria*	5.7	5.9	5.8	0.1	1.7

The data on ^{210}Po and ^{210}Pb concentrations in daily food products reported from various countries indicate that their concentrations depend on the climate- and geological- and agricultural- conditions. In Table 9 is given a summary of estimated ^{210}Po and ^{210}Pb concentrations in various food items.

Direct dry and wet deposition of ^{210}Po and ^{210}Pb on leafy plants enhances the concentration of unsupported ^{210}Po in vegetables and grass. Thus, in terrestrial food chains, the highest level of ^{210}Po is usually found in leafy vegetables.

The activity concentrations of ^{210}Pb and ^{210}Po reported in fish and seafood, however, are about 30 and 100 times higher than in terrestrial food (Beasley et al. 1969; Carvalho 1995; Lee et al. 2009; Stewart and Fisher 2003). This is due to the extremely high accumulation of these radionuclides in plankton (Fowler 2011).

Polonium in Terrestrial Environment, Table 9 Activity concentrations of ^{210}Pb and ^{210}Po in various fresh food products Bq kg^{-1} fresh weight. UNSCEAR's reference value (UNSCEAR 2000) and the values estimated in this work

	^{210}Pb mBq kg^{-1} Reference: UNSCEAR (2000)	^{210}Pb mBq kg^{-1} Average This work	SE	^{210}Po mBq kg^{-1} Reference: UNSCEAR (2000)	^{210}Po mBq kg^{-1} Average This work	^{210}Po SE
Cereals	50	138	35	60	239	78
Leafy vegetables	80	64	22	100	318	193
Root vegetables and fruits	30	40	4	40	82	18
Milk products	15	32	13	15	59	77
Meat products	80	81	11	60	70	38
Terrestrial products		71	42		154	118
Marine products Henricsson and Persson (2012)	200	2,400	750	2,000	14,500	6,800

Polonium in Terrestrial Environment, Table 10 The minimum and maximum annual dose equivalents of ^{210}Po from drinking water and various food products estimated in this work compared with the reference levels given by UNSCEAR 2000 in the rightmost column (UNSCEAR 2000)

	^{210}Po Min µSv y^{-1}	^{210}Po Max µSv y^{-1}	^{210}Po Average µSv y^{-1}	SE	^{210}Po Reference level UNSCEAR (2000)
Drinking water	0.02	3,900	13	23	3
Cereals	15	152	40	13	10
Leafy vegetables	2	150	23	14	7
Root vegetables and fruits	7	39	17	4	8
Milk products	0	10	7	4	2
Meat products	1	7	4	1	4
Terrestrial products	**25**	**358**	**91**	**70**	**31**
Marine products	1	2,160	261	122	36
Total	26	6,418	352	141	70

Radiation Absorbed Dose to Man from Intake of ^{210}Po

According to the model recommended by the International Commission on Radiological Protection (ICRP), about 10–50% of ingested ^{210}Po is absorbed by the intestine into the bloodstream and deposits mostly in the liver, kidneys, spleen, and red bone marrow (ICRP 1993). To distinguish between ingestion of the organic and inorganic forms of polonium, ICRP recommended that 10% of ingested ^{210}Po material of workers was in a form with fast or moderate absorption to the blood. But for members of the public, 50% of ingested ^{210}Po material is of a form with slow absorption to the blood. For intake by inhalation, ICRP recommended that 10% is absorbed by the blood for workers and 1% for members of the public (ICRP 1994, 1996).

The effective dose coefficient for dietary intake of ^{210}Po for adult members of the public recommended by ICRP is 1.2 µSv Bq^{-1}, considering an alimentary tract transfer coefficient of 50% for ^{210}Po (Hunt and Rumney 2007; ICRP 1993). That is the highest value for any of the natural radionuclides ^{3}H, ^{7}Be, ^{14}C, ^{23}Na, ^{40}K, ^{238}U(series); ^{232}Th(series); or ^{235}U(series) (UNSCEAR 2000). In Table 10 is given the minimum and maximum annual dose equivalent of ^{210}Po from drinking water and various food products estimated in this work compared with the reference levels given by UNSCEAR displayed in the rightmost column of Table 10 (UNSCEAR 2000).

For inhalation of ^{210}Po aerosols, assuming 10% absorption to the blood, the recommended effective dose coefficient is 3.3 µSv Bq^{-1} (ICRP 1996). The committed effective dose equivalent by breathing inhalation of natural ^{210}Po aerosols (50 µBq m^{-3}) has been estimated to be about 1.2 µSv y^{-1} (UNSCEAR 2000).

The average median daily dietary intakes of ^{210}Po for the adult world population from terrestrial products is estimated to be in the range of 20–300 Bq y^{-1} with an average of 80 ± 60 Bq a^{-1}. That corresponds to a committed annual effective dose of 91 ± 70 µSv y^{-1} for ^{210}Po from terrestrial products. The dietary intakes of ^{210}Po and ^{210}Pb from vegetarian food, however, were estimated to correspond to annual effective doses of only about 30 µSv y^{-1} and 10 µSv y^{-1}, respectively (Persson and Holm 2011).

Since the activity concentrations of ^{210}Po in seafood are significantly higher than in terrestrial food products, the world average effective doses estimated for ^{210}Po from marine products are higher (260 ± 120 µSv y^{-1}). The effective dose to populations consuming a lot of seafood is estimated to be up to four to eight times higher than this world average (Carvalho 1995; Sugiyama et al. 2009).

References

Akyil S, Aslani MA, Gurboga G et al (2002) Activity concentration of radium-226 in agricultural soils. J Radioanal Nucl Chem 254:9–14

Akyil S, Gurboga G, Aslani MA et al (2008) Vertical distribution of Ra-226 and Po-210 in agricultural soils in Buyuk Menderes Basin, Turkey. J Hazard Mater 157:328–334

Al-Masri MS, Mukallati H, Al-Hamwi A et al (2004) Natural radionuclides in Syrian diet and their daily intake. J Radioanal Nucl Chem 260:405–412

Al-Masri MS, Mamish S, Al-Haleem MA et al (2005) Lycopodium cernuum and Funaria hygrometrica as deposition indicators for radionuclides and trace metals. J Radioanal Nucl Chem 266:49–55

Al-Masri MS, Al-Akel B, Nashawani A et al (2008) Transfer of K-40, U-238, Pb-210, and Po-210 from soil to plant in various locations in south of Syria. J Environ Radioact 99:322–331

Al-Masri MS, Al-Hamwi A, Eadan Z et al (2010) Transfer factors of Polonium from soil to parsley and mint. J Environ Radioact 101:1038–1042

Beasley TM, Palmer HE (1966) Lead-210 and Polonium-210 in biological samples from Alaska. Science 152:1062–1064

Beasley TM, Osterber CL, Jones YM (1969) Natural and artificial radionuclides in seafoods and marine protein concentrates. Nature 221:1207

Berger KC, Erhardt WH, Francis CW (1965) Polonium-210 analyses of vegetables cured and uncured tobacco and associated soils. Science 150:1738

BFS (2003). Umweltpolitiek. Umweltradioaktivität und Strahlenbelastung. Bundesamt für Strahlenschutz Jahresbericht Bonn:: Das Bundesministerium für Umwelt, Naturschutz und Reaktorsicherheit

Blanchar RL, Moore JB (1970) Pb-210 and Po-210 in tissues of some alaskan residents as related to consumption of caribou or reindeer meat. Health Phys 18:127–134

Bolca M, Sac MM, Cokuysal B et al (2007) Radioactivity in soils and various foodstuffs from the Gediz River Basin of Turkey. Radiat Meas 42:263–270

Bonotto DM, Bueno TO (2008) The natural radioactivity in Guarani aquifer groundwater, Brazil. Appl Radiat Isot 66:1507–1522

Brown JE, Gjelsvik R, Roos P et al (2011) Levels and transfer of (210)Po and (210)Pb in Nordic terrestrial ecosystems. J Environ Radioact 102:430–437

Carvalho FP (1995) Po-210 and Pb-210 Intake by The Portuguese population – The contribution of seafood in the dietary-intake of Po-210 and Pb-210. Health Phys 69:469–480

Cunha IL, Bueno L, Favaro DIT et al (2001) Analysis of Pb-210 and Po-210 in Brazilian foods and diets. J Radioanal Nucl Chem 247:447–450

Desideri D, Meli MA, Feduzi L et al (2007) Po-210 and Pb-210 inhalation by cigarette smoking in Italy. Health Phys 92:58–63

Din KS (2011) Determination of (210)Po in various foodstuffs and its annual effective dose to inhabitants of Qena City, Egypt. Sci Total Environ 409:5301–5304

Dowdall M, O'dea J (1999) Comparison of point estimation techniques in the spatial analysis of radium-226 and potassium-40 in soil. Environ Monit Assess 50:191–209

Ekdal E, Karali I, Sac MM (2006) Po-210 and Pb-210 in soils and vegetables in Kucuk Menderes basin of Turkey. Radiat Meas 41:72–77

Fowler SW (2011) 210Po in the marine environment with emphasis on its behaviour within the biosphere. J Environ Radioact 102:448–461

Globel B, Muth H, Oberhaus E (1966) Aufnahme und ausscheidung der naturlichen radionuklide 21Pb und 210 Po durch den menschen. Strahlentherapie 131:218

Hasanen E (1977) Dating of sediments, based on ^{210}Po measurements. Radiochem Radianal Lett 31:207

Henricsson F, Persson BRR (2012) Polonium-210 in the biosphere: bio-kinetics and biological effects. In: Guillén Gerada J (ed) Radionuclides: sources, properties and hazards. Nova Science Publishers, Hauppauge NY, pp 1–39

Hill CR (1966) Polonium-210 content of human tissues in relation to dietary habit. Science 152:1261–1262

Holm E, Samuelsson C, Persson BRR (1982) Natural radioactivity around a prospected Uranium mining site. In: Vohra KG (ed) Natural radiation environment : proceedings of the Second Special Symposium on Natural Radiation Environment, held at Bhabha Atomic Research Centre, Bombay, India, during January 19–23, 1981 Bhabha Atomic Research Centre, Bombay 400 085, India: A Halsted Press book. ISBN 100470273305

Holtzman RB (1966) Natural levels of Lead-210 Polonium-210 and Radium-226 in humans and biota of Arctic. Nature 210:1094

Hunt GJ, Rumney HS (2007) The human alimentary tract transfer and body retention of environmental polonium-210. J Radiol Prot 27:405–426

IAEA (1994) Handbook of parameter values for the prediction of radionuclide transfer in temperate environments, Technical Report Series 364. IAEA (International Atomic Energy Agency), Vienna

Ibrahim SA, Whicker FW (1987) Plant accumulation and plant/soil concentration ratios of ^{210}Pb and ^{210}Po at various sites within a uranium mining and milling operation. Environ Experiment Bot 27:203–213

ICRP (1993) Age-dependent doses to members of the public from intake of radionuclides: Part 2 ingestion dose coefficients. Annals of ICRP 23, pp 75–84

ICRP (1994) Limits for Intakes of radionuclides by workers. Annals of ICRP 24, pp 1–413

ICRP (1996) Age-dependent doses to members of the public from intake of radionuclides: Part 5 compilation of ingestion and inhalation dose coefficients. Annals of ICRP 26, pp 1–90

Iyengar MAR, Rajan MP, Ganapathy S et al (1980) Sources of natural radiation exposure in a low monazite environment. In: Gesell TF, Lowder WM (eds) Natural radiation environment III, Proceedings of international Conference, CONF-780422 1980 Houston, 1978. Technical Information Centre, US Department of Energy, Oak Ridge TN. pp 1090–1106

Jia G, Torri G (2007) Estimation of radiation doses to members of the public in Italy from intakes of some important naturally occurring radionuclides (^{238}U, ^{234}U, ^{235}U, ^{226}Ra, ^{228}Ra, ^{224}Ra and ^{210}Po) in drinking water. Appl Radiat Isot 65:849–857

Kannan V, Iyengar MA, Ramesh R (2001) Dose estimates to the public from 210Po ingestion via dietary sources at Kalpakkam (India). Appl Radiat Isot 54:663–674

Karunakara N, Avadhani DN, Mahesh HM et al (2000) Distribution and enrichment of Po-210 in the environment of Kaiga in South India. J Environ Radioact 51:349–362

Kauranen P, Miettinen JK (1969) ^{210}Po and ^{210}Pb in Arctic food chain and natural radiation exposure of Lapps. Health Phys 16:287–295

Khandekar RN (1977) Polonium-210 in Bombay diet. Health Phys 33:149–150

Lee CW, Kang MJ, Lee WN et al (2009) Assessment of Po-210 in foodstuffs consumed in Korea. J Radioanal Nucl Chem 279:519–522

Macdonald CR, Ewing LL, Elkin BT et al (1996) Regional variation in radionuclide concentrations and radiation dose in caribou (Rangifer tarandus) in the Canadian Arctic; 1992–1994. Sci Total Environ 182:53–73

Malmer N, Holm E (1984) Variation in the C/N-quotient of peat in relation to decomposition rate and age determination with ^{210}Pb. OIKUS 43:171–182

Marsden E (1964) Incidence + possible significance of inhaled or ingested polonium. Nature 203:230

Mcdonald P, Jackson D, Leonard DRP et al (1999) An assessment of Pb-210 and Po-210 in terrestrial foodstuffs from regions of England and Wales. J Environ Radioact 43:15–29

Morse RS, Welford GA (1971) Dietary intake of Pb-210. Health Phys 21:53

Myrick TE, Berven BA, Haywood FF (1983) Determination of concentrations of selected radionuclides in surface soil in the United States. Health Phys 45:631–642

Nagaiah N, Malini S, Paramesh L et al (1995) Dependence of Po-210 to Ra-226 ratio on soil characteristics. In: Sundararajan AR, Kirshnan LV, Surya Narayana DS, Rajagopal V, Mathiyarasu R (eds) Fourth national symposium on environment, 1995 Chennai. Anna University, pp 219–222

Parfenov YD (1974) Polonium-210 in environment and in human organism. At Energy Rev 12:75–143

Persson BRR (1970a) Pb-210 atmospheric deposition in lichen-carpets in northern Sweden during 1961–1969. Tellus 22:564

Persson BRR (1970b). ^{55}Fe, ^{90}Sr, ^{134}Cs, ^{137}Cs and ^{210}Pb in the Biosphere. Radiological Health Aspects of the Environmental Contamination from Radioactive Materials in Northern Sweden. Ph.D. thesis, Lund University

Persson BRR (1972) Lead-210, Polonium-210, and stable lead in the food-chain lichen reindeer and man. In: Adams JAS, Lowder WM, Gesell TF (eds) The natural radiation environment II, 1972. Rice University, Houston, pp 347–367

Persson BRR, Holm E (2011) Polonium-210 and lead-210 in the terrestrial environment: a historical review. J Environ Radioact 102:420–429

Persson BRR, Holm E, Lidén K (1974) Radiolead (Pb-210) and stable lead in lichen Cladonia alpestris. Oikos 25:140–147

Pietrzakflis Z, Chrzanowski E, Dembinska S (1997) Intake of Ra-226, Pb-2 l0 and Po-210 with food in Poland. Sci Total Environ 203:157–165

Pietrzak-Flis Z, Suplinska MM, Rosiak L (1997) The dietary intake of U-238, U-234, Th-230, Th-232, Th-228 and Ra-226 from food and drinking water by inhabitants of the Walbrzych region. J Radioanal Nucl Chem 222:189–193

Radhakrishna AP, Somashekarappa HM, Narayana Y et al (1993) A new natural background-radiation area on the southwest coast of India. Health Phys 65:390–395

Ruberu SR, Liu YG, Perera SK (2007) Occurrence and distribution of Pb-210 and Po-210 in selected California groundwater wells. Health Phys 92:432–441

Santos RC, Gouvea IR, Dutta VA et al (1990) Accumulation of Po-210 in foodstufs cultivated in farms around the Brazilian mining and milling facilities on Pocos de Caldas Plateau. J Environ Radioact 11:141–149

Santos PL, Gouvea RC, Dutra IR (1993) Pb-210 in vegetables and soils from an area of high natural radioactivity in Brazil. Sci Total Environ 138:37–46

Santos EE, Lauria DC, Amaral ECS et al (2002) Daily ingestion of Th-232, U-238, Ra-226, Ra-228 and Pb-210 in vegetables by inhabitants of Rio de Janeiro City. J Environ Radioact 62:75–86

Seiler RL (2011) Po-210 in Nevada groundwater and its relation to gross alpha radioactivity. Ground Water 49:160–171

Siddappa K, Balakrishna KM, Radhakrishna AP et al (1994) Distribution of natural and artixcial radioactivity components in the environs of coastal Karnataka, Kaiga and Goa (1991–94). Mangalore University, Mangalore

Skuterud L, Gwynn JP, Gaare E et al (2005) Sr-90, Po-210 and Pb-210 in lichen and reindeer in Norway. J Environ Radioact 84:441–456

Smithbriggs JL, Bradley EJ (1984) Measurement of natural radionuclides in UK diet. Sci Total Environ 35:431–440

Smithbriggs JL, Bradley EJ, Potter MD (1986) The ratio of Pb-210 to Po-210 in UK diet. Sci Total Environ 54:127–133

Staven LH, Rhoads K, Napier BA et al (2003) A compendium of transfer factors for agricultural and animal products. Pacific Northwest National Laboratory, Richland, Washington

Stewart GM, Fisher NS (2003) Experimental studies on the accumulation of polonium-210 by marine phytoplankton. Limnol Oceanogr 48:1193–1201

Štrok M, Smodiš B (2010) Natural radionuclides in milk from the vicinity of a former uranium mine. Nucl Eng Design 241:1277–1281

Sugiyama H, Terada H, Isomura K et al (2009) Internal exposure to (210)Po and (40)K from ingestion of cooked daily foodstuffs for adults in Japanese cities. J Toxicol Sci 34:417–425

Tso TC, Harley N, Alexande LT (1966) Source of Lead-210 and Polonium-210 in tobacco. Science 153:880

Ugur A, Ozden B, Sac MM et al (2003) Biomonitoring of Po-210 and Pb-210 using lichens and mosses around a uraniferous coal-fired power plant in western Turkey. Atmos Environ 37:2237–2245

UNSCEAR (2000) United Nations Scientific Committee on the Effects of Atomic Radiation. Sources, effects and risk of ionizing radiation. United Nation, New York

USAEC (1980) Po-210 in soils and plants. United States Atomic Energy Commission, Washington DC

Vandenhove H, Olyslaegers G, Sanzharova N et al (2009) Proposal for new best estimates of the soil-to-plant transfer factor of U, Th, Ra, Pb and Po. J Environ Radioact 100:721–732

Vesterbacka P (2007) Natural radioactivity in drinking water in Finland. Boreal Environ Res 12:11–16

Vesterbacka P, Makelainen I, Arvela H (2005) Natural radioactivity in drinking water in private wells in Finland. Radiat Prot Dosimetry 113:223–232

Wallner G, Wagner R, Katzlberger C (2008) Natural radionuclides in Austrian mineral water and their sequential measurement by fast methods. J Environ Radioact 99:1090–1094

WHO (2004). Guidelines for drinking water quality: radiological aspects. World Health Organization, Geneva. http://www.who.int/water_sanitation_health/dwq/gdwq3/en/S

Wu Q, Zhu H, Fan T et al (2008) Re-estimation of internal dose from natural radionuclides for Chinese adult men. Radiat Prot Dosimetry 130:434–441

Polonium, Physical and Chemical Properties

Fathi Habashi
Department of Mining, Metallurgical, and Materials Engineering, Laval University, Quebec City, Canada

The discovery of polonium in uranium ores was possible only because of its radioactivity. The equilibrium concentration of ^{210}Po from uranium ores is 7.4×10^{-11} g ^{210}Po/g ^{238}U. Today, all isotopes between mass numbers 193 and 218 are known. The longest-lived isotope is ^{209}Po ($t_{1/2} = 102$ years). Nearly all the chemical and physical studies on polonium were performed with the pure α emitter ^{210}Po ($t_{1/2}=138$ day). Figure 1 shows the naturally occurring polonium isotopes.

Polonium may enter the respiratory system as a result of inhaling radon gas generated from radium, as shown in Fig. 1.

Nowadays, most ^{210}Po is isolated either from "old" radium samples or by irradiation of very pure bismuth with neutrons in a nuclear reactor. One year of irradiation of 1 kg bismuth at a neutron flux of 10^{14} n cm^{-2} s^{-1} gives 25 mg ^{210}Po (ca. 4×10^{12} Bq). The isolation of such quantities from natural sources is practically impossible. Following irradiation of bismuth, ^{210}Po is separated from molten bismuth by vacuum distillation. Final purification of polonium generally is accomplished by its spontaneous deposition on silver or copper powders. Metallic polonium is then prepared by sublimation of this deposit at about 700°C. It is also obtained by reduction of polonium sulfide with hydrogen at 500°C. Polonium sulfide itself is precipitated from an acid solution of polonium ion by gaseous H_2S.

^{210}Po is used also as an α emitter for (α, n) neutron sources. A BeO – ^{210}Po source emits up to 93 neutrons per 10^6 α particles.

Physical Properties

Atomic number	84
Atomic weight	208.95
Relative abundance in the Earth's crust, %	3×10^{-14}
Melting point, °C	~250
Boiling point, °C	962
Atomic radius, pm	164
Crystal structure	
α Po (<18°C)	Cubic, $a = 335.9$ pm
β Po ((>54°C)	Rhombohedral, $a = 336.9$ pm, α $= 98.23°$
Density, g/cm^3	
α Po	9.142
β Po	9.352
Vapor pressure (438–745°C), Pa	$\log p = (964.36 \pm 0.9064) - (716\,860 \pm 893)/T(°C)$

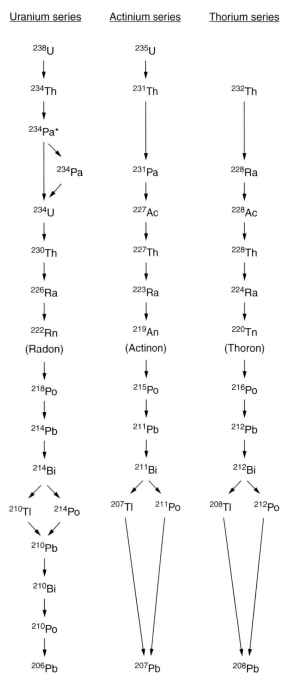

Polonium, Physical and Chemical Properties, Fig. 1 Naturally occurring polonium isotopes 210, 211, 212, 214, 215, 216, and 218

Chemical Properties

Polonium is a metalloid, chemically similar to bismuth and tellurium. The oxidation states +2, +4, +6, and −2 are known, but Po(IV) is the most stable one. Po(VI) is easily reduced to a lower oxidation state by its own radiation. Polonium is soluble in HCl and forms Po(II); in HNO_3 and aqua regia, Po(IV) is produced.

The presence of polonium in tobacco smoke has been known since the early 1960s. Radioactive polonium-210 contained in phosphate fertilizers is absorbed by the roots of plants (such as tobacco) and stored in its tissues. Tobacco plants fertilized by rock phosphates contain polonium-210, which emits alpha radiation.

References

Habashi F (1996) Pollution problems in the mineral and metallurgical industries. Métallurgie Extractive Québec, Québec City. Distributed by Laval University Bookstore "Zone". www.zone.ul.ca

Luig H et al (1997) Radioactive Metals. General. In: Habashi F (ed) Handbook of extractive metallurgy. Wiley, Weinheim, pp 1585–1598

Polonium-210

▶ Polonium in Terrestrial Environment

Polynuclear Pt Complexes

▶ Zinc, Metallated DNA-Protein Crosslinks as Finger Conformation and Reactivity Probes

Polyphosphates

▶ Polonium and Cancer

Porous Silicon for Drug Delivery

Hélder A. Santos[1], Jarno Salonen[2] and Luis M. Bimbo[1]
[1]Division of Pharmaceutical Technology, University of Helsinki, Helsinki, Finland
[2]Laboratory of Industrial Physics, Department of Physics, University of Turku, Turku, Finland

Synonyms

Drug carrier; Large surface-area-to-volume; Loading and release of drugs/proteins; Ratio material

Definition

▶ Porous silicon (PSi) refers to a material containing interconnected networks of empty holes (pores) in a Si structure, rendering the material with a large surface-to-volume ratio properties. PSi materials are classified according to pore diameter, that can vary from a few nanometers to a few micrometers depending on the fabrication parameters. Most commonly, the PSi structure is referred as mesoporous (pores with diameters between 2 and 50 nm). Sometimes the word "nanoporous" is used instead of mesopores to emphasize the nanometric dimension of the pores. The volumetric fraction of empty space (voids) in the structure of the material is called porosity. The internal surface area of PSi per volume unit can be as large as 500 m^2/cm^3 and can act as reservoirs for storing therapeutic compounds for drug delivery applications.

Fabrication Procedures and Role in Drug Delivery Applications

Si is a tetravalent metalloid that does not occur in nature in free form. However, in its combined form, mainly as silicate minerals, it accounts for nearly 25 % of the Earth's crust by mass. Si has found major applications in metallurgical alloys, especially of ▶ aluminum and iron, and in the semiconductor industry. When in porous form, Si exhibits a number of properties that make it an attractive material for controlled drug delivery applications. The electrochemical production of PSi allows the construction of tailored pore sizes and volumes that are controllable from the scale of nanometers to microns. A number of convenient chemistries exist for the modification of PSi surfaces that can be used to control the amount, identity, and the release rate of therapeutic payloads. The material can also be used as a template for organic molecules and biopolymers, and to prepare composites with a designed nanostructure. PSi is fabricated by a top-down approach, where the larger bulk Si material is processed into smaller fractions by various mechanisms, such as milling.

Manufacture of PSi

The accidental production of PSi in 1956 at Bell Labs and the discovery of its porous nature in 1970s opened up new avenues for exploring the properties of this material. Canham was one of the first to hypothesize the quantum confinement effects of PSi, as well as the pioneer to recognize the potential of PSi in biomedical applications (Canham 1997). Figure 1 shows the simpler and most common setup for PSi fabrication by electrochemical anodization (also referred as electrochemical etching) of monocrystalline Si wafer (acting as an anode) and platinum plates (acting as a cathode) both immersed in hydrofluoric acid (HF)-ethanolic solution under an applied potentiostatic (voltage-controlled) or galvanostatic (current-controlled) mode (Salonen et al. 2008; Sailor 2011). The constant current method is preferable and allows better control of the porosity and thickness than the constant voltage method, and also results in better reproducibility of the PSi properties. The porous layer is then formed on the surfaces of the anode Si during the anodization. The cathode made of platinum (Pt) ensures good conductivity and stands the harsh cathodic conditions repeatedly. Ethanol is usually used to improve the electrolyte (HF) penetration through the pores and to minimize the formation of hydrogen bubbles (which reduce the uniformity of the PSi layer). The pore diameter of PSi can be larger or smaller by decreasing or increasing, respectively, the HF concentration. Low HF concentrations and the presence of ethanol also lead to smoother and straighter pores. Depending on the etching conditions, pores can be cylindrical (or spherical), rectangular, and when low current densities are employed, firtree- or sponge-like structures are formed with interpore connections and pore branching (e.g., Fig. 1). In order to form PSi, the current at the Si side of the Si/electrode interface has to be carried

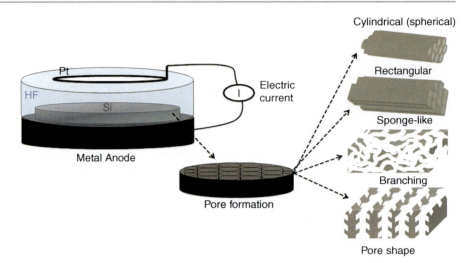

Porous Silicon for Drug Delivery, Fig. 1 Schematic representation of an electrochemical etching setup for PSi fabrication and pore formation: cylindrical (*spherical*) and *rectangular* without interconnections, sponge-like, and branching

by holes, injected from the bulk toward the interface. The current must be kept below an electropolishing threshold. The mechanism of the chemical reaction that has been suggested to better describe the anodization process of Si in fluoride-containing solutions is as follows:

$$Si + 6HF \rightarrow H_2SiF_6 + H_2 + 2H^+ + 2e^- \quad (1)$$

The morphological properties of the obtained PSi film, such as porosity, pore layer thickness, pore size, and shape, are all determined by the fabrication conditions. These conditions include current density, type and doping degree of Si wafer and resistivity, HF concentration, chemical composition of the electrolyte solution, crystallographic orientation, temperature, time, stirring, illumination intensity, and wavelength (Sailor 2011). For example, PSi with pore size of ca. 10 nm can be produced with p^+-type Si wafers (orientation: $\langle 100 \rangle$; resistivity: 0.01–0.02 W.cm), electrolyte of HF (38 %)-ethanol mixture, and constant etching current of 50 mA/cm^2. The initial type of Si wafer affects the pore morphology of PSi, and the morphologies are usually grouped into four categories based on the doping levels of substrates: n, p, n^+, p^+. As the conductivity of Si (i.e., the dopant concentration) increases, the pore diameters and interpore matter lengths also increase. Finer mesoporous structures are obtained with low doped Si substrates. The type of Si wafer also affects the pore diameters and the pore structure changes as a function of the resistivity: for low doped n- and p-type PSi, the pores are relatively randomly orientated in $\langle 100 \rangle$ directions and form spherical sponge-like networks. Sponge-like pores of PSi produced on low doped substrates have typical sizes of 1–8 nm and very large specific surface areas (up to 900 m^2/cm^3). As the resistivity of the substrate decreases, the average pore size increases, and the specific surface area decreases. In the highly doped substrates, such as n^+- and p^+-type PSi, the average pore size is 6–20 nm and the specific surface area is 100–300 m^2/cm^3. The pores are already well orientated perpendicular to the initial surface of the substrate, and in many cases, the pores are cylindrical with smooth pore walls that are not interconnected.

After the etching, the PSi film is detached from the substrate by increasing the etching current abruptly. Regarding the applicability of PSi materials in biomedical applications, microparticles are produced from the PSi free-standing film by milling and subsequent sieving. Typical values of the surface area, pore volume, and porosity obtained for microparticles vary in the ranges of 200–500 m^2/g, 0.5–2.0 cm^3/g, and 50–80%, respectively. In the case of PSi nanoparticles, a similar methodology is used as for PSi microparticles. In the simplest method, fragile, relatively porous layers are brook down into nanoparticles with sonification (Sailor 1992). But instead of using a constant current density during etching, nanoparticles can also be prepared more efficiently using a pulse etching method. In this method, three different subsequent current pulses are employed (Bimbo et al. 2010): (1) the first low current pulse is used to produce a similar mesoporous structure as in the case of microparticles; (2) the second one is shorter, with a high current pulse near the electropolishing region to produce highly porous and

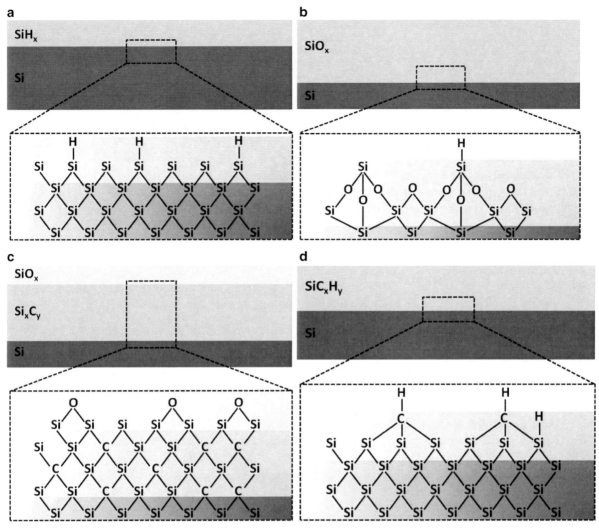

Porous Silicon for Drug Delivery, Fig. 2 Surface chemistry of the PSi materials after anodization and after surface treatments: (**a**) as-anodized; (**b**) oxidized; (**c**) carbonized; and (**d**) hydrocarbonized (Adapted with permission from Sarparanta et al. (2011). Copyright © 2011, American Chemistry Society)

mechanically fragile layers; and (3) a third zero-current pulse is used to remove the possible electrolyte concentration gradients in the pores formed during the high current pulse before the new etching cycle starts. After drying, the production of nanoparticles is performed by milling the porous film. In addition, wet milling is used in order to increase the grinding efficacy and to facilitate the collection of the nanoparticles. The final size separation is done, for example, by centrifugation or filtration.

After the electrochemical etching, the free-standing PSi film is hydrogen terminated (as-anodized, Si-H_x), and thus, the surface is relatively reactive and unstable. These Si-H_x bonds can render the Si surface prone to oxidation even in dry ambient air. Figure 2 shows how the freshly (as-anodized) made PSi surface can be stabilized to prevent unfavorable interactions with the payload molecules by converting the reactive groups to more stable oxidized, carbonized, or hydrocarbonized forms. The PSi surface can be stabilized by employing several methods which generate different surface properties to the material (Salonen et al. 2008; Sailor 2011). These methods usually involve a thermal reaction, changing the Si bond termination and stabilizing the PSi surface against oxidation. The oxygen stabilization (SiO_2) produces Si-O bonds, which have bond strength considerably higher than the Si-H or Si-Si bonds, and therefore more stable. The treatment is done under oxygen containing atmosphere at slightly elevated temperatures (around 300 °C) or chemically, rendering the PSi surface hydrophilic and with a negative surface charge (zeta-potential). The stabilization with carbon (Si_xC_y) produces Si-C bonds, which are more stable than the Si-O bonds, despite the lower bond strength. This is most likely due to the nonpolar nature of the Si-C bond, which reduces the possibility of hydrolysis. The heat treatment is done at high temperature (>700 °C) for the surface where acetylene has already been adsorbed. The surface is extremely stable, hydrophilic, and negatively charged. If the treatment is done under a continuous acetylene flush at temperatures below 700 °C, the PSi is thermally hydrocarbonized ($Si-C-H_x$). This PSi surface is hydrophobic and negatively charged.

The techniques used to assess the various structural and physical properties of PSi include various microscopy (pore diameter, microstructure, and layer thickness), gravimetric analysis (porosity and layer thickness), and sorption isotherms of gases at low temperatures, for example, Brunauer-Emmett-Teller method for pore volume determination (Salonen et al. 2008). The surface chemical composition of PSi can be probed with Fourier Transform Infrared (FTIR) spectroscopy, X-ray photoelectron spectroscopy (XPS), energy-dispersive X-ray (EDX), etc.

Properties of Drug Loading into PSi

PSi-based materials have been widely studied as drug delivery and imaging vehicles (Santos et al. 2011; Godin et al. 2011) in order to enhance the dissolution properties of poorly soluble drugs as well as to facilitate the delivery of larger molecules (Salonen et al. 2008; Lehto et al. 2012), such as peptides and proteins. From a pharmaceutical point of view, it is estimated that more than 90 % of new drug candidates suffer from poor bioavailability when administered orally due to their poor thermodynamic solubility and/or dissolution rate in the intestinal lumen, and poor permeability across the gastrointestinal wall, as well as high intestinal or hepatic first pass metabolism. Poor solubility alone is an issue for 40–60 % of the new drug candidates. Loaded molecules inside micro- or nanofabricated PSi materials could overcome these problems, because the drugs are confined in the pores (usually not much larger than the drug molecules themselves), thus retaining the drug in its amorphous (noncrystalline) disordered or nanocrystalline form (Salonen et al. 2008; Lehto et al. 2012). The increase in solubility and dissolution rate of the loaded drug is achieved as a result of the low lattice energy of the amorphous state of the molecule inside the pores (Salonen et al. 2008).

In addition to small drug molecules, also small hydrophobic peptides are readily loaded inside the PSi pores, whereas larger hydrophilic proteins accumulate mainly on the PSi surface. The loading of drugs into PSi is commonly achieved by simple immersion of the PSi particles into a loading solution (pore filling takes place via capillary forces) in which the desired drug is dissolved in a suitable solvent (usually organic or water in case of hydrophilic peptides/proteins), or by impregnation where a controlled amount of drug is added to the particles and allowed to infuse through capillary forces into the pores (Salonen et al. 2008). Typically, the loading parameters are optimized so that most of the drug molecules are to be located inside the pores without any crystalline fraction on the external surface of the particles. The common advantage of this method is that the loading can be conducted at room temperature and the drug to be loaded is not exposed to harsh chemical conditions, particularly in the case of protein and peptide delivery. The drug loading process begins with a monolayer adsorption on the surface of the carrier. Critical parameters related to optimal loading procedure are the drug–surface interactions, surface–solvent interactions, and solvent–drug interactions (Lehto et al. 2012). For an efficient monolayer adsorption, the solvent–drug interactions are minimized and the drug–carrier surface affinity is dominant and stronger. Carrier–solvent interactions may hinder the effect on the adsorption process. The main adsorption mechanism that takes place when the drug is loaded into PSi materials is physisorption, a reversible process that allows quick release of the loaded compound with minimal activation. This adsorption mechanism is used for enhancing the solubility of the payload compound. In case of a favorable strong chemical interaction between the drug molecule and the surface of the carrier (chemisorption), a sustained drug release is obtained (Lehto et al. 2012). The PSi properties such as hydrophilicity/hydrophobicity, pore size, surface chemistry together with the nature of the loaded drug (charge, chemical nature, and molecular size), spatial

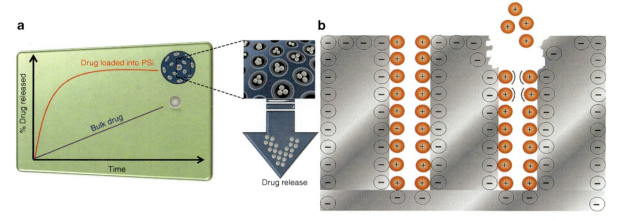

Porous Silicon for Drug Delivery, Fig. 3 (a) Schematic overview of the drug release/dissolution profiles of drug-loaded PSi materials compared to bulk drug molecules. (b) Cationic drug release from negatively charged PSi materials: combined effect of leaching and matrix dissolution

arrangement of the molecules inside the pores, and loading method all affect the drug adsorption mechanism and loading degree. For example, negatively charged PSi surface loaded with cationic drug molecules will lead to strong interactions between the PSi surface and the drug, resulting in a high loading degree and slow drug release kinetics of the molecules attached to the pore surface.

The distinction between the amounts of drug loaded inside the pores of PSi materials or adsorbed onto their external surface is achieved by employing a wide range of complementary characterization methods (Salonen et al. 2008). For example, the most commonly used method to assess the drug load is nitrogen ad/sorption isotherms. From these measurements, the pore volume, surface area, and size distribution of the PSi materials are calculated. The difference in pore volumes of the sample measured before and after loading gives a good estimation on the amount of loaded drug. Combining thermogravimetry (TG) and differential scanning calorimetry (DSC) provides a better quantification of the drug loaded into the PSi materials (Salonen et al. 2008; Lehto et al. 2012). In TG, the sample weight is monitored during constant heating rate at controlled gas atmosphere. The drug decomposes at a certain temperature and desorbs from the sample, resulting in a weight decrease of the sample. DSC is used to distinguish between the amount of drug crystallized on the surface of the PSi from that loaded inside the pores. By measuring the melting points (T_m) of the crystalline drug on the surface of PSi, the T_m of the drug confined in the pores (usually T_m is depressed to lower temperatures), or the T_m of the drug in a disordered state (no T_m is detected) is obtained. Information regarding the stability of the loaded drug is usually provided by using high-performance liquid chromatography (HPLC).

Aspects of Drug Release from PSi

PSi-based materials are also used in drug delivery to improve the ▸ physicochemical properties of poorly soluble drug molecules (Salonen et al. 2008; Santos et al. 2011; Lehto et al. 2012). The idea of enhancing drug dissolution/release of drugs loaded into PSi materials is based on the adsorbed state of the drug molecules inside the pores of the particles (usually in a noncrystalline form). As buffer solutions or body fluids come into contact with both the pores and the drug molecules, an initial drug/release dissolution burst is observed compared to the bulk drug material, followed by a very slow release (Fig. 3a). The release of the drug from the pores is usually associated with diffusion or dissolution processes as well as with combined effects of leaching and PSi matrix dissolution (Fig. 3b), which depend on the PSi surface treatment and environment conditions where the release takes place. The drug release usually follows a first-order kinetics with respect to drug concentration. Many parameters affect the drug release, such as the surface chemistry, morphology, and the pore size of the particles. All these parameters can be

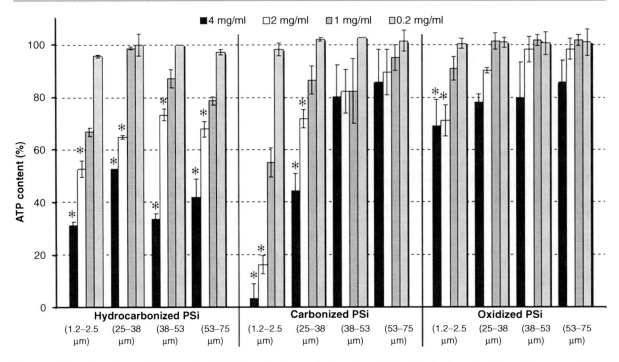

Porous Silicon for Drug Delivery, Fig. 4 ATP production (cell viability) of intestinal Caco-2 cells determined by a luminescence assay after 3 h incubation at 37 °C with hydrocarbonized, carbonized, and oxidized PSi microparticles ($*P \leq 0.05$; Modified with permission from Santos et al. (2010). Copyright © 2010, Elsevier B.V.)

fine-tuned to improve drug release/dissolution and drug permeation behavior, thus improving drug bioavailability.

Biocompatibility, Biodegradability, and Cytotoxicity of PSi

Si can be absorbed daily from food intake and has a positive role in the health of connective tissues and bone. Si does not accumulate in the body and, instead, is readily excreted into urine as orthosilicic acid [$Si(OH)_4$] which is the most natural form of Si in the environment, nontoxic, and important in human physiology by protecting against the toxic effects of aluminum. The advantages of the PSi materials are its proven bioactiveness, bio-inertness, resorbability, and bioerosion when in biological solutions (Canham 1997). For example, the bioerosion of PSi is greatly influenced by the overall porosity, pore size, shape, and surface of the PSi material, which in turn can be controlled by the choice of the material fabrication parameters. PSi with a porosity >70% dissolves in all the simulated body fluids (except gastric fluids), whereas PSi with a porosity <70% is bioactive and slowly bioerodible, degrading mainly into harmless [$Si(OH)_4$].

Cytotoxic studies are used to assess the biological activity of tested substances or materials in a wide range of different ▶ cell types, as well as their behavior and survival (viability) at different experimental conditions. The in vitro analysis is the first screening steep for understanding the mechanisms resulting from the cell–biomaterial interactions and the first assessment before moving to in vivo tests (Lehto et al. 2012). For example, the most significant parameters of several PSi-based microparticles of different size fractions and surface chemistries affecting the cyto ▶ toxicity of Caco-2 intestinal cell models (human epithelial colorectal adenocarcinoma cell line) when exposed to different doses and sizes of particles can be monitored in vitro (Santos et al. 2010). Figure 4 shows that the cytotoxic mechanism is clearly dependent on the size of the particles as well as on the concentration and surface chemistry of the particles. The most favorable particles (least

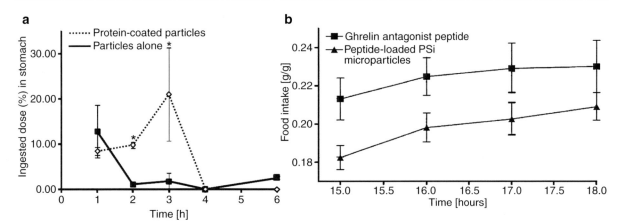

Porous Silicon for Drug Delivery, Fig. 5 (a) Gastric clearance of orally administered hydrophobin protein-coated and -uncoated radiolabeled ^{18}F-PSi nanoparticles (mean ± s.d.; $n = 3$ per time point; *$p < 0.05$; Modified with permission from Sarparanta et al. (2012). Copyright © 2012, Elsevier B.V.).

(b) The suppression effect of food intake in mice by ghrelin antagonistic-loaded PSi microparticles compared to pure ghrelin antagonists after subcutaneous administration (Modified with permission from Kilpeläinen et al. (2009). Copyright © 2009, Elsevier B.V.)

interacting with the cells) were oxidized PSi particles with non-cytotoxic doses below 4 mg/mL, regardless of the particle size. Cell exposure to hydrocarbonized and carbonized PSi particles of sizes above 25 μm induced non-cytotoxic effects below doses of 2 and 4 mg/ml, respectively.

Applications of PSi in Peptide and Protein Delivery

Peptides are promising but challenging molecules for drug delivery due to their large molecular size, chemical characteristics, and sensitivity to breakdown in harsh conditions. Typically, peptides/proteins are administered parenterally either in solutions or as particulate drug delivery systems, as their poor oral bioavailability restricts their administration routes. In addition, most peptides/proteins have short biological half-lives and undergo rapid hepatic clearance. Therefore, peptide solutions or suspensions are frequently administered as injections. In the treatment of several diseases such as ▶ diabetes and obesity, regular daily or at least weekly injections are required. In this respect, PSi-based materials have been investigated with several peptides and proteins, for example, ghrelin antagonist ([D-Lys-3]-GHRP-6, H-His-D-Trp-D-Lys-Trp-D-Phe-Lys-NH$_2$, which inhibits the action of a 28 amino acid containing peptide hormone ghrelin to increase food intake and blood pressure), PYY$_{3-36}$ (an endogenous peptide belonging to the same family of neuropeptide Y and pancreatic polypeptide, consisting of 36 amino acids secreted by enteroendocrine L-cells of the gastrointestinal tract in response to a caloric load), and hydrophobin (a family of surface active proteins of fungal origin consisting of 86 amino acids) (Kovalainen et al. 2011; Bimbo et al. 2011). The interactions between the loaded peptides/proteins and the PSi surfaces define the success of the loading and/or release of the peptides/proteins from the particles, which dictates the adsorption degree, protein structure, and bioactivity. The protein adsorption mechanisms onto PSi materials are controlled by both the protein structure and PSi surface chemistry via hydrophobic interactions and electrostatic attractions. By self-assembling a biofilm of hydrophobin proteins onto the PSi surfaces, the biocompatibility of micro- and nanoparticles is improved (Bimbo et al. 2011; Sarparanta et al. 2011). This family of surface active proteins of fungal origin modifies the surface binding properties, turning hydrophobic surfaces into hydrophilic ones, thus altering the wettability properties of PSi (Bimbo et al. 2011). The properties of such hydrophobin-coated PSi have been explored in oral drug delivery applications as the stable protein-coated PSi micro- and nanoparticles exhibit mucoadhesive and gastroretentive properties for controlled drug delivery (Sarparanta et al. 2012). Figure 5a shows an example of the mucoadhesive and gastroretentive properties of hydrophobin-coated PSi nanoparticles up to 3 h in rats after oral administration. These results can be explored in oral drug formulations to increase the efficacy and efficiency of therapeutic compounds. In this study, the particles were radiolabeled with ^{18}F isotope, which exhibits excellent in vivo stability when attached to the PSi particles and can further allow the biodistribution determination of the particles in the animal's body. In addition, Fig. 5b shows PSi microparticles loaded with ghrelin antagonist (20 % w/w) and their effect on prolonged inhibition of food intake up to 18 h in mice after subcutaneous administration (dose 14 mg/kg). The peptide activity was not affected after the sustained release, leading to a gradual decrease in the heart rate and blood pressure of the mice. These results are rather promising regarding the use of PSi-based materials in controlled peptide/drug delivery, with prolonged physiological effects in vivo.

Overview of the Potentialities and Drawbacks of PSi-Based Materials in Drug Delivery Applications

PSi is an acceptable biomaterial with several advantageous features which render it an efficient drug carrier, such as: (1) confinement of the drug/compound molecules into the pores hinders molecule crystallization and stabilizes the disordered structure against crystallization or reduces the crystal size; (2) the confinement of the payload molecules protects them from the harsh conditions of the body environment and from enzymatic degradation, ideal for peptide/protein delivery; (3) the compound release kinetics can be controlled by tuning the pore size and manipulating the interaction between the payload molecule and the pore wall; (4) hydrophilic particles aid to improve drug wettability; (5) high drug loading degrees into PSi up to 60 w% can be obtained; (6) the surface chemistry, particle and pore size, and shape and morphology of the PSi carriers can easily be controlled by the fabrication parameters producing materials with good biocompatibility and biodegradability; (7) relatively easy surface functionalization for more advanced controlled delivery or targeting; and (8) PSi production by a top-down approach enabling fast scaling up.

Although PSi has been shown to be a good carrier for protein/peptide compounds, the interactions between these compounds and the surface of the PSi materials are yet poorly understood, as well as the effect on the peptide/protein activity after delivery/release from the particles' surface. Also, from an experimental in vivo point of view, all the features of PSi in oral (or other routes) drug delivery are not yet thouroughly understood – the carrier can improve not only the dissolution behavior of the drugs but also their permeation/absorption in the intestinal wall. PSi material is easy to produce in small scale, but regarding commercialization, the production of PSi materials in large amounts is still limited. Another issue is the safety of the material especially in case of nanoparticles, which has not yet been fully evaluated. Nonetheless, PSi-based materials have many interesting properties that are useful for biomedical applications in detecting (▶ biosensors), identifying, imaging, and therapeutically delivering to tissues, organs, and cells of interest (Santos et al. 2011; Godin et al. 2011).

Cross-References

▶ Porous Silicon Wafer-Based "Lab on Chip" Sensors
▶ Silica, Immunological Effects
▶ Silicateins
▶ Silicon Exposure and Vasculitis
▶ Silicon Nanowires
▶ Silicon Transporters
▶ Silicosis

References

Bimbo LM, Sarparanta M, Santos HA et al (2010) Biocompatibility of thermally hydrocarbonized porous silicon nanoparticles and their biodistribution in rats. ACS Nano 4:3023–3032

Bimbo LM, Mäkilä E, Raula J et al (2011) Functional hydrophobin-coating of thermally hydrocarbonized porous silicon microparticles. Biomaterials 32:9089–9099

Canham LT (ed) (1997) Properties of porous silicon. Short Run Press, London

Godin B, Tasciotti E, Liu X et al (2011) Multistage nanovectors: from concept to novel imaging contrast agents and therapeutics. Acc Chem Res 44:979–989

Heinrich JL, Curtis CL, Credo GM, Kavanagh KL, Sailor MJ (1992) Luminescent colloidal silicon suspensions from porous silicon. Science 255:66–68

Kilpeläinen M et al (2009) In vivo delivery of a peptide, ghrelin antagonist, with mesoporous silicon microparticles. J Control Release 37:166–170

Kovalainen M, Mönkäre J, Mäkilä E et al (2011) Mesoporous silicon (PSi) for sustained peptide delivery: effect of PSi microparticle surface chemistry on peptide YY_{3-36} release. Pharm Res 29:837–846

Lehto V-P, Salonen J, Santos HA, Riikonen J (2012) In: Douroumis D, Fahr A (eds) Nanostructured silicon based materials as a drug delivery system for insoluble drugs. Wiley-Blackwell, Weinheim

Sailor MJ (ed) (2011) Porous silicon in practice: preparation, characterization and applications. Wiley-VCH, Weinheim

Salonen J, Kaukonen AM, Hirvonen J, Lehto V-P (2008) Mesoporous silicon in drug delivery applications. J Pharm Sci 97:632–653

Santos HA, Riikonen J, Salonen J et al (2010) In vitro cytotoxicity of porous silicon microparticles: effect of the particle concentration, surface chemistry and size. Acta Biomater 6:2721–2731

Santos HA, Bimbo LM, Lehto V-P et al (2011) Multifunctional porous silicon for therapeutic drug delivery and imaging. Curr Drug Discov Techol 8:228–249

Sarparanta M, Mäkilä E, Heikkilä T et al (2011) ^{18}F-labeled modified porous silicon particles for investigation of drug delivery carrier distribution in vivo with positron emission tomography. Mol Pharmaceutics 8:1799–1806

Sarparanta MP, Bimbo LM, Mäkilä EM et al (2012) The mucoadhesive and gastroretentive properties of hydrophobin-coated porous silicon nanoparticle oral drug delivery systems. Biomaterials 33:3353–3362

Porous Silicon Wafer-Based "Lab on Chip" Sensors

Sabato D'Auria, Antonio Varriale, Giuseppe Ruggiero and Maria Staiano
National Research Council (CNR), Laboratory for Molecular Sensing, Institute of Protein Biochemistry, Naples, Italy

Definition

Sensing devices based on the utilization of a porous silicon wafer.

It is widely demonstrated that porous materials can be a very convenient choice as substrate for the design of biochips, "lab on chip," and bio/devices.

In fact, the inner surface of the porous materials can be exploited for the immobilization of large amount of proteins, allowing the fabrication of highly sensitive bio/devices.

Among porous materials, Porous Silicon (PS) has been the object of the greatest attention of the researchers, due to its important properties, such as photoluminescence, biocompatibility, and its easiness to be integrated in the microelectronic devices. Also, many patented biotechnologies involve the use of PS substrates.

Recently, an innovative technology was proposed which implied the integration of a Scanning Probe Microscopy (SPM) technique with a porous substrate (Borini et al. 2005, 2007; Rocchia et al. 2007; D'Auria et al. 2011).

In this case, a standard Electron Beam Lithography (EBL) system permitted a direct writing process free of organic contamination. Here, porous silicon was used as a substrate. This porous silicon was easily obtained by electrochemical etching of silicon in hydrofluoric acid solution. The method is based on the electron irradiation of fresh PS in a Scanning Electron Microscope (SEM). This procedure generated the Electron Beam Activated Porous Silicon (EBAPS) regions. The three steps involved in this nano-patterning process are summarized below:

The sample was irradiated with an electronic dose of 140 mC/cm2, and exposed to a rhodamine-labeled glucose-binding protein solution for 1 h at 37°C. After the immobilization, the protein retained its functionality. In fact, the immobilized protein was able to bind to a specific analyte, for example, glucose in the case of the glucose-binding protein, even when attached to the porous substrate. Importantly, the process can be serially repeated, so that different biomolecules can be patterned on the same chip, allowing the fabrication of protein-based lab-on-chips. The EBAPS regions can be defined with sub-micrometer resolution, due to the high resolution of the electron beam. Furthermore, by varying the electron beam energy, it is possible to control the dimension of the EBAPS patterns even in depth (in the z direction).

In conclusion, the aforementioned EBAPS technology is an example of how advantageous can be an approach based on the combining together different features, such as the high resolution of SPM and the large specific surface of a porous silicon wafer, for the invention of an effective nanofabrication method for the realization of a "lab on chip."

Cross-References

▶ Silica, Immunological Effects
▶ Silicon Nanowires
▶ Silicon Transporters

References

Borini S, D'Auria S, Rossi M, Rossi AM (2005) Writing 3D protein nanopatterns onto a silicon nanosponge. Lab Chip 5:1048–1052

Borini S, Staiano M, Rocchia M, Rossi AM, Rossi M, D' Auria S (2007) Advanced nanotechnological approaches for designing protein-based "lab-on-chips" sensors on porous silicon wafer. Recent Pat DNA Gene Seq 1:1–7

D' Auria S, Borini SM, Rossi AM, Rossi M (2011) Process of immobilizing biomolecules in porous supports by using an electronic beam. United States Patent N. US 7, 964, 386 B2

Rocchia M, Borini S, Rossi AM, Rossi M, D' Auria S (2007) Direct writing of a protein micro-array: lab-on-a-chip for multipurpose sensing. In: Ultrasensitive and single-molecule detection technologies II proceedings of SPIE, San Jose, CA, US vol 6444

Post-Electrophoretic Detection of Proteins

▶ Silver in Protein Detection Methods in Proteomics Research

Potassium

▶ Potassium in Health and Disease

Potassium (K^+)

▶ Magnesium and Vessels

Potassium Channel Diversity, Regulation of Potassium Flux across Pores

Ameer N. Thompson[1,2] and Crina M. Nimigean[1,2,3]
[1]Department of Anesthesiology, Weill Cornell Medical College, New York, NY, USA
[2]Department of Physiology and Biophysics, Weill Cornell Medical College, New York, NY, USA
[3]Department of Biochemistry, Weill Cornell Medical College, New York, NY, USA

Synonyms

Gating; Ion channel; Permeation; Selectivity

Definition

Potassium channels allow for the movement of potassium ions across the cell membrane. Access to the ion permeation pathway is regulated by a variety of protein modules appended to the K^+-selective pore.

A Channel for Every Season

While all K^+ channels show strong similarities to each other in the pore region responsible for selective K^+ permeation, other regions of the channels are variable as they are tailored to respond to the diverse stimuli that regulate K^+ access to the permeation pathway. The opening and closing (i.e., gating) of ion channels can be modulated in response to several types of stimuli such as changes in membrane voltage (voltage-gated K^+ channels), binding of molecules to a particular domain of the channel (ligand-gated K^+ channels), or membrane tension (stretch-activated channels). The ability to respond to disparate types of stimuli is reflected in the existence of specialized domains for the various K^+ channels. Channels whose gating is modulated by changes in the transmembrane electric field contain voltage-sensing domains appended to the K^+ channel pore. Likewise, channels that respond to ligands contain domains that specifically bind these ligands and bias the gating equilibrium. What follows is a brief description of several types of K^+ channels. While they vary in makeup, they all contain a canonical K^+-selective pore. Attached to the respective pores are protein modules that regulate the gating of the channel.

Voltage-Gated K^+ Channels

Voltage-gated K^+ (Kv) channels play a critical role in cellular excitability and are part of the superfamily of voltage-gated channels that also includes voltage-gated Na^+ and Ca^{2+} channels. Kv channels are widely expressed in excitable cells and are activated by membrane depolarization. The Kv channels involved in the action potential open after a large Na^+-mediated membrane depolarization, returning the transmembrane voltage to the resting potential. These channels control the duration and frequency of action potential firing. Each subunit of the tetrameric complex contains six transmembrane domains, S1 (at the amino-terminus) through S6 (carboxy-terminus), with the pore region between S5 and S6 (Fig. 1a). Both a prokaryotic

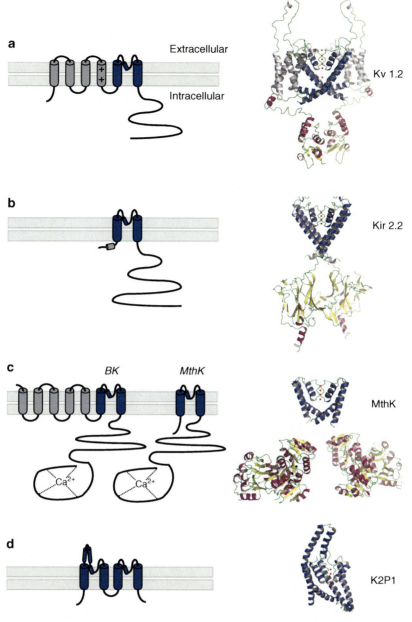

Potassium Channel Diversity, Regulation of Potassium Flux across Pores, Fig. 1 The K^+-selective pore is highly conserved. (a) Membrane topology of a typical voltage-gated K^+ channel. Each subunit is composed of six transmembrane domains, denoted S1–S6. The voltage-sensing domain consists of transmembrane domains S1–S4 (*gray*). The voltage sensor is attached to the K^+-selective pore domain (*blue*). Adjacent to the topology diagram is the structure of Kv 1.2 (pdb: 3LUT) (b) Membrane topology of inward rectifier K^+ channels. Like KcsA, Kir channel subunits are composed of two transmembrane domains. Adjacent to the topology diagram is the structure of Kir 2.2 (pdb: 3JYC) (c) Membrane topology of the Ca^{2+}-gated K^+ channels BK (eukaryotic) and MthK (prokaryotic). BK has seven transmembrane domains, S0–S6. In addition to the K^+-selective pore (*blue*) BK has an intracellular *N*-terminal calcium-binding (regulator of K^+ conductance or "RCK") domain that binds intracellular Ca^{2+}. MthK, like BK, also has a K^+-selective pore attached to soluble, Ca^{2+}-binding, intracellular RCK domains. However, MthK lacks transmembrane domains S0–S4. Adjacent to the topology diagram is the structure of MthK (a combination of pdb: 3LDC for the pore region and pdb:1LNQ for the RCK domains) (d) The membrane topology of the two-pore K^+ channel. One subunit of the channel has two P-loop domains. The crystal structure (pdb: 3UKM) reveals a helical cap between transmembrane helices one and two, which imparts a two-fold symmetry about the pore. Each structural picture shows only two of the four subunits for clarity

(KvAP (Jiang et al. 2003)) and a eukaryotic (Kv1.2 and Kv2.1/Kv1.2 chimera (Long et al. 2005, 2007)) channel were crystallized, with their structures determined to high resolution. Figure 1a shows the structure of Kv 1.2. The S4 segment (gray in Fig. 1a) is part of the voltage sensor, and it contains a series of positively charged arginine and sometimes lysine residues (Papazian et al. 1987; Aggarwal and MacKinnon 1996; Seoh et al. 1996; Long et al. 2005). The voltage-sensing domain is attached to the pore domain via a linker helix between S4 and S5 (Fig. 1a) that is believed to couple the conformational changes in the voltage sensor to the channel gate. Depolarization of the membrane causes a change in the voltage-sensing domain, which is propagated through the channel causing the gate to open.

The pore-forming region is quite similar in architecture, sequence, and structure to the pore domain of KcsA. The GYG-containing selectivity filter is conserved in this class of channels. However, unlike the KcsA x-ray crystal structure, which shows the channel in a closed (nonconducting) conformation, the Kv channels were captured with the S6 transmembrane segment in an open (presumably conducting) conformation. Eukaryotic Kv channels alone display a conserved triplet sequence (Pro-X-Pro, where X is any amino acid) at the innermost part of the S6 helix that was shown to play an important role in gating by allowing this part of the inner helix to curve outward when the channel opens (del Camino et al. 2000; Long et al. 2005).

Inward Rectifying K$^+$ Channels

In excitable cells, inward rectifying K$^+$ channels (Kir) channels serve several physiological functions such as maintenance of the resting membrane potential, control of action potential duration, and regulation of metabolism. They are expressed in a number of cells including blood cells, cardiac myocytes, endothelial cells, epithelial cells, glial cells, oocytes, and pancreatic β cells (Nichols and Lopatin 1997; Hibino et al. 2010). They are so named because they allow robust inward and very little outward K$^+$ current. Inward rectification is due to intracellular pore block by Mg^{2+} and polyamines, which bind in the ion permeation pathway when the membrane is depolarized, preventing K$^+$ from permeating the channel (Matsuda et al. 1987; Lopatin et al. 1994). Fifteen Kir genes have been identified (Hibino et al. 2010). These are classified into seven subfamilies; Kir 1.x–7.x. Both the structure of a prokaryotic (KirBac 1.1 (Kuo et al. 2003)) and a eukaryotic (Kir 2.2 (Tao et al. 2009), Fig. 1b) Kir channel verified the hypothesized structural similarity with KcsA; Kir channels are tetrameric, and each subunit is composed of two transmembrane helices (TM1 and TM2) containing the highly conserved GYG selectivity filter. Kir channels have a cytoplasmic domain that is primarily made up of carboxy-terminal residues from each of the four subunits (Kuo et al. 2003; Nishida et al. 2007). The cytoplasmic domain is believed to contain five permeant ion-binding sites (in addition to those in the canonical pore) and extends the ion permeation pathway by $\sim$30 Å (Xu et al. 2009). An amphiphilic helix (the "slide helix") was found at the cytoplasmic end of TM1, where it may interact with both the inner leaflet of the membrane as well as the cytoplasm (Fig. 1b). Kir channels are regulated by PIP2 (phosphatidylinositol 4,5 bisphosphate) anchored in the cell membrane, which is believed to associate with basic and/or neutral residues in the cytoplasmic domain. Opening of the Kir channel pore can also be controlled by the G-protein βγ subunit (Kir 3.x), phosphorylation, ATP (Kir 6.x), extracellular K$^+$, and protons (Hibino et al. 2010).

Ca^{2+}-Activated K$^+$ Channels

Calcium-activated K$^+$ (K$_{Ca}$) channels function to link changes in intracellular Ca^{2+} to the membrane potential. K$_{Ca}$ channels are also members of the S4 superfamily of voltage-gated ion channels. The large conductance Ca^{2+}-activated K$^+$ channels (BK, MaxiK, K$_{Ca}$1) are both depolarization and Ca^{2+}-activated (Fig. 1c). Ca^{2+} activation occurs via Ca^{2+} binding to multiple sites on the intracellular region of this channel and voltage activation occurs via an S4 voltage sensor domain similar to Kv channels (Cui et al. 1997; Shi et al. 2002; Xia et al. 2002). Conversely, the small conductance Ca^{2+}-activated K$^+$ (SK, K$_{Ca}$2, and 3) channels are not voltage gated (despite possessing an S4, voltage sensor-like domain) and Ca^{2+} activates them via binding to calmodulin, which is constitutively bound to the cytoplasmic portion of the channel (Xia et al. 1998). The BK channels (Fig. 1c) contain specialized Ca^{2+} binding "regulator of K$^+$ conductance" (RCK) domains appended to their pore that allow them to rapidly respond to changes in Ca^{2+}, leading to K$^+$ efflux and membrane hyperpolarization, thus dampening excitatory events. BK channels are further modulated by a family of single transmembrane-spanning β subunits that are believed to interact with S0, an N-terminal transmembrane helix particular to this channel (Fig. 1c) (Wallner et al. 1996). There is no high resolution structural information available for a eukaryotic BK (or SK) channel (with the exception of the cytoplasmic RCK-like domains (Wu et al. 2010; Yuan et al. 2010)), but the structure of a prokaryotic homolog of BK, the MthK channel was solved (Jiang et al. 2002) (Fig. 1c). MthK differs from its eukaryotic counterparts in that it has only two transmembrane domains, it lacks a voltage-sensing S4 domain, and the RCK domains assemble as an octameric complex of four intrinsic and four soluble RCK domains derived from the same sequence (Jiang et al. 2002). Functionally, MthK is different than BK and SK. Ca^{2+} activation occurs over a millimolar affinity range while all eukaryotic K_{Ca} channels are activated by micromolar Ca^{2+} concentrations (Jiang et al. 2002; Zadek and Nimigean 2006). Structurally, the selectivity filter of the MthK pore appears similar to that of KcsA. The intracellular part of the pore differs in that the inner helices (TM2) are curved outward to open this part of the pore to allow ions and blockers to enter (Jiang et al. 2002; Ye et al. 2010). The curvature occurs due to a so-called "glycine hinge" on the inner helix, a glycine conserved in most K^+ channels (Fig. 1d). It has been suggested that a similar curvature is achievable in the eukaryotic voltage-gated channels by a Pro-X-Pro motif (discussed above) (Long et al. 2005).

Two-Pore-Domain K⁺ Channels

Two-pore potassium channels (K2P) are responsible for the background leak K^+ current observed at resting membrane potentials. They are abundantly expressed in several locations including kidney, brain, lung, and heart (Lesage et al. 1996). In total, 15 K2P channels have been cloned, each belonging to one of six subfamilies: TWIK, TREK, TASK, TALK, THIK, and TRESK. Two-pore potassium channels, as their name suggests, are defined by their membrane topology. One K2P subunit is composed of two linked units, each containing two transmembrane domains (M1/M2 and M3/M4) with a signature sequence for K^+ in between (Lesage et al. 1996). As a result, K2P channels arrange as dimers instead of tetramers, thereby maintaining the same number of pore domains (4) as other K^+ channels. However, unlike other K^+ channels, K2P channels are two-fold symmetric. Each subunit has an extracellular "helical cap" composed of ~50 amino acids between transmembrane helices M1 and M2 (Miller and Long 2012; Brohawn et al. 2012). The helical cap is not present between M3 and M4. The function of this cap is unknown. The channels approach four-fold symmetry at the selectivity filter. Similar to prokaryotic voltage-gated sodium channels, the K2P structure reveals a large fenestration in the membrane plane, offering a possible clue as to how hydrophobic compounds (e.g., lipids, anesthetics, etc.) can interact with the channel and modulate its function. The open probability of these channels is largely voltage independent, though some channels display weak voltage dependence. K2P channels can be regulated by membrane tension, intracellular and extracellular pH, lipids, and serine phosphorylation (Lesage et al. 1996, 2000; Patel et al. 1998).

Conclusion

The ability to finely regulate K^+ flux through a cell membrane is critical to numerous physiological functions. Biology has developed a highly tuned mechanism to accomplish this task. There is a wide diversity of K^+ channels throughout cell types and across organisms, and they allow for rapid, selective, and highly controlled movement of K^+ across the cell membrane. The mandatory stringent K^+ selectivity and rapid flux is achieved via a highly conserved pore region. Control over the gate that regulates K^+ flux is accomplished via attaching to this pore one or several domains that are uniquely specialized for the particular channel function and the role it plays in the cell. Several decades of scientific exploration have yielded a multitude of insights into the inner workings of this biophysical ballet.

Cross-References

▶ Calcium Ion Selectivity in Biological Systems
▶ Cellular Electrolyte Metabolism
▶ Lithium, Physical and Chemical Properties
▶ Potassium Channels and Toxins, Interactions
▶ Potassium in Biological Systems
▶ Potassium in Health and Disease

- Potassium, Physical and Chemical Properties
- Potassium-Binding Site Types in Proteins
- Sodium, Physical and Chemical Properties
- Sodium-Binding Site Types in Proteins

References

Aggarwal SK, MacKinnon R (1996) Contribution of the S4 segment to gating charge in the Shaker K$^+$ channel. Neuron 16:1169–1177

Brohawn SG, del Marmol J, Mackinnon R (2012) Crystal structure of the human K2P TRAAK, a lipid-and mechano-sensitive K$^+$ ion channel. Science 335:436–441

Cui J, Cox DH, Aldrich RW (1997) Intrinsic voltage dependence and Ca^{2+} regulation of mslo large conductance Ca-activated K$^+$ channels. J Gen Physiol 109:647–673

del Camino D, Holmgren M, Liu Y, Yellen G (2000) Blocker protection in the pore of a voltage-gated K$^+$ channel and its structural implications. Nature 403:321–325

Hibino H, Inanobe A, Furutani K, Murakami S, Findlay I, Kurachi Y (2010) Inwardly rectifying potassium channels: their structure, function, and physiological roles. Physiol Rev 90:291–366

Jiang Y, Lee A, Chen J, Cadene M, Chait BT, MacKinnon R (2002) Crystal structure and mechanism of a calcium-gated potassium channel. Nature 417:515–522

Jiang Y, Lee A, Chen J, Ruta V, Cadene M, Chait BT, MacKinnon R (2003) X-ray structure of a voltage-dependent K$^+$ channel. Nature 423:33–41

Kuo A, Gulbis JM, Antcliff JF, Rahman T, Lowe ED, Zimmer J, Cuthbertson J, Ashcroft FM, Ezaki T, Doyle DA (2003) Crystal structure of the potassium channel KirBac1.1 in the closed state. Science 300:1922–1926

Lesage F, Guillemare E, Fink M, Duprat F, Lazdunski M, Romey G, Barhanin J (1996) TWIK-1, a ubiquitous human weakly inward rectifying K$^+$ channel with a novel structure. EMBO J 15:1004–1011

Lesage F, Terrenoire C, Romey G, Lazdunski M (2000) Human TREK2, a 2P domain mechano-sensitive K$^+$ channel with multiple regulations by polyunsaturated fatty acids, lysophospholipids, and Gs, Gi, and Gq protein-coupled receptors. J Biol Chem 275:28398–28405

Long SB, Campbell EB, Mackinnon R (2005) Crystal structure of a mammalian voltage-dependent Shaker family K$^+$ channel. Science 309:897–903

Long SB, Tao X, Campbell EB, MacKinnon R (2007) Atomic structure of a voltage-dependent K$^+$ channel in a lipid membrane-like environment. Nature 450:376–382

Lopatin AN, Makhina EN, Nichols CG (1994) Potassium channel block by cytoplasmic polyamines as the mechanism of intrinsic rectification. Nature 372:366–369. doi:10.1038/372366a0

Matsuda H, Saigusa A, Irisawa H (1987) Ohmic conductance through the inwardly rectifying K channel and blocking by internal Mg^{2+}. Nature 325:156–159. doi:10.1038/325156a0

Miller AN, Long SB (2012) Crystal structure of the human two-pore domain potassium channel K2P1. Science 335:432–436

Nichols CG, Lopatin AN (1997) Inward rectifier potassium channels. Annu Rev Physiol 59:171–191. doi:10.1146/annurev.physiol.59.1.171

Nishida M, Cadene M, Chait BT, MacKinnon R (2007) Crystal structure of a Kir3.1-prokaryotic Kir channel chimera. EMBO J 26:4005–4015

Papazian DM, Schwarz TL, Tempel BL, Jan YN, Jan LY (1987) Cloning of genomic and complementary DNA from Shaker, a putative potassium channel gene from Drosophila. Science 237:749–753

Patel AJ, Honore E, Maingret F, Lesage F, Fink M, Duprat F, Lazdunski M (1998) A mammalian two pore domain mechano-gated S-like K$^+$ channel. EMBO J 17:4283–4290. doi:10.1093/emboj/17.15.4283

Seoh SA, Sigg D, Papazian DM, Bezanilla F (1996) Voltage-sensing residues in the S2 and S4 segments of the Shaker K$^+$ channel. Neuron 16:1159–1167

Shi J, Krishnamoorthy G, Yang Y, Hu L, Chaturvedi N, Harilal D, Qin J, Cui J (2002) Mechanism of magnesium activation of calcium-activated potassium channels. Nature 418:876–880

Tao X, Avalos J, Chen J, MacKinnon R (2009) Crystal structure of the eukaryotic strong inward-rectifier K$^+$ channel Kir2.2 at 3.1 A resolution. Science 326:1668–1674

Wallner M, Meera P, Toro L (1996) Determinant for beta-subunit regulation in high-conductance voltage-activated and Ca($^{2+}$)-sensitive K$^+$ channels: an additional transmembrane region at the N terminus. Proc Natl Acad Sci USA 93:14922–14927

Wu Y, Yang Y, Ye S, Jiang Y (2010) Structure of the gating ring from the human large-conductance Ca($^{2+}$)-gated K($^+$) channel. Nature 466:393–397

Xia XM, Fakler B, Rivard A, Wayman G, Johnson-Pais T, Keen JE, Ishii T, Hirschberg B, Bond CT, Lutsenko S, Maylie J, Adelman JP (1998) Mechanism of calcium gating in small-conductance calcium-activated potassium channels. Nature 395:503–507

Xia XM, Zeng X, Lingle CJ (2002) Multiple regulatory sites in large-conductance calcium-activated potassium channels. Nature 418:880–884

Xu Y, Shin HG, Szep S, Lu Z (2009) Physical determinants of strong voltage sensitivity of K($^+$) channel block. Nat Struct Mol Biol 16:1252–1258

Ye S, Li Y, Jiang Y (2010) Novel insights into K$^+$ selectivity from high-resolution structures of an open K$^+$ channel pore. Nat Struct Mol Biol 17:1019–1023

Yuan P, Leonetti MD, Pico AR, Hsiung Y, MacKinnon R (2010) Structure of the human BK channel Ca^{2+}-activation apparatus at 3.0 A resolution. Science 329:182–186

Zadek B, Nimigean CM (2006) Calcium-dependent gating of MthK, a prokaryotic potassium channel. J Gen Physiol 127:673–685

Potassium Channels

- Barium Binding to EF-Hand Proteins and Potassium Channels

Potassium Channels and Toxins, Interactions

Vladimir Yarov-Yarovoy
Department of Physiology and Membrane Biology, Department of Biochemistry and Molecular Medicine, School of Medicine, University of California, Davis, CA, USA

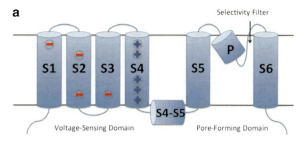

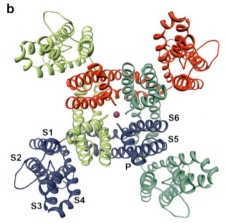

Synonyms

Ion channel structure; Peptide toxins; Voltage-gated ion channels

Definition

Voltage-gated potassium (Kv) channels are transmembrane proteins that open and close their ion-conducting pores in response to changes in membrane voltage, that is essential for repolarization of action potential in excitable cells (Hille 2001). Ion channels are the molecular targets of a variety of peptide neurotoxins from venom of certain types of spiders, snakes, scorpions, and snails. Neurotoxins targeting Kv channels paralyze the nervous system by inhibiting channel opening that causes extreme stimulation of synaptic transmission within the nervous system. This entry focuses on toxins that target the voltage-sensing domain of the voltage-gated potassium channels.

Potassium Channels and Toxins, Interactions, Fig. 1 *Structure of voltage-gated K^+ channels.* (**a**) Two-dimensional topology of Kv channels. Alpha helical transmembrane segments S1–S6 shown as *cylinders* and labeled. Positions of highly conserved positively and negatively charged residues within the voltage-sensing domain are indicated by "+" and "−" sign, respectively. (**b**) Three-dimensional structure of Kv1.2–Kv2.1 chimera channel (Long et al. 2007). View of the Kv1.2–Kv2.1 chimera channel structure from the extracellular side of the membrane. All four subunits are colored individually. Transmembrane segments S1–S6 and P-helix for the *blue*-colored subunit are labeled accordingly. Potassium ion is shown in the middle of the structure and colored *purple*

Introduction

Voltage-gated potassium (Kv) channels are transmembrane proteins that open and close in response to changes in membrane voltage – a process that is essential for repolarization of action potential in excitable cells (Hille 2001). Ion channels are the molecular targets of variety of peptide toxins from venom of spiders, snakes, scorpions, and cone snails. Toxins targeting Kv channels paralyze the nervous system by inhibiting channel opening and thereby cause extreme stimulation of synaptic transmission within the nervous system. High affinity of toxin binding to Kv channels is key for studying molecular mechanism of toxin action and for designing of therapeutic peptides and drugs targeting ion channels.

Structure of Kv Channels

Kv channels are formed by four identical subunits each containing six transmembrane alpha helical segments (S1–S6), with segments S1–S4 forming the voltage-sensing domain and segments S5–S6 forming the pore-forming domain (Fig. 1a). An alpha helix between the voltage-sensing and pore-forming domains (S4–S5 linker) plays key role in coupling of conformational changes in the voltage-sensing domain to pore-forming domain opening and closing (Fig. 1a). Short loop between P-helix and S6 segment forms the selectivity filter (Fig. 1a–b). Crystal structure of

Kv1.2–Kv2.1 chimera channel revealed that the voltage-sensing domain from one subunit interacts with the pore-forming domain from adjacent subunit in clockwise direction when viewed from the extracellular side of the membrane (Long et al. 2007) (Fig. 1b). The voltage-sensing domain is responsible for voltage sensing across the membrane, and the pore-forming domain is responsible for ion selectivity and pore gating.

The S4 segment in voltage-sensing domain contains several (usually four to five) highly conserved positively charged residues (mostly arginines) that sense changes in the transmembrane voltage within physiological range (from -70 to $+40$ mV) and carry the gating charge across the membrane (Figs. 1a, 2a). These positively charged residues are separated by two hydrophobic residues in S4 sequence and form 3–10 helix structures during gating (Long et al. 2007). During channel activation and deactivation, the gating charge carrying arginines interact with highly conserved negatively charged residues in S1, S2, and S3 segments (Figs. 1a, 2a).

Extensive alanine scan of the voltage-sensing domain of Kv channel suggested that three residues within the extracellular half of S3 (I273, F274, and E277) are important for interaction with Hanatoxin (Li-Smerin and Swartz 2000). Notably, all of these residues are oriented away from neighboring S4 segment and toward the hydrophobic lipid environment (Fig. 2b).

The extracellular part of the pore-forming domain forms the selectivity filter of Kv channel structure where backbone carbonyl oxygen atoms form octagonal shell of favorable interactions with potassium ions (Figs. 1a, 2c, d) (Zhou et al. 2001). At resting membrane potentials (between -60 and -70 mV), the pore-forming domain is closed by the intracellular gate formed by the S6 helices (Fig. 2c). At depolarized membrane potentials the S6 helices sway open to allow flow of potassium ions from the intracellular to the extracellular side of the membrane (Fig. 2d).

Structures of Toxins

The overall fold of toxins targeting the voltage-sensing domain of Kv channels is shown in Fig. 3a. Toxin structure is formed by a triple-stranded anti-parallel β-sheet and stabilized by absolutely conserved three pairs of disulfide bonds (Fig. 3a, b). Nuclear magnetic resonance (NMR) structures of several toxin structures targeting the voltage-sensing domain of Kv channels are shown in Fig. 4a–c. The toxin structure displays

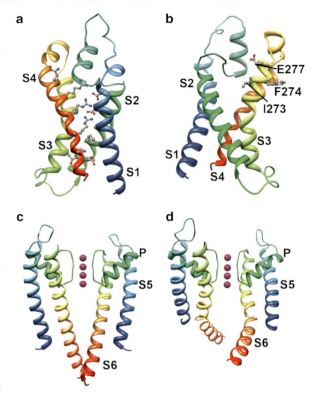

Potassium Channels and Toxins, Interactions, Fig. 2 *Structure of the voltage-sensing and pore-forming domains of Kv channels.* (**a**) Transmembrane view of a *ribbon* representation of a single subunit of the voltage-sensing domain of Kv1.2–Kv2.1 chimera channel structure (Long et al. 2007) with segments S1 and S4 on the front. Segments S1–S4 colored individually and labeled. Side chains of gating-charge-carrying arginines in S4 and negatively charged residues in S1–S3 segments are shown in *stick* representation. *Blue, red,* and *cyan* atoms in the side chains shown represent nitrogen, oxygen, and carbon atoms, respectively. (**b**) Transmembrane view of a *ribbon* representation of a single subunit of the voltage-sensing domain of Kv1.2–Kv2.1 chimera channel structure (Long et al. 2007) with segments S2 and S3 on the front. Side chains of residues important for binding of Hanatoxin are shown in *stick* representation and labeled. (**c**) Transmembrane view of a *ribbon* representation of the *closed state* of the pore-forming domain of KcsA channel (Zhou et al. 2001). Only two subunits are shown for clarity. Transmembrane segments S5–S6 and P-helix are colored individually and labeled accordingly. Potassium ions are shown within the selectivity filter and colored *purple.* (**d**) Transmembrane view of a *ribbon* representation of the *open state* of the pore-forming domain of Kv1.2 channel structure (Long et al. 2007). Only two subunits shown for clarity. Transmembrane segments S5–S6 and P-helix are colored individually and labeled accordingly. Potassium ions are shown within the selectivity filter and colored *purple*

amphipathic pattern with a number of large hydrophobic residues forming a hydrophobic patch on one side of the toxin and charged and polar residues forming the

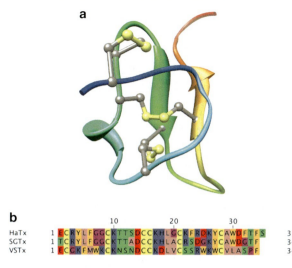

Potassium Channels and Toxins, Interactions, Fig. 3 *Structure and sequence of toxins targeting voltage-sensing domain of Kv channels.* (**a**) Hanatoxin structure of the first model in NMR ensemble (Takahashi et al. 2000). Backbone is colored by a rainbow color scheme – from N-terminus (*blue*) to C-terminus (*red*). Side chains of cysteines are shown in *stick* representation with sulfur atoms are colored *yellow* and carbon atoms are colored *gray*. Disulfide bridges (colored *yellow*) formed by conserved cysteines are shown in *stick* representation. (**b**) Multiple sequence alignment between Hanatoxin (HaTx), SGTx, and VSTx. Highly conserved cysteine residues are colored *yellow*

surface of the rest of the toxin structure (Fig. 4a–c). At least some of the hydrophobic residues on the toxin surface dip into the hydrophobic layer of the membrane where they can also interact with residues on the voltage-sensing domain of Kv channels (Phillips et al. 2005). Amino acid identities of residues in the hydrophobic patch vary between different toxins (Fig. 4a–c).

Molecular Mechanism of Toxin Action on Kv Channels

Studies of Hanatoxin inhibition of Kv channel activity using the electrophysiological recordings revealed that the toxin modifies channel gating through stabilization of one of the resting states of the voltage-sensing domain of the channel and consequently shifting channel activation to more depolarized potentials (Swartz and MacKinnon 1997; Swartz and MacKinnon 1997). In addition, Hanatoxin significantly decreases the maximum conductance of Kv channels at depolarized voltages that is

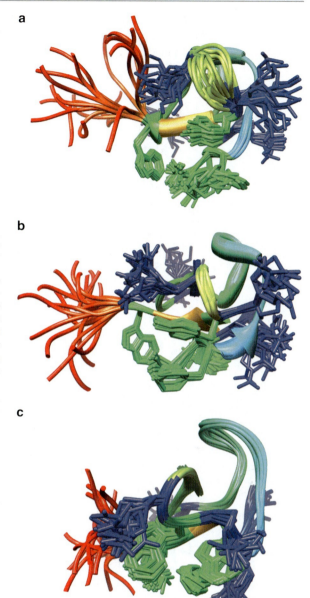

Potassium Channels and Toxins, Interactions, Fig. 4 *NMR structures of toxins targeting voltage-sensing domain of Kv channels.* (**a**) Hanatoxin from Chilean rose tarantula Grammostola spatulata (Takahashi et al. 2000). (**b**) SGTx toxin from the venom of the spider Scodragriseipes that inhibits Kv2.1 channel (Lee et al. 2004). (**c**) VSTx toxin from Chilean rose tarantulaGrammostolaspatulata that inhibits KvAP channel (Jung et al. 2005). Side chains of key residues are shown in *stick* representation. Backbone is colored by a rainbow color scheme – from N-terminus (*blue*) to C-terminus (*red*). Hydrophobic residues are colored *green* and positively charged residues are colored *blue*

attributed to an effect of the toxin on the concerted conformational change within the pore-forming domain that leads to channel opening (Swartz 2007). The toxin-bound channels can still open during channel gating if strong depolarization is applied to the membrane.

While there is no x-ray or NMR structure of a complex between a Kv channel and a toxin is available, extensive mutagenesis studies on SGTx1 toxin revealed that active surface of the toxin contains three hydrophobic residues within the hydrophobic patch on the toxin structure – L5, F6, and W30 (Wang et al. 2004) (Fig. 4b). Since two hydrophobic residues within the S3 segment of Kv2.1 channel (I273 and F274) are important for interaction with the toxin (Fig. 2b) (Li-Smerin and Swartz 2000), it is feasible that they are involved in hydrophobic interactions with the residues forming the hydrophobic patch on the surface of the toxin (Swartz 2007). Negatively charged residue at the end of S3 segment of Kv2.1 channel (E277) (Li-Smerin and Swartz 2000), that is also important for interaction with the toxin, may interact with one of the positively charged residues on the toxin surface (Figs. 2b, 4a, b). Thus, toxins targeting the voltage-sensing domain of Kv channels specifically evolved to bind to the extracellular half of the S3 segment in order to affect the movement of the adjacent S4 segment during channel gating and stabilize it in the resting, intermediate, or activated states. In addition to the protein-protein interactions between the toxin and the voltage-sensing domain of Kv channels, toxin interaction with membrane lipids have also been proposed based on experiments demonstrating membrane partitioning of Hanatoxin (Phillips et al. 2005). Therefore, toxin–lipids interactions may also contribute to the molecular mechanism of toxin action on the Kv channels. Position of the toxin at the interface between the hydrophobic and polar layers of the membrane would allow for the hydrophobic toxin residues to dip into the hydrophobic layer of the membrane and for the positively charged toxin residues to interact with the lipid phosphate groups.

Cross-References

▶ Potassium Channels, Structure and Function
▶ Sodium Channel Blockers and Activators
▶ Sodium Channels, Voltage-Gated

References

Hille B (2001) Ion channels of excitable membranes. Sinauer Associates, Sunderland
Jung HJ, Lee JY, Kim SH et al (2005) Solution structure and lipid membrane partitioning of VSTx1, an inhibitor of the KvAP potassium channel. Biochemistry 44:6015–6023
Lee CW, Kim S, Roh SH et al (2004) Solution structure and functional characterization of SGTx1, a modifier of Kv2.1 channel gating. Biochemistry 43:890–897
Li-Smerin Y, Swartz KJ (2000) Localization and molecular determinants of the Hanatoxin receptors on the voltage-sensing domains of a K(+) channel. J Gen Physiol 115: 673–684
Long SB, Tao X, Campbell EB et al (2007) Atomic structure of a voltage-dependent K + channel in a lipid membrane-like environment. Nature 450:376–382
Phillips LR, Milescu M, Li-Smerin Y et al (2005) Voltage-sensor activation with a tarantula toxin as cargo. Nature 436:857–860
Swartz KJ (2007) Tarantula toxins interacting with voltage sensors in potassium channels. Toxicon 49:213–230
Swartz KJ, MacKinnon R (1997) Hanatoxin modifies the gating of a voltage-dependent K + channel through multiple binding sites. Neuron 18:665–673
Takahashi H, Kim JI, Min HJ et al (2000) Solution structure of hanatoxin1, a gating modifier of voltage-dependent K(+) channels: common surface features of gating modifier toxins. J Mol Biol 297:771–780
Wang JM, Roh SH, Kim S et al (2004) Molecular surface of tarantula toxins interacting with voltage sensors in K(v) channels. J Gen Physiol 123:455–467
Zhou Y, Morais-Cabral JH, Kaufman A et al (2001) Chemistry of ion coordination and hydration revealed by a K + channel-Fab complex at 2.0 A resolution. Nature 414:43–48

Potassium Channels, Structure and Function

Ameer N. Thompson[1,2] and Crina M. Nimigean[1,2,3]
[1]Department of Anesthesiology, Weill Cornell Medical College, New York, NY, USA
[2]Department of Physiology and Biophysics, Weill Cornell Medical College, New York, NY, USA
[3]Department of Biochemistry, Weill Cornell Medical College, New York, NY, USA

Synonyms

Gating; Ion channel; Permeation; Selectivity

Definition

Potassium channels are integral membrane proteins that allow for selective passage of K^+ ions across the cell membrane while excluding the passage of other similar ions.

Introduction

Ion channels are proteins that allow ions to move across cell membranes, forming the basis of electrical-osmotic balance with the environment as well as electrical signaling (Hille 2001). Their proper function is of such importance that mutations in ion channel genes underlie multiple diseases including cardiac arrhythmias, muscular disorders, certain types of diabetes and epilepsy (Ashcroft 2006). Ion channels are grouped into families depending on the type of ions they pass (e.g., K^+, Na^+, Ca^{2+}) or the type of environmental variable that drives their gating (opening and closing of the channel pore) mechanisms. Ion channels display a high degree of diversity and specialization of function (Hille 2001).

Potassium (K^+) channels, integral membrane proteins that allow K^+ to flow down its electrochemical gradient, can be found in a multitude of cell types across all kingdoms of life. In excitable cells, the ability of these channels to allow selective passage of K^+ is crucial toward proper membrane repolarization after an action potential. In order to fulfill their physiological roles, K^+ channels must allow for K^+ (ionic radius = 1.33 Å) to cross the membrane at near diffusion-limited rates of up to 10^6–10^8 ions/s while contemporaneously excluding the smaller and abundant Na^+ ions (ionic radius = 0.95 Å). This leaves us with two interesting problems: (1) how do these channels allow for the selective passage of K^+ at the exclusion of other remarkably similar ions and (2) how does the channel regulate the flow of K^+ across the pore? Several decades of vigorous scientific inquiry have led to substantial advances in this field. Multidentate approaches incorporating biochemical assays, biophysical measurements, structural biology, and theory, such as molecular dynamics simulations, have contributed to our knowledge about the molecular workings of these essential proteins.

Architecture of the K^+ Channel Pore

Our first structural snapshots of the K^+ channel pore came from KcsA, a prokaryotic potassium channel from the Gram-positive actinomycete *Streptomyces lividans* (Schrempf et al. 1995; Doyle et al. 1998). Prior to the elucidation of the first K^+ channel structure, forays into the mechanism of selectivity and gating in K^+ channels with electrophysiological methods yielded a cartoon structural model consistent with the functional data (Fig. 1a). This consensus model contained elements that were confirmed and refined by the KcsA structure (Doyle et al. 1998). The channel is a tetrameric protein arranged around one central pore so that each subunit contributes to the pore formation. The pore has two main sections: a narrow selectivity filter toward the extracellular side (composed mainly of backbone carbonyl oxygens) that can accommodate dehydrated K^+, and a large hydrophobic cavity toward the intracellular side, capable of accommodating hydrated cations and large organic cation blockers (Armstrong and Hille 1972; Bezanilla and Armstrong 1972; Holmgren et al. 1997). The pore contains 5 K^+-binding sites (four in the selectivity filter and one in the cavity) arranged in single file, with at least two sites able to be occupied simultaneously (Hille and Schwarz 1978; Neyton and Miller 1988; Zhou et al. 2001).

KcsA is a minimal K^+ channel where each subunit is composed of only two transmembrane helices (TM1 and TM2), similar in architecture with the inward rectifier K^+ channels (Fig. 1d) (Schrempf et al. 1995; Doyle et al. 1998). However, the amino acid sequence is more homologous with the TM5 and TM6 (often termed the S5 and S6 helices) of voltage-gated K^+ channels (Fig. 1d). Consequently, KcsA seems to represent a valid structural model for the pore-forming regions of potentially the entire, diverse family of K^+ selective channels (Fig. 1b). At the N-terminus, TM1 initiates in the cytoplasm and traverses the membrane as an alpha helix. At the extracellular side of the membrane, the protein continues as a loop (the "turret"). The loop then reenters the membrane as an alpha helix tilted ~45° to the membrane normal, forming the pore helix. The negative charge of the pore helix dipole is oriented toward the center of the membrane, and was proposed to attract and stabilize cations in the large intracellular cavities of some K^+ channels (Roux 1995; Doyle et al. 1998). The P-loop makes a sharp turn starting at the end

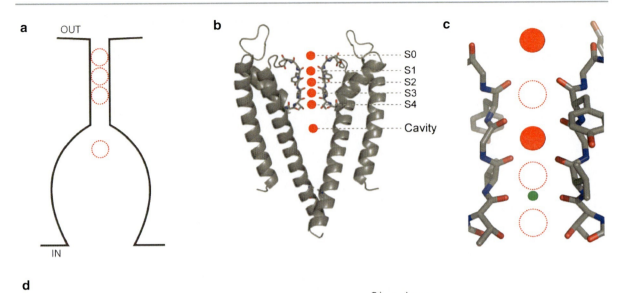

Potassium Channels, Structure and Function, Fig. 1 The K⁺ selective pore is the basic functional unit of K⁺ channels. (**a**) A cartoon model of the prestructure K⁺ channel depicting a narrow pore containing 4 (*red dotted circles*) ion-binding sites in the ion permeation pathway. (**b**) Two opposing subunits from the tetrameric KcsA K⁺ channel structure from *S. lividans* (PDB: 1k4c). Consistent with the prestructure experimental predictions (shown in **a**), the K⁺ channel consists of a narrow pore with several ion-binding sites (S0–S4 and the cavity site) in the ion permeation pathway. (**c**) The K⁺ channel selectivity filter is made of the highly conserved GYG amino acid sequence. Backbone carbonyls from the residues mimic the waters of hydration, forming cage-like binding sites for K⁺ ions (*red solid* and *dotted circles*). Small monovalent cations such as Na⁺ or Li⁺ (*green*) are suggested to bind in the plane of carbonyl atoms between two adjacent K⁺-binding sites. The *filled circles* (*green* and *red*) are supposed to indicate a potential, albeit infrequent, ionic distribution in the K⁺ channel filter when both K⁺ and Na⁺ ions are present. (**d**) Sequence alignment of representative K⁺ channels. The sequence for Kir2.2 was taken from Protein Data Bank entry 3JYC. All other sequences were taken from Uniprot Knowledgebase (KcsA-P0A334, MthK-O27564, shaker-P08510, Kv1.2-P16389, Kca1.1-Q12791, and TWIK-1-Q0IHD4). Regions in *blue* correspond to the S5, S6, and pore helices. The signature sequence is highlighted *yellow*

of the pore helix and then goes normal to the membrane toward the extracellular side. This loop contains the highly conserved GYG signature sequence for potassium selectivity. TM2 initiates on the extracellular leaflet and traverses the membrane to the cytoplasm. TM2 helices from all four subunits line the inner part of the pore below the signature sequence loop. The bottom two thirds of the channel, from the middle of the membrane and extending to the intracellular leaflet, form a large water filled cavity capable of accepting hydrated cations and large organic cations (Bezanilla and Armstrong 1972; Zhou and MacKinnon 2004). As a consequence, a K⁺ permeating the channel from the intracellular milieu can traverse a substantial portion of the membrane fully hydrated.

GYG Selectivity Filter

The structural feature responsible for the channels ability to discriminate between ions is the selectivity filter, a narrow passageway located in the top third of the protein toward the extracellular leaflet of the membrane (Heginbotham et al. 1992; Doyle et al. 1998). The filter contains the highly conserved GYG amino acid sequence (Heginbotham et al. 1992, 1994).

The pore is lined not with the side chains of these amino acids, but with the backbone carbonyl groups that project their oxygen atoms into the ion permeation pathway, forming four evenly spaced K^+ cage-binding sites, denoted S1–S4 (Zhou and MacKinnon 2003) (Fig. 1c). In order to enter the narrow selectivity filter and traverse the channel, a hydrated K^+ must shed its waters of hydration, a costly energetic process. It has been proposed that the carbonyl groups in the selectivity filter mimic the waters of hydration by matching the bulk number of coordinating ligands and coordination distance of a K^+ in solution (Doyle et al. 1998; Zhou et al. 2001). In this way, the dehydration costs are balanced by the gain of pseudo waters from the cage-binding site carbonyls, thus making the transition from hydrated ion in solution to dehydrated ion in the selectivity filter an isoenergetic step (Neyton and Miller 1988; Doyle et al. 1998; Berneche and Roux 2001; Morais-Cabral et al. 2001).

One view of selectivity posits that Na^+ is not privy to the same thermodynamically favorable binding in the filter (Neyton and Miller 1988; Zhou et al. 2001; Noskov et al. 2004; Noskov and Roux 2006; 2007). A Na^+ attempting to traverse the channel is also subject to a large energetic penalty imposed by the need to dehydrate prior to accessing the filter. It was proposed that the carbonyl oxygens forming the cage K^+-binding sites cannot move in far enough to accommodate the smaller Na^+, a conformation change thought to be required in order to reconcile the difference in coordination distance between K^+ and Na^+ (Doyle et al. 1998; Noskov et al. 2004). Therefore, Na^+ cannot readily bind in the canonical cage-like selectivity filter-binding sites. It followed then that the lack of thermodynamically favorable binding sites inside the filter drives selectivity against Na^+ in K^+ channels.

Selectivity has largely been thought of as a thermodynamically driven process. Given that K^+ channels are multi-ion pores, the possibility that there are additional factors that contribute to selectivity was also considered (Thompson et al. 2009). Recent evidence suggested that there are in fact favorable binding sites in the selectivity filter of K^+ channels for small monovalent cations like Na^+ and Li^+. These sites are not in the canonical cage-like binding sites for K^+ (S1-S4, see Fig. 1c). Rather, they are in the plane of carbonyls from two adjacent cage sites (Fig. 1c). However, it is clear that Na^+ is excluded from its binding site in the pore under physiological conditions; efflux of K^+ through K^+ channels repolarizes the cell membrane after an action potential despite a large inward driving force for Na^+ from the extracellular side. Evidence suggests that in order for Na^+ to access its binding in the plane of the carbonyls, a rearrangement of the K^+ ions in the filter needs to occur as both cage K^+ sites flanking the Na^+ plane site need to be empty for Na^+ to occupy the site. This movement of K^+ ions was calculated to be associated with a large free energy barrier (Berneche and Roux 2001; Thompson et al. 2009).

If K^+ binds with relatively high affinity in the selectivity filter, how are the high conduction rates achieved? Out of the four crystallographically identified K^+-binding sites in the filter, it was suggested that at physiological ionic concentrations, only two nonconsecutive binding sites (either S1 and S3 or S2 and S4) are typically occupied by K^+ at the same time. A third ion entering the filter lowers the free energy barriers for the existing two filter ions via electrostatic repulsion and they move up one register. In the process, the extreme most ion exits the pore at the opposite side of the filter. This knock-on kinetics, first described by Hodgkin and Keynes, is believed to be responsible for the high rate of flux through K^+ channels (Hodgkin and Keynes 1955; Morais-Cabral et al. 2001).

Conclusion

The pore domain, which contains a large aqueous pore and the selectivity filter, is the basic functional unit of the K^+ channel. K^+ channels are defined by their ability to allow for the selective passage of K^+ ions and to exclude other remarkably similar ions. This feat is achieved through the highly conserved GYG selectivity sequence. Several decades of research have revealed the thermodynamic and kinetic basis for this critical physiological process.

Acknowledgments We thank J. McCoy for help with the figures and D. Posson for critically reading the manuscript.

Cross-References

▶ Calcium Ion Selectivity in Biological Systems
▶ Cellular Electrolyte Metabolism
▶ Lithium, Physical and Chemical Properties

▶ Potassium Channels and Toxins, Interactions
▶ Potassium in Biological Systems
▶ Potassium in Health and Disease
▶ Potassium, Physical and Chemical Properties
▶ Potassium-Binding Site Types in Proteins
▶ Sodium, Physical and Chemical Properties
▶ Sodium-Binding Site Types in Proteins

References

Armstrong CM, Hille B (1972) The inner quaternary ammonium ion receptor in potassium channels of the node of Ranvier. J Gen Physiol 59:388–400

Ashcroft FM (2006) From molecule to malady. Nature 440:440–447

Berneche S, Roux B (2001) Energetics of ion conduction through the K^+ channel. Nature 414:73–77

Bezanilla F, Armstrong CM (1972) Negative conductance caused by entry of sodium and cesium ions into the potassium channels of squid axons. J Gen Physiol 60:588–608

Doyle D, Morais Cabral J, Pfuetzner R, Kuo A, Gulbis J, Cohen S, Chait B, MacKinnon R (1998) The structure of the potassium channel: molecular basis of K^+ conduction and selectivity. Science 280:69–77

Heginbotham L, Abramson T, MacKinnon R (1992) A functional connection between the pores of distantly related ion channels as revealed by mutant K^+ channels. Science 258:1152–1155

Heginbotham L, Lu Z, Abramson T, MacKinnon R (1994) Mutations in the K^+ channel signature sequence. Biophys J 66:1061–1067

Hille B (2001) Ion channels of excitable membranes, 3rd edn. Sinauer Associates, Inc, Sunderland

Hille B, Schwarz W (1978) Potassium channels as multi-ion single-file pores. J Gen Physiol 72:409–442

Hodgkin AL, Keynes RD (1955) The potassium permeability of a giant nerve fibre. J Physiol 128:61–88

Holmgren M, Smith PL, Yellen G (1997) Trapping of organic blockers by closing of voltage-dependent K^+ channels: evidence for a trap door mechanism of activation gating. J Gen Physiol 109:527–535

Morais-Cabral JH, Zhou Y, MacKinnon R (2001) Energetic optimization of ion conduction rate by the K^+ selectivity filter. Nature 414:37–42

Neyton J, Miller C (1988) Discrete Ba^{2+} block as a probe of ion occupancy and pore structure in the high-conductance $Ca2^+$ - activated K^+ channel. J Gen Physiol 92:569–586

Noskov SY, Roux B (2006) Ion selectivity in potassium channels. Biophys Chem 124:279–291

Noskov SY, Roux B (2007) Importance of hydration and dynamics on the selectivity of the KcsA and NaK channels. J Gen Physiol 129:135–143

Noskov SY, Berneche S, Roux B (2004) Control of ion selectivity in potassium channels by electrostatic and dynamic properties of carbonyl ligands. Nature 431:830–834

Roux B (1995) The calculation of the potential of mean force using computer simulations. Comput Phys Commun 91:275–282

Schrempf H, Schmidt O, Kummerlen R, Hinnah S, Muller D, Betzler M, Steinkamp T, Wagner R (1995) A prokaryotic potassium ion channel with two predicted transmembrane segments from Streptomyces lividans. EMBO J 14:5170–5178

Thompson AN, Kim I, Panosian TD, Iverson TM, Allen TW, Nimigean CM (2009) Mechanism of potassium-channel selectivity revealed by Na(+) and Li(+) binding sites within the KcsA pore. Nat Struct Mol Biol 16:1317–1324

Zhou Y, MacKinnon R (2003) The occupancy of ions in the K^+ selectivity filter: charge balance and coupling of ion binding to a protein conformational change underlie high conduction rates. J Mol Biol 333:965–975

Zhou Y, MacKinnon R (2004) Ion binding affinity in the cavity of the KcsA potassium channel. Biochemistry 43:4978–4982

Zhou Y, Morais-cabral J, Kaufman A, Mackinnon R (2001) Chemistry of ion coordination and hydration revealed by a K^+ channel-Fab complex at 2. 0 A resolution. Nature 414:43–48

Potassium Dependent Sodium/Calcium Exchangers

Paul P. M. Schnetkamp
Department of Physiology & Pharmacology, Hotchkiss Brain Institute, University of Calgary, Calgary, AB, Canada

Synonyms

K^+-dependent Na^+/Ca^{2+} exchanger; Na^+/Ca^{2+}-K^+ exchanger; NCKX; *SLC24* gene family

Definition

The *SLC24* gene family consists of five distinct genes encoding the NCKX1-5 Na^+/Ca^{2+}-K^+ exchangers. NCKX proteins are found throughout the animal kingdom and are important in Ca^{2+} extrusion in tissues that utilize fast Ca^{2+} signaling, e.g., many neurons, photoreceptors, and smooth muscle. NCKX5 is important for pigmentation of the human skin, and people of European descent carry a NCKX5 variant that is responsible for their fair skin.

Background

The most prevalent cellular signaling mechanism is mediated by a transient increase in the cystosolic free Ca^{2+} concentration. This increase is brought about by Ca^{2+} influx into the cytosol via a variety of different Ca^{2+} permeable ion channels located in the plasma membrane of cells or in intracellular organelles. Restoring Ca^{2+} to resting levels of 50–100 nM is accomplished by the combined action of ATP-driven Ca^{2+} pumps located in either the plasma membrane (PMCA) or endo(sarco)plasmic reticulum (SERCA) and Na^+-coupled Ca^{2+} transporters located in the plasma membrane.

Na^+-coupled Ca^{2+} transport or Na^+/Ca^{2+} exchange was first described in the early 1960s and has been extensively studied in various systems, including cardiac sarcolemma vesicles, squid giant axon, and the outer segments of retinal rod photoreceptors (ROS). In the late 1980s, it was discovered that Na^+/Ca^{2+} exchange in ROS differed from that found in most other tissues in that, in ROS, Na^+/Ca^{2+} exchange required the presence of both Ca^{2+} and K^+ (for a general review on Na^+/Ca^{2+} exchange including its history see Blaustein and Lederer (1999)). It is now known that the ROS Na^+/Ca^{2+} exchanger not only requires but also transports K^+ and was the first known member of the *SLC24* gene family of Na^+/Ca^{2+}-K^+ exchangers or NCKX proteins. In contrast, the Na^+/Ca^{2+} exchangers belong to the *SLC8* gene family of NCX proteins. Here, the current knowledge of the five NCKX1-5 proteins encoded by the *SLC24a1-5* genes will be reviewed.

In Situ Characteristics of Na^+/Ca^{2+}-K^+ Exchangers

The vast majority of detailed in situ studies of NCKX function have been carried out in retinal rod outer segments (ROS) since isolated ROS preparations are readily obtained and are also easily accessible for either biochemical or electrophysiological measurements (reviewed in Schnetkamp (1989)). Moreover, the NCXK1 protein is present in high concentration and is the only functional ion transporter in the plasma membrane of isolated ROS greatly simplifying its study. The most salient feature of NCKX1 is that it carries out obligatory coupled exchange of four Na^+ ions for one Ca^{2+} plus one K^+ ion, in contrast to the coupled transport of three Na^+ ions against one Ca^{2+} ion mediated by the NCX Na^+/Ca^{2+} exchangers. The second salient feature of NCKX is that the direction of transport is dictated by the direction of the ion gradients; i.e., Ca^{2+} transport can be both forward (efflux) or reverse (influx). This has the important physiological consequence that when the transmembrane Na^+ gradient is compromised, the direction of NCKX-mediated Ca^{2+} transport switches from efflux to influx. Normally, the large transmembrane Na^+ gradient drives Ca^{2+} extrusion from the cells. However, when external Na^+ is removed during experimental manipulations, the Na^+ gradient is now reversed (higher inside cells), and NCKX readily mediates a large Ca^{2+} influx even at micromolar external Ca^{2+} concentrations (see below).

In addition to Na^+-coupled Ca^{2+} and K^+ extrusion, NCKX1 in ROS also mediates self-exchange fluxes of intracellular Ca^{2+} + K^+ with extracellular Ca^{2+} + K^+ or intracellular Na^+ with extracellular Na^+ at rates that are comparable to Na^+-coupled Ca^{2+} and K^+ transport. SLC transporters like NCX and NCKX are generally thought to operate via a consecutive or alternate access mechanism, i.e., a single set of binding sites can accommodate either four Na^+ or one Ca^{2+} plus one K^+ and access to these binding sites switches from outward facing to inward facing and vice versa during the transport cycle. The observation of self-exchange fluxes is consistent with such a mechanism. Also consistent with an alternate access mechanism is that both Na^+ and Ca^{2+} were shown to compete for a common binding site on the NCKX protein.

The cation selectivity for the two different configurations of the NCKX cation binding sites has been determined as well: (1) Ca^{2+} can be replaced with Sr^{2+} in transport whereas other divalent cations compete with Ca^{2+} for binding but are not transported; (2) NH_4^+ and Rb^+, and to a much lesser extent Cs^+, are the only monovalent cations that can replace K^+ for transport; and (3) the selectivity for binding of multiple Na^+ ions is absolute as no other monovalent cation can replace Na^+, not even Li^+. As Ca^{2+}, Na^+ and K^+ share a common set of binding sites, both Ca^{2+} and K^+ concentrations affect the observed Na^+ dependency of Ca^{2+} transport and vice versa. K_m values in the absence of competing cations were found to be roughly symmetrical, i.e., the same values were observed for both internal and external K_m: (1) a sigmoidal dependence on $[Na^+]$ was observed with a Hill coefficient of

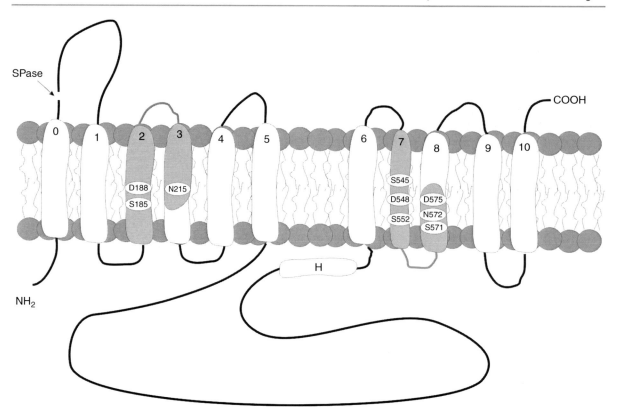

Potassium Dependent Sodium/Calcium Exchangers, Fig. 1 Current topological model of NCKX proteins. The orientation of the NCKX proteins embedded in a phospholipid bilayer membrane is shown. Hydrophobic segments of ~20 residues and capable of traversing the membrane in the form of α helices (TMS) are shown as *bars*, whereas hydrophilic stretches of proteins are shown as *solid lines*. TMS0 is a signal peptide that can be cleaved by a signal peptidase (SPase). The *bars* labeled *1–5* and *6–10* are thought to be sets of five transmembrane alpha helices each, while the *bar* labeled "*H*" is thought to be a hydrophobic segment near the cytosolic surface. The shaded segments represent the so-called alpha repeats as discussed in the text. Highlighted are also some of the residues shown to be important for cation binding and cation transport

2–3 and a K_m of 35 mM, and (2) simple hyperbolic binding curves (suggesting binding of a single Ca^{2+} and K^+) were obtained for both Ca^{2+} and K^+ with K_m values of 1–3 μM and 1–5 mM, respectively.

The SLC24 Gene Family

NCKX1 has been extensively characterized in and purified from retinal ROS and the first NCKX1 cDNA was cloned from bovine retina. Subsequently, four other NCKX genes have been cloned from mammalian species while other NCKX genes have been identified throughout the animal kingdom. NCKX1 transcripts are found in retinal rod photoreceptors, NCKX2 transcripts are found in retinal cone photoreceptors, retinal ganglion cells and throughout the brain, NCKX3 and 4 transcripts are more wide spread in brain as well as various smooth muscle tissues, while NCKX5 transcripts are found in pigmented cells such as melanocytes and retinal pigment epithelium. In addition, multiple splice variants for the various NCKX transcripts have been described although their physiological significance remains to be established. Hydropathy analysis reveals that all NCKX contain 12 hydrophobic segments of ~20 residues that could represent α helices constituting transmembrane segments (TMS) (Fig. 1). The first hydrophobic segment is located at the N-terminus and is thought to be a cleaved signal peptide or an uncleaved signal anchor. This segment is followed by a hydrophilic loop located in the extracellular space. Next are five hydrophobic segments and putative TMS with very short connecting loops followed by a large hydrophilic loop, thought to be located in the cytosol. The C-terminal part consists of six hydrophobic segments with short connecting loops and a very short hydrophilic loop at the C-terminus. Sequence identity between all five NCKX isoforms is limited to the 11 hydrophobic segments excluding the one at the N-terminus. The human NCKX2-5 range between 500 and 661 residues with most of the difference accounted for by differences in size of the large hydrophilic loop located in the cytosol. In contrast, human and other mammalian NCKX1 are considerably larger with ~1,200 residues. The large size of NCKX1 is only seen with the mammalian isoform and is accounted for by extensions of the two large hydrophilic loops. Interestingly, those extensions, in particular that of the large cytosolic loop, show relatively poor sequence homology, even among related mammalian species. The significance of this observation remains to be established.

In the current topological model of NCKX proteins, the aforementioned 11 hydrophobic segments are arranged in two sets of five TMS (thought to be α helices) (1–5 and 6–10) and one hydrophobic segment (H) not traversing the membrane and located in the cytosol (Fig. 1). The highest degree of sequence identity is found in two segments known as the alpha repeats (shaded bars in Fig. 1), thought to have arisen from an ancient gene duplication event. The alpha repeats are the only areas of (very limited) sequence conservation between members of the *SLC8* and *SLC24* gene families and contain many residues important for cation binding and cation transport (Fig. 1). For an earlier review on the *SLC24* gene family, see Schnetkamp (2004) and its bibliography.

Physiological Role of NCKX Proteins

The difference in stoichiometry observed for the NCX and NCKX proteins, respectively, implies that NCKX proteins can maintain much larger transmembrane Ca^{2+} gradients. Also, a much greater membrane depolarization and/or reduction of the transmembrane Na^+ gradient is required to reverse the action of NCKX from Ca^{2+} extrusion to Ca^{2+} influx. The physiological role of NCKX has been most clearly defined in retinal rod and cone photoreceptors. In darkness, a large and sustained inward current via cGMP-gated channels is carried by Na^+ (80–90%) and Ca^{2+} (10–20%). This causes a sustained membrane depolarization and sustained Ca^{2+} influx and results in elevated cytosolic free Ca^{2+} concentrations, or $[Ca^{2+}]$. Exposure to light initiates an enzymatic cascade, leading to hydrolysis of the excitatory messenger cGMP and a graded closure of cGMP-gated channels, resulting in membrane hyperpolarization and a drop in cytosolic $[Ca^{2+}]$. The drop in cytosolic $[Ca^{2+}]$ leads to an increase in cGMP synthesis and, as a result, leads to reopening of cGMP-gated channels. Ca^{2+} regulation of guanylyl cyclase (the enzyme responsible for cGMP synthesis) is mediated by Ca^{2+}-binding proteins of the GCAP family and is thought to be responsible, at least in part, for the process of light adaptation, i.e., the change in absolute light sensitivity caused by different levels of ambient background light (e.g., a bright sunny day outdoors versus a candlelight dinner after sunset).

The physiology of the NCKX2-4 isoforms is much less well understood. NCKX2 is widely expressed in the brain and may contribute to either presynaptic or postsynaptic Ca^{2+} clearance or both, in order to restore the resting Ca^{2+} gradient following events of Ca^{2+} influx via voltage-gated Ca^{2+} channels or neurotransmitter-gated channels, respectively. Elimination of NCKX2 in the mouse resulted in a loss of hippocampal plasticity and a loss of motor learning and memory (Li et al. 2006). However, neurons cultured from the knock-out mice showed both NCX-mediated Ca^{2+} fluxes and residual K^+-dependent Na^+/Ca^{2+} exchange suggesting that these neurons may coexpress NCX, NCKX2 as well as other NCKX isoforms, i.e., NCKX3 or 4. In the absence of specific inhibitors of NCKX, it may be difficult to unravel the specific contribution of these various NCX and NCKX proteins to Ca^{2+} homeostasis in neurons.

Expression of NCKX1-4 cDNAs in cell lines results in the functional appearance of Na^+/Ca^{2+}-K^+ exchange in the plasma membrane, suggesting NCKX1-4 normally function as plasma membrane proteins. For a more detailed survey of NCKX physiology and a full bibliography, see Altimimi and Schnetkamp (2007), Lytton (2007). Various genetic studies (search Pubmed for SLC24a*) have suggested a role for NCKX1, NCKX3 and NCKX4: a premature stop in the *SLC24a1* gene results in congenital stationary night blindness (consistent with the essential role for NCKX1 in rod photoreceptors discussed above), single-nucleotide polymorphisms (SNP) in the *SLC24a3* gene may associate with salt-dependent high blood pressure, while SNPs in the *SLC24a4* gene may correlate with blue eyes and blond hair in European populations.

The most surprising case is *SLC24a5* which has been shown to be a pigmentation gene in both the zebrafish and in humans (Lamason et al. 2005). A SNP in both alleles of the *SLC24a5* gene is associated with fair skin in people from European descent. Unlike the case for the NCKX1-4 cDNA's, NCKX5 cDNA does not result in plasma membrane expression when expressed in cell lines. In melanocytes, the NCKX5 protein also appears located in intracellular organelles, either in melanosomes or in the trans-Golgi network (TGN). No direct demonstration of NCKX fluxes mediated by NCKX5 has been shown in either melanosomes or TGN.

Structure-Function Relationships of NCKX

Whereas nearly all of the in situ characterizations of NCKX proteins are concerned with NCKX1 in retinal ROS, most of the functional studies of NCKX cDNAs expressed in cell lines are concerned with NCKX2. All the structure-function studies to date have been carried out using mutants of the NCKX2 cDNA. The stoichiometry, cation selectivity, and K_m values observed for NCKX2 expressed in cell lines are very similar to those previously reported for NCKX1 in retinal ROS. Topological studies have resulted in the topological model (Shibukawa et al. 2007) illustrated in Fig. 1 which consists of two sets of five TMS each, placing the C-terminus in the extracellular space.

Expression of NCKX2 in cell lines invariably results in two protein bands of slightly different MW reflecting full-length or uncleaved NCKX2 (higher MW) and NCKX2 from which the signal peptide has been cleaved (lower MW). Deletion of the signal peptide did not prevent protein expression, but the expressed mutant protein did not traffic to the plasma membrane, suggesting that signal peptide cleavage may be associated with intracellular transport to the plasma membrane.

Other studies have focussed on the membrane topology of NCKX resulting in the model illustrated in Fig. 1, and on residues important for NCKX function (see below). However, the functional role of the large hydrophilic loops, in particular that located in the cytosol and subject to alternate splicing, remains largely unexplored and terra incognita. Quite a bit of information has been gained by mutagenesis in which single amino acids of interest are replaced by another and the effect on NCKX function examined after the mutant NCKX2 cDNAs are expressed in cell lines. Mutagenesis studies have focused on the two alpha repeats, as these are the most conserved parts of the NCKX protein and most likely contain key residues important for cation binding and transport. Consistent with the role of NCKX as a cation transporter, the only polar residues found in the interior TMA are those suited for cation binding such as Glu, Asp, Asn, and Ser. Figure 1 shows some of the key residues involved in cation binding and transport as identified by mutagenesis studies (Shibukawa et al. 2007). Although these residues, e.g., E188 and D548, are far apart in the linear protein sequence it has been shown that they are in close proximity in the three dimensional structure of the protein and are likely to contribute coordination sites for binding cations that are transported by NCKX.

Acknowledgment Research on the *SLC24* gene family in my laboratory is funded through operating grant MOP-81327 from the Canadian Institutes for Health Research. My thanks to Dr. Lorraine Aha for her careful reading of the text and to David Schnetkamp for producing the graphic illustration in Fig. 1.

Cross-References

▶ Biological Copper Transport
▶ Calcium Homeostasis: Calcium Metabolism
▶ Na^+/Ca^{2+}-K^+ Exchanger
▶ Potassium Channels, Structure and Function

References

Altimimi H, Schnetkamp PPM (2007) Na^+/Ca^{2+}-K^+ exchangers (NCKX). Functional properties and physiological roles. Channels 1:62–69

Blaustein MP, Lederer WJ (1999) Sodium/calcium exchange: its physiological implications. Physiol Rev 79:763–854

Lamason RL, Mohideen MA, Mest JR, Wong AC, Norton HL, Aros MC, Jurynec MJ, Mao X, Humphreville VR, Humbert JE, Sinha S, Moore JL, Jagadeeswaran P, Zhao W, Ning G, Makalowska I, McKeigue PM, O'donnell D, Kittles R, Parra EJ, Mangini NJ, Grunwald DJ, Shriver MD, Canfield VA, Cheng KC (2005) SLC24A5, a putative cation exchanger, affects pigmentation in zebrafish and humans. Science 310:1782–1786

Li XF, Kiedrowski L, Tremblay F, Fernandez FR, Perizzolo M, Winkfein RJ, Turner RW, Bains JS, Rancourt DE, Lytton J (2006) Importance of K^+-dependent Na^+/Ca^{2+}-exchanger 2, NCKX2, in motor learning and memory. J Biol Chem 281:6273–8262

Lytton J (2007) Na^+/Ca^{2+} exchangers: three mammalian gene families control Ca^{2+} transport. Biochem J 406:365–382

Schnetkamp PPM (1989) Na-Ca or Na-Ca-K exchange in the outer segments of vertebrate rod photoreceptors. Prog Biophys Mol Biol 54:1–29

Schnetkamp PPM (2004) The SLC24 Na^+/Ca^{2+}-K^+ exchanger family: vision and beyond. Eur J Physiol 447:683–688

Shibukawa Y, Kang KJ, Kinjo TG, Szerencsei RT, Altimimi HF, Pratikhya P, Winkfein RJ, Schnetkamp PP (2007) Structure-function relationships of the NCKX2 Na^+/Ca^{2+}-K^+ exchanger. Ann N Y Acad Sci 1099:16–28

Potassium in Biological Systems

Sumukh Deshpande, Serdar Durdagi and Sergei Yu. Noskov
Institute for Biocomplexity and Informatics,
Department of Biological Sciences,
University of Calgary, Calgary, AB, Canada

Synonyms

Alkali cation binding sites and structure–functions relations; K^+ dependent proteins

Definition of the Subject

This entry provides a comprehensive reference of the role of potassium ions in biological system with examples from bacteria, plant, and animal kingdoms. These involve mediation of ATPase activity, selective conductance across ion channels and pumps.

General Information

Potassium plays an important role in regulation of cellular electrolyte metabolism, electric signaling in cells, transport of essential nutrients, and enzymatic analysis. In alliance with other (i.e., sodium and chloride ions), it contributes to overall electrolyte balance of virtually all living organisms. It also assists in the conduction of nerve impulses. In this entry, various roles played by the potassium ion in mammals, bacteria, and plants will be briefly reviewed.

Potassium in Bacteria

K^+ is involved in regulation of osmotic pressure, activation of intracellular enzymes, regulation of internal pH, and as a second messenger in signaling pathways. Certain halophilic bacterial species including eubacteria and archaebacteria use K^+ as a major cytoplasmic solute. Those bacterial species that possess a cell wall maintain turgor pressure (pressure exerted by solvent inside the cell against the cell wall) across the cell envelope which acts as a signal for K^+ transport across the cell membrane. This means that any rise/fall in turgor pressure may signal the cations (e.g., K^+) to move across the membrane for balancing ion concentration inside and outside the cell. To maintain a constant K^+ concentration, K^+ transport systems are required such as Trk, Kup, and Kdp transporters (Epstein 2003). Trk and KuP are low-affinity transporters which allow K^+ uptake at neutral or alkaline pH whereas Kdp is a high-affinity transporter allowing K^+ uptake at much lower K^+ concentrations. This explains how the affinity of transporter is coupled with extracellular pH changes. The most extensively studied transport systems are the members of the Trk transport system which use the electrochemical gradient of H^+ to transport K^+ across the cell membrane (Fig. 1) (Epstein 2003). On the other hand, members of the Ktr transport system use Na^+ electrochemical gradients which are created by imbalances in Na^+ and K^+ concentration between the ICF (intracellular fluid) and ECF (extracellular fluid) (Epstein 2003). These systems mediate uptake of K^+ upon reduction in turgor pressure until equilibrium is reached. A constant internal K^+ concentration is needed to maintain turgor pressure for optimal growth in different ionic conditions.

Certain symporters (transport of two or more different solutes in the same direction) such as TrkA/KtrA couple the gradient of H^+ for transporting K^+ by binding to NAD^+. Certain species such as acidophiles express proton/cation antiporters (secondary active transport of solutes in opposite direction) to avoid excessive turgor pressure and accumulation of cations (Epstein 2003). This leads to K^+ efflux mediated by efflux systems such as KefB and KefC system

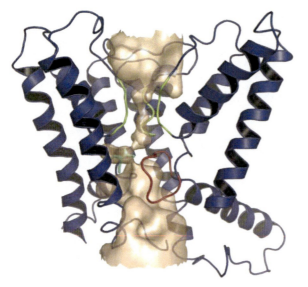

Potassium in Biological Systems, Fig. 1 Surface representation of TrkH subunit pore (Cao et al. 2011)

in *Escherichia coli*. These systems release K^+ upon activation of the transport system by toxic compounds such as methylglyoxal (Epstein 2003). Certain mutations which cause the channel to remain active in absence of toxins identify regions of protein which may be important for controlling channel activity (Epstein 2003). Bacteria also express K^+ channels which are similar to the ones expressed in eukaryotes such as KcsA and MthK channels crystallized at 3.2 and 3.3 Å from *Streptomyces lividans* and *Methanobacterium thermoautotrophicum*, respectively. These channels conduct K^+ ion rapidly across the membrane and show higher preference for K^+ over other cations. The channel displays open, closed, and inactivated conformational states for regulating ion flow across the membrane. Bacteria also express mechanosensitive channels (Msc) (Fig. 2) which are similar in function to KefB and KefC, which open upon rise in turgor pressure and high concentration of extracellular K^+, thus protecting the cell from rupture or osmotic shock (Epstein 2003).

Apart from maintaining osmotic pressure, K^+ as a major intracellular cation is also used to control pH of cytoplasm where it uses various buffer anions such as the phosphate groups on nucleic acids and charged residues in proteins for displacing protons where the antiporter exchanges K^+ for H^+, for example, the

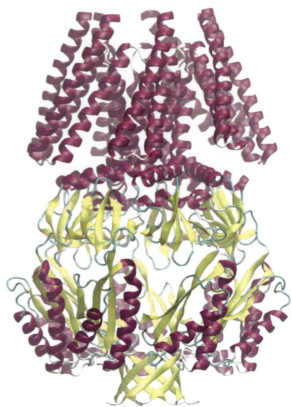

Potassium in Biological Systems, Fig. 2 Secondary structure representation of crystal structure of mechanosensitive channel of small conductance (MscS) (PDB code: 2OAU) (Helix, color *magenta*; Beta Sheets, color *yellow*, loops, color *cyan*) (Bass et al. 2002)

K^+/H^+ antiporter in *Enterococcus hirae* reduces internal pH when the intracellular concentration of NH_4^+ increases (Epstein 2003).

Potassium in Plants

Potassium is the most essential and highly absorbed nutrient in plants. It plays a major role in regulation of osmotic pressure, stabilizing metabolism, enzyme activation, neutralization of charged residues on proteins, regulation of membrane potential, etc. Potassium is highly concentrated in leaf tissues, protecting the plant from frosting and forming a defense against insects and disease (Dreyer and Uozumi 2011). A stable K^+ concentration of approximately 100 mM is required in active compartments of the plant cell for enzyme activation and protein synthesis. Besides this,

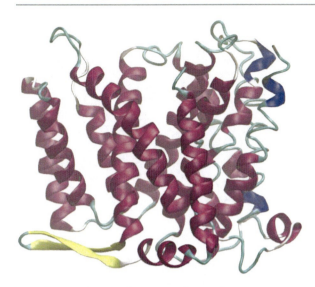

Potassium in Biological Systems, Fig. 3 Secondary structure representation of crystal structure of Nha antiporter (PDB code: 1ZCD) (Helix, color *magenta*; Beta Sheets, color *yellow*, loops, color *cyan*) (Hunte et al. 2005)

K^+ is also used as a major cation in turgor driven process such as stomatal movement, cell elongation, and phototropism. Due to variation of K^+ concentration in soil, its uptake in plants is mediated by multiple transport systems with high and low affinity for K^+ (Dreyer and Uozumi 2011). K^+ uptake in plants is driven either by channels or by electrochemical potential-driven transporters including uniporters (transport of solute across its gradient), symporters, or antiporters (Grabov 2007). Plants exhibit various cation transporters which include Trk/HKT system, KEA (K^+/H^+ antiporter), CHX (cation/H^+ antiporter) (Fig. 3), KT/HAK/KUP system, Glutamate Receptors, Cyclic Nucleotide-Gated Channels (CNGC), and *Shaker-type* K^+ channels (Grabov 2007; Demidchik and Maathuis 2010). Members of *Shaker-type* channels involving voltage-gated channels have been identified which include both inward- and outward-rectifying channels which mediate K^+ uptake by membrane hyperpolarization and depolarization respectively (Demidchik and Maathuis 2010). CNGC are ligand-gated channels (e.g., AtCNGC2) permeable for K^+ and are modulated by binding of cAMP/cGMP and calmodulin (Demidchik and Maathuis 2010). Glutamate receptor channels express putative glutamate binding subunits and uptake K^+ from extracellular environment of the cell. For example, roots of *Arabidopsis thanliana* show higher expression of AtGLR receptor which help the plant in regulating ion uptake (Demidchik and Maathuis 2010). Trk/HKT transport systems in plants share homology with Ktr and Trk systems in bacteria. HKT functions as alkali-cation transport systems which links osmotic homeostasis with soil salinity tolerance. Certain plants such as barley express HvHKT2 which is upregulated by lower K^+ concentration and overexpressed by higher Na^+ concentrations and functions as Na^+ uniporter whereas in wheat HKT1 transporter cotransports Na^+ and K^+ (Dreyer and Uozumi 2011; Grabov 2007). CHX and KEA functions as antiporters by transporting K^+ into and out of the cell. For example, AtCHX17 identified from *Arabidopsis thanliana* mediates K^+ uptake (Dreyer and Uozumi 2011; Grabov 2007). KT/HAK/KUP system is one of the diverse and high-affinity transport systems found in various plant species ranging from green algae, fungi to angiosperms and operates under low K^+ concentrations. For example, HAK1 transporter from *Sarcoscypha occidentalis* transports K^+ against extremely low nutrient concentration in the environment (Grabov 2007).

Plants are sensitive to environmental conditions such as salinity which can affect the growth of roots, shoots, and leaves and K^+ availability for the plant decreases. As NaCl dependent depolarization occurs in roots and leaves, flux of K^+ through outward-rectifying channels increases. This activates H^+-ATPases which pump H^+ by hydrolyzing ATP and returning to resting membrane potential. This further activates hyperpolarization and depolarization activated voltage-gated K^+ channels which uptake or release K^+ depending on K^+ concentration based on salinity level (Demidchik and Maathuis 2010).

Potassium in Mammals

K^+ is one of the principal monovalent cation in the intra- and extracellular milieu. Average contributions of K^+ in different cells are shown in Table 1 (Parthasarathy and Ravishankar 2007). An average human has about 3,500 mM of K^+ or 50 mM/kg in molality units (Parthasarathy and Ravishankar 2007). Its concentrations in the ICF vary from 140 to 150 mEq/L (milliequivalent per liter), while ECF displays K^+ concentrations from 3.5 to 5 mEq/L (Parthasarathy and Ravishankar 2007). Concentration of serum K^+ is maintained by orchestration of various

Potassium in Biological Systems, Table 1 Potassium concentration in body fluids (Parthasarathy and Ravishankar 2007)

	Concentration of K^+ (mmol/L)	Total K^+ (mmol)
Extracellular fluid (ECF)	5	30–70
		Plasma – 15
		Interstitial fluid – 35
Intracellular fluid (ICF)	150	2,500–4,500
		Muscle – 2,650
		Liver – 250
		Red blood cells – 250
		Others – (urine, sweat, stool, and bone)

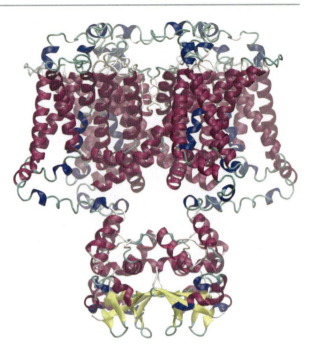

Potassium in Biological Systems, Fig. 4 Secondary structure representation of crystal structure of voltage-gated *Shaker* channel (Kv1.2) (PDB code: 2A79) (Helix, color *magenta*; Beta Sheets, color *yellow*, loops, color *cyan*) (Long et al. 2005)

transporters and channels involved in controlling ion intake and excretion for controlled biological function. Maintenance of intra- and extracellular ion distribution depends on cell membrane integrity, work of membrane proteins known as pumps, pH, osmotic pressure, concentrations of hormones such as insulin, aldosterone, catecholamines, etc. This is just to name several important players involved in electrolyte metabolism in mammals. An imbalance in K^+ concentration (Table 1) between the ICF and ECF creates an electrochemical potential, a key property of excitable membranes (Parthasarathy and Ravishankar 2007).

Neuromuscular excitability depends on local perturbation in the ICF/ECF K^+ ratio. In the heart alteration of ICF/ECF ratio due to ion flows across cell membranes plays a tremendous role in various physiological functions such as pacemaking, conducting systems, and contractile states which involves >200 genes encoding a variety of K^+ channels (Parthasarathy and Ravishankar 2007; Shieh et al. 2000). Any decrease in ECF K^+ concentrations can lead to myocardium ventricular fibrillation and catecholamine-prone arrhythmias which are characterized by abnormal conduction of K^+ across the cell caused by mutations in voltage-gated channel in response to release of catecholamines (Parthasarathy and Ravishankar 2007). With the decrease in ECF K^+ concentration, resting membrane potential becomes negative leading to hyperpolarization of myocytes, thereby decreasing muscle excitability. Increase in ECF K^+ concentration causes depolarization resulting in prolongation of refractory periods.

Regulation of K^+ concentration across the cell is mediated by various ion channels which conduct K^+ across the membrane. These include voltage-gated *Shaker-type* K^+ channels (6-TM 1 pore protein), inward-rectifier K^+ channels (2-TM 1 pore protein), and two-pore K^+ channels (4-TM 2 pore protein). Voltage-gated K^+ channels activated by membrane depolarization (Kv channels) include *Shaker-type* channels (Fig. 4), human-ether-a-go-go (hERG) channels, Ca^{2+}-activated and KCNQ channels. Inward-rectifying potassium channels (K_{ir}) allow inward flow of K^+ ions to restore the resting membrane potential (e.g., −60 mV). For example, K_{ir} channels which are regulated by G-protein coupled receptors (GPCRs) and ATP-sensitive channels are linked to cellular metabolism (Hibino et al. 2010). Two-pore domain (KCNK) channels are regulated by physical and chemical stimuli, temperature, pH and osmolarity, and homeostatic control through reflexes mediated through central nervous system. For example, TASK-1 channel is pH sensitive and is a target for important anesthetic effects in mammalian neurons (Sirois et al. 2000).

Certain hormones like aldosterone and Antidiuretic Hormone (ADH) regulate the K^+ concentration by increasing Na-K-ATPase pump activity, thereby stimulating K^+ uptake (Parthasarathy and Ravishankar 2007).

blockage of N^+-K^+-ATPase pump by drugs such as *Digitalis* that impair the channel and prevent the entry of K^+ into the cells. Increased secretion of K^+ through channels could lead to severe hyperkalemia and cause life-threatening arrhythmia. On the other hand, K^+ concentration in serum of less than 3.5 mM causes hypokalemia characterized by starvation, alcoholism, and diarrhea.

Cross-References

▶ Cellular Electrolyte Metabolism
▶ Potassium Channel Diversity, Regulation of Potassium Flux across Pores
▶ Potassium Channels, Structure and Function
▶ Potassium Dependent Sodium/Calcium Exchangers

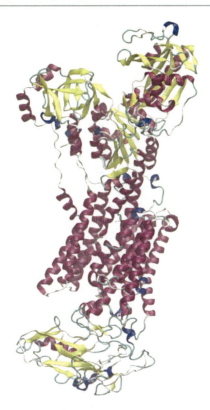

Potassium in Biological Systems, Fig. 5 Secondary structure representation of crystal structure of Na-K-ATPase pump (PDB code: 2ZXE) (Helix, color *magenta*; Beta Sheets, color *yellow*, loops, color *cyan*)

The pump works by hydrolysis of terminal phosphate bond of ATP which provides energy for pumping Na and K through conformational change in the protein. Catecholamine and insulin drives K^+ intracellularly to the muscles and liver by stimulating the activity of Na-K-ATPase (Fig. 5) (Shinoda et al. 2009). Insulin dependent cellular K^+ uptake is independent of glucose uptake caused by insulin. Insulin deficiency can cause decreased cellular K^+ uptake. However, if external K^+ is administered, it could result in hyperkalemia (high K^+ concentration). It should be noted that K^+ concentration in tissues has to be tightly controlled. For example, hyperkalemia causes respiratory and metabolic acidosis which is a common acid–base disorder produced by excessive acid production and reduced acid secretion or loss of body alkali which causes an imbalance between CO_2-HCO_3 buffer system (Parthasarathy and Ravishankar 2007). Furthermore, hyperkalemia can also be caused by either depolarizing muscle relaxants such as succinylcholine that increase permeability of muscle cells resulting in greater K^+ efflux or by

References

Bass RB, Strop P, Barclay M, Rees DC (2002) Crystal structure of *Escherichia coli* MscS, a voltage-modulated and mechanosensitive channel. Science 298:1582–1587

Cao Y, Jin X, Huang H, Derebe MG, Levin EJ, Kabaleeswaran V, Pan Y, Punta M, Love J, Weng J, Quick M, Ye S, Kloss B, Bruni R, Martinez-Hackert E, Hendrickson WA, Rost B, Javitch JA, Rajashankar KR, Jiang Y, Zhou M (2011) Crystal structure of a potassium ion transporter, TrkH. Nature 471:336–340

Demidchik V, Maathuis F (2010) Ion channels and plant stress responses. Springer, Heidelberg/New York

Dreyer I, Uozumi N (2011) Potassium channels in plant cells. FEBS J 278:4293–4303

Epstein W (2003) The roles and regulation of potassium in bacteria. Prog Nucleic Acid Res Mol Biol 75:293–320

Grabov A (2007) Plant KT/KUP/HAK potassium transporters: single family – multiple functions. Ann Bot 99:1035–1041

Hibino H, Inanobe A, Furutani K, Murakami S, Findlay I, Kurachi Y (2010) Inwardly rectifying potassium channels: their structure, function, and physiological roles. Physiol Rev 90:291–366

Hunte C, Screpanti E, Venturi M, Rimon A, Padan E, Michel H (2005) Structure of a Na^+/H^+ antiporter and insights into mechanism of action and regulation by pH. Nature 435:1197–1202

Long SB, Campbell EB, MacKinnon R (2005) Crystal structure of a mammalian voltage-dependent shaker family K^+ channel. Science 309:897–903

Parthasarathy S, Ravishankar M (2007) Potassium ion and anaesthetic implications. J Anaesthesiol Clin Pharmacol 23:129–144

Shieh CC, Coghlan M, Sullivan JP, Gopalakrishnan M (2000) Potassium channels: molecular defects, diseases, and therapeutic opportunities. Pharmacol Rev 52:557–594

Shinoda T, Ogawa H, Cornelius F, Toyoshima C (2009) Crystal structure of the sodium-potassium pump at 2.4 angstrom resolution. Nature 459:446–450

Sirois JE, Lei QB, Talley EM, Lynch C, Bayliss DA (2000) The TASK-1 two-pore domain K^+ channel is a molecular substrate for neuronal effects of inhalation anesthetics. J Neurosci 20:6347–6354

Potassium in Health and Disease

Ruth Schreiber and Daniel Landau
Department of Pediatrics, Soroka University Medical Centre, Ben-Gurion University of the Negev, Beer Sheva, Israel

Synonyms

Potassium = Kalium (K^+)

Definitions

Hypokalemia: low K^+ plasma concentration; Hyperkalemia: high K^+ plasma concentration; TTKG: transtubular K^+ gradient.

Physiology of Potassium Handling

Potassium (K^+) is the most abundant intracellular cation. High concentrations of K^+ in the cytosol are required for many normal cellular functions, including metabolism and growth, cell division, optimal enzyme function, protein and DNA synthesis, volume regulation, and intracellular acid-base balance. In addition, a steep concentration gradient of K^+ across the cell membranes is required for nerve excitation and muscle contraction. Thus, both hypokalemia (low plasma K^+ concentration) and hyperkalemia (high-plasma K^+ concentration) can result in potentially fatal muscle paralysis and cardiac arrhythmias, in part by altering conduction in skeletal and cardiac muscle.

These functions are achieved by maintaining high intracellular and low extracellular K^+ concentrations. The intracellular and extracellular fluids (ECF) are separately controlled to maintain K^+ homeostasis. Whereas homeostasis in adults seeks to maintain zero K^+ balance, growing infants and children must accumulate K^+ for growth.

The intracellular K^+ concentration, 100–150 mEq/L, far exceeds its concentration in ECF, 3.5–5 mEq/L. This transcellular K^+ gradient is maintained by the action of the enzyme Na^+-K-adenosine triphosphatase (Na^+-K^+-ATPase) in the cell membrane, which pumps Na^+ out and K^+ into the cell in a 3:2 ratio. Thus, serum potassium level provides only an indirect estimate of total body K^+ status because it represents only 2% of total body K^+ (Satlin 2009; Mount and Kambiz 2007).

Distribution of K^+ Between the Cells and the Extracellular Fluid

Regulation of the internal distribution of K^+ must be extremely efficient, since the movement of as little as 1.5–2% of the cell K^+ into the ECF can result in a potentially fatal increase in the plasma K^+ concentration to as high as 8 meq/L or more. Physiologic determinants that influence this process include catecholamines, insulin, aldosterone, exercise, Na^+-K^+-ATPase activity (which is increased by catecholamines and insulin), and the plasma K^+ concentration itself. After a K^+ load, most of the extra K^+ is initially taken up by the cells, a response that is facilitated by basal levels of catecholamines and insulin. This cell uptake minimizes the increase in the plasma K^+ concentration, pending the excretion of the excess K^+ in the urine (Rose and Post 2001).

Potassium Excretion

Although small amounts of K^+ are lost each day in stool (5–10 meq) and sweat (0–10 meq), the kidney plays the major role in the maintenance of K^+ balance, appropriately varying K^+ secretion with changes in dietary intake (normal range is 40–120 meq/day). The primary tubular segment responsible for urinary K^+ excretion is the cortical collecting duct (particularly the principal cells) in distal nephron. The main factors modulating this process are aldosterone and the plasma K^+ concentration itself. Distal flow rate and the transepithelial potential difference (which is generated primarily by Na^+ reabsorption) play a more permissive role: they do not change directly with K^+

Potassium in Health and Disease, Table 1 Causes of hyperkalemia

Etiology	Remarks/associated findings
I. Increased potassium release from cells	
1. Pseudohyperkalemia	Hemolysis, leukocytosis, thrombocytosis
2. Metabolic acidosis	
3. Insulin deficiency, hyperosmolarity	
4. Increased tissue catabolism	
5. Beta-adrenergic blockade	
6. Exercise	
7. Hyperkalemic periodic paralysis[a]	
8. Drugs	Trimethoprim-Sulphamethoxazole, pentamidine, ACE inhibitors, aldosterone antagonists, etc.
II. Reduced urinary potassium excretion	
1. Hypoaldosteronism[a]	Aldosterone deficiency
2. Pseudohypoaldosteronism[a]	Aldosterone resistance
3. Renal failure	
4. Hyperkalemic (type IV) renal tubular acidosis[a]	
5. Selective impairment of potassium secretion[a]	
6. Ureterojejunostomy	Metabolic acidosis

[a]Can be caused by genetic defects (see Table 3)

Potassium in Health and Disease, Table 2 Causes of hypokalemia

Etiology	Remarks/Associated findings
I. Decreased potassium intake	
	Will appear faster with fasting
II. Increased entry into cells	
1. Elevation in extracellular pH (Alkalosis)	Metabolic or respiratory
2. Increased availability of insulin	
3. Elevated β-adrenergic activity -	Stress or administration of beta agonists
4. Hypokalemic periodic paralysis[a]	
5. Marked increase in blood cell production	
III. Increased gastrointestinal losses	
1. Vomiting	Induces secondary hyperaldosteronism and urinary K losses
2. Diarrhea	
3. Tube drainage	
4. Laxative abuse	
IV. Increased urinary losses	
1. Drugs	Furosemide, thiazides, Amphotericin B
2. Selective tubular transport defects[a]	
With normal blood pressure	Bartter's or Gitelman's syndromes[a], renal tubular acidosis
With elevated blood pressure	Liddle syndrome[a], apparent mineralocorticoid excess[a]
3. Adrenal defects	
Acquired	Cushing syndrome, Adrenal adenoma
Genetic	Adrenal enzyme defects[a]
V. Increased sweat losses	
	Cystic fibrosis

[a]Can be caused by genetic defects (see Table 4)

balance, but relatively normal values are required for adequate K^+ secretion.

Understanding these principles can simplify the approach to patients with disorders of K^+ balance. Chronic hyperkalemia, for example, must be associated with a defect in distal K^+ secretion, since the adaptation response would normally permit excretion of the excess K^+. The two major mechanisms by which K^+ secretion might be impaired are hypoaldosteronism and decreased distal urinary flow (due to marked volume depletion or advanced renal failure). These conditions, therefore, constitute most of the differential diagnosis of persistent hyperkalemia (Table 1).

Urinary K^+ wasting and hypokalemia, on the other hand, are due to activation of the distal secretory process. This most often occurs with hyperaldosteronism (as long as distal flow is maintained), increased distal flow (as long as aldosterone secretion is normal or elevated, as with diuretic therapy), or with the delivery of Na^+ to the distal nephron with a nonreabsorbable anion (as is seen in ketoacidosis or proximal renal tubular acidosis) (Table 2) (Rose and Post 2001).

Hyperkalemia

Hyperkalemia, serum potassium (K^+) concentration higher than 5.5 mEq/L, is a rare occurrence in normal subjects, due to cellular and renal adaptations preventing significant accumulation of K^+ in the

Potassium in Health and Disease, Table 3 Genes associated with hyperkalemia

Mutated gene	Chromosomal location/Heredity	Cell involved	Phenotype	Special features
CYP11B2 (aldosterone synthase)	8q21/AR	Adrenal	Congenital adrenal hyperplasia	Adrenal enzymatic defects leading to reduced or absent aldosterone
CYP11B1 (11-β hydroxylase)	8q21/AR	Adrenal		
CYP21A2 (21 hydroxylase)	6p21.3/AR	Adrenal		
ENaC	16p13-p12	CCD	Pseudohypoaldosteronism (PHA type I)	Loss-of-function mutation. Severe phenotype. Multiorgan involvement
	12p13/AR			
MLR (Mineralocorticoid receptor)	4q31.1/AD	CCD		Mild/transient renal phenotype
WNK4	17q21/AD	DCT	Gordon syndrome (PHA type II)	Low renin hypertension. Gain-of-function missense mutations. An additional locus, referred to as PHA IIA, has been mapped to 1q
WNK1	12p/AD	DCT		
KCNJ1 (ROMK)	11q24/AR	TAL or DCT	Antenatal Bartter syndrome	Transient hyperkalemia in the neonatal period
SCNA4	17q23AD	Skeletal muscle	Hyperkalemic periodic paralysis	Point mutations encoding skeletal muscle sodium channel

AR Autosomal recessive, *AD* Autosomal dominant, *CCD* cortical collecting duct, *DCT* Distal convoluted tubule, *TAL* thick ascending limb

extracellular fluid. Clinical consequences of hyperkalemia are related to the adverse electrophysiologic effects of an altered transmembrane K^+ gradient on excitable tissues, reducing the resting membrane potential, initially enhancing but ultimately suppressing tissue excitability ("depolarizing block"). Cardiac arrhythmias (potentially fatal) generally develop when the plasma K^+ concentration rises above 7 mEq/L; however, adverse effects may be seen with lower plasma levels if hyperkalemia has developed acutely or with other metabolic derangements such as acidosis or hypocalcemia. Neuromuscular system manifestations include skeletal muscle weakness, paresthesias, and ascending flaccid paralysis.

The major three etiological categories include high K^+ intake, movement of K^+ from cells to the extracellular fluid, but mostly impairment in urinary K^+ excretion (Tables 1, 3). Unless given acutely (e.g., by the intravenous route), increased intake alone will not lead to hyperkalemia if adrenal and renal functions are intact, since the excess K^+ will be excreted rapidly in the urine (a phenomenon called *potassium adaptation*).

Investigation includes urine and plasma electrolytes and osmolality, blood gases and ECG. Calculation of the transtubular K^+ gradient (TTKG) = $[(U_K \times P_{osm})/(P_K \times U_{osm})]$ can estimate the degree of aldosterone action. A value below 7, and particularly below 5, is highly suggestive of aldosterone deficiency or resistance.

The treatment of hyperkalemia is divided into three general categories: (1) antagonism of the cardiac effects of hyperkalemia with intravenous calcium; (2) rapid reduction in serum K^+ by redistribution into cells with β2-agonists and insulin, and sodium bicarbonate when metabolic acidosis exist; and (3) removal of K^+ from the body with diuretics, enteral cation exchange resins, and dialysis (Mount and Kambiz 2007).

Hypokalemia

Hypokalemia, a serum K^+ level less than 3.6 mmol/L, is a common electrolyte abnormality encountered in clinical practice. It may be caused by decreased K^+ intake, increased translocation into the cells, or most often, increased losses in the urine, gastrointestinal tract, or sweat (Tables 2, 4). The clinical manifestations of K^+ depletion include skeletal muscle weakness and paralysis in severe cases (including the smooth

Potassium in Health and Disease, Table 4 Genes associated with hypokalemia

Mutated gene	Gene locus/heredity	Affected tubular segment	Phenotype	Special features
SLC12A1	15q15–q21.1/AR	TAL	Bartter syndrome	
KCNJ1	11q24/AR	TAL + CCD		
BSND	1p31/AR	TAL + DCT		With sensory neural deafness
CLCNKB	1p36/AR	TAL + DCT		
SLC12A3	16q13/AR	DCT	Gitelman syndrome	Hypomagnesemia
SCNN1B	16p12.2–12.1/AD	CCD	Liddle syndrome	Hypertension
SCNN1G	16p12			
CACNA1S (Type I)	1q32/AD	Skeletal muscle Na^+ channel	Hypokalemic periodic paralysis	Thyrotoxicosis
SCN4A (Type II)	17q23.3/AD			Myotonia
HSD11B1/B2	16q22/AR	Enzyme complex 11-β-HSD	Apparent mineralocorticoid excess	Defective cortisol degradation leading to hypertension
KCNJ10	1q23.2/AR	DCT + CCD	EAST syndrome	Tubulopathy with epilepsy, ataxia, sensorineural deafness, mental retardation

AR Autosomal recessive, *AD* Autosomal dominant, *HSD* Hydroxysteroid dehydrogenase, *TAL* Thick ascending limb, *CCD* Cortical collecting duct, *DCT* Distal convoluted tubule

muscle of the gastrointestinal tract), muscle cramps, rhabdomyolysis, and myoglobinuria. Electrical disturbances of conduction and rhythm generally develop when the plasma K^+ concentration falls. In patients who have primary cardiac dysfunction, hypokalemia significantly increases the risk of cardiac arrhythmia. The metabolic changes of K^+ deficiency lead to suppression pancreatic insulin release, which results in increased glucose intolerance especially in diabetic patients.

The goals of therapy are to prevent life-threatening conditions by replacing any K^+ deficit, and diagnose and correct the underlying cause. A rapid drop to less than 2.5 mEq/L poses a high risk of cardiac dysrhythmias and calls for urgent replacement. Renal function, medications, and comorbid conditions such as diabetes should be considered to gauge the risk of overcorrection. In addition to K^+ supplementation, strategies to minimize its losses such as using K^+ sparing diuretics or replacement of additional sodium losses (to abolish secondary hyperaldosteronism) should be considered. It is important to correct hypomagnesemia in all patients with hypokalemia (Mount and Kambiz 2007; Devuyst et al. 2009).

Cross-References

▶ Potassium, Physical and Chemical Properties

References

Devuyst O, Konard M, Jeunemaitre X, Zennaro MC (2009) Tubular disorders of electrolyte regulation. In: Avner ED, Harmon WE, Niaudet P (eds) Pediatric nephrology. Lippincott Williams & Wilkins, Philadelphia, pp 929–977

Mount DB, Kambiz ZN (2007) Disorders of potassium balance. In: Brenner MB (ed) Brenner and rector's the kidney. Saunders, Philadelphia, pp 556–576

Rose BD, Post TW (2001) Clinical physiology of acid-base and electrolyte disorders. McGraw-Hill, New York, pp 372–402, 822–835, 888–930

Satlin ML (2009) Potassium. In: Avner ED, Harmon WE, Niaudet P (eds) Pediatric nephrology. Lippincott Williams & Wilkins, Philadelphia, pp 185–204

Potassium, Physical and Chemical Properties

Sergei Yu. Noskov
Institute for Biocomplexity and Informatics, Department of Biological Sciences, University of Calgary, Calgary, AB, Canada

Synonyms

Alkali cations; Alkali metals; K, *kalium* (Latin); Monovalent cations

Potassium, Physical and Chemical Properties, Table 1 Physical properties of potassium

Atomic number	19
Atomic mass	22.98977 g mol^{-1}
Relative abundance on Earth	2.4%
Oxidation states	+1, 0, −1
Ionization energy (first, second, third)	418.8, 3,052, and 4,420 kJ mol^{-1}
Ionic radius	0.133 nm
Van der Waals radius	0.275 nm
Atomic radius	0.226 nm
Electronic shell	[Ar]4s^1
Standard potential	−2.94 V
Phase	Solid
Molar heat capacity	29.6 J mol^{-1} K^{-1}
Melting point	63.5°C
Boiling Point	759°C
Hydration free energy (ΔG^0)	−300 kJ/mol
CAS registry number	7440-09-7

Definitions

CAS: Chemical Abstracts Service

Alkali metal: Group 1 element

Physical Properties: Potassium (K) is a member of the group 1 of alkali cations, sometimes also referred to group IA. Its atomic number is 19 and atomic weight is 39.0983 g mol^{-1}. As other members of the alkali metals family, potassium is highly reactive. Being highly soluble in water (react with it exothermically), a monovalent (K$^+$) potassium cation is abundantly present in salt- and freshwater basins, soil, and earth crust, but in smaller amounts than sodium. It is most commonly found in complex with chloride (ordinary or table salt) but also occurs in other minerals such as polyhalite, carnallite, sylvite, and others. The principle mineral source of potassium is sylvite (KCl). Most potassium minerals are insoluble. The greatest demand for potassium compounds is in agriculture where it is being used as a potent fertilizer. A list of key physical properties of K is collected in Table 1.

Chemical and Biochemical Properties: As all other alkali metals, potassium is highly reactive compared to metals from other groups. A pure metal is usually obtained by electrolysis of its salts. In fact, potassium was the first metal to be obtained by electrolysis. It can be easily oxidized because the first ionization energy of the metal is so low; thus it is mostly found as an ion. Potassium standard potential is also very electronegative, which makes potassium a strong reducing agent. The most common oxidation state of potassium is +1. It forms stable oxides with oxygen-oxygen bonds. Most of the potassium salts are soluble in water with only a few notable exceptions. In the gas state, potassium like other alkali metals exists as either atoms or M$_2$ molecules. Potassium vapor can be easily detected with a flame test. When the potassium salts are heated strongly in the Bunsen flame, a characteristic lilac flame color is observed. Its chemical properties and reactivity are similar to those of Na.

The average hydration number for K$^+$ is 6.3, but other coordination states were also reported. The hydration enthalpy ($\Delta_{hyd}H^0(298\ K)$) for K$^+$ is −321 kJ mol^{-1}, hydration entropy ($\Delta_{hyd}S^0(298\ K)$) is −70 JK^{-1} mol^{-1}, and the corresponding free energy of hydration ($\Delta_{hyd}G^0(298\ K)$) is −300 kJ mol^{-1}, respectively. The hydration data is reported for the following reaction:

$$K^+_{(gas)} \rightarrow K^+_{(aqueous)}.$$

The oxides of potassium (K$_2$O) can be obtained by using a limited air supply. A better approach to their synthesis is a decomposition of potassium superoxide formed by oxidation in pure oxygen:

$$K + O_2 \rightarrow KO_2.$$

Potassium superoxides have a magnetic moment of 1.74 μ_B that corresponds to one unpaired electron.

Potassium forms stable complexes with cyclic polyethers including 1-, 4-, 7-, 10-, 13-, 16-hexaoxacyclooctadecane. A similar binding/coordination principle is used for a production of the complex organic ions known as potassium cryptates. A cryptand is usually a polycyclic organic ligand that can accommodate a specific metal ion into its cavity. Both sodium and potassium cryptates are excellent models for understanding of membrane transport. A popular system for membrane transport studies is a small ionophore, valinomycin. Valinomycin is a small cyclodepsipeptide that can catalyze the permeation of cations across lipid membranes. It is highly specific for K$^+$ over Na$^+$, and its three-dimensional structure ion complex with K$^+$ is available (see Fig. 1 for

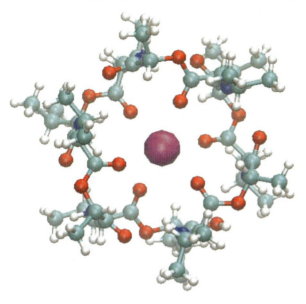

Potassium, Physical and Chemical Properties, Fig. 1 CPK-representation of the valinomycin complex with K$^+$ (shown in *magenta*). Six oxygens are forming a coordination shell for the bound ion

illustration). Several studies have put forward selective K$^+$-binding model that relies on the local covalent connectivity that controls the size of the binding cavity in valinomycin.

Although matching between ion size and cavity radius of the host is important for high specificity of binding in the guest-host model, it is also known that many other factors are involved into control of the ion selectivity of the host.

Potassium plays an important role in biological systems. Its extracellular concentration is lower that the intracellular one. Therefore, K$^+$ can leave the cell following its electrochemical gradient. Transport of K$^+$ across cellular membranes is utilized by the cell to generate the resting and action potentials in excitable membranes (nerves, muscle cells containing gated Na$^+$ and K$^+$ channels as well as a number of pumps (Na$^+$/K$^+$ ATPases) which restore resting concentrations of ion in the cell) as to control cell volume.

Cross-References

▶ Cellular Electrolyte Metabolism
▶ Potassium-Binding Site Types in Proteins
▶ Sodium, Physical and Chemical Properties

References

Dean JA, Lange NA (1998) Lange's handbook of chemistry. McGraw-Hill, New York
Haynes WM (ed) (2011) CRC handbook of chemistry and physics, 91st edn. CRC Press, Boca Raton
Hille B (2001) Ion channels of excitable membranes, 3rd edn. Sinauer, Sunderland

Potassium-Binding Site Types in Proteins

Serdar Durdagi[1], Benoît Roux[2] and Sergei Yu Noskov[1]
[1]Institute for Biocomplexity and Informatics, Department of Biological Sciences, University of Calgary, Calgary, AB, Canada
[2]Department of Pediatrics, Biochemistry and Molecular Biology, The University of Chicago, Chicago, IL, USA

Synonyms

Alkali cation binding sites and structure-functions relationships; K$^+$-dependent proteins

Definitions

Potassium: An alkali metal (K) with atomic number 19. Potassium ion (K$^+$) is essential for numerous physiological functions including participation in the action potential generation, maintenance of the osmotic balance and control of muscle contractions. Deficit of potassium in body fluids may lead to potentially lethal condition known as hypokalemia.

Potassium-binding protein: Any protein or enzyme that requires the binding of a potassium ion to its structural stability or functional activity.

Binding Site: Region in the protein where metal ions will favorably bind. The ion has to be completely or partially dehydrated with water molecules replaced by coordinating groups from protein amino-acid residues.

Coordinating Ligand: Atom or molecule that binds to an ion.

Coordination Number: Number of atoms in ligands in direct contact with an ion.

Introduction

The binding of small metal ions plays a pivotal role in the structure and function of proteins and nucleic acids. Key physiological cations, such as K^+ and Na^+, are required for proper protein folding and stability, cellular homeostasis, and electrical activity of the cell. K^+ ions are abundant inside the cells, whereas Na^+ ions are mostly found outside the cells. For example, human plasma contains ~135–146 mM Na^+ and 3.5–5.2 mM K^+ ions, while intracellular cytosol contains 25–35 mM Na^+ and 130–145 mM K^+. This unequal distribution of Na^+ and K^+ across the cell membrane provides an energy source for action potentials generated by the opening of the Na^+ and K^+ channels, but also for the movement of various substances and other ions through the membranes (i.e., transporters), and for ATP synthesis in some organisms. Thus, most living cells have evolved to tightly control the intracellular concentration of these two cations (Hille 2001). K^+-binding proteins play an important role in human health and disease. Induced or acquired mutations that affect metal binding to proteins are thought to be connected to a broad spectrum of pathological conditions, including certain types of cancer, such as breast and pancreatic cancers, cardiovascular diseases, such as acquired long-QT syndrome, and protein deposition disorders, such as Alzheimer's disease or neonatal epilepsy (Ashcroft 1999). In this entry, the structures and functions of some of the best-studied K^+-binding sites are considered.

Thermodynamics of Ion Binding and Selectivity

K^+-specific proteins allow favorable K^+ binding and discrimination of other competing cations. To ensure optimal biological function, ion binding and the ion permeation across cell membranes have to be tightly controlled. The selectivity (preference for a particular ion) of a protein site is controlled by a combination of different factors, including the chemical composition of the site, the number of coordinating ligands, and even the configuration of surrounding residues that may restrict the flexibility of the binding site and its cavity size. Proteins can achieve remarkable selectivity between Na^+ and K^+, the two chemically similar alkali cations present in the cell. For example, K^+ channels are 100–1,000 times more permeable to K^+ than Na^+. To some extent, ion binding and selectivity in K^+-specific proteins can be explained in terms of equilibrium thermodynamic processes (Neyton and Miller 1988; Di Cera and Page 2006). All small cations are strongly hydrated and binding to a protein site is possible only when the gain in interaction free energy in the site is sufficient to counterbalance the considerable loss of hydration free energy. In order to favorably bind an ion, a protein has to compensate for the dehydration penalty, i.e., the energetic price of striping water molecules away from an ion. From this perspective, ion selectivity reflects the relative change in hydration and interaction free energy for a pair of ions. This statement and ion selectivity can be expressed quantitatively in terms of free energy differences between two ions competing for a binding site, for example, in the case of K^+ versus Na^+:

$$\Delta\Delta G_{Na,K} = \left[G_{Na}^{site} - G_{Na}^{bulk}\right] - \left[G_{K}^{site} - G_{K}^{bulk}\right]$$
$$= \Delta G_{Na,K}^{site} - \Delta G_{Na,K}^{bulk}$$

where $\Delta G_{Na,K}^{site} = \left[G_{Na}^{site} - G_{K}^{site}\right]$ and $\Delta G_{Na,K}^{bulk} = \left[G_{Na}^{bulk} - G_{K}^{bulk}\right]$.

These simple thermodynamic considerations are adequate to explain the process of equilibrium binding selectivity to a single site. For example, many enzymes requiring K^+ for activation such as kinases and molecular chaperons can also be activated by NH_4^+. At the same time, they typically cannot be activated by larger cations, such as Cs^+, or smaller cations, like Na^+ and Li^+ indicating tight coupling between specificity of an ion-binding site and solvation thermodynamics. However, many of K^+-binding proteins have multiple binding sites; therefore, ion binding may depend on the competitive occupancy of the different binding sites (Derebe et al. 2011). In these systems, ion selectivity may be controlled by out-of-equilibrium kinetic processes as well.

General Principles of Organization of K^+ Sites from a Survey of Known Binding Sites

A comprehensive review of K^+-binding sites in proteins was performed by Harding (Harding 2002). The survey examined the ligand packing around K^+

ions in protein sites and highlighted some common organizational principles of biologically relevant small molecules from the Cambridge Structural Database (CSD). Although a unique coordination number can be defined for most of the divalent physiological cations (i.e., Ca^{2+}, Mg^{2+}, and Mn^{2+}), this is not the case for monovalent cations such as K^+ and Na^+. Analysis of known binding sites shows that a wide spread of coordination numbers, from 4 to 9, can be observed for these two cations. It is thought that difference in the strength of ion-ligand interactions between monovalent and divalent cations lead to a greater flexibility in the coordination geometry for Na^+ and K^+.

In the case of K^+ binding to organic molecules, eight is the most commonly observed coordination number, but six and seven are observed almost as frequently. If the coordination distance limit is set to 3.2 Å, a criteria developed from the analysis of small molecules in the CSD, the percentages of K^+-binding sites found with coordination numbers $N = 5, 6, 7, 8$, and 9 are 6.6, 20.6, 27.4, 35.4, and 10.0%, respectively (Harding 2002). Statistics regarding the most common coordinating ligands (CSD of small molecules) have shown that oxygen atoms comprise 90% of all coordinating ligands, with nitrogen, fluorine, and chlorine coming in second place. The dominant role of oxygen, an electronegative atom, is not surprising as it provides favorable interactions to bind a K^+ ion.

It is worth mentioning that many of the crystal structures show at least partial hydration of K^+ ions bound to protein sites. The distribution of K^+ coordination numbers in proteins tends to show a slightly lower number of coordinating ligands compared to that in the small molecules database. It also displays a similarly broad spread of coordination numbers, ranging from 4 to 9. Because the constraints imposed on the binding site by the protein matrix makes it difficult to achieve coordination numbers similar to that in bulk liquids, a multi-dentate coordination is often used by proteins to enhance energetically ion-ligand interaction (Harding 2002).

Examples of the K^+-Binding Sites in Proteins

The available crystal structures of enzymes in the Protein Data Bank (PDB) that contain a K^+ ion-binding site are cytochrome C peroxide, pyruvate kinase, ribokinase, and fructose-1,6-biphosphatase, as well as a number of structures for prokaryotic and eukaryotic potassium channels. The K^+ binds to most of these proteins with dissociation constants in the mM to μM range (Di Cera and Page 2006). Although progress has been made in the elucidation of the crystal structures of several K^+ channels and pumps, only a small number of examples of a potentially diverse class of enzymes are currently available. The general structural rules used to identify a tentative protein site as K^+ selective are still to be formulated. Most of the known sites, however, display some common traits. Oxygen atoms from backbone carbonyls and, less frequently, carboxylate groups from aspartic or glutamic acids are the most commonly found forming K-selective sites.

K^+-Activated Enzymes

Specific ion binding to an enzyme may lead to local and/or global conformational changes resulting in the stabilization of an active conformation. K^+ is not a primarily a cation required for catalysis, but it can facilitate substrate binding and chemical activity by lowering the energy barriers or stabilizing appropriate configurations of active site of an enzyme. This increase in catalytic activity may rely on a number of different factors. K^+ binding to a protein may potentially increase polarity of the binding pocket resulting in favorable entropic and enthalpic contributions to the protein activity. Another mechanism explaining enzyme activation by K^+ operates with the ion-dependent substrate binding to a stabilized enzyme-ion complex. That is, the entropic penalty of ordering the enzymatic binding site is offset by the previously bound cation. The substrate may or may not be directly interacting with a bound ion within this mechanism.

Crystal structures of K^+-activated enzymes have broadened our understanding of K^+ ion selectivity in soluble proteins. In particular, it was found that replacement of K^+ with other monovalent cations does not necessarily lead to significant structural changes. Although pyruvate kinase is inactive without the K^+ ion, replacement of K^+ with Na^+ does not result in significant structural changes. In dialkylglycine dehydrogenase and Hsc70, however, the replacement of K^+ with Na^+ drastically changes the geometry of the

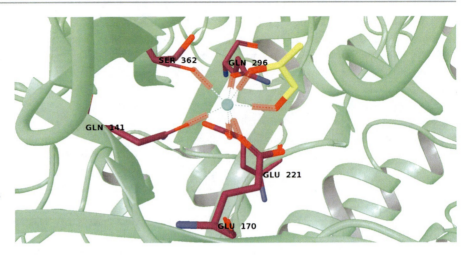

Potassium-Binding Site Types in Proteins, Fig. 1 Structural representation of type-I diol dehydratase enzyme (pdb ID: 1DIO) K$^+$-binding site. Carbon atoms of substrate (propanediol) and amino acid residues are shown with *yellow* and *maroon* colored sticks, respectively. K$^+$ ion shown with *turquoise* colored sphere in the figure. K$^+$ ion is coordinated with five residues at the binding site of the enzyme

coordination. It was suggested that these enzymes have evolved K$^+$ ion selectivity by imposing geometric constrains at the coordination shell that cannot be ordered by the Na$^+$ ion, which has a smaller ionic radius (Di Cera and Page 2006). Another important factor for K$^+$ selectivity of the protein is the chemical composition of the binding site. For example, the GXG (X = Y or F amino acids) signature sequence of the selectivity filter is conserved among all families of K$^+$ channels. The K$^+$-activation mechanisms of several enzymes can be further categorized into a cofactor-like (type-I) or allosteric regulatory mechanism (type-II) (Di Cera and Page 2006). In K$^+$-activated type-I enzymes, substrate binding to the active site of the target is mediated by K$^+$. Thus, K$^+$ binding to an enzyme is absolutely required for catalysis or substrate recognition. Examples of type-I enzymes are provided by diol- and glycerol-dehydratases. Diol-dehydratase is a coenzyme B$_{12}$-dependant protein.

The requirement of K$^+$ coordination for substrate binding can be better understood from the crystal structure of the binding pocket with bound propanediol. The ion-binding site is illustrated in Fig. 1. The K$^+$ ion is coordinated by five residues donating carboxylate oxygens from E170 and E221, the side-chain oxygens of S362 and Q141, the main-chain oxygen of Q296, as well as two hydroxyl oxygen atoms of the propanediol substrate. Thus, it follows that the absence of a K$^+$ ion renders substrate binding impossible.

In K$^+$-activated type-II enzymes, however, enzymatic activity is enhanced by conformational transitions triggered by ion binding. In this case, K$^+$ ion makes no direct contact with the substrate. A good example of such a system can be found in the type-II DNA repair enzyme MutL. The enzyme requires both Mg^{2+} and K$^+$ for its activity. A crystal structure shows that the K$^+$ ion is coordinated with four carbonyl oxygen atoms from the enzyme. This coordination is supplemented by the indirect contact between K$^+$ and the P$_\alpha$O atom of the ATP analogue. According to the crystal structure, this interaction is mediated by a stable water molecule, and is illustrated in Fig. 2.

Another example of a K$^+$-dependent molecular switch can be found in dealkylglycine decarboxylase, which is a PLP-dependent enzyme. This enzyme is a tetramer composed of identical subunits and each monomer contains a PLP binding domain. The coordination is octahedral, with the K$^+$ ion coordinated by Ser80 (O$_\gamma$), Asp307 (O$_{\delta 1}$), and the carbonyl oxygens of Leu78, Thr303, and Val305 and a water molecule at the binding site (Fig. 3). When Na$^+$ binds to the same site, the ion coordination by Thr303 and Ser80 is lost (the carbonyl oxygen atom of Thr303 and O$_\gamma$ of Ser80 are replaced by a water molecule from their coordinating positions). The O$_\gamma$ of Ser80 repositions itself and clashes with the aromatic ring of Tyr301, leading to a new conformation of the binding site in which substrate binding become unfavorable. One explanation is that the difference in ionic radius between Na$^+$ and K$^+$ ions allows for additional water molecule and may lead to this drastic rearrangement (Di Cera and Page 2006). However, proteins should be considered as soft materials to some extent. Their thermal fluctuations are typically at least a factor of two

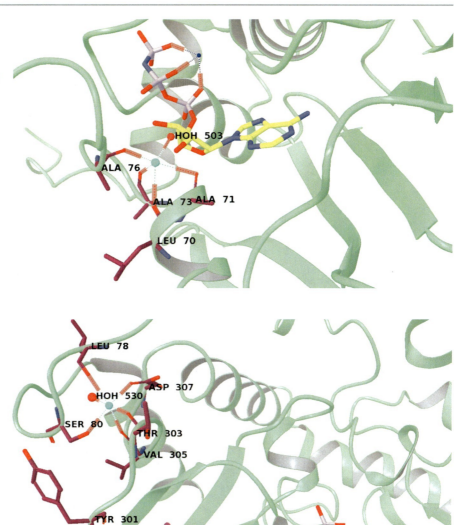

Potassium-Binding Site Types in Proteins, Fig. 2 Structural representation of type-II enzyme MutL (pdb ID: 1NHI) K⁺-binding site. Shown are substrate (CPK, C in *yellow*), relevant residues (CPK, C in *maroon*), K⁺ (*turquoise sphere*), and Mg^{2+} (*blue sphere*)

Potassium-Binding Site Types in Proteins, Fig. 3 Structural representation of type-II enzyme dialkylglycine decarboxylase (pdb ID: 1DKA) K⁺-binding site. Shown are substrate (CPK, C in *yellow*), relevant residues (CPK, C in *maroon*), K⁺ (*turquoise sphere*), and Na⁺ (*pink sphere*)

larger than the difference in ionic radii between Na⁺ and K⁺. Therefore, a mechanism for the K⁺-dependent switching in dealkylglycine decarboxylase accounting for protein flexibility will require further investigations.

K⁺ Channels

K⁺ channels selectively transport K⁺ ions across membranes. They set the resting cellular membrane potential and control the duration of action potentials. While the generation of action potentials is the best-known role of ion channels, they have many other functions. For example, they are crucial for the transepithelial transport of salt and water, the regulation of cellular volume and pH, and chemical signaling (Hille 2001). Membrane proteins have a hydrophobic exterior, which makes them insoluble in water and hard to crystallize for X-ray diffraction studies. NMR spectroscopy may potentially offer an alternative method, but has been limited to date to relatively small membrane-inserted peptides and protein fragments. For many years, selective ion channels,

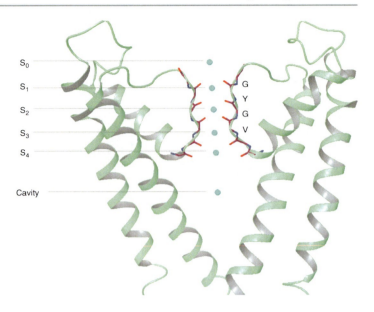

Potassium-Binding Site Types in Proteins, Fig. 4 X-ray structure of KcsA K^+ channel (pdb ID: 1BL8). The channel made of four identical subunits (for clarity only two of four subunits are shown in the figure). Cation binding sites are shown with *turquoise spheres*

a critically important class of membrane proteins, seemed to be almost beyond the reach of structural determination methods. The determination of the structure of the bacterial KcsA channel from the thermophile *Streptomyces lividans* in 1998 by Rod MacKinnon and coworkers (Doyle et al. 1998) offered a first view of the general architecture of biological K^+ selective channels at the atomic level (see Fig. 4). Although KcsA was a bacterial channel, its high sequence identity with the pore domain of important eukaryotic K^+ channels such as *Shaker* (~38%), guaranteed that the X-ray structure would help understand a large class of biologically important channels.

The membrane-inserted channel provides water-filled pathway, which enables selective K^+ permeation across the lipid membrane at a rate near the diffusion limit. This complex process is achieved by highly conserved structural element known as "Selectivity Filter" (Fig. 4). The narrow selectivity filter provides a translocation pathway for K^+ ions lined with polar carbonyl oxygens. The combination of experimental investigations and computational studies based on all-atom MD simulations led to the established of the molecular mechanism of ion permeation (Berneche and Roux 2001). At physiological concentrations, the selectivity filter is occupied predominantly by two K^+ ions in the $[S_1, S_3]$ or $[S_2, S_4]$ configurations (see Fig. 4). During an outward conduction, a third ion from the intracellular cavity enters S_4, while two ions are located at S_1 and S_3 sites in the selectivity filter. An electrostatic repulsion due to an incoming ion induces a concerted transition to a state with the three ions occupying S_4, S_2, and S_0 sites. This is a short-living intermediate followed by the rapid dissociation and departure of the outermost ions in S_0, resulting in the overall conductance of single K^+. The two ion configurations $[S_1, S_3]$ and $[S_2, S_4]$ are thought to have similar free energies and relatively low separating barriers; therefore, the exchange rate between these ion configurations is extremely rapid.

Based on the crystal structure of KscA and computational studies, it has been suggested that K^+/Na^+ selectivity relies on a number of factors, such as the chemical composition of the site (coordination by eight carbonyl oxygens), the number of ligands involved, and the flexibility of the site as well as the indirect contribution of the protein matrix surrounding the ion-binding pocket (Yu et al. 2010). It was shown that the presence of polar ligands, such as carbonyl oxygens, ensures favorable binding to the given site in the selectivity filter and efficient dehydration (Zhou et al. 2001; Berneche and Roux 2001).

Conclusion

In this entry, the thermodynamic determinants of K^+ selectivity of protein binding sites were primarily discussed. How these simple binding processes affect function can become fairly complex in real systems, and may rely on a competitive occupancy of different sites in the protein, where multi-ion correlations modulate the conformation and function of the protein (Derebe et al. 2011). More efforts will be required to better understand the complex phenomenon of ion-coupled conformation-driven enzymatic activity, as well as to provide better insight into the coupling between multiple binding sites and functional activity.

Cross-References

▶ Cellular Electrolyte Metabolism
▶ Potassium Channel Diversity, Regulation of Potassium Flux across Pores
▶ Potassium Channels, Structure and Function
▶ Potassium Dependent Sodium/Calcium Exchangers

References

Ashcroft FM (1999) Ion channels and disease. Academic, San-Diego
Berneche S, Roux B (2001) Energetics of ion conduction through the K$^+$ channel. Nature 414:73–77
Derebe MG et al (2011) Tuning the ion selectivity of tetrameric cation channels by changing the number of ion binding sites. Proc Natl Acad Sci USA 108:598–602
Di Cera E, Page MJ (2006) Role of Na$^+$ and K$^+$ in enzyme function. Physiol Rev 86:1049–1092
Doyle DA, Morais Cabral J et al (1998) The structure of the potassium channel: molecular basis of K$^+$ conduction and selectivity. Science 280:69–77
Harding MM (2002) Metal-ligand geometry relevant to proteins and in proteins: sodium and potassium. Acta Cryst D 58:872–874
Hille B (2001) Ion channels of excitable membranes, 3rd edn. Sinauer, Sunderland
Neyton J, Miller C (1988) Potassium blocks barium permeation through a calcium-activated potassium channel. J Gen Physiol 92:549–567
Yu H, Noskov SY, Roux B (2010) Two mechanisms of ion selectivity in protein binding sites. Proc Natl Acad Sci USA 107:20329–20334
Zhou Y et al (2001) Chemistry of ion coordination and hydration revealed by a K$^+$ channel-Fab complex at 2.0 A resolution. Nature 414:43–48

PP2B

▶ Calcineurin

PP3

▶ Calcineurin

Praseodymium

Takashiro Akitsu
Department of Chemistry, Tokyo University of Science, Shinjuku-ku, Tokyo, Japan

Definition

A lanthanoid element, the second element (cerium group) of the f-elements block, with the symbol Pr, atomic number 59, and atomic weight 140.90765. Electron configuration [Xe]4f^{3}6s^2. Praseodymium is composed of stable (^{141}Pr, 100%) and two synthetic radioactive (^{142}Pr; ^{143}Pr) isotopes. Discovered by C. A. von Welsbach in 1885. Praseodymium exhibits oxidation states III and II (and IV in the solid states); atomic radii 182 pm, covalent radii 204 pm; redox potential (acidic solution) Pr^{3+}/Pr -2.462 V; Pr^{3+}/Pr^{2+} -2.7 V; electronegativity (Pauling) 1.13. Ground electronic state of Pr^{3+} is $^3H_4$ with $S = 1$, $L = 5$, $J = 4$ with $\lambda = 360$ cm^{-1}. Most stable technogenic radionuclide ^{143}Pr (half-life 13.57 d). The most common compounds: PrO, Pr$_2$O$_3$, Pr$_6$O$_{11}$, PrF$_2$ PrF$_3$ and PrF$_4$. Biologically, praseodymium is of low to moderate toxicity, Pr^{3+} ions prevent from accumulation of Ca^{2+} in mitochondria and reactions of Ca^{2+} in the inner membrane of mitochondria (Atkins et al. 2006; Cotton et al. 1999; Huheey et al. 1997; Oki et al. 1998; Rayner-Canham and Overton 2006).

Cross-References

▶ Lanthanide Ions as Luminescent Probes
▶ Lanthanide Metalloproteins
▶ Lanthanides and Cancer
▶ Lanthanides in Biological Labeling, Imaging, and Therapy
▶ Lanthanides in Nucleic Acid Analysis
▶ Lanthanides, Physical and Chemical Characteristics

References

Atkins P, Overton T, Rourke J, Weller M, Armstrong F (2006) Shriver and Atkins inorganic chemistry, 4th edn. Oxford University Press, Oxford/New York

Cotton FA, Wilkinson G, Murillo CA, Bochmann M (1999) Advanced inorganic chemistry, 6th edn. Wiley-Interscience, New York

Huheey JE, Keiter EA, Keiter RL (1997) Inorganic chemistry: principles of structure and reactivity, 4th edn. Prentice Hall, New York

Oki M, Osawa T, Tanaka M, Chihara H (1998) Encyclopedic dictionary of chemistry. Tokyo Kagaku Dojin, Tokyo

Rayner-Canham G, Overton T (2006) Descriptive inorganic chemistry, 4th edn. W. H. Freeman, New York

Prevention of Recurrences in Bipolar Disorders

▸ Lithium as Mood Stabilizer

Primary Amine Oxidase

▸ Copper Amine Oxidase

Primary Amine:oxygen Oxidoreductase (Deaminating)

▸ Copper Amine Oxidase

Prion

▸ Copper and Prion Proteins

Procofactors: Procoagulant Cofactors

▸ Blood Clotting Proteins

Proline Oxidase, Proline Dehydrogenase

▸ Chromium(VI), Oxidative Cell Damage

Promethium

Takashiro Akitsu
Department of Chemistry, Tokyo University of Science, Shinjuku-ku, Tokyo, Japan

Definition

A lanthanoid element, the fourth element (cerium group) of the f-elements block, with the symbol Pm, atomic number 61, and atomic weight 145. Electron configuration $[Xe]4f^5 6s^2$. Promethium is composed of no stable and three radioactive (^{145}Pm; ^{146}Pm; ^{147}Pm) isotopes. Discovered (J. A. Marinsky, L. E. Glendenin, and C. D. Coryell) in 1947. Promethium exhibits oxidation state III; atomic radii 183 pm, covalent radii 199 pm; redox potential (acidic solution) Pm^{3+}/Pm -2.423 V electronegativity (Pauling) 1.13. Ground electronic state of Pm^{3+} is $^5I_4$ with $S = 2$, $L = 6$, $J = 4$ with $\lambda = 260$ cm^{-1}. Most stable technogenic radionuclide ^{145}Pm (half-life 17.7 years). The most common compounds: Pm_2O_3, $PmCl_3$ and Pm(OH)$_3$. Biologically, promethium is of low to moderate toxicity, and Pm3$^+$ is known to be removed from the surface of platelets and smooth muscle by both Ca^{2+} and La^{3+} (Atkins et al. 2006; Cotton et al. 1999; Huheey et al. 1997; Oki et al. 1998; Rayner-Canham and Overton 2006).

Cross-References

▸ Lanthanide Ions as Luminescent Probes
▸ Lanthanide Metalloproteins
▸ Lanthanides and Cancer
▸ Lanthanides in Biological Labeling, Imaging, and Therapy
▸ Lanthanides in Nucleic Acid Analysis
▸ Lanthanides, Physical and Chemical Characteristics

References

Atkins P, Overton T, Rourke J, Weller M, Armstrong F (2006) Shriver and Atkins inorganic chemistry, 4th edn. Oxford University Press, Oxford/New York

Cotton FA, Wilkinson G, Murillo CA, Bochmann M (1999) Advanced inorganic chemistry, 6th edn. Wiley-Interscience, New York

Huheey JE, Keiter EA, Keiter RL (1997) Inorganic chemistry: principles of structure and reactivity, 4th edn. Prentice Hall, New York

Oki M, Osawa T, Tanaka M, Chihara H (1998) Encyclopedic dictionary of chemistry. Tokyo Kagaku Dojin, Tokyo

Rayner-Canham G, Overton T (2006) Descriptive inorganic chemistry, 4th edn. W. H. Freeman, New York

Promyelocytic Leukemia–Retinoic Acid Receptor α

▶ Promyelocytic Leukemia–Retinoic Acid Receptor α (PML–RARα) and Arsenic

Promyelocytic Leukemia–Retinoic Acid Receptor α (PML–RARα) and Arsenic

Sai-Juan Chen[1], Xiao-Jing Yan[2] and Zhu Chen[1]
[1]State Key Laboratory of Medical Genomics, Shanghai Institute of Hematology, Rui Jin Hospital Affiliated to Shanghai Jiao Tong University School of Medicine, Shanghai, China
[2]Department of Hematology, The First Hospital of China Medical University, Shenyang, China

Synonyms

Acute promyelocytic leukemia; Arsenic sulfide; Arsenic trioxide; Promyelocytic leukemia–retinoic acid receptor α; RBCC domain

Definition

Translocation and the resultant fusion promyelocytic leukemia (PML)–retinoic acid receptor-α (RARα) are the most constant genomic abnormalities in APL cells. APL, the M3 subtype of acute myeloid leukemia (AML), seemed firstly to be the most fatal type of acute leukemia with a severe bleeding tendency and a fatal course of only weeks. Since the introduction of all-trans retinoic acid (ATRA) in the treatment, the 5-year disease-free survival (DFS) has been up to 75%. The use of arsenic trioxide (As_2O_3) since early 1990s has further improved the clinical outcome of refractory or relapsed as well as newly diagnosed APL. Combination of ATRA and As_2O_3 cures most patients with this leukemia.

Acute Promyelocytic Leukemia and PML–RARα

APL is a distinct subtype of AML. It was firstly described in 1957 by Hillestad, which "seems to be the most malignant form of acute leukemia." Clinically, APL had a dismal prognosis, particularly because of fibrinogenopenia and disseminated intravascular coagulation, before the introduction of anthracyclines and differentiation therapy (Wang and Chen 2008). The introduction of all-trans retinoic acid (ATRA) in 1985 and the pure form of As_2O_3 in 1994 to improve the response of these patients has led to a revolution in the prognosis of this disease. Indeed, treatment regimens combining ATRA and As_2O_3 have yielded a long-term remission in most patients, 90% of them being disease-free and off-treatment after 5 years (Chen et al. 2011; Zhu et al. 2002).

Cytogenetically, APL is characterized by a balanced reciprocal translocation between chromosomes 15 and 17 in more than 95% of patients. The t(15;17) fuses retinoic acid receptor-α (*RARα*) gene on chromosome 17q21 to promyelocytic leukemia (*PML*) gene on 15q22, yielding the PML–RARα fusion protein that initiates APL through both a block to differentiation and an increasing self-renewal of leukemic progenitor cells (Pandolfi 2011; Puccetti and Ruthardt 2004). In rare cases, *RARα* can form fusion with other partner genes such as the promyelocytic leukemia zinc finger (*PLZF*) located on chromosome 11q23. RARα is a RA-responsive transcription factor of the nuclear receptor superfamily, and PML is the organizer of nuclear subdomains that are linked to posttranslational modifications and the control of stem cell self-renewal. The PML–RARα fusion protein retains all the functional domains of RARα and PML (Fig. 1): the DNA-binding, hormone-binding, and retinoid-X-receptor (RXR)-binding domains in RARα; the RING domain, two B box motifs, and a coiled-coil domain in PML (also called RBCC domain) (de Thé and Chen 2010; Chen et al. 2011). Historically, pathogenetic models of APL are largely based on the assumption that PML–RARα transforms hematopoietic progenitors through the transcriptional repression of RARα target genes in

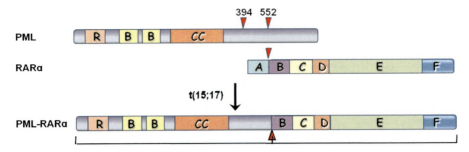

Promyelocytic Leukemia–Retinoic Acid Receptor α (PML–RARα) and Arsenic, Fig. 1 Structure of the promyelocytic leukemia (PML) and retinoic acid receptor-α (RARα) proteins, together with PML–RARα fusion. RING (*R*), B box (*B*), and coiled-coil (*CC*) domains in PML are indicated. The RARα DNA-binding domain (*C*) and hormone-binding domain (*E*) are shown. *A*, *B*, *D*, and *F* are other regulatory domains

a dominant-negative manner (Fig. 2). PML–RARα homodimerizes and forms a complex with another nuclear receptor, RXR (retinoid X receptor). The complex recruits the transcriptional corepressors, histone deacetylases (HDAC), and DNA methyltransferases to repress transcription. This whole complex could mediate the myeloid differentiation block that leads to APL. However, collective studies show that deregulated RARα transcriptional regulation alone does not sufficiently explain APL pathogenesis. The recent identification of several crucial PML–RARα features now yields a much more complex image of APL pathogenesis. In the revised model for APL pathogenesis, the PML–RARα fusion is a multifaceted protein that deregulates differentiation and self-renewal of myeloid progenitors and confers resistance to apoptosis, involving RXRα binding, corepressor and DAXX binding, sequestration of the essential myeloid regulator PU.1, PML delocalization, polycomb recruitment, p53 deregulation, binding site specificity degeneration, and AKT or PTEN deregulation (de Thé and Chen 2010).

PML–RARα as a Direct Arsenic Target

That As_2O_3 exerts drastic therapeutic effects against APL, but not other subtypes of AML, suggests a crucial link between its mechanism of action and PML–RARα. The RBCC domain retained in PML contains several cysteine-rich zinc fingers (ZFs), a RING domain, and two B box motifs, which are capable of binding metal (physiologically zinc) ions. Indeed, it has been shown that arsenic efficiently triggers the degradation of PML–RARα through its PML moiety (Fig. 2). In pharmacological concentrations, arsenic is able to competitively replace zinc in PML zinc fingers. It binds the ZFs of the PML RBCC domains through thiol groups of cysteines with the formation of arsenic-sulfur bonds, either intramolecularly or by forming cross-links between the two RBCC molecules in the homodimer. Of note, arsenic tends to coordinate with three sulfur atoms from the three conserved cysteines in PML ZFs, whereas zinc exhibits the typical tetrahedral coordination with conserved cysteine and histidine residues. Consequently, the protein can adopt different conformations upon coordination with different metals. The resultant conformational changes may facilitate further oligomerization of PML–RARα and promote SUMOylation of the protein through enhanced interaction of PML with the enzymes that catalyze this modification (such as UBC9) or through enhanced exposure of the modification sites. Recent studies indicate that RNF4, a ubiquitin E3 ligase containing SUMO interaction motifs, could recruit sumoylated PML–RARα and promote its proteasomal degradation (Chen et al. 2011; de Thé and Chen 2010; Liu et al. 2012).

In eukaryotic cells, PML proteins represent the main components of the nuclear bodies (NBs) which are large, spherical-shaped nuclear organelles essential to the regulation of apoptosis, gene transcription, and storage/modulation of certain nuclear proteins. In APL cells, NB structures are disrupted owing to the formation of PML/PML–RARα heterocomplex, leading to numerous tiny dots in immunofluorescence examination. When the cells were treated with arsenic, these PML protein-containing dots transiently aggregate to form larger

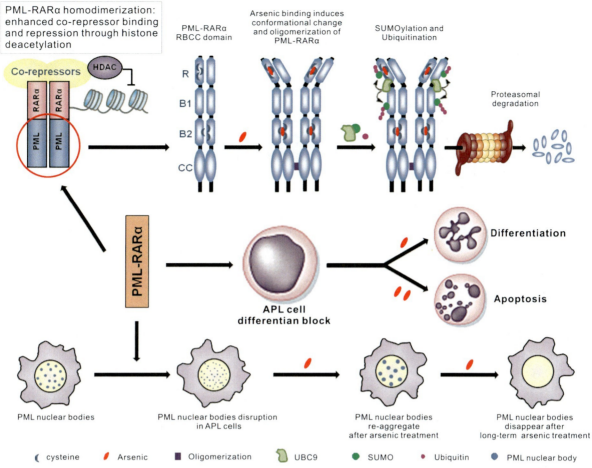

Promyelocytic Leukemia–Retinoic Acid Receptor α (PML–RARα) and Arsenic, Fig. 2 The classical model of APL pathogenesis and the effects of As_2O_3 on PML–RARα function. The PML–RARα fusion is a multifaceted protein that deregulates differentiation of myeloid progenitors. PML–RARα homodimers bind and repress RARα targets through enhanced recruitment of corepressors. It also disrupts PML nuclear bodies. Arsenic directly targets PML RBCC domain to control the fate of PML–RARα. It also restores the PML nuclear bodies which are disrupted in APL cells

particles at the nuclear matrix and then disappear, while the cells are committed to apoptosis or partial differentiation (Fig. 2) (Bernardi and Pandolfi 2007; Chen et al. 2011).

Arsenic also induces ROS production to initiate intermolecular disulfide formation. PML or PML–RARα multimers linked by disulfide associate with nuclear matrix and form PML NBs, indicating that PML oxidation regulates NB biogenesis. Interestingly, nonarsenical oxidants also induce PML–RARα multimerization, NB association, degradation, and leukemia response in vivo. However, oxidants cannot affect APL driven by PLZF–RARα, demonstrating that PML is the key target of arsenic (de Thé and Chen 2010).

Other Mechanisms of Action

Besides the direct targeting on PML–RARα, arsenic can eliminate leukemia-initiating cells (LICs) which are pluripotent, self-renewing, phenotypically primitive, and mitotically quiescent cells that have been identified in leukemia. In APL, PML–RARα is required, and even a minute amount of the oncoprotein allows LICs self-renewal in vivo. The ATRA/As_2O_3 combination rapidly clears PML–RARα[+] LICs, resulting in APL eradication in murine models and patients. However, the precise molecular mechanisms remain obscure. Arsenic can facilitate elimination of LICs through decreasing PML expression, inhibiting notch pathway, antagonizing the hedgehog pathway by reducing stability of the Gli2 transcriptional effector and repressing NFκB and β-catenin (Chen et al. 2011; Puccetti and Ruthardt 2004).

A large body of knowledge indicates that arsenic can induce apoptosis of leukemia cells via histone H3 phosphoacetylation at *Casp-10*; up-regulation of a set of genes responsible for ROS production and intracellular oxidative DNA damage; perturbation of *ATR*, *Chk2*, *p53*, *JNK*, anion exchanger 2, and *GSTP1-1*; repression of telomerase reverse transcriptase gene (*hTERT*), *C17*, and *c-Myc*; and suppression of *NFκB* and *Wt1*. Arsenic also induces autophagy by the mammalian target of rapamycin (mTOR) or MEK/ERK pathway, while autophagic degradation contributes significantly to proteolysis of PML–RARα (Wang and Chen 2008; Zhou et al. 2005).

Mechanism of As$_4$S$_4$-Containing Formula Based on TCM

The remarkable results of the realgar-indigo naturalis formula (RIF), in which As$_4$S$_4$ is the principle ingredient, have been reported. The mechanisms of action of RIF are investigated by using As$_4$S$_4$ (A), indirubin (I), and tanshinone IIA (T) as representatives of realgar, indigo naturalis, and Salvia miltiorrhiza, respectively. The results show that ATI combination exhibits enhanced therapeutic efficacies against APL in murine model and causes a much more profound differentiation of APL cells. Molecularly, ATI combination triggers potentiated ubiquitination and degradation of PML–RARα oncoprotein and stronger reprogramming of myeloid differentiation regulators in APL cells as compared to monoagents or biagents. Moreover, T and I facilitate transportation of arsenic into malignant promyelocytes by up-regulation of AQP9, which in turn intensify arsenic-mediated PML–RARα degradation and therapeutic efficacies (Chen et al. 2011; Wang and Chen 2008).

Arsenic is a natural substance and a traditional poison. It has also been used as a drug with appropriate application for over 2000 years in both traditional Chinese medicine and the Western world. However, the medical use of arsenic faces an embarrassment partly due to its low effect compared to modern therapy but mostly because of concerns about the toxicity and potential carcinogenicity. The great success of As$_2$O$_3$ on treatment of APL revives this old remedy. The underlying molecular mechanism reminds us that there is an extraordinary richness of traditional natural medicine and an ancient medicine, revived through careful clinical and biological studies in modern times, will have an even greater impact on human health.

Cross-References

▶ Arsenic and Primary Human Cells
▶ Arsenic in Nature
▶ Arsenic in Pathological Conditions
▶ Arsenic in Therapy
▶ Arsenic in Tissues, Organs, and Cells

References

Bernardi R, Pandolfi PP (2007) Structure, dynamics and functions of promyelocytic leukaemia nuclear bodies. Nat Rev Mol Cell Biol 8:1006–1016

Chen SJ, Zhou GB, Zhang XW et al (2011) From an old remedy to a magic bullet: molecular mechanisms underlying the therapeutic effects of arsenic in fighting leukemia. Blood 117:6425–6437

de Thé H, Chen Z (2010) Acute promyelocytic leukaemia: novel insights into the mechanisms of cure. Nat Rev Cancer 10:775–783

Liu JX, Zhou GB, Chen SJ et al (2012) Arsenic compounds: revived ancient remedies in the fight against human malignancies. Curr Opin Chem Biol 16:92–98

Pandolfi PP (2011) Oncogenes and tumor suppressors in the molecular pathogenesis of acute promyelocytic leukemia. Hum Mol Genet 10:769–775

Puccetti E, Ruthardt M (2004) Acute promyelocytic leukemia: PML/RARalpha and the leukemic stem cell. Leukemia 18:1169–1175

Wang ZY, Chen Z (2008) Acute promyelocytic leukemia: from highly fatal to highly curable. Blood 111:2505–2515

Zhou GB, Chen SJ, Chen Z (2005) Acute promyelocytic leukemia: a model of molecular target based therapy. Hematology 10(Suppl 1):270–280

Zhu J, Chen Z, Lallemand-Breitenbach V (2002) How acute promyelocytic leukaemia revived arsenic. Nat Rev Cancer 2:705–713

Propagermanium

▶ Germanium-Containing Compounds, Current Knowledge and Applications

Protactinium, Physical and Chemical Properties

Fathi Habashi
Department of Mining, Metallurgical, and Materials Engineering, Laval University, Quebec City, Canada

Protactinium is a radioactive metal formed in the decay chain of uranium but not thorium (Fig. 1). It exists in two crystalline forms: α-Protactinium is tetragonal body-centered ($a = 393.1$ pm, $c = 323.6$ pm) and β protactinium is cubic face-centered ($a = 501.9$ pm). Protactinium isotope ^{231}Pa ($t_{1/2} = 32{,}480$ years) is the most important of all 24 known isotopes. The isotope ^{233}Pa is of special importance because it is formed as an intermediate in the production of fissile ^{233}U from non-fissile ^{232}Th by neutron capture:

$$^{232}_{90}\text{Th} + ^{1}_{0}\text{n} \rightarrow ^{233}_{90}\text{Th} \xrightarrow[22.3 \text{ min}]{\beta^-} ^{233}_{91}\text{Pa} \xrightarrow[26.967 \text{ d}]{\beta^-} ^{233}_{92}\text{U}$$

Physical Properties

Atomic number	91
Atomic weight	201.04
Relative abundance in Earth's crust, %	8×10^{-11}
Atomic radius, pm	163
Density, g/cm^3	15.37
Melting point, °C	1,568
Boiling point, °C	4,027
Heat of fusion, kJ·mol^{-1}	12.34
Heat of vaporization, kJ·mol^{-1}	481
Electrical resistivity at 0°C, nΩ·m	177
Thermal conductivity, W·m^{-1}·K^{-1}	47

Chemical Properties

Protactinium metal is best prepared by the Van Arkel method using a PaC–I$_2$ mixture. In its chemical and physical behavior, protactinium is similar to tantalum. In the solid state and in solution, protactinium can occur in the oxidation state Pa^{4+} and Pa^{5+}. Binary compounds exist such as the tetrahalides PaX$_4$ and pentahalides PaX$_5$. Complex compounds include the sparingly soluble K$_2$PaF$_7$. The stable white Pa$_2$O$_5$ occurs in five modifications. The reduction of Pa$_2$O$_5$ with hydrogen at 1,550°C yields black PaO$_2$. A protactinium peroxide can be precipitated from dilute acids. This method is used to separate protactinium from tantalum and niobium which form soluble peroxide complexes.

Pa (V) is hydrolyzed in aqueous solution even in strong acids, if not complexed, for example, by F$^-$ ions. No Pa^{5+} exists in solution. Only partly hydrolyzed species with two positive charges (in 1 M HCl) or one positive charge (in $10^{-3} - 10^{-1}$ M H$^+$) could be identified. The partly hydrolyzed protactinium species form stable complexes with most organic ligands. Pa (V) can be extracted from aqueous solution by amines and most neutral and acid phosphoric and phosphonic extraction agents. Pa (IV) solution is obtained by reduction of Pa (V) solutions by zinc or electrochemical methods. The behavior of Pa (IV) is similar to other actinides of oxidation state IV.

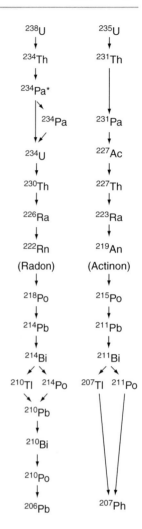

Protactinium, Physical and Chemical Properties, Fig. 1 Protactinium in the decay scheme of uranium

References

Habashi F (2003) Metals from ores. An introduction to extractive metallurgy. Métallurgie Extractive Québec, Québec City. Distributed by Laval University Bookstore, www.zone.ul.ca

Luig H et al (1997) Radioactive metals. General. In: Habashi F (ed) Handbook of extractive metallurgy. Wiley, Weinheim, pp 1585–1598

Protein Adsorption onto Ultrasmall Gold Nanoclusters

▶ Gold, Ultrasmall Nanoclusters and Proteins, Interaction

Protein Corona Formation Around Ultrasmall Gold Nanoclusters

▶ Gold, Ultrasmall Nanoclusters and Proteins, Interaction

Protein Interaction Network

▶ Proteomic Basis of Cisplatin Resistance

Protein Kinase A (PKA)

▶ Calcium Sparklets and Waves

Protein Kinase Cα (PKCα)

▶ Calcium Sparklets and Waves

Protein Phosphatase 2B or Calcineurin (PP2B)

▶ Calcium Sparklets and Waves

Protein-Gold Nanoparticle Conjugation

▶ Gold Nanoparticles and Fluorescent Proteins, Optically Coupled Hybrid Architectures

▶ Gold Nanoparticles and Proteins, Interaction

Proteins Containing 2-Selenoalanine

▶ Selenoproteins in Prokaryotes

Proteins with Mercury as "Natural" Substrate

▶ Bacterial Mercury Resistance Proteins

Proteomic Basis of Cisplatin Resistance

Juan D. Chavez and James E. Bruce
Department of Genome Sciences, University of Washington, Seattle, WA, USA

Synonyms

Protein interaction network; Resistance of platinum anticancer drugs

Definition

Cisplatin along with other platinum-based drugs including oxaliplatin, carboplatin are some of the most widely used chemotherapeutic agents. Unfortunately, drug resistance remains a major problem for the successful chemotherapeutic treatment of cancer. Current evidence suggests that resistance toward platinum chemotherapeutics is a multifactorial problem due to changes in the expression levels and activities of a wide number of proteins.

Cisplatin Resistance and Changes to the Proteome

The primary goal of cancer chemotherapy is to disrupt tumor cell growth by interfering with cell division (mitosis) or activating programmed cell death pathways (apoptosis). The powerful anticancer properties of cisplatin ((SP-4-2)-diamminedichloridoplatinum, cis-PtCl$_2$(NH$_3$)2) were first discovered by accident in a series of studies carried out in the 1960s and 1970s. The United States Food and Drug Administration (FDA) approved the use of cisplatin as an anticancer treatment in 1978, and spawned much research into the biological activities of platinum and other transition metal-based complexes. Since then 23 additional platinum-based complexes have entered clinical trials, however, at present there are only three platinum-based drugs (cisplatin, oxaliplatin, and carboplatin) used for the clinical treatment of cancer worldwide (▶ Platinum Anticancer Drugs). Three additional platinum-based drugs (nedaplatin, lobaplatin, and heptaplatin) are approved for use in Japan, China, and Korea, respectively. Chemical structures for these six platinum-based drugs under current clinical use are shown in Fig. 1.

Two major complications of cisplatin use in chemotherapy are the undesired dose limiting side effects caused by nephrotoxicity and resistance of the cancer cells to the drug. Generally, acquired resistance is the most common reason for cancer chemotherapy failure. Possible mechanisms for resistance to cisplatin include reduced intracellular concentration of cisplatin by increased drug efflux and/or decreased drug influx, inactivation of cisplatin by reaction with glutathione and other intracellular nucleophiles, increased repair of DNA damage, and altered apoptotic signaling pathways.

Proteins are the molecular mediators of nearly all cellular processes and most function is achieved through a complex protein interaction network termed the interactome. Changes in the activities and cellular concentrations of proteins allow cells to adapt and survive environmental stresses, including exposure to cytotoxic drugs such as cisplatin. Therefore, quantitative information on relative protein expression levels between drug sensitive and resistant cancer cells is important for understanding the molecular mechanisms which give rise to drug resistance. Proteins that display differential expression with cisplatin resistance are likely to be involved in functional pathways important to the mechanism of resistance and are therefore potential targets for monitoring drug treatment response as well as development of new chemotherapy drugs.

The field of proteomics has advanced greatly since the early 1990s primarily due to technological advances in liquid chromatography-mass spectrometry(LC-MS) analytical platforms. Modern mass spectrometry-based quantitative proteomics experiments are capable of identifying and monitoring the relative expression levels of thousands of proteins allowing for comparison between different biological conditions, akin to using DNA microarrays for monitoring gene expression data. Mass spectrometry has therefore become the method of choice for comparative proteomics studies, for example, comparing proteome changes between drug resistant and sensitive cancer cells (Zhang and Liu 2007). A typical mass spectrometry-based proteomics workflow is illustrated in Fig. 2. While there are many different varieties of quantitative LC-MS experiments, they can be generally classified into two types: those employing isotope or mass labeling and label-free methods. Both approaches have specific advantages and disadvantages which are not discussed in detail here, but both have established widespread applications in current proteomics research. More traditional approaches for quantifying protein expression levels such as SDS-PAGE, Western blotting, and immunoassays are generally much more limited in scope in terms of numbers of proteins that can be quantified, however, still find widespread use in monitoring specific protein targets of interest. Each of these analytical approaches has contributed valuable information to the body of knowledge we currently have about how the proteome contributes to the molecular mechanisms that give rise to drug resistance in cancer. Global quantitative proteomics experiments produce vast quantities of data making detailed discussion on each protein impractical. The following discussion will therefore focus on examples of specific proteins from varying functional classes that have been measured by one or more techniques and are well established in the scientific literature to have altered expression levels correlated with cisplatin resistance.

Cisplatin Resistance and Cellular Influx/Efflux

Cisplatin may enter cells by either passive diffusion or active transport across the plasma membrane with the aid of membrane bound ion transport proteins. Evidence

Proteomic Basis of Cisplatin Resistance, Fig. 1 Chemical structures for platinum-based drugs under current clinical use around the world

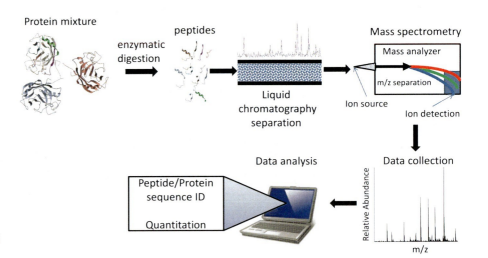

Proteomic Basis of Cisplatin Resistance, Fig. 2 Overview of LC-MS experiment

exists for both modes of transport and it is estimated that 50% of cisplatin is transported into the cell by active transport by membrane proteins (Kartalou and Essigmann 2001). Once inside the cell, the chloride ion concentration is much lower than the extracellular matrix and cisplatin undergoes spontaneous aquation with water molecules replacing the *cis*-chlorido ligands. The aquation reaction activates cisplatin making it a potent electrophile which will react with and covalently modify nucleophilic sites on biomolecules including DNA, RNA, and proteins. The cytotoxic activity of cisplatin is primarily attributed to its reaction with nucleophilic purine bases in DNA to form intrastrand DNA cross-links. A simplified diagram illustrating the mechanism of action of cisplatin is shown in Fig. 3. Proteins thought to be involved in the transport of cisplatin include sodium-potassium dependent ATPases as well as the copper transporter Ctr1 (▶ Platinum

Proteomic Basis of Cisplatin Resistance, Fig. 3 Cisplatin transport into cells

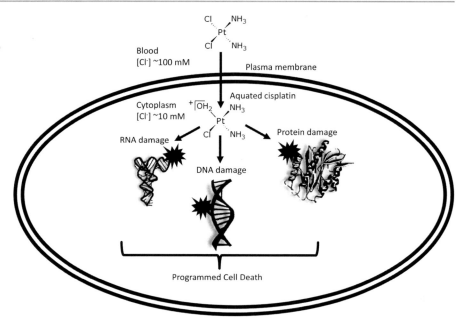

Interaction with Copper Proteins). Ctr1 is the primary transporter for copper into cells and has been shown to compete with copper for transport into cells. In certain cellular systems, decreased expression levels of Ctr1 have been shown to lower cisplatin uptake to 35% of normal levels and lead to a two- to threefold increase in drug resistance (Kelland 2007). Reduced influx of cisplatin has also been demonstrated over a wide range of extracellular concentrations implicating changes in the non-saturable process of passive diffusion of cisplatin through the cell membrane. Due to the multifactorial mechanism of drug resistance, decreased influx of cisplatin can only partially explain the observed resistance. Generally, increased drug efflux is thought to play a lesser role in resistance to cisplatin compared with other chemotherapy drugs. In part, this could be because cisplatin is not known to be a substrate of P-glycoprotein also known as multidrug resistance protein 1 (MDR1), a member of the ATP-binding cassette protein family, which is a widely implicated drug efflux pump in the multidrug-resistant phenotype. Studies have yet to yield a consensus of directional shift of expression levels of MDR1 or any of the ATP-binding cassette proteins. Other efflux proteins that have been implicated in the export cisplatin are the copper-transporting ATPases ATP7A and ATP7B. While the precise mechanism of how ATP7A/B participate in cisplatin resistance remains unclear, it has been shown that cisplatin binds to the copper-binding sites in these proteins and increased expression of these proteins tends to increase resistance. However, measurements on *trans*-membrane proteins like ATP7A/B are notoriously difficult due to their hydrophobicity and poor solubility outside of the lipid bilayer. This could in part explain why identifying and quantifying each of the specific membrane protein transporters playing a role in the influx and efflux of cisplatin remains difficult.

Cisplatin Resistance and Intracellular Nucleophiles

As mentioned above, cisplatin becomes aquated inside the cell to yield a highly reactive electrophile. Aquated cisplatin will readily react with nucleophilic sites on proteins and other biomolecules. In proteins, the sulfur-containing thiol functional groups on cysteine side chains are the most nucleophilic sites, and are frequently used by proteins to bind metals, for instance, the metal-binding metallothioneins (▶ Metallothioneins and Copper). While cisplatin has been shown to bind to metallothioneins and their overexpression has been associated with cisplatin resistance, increased expression of these proteins with cisplatin resistance does not appear to be a universal observation (Kartalou and Essigmann 2001).

One of the most abundant thiol-containing nucleophiles, reaching concentrations of ~10 mM in the cell, is the tripeptide glutathione (γ-L-glutamyl-L-cysteinylglycine, GSH). Glutathione functions as a major antioxidant in cells maintaining cellular redox homeostasis and serving to protect important biomolecules from damage from free radicals, reactive oxygen species (ROS), and electrophilic molecules. The conjugation reaction between cisplatin and GSH is a primary method of cellular inactivation of cisplatin and generally considered to be a significant factor in the mechanism of resistance. While this conjugation reaction will proceed spontaneously, it can also be catalyzed by the glutathione-S-transferase family of enzymes. Not surprisingly then, GSH-S-transferases have frequently been measured with increased expression levels associated with cisplatin resistance. Once formed the cisplatin-GSH conjugates can then be exported from the cell via an ATP-dependent glutathione-S conjugate pump. GSH-S-transferases also contribute to increase cisplatin resistance by participating in cell survival and apoptotic signaling pathways. For example, overexpression of the GSH-S-transferase GSTO1 was found to confer resistance to cisplatin primarily through activation of the phosphatidylinositol-3-kinase/serine/threonine kinase (PI3-K/AKT) pathway which promotes cell survival, as well as inhibition of the c-Jun N-terminal kinase (JNK) apoptotic pathway. Additionally, increased levels of γ-glutamyltransferase (γ-GT), a key enzyme for maintaining proper levels of GSH, has also been associated with cisplatin resistance. Cysteinylglycine, which is 10 times more reactive with cisplatin than GS is generated during the γ-GT catalyzed degradation of GSH and may explain how γ-GT participates in the mechanism of cisplatin resistance (Siddik 2003).

Cisplatin Resistance and Redox Homeostasis

Peroxiredoxins are a family of antioxidant enzymes which decompose peroxides and are functionally recycled by GSH and GSH-S-transferase enzymes. Increased levels of several members of the peroxiredoxin family, including peroxiredoxins 2 and 6, have been measured in cisplatin-resistant cells. Increased levels of antioxidant enzymes in the peroxiredoxin and glutathione families are a result of redox adaptation of the cancer cells, which is frequently observed in late stages of the disease and is associated with poor prognosis. Redox adaptation enables the cancer cells to survive under increased ROS stress and promotes resistance to many anticancer drugs, through the mechanisms of increased tolerance of exogenous stress, upregulation of survival pathways, and increased capacity for drug inactivation. Recently, the targeting of cellular redox adaptation observed in many cancers and drug-resistant phenotypes has become the focus of developing new therapeutic approaches to kill cancer (Trachootham et al. 2009).

Cisplatin Resistance and DNA Damage Response

The formation and persistence of DNA-cisplatin adducts activates cellular DNA damage response signals which are critical for the cytotoxic effects of cisplatin. Therefore, increased DNA damage repair is an important factor in the mechanism of resistance. Cells have multiple mechanisms for repairing DNA damage. Nucleotide excision repair (NER) is the primary pathway for the removal and repair of platinum-DNA adducts. NER is a highly conserved DNA repair pathway that repairs DNA lesions which alter the helical structure of the DNA molecule and interfere with DNA replication and transcription. Protein complexes carry out the important steps in this process including the recognition of site of damage to the DNA helix, unraveling and removal of the damaged section, and repair of the DNA molecule. There are at least nine major proteins involved in NER and the increased expression levels of several of them have been tied with cisplatin resistance. In particular, levels of the DNA excision repair protein ERCC1 have been correlated with response to platinum-based therapy with increased levels associated with resistance. In testicular cancer in particular, it's been observed that low levels of ERCC1 are linked with a low capacity for NER in testicular cancer cell lines and listed as a reason why testicular cancer responds very favorably to cisplatin treatment (Kelland 2007). Another DNA repair pathway is nonhomolgous end joining (NHEJ), in which DNA double strand breaks are ligated. A number of proteins participate in the NHEJ process including the Ku protein complex. Ku is a heterodimer protein complex consisting of two protein subunits Ku70 and Ku80. The Ku complex binds to the catalytic subunit of DNA-dependent protein kinase (DNA-PKc) to form the complete active DNA-dependent protein kinase complex (DNA-PK). DNA-PK plays an important role in DNA damage response with the Ku proteins binding to the broken ends of the DNA strand and the kinase activity of DNA-PKc initiating a signaling cascade which can ultimately lead to cell death. Interestingly, decreased expression levels of both Ku80 and DNA-PKc have been linked with cisplatin resistance and have also been shown to induce cell death in response to cisplatin through a cell-interdependent signaling pathway involving gap junctions (Jensen and Glazer 2004).

Cisplatin Resistance and Calcium-Binding Proteins

Several quantitative proteomics studies on cisplatin resistance have identified calcium-binding proteins to be differentially expressed in cisplatin-resistant cancer cells. Generally, Ca^{2+}-binding proteins display increased expression levels correlated with drug resistance, suggesting an important role for Ca^{2+} signaling in the mechanism of resistance. Annexins comprise a large family of bind Ca^{2+}-binding proteins and are involved in a wide variety of cellular processes including apoptosis, cell division, vesicle trafficking, and calcium signaling (▶ Annexins). Annexins are thought to play several roles in the development of cancer related to tumor invasiveness, metastasis, and drug resistance (Mussunoor and Murray 2008). In particular, annexin A4 and annexin A11 exhibit differential expression with drug resistance, and are often used as markers tied with poor prognosis in cancer patients. Another family of Ca^{2+}-binding proteins that show differential expression with cisplatin resistance is the S100 proteins. S100 proteins are a family of low molecular weight (10–12 kDa) Ca^{2+}-binding EF-hand that are known to play a diverse role in many forms of cancer (▶ S100 Proteins). Both S100A4 and S100A6 have attracted significant attention in relation to drug resistance due to their established role in the progression of metastatic tumors. There are several known protein-protein interactions between annexins and S100 proteins, for example: annexin 1 with S100A11, annexin 2 with S100A10 and S100A4, and annexins 4 and 11 with S100A6. These interactions are highlighted in Fig. 4 displaying a heat map of over 800 quantified proteins and a quantitative protein interaction network produced from data generated in a recent proteomics study on cisplatin resistance in cervical cancer cells (Chavez et al. 2011). The depth of data generated from modern proteomics studies provide researchers the powerful opportunity to monitor changes to protein expression levels across multiple families of proteins as shown for the S100 and annexin proteins. Although the mechanism of annexin and S100 mediated drug resistance is currently not fully understood, it is clear they do play an important role.

Cisplatin Resistance and Ribosomal Proteins

Ribosomal proteins function primarily in the synthesis of new proteins through the process of translation. Human cells have 80s ribosomes consisting of a small and large subunit; the 40s subunit which contains 33 proteins, and a 60s subunit which contains approximately 50 proteins. These subunits are assembled together into a large biomolecular complex with ribosomal RNA to form the complete ribosome. In addition to their critical role in translation of mRNA into protein, several ribosomal proteins perform additional functions in apoptosis, DNA repair, and transcription (Lindstrom 2009). Ribosomal proteins including RPL5, 7, 11, and 23 have been shown to induce apoptosis when overexpressed. Interestingly, several comparative proteomics studies have found altered expression levels of several ribosomal proteins to be correlated with drug resistance (Zhang and Liu 2007). One recent LC-MS-based proteomics study identified 32 proteins from the 60S ribosomal subunit and 20 proteins from the 40S ribosomal subunit with significantly decreased expression levels in cisplatin-resistant cells (Chavez et al. 2011). The role the ribosomal proteins play in the mechanism of drug resistance is unclear; however, there is the intriguing possibility that their extraribosomal functions in apopotosis and DNA repair could be contributing factors.

Outlook

Drug resistance remains a major hurdle for the successful treatment of cancer. Proteomics studies are shedding new light on the molecular mechanisms

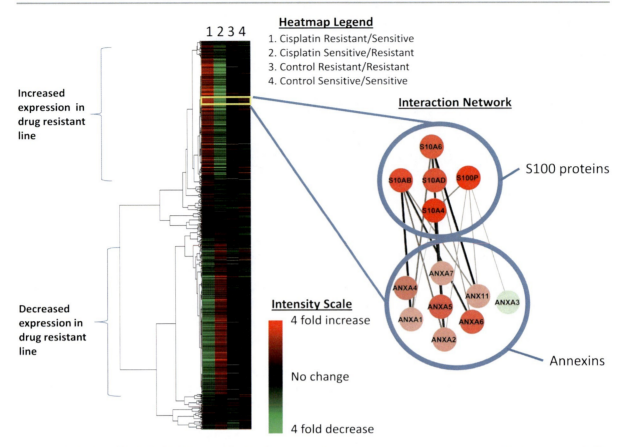

Proteomic Basis of Cisplatin Resistance, Fig. 4 Heat map and quantitative protein interaction network of annexin and S100 protein family members

involved in the drug-resistant phenotype. It is becoming more evident through proteomics studies performed to date that the drug-resistance problem is the result of changes to multiple interwoven and independent biological pathways involving hundreds of various proteins. Therefore, future treatments will be required to be multifaceted and combinatorial approaches to treatment will be necessary to circumvent the drug-resistance issue. Ultimately, the integration of quantitative proteomics experiments with dynamic measurements on the proteome and protein-protein interactions with genomic, transcriptomic, and metabolomic data will provide a more complete systems-level understanding of the drug-resistance phenomenon, which will hopefully lead to improved cancer treatment.

Acknowledgments This laboratory is grateful for support from the National Institutes of Health through grant numbers: 5R01RR023334, 1R01GM097112, 5R01GM086688, 7S10RR025107, and 1R01HL110879.

Cross-References

▶ Annexins
▶ Metallothioneins and Copper
▶ Platinum Anticancer Drugs
▶ Platinum Interaction with Copper Proteins

References

Chavez JD, Hoopmann MR et al (2011) Quantitative proteomic and interaction network analysis of cisplatin resistance in HeLa cells. PLoS One 6(5):e19892

Jensen R, Glazer PM (2004) Cell-interdependent cisplatin killing by Ku/DNA-dependent protein kinase signaling transduced through gap junctions. Proc Natl Acad Sci USA 101(16):6134–6139

Kartalou M, Essigmann JM (2001) Mechanisms of resistance to cisplatin. Mutat Res 478(1–2):23–43

Kelland L (2007) The resurgence of platinum-based cancer chemotherapy. Nat Rev Cancer 7(8):573–584

Lindstrom MS (2009) Emerging functions of ribosomal proteins in gene-specific transcription and translation. Biochem Biophys Res Commun 379(2):167–170

Mussunoor S, Murray GI (2008) The role of annexins in tumour development and progression. J Pathol 216(2):131–140

Siddik ZH (2003) Cisplatin: mode of cytotoxic action and molecular basis of resistance. Oncogene 22(47):7265–7279

Trachootham D, Alexandre J et al (2009) Targeting cancer cells by ROS-mediated mechanisms: a radical therapeutic approach? Nat Rev Drug Discov 8(7):579–591

Zhang JT, Liu Y (2007) Use of comparative proteomics to identify potential resistance mechanisms in cancer treatment. Cancer Treat Rev 33(8):741–756

Proximal Tubular Cells

▶ Mercury Nephrotoxicity

PrP^{Sc}

▶ Copper and Prion Proteins

Pt-DNA Adducts

▶ Zinc, Metallated DNA-Protein Crosslinks as Finger Conformation and Reactivity Probes

P-Type ATPase/Ion Pump

▶ Lead Transporters

Public Health

▶ Cadmium and Health Risks

Purine Hydroxylase

William Self
Molecular Biology & Microbiology, Burnett School of Biomedical Sciences, University of Central Florida, Orlando, FL, USA

The selenium-dependent molybdoenyzme purine hydroxylase (PH) catalyzes the hydroxylation of purines at either the 2 or 6 position of the purine ring (Self and Stadtman 2000). This specificity is unique among the class of molybdenum (Mo) hydroxylases as other members of this family of enzymes such as xanthine dehydrogenases catalyze hydroxylation of the 2, 6 and 8 position of the purine ring (Choi et al. 2004; Durre and Andreesen 1983; Schwarz and Mendel 2006). PH has only been studied from *Clostridium purinolyticum*, a microbe that was isolated using adenine as a primary carbon and nitrogen source (Durre and Andreesen 1983). The rationale for having an enzyme that efficiently converts hypoxanthine to xanthine but *not* xanthine to uric acid lies in the fact that xanthine is the purine intermediate that is attacked by a series of enzymes to obtain carbon (formate) and nitrogen (ammonia) through its catabolic breakdown (Durre and Andreesen 1983). The form of selenium in this enzyme, as in all other selenium-dependent molybdenum hydroxylases, is not selenocysteine, but it may be presumed to be either a ligand of the molybdenum cofactor or in the vicinity of the Mo atom as a selenosulfide (S-Se) form (Self et al. 2003).

Although the genes encoding this enzyme have not yet been cloned and sequenced, the regulation of the expression of the protein subunits was evaluated using polyclonal antibodies raised against the purified enzyme (Self 2002). The production of the enzyme was dependent on the level of selenium (selenite) or molybdenum (molybdate) in the culture medium. This is reminiscent of the regulation of the formate hydrogenlyase and nitrate reductase in *E. coli* in response to the availability of molybdate (Self et al. 1999). The level of enzyme activity paralleled the presence of the protein subunits by western blot, and the presence of the purine substrates (such as hypoxanthine) also played a role in upregulation of the enzyme (Self 2002). The molecular mechanism for regulation of this pathway remains unstudied since this organism is not likely to be genetically tractable.

The chemical form of the labile selenium cofactor, and whether it is bound to the Mo cofactor, is still unresolved for this enzyme. Electron paramagnetic analysis of the enzyme revealed the presence of the expected Mo(V) species upon reduction by hypoxanthine or 2OH-purine (Self et al. 2003). To probe whether the selenium atom was present as a ligand of the Mo cofactor (i.e., bound to molybdenum), a preparation of enzyme was prepared using ^{77}Se stable isotope. Upon reduction by hypoxanthine, the characteristic Mo(V) species was detected; however, no hyperfine coupling with Se was observed. This negative yet critical result strongly suggests that at least in the reduced form, the selenium atom in PH is not a direct ligand of the active site. Parallel x-ray absorption analysis of PH also suggests that Se and Mo are not bound in the oxidized enzyme (Robert Scott, personal communication). Thus, this enzyme and its putative reaction mechanism, as well as the catalytic requirement for selenium, are still poorly understood.

References

Choi EY, Stockert AL, Leimkuhler S, Hille R (2004) Studies on the mechanism of action of xanthine oxidase. J Inorg Biochem 98:841–848

Durre P, Andreesen JR (1983) Purine and glycine metabolism by purinolytic clostridia. J Bacteriol 154:192–199

Schwarz G, Mendel RR (2006) Molybdenum cofactor biosynthesis and molybdenum enzymes. Annu Rev Plant Biol 57:623–647

Self WT (2002) Regulation of purine hydroxylase and xanthine dehydrogenase from *Clostridium purinolyticum* in response to purines, selenium, and molybdenum. J Bacteriol 184:2039–2044

Self WT, Grunden AM, Hasona A, Shanmugam KT (1999) Transcriptional regulation of molybdoenzyme synthesis in *Escherichia coli* in response to molybdenum: modE-molybdate, a repressor of the modABCD (molybdate transport) operon is a secondary transcriptional activator for the hyc and nar operons. Microbiology 145(Pt 1):41–55

Self WT, Stadtman TC (2000) Selenium-dependent metabolism of purines: a selenium-dependent purine hydroxylase and xanthine dehydrogenase were purified from *Clostridium purinolyticum* and characterized. Proc Natl Acad Sci USA 97:7208–7213

Self WT, Wolfe MD, Stadtman TC (2003) Cofactor determination and spectroscopic characterization of the selenium-dependent purine hydroxylase from *Clostridium purinolyticum*. Biochemistry 42:11382–11390

PxIxIT

▶ Calcineurin

QM/MM, Quantum Mechanical/Molecular Mechanical

► Zinc Aminopeptidases, Aminopeptidase from Vibrio Proteolyticus (Aeromonas proteolytica) as Prototypical Enzyme

Quadrupolar NMR in Biology

► NMR Spectroscopy of Gallium in Biological Systems

Quicksilver

► Mercury Neurotoxicity

R

rAAP, Recombinant Leucine Aminopeptidase

► Zinc Aminopeptidases, Aminopeptidase from Vibrio Proteolyticus (Aeromonas proteolytica) as Prototypical Enzyme

α-Radiation Emitter

► Polonium in Terrestrial Environment

Radical Reactions

► Iron-Sulfur Cluster Proteins, Fe/S-S-adenosylmethionine Enzymes and Hydrogenases

Radical Rearrangement

► Cobalt Proteins, Overview

Radioactive Cesium

► Cesium, Therapeutic Effects and Toxicity

Radioactive Element

► Polonium in Terrestrial Environment

Radioactivity

► Polonium and Cancer

Radionuclide Cell Imaging

► Labeling, Human Mesenchymal Stromal Cells with Indium-111, SPECT Imaging

Radium

► Polonium and Cancer

Radium, Physical and Chemical Properties

Fathi Habashi
Department of Mining, Metallurgical, and Materials Engineering, Laval University, Quebec City, Canada

Physical Properties

Radium is a radioactive alkaline-earth metal that is found in trace amounts in uranium and thorium ores. One gram uranium contains 0.33×10^{-6} g Ra. Uranium 235 is present to the extent of 0.7% in uranium ores; it was once called actino-uranium because it produced actinium, hence the actinium series. Radium and its

V.N. Uversky et al. (eds.), *Encyclopedia of Metalloproteins*, DOI 10.1007/978-1-4614-1533-6,

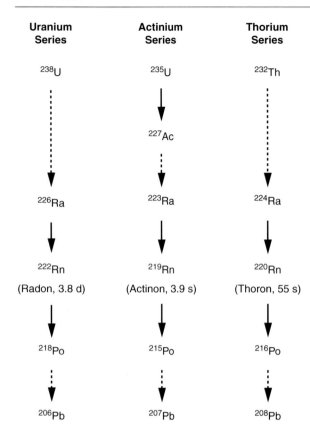

Radium, Physical and Chemical Properties, Fig. 1 Formation of radium by the disintegration of uranium and thorium. Other intermediate radioactive decay products are Bi, Pb, Tl

salts exhibit luminescence and impart a carmine color to flame. Radium emits alpha, beta, and gamma rays and was once used for cancer treatment because of its gamma rays. It produces neutrons when mixed with beryllium. Its most stable isotope, Ra 226, has a half-life of 1,601 years. Radium 223 has a half-life of 11.43 days and that of Ra 224 is 3.363 days. All three isotopes decays into radon gas which then decays to polonium which is a solid product hence the danger of respiring radon – once it enters the lungs it will never leave and may cause lung cancer because of the bombardment of the tissue with alpha particles (Fig. 1). Radium loses about 1% of its activity in 25 years, being transformed into elements of lower atomic weight with lead being the final product of disintegration.

The SI unit of radioactivity is the becquerel (Bq), equal to one disintegration per second. The Curie is a non-SI unit defined as that amount of radioactivity which has the same disintegration rate as 1 g of Ra 226 (3.7×10^{10} disintegrations per second, or 37 GBq).

Atomic number	88
Atomic weight	226
Density	5.5
Relative abundance in Earth's crust, %	1.3×10^{-10}
Melting point, °C	700
Boiling point, °C	1737
Heat of fusion, kJ·mol^{-1}	8.5
Heat of vaporization, kJ·mol^{-1}	113
Crystal structure	Body-centered cubic
Electrical resistivity at 20°C, μΩ·m	1
Thermal conductivity at 25°C, W·m^{-1}·K^{-1}	18.6

Chemical Properties

Radium has the electronic structure 2, 8, 18, 32, 18, 8, 2. When it loses its two outermost electrons, it will attain the inert gas electronic structure; hence it is a typical metal. Since radium is chemically similar to calcium, it has the potential to cause great harm by replacing it in bones. Inhalation, injection, ingestion, or body exposure to radium can cause cancer and other disorders. Stored radium should be ventilated to prevent accumulation of radon. Emitted energy from the decay of radium ionizes gases, affects photographic plates, causes sores on the skin, and produces many other detrimental effects.

Infinitesimal amounts of radium in sulfate solution can be co-precipitated with barium sulfate as (Ba,Ra)SO$_4$, when barium chloride solution is added.

References

Enghag P (2004) Encyclopedia of the elements. Wiley, Weinheim, pp 1153–55

Habashi F (1996) Pollution problems in the mineral and metallurgical industries. Métallurgie Extractive Québec, Québec City; Distributed by Laval University Bookstore "Zone". www.zone.ul.ca

Radon

▶ Polonium and Cancer

Rare Earth Elements

▶ Lanthanides and Cancer
▶ Lanthanides, Toxicity

Rare Earths

▶ Lanthanides, Physical and Chemical Characteristics

Ratio Material

▶ Porous Silicon for Drug Delivery

Ratio of Off Rate to On Rate

▶ Magnesium in Biological Systems

RBCC Domain

▶ Promyelocytic Leukemia–Retinoic Acid Receptor α (PML–RARα) and Arsenic

Reactive Oxygen Species

▶ Arsenic, Free Radical and Oxidative Stress
▶ Selenium and Glutathione Peroxidases
▶ Tellurium and Oxidative Stress

ReAsH-EDT$_2$

▶ Biarsenical Fluorescent Probes

Redox Regulation

▶ Tellurium in Nature

Reduction Potential

▶ Iron-Sulfur Cluster Proteins, Ferredoxins

Refitting

▶ Cadmium Exposure, Cellular and Molecular Adaptations

Regulation of Iron Traffic

▶ Iron Homeostasis in Health and Disease

Regulatory Calcium Binding Proteins

▶ EF-Hand Proteins

Relaxation of Cardiac Muscle

▶ Calcium in Heart Function and Diseases

Reporter Gene Cell Tracking

▶ Labeling, Human Mesenchymal Stromal Cells with Indium-111, SPECT Imaging

Reprolysin

▶ Zinc Adamalysins

Resistance of Platinum Anticancer Drugs

▶ Proteomic Basis of Cisplatin Resistance

Resistivity

▶ Cadmium Exposure, Cellular and Molecular Adaptations

Respiratory Proteins

▶ Heme Proteins, the Globins

Revascularization, Reflow, Reperfusion

▶ Rubidium in Biological Systems and Medicine

3,3′,5′-Reverse Triodothyronine, Reverse T3, RT3

▶ Selenoproteins and Thyroid Gland

Reversibly Mineralizing Protein Nanocage

▶ Iron Proteins, Ferritin

RGII – Rhamnogalacturonan II

▶ Boron, Biologically Active Compounds

Rhenium, Physical and Chemical Properties

Fathi Habashi
Department of Mining, Metallurgical, and Materials Engineering, Laval University, Quebec City, Canada

Physical Properties

Rhenium is a transition metal. Like the other members of the group it is less reactive than the typical metals but more reactive than the less typical metals. The transition metals are characterized by having the outermost electron shell containing two electrons and the next inner shell an increasing number of electrons. Because of the small energy differences between the valence shells, a number of oxidation states occur. The metal occurs in nature in association with molybdenite, MoS_2, which in turn is associated with chalocopyrite in porphyry copper ores. Molybdenite occurring in quartz veins does not contain rhenium.

Atomic number	75
Atomic weight	186.207
Relative abundance in the Earth's crust, %	7×10^{-8}
Density at 20°C, g/cm³	21.0
Metallic radius, pm	137
Ionic radius (VII), pm	53
Melting point, °C	3,180
Boiling point, °C	5,870
Heat of fusion, kJ/mol	33
Heat of sublimation ΔH_B (monatomic gas)	$+779 (\pm 8)$ kJ/mol
Enthalpy of formation of Re_2O_7, kJ/mol	−1,241
Electrical conductivity, $\mu\Omega^{-1}$ cm^{-1}	0.051
Specific resistivity at 20°C, $\mu\Omega\cdot$cm	19.3
Superconductivity, K	1.699
Tensile strength at 20°C, (annealed), MPa	12
Modulus of elasticity at 20°C, GPa	459.9
Recrystallization temperature (depending on degree of working), K	1,573–2,073

Chemical Properties

Rhenium is a heat-resistant metal, provided it does not come in contact with oxidizing agents. It is practically insoluble in hydrochloric and hydrofluoric acids. In oxidizing acids, it dissolves to form perrhenic acid ($E°$ for Re/ReO_2: 0.251 V). It forms volatile oxides with oxygen at high temperature. In air at 350°C, the heptoxide, Re_2O_7, is formed. When heated with fluorine or chlorine, the fluorides or chlorides are formed. Rhenium reacts with silicon, boron, and phosphorus at elevated temperature to form silicides, borides, and phosphides. Rhenium closely resembles its neighbors tungsten and osmium.

In the most important rhenium compounds such as perrhenic acid, the perrhenates, and dirhenium heptoxide, rhenium has the oxidation state +7.

It exhibits its lowest oxidation states (as low as −3) in carbonyl complexes. The yellow hygroscopic rhenium (VII) oxide, Re_2O_7, is the most stable oxide of rhenium. It is formed from rhenium metal powder or other rhenium oxides in dry air or an oxygen atmosphere above 350°C. It is readily soluble in water, forming the colorless perrhenic acid, $HReO_4$.

Perrhenic acid forms salts ($MReO_4$) with a tetrahedral ReO_4^- ion. Ammonium perrhenate, NH_4ReO_4, is an important starting material, which can be reduced to metal and used for the production of other rhenium compounds. When Re_2O_7 is heated with rhenium powder, the oxides ReO, Re_2O_5, ReO_2, and Re_2O_3 are formed. The black rhenium (VII) sulfide is formed when H_2S is passed into acidic solutions of perrhenates. Reduction of Re_2S_7 by hydrogen produces ReS_3. The most stable sulfide ReS_2 is formed by the thermal decomposition of Re_2S_7 or by heating the elements together.

References

Habashi F (2003) Metals from ores. An introduction to extractive metallurgy. Métallurgie Extractive Québec, Quebec City, Canada. Distributed by Laval University Bookstore, www.zone.ul.ca

Nadler HG (1997) Rhenium and rhenium compounds. In: Habashi F (ed) Handbook of extractive metallurgy. Wiley-VCH, Weinheim, pp 1491–1502

Rhodium, Physical and Chemical Properties

Fathi Habashi
Department of Mining, Metallurgical, and Materials Engineering, Laval University, Quebec City, Canada

Rhodium is a member of the group of six platinum metals which occur together in nature and is located in the center of the Periodic Table in the Iron-Cobalt-Nickel group:

Fe	Co	Ni
Ru	*Rhodium*	Pd
Os	Ir	Pt

Physical Properties

Atomic number	45
Atomic weight	102.91
Relative abundance in Earth's crust, %	1×10^{-7}
Abundance of major natural isotopes	103 (100%)
Crystal structure	Face-centered cubic
Lattice constants at 20°C	0.38031
a, nm	
c, nm	
Atomic radius, nm	0.134
Melting point, °C	1,966
Boiling point, °C	3,760
Specific heat at 25°C c_p, J g^{-1} K^{-1}	0.24
Thermal conductivity λ, W m^{-1} K^{-1}	89
Density at 20°C, g/cm^3	12.41
Brinell hardness	101
Young's modulus E, N/mm^2	379,058
Tensile strength σ_B, N/mm^2	412
Specific electrical resistance at 0°C, $\mu\Omega$ cm	4.33
Temperature coefficient of electrical resistance (0–100°C), K^{-1}	0.0046
Thermoelectric voltage versus Pt at 100°C E, V	+0.70

Chemical Properties

Because of the small energy differences between the valence shells, a number of oxidation states occur. The following oxidation states are known in the compounds of rhodium: −1, 0, +1, +2, +3, +4, +5, +6. The most common oxidation state is +3. Rhodium can be converted into water-soluble rhodium(III) sulfate at ca. 600°C by melting with potassium or sodium hydrogen sulfate, which is converted to the pyrosulfate with loss of water. Platinum, iridium, and ruthenium are not attacked.

Ammonium hexachlororhodate, $(NH_4)_3[RhCl_6]$, can be isolated from hydrochloric acid solutions such as those obtained after removal of the other platinum group metals. Ammonium chloride is added, followed by concentration and crystallization. Much of the remaining $(NH_4)_2[PtCl_6]$ and $(NH_4)_2[IrCl_6]$ is precipitated at the same time. Chlorides of the base metals mostly remain in solution. The $(NH_4)_3[RhCl_6]$ can be selectively redissolved in water at room temperature to form an almost saturated solution. This gives a purified solution of rhodium. Both $(NH_4)_2[PtCl_6]$ and $(NH_4)_2[IrCl_6]$ are recovered by filtration.

Rhodium can also be precipitated as chloropentamminerhodium (III) chloride, [RhCl(NH$_3$)$_5$]Cl$_2$, in weakly alkaline solution. This method is unsuitable for solutions that contain large quantities of base metals because, under reaction conditions, these form insoluble hydroxides that are difficult to filter.

Ammonium hexachlororhodate (III) is subjected to thermal reduction by hydrogen to yield the metal:

$$2(NH_4)_3[RhCl_6] + 3H_2 \rightarrow 2Rd + 6NH_4Cl + 6HCl$$

Wet chemical reduction with hydrazine must be carried out with a large excess of hydrazine and NaOH at about 100°C to ensure complete reaction:

$$4(NH_4)_3RhCl_6 + 3N_2H_4 + 24NaOH \rightarrow$$
$$4Rh + 3N_2 + 12NH_3 + 24NaCl + 24H_2O$$

Fission of ^{235}U in nuclear power stations produces considerable amounts of platinum group metals; 1 t of spent reactor fuel contains 1.2 kg of palladium, 0.5 kg of rhodium, and 2.3 kg of ruthenium. The most important isotopes are: ^{107}Pd ($t_{1/2}$ 7 × 10^6 years), ^{102}Rh ($t_{1/2}$ 3 years), and ^{106}Ru ($t_{1/2}$ 1 year).

Platinum-rhodium gauzes are used as catalysts for the oxidation of ammonia and platinum-rhodium alloys are used for thermocouple components.

References

Habashi F (2003) Metals from ores. An introduction to extractive metallurgy. Métallurgie extractive Québec, Quebec City, Canada. Distributed by Laval University Bookstore, www.zone.ul.ca

Renner H (1997) Platinum group metals. In: Habashi F (ed) Handbook of extractive metallurgy. Wiley-VCH, Weinheim, pp 1269–1360

Ribonucleotide Reductase

Britt-Marie Sjöberg
Department of Biochemistry and Biophysics,
Stockholm University, Stockholm, SE, Sweden

Synonyms

RNR; also see Table 1

Definition

Ribonucleotide reductase is an enzyme that catalyzes de novo synthesis of DNA building blocks by reduction of ribonucleotides to deoxyribonucleotides.

Introduction

Ribonucleotide reductase (RNR) is a fascinating enzyme family whose origin dates back more than 3.5 billion years ago. An ancestral RNR function is inferred to the time when the molecular foundations for life evolved on Earth. The contemporary RNR family is diverse and all RNRs contain metalloproteins. In addition to members with non-heme iron clusters, the RNR family also comprises members with manganese, cobalt, and zinc sites.

RNRs catalyze the formation of DNA building blocks (deoxyribonucleotides) by reduction of the corresponding ribonucleotides (RNA building blocks). As this is the sole pathway that cells can produce deoxyribonucleotides (dNTPs) de novo, RNR is an essential constituent of all organisms (Fig. 1). There action mechanism is superficially simple – substitution of the 2′-hydroxyl group in the ribose sugar of the ribonucleotide with a hydrogen. However, this is a chemically demanding reaction, and all known RNRs depend on radical chemistry and follow the same basic reaction mechanism (Fig. 2a).

Three Major RNR Classes

Three major classes of RNRs are known and denoted I, II, and III (Table 1). The essentials of the reaction mechanism are shared between the classes, but the composition of the enzymes differs, both in terms of protein components and cofactors (Nordlund and Reichard 2006; Andersson 2008). Class I RNRs consists of two components. One component (denoted NrdA or NrdE, depending on subclass) harbors the active site, while the other component (NrdB or NrdF) harbors a dinuclear metal center. In most cases, the latter contains a built-in stable tyrosyl radical close to the metal center. Class II and III RNRs consist of a single catalytic component with the active site. Class II RNRs (NrdJ) require the cobalt-containing vitamin B$_{12}$ coenzyme 5′-deoxyadenosylcobalamin (AdoCbl). The catalytic

Ribonucleotide Reductase, Table 1 Gene, subunit, and protein names of RNR enzymes

RNR class	Gene name	Polypeptide name	Quaternary structures of active holoenzyme	General and synonymous protein names
I	NrdA	α	$\alpha_2\beta_2$ (general), $\alpha_6\beta_{2\text{-}6}$ (mammals, yeast)	NrdA, R1
				M1, RR1 (mammals), RNR1 (yeast), RNR3 (yeast)
	NrdB	β		NrdB, R2
				M2, RR2 (mammals), p53R2, RR2B (p53-regulated R2, mammals), RNR2 (yeast), RNR4 (yeast)
Ib	NrdE	α	$\alpha_2\beta_2$	NrdE, R1E
	NrdF	β		NrdF, R2F
	NrdI	n.a.[a]		NrdI(flavodoxin)
Ic	NrdA	α	$\alpha_2\beta_2$	NrdA, R1
	NrdB$^{\text{Phe/Leu/Val}}$	β$^{\text{Phe/Leu/Val}}$		NrdB$^{\text{Phe/Leu/Val}}$, R2$^{\text{Phe/Leu/Val}}$
II	NrdJ	α	α or α_2	NrdJ
III	NrdD	α	α_2	NrdD
	NrdG	n.a.[a]		NrdG(activase)

[a]n.a., not applicable

Ribonucleotide Reductase, Fig. 1 The basic reaction catalyzed by RNR – reduction of a ribonucleotide (RNA building block) to the corresponding deoxyribonucleotide (DNA building block), and the importance of RNR in cell growth and proliferation as well as evolution

component of class III RNRs (NrdD) harbors a built-in stable glycyl radical in the vicinity of the active site, and a Zn-containing metal center. An iron-sulfur center carried on a separate radical SAM protein (NrdG) activates class III RNRs by generating the glycyl radical in NrdD. During catalysis, the radical-containing NrdD works independent of the activating protein.

A Crucial Transient Thiyl Radical (Cys•) in the RNR Active Site

The amino acid sequences of the active site-containing components (NrdA, NrdE, NrdD, NrdJ) from the different RNR classes are in some cases very dissimilar speaking in favor of two or three independent evolutionary origins of the enzyme function. Structurally the catalytic core of all three classes is however fully conserved. It consists of a unique structure of two five-stranded β/α half-barrels that together form a wide ten-stranded β/α barrel with the five strands in each half-barrel in opposite directions (Fig. 2b). The comparatively wide barrel allows protrusion of a so-called finger loop with an essential cysteine residue at its tip. This cysteine is the key residue of the common reaction mechanism as it transiently forms a cysteinyl radical (Cys•) that initiates the reaction mechanism by abstracting the 3′-hydrogen of the substrate (Fig. 2a). Interestingly, the enzyme thus initiates

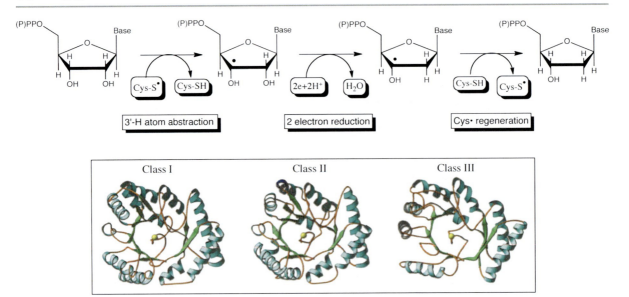

Ribonucleotide Reductase, Fig. 2 (*top*) The general RNR reaction mechanism is initiated by formation of a transient thiyl radical in a cysteine residue that then abstracts the 3′-H atom of the substrate ribose sugar. In the next step the protonated 2′-hydroxyl is released as water and the 2′ position is reduced. The reaction is completed by reintroduction of the earlier abstracted 3′-hydrogen atom and regeneration of the transient cysteinyl radical. (*bottom*) The three-dimensional catalytic core structures of *E. coli* class I, *Lactobacillis leichmannii* class II, and bacteriophage T4 class III RNRs

the reduction by a one-electron oxidation of the substrate. This enables release of the ribose 2′-hydroxyl as water, followed by the actual two-electron reduction of the ribose 2′-position. The reaction is completed by return of the earlier abstracted hydrogen to the 3′-position of the sugar, which regenerates the transient cysteinyl radical in the enzyme. All RNR enzymes, regardless of class or subclass, follow these basic reaction steps.

Generation of the Initiating Radical

A common necessity in all RNRs is to avoid storing a reactive Cys• in the relatively open active site. The problem is solved by storing an unpaired electron, or the capacity to generate a reactive unpaired electron, in a different chemical structure some distance from the active site. This way the Cys• can be generated only when it is needed for catalysis. The three RNR classes differ in the way that they store the unpaired electron (Table 2) and generate the transient cysteinyl radical and this is where most of the RNR metal sites have a crucial role.

The Stabilized Unpaired Electron in the Metal-Containing Proteins of Class I RNRs

In class I RNRs, one subunit (the homodimeric NrdA protein or α_2 [see later in this chapter for variations of the homodimeric forms]) contains the active site, whereas the radical functionality is stored in a separate homodimeric protein (NrdB or β_2) that harbors a stable tyrosyl radical close to a dinuclear metal center (Cotruvo and Stubbe 2011; Sjöberg 2010). In the canonical form of class I RNRs the dinuclear metal center is a diferric-oxo center (Fig. 3a). Both ferric ions are ligated by carboxylate and histidine side chains. An aspartic acid and a histidine ligate the Fe1 ion that is closest to the tyrosine (Tyr122 in *E. coli* numbering, Fig. 3a) that harbors the radical, and two glutamates and a histidine ligate Fe2. In addition, one glutamate and a μ-oxo-group derived from molecular oxygen bridge both irons. Each metal ion also has one water-derived ligand. The protein-derived ligands are provided by four helices that form a typical four-helix bundle also found in the ferritins, bacterioferritins, rubrerythrins, dimanganese catalases, bacterial methane monooxygenases, and fatty acid desaturases (Lundin et al. 2012). The iron ions are

Ribonucleotide Reductase, Table 2 Radical generating metal site, initiating radical, substrate phosphorylation level, and distribution of RNR enzymes

	Class I			Class II	Class III
	Ia (canonical)	Ib	Ic		
Operation	Aerobic	Aerobic?[a]	Aerobic	Oxygen independent	Anaerobic
Radical generating metal site	$Fe_2^{II/II} \rightarrow Fe^{III}\text{-O-}Fe^{III}$	$Mn_2^{II/II} \rightarrow Mn^{III}\text{-O-}Mn^{III}$	$Mn^{II}/Fe^{II} \rightarrow Mn^{IV}\text{-O-}Fe^{III}$	$AdoCbl(Co^{III}) \rightarrow Ado\bullet (Co^{II})$	$[4Fe4S]^{1+}/AdoMet \rightarrow [4Fe4S]^{2+}/Ado\bullet$ (in RNR activase)
Initiating radical	Tyr•	Tyr•	$Mn^{IV}\text{-O-}Fe^{III}$	Ado•	Gly•
Substrates	NDPs	NDPs	NDPs	NDPs or NTPs	NTPs (ribonucleoside triphosphate)
Distribution[b]	In almost all eukaryotes, abundant in bacteria, rare in archaea	Abundant in bacteria	Abundant in bacteria, rare in archaea	Abundant in bacteria and archaea, rare in unicellular eukaryotes	Abundant in strict and facultative anaerobic bacteria and archaea, rare in unicellular eukaryotes

[a]Unknown oxidant.
[b]For additional information on distribution, see the Ribonucleotide Reductase Database (RNRdb) at http://rnrdb.molbio.su.se.

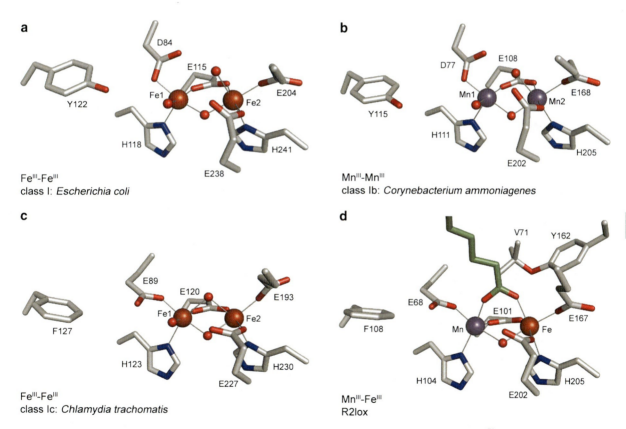

Ribonucleotide Reductase, Fig. 3 Three-dimensional structures of metal sites in class I RNRs and in the alkane oxidase R2lox. (**a**) The Fe_2^{III} center in *E. coli* class I NrdB, (**b**) the Mn_2^{III} center in *Corynebacterium ammoniagenes* class Ib NrdF, (**c**) the *Chlamydia trachomatis* $NrdB^{Phe}$ protein crystallized with a Fe_2^{III} center, and (**d**) a $Mn^{III}\text{-}Fe^{III}$ center in the related R2lox from *Mycobacterium tuberculosis*

Ribonucleotide Reductase, Fig. 4 Schematic reaction mechanisms for generation of metal sites and radicals in class I RNRs. Divalent metal ions are bound to apoproteins, the metal sites are oxidized to high-valent forms by molecular oxygen (class Ia and Ic) or a highly reactive dioxygen species (class Ib). A stable radical is generated in a nearby tyrosine residue by electron transfer to the high-valent metal site (class Ian ad Ib), or the unpaired electron is stably stored in the mixed-valent metal site (class Ic)

initially bound to the apoNrdB protein as ferrous ions (Fig. 4). In presence of molecular oxygen, a series of carboxylate shifts leads to formation of an oxo-bridged Fe^{III}-Fe^{IV} center that generates the stable tyrosyl radical by a one-electron abstraction and thereby forms the Fe^{III}-O^{2-}-Fe^{III} site. To reduce molecular oxygen to water, the protein thus provides three intrinsic electrons in addition to one solution-derived electron. Even though the NrdB protein is a tight dimer and harbors a diferric-oxo center in each polypeptide, the radical content in the isolated protein is usually not more than 1 radical per dimer, but the reason for this is currently not known.

The tyrosyl radical in the NrdF protein of the Ib subclass can be stabilized by either a diferric-oxo center or a dimanganese-oxo center in vitro. The carboxylate, histidine, and oxo ligands of the NrdF metal center are in both cases identical to corresponding ligands in the NrdB protein (Fig. 3b). Interestingly, an RNR-specific flavodoxin – the NrdI protein encoded in the same operon as the other RNR genes coding for class Ib RNR (Fig. 5) – is needed to form the dimanganese center and the adjacent tyrosyl radical. Analogously to the diiron case described above, manganese initially forms a Mn_2^{II} site that, in presence of reduced $NrdI_{hq}$ (containing the reduced hydroquinone form of FMN) and molecular oxygen, forms an oxo-bridged Mn^{III}-Mn^{IV} site (Fig. 4). The mixed-valent dimanganese center then generates the tyrosyl radical and thereby forms the Mn^{III}-O-Mn^{III} site. The ultimate oxidant of the metal site is not yet known. Neither molecular oxygen, nor hydrogen peroxide can by themselves promote the radical generation. It has instead been suggested that the $NrdI_{hq}$ generates a hydroperoxy anion that oxidizes the Mn_2^{II} site (Cotruvo and Stubbe 2011). A crystal structure of the NrdI: NrdF complex suggests that a diatomic species is found in the channel leading from the FMN cofactor in NrdI to the metal site in NrdF (Boal et al. 2010). Independent results suggest that the dimanganese NrdF is the physiologically relevant form.

The $\beta_2$protein of subclass Ic has a phenylalanine residue at the position of the tyrosyl radical in canonical class I RNRs (Fig. 3c), and the unpaired electron needed to generate the transient cysteinyl radical in the NrdA protein is instead stabilized in a Mn^{IV}-O-Fe^{III} metal center in theNrdB^{Phe}protein (Fig. 4). Biochemical results suggest that Mn^{2+} binds in the position closest to the phenylalanine (corresponding to the radical tyrosine in NrdB). However, the only three-dimensional structure for a NrdB^{Phe} protein is an inactive variant with a Fe_2^{III} site (Fig. 3c). Results from

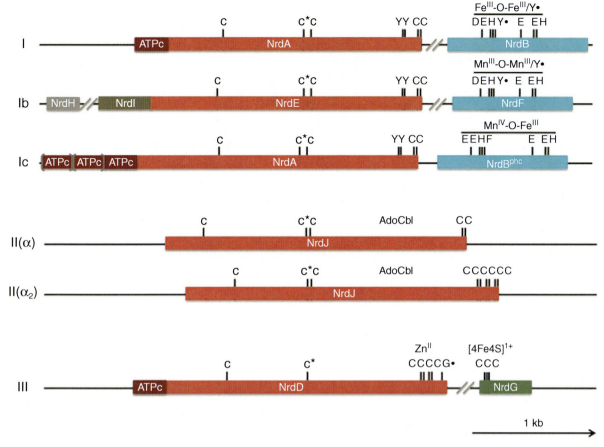

Ribonucleotide Reductase, Fig. 5 Comparison of protein components in RNR classes and subclasses, schematically represented as operon structures. Catalytically important residues and ligands of metal sites are marked. Representatives are (from *top* to *bottom*, *left* to *right*): *E. coli* NrdA (C225, C•439, C462, Y730, Y731, C754, C759), NrdB (D84, E115, H118, Y•122, E204, E238, H241); *E. coli* NrdH, NrdI, NrdE (C178, C•388, C415, Y692, Y693, C709, C712), NrdF (D67, E98, H101, Y•105, E158, E192, H195); *C. trachomatis* NrdA (C518, C•672, C687, Y990, Y991, C1043, C1046), NrdB^{Phe} (E89, E120, H123, F127, E193, E227, H230); *L. leichmannii* monomeric NrdJ (C108, C•408, C419, C731, C736); *Thermotoga maritima* dimeric NrdJ (C134, C•322, C333, Y790, Y791, C800, C803, C815, C818, C821, C826); *E. coli* NrdD (C175, C•384, C644, C647, C662, C665, G•681), NrdG (C26, C30, C33). The *dark gray* interruptions between the genes and the *dark gray* parentheses indicate common variations among the RNR classes and subclasses

advanced spectroscopic studies indicate that the Mn^{II}-Fe^{II} site in NrdB^{Phe} is oxidized by molecular oxygen to a high-valent Mn^{IV}-O-Fe^{IV} form that converts to the more stable Mn^{IV}-O-Fe^{III} form by abstraction of an externally derived electron (Fig. 4). Comparison to the three-dimensional structure of the distantly related alkane oxidizing R2lox protein that has a similar Mn/Fe site suggests that the ligands of the active Mn^{IV}-Fe^{III} NrdB^{Phe} are glutamates, histidines, and oxo ligands (Högbom 2011) as in the structure of the diferric form of NrdB^{Phe} (Fig. 3d). Recently deduced bacterial genome sequences indicate that a few related species with class Ic RNRs encode a leucine (NrdB^{Leu}) or a valine (NrdB^{Val}) at the position of the phenylalanine in NrdB^{Phe} and the tyrosyl radical in canonical NrdB.

The Transient Ado• in Class II RNRs

Class II RNRs are called NrdJ and are either monomeric (α) or homodimeric (α_2) proteins (Figs. 5 and 6) and require the AdoCbl cofactor. Homolytic cleavage of the Co^{III}-deoxyadenosyl bond of the bound cofactor generates a transient 5′-deoxyadenosyl radical (Ado•) and a Co^{II} corrin ring that remains bound to the enzyme during catalysis (Fig. 7). No other variant of cobalamin than AdoCbl can fulfill this function in class II RNRs.

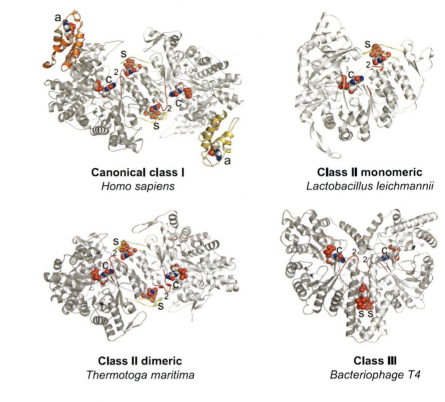

Ribonucleotide Reductase, Fig. 6 Three-dimensional structures of *Homo sapiens* class IR1, *L. leichmannii* class II monomeric NrdJ, *T. maritima* class II dimeric NrdJ (lacks 82 C-terminal residues), and bacteriophage T4 class III NrdD. Bound substrate and allosteric effector nucleotides are highlighted. The a-sites, s-sites, catalytic sites, and loop 2 are indicated by a, s, c, and 2. The ATP cones in each α subunit of the human class I enzyme is shown in *orange* or *gold*, respectively, whereas it is absent in the representatives from the other classes. Loop 1 and loop 2 are colored in *yellow* and *red*, respectively

Ribonucleotide Reductase, Fig. 7 Reaction mechanism for generation of the transient Ado• and Cys• in class II RNRs. A redox-active disulfide is formed in the enzyme during catalysis, and needs to be reduced by NADPH via thioredoxin reductase (TR), thioredoxin (Trx), and prior to the next round of catalysis. It has been suggested that the products of AdoCbl homolytic cleavage remain bound in the active site during catalysis, but not during reduction of the enzyme disulfide

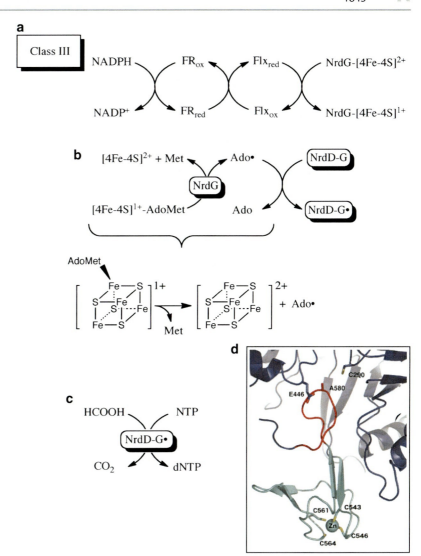

Ribonucleotide Reductase, Fig. 8 Generation of the stable Gly• in class III RNRs. (a) The iron-sulfur cluster in the NrdG activase is reduced by NADPH via flavodoxin reductase (FR) and flavodoxin (Flx), (b) the reduced iron-sulfur cluster in NrdG binds and cleaves AdoMet to generate the transient Ado• that abstracts a hydrogen atom from the conserved glycine residue in NrdD to generate the stable Gly•. (c) Gly•-containing NrdD reduces substrate using formate as reductant, and can cycle many turnovers. (d) A close-up of the active site region in Bacteriophage T4 NrdD with the transient thiyl radical cysteine (C290), the position of the stable glycyl radical residue (here as A580), and the C-terminal Zn(Cys)$_4$ motif (C543, C546, C561, C564) highlighted (From Logan (2008) in Andersson (2008))

The Stable Gly• in Class III RNRs

Class III RNRs are homodimeric (α_2) proteins that harbor a stable glycyl radical that in the three-dimensional structure of NrdD is close to the crucial cysteine in the finger loop (Fig. 6) (Nordlund and Reichard 2006). The glycyl radical is generated by a hydrogen atom abstraction by a specific activase (NrdG), a 4Fe-4S iron-sulfur protein usually encoded in the same operon as NrdD (Fig. 5). The NrdG protein belongs to the "radical SAM" protein family, a diverse protein superfamily with a C(X)$_3$C(X)$_2$C amino acid sequence motif. The 4Fe-4S cluster in the radical SAM proteins ligates S-adenosylmethionine (denoted AdoMet or SAM) to the Fe that lacks a cysteine ligand. Generation of the glycyl radical requires a radical transfer process involving NrdG and AdoMet. The [4Fe-4S]$^{1+}$-AdoMet center in NrdG is cleaved to [4Fe-4S]$^{2+}$, Ado•, and Met, and Ado• then abstracts a hydrogen atom from the glycine residue in the NrdD protein (Fig. 8). The oxidized iron-sulfur center in the activase is regenerated to the functional [4Fe-4S]$^{1+}$ by the action of a canonical flavodoxin (Flx, different from NrdI), flavodoxin reductase and NADH. When the stable Gly• has been formed, NrdD is capable of many catalytic turnovers, so in contrast to the class I proteins that require a holoenzyme complex of the α_2 and the β_2 components (NrdAB or NrdEF) during catalytic turnover, the NrdD protein is the enzyme

Ribonucleotide Reductase, Fig. 9 Radical transfer pathways in the three RNR classes. (**a**) A modeled class I NrdAB complex from *E. coli*, (**b**) class II AdoCbl binding is derived from adenosine and cobalamin structures of *T. maritima* RNRNrdJ, (**c**) class III structure shows a glycine to alanine mutant of the glycyl radical site in bacteriophage T4 NrdD (From Nordlund and Reichard 2006)

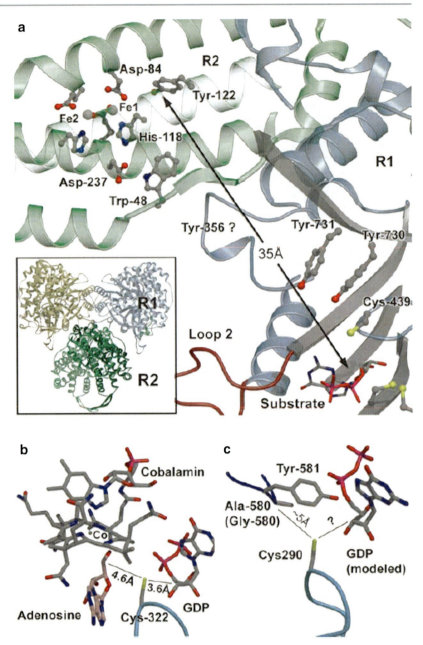

proper of class III RNRs. In the C-terminal part, the NrdD protein harbors a $Zn(Cys)_4Zn$-ribbon motif that is believed to control the formation and/or stabilization of the glycyl radical (Fig. 8).

Radical Transfer Pathways

The stabilized unpaired electron (or the capacity to form a transient unpaired electron) is stored at a site some distance from the crucial cysteine residue in the finger loop of the α – component. The radical function must hence be transferred from where it is stored to the cysteine in the finger loop near the active site when a substrate binds to the active site of RNRs (Nordlund and Reichard 2006; Cotruvo and Stubbe 2011). In its extreme, the radical transfer pathway in class I RNRs is approximately 35 Å (Fig. 9a) – a record in naturally occurring radical transfer pathways.

Formation of Cys• in Class I RNRs

When the class I NrdAB (or class Ib NrdEF) holoenzyme complex is formed and substrate is bound in the active site of NrdA (or NrdE), a radical transfer chain is initiated between the tyrosyl radical in NrdB (or NrdF) and the active site cysteine in NrdA (or NrdE). Several conserved aromatic residues participate in the transfer, which is the reason that the radical transfer can occur at all over the 35 Å distance. The transfer is proposed to involve proton-coupled electron transfer between the tyrosyl radical and the surface of NrdB/F(Tyr122, Trp48, Tyr356 in *E. coli* numbering) and probably hydrogen atom transfer between the tyrosine pair and the cysteine in the finger loop in NrdA/E(Tyr731, Tyr730, Cys439 in *E. coli* numbering) (Fig. 9a). Radical transfer between the mixed-valent metal site to the α_2 component in class Ic is believed to follow corresponding pathways as in the other class I holoenzymes, and residues involved in radical transfer in the α_2 and β_2 components are highly conserved (Fig. 5).

Formation of Cys• in Class II RNRs

Class II RNRs bind the AdoCbl cofactor close to the finger loop cysteine. The reaction mechanism is initiated by homolytic cleavage of the Co^{III}-AdoCbl bond to form the Ado• that then abstracts an electron from the active site cysteine (Fig. 9b). As the Co^{II}-corrin ring is bound tightly in the protein and shields the finger loop cysteine from contact with water, it has been suggested that the 5′-deoxyadenosine stays in the pocket during catalysis and that AdoCbl is regenerated at the end of the reaction mechanism (Fig. 7). It is interesting to note that Co, as well as Fe and Mn are redox-active metals belonging to the transition metals and have consecutive atom numbers in the periodic table.

Formation of Cys• in Class III RNRs

Class III RNRs harbor the glycyl radical on the tip of a second finger loop that meets the cysteine-containing finger loop in the center of the 10-stranded β − barrel, and it is believed that the transient Cys• is formed via a direct hydrogen atom abstraction by Gly• (Fig. 9c). One enigmatic feature of class III RNRs is the presence of a Zn-ribbon motif, in most variants with four cysteine ligands (Fig. 8). The function of the $Zn(Cys)_4$ site is unclear although mutational studies suggest a role in acquisition and/or stabilization of the glycyl radical. Class III RNRs only function anaerobically because the enzyme is cleaved at the stable glycyl radical in presence of molecular oxygen. This irreversible modification of the α − polypeptide has been used as a facile monitor of the presence of the glycyl radical in vivo, that is, a catalytically competent form of the enzyme during physiological conditions.

Physiological Reducing Systems

As described above the overall reaction mechanism is similar for all RNRs in its major steps, however mechanistic details differ between the different classes. The immediate reducing entities in class I and II RNRs are a redox-active cysteine pair partly buried in the active site. The disulfide formed during catalysis needs to be reduced again in each enzyme turnover. Small redoxin proteins (thioredoxin, glutaredoxin, NrdH-redoxin) mediate this reduction. The redoxins reduce a disulfide in the flexible C-terminal part of the α − polypeptide chain and the resulting cysteine pair in turn reduces the disulfide formed in the active site. The ultimate reductant in this cascade of disulfide interchanges is NADPH (Figs. 10 and 7). Thioredoxins require the flavoprotein thioredoxin reductase for their reduction, and glutaredoxins require glutathione and the flavoprotein glutathione reductase. NrdH-redoxins that are specific for class Ib RNRs is reduced by thioredoxin reductase. Class III RNRs on the other hand, have only one cysteine residue in the active site in addition to the radical cysteine residue, and no disulfide is formed during catalysis. The reductant in the class III reaction mechanism is instead formate that is oxidized to CO_2 during catalysis. As formate is consumed and no disulfide is formed there is no need to reduce the class III RNR prior to the next catalytic turnover (Fig. 8).

Allosteric Regulation

DNA building blocks need to be provided in balanced concentrations reflecting the base composition of the organism's DNA. Otherwise, the DNA replication will not be as accurate as required. The RNR enzymes reduce all four substrates and the appropriate supply of deoxyribonucleotides is largely controlled by RNR via an elaborate allosteric regulation (Hofer et al. 2012). Binding of dNTPs (deoxyribonucleoside triphosphate)

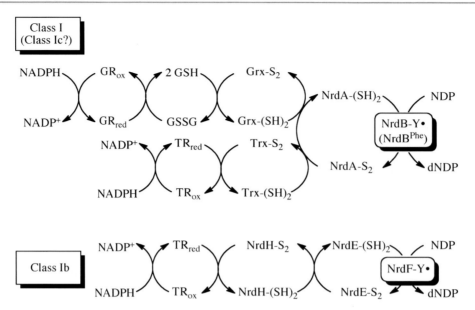

Ribonucleotide Reductase, Fig. 10 Physiological turnover cycles of class I RNRs. (*top*) A redox-active disulfide is formed in canonical class I NrdA during catalysis, and needs to be reduced by NADPH prior to the next round of catalysis; two parallel pathways involve either thioredoxin (Trx) and thioredoxin reductase (TR), or glutaredoxin (Grx), glutathione (GSH, reduced, and GSSG oxidized), and glutathione reductase. (*bottom*) The redox-active disulfide formed in class Ib NrdE during catalysis is reduced by NADPH prior to the next round of catalysis via the Ib-specific NrdH-redoxin, and thioredoxin reductase (TR)

and ATP to the allosteric specificity site induce conformational changes in the enzyme and thereby balance the four dNTP pools by dictating the substrate specificity at the active site. In this way, ATP and dATP are positive regulators for production of dCTP and dUTP, dTTP is a positive regulator for production of dGTP, and dGTP is a positive regulator for production of dATP. Almost all canonical class I RNRs and class III RNRs also have an additional allosteric site that in combination with the other allosteric site controls the overall activity of the enzyme. Only ATP and dATP bind to this site. ATP has similar affinity for both allosteric sites, whereas dATP has much stronger affinity for the specificity site compared to the overall activity site. In general terms, ATP acts as an activator and dATP as an inhibitor when bound to the overall activity site. With the two complementary allosteric sites RNR enzymes can control both the balance between the different DNA building blocks and their overall concentrations.

The allosteric regulation has been studied in great detail in a few eukaryotic class I RNRs, and these holoenzyme complexes are larger than the $\alpha_2\beta_2$ holoenzyme complex described earlier in this chapter. A seminal study of yeast and human RNRs (Fairman et al. 2011; Logan 2011) provided structural details of a dATP-inhibited $\alpha_6\beta_2$ complex where the radical transfer pathway between α and β is interrupted. It also provided insight into how the active ATP-induced $\alpha_6\beta_{2-6}$ complexes differ from the inhibited complex. In contrast, the canonical class I RNR from *E. coli* cycles between the active $\alpha_2\beta_2$ holoenzyme and an inactive dATP-induced $\alpha_4\beta_4$ complex. Clearly, the on/off regulation mediated by the allosteric overall activity site uses different structural solutions in different organisms.

Environmental Occurrences

Based on the different chemistries needed to initiate the common reaction mechanism, the different RNR classes have different environmental requirements (Table 2): class I RNRs are oxygen requiring as the formation of the tyrosyl radical requires an oxygen-dependent oxidation of the metal center, class II RNRs are oxygen-independent but are AdoCbl-requiring, and class III RNRs are oxygen sensitive, as both the glycyl radical in the catalytic component and the iron-sulfur center in the activase are destroyed by oxygen (Torrents et al. 2008). All multicellular and almost all unicellular eukaryotes carry a canonical class I RNR. The majority of bacteria (80% of sequenced species) also carry genes for class I RNRs, but class I RNR is relatively rare in archaea. Class II and III RNRs are abundant in bacteria and archaea, but very rare in eukaryotes and found only in some unicellular eukaryotes. Interestingly, two thirds of all sequenced bacteria and one third of the archaea code for more than one RNR operon. In fact, 7% of sequenced bacterial species encode at least one gene copy of each class (I, II, III), and there are also one archaeon and 3 unicellular eukaryotes that code for one gene copy of each class. Having the capacity to produce DNA building blocks, both during aerobic and anaerobic conditions, may give competitive advantages to organisms in fluctuating environmental conditions.

Origin and Evolution of RNR Classes

There are two major questions concerning the evolution of RNR: do all contemporary RNRs have a common origin, and if so when did the ancestral RNR appear. The answer to the first question became obvious when three-dimensional structures for the RNR classes were available. Although the amino acid sequence of RNR classes have diverged – in some cases beyond recognition – the high degree of structural similarity between RNR classes and the conserved reaction mechanism make a single origin of the contemporary RNR classes unquestionable (Fig. 2). The answer to the second question is intimately connected to the generally accepted view that life on Earth evolved from an RNA-world to the DNA-world we know of today. In the RNA-world, RNA played the role of both catalyst and information carrier, whereas most catalysts are proteins in the DNA-world and the genetic information is stored in DNA. As the only means to produce DNA building blocks is the enzymatic reduction of the corresponding RNA building blocks, RNR played a crucial role in the transition from the RNA-world to modern life more than 3.5 billion years ago (Fig. 1), that is, after evolution of protein synthesis and before evolution of organisms with DNA genomes.

The ancestral RNR must have been very similar to the contemporary classes in terms of reaction mechanism and tertiary structure: the ten-stranded β/α-barrel with the finger loop, possibly with an origin as a half-barrel. Furthermore, although the earliest forms of RNR probably were rather promiscuous in terms of substrate specificity, the extant allosteric regulation of specificity – conserved in all classes – was most likely present in the common ancestor of extant RNR classes. We can however not infer what cofactors or which quaternary structure the ancestral RNR had, since no transitional forms between the classes have been found. Conceivably, class I RNR is more recent than class II and III, as it is dependent on the availability of molecular oxygen.

In a more recent perspective, it is interesting to note that the repertoire of RNR classes encoded by organisms bears likely evidence of adaptation. In the extreme, some obligate intracellular parasites lack RNRs altogether, probably because they can salvage deoxyribonucleotides from their hosts. The other extreme, organisms encoding all three classes, can also be found, possibly adaptations to varying oxygen concentrations and AdoCbl availability. Not only do we see different repertoires of RNRs in organisms, but the RNR repertoire also appear to often evolve rather rapidly by horizontal transfer of RNR genes between organisms (Torrents et al. 2008). Furthermore, larger dsDNA-viruses commonly encode RNRs, supposedly a selective advantage for a virus. It also provides a possible vector for horizontal transfer of the RNR between different organisms infected by the virus.

Biomedical Importance

The essential role of RNR for DNA synthesis, and consequently cell proliferation and organismal growth has obvious biomedical consequences. They have however been moderately explored hitherto. Two antiproliferative drugs on the market are directed toward RNR; Hydrea® (hydroxyurea) that effectively reduces the radical in class I RNRs, and Gemzar® (gemcitabine) that acts as a suicidal substrate analog of RNRs. One major reason for the low exploitation of RNR as an antiproliferative, antimicrobial, and/or antiviral target enzyme is the fact that all hitherto developed assays for enzyme activity are both laborious and time consuming, and do not allow use of high-throughput screening techniques. More than half a century after the first discovery of an RNR activity, a novel assay designed for microplate format and robotics has been developed and shown to have the capacity to screen for novel antimicrobials directed toward bacterial RNRs. This may allow a renaissance for novel antibiotics as well as exploring human RNRs as targets for novel antitumorogenic drugs.

Cross-References

▶ Cobalt Proteins, Overview
▶ Cobalt-containing Enzymes
▶ Copper Amine Oxidase
▶ Iron Proteins, Ferritin
▶ Iron-Sulfur Cluster Proteins, Fe/S-S-adenosylmethionine Enzymes and Hydrogenases
▶ Iron-Sulfur Cluster Proteins, Ferredoxins

References

Andersson KK (ed) (2008) Ribonucleotide reductase. Nova Science, Hauppage

Boal AK, Cotruvo JA Jr, Stubbe J, Rosenzweig AC (2010) Structural basis for activation of class Ib ribonucleotide reductase. Science 329:1526–1530

Cotruvo JA, Stubbe J (2011) Class I ribonucleotide reductases: metallocofactor assembly and repair in vitro and in vivo. Annu Rev Biochem 80:733–767

Fairman JW, Wijerathna SR, Ahmad MF et al (2011) Structural basis for allosteric regulation of human ribonucleotide reductase by nucleotide-induced oligomerization. Nat Struct Mol Biol 18:316–322

Hofer A, Crona M, Logan DT, Sjöberg B-M (2012) DNA building blocks: keeping control of manufacture. Crit Rev Biochem Mol Biol 47(1):50–63

Högbom M (2011) Metal use in ribonucleotide reductase R2, di-iron, di-manganese and heterodinuclear–an intricate bioinorganic workaround to use different metals for the same reaction. Metallomics 3:110–120

Logan DT (2008) The anaerobic ribonucleotide reductases: recent progress in understanding. In: Andersson KK (ed) Ribonucleotide Reductase. Nova Science, Hauppage, pp 185–208

Logan DT (2011) Closing the circle on ribonucleotide reductases. Nat Struct Mol Biol 18:251–253

Lundin D, Poole AM, Sjöberg B-M, Högbom M (2012) Use of structural phylogenetic networks for classification of the ferritin-like superfamily. J Biol Chem [Epub ahead of print]

Nordlund P, Reichard P (2006) Ribonucleotide reductases. Annu Rev Biochem 75:681–706

Sjöberg B-M (2010) Biochemistry. A never-ending story. Science 329:1475–1476

Torrents E, Sahlin M, Sjöberg B-M (2008) The ribonucleotide reductase family – genetics and genomics. In: Andersson KK (ed) Ribonucleotide reductase. Nova Science, Hauppage, pp 17–77

Rigid Lamb Syndrome

▶ Selenium and Muscle Function

Rigid Spine Muscular Dystrophy (RSMD1)

▶ Selenium and Muscle Function

RNA Structure Probing

▶ Lead and RNA

RNR

▶ Ribonucleotide Reductase

Role of Metal Ions in Protein–Nucleic Acid Complexes

▶ Barium and Protein–RNA Interactions

Role of Metallothioneins in Cadmium-Induced Hepatotoxicity and Nephrotoxicity

▶ Cadmium and Metallothionein

Role of Zinc in Etiopathogenesis of Bleeding

▶ Zinc in Hemostasis

ROS: Reactive Oxygen Species

▶ Iron Proteins, Plant Iron Transporters

Rubidium in Biological Systems and Medicine

Valery V. Kupriyanov
Institute for Biodiagnostics, National Research Council, Winnipeg, MB, Canada

Synonyms

Na/K-ATPase, Na/K pump, sodium- and potassium-dependent adenosine triphosphatase; Revascularization, reflow, reperfusion

Definitions

Chemical symbols of ions (Na^+, K^+, Rb^+, Li^+) in square brackets: concentrations of free ions; *subscripts "in" and "out"*: intracellular and extracellular concentrations, respectively. *Infarct/infarction*: nonviable organ area; *ischemia*: insufficient blood supply to organ area. *NMR*: nuclear magnetic resonance, which involves *MRS* and *MRI*, nuclear magnetic resonance spectroscopy and imaging, respectively. T_1 and T_2: longitudinal and transverse relaxation times of NMR-sensitive nuclei, respectively. *Perfusion*: supply of blood or its substitute (perfusate) into the vascular system of an organ (e.g., heart, kidney, liver). *Revascularization*: opening of obstructed vessel(s) and restoration of blood flow (reperfusion).

Introduction

Rubidium (symbol Rb) is the 37th element of the 1st group of the periodic chart, closest analog of an alkali metal, potassium. Its abundance in the earth's crust is 16th among metals and 23rd among all the elements (Butterman and Reese 2003). In the earth's crust, Rb is present as a mixture of two isotopes, ^{85}Rb and ^{87}Rb, in proportion 72–28% (Table 1). Although ^{87}Rb is radioactive, its half-life ($t_{1/2}$) of 49 billion years (Holden 2004) is much longer than the age of the Universe (~14 billion years), which means that in a practical sense, Rb is a stable element with a very weak radioactive emission. This resembles a situation with naturally occurring potassium (symbol K) isotopes, which contain a small fraction of radioactive ^{40}K with a half-life of 1.3 billion years. Among many artificial radioactive isotopes of Rb, only two of them deserve consideration for biological applications. ^{86}Rb ($t_{1/2} = 18.7$ days) is manufactured in a synchrotron and emits radiation in the form of electrons (β-) and γ-photons while ^{82}Rb ($t_{1/2} = 76$ s) is produced by 82-strontium (^{82}Sr) generators and emits positrons (β+, Table 1) (Holden 2004).

Naturally occurring Rb ion (Rb^+) does not play any significant biological role as its biochemical properties are very similar to K^+ and Rb^+ to K^+ molar concentration ratio is very low in animal tissues, ca. 1/800 (20–60 ppm of dry weight; > 300 mg in average-sized human). In seawater, this ratio is even lower, ca. 1/2,000 (Relman 1956; Meltzer and Fieve 1975). Rb was found in different soil types with contents ranging from 1% to 15×10^{-3}% and plants where its content varies considerably from 1 to 100 ppm since Rb uptake depends on the Rb/K ratio in soil (Call 1963).

On the other hand, similar biochemical properties of K and Rb stimulated research of K replacement with Rb and use of Rb isotopes as K^+ tracers (Relman 1956; Meltzer and Fieve 1975; Call 1963; Kupriyanov and Gruwel 2004). The latter application was also driven by the inconvenience of using relatively short-living radioactive ^{42}K isotope ($t_{1/2} \sim 12$ h) for long-term tracer studies.

The scope of this entry covers the applications of rubidium isotopes in biological research and medical diagnostics. Research applications are mainly focused on animal and human studies of potassium ions replacement with rubidium ions (Rb^+) and use of Rb^+ as a tracer of K^+ transport routes. Four major Rb isotopes are considered: naturally occurring 85 and 87 and artificial radioactive 82 and 86. Diagnostic application involves cardiac coronary artery disease assessment using positron emission tomography (PET) imaging of ^{82}Rb distribution. Rb^+ applications for studies in plants and fungi are beyond the scope of these entry (see, e.g., Rodrigues-Navarro 2000).

Research Applications of Rb

Animal Studies

Animal research of Rb^+ can be subdivided into two large categories: (1) early studies of effects of acute and chronic K^+ replacement with Rb^+ on animal and

Rubidium in Biological Systems and Medicine, Table 1 Properties of some rubidium isotopes

Isotopes	Natural abundance	Source	Half-life	Decay	Spin n	NMR properties Sens.	Res. freq.
1. ^{87}Rb	28%	Earth crust	4.9×10^{10} y	β–	3/2	5%	1/3
2. ^{85}Rb	72%	Earth crust	Stable	n/a	5/2	0.8%	1/10
3. ^{86}Rb	Artificial	Cyclotron	18.6 days	β–, γ	2	n/a	n/a
4. ^{82}Rb	Artificial	^{82}Sr decay	76 s	β+	1	n/a	n/a

β– electron, β+ positron, *Sens.* NMR sensitivity expressed as a percentage of ^{1}H sensitivity at the same field strength corrected for natural abundance, *Res freq.* resonance frequency relative to ^{1}H at the same field strength

organ physiology including toxicology issues and (2) tracer studies focused on measurements of K^+ fluxes and distribution using Rb^+ radioisotopes (^{86}Rb and ^{82}Rb) and ^{87}Rb$^+$ as an NMR-sensitive probe (Relman 1956; Meltzer and Fieve 1975; Call 1963; Kupriyanov and Gruwel 2004). These studies were performed at different levels of biological organization: live animals in vivo, isolated organs (e.g., heart, liver, kidneys) and cells (e.g., erythrocytes), subcellular structures (e.g., membrane vesicles, mitochondria), and enzymes (e.g., Na/K-ATPase). Isolated organs can be maintained in their normal physiological state for hours by providing a sufficient vascular flow (perfusion) of oxygenated blood or artificial blood substitutes (perfusate), typically composed of blood plasma concentrations of salts, such as NaCl, KCl, $CaCl_2$, $MgCl_2$, $NaHCO_3$, KH_2PO_4, and glucose (crystalloid perfusate) with or without other oxidizable substrates (e.g., pyruvate, lactate, and fatty acids).

Studies of the effects of Rb salts on the various organisms had begun in nineteenth century and involved toxicity and physiology issues (Meltzer and Fieve 1975). Cardiac effects were tested in isolated rat hearts by S. Ringer and in dog hearts by S. Botkin. The results suggested similarity in K^+ and Rb^+ properties. At the same time, toxicology studies were performed by Richet on snails, crayfish, reptiles, amphibians, fish, birds, guinea pigs, and dogs. Behavioral effects of Rb salts were noted in rats in vivo in the next century (Meltzer and Fieve 1975). Subsequent studies in rats demonstrated the ability of animal tissues to concentrate the Rb ion in the same way as the potassium ion, such that its concentration ratio between the intracellular and extracellular compartments is similar to that of potassium, i.e., 22–30. Rb^+ accumulated in excitable tissues of dogs and rats, such as the brain, the muscles, and the heart as well as in the liver and kidneys. Rb^+ accumulation in the brain was slower due to its limited permeability (similar to K^+) across the blood–brain barrier. Gradual K^+ replacement with Rb^+ was achieved by feeding animals with a high Rb/low K diet (food and water) over weeks and months. The degrees of substitution were in the 20–50% range; 20–30% was considered safe. Higher substitution levels of 30–50% produced side effects such as increased irritability, aggressiveness, loss of body mass, and even death (Relman 1956; Meltzer and Fieve 1975). The lower degrees of K^+ replacement (<30%) did not interfere with physiological functions and ability to reproduce (Meltzer and Fieve 1975). Rb^+ has a long half-life in animal and human tissues of about 60 days and is cleared from the body by the kidneys with urine (Relman 1956; Meltzer and Fieve 1975). The absolute urine content of naturally occurring Rb^+ varied considerably; however, Rb-to-K ratio was more constant (approximately 350 ppm) (Meltzer and Fieve 1975).

Biochemical basis for K^+ and Rb^+ accumulation in animal tissues is adenosine 5′-triphosphate (ATP)-dependent K^+ (and Rb^+) uptake coupled to Na^+ extrusion by the Na/K-ATPase (Na:K:ATP = 3:2:1) located in the cell membrane of any mammalian cell (Scou 1965). This process builds up outwardly directed K^+ and inwardly directed Na^+ gradients. For example, in muscle tissue, $[K^+]$in/$[K^+]$out is ~120/5 mM and $[Na^+]$in/$[Na^+]$out ~ (5–10)/140 mM. Potassium ion gradient is important for generation of a transmembrane potential (−80 mV at rest) which is produced by the outward movement of K^+ ions through the K^+-selective voltage-gated channels (Fozzard and Arnsdorf 1986). When Rb^+ concentration in blood plasma increases above its natural level (due to RbCl injection or intake with food/water), Rb^+ competes with K^+ for binding with Na/K-ATPase, which results in increased Rb^+ and decreased K^+ influx into the cytoplasm. Rb^+ replaces K^+

on nearly 1:1 basis due to very similar kinetic properties for Na/K-ATPase transport, which is, perhaps, based on similarity of the ionic radii of Rb^+ and K^+ in crystals (149 vs. 133 pm, respectively) and water (hydrated R-values, 293 vs. 276 pm). Eventually, increased Rb^+ delivery into the cells is compensated by its increased efflux via K^+ channels and other transporters. This effect leads to the establishment of new equilibrium distribution of Rb^+ and K^+. Termination of Rb^+ delivery results in its efflux via K^+ channels and other transporters and reestablishment of the previous equilibrium. The latter forms a basis for analysis of K^+ efflux pathways, such as K^+ channels. Note that an acute Rb injection into the bloodstream should be performed with a great care as plasma (K^+ + Rb^+) concentration can be easily doubled or tripled (blood volume *plus* interstitial space is relatively small, ∼140 ml/kg body wt.). Such increase may cause arrhythmias, fibrillation, and cardiac arrest. The effect is not due to RbCl toxicity per se but is similar to that of increased K^+ which causes cell membrane depolarization and changes excitability. In the experiments with isolated perfused organs (e.g., heart (Kupriyanov and Gruwel 2004)), cells, and subcellular systems, Rb^+ + K^+ content in perfusate or incubation medium is much easier to control.

Radiotracer experiments in vivo do not impose such a problem as ^{86}Rb or ^{82}Rb concentrations are negligible relative to that of K^+. The isotopes provide a relatively easy way to follow Rb uptake kinetics and evaluate the activity of the Na/K-ATPase at the subcellular, cellular, organ, and in vivo levels using a specific inhibitor of the pump, ouabain (Kupriyanov and Gruwel 2004; Scou 1965). However, in the in vivo and isolated organ experiments, the isotope is delivered with blood or perfusate flow, which potentially can limit Rb uptake under certain conditions. When blood flow and vascular permeability is not compromised by disease, Rb extraction by cells is mostly determined by the Na/K-ATPase activity (except for brain). In the heart and the muscles, this activity increases as the rate of heart beat or muscle stimulation increases, leading to an elevated intracellular Na^+ concentration. In addition, the cellular metabolic state affects ATPase activity via ATP and adenosine-5′-diphosphate (ADP) concentrations and intracellular pH. For instance, under conditions of insufficient oxygen supply (needed for ATP synthesis), ATP/ADP ratio and pH decline, leading to the Na/K-ATPase inhibition. This condition is known as ischemia, and arises due to mismatch between blood supply and cellular energy demand caused by vascular disease (e.g., atherosclerosis). Therefore, tissue Rb uptake is reduced owing to its decreased delivery and secondary changes in the cell metabolism. Prolonged severe ischemia results in cell damage and subsequent death, such that cells become "leaky" and are unable to maintain K^+ and Na^+ gradients any longer. This condition is known as tissue infarction and is associated with a heart attack and stroke. The sensitivity of Rb^+ uptake to blood flow and metabolic state makes Rb a suitable flow and viability tracer (Kupriyanov and Gruwel 2004).

In experiments with perfused isolated organs such as the heart, ^{86}Rb isotope concentration can be maintained constant by means of continuous infusion, which gives rise to exponential accumulation curves. A bolus ("instantaneous") injection results in first-pass kinetics when the radioactivity level quickly rises and then diminishes providing information on "wash-in" and "wash-out" kinetics from which perfusion flow can be obtained. This method is used predominantly in the in vivo settings with a short-living isotope, ^{82}Rb (see section "Rb in Medical Diagnostics"). Note that for quantitative measurements of Rb^+ uptake with the radiotracers in vivo, the natural abundance of Rb in blood plasma (ca. 3 μM) should be taken into account due to dilution of the isotope with "cold" Rb and a decrease in its specific radioactivity.

Development of magnetic resonance spectroscopy (MRS) and imaging (MRI) of biologically relevant cations, such as $^{23}Na^+$, $^{39}K^+$, $^{87}Rb^+$, and $^{133}Cs^+$, opened the possibility of measurements of Rb^+ fluxes and distribution in cells, isolated organs, and living organisms (reviewed in Kupriyanov and Gruwel 2004). Despite the rather low abundance of ^{87}Rb (28%), it has a relatively high MR sensitivity (absolute = 17%; relative = 5% of ^1H) (Table 1). Furthermore, a very short longitudinal relaxation time, T_1 (determines the time needed for recovery of magnetization to equilibrium) of 2.5 ms, allows frequent pulsing without saturation effects, which provides good signal-to-noise ratios (SNR) over short periods of time when the stability of the biological system is ensured. However, very short transverse relaxation times, T_2 of <2.5 ms (determines the rate of signal decay following excitation), makes spectral lines wide (>150 Hz), which reduces SNR values and complicates imaging applications (Kupriyanov and Gruwel 2004).

Pioneering studies of the George Radda group in 1980 have demonstrated the feasibility of ^{87}Rb-MRS for assessment of K$^+$ transport in red blood cells (RBC), kidney, heart, and skeletal muscle.

Later on, ^{87}Rb-MRS and MRI were mostly used in cardiovascular research using isolated perfused rat, pig, and mouse hearts (Kupriyanov and Gruwel 2004). In rat hearts, the Rb$^+$ uptake and efflux (20% K$^+$ substitution) were assessed by applying pharmacological agents that specifically inhibit Na/K-ATPase (ouabain), Na/K/2Cl co-transporter (bumetanide), voltage-gated K$^+$, and Na$^+$ channels (4-aminopyridine and lidocaine, resp.) and stimulate Na$^+$ entry (Na$^+$ ionophore, monensin). About ~80% of Rb$^+$ uptake was accounted for the Na/K-ATPase and was dependent on the intracellular Na$^+$. Bulk of Rb$^+$ efflux could not be assigned to any studied carrier. Manipulations with intracellular pH (acidosis and alkalosis) induced by an osmotic shock, by addition of ammonium chloride or monensin, revealed the functional presence of a putative nonspecific cation/H$^+$ exchanger (Kupriyanov and Gruwel 2004).

A link between cardiac energetics and K$^+$/Rb$^+$ efflux, which is provided by ATP-sensitive K$^+$ channels (K(ATP)), was studied in great detail. These channels are silent under normal physiological conditions and are activated in response to a stress (e.g., ischemia). Inhibition of oxidative phosphorylation with cyanide or uncoupling with 2,4-dinitrophenol (documented by ^{31}P-MRS) resulted in stimulation of Rb efflux. The stimulation was reversed by sulfonylurea derivative, glyburide, specific inhibitor of K(ATP). K(ATP) opener, cyanoguanidine derivative P-1075, also stimulated Rb efflux in a glyburide-sensitive manner. Overall, these studies demonstrated the feasibility of ^{87}Rb-MRS for analysis of K$^+$ movements in living hearts under normal conditions and during metabolic stress.

^{87}Rb-MRI was applied to study the changes in Rb$^+$ distribution kinetics associated with pathological states, such as regional ischemia and infarction in pig hearts (Kupriyanov and Gruwel 2004). Three-dimensional chemical shift imaging (3D-SCI) provided fast data acquisition with a time resolution of 5 min, which enabled measurements of the kinetics of Rb accumulation. For this reason, the spatial resolution was relatively poor, 10 × 10 mm^2 in plane, enhanced to 2.5 × 2.5 mm^2 by extrapolation. In isolated perfused pig hearts, left anterior descending artery (LAD) feeding mainly left ventricular anterior wall (facing front of chest wall) was partially or completely blocked while blood supply to the opposite posterior wall was not affected. This mimics the real-life situation when blood flow is restricted by atherosclerotic plaque or a thrombus. Partial blockage of the LAD caused severe ischemia, while complete occlusion and subsequent restoration of flow (reperfusion, mimics vessel revascularization) resulted in irreversible damage and formation of an infarction area. Replacement of K$^+$ in blood-containing or crystalloid perfusate resulted in 25% or 67% substitution of the intracellular K$^+$ with Rb$^+$ in healthy tissue, respectively. Ischemic and infarction areas appeared as zones of reduced intensity on ^{87}Rb images due to reduced rates of Rb$^+$ accumulation. In case of ischemia, it was due to severe flow restrictions and associated metabolic changes (inhibition of Na/K-ATPase) while in case of infarction, it was caused by cell death and inability to maintain ionic gradients. The overall location and size of ischemic and infarct areas were correctly determined by ^{87}Rb-MRI as was confirmed by histological examination.

Thus, ^{87}Rb MR spectroscopy and imaging is a convenient research tool for studies of K$^+$ transport in isolated organs under normal and pathological conditions. In vivo applications are plausible in chronic settings when K$^+$ substitution with Rb$^+$ takes place gradually over a period of weeks by means of the Rb-enriched diet. In this case, Rb$^+$ washout can be monitored in connection with simulated pathological situations. An example of this application is recent studies of brain ischemia and infarction in rats (Yushmanov et al. 2007).

Human Studies

Soon after the first animal studies at the end of nineteenth century, RbCl (300 mEq/25 days; 1 mEq = 1 mmol) was tested on cardiac patients by Botkin who reported an increase in heart rate and blood pressure (Meltzer and Fieve 1975). One of the applications of Rb$^+$ was replacement of K$^+$ in KBr therapy of epilepsy at the doses about 200 mEq/15 days. No advantage or disadvantage was observed (Meltzer and Fieve 1975). Rubidium iodide was administered to syphilitics in place of KI, producing similar therapeutic and side effects. Overall about 60 patients received Rb salts, which were available in pharmacies across central Europe (Meltzer and Fieve 1975). In the middle of twentieth century, ^{86}Rb equilibration in the human body was studied. Later on, Rb salts were tested as antidepressants in research and small-scale clinical trials. The total administered dose was 268 mEq over a 3-month period, of which no more than 185 mEq was retained based on urinary secretion. The total body K replacement level did not exceed 12% and activity of blood enzymes such alkaline phosphatase, lactate dehydrogenase, and serum glutamate-oxaloacetate transaminase did not change. Parallel to Rb^+ accumulation, electro-encephalogram (EEG) activation was noted; however, the effect on a mood was scattered showing decrease, no change, and increase in 10 patients. Simultaneous administration of another antidepressant, lithium (Li^+), accelerated Rb^+ clearance into urine. Overall neurophysiological and behavioral effects of Rb^+ and Li^+ were considered as opposite (Meltzer and Fieve 1975). In addition, RbI was used in place of KI in treating enlargement of the thyroid gland, as an antishock agent in arsenic poisoning (Butterman and Reese 2003) and as a component of eye drops.

Rb in Medical Diagnostics

Diagnostic applications of Rb are based on use of its positronic isotope, ^{82}Rb, which is produced in ^{82}Sr generators (Table 1). Positrons emitted by ^{82}Rb decay (average pathlength in tissue of the order of ~2 mm) and collide with the tissue electrons, resulting in annihilation with formation of two gamma-photons traveling in opposite directions. This forms the basis for their detection as coincidence events in a ring positron emission tomography (PET) scanner. Spatial resolution of PET in clinical scanners is in the range of 4–7 mm, which is compensated by a high time resolution that allows monitoring first-pass tracer kinetics (Schindler et al. 2010). It is used for perfusion measurements in cardiac diagnostics although it is not inert tracer and avidly extracted by cardiac tissue (Na/K-ATPase) at 65% during first-pass measurements. By this parameter, it is closer to the deposit flow tracers which ideally should have a 100% extraction coefficient and linear flow dependence of their deposition. ^{82}Rb extraction decreases as flow increases; therefore, flow dependence is nonlinear as deposition is the product of flow, ^{82}Rb concentration, and the extraction coefficient. Under certain conditions, ^{82}Rb may also serve as a viability tracer as its uptake depends on the Na/K-ATPase activity and the "leakiness" of a cell's membrane. Furthermore, mature infarct area represents mostly the extracellular space as the contractile cardiac cells (myocytes) are replaced with the extracellular matrix and relatively small amounts of myofibroblasts.

Due to its short half-life (76 s), ^{82}Rb can be used several times during the course of examination (Schindler et al. 2010). A common example is testing ^{82}Rb kinetics under baseline (rest) conditions followed by measurements under stress and then again at baseline conditions. This sequence allows identification of "hidden" perfusion defects, which are not detectable during the rest due to insufficient narrowing of medium and large vessels by atherosclerosis. Injection of the stress agents, such as coronary vasodilators, dipyridamole, or adenosine, widens (dilates) small vessels in normal tissue and increases its perfusion several-fold. The degree of this increase is termed as a coronary flow reserve. The vascular bed in tissue downstream of the flow blockage is unable to respond to these agents as it is already fully dilated, which results in reduced coronary flow reserve. ^{82}Rb-PET detects areas which poorly respond to adenosine as zones of compromised perfusion. Termination of adenosine administration rapidly reestablishes baseline flows since adenosine half-life in the blood is very short (few seconds). Another type of stress agent is a short-living (few minutes) β-agonist, dobutamine, which in addition to vasodilation stimulates contraction of cardiac muscle, thereby increasing energy demand and matching coronary flow. In the regions with compromised perfusion, coronary flow is limited which causes a supply/demand mismatch and transient ischemia. Again flow and ^{82}Rb uptake is increased in normal areas relative to the compromised area in response to dobutamine stress. Thus, ^{82}Rb PET imaging allows quantitative functional evaluation of location and degree of coronary artery disease and detects circulation defects (ca. 90% specificity and sensitivity) in addition to routine two-dimensional coronary X-ray angiography.

In early studies, ^{82}Rb PET was tested for tumors detection. Despite initial promising results, it did not find further diagnostic applications in oncology.

Conclusion

Rb salts containing natural isotopes, ^{85}Rb and ^{87}Rb, have been used as potassium analogs in various research applications at different levels of biological organization. In humans, the applications were mainly limited to the field of psychiatry. Recent noninvasive research applications involved ^{87}Rb-MRS and MRI studies of K$^+$ transport in animal models of disease. Radioactive isotopes, ^{86}Rb and ^{82}Rb, were intensively utilized in animal and human studies of K$^+$ transport under normal and pathological conditions. ^{82}Rb is currently used in noninvasive diagnostics of coronary artery disease with PET imaging. Overall, Rb isotopes were and still remain valuable probes in biomedical research.

Cross-References

- Thallium-201 Imaging
- Cesium
- Potassium in Biological Systems
- Rubidium, Physical and Chemical Properties
- Potassium Channels, Structure and Function
- Potassium in Health and Disease
- Potassium, Physical and Chemical Properties
- Rubidium, Physical and Chemical Properties

References

Butterman WC, Reese RG (2003) Mineral commodity profiles: rubidium. US geological survey. Open file report 03-045. http://pubs.usgs.gov/of/2003/of03-045/of03-045.pdf. Accessed July 20, 2011

Call F (1963) Biological properties of rubidium. In: Mellor's comprehensive treatise on inorganic and theoretical chemistry, vol II, Suppl. III. The alkali metals, pt. 2. Wiley, New York, pp 2279–2285

Fozzard HA, Arnsdorf MF (1986) Cardiac electrophysiology. In: Fozzard HA, Haber H, Jennings RB, Katz AM, Morgan H (eds) The heart and cardiovascular system, vol 1, 1st edn. Raven, New York, pp 1–30

Holden NE (2004) Table of isotopes. In: Lide DR (ed) CRC handbook of chemistry and physics, 85th edn. CRC Press, Boca Raton, Section 11

Kupriyanov VV, Gruwel MLH (2004) Rubidium-87 magnetic resonance spectroscopy and imaging for analysis of mammalian K$^+$ transport. NMR Biomed 18:111–124

Meltzer HL, Fieve RR (1975) Rubidium in psychiatry and medicine. An overview. In: Current developments in psychopharmacology, vol 1. Spectrum Publications, New York

Relman AS (1956) The physiological behaviour of rubidium and cesium in relation to that of potassium. Yale J Biol Med 29:248–262, http://www.ncbi.nlm.nih.gov/pmc/articles/PMC2603856/pdf/yjbm00642-0078.pdf

Rodrigues-Navarro A (2000) Potassium transport in fungi and plants. Biochim Biophys Acta 1469:1–30

Schindler TH, Schelbert HR, Quercioli A, Dilsizian V (2010) Cardiac PET imaging for the detection and monitoring of coronary artery disease and microvascular health. J Am Coll Cardiol Imaging 3:623–640

Scou JC (1965) Enzymatic basis for active transport of Na$^+$ and K$^+$ across cell membrane. Physiol Rev 45:596–617

Yushmanov VE, Kharlamov A, Boada FE, Jones SC (2007) Monitoring brain potassium with rubidium flame photometry and MRI. Magn Reson Med 57:494–500

Rubidium, Physical and Chemical Properties

Fathi Habashi
Department of Mining, Metallurgical, and Materials Engineering, Laval University, Quebec City, Canada

Rubidium is a very reactive soft metal with a silvery luster and has the typical chemical properties of an alkali metal. It is more abundant in the Earth's crust than lithium and cesium. Due to the similarity in their ionic radii, rubidium is always found in potash salt deposits, e.g., carnallite, sylvite, and langbeinite, which contain 10–300 g/t rubidium oxide. Rubidium is also widely distributed in very small concentrations in spring water, rivers, lakes, and seas. Its compounds are used in psychiatry and medicine, in magneto-optic modulators, solid-state lasers, phosphors (rubidium aluminate), Rb – C molecular sieves for hydrogen absorption, paper pigments, glass hardening by ion exchange, and components of electrolytes for fuel cells (Lenk and Prinz 1997).

Physical Properties

Atomic number	37
Atomic weight	85.468
Relative abundance in Earth's crust, %	3.1×10^{-2}
Density at 18°C, g/cm^3	1.522
Density at m.p., g/cm^3	1.472
Melting point, °C	39
Boiling point, °C	696
Ionization potential, eV	4.16
Normal electrode potential, V	−2.99
Electrical conductivity at 0°C, Ω^{-1} cm^{-1}	8.86×10^4
Thermal conductivity of liquid, W m^{-1} K^{-1}	29.30
Specific heat capacity	

(continued)

Solid, J g^{-1} K^{-1}	0.332
Liquid (from m.p. to 1,093°C), J g^{-1} K^{-1}	0.368
Gas, J g^{-1} K^{-1}	0.242
Heat of fusion, J/g	25.71
Heat of sublimation at 25°C, J/g	1004.3
Heat of vaporization, J/g	887.6
Mohs hardness	0.3
Atomic radius, nm	0.243
Covalent radius, nm	0.211
Ionic radius, nm	0.149
Crystal structure	Body-centered cubic

Chemical Properties

Rubidium is a typical metal; its electronic structure is 2, 8, 18, 8, 1. When it loses its outermost electron, it will have the inert gas electronic structure. It is a monovalent highly reactive metal toward water, steam, air, and many other gases and substances. It reacts explosively with water to form colorless rubidium hydroxide. Rubidium salts are hygroscopic and must be protected from moisture during storage and transport (Schreiter 1961).

References

Lenk W, Prinz H (1997) Chapter 53. In: Habashi F (ed) Handbook of extractive metallurgy. Wiley, Weinheim

Schreiter W (1961) Seltene metalle, vol 2. VEB Deutscher Velag fuer Grundstoffindustrie, Leipzig, pp 300–306

Rubredoxin, Interaction with Germanium

Griselda Hernández[1], David M. LeMaster[1] and Janet S. Anderson[2]
[1]New York State Department of Health, Wadsworth Center, Albany, NY, USA
[2]Department of Chemistry, Union College, Schenectady, NY, USA

Synonyms

Biomacromolecular coordination of germanium

Definition

The tetracysteine metal coordination site of the rubredoxins from the eubacterium *Clostridium pasteurianum* and the hyperthermophilic archaeon *Pyrococcus furiosus* forms kinetically stable coordination with the inorganic Ge(IV) ion.

Clinical Toxicity of Inorganic Germanium Complexes

Inorganic germanium(IV) forms coordination complexes with a wide range of O, N, and S-bearing ligands (Holloway and Melnik 2001). Elsewhere in this volume, the potential utility of a small molecule germanium complex as a radioprotectant for whole body γ-irradiation treatment is considered. Unfortunately, during the past several decades, the clinical relevance of germanium ion coordination has been dominated by the marketing of elixirs containing citrate and lactate coordination complexes of inorganic Ge(IV) that are proposed to treat numerous types of severe maladies (Tao and Bolger 1997). A number of fatalities have resulted, primarily due to renal failure. In aqueous solutions, these complexes generally hydrolyze to form a highly insoluble lattice in which each Ge(IV) ion is tetrahedrally coordinated to oxygen, denoted GeO_2. This nephrotoxicity correlates with the formation of GeO_2 deposits in the kidneys of these victims (Sanai et al. 1990).

Metal Binding Site of the Electron Transfer Protein Rubredoxin

The small electron transfer protein rubredoxin contains a single Fe(S-Cys)$_4$ cluster in its active site. The ease in production and characterization of this protein has led to its extensive use in analyzing the structural basis of electron transfer and reduction potential. In addition to the physiologically relevant Fe^{3+} (Watenpaugh et al. 1979) and Fe^{2+} (Min et al. 2001) forms of the rubredoxin from the anaerobic eubacterium *Clostridium pasteurianum*, high resolution X-ray structures have been obtained in which the iron is substituted with a number of different divalent and trivalent metal ions, including the diamagnetic Zn^{2+} (Dauter et al. 1996) and Ga^{3+} (Maher et al. 2004) forms. In Fig. 1 are shown the backbone and cysteine sidechain atoms

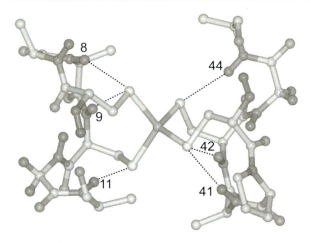

Rubredoxin, Interaction with Germanium, Fig. 1 The active site region of Ga³⁺-substituted rubredoxin from *Clostridium pasteurianum*. Residues 6, 9, 39, and 42 contribute the four cysteine sulfurs which coordinate the metal (Maher et al. 2004). Indicated are the six backbone amide protons which donate hydrogen bonds to these cysteine sulfurs in a pseudo twofold structural arrangement (Reproduced with permission LeMaster et al. 2006b)

from the metal site region which are arranged in a pseudo twofold symmetry as illustrated from the Ga^{3+}-substituted rubredoxin structure (Maher et al. 2004). The six backbone amides that hydrogen bond to the metal-coordinated sulfur atoms are numbered.

Structural Characterization of Ge(IV)-Substituted Rubredoxin

To examine if the diamagnetic series of Zn^{2+}- and Ga^{3+}-substituted rubredoxins could be extended to the Ge^{4+} form, the apoprotein was reconstituted with $Ge(OEt)_4$ at pH 8, analogous to the protocol used for the previously described metal reconstitution studies (Moura et al. 1991; Maher et al. 2004). Following isolation by ion exchange chromatography in metal-free buffer, inductively coupled plasma mass spectrometry analysis of ^{74}Ge indicated a metal binding of 0.93 (±0.04) atom per rubredoxin molecule (LeMaster et al. 2006b). The Ge^{4+}-substituted rubredoxin loses its metal at a more

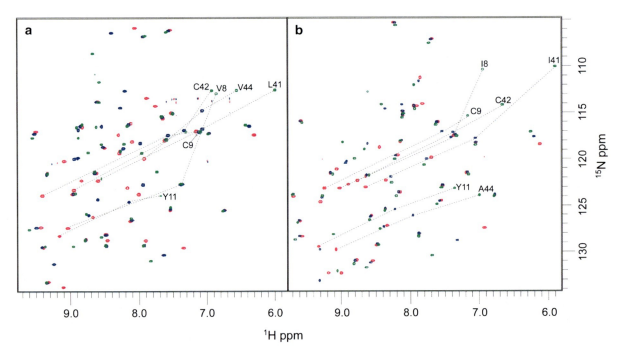

Rubredoxin, Interaction with Germanium, Fig. 2 ^1H-^{15}N 2D NMR spectra of Zn^{2+}-, Ga^{3+}-, and Ge^{4+}-substituted *Clostridium pasteurianum* (panel **a**) and *Pyrococcus furiosus* (panel **b**) rubredoxin. Active site tetracysteine coordination to Zn^{2+} (*pink*), Ga^{3+} (*blue*), and Ge^{4+} (*green*) at pH 6.0. The chemical shift migration pattern of the six amides that hydrogen bond to the coordinating sulfur atoms is indicated (Reproduced with permission LeMaster et al. 2006b)

rapid rate as the pH of the solution is raised from 4.5 to 8.5 (LeMaster et al. 2006a). However, under physiological buffer conditions the metal coordination is stable on the order of weeks.

^{1}H-^{15}N 2D NMR correlation spectra for the Zn^{2+}-, Ga^{3+}-, and Ge^{4+}-substituted rubredoxins demonstrated that for many of the backbone positions the amide crosspeaks from the three metal forms are essentially superimposed. However, a number of the other amide crosspeaks exhibit significant differential chemical shifts induced by the change in metal charge. The magnitude of these shifts correlates with the distance and orientation, relative to the metal site, consistent with an electric field–induced change in the chemical shifts (Fig. 2a).

A qualitatively similar differential chemical shift pattern is observed when these three metal ions are substituted into the rubredoxin from the hyperthermophilic archaeon *Pyrococcus furiosus*, which shares a 59% sequence identity with the *Clostridium pasteurianum* protein (Fig. 2b). When compared to the NMR spectra of *Clostridium pasteurianum* rubredoxin, the hyperthermophile rubredoxin exhibits more nearly a twofold symmetry in the metal charge-induced chemical shift effects for amide crosspeaks within the active site, consistent with the more closely preserved twofold symmetry observed in the corresponding X-ray structures of *Pyrococcus furiosus* rubredoxin.

Possible Avenues of Clinical Relevance for the Biomacromolecular Coordination of Germanium

Ge^{4+}-substituted rubredoxin is the first characterized coordination complex of tetravalent germanium with a biological macromolecule. The range of additional mechanisms of action which might contribute to the renal pathology and associated hepatic and neurological dysfunctions observed in these patients might be extended if the Ge(IV) ion was found to form stable complexes with other biological macromolecules.

Cross-References

▶ Germanium l-Cysteine Alpha-Tocopherol Complex as Stimulator to Antioxidant Defense System
▶ Germanium, Physical and Chemical Properties

References

Dauter Z, Wilson KS, Sieker LC, Moulis JM, Meyer J (1996) Zinc- and iron-rubredoxins from *Clostridium pasteurianum* at atomic resolution: a high precision model of a ZnS_4 coordination unit in a protein. Proc Natl Acad Sci USA 93:8836–8840

Holloway CE, Melnik M (2001) Germanium coordination compounds: classification and analysis of crystallographic and structural data. Main Group Met Chem 24: 681–726

LeMaster DM, Anderson JS, Hernández G (2006a) Role of native-state structure in rubredoxin native-state hydrogen exchange. Biochemistry 45:9956–9963

LeMaster DM, Minnich M, Parsons PJ, Anderson JS, Hernández G (2006b) Tetrathiolate coordination of germanium(IV) in a protein active site. J Inorg Biochem 100:1410–1412

Maher M, Cross M, Wilce MCJ, Guss JM, Wedd AG (2004) Metal-substituted derivatives of the rubredoxin from *Clostridium pasteurianum*. Acta Crystallogr D 60: 298–303

Min TP, Ergenekan CE, Eidsness MK, Ichiye T, Kang CH (2001) Leucine 41 is a gate for water entry in the reduction of *Clostridium pasteurianum* rubredoxin. Protein Sci 10:613–621

Moura I, Teixeira M, LeGall J, Moura JJ (1991) Spectroscopic studies of cobalt and nickel substituted rubredoxin and desulforedoxin. J Inorg Biochem 44:127–139

Sanai T, Okuda S, Onoyama K, Oochi N, Oh Y, Kobayashi K, Shimamatsu K, Fujimi S, Fujishima M (1990) Germanium dioxide-induced nephropathy: a new type of renal disease. Nephron 54:53–60

Tao SH, Bolger PM (1997) Hazard assessment of germanium supplements. Regul Toxicol Pharmacol 25:211–219

Watenpaugh KD, Sieker LC, Jensen LH (1979) The structure of rubredoxin at 1.2 Å resolution. J Mol Biol 131: 509–522

Ruthenium, Physical and Chemical Properties

Fathi Habashi
Department of Mining, Metallurgical, and Materials Engineering, Laval University, Quebec City, Canada

Ruthenium is a member of the group of six platinum metals which occur together in nature and is located in the center of the Periodic Table in the Iron-Cobalt-Nickel group:

Fe	Co	Ni
Ruthenium	Rh	Pd
Os	Ir	Pt

Physical Properties

Ruthenium is a transition metal like the other members of the group. They are less reactive than the typical metals but more reactive than the less typical metals. The transition metals are characterized by having the outermost electron shell containing two electrons and the next inner shell an increasing number of electrons. Because of the small energy differences between the valence shells, a number of oxidation states occur.

Atomic number	44
Atomic weight	101.07
Relative abundance in the Earth's crust	1×10^{-6}
Abundance of major natural isotopes	101 (17.0%)
	102 (31.6%)
	104 (18.7%)
Crystal structure	Hexagonal closed packed
Lattice constants at 20°C	
a, nm	0.27058
c, nm	0.42819
Atomic radius, nm	0.133
Melting point, °C	2,310
Boiling point, °C	4,050
Specific heat at 25°C c_p, J g^{-1} K^{-1}	0.25
Thermal conductivity λ, W m^{-1} K^{-1}	106
Density at 20°C, g/cm^3	12.45
Brinell hardness	200
Young's modulus E, N/mm^2	475,785
Tensile strength σ_B, N/mm^2	490.5
Specific electrical resistance at 0°C, $\mu\Omega$ cm	6.71
Temperature coefficient of electrical resistance (0–100°C), K^{-1}	0.004
Thermoelectric voltage versus Pt at 100°C E, V	+0.68

Chemical Properties

Because of the small energy differences between the valence shells, a number of oxidation states occur. The following oxidation states are known in the compounds of ruthenium: Ru: -2, 0, $+2$, $+3$, $+4$, $+5$, $+6$, $+7$, $+8$. The most common are $+3$, $+4$.

Ruthenium is converted into water-soluble $K_2[RuO_4]$ when reacted with KOH–KNO$_3$ and is readily attacked by molten Na_2O_2.

The most important industrial process for the isolation of ruthenium is distillation of the volatile ruthenium (VIII) oxide, RuO_4, from aqueous solution. This compound is formed in solutions of potassium ruthenate (VI), $K_2[RuO_4]$, and hexachlororuthenate (III), $K_3[RuCl_6]$, by oxidation with chlorine at neutral pH. Oxidation can also be carried out in dilute sulfuric acid solution with potassium permanganate, or with potassium chlorate in sulfuric acid solution, but these methods are less satisfactory for safety reasons. The volatile RuO_4 is absorbed by dilute hydrochloric acid, and is converted into water-soluble chlororuthenate complexes.

Even in the absence of potassium permanganate, chlorate, or their reaction products, the distillation of RuO_4 is hazardous. The presence of NH_4^+ salts can lead to the formation of explosive chlorides of nitrogen. Therefore, removing ruthenium at the start of a separation process is often preferable. The formation of chlorine dioxide, chlorate, and perchlorate is also possible. Moreover, RuO_4 can explode by spontaneous decomposition, especially at $>100°C$ and on contact with organic substances. To improve safety, work should be carried out under an inert gas and by avoiding the presence of large quantities.

Ammonium hexachlororuthenate (III–IV) is reduced in a hydrogen atmosphere at 800°C. Aqueous solutions of ruthenates (IV) (e.g., produced by adding RuO_4 to an alkali-metal hydroxide solution) are treated with ethanol to precipitate hydrated RuO_2, which is reduced in a hydrogen atmosphere at 800°C.

Ruthenium (VIII) oxide, RuO_4, is formed by the oxidation of aqueous solutions of ruthenates. It has a melting point of 25.5°C and is therefore a liquid at room temperature. It is toxic and explosive.

Soluble Ruthenium (IV) chloride consists mainly of $Ru(OH)Cl_3$ with some $RuCl_3 \cdot H_2O$ and is prepared by reacting RuO_4 with hydrochloric acid and evaporating the solution. The compound $(NH_4)_2[RuCl_6]$ can be precipitated by adding NH_4Cl.

Potassium ruthenate (VI), $K_2[RuO_4]$, is formed by fusing ruthenium with KOH–KNO_3 or by dissolving RuO_4 in potassium hydroxide solution.

References

Habashi F (2003) Metals from ores. An introduction to extractive metallurgy. Métallurgie Extractive Québec, Quebec City, Canada 2003. Distributed by Laval University Bookstore, www.zone.ul.ca

Renner H (1997) Platinum group metals. In: Habashi F (ed) Handbook of extractive metallurgy. Wiley-VCH, Weinheim, pp 1269–1360

Ryanodine Receptors (RyRs)

▶ Calcium Sparklets and Waves

Printed by Books on Demand, Germany